PRAISE FOR THE BOOK

This book provides a comprehensive review of the occurrence, abundance, conservation status, temporal trends, and some ecological information about odonates (dragonflies and damselflies) in Oklahoma. It is a new contribution to Odonatologysss and I learned a lot from it. The book represents the most complete collection of information about odonates in Oklahoma, with some intriguing patterns mentioned in the species accounts that will hopefully spur more research into this fascinating, beautiful, and important group of animals.

– Nancy E. McIntyre, Professor & Associate Department Chair,
Landscape & Community Ecology, Texas Tech University, USA

Dragonflies at a Biogeographical Crossroads demonstrates how these extraordinary insects can reveal for us the riches of the physical landscape and the human experience. Far more than a scholarly work about dragonflies, and covering terrain and ideas beyond Oklahoma, this is an almanac of history, ecology and conservation from a state too long dismissed as flat and dusty. Among its dragonflies, you will discover the vast biological diversity of Oklahoma and the state's rightful place in the geography of America. Let this volume set a high standard for insect discovery in other states and regions.

– Bryan Pfeiffer, President of the Dragonfly Society of the Americas, 2018-21

This book is a fantastic contribution to the field. Not only does it present a great array of new information, it even presents new ways to present the information. I have never seen a book about the flora or fauna of a state that has as much information as this book contains. I paged through it in awe of the scholarship, thoroughness and even imagination expressed in the pages. The occurrence maps are fantastic, better than any I have ever seen, as they combine specimens, photos and sight records in an easily understandable way. The historical aspect of the county maps and all the history related in the text is also unique to this book. As well, it often deals with taxonomic and other questions that far exceed the borders of the state. So many things about this book are unique! Conservation becomes a more and more important feature of our writing about odonates and other organisms, and this book treats that thoroughly. Finally, we need much more published about the natural history of odonates, and the species accounts in this book contain much of interest in that regard.

– Dennis R. Paulson, Director Emeritus, Slater Museum of Natural History, USA

Brenda Smith and Michael Patten have succeeded admirably in completing the most thorough treatment of Dragonflies and Damselflies of any state I have seen. This treatise on the Oklahoma fauna includes a series of rich and interesting introductory chapters followed by detailed accounts of each species in the state. This is an impressive compilation and analysis of an especially diverse state that will be useful to anyone wanting to explore dragonflies and damselflies in Oklahoma.

– John C. Abbott, Ph.D., Chief Curator & Director of Museum Research and Collections,
University of Alabama Museums, The University of Alabama, USA

Dragonflies at a Biogeographical Crossroads is a truly unique and engaging book, absolutely indispensable if you are interested in the odonates of Oklahoma but also of great interest to any student or lover of this amazing group of creatures. One of the themes of the book is "beyond," and it certainly exceeds even lofty expectations in that regard: it is beyond a field guide, beyond just Oklahoma, and beyond just a book about dragonflies (here the title is a little misleading as it includes damselflies as well). The chapter on the ancestors of Oklahoma odonata is simply the most complete history on this group I have read, and the other introductory chapters are equally riveting. The species accounts are like complete research papers for each and every species, with life histories, seasonality and incredible range data, packed with more information than is available almost anywhere else. The treatise on Macromia field identification alone makes buying this book worthwhile! This is a book that will, and should, find its way onto the shelf and into the lap of everyone who is interested in this fascinating group of insects.

– Giff Beaton, naturalist and author, USA

Dragonflies at a Biogeographical Crossroads

The Odonata of Oklahoma and Complexities Beyond its Borders

BRENDA D. SMITH, MICHAEL A. PATTEN
with contributions from
BRUCE W. HOAGLAND, ROY J. BECKEMEYER

CRC Press
Taylor & Francis Group
Boca Raton London New York

CRC Press is an imprint of the
Taylor & Francis Group, an **informa** business

First edition published 2021
by CRC Press
6000 Broken Sound Parkway NW, Suite 300, Boca Raton, FL 33487-2742

and by CRC Press
2 Park Square, Milton Park, Abingdon, Oxon, OX14 4RN

© 2021 Taylor & Francis Group, LLC ss

CRC Press is an imprint of Taylor & Francis Group, LLC

Library of Congress Cataloging-in-Publication Data

Library of Congress Control Number: 2020945970

ISBN: 978-0-367-44035-0 (hbk)
ISBN: 978-1-003-00756-2 (eBk)

Typeset in Janson Text
by Deanta Global Publishing Services, Chennai, India
Visit the companion website: www.routledge.com/cw/patten

~ Dedicated to the Pioneer Oklahoma Odonatologists ~

Ralph D Bird, A Earl Pritchard,

George H Bick, Juanda C Bick, and Lothar E Hornuff

and

to swan songs

CONTENTS

Contents

Preface

The ode bug has bitten, and it is not about to let go anytime soon.
—Steven Daniel, *Argia* **27(4):30**

… I really couldn't help but pay attention.
I had been a birder, so it was only natural for me to want to identify and list them.
—Kathy Biggs, *Argia* **28(2):26**

Most people do not like the idea of getting bit by a bug. But those of us who have been bitten by the *ode bug* are glad that we were its victim.

Many of us spent years overlooking the little jewels known as dragonflies and damselflies, or "odes," or odonates, or Odonata. We came to our love of nature by looking through our binoculars and sometimes our cameras—*at birds.* For years, we stood at the edges of ponds and streams and looked beyond the water, into the shrubs, trees, and reeds, desperately trying to get a good look at that bold and striking warbler or that subtle but gorgeous wren before it flew away and we missed one for our "life list" of species we'd seen. For years, we somehow *looked past* the equally bold and striking, the subtle but gorgeous dragonflies that hovered in front of our very faces, that landed at our very feet.

It wasn't until that hot summer's day, when the birds quieted down. It wasn't until we found ourselves on the verge of boredom thinking we had hours to wait for the next go around of bird activity. It wasn't until then that we finally noticed.

We finally noticed that brilliant orange skimmer hovering before our eyes, daring us, for even just a moment, to realize that the beauty that is birds *is also bugs.* Look there! That one with all the spots! Oh my! And that one, how can anything be so red, so blue, so acrobatic! How could we have missed all of this for so long?

We travelled the world looking for beauty. But it was right in our backyard all this time.

Now we can take those skills we developed and honed with birdwatching—our receptiveness to a gentle stirring, our reflexive lifting of binoculars when we sense something is there, our lightning-quick ability to zero in on a moving object—to also, or instead, find a dragon.

Now, with field guides at our disposal, we can identify dragonflies. Now, with a camera, we can document new records. Now, we can contribute. Now, we can all be part of the science we once thought was only for others. Now, as a community of nature lovers, we can move beyond simply identifying dragonflies to instead—*do what this book does*—see how history has shaped our understanding of how and why they are where they are.

For those of you reading this who are not birdwatchers, past or present, you too can contribute to and enjoy the wonderful, miraculous world of dragonflies. All are welcome. Be a part of the Ode to Joy!

Here's to wishing you get bit by the ode bug …

Authors

Brenda D. Smith has worked with Odonata for over 20 years in the United States and Central America. She is a principal investigator for the Odonata of Oklahoma Project and for various projects investigating odonate species of conservation concern, including acting as chief editor of the Dragonfly Society of the Americas odonate species richness project. A former archaeologist, historian, Native American affairs consultant, and museum registrar and collections manager, she found her true love in biology. She is currently a biogeographer and conservation biologist with the Oklahoma Natural Heritage Inventory, dealing with mammals, herpetiles, birds, fish, and various invertebrates, with special focus on tiger beetles and, of course, dragonflies.

Michael A. Patten has published extensively (nearly 250 scientific publications, including 2 previous books) on conservation biology, evolutionary ecology, and biogeography. Much of his research has focused on birds, but he has focused increasingly on dragonflies and damselflies over the past 15 years. He and Brenda Smith have conducted both intensive and extensive field surveys for odonates across Oklahoma, amassing over 55,000 records for 176 species.

Roy J. Beckemeyer is a man of many talents. He is a retired aeronautical engineer and executive who has made many contributions to odonatology and paleoentomology, including discovering 20 Paleozoic fossil insects. He is also a nature photographer, writer, and award-winning poet.

Bruce W. Hoagland is a professor at the University of Oklahoma in the Department of Geography and Environmental Sustainability and at the Oklahoma Biological Survey, where he is the Coordinator of the Oklahoma Natural Heritage Inventory. He is a jack of many trades with particular focus on plant ecology, vegetation classification, and biogeography.

If you just picked up this book, you might be asking yourself, "Why a book about dragonflies in Oklahoma?" But what you should be asking is, "*Why not* a book about dragonflies in Oklahoma?"

It always astonishes those who hear it—*Oklahoma ranks in the top ten of U.S. states in terms of the number of dragonfly and damselfly species recorded within its borders.*[1] Oklahoma ties with Georgia,[2] for example, and has only 20 fewer species than Virginia, which lies along the species-rich Eastern Seaboard and is the state ranked second in the U.S. The top-ranked state, Texas, counts 246 species, which is not surprising given its massive size and shared border with Mexico (and thus shared subtropical fauna). Oklahoma's *176 species* is >70% that of Texas' total species list, while being only *a quarter of the size* of that state. Remarkably, one of Oklahoma's counties has more documented species than 15 U.S. states and 9 Canadian provinces and territories.

You may wonder why Oklahoma, hosting more than a third of the 478 odonate species known for the United States and Canada, is so species rich. To answer that, Oklahoma must first be put into its proper geographical perspective. For starters, Oklahoma is larger than 30 U.S. states. More impressive, however, is that it is larger than 107 countries! As such, Oklahoma and its diversity call for us to reconfigure our thinking about size, geographical placement, and jurisdictional borders. Granted, richness is a complex equation, so size does not always equate to more diversity, but in the case of Oklahoma, there is a definite correlation. And yet, Oklahoma, because of where it lies in the world, continues to face a bias of an unfortunate assumption that it holds little biological relevance, which in reality is far from the truth. It is amusing to think that, were it a European country, for instance, given its size and richness, Oklahoma would be seen in a very different light (for example, Switzerland and the Netherlands are less than a quarter of the size of Oklahoma and have but a subset of its odonate diversity, yet both have multiple books about their dragonflies). This skewing of perception says much about the interplay of mental constructs of borders, biodiversity, and value placement. It is also a counterproductive paradigm because in a real sense, Oklahoma is a country unto itself, in its size, and especially in the biological significance it holds for odonates.

But size alone does not account for species richness. Oklahoma hosts so many species, too, because it lies at a biogeographical crossroads—it is where 12 ecoregions converge (Figure 1.1). That is the second highest number of ecoregions of any state in the United States (just one region behind California and tied with Texas, both states that are much, much larger). Those ecoregions span almost a quarter of the contiguous United States, but their reach, as part of two broader ecoregions, covers more than half of the contiguous United States (4.2 million km² or 1.6 million mi²). Consequently, what has been documented in Oklahoma in terms of the ecology of various species and phenotypic variation, but for two examples, is meaningful on a much larger scale, not only within that immense area where Oklahoma's ecoregions span but unquestionably beyond. Without a doubt, what has been learned of Oklahoma's odonates will have implications and applicability across much of North America.

But wait a minute you might say …isn't Oklahoma flat and boring?[3] Flat and boring it is not—not from a geological standpoint, nor so from a biogeographical standpoint, nor from a cultural standpoint. From the geological forces that shaped the land and allowed biogeographical provinces to form and thrive, to the cultural forces that have and continue to alter the natural landscape, Oklahoma is far from flat and boring.

We came to Oklahoma with that very bias in mind, yet we found so much more. It is our hope that you will do as we have done, and that through this book, you will be able to take the time to learn about Oklahoma and its dragonflies. To learn the when, where, how, and why dragons are where they are in this diverse state, and how those complexities span a great distance in both space and time. We hope as well that you will come away with a changed mindset—an impression of Oklahoma as a land so rich and varied that you will find the answer to the question, "Why not a book about Oklahoma dragonflies?"

When we began to write about how this book differs from other books about dragonflies, we kept coming back to the notion that it is "beyond a field guide." This notion was intuitive to us as long-time biologists, but we received blank stares or polite nods from colleagues and non-scientists alike when we said our book will be "beyond a field guide." We struggled for quite some time to grasp why the concept did not resonate, why people did not say (with triumphant trumpets blaring), "beyond a field guide; what an amazing idea!!" It wasn't until the day we were asked "what's a field guide?" that it dawned on us that there was a

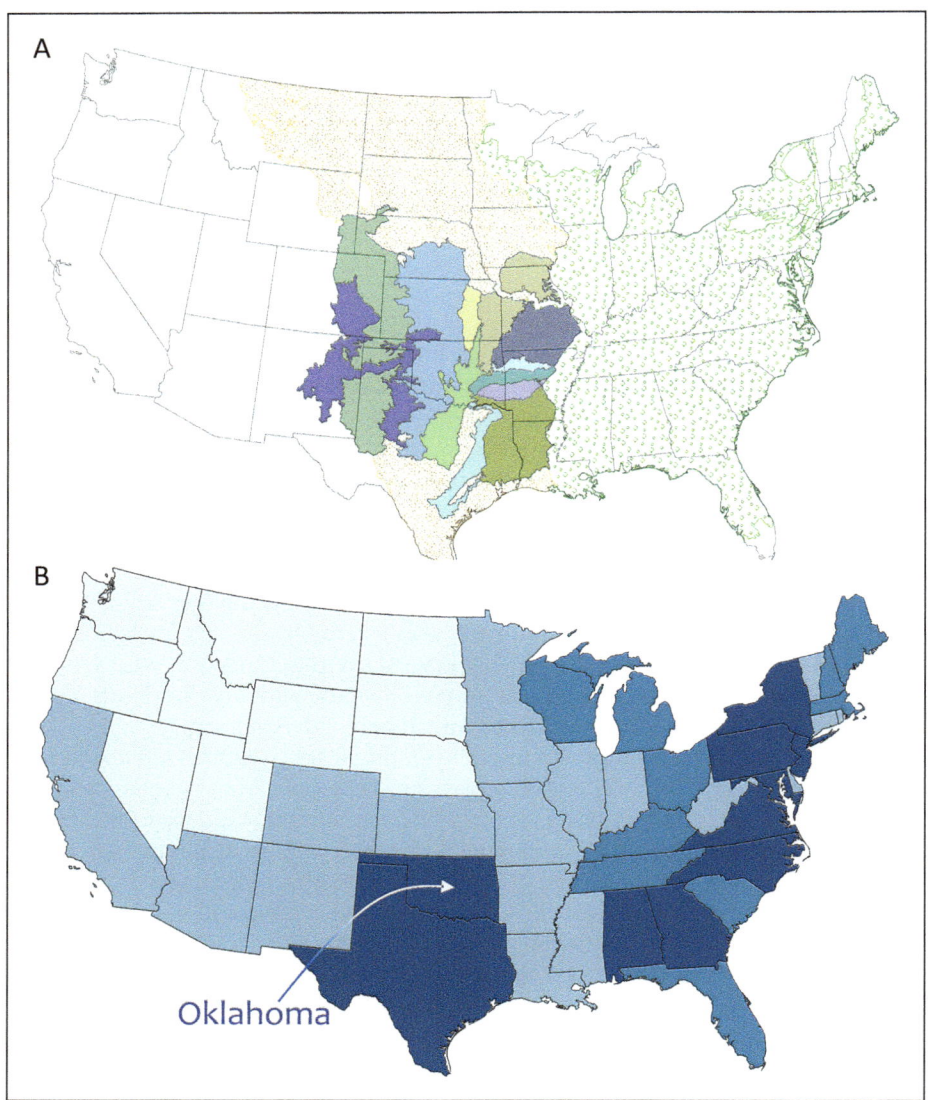

Figure 1.1 Oklahoma is at the heart of a biogeographical crossroads. A) The 12 ecoregions (solid colors) that converge in Oklahoma are also found in 11 other states and the broader ecoregions of the Great Plains (brown stippling) and the Eastern Temperate Forests (green stippling), of which Oklahoma shares affinities, covers 52% of the contiguous United States, spanning a 4.2 million km² (1.6 million mi²) area. This convergence, bringing in western, eastern, southeastern, midwestern, and subtropical odonates, is responsible for the state reigning among the top ten U.S. states in odonate species richness (B; darkest blue ≥ 174 species; lightest blue includes North Dakota, the state with the smallest list, at 68 species). Oklahoma also shares many species with areas throughout the Americas; as such, patterns and traits revealed in the state have biological significance well beyond the state's borders.

fundamental part, the foundational part, of the conversation that was lacking and, as such, what we thought was a topic of awe-inspiring proportions was in fact one with a gaping hole in it. Our *aha!* moment was falling flat because we made assumptions that everyone was familiar with field guides, and further that everyone had an idea of how natural history in general has been explored through time. We all know that making assumptions is rarely if ever a good idea, so to make sure that we are all on the same page, we will back up a bit to look at how natural history information has been conveyed and what that means in terms of how this book differs from others.

The progression of presentation in natural history has traditionally had a starting point in technical, taxonomic dichotomous keys. Quite the mouthful, yes, but simply said, keys are a series of questions, akin to a flow chart, that slowly eliminate species based on their physical features. Characteristics of an organism's body (or morphology) are run through a series of couplets—is it this type or that type, if this go to the next step, if that you have species X—until one arrives at a determination for species X or Y that is ideally species X or Y. We say ideally because one has to hope that errors were not made along the way, a hope that may leave one feeling queasy about the final species determination. These keys are gradually revised over time as new species are described to science and we notice additional morphological differences among organisms. But the downside to keys is that they tend to be jargon-filled, seemingly impenetrable, and often confusing to even those who are trained in the field. Not surprisingly then, keys scare off all but the most intrepid and dedicated of individuals. Add to that the need to kill whatever animal one is working with and you have a recipe for discouragement to say the least.

And so along came the curative—*field guides*, books that present illustrations and descriptions of a taxonomic group that aid in determining what species has been encountered.

At last there was, at least for some species, no need to obtain a specimen and no need to spend hours trying to tease apart what is meant by ambiguous couplets (species X is "yellow" and species Y is "pale yellow"). Instead a naturalist could walk along with her field guide in her pocket, see a critter, pull out said field guide, flip through the illustrations, read a bit of text, and have the coveted "*Eureka!*" moment within minutes. The value of field guides, like scholarly reference books in their own right, is immeasurable. Arguably the most value comes from their inclusivity by allowing access to the natural world to anyone who happens upon a field guide. And to many people, if they cannot name something they see, that something is inherently lacking; it remains intangible, with little chance of seizing their attention. But give that person a field guide, which allows attachment of a name and *voila!* you have a convert to a life-long commitment to studying nature.

That commitment may be as seemingly insignificant as going to a local park once a month and writing down all of the species you saw, i.e., creating a "species list." Or perhaps you took a photo of something unusual that you encountered. It may be hard to believe, but those seemingly insignificant actions add up to a book like the one you hold. Each time you went to the park and made a species list, each time you noted when and where you took a photo, you created a "record." You created a record of a given species, at a given time, at a given place. Records add up over time and form the basis of our knowledge about, for example, where species occur (= *distribution*), when they occur (= *seasonality/phenology*), with what else they occur (= *ecology*), and how rare they are (= *status*). *Your records* inform that enigma called science.

Natural history in particular would lag even more woefully had it not been for citizen scientists contributing records, studying distributions, and documenting previously unreported observations.[4] Many a well-known and respected natural historian/biologist has done so "on the side" of their non-biology profession—Carl Cook, Bill Mauffray, Karl H. Stephan, Giff Beaton, to name but a few. It is their work, *it is your work*, that can contribute to the progression of presentation by becoming data sets and leading to published articles, such as natural history "notes" or as scientific "papers" that are then used for works such as Dennis Paulson's North American field guides for damselflies and dragonflies.

But the progression of presentation does not stop there. With time comes a desire for a narrowing of geographic scope and a synthesis of more local and regional data. To illustrate, let us use the bird world as an example.

There was a time when the primary way to study birds was to shoot them, stuff them, and then closely examine them in a museum. Comparisons to other birds were made and taxonomic keys were devised. Birds remained largely in the realm of the professional ornithologist/taxonomist. But then in 1889 came Florence Merriam Bailey with her *Birds through an Opera Glass*, which was essentially the first field guide to birds, and then much later, in 1934, the modern field guide was born when Roger Tory Peterson brought out *A Field Guide to the Birds* (Stevenson et al. 2003). Once field guides were available to use for species identification, naturalists began coming into the birdwatching or "birding" world in droves. They craved more and more field guides, and as those guides got better and more detailed, birders wanted even *more detail*. No longer were they happy with the broad range maps and broad descriptions of habitat they found in North American guides; they wanted to know specifics about birds, both in their area and in someone else's area, including if there were any regional or local differences in what birds looked like or when and where to find them. And thus, the invention of state compendia on status and distribution, of bird-finding guides, and state- or region-specific field guides. Still not sated, state atlases, for both breeding and wintering birds, began to be produced. Some state rare bird records committees, for example, the California Bird Records Committee, even began to produce record-specific works that provided minute details about a suite of records.

Now, that voraciousness is turning toward *charismatic microfauna*—dragonflies and damselflies—as more naturalists discover odonates, but especially as birders become dragonfly watchers. Now that birders are becoming "oders" there is a need for the same accessible detail to which they have become accustomed. We are at the moment in the progression of presentation with odonates that we are ready to go to the next level. We have the continually revised dichotomous keys, we have the North American field guides, we have a few state- or region-specific field guides, and we have the rising person power for data collection. Now it is time to go "beyond field guides." One recent example, *The Dragonflies and Damselflies (Odonata) of Utah* by Alan Myrup and Richard Baumann (2016), is a detailed work that headed down the record-specific path of which we advocate. We continue down that path of going "beyond the field guide" by presenting details for Oklahoma and the region that, upon a first look may seem pertinent only locally or regionally, but that actually informs beyond jurisdictional borders, providing insight into species rangewide.

Facts and figures are frequently considered the *boring stuff* of both history and science. The allure and significance of those facts and figures lies not in the dry details but in the contextual wrapping. That wrapping is what convinces a reader to read on and, with any luck, be enraptured and, maybe, inspired.

But without facts and figures there is no basis, there is no skeleton to flesh out, there essentially is no need for context because there is nothing to wrap that context about. For all intents and purposes, you cannot have one without the other. You may be able to read past the facts and figures, but you must still be able to refer back to them and, if only on a subconscious level, understand that the context you are engrossed in is, in fact, *based on facts*.

At heart this is a scholarly book. We do not expect someone to sit down and read it cover to cover (although if you do, enjoy!). Instead, like a reference book, you can read sections of it at a time and refer to portions again and again as needed when a question arises about a species. We present many facts and figures, for reference and for posterity, but what we hope is that you will zero in on the context; how we tied together those facts and figures to present a story about the species of interest. Look, for instance, for the biogeographical substance and the historical factors that brought that species to that place in time and space; reserve the facts and figures for moments when some specific detail is needed.

The two key contextual themes of the book are the connection of individual records to 1), the physical landscape, and 2), the human landscape. The physicality of a place is integral to its biology. That physicality is the result of a chain of events: geological processes shape the land, which shapes habitat, which shape ecoregions. And those events resulted in an area's biogeography, or how the physical landscape and biology interacted in ways that affected the distribution of organisms. We discuss how biogeography, as evidenced by the fossil history and the contemporary landscape, plays a huge role in when and what odonate species we see today.

We explore Oklahoma's human landscape in multiple ways. The most obvious part of the human landscape—how humans alter the natural environment to suit our needs—is a fundamental aspect of species distributions and is a topic that is inescapable when speaking of species conservation. As such, we provide a chapter that summarizes some of the more noteworthy, and at times scarring, parts of Oklahoma's environmental history, and follow that with a discussion about the conservation of Oklahoma's dragonflies and damselflies.

We provide another chapter noting the relatively long history of odonatological study in Oklahoma. Our foremost reason for that inclusion is that we do not want any reader to conclude or assume that this book is the result solely of our efforts; it was built on foundations laid by our forbearers, some of whom made tremendous strides in the early stages of odonatology in general, as well as in Oklahoma. We further feel that history deserves to be documented, not just because of our personal bias of the intrinsic value of history, but because it influences how we see and characterize odonate species to this day. In a sense, because we as humans are the ones gathering data and writing about the biology, ethology, and ecology of species, our history is inextricably tied to those species. Our understanding of a species' biology is, in part, an ethological endeavor, in the human-focused sense of the word. We discover our own human character during the transformation we make while studying another species.

That transformation is why it is so vital to bring to life and acknowledge the human history behind the biological history. Without the efforts of people who were transformed by biology, we would not have biology.

ACKNOWLEDGMENTS

Not surprisingly, this book was made possible by the effort of a multitude of people, so many in fact that we hope we did not miss anyone. Two people that especially stand out are David Arbour and Bill Carrell. David and Bill started in the odonate world at roughly the same time, almost two decades ago, and instantly began submitting records; they continue to do so. They deserve immeasurable thanks for all of the time and energy they have dedicated to the Oklahoma Odonata Project (OOP). Berlin Heck, Vic Fazio, John Fisher, Jim Arterburn, and Ken Williams also deserve many special thanks for all of the years they have sought out dragons in Oklahoma. These seven guys have many county and state records, re-discoveries, and literally thousands of records (of other taxa, too!) under their belts. Our enduring thanks to you all!

There have been a handful of other record submitters that deserve special mention for the number of records submitted: Mark Dreiling, Jona Tucker, Franklin "Leroy" Alm, Colby Farquhar, Cliff and Jon Ivy, and Bryan E. Reynolds. The "Jasons"—Jason Heinen and Jason Bried—contributed many specimens, photos, and data that we are thankful for. Two other submitters, Abigail Mills and her mother Ruth, whom Brenda had the fortune of meeting after she gave a talk at a Tulsa Audubon meeting in 2014, are not only two of the sweetest people one could ever meet, but Abigail now teaches children about the wonders of nature!

Others who submitted records and to whom we send our thanks include: Aadriaan Aaronson, John C. Abbott (Odonata Central data, etc.), Randy C. Anderson, Alexandra A. Barnard, Cheryl B. Barr, Eric J. Beck, Roy J. Beckemeyer, Elizabeth A. Bergey, Wade Boys, Zachary Bragg, Jim Burns, Amy Buthod, Janalee P. Caldwell, Alex Cooper, Priscilla Crawford, Claire M. Curry, Doug Danforth, Jerrell Daigle, Mike Dillon, Bill Dobbins, Jena Donnell, Thomas W. "Nick" Donnelly, Zach DuFran, Sidney W. Dunkle, Steven Easley, Richard A. Erickson, Mark and Molly Ferguson, Kathy Furneaux, Kate Goodenough, Rick Grantham, Joe Grzybowski, Sylvia Hanson, Alex Harman, Rachel Hartnett, Adam Hasik, Ford Hendershot, Emily A. Hjalmarson, Greta and Pat Heck, Diane, Terry, and Troy Hibbitts, Bruce W. Hoagland, Bob and Hans Holbrook, Bill Horn, Sam Houston, Eric Isley, Dan Jackson, J. Harrell Johnson, Shawn Johnson, Randy Kelley, Tom Kompier, Boris C. Kondratieff, T. Kuder, Vern LaGesse, Brett H. P. Landwer, Greg W. Lasley, Heather K. LePage, Tony Leukering, Charles S. Lewallen, Brian Mannel, Trey McFall, Paul McKenzie, Brad Minson, Terry Mitchell, Gary Murphy, David Oakley, William F. Oakley, Dennis R. Paulson, Mark Peterson, George Pierson, Jay A. Pruett, Darren E. Purcell, S. Queen, Brett Roberts, Robert Sanders, Kurt Schaefer, Tim Schreckengost, Rosemary Seidler, Greg Shelton, William D. Shepard, George Sims, Don Stanley, Kent

Sowers, Bill Stark, Ryan Steiner, Kenneth J. Tennessen, Jeff Trahan, Cynthia Van Den Broeke, Laurie J. Vitt, Tim Vogt, Robert Webster, J. B. Wheatley, Lauren A. Wilkerson, Bill Wilwers, Lauren Wishard, Doug Wood, and a slew of iNaturalist users whose aliases prevent us from fully acknowledging their contributions, but thank you. Additional data were provided by Julie Craves, Bob Glotzhober, Chris Hill, Steve Krotzer, and Darrin O'Brien. Some key people who provided insight, data, and much kindness and encouragement over the years include Giff Beaton, Thomas W. "Nick" Donnelly, Rosser W. Garrison, George L. Harp, Greg W. Lasley, Bill Mauffray, Nancy E. McIntyre, Dennis R. Paulson, Kenneth J. Tennessen, and Tim Vogt. Oodles of thanks to you all, and to those who allowed us to use their photos in the book.

Many people allowed access to their personal collections, supplied collection data, or helped facilitate research. We thank Dennis R. Paulson, Rosser W. Garrison, Kenneth J. Tennessen, Bob Dubois, John C. Abbott, Nick Donnelly, Sidney W. Dunkle, and Carl Cook for access to their personal collections and help with identifications. Bill Mauffray and Paul Skelley, of the Florida State Collection of Arthropods (Mauffray also affiliated with the International Odonata Research Institute at FSCA), allowed us to work with the collection on multiple visits and Bill was exceptionally kind to pull other specimens, as needed, to assess characters, provide measurements, photos, etc., to save us from doing so ourselves. Donald C. Arnold and Richard Grantham, of Oklahoma State University, facilitated countless visits to the KC Emerson Entomology Museum. Many visits were also made to the insect collection at the University of Central Oklahoma (UCO), so thanks go to David Bass, Wayne Lord, and Lynda Loucks for providing access and loaned materials. Brenda sends her sincere gratitude to David for the extra special treat of looking though Lothar Hornuff's *Manual of the Dragonflies of North America*. Boris C. Kondratieff, of Colorado State University's Gillette Museum, provided data, reports, and loans, and Inez and Bill Prather identified specimens that the museum collected at Fort Sill in the early 2000s. We thank Jennifer C. Thomas and Zachary Falin, of the Snow Entomological Museum at the University of Kansas, who allowed access to the collection, as well as loaned materials (thank you, Jennifer, for going all the way over to the fluid collection!). We also thank Dixie Smith, Hermann Nonnenmacher, Neil Snow, and Virginia Rider for facilitating work with the collections at Pittsburg State University, Kansas. Thanks to Patricia Gentili-Poole, Karolyn Darrow, Erin Kolski, David Furth, and Gary F. Hevel, of the Smithsonian Institution, for their help with collections, including loaning materials; but special thanks are in order to Oliver S. Flint for all of the help he was through the years. Mark F. O'Brien facilitated a productive and enlightening visit to the University of Michigan, Museum of Zoology. Thank you, Mark, for your help with the collection and especially for sharing correspondence between Ralph Bird and Leonora Gloyd. When Brenda was the caretaker (2006–2011) of the Recent Invertebrates collection at the Oklahoma Sam Noble Museum of Natural History (OMNH), she was fortunate to have had some wonderful collections assistants, namely Jamie Lentz, Jessie Tanner, and William R. Winfree. In later years, Katrina Menard, Charles "Andy" Boring, and Melissa Sadir were super helpful in getting OMNH data and allowing us to examine specimens and archival materials. We wish for a favorable outcome for the collection. Additional museum staff that assisted with data and loans are: James H. Boone (Field Museum), Rob Cannings (Royal British Columbia Museum), James Cokendolpher (Museum of Texas Tech University), Douglas C. Currie (Royal Ontario Museum), Jason J. Dombroskie (Cornell University Insect Collection), Colin Favret (Ouellet-Robert entomological collection, Université de Montréal), Heath J. Garner (Museum of Texas Tech University), Christopher C. Grinter (Illinois Natural History Survey), Brad Hubley (Royal Ontario Museum), Peter T. Oboyski (Essig Museum of Entomology), M. J. Paulsen (University of Nebraska State Museum), Philip D. Perkins (Museum of Comparative Zoology, Harvard University), Raymond Pupedis (Peabody Museum of Natural History), Eddie Reese (Oxley Nature Center), and Jason D. Weintraub (Academy of Natural Sciences). Thanks to Hal White and Frank L. Carle, who helped clear up a misreported record of *Gynacatha nervosa*, and to Edward Johnson, of the Staten Island Museum, who cleared up an issue with Oklahoma *Neurocordulia virginiensis*. Also many thanks to the Arkansas Natural Heritage Commission, Missouri Natural Heritage Program and Department of Conservation, Louisiana Natural Heritage Program, and Texas Parks and Wildlife for providing data, to Jeff Casida, City Manager for the City of Hobart, Pat Gwin of The Cherokee Nation, the Arkansas Game and Fish Commission and State Parks for allowing access to properties, and to the Odonata Central vetter's list serve participants for identification help.

Various people were instrumental to historical research and deserve thanks. Sara Whyatt and Carol Jasak of the Oklahoma Historical Society (OHS) helped BS-P track down a 1926 issue of *Outdoor Oklahoma*. Sara helped, too, to find correspondence between Frank Collins and EB Williamson as well as some from A. E. Pritchard. That correspondence (and a photo) was obtained from the Bentley Historical Library, whose staff were extremely helpful and deserve many thanks. Other OHS staff, including Felecia R. Vaughn and Robert "R. J." Wilkins were helpful in researching O'Reilly Sandoz, and Rachel Mosman provided various photographs. Darrin Hill, of the Oklahoma Department of Wildlife Conservation (ODWC), was most helpful in tracking down agency history. Jo Crabtree and Debbie Neece kindly helped obtain historical photos from the Bartlesville Area History Museum. Gloria Caddell and Donna Bass, of UCO, went above and beyond to help Brenda find information and a photo of Lothar Hornuff. Sandra Thomas and Darryl J. Rainbolt, of Southeastern Oklahoma State

University, provided a photo of Lothar from their institution. Benjamin Hedges and Andrew Arterbery, of the OSU Archives, digitized a photo of A. E. Pritchard as a teenager. Erin George, of the University of Minnesota Archives, examined their collection for Pritchard's papers. Rachael Lester, of OU's Western History Collections, digitized a photo of Lois Bird. Dr. Gordon Goldsborough put BS-P in touch with Dr. Charles Durham Bird, son of Lois and Ralph Bird, who deserves our deepest and gleeful gratitude for providing several photos and additional biographical information for his parents. For geological and climatological insights, we thank Virginia L. McGuire, Shana Mashburn, and Bradley G. Illston.

Without the financial and logistical support for research that we received from colleagues in federal and state agencies, we would not have been able to accomplish what we have. One person who has gone above and beyond with his support over many years, and was and is absolutely indispensable as a colleague, is Mark Howery of ODWC. Mark is part of a group of people at ODWC who demonstrate time and again their endearing support of conservation research, including Curt Allen, David Arbour, Clay Barnes, Jena Donnell, Colby Farquhar, Matt Fullerton, Kurt Kuklinski, Alan Peoples, John Skeen, Weston Storer, Curtis Tackett, Marcus Thibodeau, Larry Wiemers, and Eddie Wilson. Special thanks to David and Colby for all of our field time together. Research that contributed to the OOP came from grants provided by ODWC: F13AF01188 (T-73-1) and F18AF00919 (T-108-R-1). We are also grateful to those in federal agencies with whom we work closely. These individuals also thankfully embody the spirit of conservation: Robert Bastarache, Shea Hammond, Daniel A. Jackson, Neil Lalonde, Melissa Lombardi, David Martinez, Paige Schmidt, Paul McKenzie, Richard Stark, Timmy Walker, David Weaver, and Amber Zimmerman. Special thanks, too, to the refuge guys who went to the field with Brenda: Levi Feltman, Glen Hensley, Steve Hodge, Scott Johnson, Shane Kasson, Daniel T. McDonald, and Barry Smart. Other folks who have been generous with their time and resources include John Drake, Dian Jordon, John Fisher, Jona Tucker, and Jay Pruett. Private landownership access was provided by a handful of kind people whom we misgivingly failed to get the names of. Some kindhearted people we were smart enough to note include Mark DeWitt, Mike Bailey, Kenny Joiner, Roger Wilson, Pete Thurmond, Larry and Raymond Moody, Jerry and Joel Alexander, and Kim Bartlett. It is people like these who make up for all the times we were asked, "May I help you?" when that person actually meant that he or she was just about to call the police on us. The world is not as scary a place as some think. We should all take a lesson from the kindness of people like Mark, Mike, Kenny, Roger, Pete, Larry, Raymond, Jerry, Joel, and Kim.

Vic Hutchison got the editorial ball rolling, which Kent Calder took and went to great lengths to shape its appeal.

Thanks, too, to Edie Marsh-Matthews, Bill Matthews, and Sandie Holguín, who provided key editorial pointers. Our undying gratitude goes to CRC Press, namely Chuck Crumly, who recommended the book to Alice Oven, who then enthusiastically saw it through to publication. Damanpreet Kaur, Marsha Hecht, and Andrew Corrigan were key to getting us through the editorial process. Alan Myrup and Robert Kirk provided guidance in publication and greatly lifted Brenda's spirits. Bryan E. Reynolds, Laura Beth Reynolds, and Bill Carrell reviewed early drafts and helped to mold our approach and the book's accessibility. Dennis R. Paulson and Nancy E. McIntyre provided extremely supportive and instrumental reviews to an earlier draft. Two anonymous reviewers also provided positive and constructive reports. Many people helped with obtaining permission to reprint articles, graphics, and photos; they include, Lisa Black, Christopher Dick, Peggy Gough, John Heppner, Christine Johnson, and Sharon Moorman. Financial support for publication was provided by the Oklahoma Biological Survey (OBS), the Office of the Vice President for Research and the Office of the Provost, University of Oklahoma and the Dragonfly Society of the Americas. We are tremendously grateful for their generous support.

The Oklahoma Natural Heritage Inventory (ONHI) and OBS provided oodles of support for the project. Jeffrey F. Kelly and Caryn C. Vaughn specifically deserve thanks. Their emotional support alone during two disruptive personal events deserves much gratitude, but they have been supportive on so many other levels as well. Lara Souza, although coming at a later date to OBS was also extremely supportive of the project, as was Bruce W. Hoagland, ONHI Coordinator. Without OBS's and ONHI's financial and intellectual support of the OOP, you would not be reading this tome. Thanks, too, to student assistants including Robin Urquhart, Alexandra A. Barnard, Emily Hjalmarson, and David Hille. And many thanks to OBS/ONHI staff, namely Todd Fagin who helped immensely over the years with data and mapping, and Trina Steil and Ranell Madding will always hold a special place in Bee's heart.

Other friends and family who have been particularly supportive of Bee through the years include Virgie Smith, Diane and Jim Hoagland, Jo and Gene Crabtree, "The Suspects," and Mary Cay and Tom Woodfin. It would have been oh so difficult to complete the book without their belief, and that of other friends and colleagues at OU, that it could be done. David Hille was a tremendous source of inspiration to Brenda. He is an extraordinary person whose accomplishments are deserving of much adulation and whose kindheartedness is beyond measure. Jutta C. Burger deserves a medal for all she has endured with both Brenda and Michael. She is simply amazing and there is no way to say thanks for all she has done. Words likewise fail to express Bee's gratitude to Bruce W. Hoagland for being a supportive partner, including helping with fieldwork, putting up with so many late nights of work and the weariness (and grumpiness) that comes along with them, bringing joy to such a sad heart, and listening to a lot of complaining.

TERMINOLOGY

Before we continue, a few words regarding terminology are useful for context and reference. Herein can be found various terms and concepts relating to taxonomy and anatomy, life stages, and behaviors of odonates. A few miscellanea terms are thrown in at the end.

Taxonomy

Biologists, in their attempt to force order upon a chaotic world, devised a hierarchical system into which all organisms are placed (Table 1.1). For example, dragonflies and damselflies are part of the animal world (Kingdom Animalia). They are also arthropods (Phylum Arthropoda), a group that includes animals such as spiders, mites, crustaceans, millipedes, and centipedes. Arthropods also include the insects (Subphylum Hexapoda, Class Insecta), of which dragonflies and damselflies are a part, specifically the "ancient winged insects" (Subclass Pterygota, Infraclass Palaeoptera). Damselflies and dragonflies are part of the insect order Odonata,[5] or the "toothed ones," due to their well-formed jaws and teeth, though it ought to be noted that many insects have well-formed jaws with "teeth." Not all insects are further divided into suborders, but in the case of Odonata there are three: Zygoptera (damselflies), Anisoptera (dragonflies), and Anisozygoptera.[6] In the last suborder, there are three species known extant, all of which are from eastern Asia; all other known species from the suborder are from fossils. The Zygoptera

and Anisoptera are typically what people mean when they refer to damselflies and dragonflies, of which there are about 6,300 described species (Paulson 2019). Each of those suborders are further broken into families; for example, Calopterygidae (broad-winged damselflies), Lestidae (spreadwing damselflies), Gomphidae (clubtail dragonflies), and Libellulidae (skimmer dragonflies). Families can be noted using their common names, i.e., their non-scientific name, or one can shorten the name by removing the *-ae* ending; for example, an individual belonging to the family Lestidae can be called a spreadwing damselfly, a spreadwing, or a lestid (always lowercase; plural is lestids). Similarly, the suborders can be shortened: anisopteran or zygopteran. (As an aside, one must be careful when using the term *common name* because it is not synonymous with *English name* nor is there always just one English name. In our case, since we discuss Odonata from Oklahoma, we use American English common names, but we all must keep in mind that elsewhere the same or similar species may have a different common name. For instance, a calopterygid is a "jewelwing" in American English but a "demoiselle" in British English and in French.)

Beyond the family name, animals and plants are assigned a unique scientific name, which in the past was often referred to as the Latin name. We no longer use the term *Latin name* because scientific names typically are (and always have been) composed of Latin or Greek roots, or a combination of roots (or occasionally roots from other languages). A scientific name comprises a genus and species and sometimes a subspecies name. Those categories are referred to as the generic name (genus, or genera, not genuses, if plural), specific name (species, the same term when used in either the singular or plural), and subspecific name (subspecies). When the first two are taken together, one has a binomial (genus and species combination), also called an epithet, or, confusingly, just a "species." When all three are combined, it is called a trinomial (genus, species, and subspecies), which can also be an epithet, or (again, confusingly) just a "subspecies." Note that only specific or subspecific scientific names are written in italics; all other taxonomic names are written in regular (non-italicized, -bolded, or -underlined) script.

Abbreviations of scientific names are sometimes used, but one should never leave a name "naked" or allow it to be confused with a name that precedes it in text. For example, take this sentence, "*Enallagma traviatum* is found at ponds whereas *E. exsulans* is found along creeks. The two subspecies, *E. t. traviatum* and *E. t. westfalli*, are both associated with ponds." The sentence is written correctly because the *E.* refers back to *Enallagma*, i.e., *Enallagma exsulans*, and the *t.* refers to *traviatum*.[7] The sentence would be incorrectly written if it read "*E. traviatum* is found at ponds whereas *E. exsulans* …" because the *E.* is left naked when there is not a preceding genus name. It would also be incorrect to write "*Enallagma traviatum* is found at ponds whereas *E. designatus* is not." In this case the *E.* incorrectly refers back

| Table 1.1 | **Taxonomic hierarchy of animals. Major divisions (left) and those specific to damselflies (middle) and dragonflies (right).** |

KINGDOM	ANIMALIA	ANIMALIA
Phylum	Arthropoda	Arthropoda
Subphylum	Hexapoda	Hexapoda
Class	Insecta	Insecta
Subclass	Pterygota	Pterygota
Infraorder	Palaeoptera	Palaeoptera
Order	Odonata	Odonata
Suborder	Zygoptera	Anisoptera
Family	Coenagrionidae	Macromiidae
Genus	*Enallagma*	*Macromia*
Species	*traviatum*	*illinoiensis*
Subspecies	*westfalli*	*georgina*

to *Enallagma*, which would make the species *Enallagma designatus*, a non-existent damselfly species rather than *Erpetogomphus designatus*, an actual dragonfly species. In these examples, the abbreviated genus is referred, correctly or not, back to a genus in that sentence. Keep in mind, though, that a reader is expected to refer back to the preceding genus even if that name is a sentence, a paragraph, a page, or a chapter behind. Mixing up generic names can cause much confusion as in the *Enallagma–Erpetogomphus* example but when an error occurs in a work as significant as the type description, or formal description of a species, mayhem can ensue. If the *Enallagma–Erpetogomphus* error were in a type description much effort would be needed to resolve the issue of mixing a damselfly genus with the specific name and description of a dragonfly. The author, or authority, of that type description would be kicking him or herself, for sure. In such a case, that would then make the correct scientific name and authority cited as *Erpetogomphus designatus* (Booboo 1847), not because authorities are cited within parentheses but because in this instance that genus name changed due to mixing of genera. To illustrate further on authority citations, let us take some examples. These cases do not involve mixing of genera as above but rather changes in taxonomy.

But first, let us take a step back to define taxonomy, which is the hierarchical structure we have been discussing. Taxonomists, or systematists (those who study systematics, or classification), are the people who classify organisms into a taxonomic hierarchy, or *taxonomy*. Within that taxonomy are distinct entities, each known as a *taxon*. An individual species is called a taxon, as can be any single group such as a family (e.g., Lestidae is a taxon, as is the genus *Lestes*) whereas multiple species or families, when discussed together, are referred to as *taxa*, the plural of taxon. For example, the single species Ebony Jewelwing is a taxon but Ebony Jewelwing and River Jewelwing, talked about together, are taxa. Again, one would speak of the single family taxon of Lestidae but the familial taxa of Lestidae and Calopterygidae.

Taxonomic changes occur for a variety of reasons. For example, a species may be found to be more closely related to another taxon than previously thought, as such the species may be assigned to a different family or a different genus. A species can also be *split* (or separated) from the genus it was originally described as part of, or *under*, or it may be *lumped* (or merged) back into a genus it was once split from. Examples of changes include when *Anax junius*, the Common Green Darner (family Aeshnidae), was originally described as a species of skimmer (Libellulidae). Because it was described under the wrong family, its family and its genus had to change; *Libellula junius* Drury, 1773 was moved from Libellulidae to become *Anax junius* (Drury, 1773) in the family Aeshnidae. Another change occurred with *Calopteryx maculata*, which was originally called *Agrion maculatum*, but it had to be split from that genus. Before the change occurred, the species and authority would have been written as *Agrion maculatum* Beauvois, 1805, indicating that

Beauvois described the species in 1805. Now, the name is written as *Calopteryx maculata* (Beauvois, 1805) to indicate that Beauvois described this species under a different genus, i.e., as *Agrion*[8] in 1805. Recall from earlier that the authority and year of description is put into parentheses when the current genus is not the same as the original.

Lingering a tiny bit longer on taxonomy, we discover the basis of some terminology that lies within the root, i.e., the Greek *odon*, of the scientific order Odonata. In more technical language Odonata can be called odonates or singularly, odonate. The study of odonates is called odonatology and those who study the subject are odonatologists or odonatists, depending upon the scientist's preference. In more common language, odonates have been traditionally called by expressive or peculiar names, including mosquito hawks, horse stingers, snake doctors, and devil's darning needles (Figure 1.2). It is now more fashionable to call them dragons (referring to just the suborder of dragonflies or to damselflies and dragonflies collectively), damsels (only used for damselflies), or odes (dragons and damsels together or as separate groups). The act of viewing or catching dragons, in the broad sense, is sometimes called dragonfly hunting and occasionally dragonflying but, to our knowledge, never dragon-ing. Others use ode-ing, odeing, oding, or ode-watching. One rarely hears dragonfliers, but dragon hunters—not to be confused with the Dragonhunter (*Hagenius brevistylus* Selys, 1854)!—is rather common and, pungent connotations aside, oders (pronounced like "odors") seems to be catching on to describe those who partake in dragonfly watching, photography, catching, or collecting. The gist is you may use whichever terms you wish. Our philosophy is that because dragon hunting is so much fun, the terms ought to be too, so we, oders, very much enjoy odeing.

Anatomy, life stages, and behaviors

At this point you may be wondering what the differences between damselflies and dragonflies are. This is a common question. The suborders are most easily distinguished in their aquatic larval form, but because most people encounter them in their terrestrial form we offer a couple of pointers here. Generally, one can tell the two suborders apart by how an individual holds its wings when at rest. Usually dragonflies hold their wings out to the sides of the body whereas damselflies hold them together behind the body (Figure 1.3). But, of course, there are exceptions, the most prominent of which are the spreadwings (Lestidae). As their name suggests, spreadwings perch with their wings in a rather un-damselfly-like fashion by holding them outward like, but not as far out as (only about halfway, actually), a dragonfly does (i.e., at an acute angle rather than horizontal with the body). Sometimes other damselflies spread their wings. For example, the Aurora Damsel (*Chromagrion conditum*) often perches with its wings partially open. Females of other species will also partially open their wings when signaling that they are not receptive to mating (i.e., performing a refusal display). Another

Figure 1.2 Dragonflies in the news. This article comes from page 4 of *The Altus Times-Democrat* on 2 September 1920.

situation that may cause confusion is when encountering a freshly emerged dragonfly. Until the body is fully formed and the individual can use its flight muscles, it will hold its wings together behind its body. The other cue used to tell damselflies and dragonflies apart is overall body size, but this too is not a hard and fast rule because there is much overlap in sizes between the suborders. However, generally damsels are small-bodied and narrow while dragons are long-bodied and relatively bulkier (Figure 3). A third general, but imperfect, way is that all damsels have the eyes widely spaced, but many dragons are *holoptic*: their eyes meet atop the head. Naturally, the gomphids (clubtails), petaltails, and some spiketails defy this "rule"; still, a holoptic ode will be a dragon.[9]

Pointing out the superficial differences between damselflies and dragonflies brings us to some terminology that is useful to know when discussing body parts of odonates. Both suborders have the main body divisions that all insects have—head, thorax (with wings and six legs attached), and an abdomen—but the terminology used to describe the individual parts of those main divisions differs between damsels and dragons. We do not go into detail here about the parts of the body, but we do provide a general guide to help navigate what you will find later in the book. For more details of anatomy, we refer to Dennis Paulson's Princeton field guides (Paulson 2009, 2011) and Bailowitz et al. (2015) for excellent depictions of odonate body parts and how they function.

The heads of damsels and dragons differ markedly in their shape but the terminology describing them is fairly consistent (Figure 1.3). The thorax of odonates comprises three parts: the prothorax, mesothorax, and metathorax, or the front, middle, and final segments. The first pair of legs are attached to the prothorax, a body part that is colloquially referred to as the neck, to which it is somewhat analogous. The second pair of legs and the forewings are attached to the mesothorax and the third pair of legs and the hindwings

are attached to the metathorax. The meso- and meta- thoraces (plural can also be thoraxes) are fused together into the pterothorax, which is what is commonly just called the thorax. For some damselflies, a plate located at the front of the pterothorax that is known as the mesostigmal plate is very useful to diagnose (determine) species. To see that plate, one much look dorsally (from the top) at the female's thorax under a microscope. Depending upon the angle at which the specimen dried, one may be able to see the plate by simply propping the specimen up on some forceps under a microscope, but sometimes the head and prothorax must be removed to see the mesostigmal plate unobscured.

The thoraces of both damsels and dragons have patterns, or thoracic striping, that are often important in determining what species one has encountered, but the terminology between the groups differs considerably (Figure 1.3). Thoracic terminology can be confusing, especially when trying to keep in mind which parts are dark and which are not. For example, with damselflies we speak of the humeral and the antehumeral stripes. Both are sometimes referred to colloquially as *the shoulder stripe* but it is more correct to say that *the antehumeral is the shoulder stripe*, as generally that is what is being discussed. But even in the scientific literature the stripes are sometimes confused and so it needs to be remembered that for damselflies the *humeral is the dark stripe* and the *antehumeral (or the shoulder stripe) is the pale stripe* (Figure 1.3). To the extreme frustration of dyslexics (like the senior author), this rule does not apply to dragonflies. *For dragons, both stripes are dark* and may or may not be used for identification depending upon the family (Figure 1.3). An additional complication to note is discussion of the mesepisternum of *Macromia* (river cruisers). That area may have a pale *mesepisternal stripe* or it may be lacking, in either case that region of the thorax, which is located behind the eyes leading up to the bases of the wings, is key to species identification (Figure 17.4). Once thoracic terminology can be kept straight, identification

Figure 1.3 Anatomy of dragonflies and damselflies. Even the tiniest structure on an odonate carries a name, but we provide here only the simplest of details. Note, too, that certain features apply only to specific genera or families. The T1, T2, etc., system of distinguishing dark thoracic stripes, for example, is reserved for the Gomphidae (the clubtails) and different terminology is used for others (e.g., mesepisternal stripe for *Macromia*, the river cruisers; see Figure 17.4). Likewise, "eye spots" appear on the rear of the bulbous compound eyes of various zygopterans, yet it is in diagnoses of *Enallagma* (the bluets) species it is used most often. The most extreme example of specialized names is the "mesotibial keel," a feature discussed not just solely for *Macromia* but just solely for the middle leg of the male.

becomes much easier, but one needs to be patient with absorbing the terms and patterns. Before leaving the thorax, we should mention that wing venation can also get complicated in a hurry, particularly with various disagreements as what to call specific veins, so we illustrate here only basic venation to which we refer in some of the species accounts (Figure 1.3).

The abdomen is divided into ten segments, even when in the larval stage (Figures 1.3 and 1.4). The segments are referred to in the shorthand as S1, S2, S3, etc., beginning with the segment closest to the thorax and ending with S10 at the end of the abdomen (or S1, anteriorly, S10, posteriorly). In common parlance the abdomen is called *the tail*, a term that can make odonatologists and entomologists cringe (but what to say about common names such as petaltails and the Taper-tailed Darner?). The abdomen is where odonate genitalia, or the sex organs, are found. Male odonates have two sets of genitalia. One set is found on the underside, or ventrally on, S2–3. The other set is found at the end of the abdomen. Attached to the end of S10 are a pair of cerci (pronounced cer-cee; singular, cercus,

pronounced like circus) and an epiproct, if a dragonfly, or paraprocts, if a damselfly. The cerci are often referred to as the superior appendages while the epiproct or paraprocts (or procts) are the inferior appendages, as an indication of which are located dorsally (on top) on the body and which ventrally (underneath). These structures are called the caudal appendages because they are located caudally, or at the "tail" end on the body; they are also called the terminal appendages or terminalia. Although located posteriorly, the term posterior appendages is not used. They are sometimes called "claspers," indicating their function of grabbing a hold of and keeping a clasp on a female during mating. For zygopterans, the claspers fit into the mesostigmal plate much like a lock and key, so much so that even when a pair is captured by a predator and the male has been eaten, the remnant of the male's abdomen may stay attached to a female for quite some time (Figure 1.4). For anisopterans, the claspers maintain a vise-like grip on the back of the female's head or on her prothorax, sometimes clasping her eyes, which can result in puncturing.

adult – hardened and "dry"

tandem pair – sperm exchange (♀ S8 joined to ♂ S2); some species stay paired after mating

ovipositing (egg laying) ♀ – lay sites vary with species, from water to plants or logs

teneral – newly emerged and expanded adult; soft and "wet" looking

emerging nymph – crawls out of water after several months to several years; "breaks out" to leave "shell" or "skin" (exuvia)

nymph – what hatches from an egg; purely aquatic (in North America)

Figure 1.4 Basic life cycle of dragonflies and damselflies.

Although not part of the terminal appendages, a ventral opening on S9 from which sperm is obtained complete the male's terminal set of genitalia, or the primary genitalia. Secondary genitalia, located ventrally on S2–3, consist of multiple structures, including a penis and a pair of hamules, all of which can be nestled within, or with some families such as the Gomphidae (clubtails), very much extruded from the genital fossa (pocket). For some families, the primary or secondary genitalia or both are helpful with distinguishing species.

As one might expect with this anatomy, mating in odonates is rather complex. Their manner of mating is also unique in the insect world. Mating begins in one of two ways: 1), sperm is transferred from the genital opening on S9 to the secondary genitalia and then the male grabs ahold of a female, or 2), a male grabs ahold of a female and then transfers sperm from S9 to the secondary genitalia. Most dragonflies will follow the former progression but some will do the latter. Damselflies typically follow the latter. Once sperm is transferred and a female is grasped, the male will maneuver her body so that her genitalia meet with his secondary genitalia (Figure 1.4). Doing so results in the characteristic mating wheel, or heart, of odonate mating.

Female genitalia are not quite as complex as males' (Figure 1.4). For starters, they are located in one area, S8–10. There are many structures that comprise the female genitalia but only a few that we mention in the text, so we will be brief. Females have cerci, but they are not typically as prominent as on males. Females can also have an epiproct or paraprocts, but they are greatly reduced or sometimes absent. The genital opening and an ovipositor *or* a subgenital plate (or the vulvar laminae) are found ventrally on S8–9. All damselflies have ovipositors, but dragonflies differ by family as to whether an ovipositor or a subgenital plate is present. The Gomphidae (clubtails), Macromiidae (river cruisers), Corduliidae (emeralds), and Libellulidae (skimmers) have subgenital plates, and although in some cases they are elongated, they are not as prominent as the ovipositors of the Petaluridae (petaltails), Aeshnidae (darners), and Cordulegastridae (spiketails). The presence of a subgenital plate versus an ovipositor indicates how ovipositing (ovipositioning or egg laying; Figure 1.4) occurs. Species with a subgenital plate deposit eggs in the water by dipping the ends of their abdomen into the water and then releasing eggs (called exophytic

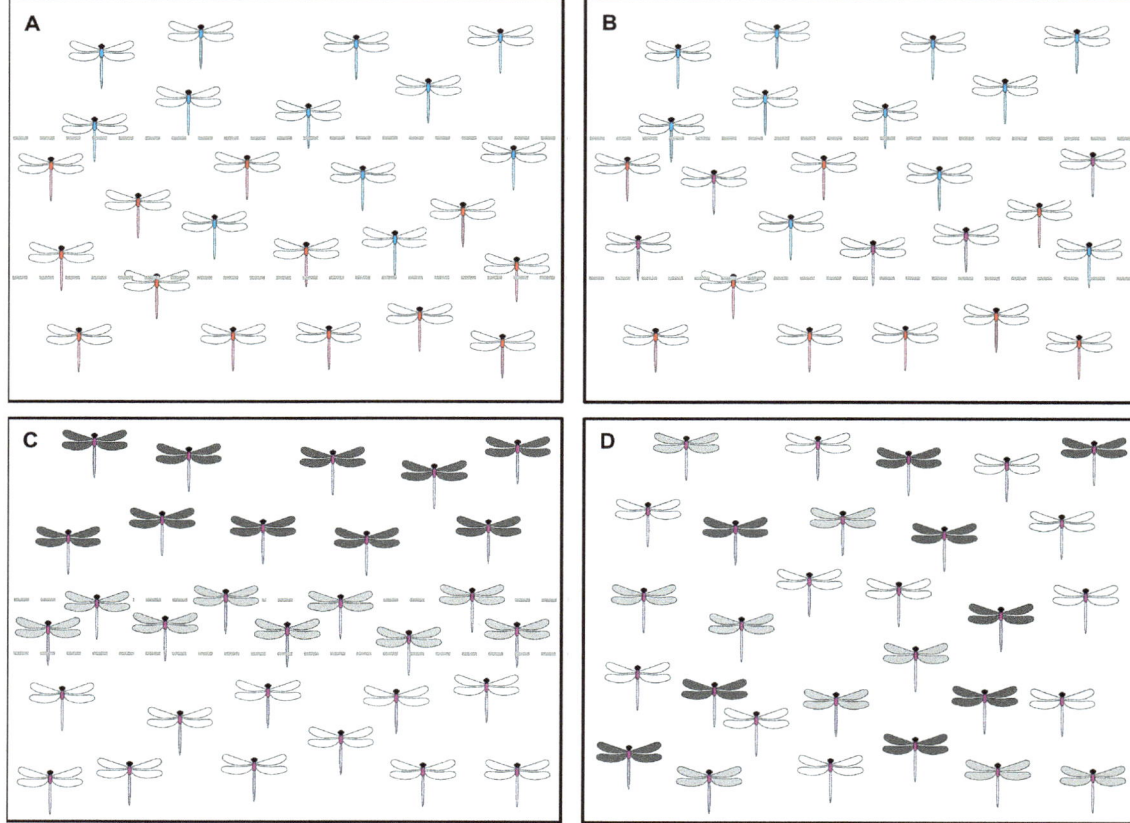

Figure 1.5 What is a species? Taxonomy is the science of classification, of how to place objects—in this case organisms—into tidy boxes. Taxonomy is predicated on the existence of such tidy boxes; for our purposes, we will assume boxes exist. Often, the burning question is "Does this set of organisms constitute a species?" This question is one of "species limits," where a subspecies—heritable geographic variation in phenotype—stops and a species starts. A long-established (since the 1940s) means to assess species limits is to determine if distinct, typically in terms of phenotype, organisms interbreed where their geographic ranges meet. A) If surveys in a contact zone (between the dashed lines) of a southern dragonfly (red) and a northern dragonfly (blue) revealed only red or blue individuals, we have no external evidence of interbreeding, and we would classify two biological species, a red one and a blue one. B) If instead a contact zone hosted varying levels of phenotypically intermediate individuals (purple ones) we might classify either two species that hybridize or a single species with diagnosable subspecies. The ultimate decision will rest not just with available data, but with an individual scientist's philosophical leanings and interpretation of prevailing species concepts. There is no simple answer, and this fuzziness spawns almost all taxonomic debate. C) Subspecies themselves always differ phenotypically and always occupy separate geographic ranges but interbreed where those ranges meet. In this example, northern individuals have their wings dark gray but are otherwise identical to southern individuals, which have their wings hyaline (clear). In a narrow contact zone—such zones are virtually always much narrower for subspecies than for species—individuals have their wings pale gray, suggesting free interbreeding and genetic exchange (i.e., gene flow). D) Another challenge taxonomists face is the existence of phenotypically distinct individuals that apparently have no distinct geographic range. Without a distinct range, such individuals cannot, by definition, constitute subspecies. They may perhaps be species if, say, individuals with dark gray wings interbred exclusively with individuals with dark gray wings, but absent any such assortative mating, distinct phenotype would be termed "morphs" (the proper technical term) or "forms."

ovipositing). The shape of the subgenital plate is often useful in species identification. Species with ovipositors will deposit their eggs within plant materials (called endophytic ovipositing) or in mud.

Ovipositing starts the life cycle of odonates—a cycle that is broken into a series of life stages beginning with eggs, then going on to prolarvae, larvae, tenerals, immatures, and finally, adults (Figure 1.4). After eggs are deposited, they can take from a few days to months to hatch, the longer time frame occurring within a single reproductive season of a species or by overwintering, as when the eggs go dormant or into a suspended state known as diapause during the winter months. When eggs hatch, they do so as prolarvae (singular prolarva), a stage that often lasts only a few minutes and the end of which results in a prolarva molting (also moulting, in British English), or shedding its skin, which is also called the exoskeleton, cuticle, integument, or epidermis. After the first molt, the prolarva enters the larval stage that is more recognizable for odonates (Figure 1.4). The larvae (singular, larva), or preferably nymphs,[10] live for as little as a month or as long as 5–6 years,[11] molting again and again, on average around 12 times, as they grow. Each stage of growth is called a stadium (plural stadia). Sometimes the term instar is used when stadium is meant, but it should be noted that, technically, usage of instar is not correct when referring to molting into the next developmental stage.

When a nymph nears completion of its metamorphosis into its next stage of life, it begins its emergence. At that time, the individual will emerge from the water and crawl up onto vegetation, a rock, sand, or some other stable surface, and then it begins its final molt (known as F-0). The back of the skin will split open and the emerging individual will push itself out and begin to fill itself with fluid until it forms into the shape we associate with an adult dragonfly or damselfly. As it is doing so, the odonate will often cling to its shed skin, or exuvia (*not* exuvium; plural, exuviae). At this stage, the newly emerged individual is called a teneral and will remain so for typically no more than 24 hours (Corbet 1999). In the teneral stage the body is soft and lacks the telltale coloration of the species. After the body hardens and the colors begin to appear, the individual is capable of flight, but will likely be rather awkward in doing so. After the teneral stage, the individual will become an "immature" that will be pre-reproductive, meaning that it will not mate. It will stay in the pre-productive stage for a couple of days up to a few months, depending upon the species (Corbet 1999). Mating occurs after the individual becomes an adult. As adults age, some odonates develop pruinescence (or pruinosity, or become pruinose or pruinescent), which causes the body, especially the abdomen, to appear frosted, typically with blue, gray, or white (e.g., as with the Powdered Dancer, *Argia moesta*). Adults live for a few weeks to a few months, depending on the species. The time between emergence and death, collectively for individuals of a given species, is known as the flight season. The flight season, or flight dates, is the species' seasonality or phenology.

It is important to keep in mind that where one finds an adult may not necessarily be where one finds nymphs (Patten et al. 2015, 2019). There are many reasons for this; for example, female odonates can often be found away from water when they are pre-productive or unreceptive to mating at a given moment. Mating, although often occurring near water, does not always correspond to where nymphs of that species will be found. A mated female can fly elsewhere to deposit eggs or eggs can wash downstream, sometimes as far as a pond or lake, and will hatch in an entirely different environment from where they were deposited. Alternatively, eggs can wash out of a pond or lake into a stream. Adults may also be found on their "feeding grounds," or places at which there are other insects that they prey upon. Sometimes while feeding odonates "swarm," or fly together in large numbers and move along with their prey items in *feeding swarms*. Feeding may occur well away from water, meaning that there may be much distance between where adults are and where nymphs are. Likewise, roosting or resting, whether for a period during the day or throughout the night, likely will not be right at the water. Habitat differences between behaviors and among life stages have yet to be teased apart for all species of odonates.

Before leaving this section, let's refer back to anatomy for a moment to mention a few directional descriptors that will be of use with odonates. We have already mentioned the terms *anterior, posterior, dorsal,* and *ventral,* or the front, back, top, and underside of the body. In addition, the term *lateral* refers to the side of the body. These terms can be combined, for example, *dorsolateral* or *ventrolateral,* referring to the side of the body nearer to the top or the underside, respectively. Other terms are used relative to another body part or portion of the body. For instance, *basal* and *proximal* refer to being at or near the base or place of attachment (e.g., basal part of the wing is where it attaches to the thorax). *Distal* refers to being away from the point of attachment, for example, wingtips are the distal portion of the wing. *Apical* refers to the *apex* of the body part being referenced which, in the case of odonates, is most often the abdomen especially when describing color patterns on a particular segment. These two terms are often mistakenly conflated, for example, when one says there is a "pale apical ring on S7," when really what is meant is that there is a pale ring on the distal portion of the segment (i.e., the portion of the segment farthest from where the abdomen attaches to the thorax), pale that "begins" at the apex and extends subapically and laterally down around the segment. Nonetheless, "apical rings" and the like are in common usage. To help keep *distal* and *apical* distinct in your mind, picture the shape of abdominal segments, which are not round tubes as often thought but instead are generally peaked on top and toward the rear. As such, the back part of the segment is the distal end, which is peaked at the top back by an apex or apical area.

Miscellanea

There are a handful of other terms that are useful to know when dealing with organisms. One that is not terribly intuitive to anyone without a background in natural history is the phrase *in the field*. Biologists will say that they are *in the field* when they go outside to conduct their research, or *fieldwork*. Because all organisms have their own seasonality, or the time they are most active (or in the case of plants, blooming or fruiting), biologists *go to the field* during that time. For adult odonates in Oklahoma, that timing is generally from mid-March to mid-November, so we call that our *field season*.

Throughout the book we make reference to the *distribution* of species. Distribution is the *geographical range* of a species, often referred to simply as the species' *range*, such as in a *range map* or textual description of where a species, as a whole, occurs, but not to be confused with use of that word in relation to the extent to which an individual can move (such as with an individual's *home range*, or movement within a given area). We discuss the latter usage rarely in the book, and always make it clear that we are discussing that concept; everywhere else, we discuss geographical range.

We also talk much about the *conservation status* (or just *status*) of species, which is essentially an assessment of how rare a species is and how likely it is to persist given known or possible threats to it directly or to the habitat with which it is associated. We discuss conservation status assessment, or *conservation ranking*, in greater detail in Chapter 7, which is dedicated to the topic of conservation.

NOTES

1. It also ranks above all of the provinces and territories in Canada. See Smith-Patten (2019) for a comparison of species list totals for the United States and Canada.
2. As of September 2019, Georgia's species total is still in dispute—likely totaling either 175 or 176 species.
3. Nope, that's Kansas (see Fonstad, Pugatch, and Vogt's 2003 paper, titled "Kansas is flatter than a pancake," in the *Annals of Improbable Research*).
4. See Sharman Apt Russell's *Diary of a Citizen Scientist: Chasing Tiger Beetles and Other New Ways of Engaging the World* (2014, Oregon State University Press, Corvallis, Oregon) for an inspiring account of how a citizen scientist transformed into an expert. Russell's own inspiration came when Dick Vane-Wright of the London Museum of Natural History said, "There is so much we don't know! … You could spend a week studying some obscure insect and you would then know more than anyone else on the planet. Our ignorance is profound." (*Here's how you can help scientists study sex, whales, and distant galaxies: Ordinary citizens are transforming science*, by Indre Viskontas, in *Mother Jones* 17 December 2014 issue).
5. Traditionally the word Odonata is broken up into four syllables, O-do-na-ta. The word's pronunciation is disputed, however, with some insisting that the first *a* is long whereas others insist it is short. Also, some insist the second *o* is long whereas others insist it is short. Generally, one will hear Odonata pronounced as either Oh-doe-NATE-a or Oh-doe-NOT-a.
6. Epiophlebioptera is considered by some odonatologists to be a replacement name for Anisozygoptera. Also note that some treat Anisozygoptera (or Epiophlebioptera) as an infraorder of Anisoptera rather than as a separate suborder. We treat it as a suborder currently containing one genus with three species, as per Paulson (2019).
7. In such a case as *E. t. traviatum*, the *t.* technically refers ahead to the subspecific name. Because this can be confusing and is inconsistent, some people (BS-P included) prefer to abbreviate subspecific names in the reverse order, for example, as *E. traviatum t.* instead of *E. t. traviatum*, but that form goes against tradition. This trinomial is known as the nominate subspecies, meaning that it shares the same species and subspecies name, whereas the other subspecies of *E. traviatum westfalli* does not.
8. The genus *Agrion* once held much of all the known damselflies and some dragonflies. Eventually *Agrion* was disposed of because it was so all-encompassing that it was impossible to determine the defining characteristics of the genus.
9. For an excellent explanation of odonate vision, see Paulson (2019).
10. There is no consensus on whether to call forms at this stage in development larvae, nymphs, or naiads. Corbet (2002) urged the use of *larva* for odonates. He argued rather strongly that neither *naiad* nor *nymph* should be used; indicating that the terms had largely been abandoned and specifically mentioning that Westfall and May (1996) and Needham et al. (2000) had adopted the term *larva*. That term is still widely used (e.g., Paulson 2009, 2011; Needham et al. 2014). However, Bybee et al. (2015) argued that *naiad* is the technically correct term for odonates based on their form of development. Others argue that *nymph* should always be used (e.g., Tennessen 2019).
11. Five to six years refers to northern North American species. At least one species, *Epiophlebia superstes*, of the Asian suborder of Anisozygoptera, is known to live for eight years as nymphs (Paulson 2019).

## CHAPTER 2

HISTORY OF OKLAHOMA ODONATOLOGY

Brenda D. Smith

Oklahoma has a relatively long history of odonatological research, dating back to at least 1907 but possibly to as early as 1877 (Table 2.1) when Oklahoma was still a territory. The earliest odonate record attributed to Oklahoma is of a *Sympetrum internum* (Cherry-faced Meadowhawk), collected by Charles Valentine Riley (USNM 487037; Figure 2.1[1]) and thought to have been taken in Indian Territory along the Red River. It is certainly an odd record, given how far south it is of this species' known range. It may be that the specimen actually was collected on the Red River of the North, often called just the Red River, which forms the border between North Dakota and Minnesota in the United States and flows into Manitoba, Canada. This Red River undoubtedly is part of the species' normal range. But until otherwise proven false, this dubious record remains part of Oklahoma's odonate history.

Oklahoma (and Indian) Territory's odonate history continued in uncertain terms due to some unfortunate events. Formal natural history investigations began in the state in 1900 under the direction of the newly created Territorial Geological and Natural History Survey (Carpenter 1990, 2000), in part, the precursor of today's Oklahoma Biological Survey.[2] The survey was a unit of the also newly created "Department of Geology and Natural History and Museum," all formed in 1899 and put under the direction of the University of Oklahoma (OU), with Dr. Albert Heald Van Vleet, Professor of Biology, named as the "Territorial Geologist and Curator of the Museum" (de Borhegyi 1957; Hollon 1961). The survey/department/museum undertook its first collecting expedition in 1900 (Gould 1932) with four members: Van Vleet, a botanist/biologist; Paul J White, a botanist; Charles N Gould, a geologist; and Roy Hadsell, the team's cook. We do not know if odonates, or any insects for that matter, were collected during this or subsequent early expeditions, although it does not seem likely (Hollon 1961). In any case, the museum that housed these early collections was destroyed by fire in 1903 and again in 1918 (Gould 1932; Carpenter 2000).[3]

We do know of a collector of odonates in this early era and, fortunately, for posterity's sake, his specimens were deposited elsewhere; as such, some early odonate vouchers still exist. In 1907, the summer before Oklahoma became a state, Edward Bruce Williamson (Figure 2.2), a banker by trade, but esteemed iris breeder and odonatologist on the side, came to what was then Indian Territory. He made his way to the territory from Texas on 2 June 1907 (Williamson

1914b) and made his first collection on the 3 June 1907. Why he chose to stop in Wister (in present-day Le Flore County) is unknown; perhaps, it simply was because he was heading back to Indiana where he lived and he was in need of respite. On 3 and 4 June he worked at an "artificial lake along [the] railroad about 1½ miles north of Wister" and on the 5th "along [a] stream west of Wister" and then he prepared to go home (Williamson 1914b:412–413). Not much of an expedition in terms of time, but it did document 25 species for Oklahoma (Table 2.2). More importantly, Williamson's short expedition resulted in the description of three new species to science. One of those species, *Tetragoneuria williamsoni* Muttkowski, 1911, later became a junior synonym to *Epitheca costalis* (Sélys, 1871), the Slender Basketail. He had better luck with another corduliid. Williamson designated the male *Neurocordulia xanthosoma* (Orange Shadowdragon) that he collected on 4 June as the holotype for *Platycordulia xanthosoma* (Williamson 1908; holotype and co-type both at UMMZ). And finally, a male *Enallagma* that he collected on 3 June and identified as *E. pollutum* (now known as the Florida Bluet), was later re-identified as *E. vesperum* Calvert, 1919 (Vesper Bluet), and designated as a paratype for the species (Calvert 1919:384).

It was on this trip that Williamson met "a boy," presumably a teenager, named Frank Collins, whom Williamson subsequently hired to collect odonates (Figure 2.2). From early August to late September Collins collected an additional 14 species along the Poteau River near Wister and around Henryetta, in present-day Okmulgee County (Table 2.2). Collins turned out to be quite the collector, taking, for example, an amazing 10 Dragonhunters between 3–5 August. Additionally, he collected a handful of other important specimens, including the co-type of *Neurocordulia xanthosoma*. Collins also collected a specimen that was initially described as *Macromia australensis* Williamson, 1909 (holotype and allotype at UMMZ), a designation that was later subsumed under *M. illinoiensis georgina*. Another re-designated species that Collins collected was *Argia intruda* Williamson, 1912, which is now a synonym of *A. moesta* (Powdered Dancer). Williamson did not designate a holotype for *A. intruda*, so RW Garrison designated a UMMZ specimen as a lectotype (i.e., replacement holotype; Garrison 1994). A fun aside to Collins' collecting is his use of "gasoline oil" to preserve specimens, a preservation technique of which I was previously unaware. In letters to Williamson he said that he was unable to procure the carbon bisulphide that was in use at the time as an insecticide (Figure 2.2).

Table 2.1 **Notable early and middle era collectors in Oklahoma odonatology. "Years active" includes only years that the person was known to have actively collected odonate records, not other taxa. Also indicated are the number of state records and specimens taken by that collector.**

	OBSERVER/COLLECTOR	YEARS ACTIVE	PRIMARY AREAS COVERED	STATE RECORDS	ODONATE SPECIMENS
Early	Charles Valentine Riley	1877	"Red River"	1	1
	Edward Bruce Williamson and Frank Collins	1907	Le Flore and Okmulgee Co.	39	583
	Theodore Huntington Hubbell	1918, 1926	SW, NW, panhandle	9	124
	Edward Bruce Williamson	1929	Tulsa, Lincoln, and Caddo Co.	4	36
OU Expeditions	Arthur Irving Ortenburger	1925–1928	statewide	10	45
	Zelmer Logsdon	1928, 1930	C, NC, SW	3	175
	Ralph Durham Bird	1929–1933	statewide	37	>5600
	Eugene Butler Webster	1930–1932	C	–	26
	Wilton Monroe Fisher	1930–1933	C, SE, SW	½*	369
	A Earl Pritchard	1930–1936	statewide	12	>488
	Lois Hazel Bird	1931, 1933	statewide	–	147
	O'Reilly Sandoz	1931–1933, 1938	C, SW, NC	–	316
	John Stankavich	1934	SE	1	1
	other OU expeditions contributors	1925–1934	statewide	2	c. 700
	RW Kaiser, WT Nailon, and Standish	1931–1941	statewide	1	95
	Albert Patrick Blair and William Frank Blair	1936, 1938	E, SW	2	122
Merry Trio	George Herman Bick	1950–1970/1991	statewide	14	c. 7200
	Juanda Claire Bick	1954–1970			
	Lothar Edward Hornuff, Jr.	1954–1978/1985			
	Thomas "Nick" W Donnelly	1954	EC, SC	–	39
	Minter J Westfall	1956, 1958	SC, SW	–	74
	Robert William Cruden	1961	SC, SW	–	89
	Donald C Arnold	1962–2007	statewide	–	38
	William D Shepard	1975–1976, 1987	statewide	–	403
	Aaron T Dossey	1977–2000	C	–	34
	William J Matthews	1981–1982	statewide	–	365
	Paul Liechti and Don Huggins	1981–1983	NW, NE	–	33
	John Nelson and Hal C Reed	1982–2008	statewide	–	53
	Karl Heinz Stephan	1982–1988	Latimer Co.	–	53
	David Bass	1986–2002	statewide	–	114
	Kenneth Stewart, John C Abbott, et al.	1991–1997	SC	–	253
	Sidney W Dunkle	1992–1993, 1996–1999	statewide	–	195
	Roy J Beckemeyer	1998–2006	statewide	–	48

C = central; Co. = County; EC = east-central; NC = north-central; NE = northeast; NW = northwest; SC = south-central; SE = southeast; SW = southwest;
* we attributed ½ to WM Fisher to account for his role with the state record of *Somatochlora ozarkensis*, the Ozark Emerald

Figure 2.1 The first odonate specimen purportedly collected in what now is Oklahoma. This meadowhawk, collected in August 1877 by Charles V Riley, was identified as *Sympetrum internum*, the Cherry-faced Meadowhawk, by TW Donnelly and confirmed by OS Flint (USNM 487037), making this the southernmost record for this species and well south of its "normal" range. Most individuals who have examined the original locality label feel it says "Red River Ind.," as in the Red River in Indian Territory, which later became Oklahoma. Nonetheless, the possibility still stands that it reads "Nd.", for North Dakota, indicating the Red River of the North.

Very little collecting of odonates is evident between 1907 and 1924. After the museum fire of 1918, Henry Higgins Lane, the head of the Department of Zoology and Embryology,[4] began amassing an entomology collection (Hollon 1961), though it is not known what it comprised. There were eight Odonata specimens collected after Williamson's time in the state and before 1925.[5] By the end of this era, there were 40 species on the Oklahoma state list.

THE FIRST ERA OF OU COLLECTING EXPEDITIONS (1925–1928)

OU's Department of Zoology began to conduct regular summer collecting expeditions in 1923 (Richards 1929; Figure 2.3), but hardcore collecting of Odonata in Oklahoma did not begin until Arthur Irving Ortenburger organized the first large-scale expedition in 1925. Ortenburger arrived in the state in 1924 when he was named faculty of OU's Department of Zoology. Prior to coming to OU, he acted as the Assistant Curator of Herpetology for the American Museum of Natural History from 1922 to 1924, so it is no surprise that the first summer of his arrival he made some minor collections of herpetiles in the Arbuckle Mountains (Carpenter 1990). Ortenburger would later come to be considered the father of Oklahoma herpetology, but first he made a worthy contribution to Oklahoma odonatology.

There are probably few sections of the United States where as little collecting has been done as in Oklahoma; and, with the possible exception of the panhandle region, the least collecting has been done in the southeastern counties of the State. For this reason and because this section is more or less burned over nearly every year, it was decided to work there first.

(Ortenburger 1926a:86)

Such was the reasoning that Ortenburger gave in 1925 to lead a collecting expedition to southeastern Oklahoma. This is the first OU expedition on which odonates are known to have been collected. Although the expedition was focused on the collection of vertebrates, they took more than 2000 insects "by only incidental collecting" (Ortenburger 1926a). The 1925 expedition lasted for seven weeks and went to Choctaw, Le Flore, McCurtain, and Pushmataha Counties (Ortenburger 1926b; Carpenter 1990); subsequent expeditions under his lead, those between 1926 and 1928,[6] covered another seven counties across the state in which odonates were collected (Ortenburger 1926c; Figure 2.3).

Ortenburger touched upon his philosophy of field expedition by saying "we do not go out for a good time or merely on a camping trip, but are out to do a definite amount of work within a certain times. [sic]" and "it is much more important to protect collections than collectors" (1926c:7). Although many students accompanied Ortenburger over

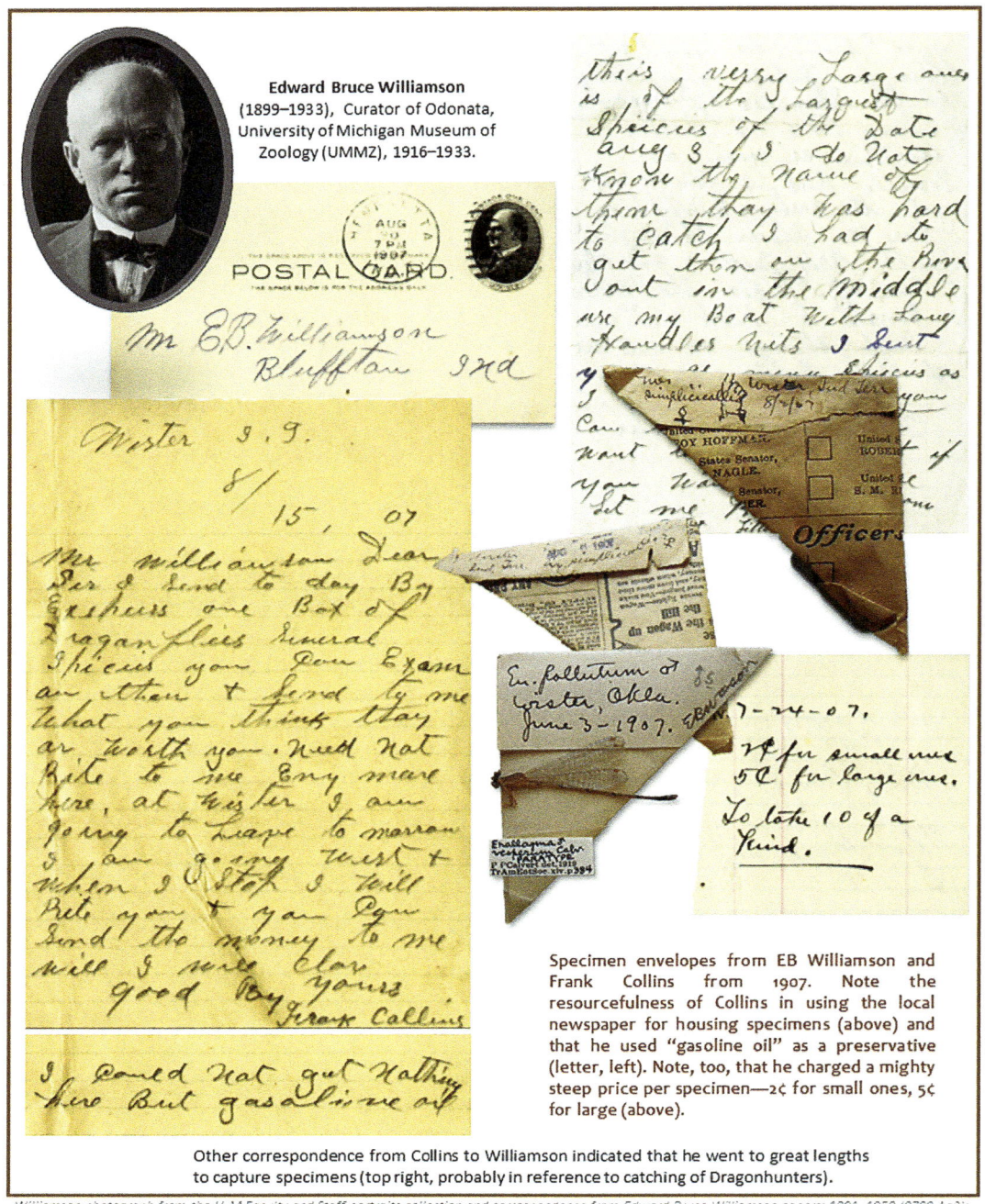

Edward Bruce Williamson (1899–1933), Curator of Odonata, University of Michigan Museum of Zoology (UMMZ), 1916–1933.

Specimen envelopes from EB Williamson and Frank Collins from 1907. Note the resourcefulness of Collins in using the local newspaper for housing specimens (above) and that he used "gasoline oil" as a preservative (letter, left). Note, too, that he charged a mighty steep price per specimen—2¢ for small ones, 5¢ for large (above).

Other correspondence from Collins to Williamson indicated that he went to great lengths to capture specimens (top right, probably in reference to catching of Dragonhunters).

Williamson photograph from the U-M Faculty and Staff portraits collection and correspondence from Edward Bruce Williamson papers: 1891–1950 (0799 Aa2); all courtesy of the Bentley Historical Library, University of Michigan.

Figure 2.2 Edward Bruce Williamson and Frank Collins were the first to collect odonates to any great extent in Oklahoma.

the years of his expeditions, few managed to avoid the near obligatory obscurity often faced by students of such men. Generally, collections have been labeled solely with Ortenburger's name, but one collector's name that survived was Anna Elizabeth Rennie (she collected dragonflies in 1926; Figure 2.3). Admittedly, her documented collection of odonates is limited but it is absolutely remarkable that she was a herpetology graduate student and part of a field team *in the 1920s.* Zelmer Logsdon (1928), another graduate student of whom I will speak shortly, was another team member who

managed to escape obscurity, in his case partly by receiving credit for three odonate state records (Appendix D, E).

TH Hubbell, who collected in the state in 1918, returned in 1926 to conduct further research with his specialty taxon Orthoptera (grasshoppers and allies). He joined the Ortenburger team that year and is the first person known to have collected odonates in the Oklahoma panhandle. His role was "to take care of all the insect collecting and have general charge of the invertebrate work" (Ortenburger 1926c:6). He assisted the expedition most admirably by

Table 2.2 **The first major collection of Odonata specimens in Oklahoma was made in 1907. On 3–5 June, EB Williamson and F Collins collected near Wister, in what was then Indian Territory, but is now Le Flore County. In August and September, Collins continued to collect for Williamson, taking an additional 14 species (*), near Wister and in Okmulgee County near Henryetta. All records are cited by Williamson (1914b); additional citations are indicated in notes.**

SPECIES	TOTAL & SEX	INSTITUTION[a]	NOTES
Zygoptera			
*Hetaerina americana**	19♂, 11♀		
*Hetaerina titia**	18♂		also Williamson (1912a)
Lestes inaequalis	2♂, 1♀	UMMZ	Williamson (1914b) had error: 1♂, 2♀
Argia apicalis	8♂, 1♀	UMMZ; 1♂	
Argia fumipennis	3♂		
Argia moesta	22♂, 17♀	MCZ, OMNH, UMMZ[d]; 13♂, 5♀	most as *A. intruda*, including UMMZ lectotype; also Williamson (1912b)
Argia tibialis	6♂, 3♀ (3 pr)	UMMZ	
Enallagma civile	7♂, 1♀ (1 pr)	UMMZ; pr	
Enallagma divagans	7♂, 1♀	UMMZ; 4♂	
Enallagma exsulans	3♂, 2♀	UMMZ; 2♂	
Enallagma geminatum	3♂, 2♀ (1 pr)	UMMZ; 2♂,1♀ (1 pr)	
Enallagma signatum	5♂, 2♀	UMMZ; 2♂	
Enallagma traviatum	40♂, 4♀	FSCA, UMMZ; 6♂	
Enallagma vesperum	1♂	UMMZ	original id *E. pollutum*; Paratype; also Calvert (1919)
Ischnura hastata	3♂, 13♀ (1 pr)	UMMZ; 3♂, 3♀;	
Ischnura kellicotti	14♂, 8♀		
Ischnura posita	12♂, 2♀	UMMZ; 6♂;	
Anisoptera			
*Anax junius**	2♂, 8♀		
*Epiaeschna heros**	1♀		
Arigomphus submedianus	3♂, 2♀	MCZ; 1♂;	also Williamson (1914a)
Dromogomphus spinosus	9♂, 1♀	UMMZ; 1♂,1♀;	
*Hagenius brevistylus**	10♂	UMMZ	Williamson (1914b) said Collins collected 10♂, but UMMZ has 1♂ and 1♀
Phanogomphus graslinellus	1♀		
*Stylurus plagiatus**	2♂, 2♀		
*Macromia illinoiensis**	7♂, 2♀	UMMZ; 1♂,1♀;	HOLOTYPE and ALLOTYPE as *M. australensis* (= *M. illinoiensis georgina*); also Williamson (1909)
Epitheca costalis	2♂, 1♀	UMMZ	HOLOTYPE, ALLOTYPE, and Paratype as *Tetragoneuria williamsoni*; paratype location is not known; also Muttkowski (1911)
Neurocordulia xanthosoma	2♂	UMMZ	HOLOTYPE and COTYPE as *Platycordulia xanthosoma*; also Williamson (1908), Byers (1937)
Erythemis simplicicollis	2♂, 23♀	UMMZ; 1♂,4♀;	
Libellula cyanea	46♂, 4♀	UMMZ; 13♂,3♀;	
*Libellula incesta**	46♂, 69♀		
*Libellula luctuosa**	2♂, 1♀	UMMZ; 1♂,1♀;	
Libellula pulchella	5♂, 2♀	UMMZ; 3♂,2♀;	
*Libellula vibrans**	17♂, 8♀		

(Continued)

Table 2.2 (Continued) **The first major collection of Odonata specimens in Oklahoma was made in 1907. On 3–5 June, EB Williamson and F Collins collected near Wister, in what was then Indian Territory, but is now Le Flore County. In August and September, Collins continued to collect for Williamson, taking an additional 14 species (*), near Wister and in Okmulgee County near Henryetta. All records are cited by Williamson (1914b); additional citations are indicated in notes.**

SPECIES	TOTAL & SEX	INSTITUTION[a]	NOTES
Pachydiplax longipennis	1♂, 17♀	UMMZ; 1♂,2♀;	
Pantala flavescens*	2♂	UMMZ	
Pantala hymenaea*	1♀	UMMZ	
Perithemis tenera	11♂, 3♀	UMMZ; 2♂;	
Plathemis lydia*	10♂, 13♀	UMMZ; 2♂,4♀;	
Sympetrum corruptum*	5♂	UMMZ; 2♂	

pr = pair; [a] Specimens listed in the previous column are all housed at the institution unless indicated otherwise by a note in this column. See Appendix B for acronyms.

collecting—*in the one field season*—almost half of the odonate specimens known to have been collected for the entire four-year period that Ortenburger led expeditions. And, he added seven species to the state list.

Regardless of the exact attribution of various specimens, there were 20 species added to the state list as a result of the early OU expeditions (Appendix D, E). With the efforts of people like Rennie, Logsdon, and Hubbell, the state list was brought up to around 60 species before Ortenburger relinquished the reins of Oklahoma odonatology to the second era of OU collectors.

THE SECOND ERA OF OU COLLECTING EXPEDITIONS (1929–1934)

The second era of OU expeditions were officially led by the Oklahoma Biological Survey. In 1929, Ralph Durham Bird[7] (Figure 2.4) came to Oklahoma as new OU faculty, taking a lead role in the expeditions.[8] Bird continued to lead the expeditions, or at least the entomological portion, until 1933, the year he left OU to return to Manitoba. Though he stayed but a short time, his expeditions were *extensive*—his teams covered 58 counties—and *productive*—collecting >7,900 specimens of 107 species. Because of his productivity and contributions, it is difficult not to think of Bird as the true father of odonatology in Oklahoma. He may not have been the first to work with odonates, but he left his mark on the study much more indelibly than any predecessor.

But, in all fairness, it was not just Bird who was responsible for the all the hard work. His wife, for one, Lois Hazel Bird *née* Gould,[9] (Figure 2.4) was on some of those expeditions and is deserving of recognition in her own right. Lois, a botanist by training, received her BS from OU in 1929, the year Ralph arrived in Oklahoma and the year she married him. While at OU she wrote a weekly article about botany for the Oklahoma City newspaper, *The Daily Oklahoman*,[10] she published two papers on the phenology of flowering plants (Bird 1972), and she pursued graduate work

(1930–1931). She has a fairly long list of insect specimens (approximately 350 records from 1929 until 1933) that were likely collected by her, not only odonates but many other insect groups too (including Blattodea, Coleoptera, Diptera, Hemiptera, Hymenoptera, Lepidoptera). I know of 147 odonate specimens specifically attributed to her for 1931 and 1933 (Table 2.1), but the amount of obscurity caused by her nominal association with Ralph is yet to be determined; I suspect that some of the specimens simply labeled as "Bird" that were attributed to Ralph could be hers. Lois' contribution appears not to have just been with collecting, as she annotated Ralph's field notes,[11] for example, to correct erroneous county names that he had written in his notebooks.

Of course, Lois was not the only person that Ralph overshadowed during these years. It has been commonplace to attribute many thousands of specimens of various groups, but especially odonates, to Ralph if those specimens had been collected during the OU expeditions that he led.[12] We will never be able to tease apart which specimens were actually collected by him and which simply added to his legend. Nonetheless the best that can be done is to mention those, like Lois, who helped Ralph through the years become the father of Oklahoma odonatology.

Besides Lois Bird, there were four major odonate collectors on the Bird expeditions. One was Zelmer Logsdon who, after having assisted Ortenburger in 1928, returned in 1930 to assist Bird.[13] Of the roughly 175 odonate specimens attributed to him, the vast majority are from July 1928 and March–May 1930 (Table 2.1). Two specimens, though, have the collection date as 6 July 1929, but they are likely from 1928 when he worked in the Wichita Mountains with Ortenburger in July of that year. Carpenter (1990, 2000) indicated that Logsdon was officially a student assistant to the OU expeditions only in 1932; but no specimens of any taxa are attributed to him in that year and he does not appear in OU's student directory.

Another student member of the Bird expeditions, O'Reilly Sandoz (Figure 2.4),[14] should, as with Zelmer Logsdon, go down in odonate history if for nothing else than his unusual

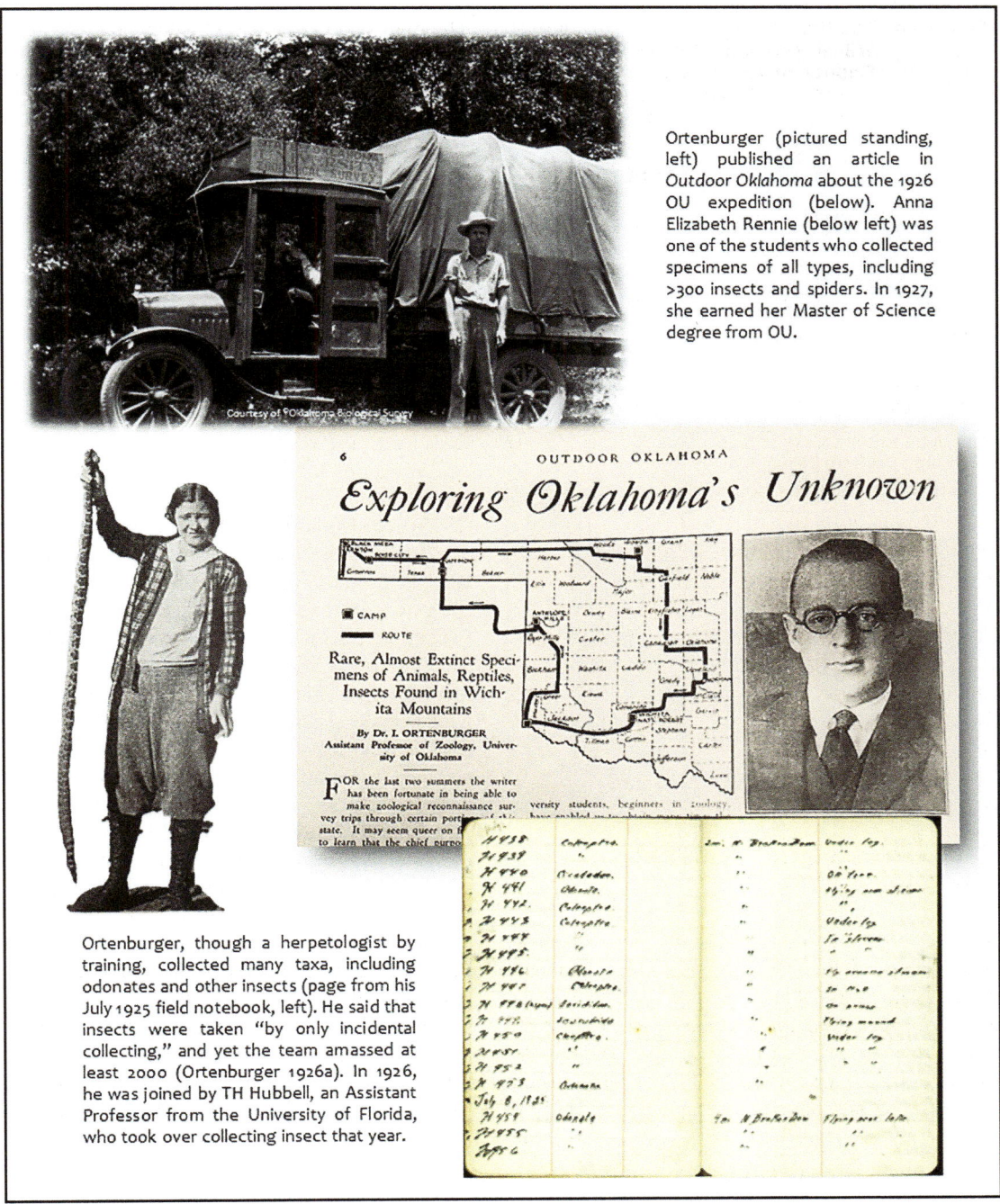

Ortenburger (pictured standing, left) published an article in *Outdoor Oklahoma* about the 1926 OU expedition (below). Anna Elizabeth Rennie (below left) was one of the students who collected specimens of all types, including >300 insects and spiders. In 1927, she earned her Master of Science degree from OU.

Ortenburger, though a herpetologist by training, collected many taxa, including odonates and other insects (page from his July 1925 field notebook, left). He said that insects were taken "by only incidental collecting," and yet the team amassed at least 2000 (Ortenburger 1926a). In 1926, he was joined by TH Hubbell, an Assistant Professor from the University of Florida, who took over collecting insect that year.

Figure 2.3 The first era (1925–1928) of University of Oklahoma expeditions was led by Arthur Irving Ortenburger.

name. But he was also a serious collector of Oklahoma odonates: 313 specimens dating 1931–1933,[15] and one from May 1938 (Table 2.1). Sandoz later became a senior biologist for the Oklahoma Department of Wildlife Conservation and served as the director of their fish hatcheries, as well as the director of the game division. Early in his more than three decades of service to the department, he contributed to the famed Duck and Fletcher map of Oklahoma (Figure 3.5). That map, known as "A Game Type Map of Oklahoma," is actually a vegetation map for the state although its purpose

was to determine the distribution and status of game birds and fur-bearing mammals across the state (Duck and Fletcher 1943; see Chapter 3 and Fuller 2011 for history and impact of the Duck and Fletcher expedition).

Wilton Monroe Fisher, an undergraduate and then graduate student at OU (BA 1932; MS 1937; Carpenter 2000), collected 369 odonate specimens starting in 1930 and continuing until 1933 (Table 2.1).[16] Fisher was responsible for collecting the holotype and allotype for the species that would be named *Somatochlora ozarkensis* Bird, 1933, the Ozark Emerald. Fisher

Ralph Durham Bird (1901–1972)
"Father of Oklahoma Odonatology"

Lois Hazel Bird née Gould
(1905–1959)
OU junior photo, 1928
botanist and insect collector

O'Reilly Sandoz
(1908–1972)
OU student
collected odonates
mostly in 1931–1933

Figure 2.4 Ralph Durham Bird was instrumental to odonatological research in Oklahoma during the second era (1929–1934) of the University of Oklahoma expeditions. His team included many students, such as O'Reilly Sandoz, as well as RD Bird's wife, Lois Bird *née* Gould. Photos used with permission: Ralph D Bird (Charles D Bird), Lois H Bird (University of Oklahoma Western History Collections), O Sandoz (Oklahoma Department of Wildlife Conservation).

eventually left the odonate world to focus on ants (Fisher 1937). As good as Fisher was with odonates and other insects, the captain of Bird's team was AE Pritchard.

Arthur Earl Pritchard (known as "A Earl"; Figure 2.5) started to collect odonates when he was merely a teenager. The first specimens that I know of Pritchard collecting were 2♂ *Sympetrum corruptum* (Variegated Meadowhawks) taken in March 1930 in Norman, Cleveland County (UMMZ). The following year, he collected 2♂ *Dythemis fugax* (Checkered Setwings) in Medicine Park, Comanche County, on 28 June

1931 (OMNH). He also collected a ♂ clubtail in Wilburton, Latimer County, on 25 April 1925 that he would describe later as a new species *Gomphus oklahomensis* Pritchard, 1935,[17] the Oklahoma Clubtail (now *Phanogomphus oklahomensis*; Figure 2.5). He was officially part of the Bird expedition in 1932 (Carpenter 2000), although he collected odonates in 1933, 1934, and 1936 as well. During these years he encountered at least 75 species, 12 of which were new for Oklahoma, and collected many hundreds of odonates in the state—we found 488 specimens, but we know from his publications that there are, or at least were, additional specimens (Table 2.1; Appendix D, E). Given how widely separated his specimens are, being at a minimum of nine institutions across the United States (CUIC, EMEC, FSCA, MCZ, OMNH, OSU, RWG, UMMZ, and USNM), it is not surprising that they have not all been located. Fortunately, the holotypes and allotypes for *Phanogomphus oklahomensis* and *Celithemis verna* (Double-ringed Pennant) as described by Pritchard (1935) are safely housed at UMMZ.

Although obviously adept at Odonata research, Pritchard later went on to work with other arthropod groups, including the one that made him most well-known, the Spider Mites (Tetranychidae). He attended Oklahoma State University for the first three years of his undergraduate education (during that time he corresponded with EB Williamson, who was apparently so taken by Pritchard's deportment that he mistook him for faculty). Pritchard then transferred to the University of Minnesota where he received his BS degree in 1936, and then went on to an MS in entomology in 1940, and a PhD in 1942 (Denning and Allen 1965). He later became faculty at the University of California, Berkeley, where he stayed until 1961. He died in 1965 as an "internationally known entomologist" (Denning and Allen 1965:807). Though by that time he was largely removed from his Oklahoma past, it should always be remembered that he got his start in entomology in Oklahoma, including his very first scientific paper (Pritchard 1935).

The Bird expeditions collected across the state, but considerable focus was spent on ten counties. In the central part of the state they spent much time collecting in Murray County, especially at Sulphur, Price Falls, and along Rock, Sand, Honey, and Falls Creeks; in Johnston County at "Oil Springs"; in McClain County at Dean's Slough (or Dean's Lake); and, of course, in Cleveland County, in and around the OU campus and Norman, with "Indian Springs" being the favorite locale.[18] They collected 800 specimens in Murray County, 252 in Johnston County, 896 in McClain County, and 1,343 in Cleveland. The counts were likely higher, given that George H Bick mentioned in his notes that he examined some that we cannot re-find. Perhaps the best, and most painful, example is the 2♂ *Amphiagrion abbreviatum* (Western Red Damsel) collected by Bird in April 1932 in Cleveland County. Bick indicated in his notes that he borrowed the specimens from OMNH, but he did

Pritchard, as shown in the 1935
Oklahoma State University yearbook.

Figure 2.5 Arthur Earl Pritchard (1915–1965) was but 17 years old when he started to work with RD Bird and correspond with the venerable EB Williamson. Days after his 18th birthday, his first scientific paper was published (Pritchard 1935, above). That paper described two new species to science, both of which were taken in Oklahoma. Photos used with permission: Oklahoma State University Archives (AE Pritchard), Greg W. Lasley (*P. oklahomensis*). Pritchard (1935) reprinted with permission from the University of Michigan Museum of Zoology.

NUMBER 319   JULY 26, 1935

OCCASIONAL PAPERS OF THE MUSEUM OF ZOOLOGY

UNIVERSITY OF MICHIGAN

ANN ARBOR, MICHIGAN   UNIVERSITY OF MICHIGAN PRESS

TWO NEW DRAGONFLIES FROM OKLAHOMA

By A. EARL PRITCHARD

DESCRIPTIONS of two new species of Odonata from Oklahoma and Georgia are given in this paper. The types are deposited in the Museum of Zoology at the University of Michigan.

Gomphus oklahomensis, n. sp.

A slender greenish species, striped with brown, the abdomen dark brown and yellow with the clubbed segments narrow.

MALE.—Labium, pale greyish, tinged with brown in the middle; anteclypeus, whitish; labrum, postclypeus, and frons, greenish yellow, except narrow posterior border of horizontal surface of frons which is dark brown; vertex dark brown with

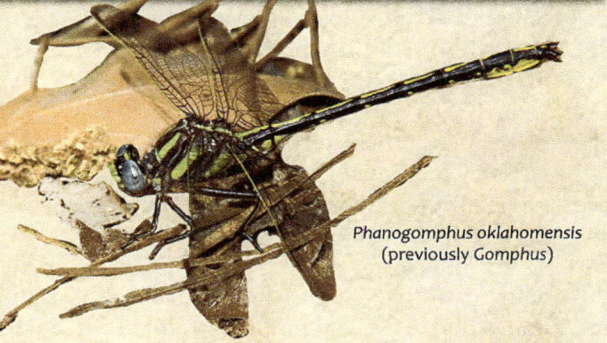

Phanogomphus oklahomensis
(previously *Gomphus*)

not indicate that he returned them, and we have not been able to relocate them. These are the only records of the species for the county and for central Oklahoma. The 3♂ and 2♀ Western Red Damsels for Comanche County are likewise an example of lost specimens that are the only documentation of the species in the county and that part of Oklahoma.

In the southwestern part of the state the team collected >1,800 specimens. They collected primarily in Comanche County, in and around what was then the Wichita National Forest and Game Preserve, present-day Wichita Mountains Wildlife Refuge (WR).[19] The team spent a minimum of 66 days in Comanche County, collecting roughly 1,150 specimens. The county with the next highest total of specimens was Cotton County, where 166 specimens were collected in only 4 days that the team spent working around Walters, Randlett, and along the Red River, the Deep Fork of the Red River, West Cache Creek, and Beaver Creek. Again, we have a disheartening example of a lost specimen taken

during this era. Bick indicated in his notes that he examined and returned to OMNH a male *Erythrodiplax minuscula* (Little Blue Dragonlet) that Bird collected on 26 April 1930 in Comanche County (Bird 1932), a specimen that has not yet turned up, and about which lies some healthy skepticism.

In the northwestern part of Oklahoma, Bird's expeditions focused on three counties—Woods, Major, and Alfalfa. The team collected in the former two counties only in 1930, but in the latter in 1930–1932. In Woods County (*n* = 260 specimens) they collected in either Waynoka, where they tended to be along Dog Creek, or at Edith. They took 202 specimens in Major County around the Glass (or Gloss) Mountains and Cleo Spring, including along Eagle Chief Creek. The main collecting localities in Alfalfa County all fall within what is now the Salt Plains National Wildlife Refuge or the adjoining Great Salt Plains State Park (SP). They took 242 specimens in this area, some of which were rather problematic for many years because they were cataloged as Cherokee County specimens because the cataloger did not realize they were taken near Cherokee, Alfalfa County.[20]

Another data error relating to this part of the state, and one that was mistakenly associated with the Bird expeditions, is credited to the cursed use of initials. Dear reader, please take the following as a cautionary tale. Over the years, the initials "EBW" brought me much aggravation and wasted time and energy. I once thought that various records with these initials referred to EB Williamson, but sometime down the road I realized there was another EBW; one Eugene Butler Webster. After much teasing apart of dates, all 1907 records were referred to Williamson because that was the only year he was known to have visited the state. All other records were assigned to Webster, since he was known to have collected insects in Oklahoma, primarily in the late 1920s and 1930s.[21] Thinking all had been resolved, I plowed ahead writing Oklahoma's odonatological history. Part of that history involved placing blame on Webster for misattributing a series of specimens (UMMZ) from October 1929 as being from Custer County when, in fact, they were from Caddo County (and when, in fact, the other EBW was responsible). Until late 2018, it was not known that Williamson had traveled again to Oklahoma. While reading correspondence between Bird, LK Gloyd, and Williamson, I discovered that Williamson had made a second short visit to the state, this time on 18–19 October 1929, when he collected 36 specimens in Tulsa ("just west of Tulsa"), Lincoln ("near Chandler"), and Caddo (*not* Custer) Counties. His localities of 7 mi E of Weatherford and 15 mi E of Weatherford should from hereon be attributed to Caddo County.[22] We all make mistakes; unfortunately for some of us, they go down in history, but it would be unfair to fault Williamson too much given that he was just passing through Oklahoma, the county line was not well-marked, and it was, after all, just a slight miscalculation of geography.[23] If only I had heeded the advice of the wise old sage EBW … 1) never assume you know who initials belong to, even if you thought you did your history homework, 2) always record primary data, and 3) *Never* use initials. With that wisdom imparted, let's return to the history of the Bird expeditions.

The Bird expeditions largely ignored the far southeastern part of the state, presumably because Ortenburger had spent so much time there in 1925. Bird et al. collected 172 specimens in Le Flore County in 1931 and 1934, and they spent ample time in Latimer County, too, in 1931, 1932, and 1934,[24] racking up 60 days in the field and taking 954 specimens. They spent virtually no time in northeastern Oklahoma.

Cimarron County in the Oklahoma panhandle gets an honorable mention given that the Bird expedition headed all the way out to Boise City and the Kenton area in June and July 1933. This must have been a rather unpleasant situation given that the region was already in an extended drought and had begun to experience the now notorious "black blizzards" of the Dust Bowl, massive dust storms that plagued the region for years on end. It is not surprising, then, that

the almost three weeks of time spent there (24 June to 17 July) produced a mere 73 specimens. It may be that when Bird was in the panhandle on 27 July 1930, although also not collecting very much material, he was left with an impression of a more verdant region than it was only three years later. And certainly, as the Dust Bowl steadily worsened, Bird must have reasoned that the area would yield little for the effort expended, so he did not return.

Collecting continued beyond the second OU era but it was not as extensive, at least relatively so. Between 1935 and 1948 there were about 325 specimens collected for 60 species (Table 2.1). All collections contribute to our knowledge of Oklahoma odonatology in at least a minor way, from the two *Calopteryx maculata* (Ebony Jewelwings) that Edith Bright Larkin, a botany MS student at OU, collected in 1936, all the way up to the impressive record of a ♂ *Gynacantha nervosa* (Twilight Darner) that TH Hubbell captured in the Ouachita NF, 1 mi NW of Page, Le Flore County, on 15 September 1935 (Kormondy 1960, OC 403468, UMMZ). That record still stands as the only one for that species in Oklahoma, and remains a major outlier from the species' normal range (Paulson 2011). Hubbell continued to collect in Oklahoma until at least 1937. Other collectors of note post-OU expeditions and pre-1950 include Albert Patrick Blair and William Frank Blair, brothers and University of Tulsa students who are known to have collected 122 specimens in 1936 and 1938; two of which, *Stylogomphus sigmastylus* (Interior Least Clubtail) and *Macromia taeniolata* (Royal River Cruiser), were first state records (Appendix E). RW Kaiser, WT Nailon, and Standish (initials unknown), whose affiliations are unclear, have 95 odonate specimens attributed to their names between 1931 and 1941. Those specimens, along with tiger beetles and other insects they collected are housed at OSU, so it may be that they were associated with that institution. One of the many specimens collected by Kaiser was the first state record of *Amphiagrion abbreviatum*, the Western Red Damsel (Appendix E).

The one and only record of *Neurocordulia virginiensis* (Cinnamon Shadowdragon) for Oklahoma was a specimen taken in 1934 by John Stankavich, an OSU student who was collecting with Pritchard on that day (Appendix E). Pritchard sent Stankavich's specimen to WT Davis, the species' describer, for confirmation. Davis compared the Oklahoma specimen to the holotype (♀[25]) and confirmed the identification (Davis 1937). Pritchard then alerted CF Byers of the find so he could include the record in his monograph covering the genus (Byers 1937). Although seemingly a small role in Oklahoma odonatology in terms of number of odonates collected, Stankavich sure nabbed a good one. The unfortunate part of the story is that it appears that the specimen has been lost.

Despite lost specimens and such, the species list for the state had grown by leaps and bounds. In correspondence with Leonora K "Dolly" Gloyd (9 May 1933), Bird indicated that with the addition of the specimen of *Paltothemis lineatipes* (Red Rock Skimmer) that she had recently determined (but which also appears lost), the state list was 121 species. In actuality, given the "clean-up" work after the fact that I, MAP, and others have done, including various taxonomic changes, finding specimens lost in collections, and re-identifications/confirmations, we believe the total was likely closer to 117 species. Nonetheless, combining all of the efforts up to 1948 brought the state list to a respectable 123 species (Appendix D).

THE MERRY TRIO ERA: THE BICKS AND LOTHAR HORNUFF (1950–1978)[26]

In the two decades between 1950 and 1970, George Herman Bick (1915–2005; Figure 2.6) spent all but five summers (1952–1953, 1965–1966, 1969) in Oklahoma. He was mainly in the state during June and July, and sometimes August. His first two summers, in 1950 and 1951,[27] were spent collecting "odonates within a twenty mile radius of the Wildlife Conservation Station of Oklahoma A & M College at Braggs, Muskogee County" (Bick 1951:178). That station is now defunct, but it holds the distinction of sparking Bick's interest in Oklahoma odonatology. Prior to Bick's first stay in Oklahoma, Muskogee County had not been collected in since the single *Ischnura verticalis* (Eastern Forktail) that RD Bird collected in the county in 1933 (OMNH 3437, 30 Apr). During his first field season, Bick also collected some in Wagoner County.

When he returned to the state in 1954, he began to work at the field station with which he is most closely associated, the University of Oklahoma Biological Station, located on Lake Texoma, Marshall County, in south-central Oklahoma, where he spent 14 summers. But George was not alone at that field station. There was an important collaborator there as well—Juanda Claire Bick *née* Bonck (1919–1999; Figure 2.6), George's wife and colleague for 55 years. It could be argued that it was Juanda who was most strongly associated with the OU's bio station, as it was she who stayed at the station most consistently so that she could study damselfly behavior. George did that, too, but he also traipsed around the state looking for distributional records, trying to get a handle on what species occurred in Oklahoma. Not that Juanda would not occasionally venture out; for instance, when she and her daughters recorded the first state record of *Anax longipes* (Comet Darner) at the Federal Fish Hatchery, Johnston County, on 29 June 1956 (Bick and Bick 1957). That was a species that George had to wait seven more years to find in the state.

During their time at the station the Bicks, according to George's notes, collected at or near the station >2,300 specimens, of which we know of only 708 that still exist (his notes indicate that he lost or discarded many). The Bicks documented 74 species at the station, including *Erythrodiplax minuscula*, the Little Blue Dragonlet (1♂, McMillan, 27 Jul 1968, FSCA), a rather rare species in the state. Near the station is also where the only *Lestes sigma*, the Chalky Spreadwing (1♂, Little City, Weno Pond, 22 Jul 1968), for the state was found by the Bicks, but unfortunately that specimen appears to be lost.

The Bicks also did a tremendous job documenting the odonates in the rest of Marshall County—documenting 78 species. As a testament to their achievement, in the almost five decades since they last worked in the county, only six additional species have been recorded. Of those six, five are spring fliers, i.e., species whose flight season is earlier than when the Bicks were in the state, and the other species is generally crepuscular and secretive.

The "Merry Trio"
Lothar E Hornuff, George H Bick, and Juanda C Bick

Lothar, George, and Juanda (left to right)

Juanda, Lothar, and George (clockwise from right)

Figure 2.6 The "Merry Trio," Lothar E Hornuff, George H Bick, and Juanda C Bick.

A second collaborator of George's that must be lauded is Lothar Edward Hornuff, Jr. (Figure 2.6). Their relationship began at Tulane University in the late 1940s when Lothar was an undergraduate student and George was an assistant professor. The pair worked on a project that produced Lothar's first publication, a study on the life history of *Aphylla williamsoni* (Two-striped Forceptail; Hornuff 1950). After serving in the Korean War (1951–1953), Lothar went on to graduate school in 1954. That summer, since he was headed to OU in Norman, and the Bicks were headed to the OU bio station in Willis, they all went together, with the Bick children in tow. Lothar worked as George's teaching assistant at the station for two summers, then as a field assistant, and eventually became George and Juanda's long-time colleague and close family friend. The three were so close that they were referred to as the "merry trio" (Donnelly 2006a; Figure 2.6), a name now used for this era in their honor.[28]

The Merry Trio's contribution was not confined to Marshall County, though. George's notes indicated that he knew of some 9,400 specimens collected across the state; almost 7,200 of which were collected by George, Juanda, and Lothar. We know of just shy of 3,600 extant specimens, but we would not be surprised if more turn up. As with Prichard's collections, the Merry Trio's collections are widely dispersed among various institutions and private collections. We know of eight collections (CCC, FSCA, INHS, JCAC, OMNH, UCO, UMMZ, USNM) that contain their specimens (as well as three specimens from the Bicks' daughters, Suzann and Patty). It appears that Bick personally recorded odonates in 42 of the 77 Oklahoma counties and Hornuff surveyed in at least 47. But, also similar to Pritchard and others on the early OU expeditions who were overshadowed by Ralph Bird, it may be that Juanda and Lothar were overshadowed by George. Not that George intentionally eclipsed his colleagues, but as with Bird, some records have been attributed to George that may have actually belonged to the others. For example, in the early years of cataloging collections at OMNH I was under the impression that, unless otherwise

clearly indicated, any specimens collected during George's time in the state were from his collection. Unbeknownst to me at the time, I was quite wrong. While cataloging I came across many specimens that had "LH" or "LEH" handwritten on the envelopes (Figure 2.7). Although I realized that the majority of the handwriting on the envelopes was not George's, he was often the determiner, and he published the records (Bick and Bick 1957). I figured that, not uncommon to professors with student field assistants, a student had labeled the envelopes and had used initials, initials I guessed were of an unnamed, never-to-be-determined assistant or at best a code of George's (later I discovered that, like most collectors, George used a variety of codes that are not decipherable by anyone but him). I felt tremendous shame when, standing one day in the FSCA collection, it dawned on me that "LEH" was Lothar Hornuff, the almost-lost-to-history, huge contributor to Oklahoma odonatology, who had always been subsumed by the towering George H Bick. Again, not that my mistake is excusable, but I am not the only person to have cataloged specimens of this era with the wrong collector attributed. Nonetheless, if anyone ever has the time to return to those specimens (and the Bird era specimens) to accurately attribute them to the proper collector, it would be a nice clean-up for history's sake. For now, all I can do is admit to my personal error and try to clean-up the history here on these pages.

As attested by the sheer number of specimens Lothar took, he was an outstanding collector of odonates. George fondly described Lothar's technique:

> He was a good collector, but never used those monstrous nets, Daigle and Dunkle style. He would make his way to one end of a pond and say, 'Just leave me alone and I'll bring them in', he insisted that a good spit would lure them close to his net.

(Bick 1996:25)

George also said, "Using his almost vicious backstroke with his surprisingly small net, he made some good captures" (Bick 1985:259). But Lothar was not only a superb collector of odonates, he was also a collector of plants. He collected at least 70 plant vouchers that are currently housed at the University of Central Oklahoma, Oklahoma State University, and the Robert Bebb Herbarium at OU (Figure 2.8). He, too, was an avid photographer of plants, landscapes, and dragonflies, including making videos of odonate behavior.[29] He was also someone whom others only have kind words and tickled memories for.

George made mention of how he and Lothar went "on extensive Western collecting trips … together, living in two vans. One van was the lab, the other the dining room. We identified, packaged and labeled everything in the field" (Bick 1996:25). Presumably he meant trips that the two went on to the Oklahoma panhandle, as well as to Nebraska, the Dakotas, Wyoming, Montana, and Colorado, states that the

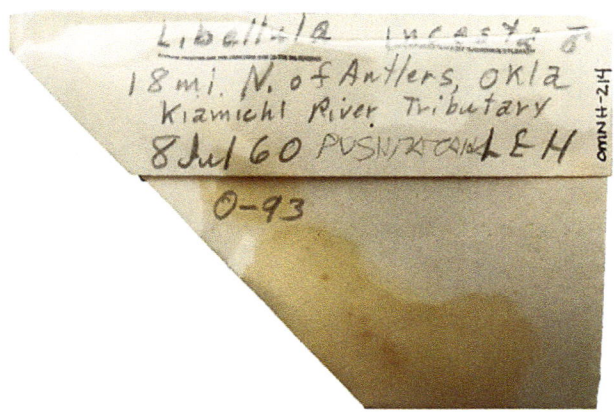

Figure 2.7 One of Lothar E Hornuff's (LEH) Odonata specimen envelopes (OMNH 214).

Figure 2.8 One of Lothar E Hornuff's plant voucher labels (Bebb Herbarium).

two and Juanda published surveys for (Bick and Hornuff 1971, 1972, 1974; Bick et al. 1977). Lothar continued working in Oklahoma well after the Bicks last collected here. He was in Oklahoma when he died in 1985, although the last known odonate specimen he collected was in 1978, thus ending the Merry Trio era, at least for Oklahoma. The three met up elsewhere and kept in contact until just a couple of weeks prior to Lothar's death.

Although not known to have visited Oklahoma again, George maintained his interest enough in the state that years later he published an update to his and Juanda's 1957 summary (Bick 1991). Seven first state records are attributed to George, and he shared two state records with Lothar and one with Juanda, making for a total of 10 state records (Appendix E). Lothar added another three, and along with one each from Juanda (and daughters) and an unknown collector, 16 new species were added to the state list by the time the Merry Trio era had come to a close. Of these, three were not confidently identified by George or Lothar at the time but turned out to be correct (*Gomphurus vastus*, the Cobra Clubtail, *Epitheca semiaquea*, Mantled Baskettail, and *Erythemis collocata*, Western Pondhawk), one was a species not yet described to science when the specimen was collected (*Gomphurus ozarkensis*, the Ozark Clubtail, but why George did not publish it in Bick 1991 is puzzling), and one was unknown to the Merry Trio (*Macromia alleghaniensis*, the Allegheny River Cruiser, an OSU specimen from 1962 misidentified as *Basiaeschna janata*, the Springtime Darner). A 16th state record also went undetected by George: a nymph of *Macrodiplax balteata*, the Marl Pennant, collected in 1986 and not found and confirmed until 2015. With these additions (Appendix E) the species total for the state was 139 by the time Bick published his final paper about Oklahoma.

LET US NOT FORGET …

Sometimes it seems as if little odonatological research was conducted outside of the OU expeditions and the Merry Trio era. But it should be remembered that there were others

working in, or at least visiting, Oklahoma from the 1950s to the early 2000s; field time that tallied >2,100 specimens. Collectors included Thomas "Nick" W Donnelly collecting 39 specimens (FSCA, USNM) in Carter, Latimer, Le Flore, Marshall, and Pittsburg Counties on 25–28 August 1954 while on his way out to California to attend graduate school. Seventy-four specimens (FSCA) from Greer, Jefferson, and Murray Counties are attributed to Minter J Westfall on 26–28 August 1956 and 14 August 1958. Robert William Cruden collected 88 specimens (UMMZ) in Bryan, Comanche, Jefferson, and Marshall Counties during his visit in June and July 1961. And then in the late 1970s and early 1980s there were some macroinvertebrate projects lead by OU, including the Grand River Dam Authority Project and the OK Springs Project, that made the first large-scale collections of odonate larvae for the state. More than 1,100 nymphs were collected in 30 counties across Oklahoma between 1975 and 1982, the majority collected in three summers by William D Shepard ($n = 387$, in 1975) and William J Matthews ($n = 365$, in 1981 and 1982). Shepard also collected 5 nymphs in 1976 and 11 adults in 1987.

Other collectors who have made noteworthy contributions include Paul Liechti and Don Huggins of the Kansas Biological Survey, who collected in 1981–1983; John Nelson and Hal C Reed, who collected during the summer of 1982 in 20 counties across Oklahoma; Karl Stephan, famed coleopterist, who collected in 1982–1988; and David Bass and his students of the University of Central Oklahoma who, among other macroinvertebrates, collected 91 nymphs between 1986–2002 in 11 counties. Oklahoma State University deserves special recognition beginning with Donald C Arnold, who collected many thousands of insects, including odonates. His odonate records date from 1962 up to 2007. Aaron T Dossey, also of OSU, collected odonates and tiger beetles from 1977 to 2000, the later years (1993 to 2000) with Richard Grantham.

In the 1960s, Kenneth Wilson Stewart studied dragonflies in the state (e.g., Stewart and Murphy 1968). Later, he and his students, including John C Abbott, collected in Oklahoma until 1997. They collected a few hundred specimens (mostly housed in JCAC), primarily near Thackerville in Love County, but also to some extent in Johnston, Murray, and Pontotoc Counties. This was also roughly the same timeframe that Sidney W Dunkle made seven visits to the state: 5–8 June 1992, 12 July 1993, 30 March 1996, 11–12 June 1996, 19–20 April 1997, 14 July 1998, and 8–10 June 1999. During those trips he collected in 11 counties (Adair, Beaver, Choctaw, Johnston, Kay, Le Flore, Love, Marshall, McCurtain, Murray, and Osage) and took some important specimens for Oklahoma, including some of the few records known for Oklahoma of *Enallagma dubium* (Burgundy Bluet), *E. daeckii* (Attenuated Bluet), and *Neurocordulia molesta* (Smoky Shadowdragon).

In the late 1990s, Roy J Beckemeyer began collecting specimens and records in Oklahoma (Table 2.1). His research resulted in a slew of publications, most of which came in the early 2000s, including the first checklist for the state (Beckemeyer 1995, 1998b, 2002a,b,c,d, 2004, 2006). Boris Kondratieff and colleagues also worked in Oklahoma during this time, primarily at Fort Sill Military Reserve (MR), Comanche County, between 2002 and 2003 (Kondratieff et al. 2004; Zuellig et al. 2006). They managed to procure a good number of important specimens, including quite a few extralimital records and one state record, *Ischnura perparva*, the Western Forktail (Table 2.3; Appendix E). Other interesting species for Comanche County that they collected include: *Hetaerina titia* (Smoky Rubyspot), *Enallagma aspersum* (Azure Bluet), *E. praevarum* (Arroyo Bluet), *E. vesperum* (Vesper Bluet), *Gomphurus ozarkensis* (Ozark Clubtail), *Cordulegaster obliqua* (Arrowhead Spiketail), and *Ladona deplanata* (Blue Corporal).

BIRDERS BECOME ODERS

In the early 2000s, Oklahoma was fortunate to have roughly half a dozen people in the state with budding interest in odonates. Almost all of these people were long-time bird watchers ("birders") and were venturing into odonate territory with the same gusto and skill they brought to birding. To our knowledge, the first to get interested in odonates was Bill Carrell. Birds have always been a focus of Bill's outdoor time, but bugs, especially odonates and tiger beetles, have also captured his heart. In 2001 Bill began photographing odonates at one of his favorite haunts, Mohawk Park in Tulsa. He photographed common species, *Argia moesta* (Powdered Dancer), *Argia apicalis* (Blue-fronted Dancer), *Celithemis elisa* (Calico Pennant), *Libellula luctuosa* (Widow Skimmer), and *Plathemis lydia* (Common Whitetail), and did not submit the records until 2011, but it was a definite start. If only he knew what those records would lead to. He has submitted >1,260 photographic records to Odonata Central, many of which are county records, and, after MAP. and myself together (*n* = 12), he holds the distinction of having the most "rediscovered" species in the state, i.e., species not seen in Oklahoma for at least 30 years (Table 2.3). His eight re-discoveries include *Lestes rectangularis*, the Slender Spreadwing, and *Argia lugens*, the Sooty Dancer. Bill was also involved with three state records—*Aeshna constricta*, the Lance-tipped Darner, *Sympetrum pallipes*, the Striped Meadowhawk, and *Tramea calverti*, the Striped Saddlebags (Table 2.3; Appendix E). He is the only person to have seen the former two in Oklahoma; at opposite ends of the state! The latter species was a co-discovery with John Fisher, an observer who supplied us with >1,100 odonate records that he tracked from various observers as well as many of his own records from across the state (J Fisher database). A fourth state record, regrettably, is on the hypothetical list because it cannot be accepted, even though MAP and I feel it is credible. On 14 May 2017, Bill saw what he identified to be a ♂ *Anax walsinghami*, the Giant Darner, at Black Mesa SP. He was unable to get documentation, without which we do not accept state records (we require a specimen or identifiable photo), even for our personal records lacking documentation, such as the ♂ *Anax amazili* (Amazon Darner) MAP identified at French Lake in the Wichita Mountains WR, Comanche County, on 26 October 2014.

Victor W Fazio III entered the Oklahoma scene in 2002. He worked primarily in the southwestern part of the state but also ventured out to some of the rest of Oklahoma, as attested by one of his three state records, *Libellula composita*, the Bleached Skimmer, being found in Texas County in the Oklahoma panhandle (Appendix E). But his other two records—*Sympetrum illotum*, the Cardinal Meadowhawk, and *Pseudoleon superbus*, the Filigree Skimmer—were found in the southwest. Vic was also responsible for four "re-discoveries" and he submitted just shy of 1,400 records to Odonata Central before he left the state in 2013 (Table 2.3). During his time working in Oklahoma he managed to get Comanche County, his primary focus area, to 98 species, at which it has sat since. We thought for sure that Comanche County would be the second county in the state to break the 100 species mark, but we were wrong. Without Vic working Comanche County so diligently, in May 2017, Atoka County took the prize instead.

David Arbour has played an integral role in Oklahoma odonatology since 2002. It is difficult to talk about David without also talking about Berlin Heck, but it was David who first became interested in dragons. He can boast seven first state records, including the first two records he submitted to Odonata Central: *Aphylla williamsoni*, the Two-striped Forceptail, and *Arigomphus maxwelli*, the Bayou Clubtail (OC 6542, 6584, the former also JCAC). He is responsible for four other state records on his own—the well-out-of-range *Tholymis citrina* (Evening Skimmer), for example. He also co-discovered two other state records (see below; Table 2.3, Appendix E). He has had three re-discoveries and been instrumental in tracking species of conservation concern that have populations, in some cases the only known Oklahoma population, at Red Slough Wildlife Management Area, such as the two gomphid species above and *Telebasis byersi*, the Duckweed Firetail, and *Tramea calverti*, the Striped Saddlebags.

Berlin Heck, though starting a few years after David, has been equally important to our knowledge of odonatology in Oklahoma. He, too, holds seven state records—six on his own and two co-discoveries (Table 2.3; Appendix E). Three of the species, *Phanogomphus lividus* (Ashy Clubtail), *Helocordulia selysii* (Selys's Sundragon), and *Cordulegaster maculata* (Twinspotted Spiketail) were, until recently, only known from the Heck property 10 km SE of Idabel, McCurtain County. One of his co-discoveries—of *Coryphaeschna ingens*—was shared by David Arbour, but it was Berlin who shot it out of the sky with his shotgun. His other shared state record was with Greta and Pat Heck when they came upon the first record of *Cordulegaster talaria*, the Ouachita Spiketail, in Oklahoma. Berlin also has one "re-discovery" under his belt, *Erythrodiplax minuscula*, the Little Blue Dragonlet. And,

Table 2.3 **Recent contributors to the Oklahoma Odonata Project along with their years active and areas covered in the state. "Years active" includes only years that the person was known to have actively collected odonate records; we explicitly did not account for years those persons collected data for other taxa. Also indicated are the number of state records and "re-discoveries", i.e., species found that went previously undetected for 30+ years, that observer is responsible for. In the case of multiple people being responsible for a state record, it was split among them (thus the half counts). The number of specimens collected in the state and the number of Oklahoma records submitted to Odonata Central are shown.**

OBSERVER/COLLECTOR	YEARS ACTIVE	AREAS COVERED	STATE RECORDS	RE-DISCOVERIES	SPECIMENS	OC RECORDS
Bill Carrell	2001–	statewide	3 ½	8	–	1362
Victor Fazio III	2002–2013	primarily SW	3	4	–	1396
David Arbour	2002–	SE	7	3	24*	284
Bryan Reynolds	2000s–	primarily central	–	–	–	16
Boris Kondratieff, et al.	2002–2003	SW	1	1	100s†	–
BD Smith-Patten and MA Patten	2003–present, statewide		6	11	1439	
Michael A Patten			4	1	694	340
Brenda D Smith-Patten			4	–	313	1257
Mark Dreiling	2005–2010	NE	–	–	–	63
John Fisher	2005–	statewide	½	–	3	37
Joe Grzybowski	2006–	statewide	–	–	–	89
Berlin Heck	2006–2014	primarily SE	7	1	53*	977
Ken Williams	2006–	primarily E	–	2	2	156
James W Arterburn	2009–	primarily E	–	–	–	390
Janalee P Caldwell and Laurie J Vitt	2009–2010	SE	–	–	51	–
Franklin "Leroy" Alm	2010–	C, SE	–	–	3	51
Jason T Bried	2012–2018	statewide	–	–	70	30
Jason R Heinen	2011–2012	primarily N	1	1	12	134
Abigail and Ruth Mills	2014–	NE, SE	–	–	4	28
Colby Farquhar	2015–	NE	–	–	5	2
Emily Hjalmarson and Brett Roberts	2015–2018	statewide	–	–	33	260
Cliff Ivy	2015–	NE	–	–	–	136
Shawn Johnson	2015–	NW	–	–	–	356
Jon Ivy	2016–	NE	–	–	–	11
Wade Boys	2017–	E	–	–	15†	4

Data to 14 September 2019.
* Heck and Arbour had seven specimens that they collected together, so the number was divided among their totals. † = unsure of exact count

although he primarily worked in the southeastern part of the state, he submitted records to Odonata Central from across Oklahoma.

During the recent era, there have been 2,876 specimens collected and about 8,000 photographic records submitted to Odonata Central (Table 2.3). There are also many thousands of photos on file, either archived with the Oklahoma Natural Heritage Inventory's Oklahoma Odonata Project (OOP) or retained by various observers. The majority of currently active participants in Oklahoma odonatology are photographers, but there are some active collectors and

there have certainly been some since 2002. We have already mentioned that Boris Kondratieff and his team collected at Fort Sill MR in 2002–2003, where they procured many hundreds of specimens. Besides Boris, et al. there were a small handful of others, including Janalee P. Caldwell and Laurie J Vitt, who collected 51 specimens on their property 3 mi E of Honobia in Le Flore County (OMNH). Berlin Heck was a principal collector in his day, with 50 specimens collected on his own and an additional eight that he collected with David Arbour. David collected about 30 other specimens. Berlin and David's specimens are housed in the collections of DRP, JCAC, or SP. Jason Heinen and Jason Bried's collection efforts are also of note. Heinen contributed some important specimens to the SP collection and, during his dissertation work, Bried collected and donated 70 specimens to SP from various sites around the state. Newish to the scene and currently active collectors (Table 2.3) include Colby Farquhar, Wade Boys, and Abigail Mills.

THE OKLAHOMA ODONATA PROJECT (OOP)

Michael Patten and I came to Oklahoma in 2003. We first lived in Bartlesville in the northeastern corner of the state, so much of our initial focus was on northeastern and north-central Oklahoma, particularly in Washington, Osage, and Tulsa Counties, but we did cover areas across the state in those early years. By 2007 we began to feel confident that we knew enough about the northeastern corner to attempt a summary of the odonates there (Smith-Patten et al. 2007). In retrospect we probably should have waited a year or so because we still had much to learn about that part of the state and have had to re-think a couple of phenological patterns that were revealed in earlier data; nonetheless, overall it proved to be a useful summary. Shortly before the paper's publication, we moved to Norman to work at the main OU campus. Over the years since, I held positions with the Recent Invertebrates collection at OMNH, as well as at the Oklahoma Biological Survey (OBS), and the Oklahoma Natural Heritage Inventory (ONHI). Michael became faculty at OBS and an adjunct with the Department of Zoology (now Biology). With our new responsibilities came the need to take on some Oklahoma-specific research and, not surprisingly, Odonata took the ticket. Coincidently, around roughly the same time, Elizabeth A Bergey, also of OBS, decided to begin assessing the conservation status of Odonata. At the time, she was the person at ONHI who was responsible for the taxon and so she began to undertake, for the first time in Oklahoma history, assignment of conservation ranks for all of the odonate species known to occur in the state at that time.

The OOP sprang out of this convergence. While technically the formation of the project began a bit earlier, we officially named 2009 as the year that the OOP really took flight. That was the year that we decided to systematically tackle the state—county by county—rather than just randomly surveying. We began the process by compiling records taken from observer field notes, photographs, published literature (e.g., Bird 1932; Bick and Bick 1957; Donnelly 2004a,b,c; Abbott 2005), and museum specimens. These records were used to calculate species totals for each of the 77 Oklahoma counties. We used these minimum counts as a baseline of effort and as a way, once plotted onto a map, to determine where to focus survey effort (Figure 2.9A). Initially we were a bit overwhelmed because some counties had all but a handful of species known from them, but we reasoned that, given a known state list at that time of around 135 species (Beckemeyer 2002a; Abbott 2005; Appendix D), each county must have at least 25 species

in it. Within a short period of time of beginning surveys, we realized that a minimum of 40 species for every county was a readily attainable goal. And, by 2013, we recalibrated to shoot for 50 species for every county (Figure 2.9B), which we reached by the end of the 2016 field season. We then chose to recalibrate again, but with a different approach in mind. The dwindling of red on the map, shifting to all white by 2016, and then to an entirely new color scheme (Figure 2.9C) indicates our mindset change, from trying to add species in order to get a handle on diversity (red, pink, white), to attaining a level of knowledge much more indicative of actual species richness across the state (green).

Through the years we continued to compile records from any sources we could find (Appendix B). A big source of data was, and continues to be, Odonata Central. Additional records came from other online portals such as GBIF (http://Global Biodiversity Information Facility, www.gbif.org), iDigBio (www.idigbio.org/portal/search), and iNaturalist (www.inaturalist.org/). As mentioned above, John Fisher's database of faunal records across the state was also an important data source. And, of course, about 12,000 records came from specimens that currently reside in museums and personal insect collections across the United States. Those records, obtained from 28 collections (Appendix B), comprise approximately 18,500 specimens, about half of which MAP and I cataloged ourselves. By finding these specimens we slowly paired literature records with their specimens, as well as matched field notes to specimen records. We also cleared up some misattributions and misidentifications, and confirmed some cautious early determinations. The process has been one akin to a roller coaster ride with many ups and downs, spirals, and even full loops.

A key misattribution and a big downer for us was when, during a visit to the University of Michigan, Museum of Zoology, we discovered a series of mislabeled specimens that necessitated removing half a dozen species from the Custer County list and adding them to neighboring Caddo County (see EBW fiasco described above). Custer County has always been species-poor compared to its neighbor, so it was disheartening to resolve this error.

Two misidentifications stand out; one compelled us to remove a species from the state list while the other added a species, so in the long run one could say all was well, but at the time we were deflated and then elated. While cataloging specimens at the International Odonata Research Institute, at the Florida State Collection of Arthropods (FSCA), we learned that the long-held notion that *Lestes congener* (Spotted Spreadwing) was part of the Oklahoma state list was in fact incorrect (Patten and Smith-Patten 2013a). The species was added to the state list based on a specimen collected in 1966 by Clarice Kerfoot in Woodward County. That specimen was published as *L. congener* by Bick (1978), but he later retracted the determination in his notes but not officially in print. We examined that specimen and confirmed Bick's later identification of *L. alacer* (Plateau Spreadwing). Another specimen, taken by RD Bird in 1932 and identified as *Argia*

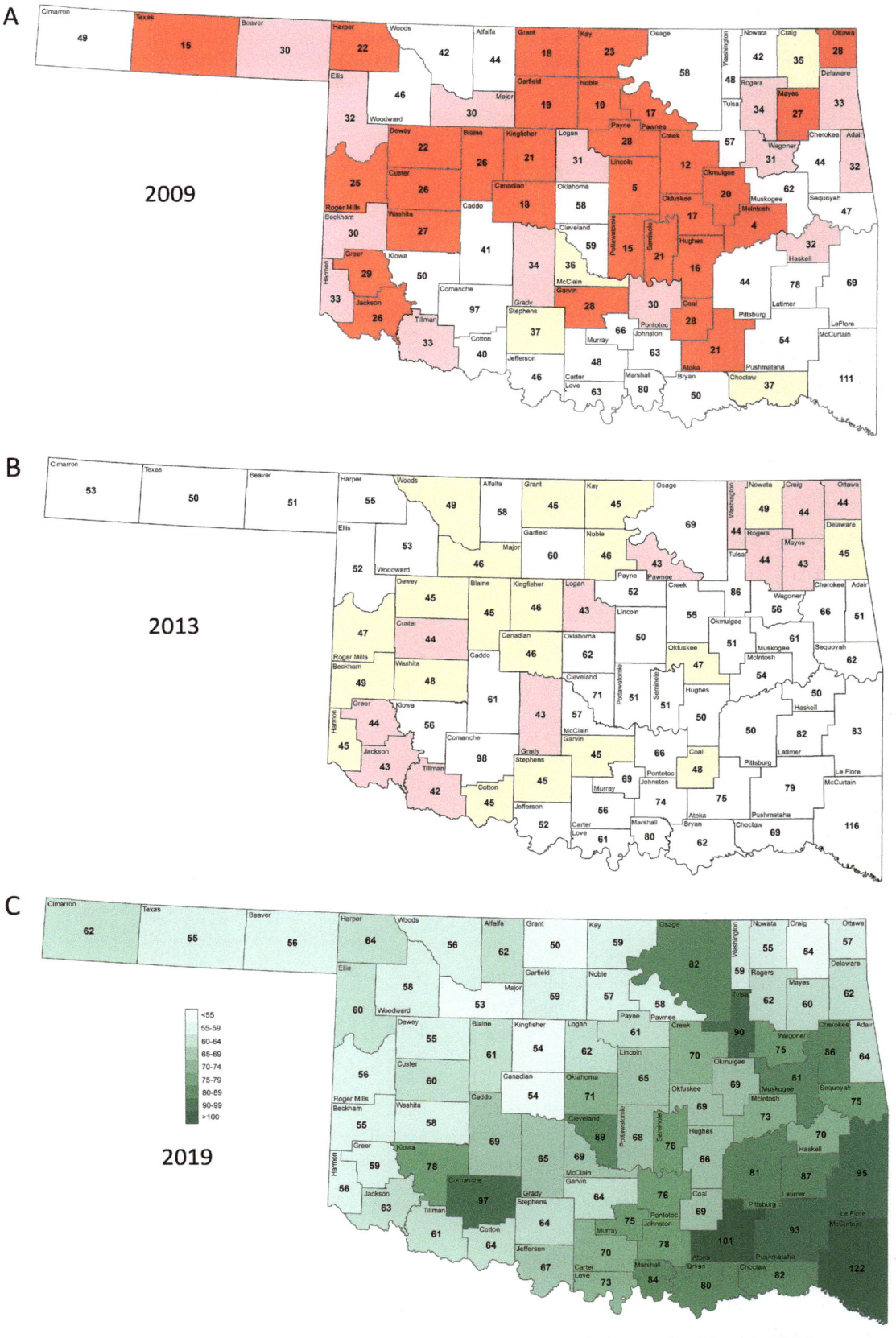

Figure 2.9 Change over time in the total of odonate species recorded in each of the 77 Oklahoma counties. Maps date from A) 2009, when the Oklahoma Odonata Project (OOP) began, B) 2013, and C) 2019. See text for details on color-coding.

sedula (Blue-ringed Dancer) was, upon our examination in 2014, determined to be *A. leonorae* (Leonora's Dancer; OMNH 3064). Recall that in Bird's time, *A. leonorae* was not yet described (Garrison 1994).

Another specimen proved to be quite exciting when I stumbled upon it while cataloging the OMNH fluid collection. I was quite shocked when I opened a vial and pulled out a female gomphid that was labeled "*Gomphus intricatus*," or what today is known as *Stylurus intricatus* (Brimstone Clubtail; OMNH 2413; Smith-Patten and Patten 2012). This specimen appeared to be a case of perhaps a little too much caution with its identification, so much so that it was lost among the collection and forgotten. Another case of an apparent cautious identification

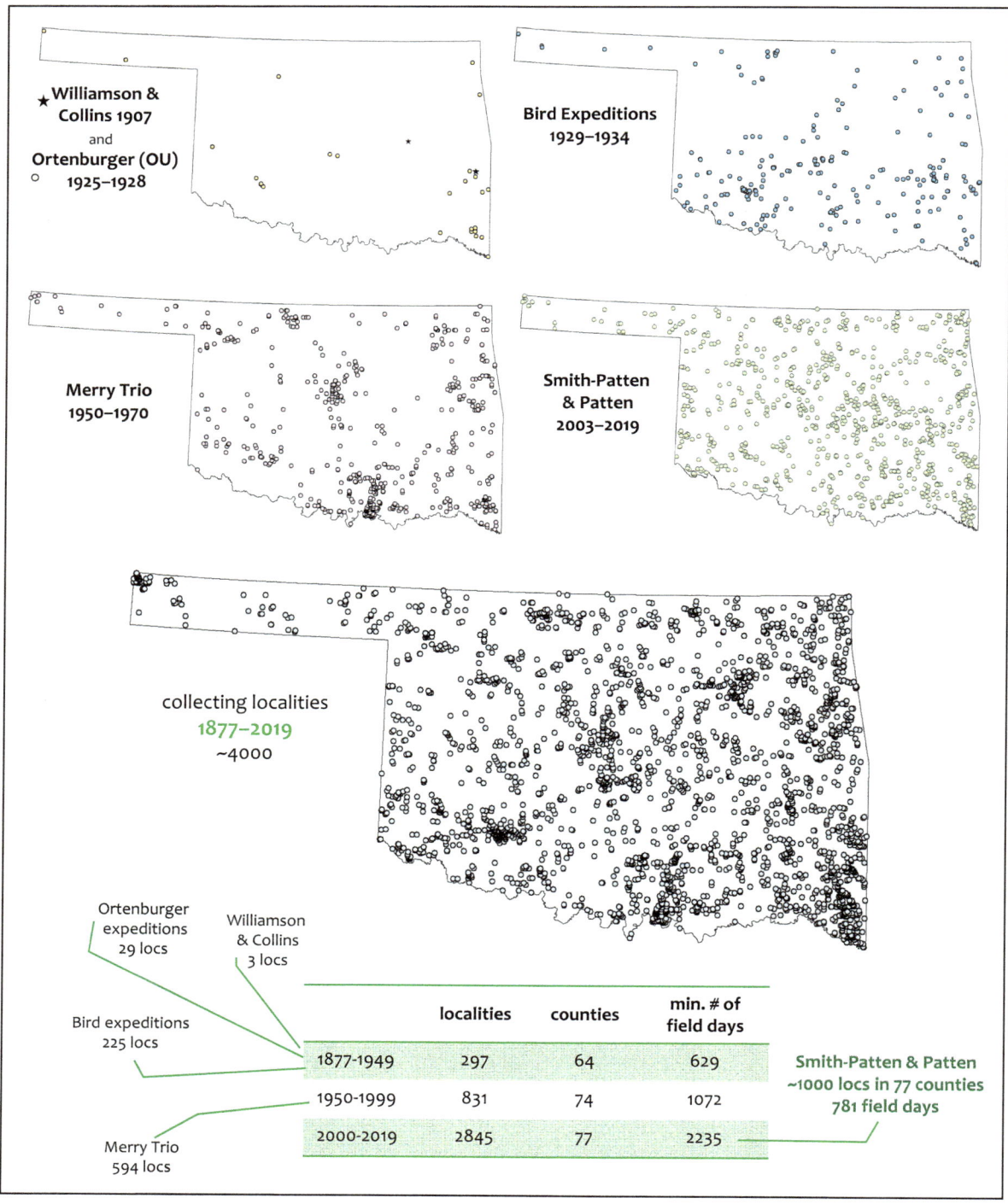

	localities	counties	min. # of field days
1877-1949	297	64	629
1950-1999	831	74	1072
2000-2019	2845	77	2235

Figure 2.10 Odonate collecting localities and effort across Oklahoma by era.

Figure 2.11 Types of occurrence records in the Oklahoma Odonata Project (OOP) database include specimens, photographs, literature citations, and notes from field surveys from all eras. Some of those records types are indicated along with the number of corresponding records in the OOP database.

is the *Erythemis collocata* (Western Pondhawk) that we came across at FSCA (Smith-Patten and Patten 2013b). GH Bick and LE Hornuff collected that male in 1970. The specimen card indicated that Hornuff had determined it to be *E. simplicicollis* (Eastern Pondhawk), but Bick's species note card indicated they had doubts and thought it was possibly *E. collocata*. When we visited the collection in 2013, we confirmed the identification as *E. collocata*. Bill Mauffray later did the same.

The above examples are but a few of the records we "cleaned-up" over the years. Our intent in doing so was not to point out errors people made, for we have certainly made quite a few ourselves over the years, but to ensure that we were working with the best possible data so that we could accurately depict the composition, ecology, status, and distribution of the odonate species in Oklahoma. An ancillary intention of our efforts, although arguably just as essential, was to determine the level of documentation for each species in each county (see Appendix C for online resource). We undertook this effort because, like the state's species list, we wanted some sort of physical documentation to support inclusion of species on county lists. Initially there was a rather low percentage of county records with physical documentation. We wished that each record could have an associated extant specimen but we realized that would mean >5,000 specimens, a large number of which we would need to collect ourselves, which is no small feat for two people covering 177,660 km² (68,595 mi²). Such an effort also meant focusing more on the mere act of collecting and less on the need to collect given that, in many instances, especially for easily identifiable species, a photograph would suffice. Presently, >96% of records are supported by a specimen or archived photograph. That small percentage of records without extant specimens or archived photographs include two species—the Chalky Spreadwing (*Lestes sigma*) and Cinnamon Shadowdragon (*Neurocordulia virginiensis*). Both of these species were reported in the literature as specimens (Bick 1978; Byers 1937; Davis 1937; also, see above), but we have been unable to locate them and fear they are lost. Nonetheless, because these specimens were confirmed by multiple renowned odonatologists, we accept them as documented species for Oklahoma.

Along with the high level of documentation of the OOP database, we can also boast of its high level of geospatial quality. I literally spent years tracking down collecting localities and pinpointing coordinates for those localities. My training as an historian proved useful in poring over primary sources including many historical maps and written accounts to find obscure, colloquial, and antiquated place names (see Appendix C for details about the gazetteer that sprang out of this research). By the end, I had only a small handful of the approximately 4,000 localities in the OOP database that I was not able to georeference.

Not surprisingly, the amount of survey effort has varied over the years (Figure 2.10). As one would expect, the current

era (post-2000) has the majority of records, with about two thirds; but we have many thousands of records dating from 1877 through the Dust Bowl era (1930s) and then close to twice as many as that dating post-Dust Bowl to the current era. Although records overlap by type (for instance, field notes may have an associated specimen, a photo of which might have been uploaded to OC; Figure 2.11), the OOP database now contains >55,000 records, accounting for hundreds of thousands of individual odonates, including almost 220,000 from just our surveys alone (Figure 2.10). As mentioned earlier, approximately 18,500 specimens have been accounted for, and there are many thousands of others that were cited in the literature or in collector's notes as being taken that may not have been retained. For instance, there are 3,761 entries for GH Bick's notes that mention >9,200 specimens that he or others collected. We have relocated but a fraction of those. Add to that >8,000 photographic records accounting for >33,000 individuals, and though neither MAP nor I are much for boasting, that makes for a tremendous amount of data collection (and for 14 state records to our names).

The OOP is archived as part of the Oklahoma Biodiversity Information System (OBIS) of the Oklahoma Natural Heritage Inventory.

NOTES

1. Oliver S Flint, of USNM, was kind enough to pull this specimen and have the labels photographed. He and others at USNM thought the collecting label read "Ind." and interpreted it as Indian Territory. MAP agreed; BS-P leans toward it reading "Nd." as in North Dakota.

2. Richards (1929) discussed the Oklahoma Biological Survey in more formal terms as an entity distinct from the Department of Zoology.

3. The fire that destroyed OU's "University Hall" on 3 January 1903 is thought to have taken with it the majority of the original natural history museum. To my knowledge, little was saved; the only reference that I have seen as to what was salvaged was a passage from the 1905 *Mistletoe*, the school's yearbook, saying, "The taxidermist opened the door under the front steps into the basement and many of the hides and mounted animals were carried out" (p. 11). Collections were later moved to the "Science Hall" and other areas as designated by H. H. Lane (Carpenter 2000). A second major fire, in 1918, was also quite destructive, consuming "a large collection of birds stored in a temporary wooden building" (Droke 2009).

OU's museum has undergone many name changes since its inception. It began as a general collection of the Department of Geology and Natural History with the three main divisions of geology, botany, and zoology. Eventually the zoology portion of the collection became the University of Oklahoma Museum of Zoology, which was the case in AI Ortenburger's early years at OU (he began in 1924). Ortenburger worked to increase the entomological portion of the collection by collecting thousands of specimens himself, as well as accessioning >2,500 insects from a donor in 1929 and 1934 (Hollon 1961). In 1943 the various geology, and natural and cultural history collections from around the campus were consolidated into a general museum called the University of Oklahoma Museum. J Willis Stovall, a paleontologist, was named director and Ortenburger was named curator of the Zoology Division. The museum was again renamed after the death of Stovall in 1953; this time in his honor as, officially, the University of Oklahoma J Willis Stovall Museum (Hollon 1961), but more commonly known as the Stovall Museum of Science and History. Various other names have been used since, officially and unofficially, namely the University of Oklahoma Museum of Natural History, the Oklahoma Museum of Natural History, from whence the coden OMNH came, the Sam Noble Oklahoma Museum of Natural History (Carpenter 2000), and now the Sam Noble Museum. The university's botanical collections are housed at the Robert Bebb Herbarium, a unit of the Oklahoma Biological Survey.

4. The Department of Geology and Natural History were broken into the Department of Botany and Department of Zoology and Embryology in 1906 (Hollon 1961). Van Vleet then became the head of the Department of Botany, and HH Lane, the head of the Department of Zoology and Embryology. Lane served as the head of the department until his departure from OU in 1920.

5. One specimen, of a ♂ *Plathemis lydia* (Common Whitetail), was taken in Ardmore, Carter County, on 1 May 1912, by an unknown collector (UNSM 43). Six of the specimens were collected by Theodore Huntington Hubbell in 1918 at Fort Sill Military Reserve, Comanche County. At the time, Hubbell was an undergraduate student at the University of Michigan, where he became a specialist in Orthoptera (grasshoppers and allies). While he was in the Wichita Mountains in 1918 looking for orthopterans (Blair and Hubbell 1938), he collected: 1 ♀ *Hetaerina americana* (American Rubyspot), 1♂ *Enallagma civile* (Familiar Bluet), 2♂ *Argia moesta* (Powdered Dancer), and 2 ♀ *Sympetrum corruptum* (Variegated Meadowhawk). All were taken on 20 October, except the Familiar Bluet (19 Oct), and all are housed at UMMZ, except one of the Powdered Dancers (FSCA specimen). The eighth specimen (OSU) is a Variegated Meadowhawk, collected in Stillwater, Payne County, on 8 September 1923, by Williamson James Brown, an Oklahoma A & M (now Oklahoma State University) then student, later instructor. Two additional specimens were purported to have been collected in 1921 (*Gomphurus externus*, ♂, Cleveland Co.: Norman, Little River, 21 April, OMNH 834; *Phanogomphus militaris*, ♂, Love Co., 13 June, OMNH 849); however, these are attributed to Ralph D Bird, who was not known to be in the state until 1929. The label for the *G. externus* clearly indicated that it was collected in 1921 by Bird, but the *P. militaris*, after re-examination of the specimen, was found to have actually been collected in 1929. The later date makes the most sense because Bird is known to have moved to Oklahoma in 1929. In 1921, he was an undergraduate student at the University of Manitoba, so it is possible that the label for OMNH 834 is in error (perhaps not an original label?).

6. Hollon (1961) indicated that Ortenburger lead OU expeditions from 1925 to 1931, but to our knowledge he did not collect odonates after 1928, except three *Anax junius* that he collected in April 1931.

7. RD Bird was born on 20 May 1901 in Arrow River, Manitoba (Bird 1972). He attended the University of Manitoba, where he received his BS in 1924 and his MS is 1926. He then went on to the University of Illinois to earn his PhD in 1929. After leaving Oklahoma he worked for the Canada Department of Agriculture Research Station until he retired in 1966. He died in Ganges, British Columbia, on 1 March 1972. If it is any indication of how esteemed Bird was to Canadian natural history, his death warranted a seven-page obituary in the *Canadian Field-Naturalist* (Bird 1972).

8. It may be, as per Hollon (1961), that Ortenburger remained the official leader of the OU expeditions until 1931. Nonetheless, as the entomologist for the Oklahoma Biological Survey, it appears that Bird was at least the lead of the entomological portions of the expeditions during his OU tenure.

9. Lois was the daughter of Charles N Gould, OU professor and head of the Oklahoma Geological Survey. Carpenter (2000) said that she was a student member of the OU expedition in 1930 and 1932. Ralph and Lois are known to have published one paper together (Bird L, RD Bird 1931. Winter food of Oklahoma Quail. *Wilson Bulletin* 43:293–305).

10. Lois published her articles between 28 March 1926 and 14 August 1927. The *Daily Oklahoman* is now just called the *The Oklahoman*.

11. Lois' handwriting was confirmed by her son, Dr. Charles Bird (*in litt*, 26 Mar 2019).

12. I am guilty of this myself. I hate to admit to it, but when I began to catalog the OMNH odonate specimens, I did not realize that Lois Bird had assisted Ralph Bird, and at the time I had no clear idea that Bird had a whole team of people collecting for him. I did find this out fairly soon upon cataloging specimens, but I know some are falsely attributed to him. I can only take solace in that I am certainly not the only one guilty of a Bird misattribute. Lessons learned …

13. It is unknown if Logsdon assisted the team in 1929. From OU's General Catalogs, it appears he was a graduate student from 1928 until at least 1930. He was listed as a graduate assistant in both 1929 and 1930.

14. His full name was Norbert O'Reilly Sandoz. He sometimes used the name N O'Reilly Sandoz, but generally he dropped the N altogether. Some, including OU's student directory, incorrectly called him O'Reilly N Sandoz. Note, too, that Duck and Fletcher (1943) misspelled Sandoz's name as O'Rielly.

15. Carpenter (1990, 2000) indicated that Sandoz was only officially apart of the OU expeditions in 1932, although he was an undergraduate and graduate student at OU for various years between 1928 and 1935 (OU directories and General Catalog).

16. As with Logsdon and Sandoz, Carpenter (1990, 2000) indicated that Fisher was a student member of the 1932 OU expedition only. OMNH has 3633 specimen records (3 reptile, 1 mollusk, and 3629 insect records) from Oklahoma and elsewhere attributed to him from 1927–1938.

17. Pritchard's first scientific publication was this description of two new species (Pritchard 1935).

18. We have yet to pinpoint the localities known as "Oil Springs," "Dean's Slough," and "Indian Springs"; as such, approximations were made for OOP database coordinates based on collector's descriptions and historical research.

19. The Wichita National Forest and Game Preserve became the Wichita Mountains WR in 1936 (O'Dell 2007).

20. Specimens were labeled only as "Cherokee," which the cataloger took as the county rather than as the town of Cherokee, Alfalfa County, which is where the Bird expedition was collecting at the time.

21. According to Carpenter (1990, 2000), EB Webster was an official student member of the OU expeditions in 1930 and 1932 only. Webster also "collected [in the] summer and fall of 1927–1929," which appears to have been independent of OU. Within the OMNH collection he has >1,000 records attributed to him from 1927 until 1940 (although there does appear to be at least one other Webster that may be confused with EB). He collected, in Oklahoma and elsewhere, a wide variety of taxa including amphibians, reptiles, birds, Hemiptera, Hymenoptera, Coleoptera (especially carabids, scarabs, and tenebrionids), and Decapoda.

22. Locality strings were changed to 1 km SSW of Hydro and 12 km E of Hydro, respectively, for inclusion in the OOP database.

23. Although we cannot say what map he would have used, none of the maps we found for the era make it look like 7 or 15 mi E of Weatherford is in Custer County. For example, the 1919 *Hammond's Map of Oklahoma* (CS Hammond & Co., New York) and both the 1921 and 1928 *Rand-McNally Standard Map of Oklahoma* show those localities are wholly within Caddo County.

24. Specimens collected in 1934 may have been collected independently by Pritchard (Latimer and Le Flore Co.) and/or by Albert H Trowbridge (Latimer County), rather than as part of an OU expedition.

25. Although tradition has it that a male specimen is designated the holotype, in the case of *Neurocordulia virginiensis*, Davis (1937) designated a female.

26. Sources of biographical information for this section were: Anonymous (2005); Beckemeyer (2002d); Bick, GH (1985, 1996); Bick, S (2006); Donnelly (2006a); DSA (1999); Dunkle et al. (2006); Hornuff (1979).

27. From specimens and field notes we gathered that Bick worked at the station from 13 June to 2 August 1950, and from 8 June to 28 July 1951. However, in his 1951 Oklahoma paper, he indicated that he only collected in June and July.

28. Although admittedly, throughout the book the term "Bick/Hornuff era" is more often used.

29. At present, there are thousands of photographic slides and other media taken by Hornuff that are housed in a wooden cabinet in a classroom at UCO. I have not had a chance to inventory or even look through the collection, but I hope to one day as I hope to see it properly curated. Many thanks to Gloria Caddell for allowing me to peek at it.

## CHAPTER 3

GEOGRAPHY AND HABITATS OF OKLAHOMA ODONATA

Bruce W. Hoagland and Brenda D. Smith

It bears repeating: from a nature perspective, Oklahoma is a rich and varied land. By way of illustration, just take a drive from the southeastern corner of the state to the far panhandle. Go from the coastal plains, where the lowest spot in Oklahoma lies, at less than 300 feet (about 90 m) where oaks, pines, rivers, bayous, and backwaters dominate and where alligators roam, to head high up (>2,600 ft, or >790 m) into the hills, where you will rise up onto ridges, and then switchback your way down and then up again through the Ouachita Highlands and their intervening valleys. In the Ouachitas you will find oak-hickory forest with scattered natural pine stands and still a lot of rivers, streams, and some forested seeps. Then back down in elevation you will go as you venture into the Cross Timbers, and Oklahoma starts to lose its feel of the American east, and you begin to taste the plains like one would taste the salty air rising from an oceanic coastline. Drive farther and those plains, with their cattle ranches, farms, and remnant grasslands, will emerge fully. This is where you will see the familiar Oklahoma, that of old Western movies, of the iconic Chisolm and the Great Western cattle trails. This is where you will plunge into the frontier imagination. You will think to yourself, "this is the real Oklahoma," but you would be wrong … there is no such thing, and there is so much more. Continue through the red dirt Southwestern Tablelands and you will think you suddenly reached New Mexico, but then you will again be in a flat wildness hidden deceptively within the apparent barrenness. And, without even noticing, you will have gradually risen thousands of feet in elevation (>4,500 ft, or >1,370 m). It seems too flat to have reached so high, but that is the magic of the High Plains. And you have seen nothing yet compared to when you reach the Southwestern Tablelands again, setting your eyes upon Black Mesa, the black rocky mass rising abruptly out of the plain. There you will have to park and walk the rest of the way of your journey—from around 4,100 feet (1,250 m) at its base, up to the highest point in Oklahoma at 4,902 feet (1,494 m). And this is just a tantalizing morsel of what Oklahoma has to offer. In the sections below (and the following chapters) we invite you to explore the physical and biological landscapes that together form Oklahoma's ecoregions, and make the state so floristically and faunistically rich—connecting it to more than half of the area that makes up the continental United States (Figure 3.1).

OKLAHOMA'S ENVIRONMENT AND LANDSCAPES

The fundamentals of landscapes, as well as classifications of vegetation communities, are climate, geology, hydrology, and plant species composition. These factors lay the foundation for how, in turn, animals are distributed. As such, we will touch upon each variable prior to discussing how researchers have used them to divide the state of Oklahoma into ecological classifications, or *ecoregions*.

Climate

Oklahoma is known as a land of meteorological extremes. For instance, one day it might be 80°F, the next there could be a snowstorm … and, believe it or not, occasionally those events happen during the same day! Drops in temperature of 30–40°F within a few hours are relatively common. Temperature extremes on record are as low as –31°F (–35°C, not counting the wind chill) and as high as 120°F (49°C), reported in 2011 and 1936, respectively. Furious super-cell storms develop within minutes and then rip through an area like a blender on its highest setting. Straight-line wind gusts can reach 150 mph (241 kmph; tornadic winds can be much higher) and hailstones, routinely the size of golf balls and sometimes softballs (the record is a 6-in [15-cm] stone!) are pelted out the sky at unbelievable speeds. Or, day after day, torrential rains will pour from the heavens and then, abruptly, there will be years with barely a drop of precipitation.

Seasonal weather patterns are strongly influenced by the Rocky Mountains to the west and the Gulf of Mexico to the southeast (Johnson and Duchon 1995), diverting cold, dry air and warm, humid air, respectively, into the state (the frequent confluence of those air masses is why Oklahoma sits within "Tornado Alley"). Temperature and precipitation are the primary variables that influence the distribution of plants and animals in Oklahoma. Both form distinct gradients: decreasing average annual precipitation from east to west, and decreasing average annual temperature from south to north (Johnson 2008; Figure 3.2). Of the two, however, precipitation has the most pronounced influence on species distributions.

Average annual temperatures tend to be highest in the counties that border the Red River, and lowest in the panhandle and a sliver of northeastern Oklahoma. For example, Harmon County in the southwest and McCurtain County in the southeast have comparable mean annual

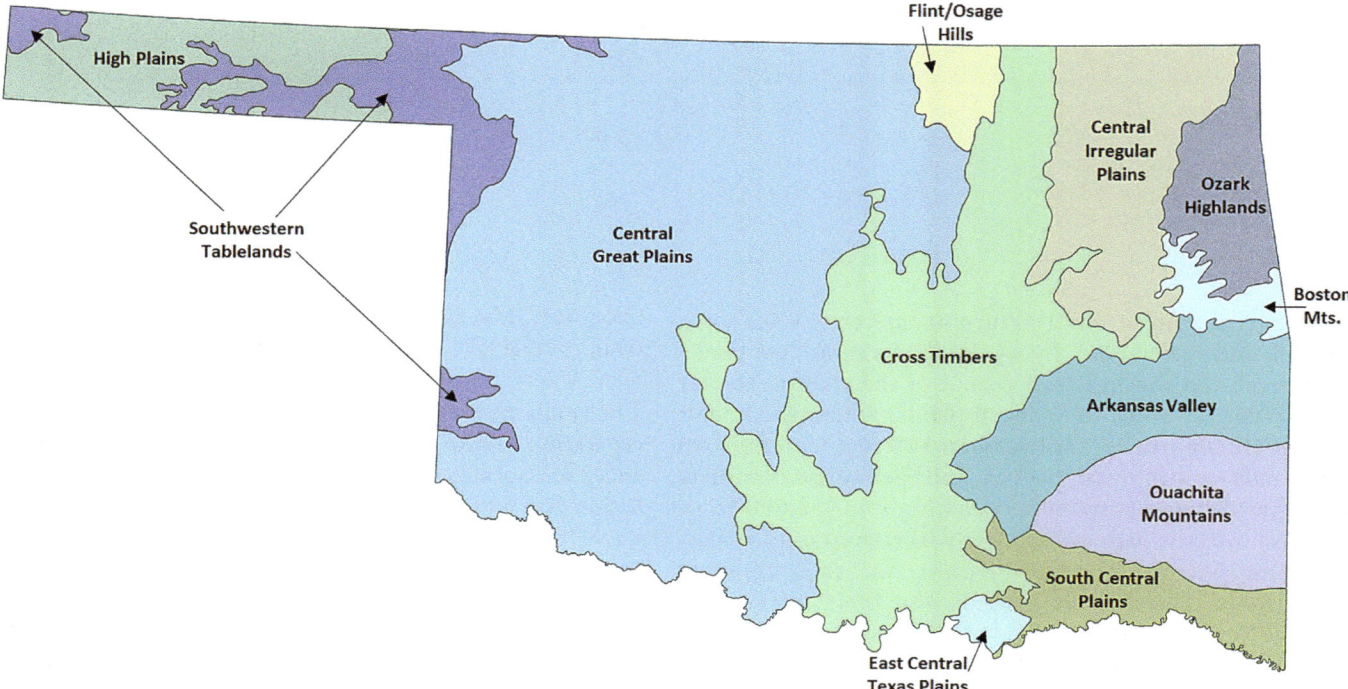

Figure 3.1 Ecoregions of Oklahoma. These ecoregions are mentioned throughout the book. They are based on the U.S. Environmental Protection Agency's Level III ecoregions map (Omernik and Griffith 2014).

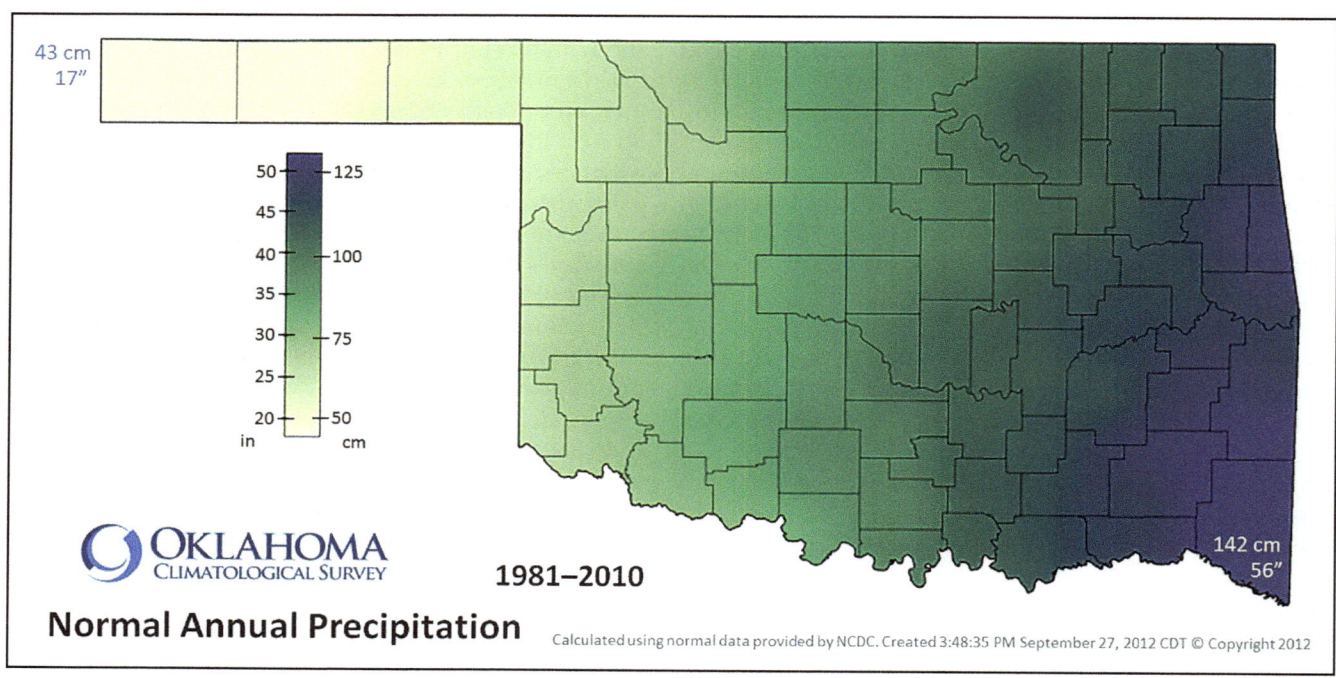

Figure 3.2 Oklahoma and the southern Great Plains have a striking east–west precipitation gradient, bested only by that found in the Pacific Northwest. Used with permission by the Oklahoma Climatological Survey.

(63°F/17°C) and average maximum (75°F/24°C) and minimum (49°F/9°C) temperatures, whereas Cimarron County, in the western panhandle and also with the highest elevation in the state, is overall much colder (56°F/13°C, 71°F/22°C, 40°F/4°C, respectively; Oklahoma Climatological Survey 2019). Not surprisingly, the southeastern counties (with the exception of the Ouachita Mountains) have the earliest dates (mid-April) for last freeze in spring and northwestern Oklahoma and the panhandle have the latest (late May; Fig. 8.1), meaning that the southeastern counties emerge from winter earlier, which means odonates begin to fly earlier in the year there than they do in the far western part of the state.

Precipitation is highest in the spring months with a secondary peak in the fall. Nevertheless, the amount of precipitation varies considerably from year to year. Droughts, periods in which precipitation is significantly below monthly and annual averages, are common. Severe to extreme drought recurs at roughly 20-year intervals (Svoboda et al. 2002). Spring rains, often heavy, can bring relief. Southeastern Oklahoma, with its subtropical humid climate (Trewartha 1968), receives the most spring precipitation (March–May), ranging from 4.5–5.8 in (11.4–14.7 cm), whereas the far panhandle receives a mere 1.2–2.6 in (3.1–6.6 cm) in the spring; annual precipitation is extreme, too, with 56 in (142 cm) in the southeast and only about 17 in (43 cm) in the far western panhandle. A sharp decrease in humidity accompanies the precipitation gradient. Plants in the panhandle experience greater moisture stress from evapotranspiration than in the southeast. Average annual relative humidity is 79% in McCurtain County and 59% in Cimarron County (Oklahoma Climatological Survey 2019).

Geology

Regional surface geology provides a template for topography, soils, hydrology, and vegetation. The physical and chemical characters of rock formations affect the types of soil that lay above, the ruggedness of the landscape (Figure 3.3), and the microclimates that result. The surface geology of Oklahoma is primarily a story of sedimentary rocks, testaments of seas and marine environments long past. But igneous formations breach the surface in the Wichita and Arbuckle Mountains, as well as within the Ozark Plateau (Johnson 2008). In this section, we will review the general features of the (primarily) surface geology of Oklahoma, starting with the buttes of far-northwestern Oklahoma and ending in the eastern mountains. We recommend that the reader refer to the excellent summary of Oklahoma geology presented by Johnson et al. (1972; http://ogs.ou.edu/docs/ed ucationalpublications.EP1.pdf) for further details, as well as to follow along with the following discussion, as we intend here to provide but a glimpse into this foundational topic.

The northwest corner of the panhandle is a land of shale and sandstone buttes. Chief among them is Black Mesa, the easternmost extent of Mesa de Maya of New Mexico and Colorado. Although the average relief at Black Mesa is but 551 ft (168 m), it is the highest point in the state at 4,902 ft (1,494 m). The name Black Mesa is derived from the cap of brownish-red to black basalt deposited in the Late Tertiary Period, which issued from Piney Mountain, an extinct volcanic vent to the west (Rothrock and Noe 1925). This corner of the state was once a land of giants. Excavations by geologist John Willis Stovall from 1935–1942 in the Jurassic Morrison Formation produced fossils of several dinosaur taxa (Moore 1953). Biologists in Oklahoma have been attracted to this corner of the state for decades because of the species present that have Rocky Mountain (or just western) affinities not found downstate. For example, although the predominant, woody plant is one-seed juniper (*Juniperus monsperma*), which co-occurs with pinyon pine (*Pinus edulis*) on exposures of the Cheyenne Sandstone, the only stand of ponderosa pine (*P. ponderosa*) in Oklahoma is found in the far panhandle (Rogers 1953). The same can be said of odonates with western affinities; for example, the Sooty Dancer (*Argia lugens*).

Beneath the remainder of the panhandle, and stretching east to portions of Ellis, Roger Mills, and Woodward counties, lay deep deposits of sand, gravel, and clay. This formation is a product of the Laramide orogeny (the uplift of the Rocky Mountains), which resulted in massive erosion. These materials were transported east by streams that were eventually overwhelmed by the quantity of material and were themselves buried. Nonetheless, the outwash continued as a broad, expansive sheet, resulting in deposits ranging from 200–600 ft (61–183 m) deep; the topography of the panhandle is level to gently rolling as a result. Since the Ice Age, water has filled the porous spaces of the gravels, filling the High Plains (Ogallala) aquifer (Hart et al. 1976).

The iconic red soils of western and central Oklahoma clearly illustrate the effect of geology on landscape development. The distinctive coloration is derived from sandstones and shales rich in iron oxides, themselves a product of incursions of oceanic waters in the Permian Period. During that time, seas receded and expanded across the region, resulting in the development of evaporite deposits in the intervals. Chief among them are gypsum, anhydrite, and halite with associated dolomite (Curtis et al. 2008). Gypsum has left a unique legacy on the natural history of western Oklahoma. First, the dissolution of gypsum over time produced a landscape pockmarked with sinkholes, caves, and springs, creating what geomorphologists refer to as karst topography. The larger caves host a variety of bats and other cave organisms. Secondly, as gypsum deposits erode, salt and other minerals are carried away by run-off into springs and streams, producing water that is high in salinity. When rivers overflow their banks following spring rains, the backwaters are flooded. But as the waters recede and moisture evaporates from the soil, a glaring white encrustation of salt emerges. Several salt flats are dotted across western Oklahoma (Gould 1900). The Great Salt Plains of northwestern Oklahoma may be the best known of these salt flats, but it is the "Jackson Salt Plains" in southwestern Oklahoma that hold key populations of the Seaside Dragonlet (*Erythrodiplax berenice*) and the Bleached Skimmer (*Libellula composita*) and, indications are, a population of the Bronzed River Cruiser (*Macromia annulata*).

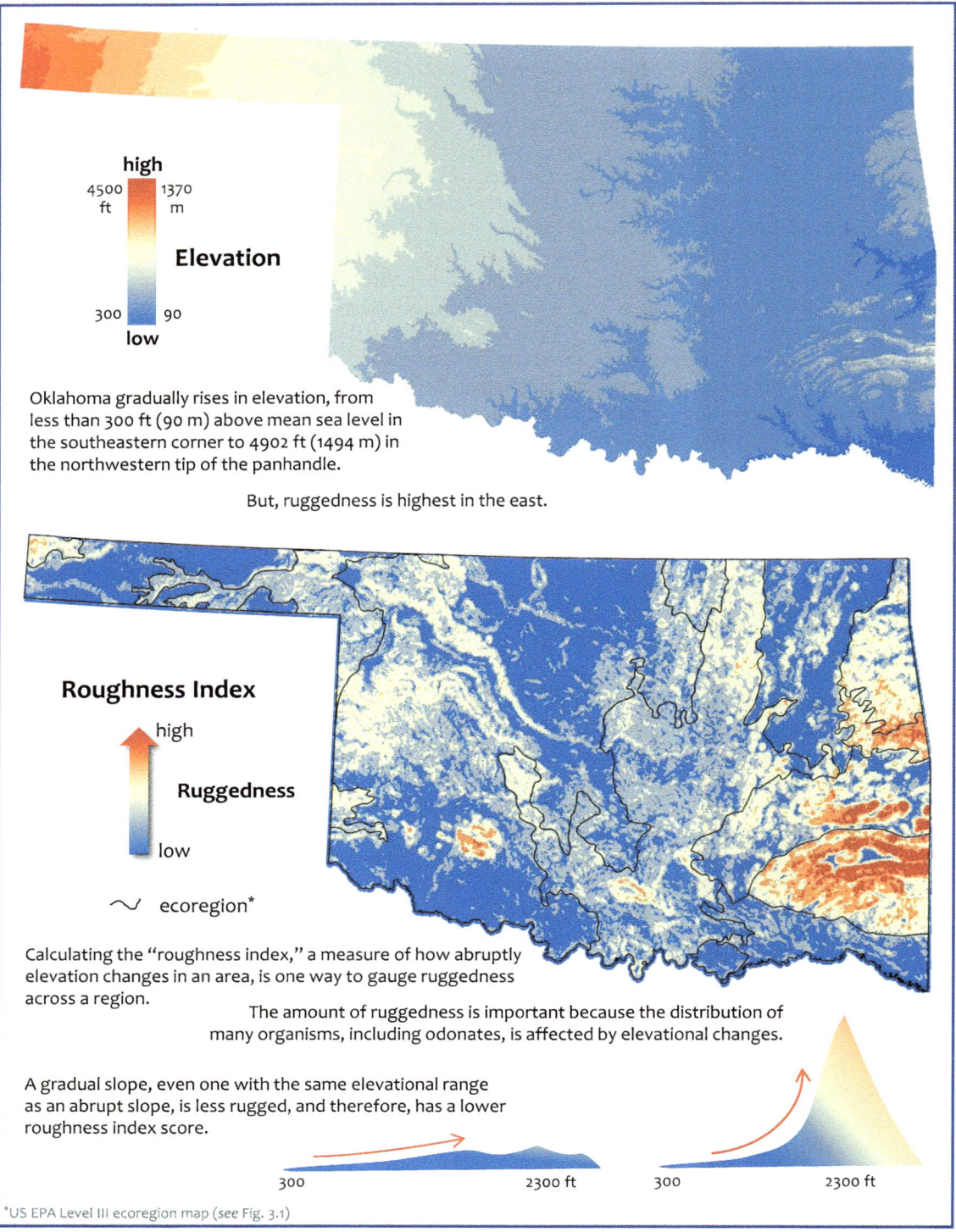

high

4500 ft   1370 m

Elevation

300   90

low

Oklahoma gradually rises in elevation, from less than 300 ft (90 m) above mean sea level in the southeastern corner to 4902 ft (1494 m) in the northwestern tip of the panhandle.

But, ruggedness is highest in the east.

Roughness Index

high

Ruggedness

low

〜 ecoregion*

Calculating the "roughness index," a measure of how abruptly elevation changes in an area, is one way to gauge ruggedness across a region.

The amount of ruggedness is important because the distribution of many organisms, including odonates, is affected by elevational changes.

A gradual slope, even one with the same elevational range as an abrupt slope, is less rugged, and therefore, has a lower roughness index score.

300   2300 ft   300   2300 ft

*US EPA Level III ecoregion map (see Fig. 3.1)

Figure 3.3 Oklahoma elevation versus ruggedness.

Jutting from the south and central portions of this rolling red landscape is a jagged terrain much older than its surroundings. The Arbuckle and Wichita Mountains, and the Ouachita Mountains to the east, were formed at roughly the same time in geological history and all hold distinct, and often disjunct, odonate populations, but the effect of these mountains on the modern landscape are quite different. The Arbuckles cover a greater extent (1,000 mi², or 2,600 km²) than the Wichitas (737 mi², or 1,857 km²), yet both were formed along an east to west axis in the Pennsylvanian Period. Both were much higher in elevation at the conclusion of the uplift (orogeny) than today, and subsequent erosion has deposited copious quantities of sediments into basins north and south. Millions of years of erosion have exposed granite, rhyolite, and gabbro in the Wichitas that date from the Cambrian Period, but the igneous rocks exposed in the Arbuckles are much older and are of the Precambrian age (Curtis et al. 2008; Johnson 2008).

As the Wichita Mountains eroded, vast quantities of rocks and sediments were deposited into what became the Anadarko Basin to the north and the Hollis Basin to the south. The action of wind and water carved canyons, crevasses, and fluted mountain sides, creating pockets of location-specific climate (i.e., climate that differs from the general conditions surrounding that location; these pockets are also referred to as microclimatic refugia) that allow non-southwestern biota to survive there. Often these species have eastern affinities, relics that stand testament to a past climate regime. The remnant sugar maple (*Acer saccharum*) populations on Mount Scott, in Hollis Canyon (Buck 1964), and in deep canyons cut from Permian Period sandstone (Little 1938) to the north are fine examples of a northeastern North American species that occurs where one would not expect.

Abutting the Wichita Mountains to the north along a Frontal fault zone are the Slick Hills, a region of sedimentary rocks dating from the Paleozoic Era. This is one of three major surface exposures of limestone in the state, the others being in the Arbuckle Mountains and the Ozark Plateau. The Slick Hills are low, gently rolling, and draped in mixed grass prairie riven by outcrops of limestone, dolomite, and conglomerate formations, many of which are exposed in the Arbuckle Mountains as well. Springs and spring-fed streams are common due to the solubility of these sedimentary formations (Curtis et al. 2008; Johnson 2008).

The Arbuckle Mountains rise out of the surrounding prairie, sloping from an elevation of 1,350 ft (411 m) in the west to 750 ft (229 m) in the east. The Arbuckles are often subdivided into two units: the Arbuckle Plain to the east and the Timbered Hills to the west. The Arbuckle Plain is predominately Cambrian granites and the topography is rolling but relatively level. The Timbered Hills are composed of limestones, shales, and conglomerates that range in age from the Ordovician to Pennsylvanian (Blome et al. 2013). These deposits were tortuously folded and faulted during the formation of the Arbuckles. The result is a landscape laced with spring-fed and travertine-lined streams (Fairchild et al. 1990). These springs and numerous artesian wells boosted an industry in mineral bath resorts in the cities of Sulphur and Bromide in the early 20th century. The majority of springs are on private land, but the Chickasaw National Recreation Area (formerly, Platt National Park) protects many of these features (Barker and Jameson 1975). The highest point (approximately 1,410 ft, or 430 m) in the Arbuckles is an unnamed peak in the Timbered Hills.

The Arbuckles have proven to be rich in natural resources. Quarries for limestone, sandstone, and granite are found throughout the range. The granite at the state capitol was quarried in the Arbuckles near Mill Creek. Even asphalt, oozing up from oil deposits within the mountains, was mined. Rich deposits of oil and gas occur on the west flank of the Arbuckles and in the Ardmore Basin to the south (Suneson 1997). The energy portfolio now includes wind turbines, which crown the south ridges of the mountains.

Thousands of feet below the surface of central Oklahoma, Permian rock formations yield to the Pennsylvanian on a contact zone, although it is not evident on the surface until well to the east. The transition is evidenced by a change in the color of surface rocks and soils from red to yellow. The Ouachita Mountains, Gulf Coastal Plain, and Ozark Plateau are the most remarkable features in this area.

The Ouachita Mountains are a rugged and beautiful landscape that support many habitats, among them, forested seeps hosting endemic and rare plants and animals such as the Ouachita Spiketail (*Cordulegaster talaria*), a dragonfly known only from these mountains. The Ouachitas consist of three provinces: the Hogback Frontal Belt, the Ridge and Valley Belt, and the Beavers Bend Hills. From the Arkansas border, the Hogback Frontal Belt and the Ridge and Valley Belt trend to the southwest (Curtis et al. 2008). The Hogback Frontal Belt consists of Pennsylvanian Period sandstone, with a relief of 500–1,500 ft (152–457 m) above the valleys below. The Ridge and Valley Belt is also composed of Pennsylvanian Period sandstone displays in long ridges (Miser 1929). Rich Mountain is the highest point (2,600 ft, or 793 m) in the Oklahoma portion of the Ouachitas and the second highest point overall, the highest being where the mountain falls within Queen Wilhemina State Park in Arkansas (2,681 ft, or 817 m).

The Beavers Bend Hills and the Potato Hills to the west are smaller in extent than the other two provinces. The Beavers Bend Hills are located in the southern Ouachitas near the Arkansas border and extend to the Gulf Coastal Plain. They consist of sedimentary formations ranging in age from the Cambrian (limestone, dolomite, and sandstone) to the Devonian (limestone and dolomite capped by a black shale folded into a series of low hills that extend to the Gulf Coastal Plain). They are clearly visible from the southern ridges of the Ridge and Valley Belt. The Potato Hills near Talihina are an east–west aligned cluster of low (200–500 ft, or 61–152 m, in relief), rolling hills that are bound by the Winding Stair fault to the north and the Octavia fault to the south. The surface rocks date from the Ordovician to the Devonian Periods. Stanley Shale provides loose footing along the steeper slopes (Arbenz 1968).

The Gulf Coastal Plain lies just south of the Ouachita Mountains and extends westward just south of the Arbuckle Mountains. It is here, where the topography is gently rolling to level, that one finds the lowest elevation in Oklahoma, at a mere 292 ft (89 m) above mean sea-level. The geology consists primarily of gravel, clays, and limestone of the Cretaceous Period (Curtis et al. 2008). The Antlers

Sand Formation follows the northern extent of the province and is characterized by sand hills that are up to 490 ft (150 m) in depth (Hart and Davis 1981). Numerous seeps and springs issue form the Antlers Sands and form what are locally referred to as bogs, though they are not technically so. These wetlands foster a number of plant species that are not found elsewhere in the state (Johnson 2008). It was in the Antlers Formation that John Willis Stovall discovered the theropod dinosaur *Acrocanthosaurus atokensis* named for Atoka County (Stovall and Langston 1950).

The Arkansas Hill and Valley Belt, rolling topography carved by the Arkansas and Canadian rivers, separate the Ozark Plateau from the Ouachita Mountains. The Ozark region (or Ozark Plateau) consists of two units, the Boston Mountains to the south and the Springfield Plateau to the north. The Boston Mountains are rugged but gently dipping hills with relief of 50 ft (15 m) above valley floors to about 2,400 ft (732 m). The surface geology is of sandstone and shale of the Pennsylvanian Period. Topography of the Springfield Plateau consists of broad plateaus that are deeply dissected at the margins and intervening broad floodplains. The geology is Mississippian Period sandstone and shale in horizontal beds. The result is a karst landscape populated by numerous springs and caves. During periods of heavy rainfall, water will seep from contacts between rock formations that are dry for most of the year (Adamski et al. 1995). Travertine is not prominent in the streams (as it is in the Arbuckle Mountains), the beds of which are often covered in deep layers of cherty rocks. Numerous threatened and endangered species are found within the Ozark Plateau, such as the Ozark cavefish (*Amblyopsis rosae*), the Oklahoma cave crayfish (*Cambarus tartarus*), and the Ozark Big-eared Bat (*Corynorhinus townsendii ingens*). Although divided by a substantial distance, the Ozarks and Ouachitas host populations of regional endemic odonates such as the Ozark Emerald (*Somatochlora ozarkensis*).

Hydrology

Oklahoma has a varied hydrological landscape that includes flowing (lotic) and still (lentic) surface waters and those harbored underground (aquifers). When most people think of heartland states, water is not generally what comes to mind, but in Oklahoma, aquifers, springs, seeps, streams, ponds, and lakes (with caveats) abound.

There are 22 major aquifers lying under Oklahoma (Johnson et al. 1972; OKWRB 2018). These hold about 390 million acre-feet of water, which is a little more than all the water contained within Lake Erie. However, about half of that water is currently too deep for extraction. The largest aquifer, the High Plains (or Ogallala) aquifer, spanning from Texas to South Dakota, underlies much of the Oklahoma panhandle and portions of the northwestern part of the state (Fig. 6.10). Oklahoma has access to 90 million acre-feet of that aquifer's storage, which would be "enough to cover the state two feet deep" (OKWRB 2018) and is equivalent to almost 40 Lake Eufaulas (or 1 Lake Nicaragua). That

apparent plentifulness has persuaded Oklahomans to drain that aquifer for all it is worth … but that is a topic for a later chapter (see Chapter 6). Not all of Oklahoma's multitude of springs and seeps originate from aquifers, but many do, and many flow into wetlands and even form streams. In recent decades, flow has been reduced from some of those springs, seeps, and streams as a result of excessive aquifer use, but that too is a topic for a later chapter.

It cannot rightly be said that Oklahoma is wanting for lotic habitat, even despite losses that have occurred in historical times (see Chapter 6). Some of the major rivers that traverse the state include the Arkansas, Neosho, Cimarron, Washita, Red, Kiamichi, Little, South Canadian, and, the state's longest at 752 miles (OKWRB 2018), the North Canadian, which is known as the Beaver River in western Oklahoma and the Oklahoma River in the Oklahoma City area. Many hundreds of smaller streams act as tributaries to those rivers. As with the rest of the United States, all rivers, streams, and other lotic waters in Oklahoma are grouped into drainage basins and watersheds that get progressively smaller.[1]

Surprising as it may be, there is great variety in the general composition of Oklahoma's lotic habitat. The same can be said broadly of its lentic habitat but, relatively, that is much more limited. To categorize streams in an orderly manner, stream ecologists evaluate several physical features, such as stream width, shoreline or bank characteristics, depth, composition of streambed, and associated vegetation. In Oklahoma, streams can be <50 cm to 100 m wide (1–300 ft), with depths of but a couple of centimeters to many meters deep (an inch to many feet), but the state's major rivers can be many hundreds of meters wide (thousands of feet) and many times deeper than its streams. With that variability in size comes differences in bank characteristics: some may be at the approximate level of the water, others gently or abruptly slope, and others may be deeply cut. Streambanks and beds can be rocky, sandy, intertwined with root masses, strewn with vegetation or leaf litter, moss-laden, tree-lined (open or closed canopy), have overhanging vegetation, be completely devoid of vegetation, or any combination of these (or a hundred other) variables.

Stream courses may be sunny or shaded, have emergent vegetation or not, be clear of debris or heavily laden with downed trees, fallen branches, and other organic matter. Substrates can be muddy, silty, sandy, pebbly, gravely, rocky, boulder-strewn (beach-ball to truck-sized), be of bedrock, or, as is often the case, vary in substrate along the stream course. Soils can be richly organic or comparatively sterile. Water can be murky or, as is the case with spring-fed streams, crystal clear. Water chemistry (e.g., pH, oxygen, nitrogen), temperature, and flow vary too. For example, spring-fed streams are often cold and at least moderately flowing, whereas slow-flowing streams are warmer. Streams may be permanent or ephemeral, with but remnant pools left in the driest parts of the year or drying completely. During wetter times of the year, some streams carry heavy loads of rainwater that severely scour their courses; others flood out into marshland or backwaters. Streams can also be impacted by non-natural occurrences, such as development, agriculture, and non-native animals (such as cows and feral pigs). And, of course, associated aquatic organisms can vary by stream, for instance, fish, crayfish, macro- and micro-invertebrates, diatoms, and algae may or may not be found. In sum, the variety of and within streams is seemingly limitless.

As we will go into in greater detail later, Oklahoma lost a lot of its lotic habitat because, in the midst of and for a few decades after the American Dust Bowl, there was a concerted effort to dam streams and rivers to provide for public and private water storage. Those dams created Oklahoma's largest lakes including, in order of size, Robert S. Kerr Reservoir, Grand Lake O' the Cherokees, Lake Texoma, and Lake Eufaula, the latter being the largest. Those lakes are between 250 and 624 miles around their shorelines and are 43,800 and 105,500 acres in size (think 43,800 American football fields on up!). It is estimated that there are now >2,300 lakes and at least 220,000 agricultural ponds in the state (Hoagland 2006a)—accounting for almost 56,000 miles of shoreline and 1,401 mi² in area, which is a larger area than the state of Rhode Island (OKWRB 2018).

All of this in a state that previously had no large naturally occurring lentic habitat, the largest sources formerly being small oxbow lakes and ephemeral playas.[2]

A final comment on Oklahoma's hydrological landscape regards the variability in streamflow and lake levels depending upon weather. It is astounding how quickly streams go from being scoured by heavy rains, to overflowing their banks, to bone dry and deeply cracked. The same can be true for lakes too. It all leaves one wondering how that affects Oklahoma's flora and fauna.

Vegetation communities and classification

Herein we describe some of the dominant vegetation communities in Oklahoma. In the ecoregion section below, we will delve further into how vegetation communities have been used alone or in combination with other factors (such as climate or soils) to define ecoregions in North America and Oklahoma. Briefly stated, like ecoregion maps, vegetation maps have been produced with both broad and fine scales in mind and everything in between. For example, Transeau (1903; Figure 3.4) presented a vegetation map of North America with a mere three communities for Oklahoma: Plains, Prairies, and Deciduous Forests. Shantz and Zon's

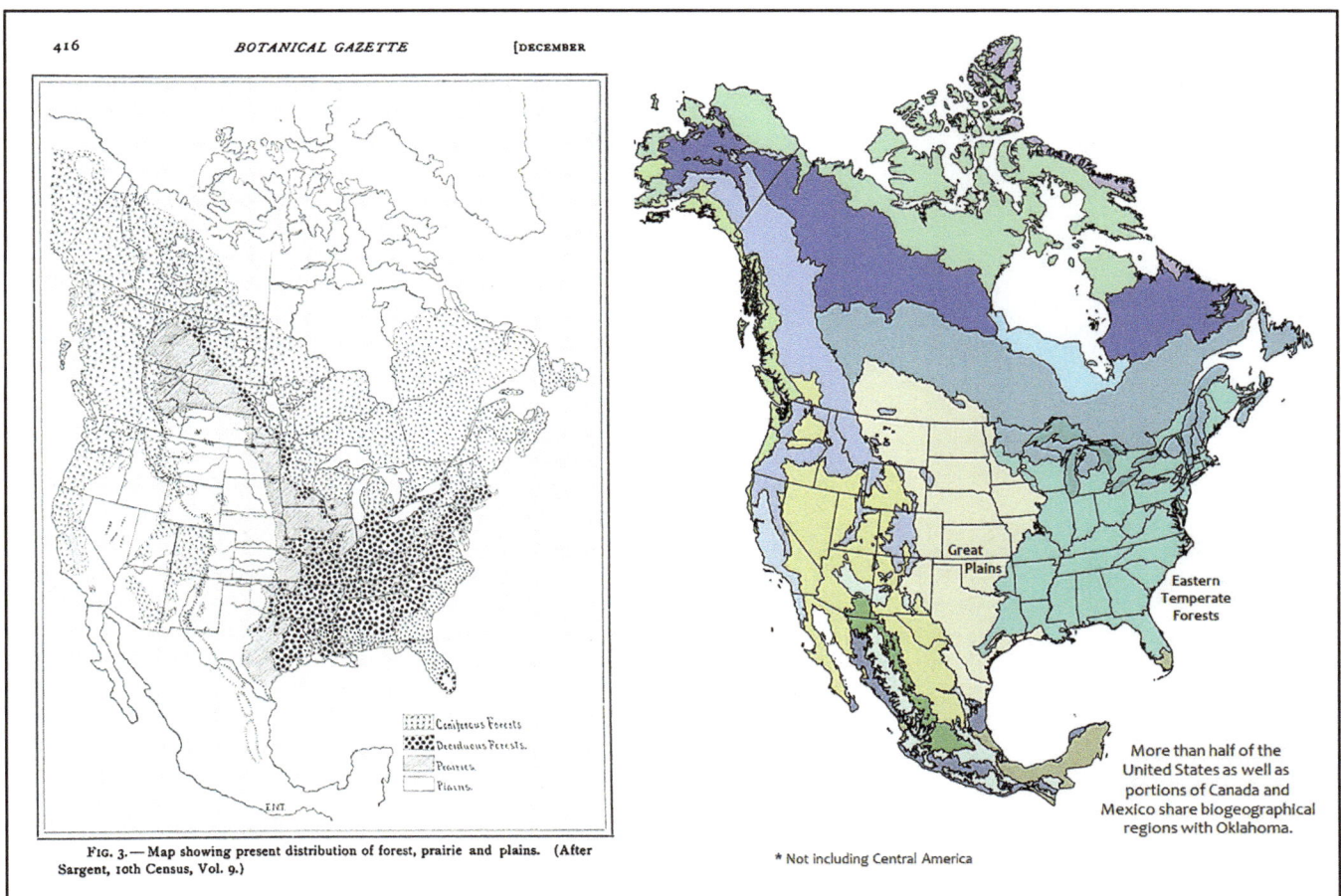

Figure 3.4 Edgar N Transeau's 1903 map showing distribution of forests, prairies, and plains in North America (left, based on Sargent 1884) compared to the U.S. Environmental Protection Agency's Level I ecoregions map of North America* (right, Omernik and Griffith 2014).

(1923) *Natural Vegetation* map is undeniably a gorgeous specimen, even if it reduced Oklahoma to but four vegetation types: Short Grass (Plains Grassland), Tall Grass (Prairies Grassland), and the Oak-Hickory and Oak-Pine Forests. Bruner (1931) offered 6 vegetation regions and Duck and Fletcher (1943) offered 15 communities (Figure 3.5). Many decades passed before Oklahoma's vegetation classification was modernized. Hoagland (2000) greatly refined vegetation communities across the state by classifying 151 associations. Hoagland's vegetation classification of Oklahoma informed the ecoregion maps and GIS layers that are most commonly used by biogeographers today for Oklahoma, including the Level III (Figure 3.1) and IV maps discussed below, as well as the Oklahoma Ecological Systems Mapping Project, which resulted in further refinement of classifications, ground truthing, fine-tuning the state's land cover map, and recognizing 165 ecological systems (Diamond and Elliot 2015; Figure 3.5).

Floral diversity in Oklahoma exhibits similar patterns to terrestrial vertebrate and invertebrates by trending with higher richness in the east. For plants, the Ouachita Highlands and the Ozark Plateau are the most floristically diverse areas in the state (Hoagland 2000). For odonates,

the Ouachita Highlands and the Ozark Plateau, in the broad sense, are species rich but are bested by the South Central Plains. Describing vegetation communities within these regions and elsewhere in the state can be difficult and take many forms. Because odonates need water for their survival, we chose to focus on descriptions of wetland vegetation communities here, in general terms, as well as dividing the state into western and eastern halves, and to relate those communities to the ecoregion map used throughout the book (Figure 3.1).

Wetland plants have unique characteristics. They are adapted to overcome diminished soil oxygen (a condition known as anoxia) from prolonged inundation and the changes in soil chemistry that ensue. Some plant species better tolerate flooding and saturated soils than other species, even within wetlands, so hydroperiod (timing and duration of flooding and soil saturation) will dictate what communities will be present (Keddy 2010). Wetlands that have experienced complete drawdown over the course of summer or winter are usually recharged or flooded during the spring. Of course, the quantity of precipitation and duration of flooding varies both interannually and with geography, and soils of high clay content tend to retain water at the surface

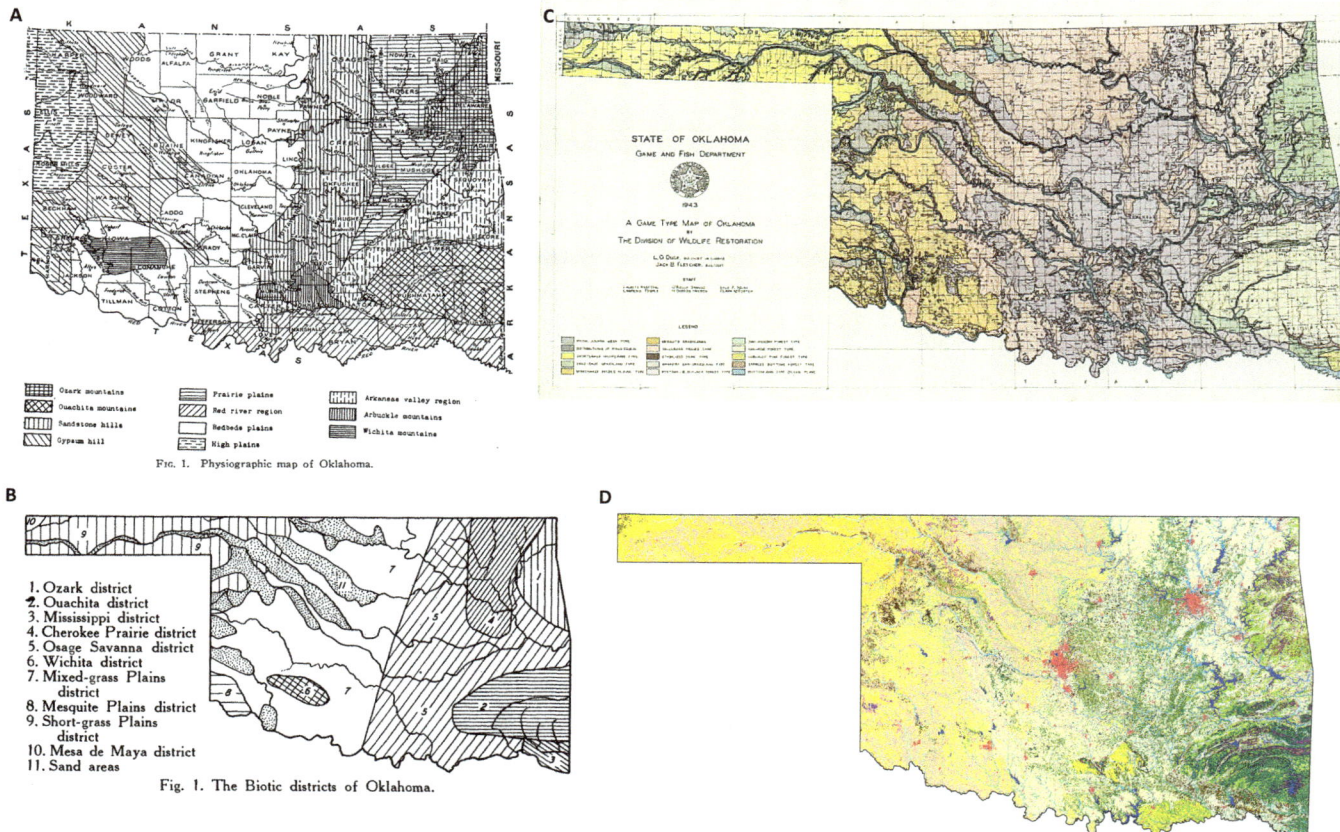

Figure 3.5 **A sampling of vegetation/ecoregion maps of Oklahoma. A) Bruner's (1931) map from *The vegetation of Oklahoma* B) Blair and Hubbell's (1938) map from *The biotic districts of Oklahoma* C) Duck and Fletcher's (1943) *A game type map of Oklahoma* D) Oklahoma Ecological Systems Mapping Project map.**

longer than soils of high sand content, so there is much variability overall. But generally, wetlands in semi-arid western Oklahoma experience low annual precipitation, low humidity, and high evaporation rates, with the opposite being true for wetlands in eastern Oklahoma.

Some wetland plants are restricted to specific ecoregions, others span multiple ecoregions or occur statewide. Among wetland plants, nodding (*Persicaria lapathifolia*) and Pennsylvania (*P. pensylvanica*) smartweeds are species of emergent wetlands that can be found in marshes and wet habitats throughout the state. More-restricted species include the yellow water buttercup (*Ranunculus flabellaris*) and the Florida mudmidget (*Wolffiella gladiata*) require stable water levels and are known only from the South Central Plains ecoregion in McCurtain County.

Cowardin et al.'s (1979) classification of wetland habitats, the most frequently adopted in the United States, employs multiple broad vegetation descriptors, four of which will be used here: forested, scrub-shrub, emergent, and aquatic beds. As one may gather, forested wetlands are characterized by the presence of trees and are most diverse and common in eastern Oklahoma. The remaining three categories can be found throughout the state and are associated with river floodplains, lakes, and unique surface features. Scrub-shrub wetlands are dominated by small trees or shrubs, emergent wetlands by herbaceous species whose stems grow above the water surface, and aquatic beds by plants that grow beneath the water surface or with floating leaves.

Western Oklahoma

Wetland vegetation does not readily come to mind when pondering the ecology of western Oklahoma, but the floodplains and associated geomorphic features of many streams are fringed with wetland and even aquatic vegetation. These wetland habitats are not restricted to one region but span the High Plains, Southwestern Tablelands, and Central Great Plains ecoregions. The common growth forms are scrub-shrub and emergent herbaceous wetlands, which often occur together. The number of plant species in these wetlands is lower than those of eastern Oklahoma.

Buttonbush (*Cephalanthus occidentalis*) is the most characteristic species of the scrub-shrub wetlands. In fact, buttonbush is the most widely distributed wetland shrub in Oklahoma, occurring in all but Cimarron and Texas counties, both in the panhandle. This shrub, which is much frequented by pollinating insects, can grow in dense stands that displace other plant species. False indigo bush (*Amorpha fruticosa*) and black willow (*Salix nigra*) share the same distribution as buttonbush and may co-occur with it in scrub-shrub wetlands. Sandbar willow (*S. exigua*), however, is a predominant species in western Oklahoma's riparian wetlands, located on floodplains, sloughs, wet depressions, or fringing impoundments.

Chairmaker's rush (*Schoenoplectus americanus*) and the inland saltgrass (*Distichlis spicata*) are two of the most common species in western herbaceous emergent wetlands. Although they may co-occur, they often form distinct vegetation communities. Soil salinity, depth, and duration of flooding dictate the occurrence and abundance of these two species. Inland saltgrass has a higher tolerance for salinity than chairmaker's rush, but not for inundation. So, in oxbow lakes and ponds, where water tends to reside for longer periods, chairmaker's rush will predominate. Both communities have the same associated plant species though: saltmarsh bulrush (*Bolboschoenus maritimus* ssp. *paludosus*), softstem bulrush (*Schoenoplectus tabernaemontani*), pale spikerush (*Eleocharis macrostachya*), seaside heliotrope (*Heliotropium curassavicum*), foxtail barley (*Hordeum jubatum*), turkey tangle frogfruit (*Phyla nodiflora*),Torrey's rush (*Juncus torreyi*), and rabbitfoot grass (*Polypogon monspeliensis*; Hoagland 2002). These species are also associated with saline-tolerant odonates, such as the Desert Forktail (*Ischnura barberi*), the Bleached Skimmer, and the Seaside Dragonlet.

Scattered across the semi-arid High Plains ecoregion in the panhandle are unique wetland habitats known as playa lakes.[3] In many ways, they do not appear to be wetlands. The hydroperiod in playas is short, due to low annual precipitation and high evaporation. But they do possess soils and vegetation characteristic of a wetland habitat. Playa lakes are basins formed on level topography by the process of deflation; the prevailing southwestern winds lift sediments from the basin and deposit them on the north rim. The lakes reside at the center of a roughly circular basin that ranges in size from a few acres to several acres. The extent of a playa wetland is defined by occurrence of Randall clay, which retains water (Hoagland and Collins 1997).

Vegetation in the playa lakes, or in mesic habitats associated with riparian zones, is dominated by Western wheatgrass (*Pascopyrum smithii*), a cool season perennial grass. In basins with an extended hydroperiod, plants such as wedgeleaf frogfruit (*Phyla cuneifolia*) and annual saltmarsh aster (*Symphyotrichum subulatum*) are found. The diminutive spotted evening primrose (*Oenothera canescens*) is known to occur only in playa lakes (Hoagland 2002). Given the short hydroperiod and interannual variation in precipitation in the panhandle, playas are often plowed and planted in winter wheat. The watersheds that surround playa lakes are often planted in row crops. Such practices result in the deposition of deep layers of sediment atop the Randall clay. In some cases, landowners excavate the center of playas to recover irrigation water and reduce surface area, thus reducing evaporation (Haukos and Smith 1992).

As noted in the geology section, western Oklahoma rivers have transported enormous loads of sand and sediment from the Rocky Mountains to the continental interior. This has created floodplains with deep alluvial deposits. Over time, the prevailing southwestern winds have blown sands and other fine sediments into the adjacent uplands, creating sand dunes in the Central Great Plains ecoregion. In the swales of these dunes, clay has accumulated, producing an impermeable layer that allows water to pool. Spring rains fill many of these ponds, which dry as summer progresses. These interdunal swale ponds range in size from a few square feet to an acre or more. Emergent vegetation differs in the swales depending upon location, size, and duration of flooding. Those with short hydroperiods are typically plowed and planted in winter wheat, just as playa lakes often are. Those with extended hydroperiods are characterized by blue mudplantain (*Heteranthera limosa*), round-leafed water hyssop (*Bacopa rotundifolia*), hooded arrowhead (*Sagittaria calycina*), and longbarb arrowhead (*Sagittaria longiloba*). Those with shorter hydroperiods foster a carpet of spike-rushes (*Eleocharis* spp.), a plant genus strongly associated with the Western Red Damsel (*Amphiagrion abbreviatum*). In either case, associated species may include scarlet tooth-cup (*Ammania coccinea*), barnyard grass (*Echinochloa crus-galli*), bearded sprangletop (*Leptochloa fusca* ssp. *fascicularis*), and hairy waterclover (*Marsilea vestita*; Hoagland 2002).

Eastern Oklahoma

Eastern Oklahoma has a diverse array of wetland vegetation communities, ranging from herbaceous, scrub-shrub, and forested types. In lentic waters, the common rush (*Juncus effusus*) forms densely with a variety of associated species, such as wingleaf primrose-willow (*Ludwigia decurrens*), marsh seedbox (*L. palustris*), swamp rose mallow (*Hibiscus moscheutos*), rice cutgrass (*Leersia oryzoides*), and swamp smartweed (*Polygonum hydropiperoides*). Emergent wetlands featuring American water willow (*Justicia americana*) are found commonly at lake margins and along the banks of rocky streams; when on streams, it often hosts odonates such as the American Rubyspot (*Hetaerina americana*) and, sometimes, the Smoky Rubyspot (*H. titia*). It occurs alone or with associates, such as rice cutgrass, smallhead doll's daisy (*Boltonia diffusa*), and ditch stonedrop (*Penthorum sedoides*), and it often grows into or across stream channels. Hornleaf riverweed (*Podostemum ceratophyllum*) is a tropical species of the riverweed family (Podostemaceae). Oklahoma is at the northwestern edge of this species' geographic range where it is found in the Ouachita Mountains clinging to rocks in streams in association with American water willow (Hoagland 2000).

Plant communities characterized by floating leaf or submerged aquatic plants are common in eastern Oklahoma. Two such communities are dominated by members of the waterlily family (Nymphaceae). Yellow pond-lily (*Nuphar lutea*; also commonly called spatterdock) can cover the surface of small, shallow waterbodies in southeastern Oklahoma. The leaves will be either floating or raised above the surface, depending upon water depth. American water lotus (*Nelumbo lutea*) leaves respond to water in a similar fashion. Of the floating leaf communities described here, examples can be found as far west as Caddo County. American white waterlily (*Nymphaea odorata*) communities are less common, although visually striking. The associated species are similar in both communities and may include humped bladderwort (*Utricularia gibba*), waterthread pondweed (*Potamogeton diversifolius*), coontail (*Ceratophyllum demersum*), longleaf pondweed (*Potamogeton nodosus*), water knotweed (*Persicaria amphibia*), and swamp smartweed. Water shield (*Brasenia schreberi*) is a common species in small ponds in much of eastern Oklahoma. It can be readily identified by its distinctive leaf shape and the gelatinous coating on the underside of the leaves and on the stems. Typical associated species are humped bladderwort and waterthread pondweed. Floating primrose-willow (*Ludwigia peploides*) fringes waterbodies with mounds of green vegetation and bright yellow flowers (Penfound 1953). This community occurs in ditches and streams with slow-flowing to still-water, and ponds. The flowers are much larger than other floating primrose species, with the exception of large-flower primrose-willow (*L. grandiflora*), which is not a species native to North America but one known from six localities in Oklahoma.

The South Central Plains harbor a unique complex of seeps and springs that are either forested or open habitats. These wetlands are associated with a sequence of sandhills that were deposited in the Cretaceous as part of the Mississippi Embayment. Locally, these wetlands are referred to as "bogs," some of which contain sphagnum and associates (e.g., Lescur's sphagnum, *Sphagnum lescurii* and common haircap moss, *Polytrichum commune*). In open, full-sun conditions, the predominant species are false nettle (*Boehmeria cylindrica*) and velvet panicum (*Dichanthelium scoparium*). Mosses and members of the sedge family (Cyperaceae) are also abundant. These wetlands harbor a number of plant species that are rare in Oklahoma (Hoagland 2000).

Although forest vegetation occurs on river floodplains statewide, those exhibiting true wetland characters (such as seasonally flooded bottomland) are limited to eastern Oklahoma. In the South Central Plains and up into the Ouachitas, a continuum of forest types exists, ranging from baldcypress (*Taxodium distichum*) swamps, which give way to overcup oak (*Quercus lyrata*) and water hickory (*Carya aquatica*), and finally willow oak (*Quercus phellos*) and blackgum (*Nyssa sylvatica*) on moist to mesic sites. These are some of the most species-diverse forests in Oklahoma. The baldcypress swamps are restricted to the Little River basin, where they occur on sloughs, backswamps, and other deepwater habitats. Baldcypress also grows along the banks of the Mountain Fork River and some tributaries, but it is not a dominant species (Little 1980). The number of associated woody species is few, but includes Virginia sweetspire (*Itea virginica*) and American snowbell (*Styrax americanus*; Hoagland et al. 1996). Baldcyress swamps rarely exhibit closed canopies. The openings vary in extent and allows for the establishment of emergent wetland and floating leaf aquatic communities, including various duckweeds. Swamp knotweed (*Persicaria glabra*), which in Oklahoma is known only from McCurtain County, can form dense mats in these openings.

Overcup oak and water hickory (*Carya aquatica*) forested wetlands occupy a geomorphic setting similar to baldcypress. In "drier" situations, there will be willow oak and blackgum forests. Both forests types usually have closed canopies, unlike baldcypress swamps, and share many of the same associated species, including American hornbeam (*Carpinus caroliniana*), parsley hawthorn (*Crataegus marshallii*), red elm (*Ulmus rubra*), water oak (*Quercus nigra*, which also can be a co-dominant), sweetgum (*Liquidambar styraciflua*), honey locust (*Gleditsia triacanthos*), planetree (*Planera aquatica*), red mulberry (*Morus rubra*), bitternut hickory (*Carya cordiformis*), and mockernut hickory (*C. tomentosa*; Blair and Hubbell 1938; Hoagland et al. 1996). In floodplain forests that have been altered or disturbed, sweetgum, and honeylocust are abundant in early successional stages.

In northeastern Oklahoma, forested wetlands characterized by pin oak (*Quercus palustris*) and pecan (*Carya illinoensis*) are common on the floodplains of the Deep Fork, Verdigris, and Neosho rivers. Pin oak does occur in southern Oklahoma, but it does not predominate stands (Blair 1938). Scrub-shrub wetland communities of eastern swamp privet (*Forestiera acuminata*) and buttonbush occur on floodplains in the northeast as well. Associated species include false indigo, halberdleaf rosemallow (*Hibiscus laevis*), and deciduous holly (*Ilex decidua*; Hoagland 2000).

ECOREGIONS OF OKLAHOMA

Logically dividing North America into ecological regions, or ecoregions, continues to be a complex task; one that is chock full of debate and feuding. As mentioned earlier, generally, there are multiple biotic and abiotic units underlying ecoregion maps. Geology and vegetation, given their mutualistic relationship, often pair up as the minimal foundation for ecoregion divisions. Climate variables and sometimes faunal boundaries (animal distributions) are also factored in when drawing regional borders. Numerous efforts, all with varying perspectives and approaches, have been undertaken to classify biogeographic provinces in North America generally and Oklahoma specifically: just a few examples include Blair and Hubbell (1938), Dice (1943), Bailey (1996), Woods et al. (2005), and Omernik and Griffith (2014). As with vegetation maps, some ecoregion renderings have been generalized while others are remarkably detailed. All have their faults and lack of precision for some regions and at some scales.

Bruner's (1931) paper titled *The vegetation of Oklahoma* and Blair and Hubbell's (1938) paper titled *The biotic districts of Oklahoma*, although ostensibly focused on flora, with the former, and fauna, with the latter, were both at heart designating ecoregions, and both were influential early on with Oklahoma biologists. The former divided Oklahoma into 6 vegetation regions, whereas the latter classified 11 districts, but biotic divisions were broadly similar (Figure 3.5) and both papers offered many details about the characteristics

that defined their biogeographical divisions, including such aspects as climate, elevation, hydrology, and, of course, vegetation. For example, Blair and Hubbell (1938:430) described their Mississippi district as "part of the Gulf coastal plain ... lying mostly between 300 and 500 feet above sea-level." They also said that it "is the warmest and most humid part of the state" and that "in contrast with the clear, swift, trellis-pattern streams of the Ouachita district, those of the Mississippi district are sluggish, south-facing consequent streams tributary to [the] Little River and the Red River. They occupy shallow valleys or mere channels in the level flood-plain, and are generally mud-banked." Equally interesting is that Blair and Hubbell related their districts to distributional divisions they noted with Oklahoma's fauna, with particular attention to their respective specialties: Blair, a mammalogist, and Hubbell, an entomologist focused on Orthoptera (grasshoppers and their allies). They further connected divisions to Oklahoma's extreme east–west precipitation gradient (see Chapter 4 for additional details).

But for many, Duck and Fletcher's (1943) map titled *A game type map of Oklahoma* (and the subsequent publication of their *Survey of the game and furbearing animals of Oklahoma* in 1945) quickly became the go-to biogeographical reference map (Figure 3.5). From the title alone one would not necessarily guess that Duck and Fletcher's map is essentially an ecoregion map of Oklahoma, but at one's first look it is obvious why it is thought of that way—the 15 vegetation communities are effectively proxies for biotic provinces. And it would be hard to believe, given the stark resemblance to some later maps, that it did not influence its successors.

Some of those successors are part of the series of ecoregions maps produced by the U.S. Environmental Protection Agency (EPA; www.epa.gov/eco-research/ecoregions; Omernik and Griffith 2014). There are four levels (I, II, III, IV) to the EPA maps, each at a different scale, becoming more detailed with each ascension. Level I displays the major ecographic regions of North America (actually just Canada, the United States, and Mexico, i.e., omitting Central America). Of the 12 Level I ecoregions in the coterminous U.S.—two occur in Oklahoma: about 80% of the state lies within the Great Plains while the other 20% is part of the Eastern Temperate Forests ecoregion (Figure 3.4). The Level II map refines Oklahoma's ecoregions slightly by presenting four regions.

But it is with the Level III and IV maps that Oklahoma's biogeography feels more tangible (Woods et al. 2005). It seems almost sacrilegious to say, but for this book and for most of what we personally do with other aspects of our research, Level IV provides too much detail—967 ecoregions for the contiguous U.S. states, with 46 of those lying within Oklahoma's borders. The Level IV map, like the Oklahoma Ecological Systems Mapping Project map, are at such fine scales that they are arguably most useful for GIS analyses rather than for an easily digestible presentation. For the latter use, our preference is the Level III map (Figure 3.1). For biogeographers, the Level III map is the Goldilocks ecoregions map of Oklahoma—not too many, not too few, but *just the right amount* of ecoregions. Twelve ecoregions are conceptually useful. The number is not overwhelming when thinking in terms of why plants and animals occur where they do. Trying to do so with the 46 ecoregions of the Level IV map is daunting, and really does take a computer to manage all of the data processing required. All of that said, when we compare Oklahoma to nearby states and those within the Great Plains, we see that Oklahoma's ecoregions are much more varied relative to the others. Most of those states harbor only 6–8 ecoregions and, at the extreme, there is North Dakota with only 4 and the colossal state of Texas with 12. Yes, Texas and Oklahoma are matched for number of ecoregions despite the tremendous difference in geographic area between the two states. Oklahoma is bested only by California, another gigantic state, which includes but one more ecoregion. Oklahoma's diverse ecoregions, coming from all directions and bringing southwestern, southern, northern, western, mid-western, Gulf coastal, eastern river valley, and eastern mountain species to the state is what makes it so species rich. Details of that richness and what biogeographical factors explain it will be explored in the following chapter.

NOTES

1. Oklahoma is divided between the Arkansas River and the Red River drainage basins, both of which eventually drain into the Mississippi River. Watershed boundaries depend upon which level of hydrological unit is being examined. The U.S. Geological Survey (USGS) established a set of Hydrological Unit Codes (HUC, pronounced like Huck) that delineate watersheds into smaller and smaller areas: e.g., 2-, 4-, 8-, 10-, and 12-digit HUCs. For more information about HUCs, explore the *Watershed Boundary Dataset* (WBD) available at the USGS website or at the U.S. Department of Agriculture's, Natural Resources Conservation Service website.

2. Oxbow lakes are formed when a bend in a river is eventually cut off by the natural tendency of rivers to flow via shorter courses over time. Those shorter courses begin as meanders, or offshoots, that, with time, close off the bend. The shorter course then becomes the river while leaving behind a small, elongated lake alongside the new river. Oxbows rarely persist for long because they typically fill with sediments, become a marsh, and ultimately stop holding water.

   Playas, or playa lakes, are shallow depressions that intermittently fill with rainwater. These areas do not hold water for long periods, but they are vital to recharging, or refilling, aquifers. For more information about playas, see the Playa Lakes Joint Venture webpages, e.g., http://pljv.org/playa-conservation/playas-are-important-source-of-water

3. See endnote above for information about playa lakes.

BIOGEOGRAPHY OF OKLAHOMA ODONATA

Michael A. Patten and Brenda D. Smith

Biogeography is the scientific study of the spatial distribution of organisms. Biogeographers seek to elucidate why species are found where they are. For millennia—in his *Historia Animalium*, Aristotle described how different species occurred in different regions—we have known that species are not distributed randomly in space, just as we have known that species richness, the number of species that occur in a given area, is not spread evenly but tends to be clumped: some areas have relatively many species and others have relatively few. "Why?" is the burning question that underlies such uneven spread. Why do some regions harbor many species whereas others harbor few? Which aspects of the environment—climatic, topographic, or hydrologic—are correlated with richness?

Patterns of odonate species richness across Oklahoma provide a fine example of uneven spread. When species richness is plotted across the state (Figure 4.1), we see that richness is highest in the southeast and generally decreases northwestward. When we look at richness in relation to the 12 ecoregions in Oklahoma and their respective geographical areas (Figure 4.2), we note that richness is highest in the South-Central Plains and the Ouachita Mountains.[1] At the county level, we see that some counties have twice the number of known species as others (Figure 4.3): for example, McCurtain County, in far southeastern Oklahoma, boasts 122 species, whereas Cimarron County, in the far-panhandle has only 62. To test if differences in county totals were due to biased survey effort, we compared documented species richness to estimated species richness, as calculated using the Chao2 estimator (Chao 1987),[2] a method to deal with "unequal catchability," or, in this case, to deal with potential survey bias. Historically, surveys were concentrated in certain areas in Oklahoma, but our surveys truly have been statewide (see Figure 2.10); nonetheless, we wanted to evaluate how closely the county lists, as they currently stand, approximate "true" species richness. Results indicated that we should have high confidence that our efforts, combined with the rest of the Oklahoma Odonata Project data, have captured the species richness in most counties across Oklahoma (Figure 4.3). In other words, county totals are highly to very highly accurate reflections of the actual species richness of the counties, as opposed to an artifact of survey effort. With survey effort bias off the table, we could then examine species richness distribution.

The uneven spread in species richness across Oklahoma suggests broader patterns; patterns that are evident despite the way richness is plotted or at what scale it is examined (Figures 4.1 through 4.2, 4.3). For instance, there is a longitudinal trend from higher richness in the eastern part of the state to a lower richness in the western part, a trend that mirrors a sharp precipitation gradient across Oklahoma (Figure 3.2), and thus leads to a hypothesis that richness is correlated with precipitation. Bick and Bick (1957) noticed that Oklahoma's east-west precipitation gradient, as well as the concomitant changes in vegetation, were correlated with odonate species richness. Specifically, they noted a relatively narrow zone of overlap between eastern and western species, and that this zone was defined by rainfall. Beckemeyer (2002f) carried forward these ideas in his analysis of odonate distributions in the Great Plains and found similar results in Oklahoma and Kansas. Bick and Bick's (1957:17) conclusions are worthy of reprint here:

> Correlated with these extremes of precipitation are the east-west changes in vegetation. In general terms the eastern part of the state is hardwood, or mixed pine and hardwood forest, the central part postoak-blackjack forest or tall grass prairie, and the western part short grass plains. The western limit in Oklahoma of eastern United States species, which do not cross the state, coincided with the principal western limit of the scrub-oak area as mapped by Duck (1943) and with the 32 inch rainfall line given by Wahlgreen (1941). Western United States species did not reach the eastern limit of the postoak-blackjack area and were absent in the forests of eastern Oklahoma. This eastern limit of western United States species was approximately the 36 inch rainfall line. Hence the 16 counties previously mentioned [Carter, Cleveland, Garvin, Johnston, Kay, Lincoln, Logan, Love, Noble, Marshall, McClain, Oklahoma, Pawnee, Payne, Pontotoc, and Pottawatomie], with 32 to 36 inches of rainfall and with vegetation neither entirely grassland nor entirely forest, constituted an area of overlap for eastern and western faunas.

> … it seems clear that odonate distribution is correlated with the precipitation gradient and not with the temperature gradient.

Odonates are not alone: orthopterans (grasshoppers and their allies), mammals, amphibians, and reptiles also follow the precipitation gradient across the state (Blair and Hubbell 1938; Costa et al. 2008). Predictably, a simple plot of precipitation against odonate richness indeed shows a significant trend, with on average of one additional species recorded for each increase in 3 cm (1.2 in) of mean rainfall (Figure 4.4).

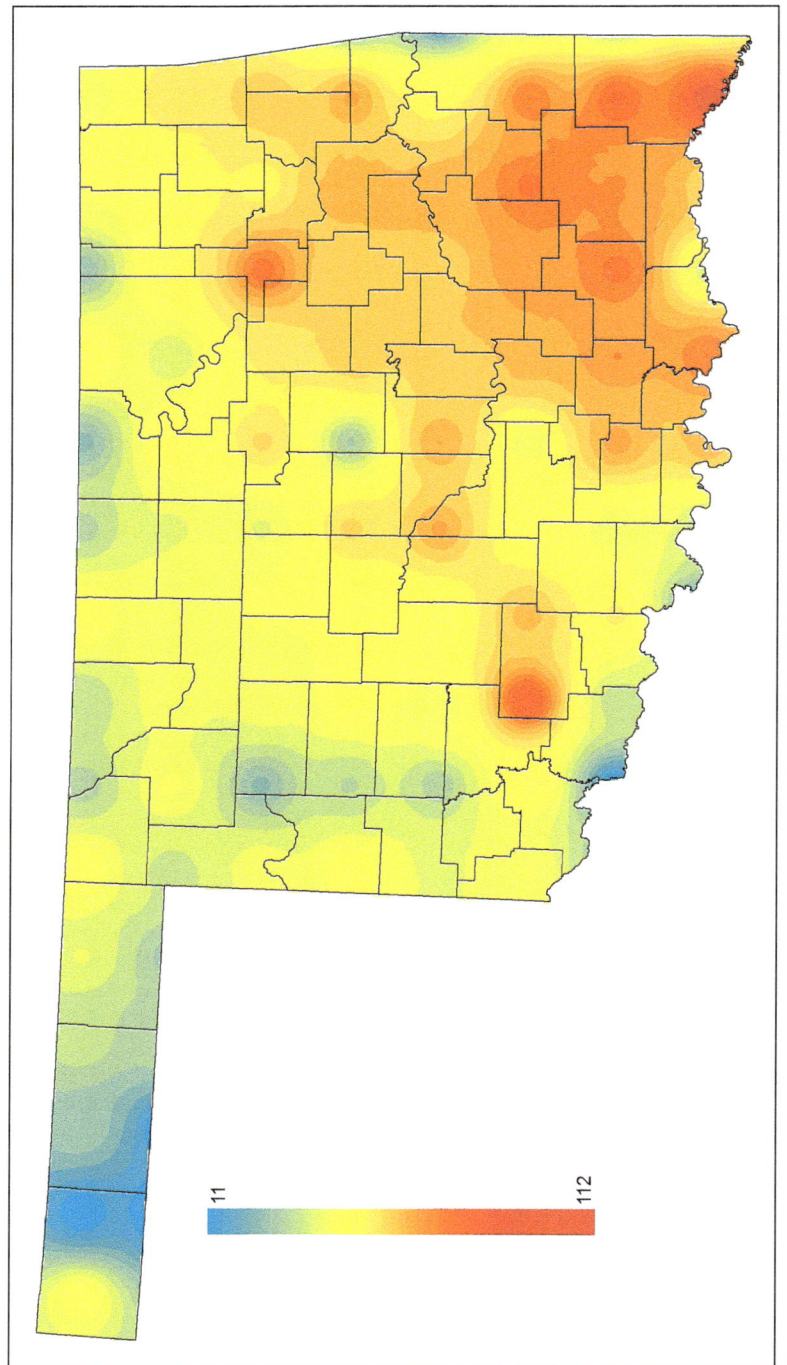

Figure 4.1 Odonate species richness in Oklahoma. Heat map of richness shows that richness is highest in southeastern Oklahoma, indicated by the red (highest) and blue (lowest) richness.

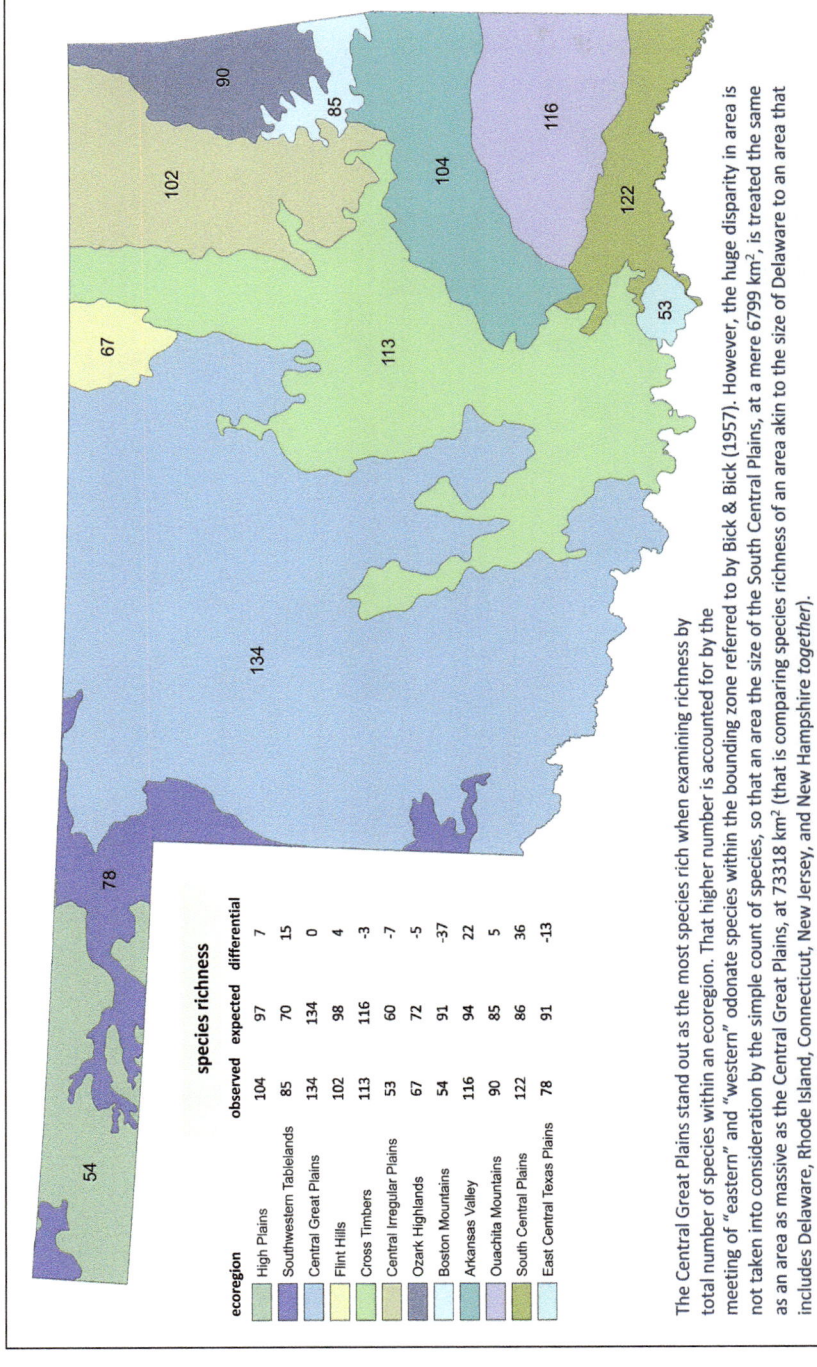

The Central Great Plains stand out as the most species rich when examining richness by total number of species within an ecoregion. That higher number is accounted for by the meeting of "eastern" and "western" odonate species within the bounding zone referred to by Bick & Bick (1957). However, the huge disparity in area is not taken into consideration by the simple count of species, so that an area the size of the South Central Plains, at a mere 6799 km², is treated the same as an area as massive as the Central Great Plains, at 73318 km² (that is comparing species richness of an area akin to the size of Delaware to an area that includes Delaware, Rhode Island, Connecticut, New Jersey, and New Hampshire *together*).

When area is accounted for, by calculating the expected richness and its difference ("differential" column) from the observed richness, we see that the Central Great Plains has a differential of 0, meaning its high richness is merely a function of its large footprint. Species richness is actually higher in eastern Oklahoma, especially in the Ouachita Mountains and the South Central Plains, where differentials are highly positive. A large negative differential, for example in the High Plains, indicates a far lower richness than expected on the basis of area.

| | | species richness | |
ecoregion	observed	expected	differential
High Plains	104	97	7
Southwestern Tablelands	85	70	15
Central Great Plains	134	134	0
Flint Hills	102	98	4
Cross Timbers	113	116	-3
Central Irregular Plains	53	60	-7
Ozark Highlands	67	72	-5
Boston Mountains	54	91	-37
Arkansas Valley	116	94	22
Ouachita Mountains	90	85	5
South Central Plains	122	86	36
East Central Texas Plains	78	91	-13

Figure 4.2 Odonate species richness across Oklahoma by ecoregion.

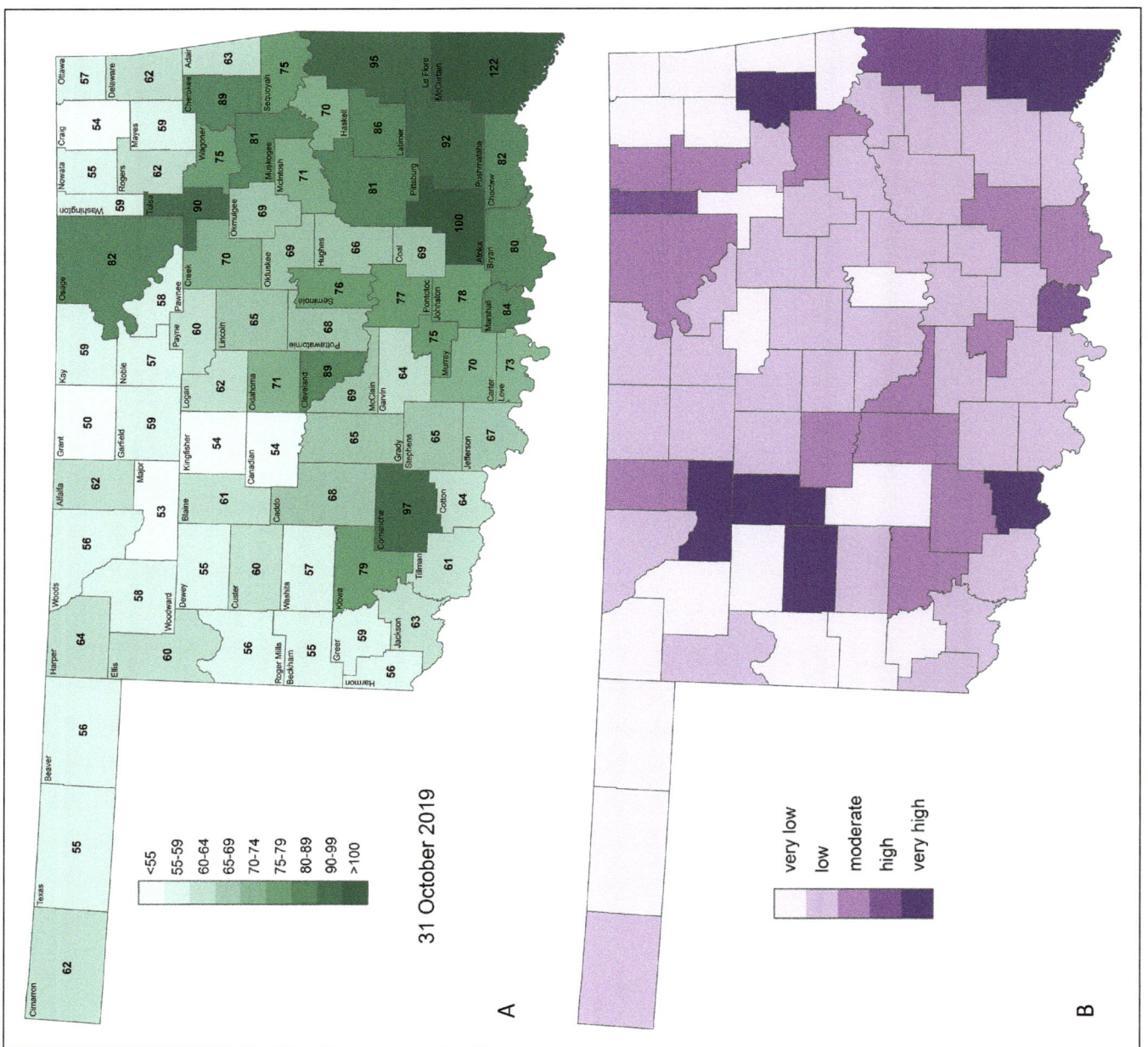

Figure 4.3 A) County-level species richness of Odonata across Oklahoma. B) A depiction of "confidence" in current species richness totals for each county in Oklahoma, by which we mean how near observed species richness is to estimated ("true") species richness.

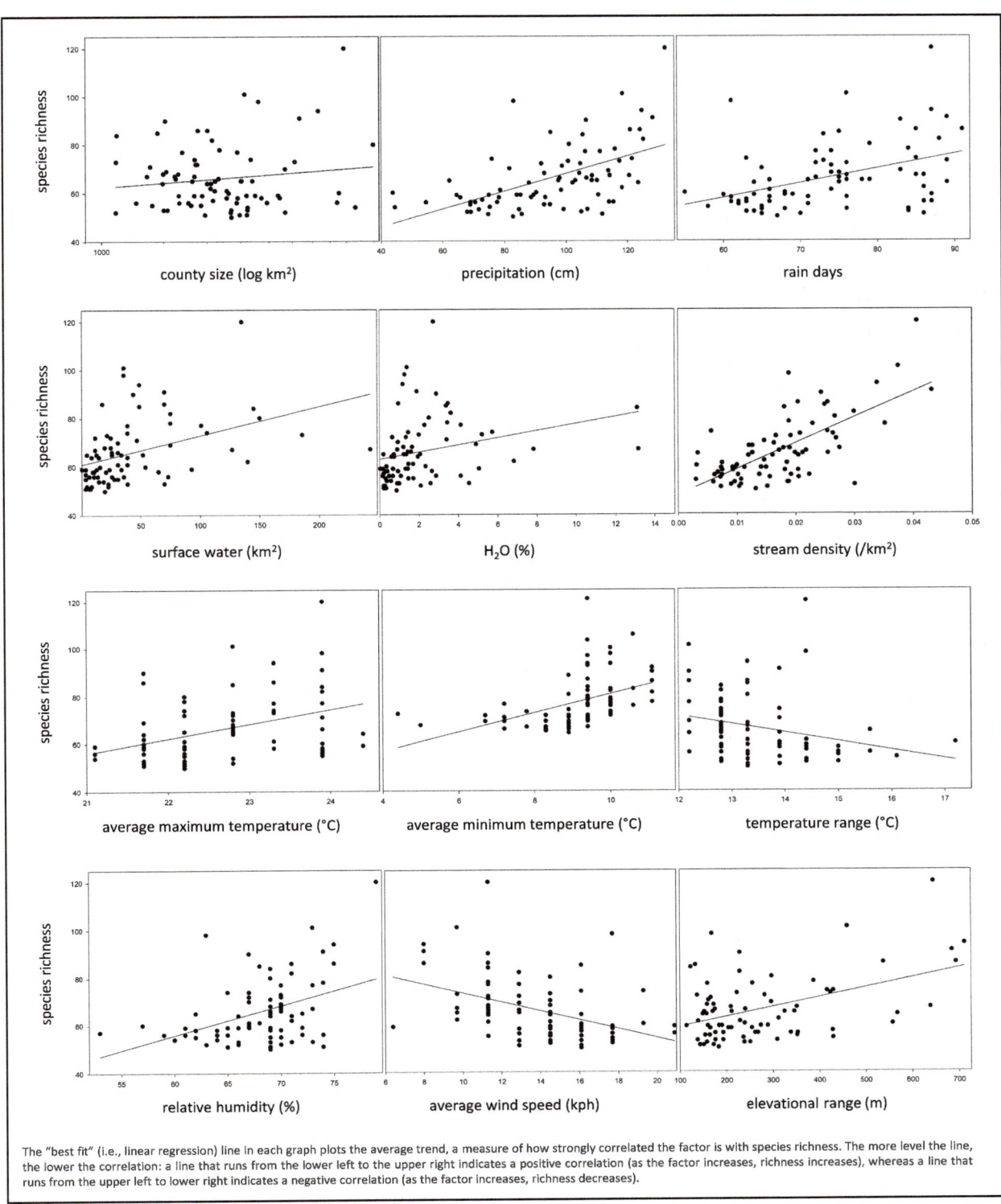

The "best fit" (i.e., linear regression) line in each graph plots the average trend, a measure of how strongly correlated the factor is with species richness. The more level the line, the lower the correlation: a line that runs from the lower left to the upper right indicates a positive correlation (as the factor increases, richness increases), whereas a line that runs from the upper left to lower right indicates a negative correlation (as the factor increases, richness decreases).

Figure 4.4 **Factors examined individually to determine correlation with odonate species richness in Oklahoma.**

Yet results of additional analyses suggest that factors beyond the amount of rainfall play an important role in determining species richness. One such factor is rain days, the number of days in a calendar year in which rain falls, which accounts for a fair amount of variation in species richness (17%) but is insufficient on its own. Richness doubtless is determined by multiple factors, including from a county's size (km^2), either the extent (km^2) or proportion (%) of surface water, the density of streams per square kilometer (/km^2), elevational range (m), maximum and minimum average temperature (°C), temperature range (°C), mean relative humidity (%) and wind speed (kph), and length of the growing season (days). We examined each factor, individually and in combination, to provide insight into why odonates occur where they do.

County size had no discernable effect on species richness, as it explained <2% of the variation in richness across the state's 77 counties. McCurtain County does not have so many species because it is relatively large, nor does Grant County have so few because it is relatively small. If county size mattered then, for example, Osage County might have a list larger than McCurtain County, or Washington and Woods Counties would not have similar numbers of species. By contrast, the extent of surface area comprising water[3] accounts for roughly 16% of the variation in richness (Figure 4.4), meaning that counties with more ponds, lakes, and streams have more species.

Interestingly, the extent of water (in km^2) correlates more highly to species richness than the proportion of water (relative to land) within a county. Even so, the explanatory power of the surface area of water is messy: there remains a good deal of variation to explain. We can begin by asking what type of surface water matters. After all, we may be inclined to think that the more water the better for aquatic species like odonates, but in our experience huge reservoirs, for example, tend to be rather sterile for odonate diversity. There may be a few species along the immediate shore—*Argia moesta*, the Powdered Dancer, is a quintessential example—but otherwise there tends to be little. The same often holds for extensive marshes or wide rivers, which themselves hold low species richness relative to the bordering vegetation. In our experience, it is the narrow, perennial creeks and small ponds, particularly those that are spring-fed, where richness abounds.

It is at that finer scale of ponds, small lakes, and, to some extent, streams where data are limited. In Oklahoma, as in many other places, ponds and small lakes tend not to be mapped or monitored in terms of size, location, and suitability for wildlife. We know such bodies of water are numerous across the state, particularly where cattle ranching is a way of life, but we currently have no feasible way to monitor them. Private property owners can restrict access to small lakes, even those built decades ago by the federal government. Landowners can build rainwater retention ponds *ad nauseum*, and springs and seeps may be ponded up, tapped, or plugged at the owner's discretion. But streams, in principle, are a different matter, because countrywide they are thought of as a public resource. Streamflow, even across private property, tends to be monitored and regulated by local, state, and federal agencies, so we have good data on where streams are located, meaning that we have good geospatial data for stream density. As it happens, stream density, which also provides an indication of the extent of natural wetland habitat in the state, explains nearly half (47%) the variation in species richness (Figure 4.4). It could be argued that the result is an artifact of availability of geospatial data, but stream density makes biogeographical sense as a factor that contributes to species richness.

Range in elevation, the distance between a county's highest point and its lowest point, likewise accounts for a fair amount of variation in species richness (21%), an understandable finding given the high topographic relief in the Ouachita Highlands in the southeastern part of the state, where richness peaks, or the relatively high richness in the east-central part of the state, coincident with the Ozark Plateau. It makes sense, too, in the Panhandle, where the northwestern corner of Cimarron County, at the tip of the state, is dominated by Black Mesa, the state's highest point at 1,494 m (4,902 ft), as well as shallow canyons and associated habitats characteristic of the Intermountain West and decidedly uncharacteristic of the Great Plains. It even makes sense given the spike in richness around the Wichita Mountains in southwestern Oklahoma, an isolated range that rises sharply from the relatively flat plain that surrounds it.

Likewise, ambient temperature correlates to some extent with odonate species richness, both in terms of average maxima (14%) and minima (20%). Curiously, richness increases as either mean maximum or mean minimum temperature increases. Odonata is a notoriously warm-weather taxon, so the correlation with increased minima makes sense, but we assumed that high maxima might have depressed richness—even odonates cannot tolerate excessive heat. These findings tell us that, although individuals may succumb to high heat at a specific time—i.e., when the critical thermal maximum (the temperature at which physiological function breaks down and death occurs) is reached—populations are able to persist under Oklahoma's average maximum temperatures, and there is no discernable effect on species richness. In contrast to these patterns, temperature range was correlated negatively, albeit somewhat weakly (7%), with richness (Figure 4.4), suggesting that increased seasonality (swings in temperature) depresses richness; an issue to consider with future climate change.

Relative humidity and average wind speed had comparable (in magnitude) but opposing associations with species richness. Each accounts for 17–19% of the variation in richness, but humidity correlates positively with richness, whereas the converse was true for wind. More dragonfly species are found in areas with higher humidity but less wind. Little research has been conducted on the effects of relative humidity on odonates; nevertheless, we know that odonates, as emerging larvae and adults, need sufficient moisture in and around their bodies to maintain normal function (Corbet 1999). Emerging nymphs in the wild, for example, may have an optimal range for relative humidity to facilitate emergence (Trottier 1973). It is not surprising, then, that having moister air, to a point, would favor odonate richness, whereas windier conditions, that would reduce humidity, would not.

Because no factor on its own explained more than 47% (i.e., stream density) of the county-level species richness we see in Oklahoma, we examined various factors together. For example, tying together humidity, rainfall, temperature, and seasonal swings (the range between minimum and maximum temperatures, on average) can be accomplished with a proxy such as growing season, a metric that in effect tells us how long plants can maintain above-ground growth. Species richness of dragonflies and damselflies correlates with growing season (18%), although, again, less variation in richness is accounted for than might be expected (or hoped) given the flexibility of such a metric.

Another complication with examining factors individually is that the various factors we considered are neither mutually exclusive—more than one could account for the prevailing pattern—nor independent—one factor may be affected by another (for example, an increase in wind or temperature by necessity will reduce relative humidity). We nonetheless can build statistical models and use a variety of techniques (we opted for an information theoretic approach[4]) to determine which of the various factors, alone or in combination, correlates strongest with species richness. For our analysis, we ran our data through 742 models. When we averaged resultant models, we found that two factors, stream density (per km^2) and elevational range (m) were included in all competing explanatory models. Every

other factor appeared in nearer to half the number of models (300–397), indicating that they were not as ecologically important in explaining species richness, even if they did contribute to some extent (Figure 4.5). We conclude that a substantial amount of the variation we see in county-level species richness is driven by variation, both in the density of streams (more is better) and the topographic relief (likewise, more is better). Relating these factors back to ecoregions, we can see why, for example, the Ouachita Mountains, with their high stream density and high roughness indices (Figure 3.3), are so species rich.

NOTES

1. We calculated the estimated species richness for each ecoregion and compared it to the observed richness, which produced a differential. We did this by means of the basic (and long-established) species–area relationship, $S = cA^z$, $\log_{10} S = \log_{10} c + z \log_{10} A$. We used SigmaPlot to fit the line to determine the slope (z) and intercept (c) to calculate the expected richness.

2. "True" species richness was determined using the Chao2 estimator (Chao 1987), a formula that is a modified ratio of the number of "singletons" (i.e., the number of species recorded only once in x surveys) against the number of "doubletons" (the number of species recorded exactly twice in x surveys). An estimate of the variance may be obtained as well, and it was the standard deviation (SD, the square root of the variance) that we used as a broad snapshot of "uncertainty" ("very low" on the map scale) or "confidence" ("very high" on the map scale) because a Chao2 estimate with low SD means there are few singletons relative to doubletons—the county has been sampled well—whereas high SD means there are many singletons relative to doubletons, an implication the county has been sampled poorly.

   At first glance, the low confidence rating of McCurtain County may catch the reader's eye. But recall that the county has a good number of species on its list that have been recorded only once. A high disparity between "singletons" and "doubletons" yields high estimates of variance in Chao's formulae. In other words, a large number of singletons relative to doubletons suggests that the current richness (122 species) is not particularly close to the "true" richness (estimated to be 145–161 species), meaning that McCurtain County, despite being well-surveyed, is highly likely to continue receiving vagrant records because it lies at or near the periphery of so many ecoregions. The other two counties that ranked low, Cotton and Rogers Counties, did so because they have not received as much attention as other counties.

3. We calculated the extent of surface area of water by determining the footprints of ponds and lakes, and adding in the lengths of streams.

4. Information theoretic approaches use empirical data to obtain a metric that informs one how close the estimate is, relatively speaking, to the "truth." Even without knowing the "truth," we can statistically obtain relative model fits. Hence, the metric allows us to rank models from best fit to worst fit. Inferences are made on the basis of the best-fit model(s) and reflects which factors are correlated with, in our case, species richness.

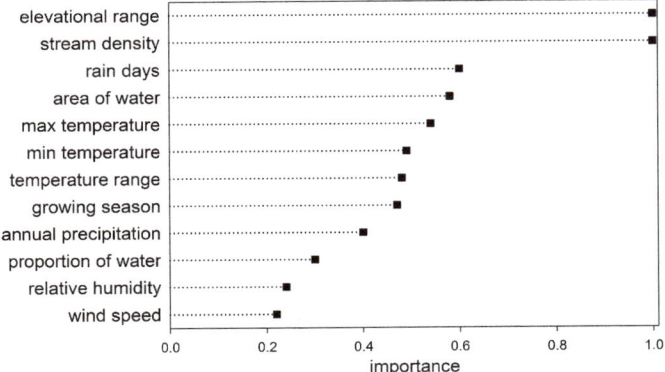

Figure 4.5 Relative importance, after building a suite of models (n = 742) for each factor individually and in combination, of each of 12 factors that putatively contribute to odonate species richness across Oklahoma's counties.

## CHAPTER 5

THE ANCESTORS OF OKLAHOMA ODONATA: THE PERMIAN GRIFFENFLIES, PROTO-DRAGONFLIES AND PROTO-DAMSELFLIES OF THE WELLINGTON FORMATION

Roy J. Beckemeyer

INTRODUCTION

The Odonata have an ancient lineage that reaches far back into the Paleozoic, at least 310–315 million years ago (MYA) (Bechly et al. 2001; Brauckmann and Zessin 1989). The Wellington Formation deposits of the Lower Permian of Oklahoma and Kansas (about 285 MYA) contain some of the most famous and important Paleozoic fossil insect remains in the world (Beckemeyer 2000; Beckemeyer and Hall 2007; Grimaldi and Engel 2005). A total of 210 species of insects in 21 orders, 53 families, and 109 genera have been described on the basis of fossils from the Kansas and Oklahoma Wellington Formation (Beckemeyer 2000; Beckemeyer and Hall 2007). Included among these are more than 20 species of dragonfly ancestors in 3 orders, 4 families, and 7 genera (Tables 5.1 and 5.2). The largest insects known to have ever flown were the giant griffenflies of Kansas and Oklahoma, *Meganeuropsis permiana* and *M. americana*.

Stratigraphically, the Wellington Formation in Kansas and Oklahoma is located within the Sumner Group of the North American Leonardian Series of Permian rocks (West et al. 2010). Dunbar et al. (1960) correlated the base of the Wellington Formation in Kansas and Oklahoma with the base of the Leonardian Series and furthermore correlated the base of the Leonardian with the base of the Russian (now the International) Artinskian Stage. Later authors (see e.g., Menning et al. 2006, Figure 4, p. 17) locate the base of the Leonardian in either the upper Artinskian or lower Kungurian. Sawin et al. (2008), the authorities cited by West et al. (2010), place the base of the Wellington and Leonardian firmly in the lower Kungurian, but with the exact boundary between the Kungurian and Artinskian not specifically placed. The latest chronostratigraphic map (GTSF 2016) dates the Kungurian from about 270 to 280 MYA, and the Artinskian from about 280 to 290 MYA, with the Permian Period dating from about 250 to 300 MYA. The Wellington Formation is thus in the Lower Permian, probably around 275–280 million years old. For a correlation of the stratigraphy of the Kansas and Oklahoma Wellington Formation beds, see Hall et al. (2016, pp. 307–308).

Figure 5.1 shows the extent of the Wellington Formation insect fossil deposits in Kansas and Oklahoma (after Hall 2004). The Elmo fossil beds of Kansas were discovered by Elias H Sellards (Sellards 1906, 1909) and studied by Robin J Tillyard and Frank M Carpenter; the "Midco" fossil insect beds of Oklahoma were discovered by Gilbert Raasch (1946) and studied by Frank M Carpenter and Paul Tasch (see Beckemeyer 2000, 2012, 2013; Beckemeyer and Hall 2007 for details and references to the original scientific papers).

Figure 5.2 is a picture showing the insect fossil-bearing layers at one of the localities in Noble County, Oklahoma. There are two rock layers that contain insect remains, and they are indicated by the black arrows. Pieces of rock removed from the site must usually be split by striking them on the edge with a rock hammer. They will split along bedding planes; if there is an insect fossil preserved, it will appear in both halves of the rock, one a positive and the other a negative impression. Most of the insect remains found in the Wellington Formation are wings or wing remnants; very few body parts or complete bodies are found (Beckemeyer and Hall 2007).

ORIGINS OF MODERN DRAGONFLIES

The dragonflies as we know them today did not appear in the fossil record until the Jurassic (about 200–250 MYA), or the age of reptiles, but better known as the age of dinosaurs. The oldest examples of insects that share a common ancestor with the dragonflies and damselflies arose in the Carboniferous (about 300–360 MYA), an era in which insects began to truly flourish. Figure 5.3 is a chart showing the relationships and "family tree" of the modern dragonflies and their extinct relatives.

DRAGONFLY ANCESTORS OF KANSAS AND OKLAHOMA

The earliest known dragonfly ancestors were the Meganisoptera (some authors refer to the group as Protodonata; Carpenter, 1960) also called "giant dragonflies," or, more recently, "griffenflies." They are known from the Upper Carboniferous through the Late Permian, but disappear from the fossil record at the Permian-Triassic extinction (Figure 5.3).

Table 5.1 **Checklist of Kansas and Oklahoma Wellington Formation fossil dragonflies and griffenflies. There are 23 species known from the formation, 15 of which have been found in the Kansas portion and 11 from Oklahoma. The state of occurrence is noted (KS = Kansas, OK = Oklahoma) for species as well as the average wing length of the species.**

Odonatoptera Martynov, 1932 [5 families, 8? genera, 22(?) sp.]

Meganisoptera Martynov, 1932 [2 families, 4 genera, 10 sp.]

Meganeuromorpha Pritykina, 1980

Meganeuridae Handlirsch, 1906 [2 subfamilies, 3 genera, 8 sp.]

Meganeurinae Handlirsch, 1906 [1 genus, 2 sp.]

Meganeuropsis Carpenter, 1939

permiana Carpenter, 1939	KS	330 mm
americana Carpenter, 1947[a]	OK	305 mm

Tupinae Handlirsch, 1919 [2 genera, 6 species]

Megatypus Tillyard, 1925 [3 sp.]

parvus Engel, 1998	KS	140 mm
schucherti Tillyard, 1925	KS	195 mm
ingentissimus Tillyard, 1925	KS	125 mm

Tupus Sellards, 1906 [3 sp.]

permianus Sellards, 1906	KS	110 mm
readi Carpenter, 1933	KS	145 mm
gracilis Carpenter, 1947	OK	145 mm

Paralogidae Handlirsch, 1906 [1 genus, 2 species]

Oligotypus Carpenter, 1931 [2 sp.]

tillyardi Carpenter, 1931	KS, OK	50 mm
sp. Carpenter, 1931	KS	100 mm

Odonatoclada Bechly, 2003 [3 families, 4? genera, 13 (?) species]

Protanisoptera Carpenter, 1931 [1 family, 1 genus, 3(?) sp.]

Ditaxineuridae Tillyard, 1926 [1 genus, 3 (?) sp.]

Ditaxineura Tillyard, 1926 [3(?) sp.]

anomalostigma Tillyard, 1926	KS	21 mm
cellulosa Carpenter, 1933	KS	27 mm
cellulosa (?) Tasch and Zimmerman, 1962	OK	27 mm

"Protozygoptera" Tillyard, 1925[b]: Archizygoptera Handlirsch, 1906 [4? genera, 10 sp.]

Kennedyidae Tillyard, 1925 [3? genera, 9 species]

Kennedya Tillyard, 1925 [4 sp.]

mirabilis Tillyard, 1925	KS, OK	34 mm
tillyardi Carpenter, 1939	KS, OK	33 mm
fraseri Carpenter, 1947	OK	40 mm

Progoneura Carpenter, 1931 [5 sp.]

grimaldii Nel et al., 2012	OK	16 mm
minuta Carpenter, 1931	KS	21 mm
nobilis Carpenter, 1947	OK	13 mm
venula Carpenter, 1947	OK	16 mm
sp. Carpenter, 1947	OK	34 mm

(Continued)

Table 5.1 (*Continued*) **Checklist of Kansas and Oklahoma Wellington Formation fossil dragonflies and griffenflies. There are 23 species known from the formation, 15 of which have been found in the Kansas portion and 11 from Oklahoma. The state of occurrence is noted (KS = Kansas, OK = Oklahoma) for species as well as the average wing length of the species.**		
Genus *incertae sedis* [1 sp.][c]		
reducta (Carpenter, 1939)	KS	35 mm
Uncertain family [1 genus, 1 sp.]		
Opter Sellards, 1909		
brongniarti Sellards, 1909[d]	KS	?

[a] Grimaldi and Engel (2005) identified *Meganeuropsis americana* as a synonym of *M. permiana*; however, they provided no data supporting the synonymy. Carpenter (1947) stated, in his description of *M. americana*, "This species is undoubtedly closely related to the genotype (*permiana*) … Since the cells are larger and the veins thicker in fragments of *permiana* than they are in *americana*, the former was probably the larger insect. More specimens of *permiana* may indicate that *americana* is identical with it, but until that is certain, I believe the two should be considered distinct." I have chosen to follow Carpenter, and to maintain the two as separate species pending publication of a detailed comparison of the available specimens of the two species.

[b] *Sensu* Nel et al. (2012).

[c] *Kennedya reducta* Carpenter, 1939 was incomplete, and some of the key characters used in the new diagnosis of the genus by Nel et al. (2012) are missing. They thus relegated this species to Kennedyidae Genus *incertae sedis*. Carpenter had expressed the opinion that if new specimens were to prove that the cross vein reduction evident in the distal portion of the wing also occurred in the basal portion, *reducta* might have to be placed in a new genus.

[d] *Opter brongniarti* Sellards, 1909 – Described in order Megasecoptera by Sellards, moved to Kennedyidae by Tillyard in 1925. Carpenter (1992) did not list the genus *Opter* with the family Kennedyidae. Nel et al. (2012) did not include it in the family because it "… is based on a very incomplete wing … Its general shape suggests affinities with the Kennedyidae, but it differs from … kennedyid genera in a very long CuA…"

Griffenflies were not all giant insects, and actually varied widely in size, from the 50-mm-long wings of *Oligotypus tillyardi* to the 330-mm-long wings of *Meganeuropsis permiana*. Figure 5.4 illustrates a life-sized model of *M. permiana* being inspected by its sculptor, Werner Kraus. This wonderful insect would have had a wingspan just a bit larger than a domestic pigeon.

Figure 5.5 shows a small fragment of the hind wing that probably came from *Meganeuropsis americana*, the giant griffenfly of Oklahoma. *M. americana* is known only from a forewing, so the figure shows the fragment against an outline based on the hind wing of *M. permiana* to show where it would have been located. It is compared with a specimen of the largest living dragonfly of North America, *Anax walsinghami*, the Giant Darner, to give a better impression of its size. The fragment and specimen are shown together in Figure 5.5a and in smaller scale against the wing outline in Figure 5.5b.

Figure 5.6 shows a photograph and drawing of the hind wing of another giant griffenfly, *Megatypus schucherti*,

Table 5.2 **Kansas and Oklahoma fossil insects originally described as Odonatoptera species, but subsequently synonymized or moved to other orders.**	
Calvertiella permiana Tillyard, 1925	Described in Protanisoptera by Tillyard. Moved to Palaeodictyoptera by Handlirsch (1937).
Campotaxineura ephialtes Tillyard, 1937	Described in Protanisoptera by Tillyard. Removed from Protanisoptera by Huguet et al. (2002), who suggested the fossil might belong to Palaeodictyoptera, although they left it in Order *incertae sedis* pending review of the type material.
Megatypus vetustus (Carpenter 1933)	Synonymized with *Tupus permianus* by Carpenter (1939).

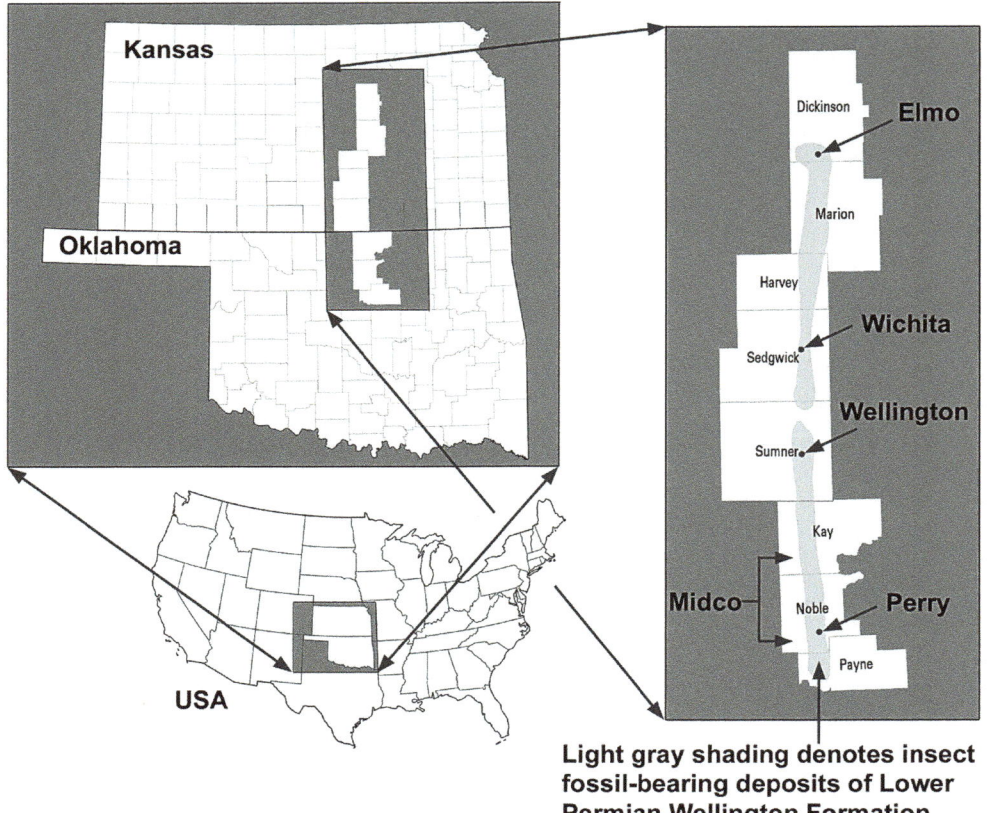

Figure 5.1 Location of the Wellington Formation fossil insect beds in central Kansas and north-central Oklahoma (after Hall 2004; Beckemeyer and Hall 2007). Fossils of extinct dragonfly relatives have been found in Kay County, Oklahoma (Tasch and Zimmerman 1959, 1962), Noble County, Oklahoma (Carpenter 1947; Tasch and Zimmerman 1962; Beckemeyer 2006), Dickinson County, Kansas (Sellards 1906; Tillyard 1925, 1926, 1937; Carpenter 1931, 1933, 1939, 1943, 1947; Klots 1944; Engel 1998), Marion County, Kansas (Tasch 1961, 1963), and Harvey County, Kansas (Tasch 1961, 1963).

Light gray shading denotes insect fossil-bearing deposits of Lower Permian Wellington Formation

from Elmo, Kansas. The wing is 195 mm long. The two halves of this wing are located in two different museums, the one shown is from the Entomology Museum at Kansas State University; the counterpart is at the Museum of Comparative Zoology, Harvard.

Figure 5.2 An example of the Midco, Oklahoma, insect fossil-bearing rock layers. Raasch 9 locality (Raasch 1946), Noble County, Oklahoma. There are two separate layers that sometimes contain insect fossils at this site; the layers are indicated by the black arrows. Photo by Roy J Beckemeyer, 2000.

Figures 5.5 and 5.6 both show the well-preserved three-dimensional sculpturing of the griffenfly wings—the corrugations of the alternating wing veins are very similar to the wings of modern dragonflies and damselflies, and provided structural stiffness. Figures 5.7 and 5.8 show specimens of two modern dragonflies perched on a fossil of a portion of the wing of the giant griffenfly, *Tupus readi*, which had a wing that was about 145 mm in length. Figure 5.7 compares *T. readi* with the modern dragonfly, *Perithemis tenera*, the Eastern Amberwing; Figure 5.8 compares *T. readi* and *Plathemis subornata*, the Desert Whitetail. In Figure 5.8, the fluting of the dragonfly and griffenfly wings can be compared.

While most griffenfly fossils are wings or wing pieces, the very first insect species described by E. H. Sellards from the Elmo, Kansas beds was the griffenfly *Tupus permianus*, and that fossil was composed of almost the entire set of four wings, and the head and a portion of the body. This specimen is in the Texas Memorial Museum in Austin, and is shown in Figure 5.9.

But not all griffenflies were giants. Figure 5.10 shows a photograph of *Oligotypus tillyardi*, the smallest Wellington Formation griffenfly. It was about the same size as the modern-day Green Darner, *Anax junius*.

Figure 5.11 is a comparison of the outlines of several Wellington Formation griffenfly wings, including the smallest and the largest. Also shown are the outlines of

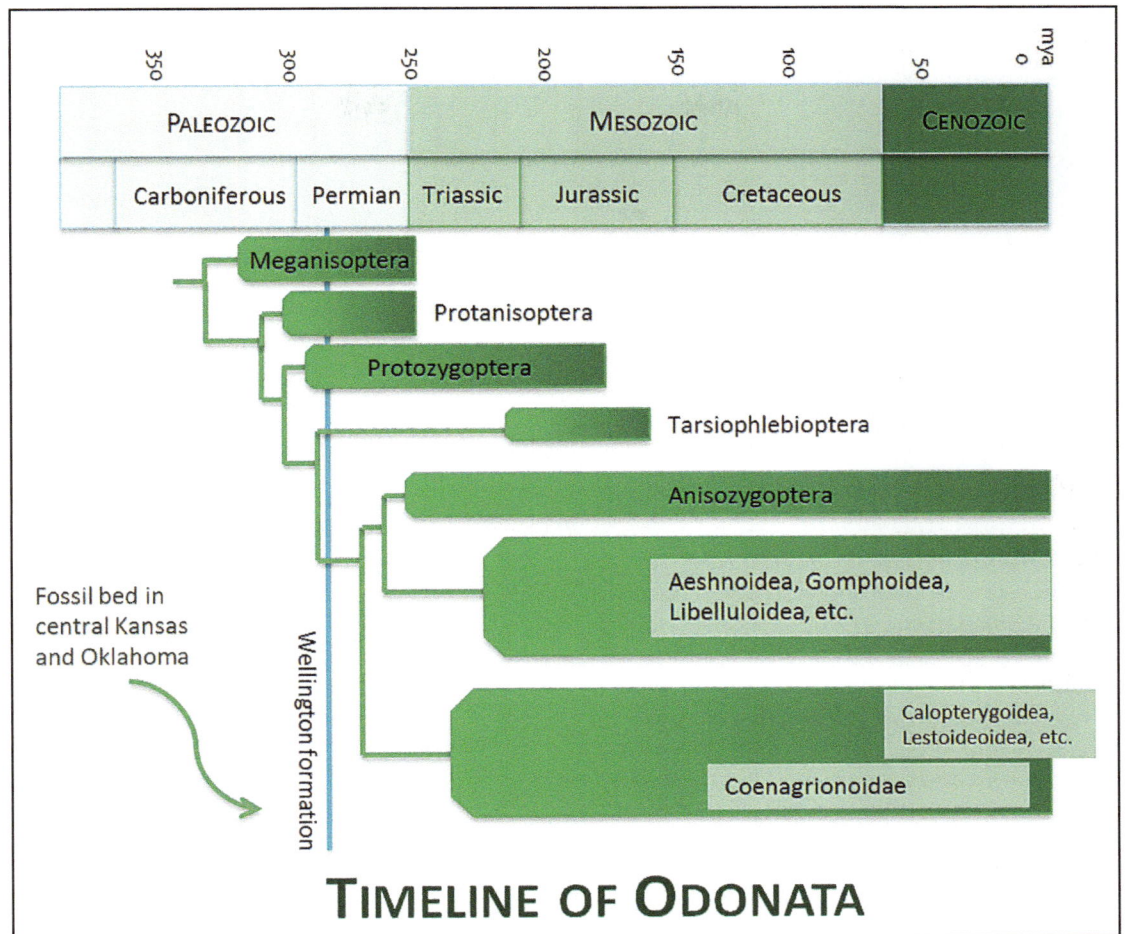

Figure 5.3 Relationships of the living Odonata and their extinct relatives over time. Modified from Grimaldi and Engel (2005), but see also Bechly (1996, 2008 for the Odonatoptera), Huguet et al. (2002, for Protanisoptera), Nel et al. (2009 for Meganisoptera), Nel et al. (2012 for "Protozygoptera") and Rehn (2003 for Odonata). The groups Meganisoptera (Griffenflies), Protanisoptera ("proto-dragonflies"), and Protozygoptera ("proto-damselflies") occurred during the Permian and are represented in the fossils of the Kansas and Oklahoma Wellington Formation.

Figure 5.4 A life-size model of the largest insect that ever flew, the giant Kansas griffenfly, *Meganeuropsis permiana*, shown with the sculptor, Werner Kraus. Image copyright by Werner Kraus, Geologie Endogene Dynamik, RWTH-Aachen, Germany, and used with permission.

one protanisopteran, *Ditaxineura anomalostigma*, and one protozygopteran, *Kennedya mirabilis*.

The Protanisoptera or "proto-dragonflies," were roughly the size of today's dragonflies. Only one genus, *Ditaxineura*, is known from the Wellington Formation. Figure 5.12 is a picture of a drawing of *Ditaxineura anomalostigma* together with a fragment of the tip of a *Ditaxineura* wing collected in Noble County, Oklahoma. One of the characteristics of Protanisoptera is that longitudinal vein RA passes through the pterostigma.

The Protozygoptera, or "proto-damselflies," were also approximately the size of today's damselflies. Two genera, *Kennedya* and *Progoneura*, are known from the Wellington Formation. Figure 5.13 is a drawing showing *K. mirabilis* (known from both Kansas and Oklahoma) and *P. nobilis* (Oklahoma).

Figure 5.5 Comparisons of one of the largest living Nearctic dragonflies (Odonata: Anisoptera: Aeshnidae: *Anax walsinghami* (Giant Darner, ♂) and a *Meganeuropsis* griffenfly. A. Specimen of *A. walsinghami* shown next to a small fragment of the anal region of the hind wing of a species of Meganeuropsis. Fossil collected from Noble County, Oklahoma, by Roy J Beckemeyer (RJB; specimen RJB2000-5a, Johnston Geology Museum). *A. walsinghami* specimen from RJB collection. Photograph by RJB Reprinted with permission from the *Bulletin of American Odonatology* (Beckemeyer 2006). B. Sketch of fossil wing venation after Carpenter's (1933) reconstruction sketch of the holotype specimen of *M. permiana*. Scan of hind wing fragment (from picture A) overlaid to scale and in approximate position relative to sketch. Wing outline estimated by author based on size and proportions of other griffenfly wings; scanned dragonfly lateral image is of same *A. walsinghami* specimen reduced to proper scale.

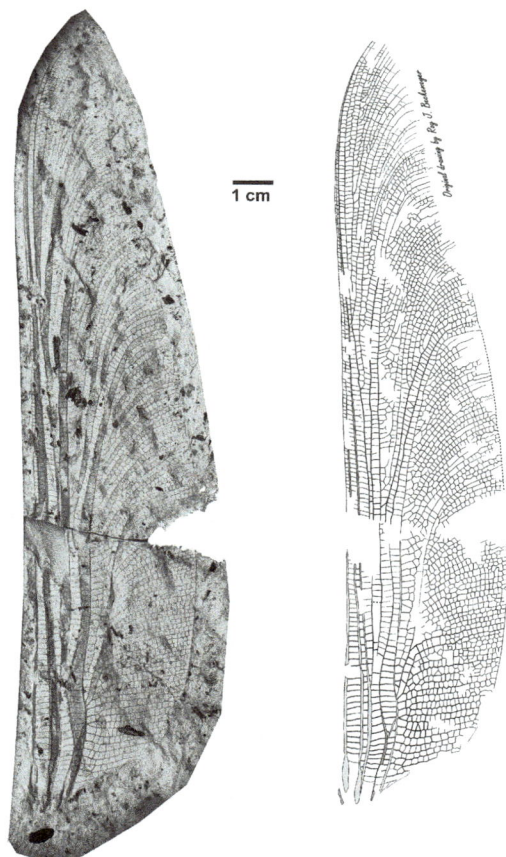

Figure 5.6 Scan and reconstruction drawing of the hind wing of the giant griffenfly *Megatypus schucherti* Tillyard, 1925. Left Image of Kansas State University (KSU) specimen scanned in 2005 by Roy J Beckemeyer with permission of R Charlton and S Ramaswamy of the KSU Entomology Department. Right Original reconstruction drawing by Roy J Beckemeyer.

SUMMARY

There was a great diversity of Odonatoptera in the Early Permian as demonstrated by the fossils of the Kansas and Oklahoma Wellington Formation. At least 22 species in 8 genera, 5 families, and 3 higher taxa are known (Table 5.1). These varied greatly in size and included the insect with the largest wingspan that we know of in the fossil record. The dragonflies and damselflies of today have a long and distinguished lineage.

Figure 5.7 Specimen of a small modern dragonfly (Odonata: Anisoptera: Libellulidae: *Perithemis tenera*, the Eastern Amberwing, ♀) placed on a fossil fragment of a forewing of the griffenfly *Tupus readi*. This gives an idea of the size of fossil dragonflies. The forewing length of *P. tenera* ranges up to about 20 mm, while the total length of the forewing length of *T. readi* was estimated at 145 mm, some 7 times the wing length of *P. tenera*. Photograph by Roy J Beckemeyer of fossil specimen FI-40105, American Museum of Natural History, collected from the Elmo fossil beds (Klots 1944). Specimen of *P. tenera* from author's collection.

Figure 5.8 Specimen of an average-sized modern dragonfly (Odonata: Anisoptera: Libellulidae: *Plathemis subornata*, the Desert Whitetail, ♂) placed on a fossil fragment of a forewing of the griffenfly *Tupus readi*. This allows comparison of the similarities in wing corrugation between the griffenflies and modern dragonflies. Note the deep fluting of both wings at the base and near the wing leading edge. The forewing length of *P. subornata* is about 35 mm, while the total length of the forewing length of *T. readi* was estimated at 145 mm, about 4 times the wing length of *P. subornata*. Photograph by Roy J Beckemeyer (RJB), of fossil specimen FI-40105, American Museum of Natural History, collected from the Elmo fossil beds (Klots 1944). Specimen of *P. subornata* from RJB collection.

Figure 5.9 The holotype specimen of the griffenfly *Tupus permianus*, the first insect from the Elmo, Kansas, fossil beds that was described by EH Sellards in 1906. This is a spectacular specimen, with a wing span of just over 200 mm, and with a portion of the body and head preserved. Specimen photographed by Roy J. Beckemeyer in 2001 at the Texas Memorial Museum with the permission of Dr. Chris Durden. Specimen 630 (EHS).

Figure 5.10 A fossil of a nearly complete forewing of the griffenfly *Oligotypus tillyardi*. (A little less than 10% of the base of the wing is obscured by rock.) This image shows that some griffenflies were the size of modern dragonflies. *O. tillyardi* had a wing length of 50 mm, comparable to that of the larger dragonflies of today such as *Anax junius*, the Common Green Darner. A penny coin (diameter 19.5 mm) is shown for scale. Specimen, from the RJB collection, was collected by Michael Montgomery from Noble County, Oklahoma, in 2003.

Figure 5.11 Size comparisons, to scale, of Wellington Formation Odonatoptera. A–C: Meganisoptera: A. *Meganeuropsis permiana* (330 mm); B. *Megatypus schucherti* (195 mm); C. *Oligotypus tillyardi* (50 mm). D. Protanisoptera: *Ditaxineura anomalostigma* (21 mm). E. Protozygoptera: *Kennedya mirabilis* (34 mm).

ACKNOWLEDGMENTS

The author wishes to thank Werner Kraus, of Geologie Endogene Dynamik, RWTH-Aachen, Germany, for allowing the reproduction here of the copyrighted photograph of him posing with his life-sized reconstruction of the giant griffenfly, *Meganeuropsis permiana*. Thanks to the American Museum of Natural History for the loan of the

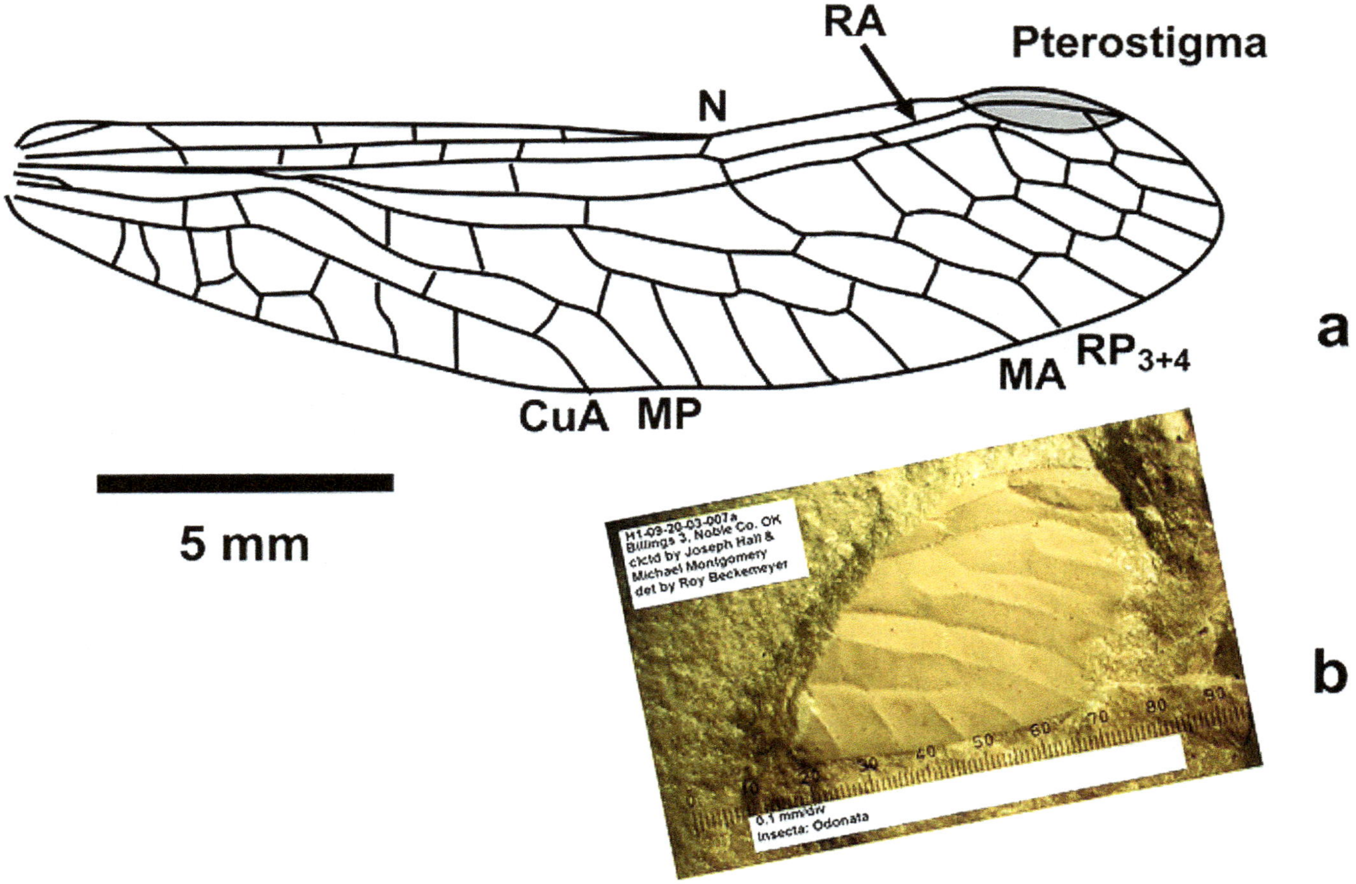

Figure 5.12 Wellington Formation Protanisoptera: Ditaxineuridae: *Ditaxineura*. A. Drawing of *Ditaxineura anomalostigma* after Carpenter (1931), but with wing veins named as in Huguet et al. (2002). B. Fragment of wingtip of a *Ditaxineura* species from Noble County, Oklahoma, showing the pterostigma. The figure is reproduced here at five times life-size.

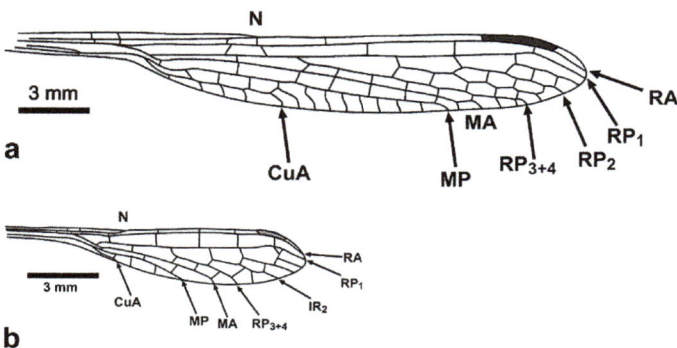

Figure 5.13 Wellington Formation Protozygoptera: Kennedyidae. A. *Kennedya mirabilis* after Carpenter (1939). B. *Progoneura nobilis* after Carpenter (1947). Notation after Nel et al. (2012).

of the Oklahoma State University KC Emerson Museum, Department of Entomology and Plant Pathology, for allowing me to access their substantial and growing collection of Wellington Formation fossil insects. Thanks to Dr. Michael Morales, director of the Museum for sponsoring my work at ESU. Dr. Sonny Ramaswamy, currently Director of the USDA National Institute of Food and Agriculture, provided access to the Kansas State University (KSU) specimen of *Megatypus schucherti* during his tenure as University Distinguished Professor and Head of the KSU Department of Entomology (1997–2004). Dr. Christopher Durden of the Texas Memorial Museum provided access to, and allowed me to photograph, some of EH Sellards' Elmo fossil material, including the type of *Tupus permianus*, at the museum in 2001. My continuing study of Paleozoic fossil insects as a Research Associate of the Division of Entomology (Paleoentomology) of the University of Kansas Natural History Museum (KUNHM) is supported by Prof. Dr. Michael S Engel. The chapter titled, *The Ancestors of Oklahoma Odonata: The Permian Griffenflies, Proto-dragonflies and Proto-damselflies of the Wellington Formation*, by Roy J Beckemeyer, is a contribution of the KUNHM Division of Entomology.

specimen *Tupus readi* (specimen FI-40105, depicted originally in Klots 1944:4, Figure 4). That loan was made through the Johnston Geology Museum at Emporia State University, Emporia, Kansas during my tenure as a Research Associate there. Thanks to Don Arnold and Dr. Richard Grantham

## CHAPTER 6

ENVIRONMENTAL HISTORY OF OKLAHOMA

Brenda D. Smith and Michael A. Patten

One cannot speak of an area's environmental history without at least cursorily addressing the human history of that area. The two cannot be decoupled, for beyond the geological and biogeographical aspects of environmental history one cannot escape the troublesome reality that humans now shape environmental history. So, we begin this chapter with a brief history of Oklahoma's human landscape and follow that with a discussion of the impacts the human landscape has had on the physical.

HUMAN GEOGRAPHY OF OKLAHOMA

The human history of Oklahoma is long and detailed, and many sources cover the subject.[1] We provide but a brief excerpt to that history here as a way of providing some context for how *settlement*, by whom and when, and *land-use*, how and by what means, has impacted Oklahoma's landscape.

Oklahoma has been peopled for many thousands of years, essentially as far back as nomadic groups are known to have come to the Americas (Brooks 2009). Various peoples have come and gone from the area that lies within Oklahoma's current border, groups of big-game hunters with familiar names such as the Clovis and Folsom cultures, for example. Other groups came later, many of whom are less familiar, such as the Dalton and Calf Creek cultures, but there were even more, those known simply as "hunters and gatherers." Starting about 4,000 years ago it appears that trading began between peoples in the region, and with trade came conflict (Brooks 2009). Such behaviors likely indicate that there was a large enough population here that groups came into contact more regularly. It is thought that Caddoan groups, including those of Spiro Mounds, the Wichita, the Pawnee, and the Caddo themselves, were in Oklahoma 1,000–2,000 years ago (Brooks 2009; Fixico 2009). Later, other groups, such as the Osage, Plains Apache, Kiowa, Comanche, Arapaho, Quapaw, and Cheyenne came to Oklahoma as part of their hunting territories, and perhaps for other reasons (Fixico 2009). But it was not until the Indian Removal Act of 1830 that Oklahoma, at the time just a portion of what was then called simply "Indian Country,"[2] saw a surge in human population. The Indian Removal era of 1830–1862 brought the Cherokee, Chickasaw, Choctaw, Creek (Muscogee), and Seminole (referred collectively and demeaningly as the "Five Civilized Tribes"), and the Alabama, Delaware, Ottawa, Quassarte, Seneca-Cayuga,

Absentee and Eastern Shawnee, Tonkawa, and Wyandotte, among others. The Indian Wars of the 1860s–1890s brought even more tribes, including other Apache groups, the Citizen Potawatomi, Iowa, Kickapoo, Kiowa, Kaw, Miami, Modoc, Otoe-Missouria, Pawnee, Peoria, Ponca, Sac and Fox, and Shawnee. Even though these eras technically brought an influx of people, the population of the state, including non-Indians, was negligible (Table 6.1), at least relative to population density and environmental impacts just a couple of decades later.

Between the 1890 census and that of 1900, "Oklahoma had the largest and Indian Territory the second largest percentage of [population] increase ... of any state or territory ..." (U.S. Census Bureau 1907). More than half a million people came to the state in those 10 years, and then another half a million came within the following 7 years, making for 1,155,520 people in less than two decades—a staggering increase in just 17 years.

Whites, in the broad sense (Germans, Russians, French, Irish, Welch, Scottish, English, Czechs, Poles, Italians, Lithuanians, other Europeans, and of course, Americans descended from those groups), were the most populous in the Twin Territories. In 1890, Whites, not Indians, were the majority (Table 6.1) in Indian Territory; this despite the fact they were not officially allowed to reside in Indian Territory. They were, however, officially allowed to participate in the land runs[3] that snatched away parcels not only in the "Unassigned Lands," i.e., Oklahoma Territory, but also those previously allocated to the Indian tribes transplanted to Indian Territory. And so, Whites came in droves.

Black people also participated in the land runs, but they were just a small proportion of those who did. That does not mean that their numbers were insignificant in that era. In 1890 and 1900, for example, Black settlers were the third most populous group in Indian Territory. By 1907 the tables had turned, and Black people outnumbered Indians (Table 6.1). In Oklahoma Territory, Black people were third in 1890, but only a decade later they began to outnumber Indians (by almost 7,000 in 1900, growing to 18,424 more than the Indian population in 1907). These numbers reflect some unique circumstances of Black settlement in the Twin Territories.

About 10,000 slaves, owned by the Cherokee, Chickasaw, Choctaw, Creek (Muscogee), and Seminole, are known to have been in Indian Territory by 1861 (Reese 2009).

Table 6.1 **Population of Indian and Oklahoma Territories in 1890, 1900, and at statehood in 1907. Note that, in 1890, there were more Whites in Indian Territory than there were in Oklahoma Territory (although it was technically illegal for them to be in Indian Territory). Also notice the number of Blacks in 1890, all of whom were either freed slaves from the post-Civil War south or from the Five Civilized Tribes.**

	INDIAN TERRITORY	OKLAHOMA TERRITORY	TOTAL POPULATION
racial grouping	1890	1890	
Whites	110,254	62,300	
Blacks	18,636	2,973	
Indians	51,279	13,177	
Asians[a]	13	25	
1890 total	180,182	78,475	258,657
	1900	1900	
Whites	302,680	367,524	
Blacks	36,853	18,831	
Indians	52,500	11,945	
Asians[a]	27	31	
1900 total	392,060	398,331	790,391
	1907	1907	
Whites	538,512	688,418	
Blacks	80,649	31,511	
Indians	61,925	13,087	
Asians[a]	29	46	
1907 total	681,115	733,062	1,414,177

[a] As "Mongolians."

By the time the Dawes Rolls were completed in 1907, 23,415 "Freedmen," as the former slaves of Indians became known, were registered. But at statehood, there were 112,160 Blacks in the former Twin Territories. So who comprised the 88,745 other Blacks in the state? Very few had migrated from the North; most were freed slaves from Texas, Arkansas, Louisiana, Kentucky, South Carolina, Alabama, Mississippi, and southeastern Missouri who had come to Oklahoma between 1890 and 1910 as part of the "Great Exodus" from the southern states. Some of these "Exodusters," as they were called, were also Black Separatists wishing to establish an all-Black state in Oklahoma (Hill 1946). Instead they were only able to join existing all-Black settlements or form a few new ones,[4] of which there were about 50, including 3 of the most famous, Red Bird (founded circa 1889, Indian Territory, Creek Nation), Langston (in 1890, Oklahoma Territory) and Boley (1903, Indian Territory, Creek Nation). Few other settlements survive.

Outside of the 3 main racial groups of Indians, Blacks, and Whites, there were tiny numbers of "Mongolians," as Asians were called then, in the Twin Territories: 38 in 1890, 58 in 1900, and 75 in 1907 (U.S. Census 1907).[5] A more substantial population at the time in the territories was Mexicans who had come to the United States principally to fill the laborer gap that Asian-exclusion laws had caused. There were more than 100,000 Mexicans in the Great Plains at the turn of the twentieth century; 10 years later that number had doubled. In the Twin Territories it is estimated that there were 138 Mexicans in 1900 and 2,645 by 1910 as a result of growth in the railroad and coal mining industries (Smith 1981). By 1930, the population had grown to 7,354 and was spread across the state, but during the Depression, a mass exodus occurred, lowering the number to a mere 1,425.

As with just about everywhere humans are found, Oklahoma's population, outside of a small dip post-Dust Bowl until about 1950, has continued to grow (Figure 6.1).

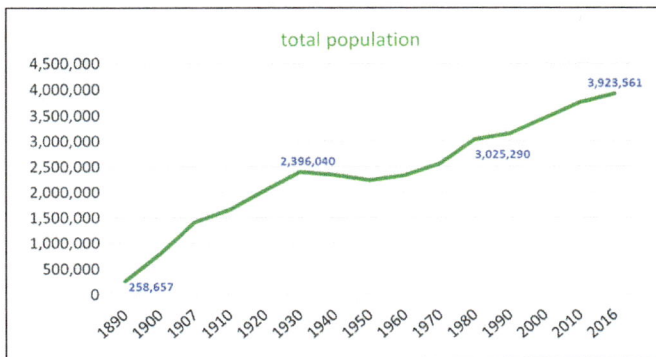

Figure 6.1 **Population of Oklahoma, 1890–2016, showing a general upward trend. Population counts for 1890 and 1900 are combined totals for the Twin Territories and the 1907 count was a special census taken for statehood (U.S. Census 1907). Other data were obtained from Morris et al. (1986) and the U.S. Census Bureau (www.census.gov/).**

The state currently has a population of almost 4 million and demographics have not changed considerably over time (Figure 6.2).

LAND-USE IN OKLAHOMA AND ITS IMPACTS

The dramatic population increase in the decades around statehood (Table 6.1) came with an attendant increase in infrastructure and changes in land-use—railroads, roads, buildings, oil rigs, farms, ranches, and such. The steady increase thereafter continued to broaden the human footprint on the Oklahoma landscape, impacting environs statewide and, consequently, odonates.

Infrastructure

Transportation systems certainly have left their mark on the Oklahoma landscape, a fine example of which is railroads. Rail lines were not legally allowed in Indian Territory until 1866. Legal battles continued to restrict railroads until June 1870, when the first track was laid. Within the first two years of railroad construction 282 miles were put down. Trackage rapidly increased from 1,939.3 miles across the Twin Territories in 1899 to 5,716.6 around statehood—that is 3,777.3 miles laid in 7 years!—until it peaked in 1928 with 6,865.4 miles (George and Wood 1943[6]; Figure 6.3).

But unlike most other infrastructure in Oklahoma, railroads began to decline by the 1930s, with more than 1,000 miles of track becoming non-operational by 1960 (George and Wood 1943; Hofsommer 1977). Now there is about half the track miles from Oklahoma's heyday still operational.[7] But non-operational track does not mean it has been removed; it just is not being used currently. As such, even defunct lines retain a footprint on the landscape, and like roads in Oklahoma, railroads take up a lot of space and make for much habitat destruction, both when laid and then subsequently maintained. Beyond the tracks themselves, there are also bridges, right-of-ways, stations, switching yards, and railroad rolling-stock and engine housing that must be considered when one thinks about the amount of habitat lost to transportation infrastructure.

The footprint of roads is more difficult to grasp due to the sheer number of them in Oklahoma. Not only are there tens of thousands of miles of paved roads in the state, but there are also many thousands of dirt or gravel county roads and "farm roads," the latter found along just about every section line in farm country if not along half- and quarter-sections as well (Figure 6.4). The landscape is literally graded along every mile

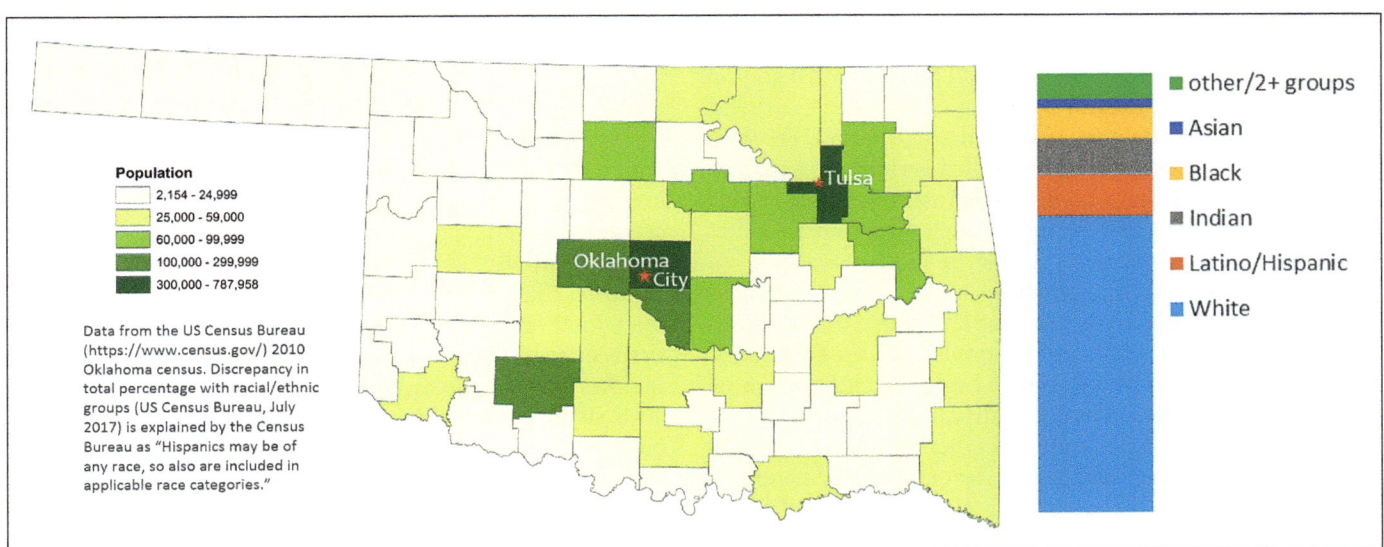

Figure 6.2 **The population distribution across Oklahoma has remained relatively stable through time, with densities highest near Oklahoma City and Tulsa and relatively few people living in western Oklahoma and the panhandle. Racial and ethnic diversity has also remained stable and wanting.**

Figure 6.3 Oil train leaving Bartlesville, Indian Territory, circa 1900. Bartlesville, Indian Territory, photo by J. Fred Hays. Used with permission by the Bartlesville Area History Museum, Bartlesville, Oklahoma.

block throughout agricultural areas in Oklahoma, whether ranching, poultry or pig farms, or growing crops. For the smaller roads there is little additional footprint beyond the road itself, but for more major roads one must also take into account medians and verges or shoulders as well as the many culverts, bridges, and "bar-ditches" (as Oklahomans call borrow ditches, which are places along major roads where fill was taken from in order to build up a road bed). That all adds up to a lot of habitat loss and fragmentation and means wildlife cannot move safely from one area to another, resulting in many, many thousands of roadkilled vertebrates each year (e.g., Smith-Patten and Patten 2008). Insects are also victims as speeding vehicles give no mercy, including to dragonflies, which die in the scores every day of the flight season.[8]

As human populations grow so does the need for buildings of various kinds: houses, apartments, outbuildings, stores, gas stations, schools, hospitals, etc. In Oklahoma, the philosophy has always been to build out rather than up, meaning that Oklahomans prefer urban sprawl over compact living. Like most Americans, Oklahomans favor building new rather than renovating or redevelopment. Abandoned box stores, malls, and houses litter the landscape, while new box stores, malls, and houses are built just down the road from the old. Also like other Americans, there is a proclivity for single-family residences with non-native lawns and landscaping. All of this means that towns and cities tend to have large footprints, containing many buildings with much asphalt and concrete within the interstices and, unsurprisingly, little native vegetation and natural water regimes.

Beyond the loss of habitat, urban environments present other challenges for odonates such as dealing with runoff of lawn chemicals into creeks and streams on which odonates depend. Some species adapt themselves to urban environments and can cope with at least moderate amounts of polluted runoff. Some can cope with loss of natural habitats as long as these are replaced with some suitable waterbody in which they can lay their eggs, but adaptable species are generalists, and so win out while specialists and more sensitive species blip out. Another urban issue is odonates laying eggs in inappropriate places. Odonates see by positive polartaxis, meaning that they are attracted to horizontal polarized light. Such light reflects off of water but it also reflects off of other surfaces such as car hoods and roofs,[9] asphalt, concrete, and puddles of oil, grease, or other chemicals. Ovipositing on such surfaces means that not only could that female odonate hurt herself (e.g., damage her ovipositor), she will also expend energy depositing eggs that are unlikely to hatch.

Industry – Natural resources extraction

Oklahoma has a handful of industries notorious for environmental degradation. These are primarily related to natural resources extraction and to agriculture, the latter sector to be discussed in the next section. Here we broach the topic of extractive industries such as mining for metals like zinc, lead, and copper, and for minerals, such as iodine, gypsum, limestone, and sand and gravel. The fossil fuel industry also has played (and largely continues to play) a huge role in the state through coal mining and oil and gas extraction. Activities of those two sectors (mining and energy) have been responsible for most of the 16 Superfund sites[10] and the >100 non-National Priority List sites in Oklahoma (www.epa.gov) as well as sites yet to be designated, such as Bokoshe, Le Flore County, that deals with coal ash (or fly ash) disposal pits. Timber extraction is another sector that has large environmental impacts. In a short summary, we cannot delve too much into the wide array of environmental issues in Oklahoma that spring from industry, but we can present a few of those issues, particularly those that we feel have had, do have, or may have negative (or even occasionally positive?) effects on odonate populations.

The most infamous Superfund site in Oklahoma is Tar Creek, centered on the former towns of Picher and Cardin in north-central Ottawa County, but it also extends southward into the towns of Commerce and North Miami and east into Quapaw.[11] The site is part of the area known as the Tri-state Mining District that included nearby southeastern Kansas and southwestern Missouri. Zinc and lead were mined in the district starting in 1891 and continued to 1970. At various times in its history, the district was the world's largest producer of zinc and lead.[12] In 1983 it was determined that the area had such high levels of zinc, lead, and cadmium that it was declared a Superfund site and residents were told to move away. Water contamination and mineshaft collapses were also problems for the area. Specific effects on odonates of the Tar Creek Superfund site are unknown, primarily because we will not survey anywhere near the area; as a result, the zone within and around the roughly 40-mi^2 Superfund site is understudied ... but an upside is that we do not have heavy metal poisoning or cancer resulting from exposure to hazardous materials. We cannot really say that about another site in Oklahoma that was responsible for

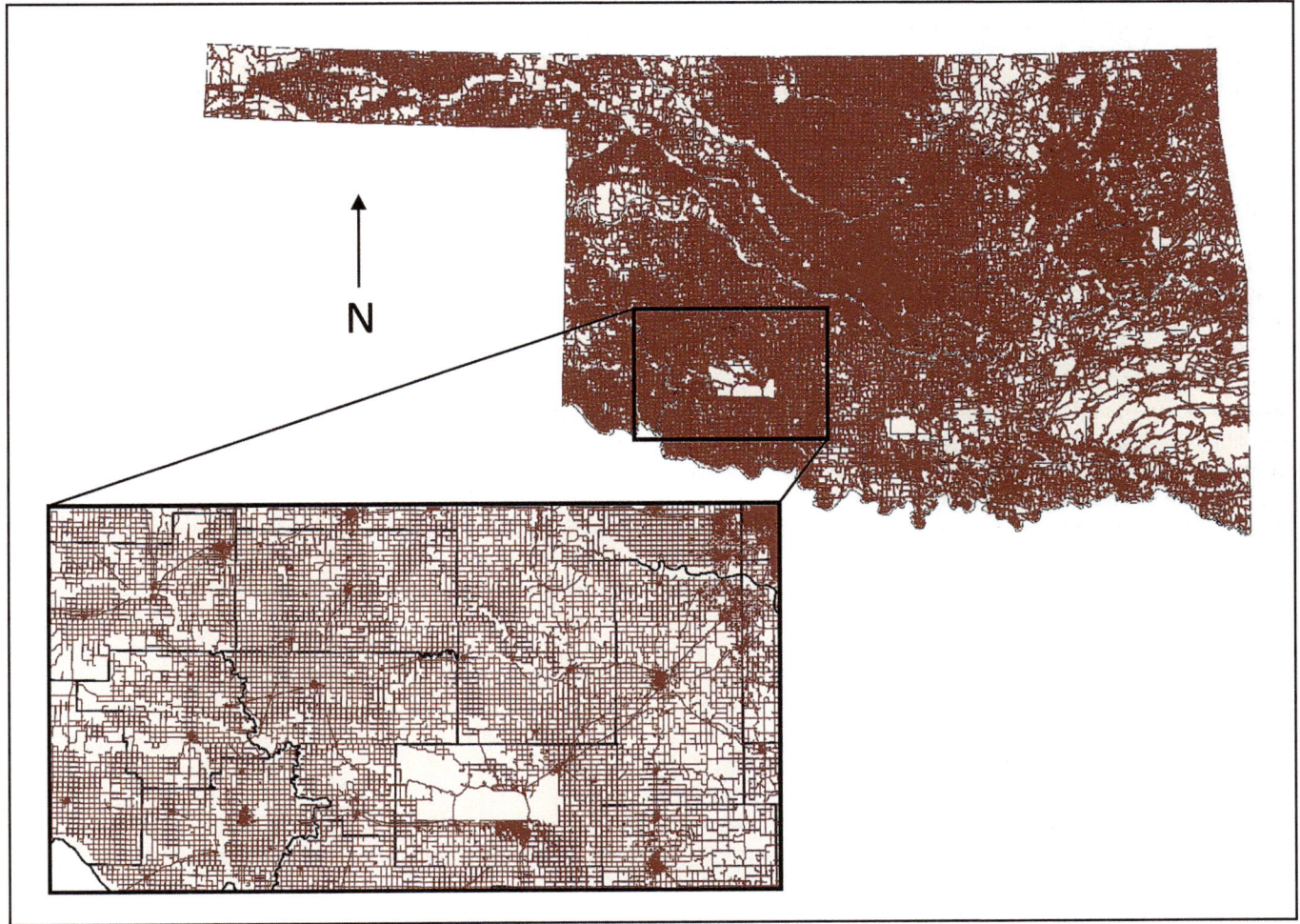

Figure 6.4 Paved and unpaved roads in Oklahoma. Inset shows an example of "farm roads" following block section lines. Large blank spot is the Wichita Mountains Wildlife Refuge and Fort Sill Military Reserve. Data source: Oklahoma Department of Transportation Roadways and Local Roads GIS files.

Bartlesville Zinc Co., Bartlesville, Okla.—14

Figure 6.5 Bartlesville Zinc Company, Bartlesville, Oklahoma, 1914. Used with permission by the Bartlesville Area History Museum, Bartlesville, Oklahoma.

zinc, lead, cadmium, and arsenic contamination because we unknowingly lived right in the middle of it when we first came to the state. That remediation site is essentially the city of Bartlesville,[13] Washington County, which was home to the National Zinc Corporation from 1907 until 1997 and the Bartlesville Zinc Company until the 1920s (May 2009; Figure 6.5). The city and a 3½-mile radius around the facilities were exposed to airborne metals, resulting in soils, sediments, and surface water being contaminated (EPA 1996[14]). Again, no studies have been conducted in the area to determine effects on odonates, but elsewhere bioaccumulation of metals and non-metal contaminates has been documented for insects and some aquatic organisms, including odonates (e.g., DeForest et al. 2007; van der Fels-Klerx et al. 2016). Fortunately, for the time being, zinc and lead mining in Oklahoma are historical pastimes.

Copper mining in Oklahoma also is historical: the last known mining operation ceased in 1975. Open-pit copper mining was centered in southern Jackson County, as such the area now has numerous small, man-made lakes. Some southern odonate species may have used these lakes to colonize, or are attempting to colonize, Oklahoma, such as *Macrodiplax balteata*, the Marl Pennant. If that is the case, then it could be argued—we will not, but someone could— that copper mining had a positive effect on odonate populations in southwestern Oklahoma. Current non-coal mining in Oklahoma is focused on limestone, gypsum, granite, sand, and gravel, and appears to principally use open-pit and hillside removal techniques, and, to a lesser extent, subsurface boring. Approximately 80 million tons of these and related materials are mined each year in Oklahoma.[15] The most obvious impact of such activities is a loss of habitat and topographic alterations; it is unclear what effects they may have on local hydrology and on aquatic organisms.

The energy industry in Oklahoma has affected the entire state in one form or another, whether that has been from mining coal in the east to drilling for oil and gas across the board. Although the state is most often associated with the oil and gas industry, Oklahoma is also part of "Coal Country," with ties that go back to at least 1829, well before the state's oil and gas boom. Coal mining is what brought many Europeans, especially eastern and southern Europeans to Oklahoma to coal towns such as McAlester; an immigration wave that ripples throughout Oklahoma's coal country today. Coal Country spans approximately a 14,550 mi[2] area in northeastern and north-central Oklahoma and is part of the Western Region of the Interior Coal Province of the United States (Doerr 1962). In the 1920s, it was estimated that there were some 79 billion tons of coal in Oklahoma, which, it was thought at the time, "at the present rate of mining is sufficient to last for 26,000 years" (Gould 1926:437).

Oklahoma's coal beds were at peak production in 1920, when close to 5 million tons were mined. Production was down to an average of 2 million tons in the 1950s; by 2016, it declined to a relatively meager sum of 670,610 tons (Doerr 1962; OK Dept. of Mines 2016 data). Early in the industry's history, subsurface shaft mines were the norm, but after World War II strip mining became more and more prevalent and is still the method that persists today. Strip mining is well known for destroying large amounts of habitat, as well as being the source of water- and air-quality issues.

The other major fossil fuels industry, that of oil and gas, has undeniably put its mark on Oklahoma's landscape, both physically and culturally. Urban and rural environments alike show the signs—pump jacks and wells are just about as likely to be found on the prairie as they are in someone's backyard or in the middle of a town (Figure 6.6) The state's

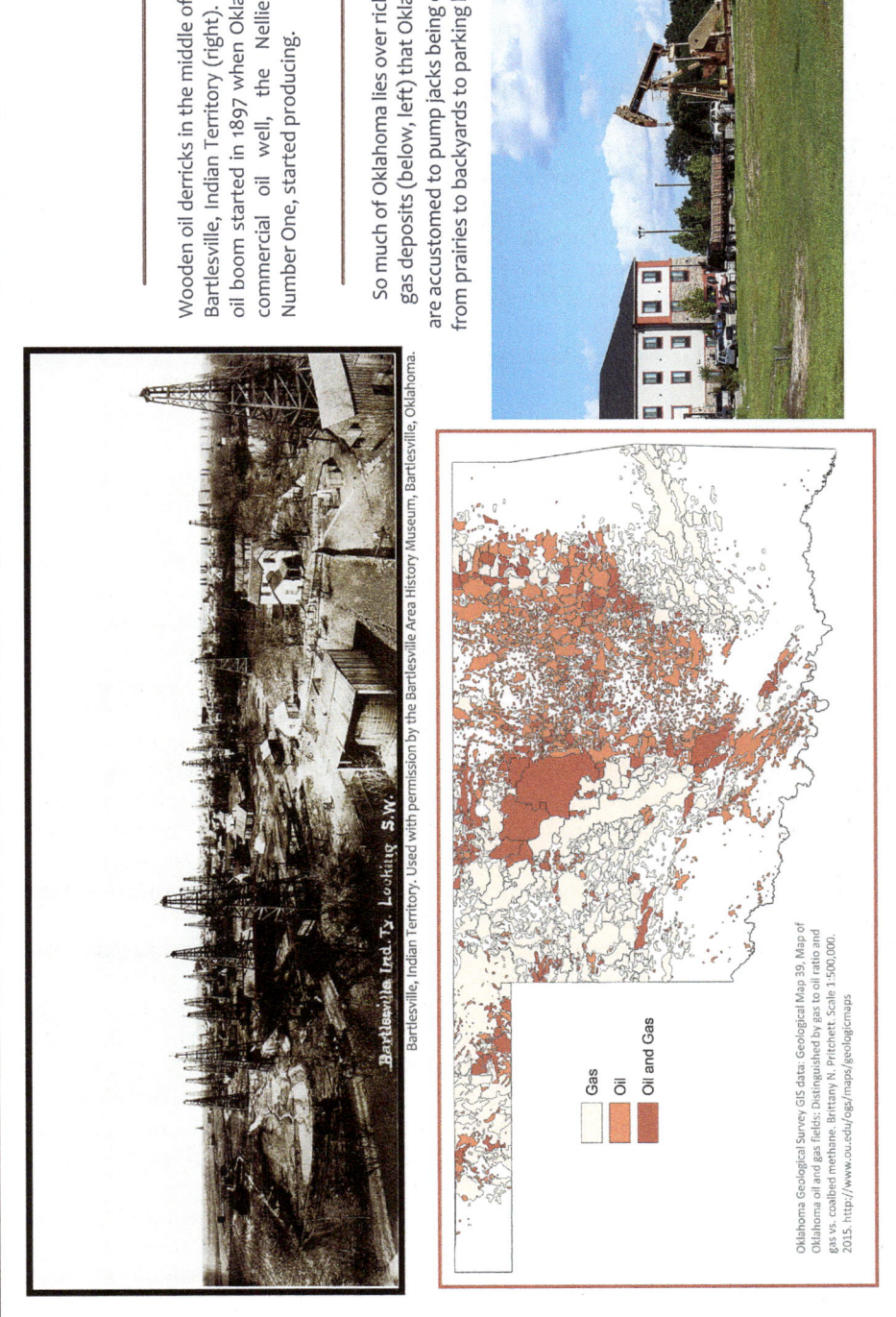

Wooden oil derricks in the middle of the town of Bartlesville, Indian Territory (right). Bartlesville's oil boom started in 1897 when Oklahoma's first commercial oil well, the Nellie Johnstone Number One, started producing.

So much of Oklahoma lies over rich oil and gas deposits (below, left) that Oklahomans are accustomed to pump jacks being everywhere, from prairies to backyards to parking lots (below).

Brenda D. Smith

Bartlesville Ind. Ty. Looking S.W.

Bartlesville, Indian Territory. Used with permission by the Bartlesville Area History Museum, Bartlesville, Oklahoma.

Gas
Oil
Oil and Gas

Oklahoma Geological Survey GIS data. Geological Map 39. Map of Oklahoma oil and gas fields: Distinguished by gas to oil ratio and gas vs. coalbed methane. Brittany N. Pritchett. Scale 1:500,000. 2015. http://www.ou.edu/ogs/maps/geologicmaps

Figure 6.6 Oil and gas in Oklahoma.

economy is so closely tied to oil and gas that a slight downturn in prices can wreak havoc and lead to a precipitous drop in production tax revenue for the state, as in recent years, that can lead to utter crisis.[16]

Oklahoma oil has been known about since at least the 1850s, but it was not developed commercially until 1897, when the Nellie Johnstone #1 oil well began to pump. That rig lead to a whole field of wooden oil derricks pumping day and night from the oil field that lay beneath Bartlesville (Figure 6.6). From there, industrial exploitation of oil ran like wildfire across a dry prairie. In 1927, 277 million barrels of oil were pumped up, setting the state's all-time record.[17] Currently about 140 million barrels of crude oil are pulled out of the ground every year and Oklahoma's 5 refineries process >511,000 barrels of petroleum per day.[18] Natural gas now plays a bigger part in the industry than oil does in the state. In 2015, Oklahoma was ranked third among states for natural gas production, producing about 2.5 million ft³ each year.[19] It is now estimated that there have been >½ million oil and gas wells drilled in Oklahoma during the industry's history.[20]

Known and potential environmental issues with oil and gas exploitation include clearing of habitat for drilling pads, wells, pump jacks, and storage facilities; habitat fragmentation; well leaks that contaminate soil, air, and water[21]; spills during transport; a slew of issues specific to fracking; wastewater disposal[22]; and water use that contributes to aquifer and streamflow depletion. Admittedly, the industry's water usage of 42,107 acre feet per year (AFY; OKWRB 2012), though nothing to sneeze at, is a drop in the bucket compared to other sectors. For example, livestock use 94,480 AFY and municipalities and industrial uses account for 601,891 AFY. Irrigation for crops consumes 745,219 AFY, which is roughly equivalent to half of the capacity of Grand Lake O' the Cherokees. These statistics lead us toward our final example of environmental impacts of land-use practices: agriculture. A tremendous amount of water is used each year for agriculture for "crops" of all kinds, whether grown crops, as with plants (primarily for livestock consumption), or raised crops, as with actual livestock. But for the moment, we will hold off from discussing the agricultural industry to first summarize the underlying theme in the land-uses reviewed above.

Up to this point in our review, we relegated water to a tangential but pervasive corollary to land-use choices made by Oklahomans. We did the same with habitat loss and fragmentation. Water, as we will revisit in the agriculture discussion, is inextricably tied to that massively landscape-altering industry. Water is also more intuitively linked to odonate conservation, but we must bear in mind that habitat loss and fragmentation, too, are serious threats to odonates. It is easy to detect a problem when, for example, we witness a wetland being drained, graded over, and paved. It is harder to notice when, say, a farmer plows over a playa lake

to plant a crop. That odonate habitat is lost for at least that season, but the more the playa is plowed, the more altered the substrate becomes until the soils can no longer function as a playa, in which case, the playa, a crucial habitat for many organisms, is lost forever.[23] Unless one knows a playa used to be in that field, the playa's demise goes unnoticed, by humans at any rate.

The upshot is that we live in a world of shifting baselines. Unnoticed is the slow fragmentation of the landscape by incrementally eliminating or altering habitat. Adult odonates are forced to fly longer distances to find suitable habitat, and larval odonates may have aquatic dispersal corridors impeded by changes in stream substrate and flow or by road culverts not designed to accommodate their needs. Siltation of streams, ponds, and lakes can result from habitat alteration or loss. Sediments that cause siltation can come from fields and areas where habitat loss has occurred such as oil well pads, open-pit mines, and construction of infrastructure. Siltation can change streamflow and put pressure on dams, not only from the buildup of sediments straining the main structure but also from clogging release mechanisms. Runoff sediments can carry chemicals (e.g., pesticides, herbicides, fertilizers, and petrochemicals) into waterbodies that negatively alter water chemistry and quality by causing eutrophication[24] and toxicity, killing aquatic organisms of all kinds, including odonates.

The moral of the story is that what seems unrelated to odonates can affect odonates. This adage applies to the next chapter of Oklahoma's environmental history we will discuss—the seemingly benign agricultural industry. It may seem like a stretch of the imagination to say that at this point in Oklahoma's history, agriculture in the broad sense is the land-use practice in the state that is most detrimental to odonates. It may seem overboard to say that in Oklahoma the worst environmental threat to odonates, especially to lotic species, is how we farm and how we raise livestock and poultry. But we hope to persuade you. Not surprisingly, water will be strategic in that persuasion.

Agriculture and its transformation of the landscape

Close to 80% of Oklahoma lies in the broad ecoregion known as the Great Plains of the United States. Because so much of the state lies within the Great Plains, in a more restrictive sense in the Southern Great Plains, it is that region on which we shall focus here. It is this region, being one of extremes—bountiful plenty and ravaged landscapes—that has had such a profound and long-lasting impact on Oklahoma's psyche and physical character, affecting both the human world and that of odonates.

It is Oklahoma's panhandle that comes to mind when we think of the state's Great Plains. We think of the great frontier wiped cleaned of its former peoples, making way for non-Indians. Mexican sheepherders came first, then White cattlemen who began to claim territory for themselves and who employed Mexican cowboys, and then … the farmers came, many of whom were German Mennonites from the Russian steppe, who brought with them winter wheat and by accident the dreaded Russian Thistle (*Salsola kali*), popularly known as tumbleweed. Lush native grasslands that once spanned thousands of miles across the entire plains were rapidly converted to highly productive ranch and crop land. That prosperity was, relatively speaking, short-lived, because poor land management transformed those lands into a desertified wasteland. Eventually much effort transformed that wasteland back into productive land, but this time instead of lush native grasslands it became America's breadbasket, or, more accurately for Oklahoma, America's livestock feed farm. Those transformations are written onto the historical landscape of Oklahoma, and that history has a clear dividing line of eras—the Dust Bowl, the period of drought and severe dust storms that ravaged the American prairies.

In the pre-Dust Bowl era (~1880 to 1930) vast tracts of grassland were plowed under at a high rate and converted to dry-farm cropland (i.e., rainfall dependent agriculture). It was also a time when much native grassland was converted to pasture for livestock. In roughly a half century, nearly 440 million acres of grassland in all of the Great Plains were converted from only 1% cropland to 31% and from a grazing intensity of 95 acres per cow, sheep, and horse to 10 acres per.[25] In Oklahoma, looking at just two crops, wheat and cotton, we see that acreage planted and bushels or bales harvested mostly increased, and did so at a terrific rate, during this era. For example, in 1896 >2.2 million bushels of wheat were harvested, but that increased to >66 million bushels by 1919 (Green 1977a); roughly equivalent to 176 million loaves of bread in 1896 versus 5.3 *billion* loaves less than 25 years later.[26] For cotton, 17,000 bales were picked in 1879 in Indian Territory, 862,383 were in 1907, and almost 1.8 million were by 1926 (Green 1977b).[27] Tractors are also an indication of how farming increased during these early years. Once gasoline-powered tractors became widely available during World War I, they began to find their way to Oklahoma: 5,789 in 1920, 10,039 in 1925, and 19,407 in 1939 (Green 1977a).

But with the boom in mechanization and productivity came eventual bust. Drought set in during the early 1930s and it held tight through the decade, affecting farms across Oklahoma. Farmers who owned land lost title, tenant farmers were forced to vacate; livelihoods were lost. Oklahomans moved in droves to Hoovervilles—homeless encampments named derisively for the then president—or became part of the mass exodus from the state; the only marked dip in population the state has ever seen (Figure 6.1). This is the iconic *Grapes of Wrath* era[28] during which "Okies" emigrated to California and other agricultural states. But that was mainly folks from and circumstances in the rest of the

state; in northwestern Oklahoma and its panhandle the situation was much worse, much worse indeed.

Over-farming of the panhandle and surrounding areas degraded once-fertile soils of the native drought-resistant grassland, leaving them exposed and susceptible to persistent high winds and gusts on the High Plains, gusts that routinely exceed 80 kph (55 mph) and sometimes 100 kph (63 mph), winds that stripped topsoil.[29] Massive dust storms rolled across the plains, at times blocking out the sun or spreading across the United States as far as the Atlantic coast[30] (Egan 2006). Although the extent of the Dust Bowl varied from year to year, its heart was in the Oklahoma panhandle near Boise City, Cimarron County (Figure 6.7). Some settlers in the panhandle, like *The Grapes of Wrath* Okies, left the state. But in the place where the Dust Bowl had its tightest and deadliest grip, the people mostly stayed. Most panhandlers, at least two-thirds, but possibly up to three-quarters, were not part of the wave of new "Exodusters" (this time the term not referring to Blacks fleeing the southern U.S. at the turn of the 19th century, but rather expropriated to Whites fleeing dust storms in the 1930s[31]).

The Dust Bowl, arguably the worst ecological disaster in the nation's history (Figure 6.7), had a huge negative economic and biological impact on the south-central United States. It convinced people on the Great Plains to take extreme measures to avoid another catastrophe. Actions in Oklahoma are a prime example of extreme measures taken. Decisions were made that would dramatically alter the hydrological landscape of Oklahoma forever. Two principal alterations in the water regime were: 1) conversion of free-flowing water (lotic habitat) to standing water (lentic habitat) via construction of reservoirs (Costigan and Daniels 2012) and 2) conversion of farming and ranching techniques, from dry-farming/ranching to irrigation-based farming/ranching that depends heavily on groundwater (Kustu et al. 2010). These conversions have not only had a dramatic effect on the physical landscape but, we argue, on the biological landscape as well.

Oklahoma is a state that lacks natural permanent lakes (there are only ephemeral playas or small oxbows). Prior to the Dust Bowl, Oklahoma had created only one lake of substantial size, Lake Lawtonka, although even its surface area was a mere 970 ha. Nineteen other reservoirs existed in the state at that time, but all were small impoundments (<280 ha). Together, these lakes were unable to sate Oklahoma's water needs during frequent cycles of drought. Amid the Dust Bowl, a frenzy of activity, spearheaded by the Army Corps of Engineers and Civilian Conservation Corps, led to the damming of dozens of rivers and major streams across the southern Plains (e.g., Costigan and Daniels 2012; Figure 6.8). Oklahoma was not immune to this frenzy and neither were just the parts of the state that were hardest hit by the Dust Bowl; instead, dams went up

For 1936 we must record another year of failure. Yet that failure might not so easily have been changed to moderate success by one good rain in the late July or August that we do not altogether despair. Will is laying up a fresh set of terraces in the hopes that next year may give us all a better chance. It seems impossible to dispense with that little word hope, even though at times we are conscious of the pain of hopes too long deferred.

– Caroline Henderson,
Eva, Oklahoma
December 8, 1936

Excerpt from Letters from the Dust Bowl (University of Oklahoma Press 2003)

1930s American Dust Bowl

Boise City, Cimarron County, Oklahoma, was the epicenter of the American Dust Bowl of the 1930s. Massive walls of dust ripped through the panhandle, devastating the landscape, covering homesteads, and choking those pioneers who stayed in the region despite the hardships they endured. The hopes of those pioneers were perhaps never so well expressed than by Caroline Henderson of *Letters from the Dust Bowl* fame.

Groundwater irrigation, the savior of farmers in the Dust Bowl, may lead to their eventual ruin with continued depletion of the High Plains aquifer.

Photos, all but first are from Cimarron County, Oklahoma (clockwise from top left): Hooker, Texas County, Oklahoma, 4 June 1937, photo by GL Risen, used with permission, Bartlesville Area History Museum; Alva, 13 April 1935, unknown photographer, Minneapolis Public Library Collection 21171-9, and farm near Felt, 3 February 1939, photo by Moore, T Bone McDonald Collection, 9319, both photos used with permission, Oklahoma Historical Society; farmer irrigating and boy, 1936, photos by Arthur Rothstein, LC-USF34-004106-E and LC-USF34-004047-E (b&w film nitrate negatives), Library of Congress.

Figure 6.7 America's worst ecological disaster to date.

across Oklahoma. Of the approximately 180 large reservoirs that now grace the state, two-thirds were created after 1934 (mid-Dust Bowl) up to the 1980s, including all of the large (>4,450 ha) ones, the first of which was completed in 1944. Overall, Oklahoma's hydrological landscape transitioned from 10,243 ha of lentic habitat prior to 1940 to 277,927 ha after 1940 (about 40 mi^2 to 1,073 mi^2; Figure 6.8), a 2,700% increase!

Damming reduced the extent of lotic habitat as streams and rivers became lakes. We estimate that as a result of the thousands of small to large impoundments created across the state after the Dust Bowl, there has been a minimum direct loss of 2,173 km (1,350 mi) of lotic habitat.[32] It is more difficult to quantify the impact to lotic habitat of another response to the Dust Bowl, a response that ushered in a new system of farming and ranching: irrigation-based agriculture, specifically center pivot irrigation, which heavily relies on groundwater. The center of irrigation-based farming in the state is Texas County

in the central panhandle, near the epicenter of the Dust Bowl (Figure 6.9). When farmers in western Oklahoma shifted from dry-farming to irrigation-based farming, they not only made farming less risky in a region that averages a mere 44.5 cm (17.5 in) of precipitation annually (Oklahoma Climatological Survey), but also began to lower the aquifer, enough so that streamflow has been reduced markedly across the Southern Great Plains (Garbrecht et al. 2004; Brikowski 2008; Patten and Smith-Patten 2013b).

Across the globe humans use a truly incomprehensible amount of water on an annual basis, estimated at 9,087 billion cubic meters, or cubic gigameters, per year[33] (Gm^3/yr) from 1996–2005 (Hoekstra and Mekonnen 2012). Irrigation consumes 75–85% of this total (Gleick 2003; UNEP 2008), and most irrigation water comes from aquifers, a source that now surpasses use of surface water (Kustu et al. 2010). In the Great Plains that groundwater source is the High Plains (Ogalalla) aquifer, which

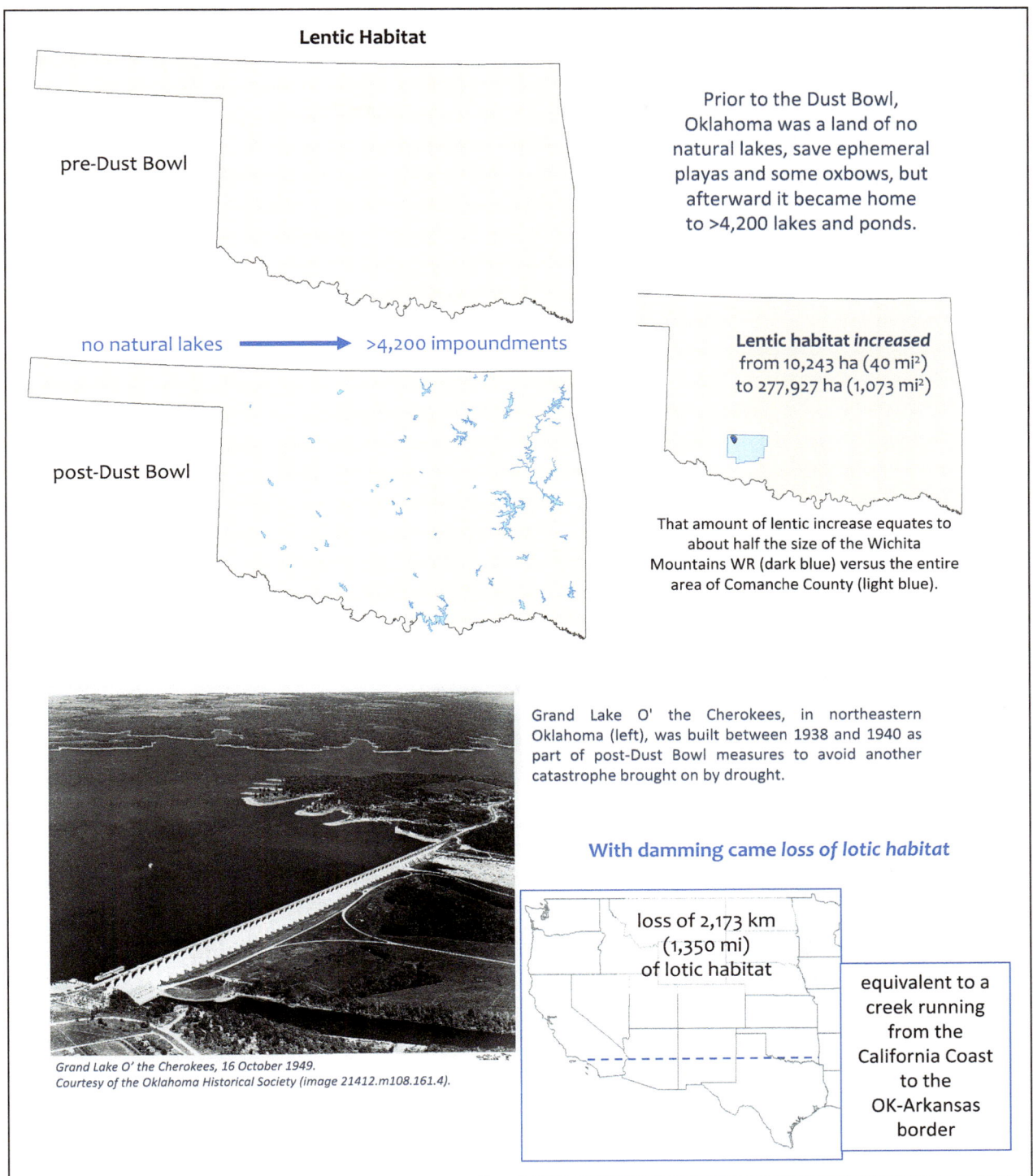

Lentic Habitat

pre-Dust Bowl

Prior to the Dust Bowl, Oklahoma was a land of no natural lakes, save ephemeral playas and some oxbows, but afterward it became home to >4,200 lakes and ponds.

no natural lakes ⟶ >4,200 impoundments

Lentic habitat *increased* from 10,243 ha (40 mi²) to 277,927 ha (1,073 mi²)

post-Dust Bowl

That amount of lentic increase equates to about half the size of the Wichita Mountains WR (dark blue) versus the entire area of Comanche County (light blue).

Grand Lake O' the Cherokees, in northeastern Oklahoma (left), was built between 1938 and 1940 as part of post-Dust Bowl measures to avoid another catastrophe brought on by drought.

With damming came *loss of lotic habitat*

loss of 2,173 km (1,350 mi) of lotic habitat

equivalent to a creek running from the California Coast to the OK-Arkansas border

Grand Lake O' the Cherokees, 16 October 1949.
Courtesy of the Oklahoma Historical Society (image 21412.m108.161.4).

Figure 6.8 Oklahoma's hydrological landscape differs strikingly pre- versus post-Dust Bowl.

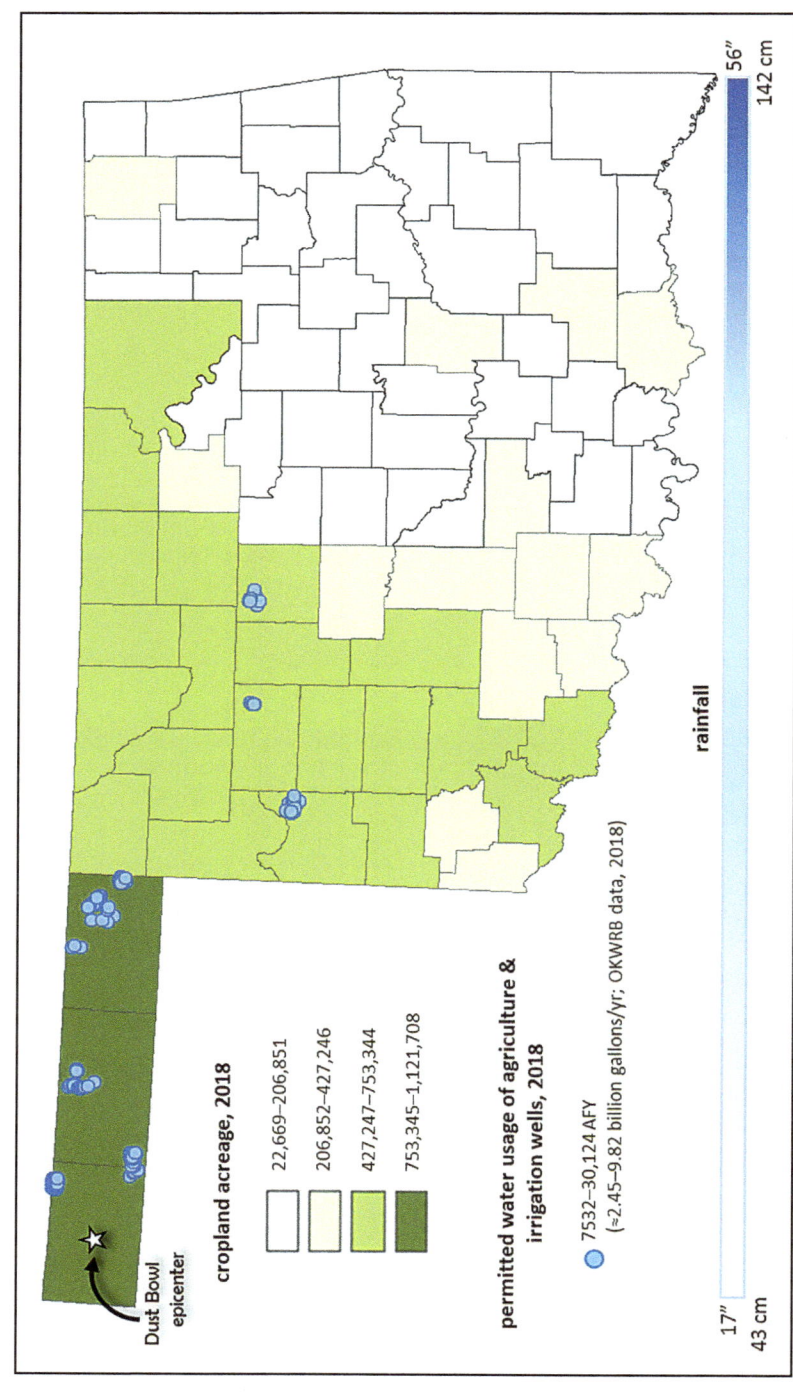

Figure 6.9 Cropland acreage in Oklahoma, by county, 2018 (USDA). Most farming is done in western Oklahoma, particularly the panhandle, where the state receives the least amount of rain. As such, there is a high dependence on groundwater for irrigation. Note the coincidence of dry areas, high levels of farming, and wells permitted to pump huge volumes of water.

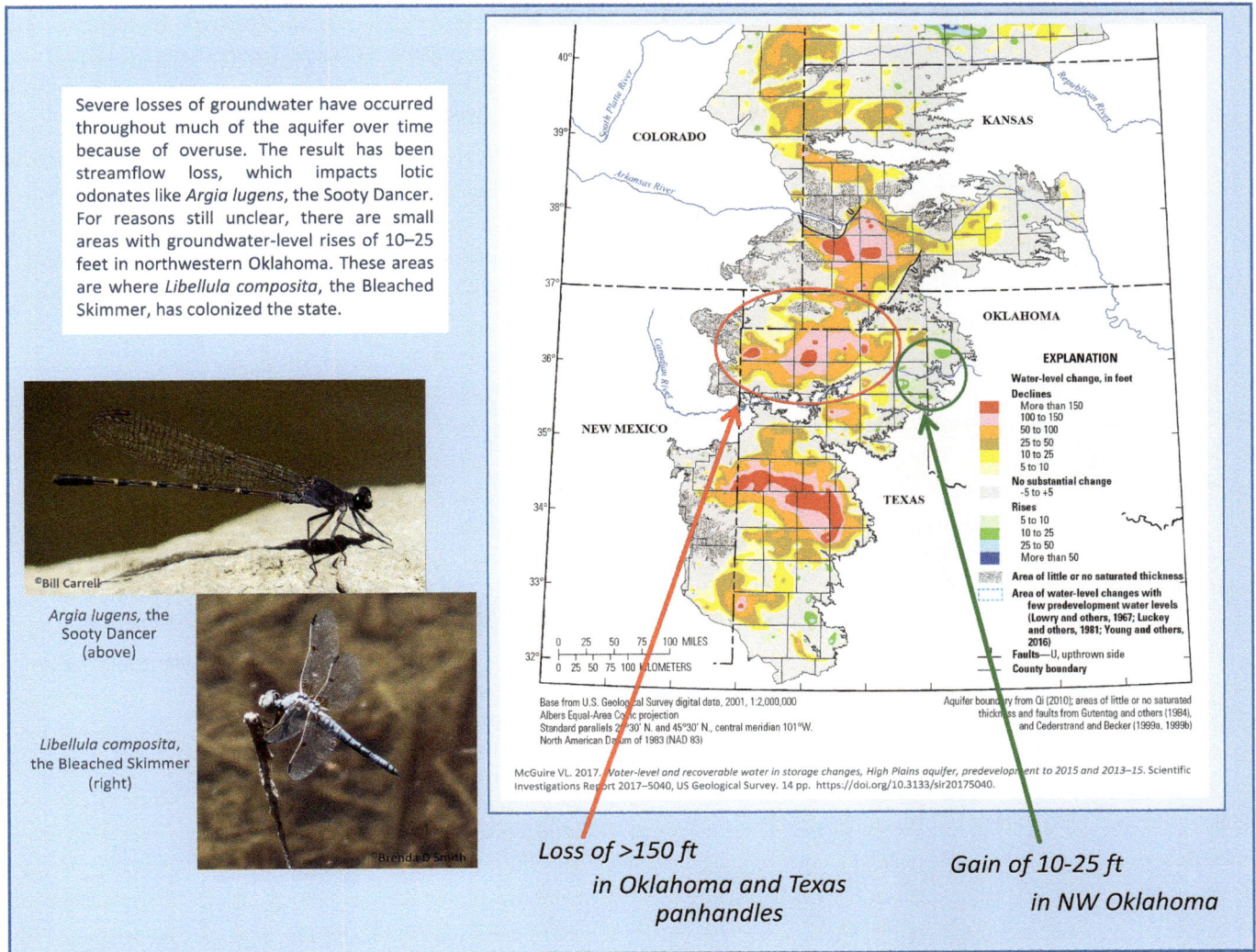

Severe losses of groundwater have occurred throughout much of the aquifer over time because of overuse. The result has been streamflow loss, which impacts lotic odonates like *Argia lugens*, the Sooty Dancer. For reasons still unclear, there are small areas with groundwater-level rises of 10–25 feet in northwestern Oklahoma. These areas are where *Libellula composita*, the Bleached Skimmer, has colonized the state.

©Bill Carrell

Argia lugens, the Sooty Dancer (above)

Libellula composita, the Bleached Skimmer (right)

©Brenda D Smith

Loss of >150 ft in Oklahoma and Texas panhandles

Gain of 10-25 ft in NW Oklahoma

Figure 6.10 Changes in water levels in the High Plains aquifer to 2015 (right, McGuire 2017).

underlies a 450,658 km² (175,000 mi²) region in eight U.S. states (Luckey and Becker 1999; McGuire 2017; Figure 6.10). Over 24.5 billion cubic meters (Gm³/day) of groundwater is pumped daily to irrigate over 6 million ha (>23,000 mi²) of land in this region (McGuire 2017), with approximately 300 Gm³ (81 trillion gallons[34]) of water "withdrawn for irrigation in Texas, Oklahoma, and Kansas from 1950 to 2005," pumping that decreased parts of the aquifer by 30–76 m "in heavily irrigated parts of Texas, Oklahoma, and Kansas" (Karl et al. 2009).

In Oklahoma, 73–80% of irrigation is drawn from groundwater (OKWRB 2011, 2018), including 335 million m³ (>88 billion gallons) of groundwater that comes from the High Plains aquifer (Gollehon and Winston 2013). The aquifer has been used at such a high rate in roughly the past half century that use outpaces recharge (Gollehon and Winston 2013). More than 330 km³ of stored volume has been depleted since 1950 (McGuire 2009; Kustu et al. 2010). The aquifer is a glacial repository with no major source. It is recharged by the slow accumulation of rainwater deposition, at a rate of 13 mm/yr in most parts of the aquifer, but a mere 0.6 mm/yr in Texas (Kustu et al. 2010), making it is impossible to compensate for present rates of use—*millimeters in but billions of cubic meters out*. Depletion of the aquifer has led to localized drying of streams, rivers, and springs throughout the Great Plains (Garbrecht et al. 2004; Brikowski 2008). Aquifer depletion is linked to greatly reduced flow in both the Cimarron and Beaver Rivers and countless streams in the Oklahoma panhandle (e.g., Wahl and Wahl 1988; Patten and Smith-Patten 2013b; Figure 6.11). In the next chapter we will see how this new hydrological landscape has impacted Oklahoma's odonates.

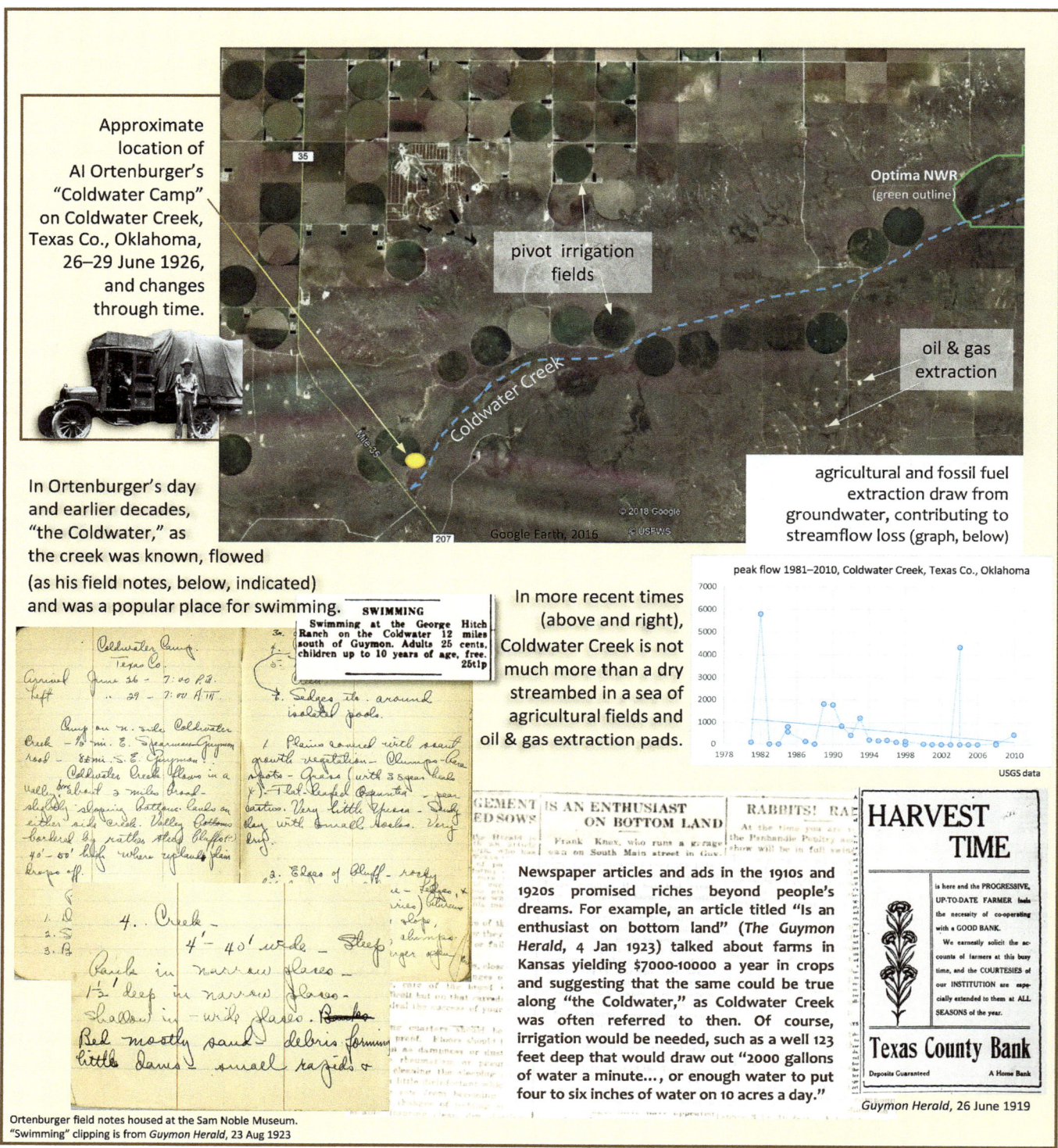

Approximate location of Al Ortenburger's "Coldwater Camp" on Coldwater Creek, Texas Co., Oklahoma, 26–29 June 1926, and changes through time.

Optima NWR (green outline)

pivot irrigation fields

oil & gas extraction

Coldwater Creek

Google Earth, 2016 © 2018 Google © USFWS

agricultural and fossil fuel extraction draw from groundwater, contributing to streamflow loss (graph, below)

In Ortenburger's day and earlier decades, "the Coldwater," as the creek was known, flowed (as his field notes, below, indicated) and was a popular place for swimming.

SWIMMING

Swimming at the George Hitch Ranch on the Coldwater 12 miles south of Guymon. Adults 25 cents, children up to 10 years of age, free.

25t1p

In more recent times (above and right), Coldwater Creek is not much more than a dry streambed in a sea of agricultural fields and oil & gas extraction pads.

peak flow 1981–2010, Coldwater Creek, Texas Co., Oklahoma

USGS data

GEMENT
ED SOWS

IS AN ENTHUSIAST ON BOTTOM LAND

Frank Knox, who runs a garage on South Main street in Guv...

RABBITS! RA

At the time you are r... the Panhandle Poultry ... show will be in full swing...

Newspaper articles and ads in the 1910s and 1920s promised riches beyond people's dreams. For example, an article titled "Is an enthusiast on bottom land" (*The Guymon Herald*, 4 Jan 1923) talked about farms in Kansas yielding $7000-10000 a year in crops and suggesting that the same could be true along "the Coldwater," as Coldwater Creek was often referred to then. Of course, irrigation would be needed, such as a well 123 feet deep that would draw out "2000 gallons of water a minute..., or enough water to put four to six inches of water on 10 acres a day."

HARVEST TIME

is here and the PROGRESSIVE, UP-TO-DATE FARMER finds the necessity of co-operating with a GOOD BANK.

We earnestly solicit the accounts of farmers at this busy time, and the COURTESIES of our INSTITUTION are especially extended to them at ALL SEASONS of the year.

Texas County Bank

Deposits Guaranteed   A Home Bank

Guymon Herald, 26 June 1919

Ortenburger field notes housed at the Sam Noble Museum.
"Swimming" clipping is from *Guymon Herald,* 23 Aug 1923

Figure 6.11   **Changes in streamflow in the Oklahoma panhandle: the example of Coldwater Creek.**

NOTES

1. A good starting point for Oklahoma history is the two volume, *The Encyclopedia of Oklahoma History and Culture*, Everett D, LD Wilson, L O'Dell, JD May, editors, 2009, Oklahoma Historical Society, Oklahoma City, Oklahoma.

2. What constitutes "Indian Country" or "Indian Territory" is a complicated matter. The gist of that history is that, at the time of the Louisiana Purchase in 1803, it was envisioned that "Indian Country" would constitute all or much of the land west of the Mississippi River. By the time of the Indian Removal Act of 1830, Indian Country was whittled down to areas west of the 95th meridian in what roughly approximates Kansas, Nebraska, Oklahoma, and part of Iowa (Everett 2009a). After the Civil War, Indian Territory, as it was generally called by then, was whittled again to present-day Oklahoma with the panhandle as the "Public Land Strip." The Land Run of 1889 took what was known as the "Unassigned Lands" and then the Oklahoma Territory Organic Act of 1890 carved out another large piece of Indian Territory by taking the Cherokee Outlet and the lands in southwestern Oklahoma that previously were assigned to the Cheyenne, Arapaho, Wichita, Caddo, Comanche, Kiowa, and Apache. The act also added to Oklahoma Territory "No Man's Land," formerly known as the Public Land Strip and now recognized as Oklahoma's panhandle.

3. The land runs lasted from 1889–1895; thereafter, a parcel bid system was imposed from 1901–1906 (Everett 2009b).

4. The leader of the Black Separatist movement, Edwin P. McCabe, continually urged southern Blacks to come to the Twin Territories in hopes of having such a large population there that the U.S. government would succumb to the pressure to form an all-Black state. Despite the rather large numbers of Blacks who came, they were never able to get congressional approval for an all-Black state; settling instead for segregated communities amongst a White population that really did not want them there (and who were willing to show it by lynching 1,902 Blacks in 1890–1909; Tolson 1970). Although dating from outside this era, it is of interest to note that the oldest all-Black settlement appears to be Tullahassee, Indian Territory (present-day Wagoner County), which was founded in 1850 (Tolson 1970). The town was probably founded so early because it was in the Creek Nation, which was less restrictive of its slaves and former slaves, including allowing inter-racial marriages. The town currently has a population of 109 people (U.S. Census Bureau). Other towns, such as Arkansas Colored (Choctaw), Canadian Colored Town (Creek), Gibson Station (Creek), and North Fork Colored (Creek) were founded before the arrival of the freed slaves from the Old South.

5. A footnote indicated that of the 1907 "Mongolian" population there were six Japanese. That tally was probably provided in accordance with the "Gentlemen's Agreement" of 1907 that was meant to purge the United States of Japanese laborers.

6. 1870–1872 mileage in the previous sentence was calculated using section IV, *Corporate History and Construction Records of Railroads in Oklahoma*, in George and Wood (1943). That section was also used to sum miles of track up to 1899 and to statehood. We used the statement "around statehood" because three of the lines (33.4 miles) that were included were started in or just before 1907 but were not completed until 1908–1909. The estimated miles of track for 1928 came from the *Railroad Map of Oklahoma* (published by the Corporation Commission of Oklahoma) of that year. Map can be accessed via the Oklahoma Digital Map Collection (https://dc.library. okstate.edu/digital/collection/OKMaps/id/8322/rec/1).

7. *2014–2015 Official Oklahoma State Railroad Map*, accessed at www.ok.gov/odot/Programs_and_Projects/Rail_Programs/index.html

8. Soluk et al. (2011) reported 2–35 individuals killed per km/day in a study conducted in Illinois. Riffell (1999) reported 10–256 kills per km/day at his study site in Michigan. Numbers like these can be detrimental to populations of conservation concern, such as with the federally endangered Hine's Emerald (*Somatochlora hineana*), for which it was estimated that 3,300 adults were roadkilled in one year in just one county (Door County, Wisconsin; Soluk and Moss 2003).

9. BS-P captured video of female *Pantala* ovipositing on car roofs (see Appendix C).

10. There is one proposed, eight active, and seven de-activated ("deleted") sites in Oklahoma on the National Priorities List.

11. Nearby Treece, Kansas, is also a part of the Environmental Protection Agency's (EPA) Tar Creek Superfund site.

12. See Everett (2009c). And, Charles N Gould (1926:437), in his article, "Oklahoma—An example of arrested development," stated that "A single county in Oklahoma, Ottawa, produces more zinc than all the rest of the United States. The amount for 1924 was 269,000 tons of zinc concentrate, valued at $35,000,000." He also said that 70,000 tons of lead was collected each year with a value of $11.5 million.

13. The EPA Superfund site is generally known as the National Zinc Corporation Superfund Site, but it is also sometimes called the National Zinc Site or St. Joe Minerals Site.

14. The EPA said the smelter operated until 1976. This site technically has been listed as a "proposed" Superfund site since 1993, but subsequent reviews of clean-up activities since 2001 "found the remedy protective of public health and the environment," as per the EPA website (11 April 2019). Also note that Bartlesville has the non-National Priority List site of the Oak Avenue Refinery, which was documented in 2007. It also had a mercury spill in 2015 and in the nearby town of Dewey there is the "Dewey Mystery Spill" of 2017.

15. Data are for 2016, grand total of minerals produced equaled 80,513,024 tons, with limestone as the largest component (48,265,197 tons), followed by sand and gravel (16,712,925 tons), gypsum (5,250,920 tons), and granite (5,028,992 tons). Data from the Oklahoma Department of Mines website (http://mines.ok.gov), accessed 25 September 2017, copy on file with BS-P

16. Oklahoma relies so heavily on revenue generated from taxes on oil and gas production that when the industry fought for and won a tax decrease from 7% to 2%, it was disastrous for the state. In 2017, Oklahoma faced a *$1.3 billion* budget shortfall. The consequences of that budget gap included taking $58 million away from public education, which resulted in many schools laying off staff and teachers, dropping classes, losing food programs for poor students, and going to four-day school weeks (and this from a state that routinely ranks in the bottom three for quality of education in the U.S.). Other sectors have not fared well either, with a collapsing health care system, overcrowding in prisons, and crumbling dams, roads, and bridges among the many problems faced by the state.

Some articles on the topic include: *Oil Companies Get a Break While Oklahoma Schools Face Budget Crisis* by Denver Nicks, 17 May 2016, *Time* (http://time.com/money); *Oklahoma Budget Crisis Spurs Battle Among Oilmen* by Sean Murphy, 23 May 2017, *US News and World Report* (www.usnews.com); and *With state budget in crisis, many Oklahoma schools hold classes four days a week* by Emma Brown, 27 May 2017, *Washington Post* (www.washingtonpost.com).

17. See Späth et al. (1998). Gould (1926:436) noted that "… during several months in 1915, 60 per cent of the high-grade oil of the world came from the Cushing field of Oklahoma. The Bartlesville sand horizon … has produced more high-grade gasoline oil than any other single geological horizon in the world."

18. Data for oil production are averaged from 2012–2017 annual totals and refinery data are from January 2017. Oklahoma is ranked as the fifth largest producer of crude oil (2016 data, not counting federal offshore drilling). All data are from the U.S. Energy Information Administration (www.eia.gov).

19. 2,513,897 million ft^3 (2017 data); ibid.

20. 532,576 estimated wells; ibid. These wells include those for more traditional oil and gas extraction and the deeper, horizontal, hydraulic fracturing injection wells (or fracking) and their associated disposal wells for wastewater, chemicals, and produced fluids. The count includes all known abandoned wells, thought to number 194,933, and active wells.

21. As but one example, see a recent article titled "750 barrels of crude oil spills into Garfield County creek; clean-up underway" (KFOR news, 30 Jan 2019).

22. Wastewater can be held on the surface, which more obviously affects wildlife, but more often than not it is injected into the ground (euphemistically called "wastewater disposal"). Not only does Oklahoma produce a lot of wastewater itself, but for many years it has actively imported wastewater from surrounding states that is then injected subsurface. It is this practice, not fracking (as per USGS), which is thought to cause the majority of Oklahoma's earthquakes, making the state have more regular quakes than California!

23. For more information on the vital function of playa lakes, see the Playa Lakes Joint Venture webpages, particularly: http://pljv.org/playa-conservation/playas-are-important-source-of-water/And: http://pljv.org/playa-conservation/playas-recharge-the-aquifer/

24. Eutrophication is a process by which water is overloaded with nutrients, such as nitrogen and phosphorus from agricultural fertilizers, causing blue-green algal blooms that result in oxygen depletion, or hypoxia, that kills aquatic plants and animals. The blooms are also toxic to humans.

25. Cropland estimates are from Cunfer (2005, Table 2.3). Cropland was estimated as 3,150,687 acres in 1880 and increased to 118,914,807 acres by 1935. Grazing intensity was calculated based on figures given in Cunfer's Table 3.2. Rather than using the cow per acre estimates provided, we calculated acreage for all livestock (cows, sheep, and horses): 4,596,294 livestock in 1880 on 436,925,448 acres of grassland compared to 25,561,957 livestock on 265,687,773 acres in 1935.

26. By 1924, Oklahoma ranked third in the nation for winter wheat production (Gould 1926).

27. See Green (1977b). By 1924, Oklahoma ranked second in the nation for cotton production (Gould 1926).

28. For those unfamiliar with Oklahoma geography, a common misconception with *The Grapes of Wrath* is that it tells the story of a family seeking refuge from the Oklahoma Dust Bowl. Actually, the family originated from near Sallisaw, Sequoyah County, in far east-central Oklahoma, not in the areas in western Oklahoma that are truly associated with the Dust Bowl. The Joads would have been affected by the drought and economic hardships of the era, but they were not a panhandle family like so many think (we did, too, before coming to Oklahoma!).

29. Mesonet wind data for Boise City, Cimarron County, Oklahoma; the National Severe Storms Laboratory classifies damaging winds as those ≥80–96 kph.

30. Such a storm hit Washington D. C. on 21 March 1935. Hugh Hammond Bennett precisely timed his testimony before Congress regarding the devastating effects of soil erosion to coincide with the storm that he knew was rushing eastward from Oklahoma. Congress and President Roosevelt yielded to the dramatic effect of Bennett's presentation—a month later, the Soil Conservation Service, which Bennett is considered the father of and which was instrumental in reversing the effects of the Dust Bowl, was born.

31. Dwight L Bolinger, in his 1941 article titled "The Revival of 'Exoduster'" (*American Speech* 16[4]:317–318), credited "Professor Louise Pound" with reviving the term for use with "dustbowl refugees in California" (Pound, cited as "LP", 1941, Folk-words, *American Speech* 16[1]:20).

32. We calculated a minimum loss of lotic habitat by overlaying current lentic habitat outlines onto an historical map of streams. Stream lengths now inundated by lakes were clipped out and added up to determine loss. This method does not account for loss of lotic habitat or for streamflow depletion resulting from groundwater use, etc.

33. This is an unfathomable number. It equates to 2.4 *quadrillion* gallons of water per year, an amount that simply cannot be grasped readily.

34. That equates to 4.5 Chesapeake Bays! (the bay holds about 18 trillion gallons of water, www.chesapeakebay.net/discover/facts)

## CHAPTER 7

CONSERVATION OF OKLAHOMA ODONATA

Brenda D. Smith and Michael A. Patten

Conservation in Oklahoma is a challenge. And not just for dragon- and damselflies. For many of the conservation challenges faced by odonates are those other organisms, including humans, must confront. We spoke in the previous chapter of various land-uses and environmental issues in Oklahoma. We spoke, too, of some of the peoples who came to what is now Oklahoma, many of whom left their indelible mark: Native Americans, those to whom Oklahoma was ancestral and those forced to come; Mexicans; Asians; Blacks, as slaves, freed slaves, and Yankees; Whites, from Europe and the United States; and many other races, ethnicities, and backgrounds. Regardless of their origins, all are now Okies,[1] and all contributed to building railroads, roads, and communities, to mining, to Superfund environmental disasters, to the timber industry, to fossil fuel extraction, to oil pipelines and spills, to methane release, to climate change, to earthquakes, to the proliferation of wind turbines, to ranching and poultry farms, to irrigation-based farming, to the Dust Bowl, to depletion of aquifers, to overuse of water, to damning of streams and changing streamflows; consequently, *to altering radically the face of Oklahoma*.

Not a one of us, as humans, can credibly argue that we have not contributed to the threats that now impact the natural world (and ourselves, as creatures embedded in that world). We are all in this together whether we like it or not. So now what? What are we going to do about it?

We suggest a shift in attitude.

We do not intend to point the finger or get preachy here, nor imply that all Oklahomans are represented by the bad behaviors we are about to mention. But we do intend to say that if such behavior can be eradicated or, at least, lessened markedly, Oklahoma and its wildlife (and its humans) will be better for it. Oklahomans (yes, after >15 years, we now must count ourselves among them) are guilty of a lot of behaviors that humans around the world are, chief among them short-sightedness and a tendency to allow fear to dictate behavior.

We humans could go a long way toward resolving environmental problems if we shed our short-sightedness. But that attitude adjustment is one humans obstinately avoid making—thus the incessant repetition of history. It is hard for us to see beyond our noses. With regard to conservation, it is not that humans aspire to apathy; we just tend to find little time and energy to be concerned about and make an effort to change our entrenched, short-term behavior despite its injurious, long-term consequences. (Quick

question: what happens to a man who depends on his farm to feed himself, but does not want to think about the amount of work he needs to do so instead gets blindingly drunk? Quick answer: eventually he runs out of liquor and hunger comes to bite him in the ass.) Human short-sightedness is not likely to be overcome anytime soon; it is seemingly engrained in our nature, as if our DNA needs a fundamental change.

But do not fear, there is hope. Hope lies in ridding ourselves of exactly that—*fear*. Fear betrays itself in varied but entangled forms. One of which is mistrust of outsiders, also called xenophobia. There is mistrust of science (Smith-Patten et al. 2015) and of government that can become vitriolic (Figure 7.1) and at times violent. We personally have not escaped this mistrust. Purely for the sin of being biologists, we have been screamed at, had dogs sicced on us, and had police summoned for the apparent violation of being on public roads. We think this mistrust stems from a fundamental false assumption (and deliberate misinformation[2]) that if a rare species is seen on your private property then the government can take away your property. That notion is particularly prevalent in western Oklahoma, where many a time we have been told something akin to, "your bug may not be endangered now, but you'll make it so, and then I'll have to stop what I'm doing on my land. No thanks! Now get outta here."

Oklahoma is a state where approximately 97% of lands are privately owned. The roughly 3% of public land finds itself habitually restricted by private fencing strung across public access points (federal and state right-of-ways, streams, county roads, etc.; Figure 7.2) and by leases to ranchers or oil and gas exploiters. That leaves little public land to survey, the little pieces of which float in what at times (rightly or wrongly) feels like a sea of paranoia and mistrust that feeds the extreme need to protect one's property and way of life. Oklahoma's circumstances are shared by more than a few U.S. states, which means that conservation biologists in other regions face the same challenges.

How do we get past this? One way is that we remind ourselves that for each person who threatens to call the police, there is a neighbor who welcomes you onto their property; thankfully, we have experienced that latter often enough to be clear-headed about the variety of views among the public. Biologists also have a deep-seated fear of the unknown. We tend to be introverted people who would rather be on the ground crawling through mud looking for bugs than to

Figure 7.1 A sign from the Oklahoma panhandle that exemplifies some of the vitriol that select landowners have toward outsiders. Photo by ©Bruce W Hoagland.

Figure 7.2 A common occurrence in Oklahoma is obstructing creeks by fencing.

speak to someone we do not know. Biologists fear talking to landowners, and landowners fear talking to biologists, but as is often the case with people, we have more in common than we think. Perhaps political ideals collide, but the goal of healthy land, air, and water is a mutual one. No matter how uncomfortable it is for either party, we need to find ways to listen to each other, to break down our barriers to listening, and let go of our preconceived ideas of what the other thinks.

How do we work with private landowners to build trust and uphold that trust over time? By being honest with one another and by finding ways that research and livelihoods are mutually beneficial. Ranchers and farmers already have the mindset that a healthy piece of land means healthy cattle and productive crops, i.e., functioning ecosystems equate to profit. If, for instance, it is explained that a healthy dragonfly

population indicates good water quality, then a landowner may be inclined to allow a species richness monitoring program on the property to help ensure that their cattle have good drinking water. Perhaps it is just a small divide forded, but one nonetheless forded.

Another area that affords hope and shuns fear is what we like to think of as "capturing the childhood spirit." We hope people will continue to grow tired of the isolating effects of our everyday lives that shackle us to the unnatural world and will continue to ween ourselves off technology while looking for tangibility, which is a quality that nature unquestionably provides. Remember as a child when you were daring and practically fearless in touching what you encountered, in experiencing opportunity? The world was not a scary place until adults told you to be afraid of it.

Get rid of that fear. Get children out into the world. Let them touch, sense, be a part of the world around them, learn respect for the world around them. Capture their spark, the spark that you will see when their eyes light up because they have a dragonfly perched on their finger for the first time, or when they see a monarch butterfly emerge from a cocoon. That spark needs to flare and be nurtured as that child grows. All too often we smother that spark by telling kids, "eww, that bug is slimy," when we have never touched one ourselves … so we don't know, or "eww, that is just gross …" Well, gross is what kids think is cool. Gross is actually not gross … it is just different … as well as a necessary part of the natural world, *the glue to our world*. We cannot stamp out that spark for nature simply because we, as adults, no longer have it. Kids need it and should hold that spark near and dear for their entire lives; that is, if we want to have a functioning world. It is vital, too, that adults reconnect with that childhood spark, that childhood fascination, that childhood daring, that childhood boldness, that lack of fear that is childhood. That spirit of understanding and connection with our world must be nurtured in our children but also in ourselves, as adults, in ourselves as a community, in ourselves as humans. Only then can society see the value in conserving the beauty we have around us.

LIVE "AS IF" YOUR ACTIONS MATTER

Conservation can be excruciatingly disheartening at times. Conservation biologists go about their days hoping that one day the world will wake up and say "Hey, from now on, let's care for the world we live in." Since that sentiment does not exactly jibe with reality, we do not hold our breath. But we do continue to encourage people to live by the adage, Do not say you want a better future for your children, *MAKE a better future* for them. One way we help ourselves to envision that future is by living "As If" (Patten and Smith-Patten 2011). *Live as if your actions matter*. In the long run, humans may destroy the planet, but if you live as if what you do on a daily basis (using less, turning off lights, upcycling, looking at bugs, whatever it is that you can do) will change the world, then, in the end, at least you know you did what you could to try and make the world a better place.

Sometimes the "As If" philosophy is all that stands between giving up entirely or trudging onward. We choose to trudge onward. We continue to collect data, we continue to provide conservation assessments and management recommendations, we continue *as if what we do will matter.*

CONSERVATION RANKING

So far, we have bandied about the term *"conservation assessment,"* but we have yet to define it. In some senses, conservation assessments are the unexciting (and less preachy) parts of conservation. They are the nitty-gritty of basic conservation biology, laying out how conservation biologists come to the conclusions we do and how we decide if a species is of conservation concern. Performing a conservation assessment provides for a *conservation rank.* There are various methods to arrive at a conservation rank. For example, NatureServe (www.natureserve.org), the umbrella organization for Natural Heritage Programs across the U.S., Canada, Latin America, and the Caribbean, uses a methodology developed to assign, at a minimum, a global rank (GRank) and individual subnational ranks (SRanks), such as for states and provinces (Table 7.1). Another commonly used method was developed by the International Union for Conservation of Nature (IUCN), which has a ranking methodology that assesses species for the "Red List of Threatened Species," usually referred to simply as the IUCN Red List (www.iucnredlist.org/). IUCN assessors generally assign one overall rank for the species, but national or regional assessments are sometimes also conducted. The NatureServe and IUCN ranking schema are broadly similar and dovetail to some extent (Table 7.2).

No matter which methodology used, conservation assessments estimate three basic factors: geographical range, rarity, and threats (terminology varies). Geographical range is generally estimated using two measurements: extent of occurrence (EOO; or range extent) and area of occupancy (AOO). EOO estimates the species' full geographical range irrespective of suitable habitat or other limiting factors to where specifically a species would occur. This type of range depicts the species' distribution, for example, as pictured on a map in a field guide. AOO is the "area of suitable habitat currently occupied" by the species (IUCN 2017:48) and can be suggestive of the amount of isolation of subpopulations. It can also suggest number of subpopulations, although it is preferable to calculate that separately. The differences between EOO and AOO (and number of populations) can be great. For example, *Quadrula fragosa,* the Winged Mapleleaf, an endangered freshwater mussel, has an estimated EOO of 242,839 km^2 but an estimated AOO of no more than 25 1-km^2 grid cells. If we knew little of this species, then that AOO would suggest there were no more than 25 populations, but in this case, because distribution of the species is rather well-studied, it is known that only 4–5 populations are extant. Therefore, this species occurs over a large geographical area (EOO), but it

Table 7.1	**NatureServe's conservation status ranking schema that is used by the Oklahoma Natural Heritage Inventory and by the authors to provide conservation ranks, or SRanks (subnational ranks for states and provinces). The scaling/categories of SRanks is the same for global and national ranks under NatureServe's methodology (GRank, NRank; e.g., G2=imperiled).**

RANK	DEFINITION
SX	**Presumed Extinct** (species) — Not located despite intensive searches and virtually no likelihood of rediscovery. **Presumed Eliminated** (ecosystems, i.e., ecological communities and systems) — Eliminated throughout its range, due to loss of key dominant and characteristic taxa and/or elimination of the sites and ecological processes on which the type depends.
SH	**Possibly Extinct** (species) or **Possibly Eliminated** (ecosystems) — Known from only historical occurrences but still some hope of rediscovery. Examples of evidence include (1) that a species has not been documented in approximately 20–40 years despite some searching and/or some evidence of significant habitat loss or degradation; (2) that a species or ecosystem has been searched for unsuccessfully, but not thoroughly enough to presume that it is extinct or eliminated throughout its range.
S1	**Critically Imperiled** — At very high risk of extinction or elimination due to very restricted range, very few populations or occurrences, very steep declines, very severe threats, or other factors.
S2	**Imperiled** — At high risk of extinction or elimination due to restricted range, few populations or occurrences, steep declines, severe threats, or other factors.
S3	**Vulnerable** — At moderate risk of extinction or elimination due to a fairly restricted range, relatively few populations or occurrences, recent and widespread declines, threats, or other factors.
S4	**Apparently Secure** — At fairly low risk of extinction or elimination due to an extensive range and/or many populations or occurrences, but with possible cause for some concern as a result of local recent declines, threats, or other factors.
S5	**Secure** — At very low risk or extinction or elimination due to a very extensive range, abundant populations or occurrences, and little to no concern from declines or threats.

Table 7.2 **Comparison of NatureServe and IUCN conservation ranks.**

NATURESERVE	IUCN RED LIST
Presumed Extinct (GX)	Extinct (EX)
Possibly Extinct (GH)	Critically Endangered (CR)*
Critically Imperiled (G1)	Critically Endangered (CR)
Critically Imperiled (G1)	Endangered (EN)
Imperiled (G2)	Vulnerable (VU)
Vulnerable (G3)	Near Threatened (NT)
Apparently Secure (G4)	Least Concern (LC)
Secure (G5)	Least Concern (LC)
Unrankable (GU)	Data Deficient (DD)

* Species that are possibly extinct are assigned CR status and are "tagged" as "possibly extinct."

is extremely limited within that area (AOO) with few populations. The advantage this species has, and a reason that both EOO and AOO are estimated, is that its *extinction risk* is lessened by its high degree of "risk spreading," meaning that it is unlikely that all populations would be extirpated by a stochastic (i.e., random) event, such as an oil spill or severe storm that might impact one population but not the others given their vast separation distance. On the flip side is a regional endemic species such as *Plethodon kiamichi*, the Kiamichi Slimy Salamander, that is limited to one mountain range (EOO) in about 500 km² of suitable habitat (AOO) in Oklahoma and Arkansas. One major wildfire in the Kiamichi Mountains could cause this species to go extinct, so it has an extremely low degree of risk spreading. Risk also can be spread depending upon the degree of motility (ability to move) of an organism. Another consideration of range is historical versus current range. If there is a difference between the two estimates, then we can discern changes in a species' range and its population dynamics, for example, *Q. fragosa* has experienced a definite range retraction: it used to occur in 15 states within roughly 1 million km², but now it is known from only five states, a loss of >850,000 km² within its historical EOO.

A species' rarity is estimated by calculating number of individuals in the overall population, as well as within subpopulations, the latter estimate getting more at a species' population density, or how common it is in a given location. As with geographical range estimates, if data are available then a comparison of historical population size to its current size can be made as a measure of population trend to determine if the species is stable, increasing, or in decline. The third major consideration in conservation assessments is gauging the severity of threats to a population. Threat level considers the number of overall threats, as well as how pervasive and how imminent they are. All environmental issues discussed above and in the previous chapter are part

of the threat assessment schema[3] used in both the IUCN and NatureServe methodologies. Some additional considerations that factor into species assessments are habitat specificity, i.e., whether a species can only survive within a specific habitat or is able to adapt to a wide array of habitats (specialist versus generalist), and population viability analyses, which examine factors such as life history traits (e.g., life span, number and survival rate of offspring) and general susceptibility to estimate how likely a given population will persist.

We trust we have conveyed a better sense of the many factors, data, analysis, thought, time, and energy that go into determining a species' conservation rank. Literally hundreds of pages are devoted to conservation ranking methodology because there are so many complicating factors to take into consideration. But we hope we have provided a summary detailed enough that you have insight into how we arrived at our SRanks for Oklahoma Odonata (Table 7.3). We discuss specific details within the species accounts.

LAND-USE REVISITED: COMPLEXITIES OF ODONATE SPECIES TURNOVER

In the previous chapter, we touched upon the human history that shaped the physical landscape of Oklahoma today. We revisit that topic here to examine some of the effects land-use changes, particularly those relating to the hydrological landscape, have had on odonates in the state.

In researching those effects, we started with the hypothesis that there would be species turnover of odonates concomitant with turnover in habitat types in Oklahoma, i.e., increased lentic (lake) habitat and decreased lotic (stream) habitat over time as a result of post-Dust Bowl era alterations of the hydrological landscape would result in increased lentic species and a decrease of lotic species. To demonstrate this seemingly intuitive notion we ran species distribution models (SDMs) for 20 odonates (12 zygopterans, 8 anisopterans) for which we had quality pre- and post-Dust Bowl data. Our results indicated both lentic and lotic odonates have shifted their ranges in Oklahoma since the Dust Bowl, with these range shifts driven by different factors, landscape alteration for one and desiccation for the other. Our initial hypotheses that lentic species would be affected positively by the increase in lentic habitat and that lotic species would be affected negatively proved to be overly simplistic (Table 12.1 and 17.1) because we failed to consider how complex Oklahoma's hydrological landscape is both above and below ground; nevertheless, the general pattern predicted by these hypotheses held.

Unsurprisingly, the turnover of lotic habitat to lentic since the Dust Bowl era was the primary variable driving positive range shifts in lentic species. These species probably relied on the slow reaches of lotic habitats during the earlier part of their historical presence in the region. It may be that although adults could adjust to such lotic habitats,

Table 7.3 **Odonate species of conservation concern in Oklahoma. Zygopterans (damselflies) on left and anisopterans (dragonflies) on right of each side of the table. Assessments made using NatureServe's methodology, as such, rankings do not necessarily reflect the species' global level of conservation concern, rather, it is meant as a measure of rarity and threat within the confines of the state's jurisdiction.**

S1 – Critically Imperiled in Oklahoma		S3 – Vulnerable in Oklahoma	
Argia bipunctulata	**Cordulegaster talaria*	*Hetaerina titia*	*Tachopteryx thoreyi*
Argia lugens	*Erythrodiplax berenice*	*Lestes inaequalis*	*Aeshna umbrosa*
Enallagma antennatum	**†Somatochlora margarita*	*Argia alberta*	*Coryphaeschna ingens*
Enallagma daeckii			**Somatochlora ozarkensis*
Enallagma praevarum			*Somatochlora tenebrosa*
Nehalennia gracilis			*Libellula semifasciata*
Telebasis byersi			*Macrodiplax balteata*
S2 – Imperiled in Oklahoma		**S3S4 – Vulnerable/Apparently Secure in Oklahoma**	
Amphiagrion abbreviatum	*Gomphaeschna furcillata*	*Ischnura denticollis*	**Gomphurus ozarkensis*
Enallagma doubledayi	*Arigomphus maxwelli*	*Ischnura kellicotti*	**Phanogomphus oklahomensis*
Enallagma dubium	*Hylogomphus apomyius*		*Erythrodiplax minuscula*
Ischnura demorsa	*Cordulegaster maculata*		*Libellula flavida*
Ischnura perparva	*Helocordulia selysii*		
Nehalennia integricollis	*Neurocordulia molesta*	**Status Unknown in Oklahoma**	
	Celithemis verna	*Lestes eurinus*	*Aphylla williamsoni*
	Libellula composita	*Argia leonorae*	*Macromia annulata*
S2S3 – Imperiled/Vulnerable in Oklahoma		*Chromagrion conditum*	*Dythemis nigrescens*
Ischnura damula	*Gomphurus hybridus*	*Enallagma carunculatum*	*Erythemis collocata*
	Phanogomphus lividus		*Miathyria marcella*
	Brechmorhoga mendax		*Micrathyria hagenii*
	Macromia alleghaniensis		*Paltothemis lineatipes*
	Macromia pacifica		*Tramea calverti*
	Plathemis subornata	* regional endemic † (S1S2?)	

the larvae, which tend to be more habitat specific, could not, and so population sizes were limited and were maintained by immigration. Once the hydrological landscape changed drastically by the introduction of close to 270,000 ha of lentic habitat, lentic species had a boon, spreading across the state with the spread of suitable habitat (and likely now have considerably higher population sizes, aided in part no doubt by buffered temperatures in lentic waters of a respectable size, which provides an advantage to lentic species in terms of overwinter survival; Danks 2007).

Whereas the pattern of range expansion of lentic species in Oklahoma was clear—and supported our prediction drawn from a hypothesized response to vast expansion of lentic habitat—the pattern for lotic species was mixed. We had predicted that lotic species would exhibit as clear a range shift, in this case a retraction, both because reservoirs have inundated substantial linear extents of streams and rivers and because streamflow has fallen to at or near

zero in many parts of the Southern Great Plains, especially in the panhandle and western one-fourth to one-third of the state, due to groundwater extraction and reduced base flow (Garbrecht et al. 2004). Part of the mixed signal was doubtless a result of our crude measure of the extent of lotic habitat, a measure that did not capture the second of these causes well nor did it capture urban environments that may mimic natural lotic habitats. Accordingly, the lotic species we modeled exhibited different "response" patterns that appeared to vary with a species' geographical range in the state. These variations reflect the severity of drought and the degree of irrigation in a given area, as well as over which part of the High Plains aquifer the locality resides (groundwater flow is from west-to-east in this aquifer). The geographical range of a typical lotic species we modeled contracted away from the area it was found in the pre-Dust Bowl era, but it also expanded its range to the north. Some species, such as *Argia sedula*, the

Blue-ringed Dancer, also spread eastward to more mesic habitats (recall the strong west-to-east gradient of increasing rainfall across Oklahoma and Kansas; see Chapter 3), moving away from the perpetually drought stricken southwest. The northward movement of *A. sedula* makes sense when one considers that although the northwestern portion of the state has also been drought stricken, the High Plains aquifer in that area is either stable or has seen an increase of 3–7 m (10–23 ft) in the water level (Kustu et al. 2010), which would keep springs and stream flowing. The eastward movement makes sense, too, because central Oklahoma is less prone to severe drought. The spread of *A. tibialis*, the Blue-tipped Dancer, northward in the eastern third of the state suggests a response to drought: although the southeastern corner of the state remains the part of Oklahoma that receives the highest annual precipitation, it has seen shortfalls of 25–43 cm (10–17 in) over the past two decades (Oklahoma Climatological Survey data). Precipitation farther north has remained stable and, in some areas, has increased. We acknowledge that northward spread also may be a predictable response to global climate change (Parmesan 2006), but we have not tested this hypothesis.

We were unable to model population declines of some lotic species, but declines can be explained by a combination of biogeography and altered hydrology. For example, although never common, four lotic zygopterans (*Enallagma praevarum*, *E. antennatum*, *Ischnura demorsa*, and *Argia lugens*, the Arroyo and Rainbow Bluets, Mexican Forktail, and Sooty Dancer) whose distributions in Oklahoma are centered on the western panhandle, were recorded somewhat regularly through to the 1980s but have been recorded only a few times since, likely as a result of extensive loss of flowing water in the panhandle (Wahl and Wahl 1988; Garbrecht et al. 2004; Brikowski 2008). *Argia lugens*, for example, was thought to be extirpated in the state—it had last been recorded in 1983—until a 2013 surge in the Cimarron River, which reached its highest flow rates in decades, brought at least three individuals to the state and, during similar times of high flow recently, has brought other individuals (see that species' account). In contrast to these zygopteran species, an anisopteran associated with spring-fed streams, *Libellula composita*, the Bleached Skimmer, appears to have colonized parts of western Oklahoma around 2011 where groundwater on the eastern side of the aquifer remains relatively high (Figure 6.10; i.e., some of the same regions where *Argia sedula* has expanded its range). Nevertheless, a more accurate means to quantify streamflow depletion, and hence desiccation of lotic habitats, is needed before we can bring a species distribution model approach fully to bear on the question of range retraction of lotic odonate species of Odonata.

Oklahoma is a political boundary within which resides a complex ecological, geological, climatological, and human history. It is a biogeographical crossroads of the American west, east, south, and mid-west (Figure 1.1). Severe weather is a mainstay of where these biomes meet, as is severe-to-extreme periodic drought that led to extensive damming of rivers and streams across the Southern Great Plains to create thousands of ponds and lakes where once there were none. As the High Plains aquifer is further depleted to unsustainably low levels (Sophocleous 2010), and as the southern plains dry further with global climate change (Rosenberg et al. 1999), the prospects for persistence of lotic species in Oklahoma, particularly those in the more arid western half of the state (Epstein et al. 1996), are not high.

All land-use changes and the rapid range shifts we documented took place in ~75 years. Odonata are top invertebrate predators and as such play an important role in trophic dynamics of arthropod communities, helping to control populations of flying pests (May 2019). We do not know the ramifications of disturbance to community structure of these predators, but it is likely that species are not fully interchangeable: as one species replaces another, subtle shifts in food web dynamics are inevitable (Covich et al. 1999). In other words, turnover from more lotic species to more lentic species has consequences for ecosystem function. We posit that the marked community shifts in the Southern Great Plains over the past 75 years has had a broad effect on ecosystem function, but only in-depth research of the community ecology of Odonata, from competition to predator–prey interactions, will elucidate to what extent the food web has been altered. Understanding these effects will undoubtedly shape our thinking on insect conservation and biogeography and its role in regional and global change dynamics.

NOTES

1. The term Okie is no longer considered derogatory because in recent years it has been re-appropriated by Oklahomans, many of whom wear the moniker proudly.
2. But one example is the article titled, "Endangered species depend on private land for habitat, so why treat landowners as the enemy?" (https://libertarianenvironmentalism.com/2017/07/31/inhabitable-habitat/), which among other falsehoods and hyperbole proclaims that the government "can designate private land as critical habitat even if it is not occupied by the species and would be inhospitable for the species." Such sentiments pass via chats with neighbors as well as at town hall meetings. These earworms reinforce the idea that conservation biologists are capricious in our decision-making and that we seek to place undue burdens on citizens. Yes, laws and regulations are complicated, but the reality of implementation of the Endangered

Species Act is that private ownership and activities are not revoked or hindered. To our knowledge, there have been few exceptions, most of which have involved landowner acceptance of federal funds such as grants and subsidies; even then, they do not necessarily make for compulsory compliance. A private landowner is much more likely to have property rights infringed upon by federal infrastructure such as highways or by national security issues.

3. www.iucnredlist.org/resources/threat-classification-scheme

## CHAPTER 8

SEASONALITY OF OKLAHOMA ODONATA

Michael A. Patten and Brenda D. Smith

Seasonality, or phenology, is the study of timing of ecological events, such as when a particular species of flower blooms or when a particular species of bird migrates. A key aspect of the biology of dragonflies and damselflies is their adult phenology,[1] otherwise known as their "flight season." That phenology spans from when adults emerge from a full-grown nymph to when adults die weeks or months later. Not surprisingly, the adult phenological cycle of odonates and other insects coincides with that of regional plant phenologies. As the number of consecutive non-freezing days hits a minimum threshold, not only do plants begin to emerge, or "greening" occurs, but early flying dragonflies and damselflies emerge, too. In Oklahoma, winter tends to loosen its grasp first on the southeastern part of the state, usually between 26 March and 10 April, so most early reports of species flying in the state originate there. Spring comes later to northwestern Oklahoma and the panhandle, averaging up to a whole month later (15–25 April), but it can be more (20 May; Figure 8.1). In some years, cold-weather snaps occur more than a month after odonates have begun to fly, such as the year when a snowstorm hit central Oklahoma in early May. During such cold snaps, many adult odonates die unless they are able to find a warm refuge. Given the frequency of extreme fluctuations in weather in this region, presumably species have evolved in such a way that allows populations to persist, but continued climate change and more erratic weather will test those limits.

Another aspect of odonate phenology to consider is that some species emerge and fly early in the season, others appear only late, whereas others split the difference to fly mid-season or across pretty much the whole of favorable conditions (in Oklahoma, "favorable conditions" coincide principally with "when it is warm enough," which in an average year means from late March to mid-November). And a few species have bimodal abundance peaks (i.e., two peaks in their annual population numbers), suggesting either a prevalence of migrants at certain seasons—for example, both *Anax junius*, the Common Green Darner, and *Sympetrum corruptum*, the Variegated Meadowhawk, are most numerous in spring (April and May) and in autumn (mid-August through October, but especially September)—or differential periods of emergence—for example, *Enallagma civile*, the Familiar Bluet, typically numbers in the thousands, especially in western Oklahoma, in early spring and in mid-autumn but can be somewhat scarce in the middle of summer.

Herein we refer to broad flight seasons of early (mid-March through May), mid-season (June and mid-August), and late (late August to mid-November), and we highlight bimodal patterns when they are clear (Table 8.1). We grounded our statements in empirical data: for each species that occurs regularly in Oklahoma, we used data we amassed for the Oklahoma Odonata Project database to create seasonality graphs. Our graph of choice is the violin plot,[2] on which seasonality is mapped across a calendar year, replete with data on relative abundance across the year (Figure 8.2). The heart of the graph contains a box plot that shows 1) the bulk (~80%) of the date range that the species has been observed in the state, indicated by the horizontal line (x-axis), 2) the part of the date range in which 50% of the occurrences fall, indicated by the rectangle, and 3) the median date,[3] or the mid-point, of all occurrences, indicated by the vertical line within the rectangle. Some of the box plots have outlying points (dots) along the x-axis that indicate that some observed dates are far enough removed from the bulk of the observed dates that they are considered outliers. We retained those data points in the graphs because they are from actual data and because today's outlier might be tomorrow's "norm." For example, a species could be found in an area that does not get adequate coverage at the beginning or tail end of its flight season and thus it may only have one or two dates recorded for the spring, not because it does not occur during that time of the year, but because the area of occurrence was surveyed only once or twice during the springtime. We also expect that climate change will make some "normal" dates become outliers and some current outlying points become part of the normal seasonal distribution.

The shape around the box plot estimates the species' relative abundance during its flight season. Plots grow "fatter" as abundance peaks. Projected early and late dates, shown as the tails on either end of the graph, are extrapolated on the basis of early and late abundances. We included extrapolations because although Oklahoma has received relatively wide coverage, geographically and seasonally, not all areas have been surveyed intensively during all seasons. The extrapolation assumes that most species will emerge at the beginning of the season in small numbers and will increase in abundance and then will likewise trail off at the end of the season. That may not be the case for every species during every flight season, such as those with mass emergences,

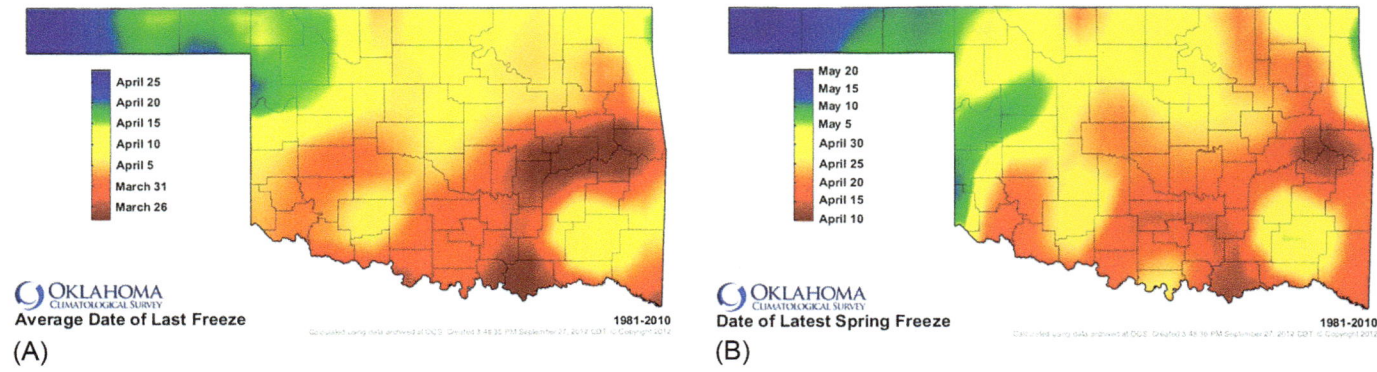

Figure 8.1 Average date of last freeze (A) and latest spring freeze (B) in Oklahoma, 1981–2010.

Table 8.1	**Examples of typical early, mid, and late fliers in Oklahoma, as well as bimodal species. A couple (*) of the "late" species occur in the panhandle beginning in spring, but appear in the rest of the state only later.**

EARLY SEASON	MID-SEASON	LATE SEASON	BIMODAL
Basiaeschna janata	*Enallagma traviatum*	*Aeshna umbrosa*	*Lestes australis*
Phanogomphus oklahomensis	*Phanogomphus militaris*	*Rhionaeschna multicolor**	*Anax junius*
Didymops transversa	*Cordulegaster obliqua*	*Libellula saturata**	*Sympetrum corruptum*
Ladona deplanata	*Macromia pacifica*	*Sympetrum ambiguum*	*Pantala hymenaea*

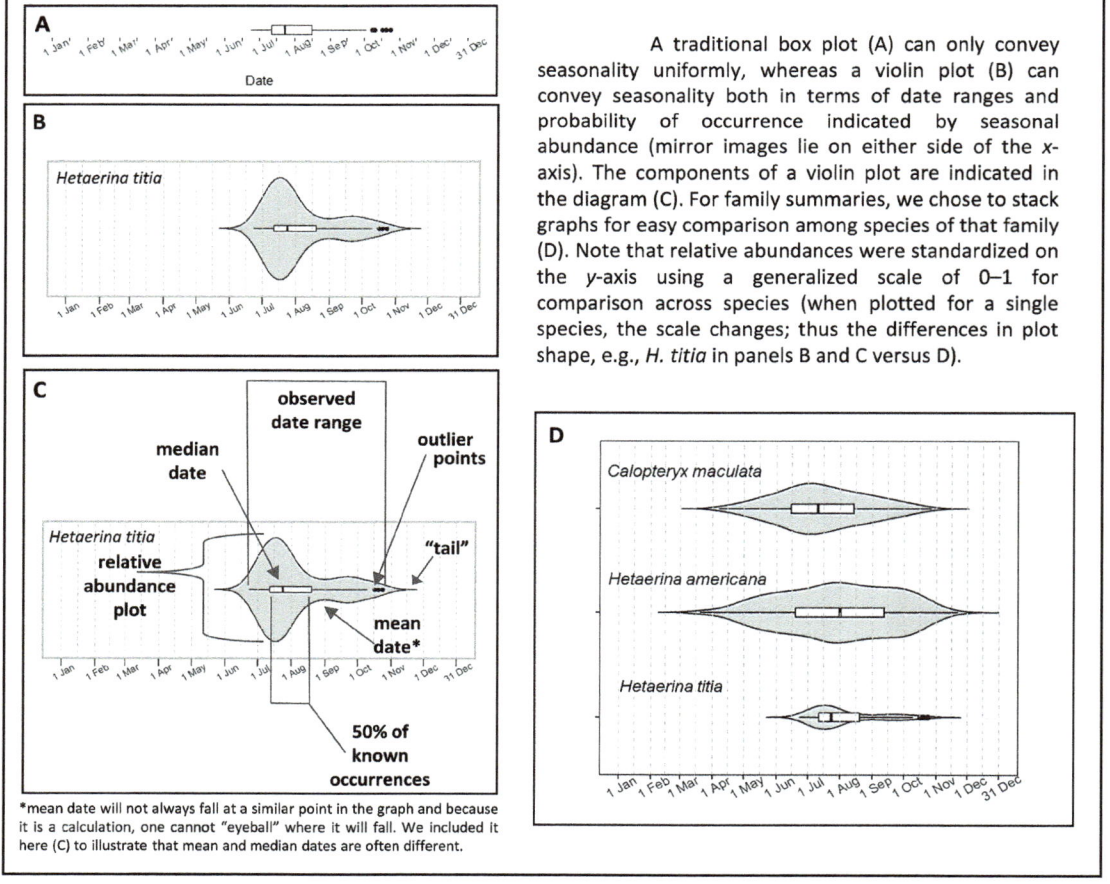

A traditional box plot (A) can only convey seasonality uniformly, whereas a violin plot (B) can convey seasonality both in terms of date ranges and probability of occurrence indicated by seasonal abundance (mirror images lie on either side of the *x*-axis). The components of a violin plot are indicated in the diagram (C). For family summaries, we chose to stack graphs for easy comparison among species of that family (D). Note that relative abundances were standardized on the *y*-axis using a generalized scale of 0–1 for comparison across species (when plotted for a single species, the scale changes; thus the differences in plot shape, e.g., *H. titia* in panels B and C versus D).

*mean date will not always fall at a similar point in the graph and because it is a calculation, one cannot "eyeball" where it will fall. We included it here (C) to illustrate that mean and median dates are often different.

Figure 8.2 Challenges (and solutions) to plotting seasonality.

but overall it was a practical and effective option for plotting seasonality data.

The focus on seasonality is more than an academic exercise: in some cases, species identification can be aided by knowledge of when it flies. For instance, the flight seasons of two broadly similar looking *Enallagma* species, *E. divagans* (the Turquoise Bluet) and *E. traviatum* (the Slender Bluet), barely overlap, with the former flying chiefly in April and May and the latter chiefly in June and July. Likewise, the two *Libellula* species in the state with bright white on the pterostigma, *L. cyanea* (the Spangled Skimmer) and *L. comanche* (the Comanche Skimmer) are mostly segregated not only geographically—the former occurs in the eastern third of the state, the latter in the western third, although their ranges meet somewhat in central Oklahoma—but phenologically as well—the former flies from late April through July, the latter from June through September.

Field identification aside, it is an open question whether seasonal segregation, as in the examples cited above, is a response to interspecific competition. What is clear is that we can never answer such a question if we do not pay careful attention to flight seasons. The same can be said about manifested phenological shifts in response to global climate change, which ought to select for earlier emergence in spring and later activity in autumn (or perhaps reduced activity or abundance during peak heat in mid-summer, possibly leading to more bi-modality). We do not know to what extent climate change already has affected seasonality in Oklahoma, but we do know that any shifts will be impossible to detect without baseline seasonal data such as that we present in the species accounts.

NOTES

1. We did not discuss larval phenology simply because there are not sufficient data to do so meaningfully, even on a generic level.

2. The name *violin plot* originates from the first time this technique was used to plot data, the shape of which is said to have resembled a violin (Hintze and Nelson 1998); however, it is rather rare, in our experience, for such a shape to appear … even with much imagination. More often, at least in BS-P's opinion, an image of lips with protruding teeth or skates/cool airplanes appear.

3. The *median date* is the mid-point of all occurrences (i.e., records/data points), meaning that it is the line that separates half of the data points on one side of the line and half on the other, whereas the *mean date* would be the average of the dates themselves. Essentially, this is the difference between determining where in time most records fall versus where the mid-point of a flight season is. Think about, for example, a species that flies for four months, say April until July, but it is mostly encountered within April (i.e., median occurrence). In such a case, we would say that it is primarily a spring flyer; we would not say that because the average occurrence date (i.e., mean occurrence) falls in early June, the species is a summer flyer. It may still be present in summer, but that season is not indicative of when the species is most encountered. Such instances result in what is called asymmetrical plots of data when more occurrences fall on one side of the plot than the other. *Hetaerina titia* (see Figure 8.2, part D) provides for a good example: 50% of data falls within the first month of the flight season and 50% falls within the remaining three, making for a median date of mid/late-July but a mean date of mid/late-August. In this case, median date is more reflective of seasonal peak in abundance during mid-summer as opposed to late. There are some species, however, such as *Calopteryx maculata*, for which the peak flight season and the mid-point date of the flight season roughly correspond (see Figure 8.2, part D), producing a more-or-less symmetrical violin plot.

INTRODUCTION TO SPECIES ACCOUNTS

The species accounts discuss the distribution and conservation status of each of the 176 odonate species reported for Oklahoma. We also discussed seasonality, habitat, and behavior specific to Oklahoma or the region. Additionally, each account presents some historical perspective on the species in Oklahoma, including noting when the species was first reported from the state.

We aimed to make the accounts readable rather than telegraphic. Because we wanted to also provide record details for posterity's sake, we included much data embedded within the text. Most data were presented parenthetically, with the idea that the reader could skim past details unless she was compelled to examine specifics. Although not strictly structured, the general flow of the accounts follows these topics:

1) distribution, habitat, and abundance
2) encounter history of the species in Oklahoma research
3) discussion of Oklahoma-specific literature
4) seasonality and a comparison to the region
5) life history and behavior
6) discussion of any glaring errors, such as misidentifications, publishing errors, etc., or records in need of clarification
7) discussion of identification issues specific to Oklahoma or the region, such as individual or geographical variation
8) mention of aberrant individuals found in Oklahoma
9) mention of particularly interesting records

The basic layout of species accounts is illustrated below and some additional details and aspects of the accounts are discussed in the following.

Family and genus summaries: species accounts are grouped by family. Each set of accounts begin with a description of the family in Oklahoma, including a comparison of seasonality of the species within the family. The comparison ought to help the reader determine seasonal species turnover, or the seasonal changes in species composition. For example, Springtime Darner (*Basiaeschna janata*) gives way to Cyrano Darner (*Nasiaeschna pentacantha*) in early May each year, when *Basiaeschna* stops flying but *Nasiaeschna* starts. Some families are so interesting biogeographically and in terms of how their populations have changed over time that we presented additional analyses and discussion. And, in some cases, we further divided summaries by genera given complexities within those genera (e.g.,

Argia, the dancers, and the *Tetragoneuria*-type *Epitheca*, the small-bodied baskettails).

Taxonomy: taxonomic order of species follows the linear phylogenetic sequence presented by Paulson (2009, 2011) rather than being alphabetical. In each account we provide the species' family, scientific name, and its American English common name. We also provided the authority citation, indicating the person who described the species and the year of description. This practice is often thought antiquated, at least with taxa such as dragonflies for which the bulk of taxonomic disputes have been resolved. Although there are still a few taxonomic issues for odonates within the United States and Canada, some of which we discuss in the species accounts, we provided authorities not as a way of tracking disputes but for context. The authority citation allows the reader to cue in on how long, or conversely how recently, a species has been thought to constitute a separate entity. We do not provide a full catalog of taxonomic changes (or *synonymy*), including name changes, splits, mergers, and the like, but we do delve into that sphere to the extent synonyms affect how a species has been dealt with specifically in Oklahoma or the region (see Appendix A). For more details about taxonomy and authority citations, see the book's Introduction (Chapter 1). See Garrison and von Ellenrieder (2019) for a detailed synonymic list.

Conservation status, timeline, and lenticity scale: each account has a set of graphics that are located below the scientific name. The first is an oval with a number. That number corresponds to the SRank we assigned to the species. The SRank, or subnational (or state) rank, is part of NatureServe's (www.natureserve.org) conservation status ranking schema (see Chapter 7), a methodology that is also used by Natural Heritage Programs across the Americas, including by the Oklahoma Natural Heritage Inventory.

The timeline graphically illustrates when the species was first reported from Oklahoma (date shown) and when it was reported in successive eras. We elaborated on the timeline within the text. The intent of providing historical perspective is to show how knowledge of a species changed as new data were gathered. The history behind discovery of a species' distribution and status helps us reflect on range expansions or contractions and thus on possible changes in population trends.

The lenticity scale indicates how strongly associated a species is with flowing (lotic) or still (lentic) water. The

Explanation of species accounts

Species accounts for regularly occurring species generally will follow this layout.

photos: when available, photos of a male and female of each species from in the field in Oklahoma were included. Sometimes a teneral, immature, or pair were also included. Photos were intended as documentation of important records, as opposed to being aesthetically pleasing or used for identification.

family
(scientific &
common name)

literature citation:
bibliographic reference (see "Literature Cited")

institutional code (coden):
code for the institution at which a specimen or photo is housed/archived (see Appendix B)

conservation rank:
based on the NatureServe's S1-S5 ranking methodology.*

sex of individuals:
♂ = male
♀ = female
U = unsexed/ not reported

scientific name and authority†

American English common name

lenticity scale:
provides an indication of how closely tied a species is to flowing (lotic) or still (lentic) water. The farther right the indicator, the more lentic a species is.

adult seasonality: known seasonality for the species in Oklahoma (top graph). Relative abundance (number of individuals counted) is indicated by height of curve. Dates indicate known flight dates: earliest (left) and latest (right). See discussion of seasonality for why graph extends beyond those dates. Regional seasonality is presented for neighboring or nearby states (bottom graph)

timeline: provides an indication of what eras the species was reported from. Also noted is the date of first report for the species.

GOMPHIDAE: CLUBTAILS   347

PHANOGOMPHUS OKLAHOMENSIS (PRITCHARD, 1935) – OKLAHOMA CLUBTAIL

Phanogomphus oklahomensis is one of two species of dragonflies A. E. Pritchard first found in Oklahoma and described to science (Figure 15.22). Pritchard first encountered this species on 25 April 1931 in, or likely north of, Wilburton, Latimer County (Pritchard 1935). The ♂ he collected that day became a paratype for the species; we have yet to locate this paratype, which is the case for another 26♂ and 7♀ paratypes he and/or C. A. Sooter collected (figure 15.23). Fortunately, we do know the whereabouts of the holotype and allotype (safely housed at UMMZ) that were collected on 28 April 1934, the second time the species was recorded in the state (Table 15.1).

Of particular note is with a specimen at the Essig Museum of Entomology (EMEC 310258) that was mislabeled as the holotype, by an unknown hand, but was relabeled as a pseudotype by RW Garrison (Fig. 15.23). Note that the label, likely not the original, has a collection date of 29 April 1934, a day that Pritchard does not indicate he collected a

specimen of *Phanogomphus oklahomensis* (Pritchard 1935); neither do we have any record of him being in the field that day. But we do know he was in the field the day before, as that is when he is known to have collected an *in copula* pair of *P. oklahomensis* he designated as the holotype and allotype (Pritchard 1935). One more data entry error of unknown origin worthy of note pertains to another EMEC specimen, this time a ♀ *P. oklahomensis* with a typewritten label, again likely not an original, shows a collection date of 12 June 1934 (EMEC 81372; Fig. 15.23). Although Pritchard was collecting at or near Wilburton on that date, we find it odd that he never mentioned collecting this specimen nor is it labeled as a paratype. Perhaps it is one of the missing paratypes collected on 12 May 1934, but we will probably never know.

The species was not recorded in the state again until 1959, when G. H. Bick collected one nymph and eight adults at Robbers Cave SP, Latimer County (18 April 7♂, 1♀, Bick

Phanogomphus oklahomensis
(Pritchard, 1935)
Oklahoma Clubtail

S3S4

25 April 1931

Figure 15.22 *Phanogomphus oklahomensis*, the Oklahoma Clubtail. Oklahoma's namesake dragonfly was discovered in and described from ... wait for it ... Oklahoma. Although the type locality is Pushmataha County in the southeastern part of the state, it is now known that the species ranges through much of eastern Oklahoma as well as into northeastern Texas, northwestern Louisiana, and western Arkansas. A) ♂, Haskell Lake, Muskogee County, 15 May 2014 (OC 422137). B) ♀, Bixhoma Lake, Wagoner County, 4 April 2012 (OC 374621). Photos by ©James W. Arterburn.

Explanation of species accounts

Maps included in the species accounts generally will be like the ones below.

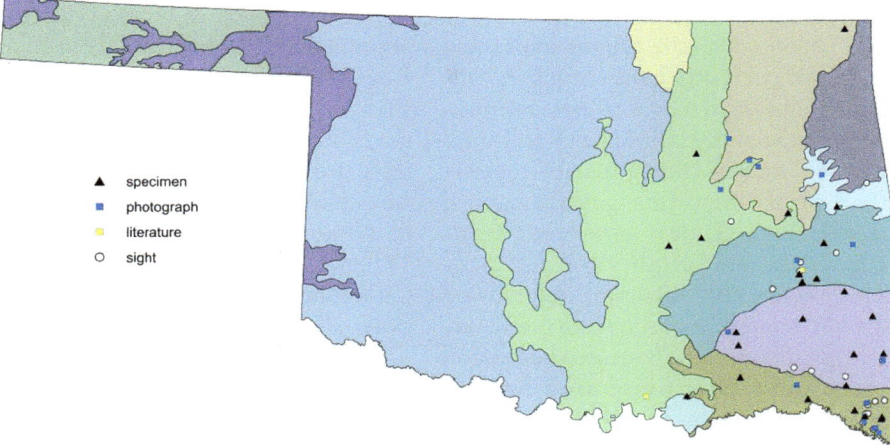

documentation level of records:
this map shows the level of support records across the species' range in Oklahoma have. The highest documentation level is "specimen" (*viz.*, physical evidence), followed by "photograph", credible "literature" report, and lastly, a "sight" record (i.e., no documentation, but we feel it is a credible record unless otherwise noted). Colored areas on the map signify the ecoregions of Oklahoma (see discussion on ecoregions in Chapter 3).

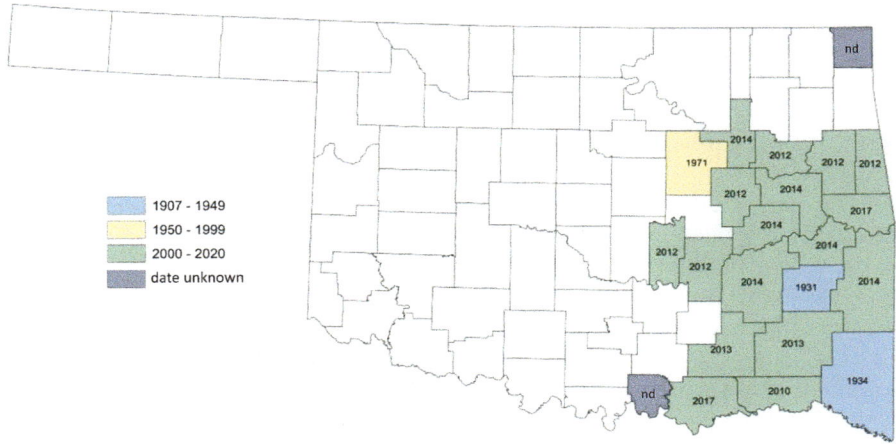

county map:
this map shows which counties the species has been recorded in as adults. Color coding indicates the era the species was first reported from. Years of first report are indicated (in some cases, years are approximated).

Endnotes

* See the Chapter 7 for more details. For general information about NatureServe, visit www.natureserve.org/ and for their conservation status assessment/ranking methodology, visit www.natureserve.org/conservation-tools/conservation-status-assessment.

† By taxonomic custom, authority names without parentheses indicate that the current genus is the genus the species was originally described as part of. For example, *Lestes sigma* was described by Calvert (1901) as a species of *Lestes* and so the scientific name and authority are written as *Lestes sigma* Calvert, 1901. Names within parentheses indicate the current genus is not the same as the original. For example, *Calopteryx maculata* was originally described as *Agrion maculatum*, so the current scientific name and authority are listed as *Calopteryx maculata* (Beauvois, 1805). Notice also the change in the ending of the species name, which was done so that the genus and species epithet would agree by gender in the Latinized name. *Agrion* (Greek) and *maculatum* (Latin) are both neuter, whereas *Calopteryx* (Greek) and *maculata* (Latin) are both feminine.

farther left the bar on the scale, the more strongly it is associated with streams. The farther right, the more strongly it is associated with ponds or lakes.[1] Species at the middle of the scale are considered to be habitat generalists in terms of their water association. The lenticity scale gauges adult odonates only because we do not currently have sufficient data for larval odonates in Oklahoma to determine associations.

Habitat: when we discuss habitat in the species accounts we do so with the intent to convey Oklahoma- or region-specific data rather than simply repeating general habitat descriptions provided elsewhere. Unless otherwise cited, habitat statements are based on our (or other's) experiences with the species in this region. If Oklahoma appears to standout with how a species' habitat has been described elsewhere, we made note of that distinction. Because the hydrological landscape is inextricably intertwined with habitat, especially for a group such as odonates, we could not avoid discussing aquatic habitat associations. As with the lenticity scale (see above), we provided habitat data for adults only because our knowledge of larval habitats is lacking.

Dates: dates are always written as day, month, year (sometimes year is omitted), such as 1 June 2010. We loathe dates written in other formats because of the confusion they cause. Admittedly, much of that confusion arises with specimen labels. For example, dates written as 6/1/10 or June 10 could be construed as 1 June 2010, 1 June 1910, 1 June 1810, 6 January 2010,[2] 6 January 1910,[3] 6 January 1810,[4] June 10th, June 2010, June 1910, or June 1810.

Seasonality: in each account, we provide a graph indicating the species' seasonality in Oklahoma (see Chapter 8 for details) and regionally. On either end of the Oklahoma graphs are the early and late flight dates, or the earliest and latest the species is known to be and expected to be active as adults in Oklahoma. The date range is good to know in terms of bounding a species' flight season, but arguably more useful to the field observer is the height of a species' season. The height of the season, also called the mid-point or peak (actually the median) in timing and abundance, is shown on the graph as a guide for optimizing encounters; while in the field you may encounter a given species any time during its flight season, but the likelihood of doing so increases the closer you are to its peak abundance.

In addition to the graphs, we discuss seasonality within the species accounts themselves. Along with presenting the records for the early and late flight dates in Oklahoma, we discuss aberrancies and provide regional context by noting and discussing known date ranges in the neighboring states of Arkansas, Kansas, Missouri, and Texas. It should be kept in mind that date ranges from Texas, although interesting from a context standpoint, will almost always be much wider than elsewhere in the region given how far south that state lies. The other three states are likely more in line with Oklahoma's timing, but Texas is the one of the four that has the most comprehensive phenological data (Abbott and Lasley 2019). We limited our inclusion of dates from Colorado and New Mexico to species that occur in southeastern Colorado and northeastern New Mexico, rather than elsewhere in those states. As a supplement to data from bordering states we included dates from Louisiana, which almost shares a border with Oklahoma. We chose to use Louisiana because Mauffray (2014) contains a wealth of data that provided a useful context for some species. On occasion, date ranges for other states or Canadian provinces were provided for context. We used Odonata Central, iDigBio, SCAN, SEMC, iNaturalist, and various publications to obtain dates (see Appendix B for these data sources). When known, we also provided date ranges for tenerals in Oklahoma (see below for further discussion about limitations of those data).

Photographs: the book is not a field guide. We nevertheless include a good deal of information about geographic and individual variation that complements that information in standard field guides (e.g., Dunkle 2000; Paulson 2009, 2011; Abbott 2011, 2015). Complementary images and text generally appear as stand-alone figures, yet we wished to let a reader know what a species looked like; hence, the header of each species account includes one or more representative photographs of each species. These photos were not selected for their perfection (although many of the photos are superb because there are several highly skilled nature photographers present in the state … the authors are not among them). No, our wish was not as mundane as to parade beauty—all damselflies and dragonflies are beautiful! Rather, we wished to highlight Oklahoma records of some geographical, seasonal, or historical interest, as well as to emphasize records from across the state. With fewer than a handful of exceptions, every photo is from Oklahoma. From this pool, we might select a photo that is nice enough aesthetically, but was of a first state record or represented a late flight date or was taken far to the west of the species' normal range. The issue of geographic spread was trickier, but there was a method to the madness. For example, two of our best and most active odonate photographers, Bill Carrell and Jim Arterburn, are based in Tulsa, meaning we could have festooned the book with photos from the Tulsa area. We tried instead to bring some attention to locales scattered across the state and its 77 counties, and in so doing we might forgo a stunning image from Tulsa for a nice enough one from somewhere else.

Breeding indicators: when records with breeding indicators were available, we noted them. Such records include those indicating pairing (including interspecies pairings), mating, mate guarding, and ovipositing. When ovipositing in Oklahoma has been reported as singly by females, in pairs, or both, we noted as such. We also noted records of tenerals, including often presenting date ranges for when tenerals were encountered. It should not be assumed that those dates always indicate the entire date range of emergence for a species because, relatively speaking, we have little data for tenerals because they are not typically noted by those in the field. Records of nymphs are generally not mentioned because of the difficulty of confidently identifying larvae to species (especially in earlier eras when larval keys were in their infancy and even now, we generally treat our own identifications with much caution given persistent issues with keys). We refrained from mentioning larval records unless the species is one that is easily identifiable at that life stage or, in a few instances, we mentioned a larval collection to bring to light a problematic record (with the hope that someone in the future can resolve the issue) or to correct a record that is clearly in error.

Breeding indicators should not be confused with the life stages and behaviors that constitute evidence of breeding or with those that define successful breeding. Although there is some disagreement as to which category a life stage or behavior should be included in (each can be included in multiple categories), it is agreed that indicators are much more broadly defined and that successful breeding ought to be rather narrowly defined. For example, pairings and mating are indicators of breeding, but they are not evidence of breeding nor of successful breeding. A male may lose hold of a female before they can mate, or a mated female may be captured by another male and he may remove the first male's sperm from the female's genitalia before it has had a chance to fertilize her eggs. Ovipositing and larvae are indicators of breeding activity and they constitute evidence of breeding, but neither is evidence of successful breeding. Exuviae and tenerals can be placed within all three categories because they indicate breeding, they are evidence of breeding, and they may document successful breeding (although some rightfully argue that pre-reproductive adults do not constitute successful breeding). Adult presence in itself is not evidence of successful breeding because adults can fly into an area from elsewhere rather than stay at their natal site (the location at which they were born). Because of the difficulty in categorization, we are advocates of using multiple variables (life stages and behaviors) when determining what constitutes evidence of breeding and successful breeding (Bried et al. 2015; Patten et al. 2019).

Morphology: when a species in Oklahoma varies in how its morphology, or body shape, size, pattern, or coloration, has been described in other references, we made note as such. We also noted individual and geographic variation of species in Oklahoma and the region.

Records and "person days": generally, a record consists of a given species, at a given location, at a given time. It would seem like breaking data into discrete species/location/date combinations would be straightforward, but in practice records can be difficult to quantify. There are many reasons, biological and geographical, why complications arise when determining what comprises a record. There are also analytical reasons to quantify records differently. For instance, we may see five Eastern Amberwings at James Collins Wildlife Management Area (WMA), which by its legal boundary undoubtedly makes it a distinct location. However, that WMA crosses the Latimer and Pittsburg county line, so you may argue that by county jurisdiction there should be two records, one for each county. Let's say we saw three individuals on the Latimer County side and two on the Pittsburg County side, so we enter data as two records. That makes biological and geographical sense if the locations surveyed were, say, half a kilometer apart, but if we were conducting an analysis with a geographic scale of one kilometer, then those two sites would be lumped into one record. That one kilometer scale is entirely reasonable if the data being compared to the site, for instance vegetative cover, also are at that scale. To complicate matters further, what if all of the amberwings were seen at one pond that straddled the county line? From a biological and geographical standpoint, both at a pond-level scale, it is the same record; as it would be from an analytical perspective. But from a county jurisdiction, it is still two records.

Other examples of complications include when a person surveys the same location for consecutive days. If we recorded Eastern Amberwings at the James Collins WMA pond on 1 and 2 June, data could be entered as few as one record to as many as four records:

- same location (county combined) with date range (1–2 June) = one record
- same location (county combined) with dates separated = two records
- location separated by county with date range = two records
- location separated by county with dates separated = four records

Again, one must think about what makes sense from a biological, geographical, and analytical perspective when determining how to quantify records. But the gist is that when trying to tally total records, these issues make what seems like an easy count become not so easy.

Another prevalent quantification issue in datasets like that of the Oklahoma Odonata Project's (OOP) is that some data types are separated out into individual records while other data types are not. For instance, specimen data is often entered as a separate line of data, or record, because each specimen has been assigned a catalog number. If two male Eastern Amberwings were collected at Lake Elmer on 1 June 1978, then they may be cataloged as SPI 989 and SPI 990 and entered into the OOP database as separate entries. In reality, they are the same record because they have the same species/location/date combination, but from a data entry perspective they are different data. When conducting any sort of analysis with these data, those records must be combined into a single record. This issue also arises with data from a source like Odonata Central. Let's say that two friends go to the field, find a new state record, they each take a photo of that individual, and they each submit a record to Odonata Central. Therefore, we have the problem of two records that are in fact just one. If your head is spinning right now, imagine dealing with these issues for >55,000 lines of data that may be >55,000 records or they may be but a fraction of that number—it all depends on how one chooses to quantify and analyze "records."

One final issue to mention is when multiple people go to field on the same day but they visit different places. This is not so much a record problem as it is a survey counting problem, specifically when trying to determine the number of surveys during a specific timeframe or the number conducted at a location throughout its history. To deal with this issue the concept of "person day" comes into play. For example, on 1 June 2015, the senior author may have been at Boiling Springs State Park in northwestern Oklahoma, while the junior author may have been at Red Slough WMA in the southeast, and Bill Carrell may have been at Mohawk Park in the northeast. Instead of saying that the state was surveyed on one day, we would say that it was surveyed on three "person days." We understand that it is a bit of a clunky way to convey that it was not just one person out on that day but three, but similar techniques are used in other fields of biology and, fortunately, we only occasionally used the term in the book.

Relative abundance: one must be careful when discussing abundance, especially in the case of disparate data such as those in the OOP database. Because the OOP's data came from a variety of sources, record types are not equal with respect to relative abundance; each record type must be considered independently. Take for example, a record that is sourced to our field notes vs. one taken from Odonata Central. When we go to the field, we record all species encountered (seen, captured, photographed, collected). We also record the number of individuals, sexes, ages, behaviors, such as pairing, mating, and ovipositing, and in what habitats species occur. Other people may only record what species they encountered or just those they photographed. Consequently, in the OOP database there might be two records that superficially indicate abundance, but in reality only one does. Hypothetically speaking, let's say that we visited a pond on 1 July and we reported 25 individuals of a given species. We entered that record into the OOP database. The next day, someone else visits that same pond and takes a photo of the same species. When he uploads that photo to Odonata Central, he makes no indication of the number of individuals he encountered. That record eventually gets imported into the OOP database as a minimum of one individual encountered. If those two records were taken at face value, it would appear that overnight the population had crashed—from 25 individuals one day to just one the next—but that likely was not the case; rather it was just a difference in data quality, or varying levels of completeness. We took such factors into account when we discussed relative abundance in the species accounts.

We also tried, for some taxa (see family and genera summaries), to take into account survey effort by era and how that might reflect relative abundance through time (i.e., increases, decreases, or stability in populations). Because we wanted to approximate trends as closely as possible, we used a raw estimate of survey days by era and known abundances of species during those days to calculate an expected average number of individuals encountered in a survey during a given era. Additionally, we attempted to correct for survey ability. We did so because correcting for survey effort is not the same as correcting for survey ability or survey bias.

It is tempting to argue that contemporary observers are more efficient record producers, and in a sense this is true, but what lies at the heart of that so-called efficiency is resource availability because we have a slew of resources that were not available previously. For instance, we have field guides and better taxonomic keys, and we now have a pretty good idea of geographic ranges for species, all of which makes species identification much faster. *Argia nahuana*, the Aztec Dancer, provides a good example of the inequities of resources through time. Record numbers and abundance indicators make it seem as if this species was not as common in the earliest era of data collection (pre-1950). We hesitate to make such a claim because it is obvious that, early on, including into the middle era (1950–1999), there were issues with confidently identifying *A. nahuana* (see that species account). Now, identification issues have been resolved and the species can be confidently identified with little effort, so more records of it does not necessarily mean is it now more common … we just have more confidence in species determination.

Earlier researchers were limited by the need to spend time capturing individuals to keep as specimens; time that we can now use looking through binoculars (or cameras) to more quickly identify species. A hypothetical comparison of 20 minutes spent in 1932 at a pond and 20 minutes spent there today might be that, in 1932, RD Bird spent that whole time capturing a *Phyllogomphoides stigmatus*, the Four-striped Leaftail, whereas we could walk up to that same pond and readily identify that species plus another 15. It is not necessarily the case that we are better field observers, but we do essentially have an unfair advantage. Such an advantage makes directly comparing record numbers between eras misleading. Abundance can also be misleading, akin to the OC examples above, because we lack complete survey notes from early researchers that indicate how many individuals they saw; instead, we know what individuals they collected as specimens and we may have supplementary data to indicate they had additional individuals *or we may not*. Our notes, on the other hand, have details about how many individuals were collected, photographed, and counted in total. Going back to that 20-minute comparison, we not only could now say that we had *P. stigmatus* and 15 other species at that pond, but we could also indicate the relative abundances of those species, whereas RD Bird may have collected one or two individuals of 5 species and those are the data we are left to work with to try and determine community composition and relative abundance. As such, it was necessary to devise a corrective index[5] in order to make any sort of logical comparisons across time.

Maps: an overwhelming majority of species accounts include two maps, one that plots individual locations coded by "highest" level of documentation (specimen, photograph, sight record, or literature report) and one that lights up a county and displays the year in which the species was first recorded in a county. The first map is straightforward, we trust. The second map may require a modicum of caveats, inasmuch as we worry that the year-of-first-record may be misinterpreted to suggest recent range expansion. Emphatically, it does not, or, at least, often enough it will not. A few species, such as Rambur's Forktail (*Ischnura ramburii*), indeed have expanded (and continue to expand) their geographic range north and west in Oklahoma, with that species reaching Kansas for the first time in mid-2019 (Isaac Fox, OC 505200), so a litany of recent year-of-first-record on the second map has that meaning. Most other species, though, have been known generally from a region for many years but may have been missed for a given county for no other reason than that county was under-surveyed during that species' flight season. A glance at the second map for the Blue-tipped Dancer (*Argia tibialis*) reveals a fine example: the species was first recorded in Muskogee County in 2019, even though it was recorded from *four* adjoining counties in the 1920s and 1930s. We have no reasons to think the species has not been in Muskogee County all along; we and others simply failed to document a record there until recently. Hence, we urge caution (and use of the species account discussion for context) when reviewing the county maps.

ABBREVIATIONS AND SYMBOLS

Collectors/observers
Some people were mentioned so often in the text, we typically abbreviated their names as:
AEP = A Earl Pritchard
BAH = Berlin A Heck
BC = Bill Carrell
BS-P = Brenda D Smith-Patten[6]
DA = David Arbour
GHB = George H Bick
JWA = James W Arterburn
KW = Ken Williams
LEH = Lothar E Hornuff
MAP = Michael A Patten
VWF = Victor W Fazio, III
Most records reported in the book are from BS-P and MAP; if someone else is not specifically cited for a given record, then the record is ours.

other collector/observer names can be found in Appendix B

Institutions/collections/data sources
The most mentioned institutions/collections/data sources are:
CSU = Colorado State University
FSCA = Florida State Collection of Arthropods

JCAC = John C. Abbott Collection
iNat = iNaturalist
OC = Odonata Central
OMNH = Sam Noble Oklahoma Museum of Natural History
OOP = Oklahoma Odonata Project
OSU = Oklahoma State University (insect collection and generically for the university)
SEMC = Snow Entomological Museum, University of Kansas
SP = Smith-Patten/Patten Collection
UCO = University of Central Oklahoma

other institutions/collections/data sources abbreviations can be found in Appendix B

Locality and measurement abbreviations
km = kilometer
mi = mile
cm = centimeter
in = inch
mm = millimeter
m = meter
E = east
N = north
S = south
W = west
jct = junction
rd = road
vic. = vicinity
Co. = County
GMA = Game Management Area
MR = Military Reserve
NA = Natural Area
NF = National Forest
NG = National Grassland
NRA = National Recreation Area
NWR = National Wildlife Refuge
PHA = Public Hunting Area
SP = State Park
TNC = The Nature Conservancy
U.S. = United States
WMA = Wildlife Management Area
WR = Wildlife Refuge (used only for Wichita Mountains WR)

Miscellaneous
cf = compare(s), as in compares to the species that the abbreviation follows (e.g., *Argia sedula*, cf)
coll. = collected by, collector, or collection (context dependent)
det. = species determined by
id = identified
imm = immature
indiv. = individual(s)
in litt = (Latin, *in litteris*) in written correspondence
n = 3 – indicates the number (n) of individuals, data points, or whatever is being discussed
ODWC = Oklahoma Department of Wildlife Conservation

pers comm = personal communication
pers obs = personal observation
USGS = United States Geological Survey
♀ = female
♂ = male
U = unsexed individual
ab = abdomen
FW = forewing
HW = hindwing
TL = total length
S = abdominal segment

NOTES

1. Our lenticity scale is quantitative rather than qualitative. First, we used data from our notes to classify an encounter with a species as being in lotic (0) or lentic (1) habitat. For example, an *Argia sedula* at Rader Park, Weatherford, Custer County, is scored 1 because its habitat is a clear, spring-fed pond, but an *A. sedula* at Moneka Park, Lake Waurika, Jefferson County, is scored 0 because we find the species on Beaver Creek, which typically flows (only during drought conditions is it reduced to isolated pools). We then calculated a weighted average of assigned scores using count as the weights. To return to the example above, if 10 individuals were encountered at Rader Park and 25 individuals were encountered at Moneka Park, the lenticity score would be 0.286; i.e., $((10 \times 1) + (25 \times 0))/(10 + 25)$. As discussed in the *Libellula saturata* species account, we are working to refine this scale to more closely capture a species true "lenticity"; we then hope to publish a paper detailing how all odonate species can be ranked for lenticity.

2. This confusion would arise if someone wrote the date in an European format, i.e. day/month/year.

3. See above.

4. See above.

5. This approach was used primarily with *Argia*; see the genus summary for more details.

6. Senior author note: I now use the name Brenda D Smith; but for the book I continued to use, Brenda D Smith-Patten (BS-P), which is the name I had during most of the Oklahoma Odonata Project and the one most people knew me by.

## CHAPTER 10

CALOPTERYGIDAE: BROAD-WINGED DAMSELS

In the United States and Canada, it is species of the genus *Calopteryx* that make one see why this family has its common name—the broad-winged damselflies. Species of *Calopteryx* do indeed have broad wings, especially female *C. maculata*. Ironically, although the genus *Calopteryx* is the namesake of the broad-winged damselflies, species in the genus are not called broadwings but rather jewelwings (in North America) or demoiselles (everywhere else).

In Oklahoma, the family Calopterygidae is represented by one *Calopteryx* and two *Hetaerina* (rubyspots) species. These two genera, as well as the family, are more speciose elsewhere, with 26 jewelwing and 37 rubyspot species worldwide, and 5 and 3 of each, respectively, in North America (Paulson 2011; estimated 167 species of calopterygids worldwide, Garrison et al. 2010). At present, only *Hetaerina titia*, the Smoky Rubyspot, is of conservation concern in Oklahoma, but because all of Oklahoma's calopterygids are found primarily on clean, clear-water streams with moderate to high flow, they should all be monitored for potential future impacts to water quality and streamflow.

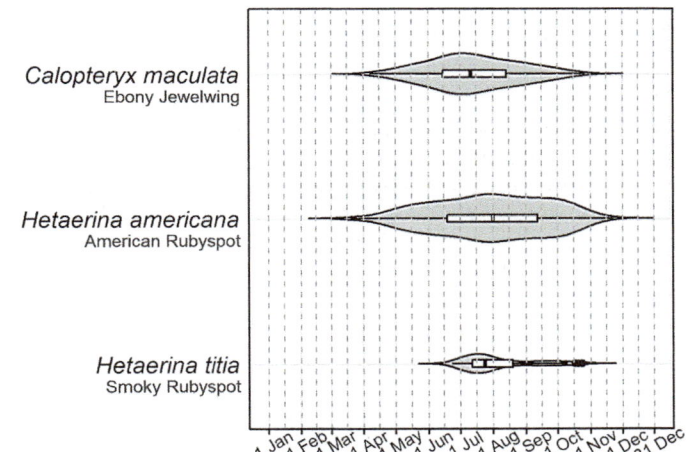

Figure 10.1 Seasonality comparison of species of Calopterygidae in Oklahoma.

CALOPTERYX MACULATA (BEAUVOIS, 1805) – EBONY JEWELWING

The Ebony Jewelwing is a lotic species typically found along narrow, shaded, clear-water creeks, usually with at least moderate flow. It occurs across much of the state, although it is more common and more widespread in the eastern half of Oklahoma. It is absent from most of the panhandle and arid southwest, which is not surprising given that both of those areas lie near the edge of the species' known range. The species is usually uncommon to locally common where it is found, being seen just a few at a time, or sometimes in the dozens and even hundreds, such as the 200 estimated individuals at Boiling Springs SP, Woodward County, 20 August 2011 (VWF, OC 331894), ~200 along Sallisaw Creek, about 8 km SW of Stilwell, Adair County, 14 July 2013, ~250 on Coal Creek, 10 km SE of Keota, Haskell County, 4 July 2014, and ~300 along Mill Creek and a nearby tributary 5 km NE

of Hanna, McIntosh County, also on 4 July 2014. The species was first recorded in Oklahoma from McCurtain County,[1] on 22 June 1925 (1♀, OMNH 1242) and has been reported regularly since.

The earliest flight date in Oklahoma for the species is 6 April (1♂, 1♀, 6U, 10 km SE of Idabel, McCurtain County, BAH, OC 374305, 374306) and the latest is 26 October (1♂, Grassy Slough WMA, McCurtain County, DA, OC 284321). Oklahoma's phenology is more-or-less in line with elsewhere in the region (AR: 2 May–25 September, OC 430665, iNat 16913022; KS: 15 May–26 September, iNat 25191611, SEMC; LA: 1 Mar–25 October, Mauffray 2014; MO: 1 May–13 November, OC 444040, INHS; TX: 12 March–1 October, Abbott and Lasley 2019). There are three Oklahoma records of ovipositing, all by lone females, between 8 May and 14 July.

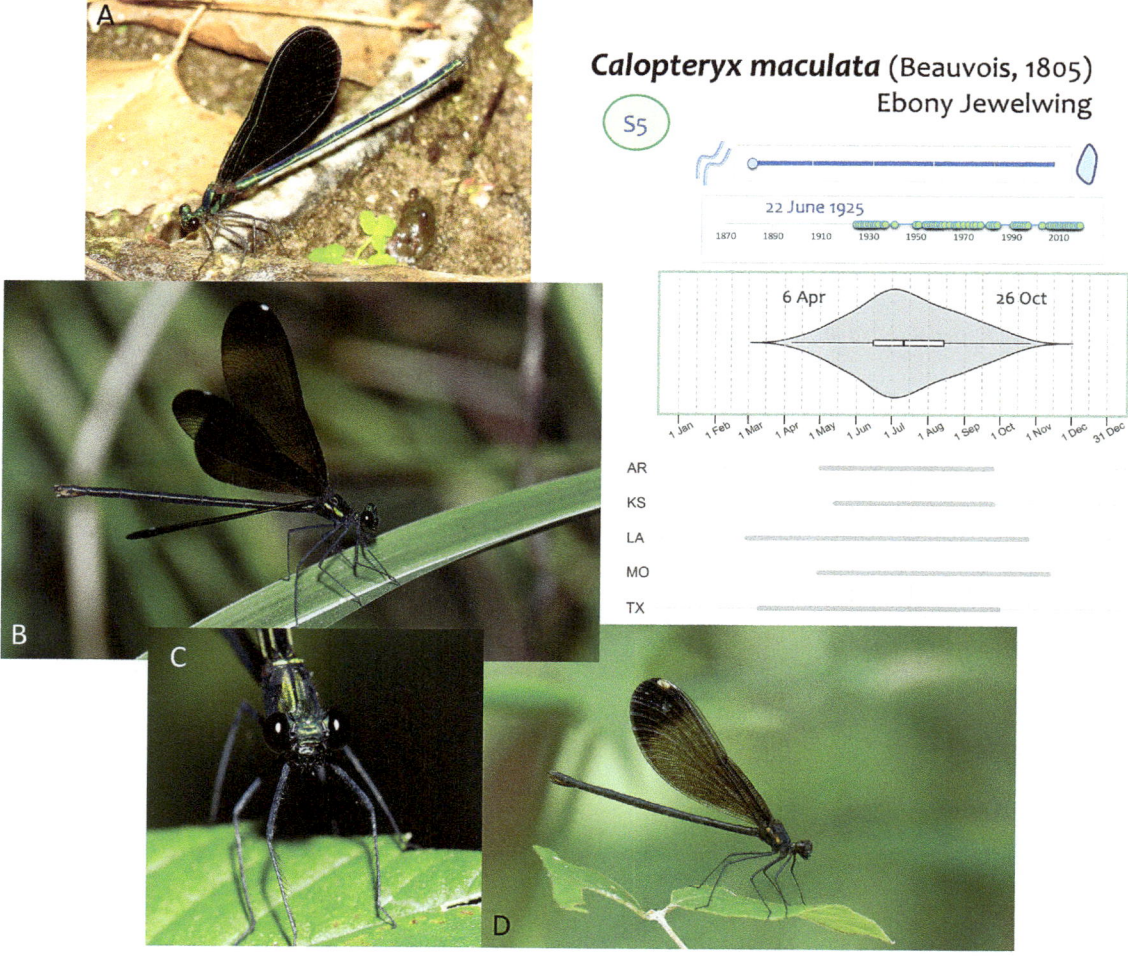

Figure 10.2 *Calopteryx maculata*, the Ebony Jewelwing. A) ♂, Doby Springs Park, Harper County, 18 September 2018 (©Bill Carrell). B) ♀ signaling whether she is receptive to copulation and C) close-up of her face, TNC Oka' Yanahli Preserve, Johnston County, 5 July 2013 (©Bryan E Reynolds). D) This ♀ at Natural Falls SP, Delaware County, 6 August 2015 (©James W. Arterburn), has a wing pattern reminiscent of *Calopteryx dimidiata*, the Sparkling Jewelwing, which occurs in northeastern Texas, but that species has narrower wings, their shape more like the wings of *Hetaerina* (rubyspots) than of *C. maculata*.

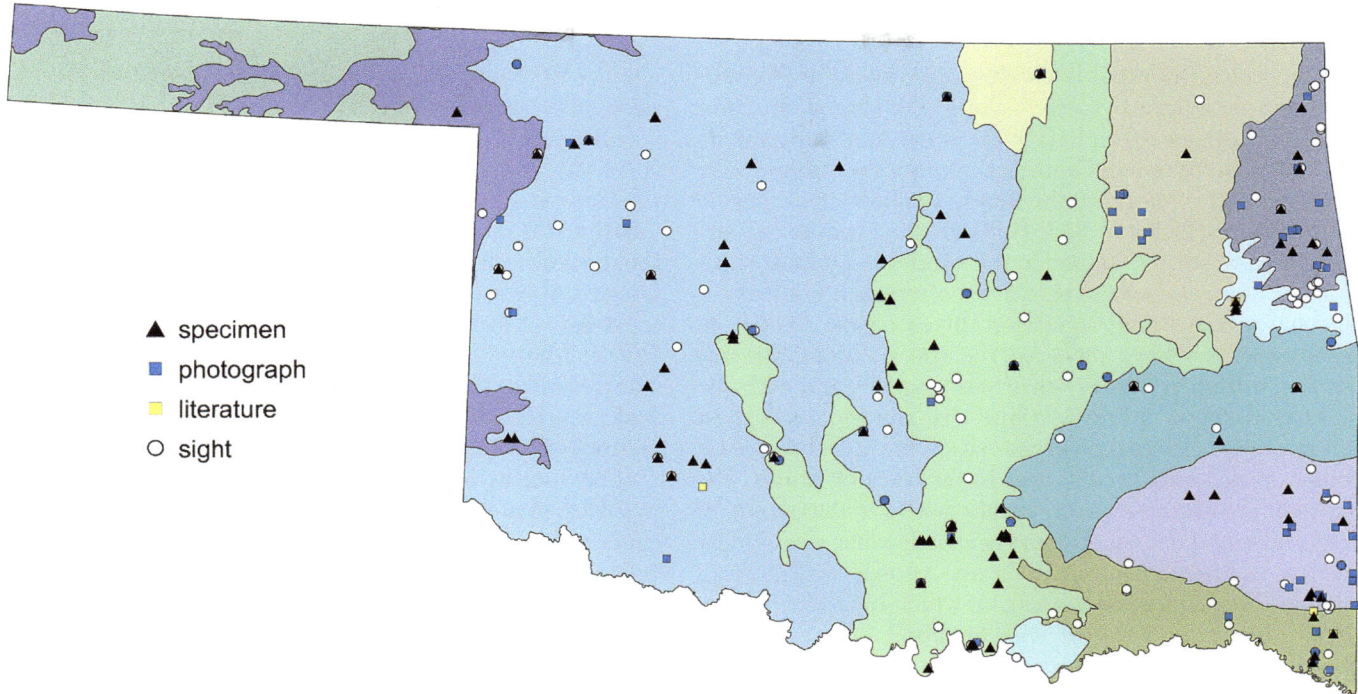

Figure 10.3 Documentation level of supporting records for *Calopteryx maculata*, the Ebony Jewelwing, in Oklahoma. Levels include specimen, archived photograph, literature reference, or sight record. Although sight records lack documentation, only those we feel are credible are plotted.

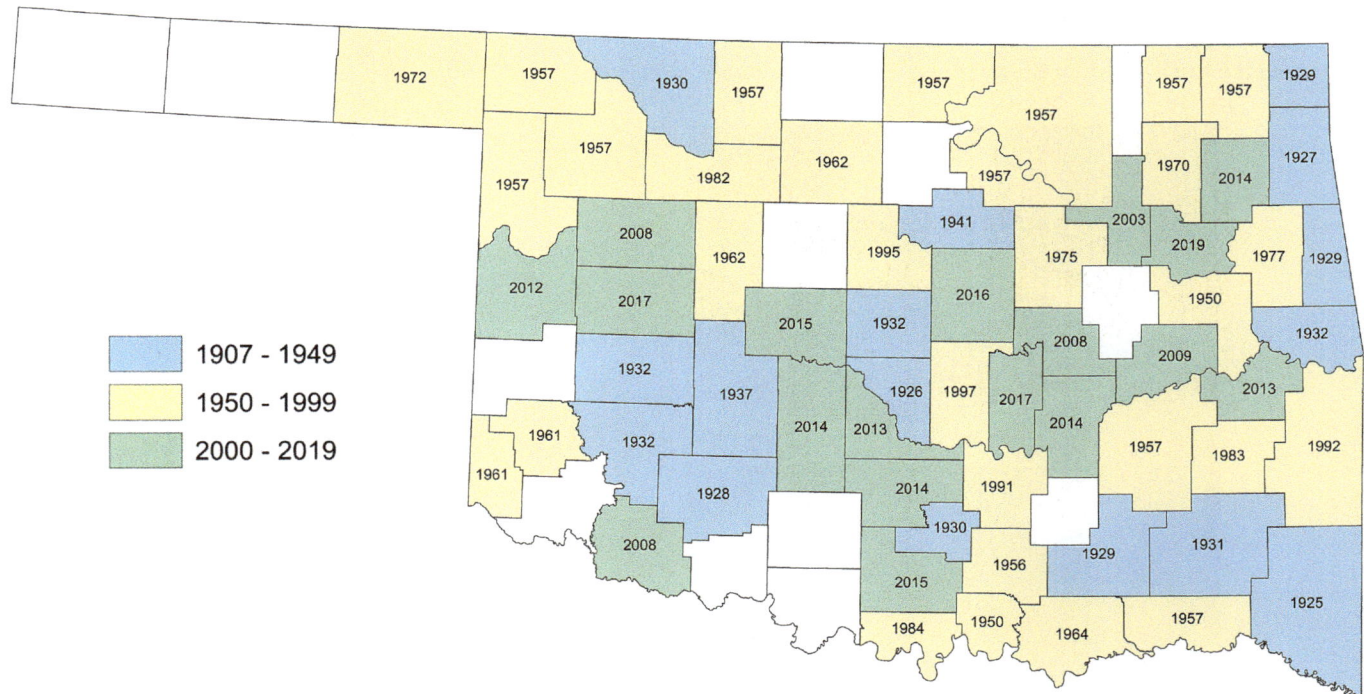

Figure 10.4 Counties in which *Calopteryx maculata*, the Ebony Jewelwing, are known to occur in Oklahoma. Year within the county is that in which the species was first reported.

HETAERINA AMERICANA (FABRICIUS, 1798) – AMERICAN RUBYSPOT

The American Rubyspot is a widespread (found in 74 counties[2]) and ubiquitous species along Oklahoma streams, especially in the western half of the state. At a study site on Cowan Creek, Marshall County, the species was found 97% of the time at a "clear, sunlit, sandy, shallow creek devoid of vegetation" (Bick and Bick 1958). In our experience, particularly in western Oklahoma, the species is found at similar types of creeks, except that more often than not there is much low, overhanging vegetation and emergent vegetation. Other observers have reported the species along streams that range from "relatively clear" to "murky" and with the substrate rocky or sandy. We find that the species is most common on clear creeks, occasional on murky, and rare on muddy creeks, and anomalously so at lakes or ponds. We routinely see the species on creeks with an abundance of rocks and emergent vegetation (especially what we jokingly refer to as "*Hetaerina* weed," better known as American water willow, *Justicia americana*), on which individuals often perch. Flow rate is a key: without good flow, American Rubyspots are generally absent.

This species has been reported regularly through all eras of collecting in Oklahoma, and it is often recorded in double digits, including the first time it was recorded in the state (19♂ and 11♀ at Wister, Le Flore County, 4 August 1907; F. Collins, Williamson 1914b). At times we have recorded the species in the hundreds, such as ~200 along Kiowa Creek near Slapout, Beaver County, 29 July 2012, and ~400 there on 12 July 2014, ~300 along the Little Washita River, 1 km N of Farwell, Grady County, 30 September 2013, and ~200 along Rock Creek in the Chickasaw NRA, Murray County, 23 August 2015.

The earliest flight season record in the state is 19 March (1932, 1♂, Johnston County, RD Bird, OMNH 3901) and the latest is 22 November (1930, 1♂, Honey Creek, Murray County, RD Bird, OMNH 3919), a shorter flight season than the species' near year-round presence in Texas (13 February–1 January, Abbott and Lasley 2019), but longer than elsewhere in the region (AR: 13 May–29 October, USNM, iNat 17951000; CO: 21 May–4 November, OC 375292, 427655; KS: 3 May–26 September, OC 327861, SEMC; LA: 2 April–2 October, Mauffray 2014; MO: 14 May–22 October, Trial 2005; NM: 10 April–8 November, OC 374425, 284357).

The species has been observed in mixed-species pairs in Oklahoma on several occasions, including inter-family/ genera and male-male pairs (Bick and Bick 1981). GH Bick collected two pairings of ♂ *Hetaerina americana* with ♀ *Argia plana* (Springwater Dancer; Marshall County, 30 June 1967, FSCA 66815, 66816, and 9 July 1967, FSCA 66811, 66812). He also collected a ♂ with a ♂ *Argia plana* (Marshall County, 5 July 1967, FSCA 66814, 66813). It is unclear from Bick's notes whether he encountered one or two ♂/♂ pairings involving ♂ *H. americana*. We cataloged FSCA 66810 that had 2♂ *H. americana* in the envelope. Bick's original envelope indicated the specimen(s) was collected on 7 July 1964 but the transcribed FSCA specimen label indicated the collection date was 6 July 1964. Bick's notes indicated that he had a tandem ♂-♂ pair on 6 July ("in tandem with

Figure 10.5 *Hetaerina americana*, **the American Rubyspot. A)** ♂, **Cimarron River, Cimarron County, 7 September 2015** (©Bill Carrell, OC 436523). **B)** ♀, **8 km E of Red Oak, Turkey Creek, Latimer County, 6 July 2014** (©Brenda D Smith).

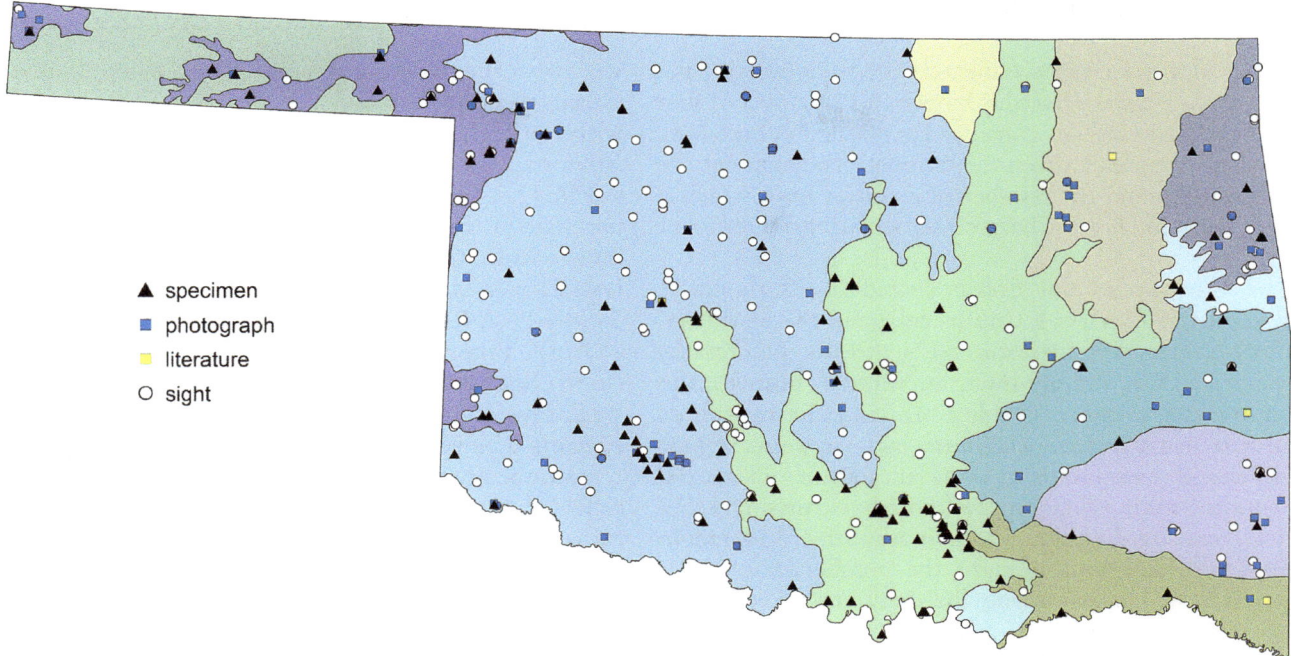

Figure 10.6 Documentation level of supporting records for *Hetaerina americana*, the American Rubyspot, in Oklahoma. Levels include specimen, archived photograph, literature reference, or sight record. Although sight records lack documentation, only those we feel are credible are plotted.

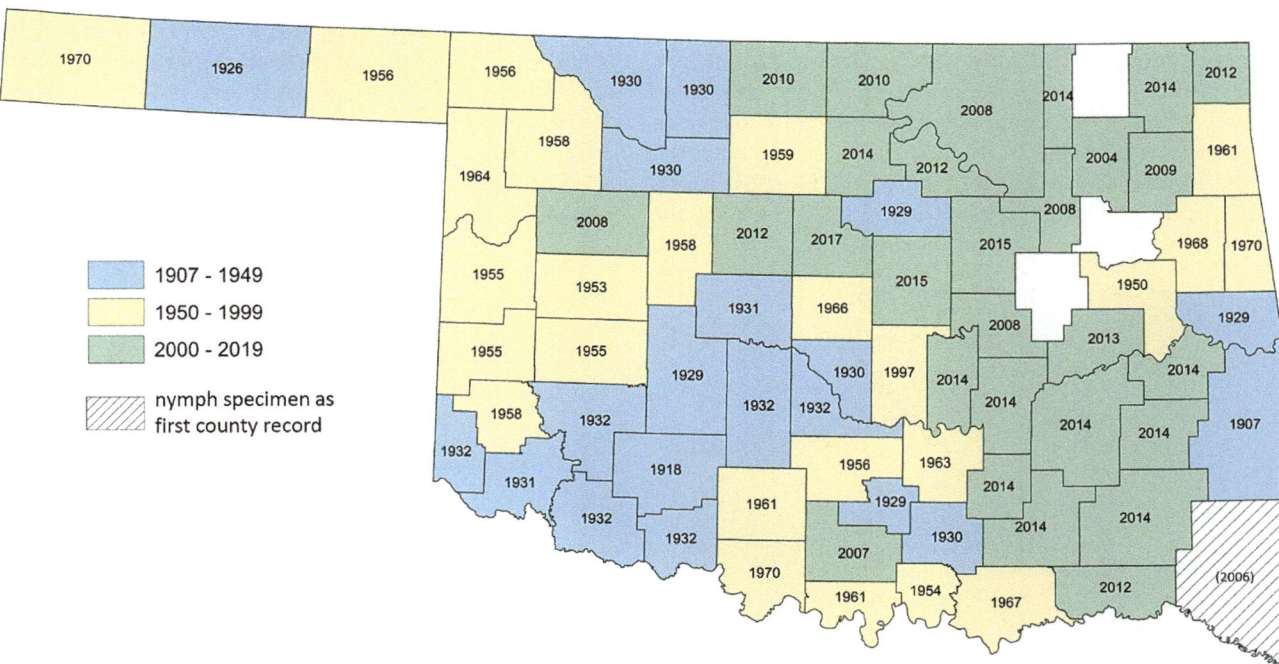

Figure 10.7 Counties in which *Hetaerina americana*, the American Rubyspot, are known to occur in Oklahoma. Year within the county is that in which the species was first reported. The first record of the species for McCurtain County (hatched) was of two nymphs collected in October 1975 (OMNH 1719). Adults were not reported from the county until August 2006.

H. amer ♂ in GHB collec") and that he had one on 7 July ("in tandem with marked *H. a.* ♂, In GHB collec"). Whether he meant that he had a pair each day or that he collected the unmarked ♂ on the 6th and the marked one on the 7th is anyone's guess. All we know is that there is one pair at FSCA. And, finally, although we have larval and teneral records in the database, as well as numerous instances of tandem pairs, we know of no ovipositing records.

HETAERINA TITIA (DRURY, 1773) – SMOKY RUBYSPOT

Hetaerina titia is far rarer and more range restricted than its congener, *H. americana*, the American Rubyspot. At first blush the species could be construed as reasonably common, or at least widespread, given the number of reports for the state (about 80) and the number of counties from which it has been reported (*n* = 31), but a closer look shows a different picture.

This lotic species was first collected in Oklahoma in early August 1907, when F Collins collected 1♂ at Wister, Le Flore County (Williamson 1912a, 1914b). The Smoky Rubyspot had not, despite intensive surveys, been seen in that county for 112 years (Table 10.1). It was next recorded in the state in 1932 when, within the last two weeks of July, the species was encountered in seven counties in central and southwestern Oklahoma (although surprisingly not from Comanche County, where it was first detected in 2003). Those additions brought the number of counties known for the species during the early eras to nine, the majority of which have not produced subsequent reports (Table 10.1). Those two amazing weeks in July 1932 also added 72 specimens (66♂, 6♀) of the species to the OMNH collection.

The Smoky Rubyspot was next collected in Oklahoma in late July 1950 by GH Bick at the "swimming hole" at Oklahoma State University Wildlife Conservation Station, Muskogee County (24 July, 2♂ and 1♀, FSCA, and 26 July, 4♂, one at FSCA). It was recorded sparingly thereafter until 1970, when it was documented in Murray County (it was reported for the county in Bird 1932, but we have not found the basis of that record; in 1970, LE Hornuff found 1♂ at 1.2 mi SE of Dougherty, 23 July, FSCA). Despite the

somewhat regular records during this era, numbers of individuals recorded were low, usually only a couple at one time, with the highest count being six (5♂, 1♀, including 1 pair, Milburn, Johnston County, 8 July 1954, GH Bick, 3♂ and 1♀ at FSCA). Those numbers are in stark contrast to the higher numbers of the previous era when it was, more often than not, recorded ≥5 at a time, including the highest count of at least 23 individuals (21♂, 2♀, Lexington, Cleveland County, 28 July 1932, RD Bird, OMNH).

At the time of our paper discussing the conservation status of Oklahoma Odonata (Patten and Smith-Patten 2013b), we had reason to believe that the Smoky Rubyspot was experiencing a probable population decline as well as a range retraction. At the time, we had no post-1970 records that reported numbers anywhere near the encounters of 1932; rather, numbers were akin to the Bick/Hornuff era. There were also 14 counties in which the species was reported historically, but where it had not been encountered for at least 40 years, and in 6 cases for 80–100+ years (Table 10.1). Moreover, most (75%) of the post-1970–2013 records were from just two locales, one in the northeast (Mohawk Park in Tulsa, Tulsa County) and one in the southwest (the Wichita Mountains area, Comanche County).

Since 2013, our perception of *Hetaerina titia* in Oklahoma has changed considerably. Although still generally encountered 1–2 at a time, we have multiple records now of >4 individuals, including 2 records for 18–19 individuals (15♂, 3♀, 3 km E of Blue, Caddo Creek, Bryan County, 24 July 2015, MAP, 1♂ as SP 1732; 18♂, 1♀, 5 km ESE of Lehigh, Muddy Boggy Creek, Coal County, 2 July 2016, MAP, OC 448362, 1 pair as SP 2003). The number of localities also

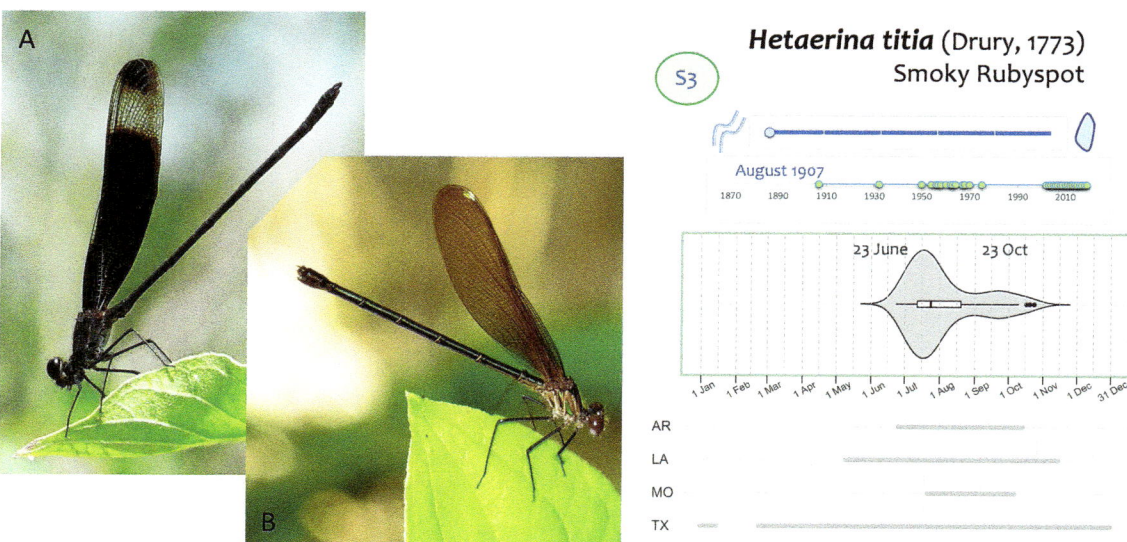

Figure 10.8 *Hetaerina titia*, the Smoky Rubyspot. A) ♂, Medicine Creek, Fort Sill Military Reserve, Comanche County, 31 July 2007 (©Victor W Fazio, III, iNat 522829). B) ♀, Oxley Nature Center, Tulsa, Tulsa County, 29 June 2018 (©Bill Carrell, OC 483349).

Table 10.1 **Oklahoma counties from which *Hetaerina titia*, the Smoky Rubyspot, was reported from 1907–1975 (top). Also indicated is the subsequent year(s) the species was recorded in the county and the years passed since the last report. New county records obtained in 2002–2019 are also provided (bottom). For some counties, for example Atoka, McCurtain, and Pushmataha, there is but a lone record despite extensive coverage.**

NEW COUNTY, 1907–1975	YEAR FIRST RECORDED	SUBSEQUENT YEAR(S) RECORDED	YEARS SINCE LAST REPORTED
Beckham	1968		51
Bryan	1954	1958, 1967, 2015	
Caddo	1932	2011	
Cleveland	1932		87
Cotton	1932	2012, 2016	
Creek	1975	2015, 2017	
Grady	1932		87
Johnston	1954	9 yrs bet. 1956–2019	
Kiowa	1932		87
Le Flore	1907		112
Marshall	1957	1963	56
McClain	1932		87
Murray	pre-1932*	1970	49
Muskogee	1950		69
Pittsburg	1954		65
Stephens	1932		87

NEW COUNTY, 2000–2017	YEAR FIRST RECORDED
Atoka	2015
Choctaw	2015
Coal	2016
Comanche	2003
Custer	2013
Garvin	2016
Jackson	2017
Latimer	2010
Love	2016
McCurtain	2016
Okfuskee	2011
Okmulgee	2017
Pottawatomie	2016
Pushmataha	2014
Tulsa	2002

* Reported by Bird (1932), but details of record are unknown. The first record for any species in the county is from 1929.

increased, and of the 15 counties in which the species was newly recorded in since 2002, 11 of those were added after 2013 (Figure 10.11). Although we can acknowledge various positive indicators of late, we still must try to reconcile the apparent blipping out of the species in 11 counties (Figure 10.11). We cannot say for sure if the species will not be found again in those counties, as it was in Bryan and Johnston Counties, where it was "rediscovered" in 2015 after having not been seen since 1967, but at this point we are not optimistic. Currently, it is unclear whether the

Figure 10.9 Variation in ♂ *Hetaerina titia* wing pattern. Such variation is present throughout the species' range (for example, see Johnson 1972b) and appears to have no geographic component given that individuals at the same location can vary from a small patch of red to fully black. (Note: Wing shape is not necessarily represented properly due to differing angles of photos. Likewise, differences in paler colors is due to differences in background of photos, not to natural differences in wing color; instead, note the darker patterns here as representative of wing variation. Photos are by ©Bill Carrell except far right, which is by ©Eric Isley.)

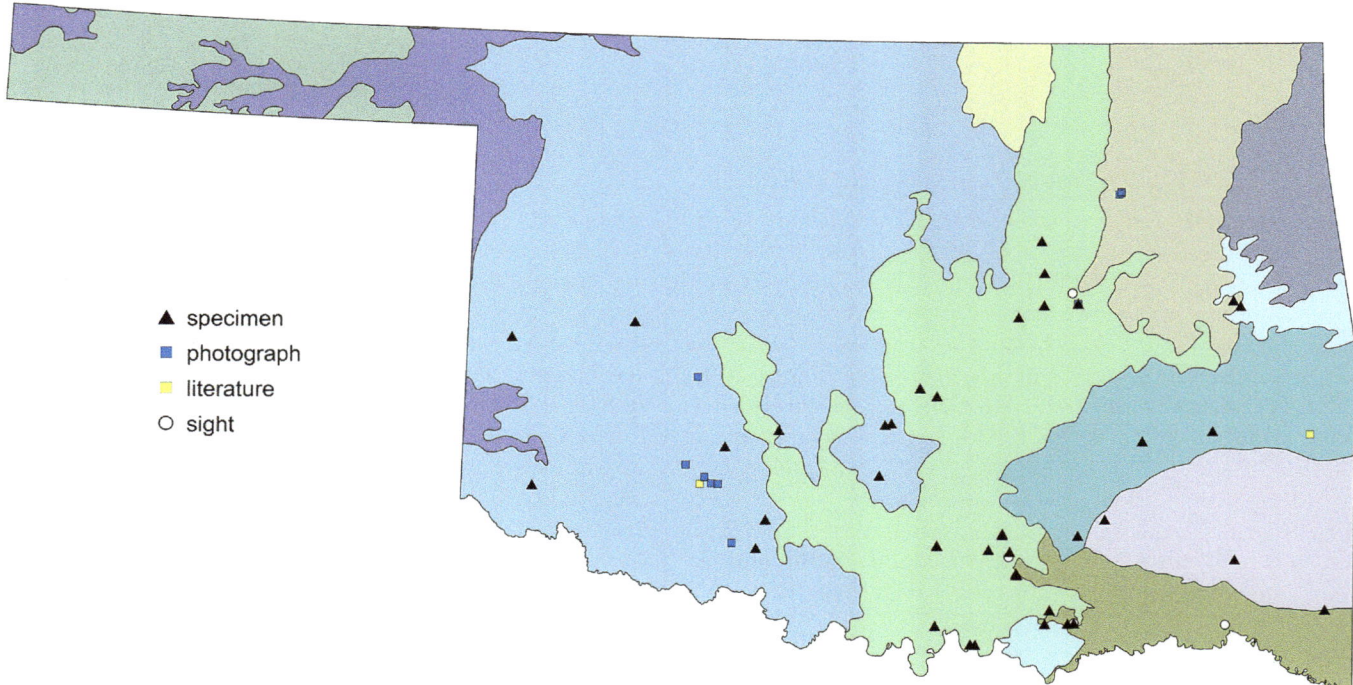

▲ specimen
■ photograph
▢ literature
○ sight

Figure 10.10 Documentation level of supporting records for *Hetaerina titia*, the Smoky Rubyspot, in Oklahoma. Levels include specimen, archived photograph, literature reference, or sight record. Although sight records lack documentation, only those we feel are credible are plotted.

Smoky Rubyspot in Oklahoma is cyclical and easily missed, is experiencing a population rebound, or we simply do not have enough data to fully characterize it. Nonetheless, given the species' possible range contraction and population decline in at least portions of its range in the state, and that it is a species of relatively low population density, we maintain the idea that this species is one of conservation concern for Oklahoma (Patten and Smith-Patten 2013b; Smith-Patten and Patten 2016), but we chose to downgrade it one category (S3).

In Oklahoma, *Hetaerina titia* is found primarily at clearwater creeks and rivers, often rocky, usually with moderate to

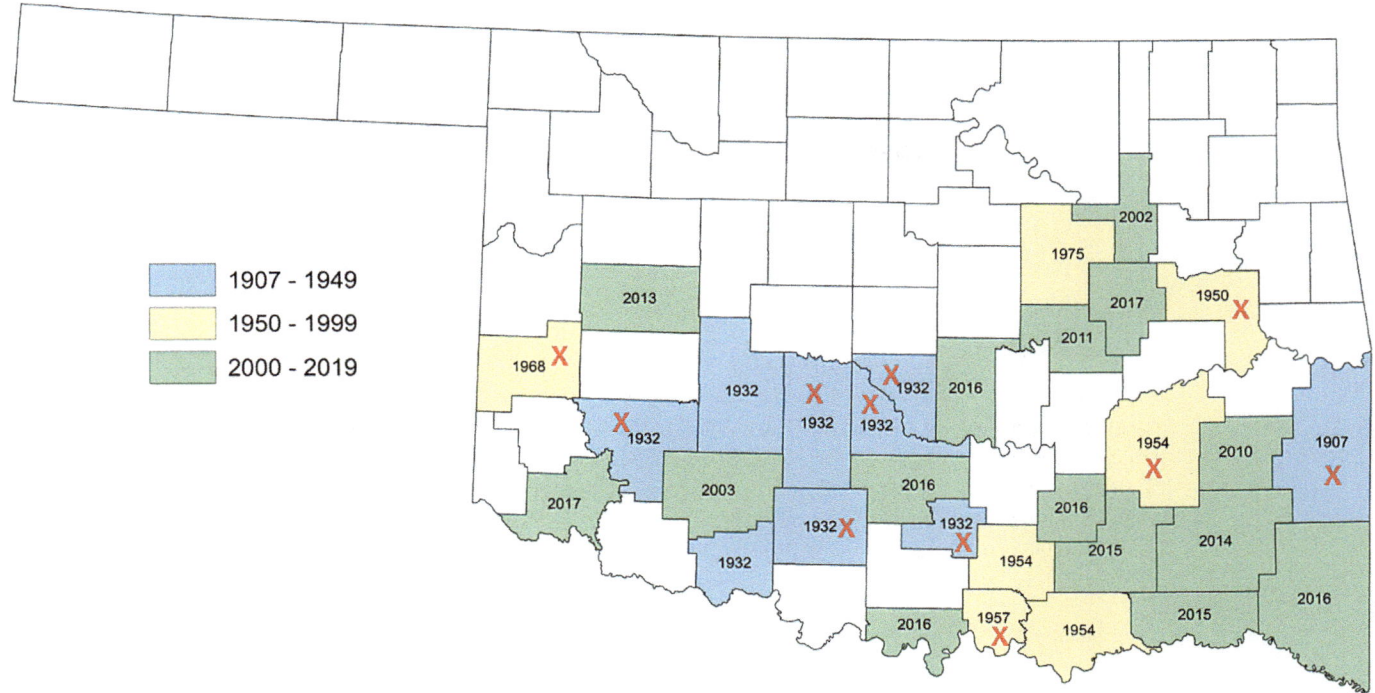

Figure 10.11 Counties in which *Hetaerina titia*, the Smoky Rubyspot, are known to occur in Oklahoma. Year within the county is that in which the species was first reported. Counties in which the species has not been reported from for approximately 50 years or more (up to 112!) are marked with a red X.

high flow, and always with at least some shade and emergent and shoreline vegetation; albeit at least once it was found at a steep-sided, mud-bottomed creek in a wooded area. In the Wichita Mountains, on Fort Sill MR, Comanche County, VWF reported it multiple times on Medicine Creek, one time counting six individuals (29 September 2007, OC 263364). The year previous he photographed a single ♂ along the creek near the North Boundary Road (2 August 2006, OC 7593). He described the location as a relatively swift creek with banks that were "thickly lined with grasses and partially shaded" and said that the ♂ was "found well within the shaded vegetation a few centimeters off the water." The last time the species was reported at Fort Sill MR, or for Comanche County, was on 20 October 2007, again on Medicine Creek, but this time near the Natural Resources Office (1♂, OC 263344). VWF speculated that the species was pushed off the creek because of the severe flooding that Fort Sill MR experienced in 2009, when many odonate habitats around the military base were obliterated or drastically changed due to the massive movement of rock and mud that scoured areas of vegetation and changed stream courses.

The early flight date for the species in Oklahoma is 23 June and the late date is 23 October (early, in 1963, 1♂, University of Oklahoma Biological Station, Marshall County, D Riggs and "MMC," "in C Harris collection," data taken from GH Bick's species notes; late, in 2016, 1♂, Sultan Park, N of Walters, Cotton County, MAP). In the region, Oklahoma's seasonality is most similar to that of Kansas and Missouri, is shorter than that in Louisiana, and much shorter than the essentially year-round presence of the species in Texas (AR: 26 June–14 October, OC 465129, iNat; KS: 2 adults and 1 nymph reported without details, SEMC, Beckemeyer and Huggins 1998 considered it rare in eastern Kansas; LA: 10 May–15 November, OC 462637, Mauffray 2014; MO: 22 July–7 October, Trial 2005; TX: 22 February–15 January, Abbott and Lasley 2019). Ovipositing has not been documented in the state, nor have tenerals. We know of only one Oklahoma larval record (15 November 1975, Bristow, Little Deep Fork, Creek County, WD Shepard, OMNH 1716).

NOTES

1. The specimen label indicated only "Broken Bow," but Ortenburger (1926b) said that the site was 2 miles N of Broken Bow, Yanubbe Creek.
2. Note that, although it has since been recorded in the county, Bird (1932) mistakenly attributed *Hetaerina americana* to Custer County, an error carried forward by Bick and Bick (1957) and Donnelly (2004c). This error was the result of EB Williamson, the collector of 7♂ and 1♀ on 19 October 1929 (UMMZ), writing the wrong county on his label; he was actually in Caddo County.

## CHAPTER 11

LESTIDAE: SPREADWINGS

Lestidae is the family of damselflies known as the spreadwings. They received this name because of their rather un-damsel-like behavior of holding their wings spread outward when they perch, behavior that is more dragonfly-like than the usual closing of the wings above the body of most damselflies.

In Oklahoma, there are 10 species of Lestidae belonging to 2 genera; 9 of Oklahoma's species belong to *Lestes*, a genus for which there are 84 species worldwide and 17 in North America (Paulson 2011). The tenth Oklahoma lestid is an *Archilestes*, a genus only having two species that occur in the United States.

Spreadwings generally are not terribly easy to detect. They tend to be small and slender and relatively inactive, allowing them to blend in well with the sedges and grasses among which they often perch. Even so, we admit we are puzzled by apparent detection differences among various eras of Oklahoma odonatology. For instance, early era researchers seemed to have little difficulty collecting *Archilestes grandis*, *Lestes alacer*, and *L. australis*, but they were not able to produce many records of *L. inaequalis*, *L. rectangularis*, *L. unguiculatus*, and *L. vigilax*. It is true that *L. australis* is the only species in the list that can be considered common across Oklahoma—it is the "default" spreadwing in the state—but the others can be locally common, and without question would have inhabited many of the areas surveyed by some of the earlier collectors. We have reservations to declare that any of Oklahoma's lestids have expanded their ranges into the state, but the pattern of early records is odd.

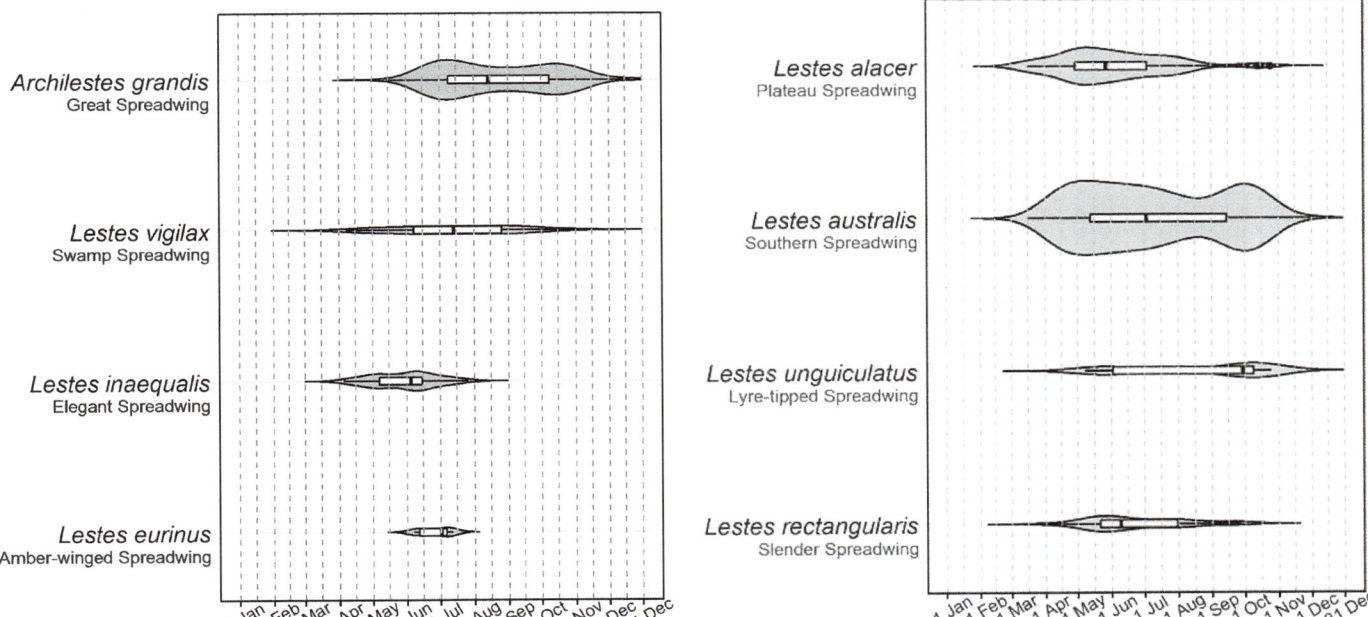

Figure 11.1 **Seasonality comparison of Oklahoma's species of Lestidae.**

ARCHILESTES GRANDIS (RAMBUR, 1842) – GREAT SPREADWING

The Great Spreadwing is uncommon in Oklahoma, although at times it can be locally common: it is usually recorded 1–3 individuals at a time, but sometimes it is found up to 20. Bick and Bick (1970) indicated that their average encounter along Cowan Creek in Marshall County, Oklahoma, was 12 adults per day.

Presently the species is widespread, which was also the case historically. It has been recorded in 46 counties, but we see little reason why it would not occur throughout Oklahoma; its apparent absence from the other 31 counties is likely happenstance, not a true reflection of reality, even if habitat needs are specific (see below). For example, the species has not been reported from the High Plains ecoregion of the Oklahoma panhandle, but it is found elsewhere in that ecoregion; for example, in Colorado and Texas. We surmise that the species is in that part of the Oklahoma panhandle, but we have yet to detect it because of the limited amount of suitable habitat we have access to within that ecoregion. The one ecoregion of exception that may not harbor the species is that of the Arkansas Valley. To our knowledge, *Archilestes grandis* has not been reported from anywhere within its boundaries. The omission is inexplicable because the species is found in neighboring ecoregions and there is no obvious reason why *A. grandis* would not inhabit the Arkansas Valley given the species' broad elevational distribution and its remarkable ability to adapt to a wide breadth of lotic habitats.

Habitat descriptions for localities in Oklahoma where *Archilestes grandis* has been found vary greatly. Descriptions include a small pool (about 1 m deep, 4 m in diameter) at the headwaters of a small creek; small (30 cm–3 m wide), clear-water, still or slow-flowing creeks with mixed or hardwood woodlands to the edge and rocky or silty substrate; marshes; sheltered roadside ditches with muddy or concrete bottoms; springs; and ephemeral or permanent ponds, sometimes with water shield (*Brasenia schreberi*) and within mixed forest. Despite the wide range of potential habitats, we have observed the species most often (i.e., >90% of our records) along narrow (sometimes only a 1 m wide), shallow, clear creeks and rills with low to moderate flow rates. For example, our most reliable locality in the state are the shallow drainage ditches around the Law School at the University of Oklahoma, Norman campus. At times these ditches are barren, at others they support many species of odonates (*n* = 34, including *Hetaerina americana*, the American Rubyspot, *Ischnura ramburii*, Rambur's Forktail, *Telebasis salva*, the Desert Firetail, *Libellula croceipennis*, the Neon Skimmer, and *Erythrodiplax umbrata*, the Band-winged Dragonlet) and non-odonates.

Not only does this species have a wide habitat breadth, it purportedly has a high tolerance to poor water quality thought to eliminate other species. In addition to habitats similar to those mentioned above, Moskowitz and Bell (1998:49), describing conditions elsewhere, included "pristine" Neotropical streams, "permanent ponds or impoundments," "temporal bodies of water in open areas," canals, and "drainage and irrigation ditches." They also described marginal or degraded habitats such as "brooks adjacent to animal pens, open bodies of water with urban and industrial contamination … ponds subject to fish kills from agricultural runoff … stream pools with high temperatures …and

Figure 11.2 *Archilestes grandis*, the Great Spreadwing, A) ♂, Black Mesa SP, Cimarron County, 10 September 2015 (©Bill Carrell, OC 436528). B) ♀, 5 km NE of Sapulpa, Polecat Creek, Creek County, 7 June 2012 (©James W Arterburn, OC 436058).

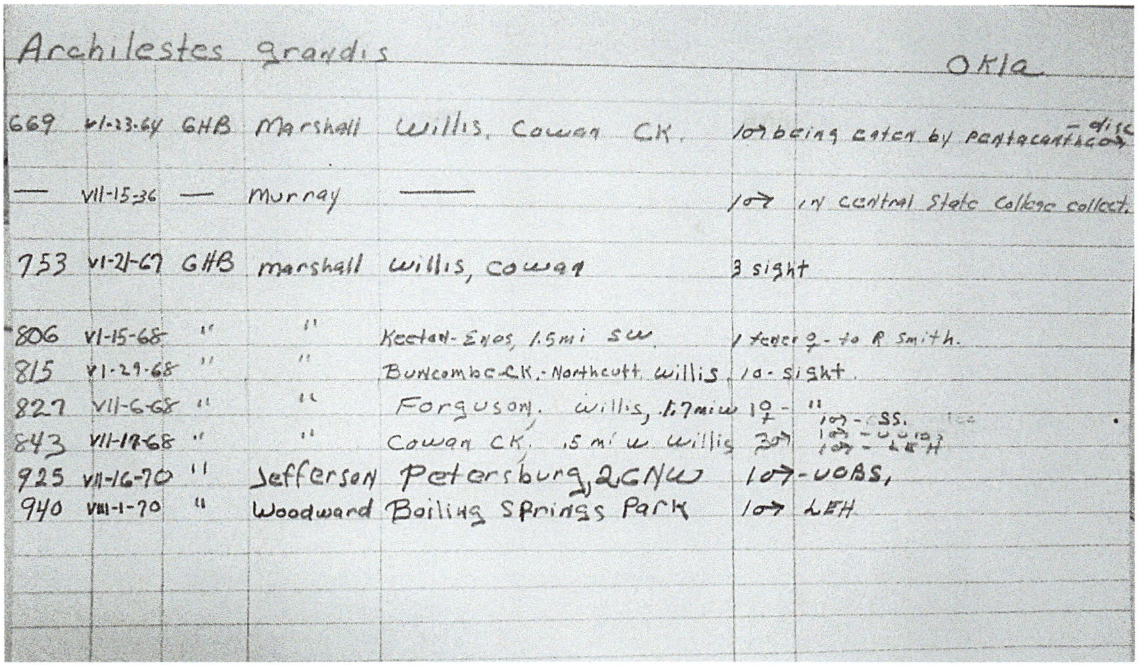

Archilestes grandis — Okla

669	VI-23-64	GHB	Marshall	Willis, Cowan CK.	1♂ being eaten by Pantacanthacon- disc
—	VII-15-36	—	Murray	—	1♂ in Central State College collect.
753	VI-21-67	GHB	marshall	willis, cowan	3 sight
806	VI-15-68	"	"	Keeton-Enos, 1.5mi SW.	1 teneral ♀ - to R Smith.
815	VI-29-68	"	"	Buncombe CK.-Northcutt. willis	1♂ - sight.
827	VII-6-68	"	"	Forguson. willis, 1.7mi w	1♀ - " 1♂ - SS.
843	VII-17-68	"	"	Cowan CK. .5 mi w. willis	3♂ 1♂ - [illegible] 1♂ - LEH
925	VII-16-70	"	Jefferson	Petersburg, 2 C NW	1♂ - UOBS,
940	VIII-1-70	"	Woodward	Boiling Springs Park	1♂ LEH

Figure 11.3 George H Bick's note card for *Archilestes grandis*, showing some of his records dating from 1964 to 1970. Also included is one record from 1936 that Bick examined at the University of Central Oklahoma (formerly Central State College).

dirty creeks around cities and towns." Williamson (1931:63) mentioned that *A. grandis* inhabited drainage ditches in which sewage and laboratory chemicals flowed, chemicals thought to have killed crawfish and goldfish, although he noted that "there were a few mayfly and stonefly larvae" present. Williamson said the spot he found *A. grandis* was a creek <1 m wide with a "very small flow of water" and "a few small willows, mint, asters, and such vegetation on the creek, but it is practically entirely open and virtually landscaped—just a little artificial-looking dab of scenery."[1]

Consequently, it is no exaggeration to say that *A. grandis* is a highly tolerant, habitat generalist. Such tolerance likely explains its extensive range (Canada south to Venezuela) and its characterization as an extremely successful colonist. Several authors have described the species' range north of Mexico as being confined to the southwestern United States until the 1920s (e.g., Abbott 2011, Westfall and May 1996,[2] 2006; Moskowitz and Bell 1998; the former two indicating the "1920s," while the latter said "early in the century"). This idea, although sometimes attributed to Gloyd (1980), appears to have originated with Donnelly (1961:4–5), who said, "…two species have probably migrated into our area from the southwestern United States in the last few decades (*Archilestes grandis* and *Enallagma basidens*) … [*A. grandis*] … spread northeastward into the mid-western States during the nineteen-thirties, but did not cross the Appalachians until the forties [10 October 1949]." By the time *A. grandis* was reported in eastern Pennsylvania (Ferris 1951) and Washington D. C. (Donnelly 1961), it was known from 16 counties across Oklahoma and was found regularly in good numbers early on—it was first recorded in Oklahoma in August 1929 (1♂, Deans Lake, McClain County, RD Bird, OMNH 1299) and was reported on at least 33 field days in 10 counties by 1936[3] (Figure 11.5).

Not only is there much Oklahoma-specific data indicating that the species was well established in the state early on, but we are also fortunate to have extremely detailed accounts of the species' breeding behavior and life history specific to the state. We have personally encountered many tandem and mating pairs in the field in Oklahoma, but we have not observed ovipositing. Bick and Bick (1970), however, studied in detail the mating behavior and ecology of *Archilestes grandis* in Oklahoma at Cowan Creek in Marshall County. During this study, they learned that adult longevity was 25 to 51 days; females were rarely at the water unless they were ready for mating; pairs almost always oviposited in tandem, perching vertically but sometimes hanging upside-down; ovipositioning in Oklahoma, unlike other areas (e.g., Williamson 1931), was common by mid-June; and it could take two hours or more for a pair to explore the habitat to find one or more oviposition sites until the deposition of eggs was complete. Pairs used a variety of plants for perching and ovipositing, but *Platanus occidentalis*, the American sycamore, was the most frequently used (55%). Oviposition tended to be in petioles that were located well above (2.5–13 m) water so that hatching larvae would fall directly into or at least near the water. Use of petioles by *A. grandis* in Oklahoma is in contrast to *A. grandis* in Ohio studied by Williamson, as they tended to use the branches, which is also the case with *A. californicus* (the California Spreadwing,

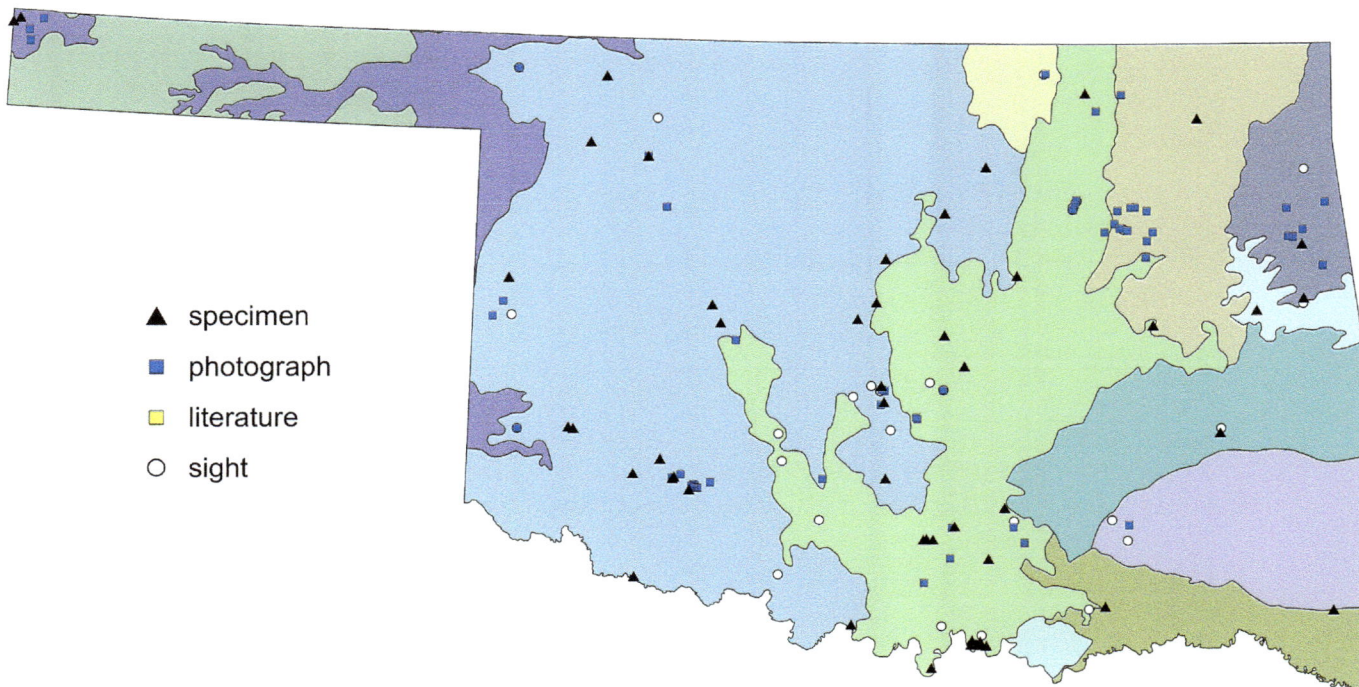

Figure 11.4 Documentation level of supporting records for *Archilestes grandis*, the Great Spreadwing, in Oklahoma. Levels include specimen, archived photograph, literature reference, or sight record. Although sight records lack documentation, only those we feel are credible are plotted.

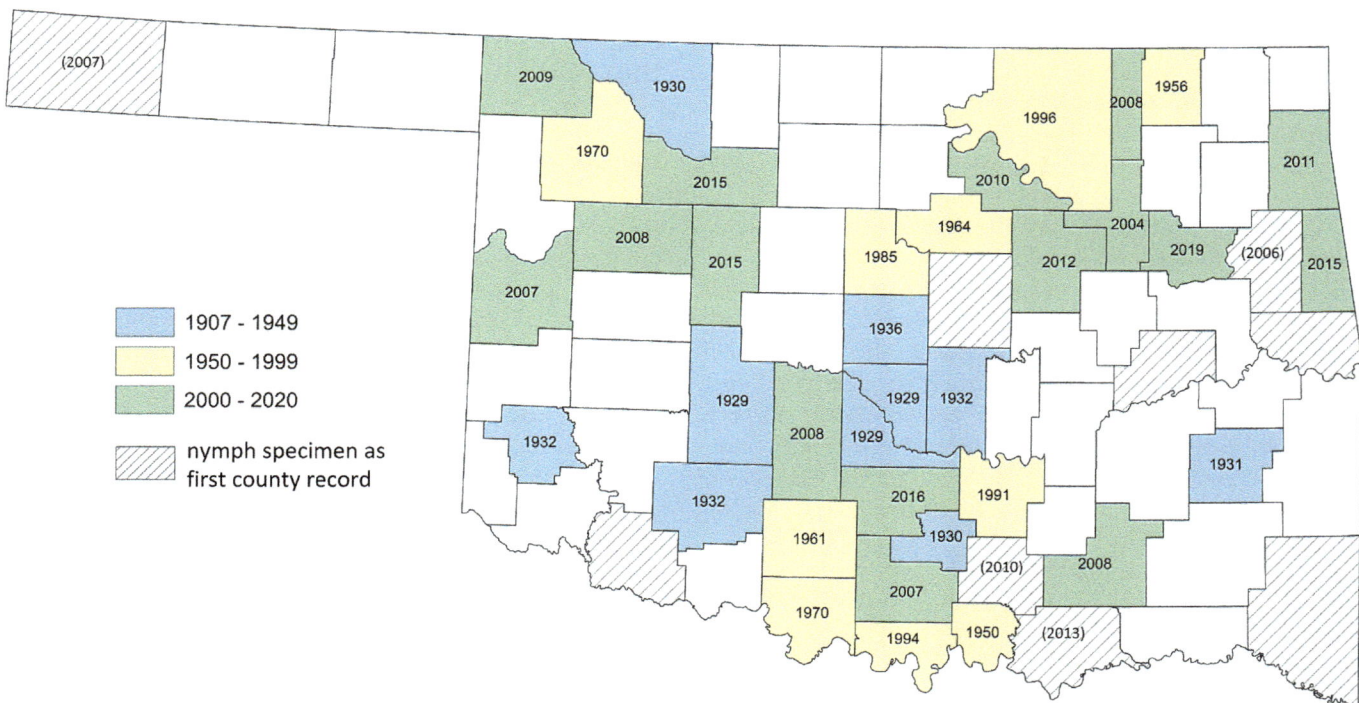

Figure 11.5 Counties in which *Archilestes grandis*, the Great Spreadwing, are known to occur in Oklahoma. Year within the county is that in which the species was first reported. Hatched counties are those with a nymph specimen as the first documentation of the species in the county. When a parenthetical year is also presented, it indicates the year of the first adult record; no year indicates the county only has larval records as support. We did not examine all of the nymph specimens of this species, but we feel the identifications are likely good given the distinctiveness of the species at that life stage.

the only other *Archilestes* in the United States). The Bicks suggested that use of petioles rather than branches indicated that *A. grandis* eggs did not overwinter in Oklahoma. Unlike *A. californicus*, Oklahoma *A. grandis* also used herbaceous non-aquatics (*Conyza canadensis*, Canadian horseweed; *Eupatorium coelestinum*, blue mistflower; *Monarda punctata*, spotted beebalm; *Rumex crispus*, curley dock; and *Verbena urticaefolia*,[4] white verbena) for oviposition. In stark contrast to many other zygopterans that frequently use *Nasturtium*, water cress, a common aquatic plant throughout the Bicks' study area, *A. grandis* would not even perch on *Nasturtium*, let alone oviposit in it. Oviposition occurred throughout the day (as early as 09:51 to as late as 15:45). The number of eggs deposited varied tremendously, with a maximum of 149 found in two *Platanus* petioles. One set of eggs that the Bicks observed closely "under almost natural conditions" hatched in 16 days, which led them to conclude that Oklahoma *A. grandis* overwinters as late instars.

As adults, *Archilestes grandis* has been recorded in Oklahoma as early as 14 May (in 1932, 1♂, Murray County, RD Bird, OMNH 2998) and as late as 21 December (in 1931, 3♂, 2♀, including 1 pair, Cleveland County, O Sandoz, OMNH 2909 and 2999). Note that 1931 was a rather warm year for December temperatures (Oklahoma Climatological Survey), which may account for this rather extreme late date for central Oklahoma, as well as for the region as a whole (OK's late date is one month beyond the next latest date of 24 November, from 1931, and 18 November, from 2007; CO: 5 May–16 November, CSU/YUHO 475, OC 457774; KS: 30 June–14 October, OC 465954, SEMC; MO: 17 May–8 November, iNat 2353493, Trial 2005; NM: 14 May[5]–19 November, USNM 362968, OC 334628; TX: 3 March–28 December, Abbott and Lasley 2019). Nonetheless, with continued warmer than "normal" winters in recent years, we may start to expect the species into December more often.

LESTES SIGMA CALVERT, 1901 – CHALKY SPREADWING

The Chalky Spreadwing is known in Oklahoma from one vagrant record. One ♂ was collected on 22 July 1968 by GH Bick at Little City ("Weno Pond"), Marshall County and was determined by LK Gloyd as *Lestes sigma* (Bick 1978). This record is the northernmost for the species and the only record north of the Red River (Abbott 2005). The Chalky Spreadwing appears to be expanding its range northward in Texas (Abbott 2011), so additional records for Oklahoma are expected, especially along the Red River (Patten and Smith-Patten 2013b). We hope for an additional record of the species not only as an indication of the species' northward expansion but also because Bick's specimen has yet to be relocated. The species is known to fly nearly year-round in Texas (Abbott and Lasley 2019).

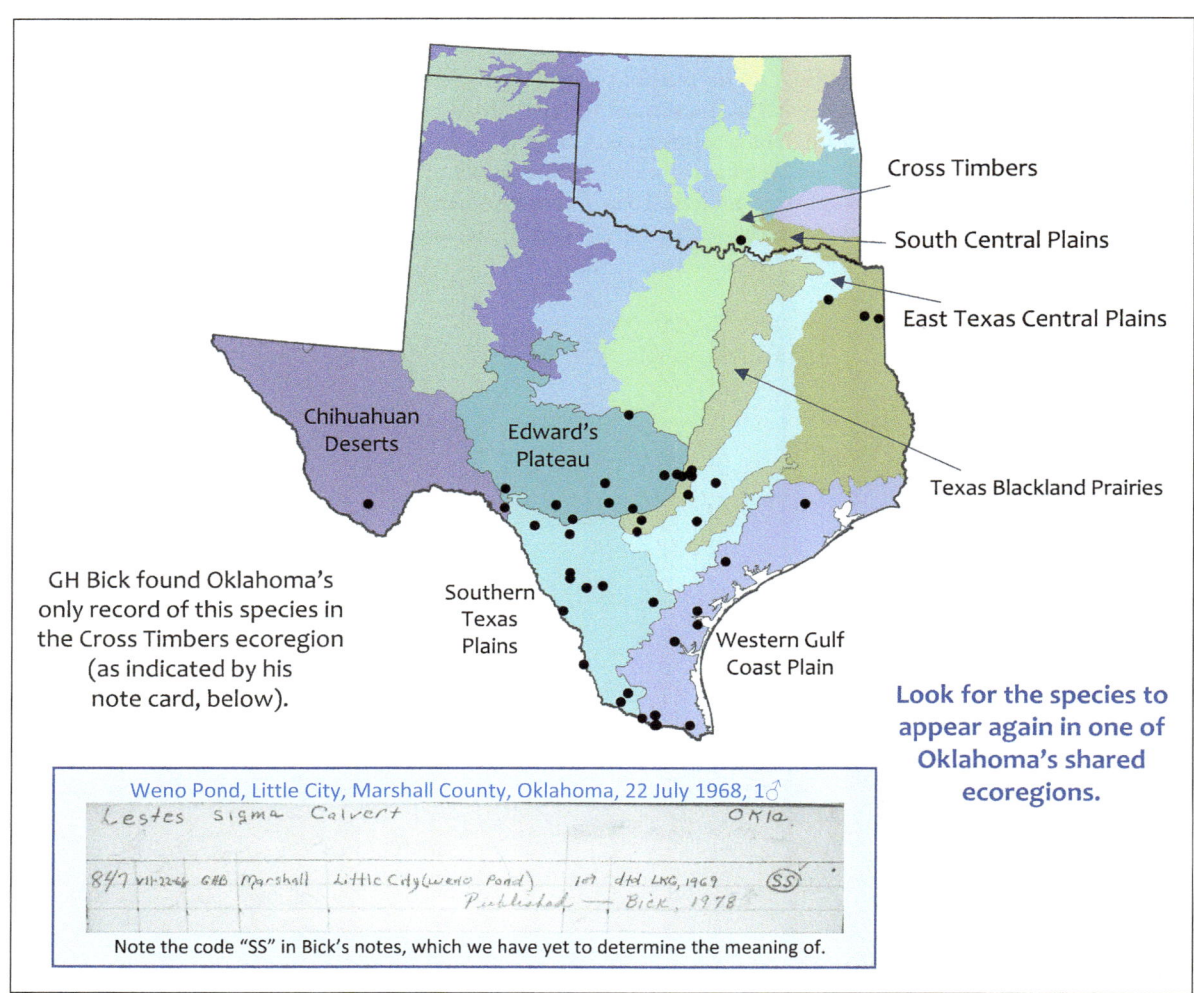

GH Bick found Oklahoma's only record of this species in the Cross Timbers ecoregion (as indicated by his note card, below).

Look for the species to appear again in one of Oklahoma's shared ecoregions.

Weno Pond, Little City, Marshall County, Oklahoma, 22 July 1968, 1♂

Note the code "SS" in Bick's notes, which we have yet to determine the meaning of.

Figure 11.6 The geographic range of *Lestes sigma*, the Chalky Spreadwing, in Texas and Oklahoma, includes eight ecoregions, three of which occur in Oklahoma—East Texas Central Plains, the South Central Plains, and the Cross Timbers.

LESTES ALACER HAGEN, 1861 – PLATEAU SPREADWING

The Plateau Spreadwing is a lentic species found throughout most of Oklahoma, being absent (or apparently so) only from the northeast, although we suspect that it occurs, too, albeit in low numbers (the recent first state record for Arkansas [24 May 2017, Ozark NF, Benton County, A. Hasik, OC 463128, Hasik and Bried 2017] argues for its presence just across the border) (Figure 11.7). Observers have mentioned finding the species in flooded fields or at ponds, naturally spring-fed and artificial, with and without fish (fish hatchery ponds and naturally populated fish ponds; but we suspect that it most often occurs at fishless ponds). Vegetation at the ponds have been described as "sedges," "tules," and "lily pads," all of which we imagine should be taken as general descriptors rather than at face value.

We mischaracterized this species (Patten and Smith-Patten 2013b) as one we thought was in decline. Prior to our paper the number of records and counties for the species, as well as its relative abundance, had all been higher in the 1930s, after which it dropped drastically during the Bick/Hornuff era (Figure 11.9). Between 2003 and 2013, the numbers did not rebound, but as if to mock us right after the publication of our paper, the number of records, counties, and individuals recorded skyrocketed. In 2014–2019, we added >45 records and 11 counties and had our highest ever count, of 45 individuals (30♂, 15♀, including 10 pair, Sandy Sanders WMA, Greer County, 10 May 2015). Generally, there are but a few individuals at any given encounter; our high count is one of a small handful to reach double digits. We now surmise the species may have been rebounding from drought or other conditions that had depressed its population in Oklahoma. Its current status indicates that it could safely be placed into an SRank

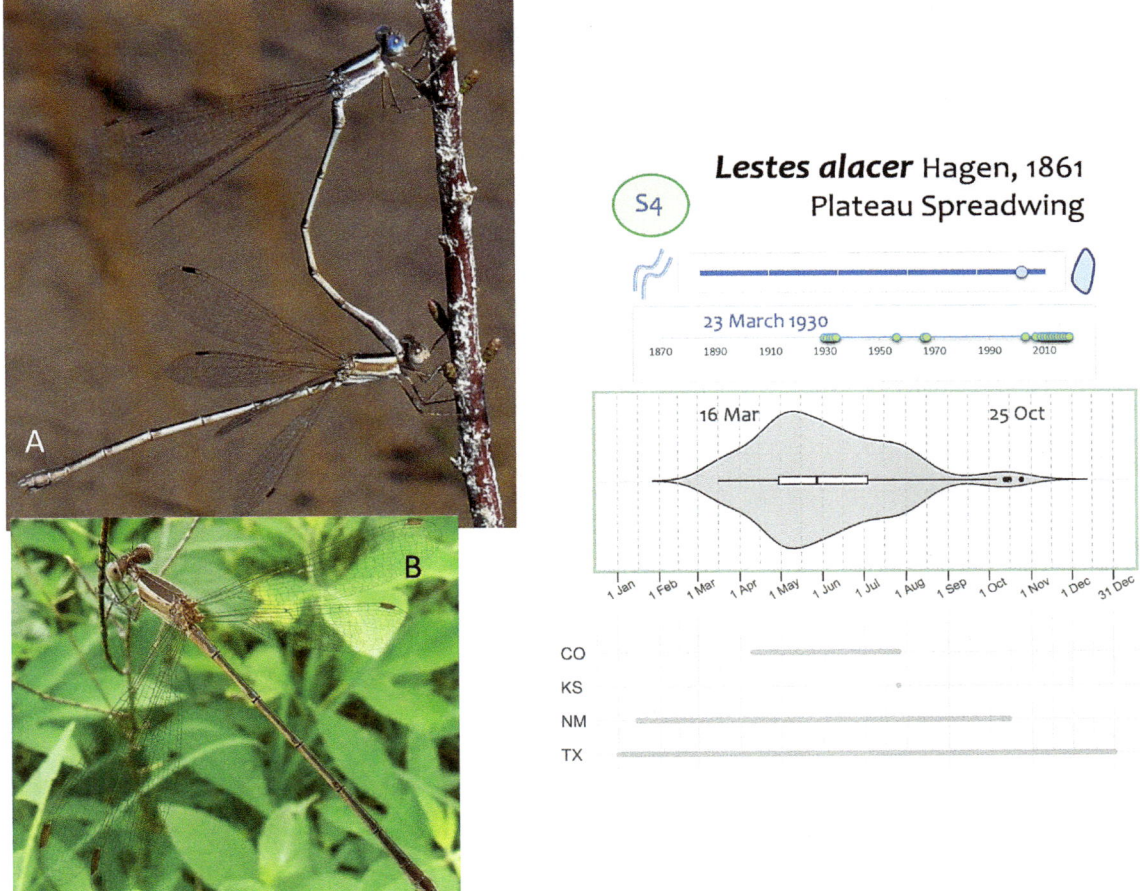

Figure 11.7 *Lestes alacer*, the Plateau Spreadwing, A) tandem pair, Mountain Park WMA, Kiowa County, 16 March 2009 (OC 312267). B) ♀, Hackberry Flat WMA, Tillman County, 16 June 2007 (OC 263638). Both photos by ©Victor W Fazio, III. Note the sharp top edge of the antehumerals (blue on the male, pale on the females).

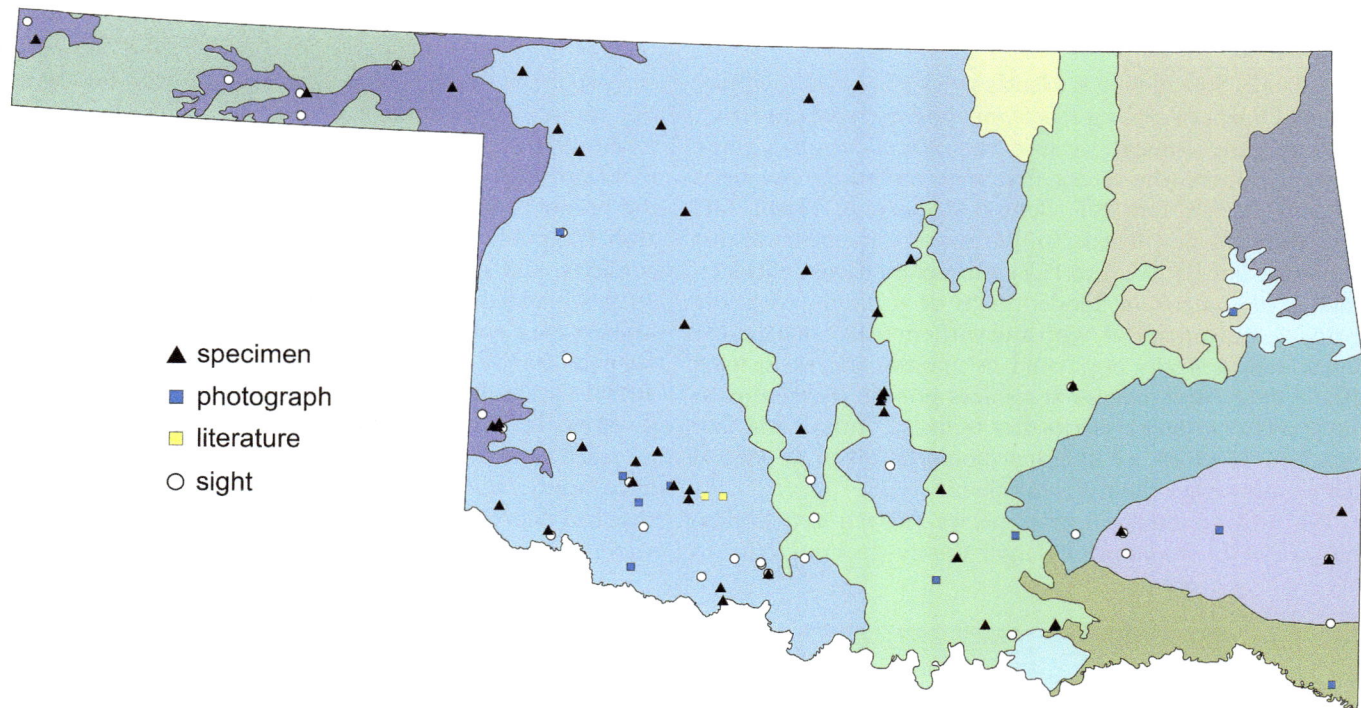

Figure 11.8 Documentation level of supporting records for *Lestes alacer*, the Plateau Spreadwing, in Oklahoma. Levels include specimen, archived photograph, literature reference, or sight record. Although sight records lack documentation, only those we feel are credible are plotted.

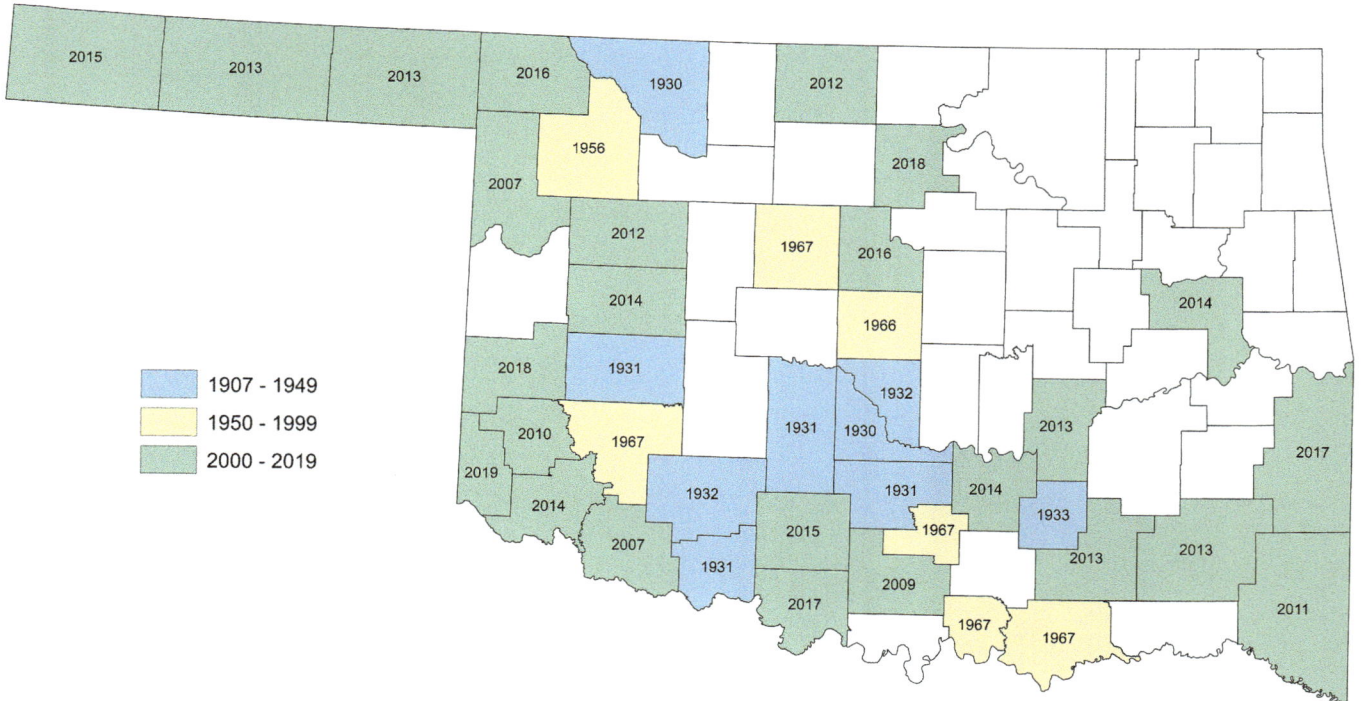

Figure 11.9 Counties in which *Lestes alacer*, the Plateau Spreadwing, are known to occur in Oklahoma. Year within the county is that in which the species was first reported.

of S4, although we suggest continuing to keep an eye on this trickster.

The Plateau Spreadwing flies in the state from mid-March to late October (16 March 2009, Great Plains SP, Kiowa County, estimated 40 individuals, VWF, OC 312267, Figure 11.7; 25 October 2015, Black Mesa SP, Cimarron County, 1♀, BC, OC 438264), but it tails off after early August. We know of no nymphs that have been collected in the state, but the handful of teneral records for Oklahoma date from 14 May to 27 June. We have two records of lone females ovipositing but no details of how and where they oviposited. Oklahoma's seasonality does not fully correspond with elsewhere in the region, as dates from other states suggest the species probably flies in Oklahoma during months from which we have yet to receive reports.

Colorado has limited adult records and as such, a limited date range (10 April–26 July, CSU); whether those dates are real or apparent is debatable. Likewise, Kansas has but two records (26 July 2019, OC). Nonetheless, the flight seasons for New Mexico and Texas are much longer (NM: January–October, Paulson 2009; TX: all year, Abbott and Lasley 2019).

Some historical records of note for this species include specimens originally identified as *Lestes congener* and *L. forficula* that were re-identified as *L. alacer* (see Appendix A). One mixed-species pair is known for this species from Oklahoma: a ♂ *L. australis* and a ♀ *L. alacer* collected in Jefferson County (Wichita Ridge Park, Lake Waurika, 5 May 2018, MAP, SP 2580).

LESTES AUSTRALIS WALKER, 1952 – SOUTHERN SPREADWING

Lestes australis is so common and widespread in the state that we consider it the "default" spreadwing in Oklahoma. By 1936, the species was known already from 35 counties and it is currently recorded in each of the state's 77 counties. High counts for the species include 70 adults (50♂, 20♀, including 15 pairs) on 10 May and 100 adults (80♂, 20♀, including 15 pairs) on 10 October; both of these records are from Great Plains SP, Kiowa County (MAP). It can be found in a wide variety of habitats, but it is most often encountered in wetlands and at ponds.

Lestes australis was first recorded in the state in early May of 1927 (2♂, Cleveland County, OMNH 68, original id as *L. forcipatus*) and has been recorded regularly since. It is known to fly in Oklahoma from mid-March until early December (16 March 2016, 1♂, DA; 2 December 2010, 1♂, BAH, OC 325124; both records from Red Slough, McCurtain

County). Because of the confusion of this species with *L. disjunctus* (and *L. forcipatus*, see that account) we chose not to provide flight dates for Colorado, Kansas, Missouri, and New Mexico, where the two species are thought to occur. Records for Arkansas may be a little more reliably called *L. australis*, as can those from Texas (AR: 20 April–27 October, OC 479314, Harp and Harp 2003, JCAC 18712; TX: 21 January–19 December, Abbott and Lasley 2019).

In Oklahoma the species is known to oviposit in submerged stems of *Eleocharis macrostachya* (pale spikerush; Bick and Bick 1961). The Bicks indicated that the species tends to oviposit in tandem and then fly away together, but females will oviposit alone (we have observed this pattern, too). They witnessed eight ovipositions in which the female was entirely submerged, the male remaining at least partially above the surface. These ovipositions "were usually

Figure 11.10 *Lestes australis*, **the Southern Spreadwing, A)** ♂, **Croton Creek in Black Kettle NG, Roger Mills County, 4 October 2014 (©Bill Carrell, OC 427359). B) immature** ♂, **Salt Plains NWR, Alfalfa County, 23 June 2012 (©Jason R Heinen, OC 376423). C) adult** ♀, **Lake Hall, Harmon County, 10 August 2014 (©Brenda D Smith). In all three individuals, note the jagged edge to the antehumeral stripes, which are expanded toward the middle and jut to a point apically. Although it is occasionally outnumbered locally, this species is easily the commonest and most widespread spreadwing in the Southern Great Plains.**

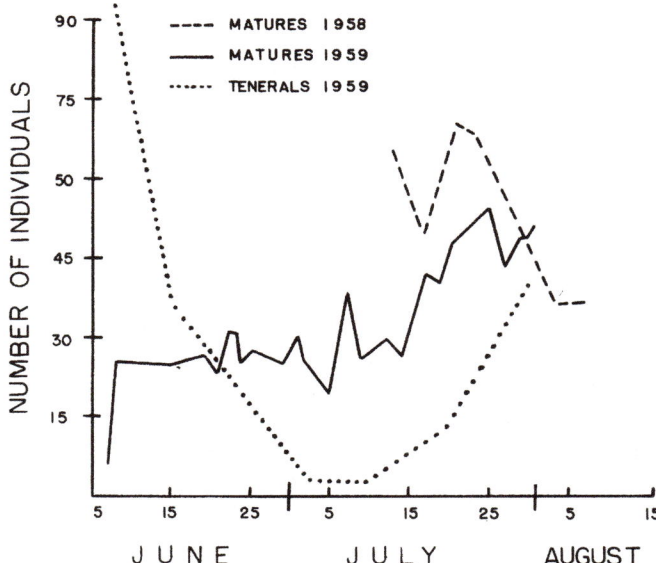

Figure 11.11 Phenological observations of *Lestes australis* by Bick and Bick (1961:114) in the summers of 1958 and 1959 at the University of Oklahoma Biological Station, near Willis, Marshall County, Oklahoma. Reprinted with permission from Southwestern Association of Naturalists.

associated with flood conditions which reduced the vertical extent of exposed *Eleocharis* stems" (Bick and Bick 1961:132). They reported mean uninterrupted time in tandem as 60.2 minutes, with a range of 12–228 minutes, and copulation lasting between 6 and 19 minutes. Individuals were not

consistent with their time in tandem, the length of which varied day-to-day. Although they estimated adult life span as roughly 3 weeks, mean reproductive time in the population they studied was 10 days (maximum of 50). More males than females were found at water, with overall totals of all adults present as well as mating highest late in the day, focused around 17:00. Males tended to stay in the same general area at the pond studied, but they were not territorial in the sense that multiple males would perch near one another (in an area <1/3 m²), although always on separate *Eleocharis* stems, and the males would rarely interact. The Bicks termed this behavior as "passive occupancy." We have observed tenerals in the state between 2 April and 15 July, but the Bicks reported them to 30 July. Their observations at a single pond in Marshall County showed that the species tends to emerge before about 08:00 in the morning, peak emergence is in June, and that teneral numbers fluctuate through the season (Figure 11.11). In their study, numbers went from >90 in early June to practically none a month later and then a rise in late July when there were about 40 tenerals present. They indicated that tenerals stayed near the pond, leaving as immatures, then returning some 13 days later for reproduction. The Bicks presented teneral data only from 1959, but adult data from 1958 and 1959. If tenerals are anything like the adult numbers they provided, then fluctuations between years ought to be expected as well (Figure 11.11).

The only known mixed-species pair including this species in Oklahoma is a ♂ *Lestes australis* and a ♀ *L. alacer* collected in Jefferson County (Wichita Ridge Park,

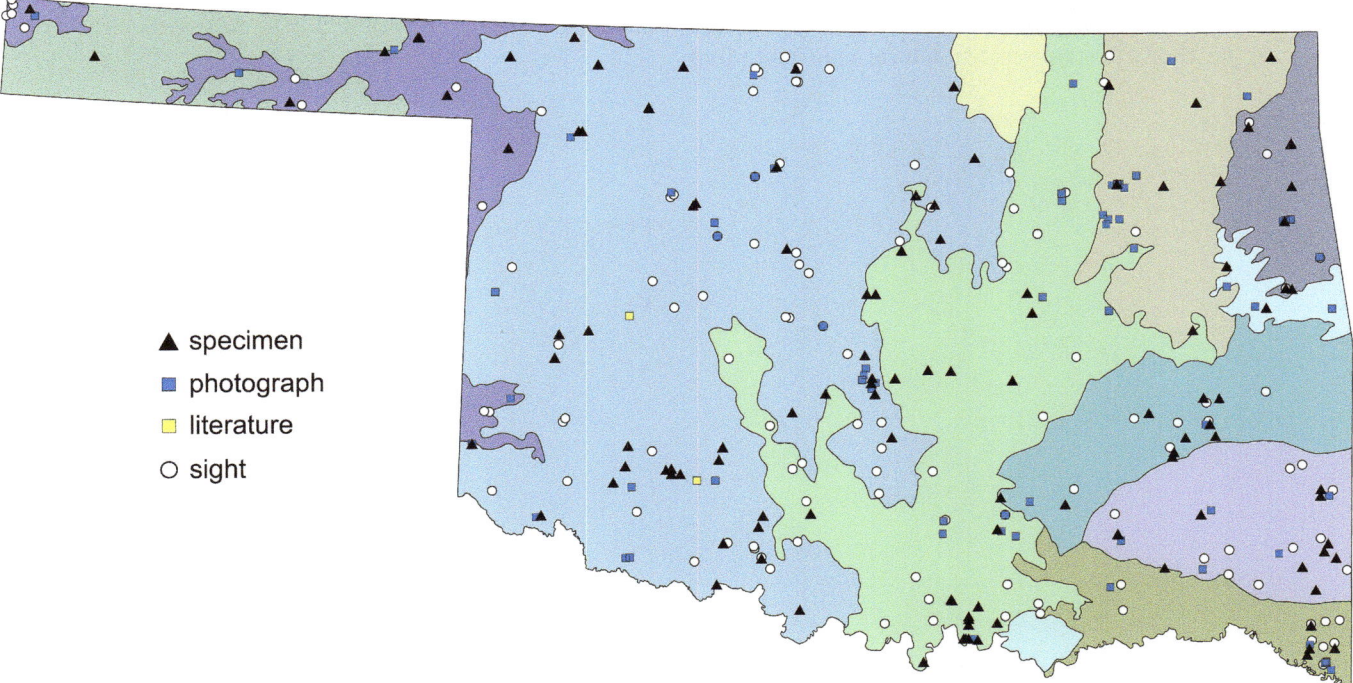

Figure 11.12 Documentation level of supporting records for *Lestes australis*, the Southern Spreadwing, in Oklahoma. Levels include specimen, archived photograph, literature reference, or sight record. Although sight records lack documentation, only those we feel are credible are plotted.

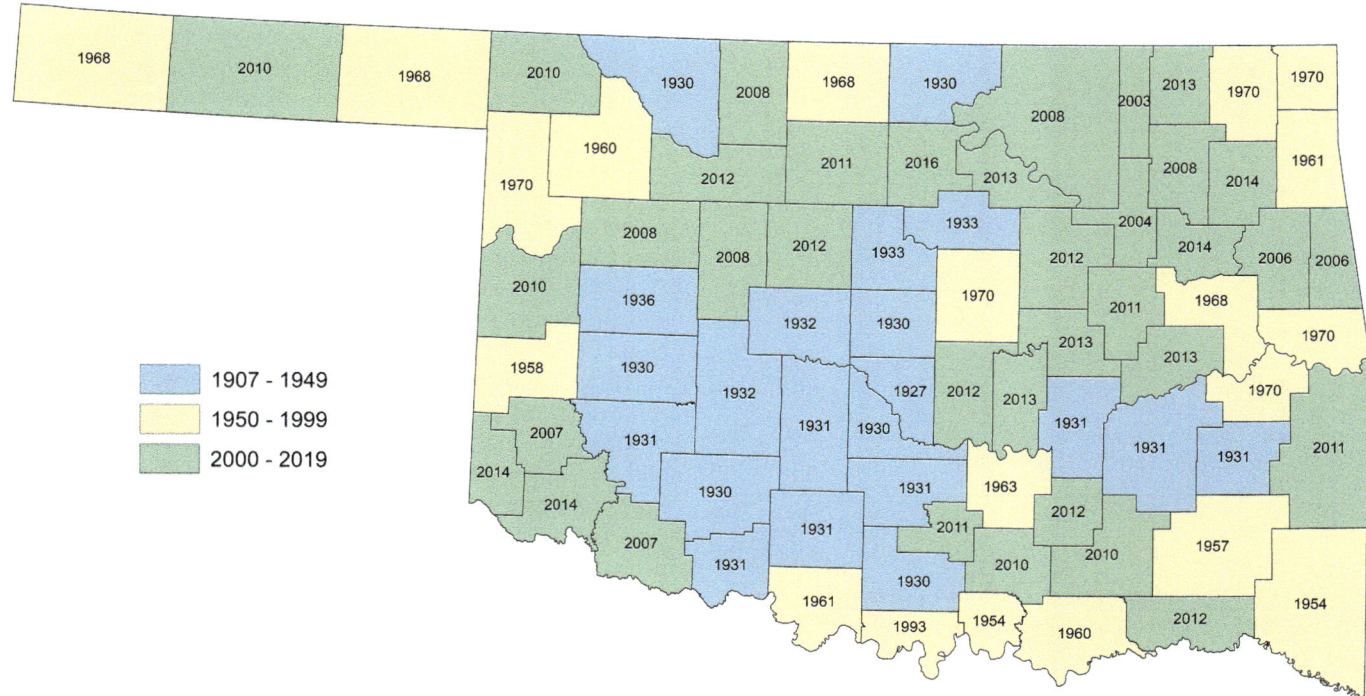

Figure 11.13 Counties in which *Lestes australis*, the Southern Spreadwing, are known to occur in Oklahoma. Year within the county is that in which the species was first reported.

Lake Waurika, 5 May 2018, MAP, SP 2580). Away from Oklahoma, Bick and Bick (1981) reported *L. australis* pairing with other species: a ♂ *L. australis* with a ♀ *Enallagma annexum* (Northern Bluet)[6] and a ♂ *L. inaequalis* (Elegant Spreadwing) with a ♀ *L. australis*. And lastly, a melanistic ♂ showing no thoracic stripe, ventrolateral pale extending anteriorly over the shoulder, and extra black on the abdomen is of note (pond near Big Hudson Creek, 12 km SSE of Smithville, McCurtain County, 21 June 2019, BS-P, SP 2788; ratio of hamule to abdomen is consistent with that presented by Donnelly 2003 for *L. australis*).

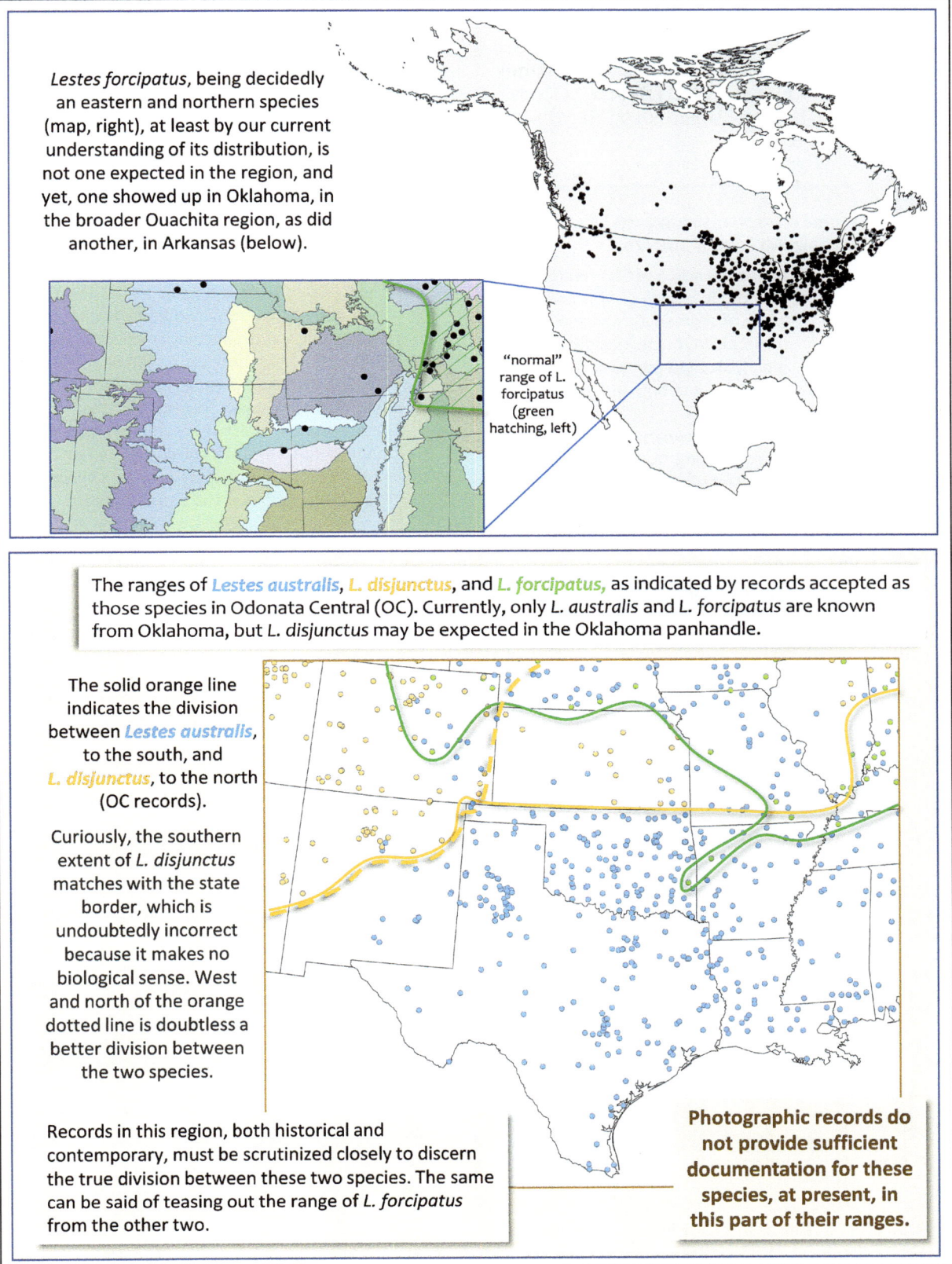

Lestes forcipatus, being decidedly an eastern and northern species (map, right), at least by our current understanding of its distribution, is not one expected in the region, and yet, one showed up in Oklahoma, in the broader Ouachita region, as did another, in Arkansas (below).

"normal" range of L. forcipatus (green hatching, left)

The ranges of Lestes australis, L. disjunctus, and L. forcipatus, as indicated by records accepted as those species in Odonata Central (OC). Currently, only L. australis and L. forcipatus are known from Oklahoma, but L. disjunctus may be expected in the Oklahoma panhandle.

The solid orange line indicates the division between Lestes australis, to the south, and L. disjunctus, to the north (OC records).

Curiously, the southern extent of L. disjunctus matches with the state border, which is undoubtedly incorrect because it makes no biological sense. West and north of the orange dotted line is doubtless a better division between the two species.

Records in this region, both historical and contemporary, must be scrutinized closely to discern the true division between these two species. The same can be said of teasing out the range of L. forcipatus from the other two.

Photographic records do not provide sufficient documentation for these species, at present, in this part of their ranges.

Figure 11.14 A closer look at the distribution of *Lestes forcipatus*, *L. australis*, and *L. disjunctus*, the Sweetflag, Southern, and Northern Spreadwings.

LESTES FORCIPATUS RAMBUR, 1842 – SWEETFLAG SPREADWING

Lestes forcipatus is known in Oklahoma from one record, a ♂ collected at a fen-like beaver dam complex below Crooked Branch Lake in the Ouachita NF, Le Flore County, 9 June 2013 (Patten and Smith-Patten 2013a; SP 670). This first state record proceeded a different first state record of another spreadwing, the Amber-winged Spreadwing (*Lestes eurinus*), by one day. As the spreadwing flies, the Oklahoma record is not far from the Franklin County, Arkansas,

---

Some characters thought useful to distinguish *L. forcipatus* are:

- the apical notch of S10 **"wide,"** (blue line in photo B, below) or 46% or more of width of segment (green line in photo B; comparisons made in posterior view)
- mostly **straight paraprocts** (also with *L. disjunctus*, but both lack curved tip of *L. australis*)
- **distance between distal and basal teeth on cerci being "long"**
- the **distal tooth of cerci blunt**

1.75 mm

A

hamule ≥1.7 mm = *L. forcipatus*

distance between distal and basal teeth = 0.60 mm

blunt tooth*

B

S10 apical notch >46%

Oklahoma's first and only *Lestes forcipatus*, the Sweetflag Spreadwing: SP 670, a ♂, collected below Crooked Branch Lake, Le Flore County, 9 June 2013.
Adapted from Patten & Smith-Patten (2013)

Donnelly (2003b) found the length of the hamule (or anterior laminae) to be a reliable character to distinguish *L. forcipatus* from the others. His hamule illustrations made it appear as if a length over about 1.7 mm would be indicative of *L. forcipatus*, with smaller hamules being *L. disjunctus* or *L. australis* (Westfall & May 2006 reproduced that figure). However, in his discussion and summary table, he urged that a direct measurement of the hamule not be taken solely, but rather in consideration of abdomen length; hence, his suggested diagnostic proportions are:

L. australis = 0.052 +/- 0.0025
L. disjunctus = 0.055 +/- 0.003
L. forcipatus = 0.065 +/- 0.004

Westfall & May (2006) followed, in part, these measurements (except note that couplet 13, p. 85 has two typos; *disjunctus* should be <0.06, *forcipatus* >0.06).

Note, too, abdominal lengths presented by these sources and resultant calculations would allow for hamule length to overlap between the species. For example, *L. australis* could, theoretically, have a hamule that measures 1.39–1.79, whereas *L. forcipatus* could range 1.76–2.15.

Paulson (2009, 2011) chose to rely on the direct measurement of the hamule, saying that ≥1.7 mm is indicative of *L. forcipatus*.

Females of *L. australis* and *L. disjunctus* are not readily distinguished, although the pale edges to the pterostigmas on *L. australis* (♂♀; also on the similar *L. unguiculatus*, the Lyre-tipped Spreadwing), but not on *L. disjunctus*, proves useful.

cercus

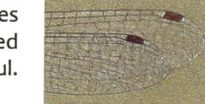

OC 436527, Cimarron Co., OK, photo by © Bill Carrell

Female *L. forcipatus* is unique among North American *Lestes* in having a bulbous abdomen tip with an extra long ovipositor that has the valve extending to or beyond the tip of the cercus.

Wharton State Forest, Atlantic County, New Jersey, 9 August 2017, photo by ©Mike Hannisian, OC 469854   valve

*some refer to this as the apical tooth, others as the distal tooth

Figure 11.15 The ease with which *Lestes forcipatus*, the Sweetflag Spreadwing, is confused with *L. australis* and *L. disjunctus*, the Southern and Northern Spreadwings, necessitates a look at features useful to distinguish these three species. However, much debate surrounds species determination, as such, sources sometimes conflict (e.g., Donnelly 2003b; Paulson 2009, 2011; Westfall and May 2006). Further complications arise because characters differ by millimeters (or *mere fractions of*) or have not been quantified at all. Moreover, there is much geographical and individual variation to account for.

record, so perhaps we should have not been too surprised to find the species in the state, even if the Arkansas record is considered a vagrant and the species' "normal" range may come no closer to Oklahoma than western Tennessee (Figure 11.14), but note that the "range" indicates accepted Odonata Central records, not the "normal" ranges of the three species discussed). If Oklahoma receives another record for the species it would be expected along its border with Missouri and Arkansas. Any further records demand careful documentation given the difficulty of distinguishing this species from *Lestes australis* and *L. disjunctus*, the Southern and Northern Spreadwings, distinctions that can be predicated on mere fractions of a millimeter (Figure 11.15). Other characters can be variable, for example, the amount of pruinosity shown by males can be useful for species determination, but Paulson (2011:69) noted that there is geographical variation in this character. In some populations of *L. australis*, pruinosity does not develop on the thorax and usually not dorsally on S8, characters that

L. disjunctus and *L. forcipatus* show. The caveat: "males in southern populations of Sweetflag may be indistinguishable from Southern. Pruinosity seems to increase with latitude, and northern populations of Southern might look like the northern species!" To complicate matters further, *L. unguiculatus*, the Lyre-tipped Spreadwing, can be confused with the other three species. Pruinosity on S2 is thought to be useful in species determination, with male *L. unguiculatus* and *L. australis* having pruinosity limited to the sides of S2, whereas *L. disjunctus* and *L. forcipatus* are mostly or entirely pruinose there dorsally (Paulson 2011). Males of all four of these species acquire pruinosity on the back of the head with age (of these species, only female *L. unguiculatus* are pale [not pruinose] on the back of the head). However, it may be that pruinosity on S2 and the head, too, vary geographically.

There are nearly 300 early specimens in the OMNH collection that were originally attributed to *Lestes forcipatus*, but actually are *L. australis* or *L. alacer* (see *L. forcipatus* in Appendix A for details).

LESTES UNGUICULATUS HAGEN, 1861 – LYRE-TIPPED SPREADWING

A species of slow-moving lotic habitats and occasionally ponds, the Lyre-tipped Spreadwing may have expanded its range in the state as more backwaters and inflows were created at the margins of reservoirs. Despite surveys across the state with his team, RD Bird encountered the species only twice: 2♂ and 1♀ in McClain County, probably at Dean's Slough, on 23 May 1930 (OMNH 101 and 1364–1365), and in Woods County, at Waynoka, on 4 July 1930, when he had

Figure 11.16 *Lestes unguiculatus,* the Lyre-tipped Spreadwing, A) adult ♂, Cedar Creek, 7 km E of Lenapah, Nowata County, 4 October 2012 (©James W Arterburn, OC 398584). Note the soft gray pruinosity, including dorsally on S8, which shows the classic V-shape of black disrupting the gray, on this older ♂. B) This younger ♂ at Optima Lake dam, Texas County, 1 August 2014 (©Bill Carrell, OC 425624), shows some pruinosity, but it has yet to spread over the segment to form the classic V, instead it forms more of a wedge shape. C) The lyre-shaped appendages for which this species is named are so distinctive that identification is straightforward even of an upside-down ♂ being eaten by a *Phidippus audax* (Bold Jumping Spider) at Skipout Lake, Roger Mills County, 4 October 2014 (©Bryan E Reynolds). D) ♀, Lake Elmer, Kingfisher County, 9 October 2014 (©Brenda D Smith, SP 1448). Note that each individual has narrow whitish bands at the proximal and distal ends of the pterostigma, a feature not unique to this species—pterostigma on, for example, *L. alacer, L. australis,* and *L. dryas,* the Southern, Plateau, and Emerald Spreadwings, have whitish edges, too—but one useful to differentiate it from *L. disjunctus,* the Northern Spreadwing.

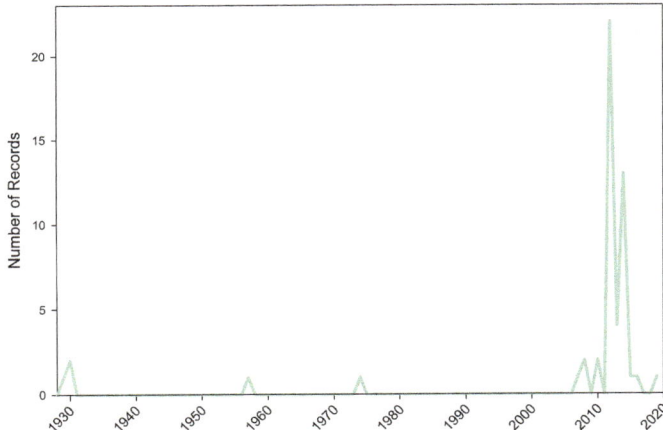

Figure 11.17 Number of reports of *Lestes unguiculatus*, the Lyre-tipped Spreadwing, fluctuate drastically annually. There was a "bloom" in the state for a few years in the mid-2010s but not much since (i.e., it has reverted back to its usual scarcity). Similar patterns have been noted elsewhere (e.g., Gregoire and Gregoire 2005, 2006).

a single ♀ (OMNH 102). EB Williamson was responsible for collecting the first record for the state, on 18 October 1929, in Chandler, Lincoln County (1♂, UMMZ). It was collected only twice during the Bick/Hornuff era,[7] but surprisingly, despite their considerable field effort, not by either collector; instead the credit goes to DR Lauck and D Knierihm.[8]

In stark contrast to the previous eras, for which there was a total of 5 records in 5 counties, the current era has seen a drastic increase in records. Between 2008 and 2016, >45 records were reported in an additional 22 counties, bringing the total for the species to 27 counties. Despite the recent boon, none were reported in 2017 or 2018 (Figure 11.17)!

Lestes unguiculatus went undocumented in Oklahoma from 1974 until 2008, when K Williams found a ♀ at the Tallgrass Prairie Preserve, Osage County, on 28 June (OC 315355) and BA Heck found a ♂ in Black Mesa SP on 29 August (OC 283758). In 2012 the species exploded across the state, being found in 11 new counties, at times in large numbers, including 75 in mid-May and 50 in early October (45♂, 30♀, including 12 pair, Canton WMA, Dewey County, 13 May, SP1241–1242; 40M, 10F, including 3 pair, Drummond Flats WMA, Garfield County, 3 October, JR Heinen, OC 382024). These records included the earliest known occurrence for the state, 7 May (OC 374809, 1♂, Salt Plains NWR, Alfalfa County, JR Heinen). The late date for the species came in 2014—also an exceptional year for the species—on 25 October, when we encountered 15♂ and 6♀, including 3 pair, at two mostly dry ponds at Great Plains SP, Kiowa County (SP 1477–1480, OC 427570). Oklahoma's seasonality is in sync with that elsewhere in the region (CO: 4 June–30 September, JCAC 23581, CSU; KS: 23 [21?] May–18 October, SEMC; MO: 2 May–8 October, Trial 2005; TX:

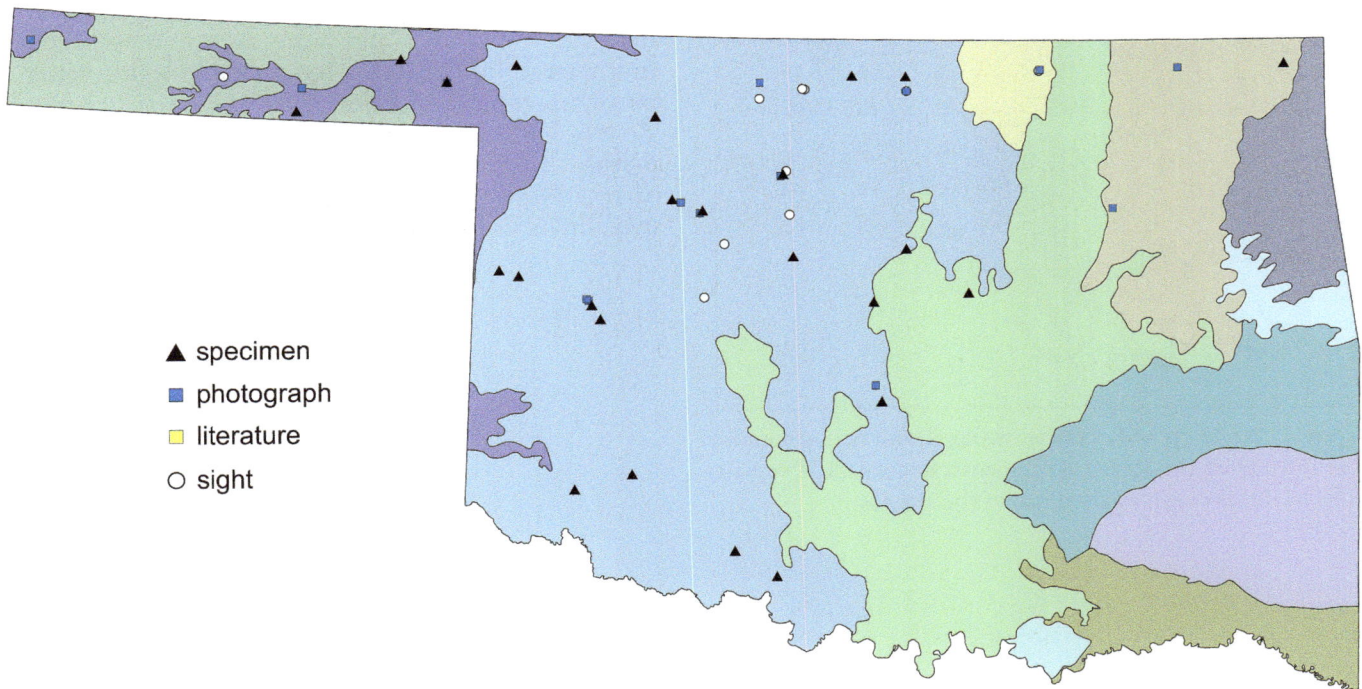

▲ specimen
■ photograph
□ literature
○ sight

Figure 11.18 Documentation level of supporting records for *Lestes unguiculatus*, the Lyre-tipped Spreadwing, in Oklahoma. Levels include specimen, archived photograph, literature reference, or sight record. Although sight records lack documentation, only those we feel are credible are plotted.

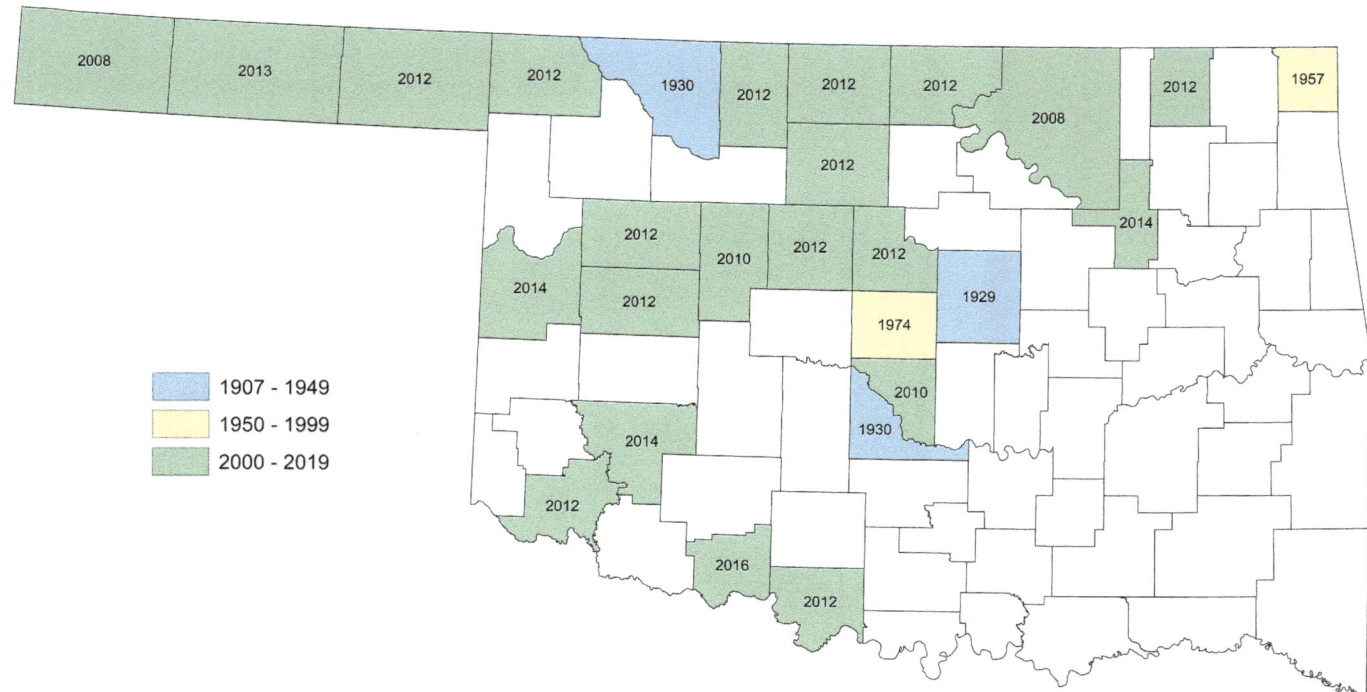

Figure 11.19 Counties in which *Lestes unguiculatus*, the Lyre-tipped Spreadwing, are known to occur in Oklahoma. Year within the county is that in which the species was first reported.

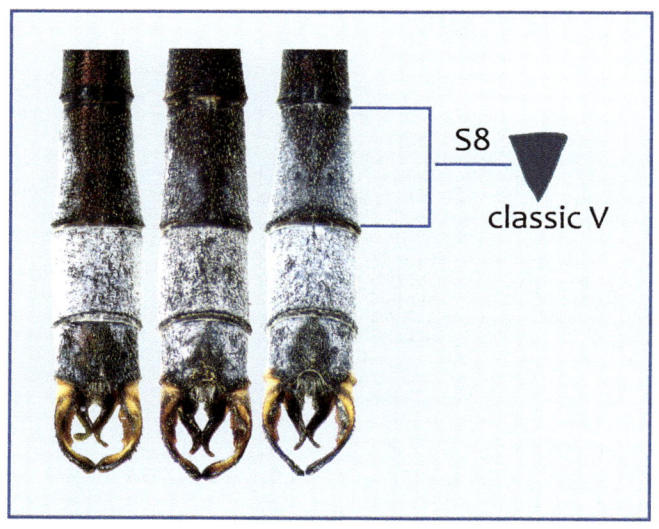

Figure 11.20 *Lestes unguiculatus* males show varying amounts of pruinosity on the terminal segments, but a black V dorsally on S8 (right) is considered typical of the species.

10 June–1 September, Abbott and Lasley 2019). Tenerals have been observed in Oklahoma from 23 May until 28 June, but we know of no larval or ovipositing records.

Some other records of note include those from Jefferson and Cotton Counties, which are not only the southernmost records for Oklahoma but some of the southernmost for anywhere in the species' range (Jefferson County: 1♂, Moneka Park at Lake Waurika, 2 October 2012, SP 449, another ♂ seen; Cotton County: 2♂, Walters, Cotton County Wetlands and Nature Preserve, 23 October 2016, one as SP 2219, MAP).

LESTES RECTANGULARIS SAY, 1839 – SLENDER SPREADWING

The Slender Spreadwing is spottily distributed and generally uncommon, although on occasion it is locally fairly common, especially in north-central or northwestern Oklahoma. For example, in two days in the summer of 1970, GH Bick and LE Hornuff collected 16♂ at Boiling Springs SP, Woodward County (1–2 August, FSCA[9]). The highest count reported for Oklahoma is our record of 50 individuals, 10 of which were teneral, at Canton WMA, Dewey County on 1 June 2013. Generally, the species is found at shaded lake inlets or flooded woodlands in the backwater areas of lakes (Figure 11.21). One outlier locality is the flooded marshy lakebed at Optima "Lake" that is not shaded by any woodland.[10]

The species was reported three times in 1930, the only year it is known to have been encountered in the state during the early era of Oklahoma odonatology. It was first found

Figure 11.21 *Lestes rectangularis*, the Slender Spreadwing, A) ♂, Lexington WMA, Cleveland County, 12 September 2017 (©Bryan E Reynolds, OC 473091). B) adult ♀, Oxley Nature Center, Tulsa, Tulsa County, 7 June 2014 (©Bill Carrell). The ♀ of this species can be tricky to identify without a specimen, but the yellow wash on the lower sides of the thorax are typical of the species, and even in photographs it can be evident that the legs are yellowish or tan (not black) and that the abdomen is proportionally longer, especially the length of S7 relative to the ovipositor (approaching twice as long many in *L. rectangularis*). C) teneral ♀, JT Nickel Preserve, Cherokee County, 6 July 2017 (©James W Arterburn, OC 466751). The warm taffy color of this individual is typical of tenerals, ♂ or ♀, of the smaller spreadwing species in the Southern Great Plains. Visible on each of these individuals is the pale vein along the edge of the wing tips, a character distinctive of this species and which both males and females show. Older adults are pruinose on the back of the thorax between the wings (as other *Lestes*) but not elsewhere on the thorax. This lack of pruinosity leaves the antehumeral always visible (Paulson 2011).

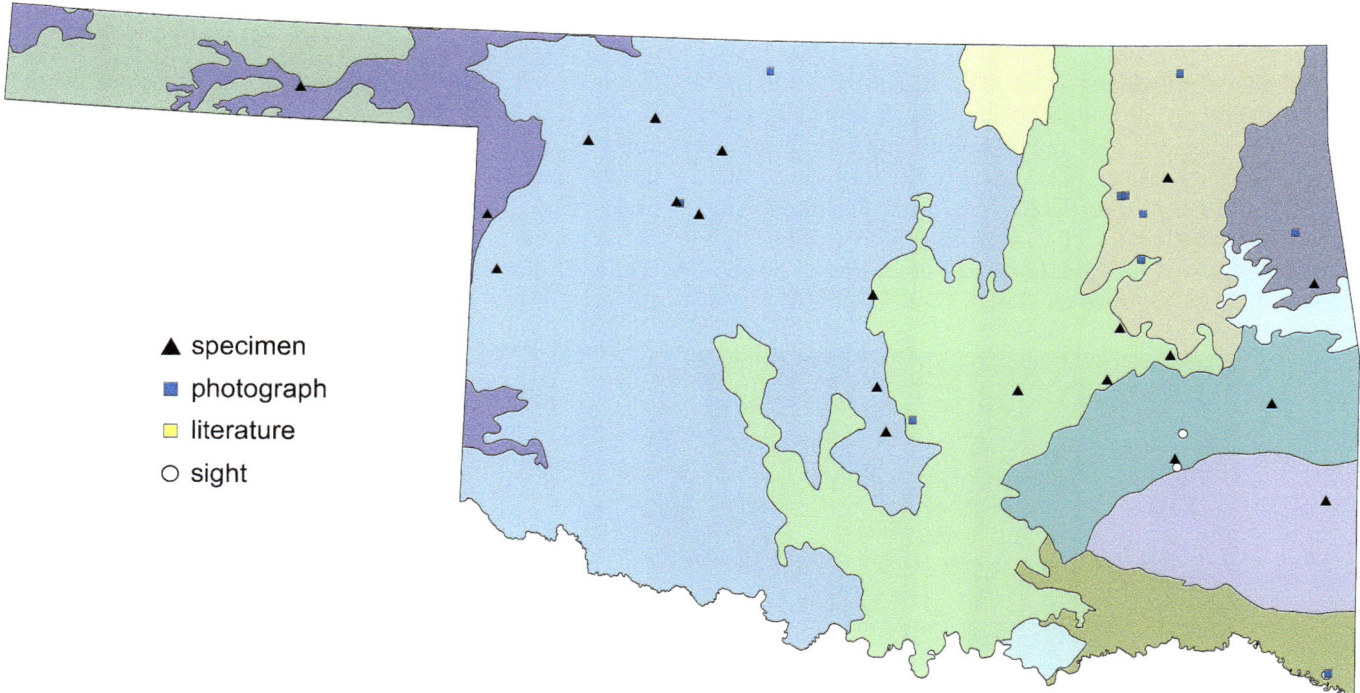

Figure 11.22 Documentation level of supporting records for *Lestes rectangularis*, the Slender Spreadwing, in Oklahoma. Levels include specimen, archived photograph, literature reference, or sight record. Although sight records lack documentation, only those we feel are credible are plotted.

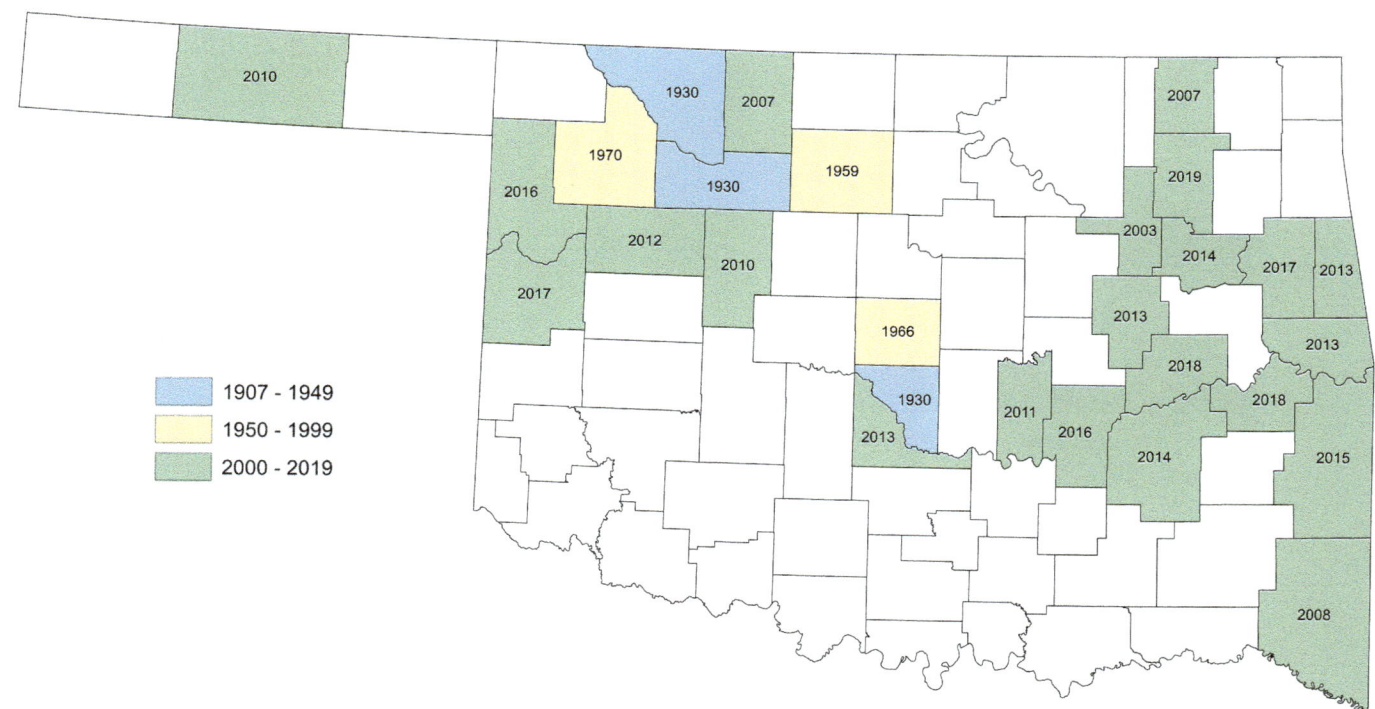

Figure 11.23 Counties in which *Lestes rectangularis*, the Slender Spreadwing, are known to occur in Oklahoma. Year within the county is that in which the species was first reported.

on 17 May in central Oklahoma (2♂, Norman, Cleveland County, RD Bird, OMNH 3823[11]). Just over a month later Bird found the species in Major County (1♂, Cleo Springs, Eagle Chief Creek, 27 June, OMNH 3826) and then a few days later in Woods County (1♀, Waynoka, 4 July, OMNH 61). It was not reported again until 1959 (2♀, Garfield County, J. F. Reinert, USNM 374500–374501). Bick's only encounter with *Lestes rectangularis* was during the two days mentioned above, but Hornuff also collected 1♂ 3 mi N of Edmond at "Spirogyra Pond," Oklahoma County (FSCA).

All early and mid-era records combined, spanning four decades and accounting for six counties, suggests a poor encounter rate relative to what is expected for *L. rectangularis* today (Figure 11.23). We know of about 35 records dating from 2003 until 2019, and in that time the species has been recorded in 22 additional counties. Currently, although there are apparent gaps in the species' range in Oklahoma, we suspect that it does occur across the northern portion of the state; with time any of those counties that have lake backwaters will have records for *L. rectangularis*. The species' absence from most of southern Oklahoma, except for McCurtain County, is probably a more accurate reflection of its distribution. We may expect the species to be found in neighboring counties to McCurtain County, but we would be surprised if it were encountered in the southwest as anything but perhaps a vagrant given the lack of what we perceive as suitable habitat. The real question with this species in Oklahoma is: was this species overlooked prior to 2003, or has it, such as we suspect with *L. unguiculatus*, expanded its range over time with the increase in lentic habitat in the state?

The long flight season of this species certainly would allow it time within a season to incrementally expand its distribution. In the region, it is known to fly from early April until late October, with Oklahoma having the earliest date and Missouri the latest (OK: 1 April–27 September, both single ♂, from Mohawk Park, Tulsa County, with early date in 2012, reported by T Mitchell, and late date in 2014 by BC, OC 374251, OC 427647, respectively; KS: early May–mid-September, Cringan 1979; MO: 10 May–23 October, Sims 2012[12]; limited records elsewhere). Only one tandem pair has ever been reported (Optima Lake, Texas County, 22 May 2010); we have no records of ovipositing or nymphs.

LESTES VIGILAX HAGEN, 1862 – SWAMP SPREADWING

The Swamp Spreadwing is a relatively scarce species in Oklahoma; we have just over 30 records. Generally it is observed no more than a half dozen individuals at a time, but our highest counts for the species were 18 individuals (6♂, 12♀; 1♀ as SP 397) on 9 September 2012 and 33 individuals (25♂, 8♀) on 14 July 2018, both times at Clayton Lake, Pushmataha County (Figure 11.24).

The species has been recorded in six counties in southeastern Oklahoma. It was first recorded in the state in Pushmataha County, probably north of Antlers, by AE Pritchard on 25 June 1932 (1♂, 2♀, OMNH 3809). He encountered the species again, between 2 May and 26 June 1934 (4♂, 7♀, EMEC 300645–300650), likely at the same location. We assume that location is the one mentioned in his 1935 paper describing *Celithemis verna* (Double-ringed Pennant; Pritchard 1935:9): "… at Antlers is a small clear lake close to the Kiamichi River … concealed by a dense growth of vegetation and trees, only the center of which

is free from prolific masses of water lilies." Such habitat is fairly common in southeastern Oklahoma and is similar to locations we have encountered *L. vigilax*. Pritchard went on to say that *L. inaequalis*, the Elegant Spreadwing, *Enallagma daeckii*, the Attenuated Bluet, *E. dubium*, the Burgundy Bluet, *Nehalennia integricollis*, the Southern Sprite, *Ischnura kellicotti*, the Lilypad Forktail, and *Ladona deplanata*, the Blue Corporal, were also encountered at the pond. Again, this associated suite of species mimics our experience with *L. vigilax*. The Swamp Spreadwing was not reported again in Oklahoma until GH Bick collected a single ♂ on 13 June 1957 (1.5 mi N of Antlers, which is in all likelihood Pritchard's locality; FSCA).

Lestes vigilax was not reported outside of Pushmataha County until SW Dunkle collected eight individuals (2♂ and 2♀ at FSCA, 4U at JCAC) at "Bokhoma Camp pond," McCurtain County, on 6 June 1992. That record was also the first time the species had been documented since Bick's 1957 collection.

Figure 11.24 *Lestes vigilax*, the Swamp Spreadwing, A) adult ♂, Lake Ozzie Cobb, Pushmataha County, 26 September 2014 (©Brenda D Smith). B) still somewhat teneral ♀—note her brown (not green) coloration and that her wings are held together over her back rather than spread, McGee Creek WMA, Atoka County, 1 May 2018, ©Bryan E Reynolds. C) adult ♀, Red Slough WMA, McCurtain County, 20 July 2011 (OC 332350, ©Tom Kompier).

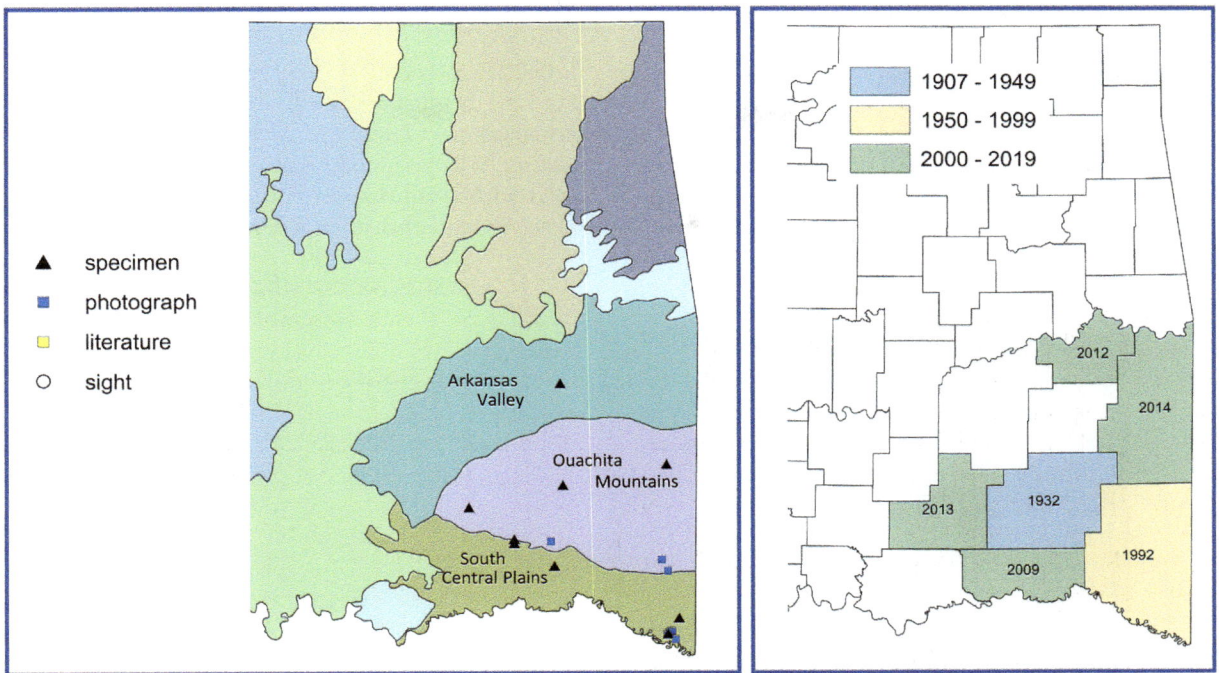

Figure 11.25 Documentation level of supporting records for *Lestes vigilax*, the Swamp Spreadwing, in Oklahoma (left). Levels include specimen, archived photograph, literature reference, or sight record. Although sight records lack documentation, only those we feel are credible are plotted. The species has been reported in six counties in eastern Oklahoma since 1932 (right).

Encounters between 2006 and 2017 rose steeply (>80% of records come from this era). It has been reported over a dozen times from Red Slough alone during 2008–2019. In 2009, it was found in a third county, when BAH reported five individuals at Schooler Lake in Choctaw County (OC 314541).

We documented the species in three additional counties. Each encounter was of a single ♂ (Atoka County: McGee Creek WMA, 13 July 2013, SP 806; Haskell County: 9 km SW of Kinta, 9 September 2012, SP 407; Le Flore County: east of Big Cedar, 24 August 2014, SP 1404). The only record subsequently in those counties is a teneral ♀ that was photographed at McGee Creek WMA on 1 May 2018 (BE Reynolds; Figure 11.24). Those records leave us unsure of the species' status there given that we regularly survey those areas but have had so few encounters.

Oklahoma falls short of the longest flight season for *Lestes vigilax* in the region by one day—218 days versus 219 for Texas (OK: 7 April–11 November, early date from 2016, 1♀ photographed by DA, OC 441058, late date from 2009, 1♂ photographed by BAH, OC 315784, both records from Red Slough WMA, McCurtain County; TX: 27 March–1 November, Abbott and Lasley 2019). Louisiana's flight season is not far behind in length, but that of Arkansas and Missouri are more limited (AR: 16 May–11 September, Harp and Harp 2003; LA: 31 March–19 October, Mauffray 2014; MO: 10 May–29 September, Trial 2005). We have no reports of pairs, ovipositing, or nymphs of *Lestes vigilax* in Oklahoma, but we do have one teneral record (10 June 2013, 1 teneral ♂ in hand and 2 adult ♀, Clayton Lake SP, Pushmataha County).

LESTES INAEQUALIS WALSH, 1862 – ELEGANT SPREADWING

This large spreadwing was, until 2014, thought to occur in Oklahoma only in the southeastern corner of the state, yet between 2014 and 2016, *Lestes inaequalis* was found in east-central Oklahoma and then, to our surprise, in northeastern Oklahoma (Haskell County: 9 km SW of Kinta, 6 July 2014, 2♀, 1 as SP 1319; Delaware County: Lake Eucha SP, 27

July 2014, 1♂, OC 425333; Figure 11.26). Really, we should not have been surprised by the first record of the species in northeastern Oklahoma. Although apparently rare in neighboring Arkansas, northeastern Oklahoma should be considered within the species' range, albeit on its western edge. A second northeastern Oklahoma record was reported in

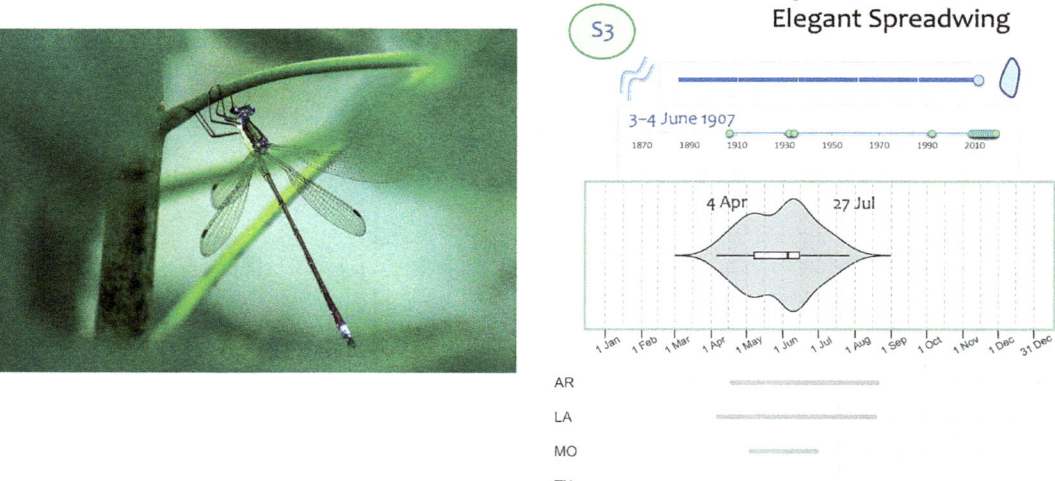

Figure 11.26 *Lestes inaequalis*, the Elegant Spreadwing, adult ♂, Lake Eucha SP, Delaware County, Oklahoma, 27 July 2014 (©Brenda D Smith), representing by far the northernmost record for the state, a curious happenstance given how far north the species occurs farther to the east. For a ♀, see Figure 11.28.

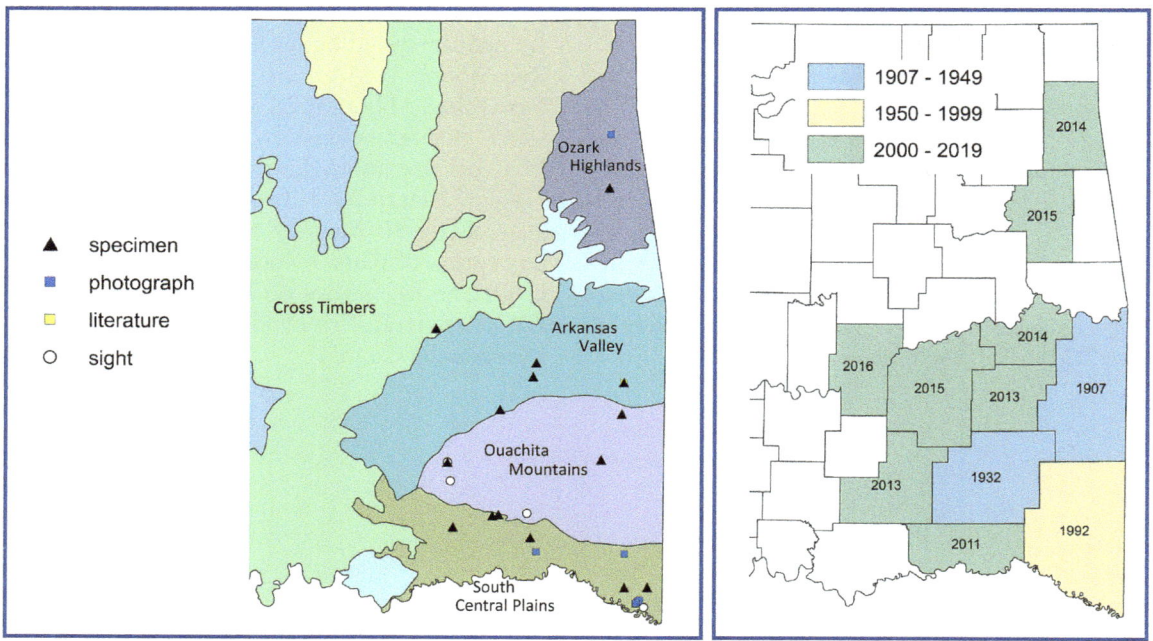

Figure 11.27 Documentation level of supporting records for *Lestes inaequalis*, the Elegant Spreadwing, in Oklahoma (left). Levels include specimen, archived photograph, literature reference, or sight record. Although sight records lack documentation, only those we feel are credible are plotted. The species has been reported in 11 counties in eastern Oklahoma since 1907 (right).

2015, this time from Cherokee County (TNC's JT Nickel Preserve, 1♂, 3 July, BS-P, BW Hoagland, SP 1691). What is odd about these records, as well as those from Hughes (2016) and Pittsburg (2015) Counties, and, for that matter, Latimer County (2013), is that all six of these counties have only had one record each. But if Le Flore County presents a lesson for patience, then we learned that sometimes one just has to wait: it took >100 years for a second encounter to be reported from that county, and there have been several more after that "rediscovery."

Given the overall rarity of this species in the state, it is unsurprising that encounters are infrequent. We tend to find a lone individual, occasionally up to three, at a time. Again, considering Le Flore County, the highest count of individuals in the state comes from Williamson's 1907 record, the first for Oklahoma, when he collected three[13] and said that "several [were] seen, but they were very wary and active, escaping in vegetation" (Williamson 1914b:414). A soaring total of three individuals has been matched just once since (1♂, 2♀, Red Slough WWA, McCurtain County, 7 May 2017, MAP, DA). *Lestes inaequalis*' rarity in Oklahoma is considered the norm for the species in the region and elsewhere (Abbott 2011, Paulson 2011; also indicated by the limited OC records for nearby states). That said, suitable habitat is not always the easiest to gain access to in that part of Oklahoma, and when one does gain entry, surveying the mucky, debris-choked areas this species often inhabits is tedious, daunting, and at times perilous, so one always wonders how thorough a given survey can be. Occupied areas tend to about sloughs and smaller lakes in heavily wooded and well-shaded areas (also see *L. vigilax* account for a habitat description and species associations). It has also been reported at wooded beaver dam complexes like the one below Crooked Branch Lake, Le

Flore County, where *Lestes inaequalis* is as regularly encountered (recorded there 4 times) as one can get in Oklahoma away from Red Slough (reported >15 times).

Encounters with *Lestes inaequalis* in Oklahoma exhibit a similar pattern to *L. unguiculatus*, *L. rectangularis*, and *L. vigilax* in reports by era. Between 1907 and 1934, it was seen three times at two locations (Williamson's 1907 record and two AE Pritchard records dating 1932–1934 from north of Antlers, Pushmataha County). It was not reported by GH Bick or LE Hornuff; SW Dunkle's 1992 records (both from McCurtain County: 1♂, 5 June, Kulli Lake and 4♂, 6 June, "Bokhoma Camp pond," all 5♂ at FSCA) being the only one from the middle timeframe of Oklahoma odonatology. Starting in 2009, the species began to be reported as regularly as a rare species can be expected: at least once a year. We now have >30 records in less than a decade.

Lestes inaequalis has a rather short documented flight season in Oklahoma (4 April–27 July, early date is from 2012, 1♂, Red Slough WMA, McCurtain County, DA, OC374288, late date is the Lake Eucha SP record mentioned above). We expect that eventually its recorded flight season in the state will be expanded to coincide with most of the nearby states that have the species (AR: 19 April–20 August,[14] OC 441224, OC 6533; LA: 7 April–18 August, Mauffray 2014; MO: 5 May–30 June, Trial 2005; TX: 21 April–13 August, iNat, Abbott and Lasley 2019). There are no known records of pairs, ovipositing, or nymphs in Oklahoma. Two tenerals, a lone ♀ collected at TNC's Boehler Seeps Preserve, Atoka County, on 1 May 2015 (SP 1532), and a ♀ examined in hand at Lake Ozzie Cobb, Pushmataha County, 4 May 2019 (MAP), are the only ones known for the state.

LESTES EURINUS SAY, 1839 – AMBER-WINGED SPREADWING

This eastern species was discovered in Oklahoma when a lone ♀ was found at a small clear-water pond nestled within a pine woodland at Pushmataha WMA, Pushmataha County, 10 June 2013 (Patten and Smith-Patten 2013a, b; SP 688, OC 400672). We considered this record to be of a vagrant because it was well to the southwest of the species' known geographical range, and so we were surprised when the species was found a year later at a small pond on Pumpkin Flats at TNC's JT Nickel Preserve, Cherokee County, where JW Arterburn found 6♂ on 13 June (OC 423037), B Carrell found 2♀ on 5 July (OC 424308), and then Arterburn and K Williams spotted the species again on 11 July (1♀, OC 424690). The Nickel Preserve hosted the species again in 2015, when Carrell encountered 2♂ on 7 June, again on Pumpkin Flats (OC 431433). We and others have

since visited that pond but have not been fortunate enough to re-find the species (although the *Micrathyria hagenii*, the Thornbush Dasher, record from the pond on 3 July 2015 was a nice consolation prize when BS-P went looking for *Lestes eurinus* there). Also in 2015, the species was found in mixed forest at a small pond with water shield (*Brasenia schreberi*) at Cookson WMA, Adair County, where up to 5♂ were present 2–12 July (1♂ SP 1688 on 2 July). In light of these seven records, it would appear that *Lestes eurinus* has expanded its overall geographic range southwestward and now occurs in small numbers in highlands in the eastern fourth of Oklahoma. Elsewhere in its range, *Lestes eurinus* flies from May to September (Paulson 2011), so we may expect to get additional records during that time of year.

Lestes eurinus Say, 1839
Amber-winged Spreadwing

10 June 2013

1870   1890   1910   1930   1950   1970   1990   2010

7 Jun   12 Jul

elsewhere

1 Jan   1 Feb   1 Mar   1 Apr   1 May   1 Jun   1 Jul   1 Aug   1 Sep   1 Oct   1 Nov   1 Dec   31 Dec

Figure 11.28 Oklahoma's first record of *Lestes eurinus*, the Amber-winged Spreadwing (A, top), in comparison with *Lestes inaequalis*, the Elegant Spreadwing (A, bottom). The *L. eurinus* ♀ was collected at Pushmataha WMA, Pushmataha County, on 10 June 2013 (SP 688). Note the yellow wings, wavy dark stripes across the lower thorax, and thicker, shorter abdomen relative to the *L. inaequalis* (♀, Robbers Cave WMA, Latimer County, 10 June 2013, SP 680). B) The second state record was 5♂, one of which is pictured here (JT Nickel Preserve, Cherokee County, Oklahoma, 13 June 2014, ©James W Arterburn, OC 423037).

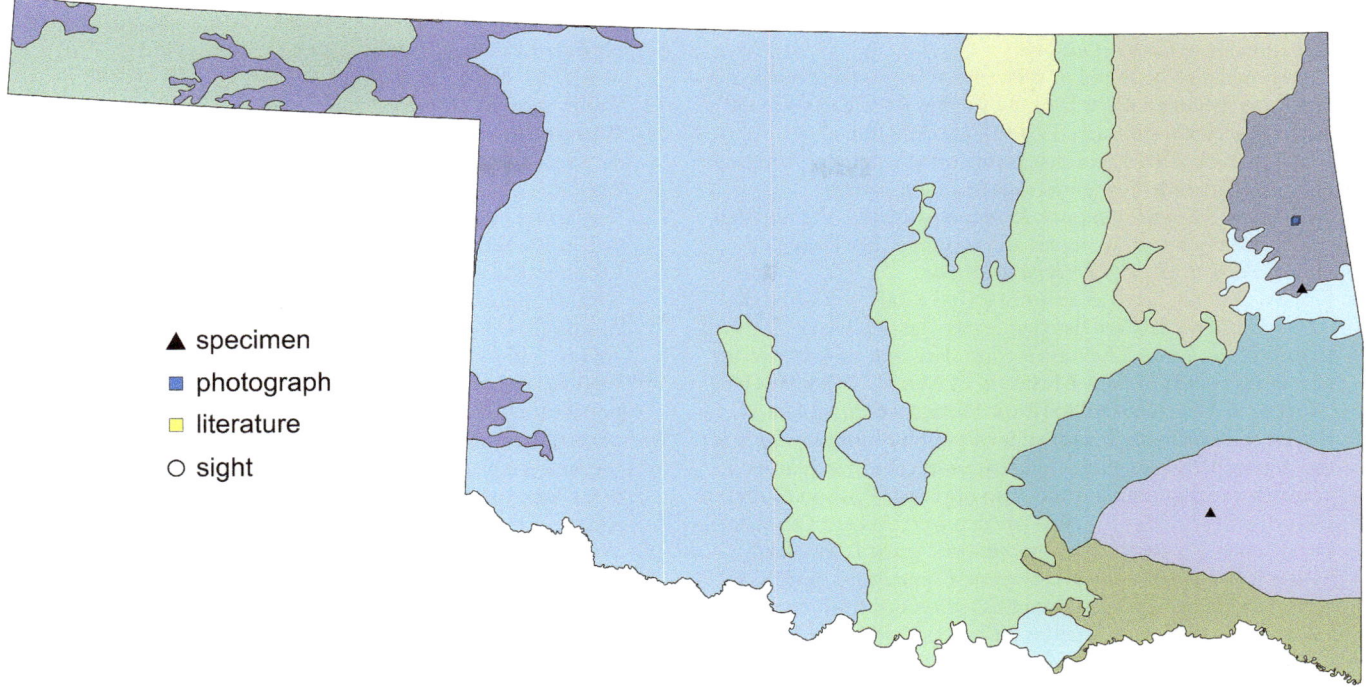

▲ specimen
■ photograph
□ literature
○ sight

Figure 11.29 Documentation level of supporting records for *Lestes eurinus*, the Amber-winged Spreadwing, in Oklahoma.

NOTES

1. Williamson (1931:64) went on to say that "It is altogether the most surprising find I have ever made in collecting dragonflies." Given his career, even to that point, this was quite the declaration. The spot he collected at was at "Western College" [for Women; now Miami University], in Oxford, Ohio. He collected 3♂ and 1♀ and saw about 12 others on 25 September 1927. On 11 October he collected 4♂ and 1♀ and "saw several" others, including "some [that] were ovipositing." He said that the majority of the individuals were found at the polluted, landscaped area, as opposed to a slightly more natural portion of the creek that was below an artificial pool along the creek; however, his description does not account for the relative water quality of the two locations.

   Following Williamson's article in the same issue of *Entomological News* was a note that indicated the first "*Archilestes*, (*A. grandis*?)" nymph for Ohio was collected in Dayton (Cotterman 1931). The author said that JG Needham verified the record. *A. grandis* was then recorded in western Pennsylvania in 1934 (Ahrens 1935) and then eastern Pennsylvania in 1951 (Ferris 1951), before reaching the east coast in 1949 (Donnelly 1961).

2. Westfall and May (1996:110 and 2006:78) stated, "Until the 1920's the species was reported in the United States only from the southwest, but since then it has undergone a remarkable range expansion (Gloyd, 1980), now occurring as far northeastward as western New England." We feel that this statement is a slight misinterpretation of Gloyd's (1980) work in which she subsumed *Superlestes* and *Cyptolestes* under *Archilestes*. She said:

   Of the five species of *Archilestes*, *A. grandis* has the most extensive range of distribution and a record of recent

expansion. For many years, in the United States it was known only from the Pacific Coast states of Washington and California, the southwestern states of Arizona, Utah, and east to Texas, Oklahoma and Kansas. The first record east of the Mississippi River was in 1927 when it was collected in Ohio by Mr. EB Williamson (1931) and it has now extended its range to states bordering the Atlantic coast, and to South Dakota.

   Had Gloyd meant to characterize the species' range as southwestern, then she included much more territory than is traditionally designated as the American Southwest. Generally, Oklahoma, Texas, and California are not considered Southwestern, and certainly Kansas and Washington never are, whereas parts of Utah and all of Arizona definitely are. We interpret Gloyd's wording as not saying the species was confined to the southwest. Moreover, she did not state that the species had not expanded its range until after the 1920s. And we believe it is unlikely that she would have interpreted the Oklahoma distribution and abundance, both of which she surely knew about at the time, as reflective of a recent colonizer. We therefore caution characterizing the species as one that was confined to the southwestern U.S. until the 1920s.

3. By 1936, *Archilestes grandis* was recorded in Caddo, Cleveland, Comanche, Greer, Latimer, McClain, Murray, Oklahoma, Pottawatomie, and Woods Counties. The species was mistakenly reported by Bird (1932) to have occurred in Custer County during his era. It may indeed occur in the county, but this specific record was based on UMMZ specimens (2♂, 1♀) that were mislabeled as Custer County. The locality for those specimens was 15 mi E of Weatherford, which places it squarely in Caddo County. But the collector, EB Williamson, was from out of state so he did not recognize his error. We

discovered the error during our 2013 visit to UMMZ and subsequently removed the species from the Custer County list and added it to Caddo County.

It should also be noted in reference to regional records that there are specimen records of the species from Kansas as early as the late 1960s and early 1970s (PSU, SEMC).

4. Bick and Bick (1970) said that plants on which ovipositioning took place included: *Erigeron canadensis*, *Eupatorium coelestinum*, *Monarda punctata*, *Rumex crispus*, and. We updated the taxonomy for these names and assumed that *Verbena urticaefolia* was actually *Verbena urticifolia*.

5. USNM 362968, listed as an adult ♂, has a date of 26 March. The specimen's data are listed as legacy data, so we are inclined to call this an error. The next earliest date that we know of for *Archilestes grandis* in New Mexico is 14 May (USNM 386199, 386200). A March record for New Mexico would be of a similar seasonality to Texas and is out of line with elsewhere in the region. Whereas a mid-May date is more consistent with elsewhere within that latitude. Also note that Paulson (2009) indicated flight dates for New Mexico were May to October.

6. Reported as a ♂ *Lestes disjunctus australis* (now *L. australis*) with a ♀ *Enallagma cyathigerum* (now *E. annexum*). Bick & Bick (1981) indicated it was an unpublished record from Shiffer.

7. 1♀, 13 June 1957, at a "shallow rush pond," Miami, Ottawa County, DR Lauck (UMMZ), and 1♂, 18 September 1974, Edmond, Oklahoma County, D Knierihm (UCO 10067)

8. DR Lauck collected 1♀ on 13 June 1957 in Miami, Ottawa County (UMMZ). D Knierihm collected 1♂ on 18 September 1974 in Edmond, Oklahoma County (UCO 10067).

9. Bick's species' notes indicated there were only 10♂.

10. SP 112, 1♀, was collected at the marshy area west of the Optima Lake dam, Texas County, on 26 September 2010 (OC 399255).

11. And it was not reported again from Cleveland County until Bryan E Reynolds photographed a ♂ at Lexington WMA on 12 September 2017 (OC 473091; Figure 11.21).

12. Abbott (2011) indicated there are two populations in Texas, one in the panhandle in Hemphill County and one in Franklin County, which is close to the border with Oklahoma near McCurtain County. The flight season for Texas is reported as 23–24 May (Abbott and Lasley 2019). Although well within the expected range for, we were only able to find four records for Arkansas, only one of which had a date associated (17 May, OC 7677). There are undoubtedly populations in Arkansas. Colorado has a fairly limited number of records, which is not surprising given that it is probably at the western extreme of the species' range. Nonetheless, flight dates span roughly two months, from 4 June (JCAC) until 1 August (OC 367317).

13. Williamson (1914b:414) reported that he collected 1♂ and 2♀ ("Wister, June 3 and 4, 1♂, 2♀") but there are 2♂ and 1♀ at UMMZ.

14. iNaturalist has photographs identified as this species from Arkansas that date from 3 September to 18 October. We do not feel comfortable confirming these given the photo quality. These dates, being out of sync with the region and more in line with *Lestes vigilax*, also gave us pause.

## CHAPTER 12

COENAGRIONIDAE: POND DAMSELS

The second-most speciose family of odonates in Oklahoma is the Coenagrionidae, the pond damsels. With 45 species recorded, it trails only the Libellulidae, the skimmers, of which 58 species have been recorded. Even so, the pond damsels account for just over a fourth (25.4%) of all odonate species known from Oklahoma. Species richness within the family is dominated by three genera: (Figure 12.1), *Enallagma* (the bluets), *Ischnura* (the forktails), and *Argia* (the dancers), enough so that we provide separate introductions for each.

There are but six species recorded in the state that are not classified in these genera (Figure 12.2), only one of which, the Desert Firetail (*Telebasis salva*), is at all common and, with a distribution that covers much of the western two-thirds of the state, at all widespread. At the other extreme, the Duckweed Firetail (*T. byersi*) and the Sphagnum Sprite (*Nehalennia gracilis*) are each known from a single site in Oklahoma (Red Slough WMA in McCurtain County and Boehler Seeps in Atoka County, respectively). Another, the Aurora Damsel (*Chromagrion conditum*), is known from just a couple of seeps and ponds in the vicinity of Pine Mountain Spring, McCurtain County. The remaining two species, the Western Red Damsel (*Amphiagrion abbreviatum*) and Southern Sprite (*N. integricollis*) are localized, the former as a denizen of springs and seepy creeks in the northwestern part of the state (including, spottily, the panhandle), the latter as an easily overlooked denizen of shaded ponds with emergent vegetation embedded in coniferous and mixed hardwood forest and woodland in the southeastern corner of the state.

ENALLAGMA

Fifteen species of bluets (*Enallagma* sp.) have been recorded in Oklahoma, although the genus truly is dominated by one species, the widespread, abundant, and aptly named Familiar Bluet (*E. civile*); it and three other species (Figure 12.3) account for three-fourths (75.4%) of all bluet occurrences in the state. At

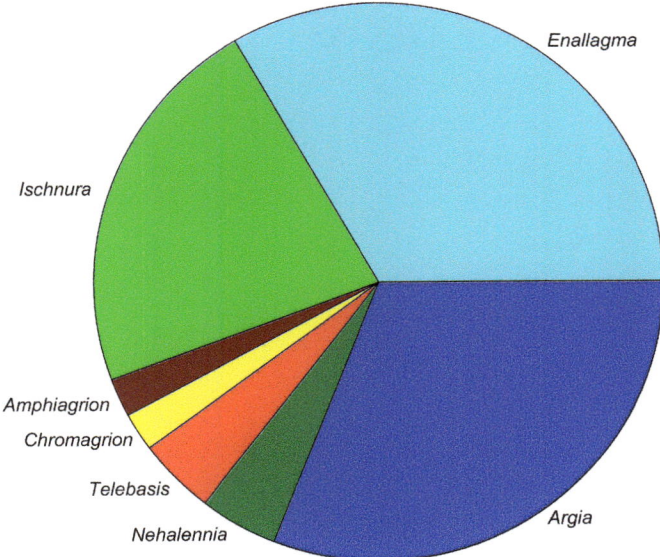

Figure 12.1 A breakdown of species richness by genus in the damselfly family Coenagrionidae, the pond damsels. Number of species in each genus is indicated.

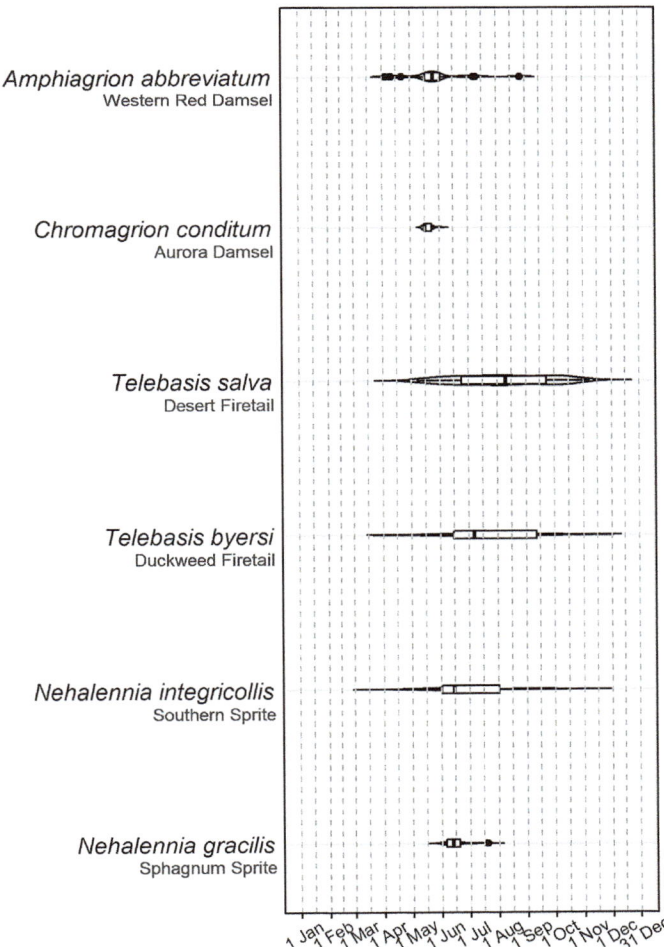

Figure 12.2 Flight seasons in Oklahoma of species in the family Coenagrionidae that are in the genera *Amphiagrion* (red damsels), *Telebasis* (firetails), and *Nehalennia* (sprites).

the other extreme, six species barely have a presence in the state. Three, the Arroyo Bluet (*E. praevarum*), Tule Bluet (*E. carunculatum*), and Rainbow Bluet (*E. antennatum*), are effectively confined to the panhandle region, where neither is recorded annually. The other three, the Atlantic Bluet (*E. doubledayi*), Attenuated Bluet (*E. daeckii*), and Burgundy Bluet (*E. dubium*), essentially each are confined to roughly a handful of known sites in the Ouachita highlands of southeastern Oklahoma.

Bluets can be found almost anywhere in the state, although most species tend to the lentic end of the water-habitat spectrum (Figure 12.4). Some species, notably the Turquoise Bluet (*E. divagans*), have a flight season restricted to the spring, but the majority of species more-or-less share a peak flight season in mid-summer (Figure 12.4). A "lookalike" of this species (and, yes, we know that many would assert that all bluets look alike), the Slender Bluet (*E. traviatum*) both flies later and is decidedly a lentic species, favoring ponds and small lakes with abundant emergent vegetation near the shore.

Apart from the Arroyo Bluet, which, as its English name might suggest, truly is a lotic species, only three other species that occur in the state are lotic or strongly tend toward the lotic end: the Stream Bluet, Rainbow Bluet, and Turquoise Bluet (Figure 12.5). Indeed, the sole reason that the Stream and Turquoise Bluets do not "score" lower on our quantitative lenticity scale is that each can be found, at times, in linear coves and inlets of lakes and even at small ponds. Even then, such occupied sites tend to be near streams or creeks, both of which are more reliably locations for either species.

The most notorious lookalikes, the four species dubbed the "*civile* group"—the Arroyo Bluet, Atlantic Bluet, Familiar Bluet, and Tule Bluet (the Alkali Bluet, *Enallagma*

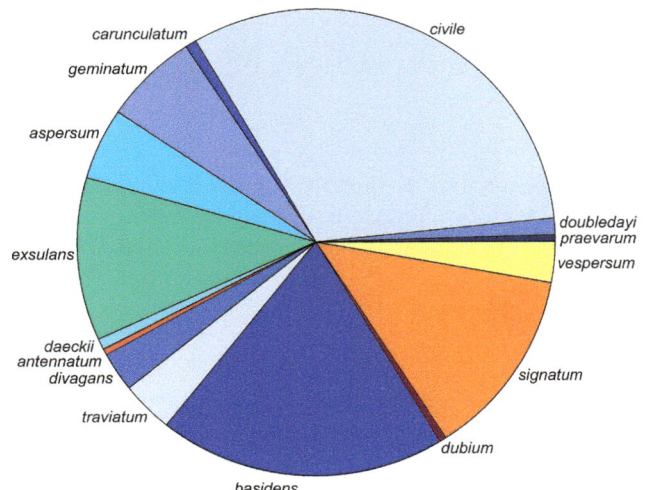

Figure 12.3 Relative commonness in Oklahoma of 15 species of bluets (*Enallagma*). Four species dominate the scene: the Familiar Bluet (*E. civile*) accounts for nearly a third (31.8%) of occurrences, and the Stream Bluet (*E. exsulans*), Double-striped Bluet (*E. basidens*), and Orange Bluet (*E. signatum*) each accounts for 11–20% of occurrences.

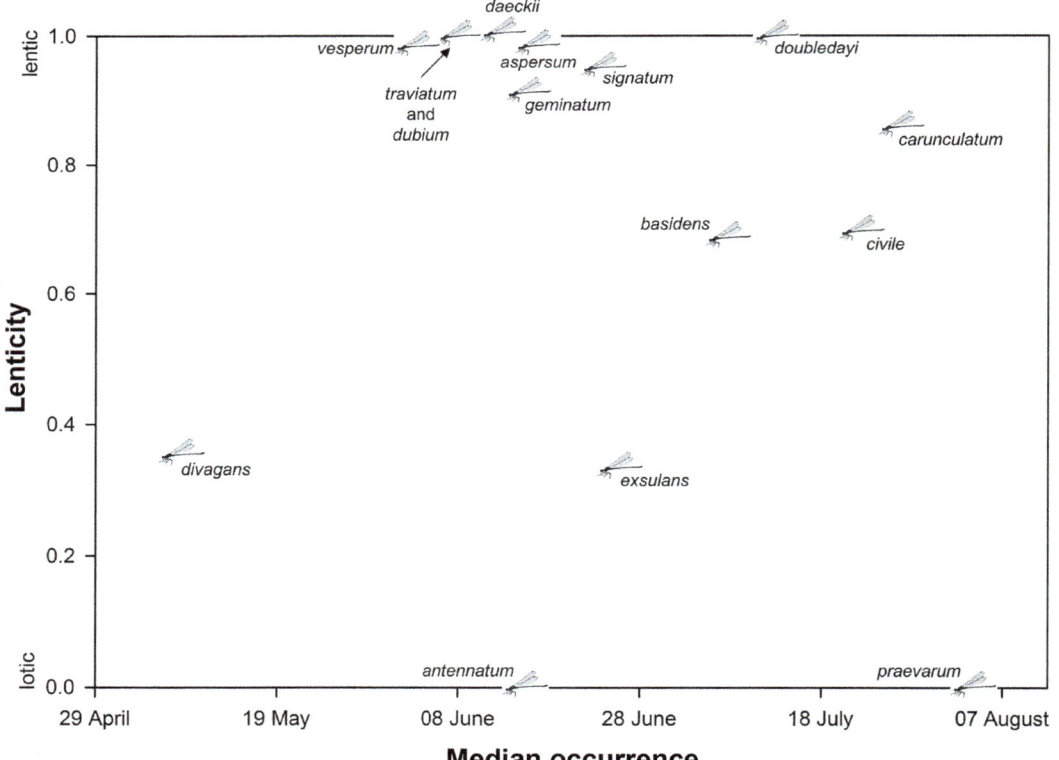

Figure 12.4 A scatter plot of the midpoint of seasonality against lenticity for fifteen Oklahoma species in the genus *Enallagma*, the bluets. Only the Turquoise Bluet (*E. divagans*) is an early-season species. It is replaced in lotic habitats by the Stream Bluet (*E. exsulans*) as the season progresses (Figure 12.5).

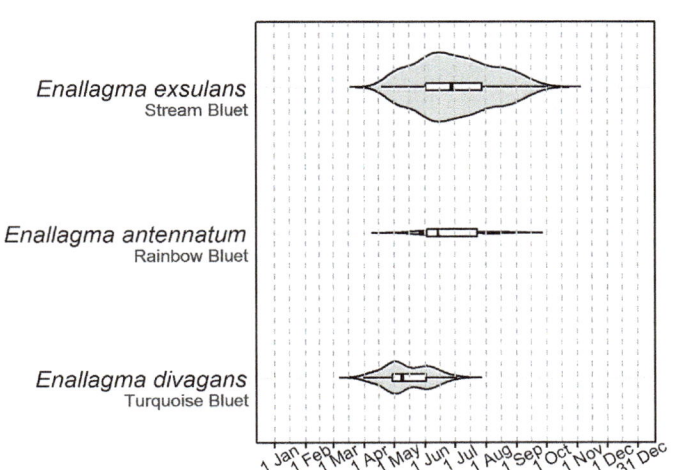

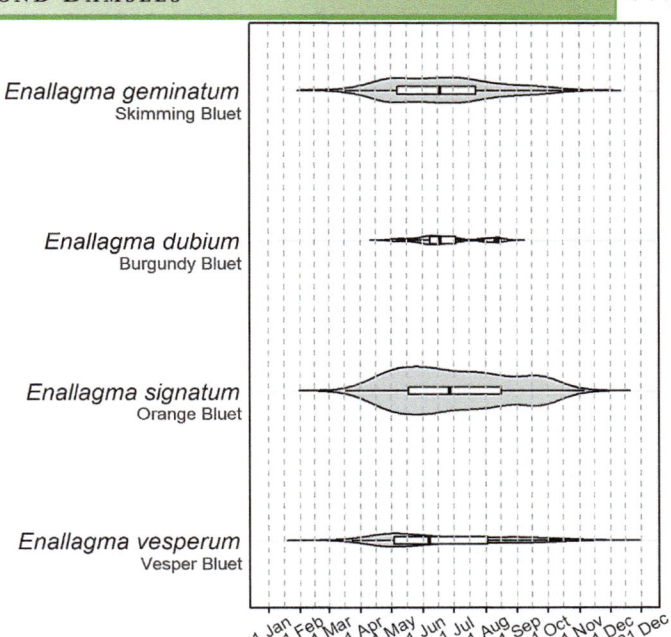

Figure 12.5 Flight seasons in Oklahoma of *Enallagma* species that chiefly occupy lotic habitats in Oklahoma (another lotic species, *E. praevarum*, the Arroyo Bluet, is treated in a different grouping; see Figure 12.6).

Figure 12.7 Flight seasons in Oklahoma of four lentic *Enallagma* species that roost typically on emergent or floating vegetation.

clausum, would join this group, if it is found in Oklahoma, as expected)—vary in habitat choice but have similar (long) flight seasons, with, as mentioned, the Familiar Bluet holding an insurmountable lead in relative abundance (Figure 12.6). (Incidentally, in erecting this "group" we imply no phylogenetic relatedness. The Arroyo Bluet may be distantly related to the other three species, which, on the basis of cerci structure, likely are in the same evolutionary clade. We group them solely because of strong phenotypic resemblance.[1])

Four lentic species—the tiny, scarce Burgundy Bluet, the widespread Orange Bluet (*E. signatum*), the crepuscular Vesper Bluet (*E. vesperum*), and the only blue bluet of

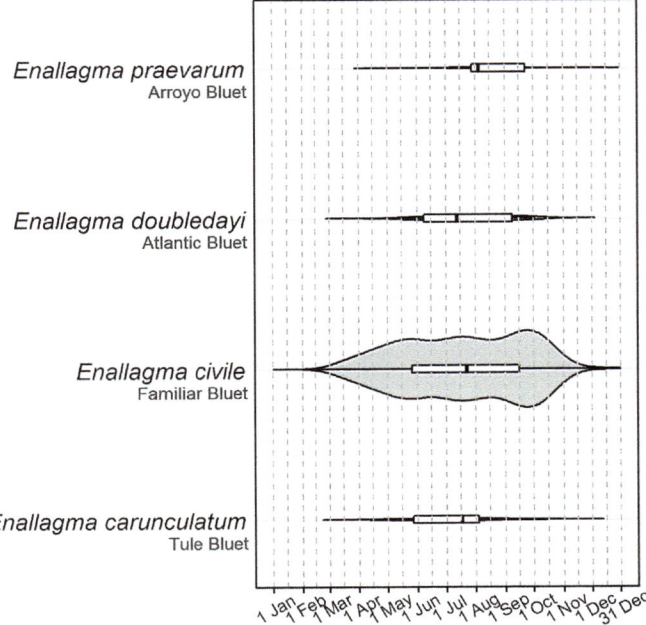

Figure 12.6 Flight seasons in Oklahoma of *Enallagma* species clustered in what we call the *"civile* group," meaning only that they strongly resemble the ubiquitous Familiar Bluet (*E. civile*).

this group, the Skimming Bluet (*E. geminatum*)—are active simultaneously during the year (Figure 12.7) and occur most often at ponds or small lakes with floating and emergent vegetation; all four species roost on water lilies, algal mats, or stems. As mentioned above, the Burgundy Bluet has but a slight presence in the state—it is known from only seven locations. And the Vesper Bluet may be more widespread than records indicate; its shy habits decrease its detection probability enough that it is often missed.

The remaining four species are a scattershot lot, grouped herein only as "other," although each tends to lentic habitats and a mid-summer flight season (Figures 12.4, 12.8). One of these species, the gangly Attenuated Bluet, not only looks unlike anything else but, like the Burgundy Bluet, barely has a toehold in the state, being known, ever, from a mere four sites in the Ouachita highland region of the southeast. Another, the Slender Bluet, has a short flight season (Figure 12.8) but typically is common in that short window.

ISCHNURA

The tiny forktails (*Ischnura*) are a familiar part of Oklahoma's damselfly fauna, even with three of the recorded species rare and localized and another ecologically specialized and, hence, virtually never encountered away from its preferred habitat—we speak, of course, of the aptly named Lilypad Forktail (*I. kellicotti*), which, as far as we know, has never been found in the state away from water lilies. Another species, the Desert Forktail (*I. barberi*), ought to be classified as a habitat specialist as well: in Oklahoma, it does not occur away from saline habitats, especially where soils have extruded salts and that support salt grass (*Distichlis spicata*) and pickleweed (*Salicornia* sp.). Curiously, although the similar, in terms of coloration and size, Rambur's Forktail

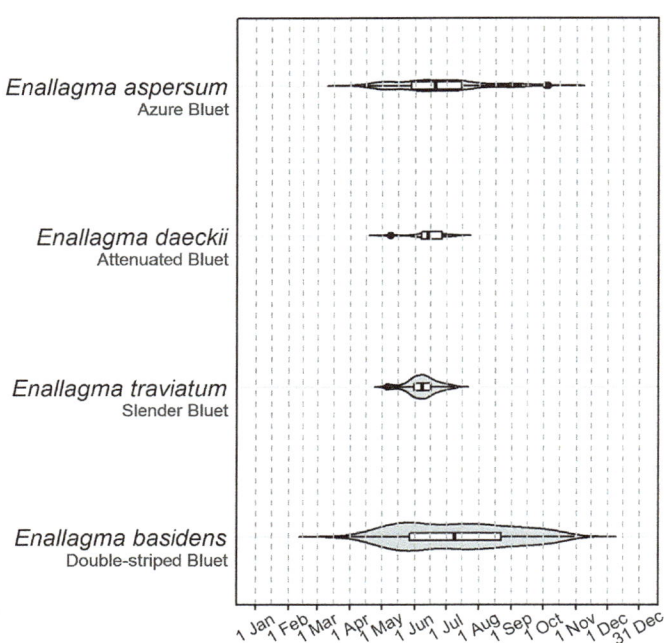

Figure 12.8 Flight seasons in Oklahoma of the remainder of *Enallagma* species, all of which tend toward lentic habitats, and only one of which—the Double-striped Bluet (*E. basidens*)—is truly abundant and widespread (it is known from every one of the state's 77 counties).

(*I. ramburii*) is a saline species elsewhere in its range, but in Oklahoma the Desert Forktail seems to exclude it from such habitats, such that they seldom co-occur.

The three rare species, which together account for just 2.1% of *Ischnura* occurrences (Figure 12.9), are the Plains Forktail (*I. damula*), Mexican Forktail (*I. demorsa*), and Western Forktail (*I. perparva*). Each is confined to the western fringe of the state, where the bulk of records are from the panhandle, and the last two species are strictly lotic,

occurring on shallow, clear, typically perennial streams, habitat that has grown scarcer in western Oklahoma over the decades with lowering of the High Plains/Ogallala aquifer and other water-use issues (Patten and Smith-Patten 2013b). The Plains Forktail occurs in such lotic habitats, but it also may be found at vegetated lake shores and ponds.

At the other extreme, the five commonest species in the state account for a remarkable 88.6% of *Ischnura* occurrences (Figure 12.9), with the Fragile Forktail (*I. posita*) alone accounting for just under one in every five. That species is absent from the western half of the panhandle but otherwise is found in every county, a tally bested only by the Citrine Forktail (*I. hastata*), which has been recorded in all 77 counties.

Forktail assemblages vary across the state. Despite its common name, the Eastern Forktail (*I. verticalis*) is commoner in the west, whereas the Fragile and Citrine Forktail numerically dominate assemblages in the east, with the former particularly prevalent and the latter gaining in prevalence from north to south (Figure 12.10). Rambur's Forktail is a southern species, whereas the Black-fronted Forktail (*I. denticollis*) is a western species that, in its proper habitat of clear, sedgy, shallow marshes or creeks, may be the most abundant forktail present. Likewise, in proper saline habitats, the Desert Forktail easily outpaces its congeners.

Seasonality does not differ that much among species (Figure 12.11), although most of the few records of the Western Forktail (*I. perparva*) are concentrated in late spring, and abundance of both Rambur's Forktail and the Desert Forktail tends to be later than others (Figure 12.11). Assemblages and seasonality may change given that Rambur's Forktail appears to have expanded (and continues to expand) its range northward into Oklahoma: in the past couple of decades the species has gone from a few strongholds along the southern fringe of the state to having established populations north to near the Kansas border. We cannot say if this range expansion has come at the expense of the Eastern Forktail, but data suggest that numbers of that species have declined, especially (and ironically) in the eastern half of the state.

Phenotypic variation among some species of forktails is extreme in Oklahoma, so much so that individuals can look superficially like one species but in fact be another (Figure 12.12). The ♂ Eastern Forktail, is particularly exemplative, having antehumeral stripes ranging from *damula*-like dots to a *posita*-like exclamation mark to the full green stripes "typical" of the species (see *Ischnura verticalis* account for details). Females, too, vary a great deal in coloration, with young differing from old and some individuals being andromorphs, or colored more like the male.

ARGIA

The arrangement worked out less than perfect, but it appears that the dancers (*Argia*) and the bluets (*Enallagma*) divvied up lotic and lentic habitats in Oklahoma, with the dancers far more numerous and speciose at creeks, springs, rivers,

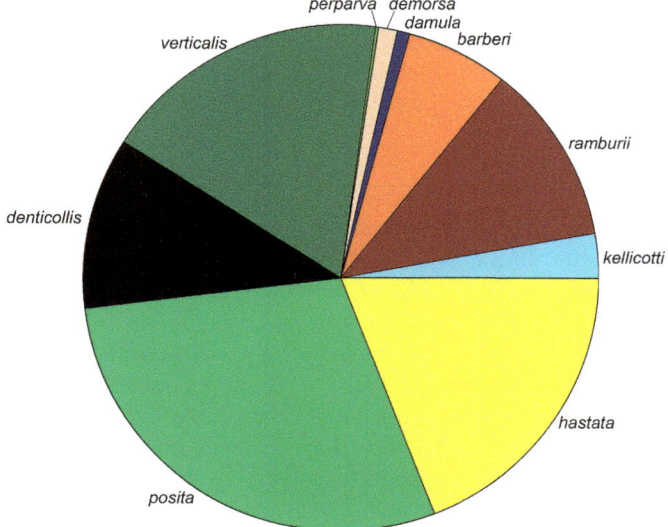

Figure 12.9 Relative commonness in Oklahoma of ten forktail species (*Ischnura*).

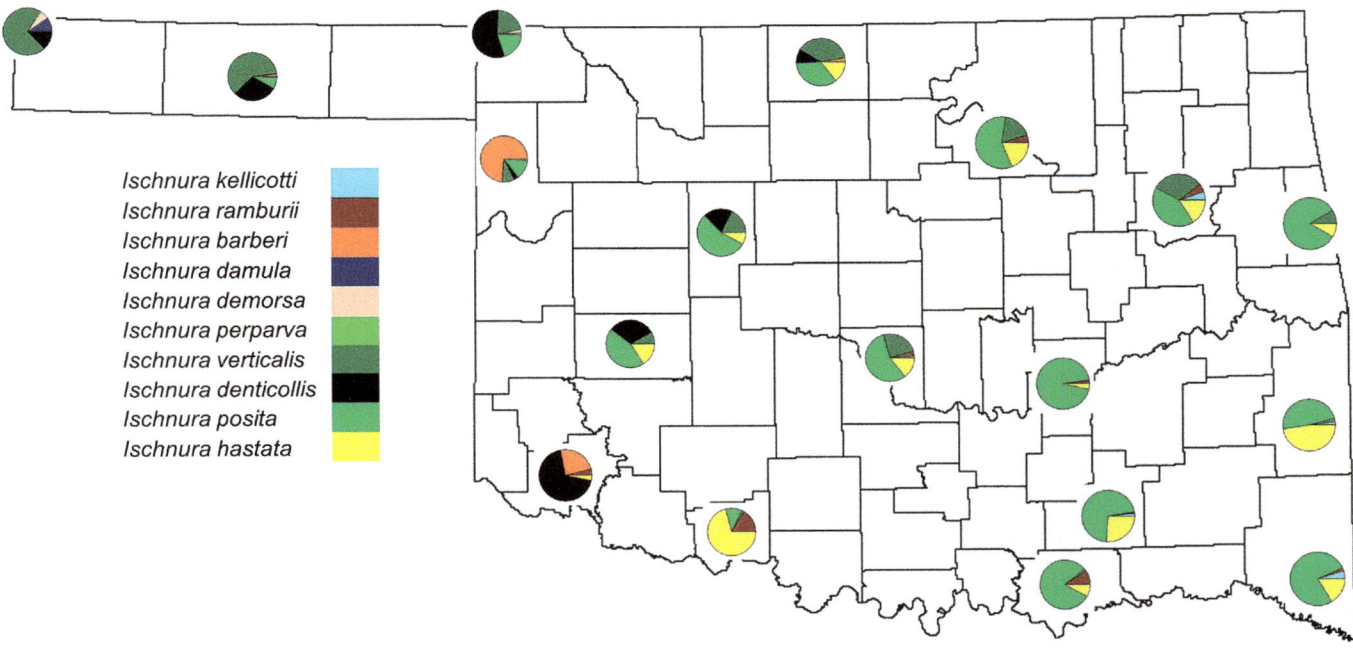

Figure 12.10 Relative composition of assemblages of forktails (*Ischnura*) across Oklahoma.

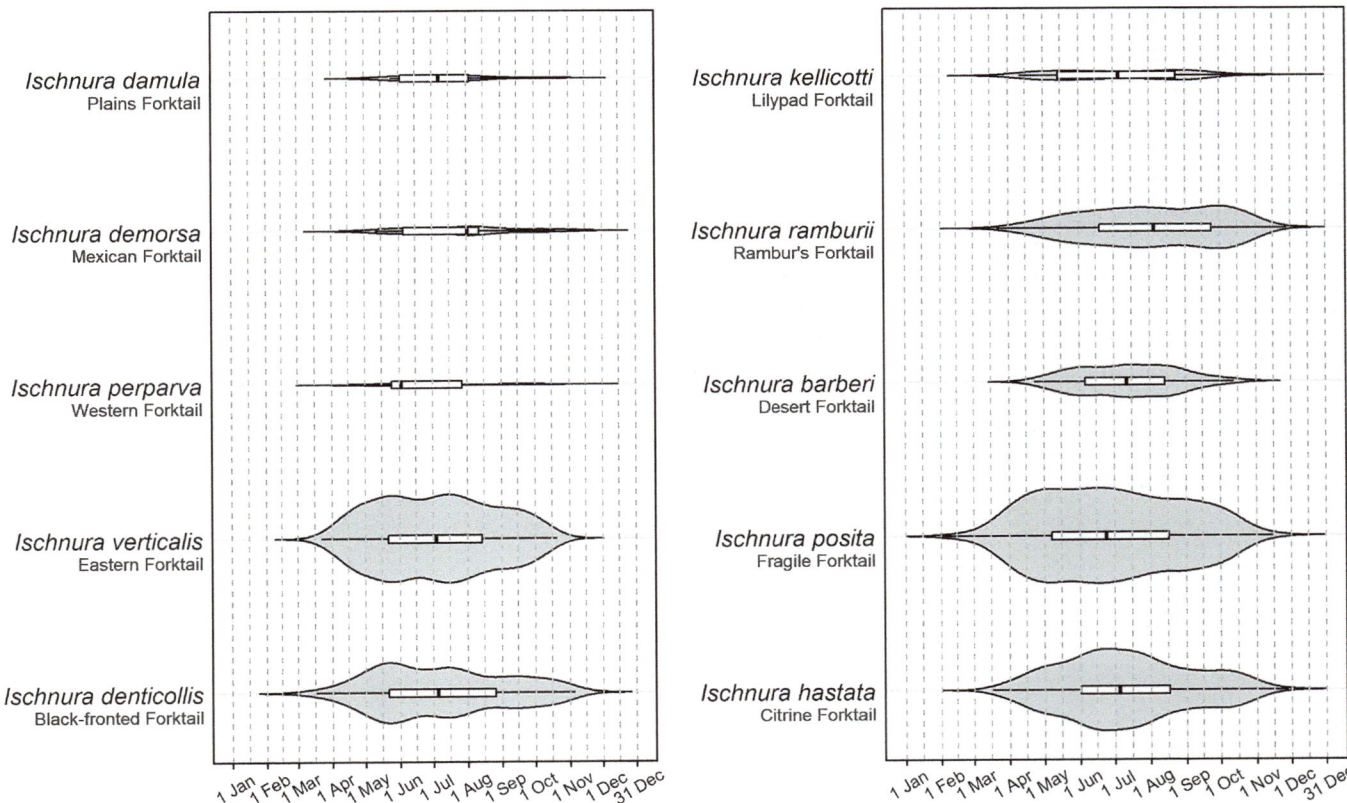

Figure 12.11 Flight seasons in Oklahoma of forktail species (*Ischnura*), parceled out by A) those species in what we dub the "Eastern Forktail (*I. verticalis*) group," which includes three scarce western species, and B) all other species, including the two most common and widespread ones, the Fragile Forktail (*I. posita*) and the Citrine Forktail (*I. hastata*), and two ecological specialists, the water lily-obligate Lilypad Forktail (*I. kellicotti*) and the saline Desert Forktail (*I. barberi*).

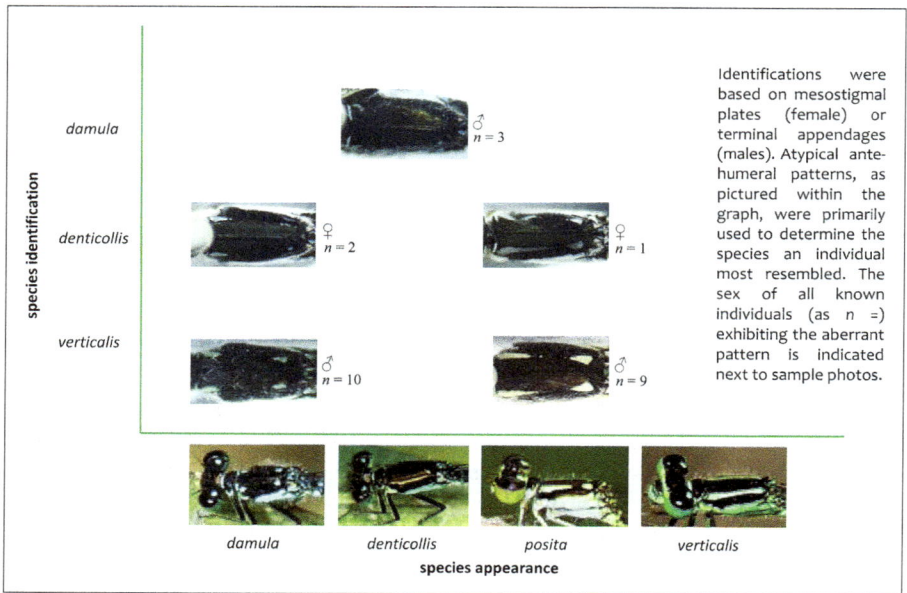

Figure 12.12 Aberrant *Ischnura* that are identified as one species but that look like another.

and other moving waters (Figure 12.13) and the bluets far more numerous and speciose at ponds, lakes, impoundments, and other still waters. The division would be cleaner were it not for the Seepage Dancer (*A. bipunctulata*), whose occurrences we classified as lentic because most are from marshy ponds formed from seeps, yet recognizing that some would classify seeps themselves as lotic, and the Blue-fronted (*A. apicalis*) and Powdered (*A. moesta*) Dancers,

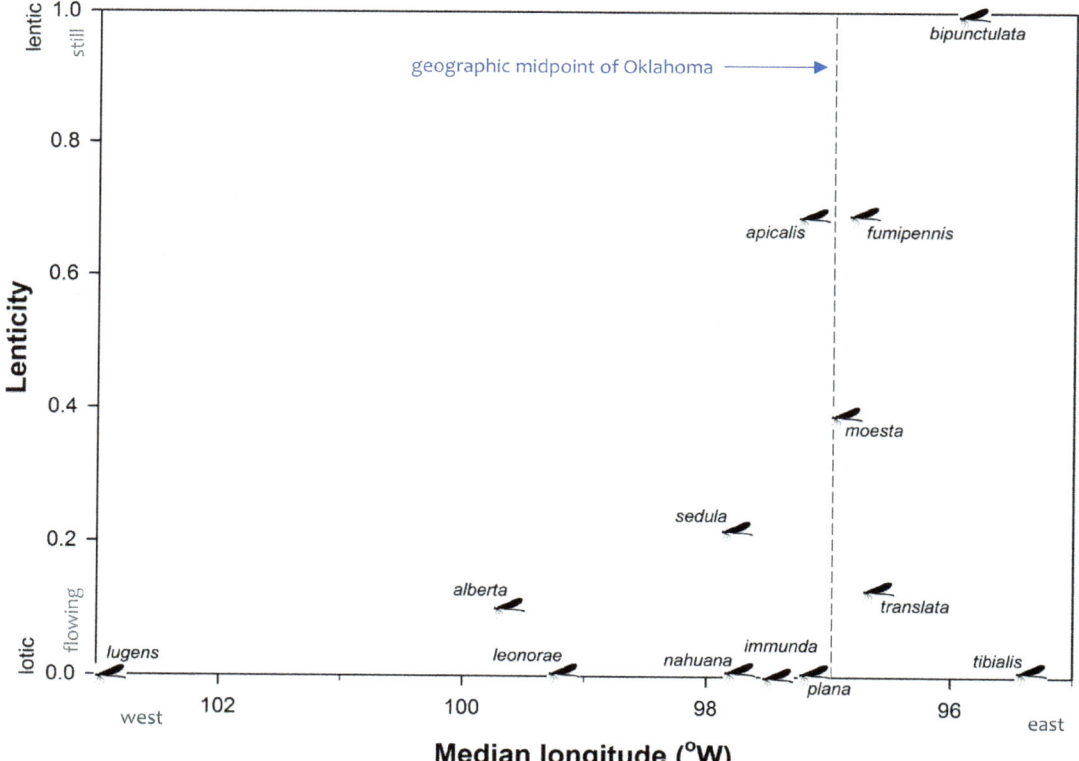

Figure 12.13 A scatter plot of midpoint (i.e., median) longitude of occurrences against tendency to occur in lotic vs. lentic habitats of 11 species of *Argia* that occur regularly in Oklahoma, along with two other species, Leonora's Dancer (*A. leonorae*), whose status is unclear but has been recorded 3 times, and the Sooty Dancer (*A. lugens*), which seems to appear in the western tip of the panhandle only in years with favorable hydrologic conditions. Note that the eastern edge of the panhandle is at the 100th meridian.

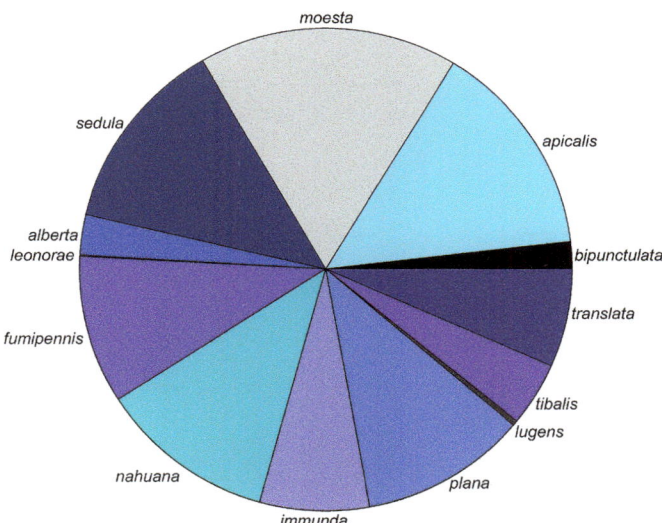

Figure 12.14 **Relative commonness in Oklahoma of 13** *Argia*. **An additional species, the Tezpi Dancer (*A. tezpi*), has occurred just once as a far-flung vagrant.**

whose abundance (Figure 12.14) and promiscuous tastes conspire to allow them to occur wherever they wish, such that both may be encountered frequently in lentic habitats (Figure 12.13), even large reservoirs. The Variable Dancer (*A. fumipennis*) also is most numerous around well-vegetated lakes and ponds. Curiously, the Blue-ringed Dancer (*A. sedula*) is most abundant on clear creeks and streams, but occurs often at spring-fed pools and ponds yet seldom at other lentic waters. Otherwise, the dancers are a strongly lotic group distributed chiefly in the western part of Oklahoma—only two species, the Seepage Dancer and Blue-tipped Dancer (*A. tibialis*), have their geographic centers of abundance distinctly in eastern Oklahoma (Figure 12.13).

In general, flight seasons of the various dancers are long (e.g., April to October) and peak in mid-summer (Figure 12.15). An exception is provided by the Seepage Dancer, which really is a spring and early summer species but whose seasonal distribution is skewed by late-season records from a single site in McCurtain County, where the species was found as late as early November (Figure 12.15A)! No other locale in the state has a record later than mid-August. Another exception is provided by the Paiute Dancer (*A. alberta*), a generally scarce species of marshy springs and shallow, clear creeks; it peaks in late spring or early summer, and most have disappeared by early September, although there are a handful of records up to mid-October (Figure 12.15B). Yet another example is provided by the Blue-tipped Dancer, which has a comparatively short flight season. By contrast, the Dusky Dancer (*A. translata*), a more widespread species with which the Blue-tipped often co-occurs, has a long flight season and peaks well after the Blue-tipped does (Figure 12.5C). We cannot say much about flight seasons for two species. The status of Leonora's Dancer (*A. leonorae*) is unclear: it has been recorded in the state just three times. We do not know if—but we expect

there is—a small, geographically disjunct population in southwestern Oklahoma. The huge Sooty Dancer (*A. lugens*) has been found in summer only in the western tip of the panhandle and then only in years with good flow in the various creeks and rivers.

In most cases, the number of individuals of a given species one would expect to encounter in an era has been stable, that is, the number encountered in the early era (before 1950) is about the same encountered in the middle (1950–1999) and current eras (2000–present; Table 12.1). Notable exceptions are the Aztec Dancer (*Argia nahuana*), with a sharp increase from 1.9 individuals early on to 14.6 individuals currently (see that species account for discussion). On the other end of the spectrum is the Blue-tipped Dancer, which appears to have experienced a decline in its numbers over time. The Paiute Dancer, also shows an apparent decline in numbers, but we suspect that is due to the near absence of records of the species in the mid-era, an omission we infer was from problems with conclusive species determination.

Our final word about dancers in Oklahoma is a word of caution: variation is high, at least in some species. For instance, we have collected a ♂ Variable Dancer on the Cimarron River at the tip of the panhandle that had an unforked humeral stripe, only slight contrast in color in segments 8 and 9 of the abdomen, and lacked a black stripe laterally on those segments (SP 1611). Outwardly, this individual easily could have been passed off as an Apache Dancer (*A. munda*); to be sure, it matched almost perfectly the photograph of a ♂ Apache Dancer included in the color supplement to Westfall and May's (2006; May and Dunkle 2007) handbook of North American Zygoptera. Nevertheless, this individual had the terminal appendages of a Variable Dancer. Likewise, we collected a strikingly odd ♂ Springwater Dancer (*A. plana*; SP [RWG] 1658) at Lake Hall, Harmon County, that had smoky amber wings, a boldly forked humeral stripe, and an extensively black abdominal pattern that could have passed as a Golden-winged Dancer (*A. rhoadsi*), yet the terminal appendages were nothing like the distinct ones of that more southerly species. In both instances, we worry—openly, here in print—that photographs of an Apache or Golden-winged Dancer submitted with good intentions to an archival site such as Odonata Central could have been accepted without knowing that the identification was incorrect.

Even a specimen will not help always. We collected a curious ♀ dancer on Pond Creek 3 km WSW of Lamont, Grant County, 30 August 2014 (SP 1408) that neither we nor the esteemed Rosser W. Garrison, the world's authority on the genus, can diagnose. There is no apparent hybrid combination that would lead to the characters this individual exhibits, so at present its identity remains a mystery. Likewise, variation in the Springwater Dancers of central Oklahoma is bewildering, with some individuals having wide black humeral forks or abdominal segments so black that there is almost no blue visible (see that species account for details).

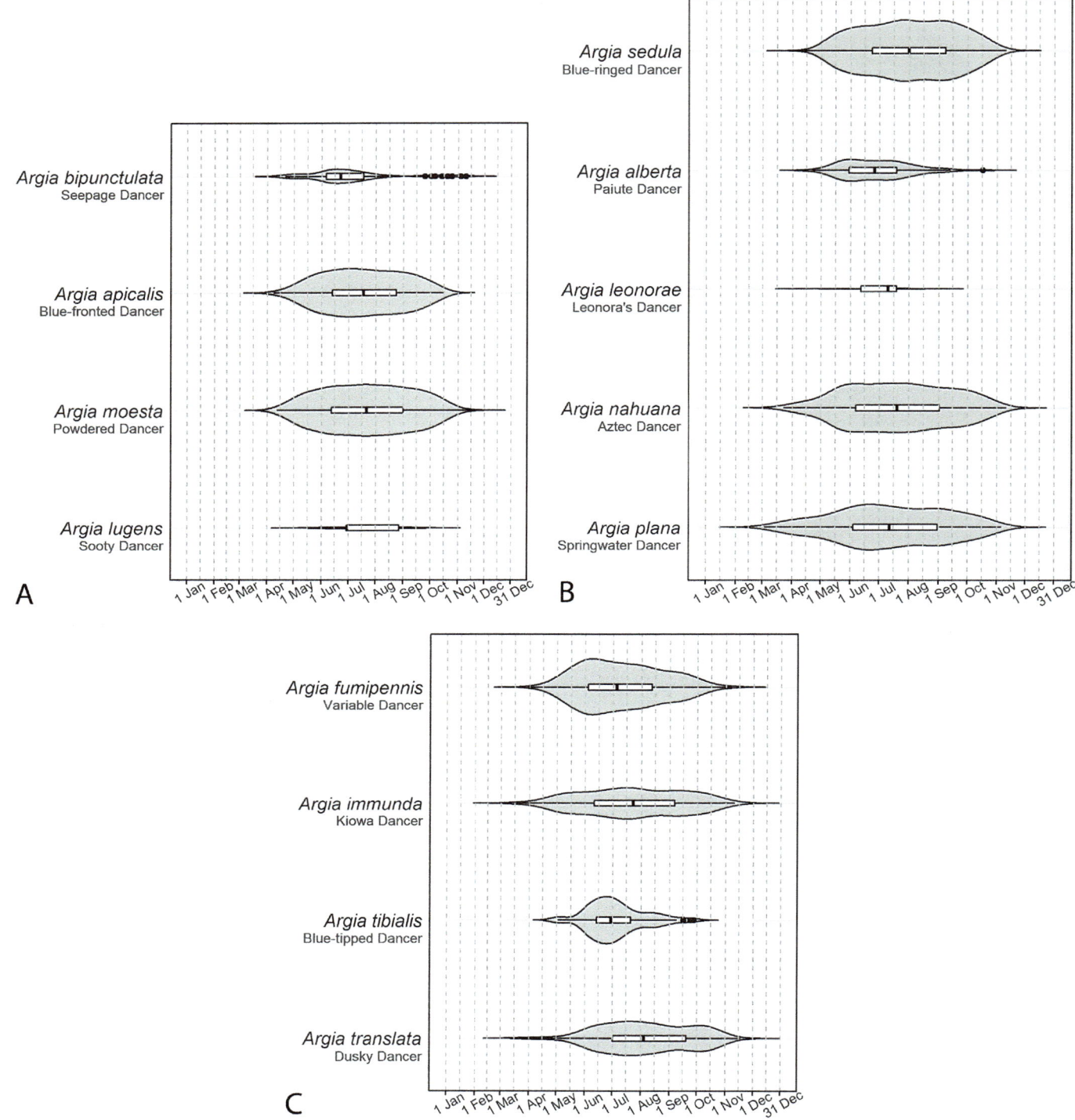

Figure 12.15 Flight seasons in Oklahoma of *Argia* species, grouped as A) widespread or localized species, B) blue species, and C) violet species. A fifth species in the violet "group," the Tezpi Dancer (*A. tezpi*), has been recorded just once, on 14 May.

Table 12.1 **Lenticity and expected number of individuals encountered by era for the twelve regularly occurring *Argia* species in Oklahoma. The lenticity scale ranges from purely lotic (=0, lighter blue) species to those that are purely lentic (=1, darker blue). Number of individuals were standardized by the number of survey days[a] in each of the three periods: early (before 1950), mid (1950–1999), and late (2000–present). Encounter rate[b] for most species has been roughly stable (yellow) but note sharp increases in encounters for some species (*A. nahuana*, *A. sedula*; darker the green, the higher the increase). Two species show apparent declines (*A. alberta*, *A. tibialis*; light and dark orange), but only that for *A. tibialis* may be a true reflection of reality.**

	LENTICITY	EARLY	MID	LATE	TREND (*C*)
Argia bipunctulata (Hagen, 1861) – Seepage Dancer	0.997	0.6	1.0	1.2	0.721
Argia apicalis (Say, 1839) – Blue-fronted Dancer	0.691	11.6	21.0	34.0	1.365↑
Argia moesta (Hagen, 1861) – Powdered Dancer	0.392	10.5	15.4	40.6	1.662↑
Argia sedula (Hagen, 1861) – Blue-ringed Dancer	0.218	4.8	7.1	26.4	2.507↑
Argia alberta Kennedy, 1918 – Paiute Dancer	0.105	1.9	0.1	3.4	-0.073↓
Argia fumipennis (Burmeister, 1839) – Variable Dancer	0.693	9.2	10.4	14.6	0.354
Argia nahuana Calvert, 1902 – Aztec Dancer	0.008	1.9	10.7	14.6	5.620↑
Argia immunda (Hagen, 1861) – Kiowa Dancer	0.000	4.5	6.2	9.9	0.796
Argia plana Calvert, 1902 – Springwater Dancer	0.013	11.1	10.3	15.5	0.157
Argia lugens (Hagen, 1861) – Sooty Dancer	0.000	0.2	0.2	0.2	0.196[c]
Argia tibialis (Rambur, 1842) – Blue-tipped Dancer	0.012	6.5	2.9	4.6	-0.429↓
Argia translata Hagen, 1865 – Dusky Dancer	0.129	2.1	4.0	9.7	2.315↑

[a] Standardization is straightforward: for each species, the number of individuals across a given survey period was divided by the number of survey days in that period, with the quotient multiplied by 100.

[b] We calculated the proportion of increase or decrease between 1) early and mid and 2) early and late and averaged those two results, the idea being to provide a single metric, *C*, that would provide a good gauge of change in encounter rate. We classified any species with $C < 0$ as declining, any species with $0 < C < 1$ (more or less) as stable, and any species with $1 < C$ as increasing (in other words: less than zero = declining, greater than zero but less than one = stable, and great than one as increasing).

[c] Actually numbers by era are 0.16, 0.19, 0.19, early, mid, and late; all round to 0.2.

ENALLAGMA PRAEVARUM (HAGEN, 1861) – ARROYO BLUET

The Arroyo Bluet was once thought to be restricted to the western panhandle because prior to 2002 it was known in Oklahoma from only six records in the Black Mesa area of Cimarron County. One would have guessed that another panhandle county would have been the second to get a record of this species, but, surprisingly, the next came from Comanche County: 1♂, a probable vagrant, was collected on Blue Beaver Creek on the Fort Sill MR on 11 October 2002 (CSU; Zuellig et al. 2006). Previously the species had only been collected in 1930 by RD Bird (on 27 July, specific locality unknown), the first record for the state and the only time it was reported prior to the Bick/Hornuff era, and from 2–11 August between 1968 and 1970 when GH Bick and LE Hornuff collected in the Black Mesa area. The specific collecting localities indicate that Bick and Hornuff collected near where North Carrizo Creek meets with the Cimarron River (2.4 mi NE of Kenton) and along Carrizozo Creek (2.1 mi W of Kenton) as well as at Lake Carl Etling. These localities mostly jibe with the idea that this is a species of slow-moving streams and lake margins. This is also the area where the species has been encountered in recent years.

In our paper discussing odonate species of conservation concern in the state (Patten and Smith-Patten 2013b), we treated the species as possibly extirpated from Oklahoma. At the time, outside of the Comanche County vagrant record, the species had not been detected in the state for 35 years despite many visits to the area during those years. Within the next few years after the paper was published, *Enallagma praevarum* was encountered on at least three occasions (3 June and 26 October 2015,[2] which are the early and late date for the species in Oklahoma; the third record was on 7 October 2017, 1♂, SP 2519; and possibly two other times.[3]). To provide context for these records as well as for

the reason we leaned toward the idea of extirpation was that between 2003 and 2017, Cimarron County had been visited by odonate watchers at least 52 days. Some of those observers are quite skilled zygopteran surveyors, such as VWF, who visited three times, and BC, who was there for 17 days. We surveyed the county on 12 days. An encounter rate of 3 out of 52 (or 32, if we wanted to limit records to ours, BC, and VWF), is poor. Conversely, GH Bick and LE Hornuff surveyed Cimarron County on 10 days between 1960 and 1978, during which they encountered the species 5 times.[4]

In Oklahoma, *Enallagma praevarum* is known to fly from early June to late October, flight dates that are much truncated compared to New Mexico and Texas, but right in line with neighboring Colorado (OK, all records from 2015 in Cimarron County: early date is 3 June, 2♂, 10 km N of Kenton, North Carrizo Creek, 1♂ as SP 1608 and 2♂, 4 km NNE of Kenton, Cimarron River., 1♂ as SP 1604, and late date is 26 October, 1♂, "Watson's Crossing," BC, OC 438265; CO: 5 June–17 October, CSU; NM: 13 March–9 November, OC 430260, OC 427829; TX: 14 February–17 November, Abbott and Lasley 2019). Note that there is but one confirmed Oklahoma record for early June (see previous endnote); all others are from late July onward, which makes us wonder about the possibility that this species is a seasonal migrant to the state, i.e., in particularly productive years, it ventures eastward beyond its "normal" range later in the season. In this sense it would be perhaps a colonizer or an occasional breeder. We have no known records indicating breeding (pairings, ovipositing, nymphs, tenerals) in Oklahoma; in fact, we only have two females ever reported. But it may be that the species' status in Oklahoma is like that in Kansas,[5] where it is known from one adult and

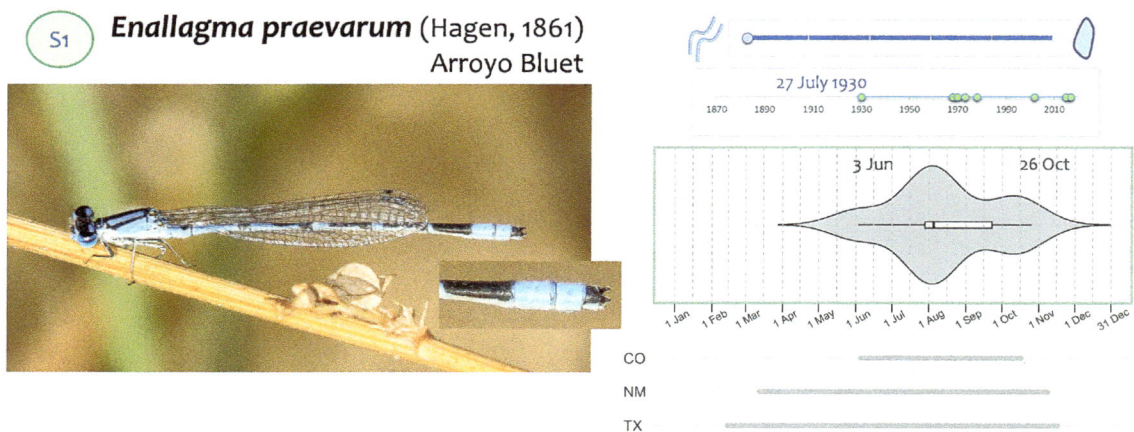

S1 ***Enallagma praevarum*** (Hagen, 1861)
Arroyo Bluet

27 July 1930

3 Jun   26 Oct

CO
NM
TX

Figure 12.16 *Enallagma praevarum*, the Arroyo Bluet. This species can be confused with both *E. civile*, the Familiar Bluet, and *E. carunculatum*, the Tule Bluet, making in-hand examination nearly always vital. An exception is when the terminal appendages of a ♂ can be seen in a good photograph, as with this one on the Cimarron River 13 km E of Kenton, Cimarron County, 26 October 2015 (©Bill Carrell, OC 438265). In general, the ♂ can be identified by its smaller size, blacker abdomen, and utterly different terminal appendages.

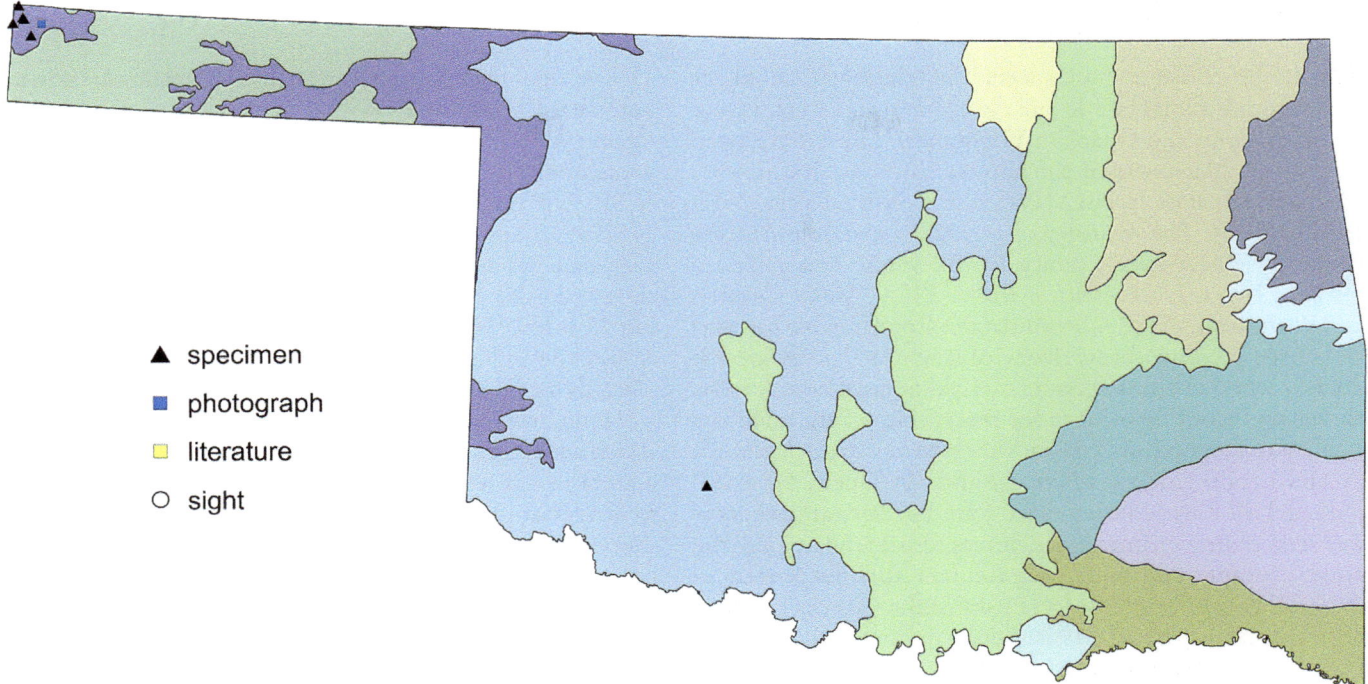

Figure 12.17 Documentation level of supporting records for *Enallagma praevarum*, the Arroyo Bluet, in Oklahoma.

three nymphs. The Kansas records suggest that the species is an occasional breeder there and may at some point colonize. However, for both Oklahoma and Kansas, only time will tell what the status is. It may be that there are detection or survey effort issues not allowing for a clear picture. Alternatively, it may be that a simple vagrant status is more appropriate. For now, we suggest carrying forward our S1

conservation status recommendation (Patten and Smith-Patten 2013b) because no matter how one looks at the situation, the species is rare in Oklahoma, it has a low population density, habitat in the panhandle is under threat given streamflow loss (see Chapter 6), and current indications are that the species has declined.

ENALLAGMA DOUBLEDAYI (SÉLYS, 1850) – ATLANTIC BLUET

The Atlantic Bluet is a rare species in Oklahoma that is limited to ponds (both tannic and clear, but usually fishless) in pineywoods in the Ouachita Highlands. There have been just shy of 20 reports of the species since we first discovered it in the state in 2012 (Patten and Smith-Patten 2012; Figure 12.18). The majority of records come from McGee Creek WMA, Atoka County (*n* = 8), where a population resides. Two other localities, Atoka PHA, Atoka County, and Pushmataha WMA, Pushmataha County, have had two and three records, respectively, of single males, making it unclear whether those spots truly hold populations. A more promising breeding locality for the species in the state is a small quarry pond off Union Valley Trail (wide dirt road) in McCurtain County where 5♂ and 1♀ (1 pair; 2♂ as SP 2745 and SP 2746, 24 May) were found in 2019 and where it was seen multiple times since. It was also found at a nearby pond. Another site with promise was also discovered in 2019: Honobia Creek WMA, Pushmataha County (5♂, 1 as SP 2780, 1 June, MAP). This species undoubtedly has been encountered at other times in Oklahoma but was passed off as *Enallagma civile*, the Familiar Bluet. The Oklahoma records are the farthest north and most inland population of the species in the Southern Great Plains.

The first record for the species in Oklahoma was at a number of small ponds within the pineywoods of McGee Creek WMA. Early in the evening (~18:30 CDT) on 2 September 2012, we encountered 30♂ and 2♀ (3♂ collected as SP 390, 391, and 392, last two donated to JCAC, and 1♀ as SP 393). Approximately 60♂ were counted on our second visit, on 8 September 2012, but no females were detected (2♂ collected as SP 394, 395, the former was donated to JCAC). On 13 July 2013, we made a high count at the pond, which is also the highest for the state, of 75 individuals (7♂, 6♀, 62U, 1♂ as SP 807), including 6 pairs. *Enallagma doubledayi* was seen every year (2012–2017[6]) at the WMA ponds, all of which are artificial and located within a few meters of a well-maintained dirt road. Vegetation at the ponds was limited to sedges and grasses of low stature along the immediate shoreline; there were no lily pads or emergent vegetation, both of which were prevalent at other ponds in the WMA. The ponds appeared to be fairly shallow throughout, with mostly clear water around the edge that was more tea-colored (likely acidic) toward the center (Figure 12.19). We encountered *E. doubledayi* in association with species such as *Nehalennia integricollis* (Southern Sprite), a first for the county, *Enallagma aspersum* (Azure Bluet), *Anax*

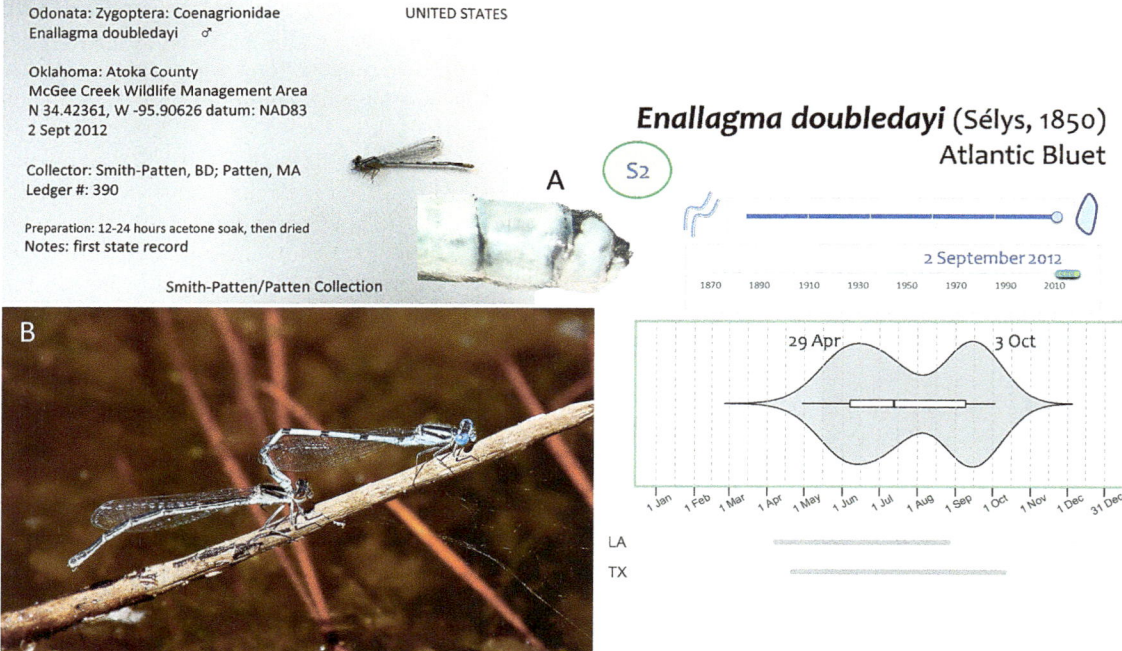

Figure 12.18 *Enallagma doubledayi*, the Atlantic Bluet, A) ♂, McGee Creek WMA, Atoka County, 2 September 2012 (SP 390, OC 381566), one of 30♂ and 2♀ at a clear, acidic pond embedded in piney woodland that were the first found in Oklahoma (Patten and Smith-Patten 2012). B) tandem pair, from the same locale on 24 September 2014 (©James W. Arterburn). Regrettably, females are often identified using a guilt-by-association method, but we must always keep in mind that mixed-species pairings are possible albeit relatively rare. The ♀ of this pair thus is likely to be *E. doubledayi* even though her antehumeral stripe is curiously shortened.

Figure 12.19 Pineywoods pond typical of where *Enallagma doubledayi*, the Atlantic Bluet, has been found in Oklahoma. McGee Creek WMA, Atoka County. Photo by ©Brenda D Smith.

junius (Common Green Darner), *Tramea lacerata* (Black Saddlebags), and *Tramea carolina* (Carolina Saddlebags).

In April 2013 we encountered the species at a second Oklahoma site, that of Atoka PHA, Atoka County (29 April, 1♂, SP 544), about 12 km north of McGee Creek WMA. We found another lone ♂ at this same pineywoods pond on 13 July 2013 (captured and released). The species has not been seen again at the PHA despite seven other visits. At Pushmataha WMA, Pushmataha County, we encountered single males on three separate dates at three different ponds (10 June 2013 [SP 690], 7 June 2015 [see below], and 3 June 2018 [SP 2614]). The ♂ in 2015 was found in tandem with a ♀ *Enallagma aspersum*; we collected the pair (SP 1644). To our knowledge, mixed-species pairs that include *E. doubledayi* have not been documented previously; *E. aspersum* has been documented as a mixed-species pair, but it was a ♂ *E. aspersum* with a ♀ *E. geminatum*, the Skimming Bluet (Bick and Bick 1981).

Enallagma doubledayi was not known to occur west of the Mississippi River until SW Dunkle discovered it in Collin County, Texas, in the 1990s. Dunkle found and collected only a single ♂ that day despite examining many *Enallagma* bluets; all other individuals proved to be *E. civile* (*in litt*). *E. doubledayi* has since been recorded in Anderson and Bastrop Counties, Texas (Abbott 2001, 2011), including a healthy population at Gus Engeling WMA, Anderson County, discovered in 2011 by M. Reid and GW Lasley. Louisiana is the only other state west of the Mississippi River in which established populations have been located;

in 2004, G and J Strickland discovered the first record for Louisiana in Natchitoches Parish (Ellzey 2004). The species has been found since in four other parishes: Grant, Rapides, St. Tammany, and Union (OC, Mauffray 2014). There is one record for Arkansas, from Union County, of a ♂ that was photographed and about 10 other individuals, found on 11 June 2016 by R and A Baldwin (OC 446603). It is unknown if this is an established population, but number of individuals may indicate as such. At this point it is unclear if the species expanded its range in the 1990s or the species was simply mistaken for *E. civile* previously. Dunkle's encounter would indicate the former; but the latter cannot be discounted.

In Oklahoma, *Enallagma doubledayi* is known to fly from late April until early October, which is similar to the seasonality for Texas (OK: 29 April 2013, as above, and 3 October 2015, McGee Creek WMA, Atoka County, 4♂, MAP, 1♂ as SP 1805; TX: 22 April–11 October, OC 443985, Abbott and Lasley 2019). Louisiana has a shorter known flight season (8 April–26 August, Mauffray 2014, OC 332116) than those of Oklahoma and Texas, but one would expect that in reality it is not different.

Three morphological features of note are from 3 of the 16 specimens (15♂, 1♀, SP, JCAC) collected in the state. One ♂ has a black splotch dorsally on S4–5 (SP 2746), another ♂ has a sooty, smudgy extension on the dorsum of S6 rather than the more typical black rocket (or dart) shape (SP 807), and another ♂ has a broken antehumeral on its left side (SP 392, now JCAC 49307).

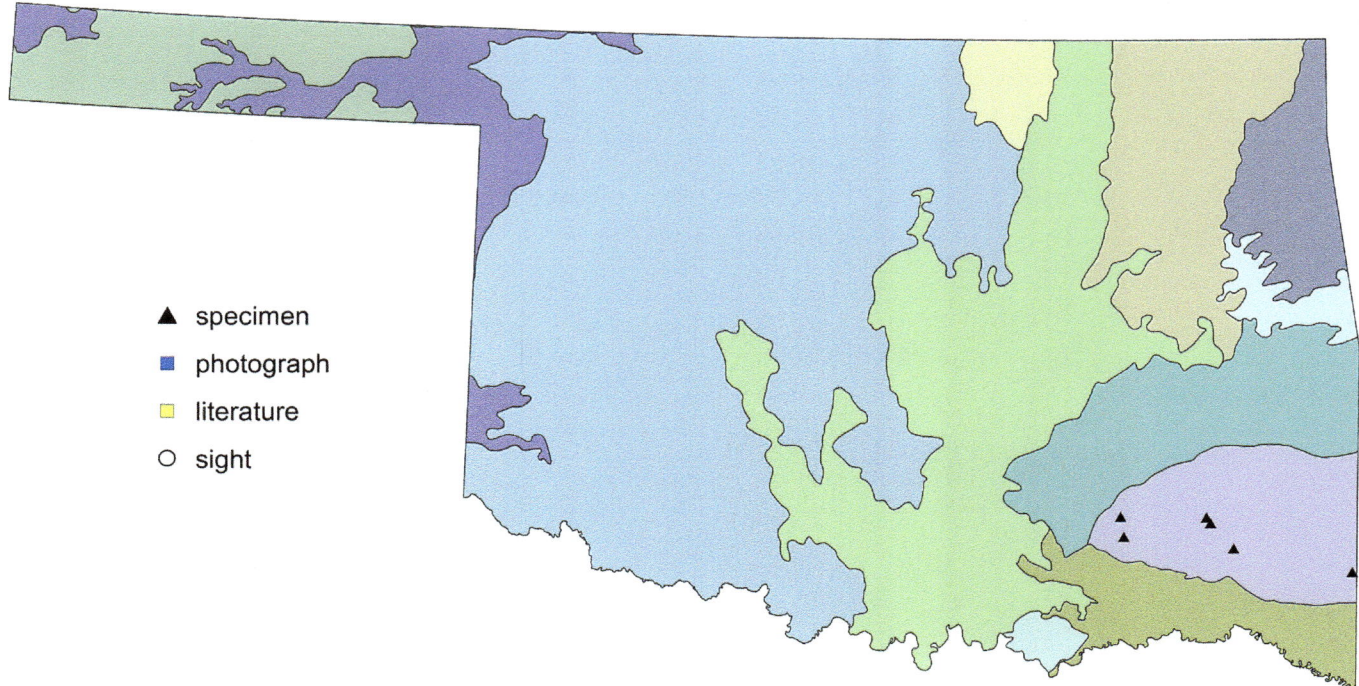

Figure 12.20 Documentation level of supporting records for *Enallagma doubledayi*, the Atlantic Bluet, in Oklahoma. Note that all records are from the Ouachita Highlands and all localities are supported by specimens.

ENALLAGMA CIVILE (HAGEN, 1861) – FAMILIAR BLUET

Just as it is elsewhere, the Familiar Bluet is aptly named for its presence in Oklahoma. It can be excruciatingly tedious and frustrating teasing out what else may be present in the swarms of *Enallagma civile* that one can encounter—it is routinely seen in the many hundreds and sometimes into the thousands, with higher numbers found in western Oklahoma (Figure 12.22). Our highest estimate is 3,000 individuals (estimated 500 pairs) on 27 September 2013 at

Figure 12.21 *Enallagma civile*, the Familiar Bluet, A) typical ♂, Cleveland Lake, Pawnee County, 9 August 2014 (©Bill Carrell, OC 425901), B) ♂ with an odd pattern of black on the abdomen at Purcell Lake, McClain County, 18 June 2016 (©Emily A Hjalmarson, OC 474238). C) Odder still, some ♂ can have extensive black on the abdomen, such as this ♂ (left) from the Black Mesa, Cimarron County, area on 25 May 2019, with a ♂ *E. carunculatum*, the Tule Bluet (right) for comparison (©Michael A Patten). The ♀ of *E. civile* can vary in color, from tan to andromorphs that are blue D) Osage Hills SP, Osage County, 2 June 2013 (©James W Arterburn, OC 427971). E) Fort Cobb SP, Caddo County, 21 July 2010 (©James W Arterburn, OC 437594).

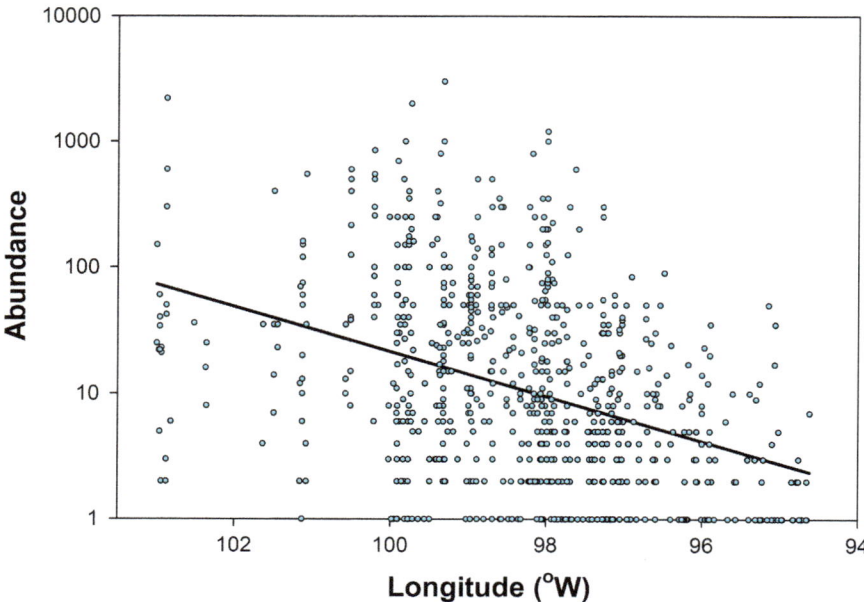

Figure 12.22 Abundance (on a log scale) in Oklahoma of *Enallagma civile*, the Familiar Bluet, as a function of longitude. The steep trend line emphasizes this species' increase in abundance from east-to-west across the state. This same pattern of higher abundance in western Oklahoma holds for several other species widely distributed in eastern North America, including *Ischnura verticalis*, the Eastern Forktail, and *Libellula pulchella*, the Twelve-spotted Skimmer.

Altus city lake, Jackson County, and we would not be surprised if our count was a gross underestimate. At that same locality, on 16 July 2009, VWF estimated 10,000 individuals (OC 314080). Huge numbers such as these are encountered elsewhere in the species' range, too.

Enallagma civile was first reported in Oklahoma by EB Williamson when he collected a pair at Wister, Le Flore County, on 3 June 1907 (UMMZ). It has been reported consistently since; on >850 days since its discovery in the state. The species also is widespread in Oklahoma: it is recorded in all 77 counties. And it is found anywhere and everywhere: creeks, streams, roadside ditches, drainage ditches, dry creek beds, springs, seeps, ponds, lakes, puddles, stock tanks, backwaters, flooded fields, pastures, muddy roadside trickles, parking lots, and sidewalks. Water can be clear or turbid, flowing or not, shallow or deep, alkaline or not. Creeks can be narrow to wide (<2–30 m), with muddy or rocky substrate. Waterbodies of all types can either have emergent or shoreline vegetation or not. The one common thread seems to be that the species rarely is found in shaded areas, seemingly preferring exposed, open spots.

Enallagma civile is known to fly almost year-round in Oklahoma, just as it does in Texas (Abbott and Lasley 2019). At their summer study site in Marshall County, the Bicks showed that adults were present at water from 18 June to 3 August, and the number increased dramatically in late July (Bick and Bick 1963). We assume they did not intend for those dates to be taken as the species' full seasonality for Oklahoma as they surely knew of records having dates beyond those they reported for the Marshall County site. We also doubt that they thought of those dates as being locally specific, even though they probably realized, as we do now, that seasonality within Oklahoma appears to vary. For example, in the panhandle the species is known to fly from

early April to late September, whereas in central Oklahoma it flies longer, from at least early March to late November. In the southeastern corner of the state it flies longer still (early March to early January). Elsewhere in the region, seasonality is more similar to those for central Oklahoma, being from spring to late autumn (AR: 6 April–20 November, OC 327645, OC 474888; CO: 16 May–24 October, CSU, OC 438577; KS: 21 April–13 November, SEMC 1347575, OC 315718; MO: 16 April–15 November, Sims 2012; NM: 13 March–24 October, OC 430259, OC 429153). Tenerals have been reported in Oklahoma from 24 April until 1 November.

Bick and Bick (1963) reported that *Enallagma civile*'s oviposition behavior varied highly, suggesting adaptability to a wide range of substrates. At their study pond, ovipositioning occurred principally in rootmats of *Salix* (willow), but also in young willow stems and leaves, *Eleocharis* (spikerush), *Bacopa* (water hyssop), "Bermuda"[7] (Bick and Bick 1963), and *Typha latifolia* (common cattail, Moss 1992). Ovipositioning often occurred beneath the surface (Bick 1963, Bick and Bick 1963, Moss 1992). We have observed ovipositing by single females and as pairs. Likely due to the difficulty with positive identifications, there have only been two collections of larvae in Oklahoma that have been identified as *E. civile* (14 nymphs in Marshall County, in June and July 1955 by GH Bick, specimens not re-located; because we are not proverbial gluttons for punishment, we rarely try to identify *Enallagma* nymphs to species).

Males with odd abdominal patterns have been found across the state and, at least in one instance, an aberrant thoracic pattern has been noted (broken antehumeral, Smith-Patten and Patten 2010, SP 92 [now OMNH 2000]). It is rather common to see males with at least a little extra black on their abdomens. The extra bit of black includes a tiny dot here or there, splattered spots, a random splotch, and

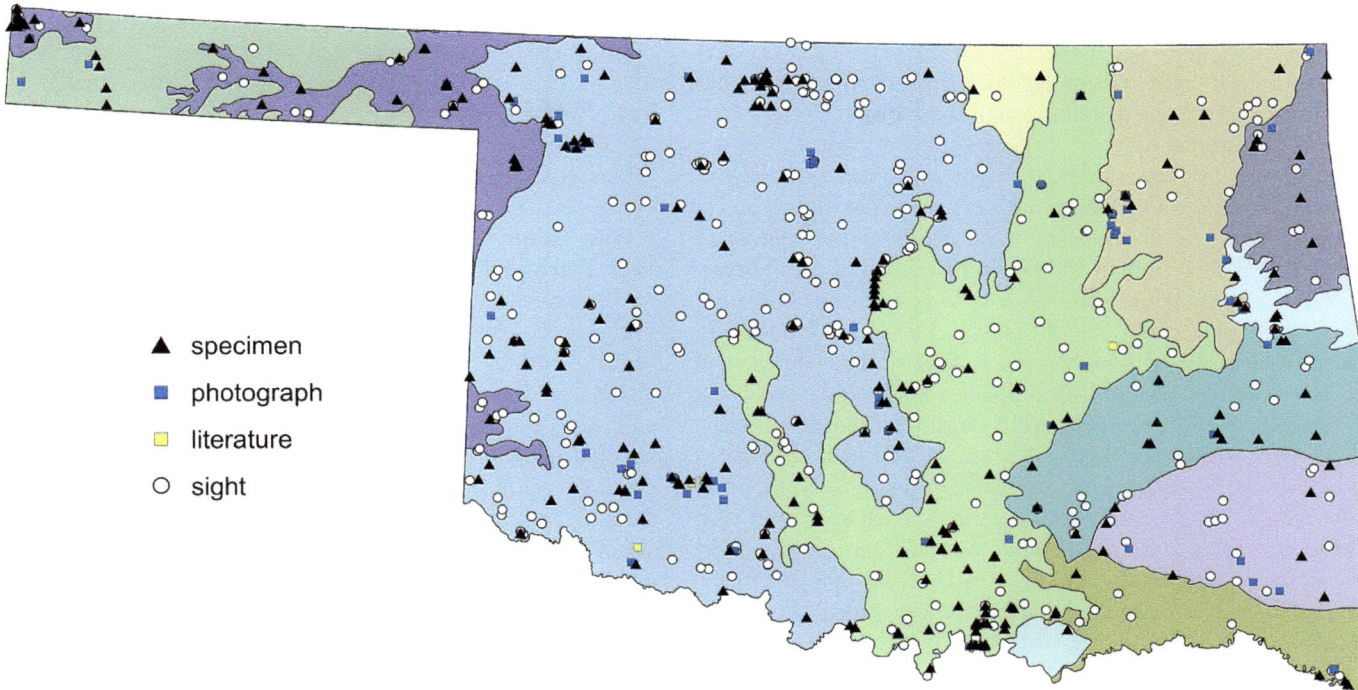

Figure 12.23 Documentation level of supporting records for *Enallagma civile*, the Familiar Bluet, in Oklahoma. Levels include specimen, archived photograph, literature reference, or sight record. Although sight records lack documentation, only those we feel are credible are plotted.

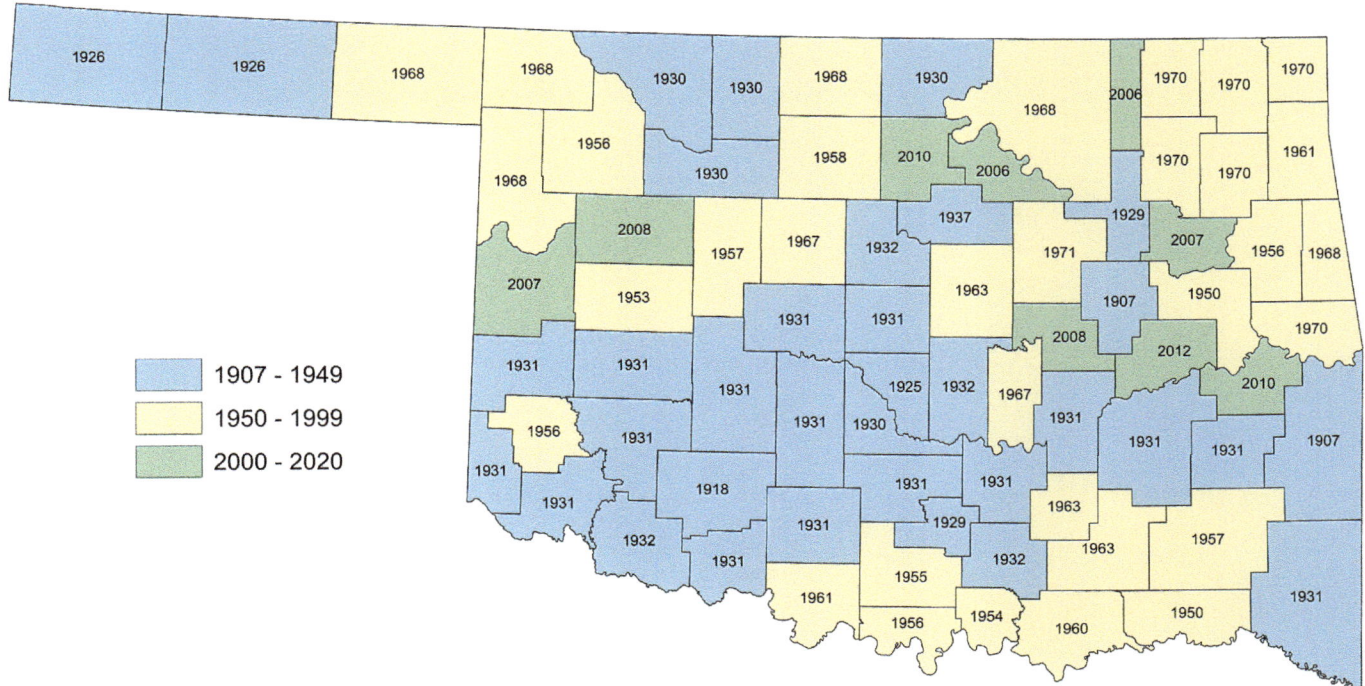

Figure 12.24 Counties in which *Enallagma civile*, the Familiar Bluet, are known to occur in Oklahoma. Year within the county is that in which the species was first reported.

extensions of the banding. Most do not appear to indicate interbreeding with other species, but there are a few encounters that have made us wonder if hybridization occurs in the region, although the distribution of these records is puzzling.

SP 441 and 442 were collected at Perry Lake, Noble County, on 29 September 2012 (the same day that we collected SP 443, an *E. carunculatum*, the Tule Bluet, at the lake; see that account below). Both exhibit some aberrancy, but SP 442 is the most interesting because of its aberrant dorsal abdominal pattern. S5–6 of this male has considerable extensions anteriorly of black and S3–4 has some additional black, all of which gives a dart-shaped appearance to the banding resembling a species such as *E. anna*, the River Bluet (SP 778, collected in Cimarron County is similar). Two other specimens of note are SP 1465 and 1466, which we collected at Altus city lake, Jackson County, on 21 October 2014. The first specimen has black splotches dorsally on S8–9, which, predictably, was confusing in the field. Even more confusing (and frustrating) was SP 1466 that we initially mistook for *E. carunculatum* given the extensive black dorsally on the abdomen, but that ♂ turned out to be an atypical *E. civile* by abdominal pattern but a typical *E. civile* by its terminal appendages (see *E. carunculatum* species account). A ♂ collected on the Cimarron River 4 km NNE of Kenton, Cimarron County, 25 May 2019 (SP 2773) likewise had extensive black dorsally, so much so that when MAP snagged it he felt certain it would prove to be other than *E. civile*, but the cerci do not lie. SP 2928, from Kiowa County, is yet another male with mixed characters that clearly call for further investigation.

We know of no mixed-species pairs involving *E. civile* in Oklahoma, but males of the species are known to attempt mating with *E. carunculatum*, *Argia moesta*, and *A. vivida* (Bick and Bick 1981). Williamson and Calvert (1906:148) said of a small pond in Indiana that, "I [Williamson] have studied probably 500 males [*E. civile* and *E. carunculatum*] from this pond, and I have found a few specimens (20–30 possibly) which were clearly intermediate." It would not be shocking then that anywhere *E. civile* and *E. carunculatum* come into contact there may be interbreeding producing atypical forms not easily determined to either species.

ENALLAGMA CARUNCULATUM MORSE, 1895 – TULE BLUET

The Tule Bluet is rarely encountered in Oklahoma—there are only 17 records—and when it is found it tends to be in low numbers. It is somewhat widespread, being found in five counties in the Oklahoma panhandle and the northwest. One record, for Noble County, may be an eastern vagrant rather than part of the Oklahoma population (see below). In Oklahoma, we have encountered it mostly at the margins of small to medium lakes, often within the vegetation consisting of *Scirpus* (bulrush), *Typha* (cattail), or grasses, habitat mostly consistent with Walker's (1953:204) description: "In general the favourite haunts of *E. carunculatum* are the beds of tall rushes, cat-tails, or other emergent plants that border lakes and rivers. It is however, by no means confined to such situations." The later part of that statement reveals why we should not have been surprised to find the species at a stark, roadside stock tank and overflow pond that completely lacked vegetation except for dead "tumbleweeds" (*Salsola*) that had tumbled into the water: upon spotting a ♂ bluet 10 km S of Conrad, Cimarron County, 26 May 2013 (SP 590) we convinced ourselves that we had found the long-awaited first state record of *Enallagma clausum*, the Alkali Bluet (shown for Oklahoma in error by Paulson 2009 and Abbott 2011), but, alas, upon capture we saw we had "only" found a ♂ (and ♀) *E. carunculatum*. This habitat was antithetical to what we thought the Tule Bluet should inhabit. The species is also known elsewhere to sometimes inhabit somewhat eutrophic as well as saline lakes, but the species is not "as characteristic of extreme environments as Alkali Bluet" (Paulson 2009:83).

Enallagma carunculatum was first discovered in Oklahoma on 8 July 1926, when TH Hubbell collected a ♀ 3 mi N of Kenton, Cimarron County (UMMZ). The only other time it was encountered in the early era of Oklahoma odonatology was on 27 July 1930, also in Cimarron County, but we are unsure where. The record was published without details by Bird (1932) but was elaborated on by the Bicks (Bick and Bick 1957, Bick note cards), who said that they examined a specimen from OMNH (probably OMNH 4351).[8] The next time *E. carunculatum* was collected in Oklahoma was in August 1970, which was the only time it was reported in the middle era of Oklahoma odonatology (Figure 12.26). In

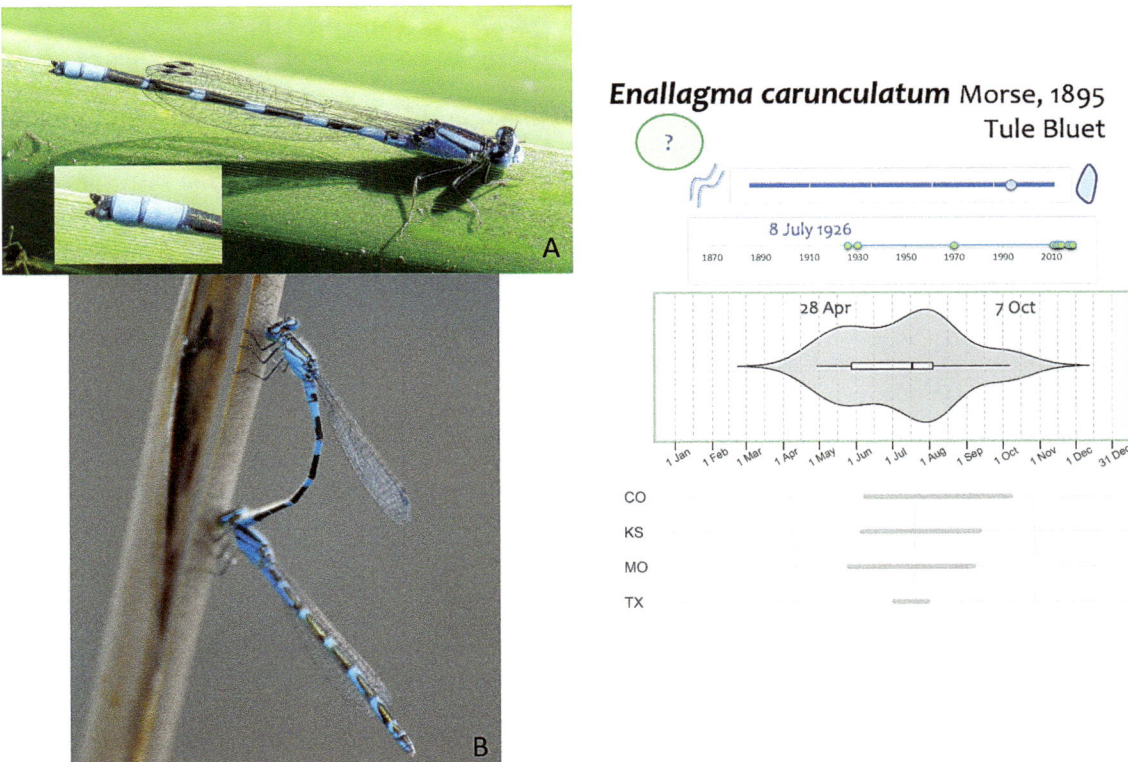

Figure 12.25 *Enallagma carunculatum*, the Tule Bluet, A) ♂, Dead Warrior Creek in Black Kettle NG, Roger Mills County, 4 October 2014 (©Bill Carrell, OC 427371). Extensive black on the abdomen is typical of this species and highly atypical of *E. civile*, the Familiar Bluet (but see Figure 12.21). If in doubt, check the cerci (inset). B) tandem pair, 25 km ESE of Buffalo, Harper County, 28 April 2012 (©Jason R Heinen, OC 374654). For all practical purposes it would be impossible to identify a lone ♀ without in-hand examination.

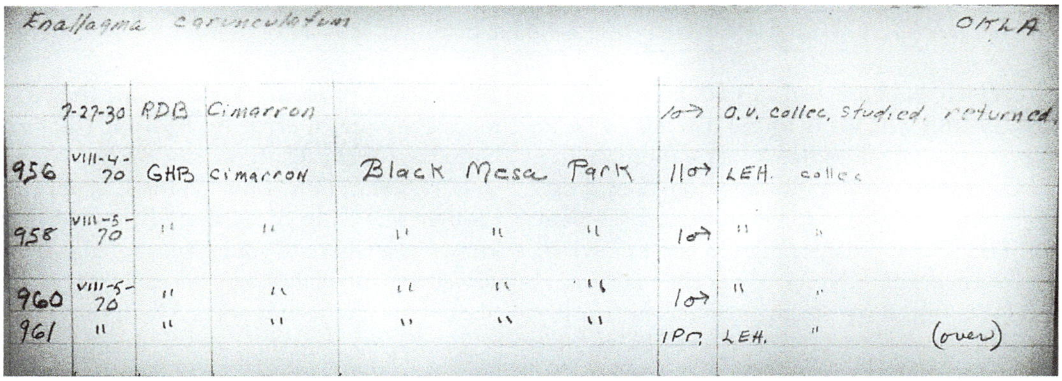

Figure 12.26 According to GH Bick's note, pictured here, the only time he (and LE Hornuff?) encountered *Enallagma carunculatum*, the Tule Bluet, in Oklahoma was on 4–5 August 1970 at Black Mesa SP, Cimarron County, where the species still occasionally occurs.

early August, GH Bick and LE Hornuff visited Black Mesa SP in Cimarron County where, according to Bick's note card, 14♂ and 1♀, including one pair, were observed during a two (or three[9]) day period. On 4 August, they had the high count for the state, with 11♂ (our highest count is 3♂, 2 June 2012, south end of Lake Evans Chambers, Beaver County, 1♂ as SP 260).

Another 41 years elapsed before the species was again found in the state (19 August 2011, VWF, Avard Lake, Woods County, OC 331879). The following year it was recorded in four additional counties—Harper (1 pair, 28 April, 25 km ESE of Buffalo, small reservoir [#5916], JR Heinen, OC 374654; 2♂, 12 May, Doby Springs, 1♂ SP 234), Beaver (3♂, 2 June, Lake Evans Chambers, 1♂ SP 260), and Noble (1♂, 29 September, Perry Lake, SP 443). And then in 2014, it was added to Roger Mills County (1♂, 4 October, Black Kettle Lake, Black Kettle WMA, BC, OC 427371), as the sixth county from which *Enallagma carunculatum* is known in Oklahoma.

Interestingly, although Oklahoma lies at the edge of the species' range and there are relatively few records, the state has the longest flight season of all of its neighbors, even those well within the species' range and having many more records. The earliest flight date for Oklahoma, of late April (see above), is a month earlier than the start date in Missouri, which is the second earliest for the region. Its late flight date in Oklahoma matches the latest date for the region (7 October), reported for Colorado; nonetheless, Oklahoma's 162 days of adult activity is 44 days longer than the season in Colorado (OK: 28 April–7 October, see Harper County record above, and 7 October 2017, Lake Carl Etling, Cimarron County, 1♂, SP 2518; CO: 8 June–7 October, OC 401417, OC 437915; KS: early June[10]–11 September, SEMC 1348708, SEMC 1347384; MO: 25 May–6 September, Sims 2012; TX: limited records, panhandle only, 2–30 July, Abbott and Lasley 2019). We have no records of teneral *Enallagma carunculatum* for Oklahoma; nor are there records of ovipositing or nymphs.

Of note, SP 443, the lone ♂ we collected at Perry Lake, Noble County, in late September 2012, clearly is *E. carunculatum*, yet its appendages led us to speculate it strayed from an eastern population. The reason for this speculation is the shape of the cercus, specifically of the dorsal extent of the dark portion (or sclerotized dorsal arm, per Westfall and May 2006). That dorsal arm sits atop a distal pale tubercle and generally that arm is anteriorly situated, leaving an obvious projection of the tubercle, which is considered so distinctive of *E. carunculatum*. In some *E. carunculatum*, that dorsal arm extends a little farther posteriorly, not projecting over the tubercle to form a "hood," as with *E. civile*, the Familiar Bluet, but extending enough to form a partial hood, akin to specimens we have from Pennsylvania and Ohio. SP 443 has such a partial hood, which we think is more akin to the eastern populations than it is to, say, a hybrid *E. carunculatum* × *E. civile*. Between the hood shape, the autumn date, the collecting locality being so far east from the Oklahoma population, and that the record is of a single ♂, it is plausible that the individual was a vagrant blown in from the east.

A second specimen (SP 1466, collected on 21 October 2014 at Altus city lake, Jackson County) is of note because it actually may be an *Enallagma carunculatum* × *E. civile* hybrid. When we first saw this extensively black ♂ we identified it as *E. carunculatum*. We caught it, stowed it away, and wrote in our notes that we had a first county record of *E. carunculatum*, one well removed geographically from other records. It was not until the next day, when we put it under the microscope and took a close look at the appendages, that we realized it was actually an aberrant *E. civile*. This individual has the abdominal pattern of *E. carunculatum*, but the terminal appendages that are unequivocally *E. civile*. We should not be at all surprised by such a hybrid, as *E. carunculatum* is known to hybridize with both *E. civile* and *E. anna*, the River Bluet (Bick and Bick 1981; Paulson 1974; Barnard et al. 2017).

Previously (Patten and Smith-Patten 2013b), we ranked this species as S3 (vulnerable). That rank may still hold, and we do continue to feel that the species is of conservation

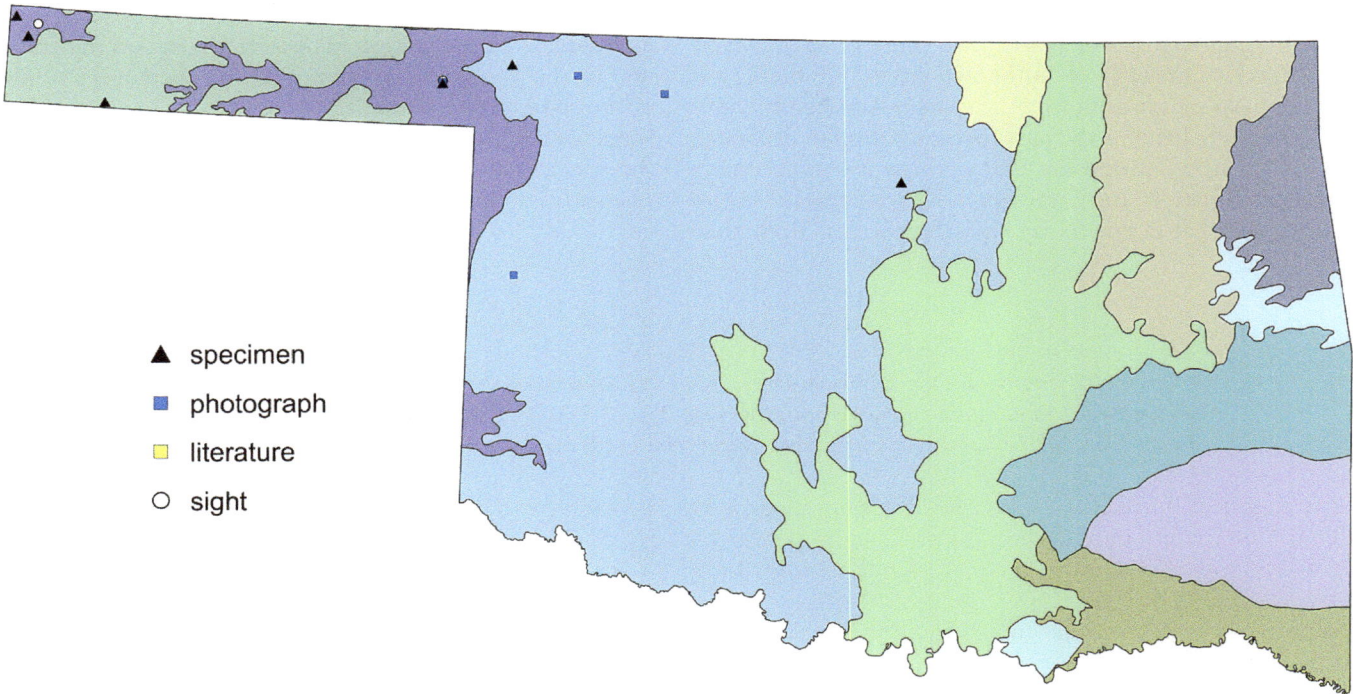

Figure 12.27 Documentation level of supporting records for *Enallagma carunculatum*, the Tule Bluet, in Oklahoma. Levels include specimen, archived photograph, literature reference, or sight record. Although sight records lack documentation, only those we feel are credible are plotted.

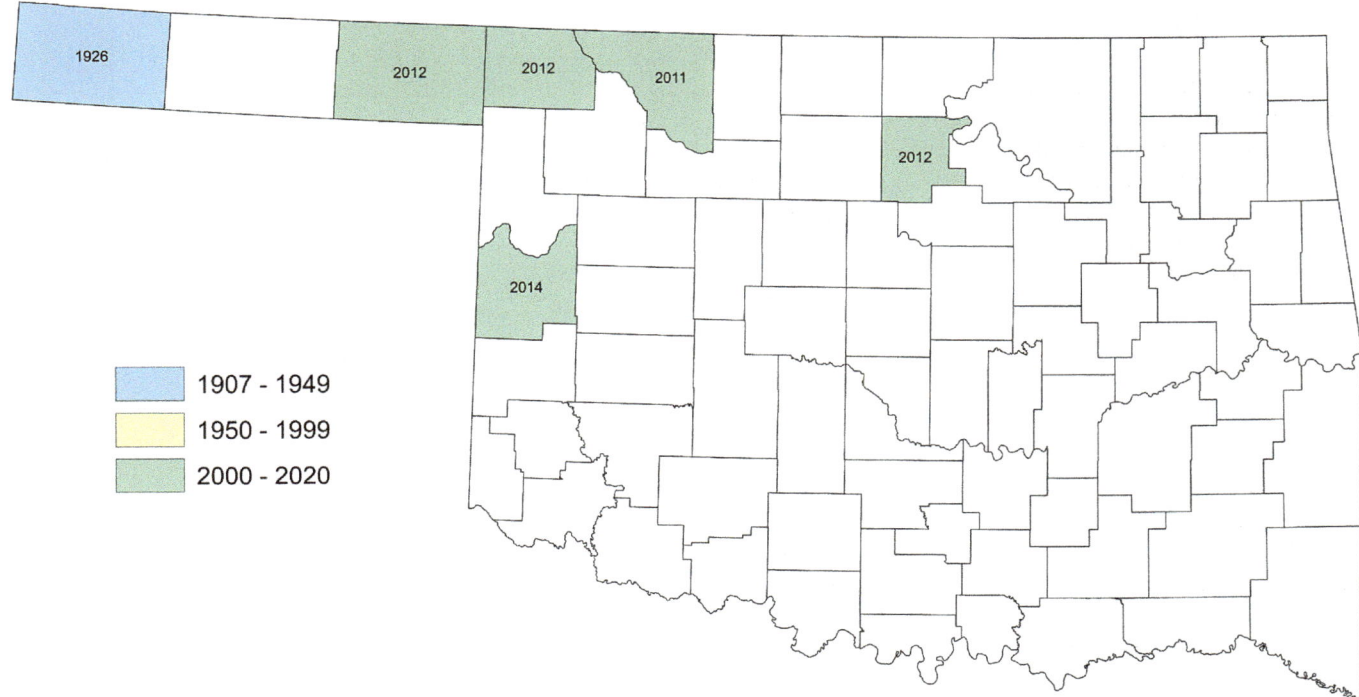

Figure 12.28 Counties in which *Enallagma carunculatum*, the Tule Bluet, are known to occur in Oklahoma. Year within the county is that in which the species was first reported.

concern within the state, but we now think that its true status is difficult to discern; as such, we chose to recategorize it as a species currently of "unknown status" (Table 7.3). If we were to speculate as to why the species has been encountered relatively little in the past few years despite dedicated surveys for it, we would say that perhaps it has become a victim of climate change and has moved northward out of Oklahoma. That is not to say that the species, from time to time, might still show itself in the state, but we fear that its residency in the state may be in the past. Alternatively, it may be that the species is not now or has been ever a resident of the state, rather there are "boom" years in which it ventures south. We hope that future data will not only clarify the species' status in the state but also dispel our present suspicions of impending loss from the state's odonatofauna.

ENALLAGMA GEMINATUM KELLICOTT, 1895 – SKIMMING BLUET

This species is part of the first series of odonate specimens ever collected in the state: EB Williamson collected 5 Skimming Bluets (3♂, 2♀, including 1 pair) on 3 June 1907 in Wister, Le Flore County (1♂ and 1 pair at UMMZ; Williamson 1914b). Since then this widespread but generally uncommon species has been reported roughly another 200 times; about 85% of those records reported between 2005 and 2018. It is now found in 58 counties (59, if one counts two sight records for Sequoyah County), primarily within the eastern half of the state. The species is generally reported fewer than a dozen at a time, although there are reports of a couple dozen at once up to 60 or maybe even around 75 individuals.[11] More often than not the species is found at clear ponds or along sandy lakeshores; occasionally it can be seen along slow reaches of streams. Regardless of location, a key component of habitat is low emergent vegetation and vegetative matter floating on or just below the water's surface.

In Oklahoma, *Enallagma geminatum* is known to fly as early as 16 March (in 2016, 1♂, Red Slough WMA, McCurtain County, DA sight record) and as late as 26 October (in 2014, 1♂, French Lake, Wichita Mountains WR, Comanche County, OC 427567), which is the early and late date for the region outside of Texas.[12] We know of no larval records for Oklahoma. Ovipositing has also not been reported but tandem pairs have been observed. There are half a dozen teneral records ranging from spring to mid-summer (31 March–15 July).

Figure 12.29 *Enallagma geminatum*, the Skimming Bluet, A) ♂, Rogers State University in Claremore, Rogers County, 29 June 2014 (©Bill Carrell, OC 423924). B) tandem pair, Lake Ponca, Kay County, 9 August 2015 (©Bill Carrell, OC 434877).

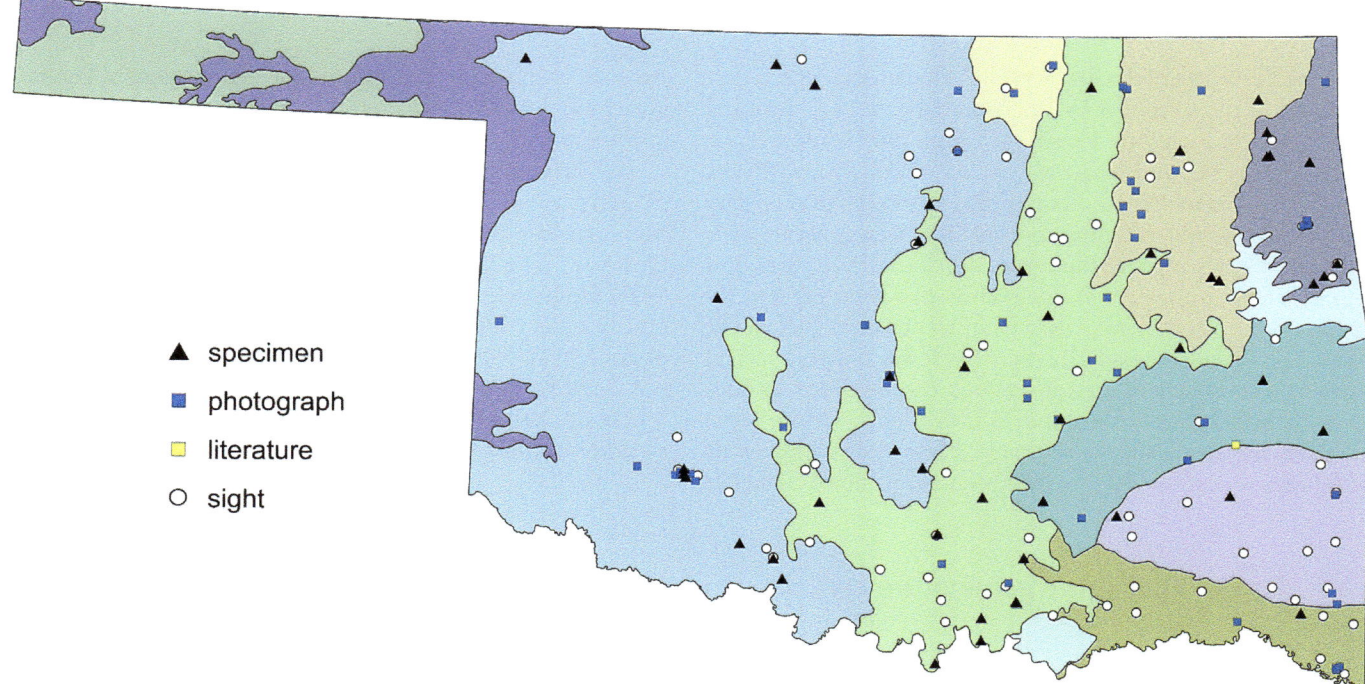

Figure 12.30 Documentation level of supporting records for *Enallagma geminatum*, the Skimming Bluet, in Oklahoma. Levels include specimen, archived photograph, literature reference, or sight record. Although sight records lack documentation, only those we feel are credible are plotted.

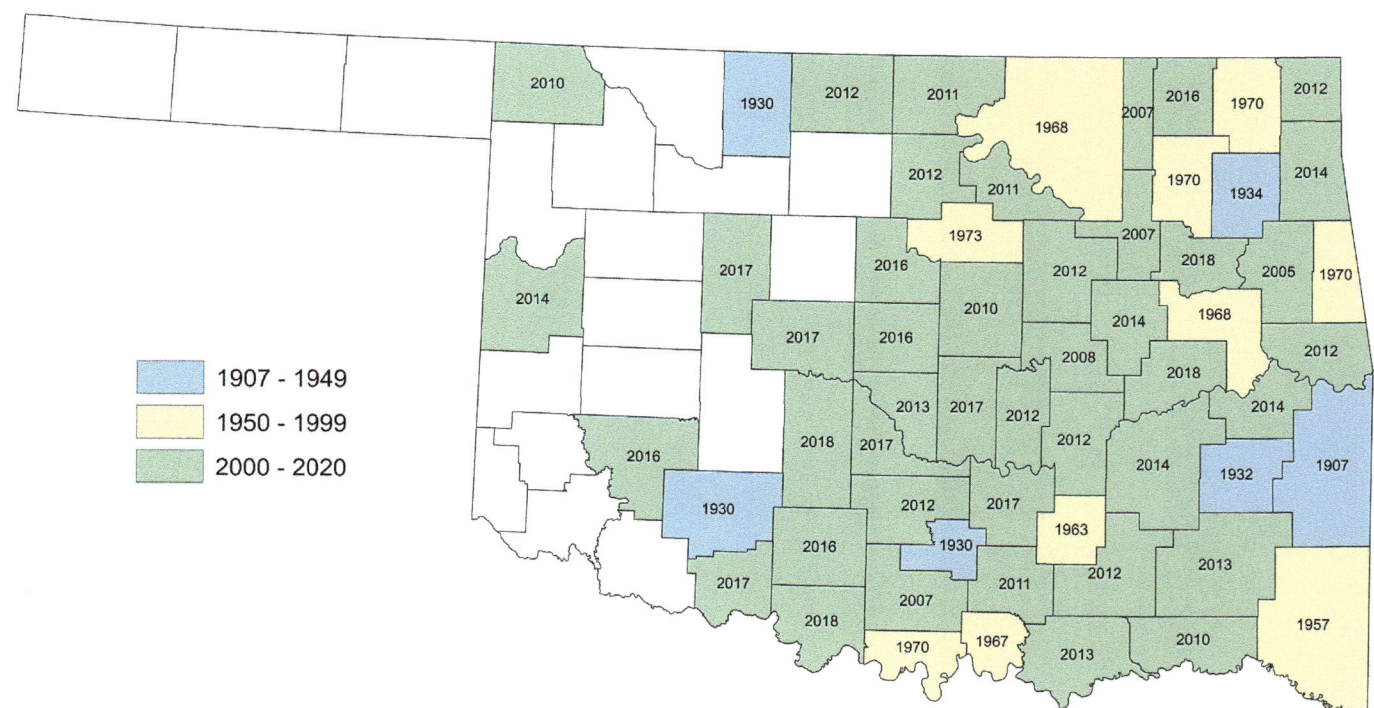

Figure 12.31 Counties in which *Enallagma geminatum*, the Skimming Bluet, are known to occur in Oklahoma. Year within the county is that in which the species was first reported. Sequoyah County is marked with hatching because it has only two records, both of which were not confidently (but probably) the species.

ENALLAGMA ASPERSUM (HAGEN, 1861) – AZURE BLUET

Enallagma aspersum is a lentic species almost exclusively of small ponds (tannic or clear water), often with water shield (*Brasenia* sp.) and almost always lacking fish. The ponds are typically within pineywoods, but occasionally some hardwoods are present. The species was first reported on 11 June 1931, when RD Bird collected 25♂ and 5♀ in Latimer County (OMNH 3262, 3266, 3267). Along with indicating that he examined 16♂ and 3♀, GH Bick mentioned that there were "numerous" individuals collected in "June 1931." Bick's notes are unhelpful in determining how many individuals were collected or if Bird had the species on more than one day during that month. Regardless, this record constitutes the sole report for that early era. The second time the species was recorded in Oklahoma was on 13 June 1957, when DR Lauck is known to have collected 1♂ in Miami, Ottawa County (UMMZ). Bick and LE Hornuff had the species on nine days between 1961 and 1970. It is hard to understand why they had so few records, but it is confounding why Bird did not have it multiple times, especially given his extensive coverage of counties such as Cleveland, Murray, and Comanche, where it was not reported until 2016, 1967, and 2003, respectively. The species' near-absence in the early era makes us wonder if the species has expanded its range in Oklahoma through the eras, perhaps as a result of the proliferation of small, manmade, post-Dust Bowl retention ponds. Since 2003, the species has been seen regularly in eastern and central Oklahoma, where it may be common, with individual counts often >50 at a time, including the highest count of 175 (150♂, 25♀, including 2 pair, 5 July 2015, JT Nickel Preserve, Pumpkin Flats, MAP). It is now known from 37 counties and seems to be pushing farther and farther west each year.

Oklahoma has the longest flight season for anywhere within the region.[13] The early flight date is 12 April (in 2016, 2♂, 1♀, including 1 pair, 4 km N of Braggs, Muskogee County, JWA, OC 441220) and the late date is 3 October (in 2015, 1♂, McGee Creek WMA, Atoka County, MAP, SP 1806). There are no larval records, but there are four records of teneral, dating from 3 June to 4 September. Ovipositing has been reported twice, once of a lone ♀ and once of three pairs.[14] We have encountered one mixed-species pair, with *Enallagma doubledayi*, the Atlantic Bluet (see that species account).

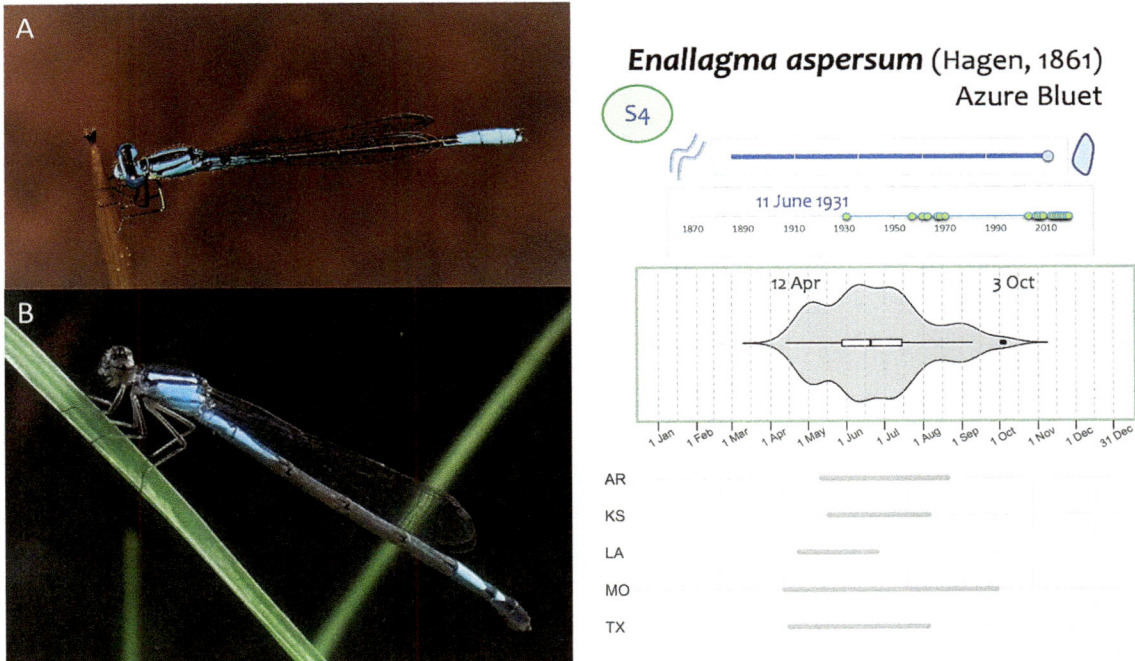

Figure 12.32 *Enallagma aspersum*, the Azure Bluet, A) ♂, Lexington WMA, Cleveland County, 11 June 2017 (©Kate Goodenough, OC 464077). B) ♀, 3 km N of Bailey, Grady County, 11 July 2008 (©Victor W Fazio III, OC 282987). Each of these records is west of what was thought to be this eastern species' geographical range, but it may have spread westward with proliferation of small, fishless ponds.

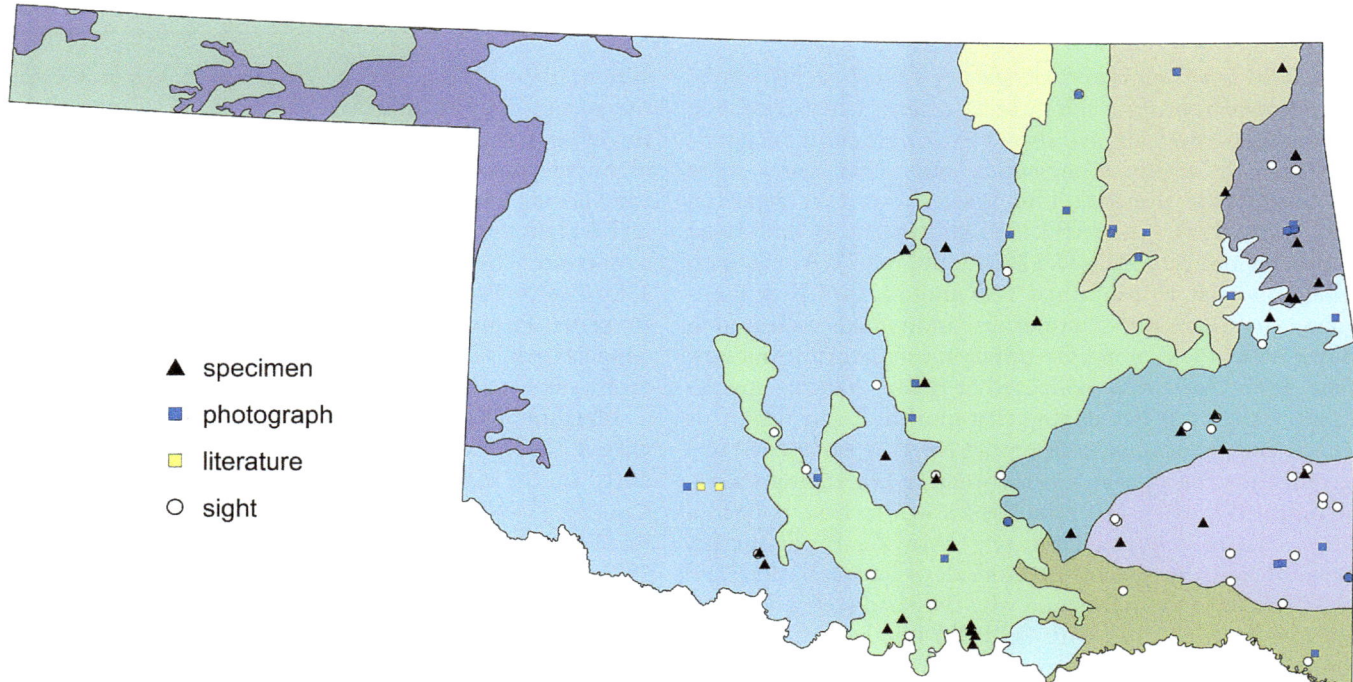

Figure 12.33 Documentation level of supporting records for *Enallagma aspersum*, the Azure Bluet, in Oklahoma. Levels include specimen, archived photograph, literature reference, or sight record. Although sight records lack documentation, only those we feel are credible are plotted.

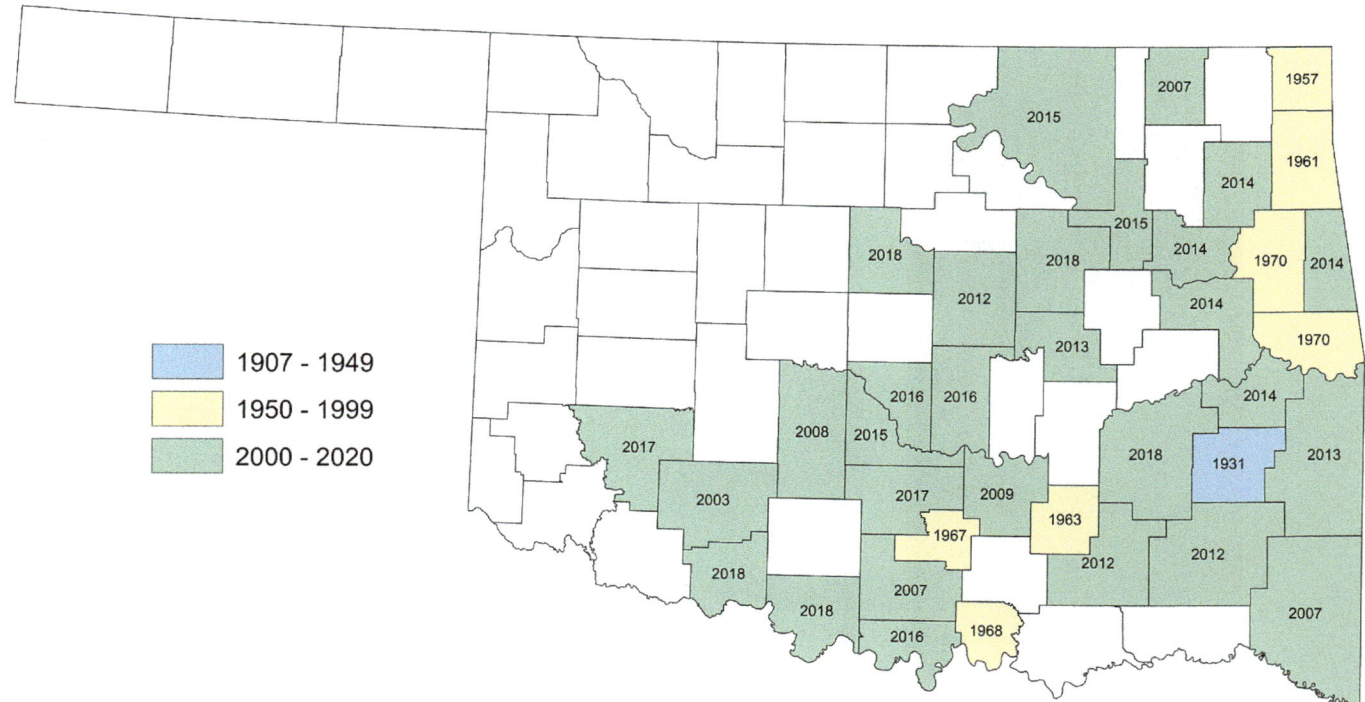

Figure 12.34 Counties in which *Enallagma aspersum*, the Azure Bluet, are known to occur in Oklahoma. Year within the county is that in which the species was first reported.

ENALLAGMA EXSULANS (HAGEN, 1861) – STREAM BLUET

The Stream Bluet was one of the first species to have been collected in the state. EB Williamson collected 3♂ and 2♀ in Wister, Le Flore County, on 3–4 June 1907 (2♂, UMMZ; Williamson 1914b). It was recorded regularly in all eras afterward (>400 records). During the early era it was recorded in an additional 12 counties and now is found in 63, ranging across the state except for the far northwest and the panhandle. As its name implies, it is a lotic species, usually being found on creeks with low to moderate flow. Creeks vary in width and depth and from muddy to rocky

and with murky to clear water. Generally woodland or shrubland is found along the creek shoreline, providing at least modest amounts of shade (but considerable shade is more characteristic), and there are emergents or overhanging vegetation on which individual bluets perch.

To encounter 5–15 individuals at any given time is typical, but it is rather regularly seen >25–50 along even short stretches of creek. It has rarely been encountered in higher numbers, but once we estimated a total of 300 individuals (50♂, 50♂, 200U, including 50 pair) on the east side of Lake

Figure 12.35 *Enallagma exsulans*, the Stream Bluet. A) ♂, Joe B Barnes Regional Park in Midwest City, Oklahoma County, 5 September 2015 (©Emily A Hjalmarson, OC 436966). Note the malformed wingtips. B) young ♀, Government Springs Park in Enid, Garfield County, 14 May 2017 (©Bill Dobbins, OC 463980). C) tandem pair, Baron Fork River 7 km SE of Tahlequah, Cherokee County, 25 June 2018 (©CA Ivy, OC 483107).

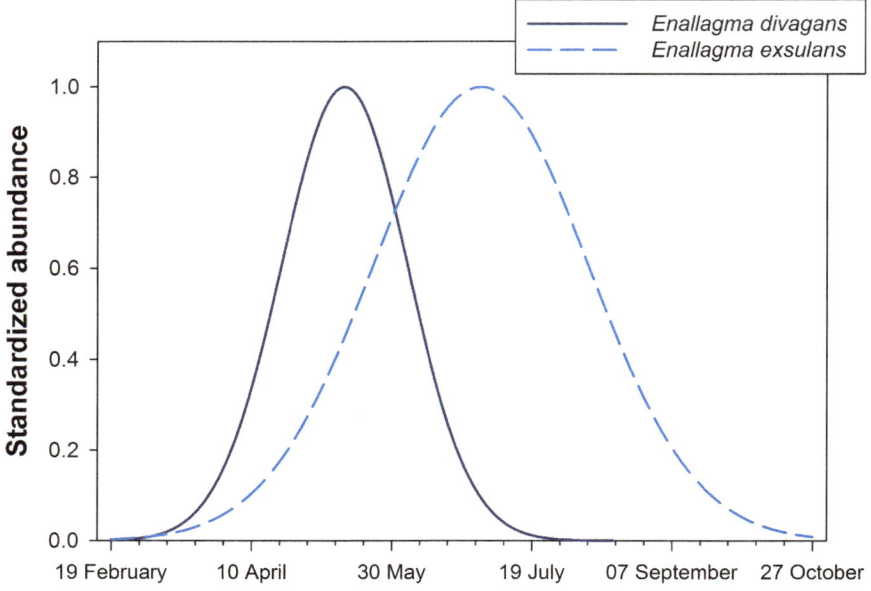

Figure 12.36 Comparison of *Enallagma exsulans* and *E. divagans* seasonal peaks in Oklahoma. The pattern of species turnover recognized by Pritchard (1935) still holds.

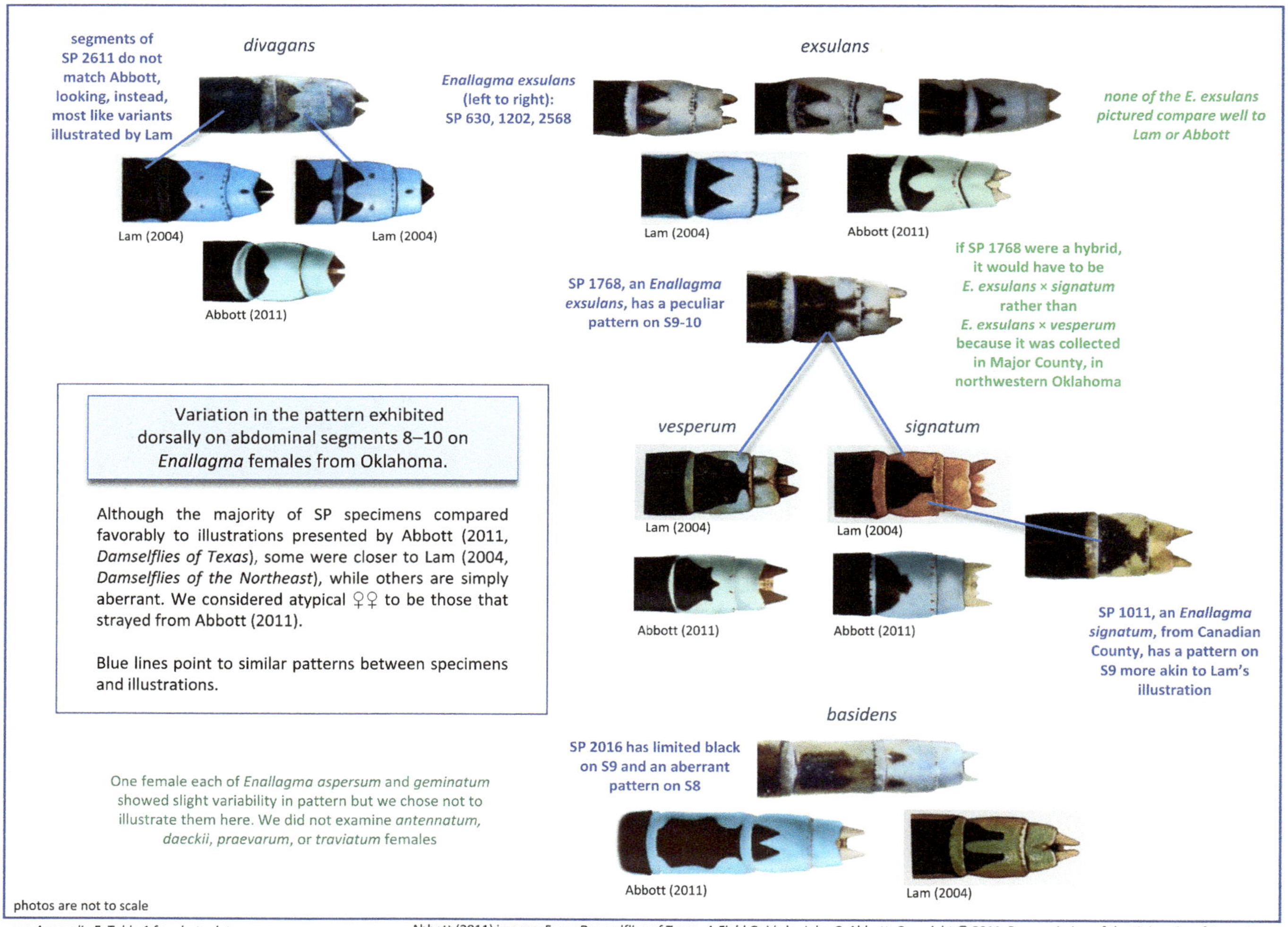

---

divagans

segments of SP 2611 do not match Abbott, looking, instead, most like variants illustrated by Lam

exsulans

Enallagma exsulans (left to right): SP 630, 1202, 2568

none of the *E. exsulans* pictured compare well to Lam or Abbott

Lam (2004)

Lam (2004)

Lam (2004)

Abbott (2011)

Abbott (2011)

SP 1768, an *Enallagma exsulans*, has a peculiar pattern on S9-10

if SP 1768 were a hybrid, it would have to be *E. exsulans × signatum* rather than *E. exsulans × vesperum* because it was collected in Major County, in northwestern Oklahoma

vesperum

signatum

Lam (2004)

Lam (2004)

Abbott (2011)

Abbott (2011)

SP 1011, an *Enallagma signatum*, from Canadian County, has a pattern on S9 more akin to Lam's illustration

> Variation in the pattern exhibited dorsally on abdominal segments 8–10 on *Enallagma* females from Oklahoma.

Although the majority of SP specimens compared favorably to illustrations presented by Abbott (2011, *Damselflies of Texas*), some were closer to Lam (2004, *Damselflies of the Northeast*), while others are simply aberrant. We considered atypical ♀♀ to be those that strayed from Abbott (2011).

Blue lines point to similar patterns between specimens and illustrations.

One female each of *Enallagma aspersum* and *geminatum* showed slight variability in pattern but we chose not to illustrate them here. We did not examine *antennatum, daeckii, praevarum,* or *traviatum* females

basidens

SP 2016 has limited black on S9 and an aberrant pattern on S8

Abbott (2011)

Lam (2004)

photos are not to scale

see Appendix F, Table 1 for photo data   Abbott (2011) images: From *Damselflies of Texas: A Field Guide* by John C. Abbott, Copyright © 2011. By permission of the University of Texas Press.

Figure 12.37 Variation in female *Enallagma* (bluets).

Claremore, Rogers County, on a visit on 22 May 2014. Not only is that record an outlier in terms of the number encountered but for where we encountered them. The location was essentially a back cove of the lake, at the mouth of a small drainage, but there was no flow, so we assumed that the lake level was low enough that these individuals were forced up from the drainage to the edge of the lake. Although fairly rare, we have encountered the species in similar circumstances at other lakes. Nonetheless, we characterize the species as lotic, probably not likely to venture into lakes any farther than the meeting point of back coves and drainage mouths that physiographically mimic streams. That said, on a rare occasion, the species is found at woodland ponds, such as when 3♂, one of which was a teneral, were observed in eastern McCurtain County (11 June 2019, BS-P).

In Oklahoma, *Enallagma exsulans* is known to fly as early as 19 April and as late as 3 October (early: from 1930, 1♂, 1♀, likely as a pair, probably from Oil Springs, Johnston County, RD Bird, OMNH-3940; late: from 2015, 1U, Osage Hills SP, Osage County, BC sight record). These dates are in line with elsewhere in the region.[15] Pritchard (1935:5) observed that by early June *E. exsulans* had replaced *E. divagans* in the areas he worked; this pattern still holds (Figure 12.36). We know of no larval records and, while pairs are often noted, ovipositing has not been recorded. Eight teneral records date between 3 May and 15 July.

We have noted some variation in the dorsal abdominal pattern of segment 9 and 10 on female *Enallagma exsulans* in the SP collection (Figure 12.37). In comparison to illustrations presented by Abbott (2011), we found that four of the seven specimens in the collection would be considered atypical. One specimen, SP 2568, is especially aberrant, looking more akin to an eastern *E. vesperum* or *E. signatum*, the Vesper and Orange Bluets (as per illustrations by Lam 2004). Although the S9–10 dorsum exhibits a strange pattern, the specimen's mesostigmal plate and humeral stripe are characteristic of *E. exsulans*.

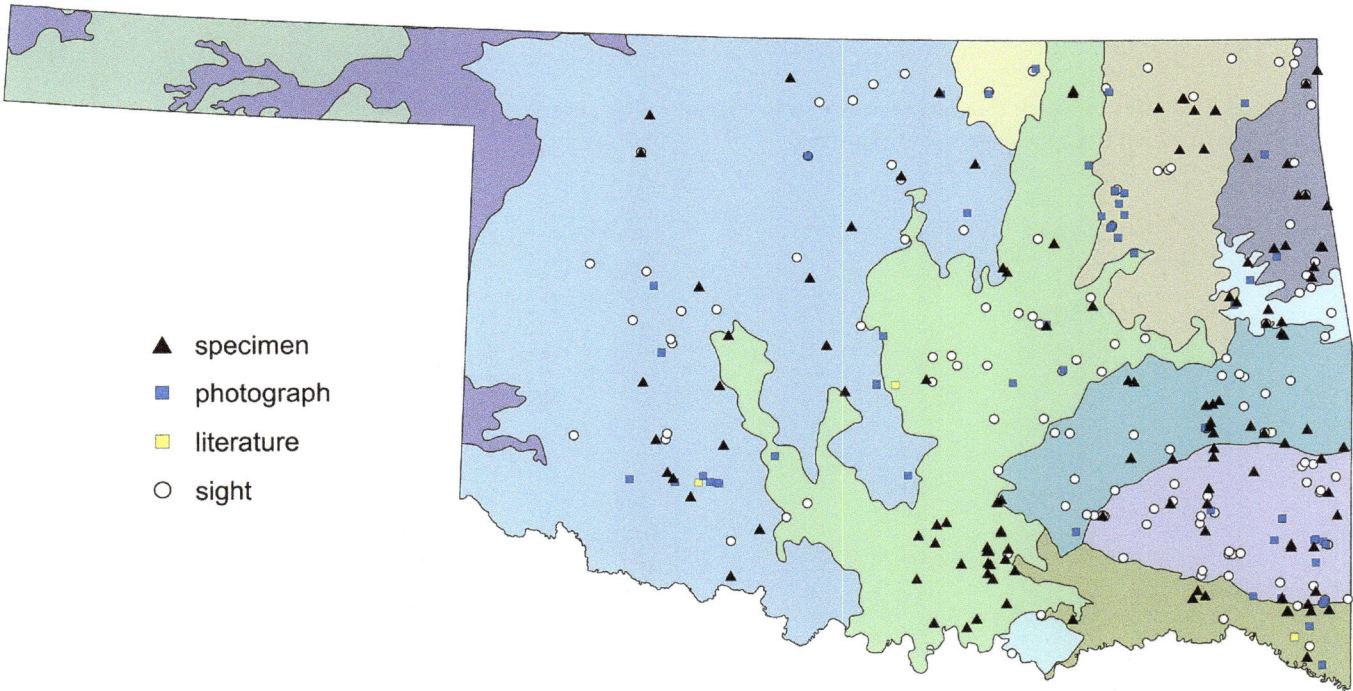

Figure 12.38 Documentation level of supporting records for *Enallagma exsulans*, the Stream Bluet, in Oklahoma. Levels include specimen, archived photograph, literature reference, or sight record. Although sight records lack documentation, only those we feel are credible are plotted.

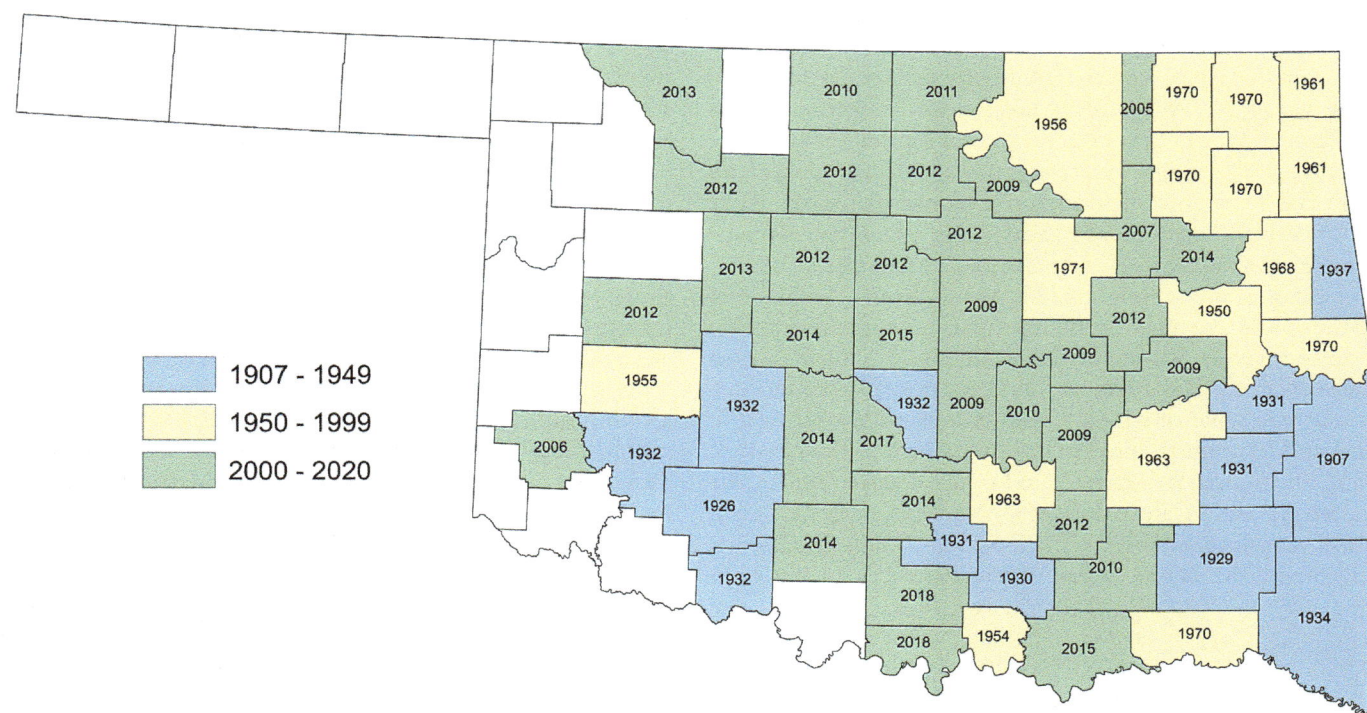

Figure 12.39 Counties in which *Enallagma exsulans*, the Stream Bluet, are known to occur in Oklahoma. Year within the county is that in which the species was first reported.

ENALLAGMA DAECKII (CALVERT, 1903) – ATTENUATED BLUET

The notably long-bodied, yet inconspicuous, Attenuated Bluet is rare in Oklahoma, where it has been recorded in four locations in three counties, making for <12 records. All but one record has been in the Ouachita Highlands, the exception being SW Dunkle's McCurtain County record (see below). Elevations for the species have ranged from about 135 m to 530 m.

AE Pritchard first found the species in Oklahoma near Antlers, Pushmataha County, probably on 16 June 1934.[16] Twenty-three years later GH Bick collected the species again in the area (1♂, 2♀, 1.5 mi N of Antlers, 13 June 1957, FSCA). *Enallagma daeckii* was not seen again in the state until 1992, when SW Dunkle discovered it at "Bokhoma Camp pond" in McCurtain County (2♂, 6 June, FSCA; Donnelly 2004c). We discovered a third population, this time in Le Flore County, at Crooked Branch Lake in the Ouachita NF (in 2013: 2♂, 8 June, one as SP 669; 3♂, 1♀, including 1 pair, 9 June; 2♂, 15 June). Surveys the following year at Crooked Branch Lake proved unfruitful, but the species was found there again in 2015, when 2♂ were encountered on 6 June (one SP 1630, MAP). No reports were received in 2016, but in 2017, MAP discovered a new population at a small pond 7 km NE of Fewell, Pushmataha County (3♂, 4♀, all individuals were teneral, 1♂ as SP 2303, 9 May; 19♂, 2♀, including 2 pair, 1♂ as SP 2442, 3 July; also 10♂, 26 May 2018, EA Hjalmarson, B Roberts, OC 485782). That location produced the early and late flight dates for *Enallagma daeckii* in Oklahoma as well as the only teneral record for the state. The Oklahoma late date is the latest for the region (AR: 2 records, 16 May–24 June, Harp and Harp 2003, OC 447441; LA: 13 April–8 June, OC 282004, Mauffray 2014; TX: 6 April–15 June, Abbott and Lasley 2019, JCAC 44309). Not surprisingly, there are no larval or ovipositing records for Oklahoma.

In Texas, *Enallagma daeckii* is described as inhabiting "margins of shady, often heavily vegetated ponds, lakes, and stream backwaters" (Abbott 2011:181) and in general it is thought to inhabit "shrubby borders of wooded sand-bottomed lakes and swamps. Always associated with woodland" (Paulson 2011:109). Those descriptions mostly fit our experience in Oklahoma—small to large woodland ponds with shorelines moderately to heavily vegetated, but we cannot comment on the substrate of the ponds at which we have encountered the species. Pritchard (1935) described the pond near Antlers as being choked with "water lilies" to all but its center (he failed to mention the species of "water lilies," a broad category in common parlance). When we visited what we thought was the site, it was choked with American or yellow lotus, *Nelumbo lutea*. Pritchard also mentioned that he had *Enallagma dubium*, the Burgundy Bluet, *Nehalennia integricollis*, the Southern Sprite, *Ischnura kellicotti*, the Lilypad Forktail, *Lestes vigilax*, the Swamp Spreadwing, *L. inaequalis*, the Elegant Spreadwing, *Ladona deplanata*, the Blue Corporal, and *Celithemis verna*, the Double-ringed Pennant at the lake. At Crooked Branch Lake we have encountered *E. daeckii* directly associated with *Lestes forcipatus* (the state's only record) and *L. inaequalis*, Sweetflag and Elegant Spreadwings, *Ischnura posita* and *I. hastata*, Fragile and Citrine Forktails, *Nehalennia integricollis*, *Plathemis lydia*, the Common Whitetail, *Libellula luctuosa*, *L. cyanea*, *L. incesta* and *L. vibrans*, Widow, Spangled, Slaty, and Great Blue Skimmers, *Perithemis tenera*, the Eastern Amberwing, *Erythemis simplicicollis*, the Eastern Pondhawk, and *Pachydiplax longipennis*, the Blue Dasher. At the shaded, woodland pond 7 km NE of Fewell, the species was associated with *Enallagma dubium*, *Ischnura hastata*, *I. posita*, *Ladona deplanata*, *Pachydiplax longipennis*, and *Perithemis tenera*.

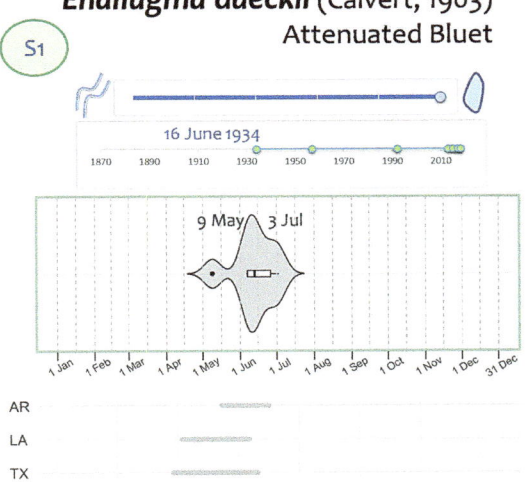

Enallagma daeckii (Calvert, 1903)
Attenuated Bluet

Figure 12.40 First record of *Enallagma daeckii*, the Attenuated Bluet, in Oklahoma for >20 years. The previous record was from 1992 (see below). This ♂ was photographed on 8 June 2013 at the beaver dam complex below Crooked Branch Lake, Le Flore County (©Brenda D Smith, OC 400667). This is one of only two known sites for the species in Oklahoma.

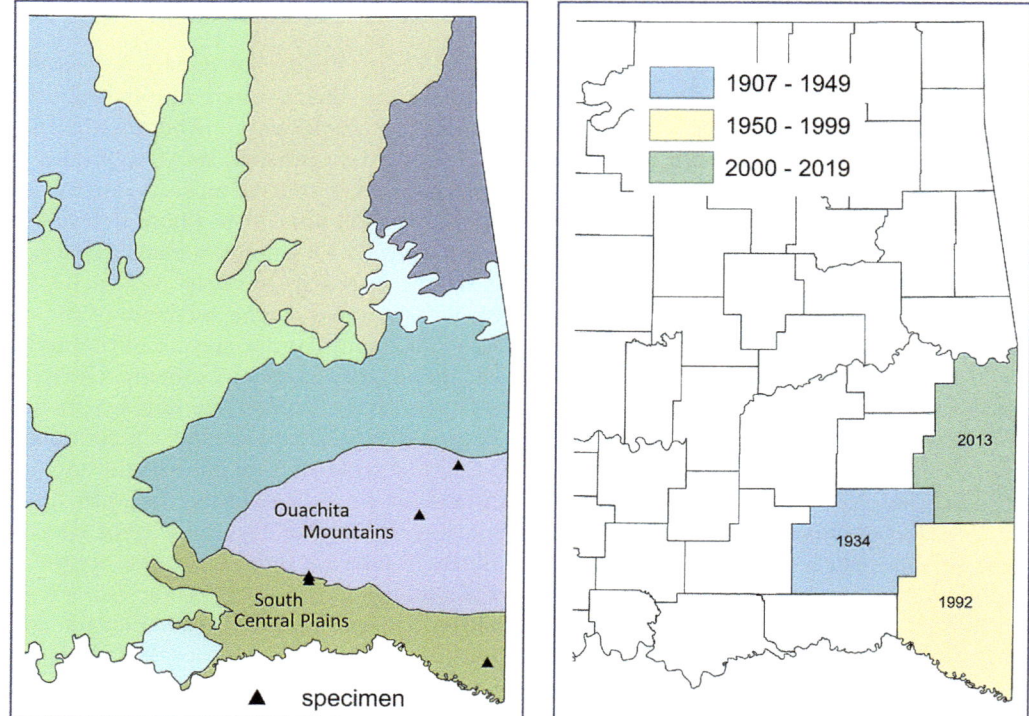

Figure 12.41 Documentation level of supporting records for *Enallagma daeckii*, the Attenuated Bluet, in Oklahoma (left). The species has been reported in only three counties in eastern Oklahoma (right) and currently, there are but two known sites for the species in Oklahoma.

ENALLAGMA ANTENNATUM (SAY, 1839) – RAINBOW BLUET

The Rainbow Bluet is known from three counties in Oklahoma: Alfalfa, Cimarron, and Texas, although it likely is regular only in Cimarron. It was first recorded in the state when RD Bird collected specimens in Alfalfa County near what is now Salt Plains NWR (15–16 June 1930, 5♂, 1♀, OMNH 3230, 3240, 3243[17]). The species was not reported again for 40 years, until GH Bick and LE Hornuff encountered it 2.4 mi NE of Kenton, Cimarron County (3♂, 1♀ on 5 August 1970, FSCA, Bick 1991). According to Bick's notes, Hornuff saw 1♀ on 10 August 1973 in Cimarron County, this time at "Boise City, 30.5 mi NW." It is odd that Bick did not note that Hornuff collected 1♀ the next day at 2.1 mi W of Boise City (FSCA). It is easy to think that perhaps Bick's notes had the day and locality a bit off or that maybe the specimen label did. Regardless, Bick and Hornuff reported the third and perhaps the fourth record of *Enallagma antennatum* for Oklahoma. Another 40 years passed before *E. antennatum* was again reported from the state: on 26 May 2013 we collected a teneral ♂ at a spring-fed pool on North Carrizo Creek, 7 km N of Kenton, Cimarron County (SP 587). That is the only teneral record for Oklahoma. The next day we found another

♂, this one an adult, at a spring-fed creek at Schultz WMA (SP 594). That was the first (and remains the only) county record for Texas County. We saw the species again on 3 June 2015 at a couple of locations in the Black Mesa area, Cimarron County, but mostly along the Cimarron River, 4 km NNE of Kenton, where the river was about five meters wide, the water was slow moving, and there was much, low, overhanging bunch grass (12♂, 1♀, including one pair that we collected as SP 1606, 1♂ photographed as OC 431500; Figure 12.42). The state's eighth record came two years later, also on 3 June and at the same spot along the Cimarron River (3♂, 1♀, MAP).

As the records above indicate, the species flies from late May until mid-August in Oklahoma. Elsewhere in the region and throughout its range, *Enallagma antennatum* flies at roughly the same time of year (CO: 5 June–27 August, OC 367206, 367406; KS: 20 May–15 August, SEMC 1348677, 1347234; MO: 16 June–28 August, Trial 2005; TX: 1 record, 1 September, OC 303190; Paulson 2009, 2011). Only one pair and one teneral have been reported for Oklahoma (see above for both records), but we have no reports of ovipositing or nymphs.

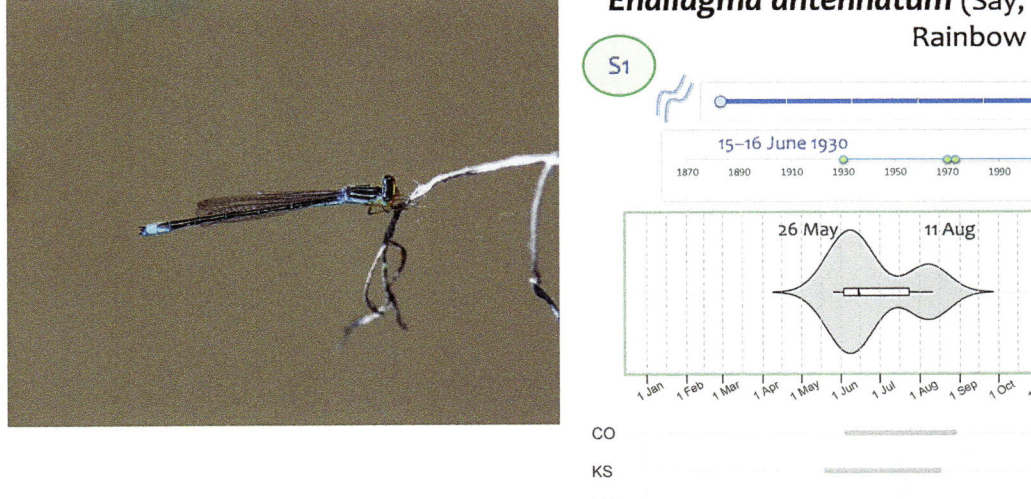

Enallagma antennatum (Say, 1839)
Rainbow Bluet

S1

15–16 June 1930

1870 1890 1910 1930 1950 1970 1990 2010

26 May   11 Aug

1 Jan 1 Feb 1 Mar 1 Apr 1 May 1 Jun 1 Jul 1 Aug 1 Sep 1 Oct 1 Nov 1 Dec 31 Dec

CO
KS
MO
TX

Figure 12.42 *Enallagma antennatum*, the Rainbow Bluet. This species' status in Oklahoma is unclear. This ♂ along Cimarron River 4 km NNE of Kenton, Cimarron County, 3 June 2015 (©Brenda D Smith, OC 431500) was one of a dozen individuals recorded that day, including a mated pair, but the species has been found only once since, and our four records from 2013–2017 are the only ones in three-and-a-half decades. We posit that disappearance of lotic habitat in the panhandle and environs has led to regional population decline.

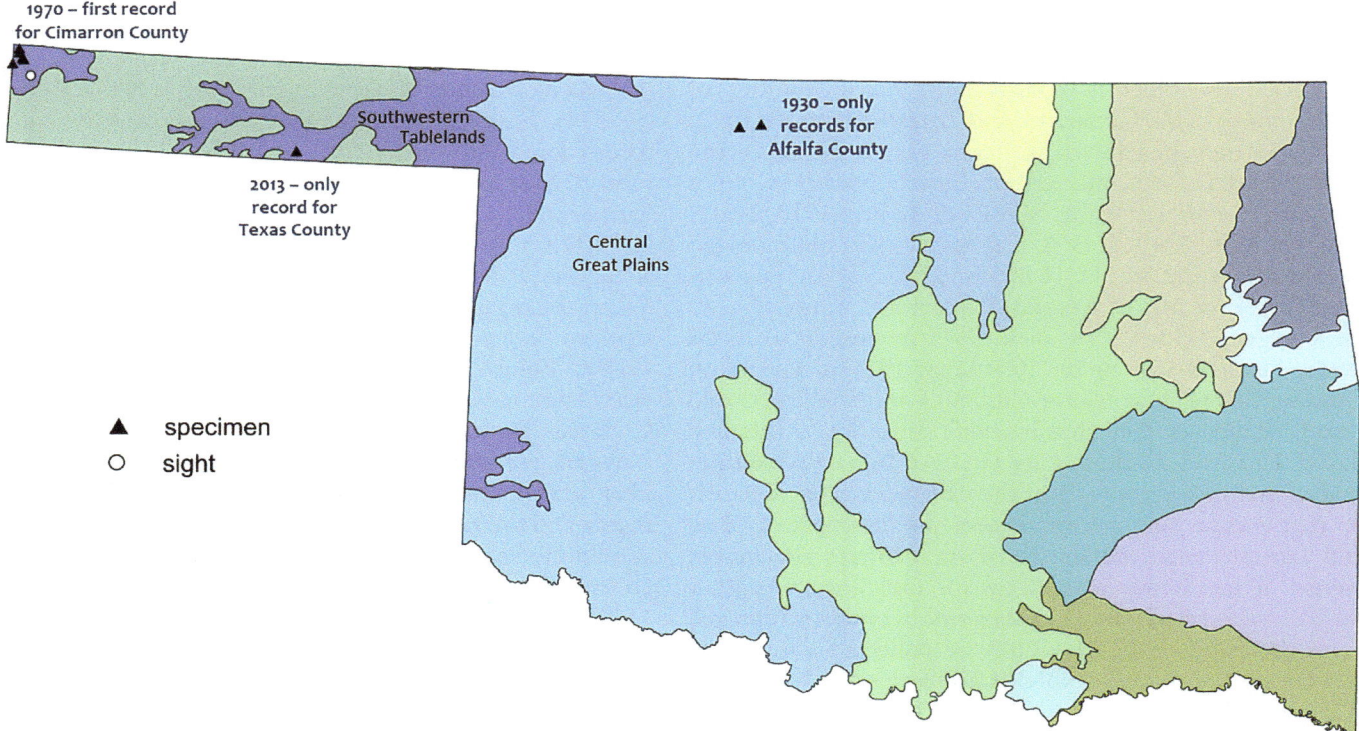

Figure 12.43 Documentation level of supporting records for *Enallagma antennatum*, the Rainbow Bluet, in Oklahoma. The species has been reported in only three counties in western Oklahoma and its panhandle.

ENALLAGMA DIVAGANS SÉLYS, 1876 – TURQUOISE BLUET

In Oklahoma, the Turquoise Bluet tends to be a species of wooded areas with ponds, beaver dam complexes, sloughs, or creeks with slow to moderate flow. It was first recorded in the state as part of the earliest specimens collected in Oklahoma (7♂, 1♀, 3 June 1907, Wister, Le Flore County, EB Williamson, Williamson 1914b, 4♂ at UMMZ). It was recorded seven times between 1931 and 1939. GH Bick and LE Hornuff contributed four records, and SW Dunkle contributed two between 1992 and 1999. Those 14 records, from 8 counties, are all that are known for the species in Oklahoma prior to the 2000s. Since 2004, there have been nearly 70 reports from an additional 15 counties. Early and mid-era records also reported one to a few individuals at a time, whereas now, even though the species is generally seen <10 at a time, it is occasionally seen >10, including the highest count of 60 individuals (50♂, 10♀, including 8 pair, Three Rivers WMA, McCurtain County, 3 May 2015). The difference between eras for record, county, and individual numbers may indicate that Oklahoma records early on were more-or-less extralimital and that the species has become more common and more widespread in the state since.

Enallagma divagans is an early season species in Oklahoma (bulk of records are from April and May, although the bracket dates are beyond that; early: 31 March 2018, Grassy Slough WMA, McCurtain County, 3♂, one in hand, MAP; late: 3 July 1958, Briar Creek, Marshall County, 1♂, GHB.'s notes, in which he said he discarded the specimen). AE Pritchard (1935:5) observed that by early June *E. divagans* "was then replaced by *E. exsulans*; the other spring dragonflies were likewise gone," a pattern that holds true (Figure 12.36). Oklahoma's seasonality is entirely consistent with the region.[18] We know of no larval records but we have six teneral records, dating 8 April to 12 May, the latter day having 19 tenerals recorded. We have observed ovipositing only once, of three pairs, at Lake Ozzie Cobb, Pushmataha County, on 2 May 2015 (25♂, 6♀, including five pairs present).

Of final note is a ♀ specimen (SP 2611) that shows a dorsal abdominal pattern on S8–9 that does not match that presented by Abbott (2011). Lam (2004) presented two illustrations of pattern variations on the terminal segments of females: S8 and S9 of SP 2611 do not resemble either illustration in their entirety, rather the pattern is more of a mixture of the two (Figure 12.37).

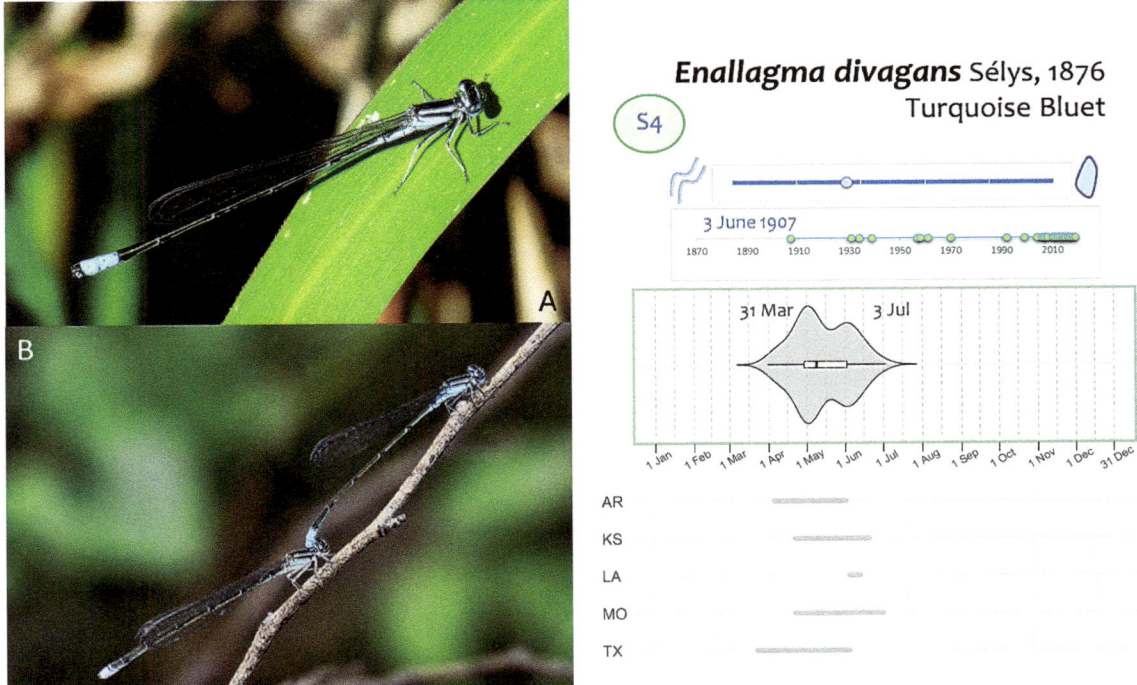

Figure 12.44 *Enallagma divagans*, the Turquoise Bluet, A) ♂, Bixhoma Lake, Wagoner County, 5 May 2014 (©Bill Carrell, OC 422015). B) tandem pair, Mountain Fork Park, McCurtain County, 27 May 2011 (©Berlin A Heck, OC 328096).

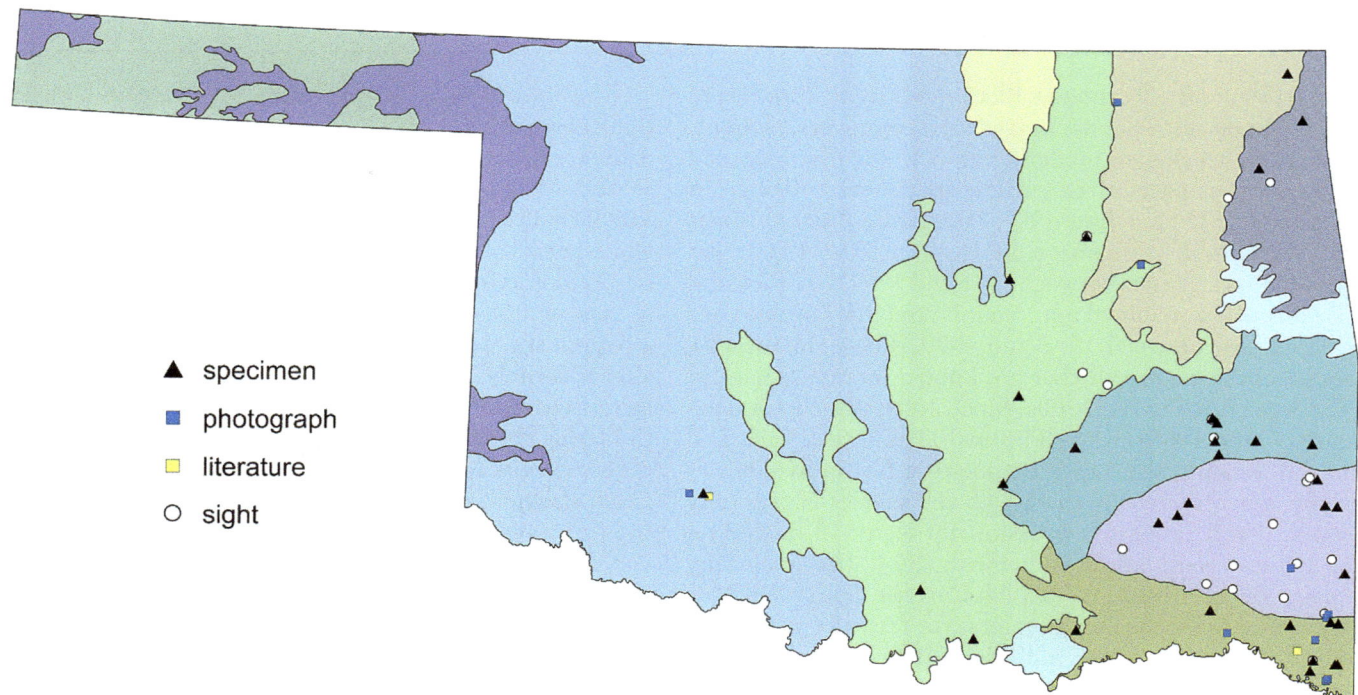

Figure 12.45 Documentation level of supporting records for *Enallagma divagans*, the Turquoise Bluet, in Oklahoma. Levels include specimen, archived photograph, literature reference, or sight record. Although sight records lack documentation, only those we feel are credible are plotted.

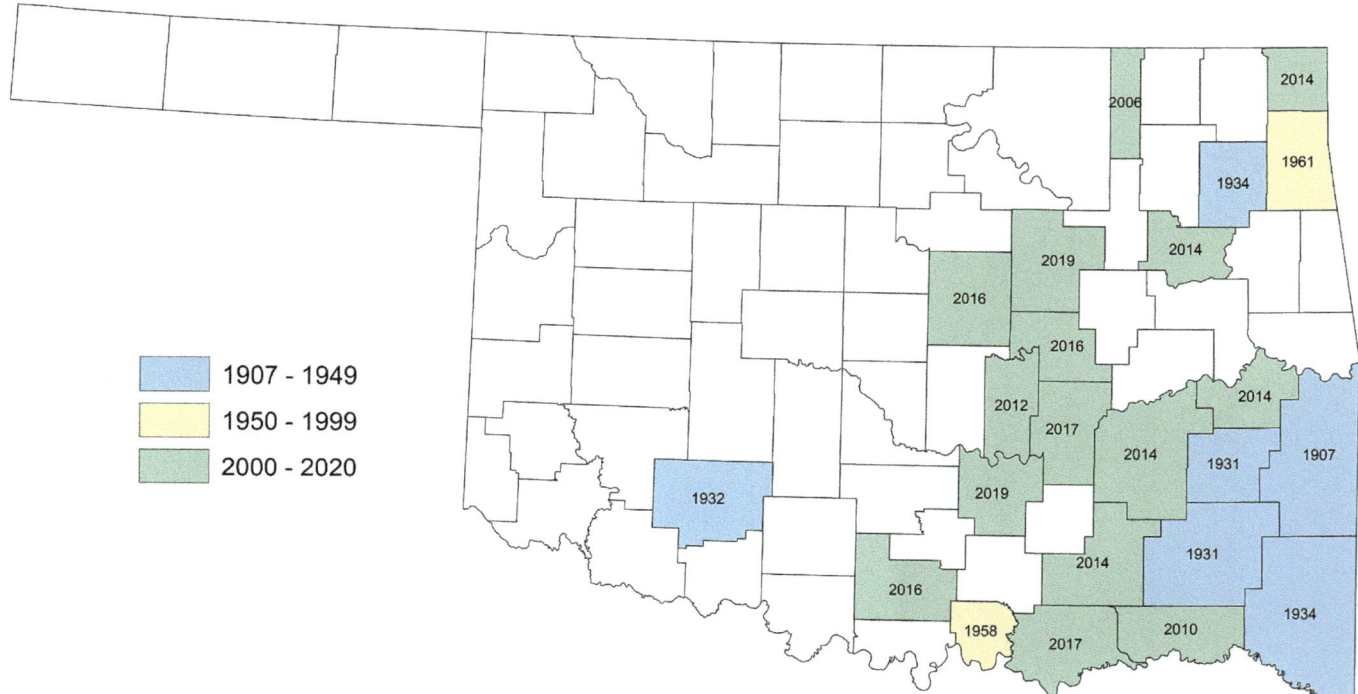

Figure 12.46 Counties in which *Enallagma divagans*, the Turquoise Bluet, are known to occur in Oklahoma. Year within the county is that in which the species was first reported.

ENALLAGMA TRAVIATUM SÉLYS, 1876 – SLENDER BLUET

Enallagma traviatum is a species of ponds, including beaver pond complexes, and back coves of lakes, with considerable low emergent forbs. Its habitat preference may explain why it has apparently expanded its range in the state (see Chapter 6). Records of the species have greatly increased over time, as have the number of counties in which it has been found.

The species was first reported by EB Williamson at Wister, Le Flore County, in early June 1907. He reported 40♂ and 4♀ "flying over the water at source of outlet of artificial lake in numbers, pairing."[19] It was next reported by RD Bird in Latimer County in the 1930s,[20] followed by GH Bick and LE Hornuff encountering the species on seven days at eight localities in Haskell, Latimer, Marshall, and

Figure 12.47 *Enallagma traviatum*, the Slender Bluet, A) ♂, Haskell Lake, Muskogee County, 28 May 2014 (©James W Arterburn, OC 422400). B) ♀, Rush Lake in the Wichita Mountains WR, Comanche County, 22 May 2008 (©Victor W Fazio III, OC 284888), showing the extensive blue on the distal abdominal segments typical of the species. The Wichitas hosts a geographically outlying population of this otherwise eastern species, a pattern repeated over and over again across the various odonate families. C) tandem pair, Rogers State University in Claremore, Rogers County, 29 June 2014 (©Bill Carrell).

Pushmataha Counties between 1957 and 1970. In the 1990s, it was reported on two days by SW Dunkle (he added the species to McCurtain County) and once by R. J. Beckemeyer. But, between 2004 and 2019 it was reported >75 times and added to 27 counties. From its humble beginnings in Le Flore and Latimer Counties in southeastern Oklahoma, it is now found in 33 counties across southeastern, northeastern, central, and some of southwestern Oklahoma. Even though it has likely expanded its range, its abundance appears to be roughly the same: Williamson's 44 individuals is completely in line with counts we and others have had in recent years, although 1–2 at a time is more typical. We know of only twice that the species has been recorded in numbers >50.[21]

Enallagma traviatum is a spring to mid-summer flyer in Oklahoma (early May to early July, with the bulk of records from mid-May through June) and the region,[22] which is in rather stark contrast to elsewhere in its range where it can fly until October (Paulson 2011 for Ohio). We know of two possible larval records (SEMC; see also "*Telleallagma* sp. [sic? *Teleallagma*]" entry in Appendix A) and three teneral records (7 May–1 June) for Oklahoma. Ovipositing has been reported once, via a photograph submitted to Odonata Central of a pair apparently ovipositing on water shield (*Brasenia schreberi*; OC 446029). Also note that *E. traviatum westfalli* is the western subspecies of *E. traviatum* (see *Enallagma westfalli* in Appendix A for details).

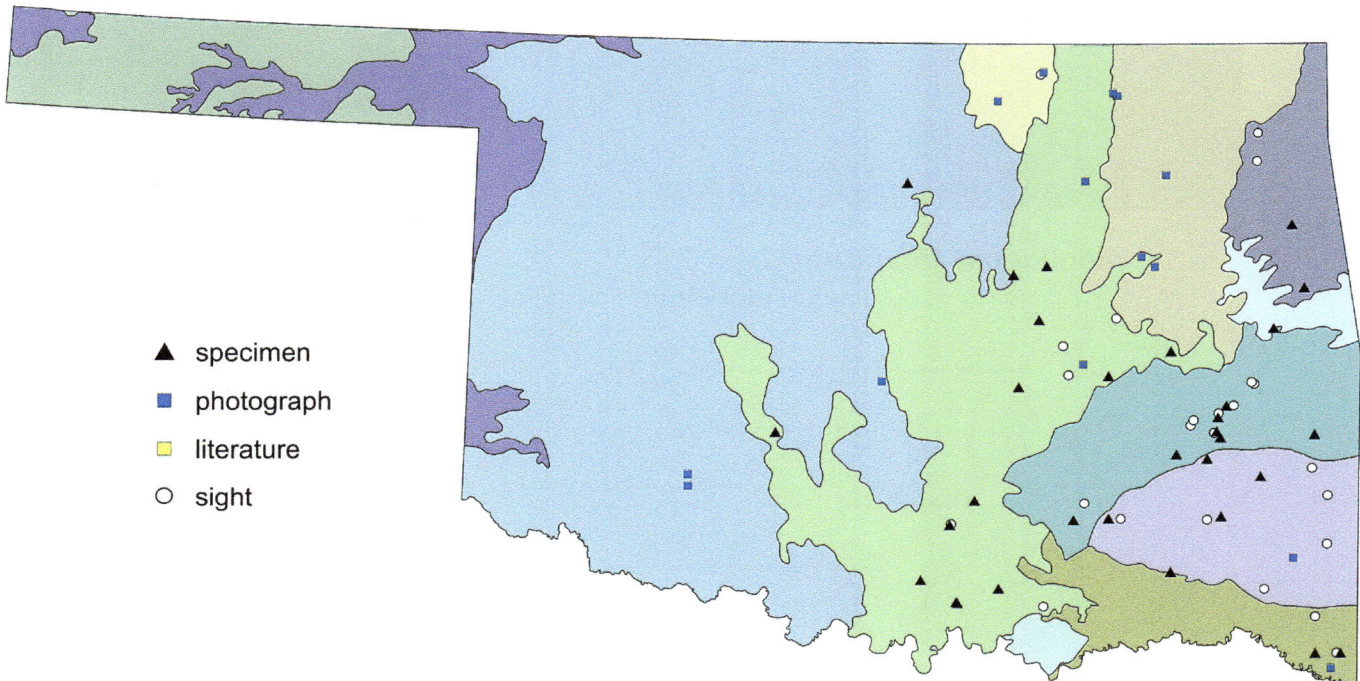

Figure 12.48 Documentation level of supporting records for *Enallagma traviatum*, the Slender Bluet, in Oklahoma. Levels include specimen, archived photograph, literature reference, or sight record. Although sight records lack documentation, only those we feel are credible are plotted.

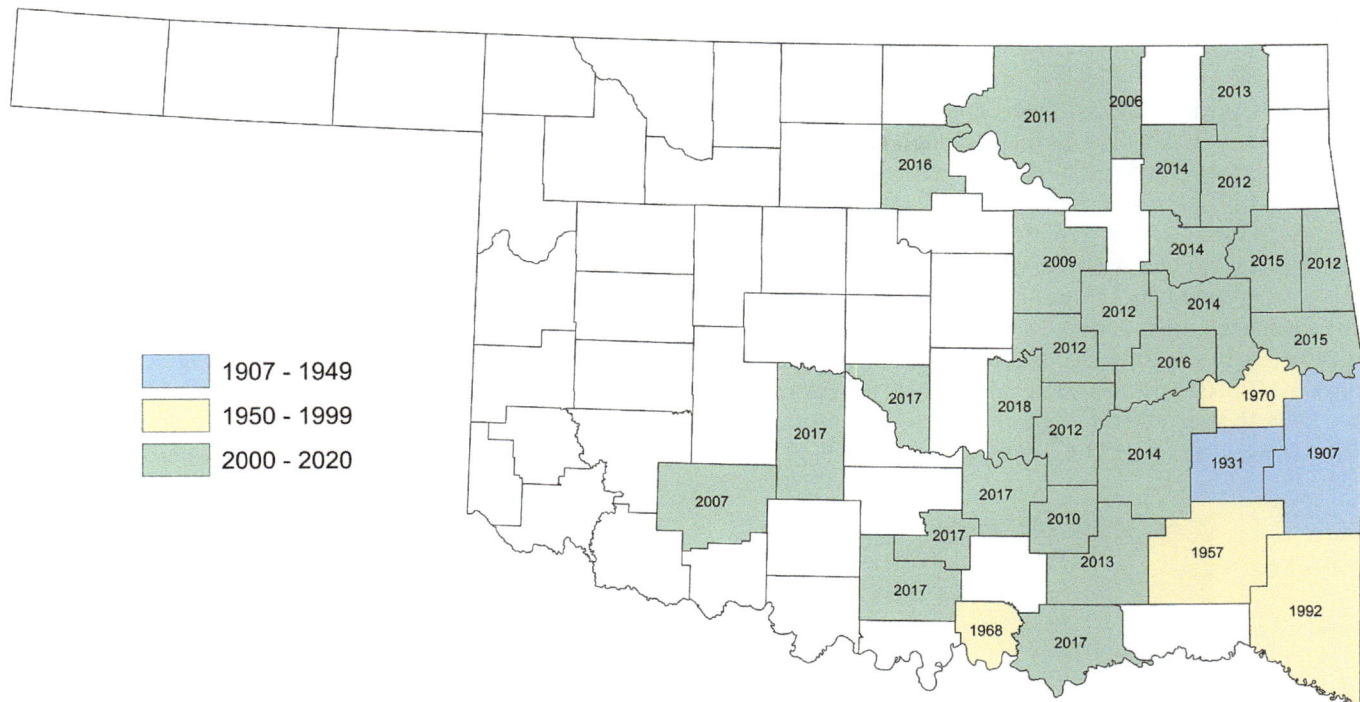

Figure 12.49 Counties in which *Enallagma traviatum*, the Slender Bluet, are known to occur in Oklahoma. Year within the county is that in which the species was first reported.

ENALLAGMA BASIDENS CALVERT, 1902 – DOUBLE-STRIPED BLUET

The Double-striped Bluet is a common (and, at times, abundant) species that has been encountered >850 times and reported in all 77 counties in Oklahoma. Although most typical of ponds and lakes, it is essentially a habitat generalist in Oklahoma, being found also along creeks, in sloughs, in ditches, and in parking lot or sidewalk puddles. Waterbodies can be with or without vegetation, whether emergent or along shorelines, and water can be spring-fed, clear, or murky. It is recorded regularly in double digits and fairly often in counts >50, including at least seven times between 100–210 individuals. All reports of triple digits have been from western Oklahoma (Harmon, Greer, Roger Mills, Beaver, Texas, and Cimarron Counties), and almost all counts >50 have been in central or western parts of the state.

Enallagma basidens was first recorded in Oklahoma in 1926, when TH Hubbell collected 4♂ at "Boulder Camp" in what is now the Wichita Mountains WR, Comanche County (9 June, UMMZ). It has been recorded regularly in Oklahoma ever since, accounting for many hundreds of days on which the species has been encountered (about 580). Between 1926 and 1939 there were 58 days on which the species was reported from 18 counties, from 1950–1991 it was reported on 131 days and added to 37 counties, but since 2001 an additional 22 counties were added, along with at least 425 encounter days. It is numbers like these

Figure 12.50 *Enallagma basidens*, the Double-striped Bluet, A) ♂, Crystal Beach in Woodward, Woodward County, 3 June 2013 (©Shawn Johnson, OC 445438). B) Note the pinkish hue of this immature ♂, photographed at Lake Murray SP, Carter County, Oklahoma, on 23 April 2018. Photo by ©Bryan E Reynolds. C) ♀, Medicine Park, Comanche County, 28 July 2011 (©Tom Kompier, OC 333114), showing the prominent blue "tail light" typical of females of this species.

that have kept odonatologists insisting that *E. basidens* has expanded its range since the 1920s (e.g., Montgomery 1942, Donnelly 1961, Huggins 1978, Cannings 1989, Westfall and May 2006). Needham and Heywood (1929:339) described the species' range as "Southwest into Tex." and said that it "... has its home primarily in the state of Texas, but it is probably to be found all along the Mexican-United States border." Needham and Heywood seemingly were unaware of Oklahoma's first record as well as Kennedy's (1917) report of the species for southeastern and northwestern Kansas (Labette, Gove, and Cheyenne County). Indiana's and Missouri's first records came in 1929, it was in North Carolina by 1940, Pennsylvania by 1959, Arizona by 1956, California by 1974, and Canada's first record, from southern Ontario, came in 1985 (see Huggins 1978 and Cannings 1989 for additional dates indicating range expansion).

Oklahoma has the longest flight season—spanning 222 days from late March to late October—in the region outside of Texas.[23] Oklahoma's flight season is about a month longer than the next longest known season in the region of 193 days for New Mexico. There is one possible larval record (UCO 5242) and about 30 teneral records for Oklahoma, the latter dating from 2 April until 18 September. Six records of ovipositing indicate that the species typically oviposits in pairs (only one of the records is of a lone ♀).

Note that a ♀ *Enallagma basidens* from Muskogee County that was erroneously published as the only record of *Ischnura ramburii* for the county (Bick and Bick 1957; Donnelly 2004; see *I. ramburii* species account for details). A more interesting note for *E. basidens* regards description of the nymphal form of the species. Bird (1931) published the initial description of the nymph, likely from an Oklahoma specimen, but he did not say which one he used. A second description of the nymph was done by Ferguson (1944). Huggins (1978) found both of those descriptions lacking in detail, particularly for individual variation, so he re-described and re-figured *E. basidens* nymphs. Indeed, Huggins' description and Mary Makepeace's and Leroy Johnson's figures are much more detailed. Perhaps of lesser note, but nonetheless interesting, is an atypical dorsal pattern on S8–9 of SP 2016 (Figure 12.37). And lastly, we know of two mixed-species pairings, including one ♂–♂ pairing, of *E. basidens* in Oklahoma. On 20 July 1963, G. H. Bick collected a ♂ *E. basidens* (FSCA 66809) that he found paired with a ♂ *Ischnura posita*, the Fragile Forktail (FSCA 66808), at Stonewall in Pontotoc County. And on 19 June 1964, Bick and his students collected a pair that had a ♀ *E. basidens* in tandem with a ♂ *Argia sedula*, the Blue-ringed Dancer (Bick, as per his notes, discarded the pair).

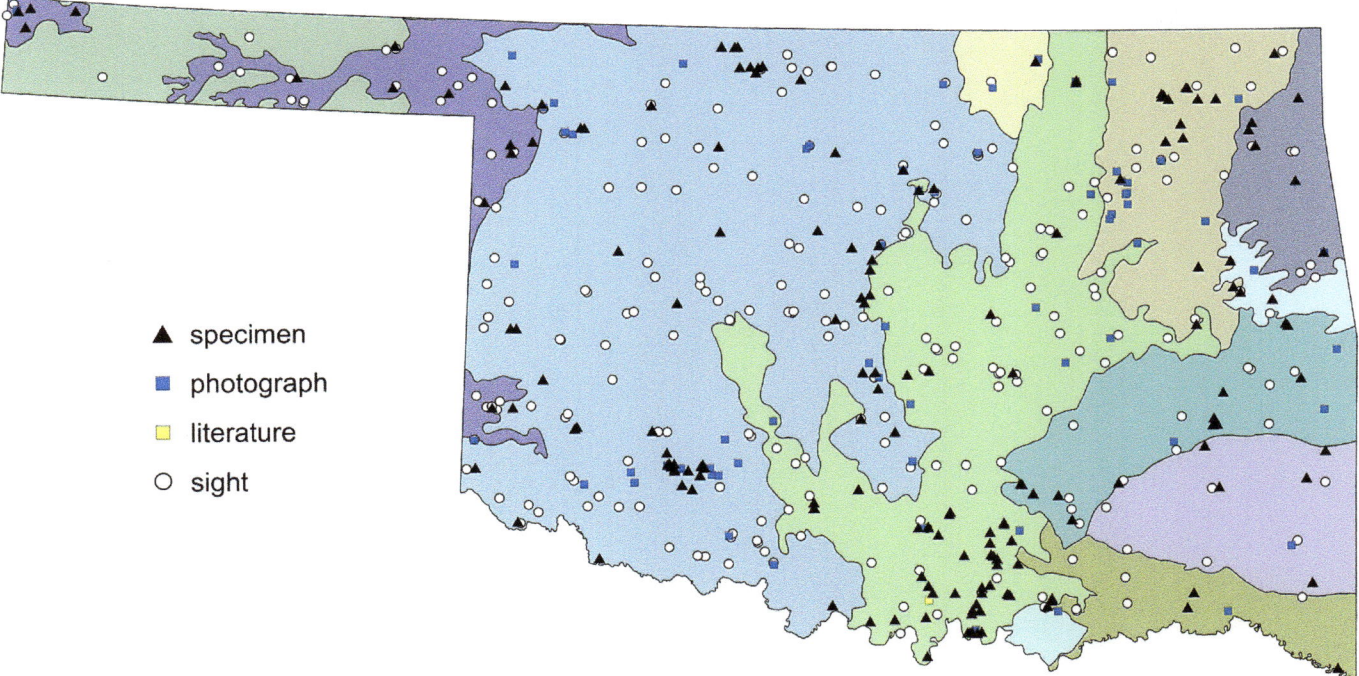

▲ specimen
■ photograph
□ literature
○ sight

Figure 12.51 Documentation level of supporting records for *Enallagma basidens*, the Double-striped Bluet, in Oklahoma. Levels include specimen, archived photograph, literature reference, or sight record. Although sight records lack documentation, only those we feel are credible are plotted.

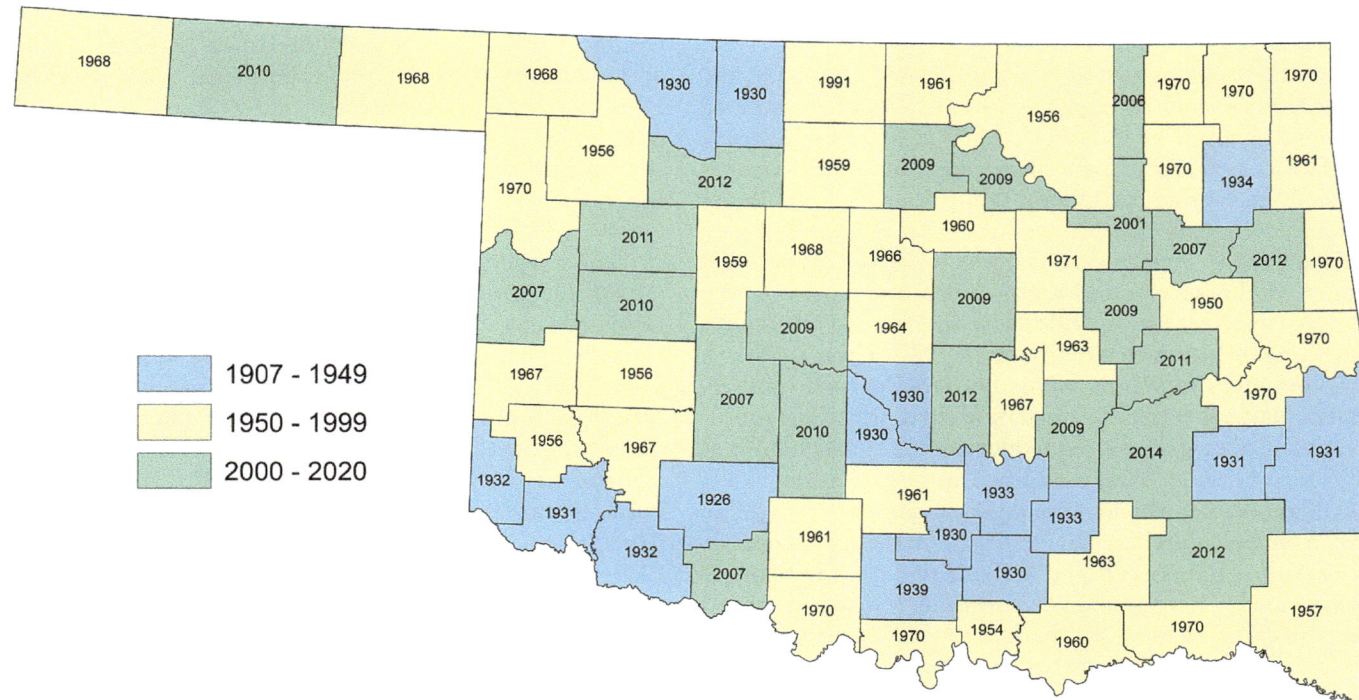

Figure 12.52 Counties in which *Enallagma basidens*, the Double-striped Bluet, are known to occur in Oklahoma. Year within the county is that in which the species was first reported.

ENALLAGMA DUBIUM ROOT, 1924 – BURGUNDY BLUET

In Oklahoma, the Burgundy Bluet is known from 17 records from 7 localities in 2 counties. It was first found in the state by AE Pritchard in Pushmataha County, north of Antlers (Pritchard 1935). We do not know if he found a population, as he gives no details, nor do we know the exact date, but it was likely in spring 1934,[24] as he collected extensively around Antlers in April, May, and June of that year. Although the majority of Pritchard's Oklahoma Odonata specimens reside at the Essig Museum of Entomology, we have yet to find an *Enallagma dubium* specimen attributed to him.

As far as we know, the earliest extant specimen of *Enallagma dubium* for Oklahoma is that collected by SW Dunkle at Kulli Lake, McCurtain County, on 5 June 1992 (2♂,

FSCA). BAH visited the site various times and we visited on 8 June 2013; all of which were fruitless. Miraculously, later in the day on 8 June 2013, we found what we thought was a healthy population (25♂, 2♀, including 2 pairs; 1♂ SP 656) in the northernmost cove of the small lake in the city park in Broken Bow, McCurtain County. DA and BAH visited the park two days later and photographed a ♂ (OC 400474), but they did not indicate how many individuals they saw. The following year the species was seen only once in the state, the only time being again at the Broken Bow city park on 14 June (1♂, MAP). Despite numerous visits by multiple people to the park in subsequent years, the species has not been reported again. In some senses, it is not surprising that the

Figure 12.53 *Enallagma dubium*, the Burgundy Bluet, A) ♂, Broken Bow, McCurtain County, 11 June 2013 (©James W Arterburn). B) tandem pair, 24 km NNE of Eagletown, McCurtain County, 22 June 2019 (©Bill Carrell, OC 497753). C) A ♂ at Broken Bow, McCurtain County, 12 June 2013, hanging out with a ♂ *E. geminatum*, the Skimming Bluet, another species that favors perching on water lilies (©James W Arterburn). (Both *E. signatum*, the Orange Bluet, and *E. vesperum*, the Vesper Bluet, also like to perch on water lilies.)

species has not been found there again given the management scheme of apparent pesticide use and trying to keep the lake and its surroundings manicured.

A more hopeful spin on the species' survival in the state was finding two new locations in 2017. At what was to be a brief stop at a small, wooded pond alongside the "Indian Highway" about 7 km NE of Fewell, Pushmataha County, MAP found 2♂, both tenerals, on 9 May (1♂ as SP 2304). He encountered 8♂ and 1♀ adults at the pond on a return visit on 3 July. In between those two visits, on 25 June and also on a lark, he pulled over to the side of OK-Highway 3 in McCurtain County, where he again found a pond/small lake holding *E. dubium* (at Tucker Lake, 5 km ESE of Ringold, 3♂, 1♂ as SP 2404). The site near Fewell has had one report since: 2♂, 26 May 2018, EA Hjalmarson, B Roberts. Tucker Lake[25] produced 4 other records (all BS-P, in 2019): 4♂ each visit on 10 and 22 June, 6♂ and 3♀, including 3 pairs (one ovipositing) on 2 August, and 4♂ (1 imm) on 14 August. 2019 proved to be a good year for the species because it was reported the most ever (8 times!) and added to 2 new localities, both in McCurtain County. It, too, was reported in relatively high numbers (second highest ever in Oklahoma; 10♂, 4♀, including 4 ovipositing pairs, 22 June, BS-P, BC) and its late flight date was extended by a month and half (15 August, one ovipositing pair, BS-P); both records were from 23 km NNE of Eagletown, McCurtain County.

Pritchard (1935:9) described the spot he collected at as "a small clear lake close to the Kiamichi River, a veritable dragonfly paradise concealed by a dense growth of vegetation and trees, only the center of which is free from prolific masses of water lilies." Here he found the species with *Lestes vigilax*, the Swamp Spreadwing, *Lestes inaequalis*, the Elegant Spreadwing, *E. daeckii*, the Attenuated Bluet, *Nehalennia integricollis*, the Southern Sprite, *Ischnura kellicotti*, the Lilypad Forktail, *Ladona deplanata*, the Blue Corporal, and *Celithemis verna*, the Double-ringed Pennant. Dunkle described his collecting locality as a "trickle outlet, beaver pond." As mentioned above, the Broken Bow city park is rather well manicured, but the lake does hold "lily pads" and sparse emergent vegetation in the area we found *E. dubium*. That portion of the lake is, except early or late in the day, exposed to direct sun, as are all of the locations where the species has been found. The pond NE of Fewell is currently the most forested, but all of the sites undoubtedly were more forested, with either evergreens or mixed forests, in the past. Currently most of the sites are mosaics of native stands of evergreens or mixed forest with farmland, pine plantations, and urban development interspersed. All of the ponds/lakes are <3 ha (<7 acres) in area and are within 2.5 km (1.5 mi) of a perennial stream. And all are manmade except the spot where Pritchard was likely to have had the species, which is a lake that looks more like a natural oxbow.

The species is known to fly from 6 May (Ouachita NF, 10 km NE of Broken Bow, OC 494656) until 15 August

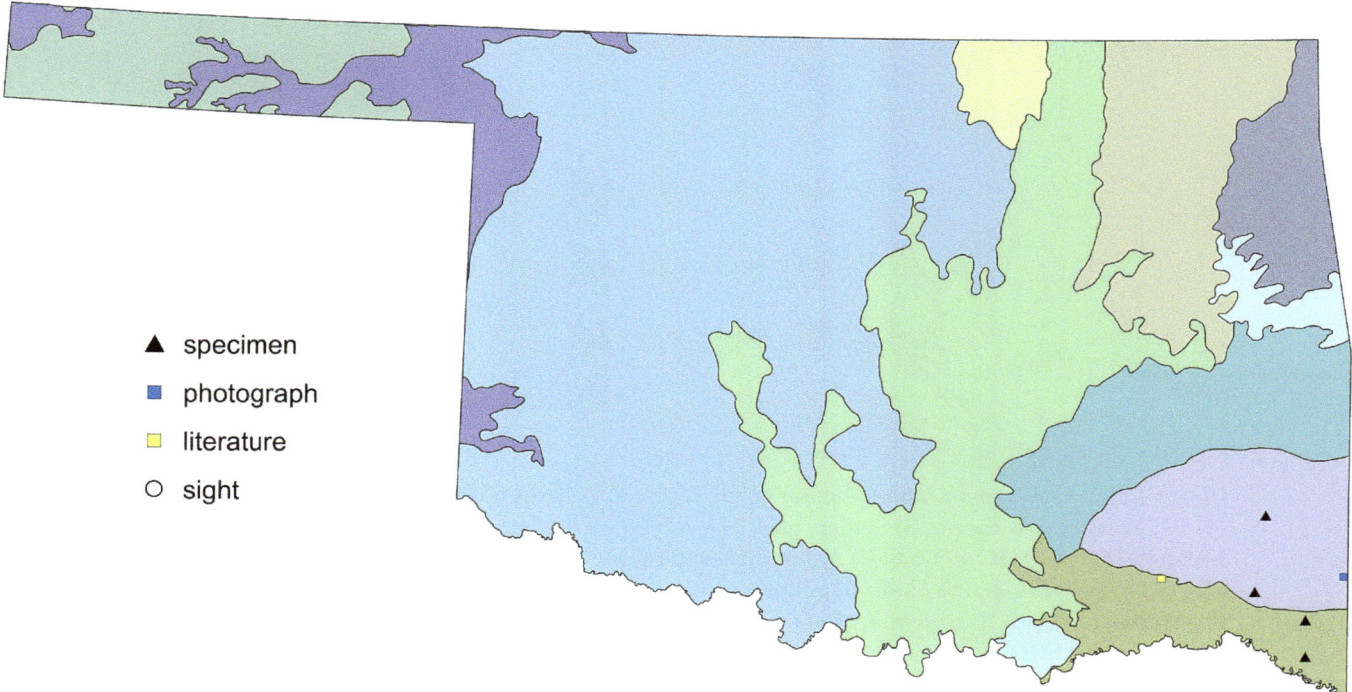

Figure 12.54 Documentation level of supporting records for *Enallagma dubium*, the Burgundy Bluet, in Oklahoma. Levels include specimen, archived photograph, literature reference, or sight record. Although sight records lack documentation, only those we feel are credible are plotted.

(Tucker Lake) in Oklahoma. Flight dates for Louisiana and Texas indicate the species can fly much longer (LA: 9 April–10 September, Mauffray 2014; TX: 9 April–5 October, Abbott and Lasley 2019). Although dates in Arkansas are limited because of only two known records of the species there (24 May to 5 June, OC 446295, 445675; both from 2016, from White Oak Lake, Nevada County), habitat notes are of interest (dominant aquatic vegetation was *Juncus effusus*, common rush, and *Utricularia inflata*, swollen bladderwort). In Oklahoma, it has been reported generically on "water lilies" and emergent near-shore vegetation and specifically on water shield (*Brasenia schreberi*) and, for example at Tucker Lake, on yellow water-lily, also known as Spatterdock (*Nuphar lutea*). In our opinion, the known flight season for Oklahoma and Arkansas indicate two things: 1) southeastern Oklahoma and southwestern Arkansas are still relatively understudied and 2) these areas almost certainly lie at the species' edge of range, so it will probably remain comparatively rare in the region. Still, taking into consideration recent data, we feel this species' conservation status is akin to that of *Enallagma doubledayi*, the Atlantic Bluet, as such we refined its rank (S2).

ENALLAGMA SIGNATUM (HAGEN, 1861) – ORANGE BLUET

The Orange Bluet is the third most widely distributed bluet in the state, behind *Enallagma basidens* and *E. civile*. It has been recorded in 75 counties; the two remaining counties are on opposite ends of the state: Ottawa in the far north-eastern corner and Jackson in the far southwestern corner. As adults, the species is primarily lentic, being most often found around ponds or lakes with clear water, sandy or muddy substrates and emergent vegetation, and some-times with lily pads or floating vegetation, on which they will perch. Tenerals have been found on spring-fed creeks, making us suspect that larval habitat may be lotic. Where the adults are found, they are usually in small numbers, often <10, but large numbers have been found. The highest estimated count that has been reported is 150 individuals (26 July 2011, Quanah Parker Lake, Wichita Mountains WR, Comanche County, T Kompier, OC 333116). That record is the only one for the species being in triple digits in Oklahoma. The next highest count is 80 individuals (including 6 pair) that MAP encountered on 10 April 2016 at Lake Durant, Bryan County.

The species has been recorded regularly and with greater frequency since it was first found in the state on 3–4 June 1907 by EB Williamson in Wister, Le Flore County (5♂, 2♀, of which 2♂ are at UMMZ; Williamson 1914b). During the early 1930s, it was reported on 17 days in 10 counties. During the Bick/Hornuff era, it was noted on 34 days and there were an additional 5 days noted between 1992 and 1997. Those records added an additional 16 counties to the species' known distribution in the state. All of that sounds impressive until one realizes that between 2003 and 2019 *Enallagma signatum* was reported on about 350 days and was added to 49 counties across Oklahoma, making one wonder about range expansion associated with increased lentic habitat post-Dust Bowl.

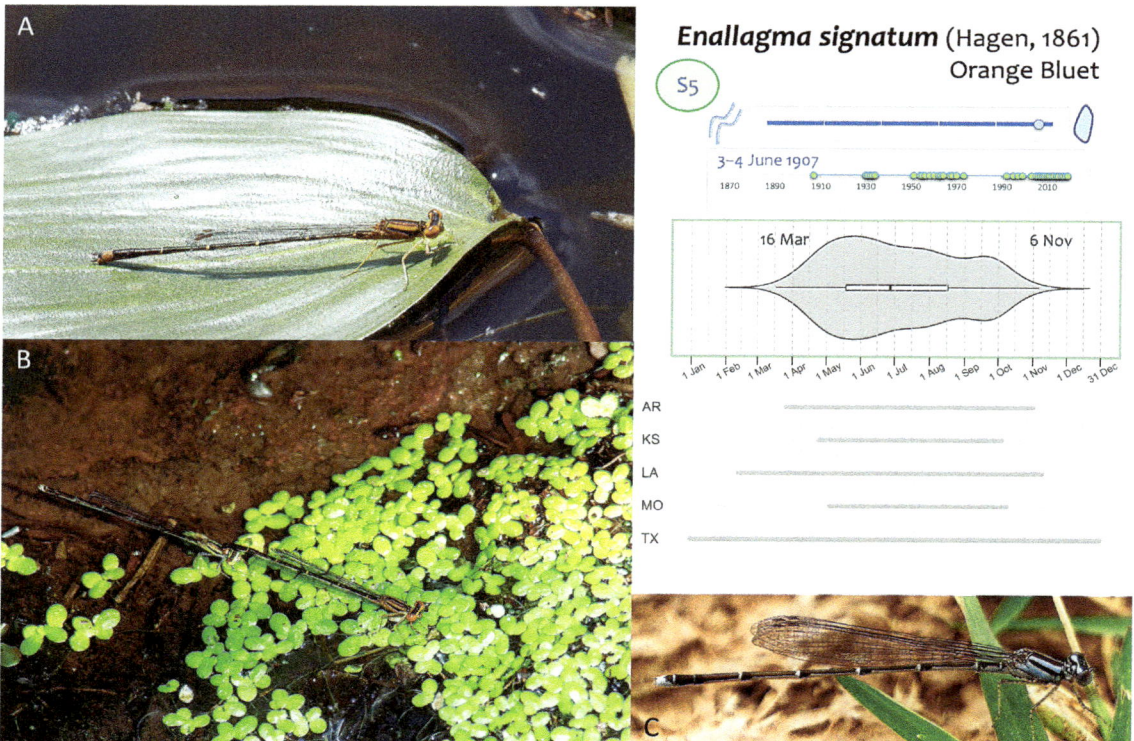

Figure 12.55 *Enallagma signatum*, the Orange Bluet, A) ♂, Haskell Lake, Muskogee County, 15 May 2014 (©James W Arterburn, OC 427983). B) tandem pair perched on duckweed, Will Rogers Park in Oklahoma City, Oklahoma County, 30 August 2006 (©Randy C Anderson, OC 7323). C) ♀, Griffin Community Park in Norman, Cleveland County, 13 April 2017 (©Emily A Hjalmarson, OC 467572). Note the blue coloration of this immature relative to the yellow of the adult ♀ in the tandem pair.

Enallagma signatum can be found in Oklahoma from early spring to late autumn, which is consistent with the rest of the region.[26] Tenerals have been recorded throughout the entire adult flight season in Oklahoma (tenerals: 25 March–1 November; adults: 16 March–6 November). GH Bick collected as least 21 nymphs in Marshall County during the summer of 1955, but those specimens were likely discarded. He said that he reared 1♀ from those collections.

Pairs are routinely seen and ovipositioning has primarily been observed while individuals are in tandem.

One specimen of note is a ♀ captured in Canadian County at Wildhorse Park in the town of Mustang (14 September 2013, SP 1011). This individual has extensive black on the dorsum of abdominal segment 9, a pattern more akin to eastern forms of this species than what we normally see in this region (Lam 2004; Figure 12.37).

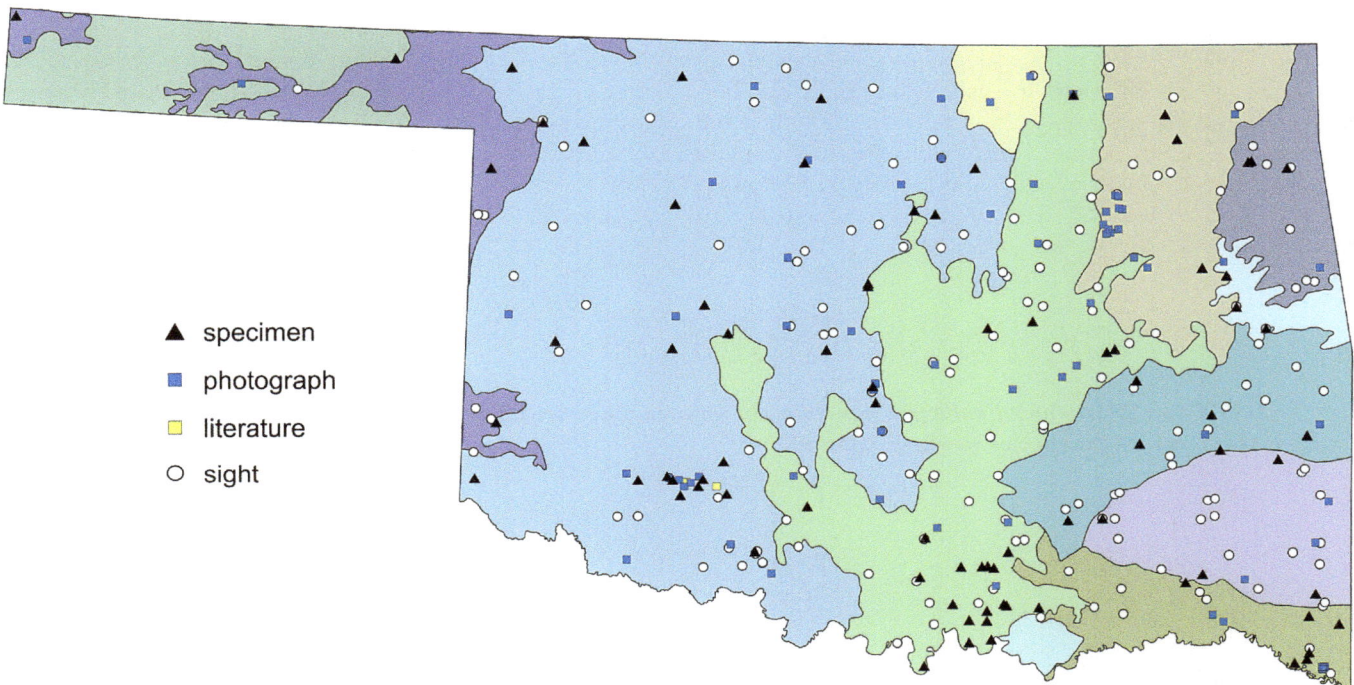

Figure 12.56 Documentation level of supporting records for *Enallagma signatum*, the Orange Bluet, in Oklahoma. Levels include specimen, archived photograph, literature reference, or sight record. Although sight records lack documentation, only those we feel are credible are plotted.

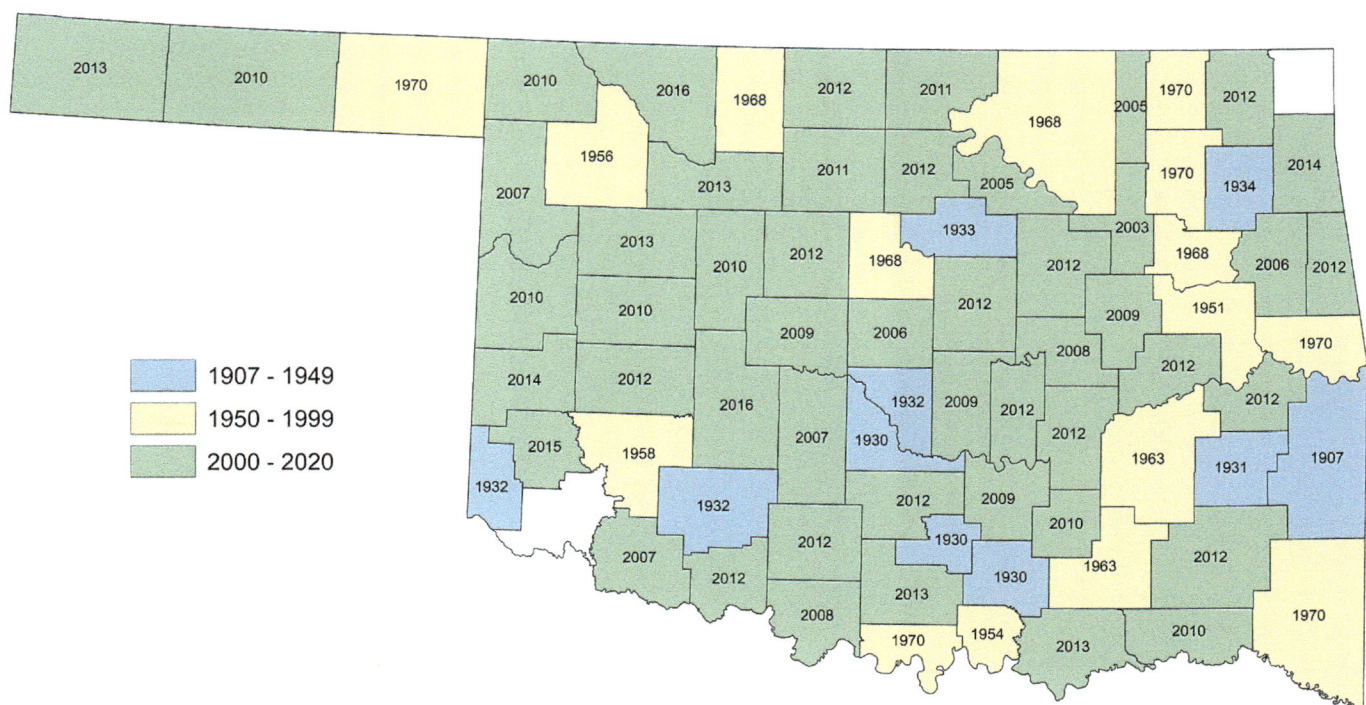

Figure 12.57 Counties in which *Enallagma signatum*, the Orange Bluet, are known to occur in Oklahoma. Year within the county is that in which the species was first reported.

ENALLAGMA VESPERUM CALVERT, 1919 – VESPER BLUET

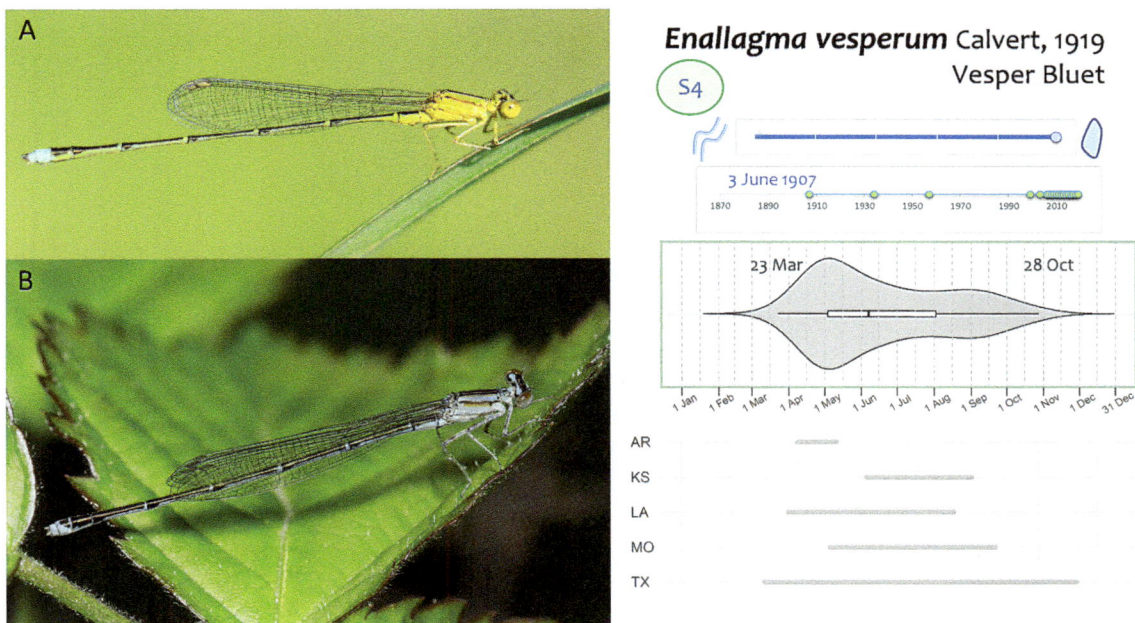

Figure 12.58 *Enallagma vesperum*, the Vesper Bluet, A) ♂, Lake Elmer, Kingfisher County, 8 September 2012 (©Jason R Heinen, OC 382129). B) ♀, JT Nickel Preserve, Cherokee County, 11 May 2015 (©Bryan E Reynolds, OC 430902).

The elusive Vesper Bluet haunts ponds and lakes that have considerable emergent or floating vegetation where, as its translated name—*evening star*—suggests, it is rarely seen unless you happen to be looking for them around dusk. Actually, it can also be found in the early morning hours and every so often during the rest of the day, if it is overcast, or if you search in heavy shade, where individuals roost during the day's heat. It is usually seen 1–2 at a time and occasionally in numbers around a half dozen. Twice it has been reported as high as 22 individuals.[27]

EB Williamson (1914b) was the first to report the species in Oklahoma. He identified the single ♂ he collected on 3 June 1907 at Wister, Le Flore County (UMMZ), as *Enallagma pollutum*, the Florida Bluet (Figure 12.59). Calvert (1919) later determined Williamson's identification to be in error and so he designated Williamson's specimen as a paratype for the newly described *E. vesperum*. Bick and Bick (1957:11) said: "Williamson (1914[b]) records *pollutum* Hagen from Le Flore and Bird (1932) gives this Williamson record under *laurenti* Calvert. Montgomery (1942) lists *vesperum* Calvert from Oklahoma. We could not find specimens of *laurenti-pollutum-vesperum* in the OU collection but judge that the above record pertains to *vesperum*." We do not profess to understand why Bird chose to call Williamson's record *E. laurenti* when Calvert clearly stated that the Williamson specimen was re-identified as *E. vesperum*.

Although *Enallagma vesperum* was not recorded again in Le Flore County until 2011, it was next recorded in

Oklahoma in 1934. AE Pritchard collected the species on 19 June (and perhaps on the 17th as well) at Broken Bow, McCurtain County (10♂, 4♀, EMEC 331120–331125 and 1♂, OSU). That was the only time it was recorded in the

Figure 12.59 The first *Enallagma vesperum* collected in Oklahoma. EB Williamson collected this ♂ on 3 June 1907 at Wister, Indian Territory (now Le Flore County, Oklahoma). He originally identified it as *E. pollutum* but it later became a paratype of *E. vesperum*, described new to science in 1919 by PP Calvert. Specimen is housed at UMMZ; Photo by ©Brenda D Smith.

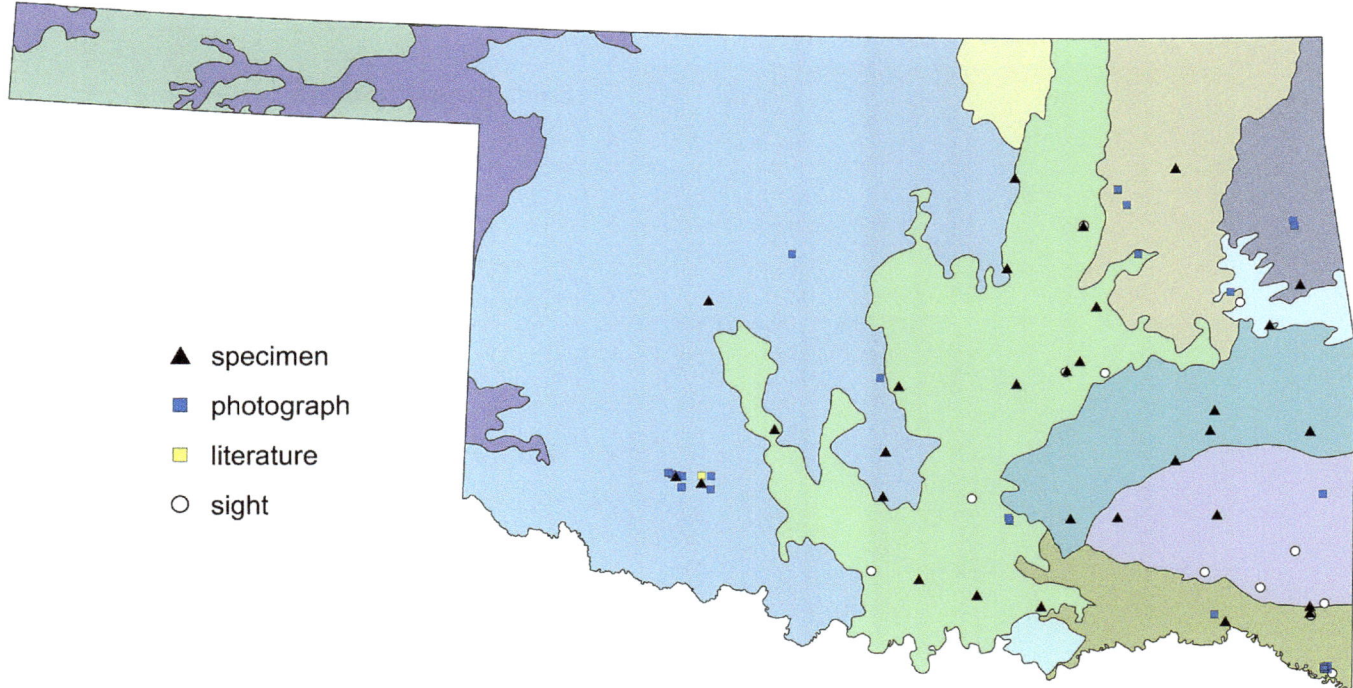

Figure 12.60 Documentation level of supporting records for *Enallagma vesperum*, the Vesper Bluet, in Oklahoma. Levels include specimen, archived photograph, literature reference, or sight record. Although sight records lack documentation, only those we feel are credible are plotted.

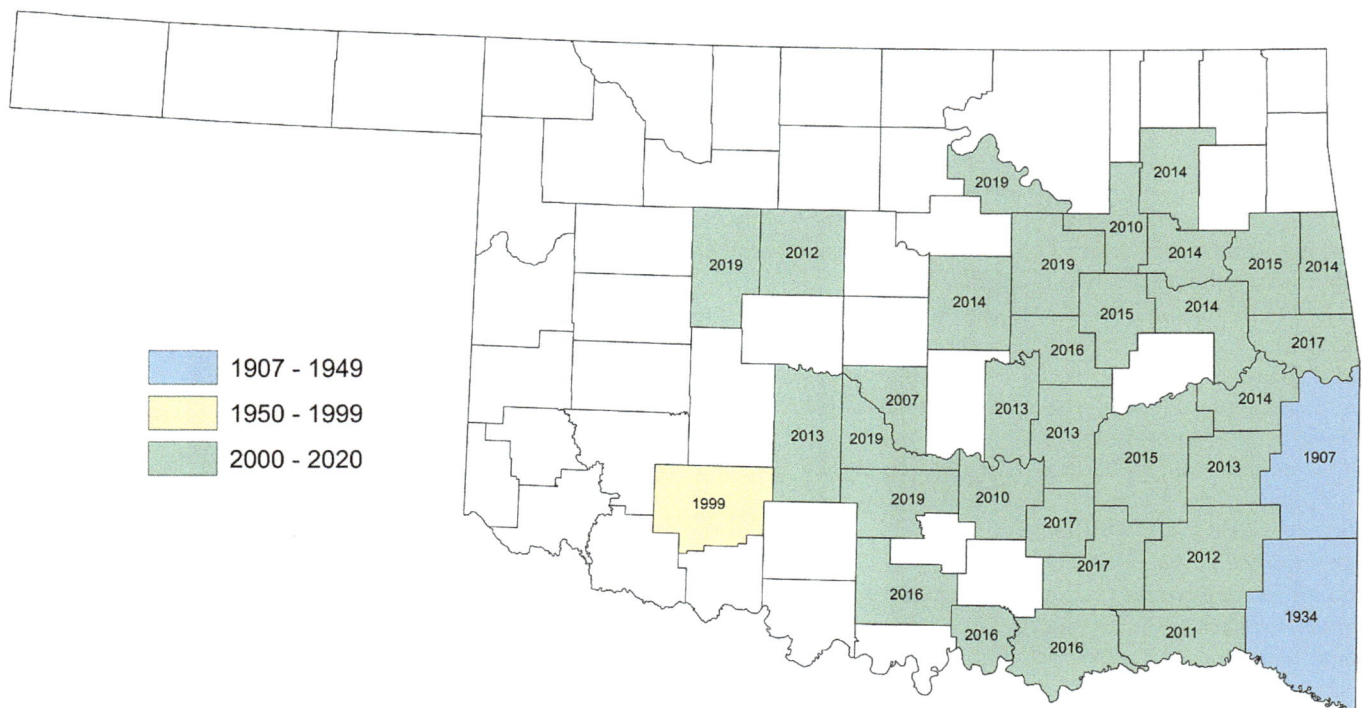

Figure 12.61 Counties in which *Enallagma vesperum*, the Vesper Bluet, are known to occur in Oklahoma. Year within the county is that in which the species was first reported.

state during the 1930s. GH Bick was next to find the species in Oklahoma, on 12 June 1957, when he collected 1♀ in McCurtain County, 2.5 mi N of Broken Bow (FSCA).

Distribution by era suspiciously points toward the species' range expansion in Oklahoma. Essentially there were three records of the species in Oklahoma, confined to two counties in southeastern Oklahoma, prior to the current era. Since 1999, there have been roughly 80 reports at >40 localities from 32 additional counties. It is hard to imagine that the earlier collectors were missing it during their intensive surveys of areas we now know it occurs, but perhaps it was simply that elusive (although MAP has a preternatural nose for it).

Vesper Bluets fly in Oklahoma from early spring until late autumn. The state is bested only by Texas in terms of length of flight season, but Oklahoma's season is much longer than everywhere else in the region (AR: 33 days; KS: 88 days; LA and MO: 139 days; OK: 219 days; TX: 262 days).[28] There are a remarkable 25 records of tenerals for Oklahoma, spanning from 10 April until 9 September, but we know of only one sighting of a pair, and there are no known ovipositing or larval records for the state.

ISCHNURA KELLICOTTI WILLIAMSON, 1898 – LILYPAD FORKTAIL

The specialized habitat—ponds or coves adorned with lily pads[29]—and slothful behavior—it sits for long periods atop lily pads (occasionally on emergent vegetation)—of this species help to explain its odd history in the state (Figure 12.62). The Lilypad Forktail was first found in Oklahoma in 1907 when EB Williamson collected 14♂ and 8♀ in Wister, Le Flore County (3–4 June; all females were "heterochromatic"; Williamson 1914b). RD Bird apparently never saw the species, nor did the Bicks. AE Pritchard briefly mentioned the species in one of his publications (Pritchard 1935), and LE Hornuff recorded it only once (18 July 1970, 2♂, 2♀,[30] Sequoyah County: 7.8 mi N of Sallisaw, FSCA). It seems odd that of all of these early collectors, the species was recorded only three times.

In contrast, since 2006 the species has been reported about 50 times and recorded at >20 additional localities in 11 counties, 10 of which were new for the era, as well as being seen again in the spot that Pritchard likely recorded it. That is such a contrast that it makes one wonder if the Lilypad Forktail is another example of range expansion as a result of increased lentic habitat in Oklahoma since the 1930s. Even though many landowners find water lilies of various types to

be a nuisance, humans have an aesthetic propensity to introducing them into public ponds and lakes; it is not surprising, then, that the Lilypad Forktail may be becoming more widespread in the state and probably the region as a result. We hypothesize that water lily introductions explain apparent range expansion of the Lilypad Forktail in the state and in particular the disjunct population in central Oklahoma at Kitchen Lake (first reported there in 2006 on 6 August by B Horn, 1♂ and 1♀ photographed, OC 7207) (Figure 12.63).

Lilypad Forktails have been reported elsewhere to occur on three types of water lilies (Harp 2005): *Nuphar* (waterlily or spatterdock), *Nymphaea odorata* (sweet-scented or fragrant water lily), and *Brasenia schreberi* (water shield). Williamson (1914b:446) reported the species in Oklahoma "on white waterlily leaves, identically as observed and described from Indiana," which we infer were *Nymphaea odorata*. We have seen Lilypad Forktails on all three water lilies in Oklahoma, but generally we and others uninformatively report the species as found on "lily pads." Although Harp (2005) thought that perhaps the rigidity of *Nelumbo lutea*, the American Lotus, might prohibit its use by Lilypad Forktails, we received a sight record of 1♂ at

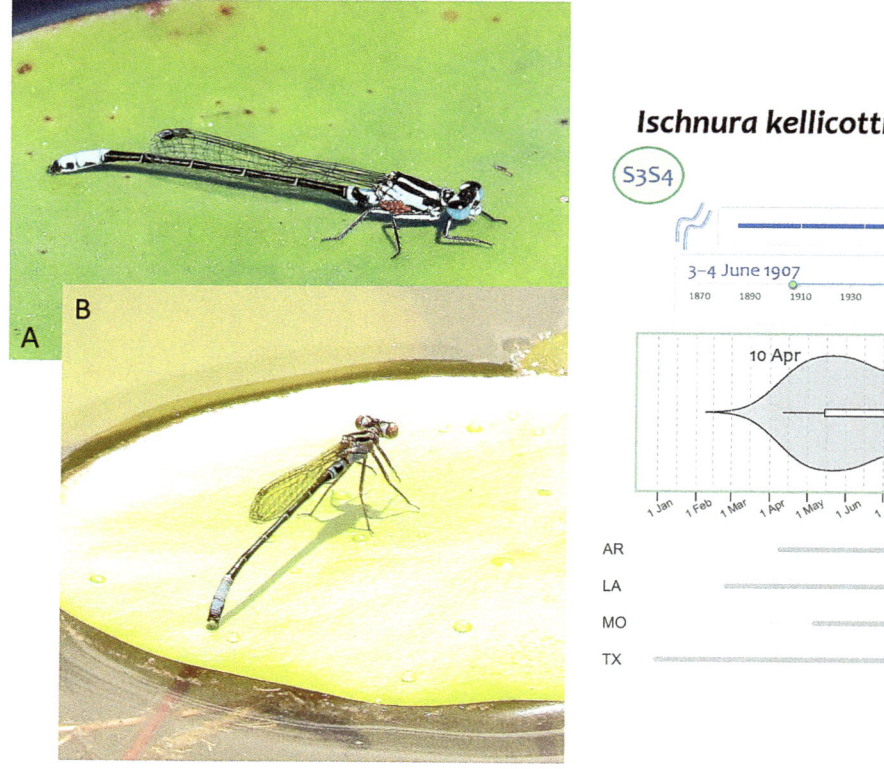

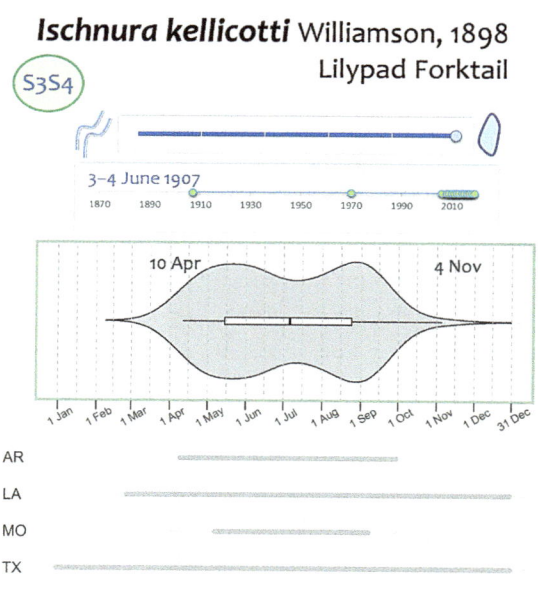

Figure 12.62 *Ischnura kellicotti*, the Lilypad Forktail. A species renowned for its botanical fidelity, one seldom sees it other than perched atop a water lily with its abdomen tip bent downward, as exemplified by both this A) ♂, Big Cedar, Le Flore County, 20 July 2011 (©Tom Kompier, OC 332367), and B) ♀, 8 km S of Smithville, McCurtain County, 5 June 2009 (©Berlin A Heck, OC 313160). This behavior obviously is innate: even a pinkish immature ♂, likely emerged no more than a day or so, acts the same. Note the amber wings on this individual, photographed 24 km NNE of Eagletown, McCurtain County, 22 June 2019 (©Bill Carrell). For the ♀, see Figure 12.63 (©Jeff Trahan, OC 463328).

Figure 12.63 When an orange damselfly is encountered, one might be quick to assume the species is either *Enallagma signatum*, the Orange Bluet, or perhaps *Ischnura hastata*, the Citrine Forktail. But, the young ♀ of many forktail (*Ischnura*) species is orange. In the case of A–C) *I. kellicotti*, the Lilypad Forktail, the A) immature ♀ is vivid orange, but the ♀ of it and all *Ischnura* species darkens within weeks, first to a B) more patterned black and green, blue, or pink, later to C) a pruinose gray, especially on the abdomen. Other examples of orange immature ♀ forktails are D) *I. ramburii*, Rambur's Forktail, E) *I. verticalis*, the Eastern Forktail, and F) *I. hastata*. The extent of orange on the abdomen is a good cue to species determination; however, with some species, one must examine the female under a microscope to confidently identify it. A) Kitchen Lake, Cleveland Co., 6 Aug 2006, ©Bill Horn, OC 7207; B) Broken Bow, McCurtain Co., 18 Aug 2013, ©Bill Carrell; C) Muskogee Co., 15 May 2014, ©James W Arterburn; D) Sequoyah NWR, Sequoyah Co., 24 Aug 2013, ©Victor W Fazio III, OC 434884; E) Wichita Mountains WR, Comanche Co., 3 Sept 2006, ©Bill Horn, OC 7336); F) Lake Stanley Draper, Cleveland Co., 28 Aug 2016, ©Emily A Hjalmarson, OC 462937

McClellan-Kerr WMA, Chouteau Unit, Wagoner County, at a "big riverine backwater wetland dominated by duckweed, buttonbush, and water lotus (*Nelumbo*)" (17 May 2012, JT Bried).

Generally, the species is found one or two handfuls at a time. The highest count is 75 individuals from a fall record from Clayton Lake SP, Pushmataha County (2♂, 2♀, 71U, including 2 pair, 9 September 2012). Williamson (1914b) reported that he collected 22 individuals in 2 days in 1907, an impressive feat given the challenge to capture the species without getting soaked. Two of those individuals (1♂, 1♀) were reported as identified by Calvert and preserved in alcohol, but we have yet to find any individuals from that collection. We know of only 8 extant specimens of the species taken in Oklahoma (2♂, 2♀, FSCA; 3♂, 1♀, SP 372, 378, 545, 1789).

The seasonality of the species in Oklahoma and Arkansas is generally mid-April (10 April 2012, DA and BAH, McCurtain County: Red Slough WMA, OC 374376) to late September (27 September 2014, Red Slough WMA), but it was once found as late as 4 November (2008, BAH, McCurtain County: Red Slough WMA, OC 284440; AR: 10 April–30 September, OC 327713, iNat 819506). Missouri has a shorter season, being known only from early May to early September (8 May–8 September, Trial 2005). It occurs in Texas year-round (Abbott and Lasley 2019, iNat) and almost year-round in Louisiana (26 February–30 December, iNat 274953, 170706). We know of no records of ovipositing but there are three reports of pairs. No nymphs are known to have been collected. There are two records of single tenerals, both of which caused a bit of a ruckus by momentarily confusing observers. The first time was when a ♂ was first spotted by DA on Bittern Lake in the late afternoon on 7 May 2018 he asked BS-P why it looked so reddish-purple. Upon looking through binoculars, discussion began about whether the individual was a Burgundy Bluet (*Enallagma dubium*), a much rarer species in Oklahoma. By that time GW Lasley had seen the individual and was wondering the same thing. We were all puzzled though by the amber wings. Because we were leading a dragonfly-watching tour at the time, a number of people whipped out their cameras to photograph the individual. When everyone was satisfied with their photos, BS-P caught the male, looked at its abdominal pattern more closely and examined its terminal appendages, and determined it to be a Lilypad Forktail. Mystery solved, but lesson not learned. The next summer (22 June 2019), BS-P and BC encountered a pink-blue damsel with amber wings that was too skittish to catch. Fortunately, BC was able to get telephoto shots that, when blown-up, revealed a teneral *I. kellicotti* (Figure 12.62).

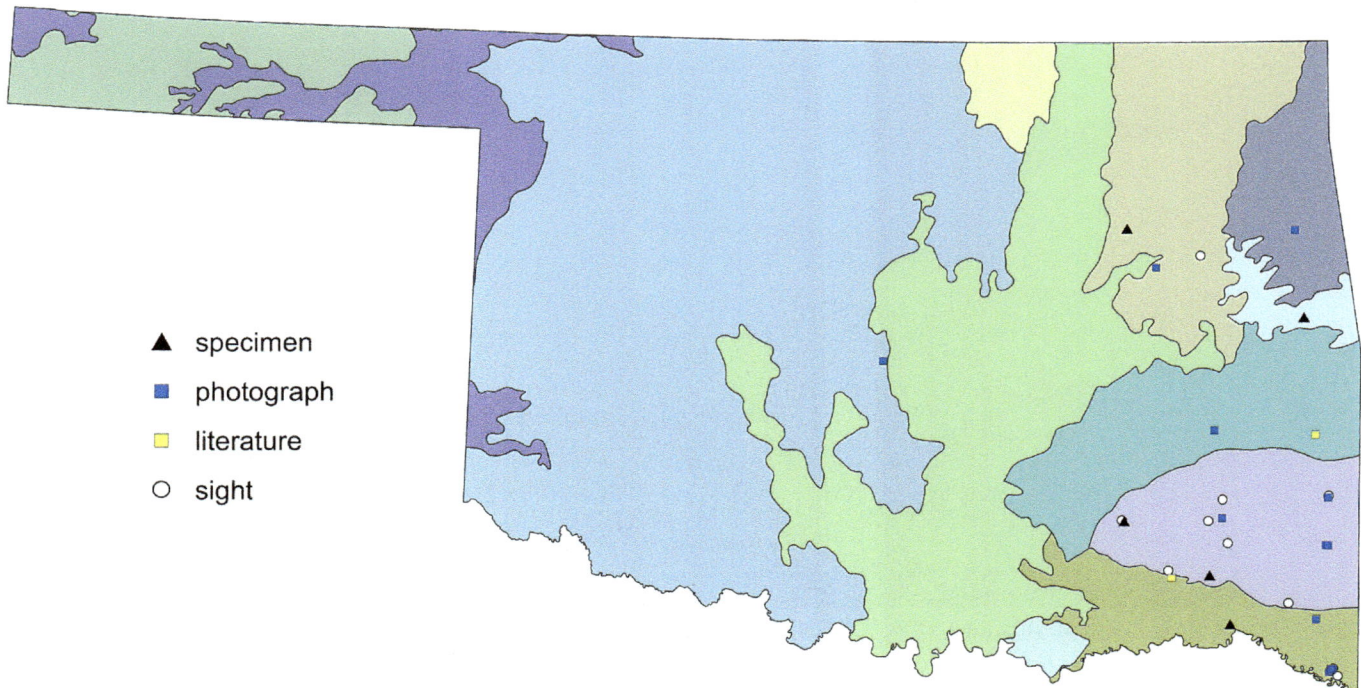

Figure 12.64 Documentation level of supporting records for *Ischnura kellicotti*, the Lilypad Forktail, in Oklahoma. Levels include specimen, archived photograph, literature reference, or sight record. Although sight records lack documentation, only those we feel are credible are plotted.

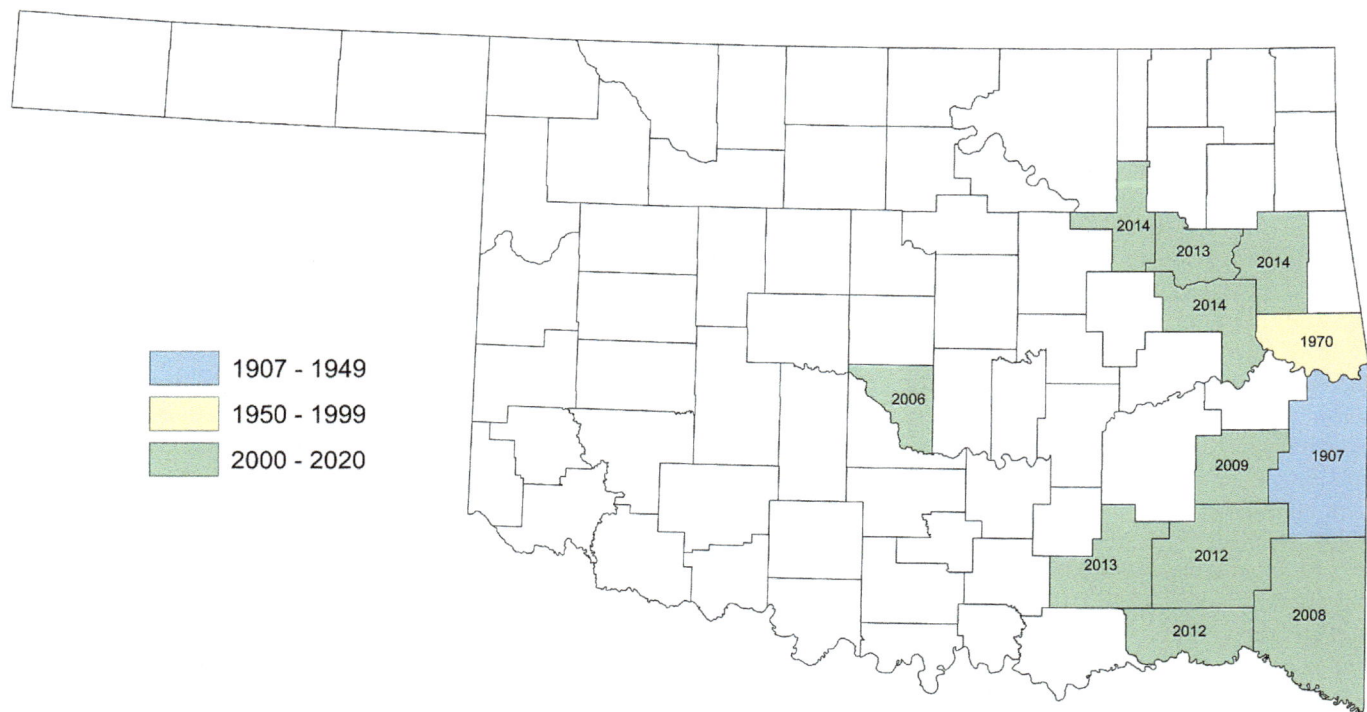

Figure 12.65 Counties in which *Ischnura kellicotti*, the Lilypad Forktail, are known to occur in Oklahoma. Year within the county is that in which the species was first reported.

ISCHNURA RAMBURII (SÉLYS, 1850) – RAMBUR'S FORKTAIL

The Rambur's Forktail is a robust forktail found at road-side ditches, clear-water lakes with mud bottoms, swampy pond edges, and heavily vegetated backwaters with grasses and cattails (Figure 12.66). It is a fairly widespread and common species: encountered >300 times in 57 counties[31] in Oklahoma. The species is most abundant in the south-central and southwestern parts of the state, but it has been recorded throughout much of the state (*sans* panhandle), with

Figure 12.66 *Ischnura ramburii*, Rambur's Forktail. A) Odonates are top predators and will even eat other odonates. This ♀ Rambur's Forktail is a fine example … she is munching on a ♂ *Enallagma exsulans*, the Stream Bluet (Mingo Creek, Tulsa, Tulsa County, Oklahoma, on 28 July 2013, OC 402525, ©Bill Carrell). Note that these two species are comparable in size: 27–36 and 31–37 mm, respectively (Paulson 2011). B) ♂, Lake Ponca, Kay County, 17 October 2014 (©Bill Carrell, OC 427483). C) ♀, Keystone Dam, Tulsa County, 2 July 2016 (©Bill Carrell, OC 448358). See Figure 12.63 for an immature ♀. D) male mimic (andromorph ♀), Cache, Comanche County, 3 September 2009 (©Victor W Fazio III, OC 314989).

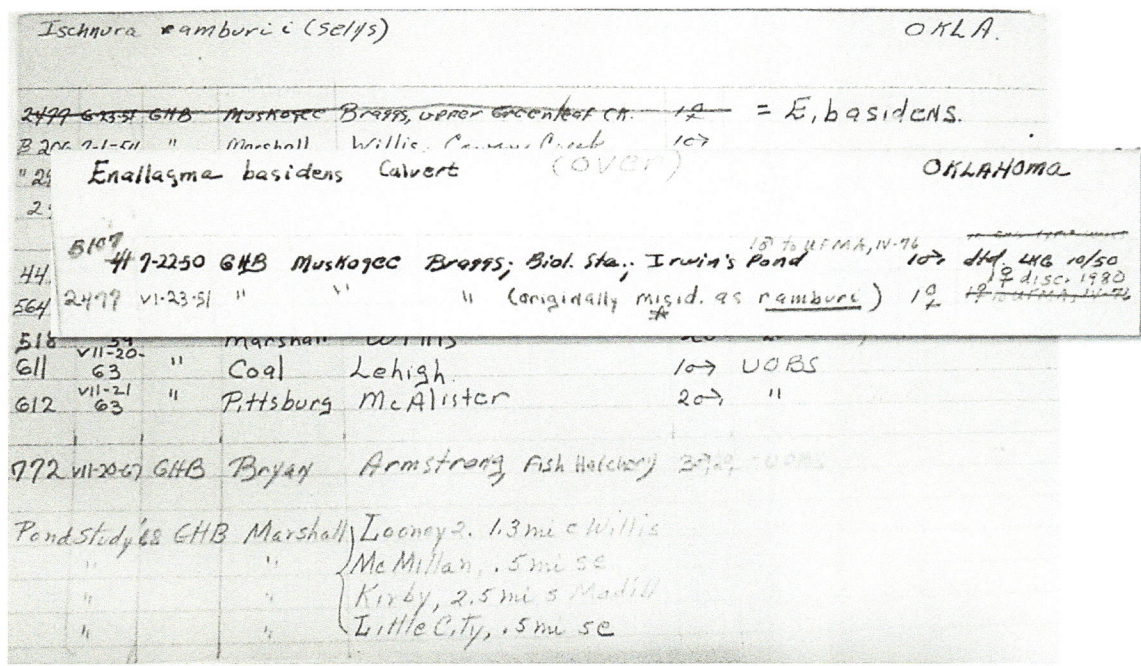

Figure 12.67 GH Bick's note cards showing how what was claimed as the first state record of *Ischnura ramburii* for Oklahoma (Bick and Bick 1957) was a misidentification of a ♀ *Enallagma basidens*, the Double-striped Bluet.

scattered records up to the Kansas border. It is fairly regularly encountered in the double digits (10–80); on two occasions "several hundred" have been reported (30 September 2007, Lake Helen, Lawton, Comanche County, VWF, OC 263583; 22 June 2012, Metcalf Pond, Tulsa, Tulsa County, BC, OC 377036).

Ischnura ramburii is another predominantly lentic species that apparently has expanded across Oklahoma with the increase in lentic habitat post-Dust Bowl. The pattern of discovery is suspicious if this is not the case: the OU expeditions never reported the species, despite working in areas in which the species now occurs commonly, and Bick and Hornuff recorded it in only seven counties, most of which were in the southern tier of the state. Yet between 2003 and 2019, the species was added to an additional 50 counties. The pattern of records strongly suggests its relatively recent spread northward (including a first state record for Kansas in 2019; OC 505200). In that northward expansion signs indicate that it is replacing *I. verticalis*, the Eastern Forktail, in areas in which that species previously occurred (see *I. verticalis*' species account).

The first record of Rambur's Forktail in Oklahoma needs clarification. A ♀ specimen collected on 23 June 1951 near Braggs, Muskogee County, was published as *Ischnura ramburii* (Bick and Bick 1957) but later was re-identified as *Enallagma basidens*, as per Bick's notes. This misidentification was not corrected in any of Bick's subsequent publications; as such, it was carried forward by the Dot Map Project (Donnelly 2004c and OC 207603, we unconfirmed the OC record). Bick's notes indicated that the specimen was discarded (Figure 12.67). As this was the only known record

for the species for Muskogee County, Rambur's Forktail was taken off the county list when we discovered the re-identification but was later re-added when it was documented (SP 2255) in the county in 2017. Therefore, the first erifiable record of the species in Oklahoma was when GH Bick collected 1♂ on 1 July 1954 at Willis, Marshall County (FSCA; Bick and Bick 1957).

The early and late flight dates for Rambur's Forktail in Oklahoma span from late March to late November, both from Red Slough WMA in McCurtain County (21 March 2012, andromorph ♀, DA, OC 374144; 28 November 2009, ♂, BAH, OC 315788). Oklahoma's date range is most in line with that of Arkansas (9 April–2 November, OC). New Mexico is not far behind (April–mid-October, Paulson 2009, OC). There are limited records for Missouri, so known dates likely do not reflect the actual flight season there (9 August–14 September, iNat, Sims 2012). The species is found in Texas year-round (Abbott and Lasley 2019).

Pairs have been reported many times in Oklahoma but ovipositing has only been reported once: a lone ♀ at Whites Lake, 12 km NW of Cookietown, Cotton County, on 3 August 2014 (3♂, 1♀ present). The only nymph reported is implied in Bick's notes: he reared 1♂ that he collected on 1 July 1955 from Cowan Creek, Willis, Marshall County. That ♂ emerged on 9 July 1955. Bick indicated that the specimen was sent to FSCA, but we did not locate it there when we cataloged the Oklahoma portion of the odonate collection. Two teneral records are known, both from the fall (23 October 2016, andromorph ♀ in hand, Waurika WMA, Cotton County, MAP; 1 November 2016, 2U, Red Slough

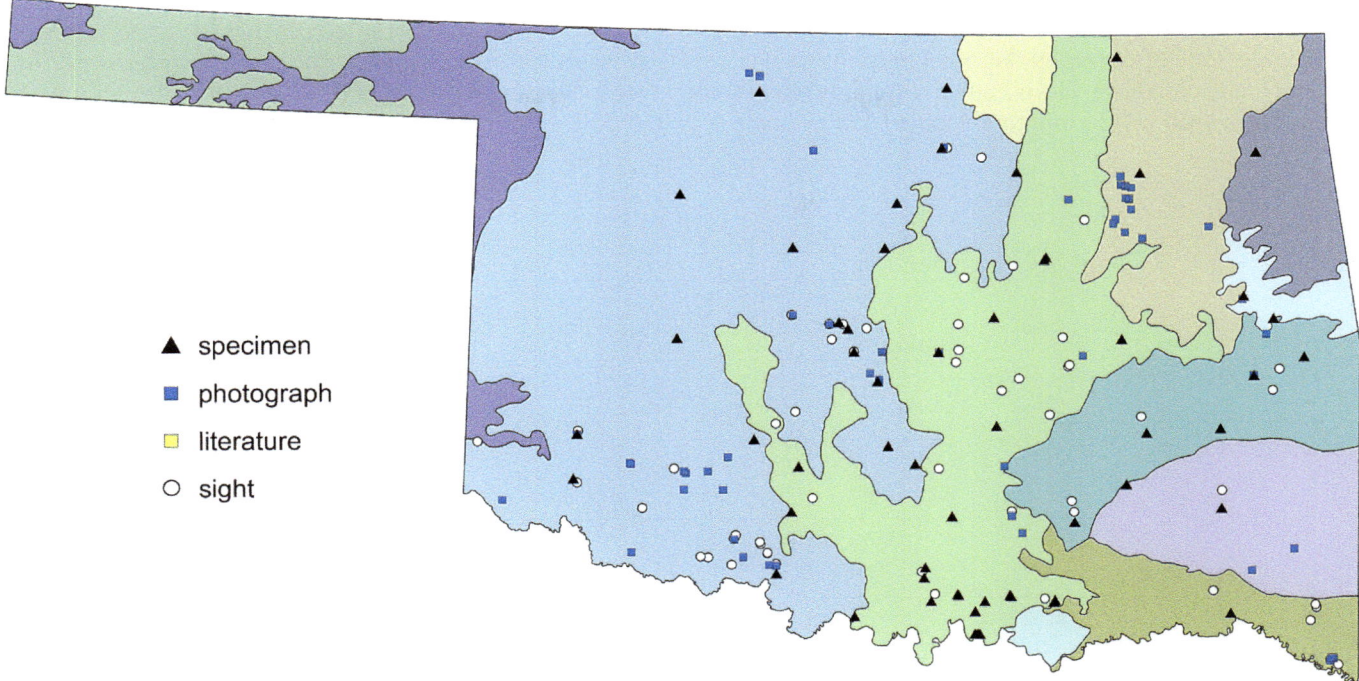

Figure 12.68 Documentation level of supporting records for *Ischnura ramburii*, Rambur's Forktail, in Oklahoma. Levels include specimen, archived photograph, literature reference, or sight record. Although sight records lack documentation, only those we feel are credible are plotted.

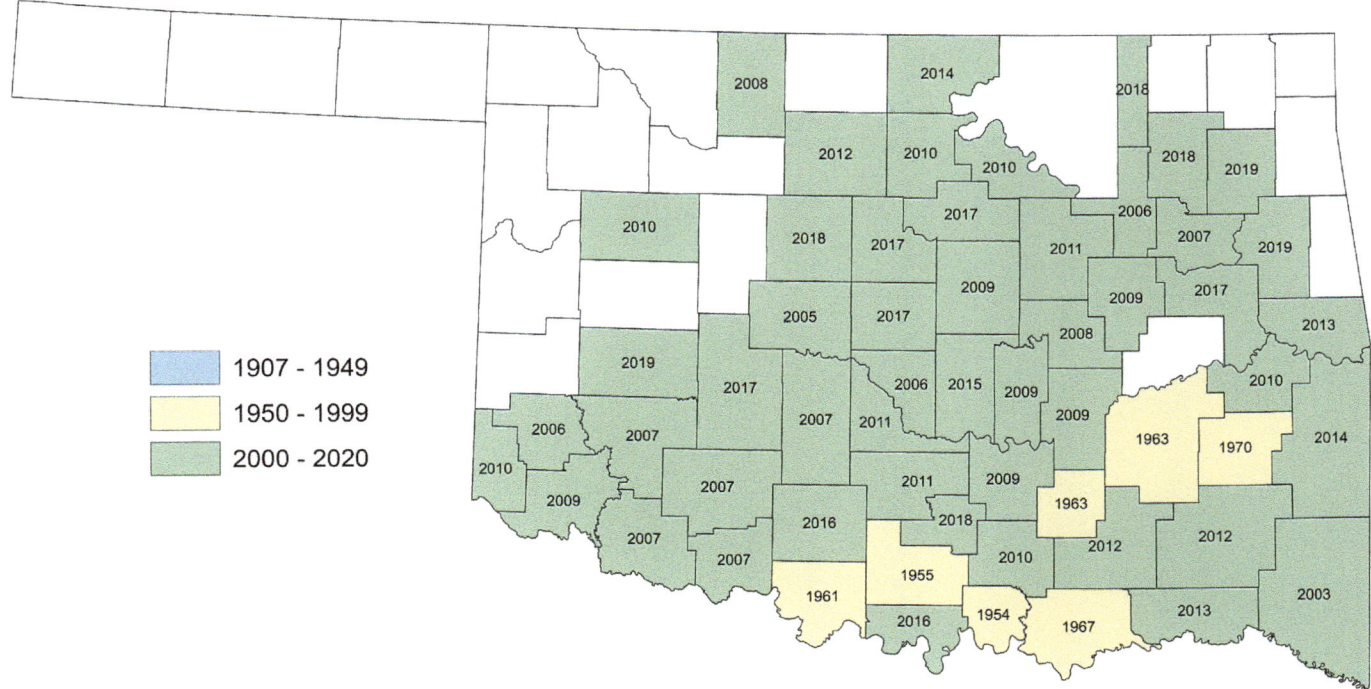

Figure 12.69 Counties in which *Ischnura ramburii*, Rambur's Forktail, are known to occur in Oklahoma. Year within the county is that in which the species was first reported.

WMA, McCurtain County, DA). An andromorph female was photographed by VWF in Cache, Comanche County, 3 September 2009 (OC 314989; Figure 12.66).

On a final note, the species is known to eat prey roughly the same size as itself (27–36 mm total length). For example, a ♀ was photographed eating an adult ♂ Stream Bluet (*Enallagma exsulans*, 31–37 mm total length; Figure 12.66). Not quite as impressive is a record of an *Ischnura ramburii* (sex unrecorded) eating a ♂ *I. posita* (OMNH 9027, formerly SP 11).

ISCHNURA BARBERI CURRIE, 1903 – DESERT FORKTAIL

This alkaline-loving species has been found in 15 counties in the western part of the state. At times it may be abundant, with numbers in the hundreds and even into the thousands (~2,400 adult males and one pair photographed at Crystal Beach Park, Woodward, Woodward County, 22 August 2011, VWF, OC 331978; ~1,000, including 100 pairs, at Artesian Beach Park, Gage, Ellis County, 25 August 2013). The area around the Salt Plains NWR, Alfalfa County, and the "Jackson Salt Plains" west and south of Eldorado, Jackson County, are two other well-known spots for the species. The refuge is where most of the Oklahoma specimens of the species have been taken. The RD Bird expedition visited there 11–15 June 1930, during which they collected 12♂, 2♀, and 4U specimens (USNM 373870–373887). Bick also collected there on 5 June 1956[32] and 13–14 July 1968 (25♂, 19♀; FSCA). To this day the salt flats around the Sandpiper Trail remain a reliable location for the species. This locality has been the center of much confusion over the years as it was once erroneously interpreted as being in Cherokee County rather than relative to the town of Cherokee, Alfalfa County, and surrounding areas, i.e., what is now the Salt Plains NWR. The confusion arose over a cataloging error, made in the early 2000s, of Bird's collection. Although Bird (1932) and the Bick's (1957) did not indicate that *Ischnura barberi* was found in Cherokee County, which is well east of the species' range, the error persisted for roughly a decade until we were able to track down its source—specimen labels with just "Cherokee" on them.

The locality of Jackson Salt Plains was equally problematic because its colloquial name was not traceable on printed maps. It was not until BS-P was able to piece together through various sources that the general location became clear. Prime habitat for the species includes areas with spring-fed ponds, lakes, creeks, and alkali plants such as *Tamarix* (salt cedar) and *Distichlis* (salt grass), but roadside flooded areas also hold the species, at times.

It is unclear if the species' range has expanded in the state or if more extensive surveys of the past decade or two have resulted in more records. Certainly in the areas we would have expected RD Bird's team to have had the species, they did, and where they worked and did not find records, we still do not have records there, for example, Comanche County.[33] But a comparison of eras—14 records in 6 counties in 1930–1970 but about 75 records with an addition of 9 counties in 2008–2019—makes it tempting to conclude a range expansion; however, we refrain from doing so.

The species flies in Oklahoma similarly to elsewhere in the region, from the spring well into the fall (OK: 22 April 2016, 1♀ in hand, Beaver River WMA, Beaver County, MAP, and 13 October 2015, 4♂, Salt Plains NWR, Sandpiper

Figure 12.70 *Ischnura barberi*, the Desert Forktail, A) ♂, the "Jackson Salt Plains" 6 km S of Eldorado, Jackson County, 11 October 2013 (OC 420221). B) ♀, Crystal Beach in Woodward, Woodward County, 22 August 2011 (OC 331979). Photos by ©Victor W Fazio III.

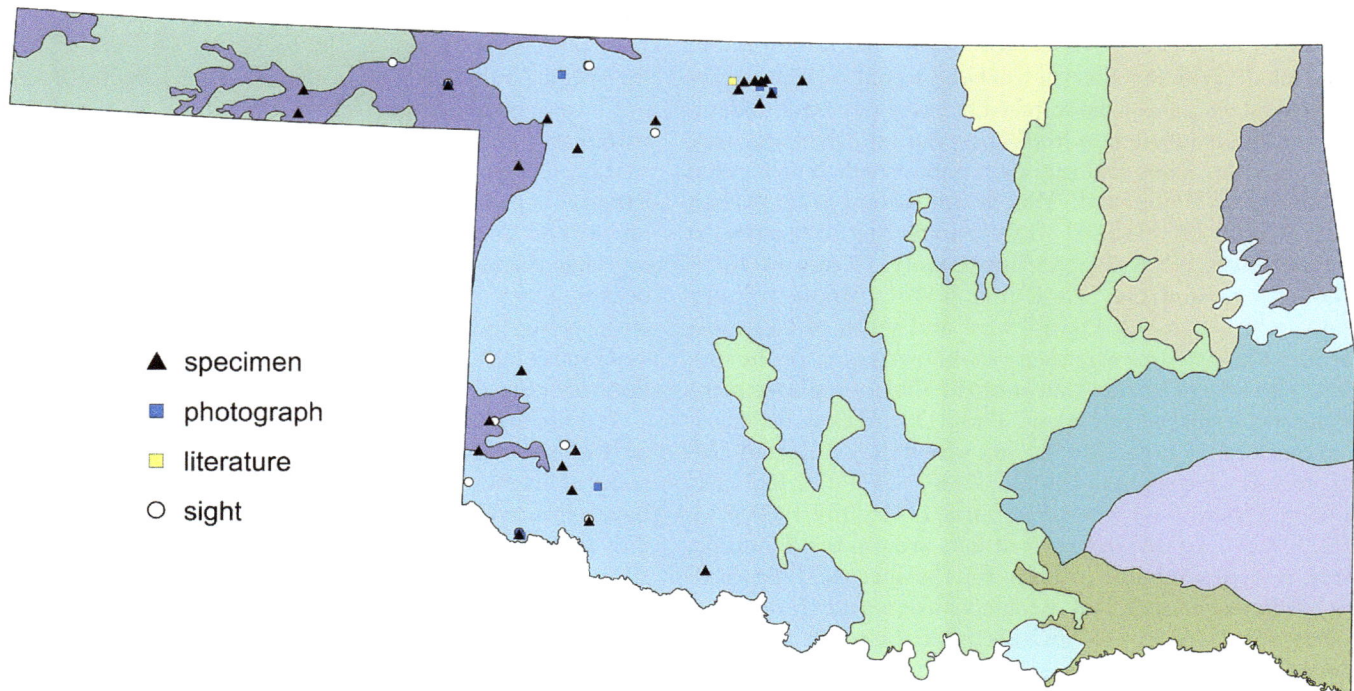

Figure 12.71 Documentation level of supporting records for *Ischnura barberi*, the Desert Forktail, in Oklahoma. Levels include specimen, archived photograph, literature reference, or sight record. Although sight records lack documentation, only those we feel are credible are plotted.

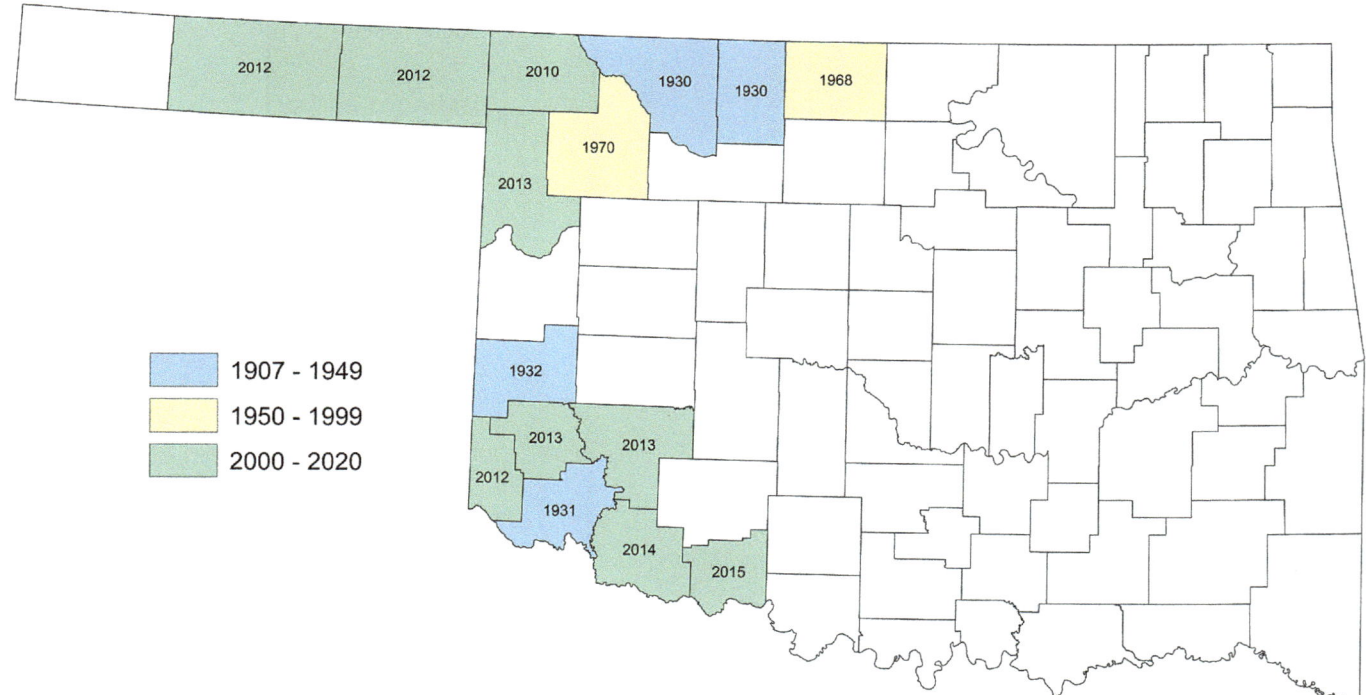

Figure 12.72 Counties in which *Ischnura barberi*, the Desert Forktail, are known to occur in Oklahoma. Year within the county is that in which the species was first reported.

Trail, Alfalfa County, BS-P, photos; KS: limited records, 7 June–1 August, SEMC, OC; NM: March–November, Paulson 2009; TX: 8 April–22 October, OC, Abbott and Lasley 2019). Many pairings have been reported (20–100 at a time), but we know of only one report of ovipositing: 15♀, at Crystal Beach, Woodward, Woodward County on 17 May 2015 (20♂ and 20♀, BS-P). That same location provides evidence of mass emergence of thousands of individuals, as does the nearby site of Artesian Beach Park in Gage, Ellis County (see above for details). Despite those records, no nymphs and only one teneral has been reported: 1♂, 7 km W of Tipton, North Fork of the Red River, Tillman County, 16 May 2014 (SP 1184).

A couple of final notes. On 5 August 2018 at Altus city lake, Jackson County, MAP collected an immature ♀ andromorph, (SP 2689), an occurrence considered to be rare (Westfall and May 2006). And, similarly to *Ischnura ramburii*, the other robust species of forktail in the state, *I. barberi* has been observed eating prey as large as itself, in this case when a ♀ *I. barberi* was seen eating a ♀ *Enallagma civile*.

ISCHNURA DAMULA CALVERT, 1902 – PLAINS FORKTAIL

In Oklahoma, the Plains Forktail is largely confined to Cimarron County, where it is generally rare but can be locally common—we have seen as many as 22 individuals in a single day. Elsewhere in Oklahoma, it is probably only a vagrant. It was first found in the state in early July 1926 when TH Hubbell collected 1♂ and 1♀ at 3 mi N of Kenton, Cimarron County (UMMZ). The second record for the species was when AE Pritchard found it in Kenton, Cimarron County, on 7 July 1933 (2♂, 1♀; USNM 374680–374682). The species was not reported for the state again until LE Hornuff and GH Bick encountered it in early August of 1968, 1970, and 1973 in Cimarron County and once in Woodward County (2 August 1970, 1♂, Fort Supply, FSCA).

Forty years elapsed before *Ischnura damula* was recorded again in Oklahoma: we encountered it on 26 May 2013 at three locations in Cimarron County—Black Mesa SP, 7 km N of Kenton, and at a cattle pond in Conrad (13♂, 9♀; SP 612–613, 623–628[34]). We had seven other Cimarron County records, between 2013 and 2019. The sole record for Texas County is from 12 km N of Goodwell, at a marshy spot on the Beaver River (3 June 2015, 2♂ [1 as SP1613], MAP). More recent records came from EA Hjalmarson and B. Roberts, when they captured and photographed an andromorph ♀ on North Carrizo Creek, 5 km NNE of Kenton, on 19 May 2018 (photos on file, OOP), and MAP, who captured and photographed a ♂ on the Cimarron River, 4 km NNE of Kenton, on 25 May 2019 (OC 495611). All records put together account for <20 at 15 localities in 3 counties.

One specimen of interest is an andromorph ♀ that we collected at Clinton Lake, 5 km SW of Foss, Washita County, 25 September 2010 (SP 96). We and JC Abbott originally identified this specimen as *Ischnura damula*, but MAP and BS-P later realized that whereas this individual's overall pattern closely resembled an *I. damula* her mesostigmal plate is that of *I. denticollis*, the Black-fronted Forktail. We have collected other aberrant individuals, all of which came from Cimarron County and have been confirmed as *I. damula* (most by RW Garrison). For example, lavender females with varying antehumeral patterns from those with full antehumerals that narrow slightly centrally (SP 625) to those that are more andromorph in character, having broken antehumerals: broken on one side only (SP 628, right antehumeral slightly broken, left side complete) to those with just lilac dots (SP [RWG] 627). Another female we collected is more of a "normal" andromorph in overall color but with blue-green elongated spots for antehumerals (SP [RWG] 624). Green females include one with broken antehumerals (SP 1601) and another with full antehumerals but they are narrow and there is a wide humeral (SP 2520). Notably, three ♀ specimens from Conrad and Black Mesa SP essentially lack the raised "nipples" on the prothorax characteristic of the species (SP [RWG] 627 and 1600, SP 628; RW Garrison, *in litt*). This condition is apparently rare.[35]

Males are not immune to aberration, as we have one with small green, rather than blue, dots (SP [RWG] 626) and three that at first glance could be mistaken for large *I. denticollis* because they lack the anterior dots and the posterior dots are reduced, making the thorax look all black dorsally (SP 613, SP 1610, SP 1613). Interestingly, a specimen BS-P captured in Montezuma County, Colorado, in the southwestern corner of that state has a similar thorax

Figure 12.73 *Ischnura damula*, the Plains Forktail. A) ♂, North Carrizo Creek 7 km N of Kenton, Cimarron County, 7 October 2017 (©Michael A Patten, OC 473885). B) tandem pair, Brett Gray Ranch, Lincoln County, Colorado, July 2011 (©Greg W Lasley), with the ♀ decidedly green; females may be pink, too, or even a male-like blue.

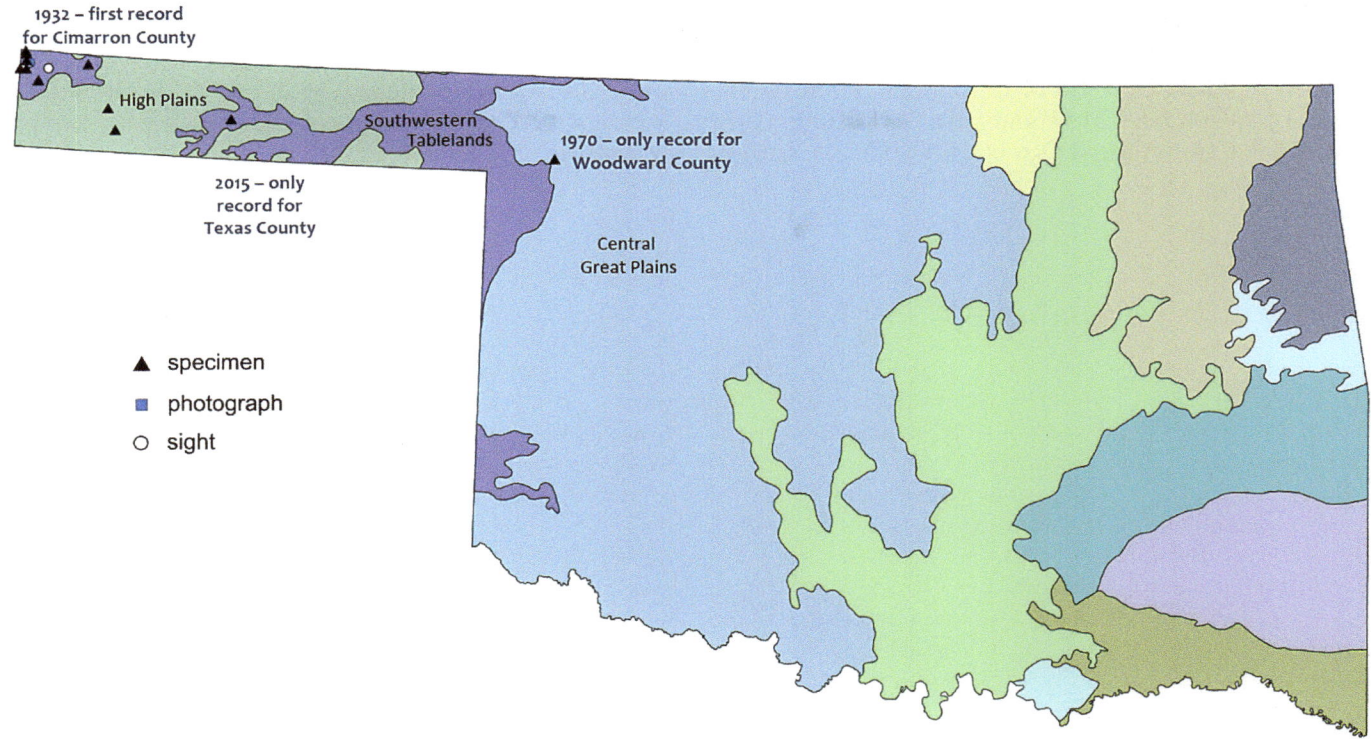

Figure 12.74 Documentation level of supporting records for *Ischnura damula*, the Plains Forktail, in Oklahoma. The few plotted sight records are those we feel are credible because of in-hand examination. The species has been reported in only three counties in northwestern Oklahoma and its panhandle.

(SP 2043, missing anterior dots, posterior dots are reduced but slightly elongated), whereas specimens we collected nearer to the Oklahoma panhandle do not exhibit the *I. denticollis*-like pattern (SP 752 and SP 2031, both from Las Animas County, Colorado).

One a final note of aberrancy regards OC 381715 that was submitted as *I. damula* but could not be confirmed as such. This ♂ was photographed at Doby Springs, Harper County, on 1 September 2012, by B. C. It looks much like *I. damula* in pattern but not coloration. Initially it was not confirmed because of what was thought then to be the possibility of *I. cervula*, the Pacific Forktail, one day showing up in northwestern Oklahoma. Those two species are generally bluer on the thorax than *I. verticalis* yet are not always confidently identifiable by photograph without close-up views of the terminal abdominal structures for the male or the prothorax and mesostigmal plates of the females. All of that said, we visited the site a couple of weeks later in hopes of finding the first state record of *I. cervula* or at least a new county record of *I. damula* but our hopes were dashed when we instead found aberrant *I. verticalis* with broken antehumeral stripes, one of which had elongated dots (SP 420) similar to that photographed by B. C. We suggest that OC 381715 is actually an aberrant *I. verticalis*. That encounter set us on

the (so very frustrating) path of trying to determine how much overlap in thoracic pattern there is between *I. damula* and *I. verticalis* (further discussion in the latter account). Years later we also learned that although once thought to range rather near to the Oklahoma panhandle, *I. cervula* is not likely to be found in Oklahoma because specimens previously identified as the species in eastern Colorado and New Mexico are now known to be *I. damula* (JN Stuart and B. Prather, *in litt*).

Paulson (2009:117) described *Ischnura damula* as inhabiting "dense vegetation beds at lake margins, ponds, and slow streams and ditches, and hot springs." We have found the species in all but the last habitat. The high count for the species in Oklahoma came from a well-water cattle pond thick with sedges (10♂, 8♀, Conrad, Cimarron County, SP 623–627).[36] The pond obviously had not been visited by cattle for quite some time. We have not been able to re-visit the pond again, but we have often wondered how the forktails are faring, especially if cattle began to visit. Such a high count was unexpected because generally we encounter the species one or two at a time.

Oklahoma's documented flight season for the Plains Forktail is quite a bit shorter than some of the surrounding states, being 9 days shorter than New Mexico's, 39 days shorter than Texas's, and 42 shorter than Colorado's 183 days (OK: early and late from Cimarron County, 19 May, see above, 7 October, 7 km N of Kenton, North Carrizo Creek, 1♂, OC 473885, MAP and Lake Carl Etling, 1♀, SP 2520, MAP; CO: 8 April–8 October, OC; NM: 11 April–16 September, OC; TX: 24 March–20 September, Abbott and Lasley 2019). Only one pair has been reported (5 mi E of Boise City, Cimarron County, 4 August 1970, GHB, LEH, Bick notes). There are no records of ovipositing, nymphs, or tenerals.

ISCHNURA DEMORSA (HAGEN, 1861) – MEXICAN FORKTAIL

The Mexican Forktail is a rare to uncommon species in Oklahoma that can be found at vegetated, shallow, clear, and flowing, albeit sometimes sluggishly, streams in western Oklahoma and its panhandle. We encountered the species once at a pond, in Elk City, Beckham County (lone ♂, 1 September 2014, SP 1421), but otherwise it is strictly lotic. We know of >30 records at 20 localities in 7 counties in Oklahoma that have been reported on only 22 days, a painfully low encounter rate given all of the time observers have spent within the range of this species in the state. Numbers found also tend to be small, usually 1–2 at a time, but higher numbers have been reported. Two of those times are from the first day that the species was reported for the state[37]—on 5 June 1956 when GH Bick found it in Harper and Woodward Counties. On that day he collected 4♂ and 4♀ at 6 mi W of May in Harper County and 4♂ and 10♀ at 3 mi E of Fort Supply in Woodward County.[38] Bick and LE Hornuff also recorded nine individuals once (9♂, 8 at FSCA, 1.4 mi N of Elmwood, Beaver County, 3 August 1970). These larger counts by Bick and Hornuff deceptively appear to indicate that numbers they encountered were overall higher than those encountered in recent years, but their median number was two individuals, which is similar to that between 2012 and 2017 (1.5 indiv). Bick and Hornuff's numbers likely only indicate that they were lucky enough to be in the panhandle on days where a recent emergence occurred. MAP may have been so fortunate when he recorded 45 forktails of the *Ischnura demorsa/perparva/verticalis* (Mexican/Western/Eastern) persuasion on 3 June 2017 along the Cimarron River 4 km NNE of Kenton, Cimarron County. Of those 45 individuals (25♂, 20♀), he was able to confirm 4♂ as *I. demorsa*, 2♂ as *I. perparva*, and 9♂ and 3♀ as *I. verticalis* (1♂ of each species as SP 2376–2378).

Part of perceived rarity of this species may be difficulty of identification. For example, 5 individuals, collected on 19–20 September 2003, were reported as *I. demorsa* from Fort Sill MR, Comanche County, (Kondratieff et al 2004, Zuellig et al. 2006): we re-identified 3♀ as *I. hastata* and 1♂ and 1♀ as *I. perparva* (CSU, see *I. perparva* account). An additional record that knit brows is of a photograph taken by J. Heinen along South Carrizo Creek at Black Mesa SP, Cimarron County, on 20 May 2012 (OC 383423). Heinen submitted the record as *I. demorsa*, but we felt that the male in the photo could be either *I. demorsa* or *I. perparva* (not *I. verticalis* because of the clearly visible green spot on the side of the prothorax). In redemption of that record, *I. demorsa* was recorded at that location on 3 June 2015 (1♂, SP 1602) and again on 25 May 2019 (1 teneral ♂ gingerly examined in hand, MAP), whereas *I. perparva* has yet to put in an appearance there.

The Mexican Forktail has been recorded flying in Oklahoma as early as 10 May and as late as 21 October (OK: early date from 2016, Beaver Dunes SP, Beaver County, BS-P,

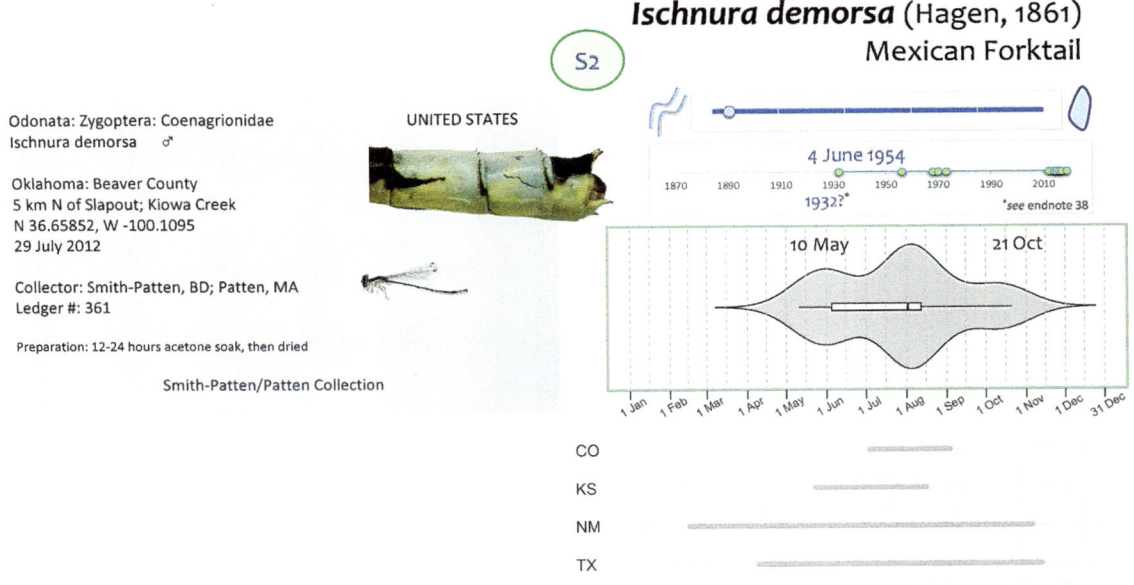

Figure 12.75 *Ischnura demorsa*, the Mexican Forktail, had not been reported from Oklahoma for almost 40 years until 2012, when we encountered this ♂ (SP 361) and a ♀ along Kiowa Creek in Beaver County. Inset shows the terminal appendages of this ♂, which are indicative of this species. The difficulty to distinguish this species, even under a microscope at times, from *I. perparva*, the Western Forktail, guarantees a need for specimens or careful in-hand examination of ♂ terminal appendages or the ♀ mesostigmal plate. A further complication is that *I. demorsa* also superficially resembles the much more numerous *I. verticalis*, the Eastern Forktail.

1♂, SP 1886, also present were 10♂ and 2♀ identified as *I. verticalis/demorsa*, sight only; late date from 2014, in a flowing, shallow ditch at Altus city lake, Jackson County, 1♂, SP1467). *I. demorsa* is known to fly longest in New Mexico with Texas not far behind, but both Colorado and Kansas have few records, which likely bias what is known about the seasonal range there (NM: February–5 November, Paulson 2009, OC; TX: 9 April–12 November, Abbott and Lasley 2019; CO: 3 July–3 September, CSU; KS: 23 May–16 August, SEMC). No nymphs have been reported for Oklahoma, but there is one record of a pair (SP 2907, 7 km N of Kenton, North Carrizo Creek, 1 September 2019, MAP) and one ovipositing record

(2♀, same location as SP 2907, 7 October 2017, MAP, 1♂ also present) and one teneral record (see above).

We classified the species as an S2 species of conservation concern because of the limited encounters and distribution in the state (Patten and Smith-Patten 2013b). We were then—and are now—greatly concerned about water sources in the panhandle and parts of northwestern Oklahoma drying up, leaving this species without suitable habitat. If continued drying occurs from persistent drought or from manmade causes, such as the overuse of aquifers and pulling water from streams and ponds for agricultural and industrial uses, then this species ought to be raised to S1 status.

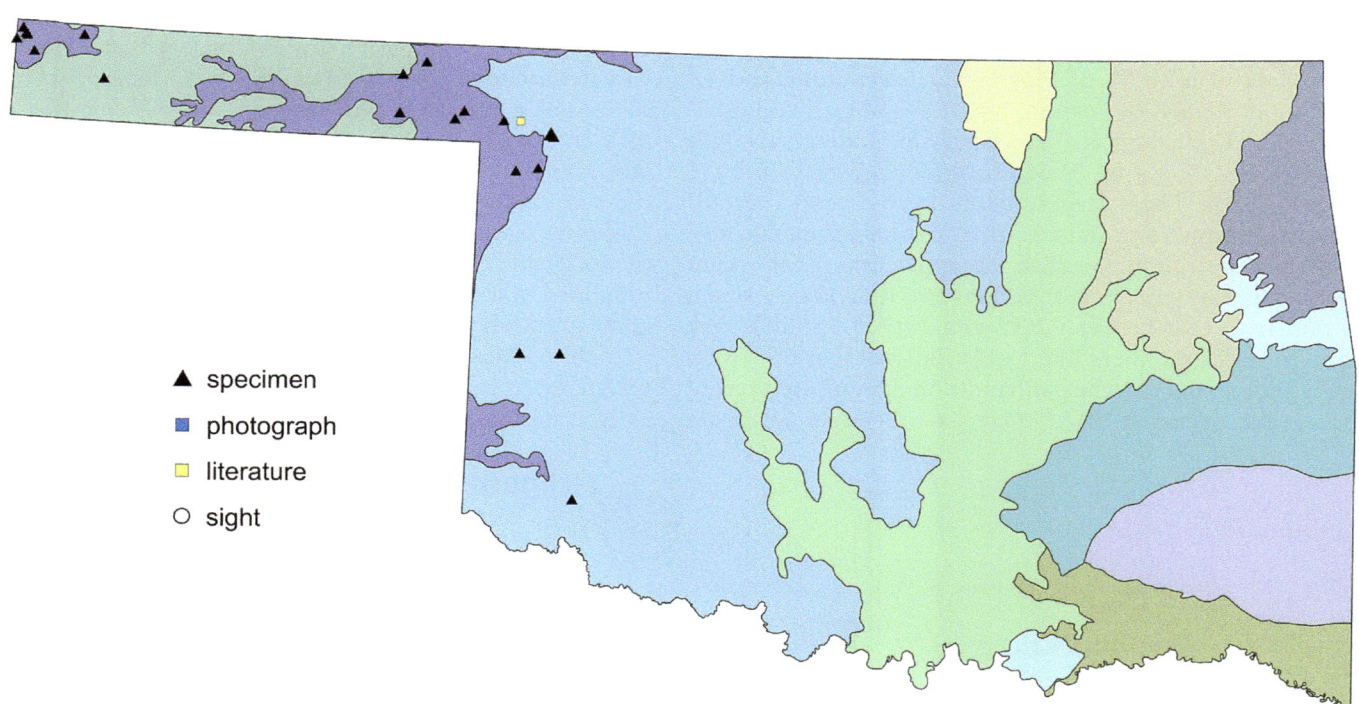

▲ specimen

■ photograph

□ literature

○ sight

Figure 12.76 Documentation level of supporting records for *Ischnura demorsa*, the Mexican Forktail, in Oklahoma.

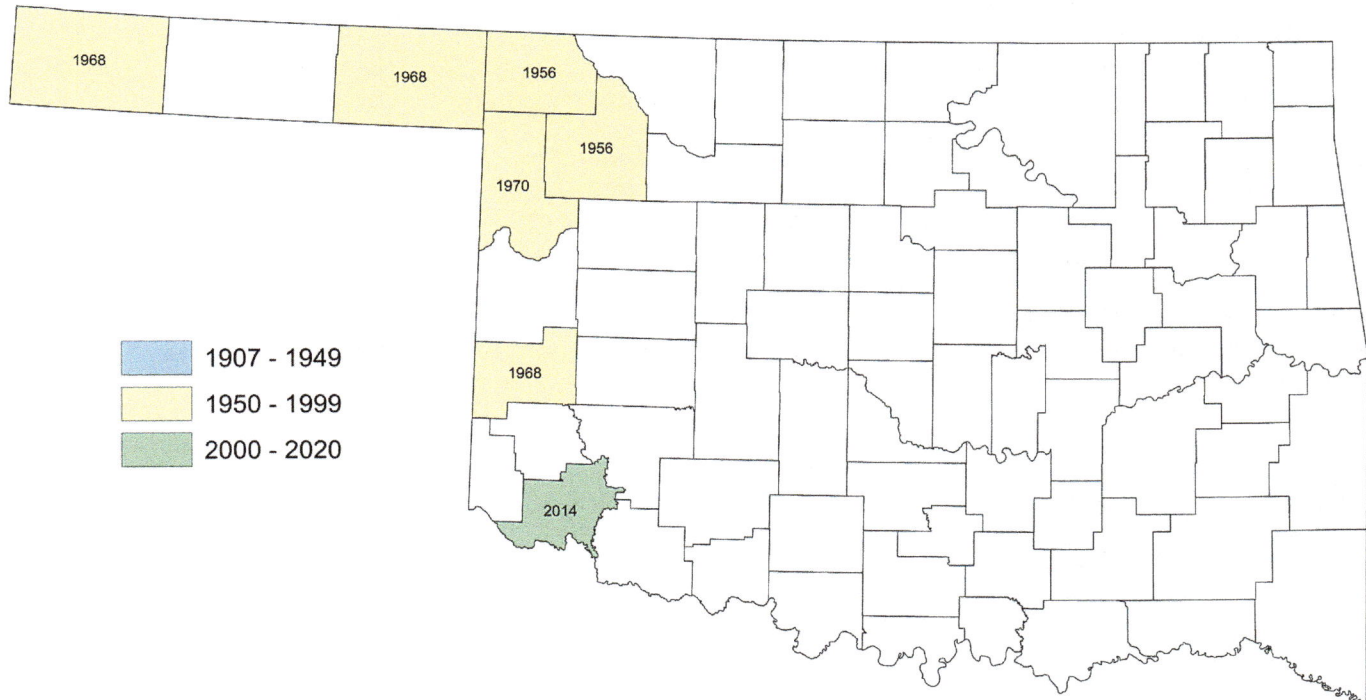

Figure 12.77 Counties in which *Ischnura demorsa*, the Mexican Forktail, are known to occur in Oklahoma. Year within the county is that in which the species was first reported.

ISCHNURA PERPARVA MCLACHLAN, 1876 – WESTERN FORKTAIL

The Western Forktail probably has been encountered a little more often than we are aware of but because observers, including ourselves, have not always captured forktails of this type within this species' range to confirm identifications, we are left with only four confirmed records at three localities. The first state record is of a ♂ from Pratt Hill in the Lake Elmer Thomas Recreation Area (LETRA), Fort Sill MR, Comanche County, 20 September 2003 (Kondratieff et al. 2004, Zuellig et al. 2006; CSU). In 2014 we examined that specimen and confirmed its identification (OC 381755). We also examined a ♂ and ♀ from the same locality and date that were identified originally as *Ischnura demorsa*, the Mexican Forktail, and re-identified them as *I. perparva*, the male positively, the female only could be assumed by association because she was too smashed for definite identification. To our knowledge these three are the only specimens of *I. perparva* for Fort Sill MR and Comanche County, yet Zuellig et al. (2006) published three localities for the species on Fort Sill: 1) on Quanah Range at a "small temporary pond near Pottawatomie Pond," 2) on West Range "near the junction of Blue Beaver Valley Road and Deer Creek Road," and 3) LETRA. Kondratieff et al. (2004) only listed LETRA for *I. perparva*, which was the only specimen that was in the CSU collected as that species at the time specimens were loaned to us in 2014. Lastly, it should also be noted that a female forktail captured at Lake George, Fort Sill, on 11 September 1999 was identified originally (we know not by whom) as *I. perparva* but proved to be *I. verticalis*, the Eastern Forktail (SP 2641).

The other records for the state are from extreme northwestern Cimarron County, at the tip of the panhandle. The second state record of Western Forktail came in 2013, when we captured a single ♂ at a small pool in the drying bed of Carrizozo Creek at Oklahoma's border with New Mexico (2 km W of Kenton, 26 May, SP 584). We recorded another male at that location and 8♂ on North Carrizo Creek that we left identified as *Ischnura demorsa/perparva*. The state's third record came in 2017 when MAP encountered 2♂ along the Cimarron River, 4 km NNW of Kenton (3 June, 1♂ as SP 2377).[39] During that encounter, he estimated 45 forktails of the Mexican/Western/Eastern persuasion, only some of which he was able to identify as *I. demorsa* and *I. verticalis* (see the *I. demorsa* account for more details; also see that account for unconfirmed but possible records of *I. perparva*). This same location recently produced the state's fourth record (2♂, 25 May 2019, MAP, OC 495612).

Despite limited records for Oklahoma, the known flight season for the state falls perfectly within its phenology elsewhere in its range (OK: 25 May–20 September). For example, it is also known to fly from May to September in Colorado, New Mexico, and Montana, from April to September in British Columbia, and from April to October in Washington and Oregon (CO: 3 May–29 September, CSU; NM: May–1 September, Paulson 2009, iNat; all others, Paulson 2009).

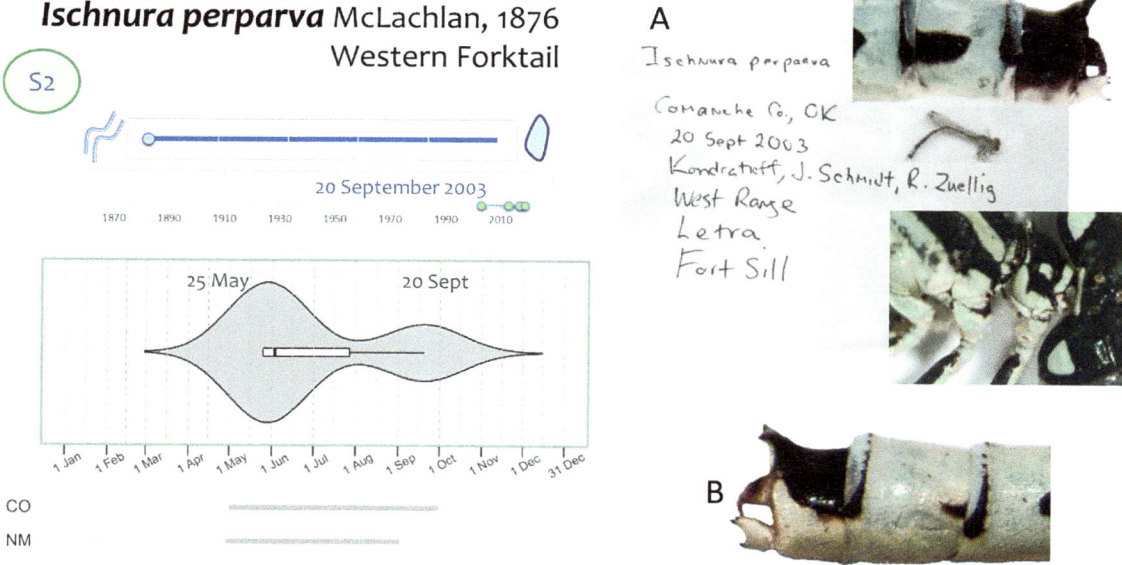

Figure 12.78 **A)** Oklahoma's first record (and inexplicable geographical outlier) of *Ischnura perparva*, the Western Forktail. This ♂ was collected at Fort Sill Military Reserve, Comanche County, on 20 September 2003 (CSU). Pictured are the specimen card (middle), the terminal appendages (upper), and the prothorax (bottom). **B)** A ♂ we collected on Carrizozo Creek 2 km W of Kenton, Cimarron County, on 26 May 2013 (SP 584), was the second to be found in the state, and was at a less geographically surprising locale given records not far to the west in northern New Mexico and southern Colorado.

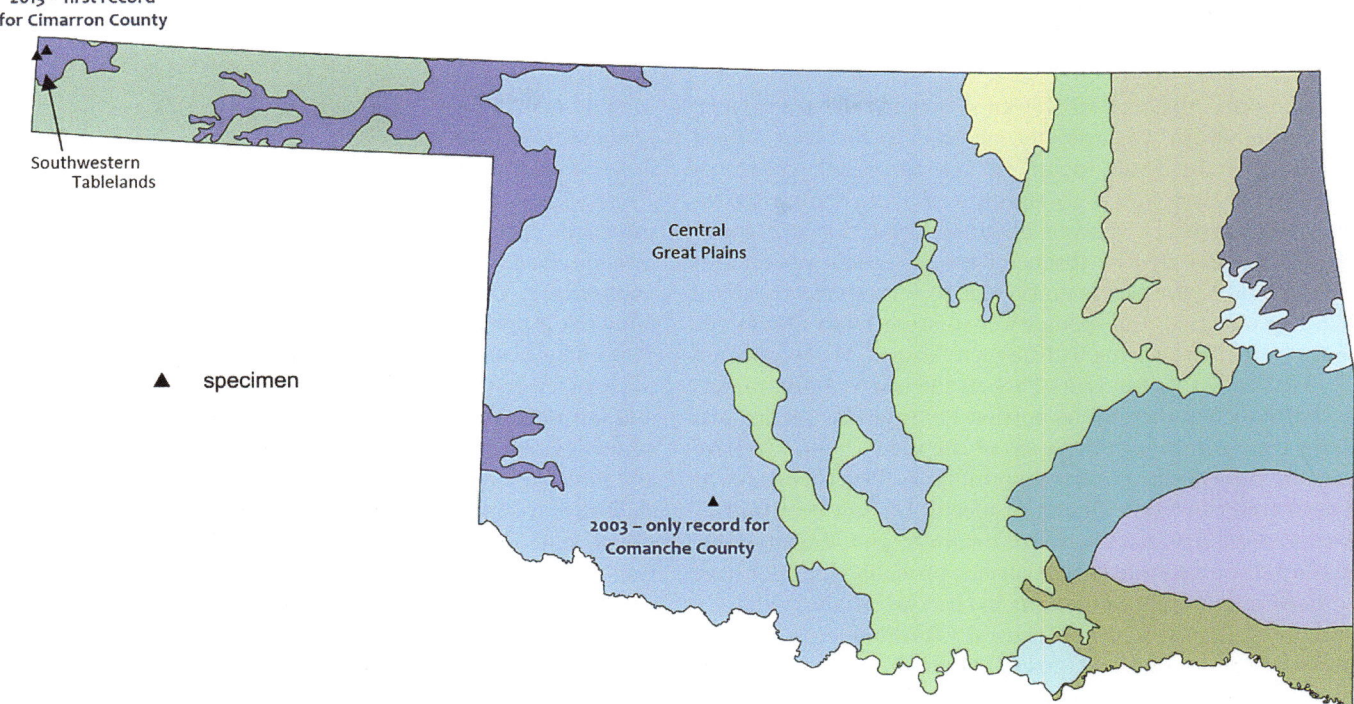

2013 – first record
for Cimarron County

Southwestern
Tablelands

▲   specimen

Central
Great Plains

2003 – only record for
Comanche County

Figure 12.79 Documentation level of supporting records for *Ischnura perparva*, the Western Forktail, in Oklahoma. The species has been reported in only two counties in Oklahoma—Comanche and Cimarron—at a great distance from one another. Such a distribution may indicate vagrancy or, perhaps more likely, that the species is one with small populations in the Texas panhandle and western Oklahoma that have gone undetected.

The species is known from northern Kansas from just a few records so the known flight season is limited (4 May–13 June, SEMC). No pairings, ovipositing, nymphs, or tenerals have been reported for Oklahoma.

The conservation status of *Ischnura perparva* in Oklahoma remains unclear. We had much hesitation when designating this species as a putative vagrant when we reviewed the species' status previously (Patten and Smith-Patten 2013b). Subsequent records have firmed our suspicion that the species has a low-density population in the Black Mesa area of the state. Our leanings in that direction are strong enough that we now feel a conservation rank akin to that of *I. demorsa* is appropriate. Although *I. perparva* is not as well documented as *I. demorsa*, it shares the same limited resources in western Oklahoma—clear, shallow, and well-vegetated creeks and rivers—as well as the same threats and thus it shares the same conservation fate.

ISCHNURA VERTICALIS (SAY, 1839) – EASTERN FORKTAIL

In Oklahoma, the Eastern Forktail is quite the contrarian. Despite its name, it is a western species in the state—most abundant in the west, including the panhandle, becomes scarce in the east, and is absent entirely from the far southeast. It is currently missing documentation for five counties. We suspect that its absence from two of those counties (Haskell and McIntosh) is simply due to survey effort and bad luck. For instance, the species was missing from Okmulgee County until late spring 2018, when Abigail Mills captured a ♂ (SP 2598) at Kiddy Lake Park in the town of Okmulgee, to finally record the species in that county; this record was a long time coming, in defiance of the many hours we have spent in the county conducting surveys. The three southeastern counties from which the species has not been definitively recorded (Choctaw, McCurtain,[40] and Pushmataha) are likely true absences as the species probably does not have regularly occurring populations in the Gulf Coastal Plain of the United States. In south-central Oklahoma and perhaps elsewhere in the state, we posit that the Eastern Forktail is being pushed out by the larger Rambur's Forktail (*Ischnura ramburii*), which seems to be expanding northward as the Eastern Forktail's range contracts northward. The range expansion and contraction may be independent of one another, yet one wonders why, for example, that in the seven counties in which Rambur's Forktail was first reported for the state, the Eastern Forktail evidently has disappeared (Figure 12.80).

Where it is found, *Ischnura verticalis* inhabits a wide range of habitats, including creeks, ponds, lakes, and roadside ditches, all ranging from having no vegetation to being heavily vegetated. Creeks have been described as being narrow to wide (<2 to 30m), flowing or not, permanent or intermittent, and sometimes spring-fed. The species has also been noted at spillways, including at a hillside seep above a dam spillway. When Eastern Forktails are found in the western part of the state they can be quite numerous, typically in the double digits. For example, during a 3-day period, 2–4 June 2015, we encountered the species in numbers around or >100 on 3 occasions (Appendix F, Table 2). On 2 June we visited Optima Lake, where we recorded about 100 individuals, and when we returned on 4 June we recorded 130 individuals. While surveying at various spots in the Black Mesa area on 3 June we recorded about 100 total individuals. We also had high counts of Eastern Forktail at three locations we surveyed in the Black Mesa area on 26 May 2013 (total of all locations was 138 individuals).

Given the species' commonness in the Oklahoma panhandle, it is not surprising that is where it was first recorded in the state (8 mi SE of Guymon, Texas County, 2♂, 5♀, 27 June 1926, TH Hubbell, UMMZ). The species has been regularly reported throughout most of Oklahoma ever since, accounting for many hundreds of records. Species associations in the western part of *Ischnura verticalis*' range in Oklahoma include *Lestes alacer* and *unguiculatus*, *Enallagma praevarum*, *carunculatum*, *antennatum*, and *basidens*, *Ischnura damula*, *demorsa*, *perparva*, and *denticollis*, *Amphiagrion abbreviatum*, *Argia alberta*, *fumipennis*, *nahuana*, *plana*, and *lugens*, *Plathemis subornata*, *Libellula saturata*, *pulchella*, and *nodisticta*, *Erythemis collocata*, *Pseudoleon superbus*, *Sympetrum internum*, *pallipes*, and *semicinctum*, *Dythemis nigrescens*, and *Brechmorhoga mendax* (also see Appendix F, Tables 2, 3 and genus summary for more details and discussion). Many of these species are tied to clean, clear-water, flowing streams and associated marshes, and several are of highest conservation priority in Oklahoma, which is in stark contrast to the S5 status of *I. verticalis* (Patten and Smith-Patten 2013b); others are of uncertain status in the state due to limited records. Species associations elsewhere in the state are not as closely tied to odonates of conservation concern arguably because there is much more habitat to choose from in other parts of *I. verticalis*' Oklahoma range.

In the region, the species is known to fly the longest in Oklahoma and Missouri, from late March until late October (OK: 213 days long, from 22 March 1930, Norman, Cleveland County, 2♂, OMNH 3830, to 21 October 2014, Altus, Jackson County, 1♀, SP 1468; MO: 209 (219) days long, 29 (or 19, which may be a typo) March–24 October, Trial 2005, Sims 2012). Texas, New Mexico, Kansas, and Colorado follow fairly close behind, but surprisingly Arkansas lags far in the rear (TX: 190 days, 6 April–13 October, Abbott and Lasley 2019; NM: May–October, Paulson 2009; KS: 170 days, 9 April–26 September,[41] OC, EMEC; CO: 136 days, 8 May–21 September, USNM, CSU; AR: 93 days, 24 May–25 August, USNM, EMEC, iNat).

Pairs of Eastern Forktails are often seen but rarely reported (a regrettable but routine omission for common species). Likewise, there are only two records of ovipositioning reported (both records of lone females in 2014, on 10 May and 4 October), although we have seen it many more times than that but did not record it in our notes. No nymphs are known from collections. One teneral record is in the OOP database. Again, we have undoubtedly seen teneral Eastern Forktails but have not recorded them in our notes, partly because we tend not to try to identify teneral damselflies because they are so delicate and must be examined in hand unlike many teneral dragonflies that can be identified using binoculars. The sole teneral record came from Lake Evans Chambers, Beaver County, on 10 May 2016 (1♂, BS-P, SP 1885).

Aberrant or "atypical" *Ischnura verticalis* have been documented from Oklahoma and adjacent areas. We, by no means, have conducted a thorough study of anomalous Eastern Forktails in Oklahoma but we have noted over 40 such individuals in the state, the vast majority of which are males (*n* = 37). Additionally, we have noted six possible aberrant males from Oklahoma, as well as ten other definite aberrant individuals from Colorado, Nebraska, and

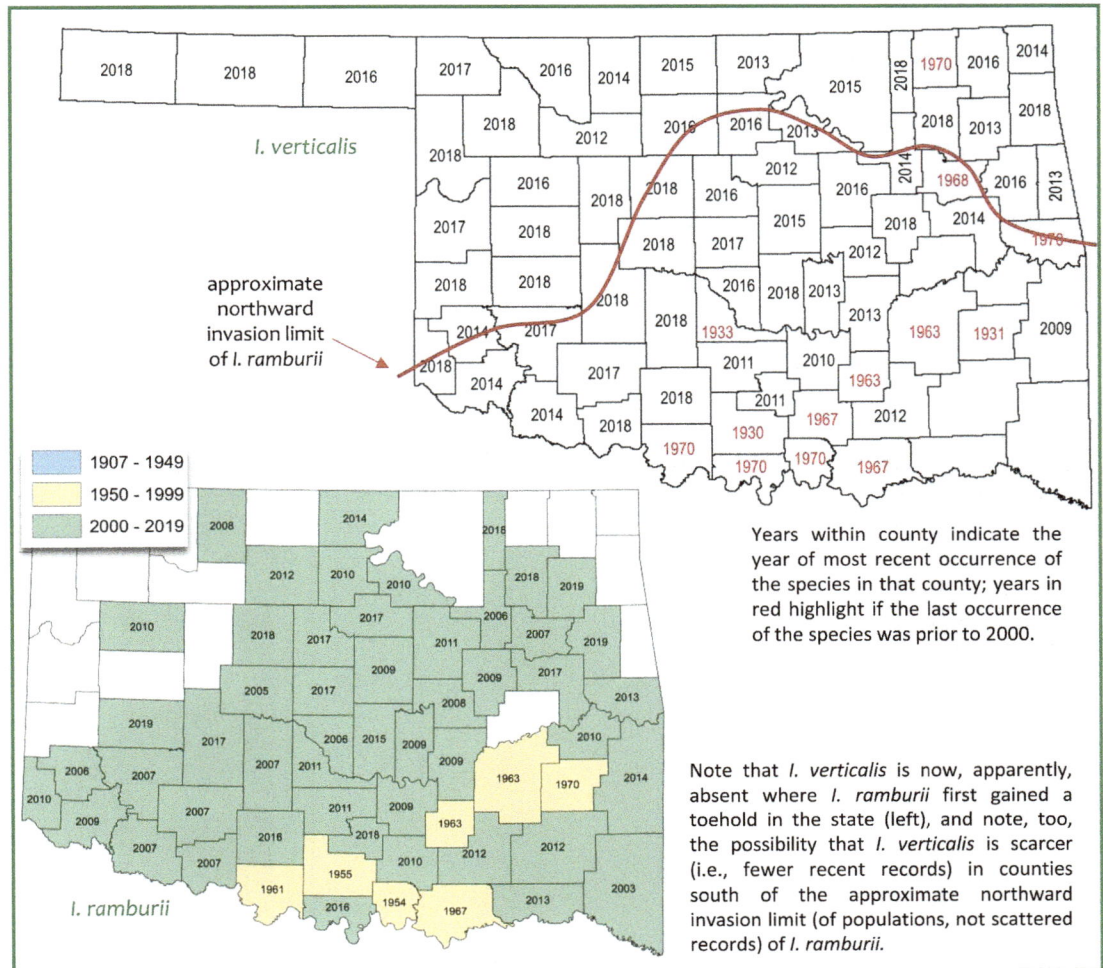

Figure 12.80 *Ischnura verticalis*, the Eastern Forktail, distribution in Oklahoma by county (it is known from 72 counties; top map).

Texas (Appendix F, Table 4 and below; 9♂, 1♀). The primary aberrancy noted with males is broken antehumerals that can make individuals resemble *I. posita* or *I. damula*, but a detailed study of other characteristics, namely the terminal appendages and genetic markers, is definitely needed.

The variability of *Ischnura verticalis'* antehumeral stripe has been mentioned previously, mainly in regard to some constriction appearing about 2/3 along the length, and being only rarely broken in such a way as to resemble the exclamation point antehumerals generally characteristic of *I. posita* (Westfall and May 2006, Paulson 2011), but the range of variability has not been well described or quantified. The overwhelming majority of *I. verticalis* we encounter in Oklahoma have the antehumerals continuous (or "full"), of subequal widths along their lengths, and are symmetrical; however, some males show slight constriction centrally. We consider both of these forms to be "typical" *I. verticalis* (Figure 12.81, A–B). Other males show much more constriction, making the stripe begin to appear or actually appear like two blobs at either end with a thin line connecting them (Figure 12.81, C). Sometimes the antehumeral

is broken or almost so. Breaks can be slight or entire (in one or more places) and lacking a distinct pattern or they can grade into various iterations of exclamation points (= *I. posita*-like)—with splattered color in the break or exhibiting clean punctuation (Figure 12.81, D–G). The antehumeral pattern that has been documented with less frequency is an *I. damula*-like pattern—two elongated spots or two distinct dots, one anteriorly and one posteriorly (Figure 12.81, H–I). Although the pattern is much like that of *I. damula*, the coloration tends to be more green than blue; nonetheless, superficially these *I. damula*-like individuals are easily mistaken for the wrong species and cannot be conclusively determined from photographs. A final note about atypical male antehumerals is that they can be asymmetrical: such as the aberrant male from Baca County, Colorado, which has elongated dots on the left side and a splattered pattern on the right or, in the case of two males, the left antehumeral is broken, yet the right side is complete and shaped like a "typical" *I. verticalis* (Figure 12.81, J–K).

In Oklahoma, aberrant males are found primarily in the panhandle and in neighboring Harper County, but there are

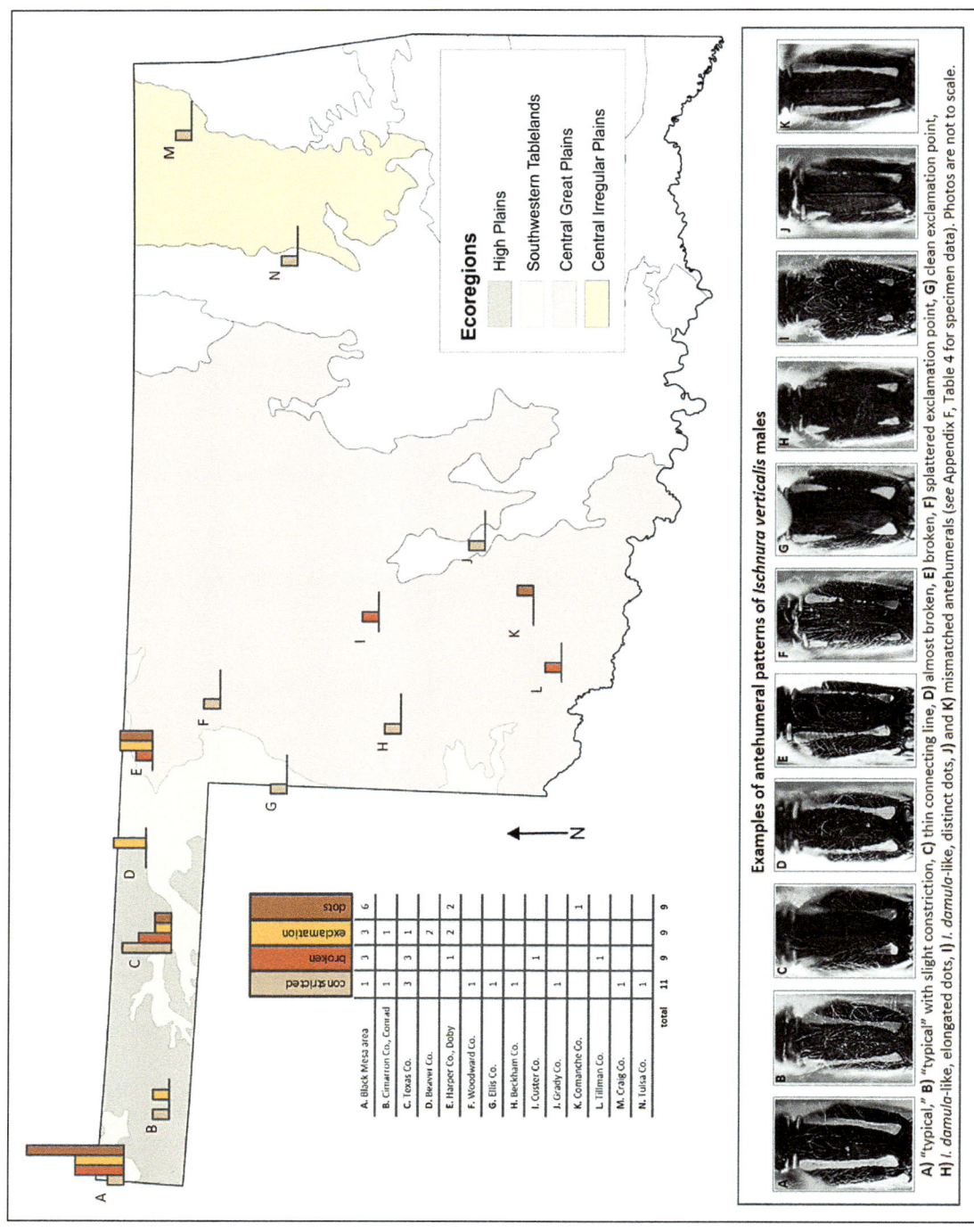

Figure 12.81 Aberrant ♂ *Ischnura verticalis* in Oklahoma and adjacent Baca County, Colorado. Data are from Smith-Patten/Patten collection (SP) specimens, Odonata Central (OC) records, and photos on file with the Oklahoma Odonata Project (OOP; record details can be found in Appendix F, Table 1). Ecoregions containing records are accented.

A) "typical," **B)** "typical" with slight constriction, **C)** thin connecting line, **D)** almost broken, **E)** broken, **F)** splattered exclamation point, **G)** clean exclamation point, **H)** *I. damula*-like, distinct dots, **I)** *I. damula*-like, elongated dots, **J)** and **K)** mismatched antehumerals (*see* Appendix F, Table 4 for specimen data). Photos are not to scale.

Examples of antehumeral patterns of *Ischnura verticalis* males

	con stricted	broken	exclamation	dots
A. Black Mesa area	1	3	3	6
B. Cimarron Co., Conrad	1		1	
C. Texas Co.	3	3	1	
D. Beaver Co.			2	
E. Harper Co., Doby	1	2	2	
F. Woodward Co.	1			
G. Ellis Co.	1			
H. Beckham Co.	1	1		
I. Custer Co.		1		
J. Grady Co.	1			
K. Comanche Co.				1
L. Tillman Co.	1		1	
M. Craig Co.	1			
N. Tulsa Co.	1			
total	11	9	9	9

Ecoregions

- High Plains
- Southwestern Tablelands
- Central Great Plains
- Central Irregular Plains

scattered records in southwestern, and, surprisingly, northeastern Oklahoma (Figure 12.81). All but one of the Tulsa County examples exhibit relatively minor constriction of the antehumerals (admittedly, it could be argued that three may fall within the confines of "typical"). The Craig County example (OC 401675) is undoubtedly aberrant in the amount of constriction shown on both sides (Figure 12.82). Whereas constricted antehumerals have been found in multiple regions in the state, *I. posita*-like males (*n* = 9) have only been encountered in the panhandle and nearby Harper and Woodward Counties. All but one male exhibiting antehumerals as dots, elongated or distinct, have also been found in that part of Oklahoma. The exception is a male from Comanche County that was photographed on French Lake in the Wichita Mountains WR on 17 April 2008 by VWF (OC 282024; Figure 12.82). That male has, on its right side, a slightly elongated anterior dot and a small posterior dot (the left side is not clearly visible, but the pattern appears to match). The other *I. damula*-like males were taken at Doby Springs, Harper County (*n* = 2), Optima Lake, Texas County (*n* = 1), or in the Black Mesa area (*n* = 6), including one that was collected in Baca County, Colorado, along North Carrizo Creek, a creek on which aberrant male *I. verticalis* have been noted in Oklahoma.

Most aberrant males have been found in the Southwestern Tablelands and the Central Great Plains, whereas others have been documented in the High Plains and, to some extent, the Central Irregular Plains (Figure 12.81). A similar pattern occurs within the region (Figure 12.83); the Nebraska record is the only one outside of those ecoregions being, instead, in the Western Corn Belt. We have yet to discover records of aberrant males from Kansas or New Mexico, but we know of some scattered ones from Indiana, Michigan, Wisconsin, and possibly North Dakota (OC; GW Lasley, *in litt*; RB DuBois, *pers comm*; Paulson 2011). At present, *I. damula*-like individuals are known only from Oklahoma.

It may be that in the Oklahoma panhandle and surrounding area, where *Ischnura verticalis* meets *I. damula* and *I. posita*, some hybridization occurs. Alternatively, because scattered records of aberrant antehumerals have been reported outside of the southern Great Plains, it may be that aberrancies are just individual variation, variation that is, for some reason, concentrated in this region.

Without a dedicated study we cannot say how prevalent these aberrancies are but accumulated relative abundance data for Cimarron, Texas, Beaver, and Harper Counties, Oklahoma, and Baca County, Colorado (Appendix F, Table 2), indicate that 1–4% of the male population in that part of the state exhibit antehumeral variation (or an encounter rate of 19.35%).[42] If pressed we would say that the

Male *Ischnura verticalis* from northeastern and southwestern Oklahoma showing aberrant antehumerals.

A) Tulsa County: exhibits the most constriction of photographs and specimens examined

B) Craig County: much constriction (glare makes stripe look wider)

C) Comanche County: elongated dots

see Appendix F, Table 1 for photo data

Figure 12.82 Male *Ischnura verticalis* from northeastern and southwestern Oklahoma showing aberrant antehumerals.

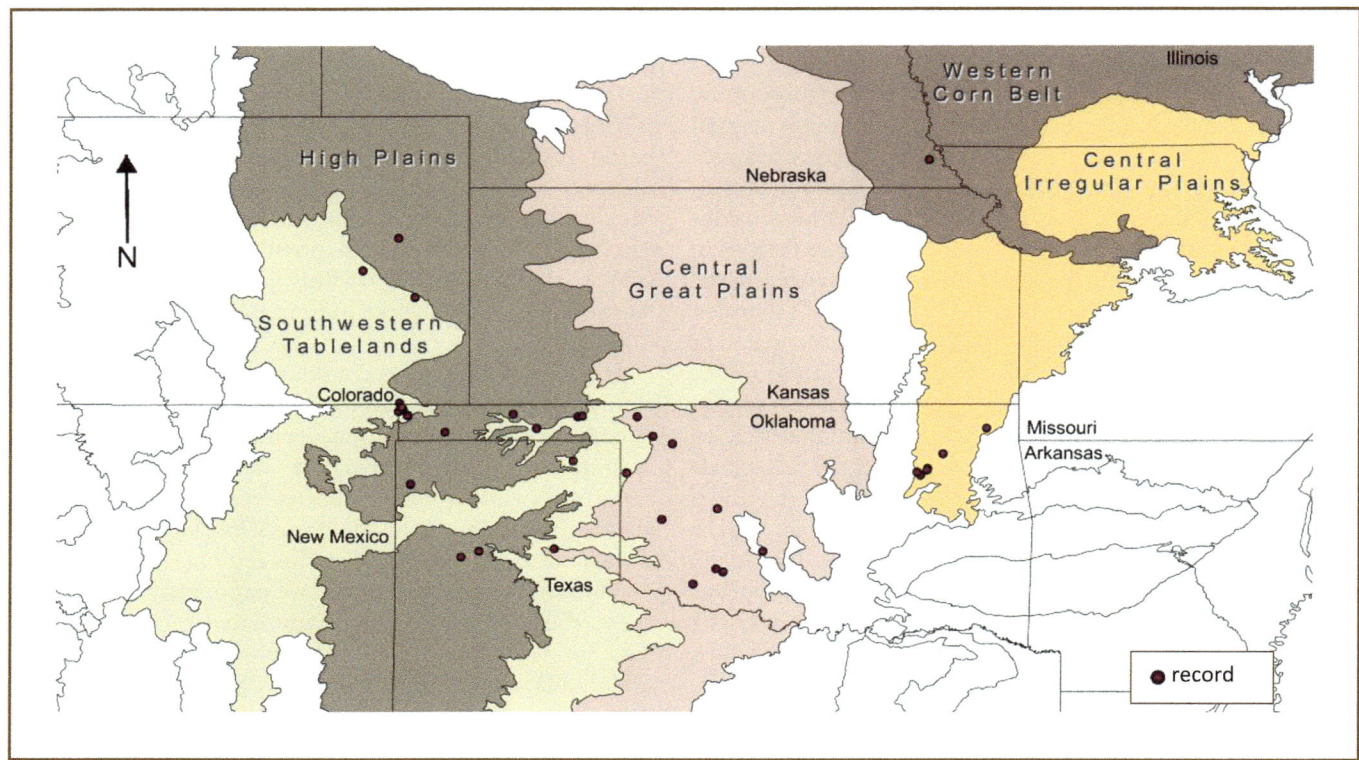

Figure 12.83 **Distribution of known aberrant** ♂ *Ischnura verticalis* **in the southern Great Plains. Locations are indicated by dots. Ecoregions (e.g., High Plains) containing records are accented. Record details can be found in Appendix F, Table 2.**

calculation is an underestimate of how much of the population is "atypical." It can be quite difficult in the field to spot constriction and breaks in the antehumeral unless they are toward the *posita/damula*-like end of the scale so we feel the possibility of undercounting is real for males, and especially so for females, which can be difficult to diagnose to species let alone to pick out aberrant characters. Nonetheless, the encounter rate is high enough for aberrant males that they ought to be on the radar of anyone visiting the Oklahoma panhandle, so we expect more records in the future.

We fear that to unearth variation in females, if it exists beyond occasional individual variation, will take considerable dedication and time. That said, there are ten Oklahoma records of atypical females deserving of note. SP 195, collected in Kiowa County has a broken antehumeral on the left side but a full one on her right (Figure 12.84). Although females have varying widths of both their antehumerals (moderately to very wide) and humerals (hairline to very wide), SP 195 is the only female we have seen exhibiting a break in the antehumeral.

Five other females are atypical by being andromorphs.[43] Although the notion of that form's rarity (Westfall and May 2006) is undoubtedly true, we imagine andromorphs are often overlooked. Paulson (2011:120) singled out immature andromorphs as being "very rarely seen," so we were rather surprised upon re-examining the SP collection to find two immature andromorphs (SP 1179, 1887), two that were transitioning from immaturity (SP 588, 1922), and one full adult (SP 1614; Figure 12.84). Five other females are atypical in their measurements, being smaller than those previously documented (all measurements in mm [ab = abdomen, TL = total length] and are from Westfall and May 2006: TL = 24.5–33.0, ab = 19.5–26.0; SP 2430, TL = 22.0, ab = 16.7; SP 17, TL = 22.5, ab = 17.4; SP 2018, TL = 22.8, ab = 17.5; SP 1422, TL = 23.4, ab = 17.8; SP 18, TL = 25.2, ab = 19.1 ab). A final note regarding *I. verticalis* females in Oklahoma: a forktail captured at Lake George, Fort Sill MR, Comanche County, on 11 September 1999, that was originally identified as *I. perparva*, proved to be *I. verticalis* (SP 2641).

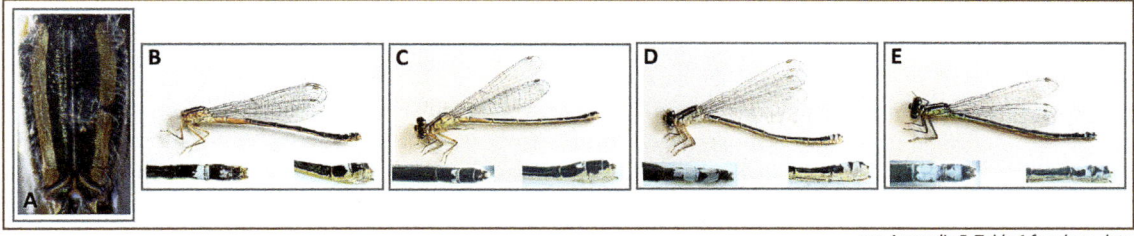

see Appendix F, Table 1 for photo data

Figure 12.84 Atypical female *Ischnura verticalis* from Oklahoma. A) SP 195, aberrant antehumeral (triangular divot on left side is not physical damage), Kiowa County. Andromorphs (insets are abdominal segments 8–10 = S8–10, left dorsal, right lateral): B) SP 1179, immature orange female with much pale blue dorsally on S8–9 and dorsolaterally on S8, Noble County. C) SP 1922, immature transitioning to adult, overall orange and green, blue dorsally on S8, Greer County. D) SP 588, almost full adult, much pearly blue dorsally S8–9, laterally S8–10, Cimarron County. E) SP 1614, full adult, most male-like of SP andromorph specimens, Texas County. Photos are not to scale.

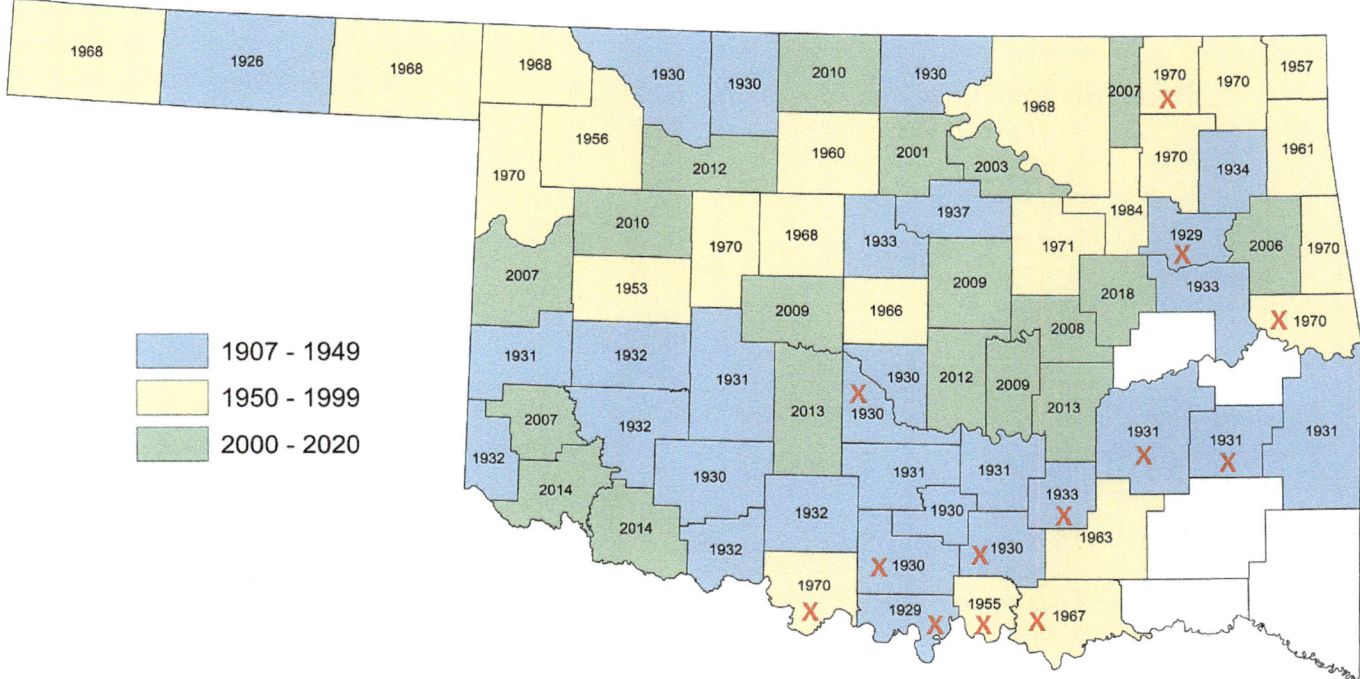

Figure 12.85 Counties in which *Ischnura verticalis*, the Eastern Forktail, are known to occur in Oklahoma. Year within the county is that in which the species was first reported. Counties with red X's indicate that the species has not been encountered in that county for about 50 years or more (see also Figure 12.80).

ISCHNURA DENTICOLLIS (BURMEISTER, 1839) – BLACK-FRONTED FORKTAIL

The Black-fronted Forktail is a western species of spring-fed, or spring-like, shallow marshes, having generally low sedges, in the generic sense, or grasses. It is most often found in areas carpeted with *Eleocharis* (spikerushes), such as those frequented by other marsh damsels such as *Amphiagrion abbreviatum* (the Western Red Damsel) and *Argia alberta* (the Paiute Dancer). This tiny forktail has virtually always been recorded at any site at which the other two have been found, although its presence does not always indicate their presence, for we have found *Ischnura denticollis* at many locales where the other two were not encountered (see *Amphiagrion abbreviatum*'s account for more details). When *I. denticollis* is encountered, double and even triple digits are easily had (it has been reported seven times >100 individuals), but abundance varies geographically.

As with *Amphiagrion abbreviatum* and *Argia alberta*, we wondered if *Ischnura denticollis* experienced a range retraction westward (Patten and Smith-Patten 2013b). But it may be that its distribution is subject to the vagaries of relative abundance from year to year, as well as the influences vagaries can have on annual geographic distribution. Records from central and south-central Oklahoma in particular made us initially consider range retraction. Two of the counties—Cleveland and McClain—really stick out because the species is known from each only from the 1930s. In Cleveland County, RD Bird collected 1♂ and 1♀ near Norman in the fall of 1932 (9 October, OMNH 3610 and 5 November, OMNH 3596, respectively) and in McClain County he collected 5♂ in April 1930 (probably at Deans Lake, OMNH 3467, as the first record for the

Figure 12.86 *Ischnura denticollis*, the Black-fronted Forktail, A) the species had not been encountered in Oklahoma County since the 1930s until this ♂ was found at Lake Hefner on 27 September 2015 (©Emily A Hjalmarson, OC 437347). B) tandem pair, Altus, Jackson County, 16 July 2009 (©Victor W Fazio III, OC 314065). C) ♀, Drummond Flats WMA, Garfield County, 17 August 2011 (©Jason R Heinen, OC 331568).

state). *Ischnura denticollis* has not been reported for either county since. Because the exact locations of those records are unknown, we cannot re-visit sites to check for persistent populations, yet it is odd that we and others have yet to find the species in the counties despite regular surveys. It may be that we have been unlucky in finding suitable habitat and that eventually what happened in Oklahoma County will happen in Cleveland and McClain Counties. After all, *I. denticollis* had not been reported from Oklahoma County since the 1930s (Bird 1932, no details known) but was found in the county again in 2015 (27 September, Lake Hefner, EA Hjalmarson, OC 437347; Figure 12.86). The other central Oklahoma county with a record of the species is Canadian, where it was recorded in 2013 for the first and only time (14 September, 1♀, SP 1014). South-central Oklahoma presents a similar puzzle with two counties there each having only one record—Jefferson (1970) and Stephens (1961)—despite many subsequent surveys.

Looking at overall distribution of the species in the state we see that county distribution is deceiving. Of the total of 29 counties where the species has been reported, more than half have only 1–3 records, most of which are of <5 individuals each. Fewer than 10 counties have sites that can be considered reliable localities for the species. The reality is that although county distribution makes the species appear widespread in the state, really it is only within the

far west and the panhandle that the species occurs regularly (Figure 12.87).

Ischnura denticollis is known to fly in the region primarily from the early spring to the late fall. In Oklahoma, for example, it flies from mid-March to early November (16 March 2009, Altus city lake, Jackson County, 8 indiv., VWF, OC 312274 and late date, as above), which is similar to Texas but not as long as New Mexico (TX: 11 March–26 October, Abbott and Lasley 2019; NM: year round, Paulson 2009). There are limited records reported from Colorado and Kansas (CO: 4 June–1 November, CSU, OC; KS: 7 May–13 August, SEMC). Tenerals have been reported in Oklahoma from 24 April to 25 September, but no nymphs are known to have been collected, and ovipositing has not been reported despite regularly reports of pairings.

We have noted quite a bit of variation in body pattern of females we have collected in the state. Some of the differences pertain to comparison of andromorphs and more typical female forms (heterochromes), but others appear to be individual variation. Although we have not conducted a full analysis of differences, we have noted variation on the prothorax and antehumerals. (To our surprise, unlike some other *Ischnura* females, we noticed little variation in the pattern and extent of pale on abdominal segments 8–10.) Along the rear edge of the prothorax's dorsum there is typically a pale spot. That spot varies in shape from a dot to a rectangle

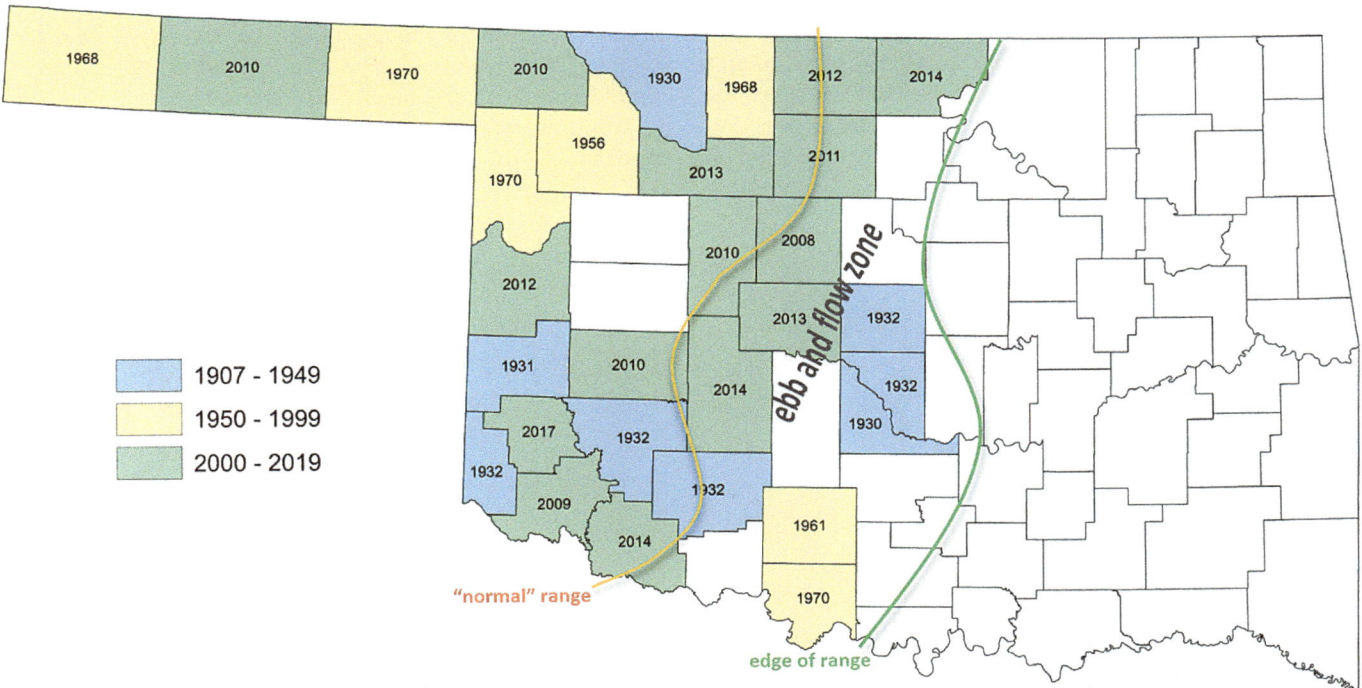

Figure 12.87 Oklahoma counties in which *Ischnura denticollis*, the Black-fronted Forktail, have been recorded. The green line indicates the documented edge of the species' range in Oklahoma; however, the orange line indicates the eastern extent of the more "normal" range of the species in the state. Given the relative rarity of *I. denticollis* in central Oklahoma and that it was last recorded in Jefferson County in 1970, Stephens in 1961, Cleveland in 1932, and McClain in 1930, it is likely that the area between the orange and green lines is an ebb and flow zone where the species occurs only occasionally.

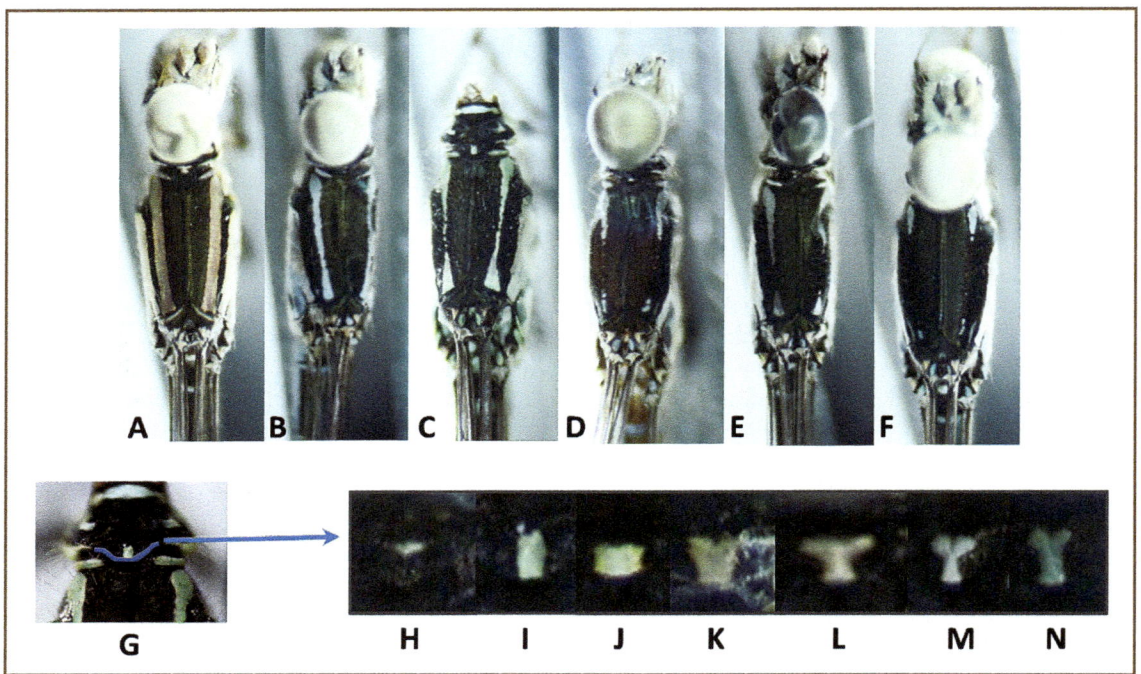

Photos are not to scale. *See* Appendix F, Table 1 for specimen data.

Figure 12.88 Variation in antehumerals (A–F) and pale spot at the rear edge of prothorax (G, blue line; variation, H–N) with female *Ischnura denticollis*.

(width- or lengthwise), to a flat-capped mushroom to bunny ears (Figure 12.88). Antehumerals range from those that are full and unconstricted toward the middle, such as those of heterochrome females (by far the most common morph), to those that are constricted centrally, broken, spattered, exclamation point-like, or have *Ischnura damula*-like dots; all of the latter marks being consistent with andromorphs (Figure 12.88). As a gentle reminder, careful consideration should be made when designating females as andromorphs. As per Westfall and May (20006:358) the pale color should be blue, not green (as pictured in Paulson 2009:121), and the antehumeral, when present, is constricted near the middle or sometimes present as "anterior and posterior spots or streaks or only with anterior spot."

ISCHNURA POSITA (HAGEN, 1861) – FRAGILE FORKTAIL

The Fragile Forktail is Oklahoma's most common and almost the most widespread forktail species. There are about 1,300 records of the species in the OOP database. It has been recorded in all but one county in Oklahoma—Cimarron. As it is more abundant in the east (similar to Citrine Forktail, *Ischnura hastata*, and opposite that of Eastern Forktail, *I. verticalis*), it is not surprising that it has not been found in the distant end of the panhandle. Double and triple digits are common, with the highest count that we know of being 300 individuals. It is a primarily lentic species, preferring ponds or non-flowing wetlands, although it is found at or near flowing creeks, too. At a pond described as "partially shaded … heavily vegetated, clear, shallow, [and] mud-bottomed," GH and JC Bick found 85% of the Fragile Forktails they encountered during their study of odonates in Marshall County (Bick and Bick 1958).

The species was first recorded in the state on 3 June 1907, when EB Williamson collected 12♂ and 1♀ near Wister, Le Flore County (6 specimens are at UMMZ). It has been seen regularly in all eras of research since. It is found in the state almost year-round, flying for 288 days (1 February–16 November, Red Slough, McCurtain County; OC 459504

Figure 12.89 *Ischnura posita*, the Fragile Forktail, A) ♂, Greenleaf SP, Muskogee County, 6 October 2018 (©Bill Carrell, OC 491279). B) ♀, Caney Creek 6 km NW of Stilwell, Adair County, 13 May 2018 (©CA Ivy, OC 479702). Even old, pruinose females such as this one typically retain a hint of the species' classic "exclamation point" thoracic stripe. Curiously, although this species, *I. verticalis*, and *I. hastata* often are encountered in large numbers, tandem pairs, such as this one C) at Bixhoma Lake, Wagoner County, 25 April 2013 (©James W Arterburn, OC 427970), are encountered rarely, yet tandem pairs of *I. ramburii* and *I. denticollis* are commonplace.

and OC 324276, respectively), and it is often the first odonate seen in the state in the spring (it competes with *Anax junius* for that title). It does fly in Texas year-round and in Louisiana almost as long, but elsewhere in the region flight dates are more limited (AR: 240 days, 1 March–27 October, OC; KS: 167 days, 11 April–25 September, USNM, SEMC; LA: 330 days, 4 February–31 December, iNat, Mauffray 2014; MO: 247 days, 14 March–16 November, Trial 2005; TX: all year, Abbott and Lasley 2019).

Fifteen reports of tenerals date from 26 February to 14 September. Many pairings have been reported, and 18 ovipositing records indicate the species tends to oviposit as lone females. The only known description of *Ischnura posita* ovipositing in Oklahoma is from Bick and Bick (1958), who observed a lone female on 19 June 1955 perched on duckweed (*Spirodela*) and curving her abdomen below

water to deposit eggs. They were not able to determine on what part of the plant the eggs were deposited. The female sat for minutes at a time while ovipositing. In a study of the species in Louisiana, females were observed depositing in a succulent plant above the water (Bick 1957). GH Bick's notes indicate that he collected 57 nymphs and reared 14 (6♂, 8♀). His notes say that he sent most of those to FSCA as specimens in alcohol. We did not find those specimens at FSCA, but we were not able to examine the fluid collection fully. Nymphs of the Fragile Forktail were almost always found by the Bicks to be clinging to *Spirodela* or other duckweed (*Lemna*) roots. There has been one mixed-species (male-male) pairing documented for *I. posita*. GH Bick collected a ♂ *I. posita* (FSCA 66808) with a ♂ *Enallagma basidens* (FSCA 66809) on 20 July 1963 in Pontotoc County at Stonewall.

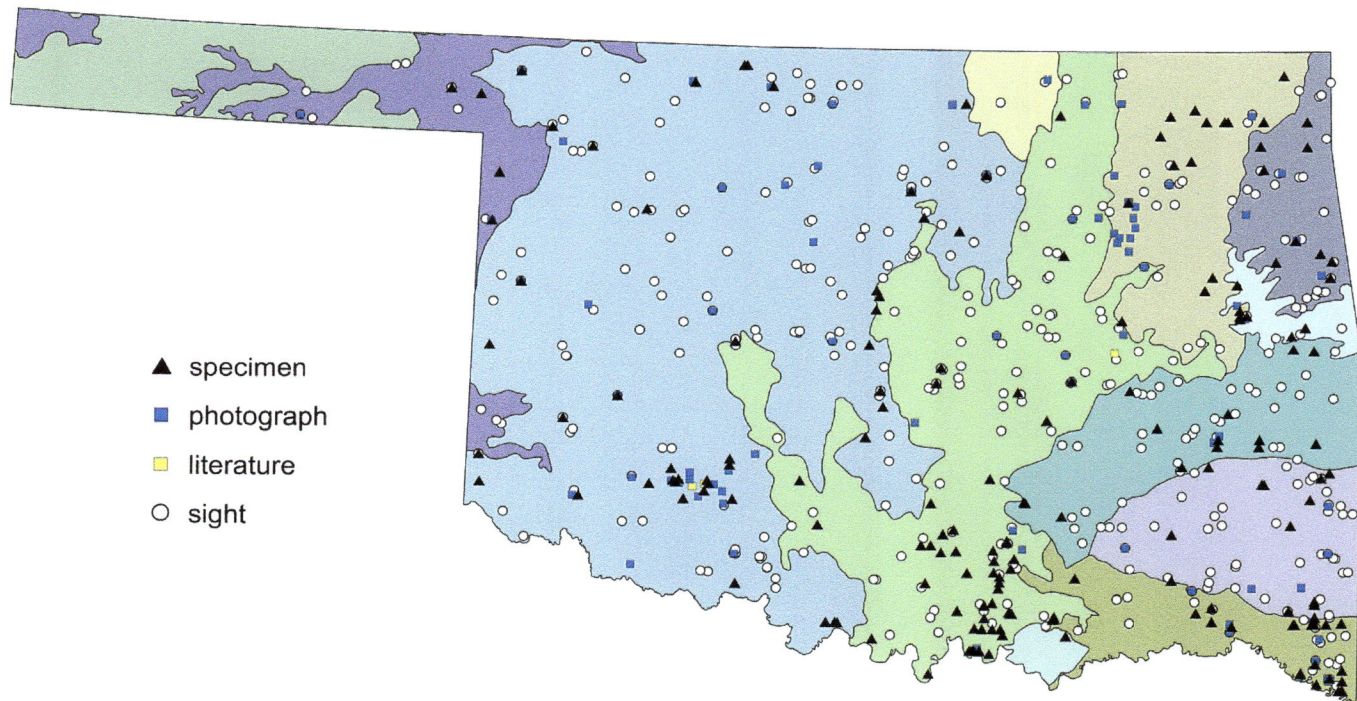

▲ specimen
■ photograph
□ literature
○ sight

Figure 12.90 Documentation level of supporting records for *Ischnura posita*, the Fragile Forktail, in Oklahoma. Levels include specimen, archived photograph, literature reference, or sight record. Although sight records lack documentation, only those we feel are credible are plotted.

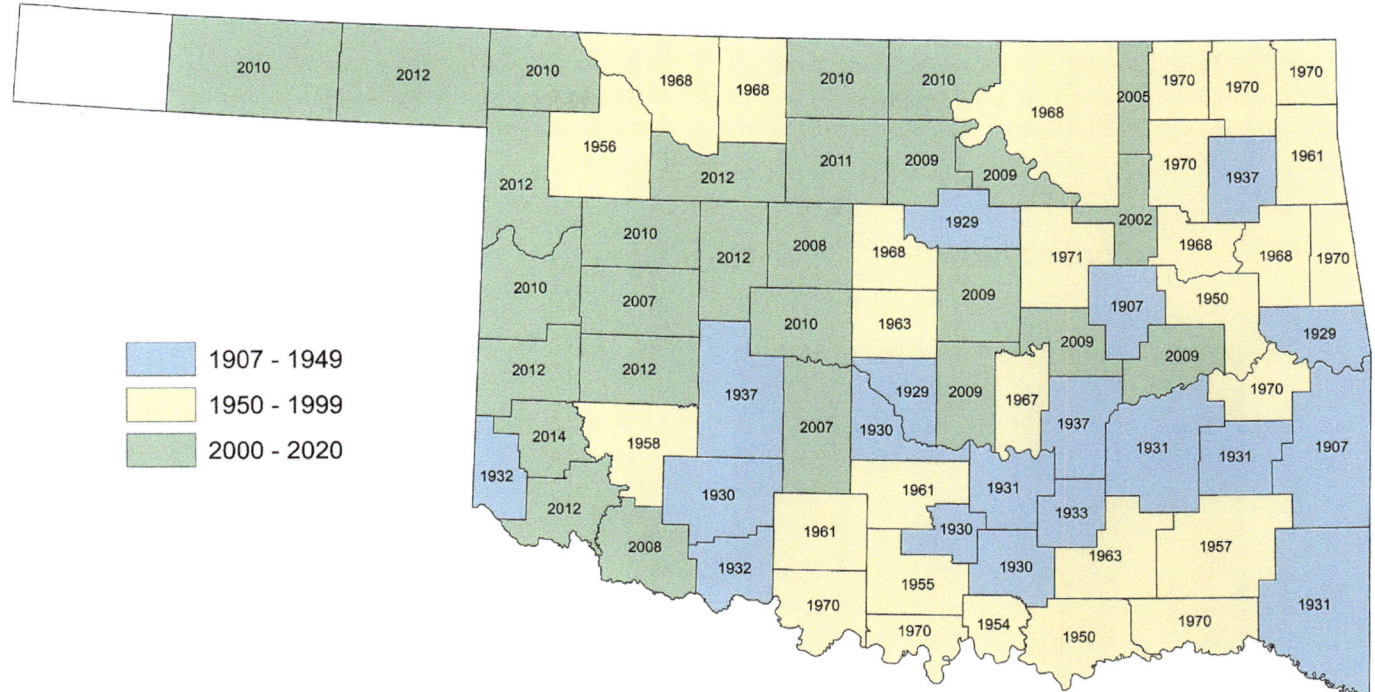

Figure 12.91 Counties in which *Ischnura posita*, the Fragile Forktail, are known to occur in Oklahoma. Year within the county is that in which the species was first reported.

ISCHNURA HASTATA (SAY, 1839) – CITRINE FORKTAIL

The Citrine Forktail is found across the state, although it decreases in abundance as one goes west toward the panhandle and also generally from south to north. It can be difficult to spot until one gets accustomed to detecting its slow and seemingly methodical movements, which are similar to that of red damsels (*Amphiagrion* sp.). Even though up

Figure 12.92 *Ischnura hastata*, the Citrine Forktail, A) ♂, Arkansas River 4 km W of Braggs, Muskogee County, 2 May 2014 (©James W Arterburn, OC 422009). B) ♂, Teal Ridge Wetlands in Stillwater, Payne County, 11 July 2011 (©Jason R Heinen, OC 329767). Note the tiny size of this species, Oklahoma's smallest odonate. C) ♀, Fry Creek in Tulsa, Tulsa County, 23 June 2011 (©Ken Williams, OC 331024); see Figure 12.63 for an immature ♀.

close one can see that males of the species are bright yellow, from a human's height looking down into the grass below, a Citrine Forktail can look little different from a dried piece of grass floating in the breeze. But once familiarized with that image, Citrine Forktails can be easy to spot and should be expected when within grasses and sedges at or near ponds and marshes.

Ischnura hastata was first recorded in Oklahoma on 3 June 1907 by EB Williamson at Wister, Le Flore County (3♂, 3♀, including one pair, UMMZ; Williamson 1914b reported as 3–4 June but all specimens are 3 June). Later that year, F Collins reported 10♀ from Henryetta, Okmulgee County between August and September (Williamson 1914b). Since then, *I. hastata* has been reported regularly. There are currently >600 records known and the species has been recorded in all 77 Oklahoma counties.

Citrine Forktails are often reported in good number, including at least nine times reported in numbers >100, with the highest being an estimated 500 individuals (250♂, 250♀, 12 km NW of Cookietown, Whites Lake, Cotton County, 4 July 2015, MAP). The species flies in the region generally from early spring to late fall, but it is known year-round or almost year-round in some states (OK: 16 March–30 November, both reported by DA from Red Slough, McCurtain County, 2016 sight record and 2017 OC 474961, respectively; AR:19 March–23 October, iNat; KS: 28 May–20 October, SEMC, OC; LA: year-round, iNat; MO: 18 April–23 October, OC, Sims 2012; NM: Jan–October, Paulson 2009; TX: year-round, Abbott and Lasley 2019). Pairs are rarely reported, and we know of no records of ovipositing. Tenerals have been reported in Oklahoma from 22 May to 15 October.

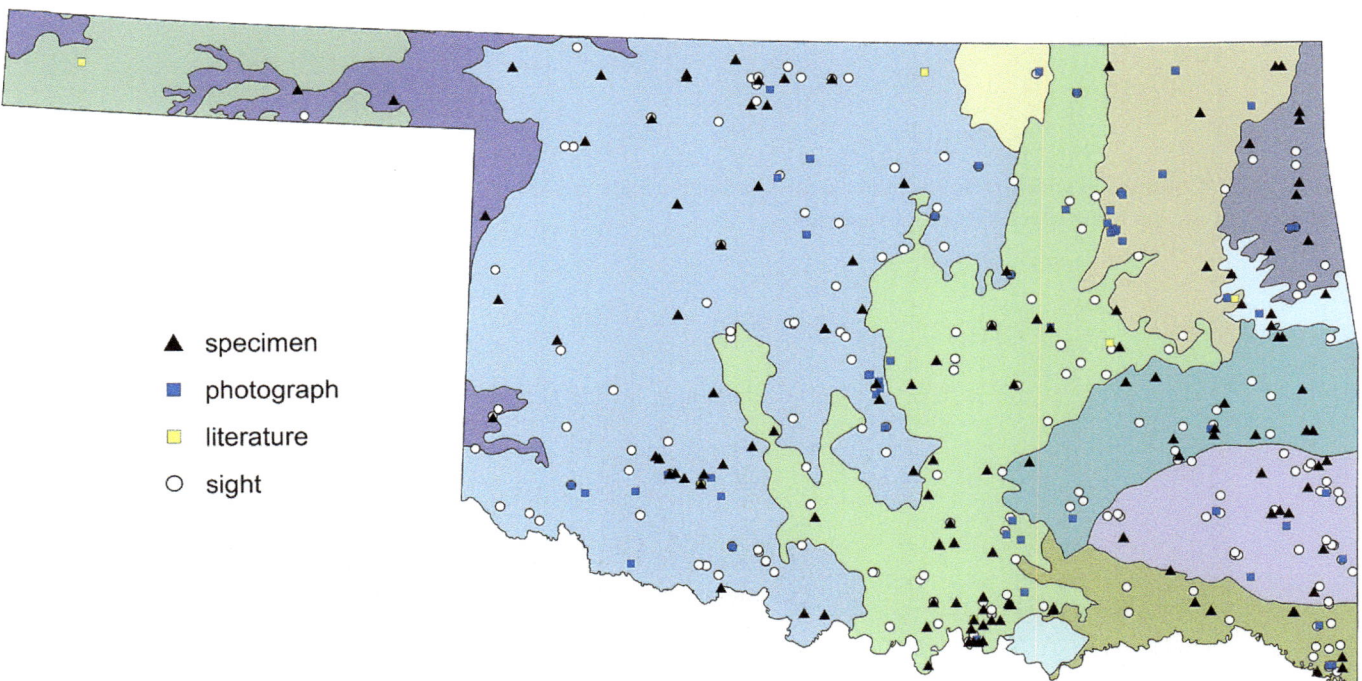

Figure 12.93 Documentation level of supporting records for *Ischnura hastata*, the Citrine Forktail, in Oklahoma. Levels include specimen, archived photograph, literature reference, or sight record. Although sight records lack documentation, only those we feel are credible are plotted.

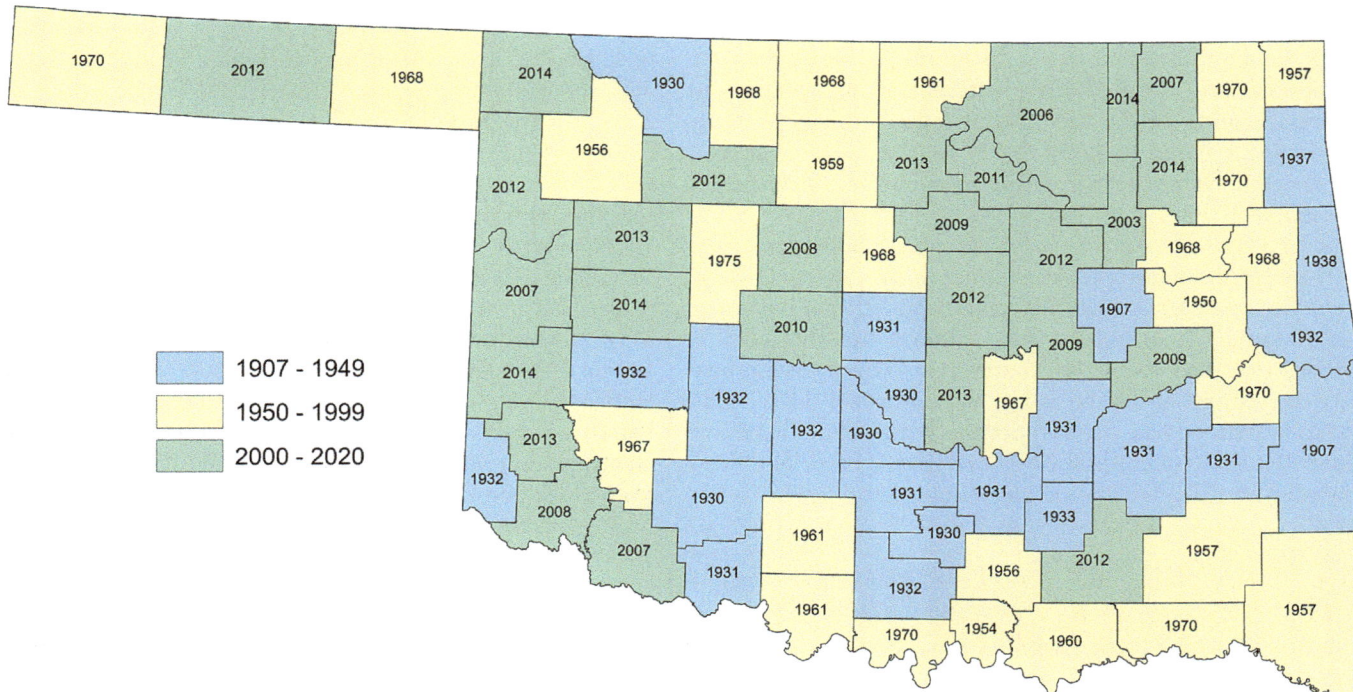

Figure 12.94 Counties in which *Ischnura hastata*, the Citrine Forktail, are known to occur in Oklahoma. Year within the county is that in which the species was first reported.

AMPHIAGRION ABBREVIATUM (SÉLYS, 1876) – WESTERN RED DAMSEL

In Oklahoma, knowing what species of red damsel you have when you catch one historically has been problematic. Three species names and one form designation—*saucium, abbreviatum, mesonum,* and "intermediate"—have been applied to Oklahoma *Amphiagrion* since red damsels were first found in the state in 1932 (Figure 12.95).

Amphiagrion saucium (Burmeister, 1839) is what all *Amphiagrion* were once thought to be; as Needham and Heywood (1929:304) remarked, "In our fauna there is a single variable species." It is that variability that has caused much consternation—is the extreme variability seen between individuals from the Atlantic seaboard and those from the western United States and Canada indicative of one species with extreme characters at the edges of its range or does it make for two (or three?) species? The former idea was propagated for some time by relegating *A. abbreviatum* (Sélys, 1876) as merely a "variety" of *A. saucium,* despite *A. abbreviatum* originally being described as a separate species. Eventually, the latter idea caught on and the two

were again considered full species (Garrison 1991). Problem solved, right? Wrong!

The beloved LK Gloyd muddied *Amphiagrion* waters by becoming convinced that a third species existed (Figure 12.96). She examined thousands of specimens and concluded that those from parts of the Great Lakes region, the northern and central Great Plains (including Oklahoma), and some parts of the southwest constituted a separate entity. If valid, the range of this species would more or less fall in line with those states listed by Westfall and May (2006:132) as having specimens that they described as "not clearly assignable to either *A. abbreviatum* or *A. saucium*": Arizona, Colorado, Illinois, Indiana, Kansas, Michigan, Nebraska, New Mexico, North Dakota, Oklahoma, South Dakota, and Wisconsin. Gloyd never published her manuscript, yet her provisional name, *A. mesonum,* has appeared in print in reports (e.g., Illinois DNR 1998), checklists (e.g., Beckemeyer 2002b; perhaps indicative of an evolution of opinion from the use of "*Amphiagrion* sp." in Beckemeyer

Figure 12.95 *Amphiagrion abbreviatum,* the Western Red Damsel. Adult red damsels were unreported in Oklahoma for 56 years until A) Jason R Heinen found the species at Drummond Flats WMA, Garfield County, 9 April 2012, when he documented this ♀, among others (©Jason R Heinen, OC 374349). B) ♂, Lake Carl Etling, Cimarron County, 14 May 2017 (©Bill Carrell, OC 462718).

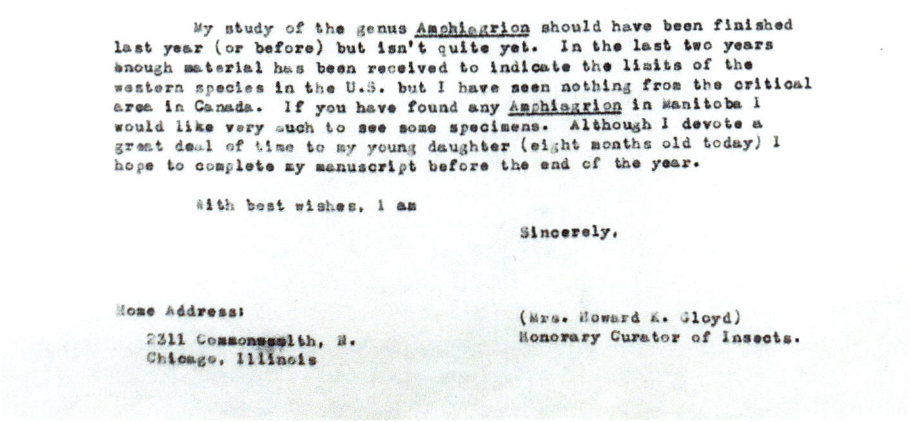

My study of the genus Amphiagrion should have been finished last year (or before) but isn't quite yet. In the last two years enough material has been received to indicate the limits of the western species in the U.S. but I have seen nothing from the critical area in Canada. If you have found any Amphiagrion in Manitoba I would like very much to see some specimens. Although I devote a great deal of time to my young daughter (eight months old today) I hope to complete my manuscript before the end of the year.

With best wishes, I am

                          Sincerely,

Home Address:
    2311 Commonwealth, N.   (Mrs. Howard K. Gloyd)
    Chicago, Illinois   Honorary Curator of Insects.

Figure 12.96 Carbon copy of a letter from LK Gloyd to RD Bird, 1 July 1940. She was hopeful that she would soon publish her study of *Amphiagrion* that described a new species, but the study remains unpublished.

and Huggins 1998), and articles (e.g., O'Brien 1999, 2008), and online numerous times. Without a proper type description, *Amphiagrion mesonum* is unavailable taxonomically; i.e., it is a *nomen nudum*, a "naked name" because there is no acceptable type description.[44]

The name "*Amphiagrion* intermediate" has also been applied to specimens from Oklahoma and the region (e.g., Donnelly 2004c; OC currently has 236 records as *A. intermediate*). This term has been used in two ways: 1) as a placeholder for later description as a third species, i.e., *A. mesonum* or 2) to indicate individuals that have characters that lie within the clinal variation between the two extremes of *one* species, *A. saucium* (i.e., intermediate to the varieties, not the species of, *abbreviatum* and *saucium*). The debate rages on (e.g., Paulson 2009, 2011[45]; Daigle and Pilgrim 2014). We and others (e.g., KJ Tennessen, JJ Daigle, JC Abbott) have talked for years about resolving the issue but none has been able to do so as yet given the immense amount of time and energy the task will take. For now, like Bick and Bick (1957), we have chosen to designate Oklahoma specimens and other records (including those on OC) as *A. abbreviatum* as we feel the Oklahoma populations most closely resembles that taxon. This is in line with Westfall and May's (2006) choice of listing Oklahoma as part of *A. abbreviatum*'s range rather than just having it as part of the specimens not readily assigned to species, as they did in their earlier treatment (Westfall and May 1996). However, we are open to accepting taxonomic changes once a thorough investigation has been completed. Now, with all of that context provided, we would like to delve into the specific history of *Amphiagrion* in Oklahoma.

The first Oklahoma specimen, a ♂ collected in Cleveland County by RD Bird on 4 April 1932, was labeled as *Amphiagrion saucium*, as was the second specimen, another ♂ Bird collected in Cleveland County, this time on 10 April 1932 (Bird 1932; data from GH Bick notes). These specimens were later examined by GH Bick and identified in his notes as "probably *mesonum*" although he published them as *A. abbreviatum* (Bick and Bick 1957). Bick did likewise for the 3♂ and 2♀ specimens Bird collected in Comanche County on 9 July 1932. We have been unable to find any of Bird's specimens, which were in the OMNH collection at the time Bick examined them. Outside of Bird's records there were three other collections made in the 1930s. The first two were collected at Cherokee, Alfalfa County: 7 July 1932, 4♂, 3♀, AE Pritchard, and 4 May 1937, 1♂, RW Kaiser (probably). The third is from 5 June 1937, 1♂, Cleo Springs, Major County, Standish and RW Kaiser. All of these specimens are at UMMZ in one of LK Gloyd's drawers labeled as "*Amphiagrion* sp."

Bick encountered *Amphiagrion* only once in the state: in Woodward County in 1956. Note that Bick and Bick (1957) had a typo indicating that he collected 5♂ and 1♀ in Woodward County on "23 June 1956," which was actually 2–3 June 1956 (3 mi W of Woodward, 2 June, 1♂, 1♀, FSCA; Boiling Springs SP, 3 June, 4♂ [2♂ FSCA, 2♂ UMMZ]). Curiously, we found the 2♂ housed at UMMZ in Gloyd's *A. abbreviatum* drawer instead of in her *Amphiagrion* sp. drawer, as other Oklahoma specimens were. The only other records known of the species for the middle era of collecting in the state are larval records. Two nymphs were collected in Alfalfa County, one by WD Butcher on 27 September 1975 (OMNH 1621, no location given) and the other by K Larson and D Gettinger on 24 September 1977, along Oklahoma highway 11, 13 mi E of Cherokee, probably at what is known as "Travellers Well" (OMNH 1622).

The species, as adults, went unrecorded in Oklahoma for over 50 years, until 9 April 2012, when, during a survey of birds at Drummond Flats WMA, Garfield County, JR Heinen discovered the first red damsel adults seen in the state since 1956 (Table 12.2; Figure 12.95). He reported the species again the following day, and then about a week-and-a-half later we were able to go to the spot and collect some specimens. The year 2012 proved to be great for the species in Oklahoma: Heinen confirmed the species' presence in Alfalfa County, and we discovered new populations in Ellis and Harper Counties. We also encountered the species along Mexico Creek below the dam at Lake Evans Chambers in Beaver County.[46] *Amphiagrion* have been reported each year since except 2014 (Table 12.2), which was a year that we

Table 12.2 Records of *Amphiagrion abbreviatum* in Oklahoma, 2012–2019.

DATE	COUNTY	LOCALITY	INDIVIDUALS	OBSERVER	SOURCE	NOTES
9 Apr 2012	Garfield	Drummond Flats WMA	2 (1♂, 1♀)	Heinen, JR	OC 374349	
10 Apr 2012	Garfield	Drummond Flats WMA	3 (1♂, 2♀)	Heinen, JR	OC 374650	
21 Apr 2012	Garfield	Drummond Flats WMA	13 (9♂, 4♀, inc. 4 pair)	MAP; BS-P	BS-P/MAP field notes; SP/JCAC	SP 143 (1♂) and SP 144 (1 pair) donated to JCAC
12 May 2012	Alfalfa	vic. Salt Plains NWR, "Travelers Well"	1♂	Heinen, JR	OC 374926	
12 May 2012	Ellis	Ellis Co. WMA	3 (1♂, 2♀)	MAP; BS-P	BS-P/MAP field notes; SP	dam spillway; SP 224 (1♂) and SP 225 (1♀)
12 May 2012	Harper	Doby Springs Park	5♂	MAP; BS-P	BS-P/MAP field notes; SP	SP 226 (1♂)
2 June 2012	Beaver	Lake Evans Chambers, Mexico Creek	1♀	BS-P; MAP	BS-P/MAP field notes; SP	SP 259 (1♀)
25 May 2013	Harper	Doby Springs Park	1♂	BS-P; MAP	BS-P/MAP field notes; SP	SP 577 (1♂)
27 May 2013	Texas	Schultz WMA	1♀	BS-P; MAP	BS-P/MAP field notes; SP	SP 593 (1♀)
27 Aug 2013	Alfalfa	vic. Salt Plains NWR, "Travelers Well"	1♂	BS-P; MAP	BS-P/MAP field notes; SP	SP 975 (1♂)
17 May 2015	Harper	Doby Springs Park	4♂	BS-P	BS-P field notes; SP	SP 1573, SP 1574 (2♂)
18 May 2015	Harper	Doby Springs Park	8 (6♂, 2♀, incl. 1 pair)	BS-P	BS-P field notes; SP; OC	SP 1582 (1♂), SP 1583 (1♀); photo (1 pair, OC 430903)
2 June 2015	Harper	Doby Springs Park	1♂	BS-P; MAP	BS-P/MAP field notes	
21 May 2016	Harper	Doby Springs Park	25 (18♂, 7♀, incl. 1 pair)	MAP	BS-P/MAP field notes	
28 May 2016	Ellis	Ellis Co. WMA	150 (100♂, 50♀, incl. 10 pair)	MAP	MAP field notes; SP	SP 1911 (1 pair); photo (1♂, OC 445176); all individuals were at the large marsh below Lake Vincent dam
14 May 2017	Cimarron	Black Mesa SP, Lake Carl Etling	2♂	BC	OC 462718	at dam spillway
16 May 2017	Ellis	Ellis Co. WMA	5 (4♂, 1♀)	BS-P	BS-P field notes; SP	SP 2317 and SP 2318 (2♂) collected below Lake Vincent dam (36.0562, -99.9148) and SP 2319 (1♀) collected at dam spillway at impoundment south of Commission Creek (36.0485, -99.9596)
21 May 2017	Harper	Doby Springs Park	10 (8♂, 2♀)	MAP	MAP field notes	
3 June 2017	Cimarron	Black Mesa SP, Lake Carl Etling	25 (22♂, 3♀)	MAP	MAP field notes; SP	SP 2374 and SP 2375 (2♂); at dam spillway
19 May 2018	Cimarron	Black Mesa SP, Lake Carl Etling	1♂	Hjalmarson, EA; Roberts, B	Hjalmarson field notes	photo (on file, BS-P); at dam spillway
26 May 2019	Harper	Doby Springs Park	110 (80♂, 30♀)	MAP	MAP field notes; OC	photo (1♂, OC 495623)

Individuals: incl. = including; Locality: NWR = National Wildlife Refuge, SP = State Park, vic. = vicinity, WMA = Wildlife Management Area; Observer, source, notes: BC = Bill Carrell, BS-P = Brenda D Smith-Patten, JCAC = John C Abbott Collection, MAP = Michael A Patten, OC = Odonata Central, SP = Smith-Patten/Patten Collection

were not able to conduct intensive surveys during the flight season at known locations for the species.

Most of the Oklahoma records are of 1–5 individuals at a time, but there have been counts of >10 individuals 6 times (out of 28 records of adults). Four of those records are of 10–25 adults, but one was a remarkable 150 adults (100♂, 50♀, including 10 pairs; Table 12.2). Ellis County WMA, the site of that high count for the state, has seen the species in numbers as few as 3 (actually 0 if we count negative surveys), to as high as 150. The most consistent site for the species, Doby Springs, Harper County, has negative reports and those ranging from 3–110 adults. So, it appears that much luck is needed with timing of surveys. Not only is this species difficult to find because of its specialized habitat— it is strongly associated with springs, both pooled (natural and artificial) and flowing, where spikerush (*Eleocharis* sp.) is found—but it may also experience mass emergences and quick reversals in numbers, as reported for its sister species, *A. saucium*.[47] Also, as reported for *A. saucium*, *A. abbreviatum* is likely localized in where it does occur.[48] If we hope to fully determine the status of this species in the state, we are faced with, in the most arid and privatized parts of Oklahoma, finding accessible springs with associated marshes of *Eleocharis* and then be lucky enough to arrive on the perfect day to spot this surprisingly unobtrusive (considering how red individuals are) species. This leaves little room for error. So … not too much to ask. Ha!

The difficulty in detectability with the species brings us to the question of whether this species has experienced a range retraction and/or a population decline. Previously we speculated these points (Patten and Smith-Patten 2013b). Despite a few more years of data under our belts, we still cannot say whether either of those possibilities can be quantified. What we do know is that there are four counties in which the species was reported earlier but not since: Cleveland (1932), Comanche (1932), Major (1937), and Woodward (1956). It may be that we have not been fortunate enough to find suitable habitat or at least the exact right spot for the species in these counties, even though we and others have surveyed intensively directly at some of the known sites (e.g., Sulphur Spring, Comanche County, and Boiling Springs SP, Woodward County). Perhaps that means the species' range has retracted or the population has declined or perhaps, more benignly, it is just bad luck. On the flip side, since early April 2012 the species was found in six counties where it was unknown previously, and it was refound in one county where adults had not been reported since 1937. In spite of these mixed signals, we feel that our assessment of the species as an S2 species of conservation concern remains warranted given its low population density, localized distribution, specialized habitat that is rare and at high risk of loss in northwestern Oklahoma and the panhandle, sensitivity to land use changes, and putative range retraction.

Amphiagrion abbreviatum is known to fly in Oklahoma from early April until late August, with the bulk of records from May. The overall flight season jibes well with that of nearby states (OK: 4 April–27 August; CO: 24 April–1 September, OC, YPM; KS: 4 May–14 August, SEMC;

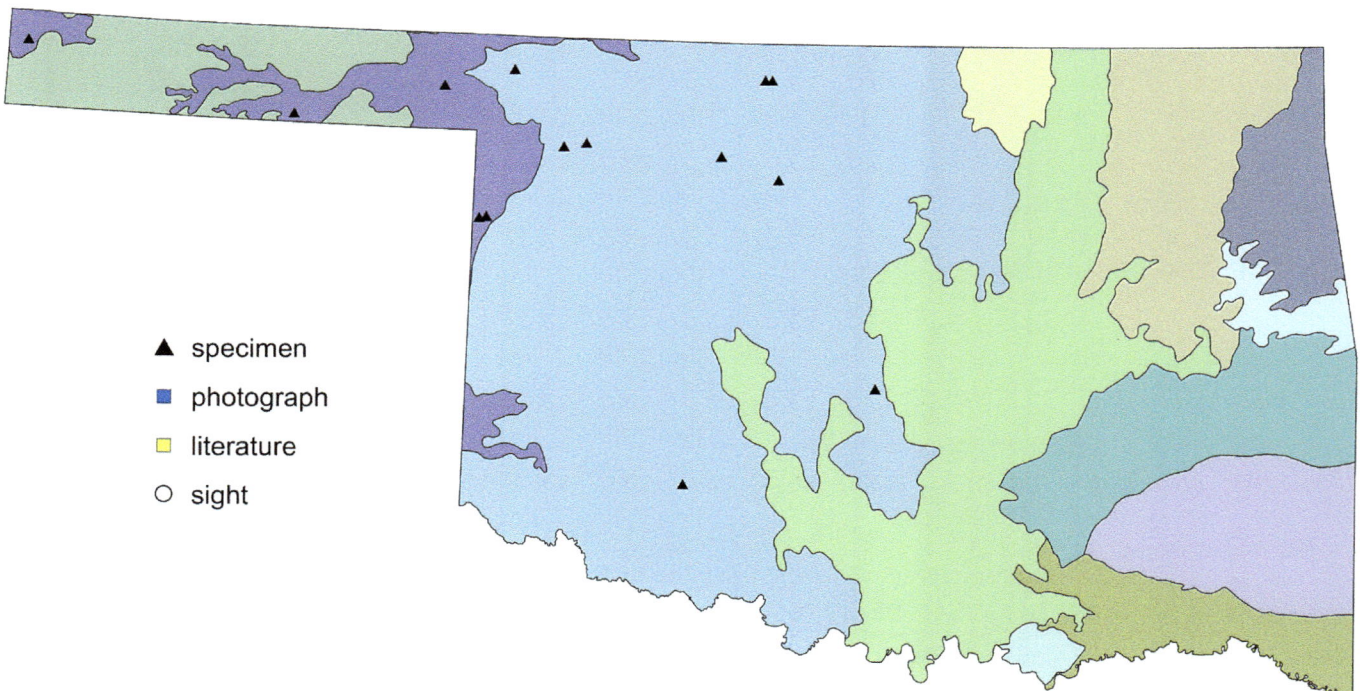

▲ specimen

■ photograph

□ literature

○ sight

Figure 12.97 Documentation level of supporting records for *Amphiagrion abbreviatum*, the Western Red Damsel, in Oklahoma.

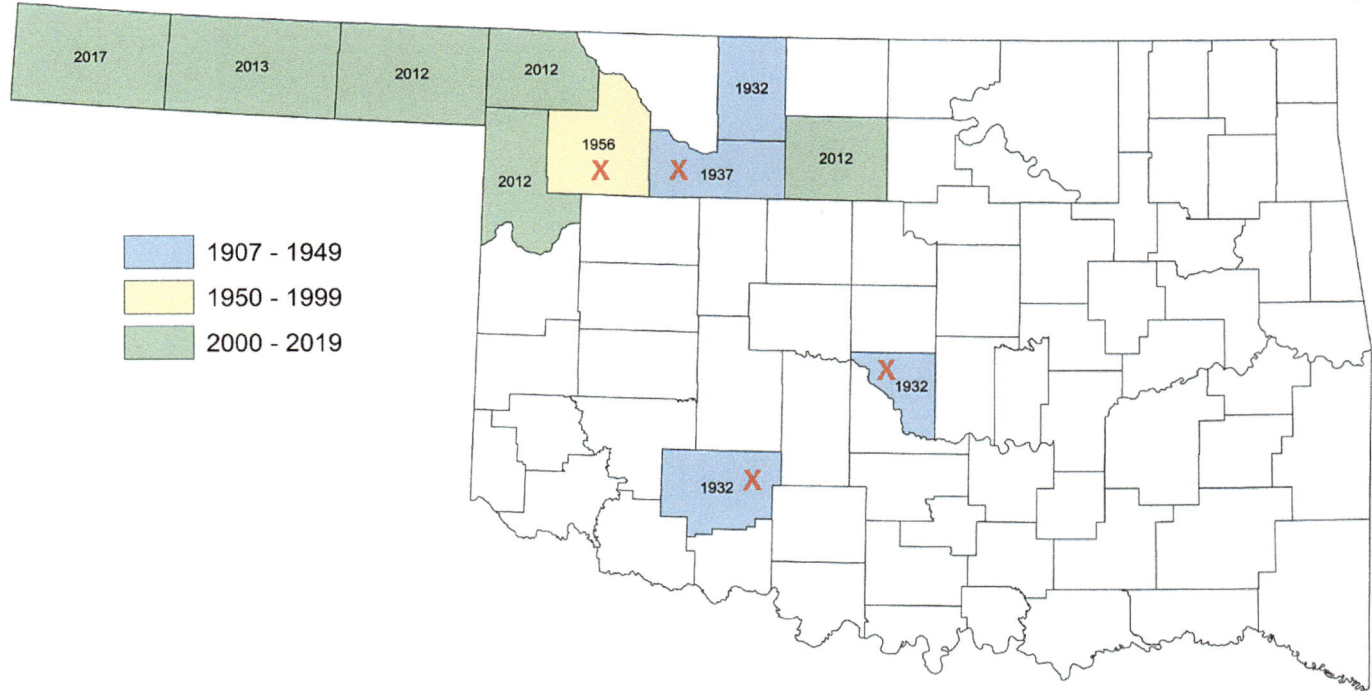

Figure 12.98 Counties in which *Amphiagrion abbreviatum*, the Western Red Damsel, are known to occur in Oklahoma. Year within the county is that in which the species was first reported. Counties with red X's indicate that the species has not been encountered in that county since the year it was first reported there.

NM: May–September, Paulson 2009). These dates are also mostly in line with elsewhere within the species' range (Paulson 2009), although it is known to fly into October in California and Utah (CA: 14 October, UCBME 51172; UT: 4 October, Myrup and Baumann 2016). The species has been reported only once in Texas, at Lake Marvin, Hemphill County, 21 May–1 June 2019 (up to 5♂ and 4♀, including 3 tandem pairs; M Dillon, OC 495053, M Reid,

OC 495855). There are no reports of tenerals in Oklahoma nor are there accounts of ovipositing. There are sufficient reports of species associations that indicate *Amphiagrion abbreviatum* is strongly associated with *Ischnura denticollis*, the Black-fronted Forktail, and *Argia alberta*, the Paiute Dancer; it has only been reported twice without either *I. denticollis* or *A. alberta* present (Table 12.3).

Table 12.3 Records of adult *Amphiagrion abbreviatum* in Oklahoma and association with other odonate species.

Species	COMANCHE	GARFIELD	ALFALFA	WOODWARD 1	WOODWARD 2	ELLIS 1	ELLIS 2	ELLIS 3	HARPER 1	HARPER 2	HARPER 3	HARPER 4	HARPER 5	HARPER 6	HARPER 7	BEAVER	TEXAS	CIMARRON
Amphiagrion abbreviatum	5	13	1	2	4	3	150	5	5	1	4	8	1	25	10	1	1	25
Lestes alacer	7																1	
Lestes rectangularis							11											
Lestes unguiculatus									1							1	1	
Argia alberta	3	4				5	40	3			4	1	5	1	4	57	40	1
Argia fumipennis						33			25	13	6	2	3	6	4	1	50	
Argia nahuana	4		9	1	6		8		8	75	75	3	75	3	23	14	25	
Argia sedula							2	1			2				3	20		
Argia plana		3	3		18	1		1	100	60	14	2	11		43	3		
Argia immunda															1			
Argia leonorae	1																	
Ischnura denticollis	7	500		7		2	4		10	23	30	20	9	150	6	43	80	11
Enallagma antennatum																	1	
Enallagma carunculatum									2							1		
Telebasis salva						1												
Plathemis subornata							1		1		5		2		3	44	3	

Site name and date[a]

Comanche = Sulphur Spring, 9 July 1932
Garfield = Drummond Flat WMA, 21 April 2012
Alfalfa = Travellers Well, 27 August 2013
Woodward 1 = Boiling Springs SP and vic., 2 June 1956
Woodward 2 = Boiling Springs SP and vic., 3 June 1956
Ellis 1 = Ellis Co. WMA, 12 May 2012
Ellis 2 = Ellis Co. WMA, 28 May 2016
Ellis 3 = Ellis Co. WMA, 16 May 2017
Harper 1 = Doby Springs, 12 May 2012
Harper 2 = Doby Spring, 25 May 2013
Harper 3 = Doby Spring, 17 May 2015
Harper 4 = Doby Spring, 18 May 2015
Harper 5 = Doby Spring, 2 June 2015
Harper 6 = Doby Spring, 21 May 2016
Harper 7 = Doby Spring, 21 May 2017
Beaver = Lake Evans Chambers, 2 June 2012
Texas = Shultz WMA, 27 May 2013
Cimarron = Black Mesa SP, 3 June 2017

WMA = Wildlife Management Area, SP = State Park, vic. = vicinity.
[a] First part of site name is its county.

CHROMAGRION CONDITUM (SELYS, 1876) – AURORA DAMSEL

We thought that when we discovered the Aurora Damsel in Oklahoma, we would recognize it instantly. After all, our experience with it in New England told us that it, could be secretive but that it was easily identified. But when the day came, recognition crept in slowly (much as "The Blob" seeking its next victim; for those unfamiliar with this reference, see the 1958 classic film of the same name). On 16 May 2019, BS-P and DA were hunting for *Cordulegaster talaria*, the Ouachita Spiketail, in eastern McCurtain County. After finding a nearby site ("Arbour Seep") supporting that species, they continued on to Pine Mountain Spring, where *C. talaria* was found in 2018 (see that species'

Figure 12.99 *Chromagrion conditum*, the Aurora Damsel, A) this ♂ was one of five found at Pine Mountain Spring, McCurtain County, on 16 May 2019 for Oklahoma's first state record of the species (USDA Forest Service; David Arbour, OC 494957). The following day an astounding 41♂ and 13♀, including 11 pairs, were discovered at a site 3 km east, for the second state record (MAP, OC 494902). One pair included B) a ♀ andromorph ("male like"), another was a heteromorph ("female like"), colored much like C) a ♀ in copula that was photographed on 24 May 2019 at the same locality (©Mike Dillon).

account). While they hiked around looking for seeps, the two came across a small, woodland impoundment (perhaps 10 m in diameter, fairly shallow, mostly clear water, with limited emergent and shoreline vegetation). BS-P stayed at the pond while DA investigated the pond's headwaters. At first it appeared that there were only a handful of Common Whitetails (*Plathemis lydia*) and a few Ebony Jewelwings (*Calopteryx maculata*) flitting about, but as she watched the whitetails tussle, two damsels caught her eye. They strayed occasionally from the edge of the pond but rarely far from it or from sunny patches, which gave them a washed-out appearance, making it difficult to discern color or pattern. That, along with alighting only briefly, made identification problematic. Being only her first full day in the field for the year, she kept saying to herself, "You know this bugger, but boy, are your id skills rusty! What are these?" She tried to cram them into a species box of one of the 44 previously recorded pond damsels (Coenagrionidae) for Oklahoma, but to no avail. Finally, just as DA was returning to the pond, one of the males landed in a shadier spot and, as the flash of ventrolateral yellow came across BS-P's eyes, so too did a flash of memory—*Chromagrion*!! It took quite some time for either male to co-operate enough for DA to get photos (OC 494957) but once he did, BS-P snapped in and caught the first specimen for the state (SP 2738). While walking back to the truck, three more males were seen along a narrow (<50 cm), shallow (<5 cm), clear-water spring run, for a total of 5♂. On a dragonfly high, the rest of the day was gravy. This was a day every ode hunter hopes for (and thrives on)— within about 2 hours, the first ♂ specimen of the Ouachita Spiketail and a new state record of another much-coveted species were had (175th species for the state and McCurtain County's 122nd!).

Upon hearing of such a spectacular day, MAP, could not help but rush down, in as much as a 5-hour drive can be rushed, to nab records for himself. And he did. The next morning, at a part of "Arbour Seep" (and an associated creek) not investigated the day prior, MAP encountered a remarkable 41♂ and 13♀ *Chromagrion*, including 11 pairs (1 pair as SP 2770); 2 days later, David Oakley made a 4-hour drive to "Arbour Seep" in inclement weather (such effort certainly attests to the perceived rarity of this species in this part of its range) to photograph 1♂ (OC 495004). Conditions were better for a return visit to "Arbour Seep" by DA and BS-P

on 24 May, when another healthy count was made (16♂, 5♀, including 4 pairs, one of which was ovipositing; all individuals, except one ♂, were at a roadside woodland stream that was about 2 m wide, spring-fed, clear-water, rocky; the lone ♂ was seen at the mucky seep run where the spiketails were previously found). The documented flight season for the species in Oklahoma was topped off by MAP's 2 June visit to the site, when he photographed the only adult *Chromagrion conditum* he encountered (1♂, OC 495965). Later visits to "Arbour Seep," Pine Mountain Spring, and the general area that year, were fruitless for the species.

The Aurora Damsel is known from just across the Oklahoma border in Arkansas from Montgomery and Scott Counties as well as in the north-central and northeastern parts of the state from Marion and Sharp Counties, the latter county from where that state's first record came (spring seep on Rock Creek, "3 ½ mi N of Sitka," 3 June 1990, ML May, GL Harp Collection; Harp and Harp 1996). There are two records known from Montgomery County, both from 1992 (spring seep, 18 May, KJ Tennessen [no additional details provided], and 1♂ at "a temporary woodland pond 8 mi W, 1 mi N Pearcy" on 7 June, GL Harp; Harp and Harp 1996). Scott County has but one record (11 May 2002, from a "wetland at NE corner of jct. US Hwy 270/FS 929," Harp and Harp 2003) as does Marion County (Moccasin Fen, 11 May 2004, PM McKenzie). A purported record for Miller County reported by Donnelly (2004c) and attributed to the Harps, is in error (GL Harp, *in litt*) and is in an incongruous ecoregion, as such we "unconfirmed" the record on Odonata Central (OC 164235). To our knowledge, the species has not been encountered in Arkansas since these early 2000s records, the dates of which indicate the species flies from mid-May to early June, as now documented in Oklahoma (16 May–2 June). Records from south-central and southeastern Missouri[49] suggest a longer flight season, extending into late June (7 May–30 June, with one report of a nymph from Iron County, on 15 October; Trial 2005). Elsewhere, the species is known to fly a little longer, generally from May to July (Paulson 2011), but a biogeographical difference in phenology is expected, so records from these states may remain bounded to May and June. We refrain from suggesting a conservation rank for this species for Oklahoma and Arkansas until further investigations of its status can be conducted.

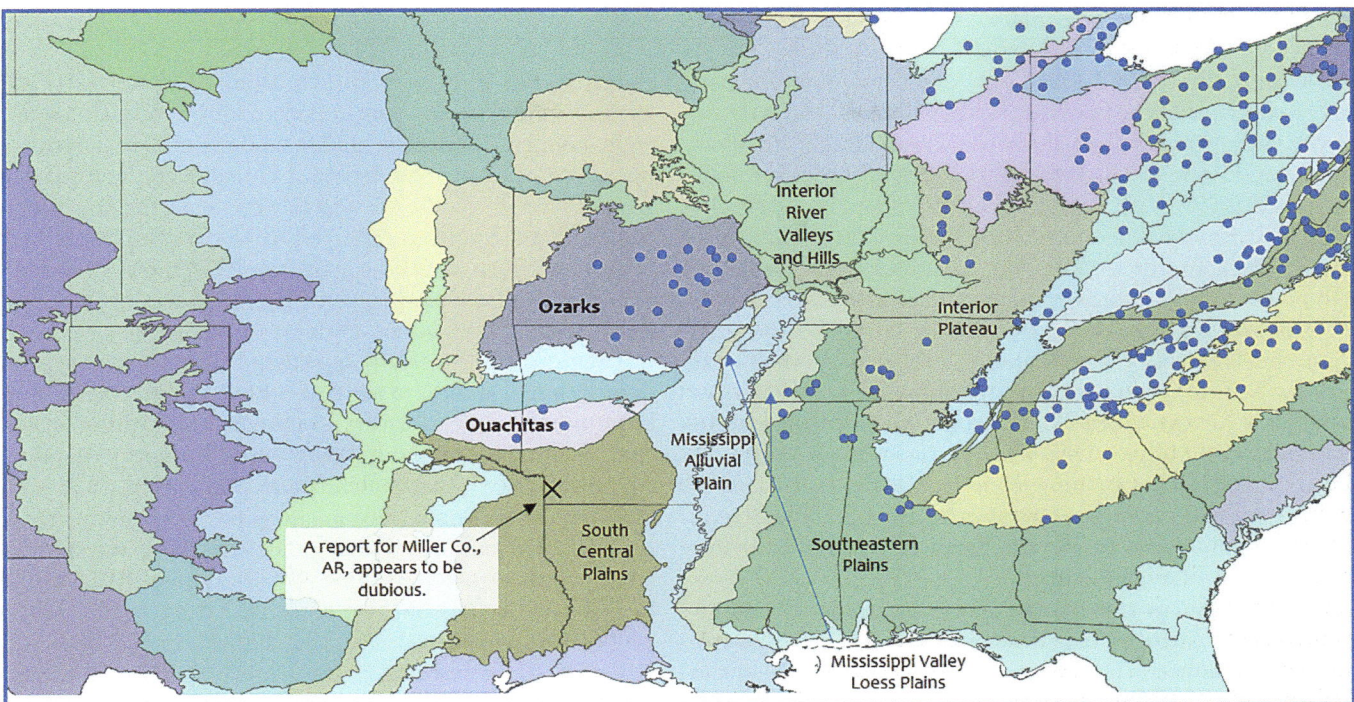

The distribution of this species very much mimics the combined range of *Stylogomphus sigmastylus* and *S. albistylus*, the Interior and Eastern Least Clubtails, in that it is found in the Ozark and Ouachita Highlands in Oklahoma, Arkansas, and Missouri, but it skips over the lowlands of the Mississippi River Valley, including the ecoregions of the Interior River Valleys and Hills and the Mississippi Alluvial Plain and most of the Mississippi Valley Loess Plains, to resume inhabiting the Southeastern Plains and the Interior Plateau and beyond.

Figure 12.100 **The Aurora Damsel is a species of eastern Canada and the United States. Oklahoma and southern Manitoba, Canada, make for the westernmost records of the species.**

TELEBASIS SALVA (HAGEN, 1861) – DESERT FIRETAIL

The Desert Firetail is a local, but relatively common, species of ponds, shallow marshes, watery ditches, and narrow, slow-moving streams. In a study on Cowan Creek, Marshall County, GH and JC Bick found the species to be strongly associated (65%) with a pond they described as "partially shaded … heavily vegetated, clear, shallow, [and] mud-bottomed" (Bick and Bick 1958). We have found the species at such ponds as well as beaver ponds and even sunlit, cement-lined ditches with shallow water and little vegetation, such as on the University of Oklahoma Norman campus near the Law School, where we have recorded large numbers of individuals for the species in the state, including the highest count of 50 individuals (30♂, 20♀, including 12 pairs, 11 August 2009). Other manmade environments that the species appears to do well in are at lakes having trickle spillways and/or marshy areas below the dams, such as that below Stroud Lake in Creek County.

The increase in the human footprint across the landscape, in the form of urban settings and dam spillways, may have benefitted this species. We say this because of the apparent spread of *Telebasis salva* across Oklahoma in recent times, where it is now known from 52 counties (Figure 12.102). The species was reported only nine times from eight counties between 1929,[50] when it was first reported, up to 1937. During the Bick and Hornuff era, records came from 14 counties, 11 of which were new for the species. Since 2002 it has been reported in 46 counties, 33 of which were new. Even so, the species remains most common in central Oklahoma and to some extent in south-central and southwestern Oklahoma. Outside of that more-or-less central region, the species is found sporadically and typically only in small numbers (one to a handful at a time). Areas lying on the species' overall current range limits, such as the Oklahoma panhandle, may only see vagrants or may currently only hold intermittent populations. Cimarron County, for example, despite multiple surveys during the Bick and Hornuff era and regular surveys in recent times, has only one record, a lone ♀[51] (4 August 1970, FSCA). Beaver County has but two records: 3 August 1970 (1♂, 1.4 mi N of Elmwood, GHB, LEH, FSCA) and 15 September 2012 (5♂, 1♀, including 1 pair, 13 km NE of Slapout, Kiowa Creek). Other counties in northwestern Oklahoma have

Figure 12.101 *Telebasis salva*, the Desert Firetail, A) ♂, University of Oklahoma Biological Station, Marshall County, 1 October 2016 (©Bill Carrell, OC 456614). B) tandem pair, Jenks, Tulsa County, 15 August 2018 (©James W Arterburn, OC 489778).

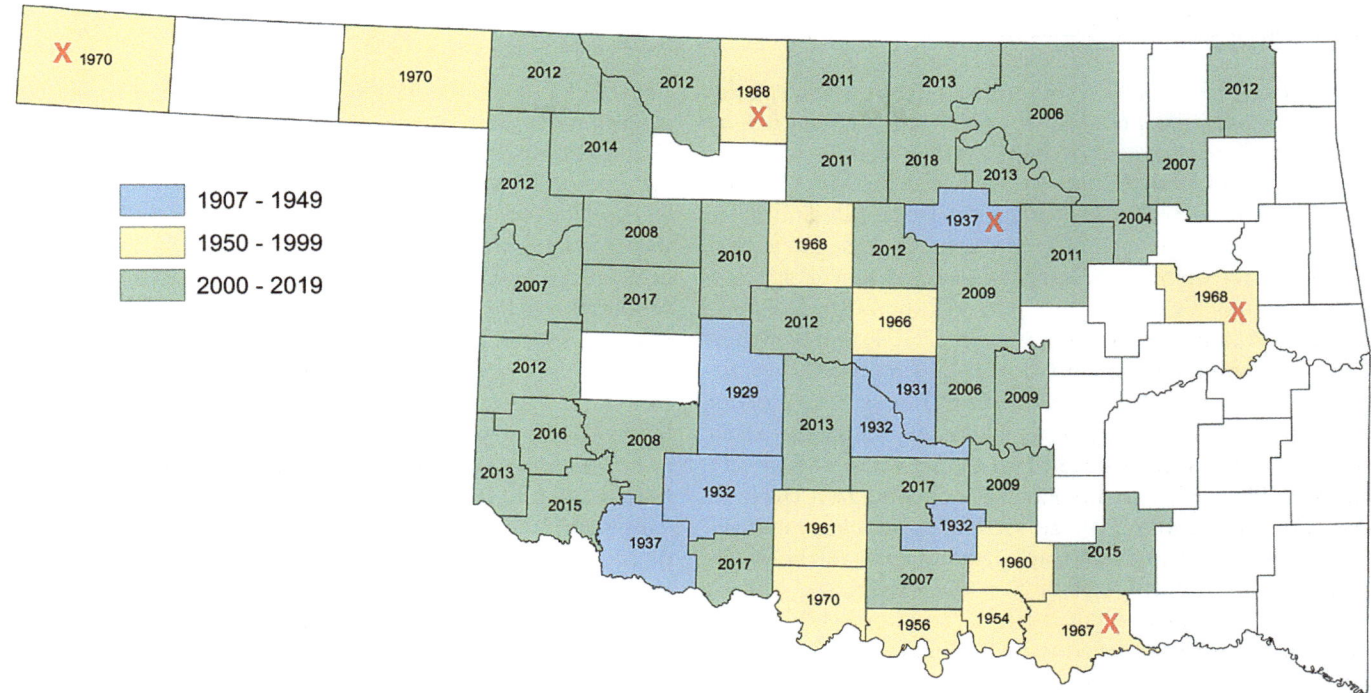

Figure 12.102 Counties in which *Telebasis salva*, the Desert Firetail, are known to occur in Oklahoma. Year within the county is that in which the species was first reported. There are 5 counties in which the species has not been reported for about 50 years or more (red X's).

similarly few records, as do those across the northern border of the state across to the northeast. At this point it is difficult to determine if these scant records indicate vagrancy, rarity, survey effort bias, or range expansion, but we have a niggling feeling that the species' range is pushing outward.

Telebasis salva is known to fly in Oklahoma and New Mexico from early summer to mid/late fall but is known year-round in Texas (OK: 6 May 2012, 15♂, 3♀, Stroud Lake, Creek County and 8 November 2013, 1♂, Hunter Park, Tulsa, Tulsa County, B. C., OC 411647; NM: 27 May–19 October, OC, iNat; TX: Abbott and Lasley

2019). In Kansas, it is known only from limited records (19 July–11 September, SEMC). Reports of tenerals in Oklahoma have not been forthcoming, the only record being from the Bicks of four tenerals that they collected on 19 June 1955 along "Croceipennis Creek" in Willis, Marshall County, but which they discarded. We have most certainly seen tenerals but have regrettably failed to record them in our notes. On the other hand, pairs are reported regularly, and ovipositing has been reported somewhat regularly, both as single females or (most often) tandem pairs.

TELEBASIS BYERSI WESTFALL, 1957 – DUCKWEED FIRETAIL

The Duckweed Firetail, as its name suggests, is a species of swampy areas carpeted with duckweed (such as *Lemna minor*). It is not surprising then that the species was first reported from Red Slough WMA, McCurtain County, where there are shady, still, duckweed-filled ditches (Figure 12.103). Perhaps more surprising, though, is that a decade after discovery Red Slough remains the species' only known locale in the state. On 14 June 2010, DA photographed a pair (OC 320802; Figure 12.104) and captured a lone ♂ (JCAC). The following year, DA reported the species again: he saw a single ♂ on 2 June. The species was not seen again until 26 June 2014, when DA reported seeing four adults. We visited the site on 1 July, when we saw and captured 1♂ (SP 1304). DA saw the species again twice in 2014, the last time on 10 July. In 2015, the region experienced severe flooding that washed away most of the duckweed from the Red Slough ditches the species was known to occur at and presumably pushed the known Oklahoma population elsewhere because DA, despite regular monitoring of the site, did not refind the species until 2017, when he reported it from 15 June until 13 July. On the earlier

date, MAP collected 1♀ (SP 2393) from the 4 individuals seen that day (2♂, 2♀, including one pair). In 2018, the species was reported multiple times, the first on 31 May and the last on 11 October; the latter (1♂, OC 492369) stands as the late date for the species' seasonality in Oklahoma, whereas the early date is of 9 individuals, including a teneral, on 14 May 2019 (OC 494769). Later that year, MAP reported the high count for the species in the state of 23 adults (21♂, 2♀, including 1 pair, 15 June, MAP).

In nearby states, *Telebasis byersi* is known to have a shorter flight season (AR: 21 June–2 October, OC; LA: 10 May–9 September, OC; TX: 14 May–28 September, Abbott and Lasley 2019). There is but one record of a teneral for Oklahoma (see above), and there are two records of ovipositing that indicate that the species will do so as tandem pairs or as a single female. Although it is relatively clear now that the species breeds regularly in Oklahoma, it is less clear whether a population will persist here given that Oklahoma lies on the very outskirts of the species' overall range. However, if suitable habitat is maintained at Red Slough and nearby areas, we suspect the species will persist in the state.

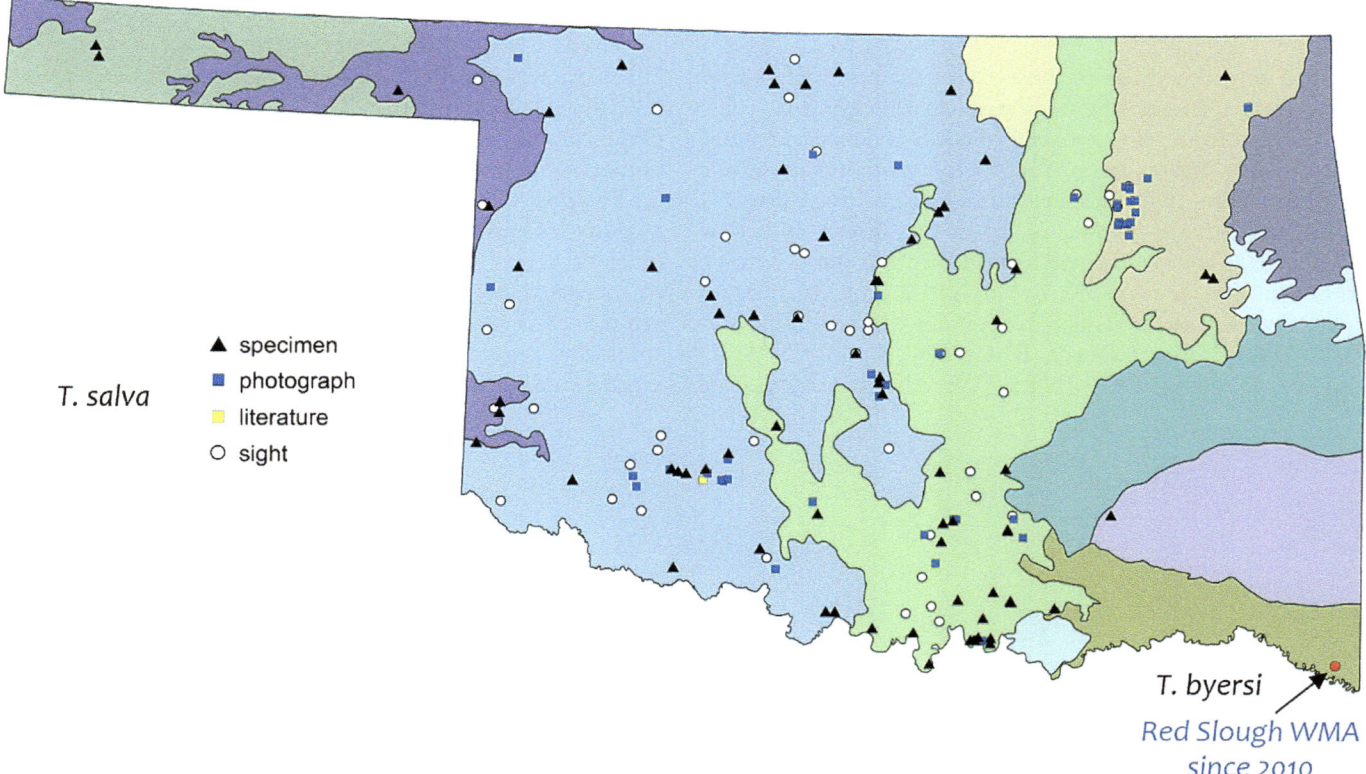

T. salva

▲ specimen
■ photograph
■ literature
○ sight

T. byersi

Red Slough WMA
since 2010

Figure 12.103 Documentation level of supporting records for *Telebasis salva*, the Desert Firetail, in Oklahoma. Levels include specimen, archived photograph, literature reference, or sight record. Although sight records lack documentation, only those we feel are credible are plotted. Also pictured is the sole location known (red dot) for *Telebasis byersi*, the Duckweed Firetail, in Oklahoma. Unlike, *T. salva*, which is found throughout much of the state, *T. byersi* is limited to the South Central Plains ecoregion.

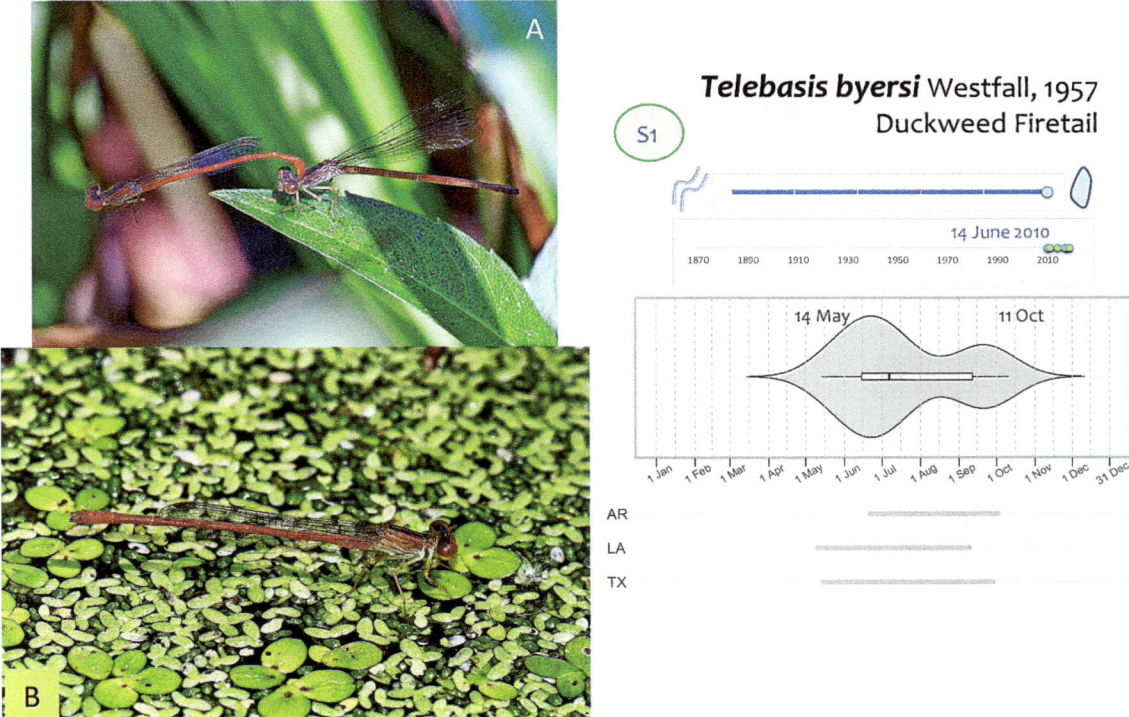

Figure 12.104 Oklahoma's first record of *Telebasis byersi*, the Duckweed Firetail. This tandem pair was photographed on 14 June 2010 at Red Slough WMA, McCurtain County, (USDA Forest Service; David Arbour, OC 320802). Red Slough harbors a small, persistent population, with numerous other documented records for the site, such as B) this ♂ on 17 September 2018 (USDA Forest Service; David Arbour, OC 490600), yet this species remains unknown from anywhere else in the state.

NEHALENNIA INTEGRICOLLIS CALVERT, 1913 – SOUTHERN SPRITE

The Southern Sprite is known in Oklahoma from just over 20 records from 9 locations in 5 southeastern counties. It was first discovered by AE Pritchard on 28 April 1934[52] at Antlers, Pushmataha County (1♂, 1♀, EMEC 331256, which were mislabeled as "*Enallagma* sp."). This date still stands as the early date for the species. Later that same year, on 16 June, Pritchard collected an additional 2♂ and 5♀, also at Antlers.[53] It appears that neither Bick nor Hornuff ever saw the species in the state, as it was not reported again until 1998 when RJ Beckemeyer collected it at Clayton Lake SP, Pushmataha County, on 26 May (Beckemeyer 1998b). He did not indicate how many he saw.

The species was next encountered in 2009, when it was recorded in McCurtain County for the first time. On 5 September, BAH found a ♂ at Teal Lake, Red Slough WMA (OC 315026). A few days later, DA had three individuals, one of which he collected (8 September, OC 315084, JCAC 24071). BAH had the species again at least 3 other times that month, including on the 30th (OC 315615), which stands as the late flight date for the species in Oklahoma. DA had the species again in 2010 and 2011 (OC 319926 and 327951; GL Harp collected a specimen

in 2010), both times in May. He did not have the species in 2012, despite some surveys. One ♂ was found at Teal Lake on 20 June 2013 (MAP), which was the last time the species was reported from Red Slough or anywhere in McCurtain County.

On 2 September 2012, we had a fine field day when we encountered *Nehalennia integricollis* at two localities in two new counties: Choctaw (3♂, 2♀, Schooler Lake, SP 385 [pair], SP 386 [♀ found dead]) and Atoka (6♂, McGee Creek WMA, SP 387 [1♂]). Le Flore County's first record came on 15 June 2013 when we found 1♂ and 1♀ at Crooked Branch Lake below the dam (♂ as SP 699). The only teneral record for the species in Oklahoma also came from Crooked Branch Lake in the following year, when we encountered three teneral ♂ on 1 June along with three adults (2♂ and 1♀, one each as SP 1228 and SP 1229). Pairs have been reported five times, but we know of no instances of ovipositing.

Generally the species has been recorded one to a handful of individuals at a time, but we have two rather high counts of 11 and 19 individuals (8♂, 3♀, including 2 pairs, Pushmataha WMA, Pushmataha County, 3 June 2018, MAP; 16♂, 3♀,

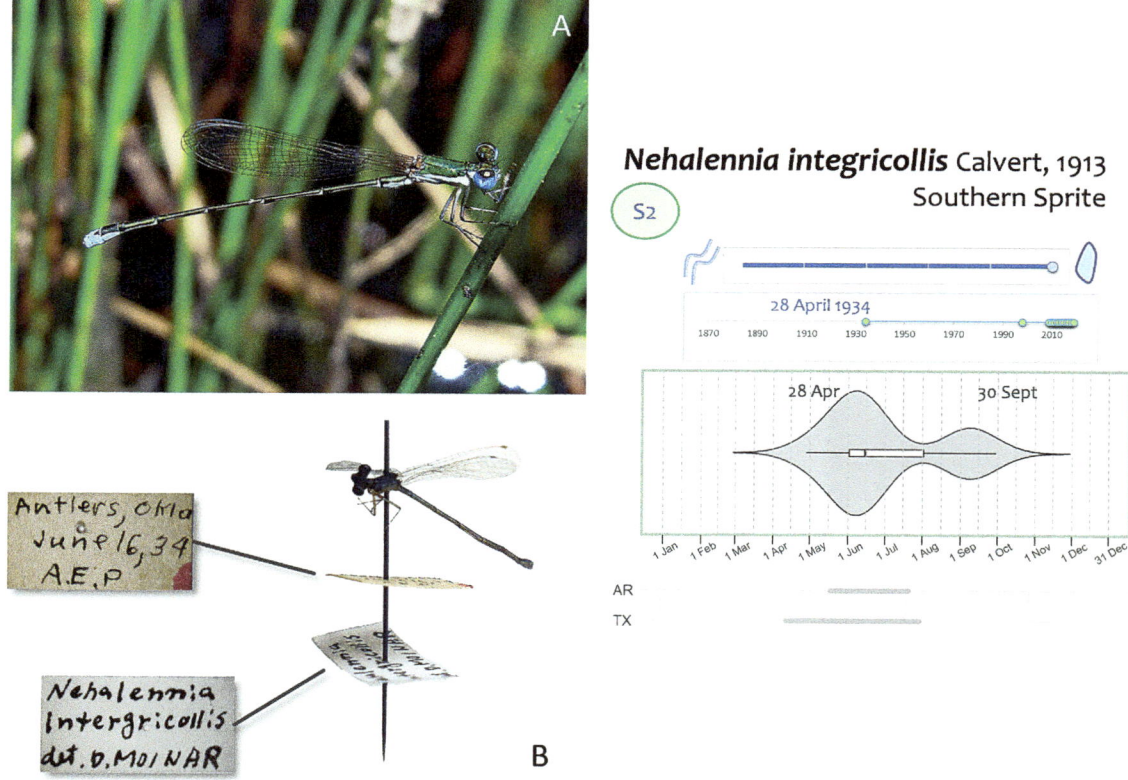

Figure 12.105 *Nehalennia integricollis*, the Southern Sprite, A) ♂, Red Slough WMA, McCurtain County, 8 September 2009 (USDA Forest Service; David Arbour, OC 315084). B), ♀, Antlers, Pushmataha County, 16 June 1934, one of AE Pritchard's specimens (OSU) dating way back to when he discovered that this dinky species ranges into southeastern Oklahoma.

McGee Creek WMA, Atoka County, 1 July 2017, MAP). Locations tend to be in boggy areas, particularly those associated with beaver pond complexes that have emergent vegetation or snags. Water can be fairly deep, knee-deep to quickly becoming chest deep or more, and ponds can have bog-like qualities such as floating mats of vegetation that one can walk upon; both of these characteristics are present at Schooler Lake, for example.

The species' flight season in Oklahoma (late April–late September) is more consistent with the timing in areas in the eastern part of the species' range (Paulson 2011), than it is to neighboring states (AR: 18 May–21 July, Harp and Harp 2003, OC; TX: 12 April–30 July, Abbott and Lasley 2019), which is unsurprising given the relative paucity of nearby records.

NEHALENNIA GRACILIS MORSE, 1895 – SPHAGNUM SPRITE

The Sphagnum Sprite[54] is known from <10 records, all from one locality in the state (Figure 12.106). It was first found on 3 June 2014 at TNC's Boehler Seeps and Sandhills Preserve, at Boehler Lake, Atoka County, when BS-P and JA Tucker spotted two ♂, and BS-P collected one (SP 1245; Smith-Patten and Tucker 2014) during a short visit to the Preserve. BS-P and MAP re-visited the site for a full survey on 6 June and found 8♂ and 5♀ (1 pair photographed, OC 422884, Figure 12.107; 2♀ as SP 1252–1253). Another visit by MAP on 13 June produced 7♂ and 1♀. The species was last seen that year on 20 July, when a single ♀ was reported. The Sphagnum Sprite has been reported from this location every subsequent year[55] in June or July with 3–5 individuals seen each time.

This sprite is an eastern species whose range does not extend terribly far inland from the eastern seaboard or south from the Great Lakes region. Until it was found in Oklahoma there were just three occurrences of the species known west of the Mississippi River—all disjunct populations, one each for Missouri, Texas, and Louisiana (see Smith-Patten and Tucker 2014 for details). It is now known from one additional population in Louisiana (OC). There are likely additional populations in the region, but because the species is habitat specific and the regional flight season is limited (OK: 3 June–20 July; LA: 3 April–18 May, OC, Mauffray 2014; TX: 12 April–15 June,

Abbott and Lasley 2019), they will, to say the least, be difficult to find.

In both Oklahoma and Texas, the Sphagnum Sprite has been found in forested areas with beaver ponds containing sphagnum. In Texas the edges of the ponds were described as "liberally covered" in sphagnum moss (Lasley and Abbott 2009). The Boehler Seeps and Sandhills Preserve is a 235-ha site that contains marshes, streams, sandhills, acidic hillside seeps, and two beaver-formed, shallow lakes: Hassell Lake in the north part of the preserve and Boehler Lake in the south (Clark 2011). Boehler Lake, classified as a semi-permanently flooded palustrine environment with broad-leaved deciduous scrub-shrub and persistent emergent vegetation (USFWS National Wetland Inventory), is found within the watershed of the Muddy Boggy and Clear Boggy Creeks, and has a surface area of <3 ha (McKnight et al. 2012). The lake is open at its center, but it has "dense stands of emergent and floating-leaved species" at its edges; the "dominant taxa include *Typha angustifolia* [narrowleaf southern cat-tail], *Nuphar lutea* [yellow water lily], and *Nymphaea odorata* [fragrant water lily]. Often quite abundant, free-floating species are *Azolla caroliniana* [Carolina mosquito fern] and *Utricularia biflora* [two-flowered bladderwort] (Clark 2011: 5)." The sponginess one feels when walking in the lake is from the carpets of vegetation that are dominated by *Sphagnum lescurii* (Lescur's sphagnum)

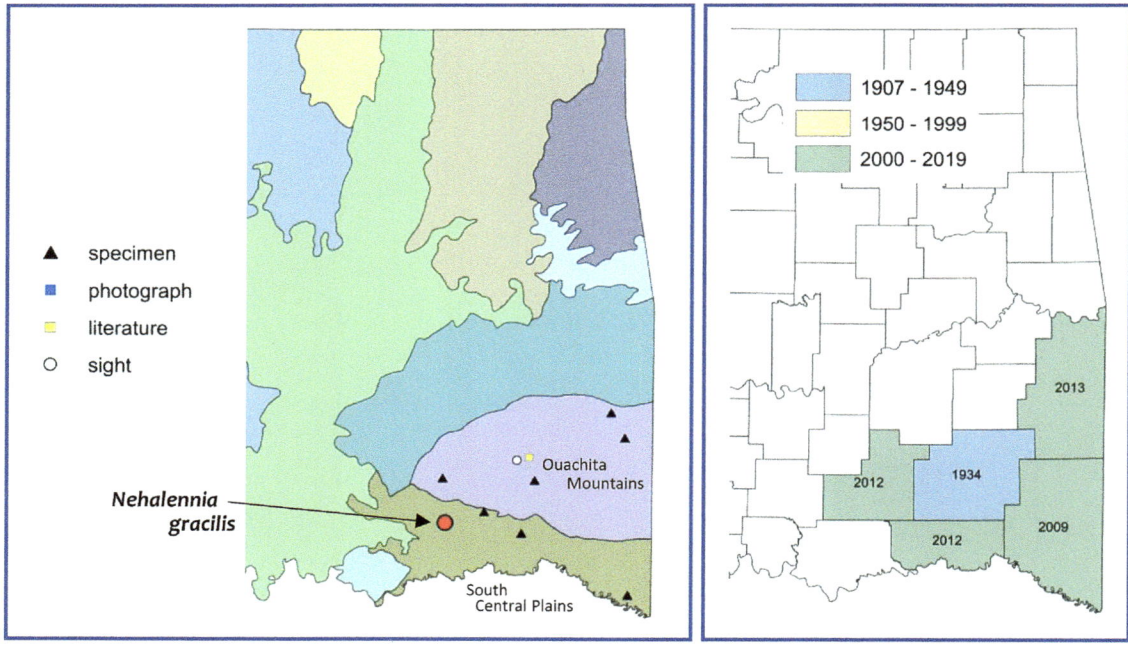

Figure 12.106 Documentation level of supporting records for *Nehalennia integricollis*, the Southern Sprite, in Oklahoma (left). Levels include specimen, literature reference, or sight record. Although sight records lack documentation, only those we feel are credible are plotted. Also shown is the sole known location (red dot) in Oklahoma for *Nehalennia gracilis*, the Sphagnum Sprite, a species that was first encountered in the state in 2014. The Southern Sprite has been reported from five counties in southeastern Oklahoma since 1934 (right; year in county is first encounter).

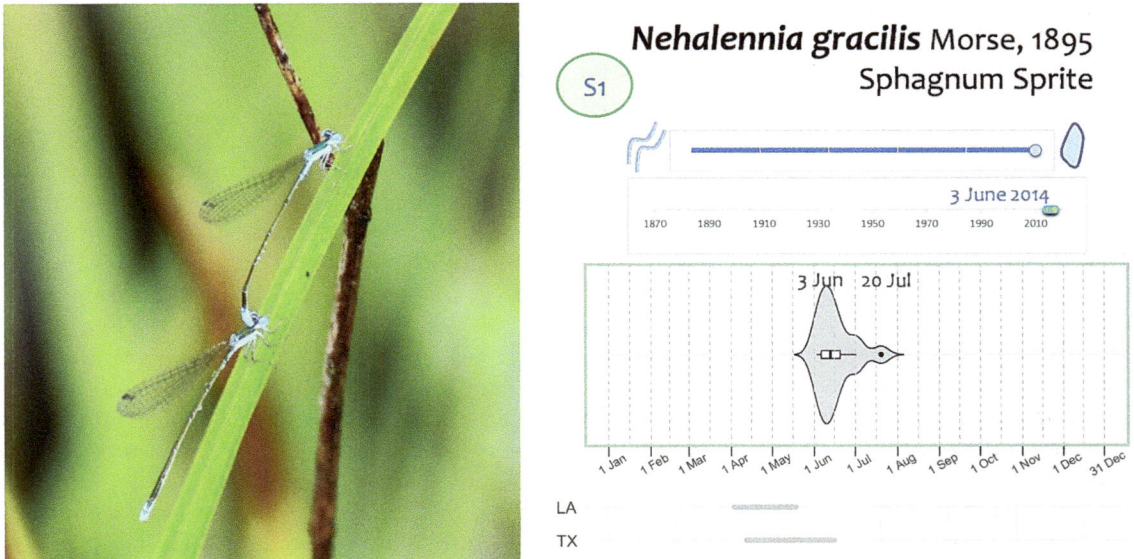

Figure 12.107 *Nehalennia gracilis*, the Sphagnum Sprite. This photograph of a tandem pair at Boehler Seeps, Atoka County, 6 June 2014 (©Michael A Patten, OC 422884), was taken just three days after the species was first discovered in Oklahoma (Smith-Patten and Tucker 2014). Boehler Seeps remains the sole sight from which the species in known in the state.

and *Polytrichum commune* (common haircap moss; Clark 2011). Sprites have been found solely on the eastern edges of Boehler Lake, where there is dense vegetation and extensive floating carpets. We have yet to detect the species at nearby Hassell Lake.

Other rare or relatively rare odonate species in Oklahoma that co-occur with *Nehalennia gracilis* at Boehler Seeps include the: Seepage Dancer (*Argia bipunctulata*), Azure Bluet (*Enallagma aspersum*), Gray Petaltail (*Tachopteryx thoreyi*), Regal Darner (*Coryphaeschna ingens*), Arrowhead Spiketail (*Cordulegaster obliqua*), Yellow-sided Skimmer (*Libellula flavida*), and Golden-winged Skimmer (*L. auripennis*). This suite of species attests to the high quality of the site and provide hope that *Nehalennia gracilis* will be found elsewhere in South Central Plains region of Oklahoma where similar odonate communities occur.

ARGIA BIPUNCTULATA (HAGEN, 1861) – SEEPAGE DANCER

The elusive Seepage Dancer is distributed spottily throughout the eastern part of the state, where it is reported from only ten counties. It was first discovered in Latimer County, where RD Bird collected a whopping 27♂ and 6♀ on 8 June 1931.[56] He returned to the county a couple of weeks later and collected an additional 45 individuals.[57] In July, he and a student, WM Fisher, again returned to the county, this time collecting a *mere* 11♂ and 15♀.[58] In under two months, all told, they collected 104 adults![59] That was the last time the species was reported in Latimer County. Coincidence?

Argia bipunctulata was next seen and collected in Bryan County at the Southeastern State College in Durant (2♂, 1 July 1950, WH Silver, FSCA). In the following year, GH Bick collected the species in Muskogee County at 3 mi S of Braggs, which was probably at the former Oklahoma State University Wildlife Conservation Station, where he collected when he began to work in the state (23 and 30 June, 4♂, 1♀, FSCA). Between 1954 and 1968 Bick recorded the species seven times in Marshall County in the area around the University of Oklahoma Biological Station (FSCA; OMNH 596; USNM 357555, 357556) and, in 1970, he collected it once in Choctaw County (1♂, 2.1 mi N of Sawyer, 27 June, FSCA). As with Latimer County, the records for

Bryan (1950), Muskogee (1951), and Choctaw (1970) are the only ones for those counties.

Contemporary records of the Seepage Dancer include a handful of sightings and one specimen from McCurtain County. Roughly half of these records are from one seep ("Heck Property") located 10 km SE of Idabel. This seep produced the early and late flight dates for Oklahoma (see below) and was productive for the species until 2011 (seen in 2007 on 3 visits, 2008 on 15, 2009 on 3, 2010 once, and 2011 twice), when it was last seen there on 26 May (photo of pair, on file OOP). In recent years the seep has dried. The only other location in McCurtain County the species is known from is the northeastern portion of Grassy Slough WMA, where we collected a ♀ on 19 June 2013 (SP 720) after the species had not been seen there for years (14 June 2007, 1♂, 1♀, DA, OC 7761), but it has not been seen since.

Two localities in Tulsa County—at a seep at an apartment complex in Tulsa proper and at the Keystone Dam—produced the species in 2011 and 2012 (OC 331508, 375621). *Argia bipunctulata* was reported for Le Flore County by Donnelly (2004b), but we have found no documentation for this record, so it may be erroneous; even so there certainly is habitat in the county that could support the species, and

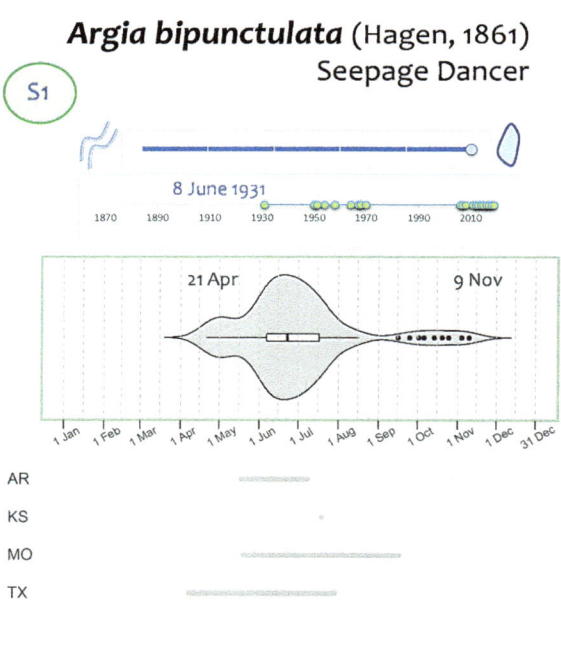

Argia bipunctulata (Hagen, 1861)
Seepage Dancer

Figure 12.108 *Argia bipunctulata*, the Seepage Dancer. This species is scarce in Oklahoma: we know of only two strongholds, one at Boehler Seeps, Atoka County, and one at Wewoka Woods, Seminole County (Figure 12.109), where A) this ♂ was photographed on 9 July 2019 (©Hans Holbrook, iNat 29446317). Away from these strongholds, the species is encountered sporadically and infrequently, as was B) this ♀ at Keystone Dam, Tulsa County, 14 August 2011 (©Ken Williams, OC 331508).

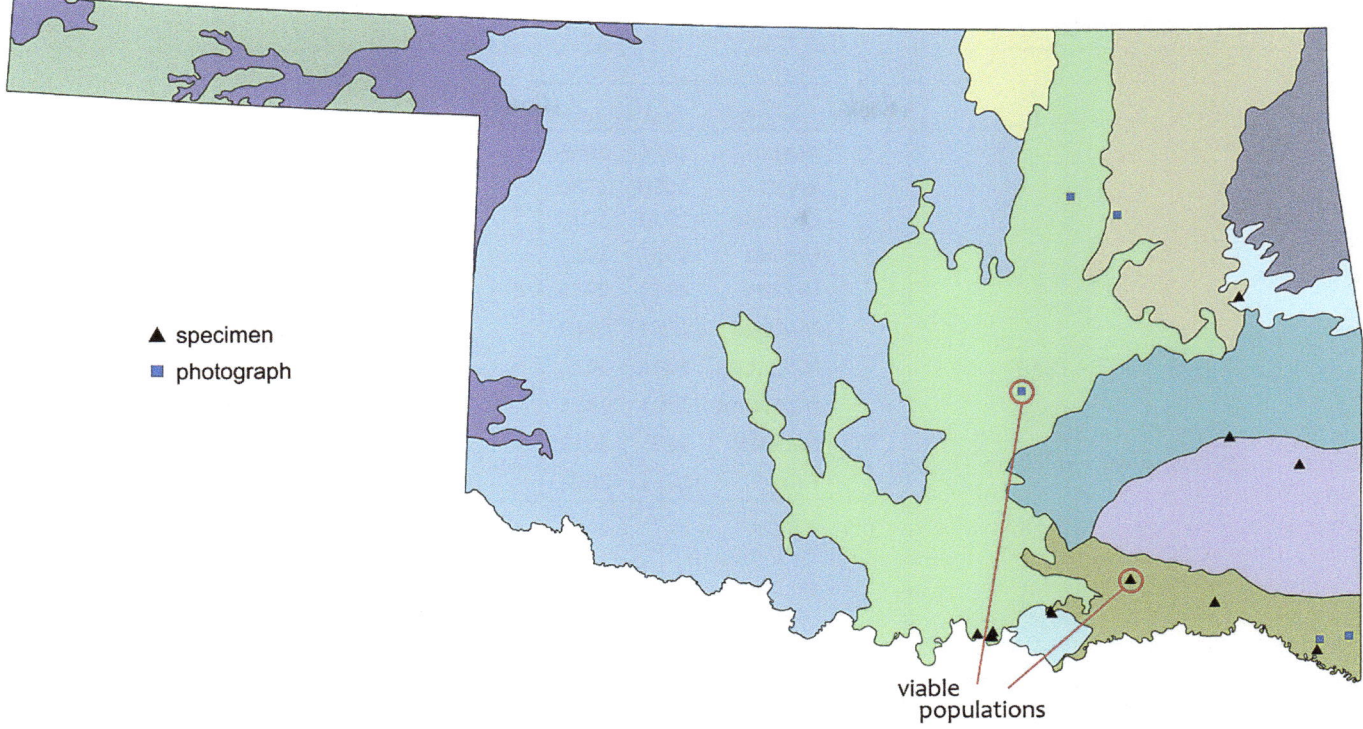

Figure 12.109 Documentation level of supporting records for *Argia bipunctulata*, the Seepage Dancer, in Oklahoma. We know of but two viable populations in the state, others, such as that known from the Heck property in McCurtain County, appear to have blipped out with the drying of the property's seep.

in 2016, MAP documented a lone ♂ at a pond along the Talimena Highway (18 km E of Talihina, SP 2010). A report for Creek County, from a beaver dam complex at Deep Fork WMA, could represent another recent record for *A. bipunctulata* but it remains a tentative sighting because the observer was not fully confident (JT Bried, 2012). But a photo of an adult ♀ documented the species in Seminole County (Wewoka Woods Camp, 11 August 2006, H Holbrook, iNat 9666550).

The species would seem to hang by a thread in the state, were it not for Atoka County, which holds hope for the species' persistence in Oklahoma. The species was added to the county in 2014, when it was found at Boehler Seeps Preserve (1♂, 3 June, BS-P and JA Tucker, SP 1246). Currently, that site is the stronghold for the species in the state: it has been seen there >10 times, in numbers estimated between 36–78 individuals, including reports of pairs. Wewoka Woods, in Seminole County, ought to be considered another stronghold given that the species has persisted there for 14 years and large numbers have been reported (30–40 adults, including at least one pair, on 9 July 2019, H Halbrook, iNat 29446317) (Figure 12.108).

The Oklahoma flight season (late April–early November) is much longer than that known for neighboring states (OK: 21 April, 1♂, 4U, all teneral, OC 282046 and 9 November, 2♂, 1♀, including one pair, OC 284439, both records from 10 km SE of Idabel in 2008 by BAH; AR: 18 May–8 July,

JCAC, USNM; KS: 19 July, SEMC; MO: 20 May–17 September, Trial 2005; TX: 8 April–30 July, Abbott and Lasley 2019). The length of the season in Oklahoma, being 202 days, compares to nearby Louisiana, with 200 days, but it is known there to start flying much sooner (14 March–30 September, OC). Elsewhere in its range, the Seepage Dancer flies principally from May to September (Paulson 2011). There are only a small handful of late flight dates known from anywhere in the species' range that are within October and there are none from November (OC, iDigBio, SCAN, iNat, Paulson 2011) except from Oklahoma.[60] Admittedly, the truncated flight seasons elsewhere could be a product merely of the onset of autumn, a time when most of us give up going to the field because of the prevalence of cold, rainy, unproductive days. Tenerals have been reported in Oklahoma from late April until late July. Pairs have been reported fairly regularly, although no one has reported seeing ovipositing.

This species remains one of conservation concern in Oklahoma (Patten and Smith-Patten 2013b). There are several reasons for this designation: scant records, possible blipping out of the species at some locations, habitat loss from drought and water use, and that there currently are but two reliable locations for the species. We recommend that this species be monitored closely where it is known to occur, and we call for intensive surveys to determine if additional sites harbor populations of the species.

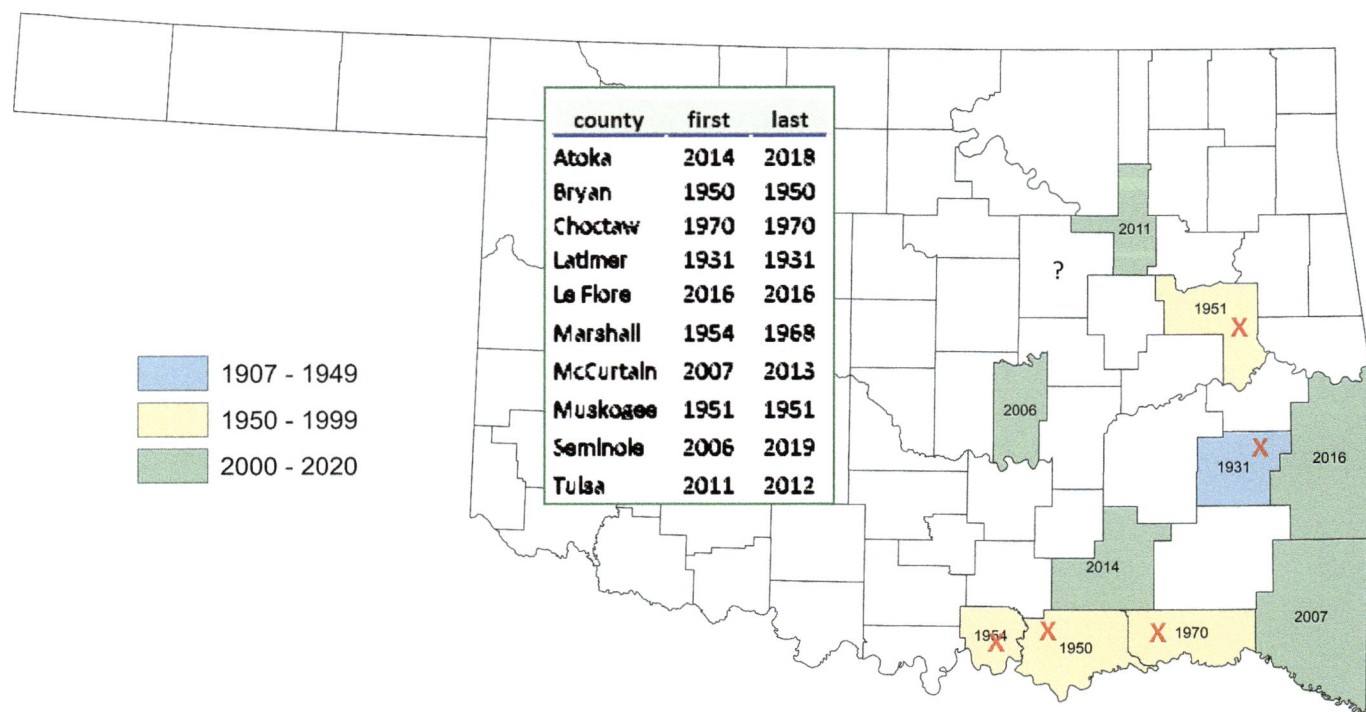

county	first	last
Atoka	2014	2018
Bryan	1950	1950
Choctaw	1970	1970
Latimer	1931	1931
Le Flore	2016	2016
Marshall	1954	1968
McCurtain	2007	2013
Muskogee	1951	1951
Seminole	2006	2019
Tulsa	2011	2012

1907 - 1949

1950 - 1999

2000 - 2020

Figure 12.110 Distribution by county of *Argia bipunctulata*, the Seepage Dancer, in Oklahoma. Year within the county is that in which the species was first reported. The report for Creek County is a tentative sighting (?) from 2012, so it is not counted among the 10 counties in which the species is documented. In McCurtain County, once thought to be the stronghold for the species in Oklahoma, it has not been seen since 2013. In 5 counties, the species has not been encountered in 50 or more years (red X's).

ARGIA APICALIS (SAY, 1839) – BLUE-FRONTED DANCER

The Blue-fronted Dancer has been recorded throughout the entire state, the first time in 1907 when EB Williamson and F Collins collected 8♂ and 1♀ at Wister, Le Flore County (1♂, UMMZ, Williamson 1914b). It has been reported regularly ever since—reports are nearing the 1,000 mark. GH Bick and JC Bick said the species was found "throughout the state and was the most frequent *Argia* [encountered] in Oklahoma" (Bick and Bick 1957:9). This supposition is debatable, but *A. apicalis* certainly is a common species, routinely seen >50 individuals at a time (highest count is of about 200,

Figure 12.111 *Argia apicalis*, the Blue-fronted Dancer, A) ♂, Lake Hall, Harmon County, 3 July 2016 (©Bill Carrell, OC 448266). Oddly, when the ♂ of this species is chilled or in tandem with a ♀ (Why would those phenomena be the same?), he becomes a muddy purple rather than a vibrant sky blue: B) ♂, Enid, Garfield County, 3 June 2017 (©Bill Dobbins, OC 478258). C) An occasional ♂ has excess black on the thorax, as does this one at Skipout Lake, Roger Mills County, 4 October 2014 (©Bill Carrell, OC 427361). D) ♀, 6 km SW of Ames, Major County, 10 July 2010 (©James W Arterburn, OC 427306). The ♀ typically is brown or tan, rarely blue. In our experience, a blue ♀ of this ilk on the southern Great Plains is more likely to be *A. moesta* (see Figure 12.114C).

including 15 pairs, at Mudeater Bend, Neosho River, Ottawa County, 20 June 2016). Moreover, data suggest that the number of individuals expected during an encounter with the species has increased over time, although it has been outpaced by *A. moesta*, the Powdered Dancer and the species that currently vies with (and likely beats) *A. apicalis* as the most common *Argia* in the state (Table 12.1; Figure 12.14)

The Bicks' experience with *Argia apicalis* at Cowan Creek, Marshall County, indicated that it was rather strongly associated (77%) with a muddy, turbid impoundment having little vegetation (Bick and Bick 1958). We also have encountered it in such situations but the species inhabits a larger variety of habitats including sloughs, oxbows, lakes, and creeks with clear or murky water, flowing or not, with and without vegetation. Being such a habitat generalist undoubtedly has contributed to this species being so widespread and common.

Pairings are commonly noted, and one mixed-species pair was reported, with an *Argia fumipennis* (see that species account). Ovipositing is done both in tandem and alone. The Bicks noted that ovipositing took place at "willow root mats at the surface, horizontal boards and sticks encrusted with periphyton …, and horizontal *Helenium* lying in the water and still green" (Bick and Bick 1965:467). These sites

were described as being "in order of frequency of utilization" and use partly depended on water level, as it appears that when water levels were high oviposition tended to be on boards and sticks but later in the season willow roots were preferred. Males of various *Argia* species darken when in tandem with a female, but perhaps none as strikingly as male *A. apicalis*, whose thorax can appear to be a grayish violet rather than vibrant blue and thus innocently send an unwary observer down a path toward misidentification. Also, some individuals are blacker than others, such as an aberrant ♂ photographed by BC at Skipout Lake, Black Kettle NG, Roger Mills County (4 October 2014, OC 427361; Figure 12.111).

As expected with this species, it flies in the region for the bulk of the odonate season, from spring to autumn (OK: 8 April 2007, Tishomingo NWR, Bell Creek, Marshall County, 1♂, 1♀, MAP and 16 October 2016, Hickory Creek WMA, Love County, 3♂, 2♀, including 1 pair, MAP; AR: 2 May–8 October, OC, iNat; CO: 8 June–22 September, CSU; KS: 14 May–27 September, OC, SEMC; LA: 9 April–24 November, iNat, Mauffray 2014; MO: 14 May–15 October, Trial 2005; NM: March–October, Paulson 2009; TX: year round, Abbott and Lasley 2019). Tenerals are known in Oklahoma from early May to late August.

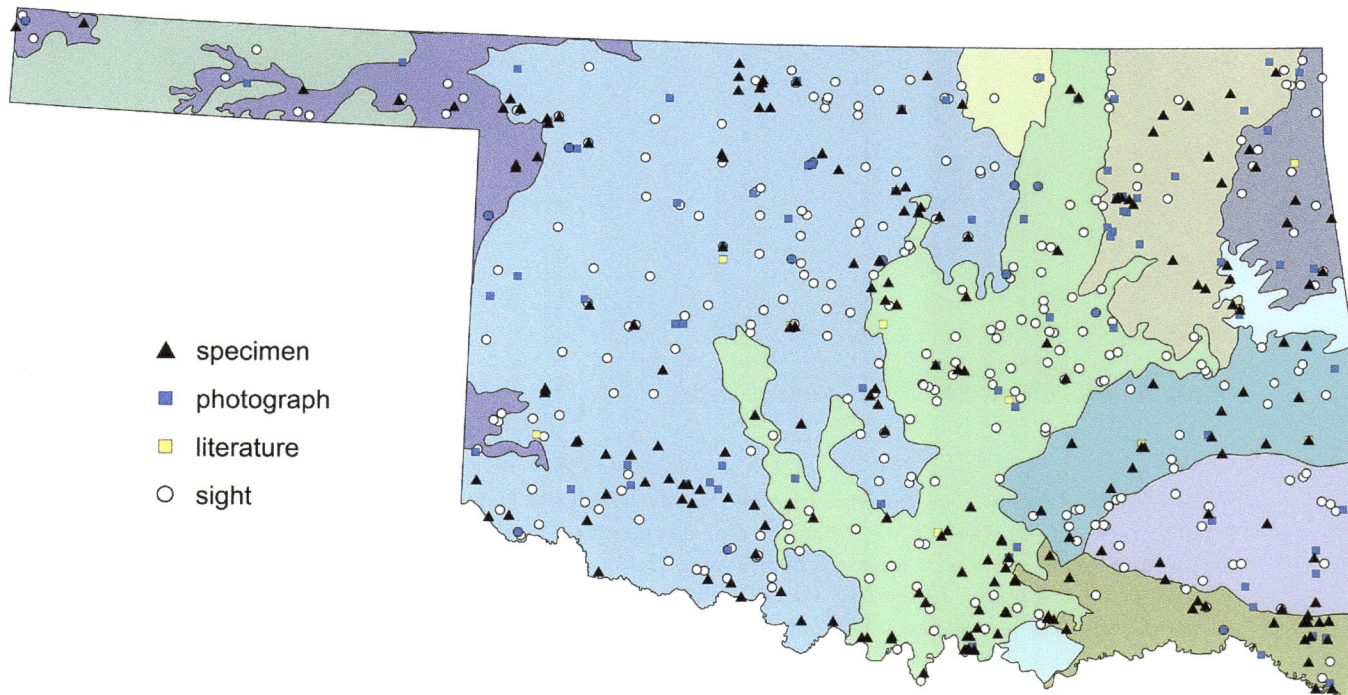

▲ specimen
■ photograph
□ literature
○ sight

Figure 12.112 Documentation level of supporting records for *Argia apicalis*, the Blue-fronted Dancer, in Oklahoma. Levels include specimen, archived photograph, literature reference, or sight record. Although sight records lack documentation, only those we feel are credible are plotted.

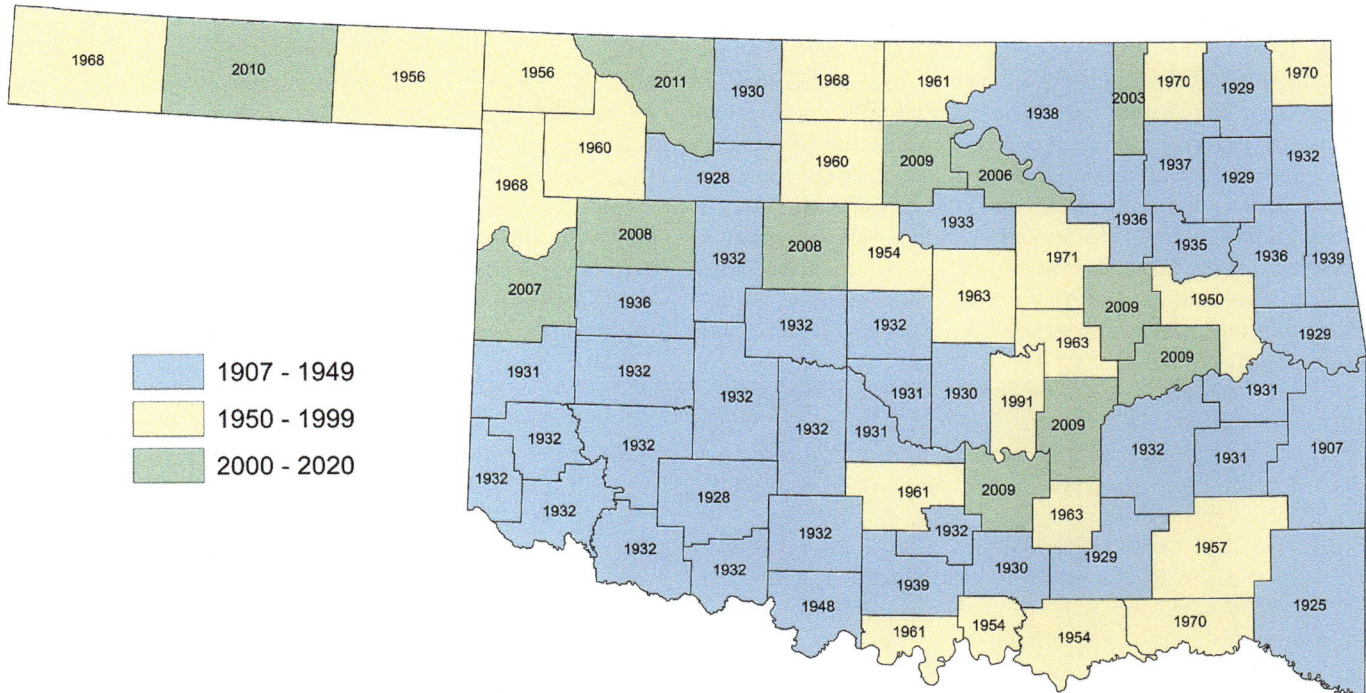

Figure 12.113 Counties in which *Argia apicalis*, the Blue-fronted Dancer, are known to occur in Oklahoma. Year within the county is that in which the species was first reported.

ARGIA MOESTA (HAGEN, 1861) – POWDERED DANCER

The Powdered Dancer is a species of rocky streams and lakeshores. It is found throughout Oklahoma, its distribution, and perhaps abundance, likely increasing along with the explosion of artificial lakes in the state after the Dust Bowl. Although creating some of these lakes resulted in loss of rocky lotic habitat, the appearance of riprap shorelines so common on the lakes' dams and spillways seems to function as surrogate habitat and may actually be preferred by this adaptable species. Our experience with the species contrasts with GH and JC Bick as at their study site at Cowan Creek, Marshall County, they encountered *A. moesta* at a "clear, sunlit, sandy, shallow creek devoid of vegetation" 95% of the time (Bick and Bick 1958).

This species was first noted in the state in 1907 when EB Williamson first visited Oklahoma and collected 1♂ on 3 June at Wister, Le Flore County (UMMZ; Williamson 1912b, 1914b). Like *Argia apicalis*, *A. moesta* has been reported regularly ever since, with >1,000 records known.

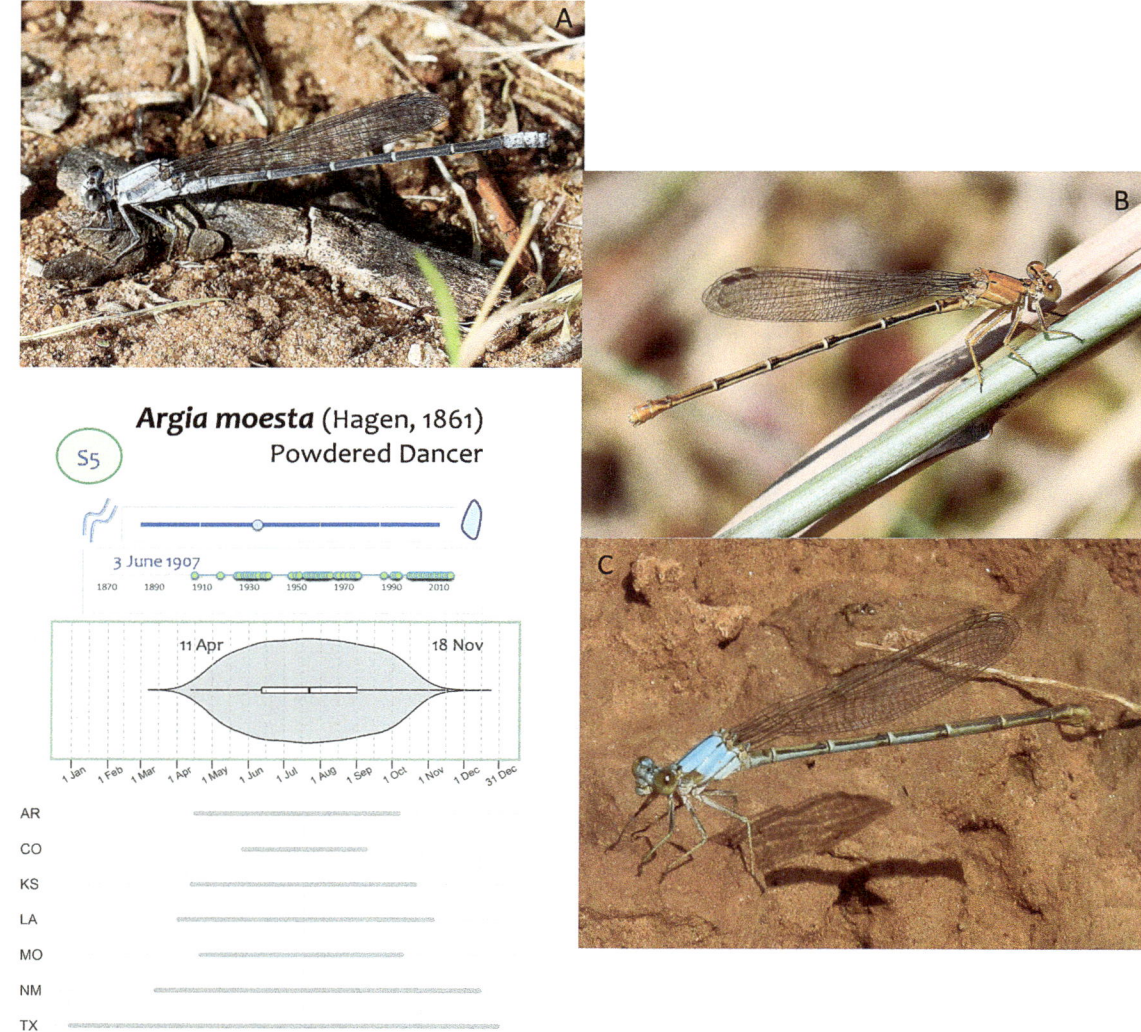

Figure 12.114 *Argia moesta*, the Powdered Dancer, A) ♂, Fort Supply Lake, Woodward County, 8 September 2017 (©Jeff Trahan, OC 472638). The ♀ comes in two "flavors," being either B) orange, as at Great Salt Plains Lake, Alfalfa County, 11 September 2017 (©Jeff Trahan, OC 472636), or C) blue, as at Beaver Creek 1 km S of Mulhall, Logan County, 21 May 2016 (©Bill Dobbins, OC 445423). Orange (or tan, when older) individuals predominate, but blue individuals are common, too.

It also appears to have increased some in its numbers over time (Table 12.1). The species is now often recorded in triple digits (high count is about 600): the first report of such high numbers came in 2006 (two reports of 100 in September by J Fisher and J Nelson).

The species' flight season in Oklahoma corresponds with elsewhere in the region—spring to late autumn (OK: 11 April–18 November, Smith-Patten et al. 2007, OC 263849; AR: 17 April–7 October, OC; CO: 28 May–9 September, OC; KS: 14 April–21 October, OC, SEMC; LA: 3 April–5 November, Mauffray 2014, iNat; MO: 22 April–10 October, Sims 2012; NM: March–December, Paulson 2009; TX: year round, Abbott and Lasley 2019). Tenerals in Oklahoma are known from at least 25 records, all of which are from late April to late September. Pairs are reported routinely (including up to 200 at a time), and ovipositing has been reported by single females and in tandem pairs. Bick and Bick (1972) reported that *Argia moesta* primarily oviposited in *Salix* (willow) roots, but they also noted several other places at which the species chose to oviposit.

Two final notes with *Argia moesta* in Oklahoma involve taxonomy and pairings. Ortenburger (1926b) used the name *A. putrida* in his species list, and Bird (1932) used both *A. intruda* and *A. moesta* in his accounting of Oklahoma odonates; all of those names relate to *A. moesta*. Lastly, there are two reports of mixed-species pairs, both with *A. plana* (*A. moesta* ♀ with *A. plana* ♂, Cowan Creek, Marshall County, 10 and 27 June 1964, Bick notes, specimens were discarded or lost).

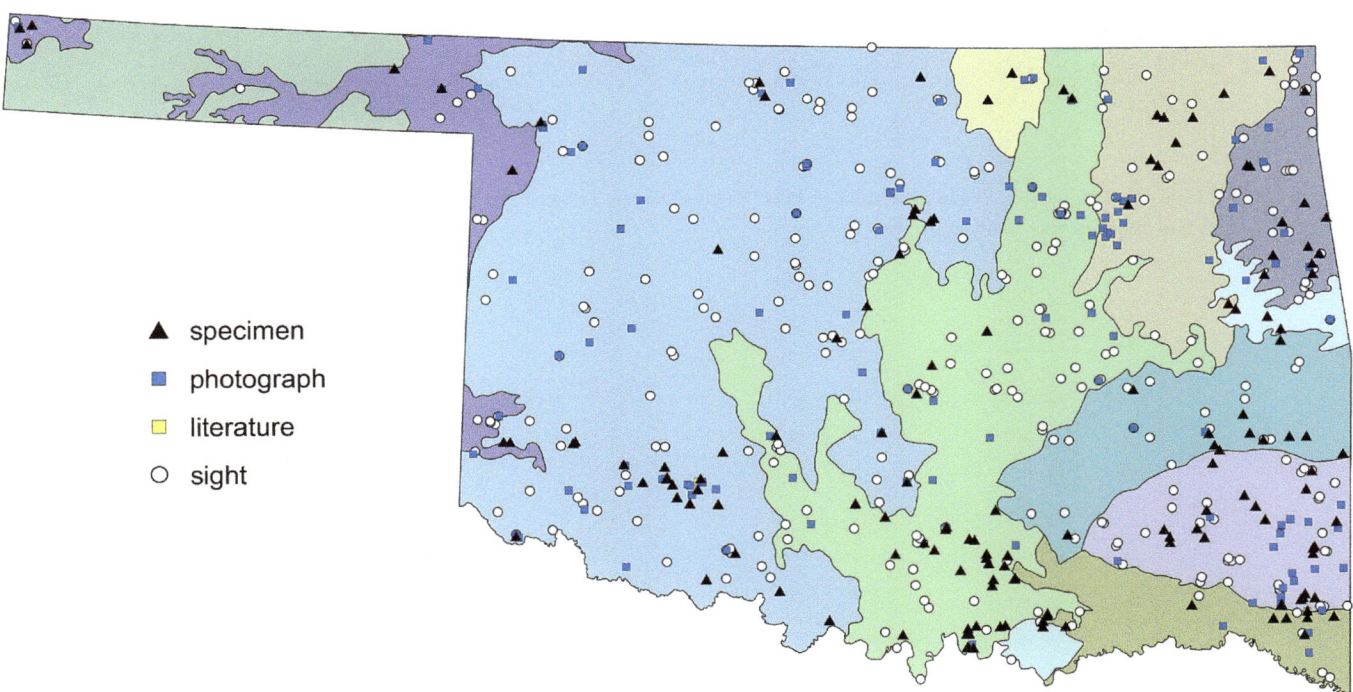

Figure 12.115 Documentation level of supporting records for *Argia moesta*, the Powdered Dancer, in Oklahoma. Levels include specimen, archived photograph, literature reference, or sight record. Although sight records lack documentation, only those we feel are credible are plotted.

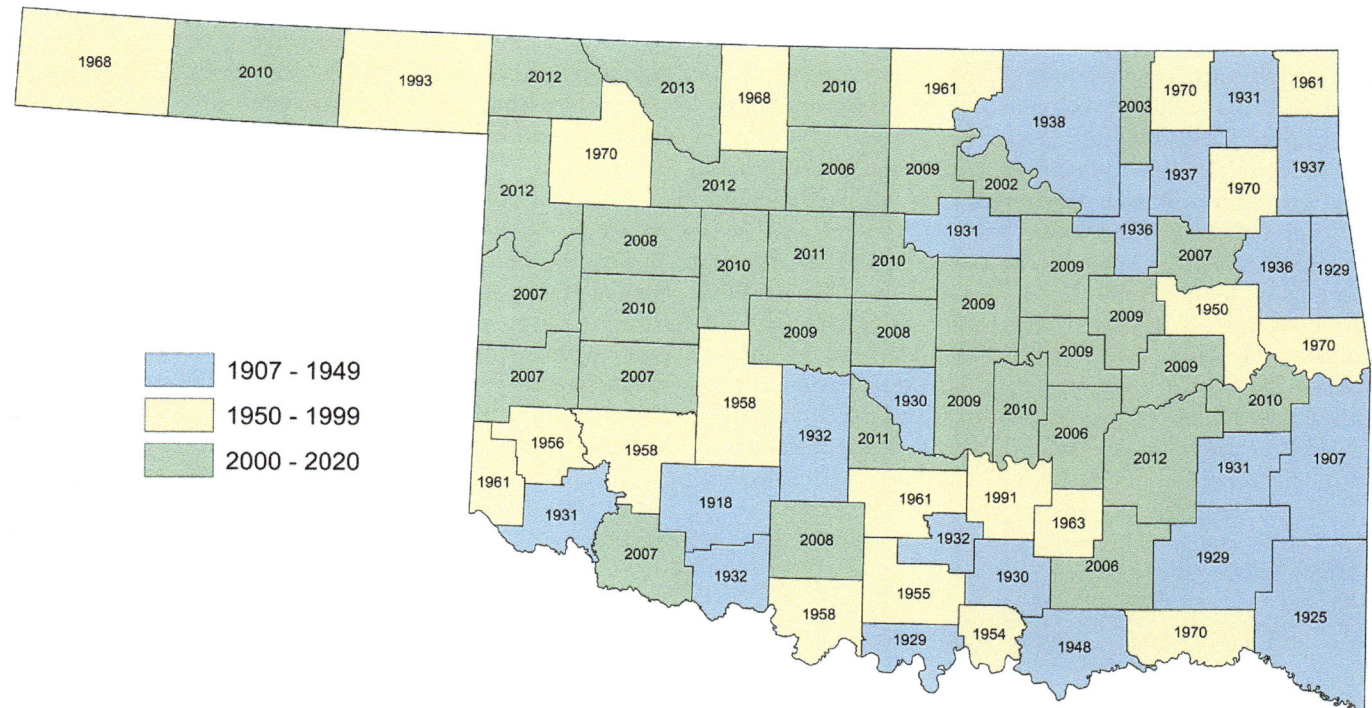

Figure 12.116 Counties in which *Argia moesta*, the Powdered Dancer, are known to occur in Oklahoma. Year within the county is that in which the species was first reported.

ARGIA SEDULA (HAGEN, 1861) – BLUE-RINGED DANCER

The Blue-ringed Dancer is primarily a lotic species, typically found perched in vegetation overhanging water (often clear and at least slightly flowing) along creeks throughout the state; although, it sometimes is found at lakeshores, ponds, and marshes if water is clear. It is rather common and widespread, although abundance declines from west to east. There are >650 records known for Oklahoma. Numbers are regularly in double and sometimes in triple digits—the high count is 360 individuals (200♂, 160♀, including 150 pairs, 11 km WNW of Thomas, Deer Creek, Custer County, 2 September 2017, MAP). That high count ranks the species second in terms of top high counts of *Argia* in Oklahoma, falling only behind *A. moesta* with its highest high count of 600 individuals. *Argia sedula* also ranks second in terms of highest number of counties it is found in: 74 (behind *A. apicalis* and *A. moesta*, which are tied for first with 77 counties each). The three

missing counties—Nowata, Ottawa, and Pittsburg—can be chalked up to the species being less common in eastern Oklahoma and may not be true absences.

The species was first recorded in Oklahoma by AI Ortenburger on 23 June 1925 in McCurtain County, 10 mi SE of Broken Bow on the Mountain Fork River (Ortenburger 1926b). No specimens associated with this record have been found, and he provided no details except to say that the river was "one of the large relatively clear streams of this section of the state" (Ortenburger 1926b:219). The earliest specimens (24♂, 9♀) known for the Blue-ringed Dancer in Oklahoma are those collected by the OU expedition in Johnston County, some at Oil Springs, others with no locality data. The majority of the specimens have 27 May 1930 as the collection date, but one has 22 May 1930, which may be an error.[61] Another error from the RD Bird-era collection is an adult ♂ originally identified as *Argia sedula* that

Figure 12.117 *Argia sedula*, the Blue-ringed Dancer. A) Typical view of this species, i.e., perched over clear, flowing water (♂, Weatherford, Custer County, 25 August 2013, ©Brenda D Smith). B) ♂, Caney Creek, 6 km NW of Stilwell, Adair County, 3 July 2017 (©CA Ivy, OC 465840). C) immature ♂, Ellis County WMA, Ellis County, 16 May 2017, SP 2320, ©Brenda D Smith. D) ♀, Mountain Fork Park, McCurtain County, 19 July 2011 (©Tom Kompier, OC 332305).

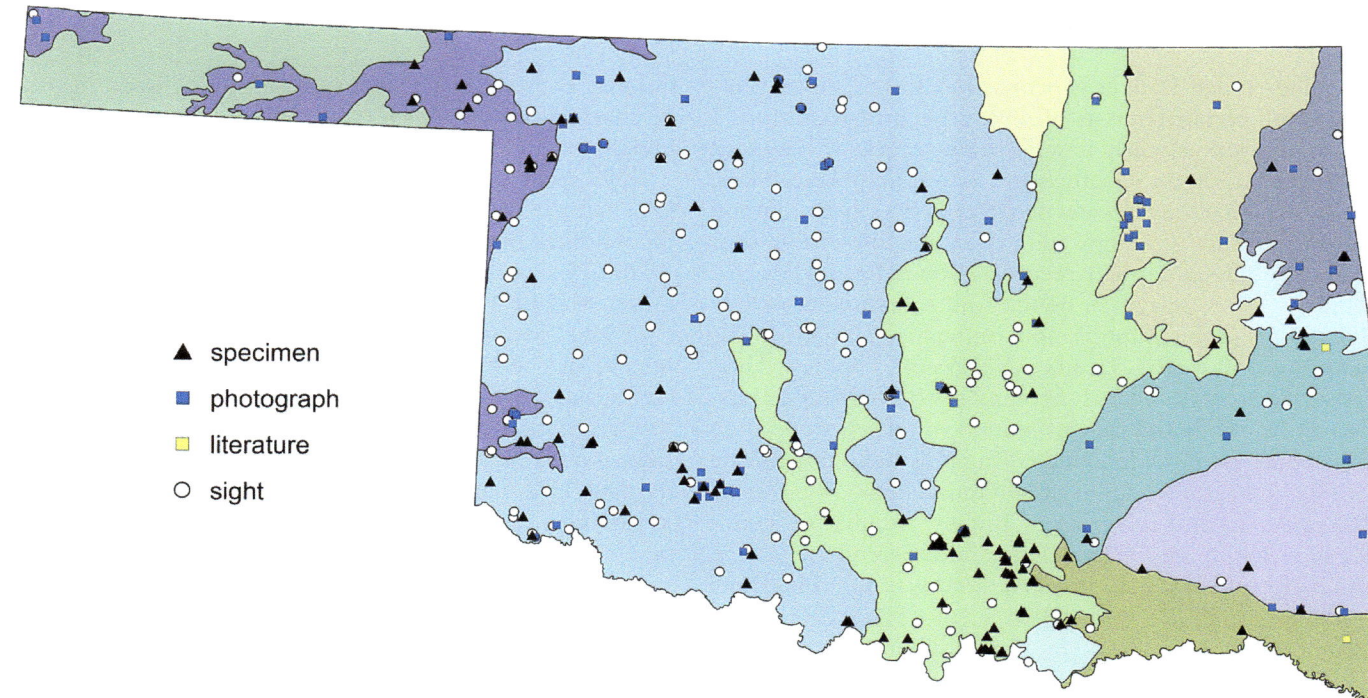

Figure 12.118 Documentation level of supporting records for *Argia sedula*, the Blue-ringed Dancer, in Oklahoma. Levels include specimen, archived photograph, literature reference, or sight record. Although sight records lack documentation, only those we feel are credible are plotted.

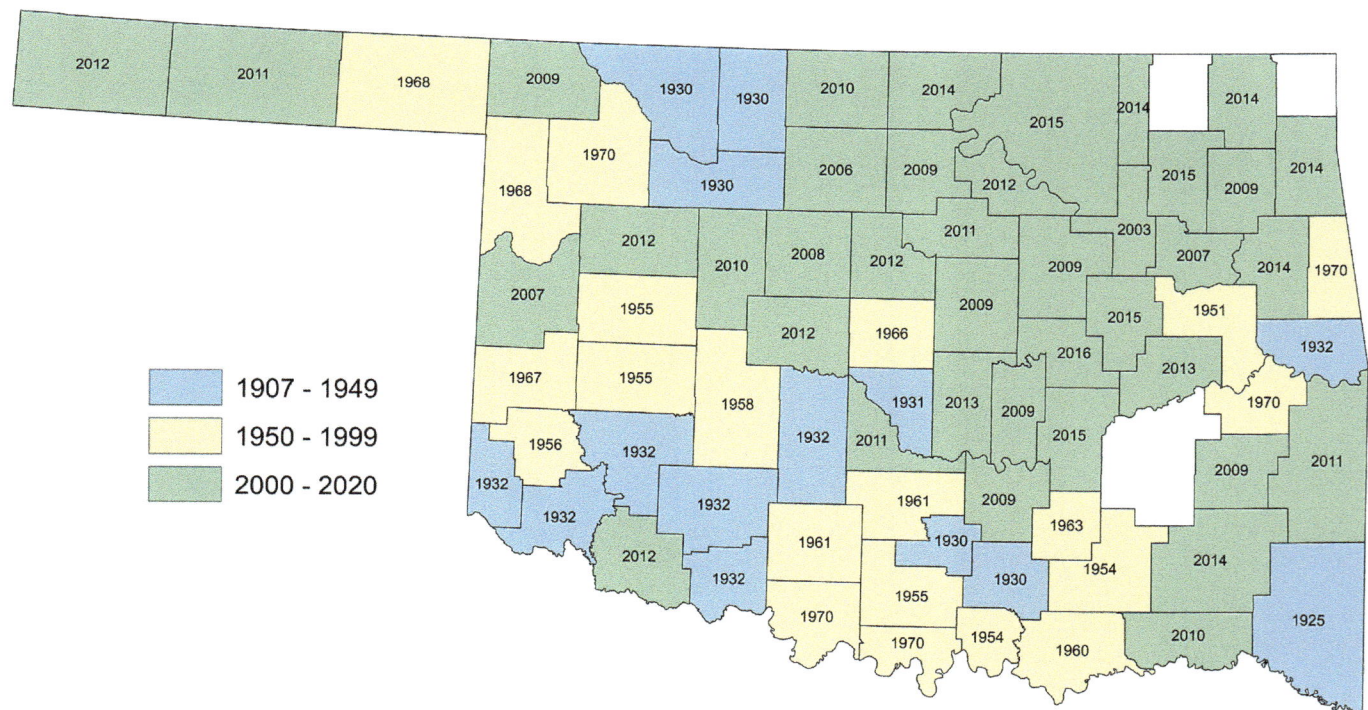

Figure 12.119 Counties in which *Argia sedula*, the Blue-ringed Dancer, are known to occur in Oklahoma. Year within the county is that in which the species was first reported.

proved to be *A. leonorae* (see that species account). Given that Leonora's Dancer was not described until the mid-1990s (Garrison 1994), it is not surprising that this error occurred, even if these two species differ considerably.

Encounters with this species may indicate that it has become more common in the state over time. The number of records per era roughly doubled between the early (pre-1950) and mid (1950–1999) eras, and then increased fivefold between the mid to late (2000–2018). Because those are just raw counts of records, it could be argued that they are too biased for trend analysis. When we accounted for survey effort in each era to approximate trend over time, we see that the expected average number of individuals encountered by era has increased: 4.8 individuals in the early era, 7.1 in mid, and 26.4 in late (Table 12.1). Additionally, species distribution modelling indicates the species has expanded its range from the southwest towards the northwest and northeast (see Chapter 7).

Oklahoma's flight dates for *Argia sedula* are similar to elsewhere in the region—generally from spring to autumn (OK: 10 April 1955, 1♀, Custer County, source is GHB notes from an unknown collector with the initials of BFS and 9 November 2012, 2♂, 1♀, including one pair, Government Springs Park, Enid, Garfield County, JR Heinen, OC 383419; AR: 13 May–18 September, OC; KS: 7 June–4 October, SEMC, OC; LA: 17 April–October, Mauffray 2014, Paulson 2011; MO: 3 May–23 September, Sims 2012; NM: March–November, Paulson 2009; TX: year round, Abbott and Lasley 2019). Pairs are commonly reported, usually 1–25 at a time, but once as high as 150 were estimated (see species high count above). One record was of a mixed-species pairing: GH Bick and his students collected a ♂ *A. sedula* in tandem with a ♀ *Enallagma basidens* in Marshall County on 19 June 1964. This pair was discarded by Bick, according to his notes. Only one record of ovipositing is known for Oklahoma: Bick reported in his notes that there was "abundant egg laying" on a day that he recorded 4♂ and 4♀ at 3 mi N of Tishomingo, Johnston County (8 July 1954). He did not report whether ovipositing was done in pairs or as single females nor was there mention of depositing substrate. Teneral records for Oklahoma are known from 11 May to 20 August.

ARGIA ALBERTA KENNEDY, 1918 – PAIUTE DANCER

The Paiute Dancer is strongly associated with flowing springs and associated marshes, which are typically dominated by spikerushes (*Eleocharis* sp.). It often co-occurs with the Black-fronted Forktail (*Ischnura denticollis*) and the Western Red Damsel (*Amphiagrion abbreviatum*). It is an uncommon species, normally encountered only in small numbers (1–5 at a time), but there are 10 instances of counts of 11–95 individuals. One other record also has a large count—an astonishing 275 individuals estimated. That encounter, on 18 June 2014, was at Drummond Flats WMA, Garfield County, a location where *Argia alberta* has been seen reliably, including two other times in high numbers (60 and 75 indiv.). BS-P surveyed the marsh that day for 1 hour and 15 minutes, and recorded 20♂ and 20♀ (20 pair, one as SP 1283) and estimated there were about 235 other individuals, about 75 of which were teneral.

The first time *Argia alberta* is known to have been encountered in Oklahoma is when RD Bird collected 1♂ at Cherokee, Alfalfa County, on 11 June 1930 (USNM 354688). Since then, >85 records have been reported from 25 counties, but the vast majority of those records have come recently, from 2010–2019 (about 75 records in 21 counties, 17 of which were new counties for the species). The distribution of records across the years begs the question, why is there such a disparity across the eras of data collection?

In his notes, Bick indicated that he and others (LE Hornuff and OM King) had some early difficulties distinguishing this species; he changed 17 records, dating between 17 June 1954 and 12 May 1956 and originally labeled as *Argia alberta*, to *A. immunda*. Bick himself encountered the species only once in Oklahoma (we confirmed his 3♂, 16 July 1970, Jefferson County: 4.5 mi N of Grady; FSCA). Bick's confusion is not surprising given that *A. alberta* and *A. immunda* can be difficult to tell apart until one has much experience with the two. Bird was also guilty of having trouble distinguishing the species. We re-identified 3♂ (OMNH 3070, 3122; from 9 July 1932, probably at Sulphur Springs, Wichita Mountains WR, Comanche County) as *A. alberta* that were originally identified as *A. sedula*. Bird and his team certainly did encounter the species including on the same date and locality as the misidentified *A. sedula*. Between 1930 and 1933, about 40 specimens were taken in Alfalfa, Woods, Cimarron, Harmon, and Comanche Counties by the OU expeditions.[62] Bird (1932) also noted that the species had been recorded in McClain and Murray Counties, but we do not know what documentation there was of these records, as we have not, nor does it appear that Bick had, found specimens for those counties, and those counties have no subsequent verifiable records. Bick also had not relocated the 9 July 1933 specimens that were collected in Cimarron

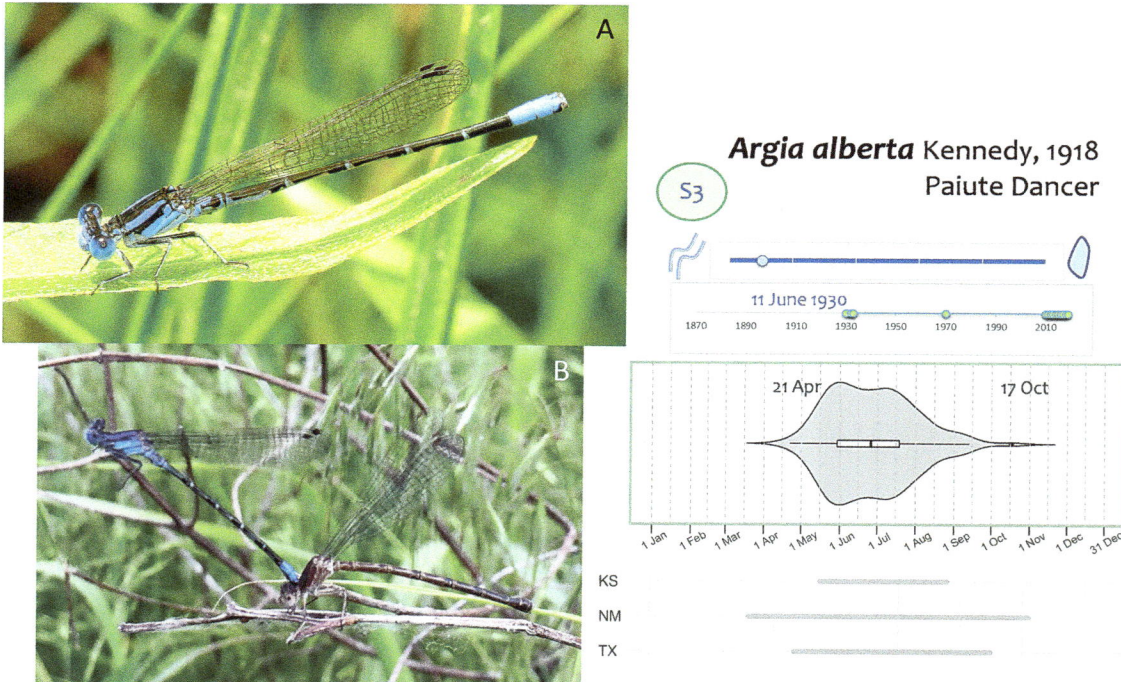

Figure 12.120 *Argia alberta*, the Paiute Dancer, A) ♂, JT Nickel Preserve, Cherokee County, 12 July 2016 (©Bill Carrell, OC 449154). This record is the easternmost in the Oklahoma, but this western species recently has been found in both Indiana and Ohio. B) tandem pair, Doby Springs, Harper County, 21 May 2017 (©Michael A Patten, OC 463133), a much more typical locale: the species is seen most often at clear, shallow, marshy creeks in the western third of the state.

County, in or near Boise City, by AE Pritchard. Those specimens were not in the OMNH collection when Bick examined specimens from the collection (we found the 4♂ at MCZ during our visit in March 2012), as such he did not include them in his notes for *Argia alberta*.

At this point, we would like to think we have managed to gather the vast majority of early records and have worked out most misidentifications and data entry errors. If that is the case, then we are left with the niggling issue of why a lotic species seems to have become more common and widespread (especially when considering the far-flung Cherokee and Osage County records, OC 449154 and OC 422477, respectively; both single ♂) while some odonate species associated with similar habitat appear to have done just the opposite. Our answer to that is … we do not know. Some have hypothesized that the recent spate of records that have extended the species' range to the east, as far as Ohio (!), may indicate the species is pushing eastward. But zooming back into Oklahoma, we must confront the confounding records presented by Bird (1932). His 1930 record for Alfalfa

County is the only one from the county and if the McClain and Murray County records are valid, then the species may have blipped out of those counties, too. That is more the pattern we would expect from a lotic damselfly in the state. It appears there is no easy answer at this point—the mystery continues.

Less of a mystery is the flight season in Oklahoma, which is equivalent to surrounding states where *Argia alberta* flies from the spring into the autumn (OK: 21 April 2012, 3♂, 1♀, including one pair as SP 145, Drummond Flats WMA, Garfield County, and 17 October 2015, 1♂, Altus city lake, Jackson County, MAP, SP 1813; CO: 29 May–29 August, OC; KS: 17 [possibly the 7th] May–27 August, SEMC; MO: 18–20 July, Trial 2005; NM: March–November, Paulson 2009; TX: 26 April–1 October, OC). Pairs have been reported regularly (1–20 pairs at a time), but only one instance of ovipositing has been reported (a pair at Sandy Sanders WMA, Beckham County, 10 August 2010, VWF, OC 322002). We know of one teneral record (see high count record above).

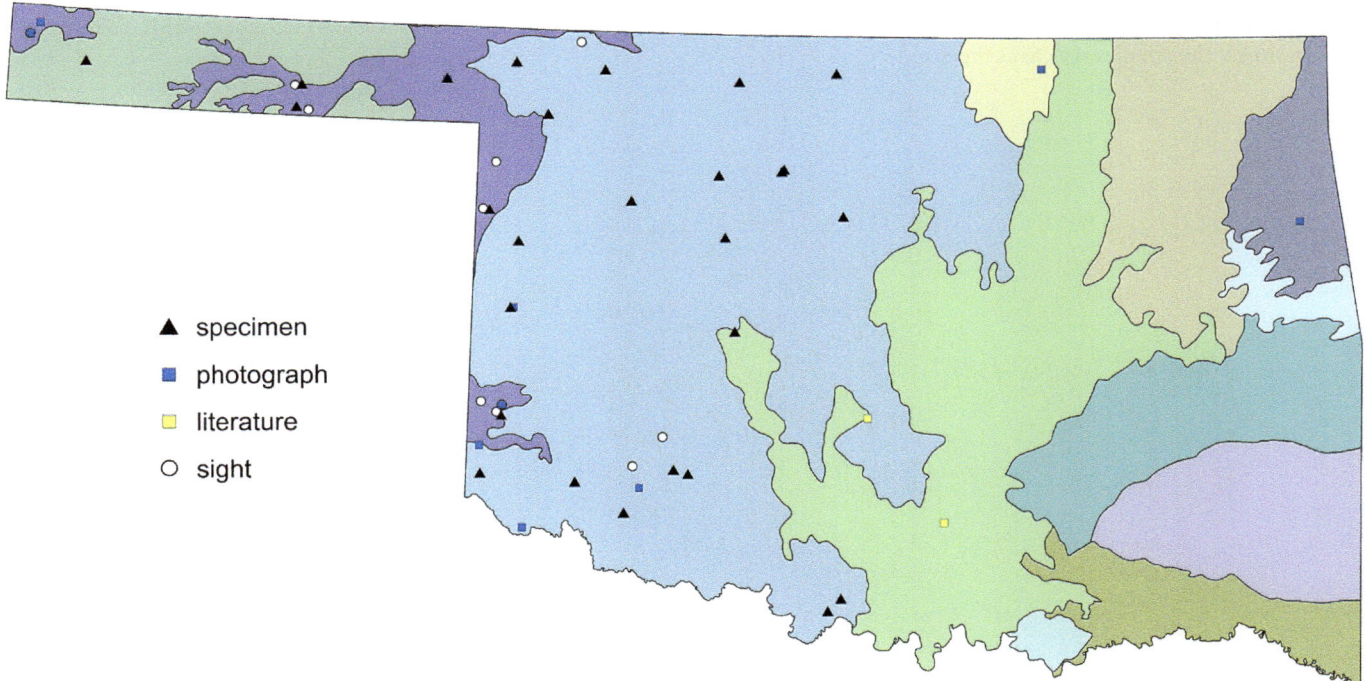

Figure 12.121 Documentation level of supporting records for *Argia alberta*, the Paiute Dancer, in Oklahoma. Levels include specimen, archived photograph, literature reference, or sight record. Although sight records lack documentation, only those we feel are credible are plotted.

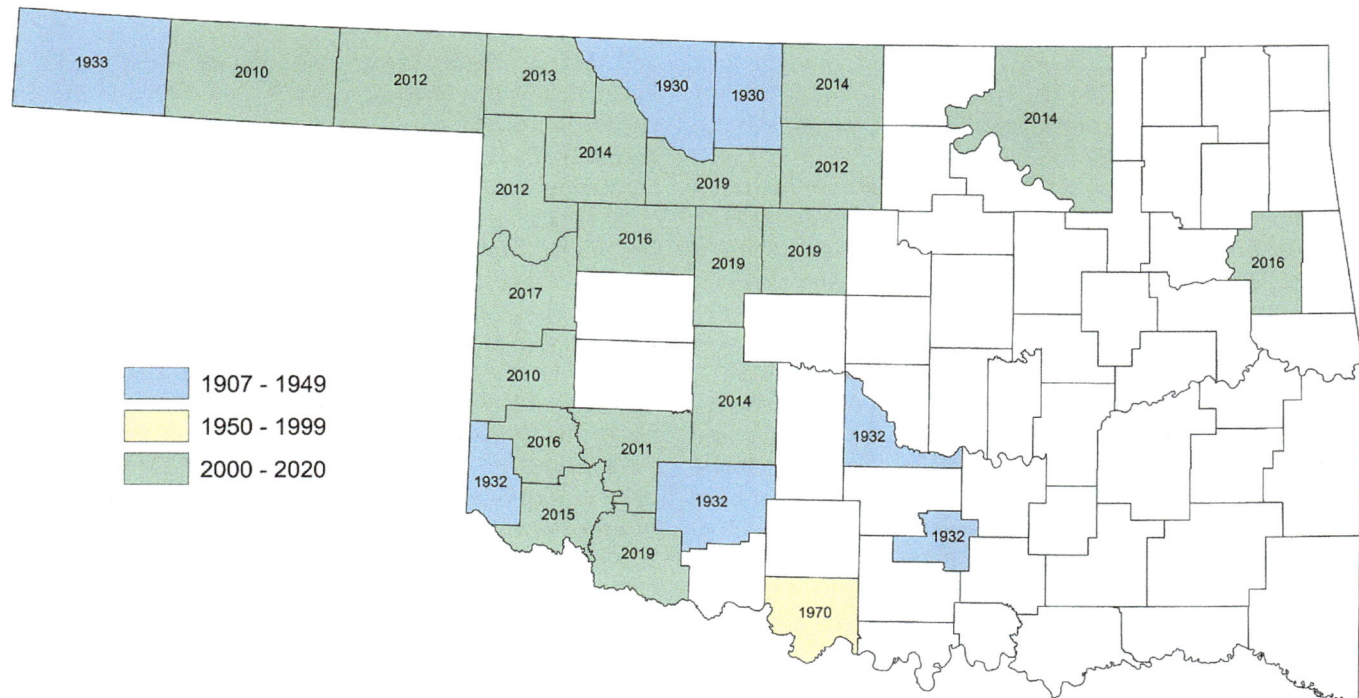

Figure 12.122 Counties in which *Argia alberta*, the Paiute Dancer, are known to occur in Oklahoma. Year within the county is that in which the species was first reported.

ARGIA LEONORAE GARRISON, 1994 – LEONORA'S DANCER

Leonora's Dancer is enigmatic in Oklahoma, and not just because the state is far (>400 km) from the species' known range in the Hill Country of Texas and adjacent tablelands of northern Mexico. There are three records for the state, all of which were essentially stumbled upon. The first publicly known record was submitted by a visitor to the state: T. Kompier photographed a single ♂ at a roadside ditch along Highway 62 south of Snyder, Kiowa County, on 28 July 2011 (OC 333094). Subsequent searches of that area proved unfruitful. We had our suspicions, however, that the species may be found within some early specimen collection given that it was not described until 1994 and is easily confused with *Argia nahuana*, especially, but also occasionally with *A. sedula*. Our suspicions were confirmed when in 2014, while looking though specimens collected in the Wichita Mountains area, we came across an adult ♂ *A. leonorae* that was originally identified as *A. sedula* (9 July 1932, Sulphur Springs, Wichitas Mountains WR, coll. and det. RD Bird, OMNH 3064). Multiple surveys have been conducted at

Sulphur Springs recently, with no luck (nor with *Amphiagrion abbreviatum*, a species that RD Bird also collected at that locale but that has not been seen there since). The third Oklahoma record, also of a lone ♂, was found in the "north ditch"—a shallow, typically flowing, clear-water channel—at the city lake in Altus, Jackson County on 14 May 2017 (MAP, SP 2327, OC 462694).

The species flies in Texas from late March to late November (19 March–21 November, Abbott and Lasley 2019), but is unknown from other states neighboring Oklahoma, except for one record in southeastern New Mexico at Rattlesnake Springs, Carlsbad Caverns National Park, Eddy County, from May 2003 (JCAC 19552). We suspect the species will be found more regularly in the state with increased climate change, but given all of the surveys we and others have conducted in the areas where we would expect to find the species in Oklahoma we infer that either a breeding population is not established as yet or it is so small that encounters are few.

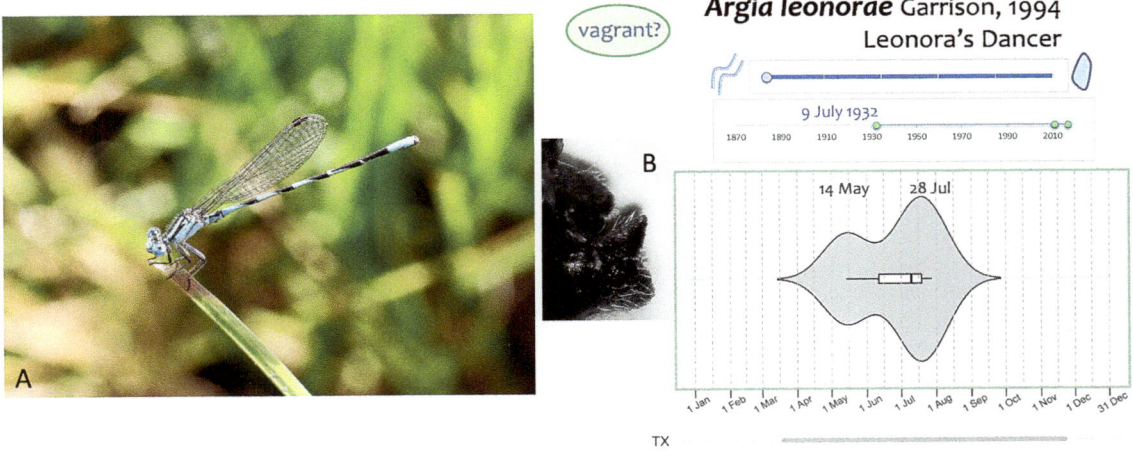

Figure 12.123 *Argia leonorae*, Leonora's Dancer. A) ♂, 3 km SE of Snyder, Kiowa County, 28 July 2011 (©Tom Kompier, OC 333094). This record was thought to represent a first for Oklahoma until we discovered B) a ♂ specimen collected by Ralph D Bird at Sulphur Spring, Comanche County, 9 July 1932 (OMNH 3064). There is but one additional record, from 2017, so this species' status in the state remains a mystery.

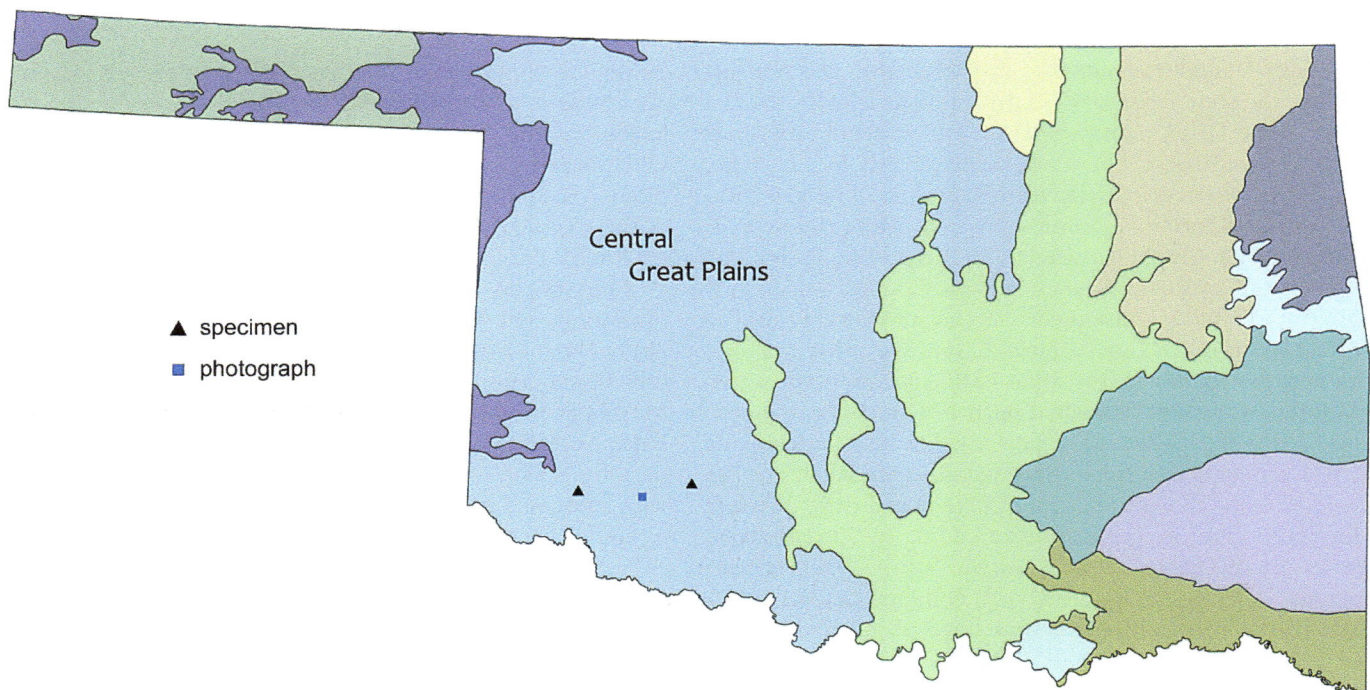

Figure 12.124 Documentation level of supporting records for *Argia leonorae*, Leonora's Dancer, in Oklahoma. Note that all records are from the Central Great Plains ecoregion.

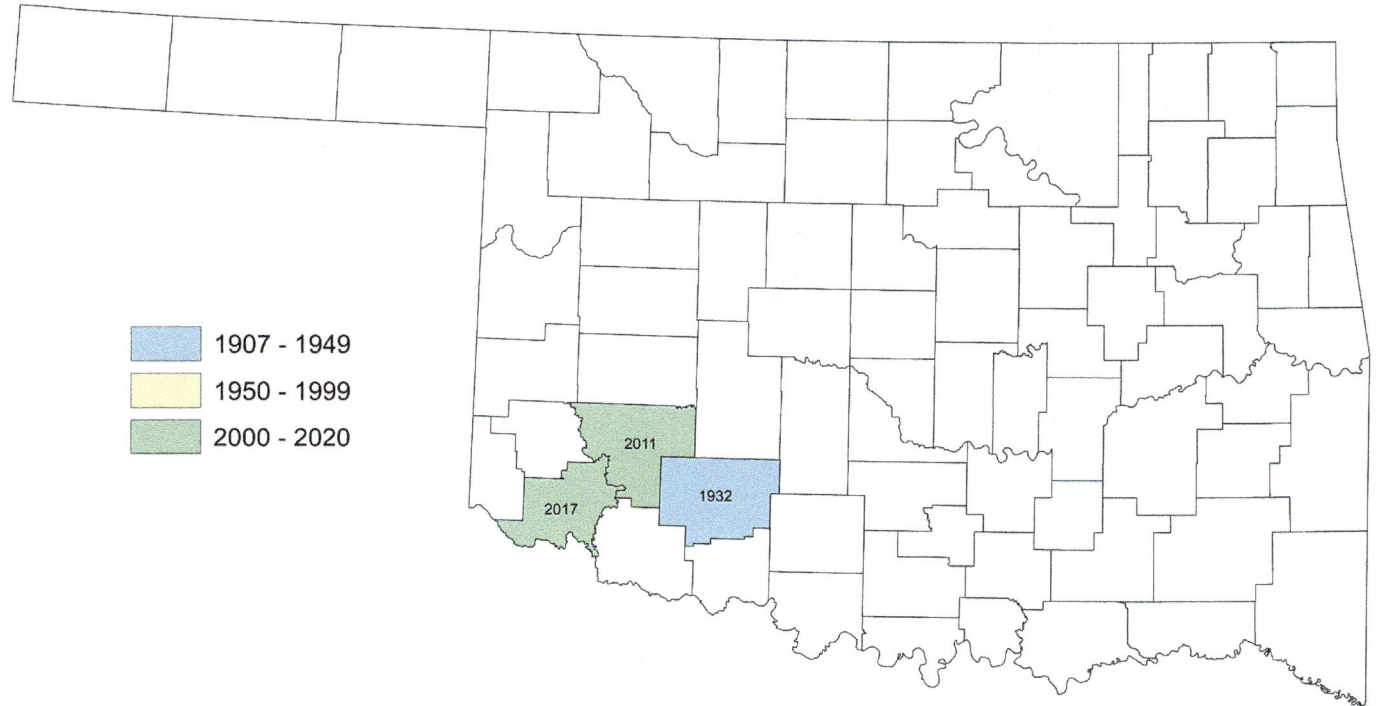

Figure 12.125 Counties in which *Argia leonorae*, Leonora's Dancer, are known to occur in Oklahoma. Year within the county is that in which the species was first reported.

ARGIA FUMIPENNIS (BURMEISTER, 1839) – VARIABLE DANCER

The Variable Dancer is a fairly common to common species of ponds, lake margins, and creeks. It was first recorded in the state by EB Williamson at Wister, Le Flore County, on 3–4 June 1907 (3♂, Williamson 1914b) and has been recorded regularly since. It is found across the state and reports of it are nearing 500. Double digits are rather commonly reported, including the highest count of 80 (50♂, 30♀) that included 65 adults and 15 tenerals seen at Bixhoma Lake, Muskogee County, on 13 May 2018 (MAP).

Argia fumipennis is a wide-ranging species, and as its common name suggests, its appearance is variable; hence,

it has three named subspecies: *atra*, *fumipennis*, and *violacea*. The most widespread subspecies, *A. fumipennis violacea*, is the one found in Oklahoma. The subspecies was considered a full species by early collectors such as Williamson (1914b), Ortenburger (1926b), and Bird (1932), each of whom called it *A. violacea*. Males tend to be strongly violet except for the telltale two-toned abdomen tip that is usually purple dorsally and anteriorly on S8, grading to blue laterally and posteriorly and on S9–10; but S8–10 can have varying violet-blue gradations or be all blue. Both males and females have post-quadrangular antenodal cells in the forewing numbering 3 or 4; Westfall and May (2006) indicate there should be

Argia fumipennis (Burmeister, 1839)
Variable Dancer

Figure 12.126 *Argia fumipennis*, the Variable Dancer. Generally the ♂ of this species, such as this one A) from Osage Hills SP, Osage County, 4 July 2015 (©Bill Carrell, OC 432706), does not vary much across the southern Great Plains, but we have noticed that individuals in the Black Mesa region of northwestern Cimarron County, at the tip of the panhandle, can be larger, blacker, and lack a forked antehumeral. The last two traits are evident on B) this ♂ from North Carrizo Creek 7 km N of Kenton, Cimarron County, 2 September 2013 (©Bill Carrell, OC 409999). The ♀, such as C) this one from Big Cedar, Le Flore County, 27 June 2013 (©James W Arterburn, OC 427982), varies little but can be impossible to diagnosis in the field relative to ♀ *A. nahuana*, the Aztec Dancer.

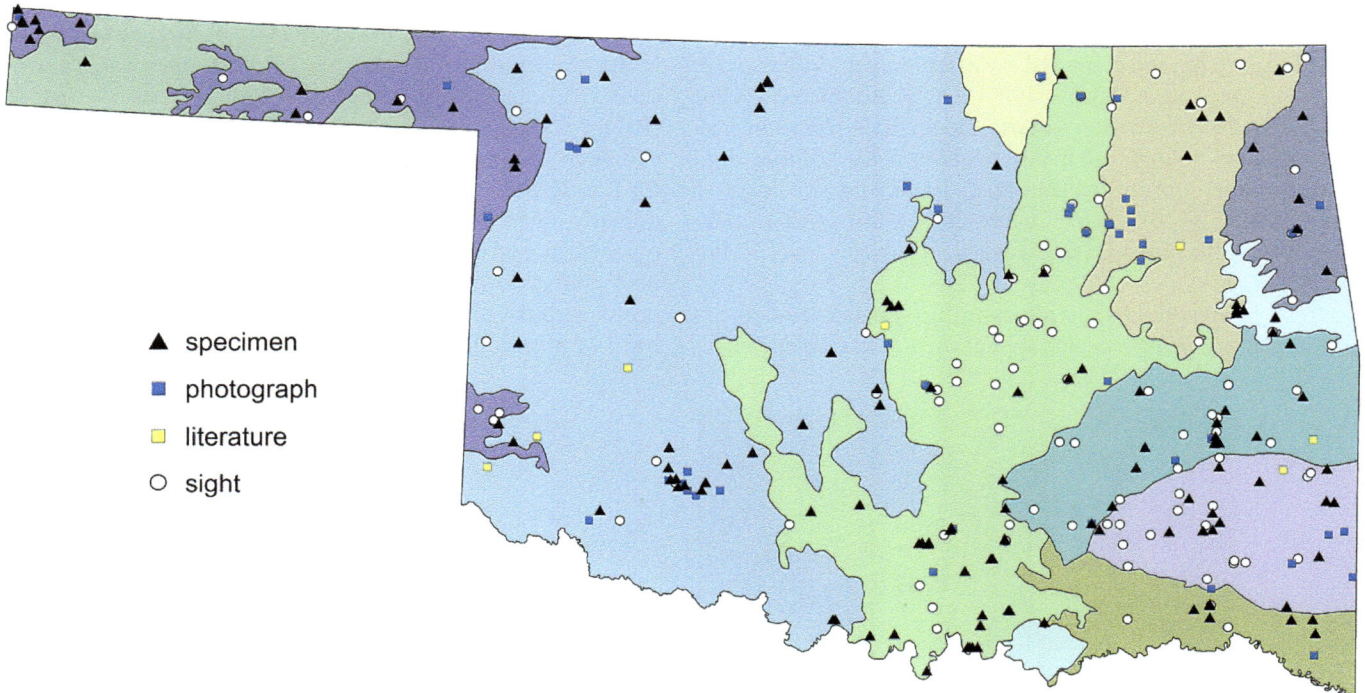

Figure 12.127 Documentation level of supporting records for *Argia fumipennis*, the Variable Dancer, in Oklahoma. Levels include specimen, archived photograph, literature reference, or sight record. Although sight records lack documentation, only those we feel are credible are plotted.

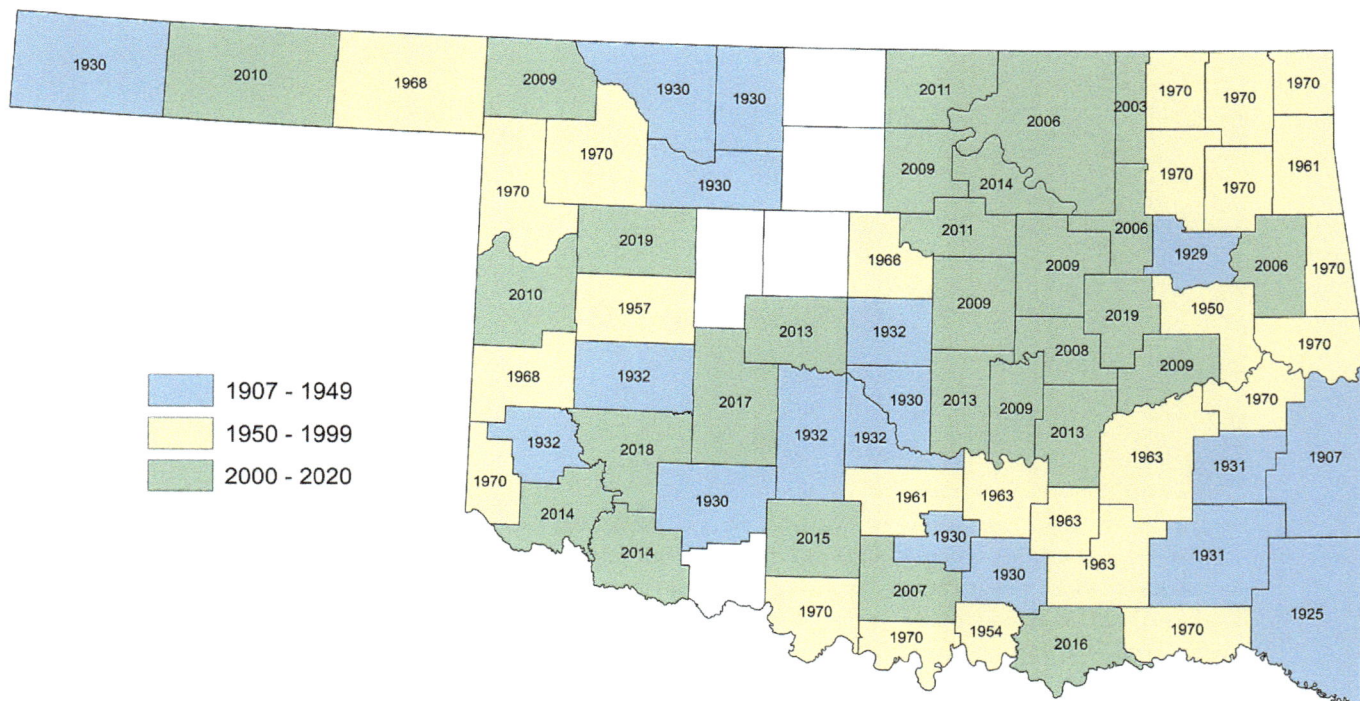

Figure 12.128 Counties in which *Argia fumipennis*, the Variable Dancer, are known to occur in Oklahoma. Year within the county is that in which the species was first reported.

four cells present, whereas Garrison (1994) indicated there are usually four but can be three.

Like many other *Argia* in the state, *A. fumipennis* flies in Oklahoma and the region starting in the spring and continues to fly well into the autumn (OK: 29 March 2012, 1U, Oka' Yanahli Preserve, Johnston County, F Alm, and 6 November 1932, 1 pair, Indian Springs, Cleveland County, RD Bird, OMNH 3386; AR: 14 May–16 September, Harp 2007, Harp and Harp 2003; CO: 2 June–1 September, CSU, OC; KS: 2 May [possibly 20 April]–5 October, SEMC, OC; LA: 26 March–4 October [nominate subspecies; 1 September for *violacea* subspecies], Mauffray 2014; MO: 21 April–10 October, Trial 2005; NM: 10 April–November, OC, Paulson 2009; TX: 29 March–17 November, Abbott and Lasley 2019). Pairs are fairly often reported. We collected a mixed-species pair of a ♂ *A. fumipennis* with a ♀

A. apicalis (SP 57, det. RW Garrison) at Perry Lake, Noble County, on 2 August 2009. We noted 50 *A. fumipennis* (40♂, 10♀, including 5 pair) and 25 *A. apicalis* (19♂, 6♀, one ♂ as SP 48) at the lake that day. We know of no records of ovipositing but there are a couple handfuls of teneral records between 11 April and 6 June.

One specimen whose patterning ought to be noted: SP 1611, a ♂ collected on the Cimarron River, 4 km NNE of Kenton, Cimarron County, on 3 June 2015. This male is slightly larger than others in the SP collection, has an unforked humeral stripe, and lacks the black stripe laterally on S8 and S9 (has only slight contrast in color on those segments and has a series of dots there instead; similar to the Lavender Dancer, *A. hinei*), features that in combination give the outward appearance of a ♂ Apache Dancer (*A. munda*), yet the cerci are those of *A. fumipennis violacea* (det. RW Garrison, 2018).

ARGIA NAHUANA CALVERT, 1902 – AZTEC DANCER

The Aztec Dancer is a strongly lotic species, preferring, as Bick and Bick (1958:248) noted, a "clear, sunlit, sandy, shallow creek devoid of vegetation." During their study of the odonates on Cowan Creek, Marshall County, the Bicks noted that 97% of the adult *A. nahuana* they encountered were at this type of habitat, although most of the species' nymphs they found were at a spring upstream. We have similar experiences with adult *A. nahuana*, but we also have recorded the species at cement-lined ditches with minimal water, little vegetation, and lots of sunlight, such as one we often check on the University of Oklahoma campus near the Law School. Such ditches may mimic what the species prefers in a natural environment.

Argia nahuana is common, especially in western Oklahoma, and widespread in the state: there are >420 records in 59 counties, chiefly in the western two-thirds of the state. In the west, counts in the double digits are common, and numbers of >100 have been reported half a dozen times, the highest being 275 individuals (200♂, 75♀, including 50 pairs, Altus city lake, Jackson County, 19 July 2015, MAP). In the eastern third of its range, the species is generally scarce, generally counted only in small numbers, and never reported >40 at a time (our highest count is only 20) until an incredible record of 225 adults came in 2019 (125♂, 100♀, including 95 pair with >20 pairs ovipositing, one pair collected as SP 2890, Posey Park, Eufaula, McIntosh County,10 August 2019, MAP; for reference, the only other times the species was recorded in the county, at the same park on 10 June 2018 and 29 June 2019, there was only 1♂ and 5♂, respectively). Of the 12 regularly occurring *Argia* in the state, *A. nahuana* appears to have increased the most in numbers encountered (Table 12.1). That trend may be because the species has adapted well to urban environments; however, some inflation may have arisen from confusion about species determination (see discussion below).

Figure 12.129 *Argia nahuana*, the Aztec Dancer. Generally, the ♂ has a distinctly forked humeral, such as A) at Buncombe Creek Public Use Area, Marshall County, 1 October 2016 (©Bill Carrell, OC 456599), but at times the humeral fork can be weak, as with this ♂ B) at Bull Creek in Vinita, Craig County, 7 July 2013 (©Bill Carrell, OC 401663). C) Tandem pairs, such as at Bixhoma Lake, Wagoner County, 14 May 2012 (©James W Arterburn, OC 375368), are a common species wherever the species occurs.

The first known encounter with *Argia nahuana* in the state was with a single ♀ (UMMZ) that EB Williamson collected on 19 October 1929. He collected the specimen at 15 mi E of Weatherford, a locale in Caddo County, but his label has the county as Custer, home of Weatherford. Bird (1932) did not publish this record, likely because the letter from which he obtained data purported to be from Custer County did not include the species (letter from LK Gloyd to RD Bird, dated 28 April 1932). Bick also did not know of this specimen, but it is likely the basis for the Dot Map Project record for *Argia nahuana* in Custer County (Donnelly 2004c), unless Bick told Donnelly of a 1955 record that Bick had in his notes. Nonetheless, there are now documented records for the species from Custer County.

Befuddlement regarding the status of *Argia nahuana* in Oklahoma extends to a taxonomic split and conflation with another species. Earlier researchers lumped *A. nahuana* with *A. agrioides*, the California Dancer. They did so under the names *A. agrioides*, *A. agrioides nahuana*, or *A. agrioides* var. *nahuana*. To befuddle matters further, early collectors in Oklahoma inadvertently misidentified *A. immunda* as *A. agrioides* (thus all previously identified *A. agrioides* cannot be attributed readily to *A. nahuana*). And, as if that did not make for sufficient befuddlement, there is the lingering possibility that *A. leonorae*—Leonora's Dancer, described in 1994—could be misidentified as *A. nahuana*. As such, over the years much befuddling has occurred. We are hopeful that the issues have been resolved, but we remain diligently aware of possible resurgence. In their attempt to correct the situation, said

specimens from Comanche (OU) labelled [sic] *agrioides* by Bird and so determined by Gloyd, were determined as subspecies *nahuana* by the writer. Numerous unidentified specimens (OU) from Cleveland as also *nahuana*. On the other hand, specimens (OU) determined by Bird as *agrioides*, from Cleveland, Jackson, Latimer,[63] and Pushmataha are *immunda* (Hagen). We did not see specimens from Murray or Wagoner in the OU collection. We believe that all county records by Bird (1932) of *agrioides* pertain to subspecies *nahuana* but I was able to verify specimens from only Cleveland and Comanche.

(Bick and Bick 1957:10)

Confusion about how to conclusively identify *A. nahuana* may be the reason that the encounter rate early on appears to be so low (Table 12.1). Alternatively, the species may have expanded its range (from the 4 counties known in the early 1930s to the 59 now known), though we suspect that is not the case.

The species is known to fly from late March to early November in Oklahoma, which is similar to the region (OK: 23 March 1933, 1♂ and 1♀ as 1 pair, Norman, Cleveland County, R Mac, OMNH 2972, and 11 November 2013, 1♂, Hunter Park, Tulsa, Tulsa County, BC, OC

411649; CO: 7 May–4 November, OC; KS: 16 April–20 November, SEMC; MO: limited records, 19 April–7 September, Sims 2012; NM: 19 March–November, OC, Paulson 2009; TX: 24 February–3 December, Abbott and Lasley 2019). Pairs are seen frequently and a dozen or so teneral records indicate emergence occurs from at least 6 April to 15 September. The Bicks noted that during their summer study in Marshall County, peak abundance of *A. nahuana*, as with the equally lotic *A. immunda* (Kiowa Dancer), was 15 July, when they had 113 adults with 14 pairs of *A. nahuana*. Numbers decreased rather dramatically by 28 July. This pattern does not hold currently (see seasonality plot above).

The Bicks were the first to describe the oviposition behavior of *A. nahuana* (Bick and Bick 1958). They observed a tandem pair on 19 June 1955 on Cowan Creek where the water was about 2.5 cm (1 in) deep. They did not indicate the species of grass on which the pair landed to deposit eggs, but they did say that the male remained vertical, holding onto the grass blade above the water while the female's abdomen

was bent at a sharp angle and its tip touched the plant one half inch below the water surface where eggs were apparently deposited. She probed for a few seconds with the tip of her abdomen, remained motionless for two and one half minutes, probed briefly and remained motionless for five minutes.

(Bick and Bick 1958:250)

The pair proceeded to other grass blades, but they did not linger at any. When they flew off, they were collected. Outside of the Bicks' study, we know of three records of ovipositing, all of which were by pairs (up to 50) ovipositing.

This species' humeral shape varies to some extent, as does overall body coloration, and dorsal pattern of abdominal segment seven. The first character varies in both males and females, whereas we have noted the other two only on males (caveat: we have not conducted a formal assessment). Humeral shape varies from a full fork, as one expects of the species, to faint or partial forks,[64] to being unforked.[65] Overall body coloration varies from the more typical medium blue to those that are strikingly pale to those strikingly dark. It may be that some of that color variation is due to age, body temperature, or in the case of specimens, preservation; as such, we do not place much emphasis on that character, but it could prove of interest to other researchers. Of more interest to us is that some males show blue dorsally on S7[66] instead of the more "normal" black. Our impression is that abnormalities of the humerals and S7 are not especially common in Oklahoma and likely result from individual variation, as we have not been able to ascertain a geographical component to it.

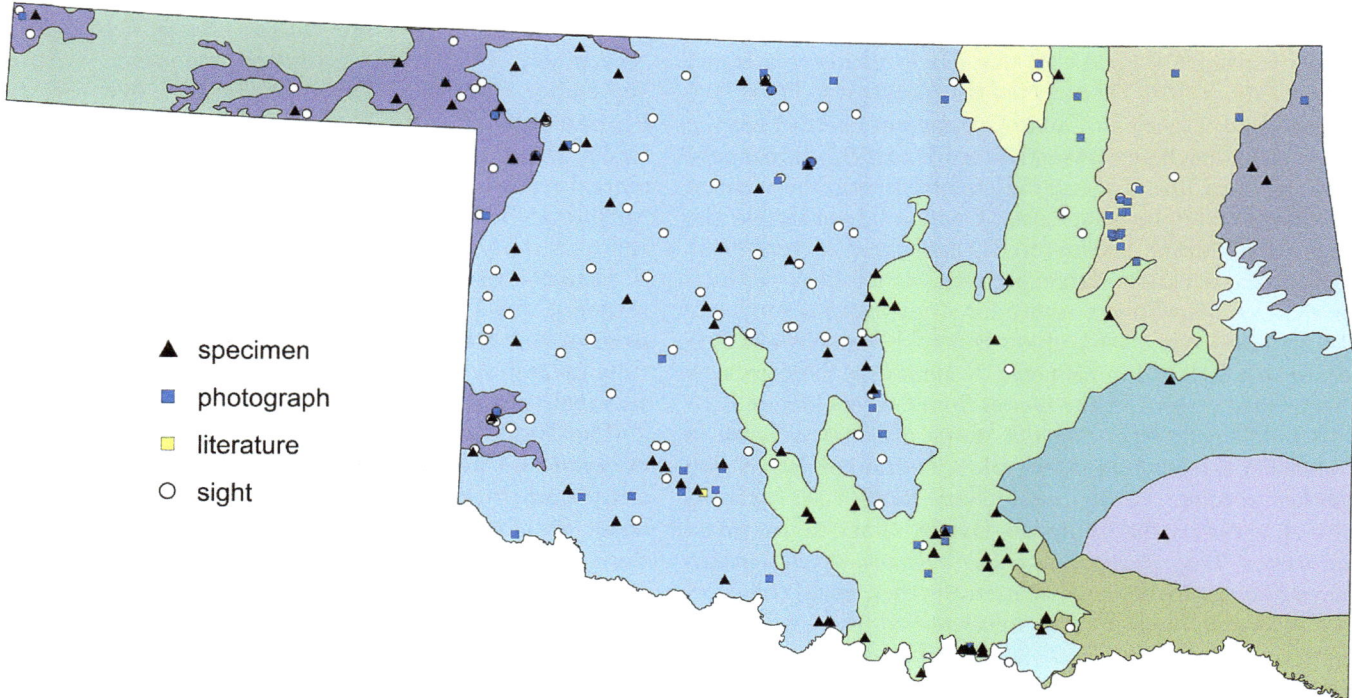

Figure 12.130 Documentation level of supporting records for *Argia nahuana*, the Aztec Dancer, in Oklahoma. Levels include specimen, archived photograph, literature reference, or sight record. Although sight records lack documentation, only those we feel are credible are plotted.

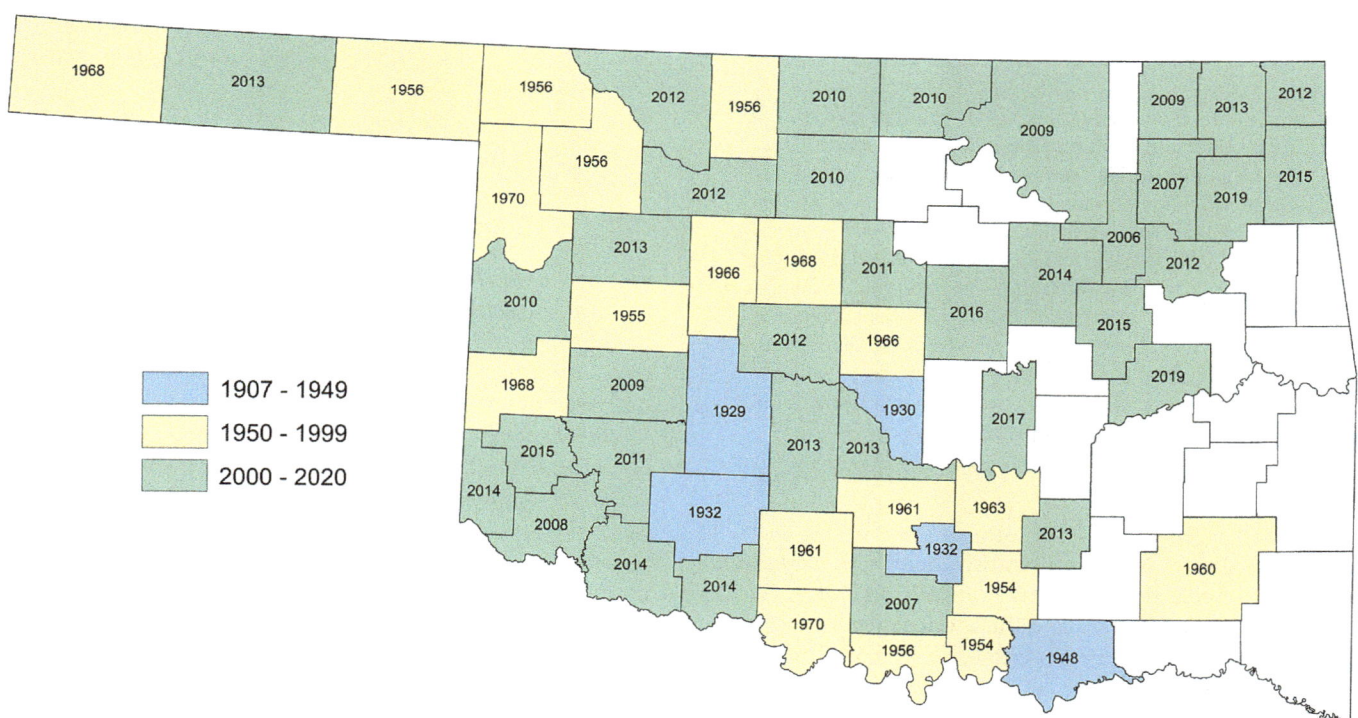

Figure 12.131 Counties in which *Argia nahuana*, the Aztec Dancer, are known to occur in Oklahoma. Year within the county is that in which the species was first reported.

ARGIA IMMUNDA (HAGEN, 1861) – KIOWA DANCER

The Kiowa Dancer is a lotic species strongly associated with clear, sandy, and shallow creeks that are sunlit and have extruding rocks and often little vegetation (Bick and Bick 1958; our *pers obs*). The species was first recorded in the state in April 1930.[67] It has been reported about 300 times since its discovery in Oklahoma. Bick and Bick (1957:10) said that *Argia immunda* "... was frequent and locally abundant in southern Oklahoma but are absent in the northernf [sic] half of the state." Although the species is a bit more common in southwestern Oklahoma, it is not absent from the northern part of the state as once thought. It is now known from 58 counties across the state. Typically, it is encountered <10 individuals at a time, but our high count is 30 individuals (25♂, 5♀, including 3 pairs, one as SP 1099, Raymond Gary SP, Choctaw County, 11 October 2013).

GH Bick appears to have struggled some initially with the identification of *Argia immunda*, as evidenced by one of his note cards from June 1954 to May 1956 that was originally labeled as *A. alberta* then changed wholesale to *A. immunda*. Bick had two other note cards that were initially labeled as *A. immunda* but later changed to *A. nahuana*. These cards included his notes from 17 June 1954 to 5 June 1956 as well as the 13 May 1930 ♂ specimen that was the first record of *A. nahuana* for the state (OMNH 534). By July 1956 Bick must have gained confidence in his ability to diagnose *Argia* because he identified the first state record (Murray County, see above) and the first county record for Comanche County as *A. immunda*. He also re-identified a series of RD Bird's specimens from *A. agrioides* to *A. nahuana* or *A. immunda* (see *A. nahuana* account for details). These included the first county records of *A. immunda* for Jackson and Pushmataha Counties. The previously undetermined specimens and those identified as *A. agrioides* indicate that Bird also struggled to identify *A. immunda*. Bird's hesitancy explains why only Cleveland County showed for the species in his 1932 publication.

Argia immunda has a long flight season in Oklahoma and the region, lasting at least 238 days, making it tied with *A. translata* as the second longest flying *Argia* in the state, both of which are just 10 days shy of *A. plana*. It is known to fly from mid-March to mid-November in Oklahoma and New Mexico, year-round in Texas, and there is one record each for Arkansas and Kansas (OK: 18 March 2009, 2♂, 1♀, including one pair, Fort Sill, Comanche County, VWF, OC 312297, and 11 November 2013, 1♀, 1U, Hunter Park, Tulsa, Tulsa County, BC, OC 411650 and 411651; AR: DMP, no date; KS: 27 April, USNM; NM: 13 March–13 November, OC; TX: Abbott and Lasley 2019). Pairs are reported rather regularly in Oklahoma, but we lack reports of tenerals. Its peak abundance was observed as 15 July by the Bicks when they recorded 33 adults and 5 pairs at Cowan Creek, Marshall County (Bick and Bick 1958); numbers were well down by 28 July. This pattern is not the same today (see seasonality plot above).

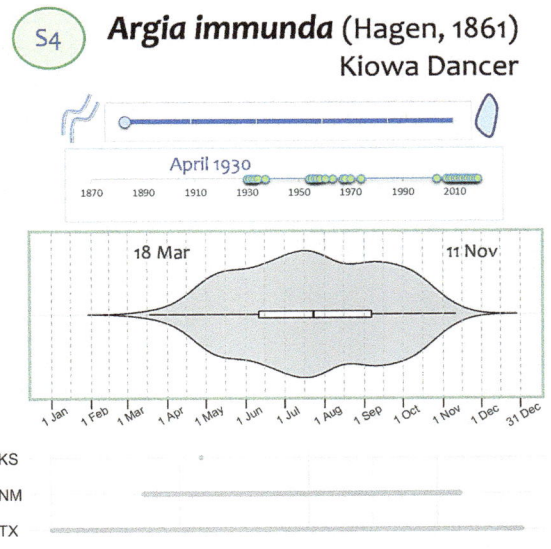

Figure 12.132 *Argia immunda*, the Kiowa Dancer, A) ♂, Government Springs Park in Enid, Garfield County, 24 July 2012 (©Jason R Heinen, OC 377957). B) ♀, Hunter Park in Tulsa, Tulsa County, 11 November 2013 (©Bill Carrell, OC 411650).

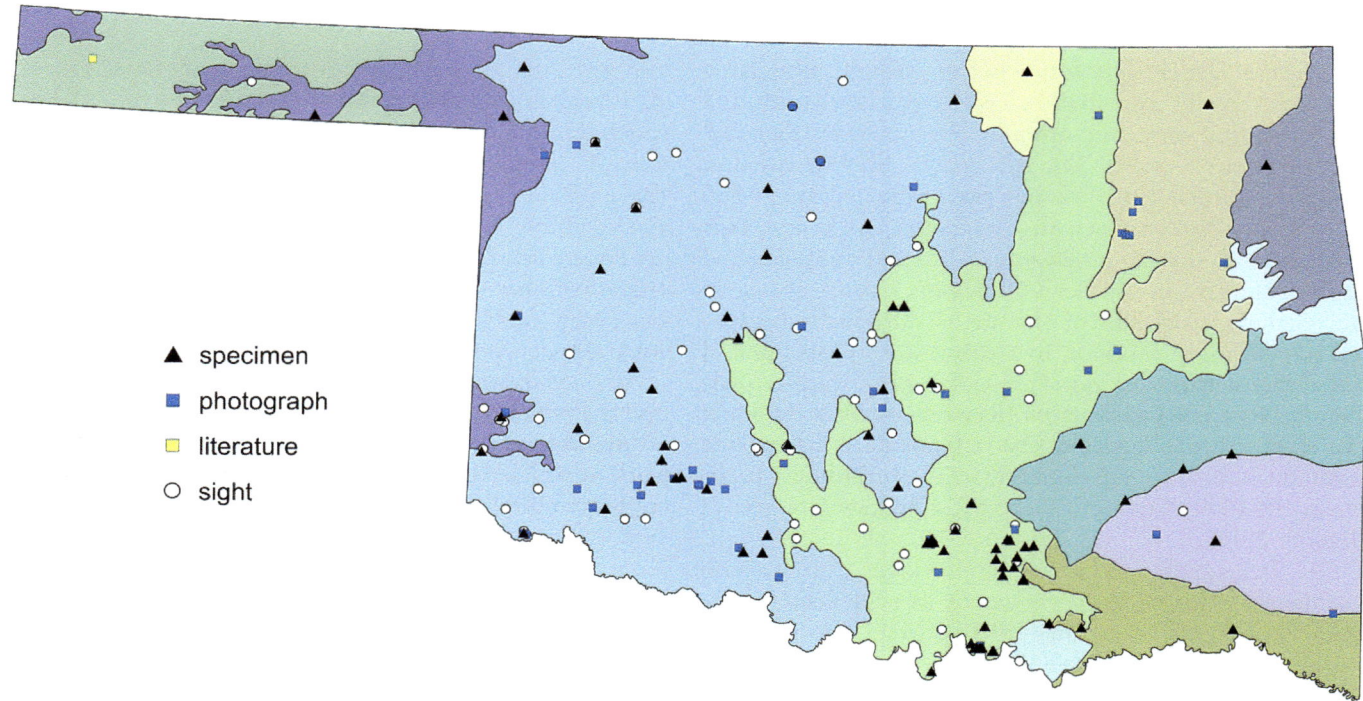

Figure 12.133 Documentation level of supporting records for *Argia immunda*, the Kiowa Dancer, in Oklahoma. Levels include specimen, archived photograph, literature reference, or sight record. Although sight records lack documentation, only those we feel are credible are plotted.

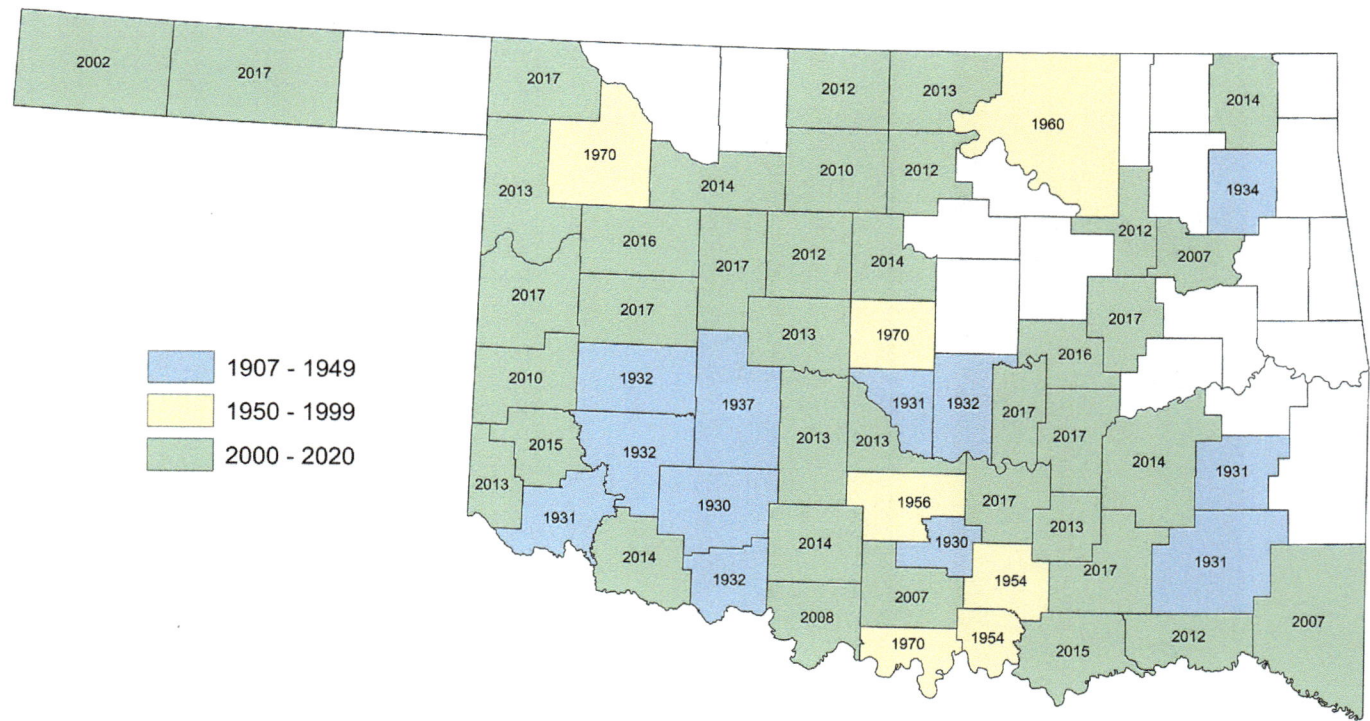

Figure 12.134 Counties in which *Argia immunda*, the Kiowa Dancer, are known to occur in Oklahoma. Year within the county is that in which the species was first reported.

ARGIA PLANA CALVERT, 1902 – SPRINGWATER DANCER

Argia plana, the Springwater Dancer, is a strongly lotic species commonly found across Oklahoma, although it is commoner in the western two-thirds of the state. We know of approximately 500 records of the species in Oklahoma. It was first discovered in the state in 1928 (4, 6, 8, and 9 July, Time-O-Day Spring, Elm Island, and a tributary of Panther Creek, Wichita Mountains WR, Comanche County, 12♂, 4♀, OMNH 1513–1528), when it was known as *A. vivida* (e.g., Bird 1932). Later it became *A. vivida plana* and later still *A. plana*. During those name transitions and ever since, it has been reported consistently (Table 12.1) and although generally encountered a dozen or fewer at a time, larger numbers have been reported. Our highest count for the species is 100 (40 adult ♂, 40 adult ♀, 20 teneral U, Doby Springs, Harper County, 12 May 2012, 1♀ as SP 235).

Argia plana was the most common species at a spring that GH and JC Bick studied along Cowan Creek, Marshall County, in the mid-1950s (118 of the 136 adult damselflies at the spring during their summer study; Bick and Bick 1958). The Bicks found 91% of *A. plana* adults at the spring that they described as a "shaded, shallow, cold-water spring with abundant vegetation." The majority (n = 33) of *A. plana* nymphs they found were also at the spring proper, but they did find a large number at a creek downstream (n = 22) as well, which they hypothesized as probably having been washed down from the spring rather than being oviposited there. Williamson (1932:16, as *A. vivida*), too, reported a strong association with springs. He was so confident of the association that he said "so dependent is it on springs that its presence anywhere may be taken as positive proof of adjacent spring water." However, we encounter the species also

Figure 12.135 *Argia plana*, the Springwater Dancer. A) ♂, Osage Hills SP, Osage County, 8 July 2018 (©Bill Carrell, OC 484489). B) ♀, Wichita Mountains WR, Comanche County, 3 June 2017 (©Jeff Trahan, OC 469938). C) andromorph ♀, Natural Falls SP, Delaware County, 24 May 2014 (©CA Ivy, OC 434861). Many males in the Cross Timbers ecoregion outwardly resemble *Argia plana* only obliquely. D) This ♂ at Keystone Dam, Tulsa County, 14 June 2011 (©Ken Williams, OC 331025) is a fine example. Curiously, aspects of the cerci and wing venation differ, too—see the text for our musings (and head clutching) on the matter.

in urban settings, such as at cement-lined drainage ditches that are not spring-fed. It could be that those settings mimic well enough spring flows and so *A. plana* will use them in addition to more pristine natural areas.

The Springwater Dancer has the longest flight season of all of the *Argia* in Oklahoma—a hearty 266 days (3 March 1930, 2♀, Indian Springs, Cleveland County, Z Logsdon, OMNH 3664, and 24 November 2016, 1♂, Woodcreek Park, Norman, Cleveland County, EA Hjalmarson, OC 478314). Oklahoma's season is much shorter than that of Texas, equivalent to that of New Mexico, and much longer than elsewhere in the region (AR: 9 May–14 September, OC, iNat; KS: limited records, 2 June–26 August, OC; MO: 5 May–29 October, Trial 2005, SEMC; NM: 13 March–12 November, OC, iNat; TX: roughly year round, Abbott and Lasley 2019). Teneral records have only been reported from 20 March to 23 July. The Bicks, at their study site in Marshall County, observed that peak abundance was from 19 June to 1 July, plummeting by 8 July (Bick and Bick 1958; but see seasonality plot for an updated pattern). They also reported collecting 57 nymphs in 1955–1956.[68] The other credible report of nymphs that we know of is 10 nymphs collected on 25 July 1969 by RE McKinley from Travertine Creek, Platt National Park (now Chickasaw NRA), Murray County; specimens were confirmed by MJ Westfall, Jr. (1990).

Also at Cowan Creek, the Bicks observed the first known description of *A. plana* ovipositing (Bick and Bick 1958). They observed eight pairs on 1 July 1955. Seven of the pairs deposited in tandem while the female held onto a twig, the male apparently supported upright by the female's prothorax. The females then deposited eggs on or in clay below the water. The female (alone?) of the eighth pair deposited also in clay, but outside of the water where the clay was still damp. All pairs were observed "in deep shade a few inches from where the clear spring water emerges from a steep bank and runs over the clay bottom at a depth of not more than one half inch" (Bick and Bick 1958:250). Almost all of the other ovipositing records we know of in the state also report that the species tends to oviposit in tandem.

Pairs have been reported frequently, including five mixed-species (mixed-family/genera) pairings, one of which was a ♂–♂ pairing. These include 2♀ *Argia plana* pairing with ♂ *Hetaerina americana*, 1♂ *A. plana* pairing with a ♂ *H. americana*, and 2♂ *A. plana* pairing with ♀ *A. moesta* (see the *H. americana* and *A. moesta* accounts for details). At least as rare as mixed-species pairs, are encountering andromorph female *A. plana*. We know of only three andromorph records of this species for Oklahoma: Delaware (BugGuide 927251), Harper (SP 1576), and Osage (SP 1946) Counties.

Argia plana is bewildering in its thoracic and abdominal patterning in the Cross Timbers region and nearby areas. This is especially true of its humeral stripe. A "typical" *A. plana* has a humeral stripe that is fairly wide towards the front of the thorax, narrowing, about half way along the thorax,

to a fine line that leads back towards the rear of the thorax, where it widens again into a small square or rectangle. More often than not the humeral is continuous, but it can be broken slightly along that thin line (e.g., SP 2138, New Mexico, Sandoval County). Occasionally, too, the humeral is continuous but thickened (e.g., SP [RWG] 1452). A much weirder phenomena for this species is a forked humeral. In approximately the area at which the humeral would begin to narrow anteriorly, there can be a bifurcation ("forking") of the stripe. The lower, secondary line then extends back, often about ⅔ or ¾ the way to the posterior margin of the thorax, but occasionally it reaches all the way to the margin to enclose a shape akin to a pale triangle (e.g., SP [RWG] 1029 or 1658). Such long and bold forks are easily visible with the naked eye as are others that do not extend back as far (e.g., Doby Springs, Harper County, 8 October 2017, BC photo). We have also noted "pseudo-forks," meaning that only a partial secondary line is present, a line that may form a rectangle showing below the main humeral line, a smudge, or a dot (SP 1952, 1953, 2197, respectively). A fork or pseudo-fork can differ from side-to-side; for example, SP 1953 has a smudgy pseudo-fork on its left side and a dot on its right, and SP 1951 has a dot on the left side but a "normal" unforked humeral on its right. Lastly, note that individuals at the same location can be "typical" or forked (e.g., 2♂, Noble Research Institute Red River Ranch, Love County, at a spring, 2 August 2019; SP 2883 with pseudo-fork but the other ♂ did not).

We have examined less closely abdominal patterns of *Argia plana* but have discovered some individuals blacker (at times remarkably so) throughout the abdomen than is typical. We have especially noticed the characteristic lateral triangles of the species being smudged and/or extended posteriorly (e.g., SP 1825, 2197, and photo of ♂ from Natural Falls SP, Delaware County, 15 September 2015, JWA). OC 331025 is a particularly interesting example of variation because it exhibits both a smudged abdominal pattern and a thickened, humeral fork (SP 1791, coll. Doby Springs, Harper County, is similar).

There may be sex differences in presentation of aberrant patterns but that cannot be said conclusively without further study. We have noted humeral forking primarily on males but the female from a pair captured in Murray County shows a small spot along the secondary line on her left side (SP 1760). Females do exhibit some smudging and other slightly aberrant patterning on the abdomen but generally aberrancies have been noted for males. We do not know if that is due to male-biased sampling given that we have not tried to conduct a formal investigation of variation in this species.

Although aberrancies occur primarily in the Cross Timbers region of the state, they have been reported elsewhere, including once in the southwestern corner (Lake Hall, Harmon County, SP [RWG] 1658) and twice in northwestern Oklahoma (both at Doby Springs, Harper

County, SP 1791, and record above). Additionally, we know of one record of a male exhibiting a forked humeral in central New Mexico (Bernalillo County, SP 2146) and at least three records of males in southwestern Texas exhibiting pseudo-forks or fully forked humerals (Presidio County, OC 403493, JCAC 2195, and at least one other male photographed by GW Lasley on 19 September 2004). The specimen from Lake Hall is of especial note. With its wide, forked humeral, extensive black dorsally on the abdominal segments, and distinctly amber wings (!), it could have been mistaken for *Argia rhoadsi*, the Golden-winged Dancer, a species recorded no closer to Oklahoma than >500 km south of the Red River. Outward appearances aside, its cerci are (more-or-less) typical for *A. plana* and nothing like the odd cerci of *A. rhoadsi*.

Before leaving the topic of variation, note that both Garrison (1994) and Westfall and May (2006) indicated that typically male and female *Argia plana* have four post-quadrangular antenodal cells (pq) in the forewing (FW). Westfall and May (2006) further noted that males can, albeit rarely, have three or five cells; they made no mention of variation with females. Of the male and female *A. plana* specimens we examined from the SP collection ($n = 92$), just over 70% had 4 pq cells in each FW. These "normal" specimens are from Oklahoma ($n = 54$), New Mexico ($n = 9$), and Missouri ($n = 1$). Some other specimens had 4 pq cells on one FW, but the cell count in the other FW did not match: e.g., SP 1759 (♀) had 3 cells in the left FW but 4 in the right and SP [RWG] 1658 (♂) had 5 cells in the left and 4 in the right (SP [RWG] 1658 also has smoky amber wings). One specimen had 3 pq cells in each FW (SP 1950, ♂). More difficult to assess wings had veins that had grown in partially or were branched into a Y or an upside-down Y or where the cell overshot the nodal vein, making it unclear if that should be counted as part of the cells considered prenodal. As such, there were FWs

that could be considered as having 3–4 cells, for example, or 4–5 (e.g., SP [RWG] 985, ♂ and SP 2418, ♀, respectively). Interestingly, most of the individuals with mixed cell counts are from the Cross Timbers ecoregion.

Also diverging from Westfall and May (2006) are some measurements of males that we captured at Lake Ponca, Kay County, in August and September 2013. Five specimens have hindwing lengths between 18.0–21.5 mm,[69] all of which are smaller than those presented by Westfall and May (2006; 22–25 mm), but one falls within measurements given by Garrison (1994; 21–24 mm). Two of those specimens have smaller abdomens (23.1 and 26.2 mm, SP [RWG] 1026 and 948, respectively) than previously reported for the species (27–33 mm, Westfall and May 2006). One of the specimens, SP [RWG] 1026, is small all around, including being >4 mm smaller (29.6 mm) than the lowest total length measurement present by Westfall and May (2006; 34–40 mm). That male was also lavender in life rather than the typical blue of Oklahoma *A. plana* males. Three of the males in this series have the humeral forked (SP [RWG] 948, 1029, 1030), one (SP [RWG] 1027) has a malformed paraproct, and one (SP [RWG] 1030) utterly lacks the lower process on the paraproct that gives *A. plana*'s paraprocts a branched appearance. Despite their aberrancies, we are confident that all of these specimens are indeed *A. plana* because they were determined as such by RW Garrison.

We can offer little to explain what appears to be a geographical component to "aberrancy," other than the possibility of gene flow with other *Argia* species, perhaps especially with *A. sedula*, the Blue-ringed Dancer. Does small size of some males support a hypothesis of hybridization with *A. sedula*? Or is prevalence of forked humerals and extensive black on the abdomen harbingers of evolutionary change in the Cross Timbers? In other words, might we bear witness to (sub-) speciation in action?

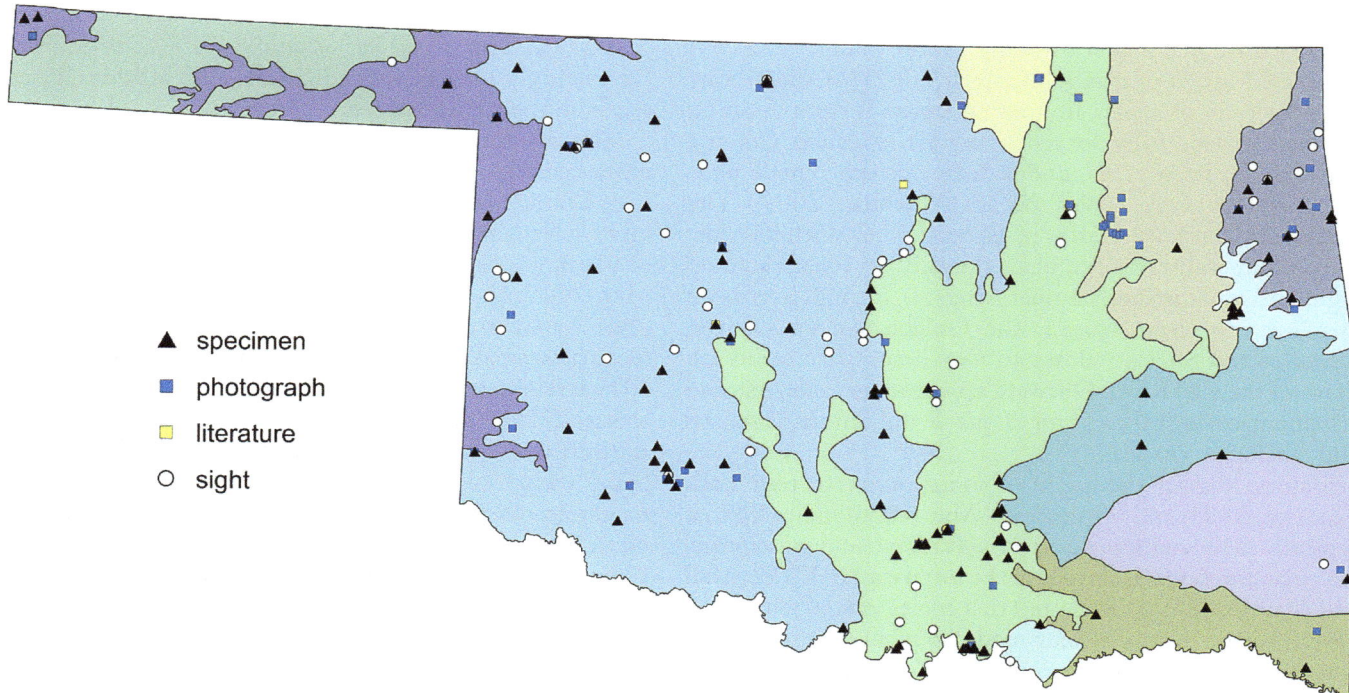

Figure 12.136 Documentation level of supporting records for *Argia plana*, the Springwater Dancer, in Oklahoma. Levels include specimen, archived photograph, literature reference, or sight record. Although sight records lack documentation, only those we feel are credible are plotted.

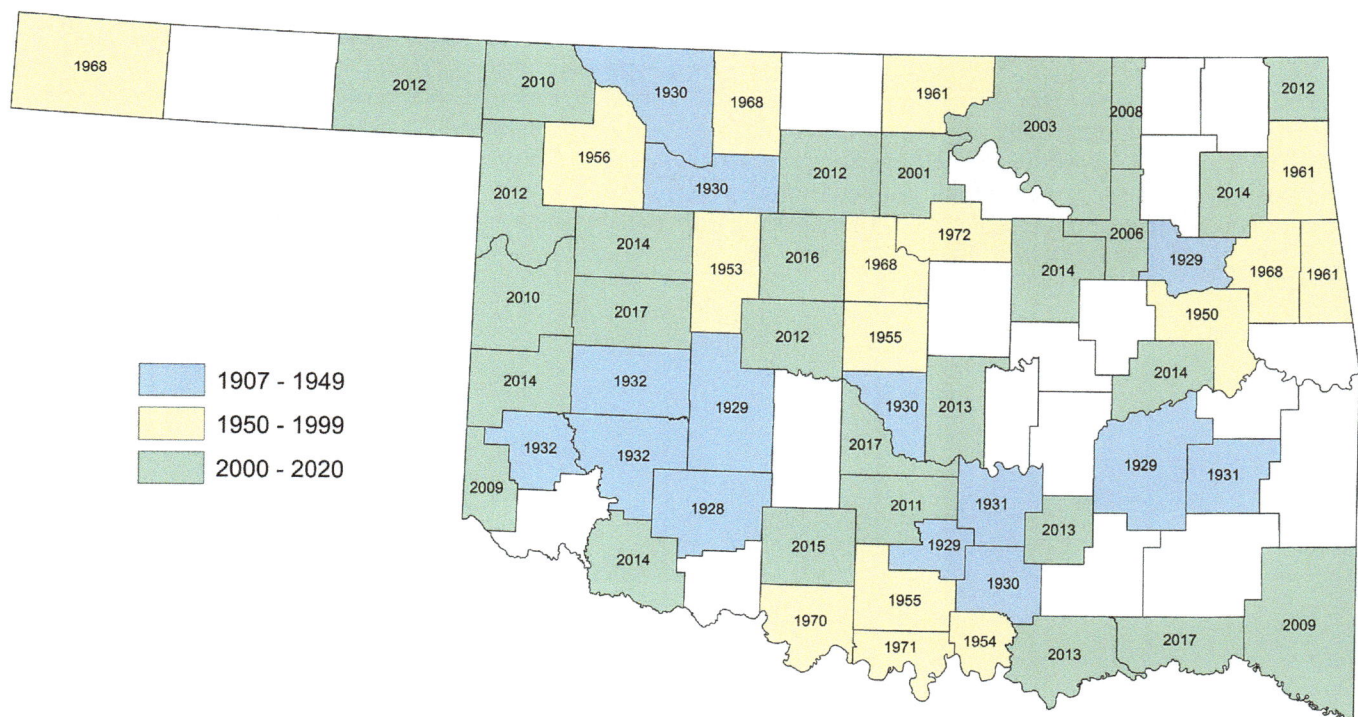

Figure 12.137 Counties in which *Argia plana*, the Springwater Dancer, are known to occur in Oklahoma. Year within the county is that in which the species was first reported.

ARGIA LUGENS (HAGEN, 1861) – SOOTY DANCER

In Oklahoma, the Sooty Dancer is known only from Cimarron County near Kenton and Black Mesa SP, which unsurprising given that the Oklahoma panhandle is at the edge of the species' range. This is a species of rocky flowing streams and rivers, a rare habitat in the panhandle, and the panhandle is a place that experiences routine droughts; even though annual rainfall is decent (about 43 cm, or 17 in, per year), it often falls in buckets, resulting in massive flooding and scoured water courses not necessarily amenable to this species.

Argia lugens has been recorded only seven times in Oklahoma. The first was when AE Pritchard collected 8♂ and 3♀ on 29–30 June 1933 near Kenton.[70] Thirty-seven years later, GH Bick and LE Hornuff had the species but once: 1♂, 2.4 mi NE of Kenton, 5 August 1970 (FSCA). The next time it was seen was on 8 June 1983, when HC Reed and J Nelson collected 3♂ and 1♀ at Black Mesa SP. These specimens went unidentified until 2011, when we found them amongst unidentified specimens at OSU. The Sooty Dancer went unseen again for another 30 years before BC saw 3 adults. He photographed a ♂ along the Cimarron River near its junction with North Carrizo Creek (3 September 2013, OC 410018; Figure 12.138). He saw the species again there 2 years later, this time noting 6♂ (7 September 2015, OC 436524). MAP also encountered *A. lugens* in 2015, this time along North Carrizo Creek, 7 km N of Kenton (8 August,

1♂, SP 1743). He found another lone ♂ along the Cimarron River, 4 km NNE of Kenton, on 3 June 2017, which was the last report for Oklahoma (SP 2379). The only other claim of this species was a nymph from Chisholm Creek in Edmond, Oklahoma County, collected on 27 March 1986 by D. Bass. This specimen (UCO 5238) was re-examined by BS-P and determined not to be *A. lugens*, although it was left undetermined to species.

For such a rare species in the state, it is almost embarrassing to admit the number of specimens in collections. When tallying records for this species we were surprised that of 27 individuals ever known to have been encountered in Oklahoma, 18 are in collections (EMEC, FSCA, OSU, SP, USNM). No nymphs, tenerals, or pairs have been collected or reported. Known flight dates for Oklahoma are similar to those of Colorado but shorter than those for New Mexico and Texas (OK: 3 June–7 September, see above; CO: 18 June–30 September, OC; NM: 28 May–19 October, OC; TX: 6 May–30 October, JCAC 39952, iNat).

We consider *Argia lugens* to be an S1 species of special conservation concern in Oklahoma. We did so previously because of an indication there was a possible population decline and range retraction (Patten and Smith-Patten 2013b). Before our assessment came out the species had been known from three records, two of which indicated a probable healthy population. By all accounts Pritchard was

Figure 12.138 *Argia lugens*, the Sooty Dancer, A) ♂, Cimarron River 4 km NNE of Kenton, Cimarron County, 3 September 2013 (©Bill Carrell, OC 410018), the day the species was rediscovered in the state after 30 years of not being seen. A ♀ has not been recorded in Oklahoma in nearly four decades, yet as evidenced by this individual B) from the Grant River Preserve, Grant County, New Mexico, 14 August 2017 (©Bill Carrell), the ♀ is distinctive. Watch for it.

an excellent field biologist, but it is likely that he collected only a portion of the *A. lugens* he saw, as the species is pestiferous in its wariness, and its preferred habitat makes it difficult to catch. Because his record indicated a sizable population, and the 1983 record (from lepidopterists) also indicated good numbers, it was hard to understand why Bick and Hornuff had only one encounter and why we and others, despite many surveys at those same locations, had not been able to refind the species. This latter point made us wonder about possible range retraction. It may be that we jumped the gun a bit with our earlier assessment, yet we still feel an S1 rank is on the mark. The population, though essentially resident, may be one that moves in and out of the state depending on habitat availability in a given year. Specialized habitat of the species and its sensitivity to land-use pressures also argues for concern. We feel that the species is highly threatened with extirpation from the state (and region, in general) given the strain on water sources in the panhandle as a result of agricultural use that is causing a severe loss of streamflow in the region. Concomitant with human water-use pressures, the region is likely to experience continued drought, the severity of which, although debatable (Basara et al. 2013), may be more common and persistent.

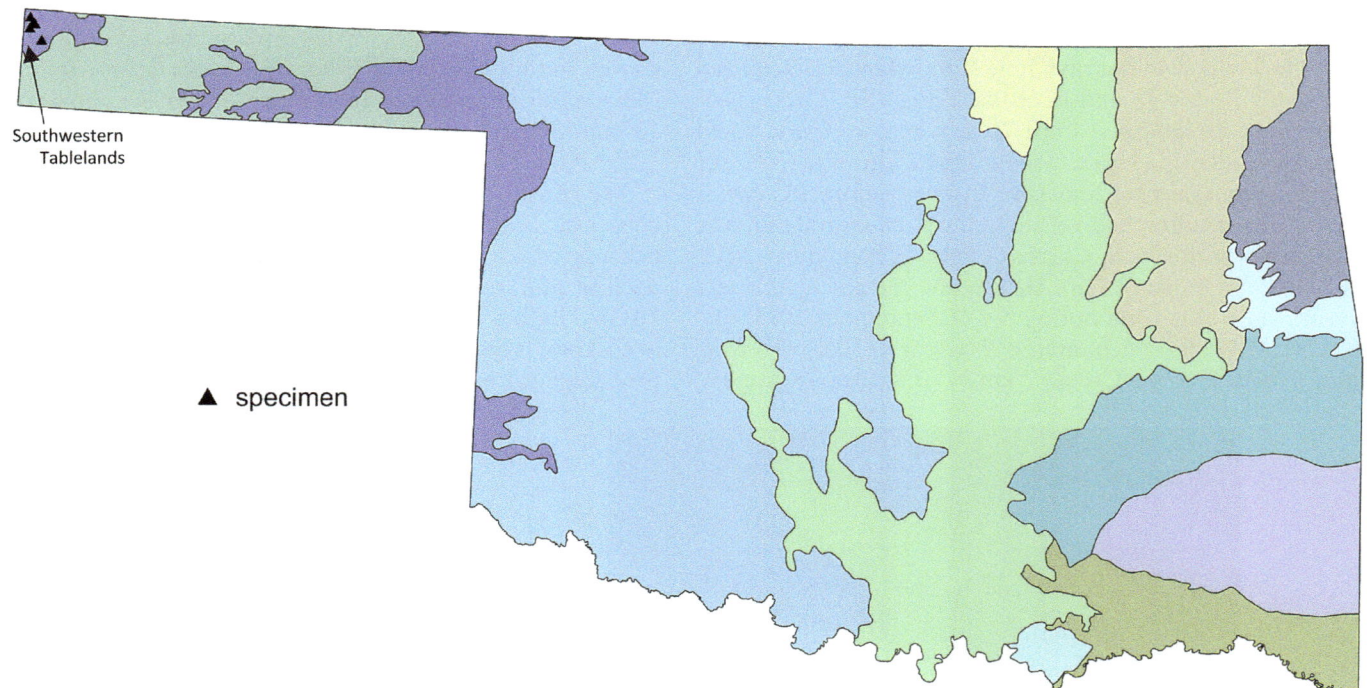

Southwestern
Tablelands

▲ specimen

Figure 12.139 **Only know from the Oklahoma panhandle, *Argia lugens*, the Sooty Dancer, is well documented by specimens.**

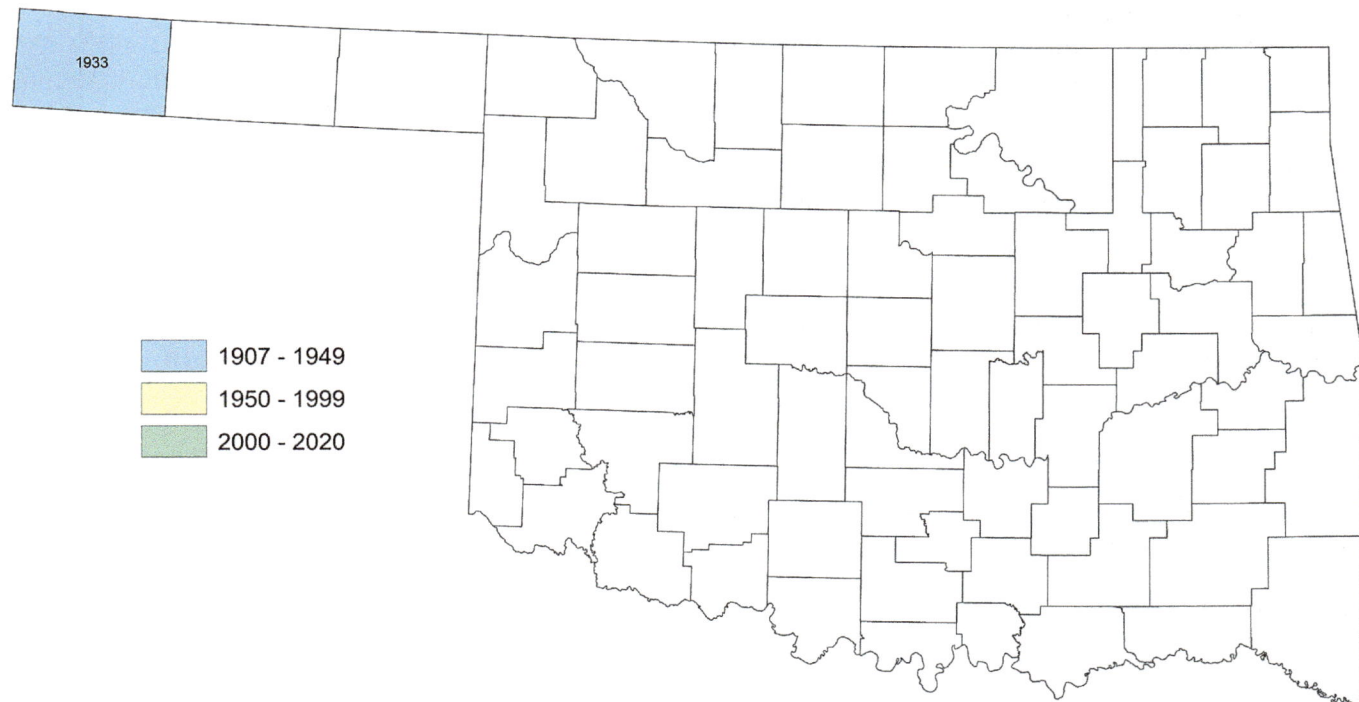

1907 - 1949
1950 - 1999
2000 - 2020

Figure 12.140 *Argia lugens*, the Sooty Dancer, is known from only one county in Oklahoma. It was first recorded in Cimarron County when AE Pritchard collected the species at or near Kenton on 29–30 June 1933.

ARGIA TIBIALIS (RAMBUR, 1842) – BLUE-TIPPED DANCER

The Blue-tipped Dancer is an uncommon lotic species of rocky streams and creeks. It was first recorded in Oklahoma in Le Flore County at Wister in early June 1907. Williamson (1914b:414) reported the record as "Oklahoma: Wister, June 3 and 4, and Aug. 2 (Collins), 6♂, 3♀; 3 pairs and 3♂ in alcohol, det. Calvert." We found 6♂ and 3♀ at UMMZ, all labeled as collected on 3 June 1907. The species was next reported in June and July 1925 (Ortenburger 1926b) and was seen somewhat regularly through the 1930s (41 records in 17 counties, 1925–1939). The species was also more-or-less regularly reported in the Bick/Hornuff era, with 31 known records for 11 counties. Currently, it is rare to go to eastern Oklahoma in the summer and not note the species at least once during an outing that includes creek walking. All told there are at least 165 records of *Argia tibialis* in about 35 counties in the state, but there is some indication that the species may be encountered in smaller numbers now than in previous eras (Table 12.1).

Argia tibialis' highest abundance and most of its distribution is in eastern Oklahoma. It is usually reported <10 individuals at a time, but there are a hearty handful of records of >20 individuals, all of which are from eastern counties, including the high count of 40 (8♂, 8♀, 24U, including

8 pairs, along Mill Creek, 5 km NE of Hanna, McIntosh County, 4 July 2014). Records from central and southwestern Oklahoma may be anomalous or at least are merely extralimital. For example, the Logan County records, both from Fitzgerald Creek, 2 km N of Langston (5♂ [1 as SP 306] and 1♀ on 4 July 2012, and 1♂ on 9 September 2017) are conceivably at the western extent of the species' "normal" range. Murray County provides a better example of why we feel the species is at its range's limit in central Oklahoma: *A. tibialis* was recorded in the county in 1931 and 1954[71] but not since, even though we have spent many hours walking creeks in the county. We feel the species may still occur in the county on occasion, but it probably does not have a stronghold there. Likewise, the southwestern records may reflect the species pushing westward in more productive years and retracting in those less so. Alternatively, it could be that the species has made attempts to colonize the Wichita Mountains area as other eastern species have done (or vestigial populations remain from a time when suitable habitat was contiguous). At this point it is difficult to determine cause and effect, but we do know that *A. tibialis* was first recorded in Caddo County in 1937 but not seen there since 1939[72] and it was only ever recorded in Comanche

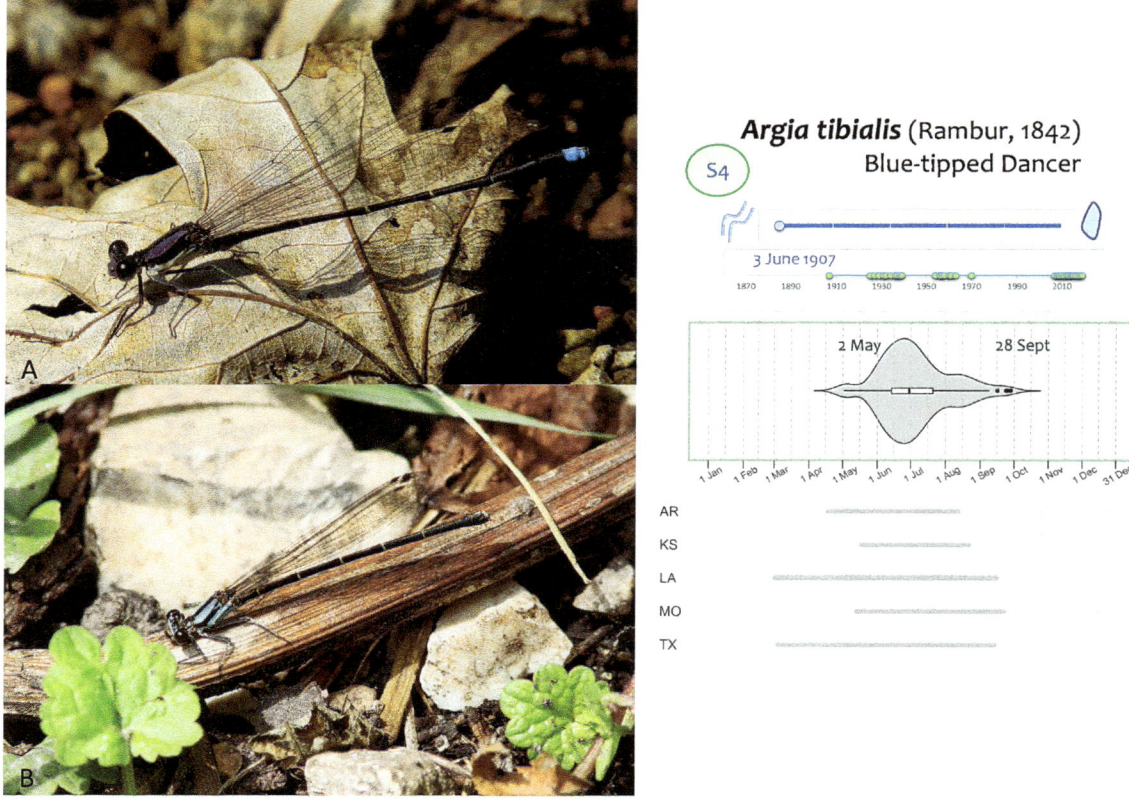

Figure 12.141 *Argia tibialis*, the Blue-tipped Dancer, A) ♂, Little Cabin Creek 6 km NE of Vinita, Craig County, 7 July 2013 (©Bill Carrell, OC 401672). B) ♀, Natural Falls SP, Delaware County, 16 July 2013 (©Ken Williams, OC 403373).

County in 1932.[73] Perhaps the species will be found in those counties again, as it was in Washita County, where it was first recorded in 1932 (2♂) but not recorded again until 2014 (1♂). On the flip side, Kiowa County, the only other southwestern county where *A. tibialis* is known, got its first record in 2018 (12 May, 3♂, 1♀, including a pair, Saddle Mountain Creek, 3 km NW of Saddle Mountain, MAP, SP 2594).

But it is not just in the southwestern part of the state where the species was reported early on and not seen again. Four other counties have similar histories: Cherokee (only reported in 1936), Mayes (1929 and 1970), Ottawa (1961), and Wagoner (reported only by Bird 1932). We thought the same of two other counties but recent records have shown that to be incorrect: Adair (seen in 1937 but not again until 2012–2016) and Haskell (1931 and again 2013–2016). Detectability may play a role in these mixed signals, but we think of this species as not terribly difficult to find.

Argia tibialis' seasonality in Oklahoma does not extend quite as long as other dancers, being confined to the window of early May to late September (2 May 2015, Raymond Gary SP, Choctaw County, 10♂, 5♀, OC 430735, and 28 September 2009, 7 mi SE of Smithville, Hudson Creek, McCurtain County, 1♂, BAH, OC 315567). Arkansas, Kansas, and Missouri have shorter reported seasons, but there is no reason to think they would not be as long as Oklahoma's though none of these states ought to be expected to have as long a season as Louisiana and Texas does (AR: 18 April–13 August, OC, JCAC; KS: 31 [18?, SEMC] May–22 August, OC; LA: 2 March–16 September, USNM, Mauffray 2014; MO: 14 May–22 September, Trial 2005; TX: 5 March–14 September, iNat, Abbott and Lasley 2019). Fewer than 20 pairs have been reported, there are no accounts of ovipositing, and no teneral records are known.

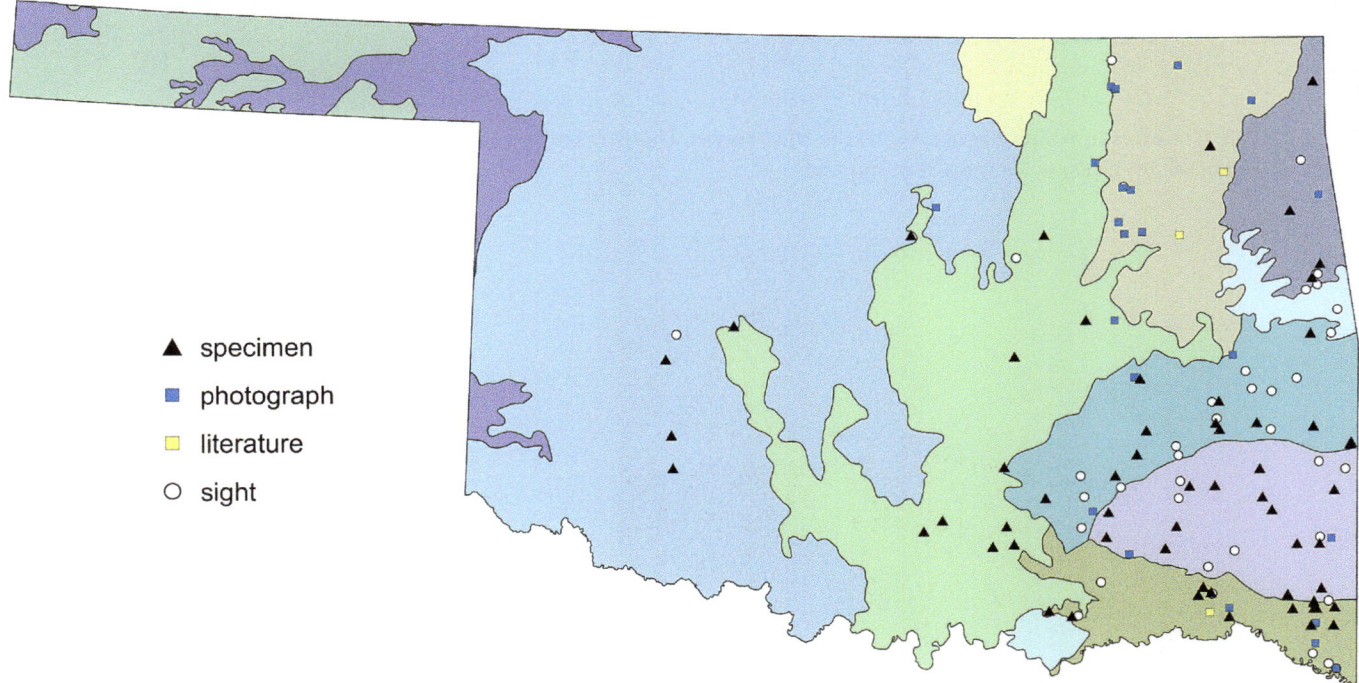

Figure 12.142 Documentation level of supporting records for *Argia tibialis*, the Blue-tipped Dancer, in Oklahoma. Levels include specimen, archived photograph, literature reference, or sight record. Although sight records lack documentation, only those we feel are credible are plotted.

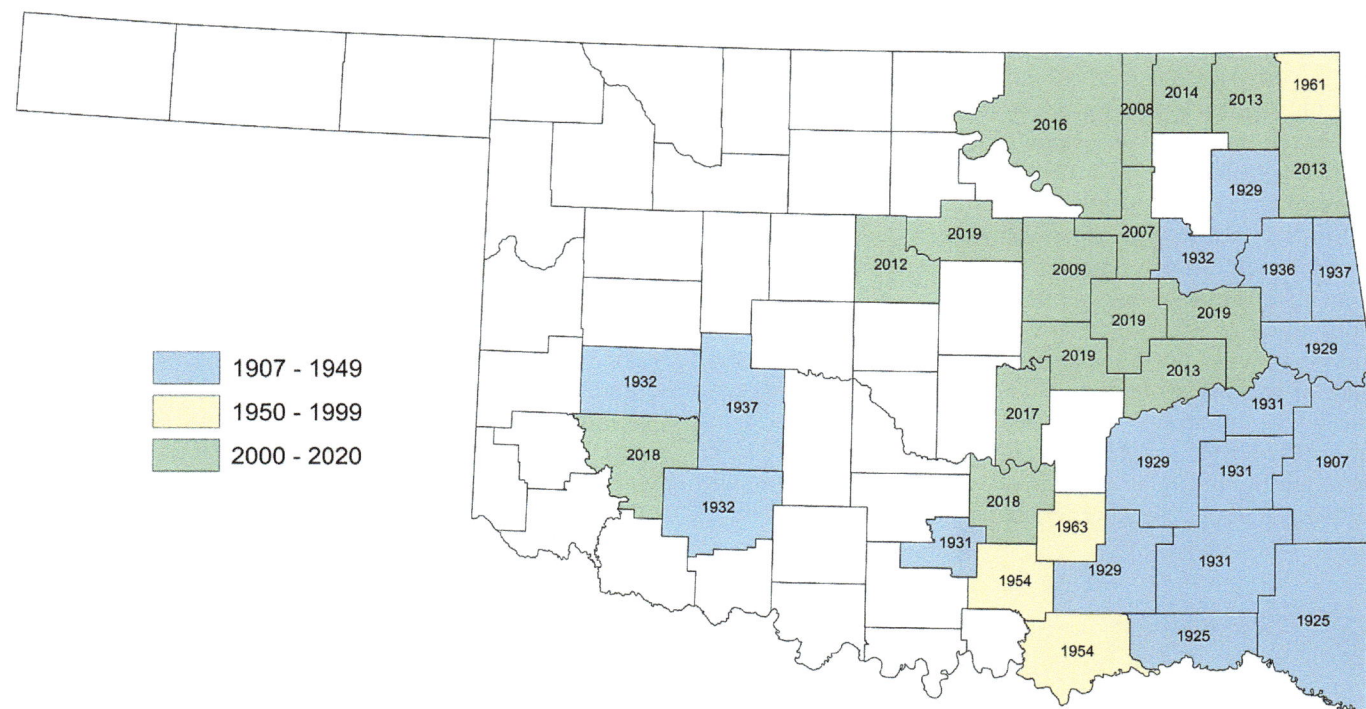

Figure 12.143 Counties in which *Argia tibialis*, the Blue-tipped Dancer, are known to occur in Oklahoma. Year within the county is that in which the species was first reported.

ARGIA TRANSLATA HAGEN, 1865 – DUSKY DANCER

Looking at record distribution over time for *Argia translata* makes it difficult not to think this highly lotic species has had a population explosion in the state both in terms of its abundance and geographic range. The species was first discovered in the state by RD Bird on 29 September 1929 in Murray County (specific location unknown), where he collected 4♂ and 1♀ (OMNH 1508–1512). It was recorded only a dozen more times and in four other counties in the state before the end of the 1930s. During the Bick/Hornuff era, *A. translata* became known from five more counties, bringing the total to ten and upping the number of records by a little more than 40. During both of these eras, the number of individuals reported was generally <5 at a time. Indeed, of the 28 records of the species ever numbering >10, only 2 were before 2006. Those 2 records were both counts of 18 individuals: 27 May 1930, Oil Springs, Johnston County (10♂, 8♀, including 2 pairs, OMNH 1498–1507, 3092, 3098, USNM 366248–366251), and 8 August 1954, Beavers Bend SP, McCurtain County

(GHB, FSCA 9♂, 3♀ of the 13♂, 5♀ reported in GHB's notes). In contrast, since 2002, there have been an additional 200 records, 43 counties have been added, and there have been 9 records with >20 individuals, including the high count of 75 individuals (50♂, 25♀, including 20 pairs, Camp Simpson, Johnston County, 12 October 2013). Expected abundance corrected by survey effort per era (Table 12.1) likewise indicates that *A. translata* has become more common over time. All of this makes a good case to argue the species has expanded its range in Oklahoma and has become more abundant. But why would a lotic species, even one that is known to inhabit a variety of streams (for example, <1m to 10m wide, but usually wooded), have spread across the state?

Argia translata has an overall odd distribution through the United States. Instead of the more typical blanketing of the eastern United States that some other damsels do, including other *Argia*, *A. translata*'s range weaves and bobs like a river. As Paulson (2011) noted, this species skips over

Figure 12.144 *Argia translata*, the Dusky Dancer, A) ♂, Cimarron River 4 km NNE of Kenton, Cimarron County, 7 September 2015 (©Bill Carrell, OC 436525), an individual far from other regional records. B) teneral ♀, Barren Fork Creek at Welling, Cherokee County, 25 June 2018 (©CA Ivy, OC 483103).

much of the southeastern lowlands of the U.S., although it inhabits low-lying areas in the neotropics, which makes little intuitive sense. That skipping is even more perplexing because in Oklahoma *A. translata* seems to have spread northward from the part of the state most like the southeastern U.S. Since 2010, it appears to be spreading into the decidedly more arid southwestern part of Oklahoma, too. While on the topic of apparent distributional oddities, note the single record of *A. translata* from Cimarron County—1♂ photographed by BC at "Watson's Crossing" on the Cimarron River, 13 km E of Kenton, 7 September 2015 (OC 436525). On a map of Oklahoma the record looks well out of range but change the scale of the map and, at least as the ode flies, the San Miguel County, New Mexico, record (OC 154210) is not far. Both of these may represent vagrant records or they may be an extension of the species' range that has not been well documented. Only time will tell with this inscrutable species.

Argia translata is known to fly in Oklahoma for an exceptional 238 days (25 March 2017, 1♂, Carl Albert Park, Durant, Bryan County, MAP., and 18 November 2007,

1♂, 1♀, Wichita Mountains WR, Comanche County, VWF, OC 263850, ♀ was ovipositing alone). In comparison with other Oklahoma *Argia*, that ranks *A. translata* as tied for second with *A. immunda* and behind *A. plana* for longest documented flight season. In the case of *A. translata*, it is in comparison to surrounding states that makes its flight season most interesting. Not surprisingly, the species flies in Texas year round (Abbott and Lasley 2019). But if we eliminate Texas because of its unfair advantages, we see that Oklahoma's season is by far the longest season in the region, being 53 days longer than New Mexico's, which has the next longest known flight period (AR: about 123 days, 18–21 May–18 September, Harp 2006 unpublished report without specific date, OC; KS: 112 days, 14 June–4 October, SEMC, OC; MO: 110 days, 7 June–25 September, Trial 2005; NM: 185 days, 10 April–12 October, OC). Pairings are reported regularly and there may be a tendency toward ovipositing as pairs, as we know of slightly more records involving pairs rather than single females. No teneral records have been reported.

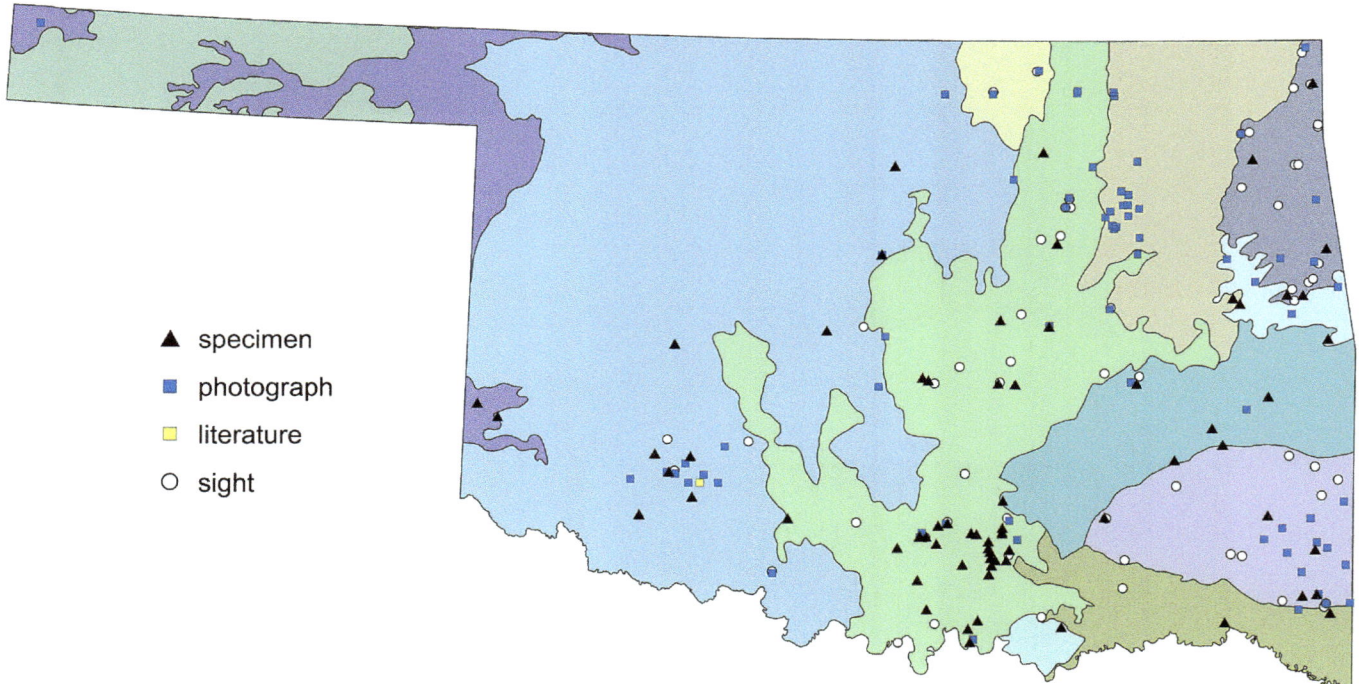

Figure 12.145 Documentation level of supporting records for *Argia translata*, the Dusky Dancer, in Oklahoma. Levels include specimen, archived photograph, literature reference, or sight record. Although sight records lack documentation, only those we feel are credible are plotted. Note the outlying record of the species in the far panhandle.

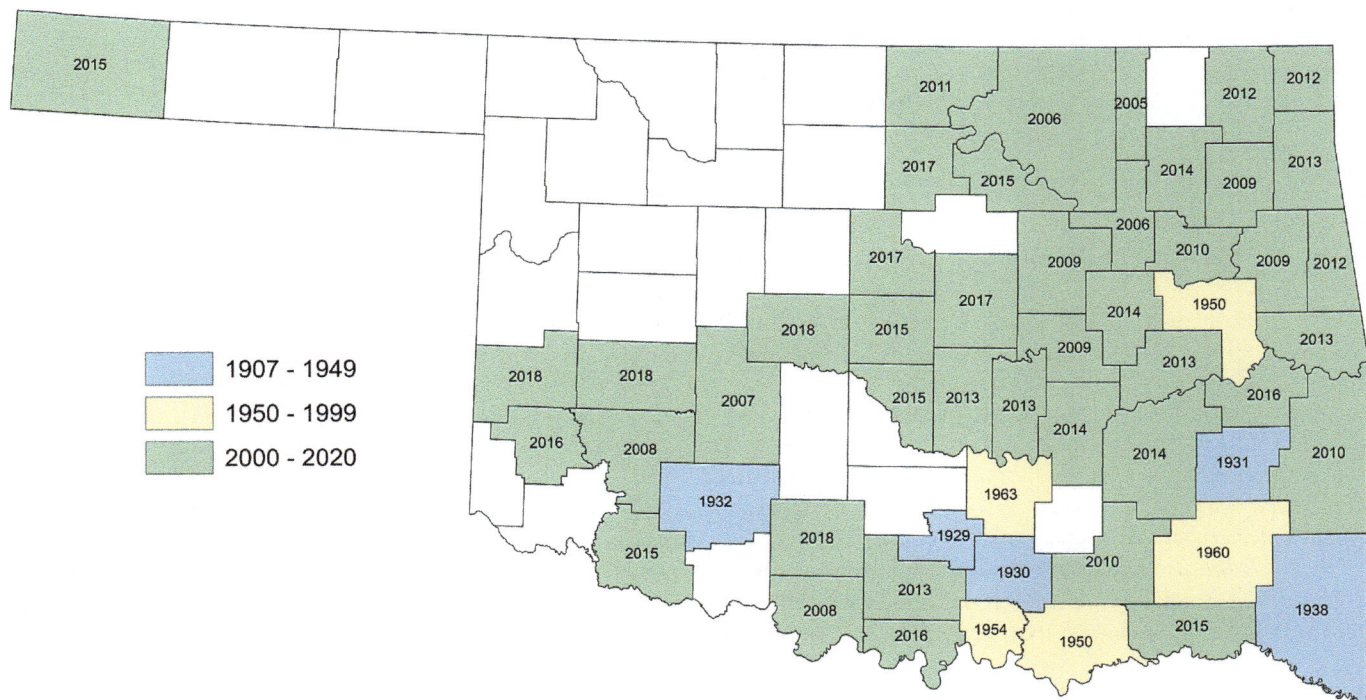

Figure 12.146 Counties in which *Argia translata*, the Dusky Dancer, are known to occur in Oklahoma. Year within the county is that in which the species was first reported.

ARGIA TEZPI CALVERT, 1902 – TEZPI DANCER

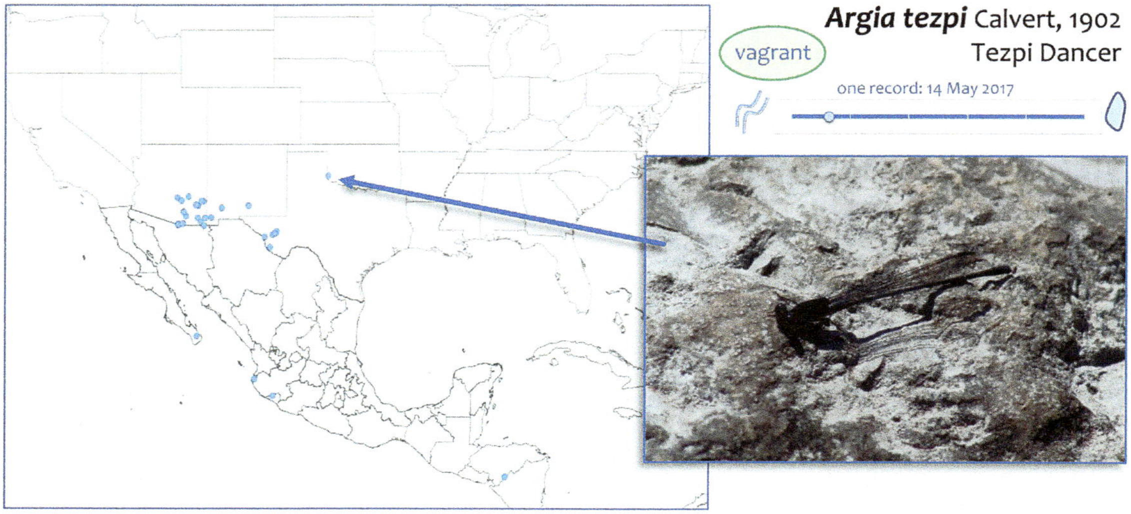

Figure 12.147 Oklahoma's only record of *Argia tezpi*, the Tezpi Dancer, was an utterly lost lone ♂ (blue dots on map are other *A. tezpi* records). This ♂ was collected at Lake Hall, Harmon County, on 14 May 2017 (©Michael A Patten, SP 2328, OC 462696).

The Tezpi Dancer is a subtropical species that barely ranges into the United States. It has been encountered most often in Arizona and New Mexico, but there are about a dozen records for Texas. There is one record for Oklahoma, and although that record was essentially along the border with Texas, it is some 600 km from the nearest record for that state and roughly the same distance from the nearest New Mexico record.

On 14 May 2017 MAP collected a lone ♂ (SP 2328; Figure 12.147) at Lake Hall in Harmon County. This record is not likely to be repeated, but one never knows. It may be that high winds brought the ♂ here, perhaps due to shifting currents from Tropical Storm Adrian that was in Central America on 9–10 May, but we are wary of that idea. If the species were to show again it arguably could be anytime during its known flight season for the U.S. (AZ: 17 April–1 November, OC; NM: 28 April–13 July, OC; TX: 20 May–24 June, OC).

NOTES

1. Or, in plainer English, we grouped these species because they superficially look alike. The five species mentioned are not necessary closely related; the Arroyo Bluet (*E. praevarum*) and perhaps the Alkali Bluet (*E. clausum*) are separated from the others by differences in the general structure of the terminal appendages (see, for example, Byers 1927 who broke these species into groups based on appendages: Group A included *E. praevarum*, Group B *E. civile*, *E. carunculatum*, and *E. doubledayi*, Group C with *E. clausum*.

2. Note that 2015 was a particularly wet year for much of Oklahoma, including the panhandle. It was also the most productive year in decades for *Argia lugens*, the Sooty Dancer, another panhandle species of conservation concern.

3. BC submitted a photograph of a ♂ that he saw at "Watson's Crossing" on the Cimarron River, on 8 September 2014 (OC 426996). Much debate transpired over this record (six people, including Nate Kohler, Dennis R. Paulson, and Bill Prather, chimed in) but no consensus could be reached. It was felt that the photograph, lacking details of the terminal appendages, could not conclusively be said to be *Enallagma praevarum*, as *E. carunculatum*, the Tule Bluet, given the rather wide variation in that species' abdominal pattern, could not be eliminated with certainty. For that matter, *E. civile*, the Familiar Bluet, with its *extreme* variation in patterning, could not be eliminated either, nor could the possibility that the record was the first for Oklahoma of *E. clausum*, the Alkali Bluet. In the end, BS-P, despite her leaning toward *E. praevarum*, decided to leave the record as "unconfirmed" given that it would have been the first record for Oklahoma for 35 years. A second record, that MAP chose to label as *E. praevarum/carunculatum*, was of a ♂ he saw on 3 June 2017, 4 km NNE of Kenton on the Cimarron River. His designation indicated that he leaned toward identifying it as *E. praevarum* but did not feel he could confidently rule out *E. carunculatum*.

4. GH Bick and LE Hornuff's records for Cimarron County between 1968 and 1978 are: 2 August 1968, Lake Carl Etling, LEH, 1♂ at FSCA; 3 August 1970, Black Mesa SP, GHB, LEH, 1♂ at FSCA; 5 August 1970, 2.4 mi NE of Kenton, GHB, 4♂ at FSCA; 11 August 1973, 2.1 mi W of Kenton, LEH, 3♂ and 1♀ at FSCA; 5 August 1978, Black Mesa SP, GHB, LEH, 2♂ at FSCA.

5. All specimens are from SEMC. There is one nymph from Barber County, which is in southcentral Kansas, on the border with Oklahoma. Cheyenne County, in northwestern Kansas, is reported to have one adult and two larval specimens. None of these specimens have been confirmed by us but they were determined by David Huggins

6. We and others were not able to check the ponds in 2018 and 2019.

7. It seems unlikely the Bicks would have meant Bermuda grass, but we cannot come up with anything else to which this reference would relate.

8. Note that the original label for the *Enallagma carunculatum* that is cataloged as OMNH 4351 had a "write-o" that indicated the specimen was collected in 1939 rather than 1930, but we believe it is the specimen that matches the 27 July 1930 record from Bick's note cards and Bick and Bick (1957). The original label matching this specimen had: "Enallagma/carunculatum/Morse/Det. RD Bird 1931; Cimarron Co. Okla./27.VII.39 [sic, 1930]. RD Bird." Although we think it in no way challenges the validity of this record, that there were two labels and two specimens in the one vial that OMNH 4351 came from could present a problem. It appears this was a quick fix at some point in OMNH's history when one of the specimens was found in a vial that had dried out so the specimen and its label was put into a vial with a different species (original label for the second specimen is: Enallagma/basidens/Det. RD Bird 1931; Murray Co. Okla./18 IV 1930/RD Bird Falls Creek). The main reason to explicate this mix up is because currently both of these specimens are cataloged under the same number.

9. There is some confusion as to the dates Bick and Hornuff were at Black Mesa SP. Bick's note cards indicate they were there on 4–5 August 1970 (Figure 12.26), but specimens at FSCA (12♂, 1♀) indicate the 3rd and 5th of August.

10. SEMC 1348708 has date range of 4–8 June.

11. MAP reported 20♂ at a roadside pond 14 km ENE of Nowata, Nowata County, on 22 July 2016 (OC 450988). J. W. A. estimated 60U at Haskell Lake, Muskogee County, on 15 May 2014 (OC 422128). And VWF said the species was "fairly common" (on our standardized estimation scale, that warrants a minimum count of 76 individuals) in the Wichitas, Comanche County, on 4 May 2008 (OC 282128).

12. AR: 7 April–22 July, OC 398651, Harp and Harp (2003); KS: 8 April–17 September, OC 374334, SEMC; LA: 1 March–18 August, OC 478522, Mauffray (2014); MO: 21 April–10 October, Trial (2005); TX: 6 February–6 December, Abbott and Lasley (2019).

13. AR: 11 May–20 August, Harp and Harp (2003), OC 6534; KS: 17 May–5 August, OC 322835, 378450; LA: 23 April–24 June, OC 462216, 313386; MO: 11 April–28 September, OC 374387, Trial (2005); TX: 15 April–4 August, OC 462795, Abbott and Lasley (2019).

14. The lone ♀ was seen on 11 July 2015 by MAP at Atoka PHA, Atoka County (also 2♂ present). The pairs were seen ovipositing in Honobia Creek WMA, 7 km NE of Corinne, Pushmataha County, on 2 May 2015.

15. AR: 15 April–17 September, USNM 375977, Harp and Harp (2003); KS: 20 April–20 September, SEMC 1348785, although SEMC 1348788 is labeled as "19 April–3 May 1976," late date is from OC 382087; LA: 1 April–10 September, OC 483763, Mauffray (2014); MO: 17 April–25 September, OC 374986, 374436, Trial (2005); TX: 20 February–28 November, iNat 2709197, 144632.

16. Pritchard (1935) casually mentioned "*Teleallagma daeckii*" as being at "Antlers Lake" north of Antlers. He provided no other details, so we do not know how many times he saw the species or in what numbers. Fortunately, there are specimens that provide us with some indication. EMEC (331092, 331093) has 2♂ and 1♀ with the date of 16 June 1934 and 1♂

with the date as 26 June 1934 (EMC 331094). There are 2♂ at MCZ with the day and month as 26 June but without a year, which we assume is 1934.

17. Bick and Bick (1957) said they examined 3♂ and 1♀ that were collected on 15 June 1930, but Bick's notes indicated there were only 2♂ and 1♀. For 16 June 1930, Bick's notes indicated they examined 4♂, but we have only located 3♂ (OMNH-3240).

18. AR: 5 April–1 June, OC 374291, USNM 375370, 375371; KS: 20/22 April–19 June, SEMC, OC 322815; LA: 3 March–12 June, USNM 375399, Mauffray (2014); TX: 22 March–3 June, Abbott and Lasley (2019). Dates for Missouri as presented by Trial (2005) are problematic: 21 April–31 August. Given that the latest date known for the state from other sources is 23 June (CMNH, USNM), which is much more in line with the species' seasonality elsewhere, we suggest that the July and August dates presented by Trial's (2005) were actually meant to be for nymphs, not adults. Trial's (2005) latest June date was the 24th, which we are willing to accept as Missouri's late date.

19. Williamson (1914b:445). We found 2♂ at UMMZ and 4♂ at FSCA. Donnelly (1973) said he examined 1♂ from MCZ, but we did not find that specimen when we visited in February 2012.

20. Dates for the Bird records are confusing. It is likely that one pair was collected on 18 June 1931 (OMNH 3484) and a ♀ (OMNH 3469) was collected on the 9th of either June or July 1931. However, Bick and Bick (1957) and Bick's notes indicate that the pair was collected on 18 June 1932 and that the ♀ was collected on 9 July. The pair's specimen label has "18 VI 1931" on it, so Bick may have mistakenly written down the year of determination, 1932. The primary specimen data as transcribed in the OMNH 3469 record has "9.VI.1931"; we were not able to double check the label.

21. MAP estimated there were 90 individuals (80♂ and 10♀, including 8 pairs) at Weleetka Lake, Okfuskee County, on 29 May 2016 (2♂ in hand, 1♂ as OC 445179). Our highest estimate was of 500 individuals (150 pairs and 200U) at Vian Lake, Sequoyah County, on 6 June 2015 (1 pair as SP 1625).

22. Oklahoma's early date is 5 May (from 2012, Bristow, Creek County, 1♂, JT Bried, SP 472). If OMNH 3469 is from 9 July (see endnote #20 above), then that is the late date for Oklahoma. If however, that specimen is from 9 June, then the late flight date record belongs to DA (7 July 2017, Red Slough, McCurtain County, 1♂, OC 466236). AR: 5–12 June, OC 446040, 469060; KS: 28 May–7 July, OC 319341, 498662; LA: 17 April–24 May, Mauffray (2014), OC 481871; MO: 21 April–18 July, Sims (2012); TX: 11 April–16 August, Abbott and Lasley (2019), OC 501858.

23. Oklahoma's early flight date is taken from 22 March 1931 (12♂ and 10♀, Murray County, RD Bird, OMNH 3555–3562, 3593–3595). The late date in Oklahoma was recorded by VWF, on 30 October in 2007 and 2008 at Fort Sill MR, Comanche County (OC 263514, 4–5U; OC 284895, 1U). AR: 24 April–15 September, OC 443908, 495631; CO: 16 May–29 August, OC 375001, 489348; KS: 22 April–16 October, Huggins (1978); LA: 4 April–5 October, Mauffray (2014); MO: 9 April–14 October, OC 374411, Trial (2005); NM: 13 March–22 September, OC 430253, USNM 386223; TX: 7 February–30 December, Abbott and Lasley (2019).

24. The summer of 1932 is also a possibility as evidenced by a letter from Pritchard to EB Williamson, dated 30 November

1932 (Figure 15.60). In that letter, Pritchard said that he added *Enallagma dubium* to the state's list. He gave no details and for all we know he may have backtracked on that identification, so we have left 1934, when we know he collected the species in the state, as the first record for Oklahoma.

25. Access to this small lake (1.6 ha or 4 acres) is limited to about 20 m (70 ft) of shoreline, so visits have been relatively short (≤30 mins). We suspect the lake holds a good-sized population, but one would need to kayak around it in order to survey fully.

26. Oklahoma's early flight date is from 16 March 2016, Red Slough WMA, McCurtain County, DA sight record of 1♂. The late flight date is from 6 November 2009, also from Red Slough, but of 1U reported by BAH (OC 315791). AR: 27 March–2 November, OC 374197, OC 439017; KS: 22 or 28 April–5 October, SEMC, OC 333979; LA: 13 February–10 November, Mauffray (2014); MO: 5 May–10 October, OC 374732, Trial (2005); TX: year round, Abbott and Lasley (2019).

27. 20♂ and 2♀, including one pair, 2 May 2015, Broken Bow city park, McCurtain County. 20 adults (17♂ and 3♀) and 2U tenerals, 6 May 2017, Atoka PHA, Atoka County, MAP.

28. AR: 8 April–11 May, OC 441073, OC 7661; KS: 5 June–1 September, SEMC 1347752, OC 436057; LA: 31 March–17 August, Mauffray (2014); MO: 5 May–21 September, Trial (2005); TX: 11 March–28 November, Abbott and Lasley (2019), iNat 4666648.

29. The only exception is a side channel of the Glover River that is essentially a pond but it is technically a pool within a stream bed (it is likely that during heavy flows of the river this side channel floods). In recent years, the channel has been holding "lily pads" and on 6 May 2018, BS-P encountered 1♂ and 1♀ there (photos of ♂, on file).

30. G. H. Bick's notes indicated only 1♂ and 1♀, but we cataloged 2♂ and 2♀ at FSCA.

31. Our report of the species from Beckham County may account for another in the county list (1♂, Lake Elk City, 5 km SSE of Elk City, 25 September 2010), yet, years later, although we cannot say exactly why, we felt queasy about this record. Because we obtained no documentation or took any notes on the characteristics of this individual, we decided it best to not validate the record even though the county is entirely reasonable for the species, just as it is for *Ischnura barberi*.

32. There are discrepancies in the individual counts for this record (exact locality was 4 mi E of Ingersoll, Alfalfa County). Bick's notes indicated there were 35 indiv. (17♂, 18♀) but Bick and Bick (1957) reported only 31 indiv. (13♂, 18♀). Furthering the confusion is that FSCA has only 30 specimens (13♂, 17♀) while Bick's notes indicated that 3♂ went "to Kiauta" and 2♂ and 2♀ were "SS," an abbreviation we have yet to decipher.

33. Of note about Bird's (1932) reports of this species is that he reported it for four counties: Alfalfa, Jackson, Woods, and Woodward. Although the species has since been recorded in Woodward County, we do not know the basis of Bird's record for that county.

34. SP 624, 626, and 627 are now part of the RWG collection.

35. As an indication of the rarity, we point to RW Garrison, in reference to SP 628 and SP [RWG] 1600, saying "The strongly elevated posterior margin of the mesostigmal plates (much higher and more prominent than in *I. verticalis*) and

the broadly projecting medial lobe of the hind lobe of the prothorax (much reduced or gently convex in *I. verticalis*) separates females of these two species [*damula* and *verticalis*]. What I was unaware of was the interesting condition of the pronotal tubercles that are normally present. Your two females essentially lack these and … [SP, RWG] 627 … [also lacks] pronotal tubercles … I checked my entire series of females of … *I. damula* … and all have the tubercles though some females have these structures reduced although not almost absent as in your females. Apparently this condition has been acknowledged…[by Westfall and May (2014:355)] … under the account for *Ischnura damula*: "Female. - Middle lobe of pronotum with prominent, nipple-like (or rarely broadly rounded) process on each side."

36. SP 624, 626, and 627 are now part of the RWG collection.

37. It may be that the species was actually first recorded in 1932, as indicated by a letter from AE Pritchard to EB Williamson, dated 30 November 1932 (Figure 15.60). In that letter, Pritchard said that he had added *Ischnura demorsa* to the state's list. He gave no details for the record and we have not found specimens. Because we do not know if Pritchard later re-identified his specimen(s), we chose to leave the first date of discovery in Oklahoma as Bick's record.

38. There are 2♂ and 3♀ from Harper County at FSCA. There are 14 individuals (4♂ and 10♀) from Woodward County at FSCA; however, Bick and Bick (1957) indicated GH Bick collected 13 (4♂ and 9♀) and Bick's note indicated he discarded 1♀.

39. Two years previous, to the day, we collected a male (SP 1607) at this location that we originally identified as *Ischnura perparva*, but which, on later inspection, actually was *I. demorsa*. The male's abdomen is smashed, but he clearly has a high, narrow fork dorsally on S10 and has no teeth on the paraprocts; however, the shape of the paraprocts, being more of a subequal C, is comparable to *I. perparva*. We feel that the apparent incongruity of features is a result of damage to the abdomen, not mutation or hybridization.

40. Raney (2002) reported *Ischnura verticalis* five times for Red Slough WMA, McCurtain County, in 2002. His report indicated that he had a total of 36 individuals: 24 on 20 April, 8 on 11 May, 2 on 15 June, 1 on 13 July, and 1 on 27 July. No specimens or photos were taken as documentation. *I. verticalis* had never been reported for the WMA nor has it been since despite the hundreds of surveys conducted there over the years. If only one or two individuals had been reported we may chalk up the record(s) to vagrants, but the large numbers reported indicate something else happened.

First, it should be noted that no field guides for damselflies existed in 2002, so identifications had to be made using outdated keys and descriptions and distributions of odonates in general were not well known. As such, there are several possibilities of what may have happened during the 2002 surveys. *I. kellicotti*, the Lilypad Forktail, a fairly common species at Red Slough, was not noted during the surveys, but we doubt that Raney, a skilled field observer, would have mistaken *I. kellicotti* and *I. verticalis*. Although he reported *I. ramburii*, Rambur's Forktail, our best guess is that he did not recognize how similar andromorph females can look to *I. verticalis* males. Confusion can occur because observers often forget that andromorph *I. ramburii*, although a mimic in other ways, lack the telltale notched S9 dorsum of male *I. ramburii* and thus can be mistaken for *I. verticalis* males. Our next best guess is that there was confusion with *Nehalennia integricollis*, the Southern Sprite, which occurs at Red Slough but was not reported during the 2002 surveys. Raney hypothesized the eventual appearance of this species at the slough, but it could be that it was one with which he was not terribly familiar and did not realize their general similar appearance to forktails. The last option that we can think of is an outside chance of *I. prognata*, the Furtive Forktail, showing up in that part of Oklahoma. That species would likely not have been on his radar as a possibility for the state and really unless a major storm blows a small population in, it is not extremely likely to ever be recorded in the state. The climate reports for early 2002 did not indicate any weather that would have caused such vagrancy (http://climate.ok.gov/index.php/climate/summary/reports_summaries).

41. There are some SEMC specimens that were entered into their database with a date of 6 or 30 November, but those appear to all be in error. Some were actually collected in June and others had only a year on the specimen label.

42. The 14% calculations were made by 1) averaging typical and atypical *I. verticalis* in northwestern Oklahoma and the panhandle (i.e., all atypical individuals/all individuals) = 4.07% and 2) limiting records to just those containing atypical *I. verticalis* ($n = 12$) and then taking a weighted average = 0.73%. Encounter rate was calculated by dividing the number of encounters with atypical individuals ($n = 12$) by the total number of encounters with atypical and typical *I. verticalis* ($n = 62$) = 19.35%.

43. There may be an additional record of an andromorph *Ischnura verticalis* in this region. There is one record from the Texas panhandle (OC 281550, Hartley County, 3 July 2004, M Reid) of a female with blue dorsally on S8 that was identified as *I. verticalis*. However, we question the id because of the pink thorax and the blue dorsally and laterally on S3. This female is in the JCAC collection, but we were not able to confirm the identification.

44. According to the International Code for Zoological Nomenclature (ICZN), which sets rules and principles for taxonomic names of all animals, a *nomen nudum* is "a Latin term referring to a name that, if published before 1931, fails to conform to Article 12; or, if published after 1930, fails to conform to Article 13." Per Article 13.1, "To be available, every new name published after 1930 must satisfy the provisions of Article 11 ["name or nomenclatural act must have been published, in the meaning of Article 8 {i.e., in a "public and permanent scientific record"}] and must … be accompanied by a description or definition that states in words characters that are purported to differentiate the taxon …" By ICZN rules, a *nomen nudum* is not available, meaning it has not met basic criteria as outlined above and thus is not a usable scientific name; however, "availability" under the ICZN may change: the same name may become available later should a proper (i.e., one that adheres to the ICZN) type description be prepared and published, at which time authorship of the name would be the person who published the later work, not the name's original usage.

The gist is that if a scientific name is not published with a written description of its diagnostic characters and it is not published in a scientific journal, that is publicly accessible and archived, then that name becomes a *nomen nudum*. The name remains as such until someone publishes a type description. In the red damsel case, were Jane Doe to publish a type description of a mid-continent species, it would be referred to as *Amphiagrion mesonum* Doe, 2018, not as *Amphiagrion mesonum* Gloyd, 1940.

45. Paulson's (2009) range map has Oklahoma's *Amphiagrion* as *A. abbreviatum* but Paulson (2011) has as *A. saucium*.

46. This part of the creek is private property, which we did not realize upon our first visit to the lake. As such, we have not visited the creek since to determine if the population of *Amphiagrion abbreviatum* persists there.

47. For *Amphiagrion saucium*, Needham and Heywood (1929:304) said "Whedon ('14, p. 92) discovered this species in 'small numbers and in teneral condition along a very small stream leading from the 'slough' to the Minnesota River at Mankato on June 11,'13. A few days after, thousands of the fully colored individuals were copulating and ovipositing in the shallow water among the sedges and the Sagittarias [sic, *Sagittaria*]. A week later their numbers began to reduce and by July 7 but an occasional specimen could be found. During the whole period their distribution was limited to an area of 200 yards along this little rivulet, so narrow that one could easily leap across it anywhere, and but a few inches deep. Such a localization is not what would be expected of a species distributed from the Atlantic to the west."

48. ibid.

49. Trial (2005) indicated the species had been recorded in Missouri from 12 counties, some records of which must not have been conveyed to Donnelly (2004) because he reported only seven counties (Carter, Dent, Howell, Iron, Reynolds, Shannon, Washington; Crawford, Dallas, Phelps, Pulaski, and St. Francois were missing from the list). In 2006 and 2014, Shannon, Reynolds, and Washington Counties received an additional record each and the species was added to Ozark County (OC).

50. The first record of the species was on 19 October 1929 when EB Williamson collected a lone male (UMMZ) from the infamous 15 mi E of Weatherford, Custer County locale. Recall from other entries that this locality was incorrectly reported for Custer County because it actually lies within Caddo County and would be better described as 12 km E of Hydro, a town lying within Caddo County. This record was initially reported by Bird (1932) for Custer County.

51. We consider this to be one record despite Bick's notes reading as one female taken at "Boise City, 5 NE" (we interpreted as 5 mi NE of Boise City) and the FSCA specimen card for a female having 5 mi E of Boise City. Our assumption is that there was a typo along the way but these refer to the same record.

52. It may be that the species was recorded earlier. In a letter dated 30 November 1932, Pritchard wrote to EB Williamson saying that he had added *Nehalennia integricollis* to the Oklahoma state list (Figure 15.60). We have not found a specimen associated with that claim and we do not know if Pritchard had backtracked on the id; as such we decided to keep the 1934 record as the first for the state.

53. 3♀ EMEC 331257; 1♂, 1♀ OSU; 1♂, 1♀ FSCA 47785.

54. Some of this account is taken verbatim or very close to verbatim from Smith-Patten and Tucker (2014).

55. Except for 2018 and 2019 because we were unable to conduct surveys for the species those years.

56. OMNH 4176, 4185, and 4188

57. They collected 3♂ and 6♀ on 22 June 1931 (OMNH 4180, 4186). Note that OMNH 4187 (9♂ and 1♀) was originally identified as *Argia bipunctulata* from this series, but we re-identified those as *Argia plana*.

58. Collected on 19 and 21 July 1931 as OMNH 4177, 4178, and 4181. Note that OMNH 4179 (6♂ and 7♀) was originally identified *Argia bipunctulata* from this series, but we re-identified those as *Argia apicalis*.

59. For many years we thought that collecting to such an extreme degree poorly reflected upon Bird and Webster. But then, BS-P discovered a letter, dated 14 December 1929, in which Bird grumbled to EB Williamson saying, "I want to thank you very much for your nice chatty letter about the Oklahoma Dragonflies. You express my sentiments in the emphasis in collecting good material. At present I seem to be fighting a solitary battle with the rest of the department, because both Drs. Richards and Ortenburger beem [sic] to think it is more important to collect for quantity than quality." Hopefully one day, Williamson's previous letter will turn up so that we may know what he said, but it is clear that Bird was unhappy with being compelled to overcollection. We concur.

60. Four records showed in the SCAN database as having been collected on 6 November 1934. The correct collection date is 11 June 1934 (confirmed by Colin Favret, *in litt*). The specimens were collected by DJ Borror at Cedar Swamp, Urbana, Champaign County, Ohio. They are housed at the Ouellet-Robert entomological collection (QMOR 4652.001, 4652.002, 4653.001, and 4653.002) at the Université de Montréal.

61. Specimens collected in the series from Johnston County taken on 27 May 1930 are housed at OMNH (554–555, 1462–1477, 3046, 3075, and 3111) and USNM (365592–365593, 365595). OMNH 3046 is the specimen that likely has a data entry error that indicates a collection date of 22 May 1930 instead of 27 May.

62. A clear count is difficult to ascertain for these records principally because of some conflicting records, missing dates, and discarded/lost specimens. In trying to reconcile GH Bick's notes with specimens we have found in collections over the years, it appears that there are (or at least were) a minimum of 23 specimens collected in 1930–1933, but there could be as many as 40. That last count includes eleven individuals (OMNH 3882) said to have been taken at Wichita Mountains WR, Comanche County, on 11 June 1932. We have not re-examined those specimens. It may be that neither did Bick, unless his notes of 3♂ from that location are part of that series. The problem with that idea is that Bick's notes indicate those specimens were collected on 11 *July* 1932.

63. We wonder if the report of *Argia nahuana* for Latimer County (Donnelly 2004) is an omission of Bick's correction in species determination. As such, we consider the Latimer County report for this species to be dubious.

64. Males in the SP Collection with faint and/or partially forked humerals are from Roger Mills (SP 179), Texas (SP 597), Blaine (SP 919), and Harmon (SP 1388) Counties. Females are from Harper (SP 1575, SP 1578, SP 1591), and Okmulgee (SP 1679) Counties. One male has a faint main line of the humeral: SP 597 (Texas County). There are females with the main line of the humeral that is thin (SP 1461, 1470, Jackson County), extremely thin (SP 1575, 1578, 1622, Harper and Blaine Cos.), or faint/partial (SP 1591, Harper County).

65. Males in the SP Collection with unforked humerals are from Woods (SP 347), Jackson (SP 1461), and Harper (SP 1575) Counties. Females are from Garfield (SP 118), Jackson (SP 1461, 1470), and Blaine (SP 1622) Counties.

66. Males in the SP Collection with blue dorsally on S7 are from Beaver (SP 268), Blaine (SP 919), Dewey (SP 606), Harper (SP 270), and Texas (SP 597, 598) Counties. Westfall and May (2006:203) note that S7 is "usually black dorsally except for transverse blue basal ring, but sometimes with blue middorsal streak in basal 2/3 …"

67. On 12 April, according to GH Bick's notes, but 18 April according to OMNH 2963: 1♂ collected by RD Bird at Falls Creek, Murray County, det. by GHB, July 1956.

68. Bick's notes for "Willis, Croceipennis Ck." indicated they collected at least 4 exuviae and 61 nymphs, six of which they reared. These records are perplexing for two reasons. One, if these are the same records the Bicks reported in Bick and Bick (1958), then why did they indicate they had 57 nymphs, not 61? Perhaps it is because four of the nymphs were so close to emergence that the Bicks counted them as tenerals whereas we counted them as nymphs? The second puzzling aspect is that in the notes the locality is Croceipennis Creek but the publication indicates it was Cowan Creek. These records all seem to match up, so if that is the case, then is Croceipennis and Cowan Creek one in the same? We have long wondered what the Bicks meant by "Croceipennis Creek," so maybe that puzzle is solved.

69. Hindwing measurements (mm) are: 18.0, SP [RWG] 1026; 19.2, SP [RWG] 948; 21.0, SP [RWG] 1027; 21.4, SP [RWG] 1030; and 21.5, SP [RWG] 1029.

70. Males: EMEC 300724 (2♂, 30th), EMEC 300725 (1♂, 29th), EMEC 300741 (1♂, 29th), FSCA (1♂, 29th), and USNM 364354–364355 and 364357 (3♂, 30th). Females (one each): EMEC 300726 (30th), FSCA (29th), and USNM 364356 (30th). The Bicks (1957) apparently knew of only 2♂ from 29 June. They indicated that they had examined the specimens at OSU, but we never found them in the collection. If there were two specimens at OSU, then it is possible that Pritchard collected a minimum of 13 individuals during his two days of collection *Argia lugens*.

71. 1931: June to July, collected by A Ireland, det. RD Bird, OMNH 3173, 1♂, 3♀, no specific location given. 1954: 17 June, collected by GH Bick at Turner Falls, 3♀, Bick's note cards.

72. There are two 1937 records, both of which are from Hinton. One adult is in the OSU collection with data of: 13 June, collected by Standish and RW Kaiser. One ♀ was in GH Bick's notes indicating it was collected on 18 July, also by Standish and Kaiser. The 1939 record, also from Hinton, is of 2♂ collected on 5 June by Kaiser and WT Nailon (OSU).

73. 1♂ collected by RD Bird, probably in the Wichita Mountains area, on 13 July 1932 (OMNH 3130).

The small dragonfly family Petaluridae, the petaltails, comprises a mere 11 species worldwide, one of which, the Giant Petaltail (*Petalura ingentissima*) of rainforests of northeastern Australia, is the world's largest dragonfly, with a length of >100 mm and a wingspan of ~160 mm (a little bigger than a 4×6 inch index card!). Two species occur in the Americas, the Black Petaltail (*Tanypteryx hageni*) of the Pacific Northwest and the Gray Petaltail (*Tachopteryx thoreyi*) of the Southeast, whose range extends westward into eastern Oklahoma.

Despite its large size—adults reach ~80 mm in length, with a wingspan >110 mm—the Gray Petaltail, which in some circles goes by the evocative name of Thorey's Grayback, is elusive. It flies little relative to other large dragonflies, and its mottled gray and black coloration allows it to blend into the bark on which they alight (Figure 13.1).

A better way to find the species is to get lucky and have one come to you. Many a person has been startled when one of these giants landed on their light-colored shirt, which the species does from time to time (Figure 13.2). They are also known to land on dirt roads and sometimes on vehicles.

Boehler Seeps Preserve, Atoka County, Oklahoma, 20 July 2014.

... or on your clothing

Boehler Seeps Preserve, Atoka County, Oklahoma, 5 June 2016

©Bill Carrell

Figure 13.1 Were you to stroll through its preferred seep-laden woodland habitat, there is a good chance you would pass by a well-camouflaged Gray Petaltail perched vertically on a tree ... or on your clothing.

TACHOPTERYX THOREYI (HAGEN, 1858) – GRAY PETALTAIL

This surprisingly (because it is so large) cryptic species, the only one in the genus *Tachopteryx*, is found in Oklahoma's eastern forests. Its uncommonness likely extends beyond crypticity: it is a habitat specialist that, in Oklahoma, generally prefers mature hardwood forests around seeps (especially), small springs, and partly flooded bottomland.

It is known from thirteen counties in Oklahoma, including Latimer County, where it was first reported for the state (Wilburton, 8 July 1948, 1♀, collector unknown). Data for this specimen came from GH Bick's notes, in which he said he borrowed the specimen from "SESC, thru Kilpatrick." We have yet to locate this specimen, but we located the specimen that constitutes the second record for the state, which was collected in Muskogee County by GHB on 23 June 1951 (1♀, 3 mi S of Braggs, FSCA). Bick had another specimen note, this one for a ♂ collected in Happy Hollow near Enos, Marshall County, by "SM" on 24 July 1964 (GHB discarded the specimen). We have an additional 6 records between 1948 and 1996 and about 40 between 2005 and 2019, with just under half of the 11 specimens collected from the state coming from the latter years (SP 662, 1285, 2657, 2658, and one reported by W Boys).

Early dates for the region include 11 March for Texas (OC 374089), 15 March for Louisiana (OC 478405), 17 May for Oklahoma (a teneral ♀ perched on her exuviae, 2019, MAP, OC 494903), 14 May for Arkansas (iNat 8602717), and 24 May for Missouri (Sims 2012, Enns). The latest record for Oklahoma, of two individuals seen in Tulley Hollow at the JT Nickel Preserve, Cherokee County, on 30 August 2006 by BC, is an anomaly for the region but not for the species. This record verges on two months beyond the late date for Texas (10 July, iNat 31159371), one and a half months beyond Louisiana (15 July, Mauffray 2014), one month past Missouri's late date (31 July, Trial 2005), a few weeks beyond Arkansas (7 Aug, iNat 15275078), and one month beyond the next latest documented record for Oklahoma (31 July 2009, McCurtain County: 4 mi NNW of Hochatown, BAH, 1 indiv., OC 314709). But an August date is consistent with the species' flight season farther east and northeast, including Kentucky, New Jersey, and Ohio (Paulson 2011) and North Carolina, where the species has occurred as late as 12 September (OC 341438).

An interesting natural history note for the species came when BAH found a Polyphemus Moth (*Antheraea polyphemus*) flopping around on a gravel road. He initially thought it was somehow impaled with a stick, but on closer inspection he realized that the stick was in fact a ♂ petaltail eating the moth's head (Figure 13.2; OC 321148). He watched for about eight minutes as the moth continued to flap its wings, pulling the petaltail to and fro. When rain started falling, the petaltail picked up the moth and "flew about 6 feet up into a small maple tree where it perched and continued to eat." Quite the feat for a dragon with a wing span of up to 110 mm to take a moth with a wing span up to 150 mm.

Tachopteryx thoreyi (Hagen, 1858)
Gray Petaltail
S3

8 July 1948

17 May   31 Jul

AR
LA
MO
TX

Figure 13.2 Gray Petaltails will alight on the ground, including when they take prey (A, B), such as this ♂ that tackled a Polyphemus Moth (*Antheraea polyphemus*), vic. Lynn Mountain, Le Flore County, 26 July 2010 (OC 321148). They will also perch on vegetation (C), such as this ♀ (10 km SE of Idabel, McCurtain County, 19 June 2008, OC 282494). Photos by ©Berlin A Heck.

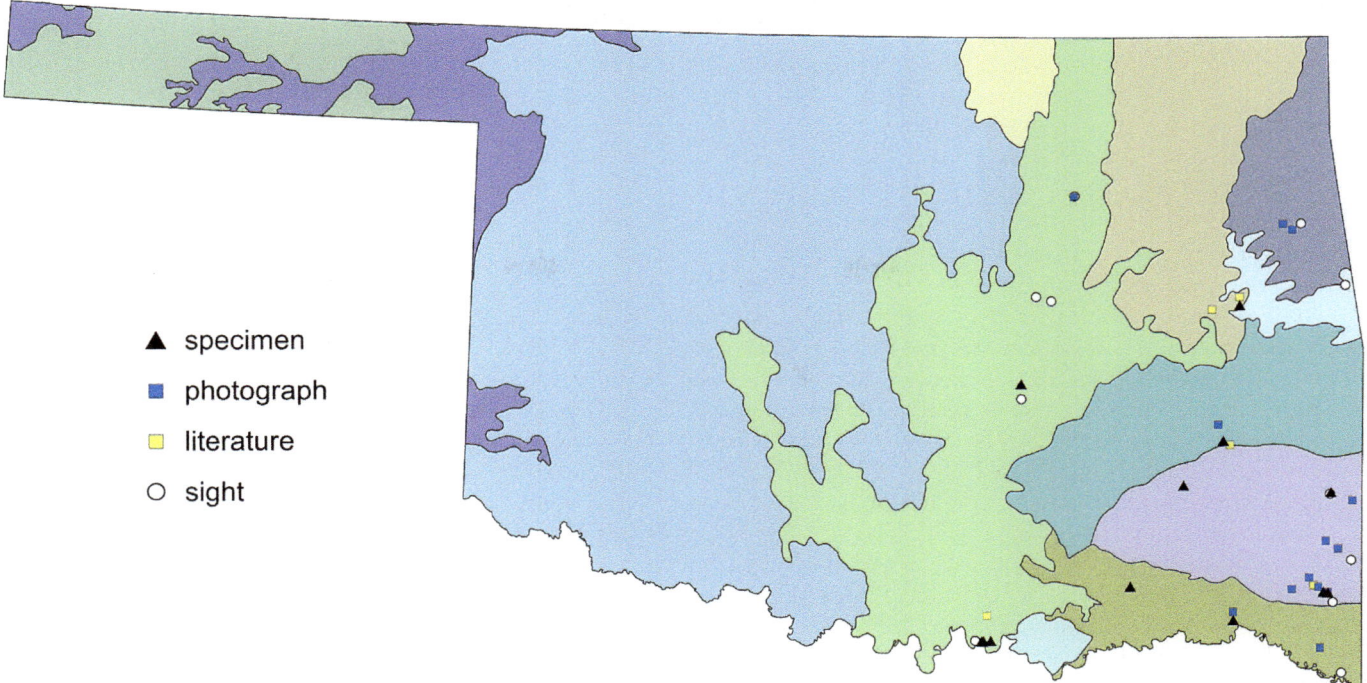

Figure 13.3 Documentation level of supporting records for *Tachopteryx thoreyi*, the Gray Petaltail, in Oklahoma. Levels include specimen, archived photograph, literature reference, or sight record. Although sight records lack documentation, only those we feel are credible are plotted.

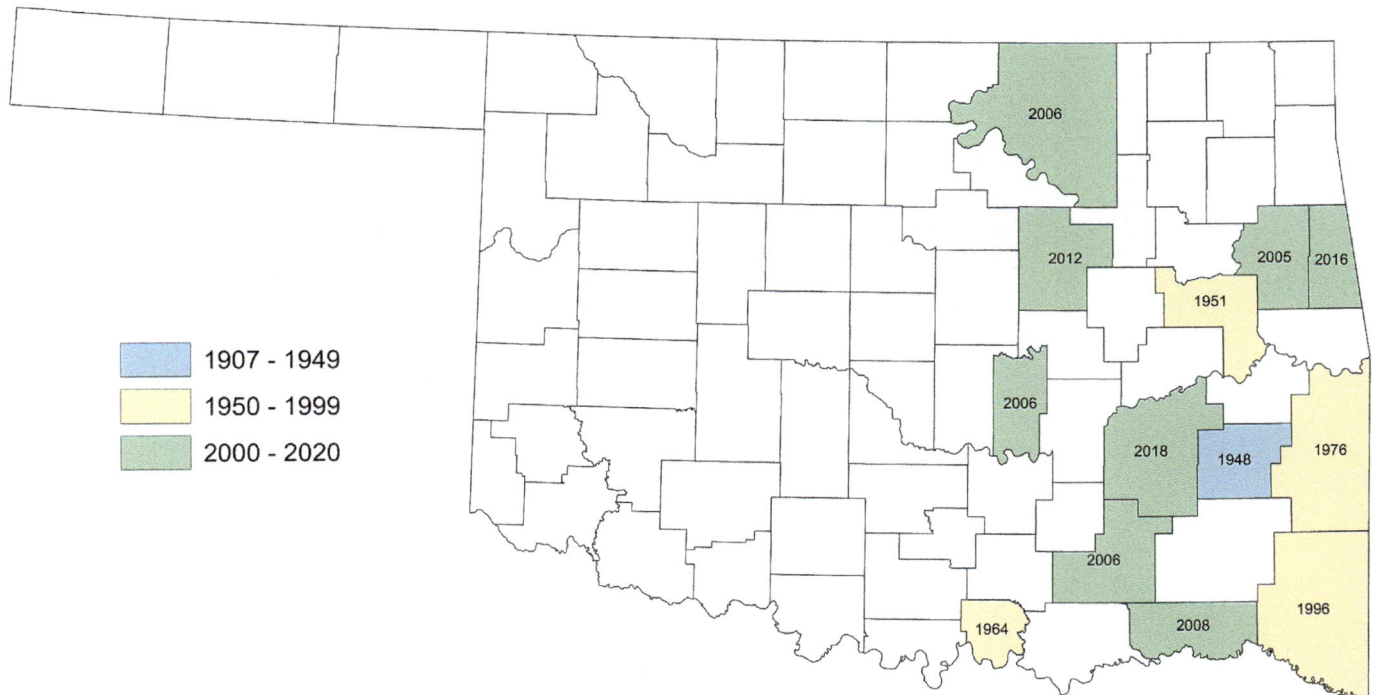

Figure 13.4 Counties in which *Tachopteryx thoreyi*, the Gray Petaltail, are known to occur in Oklahoma. Year within the county is that in which the species was first reported.

## CHAPTER 14

AESHNIDAE: DARNERS

Oklahoma hosts a moderate variety of darners, from the huge—the Swamp Darner (*Epiaeschna heros*) and Regal Darner (*Coryphaeschna ingens*) are two of the largest dragonflies that occur in the state—to the diminutive—the Harlequin Darner (*Gomphaeschna furcillata*) is just a touch larger than a small baskettail (*Tetragoneuria*-type *Epitheca*), and with the penchant for the male to hover when territorial, it can easily be mistaken for one. These species frequent a range of habitats (Figure 14.1), from purely lentic (e.g., the Regal Darner) to purely lotic (e.g., the Fawn Darner, *Boyeria vinosa*) to near-perfect tweeners (e.g., the Springtime Darner, *Basiaeschna janata*), as well as a range of light conditions, from heavily shaded (e.g., the Fawn Darner) to brightly lit (e.g., the Common Green Darner, *Anax junius*).

The full range of this diversity can be difficult to experience, however, because Oklahoma's aeshnid fauna is so thoroughly dominated by a single species, the Common Green Darner. On the basis of our own field work, 84% (5133 of 6115) of the darners we have encountered in the state have been this species. And not only is the Common Green the most numerous darner, but it has—by a comfortable margin—the longest flight season of any darner species in Oklahoma, a season that spans just about the whole of the calendar (Figure 14.2). As if its small size and tenebrous habits were not enough, the Harlequin Darner, by contrast, hinders efforts to find it by flying a mere five or six weeks in early spring (Figure 14.2).

Confounding efforts to find each of the ten regularly occurring species—two, the Twilight Darner (*Gynacantha*

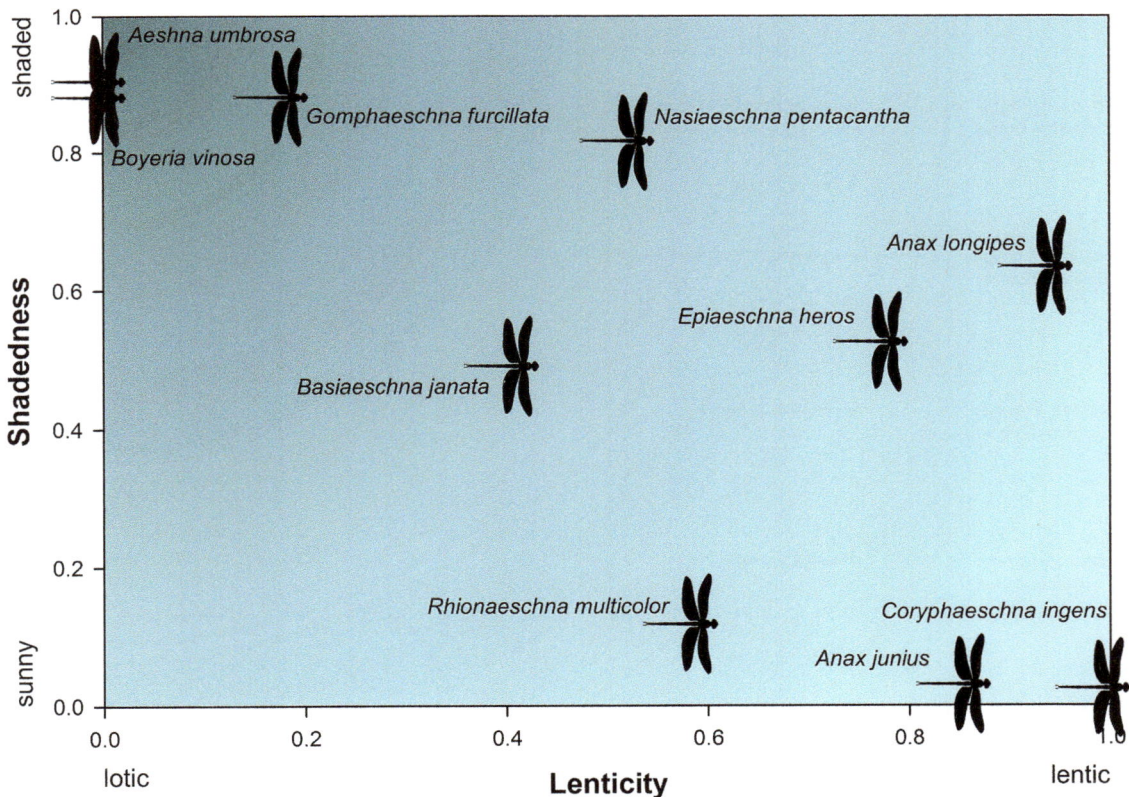

Figure 14.1 A scatter plot on two habitat axes of Oklahoma species in the family Aeshnidae, the darners. Both the lenticity score and the shadedness score were estimated directly from our field data, with each score being a proportion of individuals encountered in one habitat or the other.

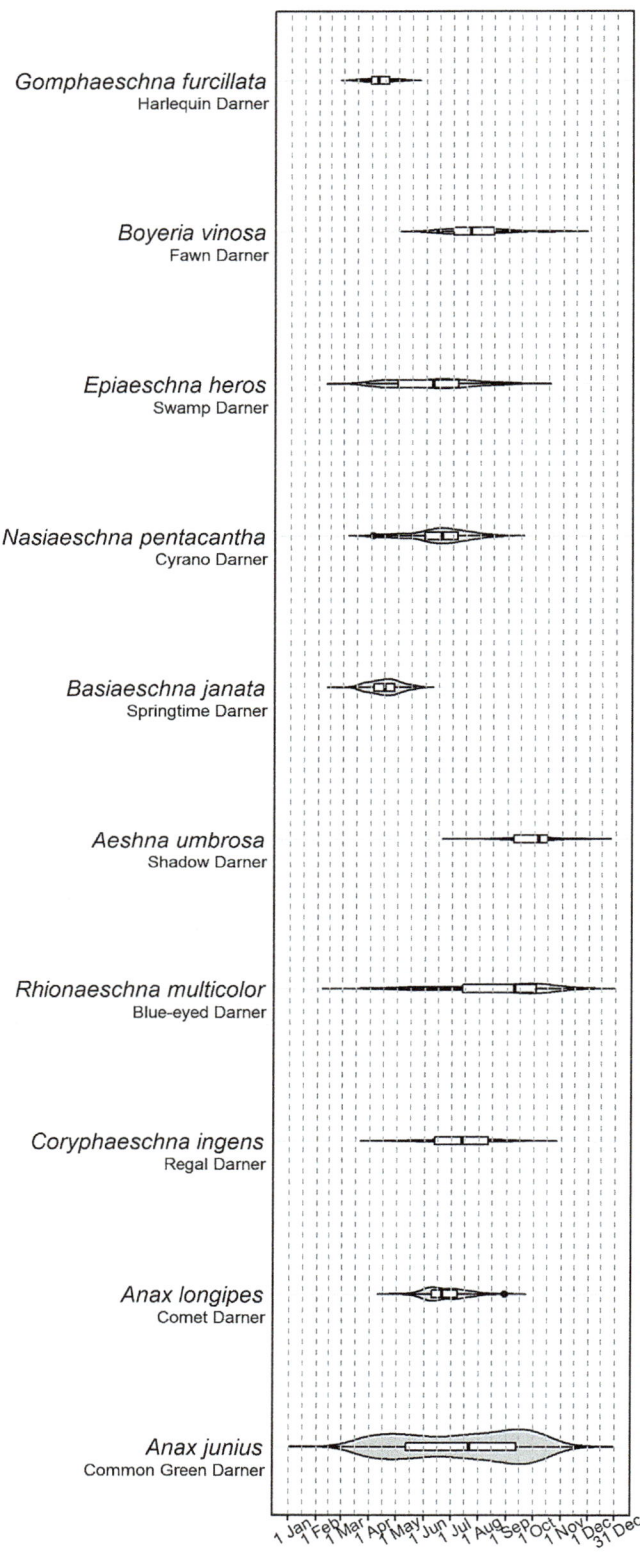

Figure 14.2 Seasonality of Oklahoma species in the family Aeshnidae, the darners. Note the long flight season of the Common Green Darner (*Anax junius*), the short spring-only seasons of the Harlequin (*Gomphaeschna furcillata*) and, the aptly named, Springtime (*Basiaeschna janata*) Darners, and the relatively short autumn-only season of the Shadow Darner (*Aeshna umbrosa*).

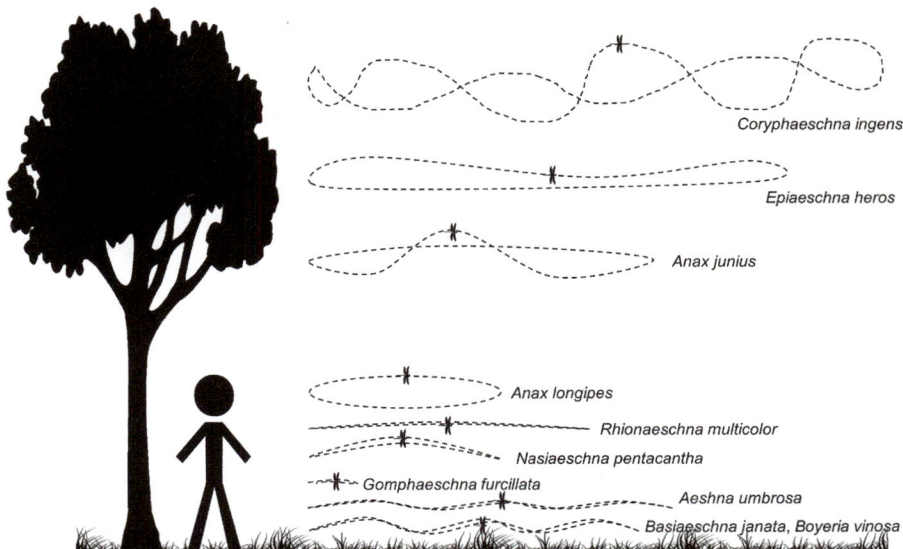

Coryphaeschna ingens

Epiaeschna heros

Anax junius

Anax longipes

Rhionaeschna multicolor

Nasiaeschna pentacantha

Gomphaeschna furcillata

Aeshna umbrosa

Basiaeschna janata, Boyeria vinosa

Figure 14.3 A schematic illustration of typical flight style—in terms of both height above the ground and length of regular "beat" flown—of Oklahoma species in the family Aeshnidae, the darners. The bottom three species tend to fly fairly low over water, each hugging the contours of the shoreline. The Comet Darner (*Anax longipes*) also tends to hug the contours of water, but it is found predominantly at clear forest pools, which it circles tirelessly. The Harlequin Darner (*Gomphaeschna furcillata*) is the only species that hovers regularly, and even without such behavior, a male's territorial beats tend to be short, generally only 1–2 m in length.

nervosa) and Lance-tipped Darner (*Aeshna constricta*), have been recorded but once each, and only the latter is likely to be found again anytime soon—is that darners fly frequently (and often rapidly) and roost inconspicuously, often hanging in shaded spots screened laterally by leaves, roots, or grass. An overwhelming percentage of encounters will be with an individual on the wing. In addition to clues offered by body size, coloration, habitat (e.g., Figure 14.1), and flight season (Figure 14.2), the intrepid observer would do well to learn where and how each species tends to fly. The massive Swamp and Regal Darners, for example, tend to fly at or above the canopy, with each species flying a long beat or circuit, whereas the Fawn, Springtime, and Shadow Darners fly at or below knee height, typically on moderate beats in which they closely follow the contours of the shoreline, whether at a stream or lake (Figure 14.3). The Cyrano Darner (*Nasiaeschna pentacantha*) might look outwardly like a Swamp or Shadow Darner, but its distinctive habitat of fluttering back-and-forth on relatively short beats in shaded areas, whether narrow backwaters of lakes or along slow-moving, often murky streams, makes it easy to identify even at a distance.

GOMPHAESCHNA FURCILLATA (SAY, 1839) – HARLEQUIN DARNER

This early season flyer is known from the state from later in March to early May, which is roughly the same flight period as in the nearby states of Arkansas, Louisiana, and Texas (AR: 1 April–4 May, OC 374336; LA: 3 February–15 April, Mauffray 2014, OC 318373; TX: 17 March–10 April, OC). The species was first documented in Oklahoma on 7 April 2008 at Grassy Slough WMA, McCurtain County, where BAH photographed a mating pair (OC 281940). Nevertheless, the species was likely photographed the day before by DA and M White, who snapped a suggestive but blurry and ultimately inconclusive photo (OC 281933). Since 2014, Harlequin Darners have been seen at Grassy Slough on nearly every visit during the species' flight season. Although tenerals have not been reported at Grassy Slough (or elsewhere in the state), paired adults have been reported twice, including 2 tandem pairs on 31 March 2018, when the state high count of 12♂ and 2♀ was achieved.

The species is now documented from four localities in addition to Grassy Slough and has been seen at one other (Figure 14.5). All records are from just two counties—McCurtain and Atoka. Elsewhere in McCurtain County, it is known from the nearby Red Slough WMA (8 April 2009, 1♂, DA, OC 312505), the only time it has been documented at that well-worked location, and there is a sight record at BAH's property, located about 10 km SE of Idabel (from April 2009, DA, *pers comm*). More recently it has been documented from two locations within Little River NWR, which spans almost 30 km from its western to its eastern

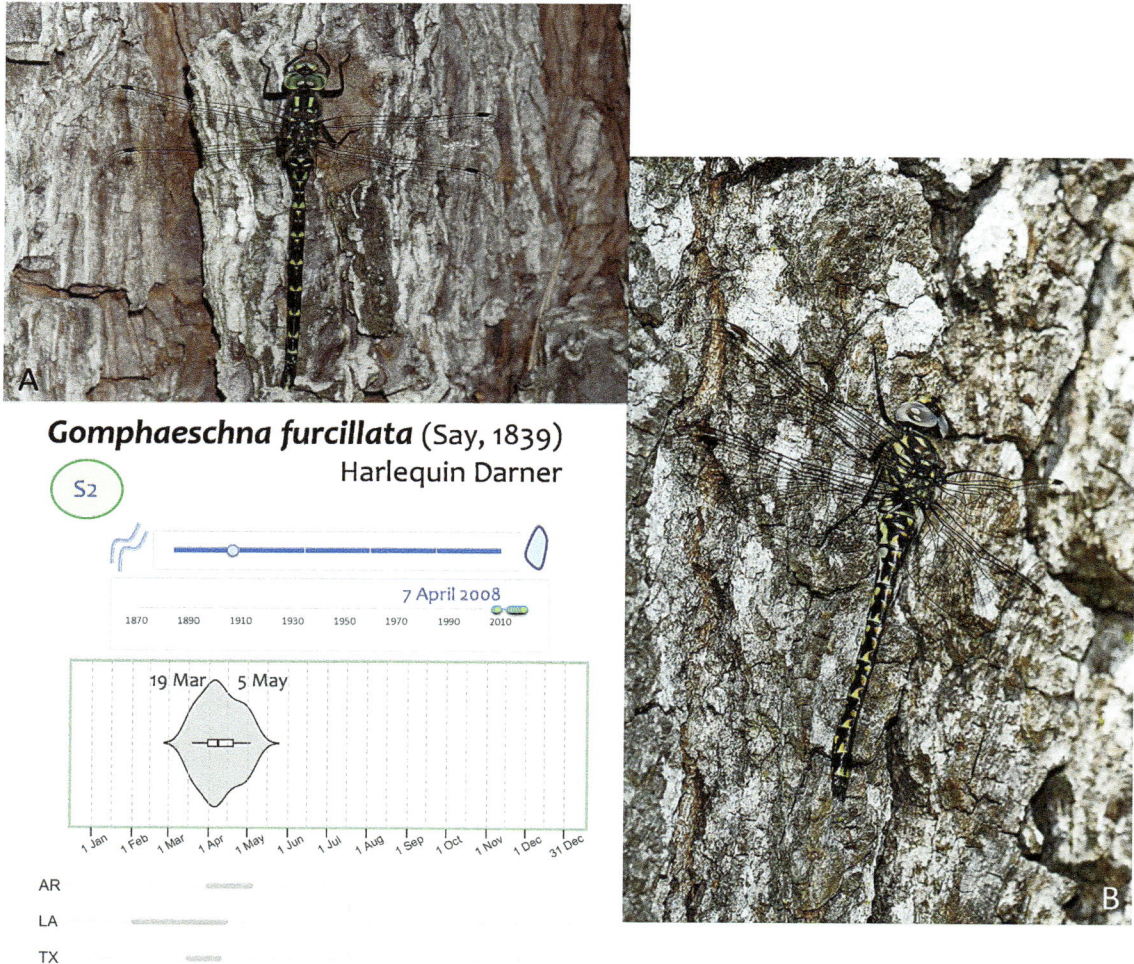

Figure 14.4 *Gomphaeschna furcillata*, the Harlequin Darner, A) ♂, Little River NWR, McCurtain County, 12 April 2018 (©Bryan E Reynolds). B) ♀, Grassy Slough WMA, McCurtain County, 19 March 2018 (USDA Forest Service; David Arbour, OC 478582). This tiny (for a darner) species was unknown in Oklahoma until a decade ago but has proven to occur regularly in early spring in the far southeast.

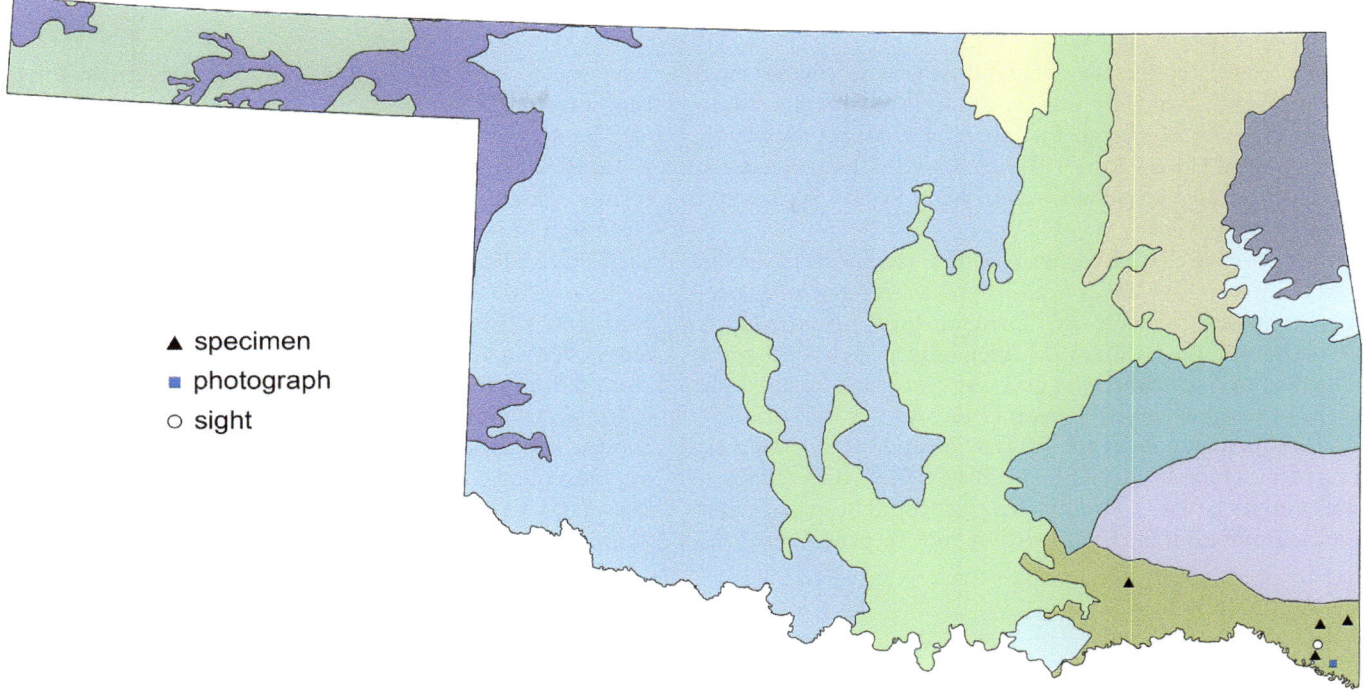

Figure 14.5 Documentation level of supporting records for *Gomphaeschna furcillata*, the Harlequin Darner, in Oklahoma. In Oklahoma, the species is confined to the South Central Plains, being known from only two counties: Atoka, where it was encountered twice in 2015, and McCurtain, where, since 2008, it has been documented at four localities and seen at another.

extents. Three individuals were seen at Forked Lake, on the eastern end of Little River NWR (1♂ as SP 1847, BS-P) on 3 April 2016, which is the only time the species was reported in 2016. On the other side of the refuge, the Harlequin Darner was first found patrolling a dirt road on 18 April 2009 (BAH, OC 312574). It was seen again there in 2014 on 3 May (2♂, one as SP 1165), also patrolling a dirt road, this time at dusk and in 2015 on 18 April (DA sight record, 1 indiv.). There was one report of the species in 2018 from the west side of the refuge, 12 April, 1♂, BE Reynolds photo, on file, OOP, and it was found there again in 2019, when a ♂ was photographed in hand (OC 494655) on 5 May, the latest date known for Oklahoma.

Harlequin Darners were first found in Atoka County on 25 April 2015 by MAP (Boehler Seeps and Sandhill Preserve, south end of Hassell Lake, 3♂, 1 as SP 1562, OC 430571). We returned to the site on 1 May, when we saw and collected 1♂ (SP 1533). We made four visits

during 2016, but only one (10 April) was within the species' flight season in Oklahoma; even so, the species was not seen that day.

Prior to our collections we doubted that the species was a resident of Oklahoma, but given the number of sightings and that mating pairs have been encountered, the species is established in McCurtain County probably Atoka and intervening counties. The swampy/boggy habitats (i.e., seasonally flooded large river bottomland) this species shows a preference for presumably confine it biogeographically to the South Central Plains where such habitats are only found in the state. Those habitats and the species' short flight season currently hinder confirmation of an established breeding population, but we remain convinced there is one in the state. We are inclined to rank this species as S3, in line with other low-density species (e.g., *Coryphaeschna ingens*, *Libellula semifasciata*) of similar priority conservation concern in southeastern Oklahoma.

BOYERIA VINOSA (SAY, 1839) – FAWN DARNER

AE Pritchard's collection of *Boyeria vinosa* contributed much to his legend. Not only was he the first to find the species in the state (at Nashoba, Pushmataha County), but in the 2-day period (14–15 June 1934) that he did, he collected an astonishing 26 individuals![1] And, he collected six more in the following week.[2]

It is possible that an earlier record of the species exists for Oklahoma. In GH Bick's notes there is reference to a specimen that was collected by "Ortenburger," presumably AI Ortenburger, on 21 June 1925. Bick indicated that the specimen was in poor condition and was "returned labelled [sic] *grafiana*[3] & checked by Calvert?" He did not indicate from where this specimen came. It is curious that it was not mentioned by Ortenburger (1926b), Bird (1932), Bick (1951), or Bick and Bick (1957), although the last could be explained by Bick's annotation on the record of "not in paper," probably

referring to Bick and Bick (1957). We have yet to relocate that specimen.

The next time the species was collected in the state was when Bick arrived in Oklahoma and began working at the former Oklahoma State University Wildlife Conservation Station in Muskogee County. He collected 21 nymphs from Little Greenleaf Creek and Greenleaf Lake between 19 and 26 June 1950 (Bick 1951). All of these specimens were discarded in 1956 (according to his notes). His first and only encounter with adult Fawn Darners was a copulating pair he collected on 20 July 1956 in Johnston County, on the Blue River near Milburn (FSCA). Between 1958 and 2003 only exuviae and nymphs were documented, beginning with two exuviae collected by HP Brown on 19 May 1958 in Norman, Cleveland County (OMNH 1765), and ending with one nymph collected and identified by EA Bergey on 25 July

♀ hiding amongst the dense, dead vegetation

Figure 14.6 *Boyeria vinosa*, the Fawn Darner. For a large dragonfly, this species is seldom seen in the Southern Great Plains because it favors heavily shaded regions of creeks and tends toward the crepuscular. A) ♂, Osage Hills SP, Osage County, 20 July 2018 (©Bill Carrell, OC 485516). B) ♀, Little Robe Creek 10 km NW of Oakwood, Dewey County, 15 August 2015 (©Michael A Patten, OC 435072), one of few ever found in western Oklahoma. C) tandem pair, Pigeon Creek at Beech Creek National Scenic Area, Le Flore County, 7 August 2010 (©Berlin A Heck, OC 321688).

2003 in Coal County at Wide Spring (OBS). Nymphs were also collected in 1975, 1977, and 1997 (1975: 5 September, Bidding Creek, Adair County, 1 indiv., OMNH 1745 and 1 November, Lake Burford, Wichita Mts. WR, Comanche County, 2 indiv., OMNH 1746, all coll. by WD Shepard; 1977: 8 October, Flint Creek, Delaware County, P. Stansly, 1 indiv., OMNH 1744 and 22 October, Medicine Park Creek, Comanche County, EA Bergey, 3 indiv., OMNH 1747; 1997: 30 June, Hanson Park, Choctaw, Oklahoma County, unknown coll., 1 indiv., UCO 5201). The Adair, Cleveland, Comanche, and Delaware County specimens were confirmed by BS-P.

Adults were not documented again in the state until 5 September 2005 when a Fawn Darner was collected in Custer County, 5 mi N of Foss on "Turkey Creek" (probably actually Oak Creek, G Warwick, 1U, JCAC 20788). This record remains the westernmost record for the species in Oklahoma and is one of the farthest west for the species anywhere. The next westernmost record in Oklahoma is one from Little Robe Creek 10 km NW of Oakwood, Dewey County, on 15 August 2015, when MAP encountered 1♂ and 2♀, taking one of the females as a specimen (SP 1749, OC 435072). Given the paucity of records, we do not know how common or rare this crepuscular species is in the western half of the state. And even in the eastern half of the state its status is unclear given the long stretches during which adults have gone undetected. The circadian rhythm of this species (active at dawn and dusk) coupled with both its affinity for shaded streams and rivers with low flow and steep, earthen banks strewn with roots and its erratic, moth-like flight low over water make it difficult to find. Even so, *Boyeria vinosa* has been documented in 18 counties, 10 of them concentrated on the eastern fringe of the state and the other eight scattered across the central and western parts of the state.

The 1958 exuvial specimens mentioned above may indicate that the species flies in Oklahoma as early as 19 May, but adults are only documented as early as 11 June (2006: TNC Tallgrass Prairie Preserve, Osage County, BC sight record of 1 indiv.; 2012: Mountain Fork River, 11 km NE of Broken Bow, McCurtain County, 1♂, TD Hibbitts, OC 376215). The late flight date in the state for the species is 20 October (2010, same location as 2012 early date record, 1♀, BAH, OC 323675 and JCAC 39980). These dates differ somewhat from neighboring states (AR: 5 May–27 September, Harp and Rickett 1977, JCAC 36665; LA: 3 May–5 November, JCAC 39008, Mauffray 2014; MO: 22 June–5 October, Trial 2005, Sims 2012; TX: 14 April–27 July, Abbott and Lasley 2019).

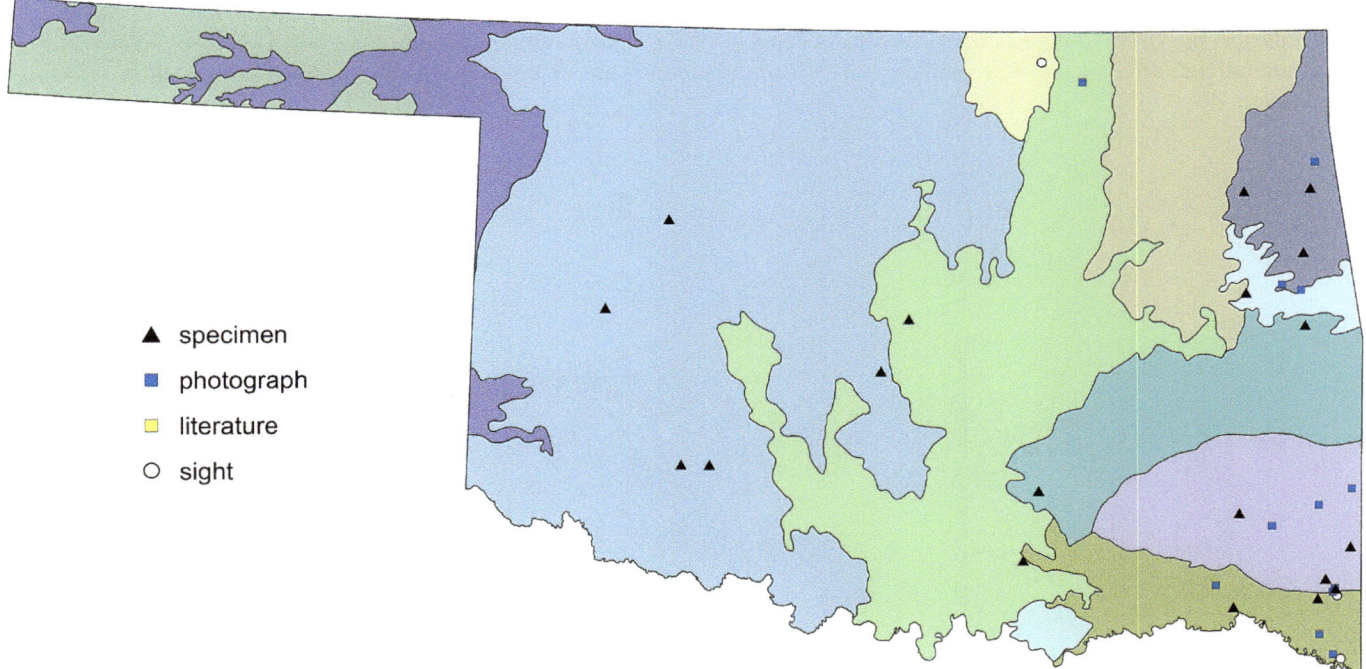

Figure 14.7 Documentation level of supporting records for *Boyeria vinosa*, the Fawn Darner, in Oklahoma. Levels include specimen, archived photograph, literature reference, or sight record. Although sight records lack documentation, only those we feel are credible are plotted.

- ▲ specimen
- ■ photograph
- ▢ literature
- ○ sight

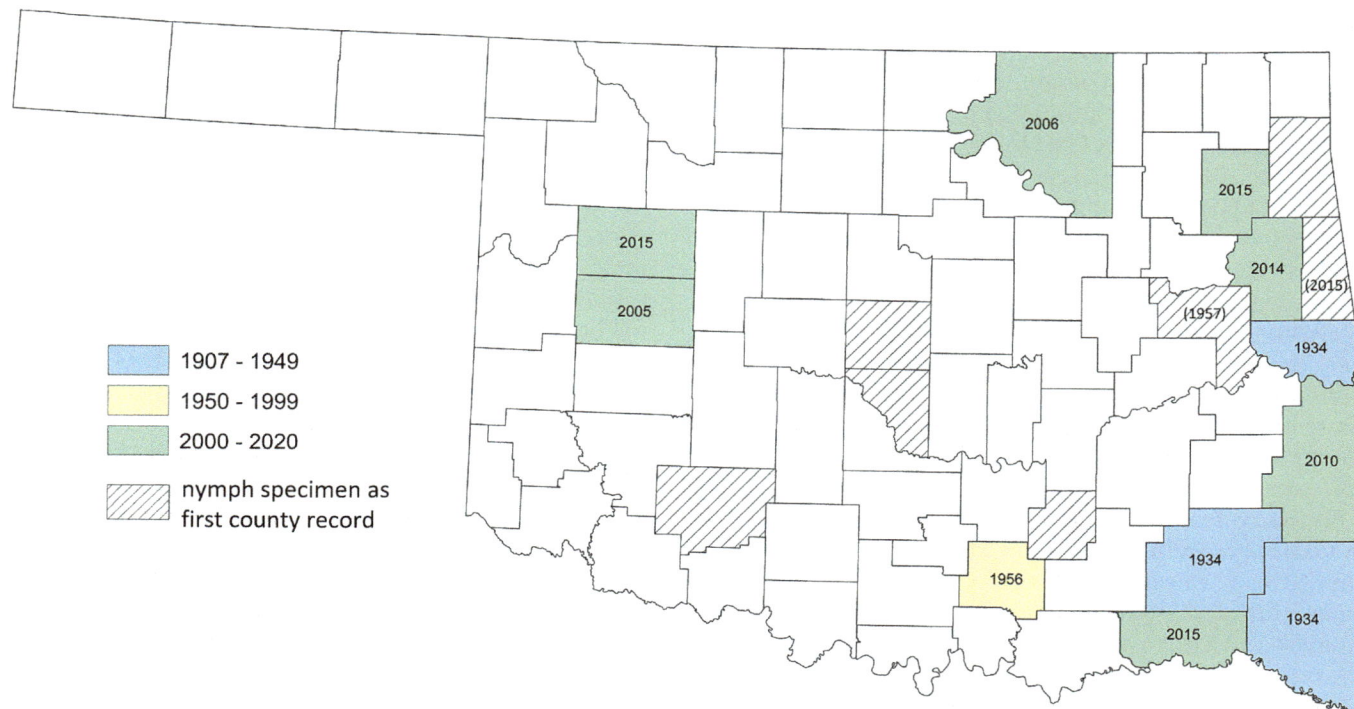

Figure 14.8 Counties in which *Boyeria vinosa*, the Fawn Darner, are known to occur in Oklahoma. Year within the county is that in which the species was first reported. Hatched counties are those with a nymph specimen as the first documentation of the species in the county. When a parenthetical year is also presented, it indicates the year of the first adult record; no year indicates the county only has larval records as support. We are fairly confident about the Adair, Cleveland, Comanche, Delaware (all det. BS-P), and Muskogee (det. GHB) County records, but we did not examine the Coal and Oklahoma County specimens.

EPIAESCHNA HEROS (FABRICIUS, 1798) – SWAMP DARNER

The Swamp Darner is an uncommon to locally fairly common species throughout eastern and central Oklahoma, with a few documented records for the northwestern corner of the state and two recent sight records for the southwestern corner. It was first recorded in Oklahoma in 1907 (Williamson 1914b), when F Collins collected a lone ♀ near Wister, Le Flore County. We have not been able to locate this specimen, so the earliest physical documentation we know of for the species is a specimen collected on 29 June 1925, presumably by AI Ortenburger, in McCurtain County near Broken Bow (likely locality 2 mi N of Broken Bow, Yanubbe Creek, as per Ortenburger 1926b; OMNH 825). It seems odd that outside of these two records, we have only one other early period record: RD Bird collected a ♀ on 4 July 1930 in Woods County, likely at Waynoka (OMNH 3019), which remains the only time the species has been reported from that county. Early on we wondered if this record involved a misidentification because we thought it too far from the species' geographic range and typical habitat of bottomland forest (often partly flooded or with pools of stagnant or slow-moving water), yet a subsequent record in the neighboring county (Woodward County, Fort Supply Lake, 18 July 2014, JT Bried, 1♀, SP 1492) and two Texas panhandle records from GW Lasley

(Hemphill County, 14 July 2009, OC 321650; Lipscomb County, 14 June 2016, OC 447949) attest to the Woods County record not being a geographical and phenological anomaly. Documented records are lacking for the southwestern part of the state, but on consecutive days in 2019 lone ♂♂ were observed repeatedly (MAP) at Great Plains SP, Kiowa County, on 8 June, and along Sandy Creek 7 km SSE of Eldorado, Jackson County, on 9 June. In each case cerci were seen well enough to sex the individuals, as were the thoracic and abdominal patterns to eliminate other huge darners (e.g., various species of *Coryphaeschna*, including *C. ingens*, the Regal Darner, which has been documented in Kiowa County), yet each eluded extended efforts at capture and dodged identifiable photographs.

We have no ready way to account for survey bias, but the sheer numbers of reported records of the Swamp Darner between the historical and contemporary eras point to the species being more consistently seen in the current era. But why? Between 1907 and 1930 there were only three records, and the record numbers increased (*n* = 14) between 1955, when GH Bick saw his first in the state, in McCurtain County, and 1997, when SW Dunkle collected a ♂ in Love County (FSCA). Across those 90 years, then, adults were reported just 16 times (and one nymph was collected, on

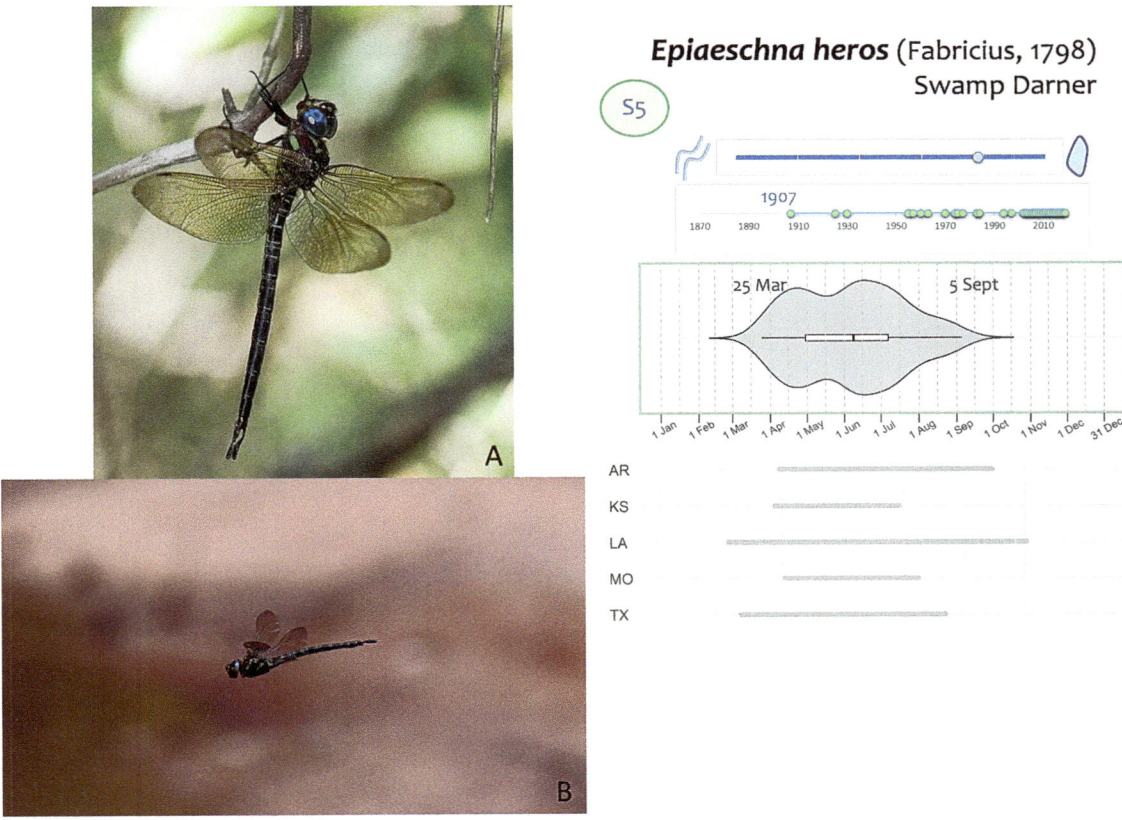

Figure 14.9 *Epiaeschna heros*, the Swamp Darner, A) ♂, Mohawk Park in Tulsa, Tulsa County, 28 July 2011 (©James W Arterburn). B) ♀, Spencer Creek 1 km S of Corinne, Pushmataha County, 2 May 2015 (©Brenda D Smith, OC 430733).

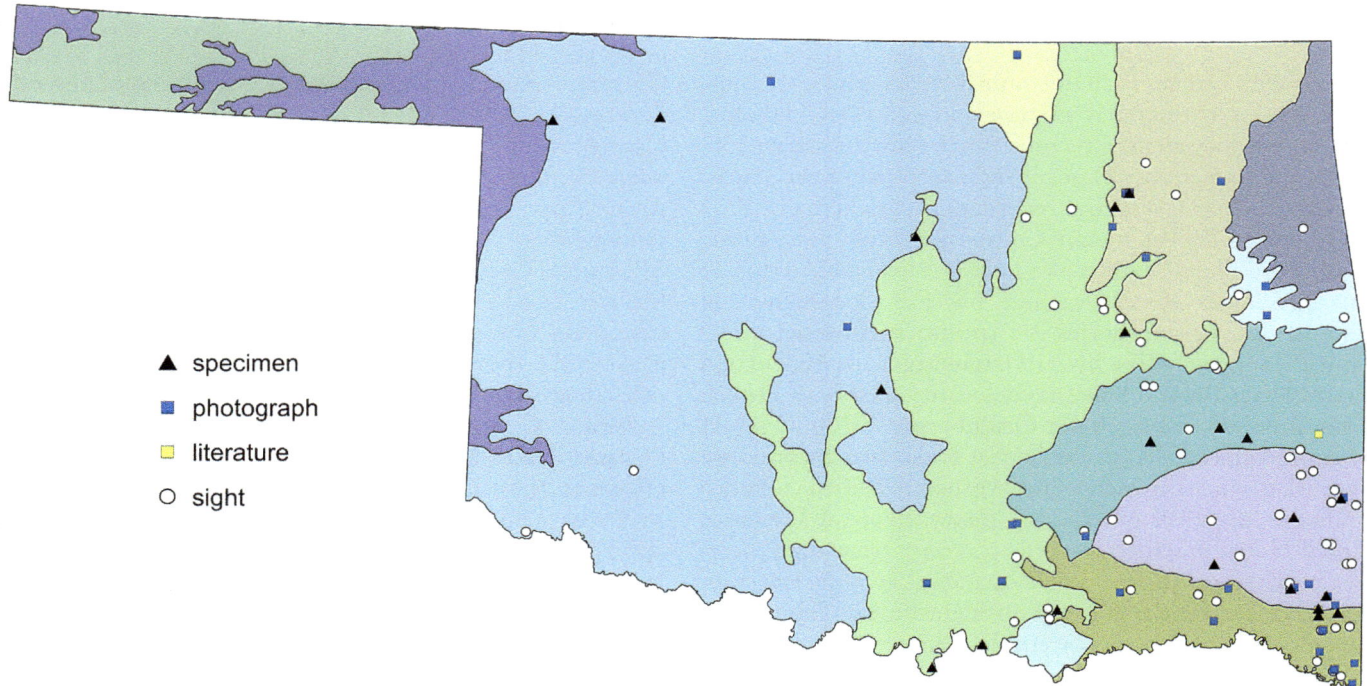

Figure 14.10 Documentation level of supporting records for *Epiaeschna heros*, the Swamp Darner, in Oklahoma. Levels include specimen, archived photograph, literature reference, or sight record. Although sight records lack documentation, only those we feel are credible are plotted.

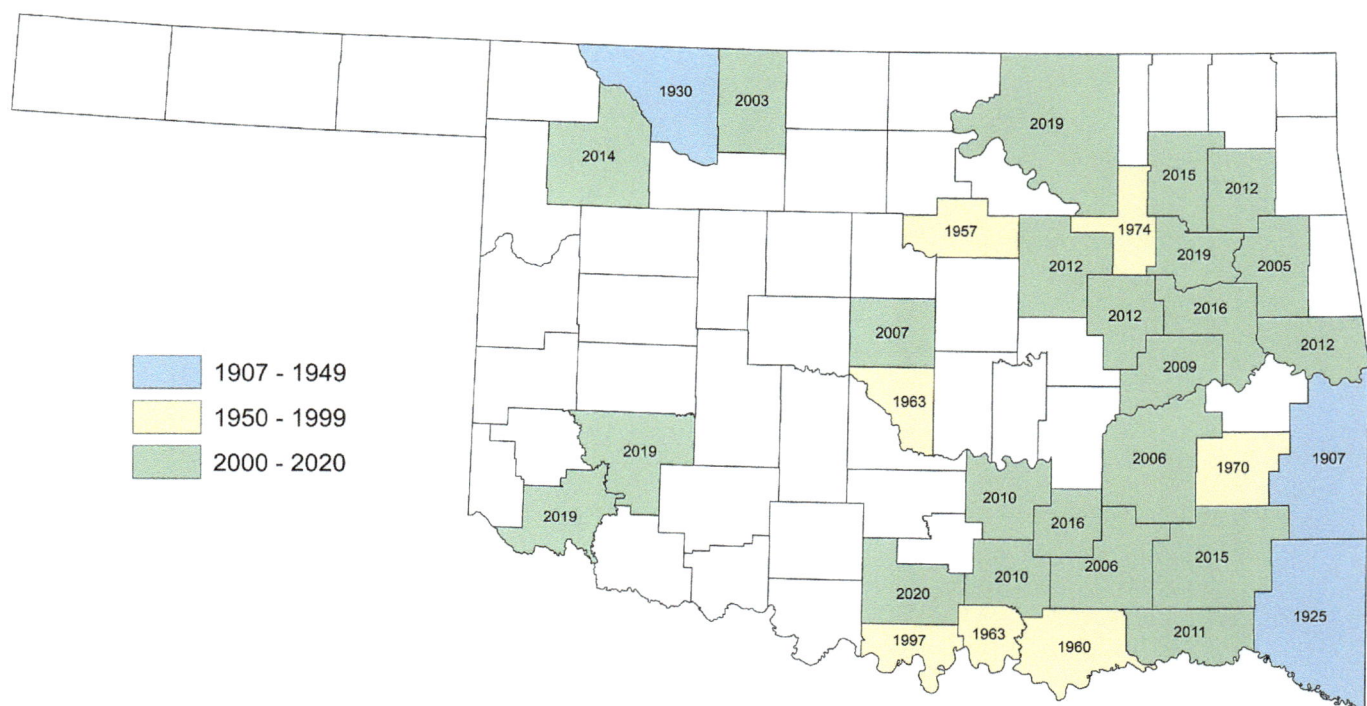

Figure 14.11 Counties in which *Epiaeschna heros*, the Swamp Darner, are known to occur in Oklahoma. Year within the county is that in which the species was first reported.

26 October 1975, Oliver's Woods in Norman, Cleveland County, coll. from the bottom of a small pool by D Butcher, det. by WD Shepard, 1 indiv., OMNH 1753). In stark contrast, in recent years (2002–2020), there have been >100 reports in 31 counties, some of which have indicated feeding "swarms" of 6 to over 10 individuals at a time, with a spectacular concentration of 75 individuals (4♂, 2♀ in hand) at Red Slough WMA, McCurtain County, 5 May 2019 (MAP, DA, JC Abbott, et al.). We posit that the apparent uptick in records is an artifact consequential of two factors, the first being the species' behavior, the second, available accoutrements. The species flies high, often at treetop level and thus generally well above the reach of a typical insect net. In short, it is easy to spot, but difficult to catch. The surfeit of recent records can be laid at the feet of modern field guides, binoculars, and cameras—all effecting an increased ability to identify the species on the wing, a distinct advantage over earlier odonatologists.

The Swamp Darner has been seen in Oklahoma as early as 25 March (2017 MAP sight record, 1♂, Carl Albert Park, Durant, Bryan County) and documented as early as 26 March (2017, MAP in hand photo, OC 461421, 1♀, ovipositing, 7 km E of Big Cedar, Le Flore County). The late flight date in the state is more than five months later on 5 September (in 2005, Salt Plains NWR, Alfalfa County, 1U), although that record, as well as two late records (from 2 September and 30 August) are only sight reports. The latest record supported by documentation is from 27 August (from 2010, mating pair photographed by T Mitchell, Mohawk Park, Tulsa County, OC 332378). As is often the case, Texas has earlier records than Oklahoma, as early as 26 February (OC 493419), which is not surprising given its latitude (also true of LA's early flight date of 22 February; Mauffray 2014). The early dates for Arkansas, Kansas, and Missouri (AR: 8 April, OC 441089; KS: 4 April, SEMC 1349052; MO: 12 April, Sims 2012) are more in line with Oklahoma. It is known to fly as late as 28 October in Louisiana (Mauffray 2014), 30 September in Arkansas (Harp and Rickett 1977), into August in Texas (22 Aug, Abbott and Lasley 2019), and into July in Kansas and Missouri (KS: 15 Jul, OC; MO: 31 Jul, iNat 29947074).

NASIAESCHNA PENTACANTHA (RAMBUR, 1842) – CYRANO DARNER

The Cyrano Darner is uncommon to fairly common at creek backwaters, swampy areas with open water, and sometimes small ponds. It favors shaded reaches of slow-moving creeks or coves, where it flies back-and-forth on short beats at chest height. A curious aspect of *Nasiaeschna pentacantha* is the tendency of other odonates to harass it, behavior akin to mobbing of birds of prey by crows, kingbirds, jays, and other species, whereby one or two birds fly in to annoy a hawk, *just to annoy it*, although the ultimate goal is to annoy the hawk into leaving. MAP has taken note of similar odonate

behavior on several occasions, usually involving *Libellula luctuosa*, the Widow Skimmer, and *L. incesta*, the Slaty Skimmer, and, to a lesser extent, *Erythemis simplicicollis*, the Eastern Pondhawk. One can see why other odonates do not care for *Nasiaeschna*'s company given that it is an impressive hunter, taking prey equally as large as itself, such as *Epitheca princeps*, the Prince Baskettail, and anything else on down the line. On the flip side, *Nasiaeschna pentacantha* is not shy about harassing other species, even if they are as large or markedly larger. For example, in an encounter with *Nasiaeschna* at Grassy

Figure 14.12 *Nasiaeschna pentacantha*, the Cyrano Darner, A) ♂, Bartlesville, Washington County, 20 June 2006 (©Mark Dreiling, OC 7066). B) A ♀ at the same locale on 17 June 2008 (©Mark Dreiling) underlined the species' penchant to lay eggs in rotting logs, and C) a ♀ at Field Station Lake, Woodward County, on 14 July 2018 (©Shawn Johnson, OC 485163) underlined the species' penchant to eat other large dragonflies, like this ♂ *Dromogomphus spoliatus*, the Flag-tailed Spinyleg, she is devouring. The Woodward County record is as far west as the species has been recorded on the Southern Great Plains.

Slough, BS-P watched as a territorial ♂ Cyrano Darner harassed three other aeshnids within just a handful of minutes. When she came up to the small, creekside, sunny pool she noticed the *Nasiaeschna* patrolling and then saw a ♀ *Anax junius* trying to oviposit but being continually harassed by the *Nasiaeschna* until she flew away. Then, what appeared to be a *Basiaeschna janata*, flew in briefly until it too was chased away. A couple of minutes later an *Epiaeschna heros* flew in, tussled with the *Nasiaeschna* (unsure who harassed whom), and then fled. The *Nasiaeschna* remained as the victor.

The Cyrano Darner is widespread in the eastern half of the state but decidedly scarcer farther west. It is never taken or seen in large numbers, although it has been recorded rather consistently in all eras of odonatological research in the state. Before 2000 the species was reported from 25 counties; in 5 of those counties it has not been seen since the 1930s (Caddo, Latimer, Major, Payne, Woods) and not since the 1960s or 1970s for 3 others (Johnston, Mayes, Murray). Conversely, the species was added to 34 counties between 2003 and 2019, such that it is now known from 59 of Oklahoma's 77 counties.

The Cyrano Darner was first collected in the state in mid-summer (15 June 1929, 1♂, Love County, RD Bird, OMNH 826), and it has since been reported from early April to late August (3 April 2017, 2U, DA sight record, Red Slough, McCurtain County; 22 August 2015, 1♀, MAP sight record, Wewoka Creek, south end of Seminole, Seminole County). We reconsidered a later report, from 6 October 2006 at TNC's Tallgrass Prairie Preserve, Osage County, of a lone ♂ seen by J Fisher, that we published (Smith-Patten et al. 2007); given the date and location, the species in question was more likely to be the Shadow Darner (*Aeshna umbrosa*). The Oklahoma late date of 22 August fits comfortably within those reported for Louisiana and Texas, but is beyond the flight seasons for Arkansas, Kansas, and Missouri (AR: 3 April–29 July, OC 440993, CLO-ML 441468–441469; KS: 7 April–29 July, SEMC 1348912, 1348916, 1348923, Allison 1921 indicated it flies until late October, which we question; LA: 24 February–12 October, OC, Mauffray 2014; MO: 13 May–8 August, OC 374957, Sims 2012; TX: 23 February–21 September, Abbott and Lasley 2019).

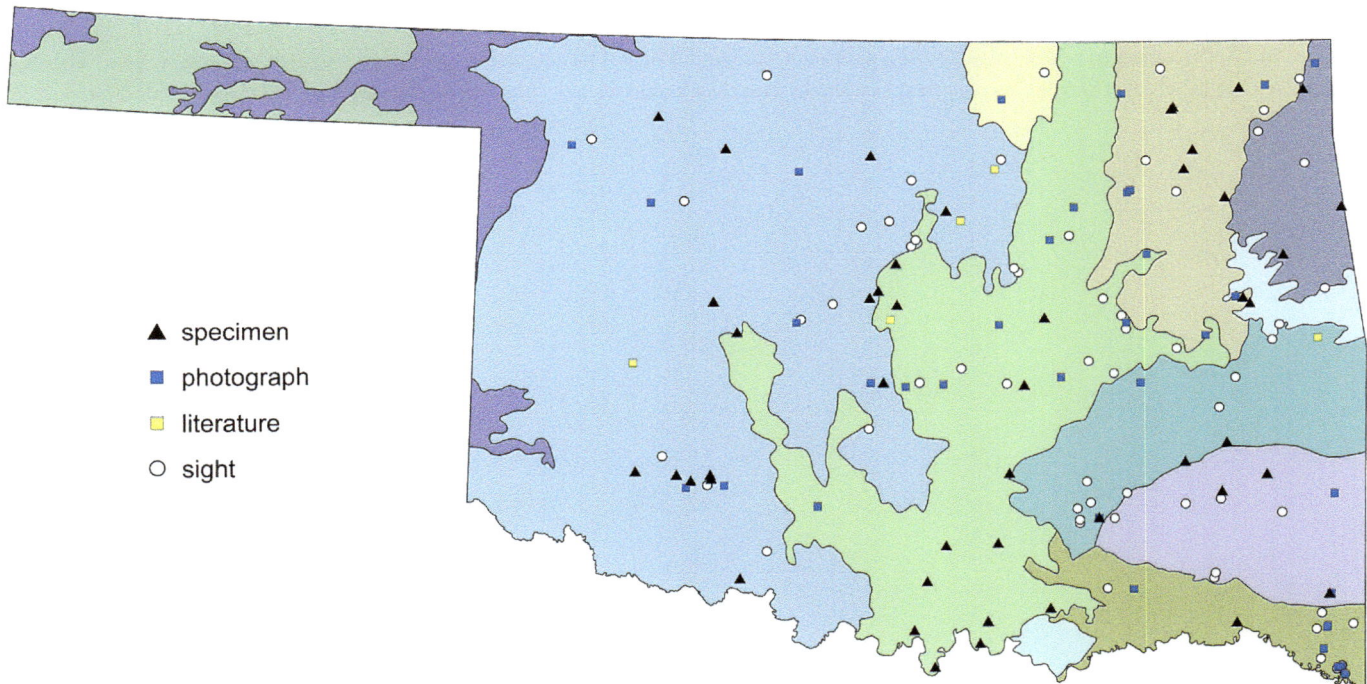

▲ specimen
■ photograph
□ literature
○ sight

Figure 14.13 Documentation level of supporting records for *Nasiaeschna pentacantha*, the Cyrano Darner, in Oklahoma. Levels include specimen, archived photograph, literature reference, or sight record. Although sight records lack documentation, only those we feel are credible are plotted.

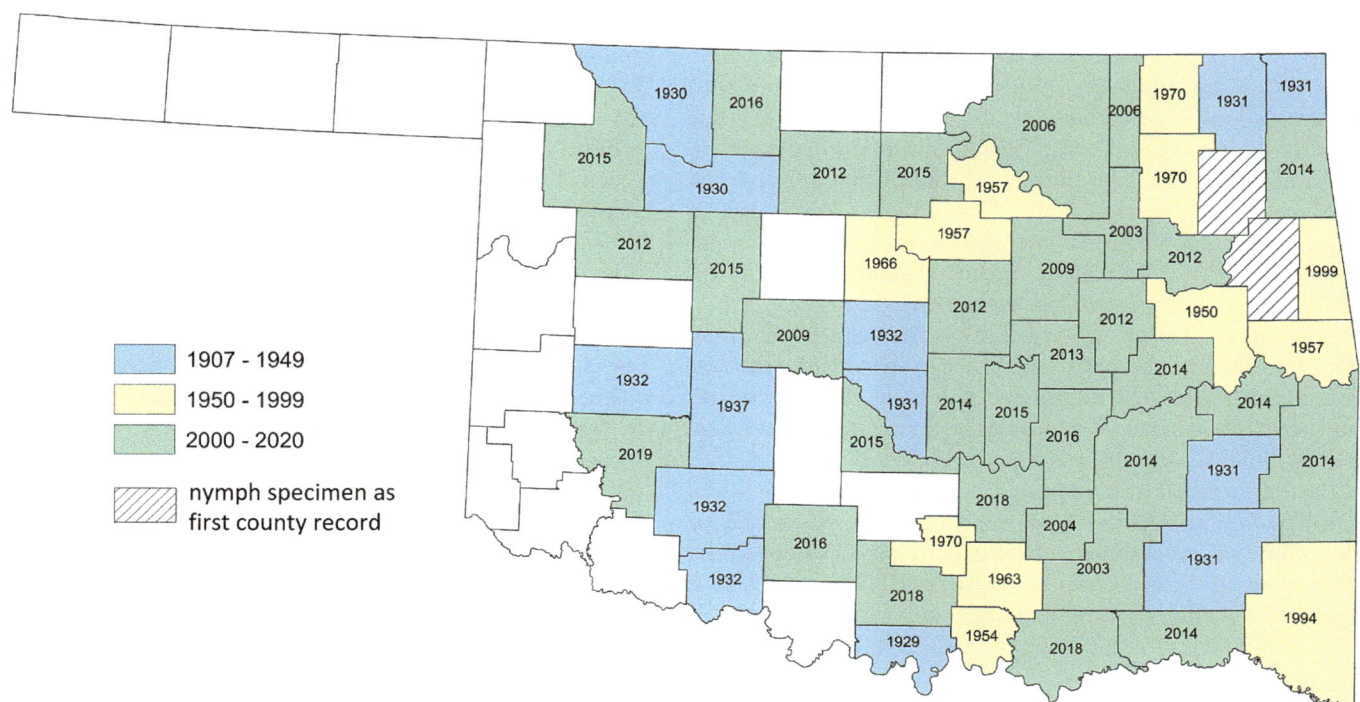

Figure 14.14 Counties in which *Nasiaeschna pentacantha*, the Cyrano Darner, are known to occur in Oklahoma. Year within the county is that in which the species was first reported. Hatched counties are those with only nymph specimens as support.

BASIAESCHNA JANATA (SAY, 1839) – SPRINGTIME DARNER

As its name implies, the Springtime Darner is a … *wait for it* … springtime dragon in Oklahoma. Its flight season is just under three months long (1 March–22 May, OC 461330, SP 1201). Arkansas's flight season is known to end at the same time, but its documented early date is much later (22 March–23 May, OC 312304, 481940). Dates for Missouri touch upon the summer (MO: 30 March–6 June, OC 376268, Trial 2005[4]), but the species has a much shorter documented flight season in Louisiana (2 March–24 April, OC 478061, OC 441320). Although the species appears to fly not as late in Texas (6 May, Abbott and Lasley 2019), that state has the earliest flight date for the species rangewide (Paulson 2009, 2011), beginning on 7 February (OC 398441).

The species is locally common to fairly common at clear, slow-flowing streams and creeks, and along the shores of clear lakes or large ponds. Regardless of habitat, it flies near to shore at or below knee height, all the while weaving in and out of emergent or overhanging vegetation (as if gleaning). It is not surprising then that it was first collected in the state along two creeks in Murray County. RD Bird collected 3♂ on Falls Creek (OMNH 2107, 2109, and 2113) and 3♂ on Sand Creek (OMNH 2110) on 18 April 1930. One would imagine that he had some help the next day when 15♂ were collected—2♂ on Sand Creek (OMNH 810 and 824, the later originally labeled as *Boyeria vinosa*) and 13♂ taken at Oil Springs in Johnston County (OMNH 811–817,

2106, 2111, 2115, USNM 333333–333334, and UMMZ). Bird (1932) reported the species from seven other counties.

The species was collected consistently during the early 1930s OU expeditions, but it was rarely reported in the 1950s–1970s. Bick apparently did not encounter the species, with good reason: his annual summer sojourns to the state came after the species' flight season. His notes nonetheless listed a handful of encounters by LE Hornuff in 1955–1956 from Johnston and McCurtain Counties, and we know of one LEH specimen from 1968 (1♂, Osage County, Osage Hills SP, Sand Creek, 4 May, FSCA). By the 2000s, the species again was reported regularly.

Of the 44 counties from which the species has been reported there is one county for which we have been unable to find documentation: Donnelly (2004a) reported the species for Love County, but neither Bird (1932) nor Bick and Bick (1957) did, nor can we find any other sources for the record. Hence, its documented presence in the county is questionable. Even so, given the species' distribution in the state—everywhere except the panhandle and the western fringe—although decidedly more numerous in the eastern half of the state—it likely occurs in Love County. Nymphs have been reported from Adair, Bryan, Cherokee, Comanche, Johnston, McCurtain, and Ottawa Counties, and both tandem pairs and ovipositing females have been reported a number of times.

Figure 14.15 *Basiaeschna janata*, the Springtime Darner, A), ♂, Blue Beaver Creek on the Wichita Mountains WR, Comanche County, 4 May 2008 (©Victor W Fazio III, OC 282134). B) ♀, Bartlesville, Washington County, 5 April 2006 (©Mark Dreiling, OC 6924).

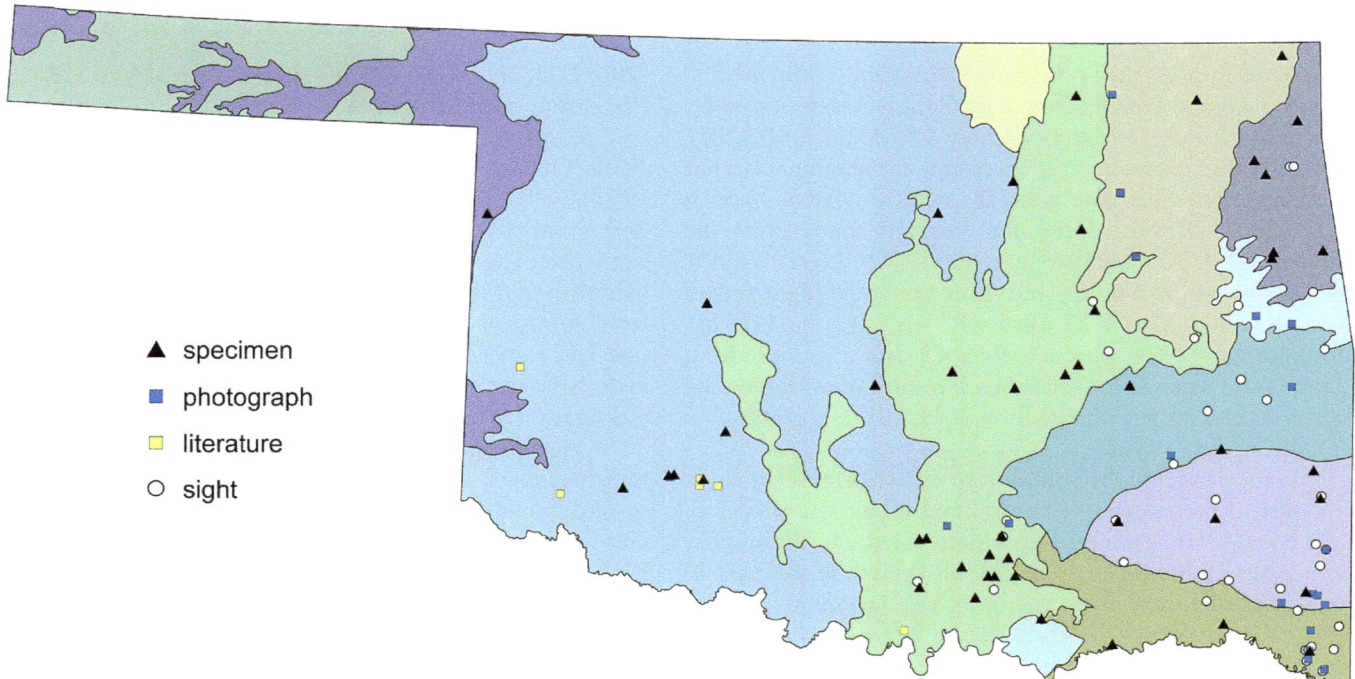

Figure 14.16 Documentation level of supporting records for *Basiaeschna janata*, the Springtime Darner, in Oklahoma. Levels include specimen, archived photograph, literature reference, or sight record. Although sight records lack documentation, only those we feel are credible are plotted.

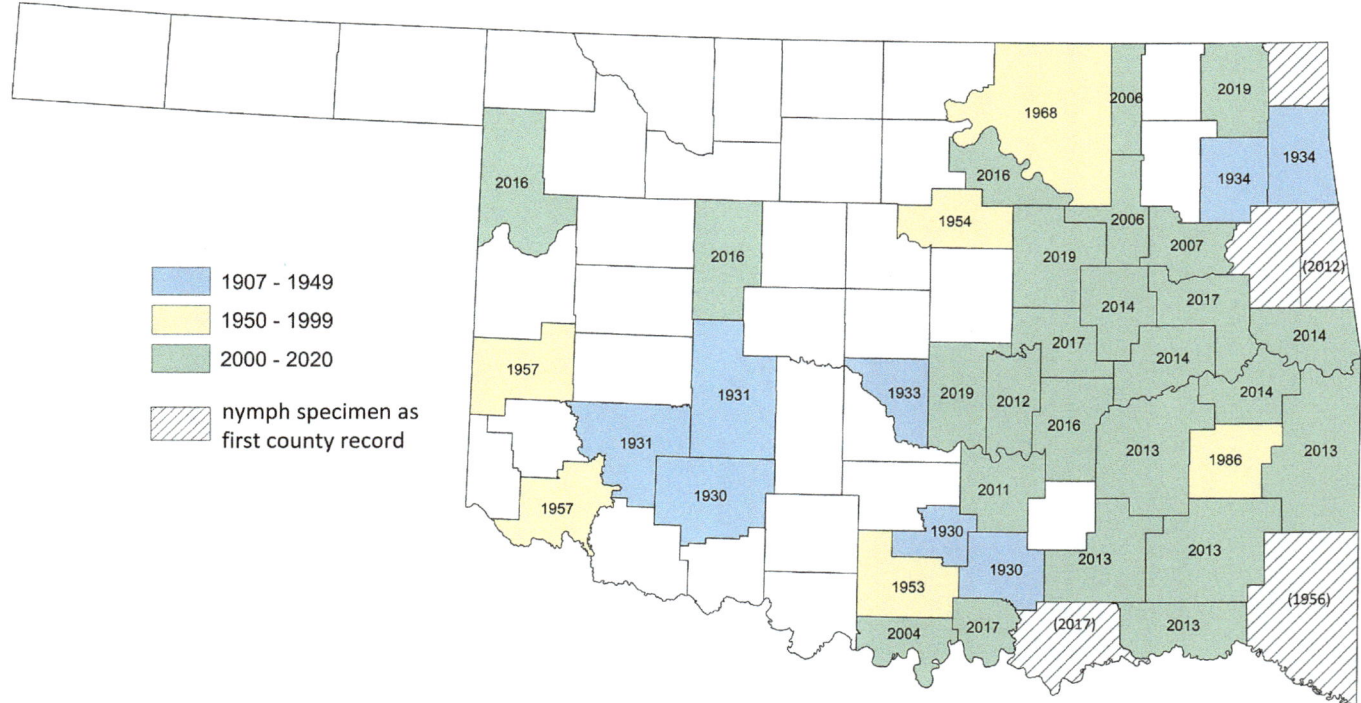

Figure 14.17 Counties in which *Basiaeschna janata*, the Springtime Darner, are known to occur in Oklahoma. Year within the county is that in which the species was first reported. Hatched counties (*n* = 5) are those with a nymph specimen as the first documentation of the species in the county. When a parenthetical year is also presented, it indicates the year of the first adult record; no year indicates the county only has larval records as support. We are fairly confident about the Cherokee County record (BS-P confirmation) but the specimens from the other four counties have not been examined by us.

GYNACANTHA NERVOSA RAMBUR, 1842 – TWILIGHT DARNER

The Twilight Darner, a crepuscular Neotropical species, has been recorded in five U.S. states. It was originally thought to occur in Florida and California (Williamson 1923, Kormondy 1960). It occurs regularly in Florida, but eventually it was realized it did not occur in California. It is somewhat regular, albeit rare, in extreme southeastern Georgia (Beaton 2007). There are single records for Alabama (Lee County, Donnelly 2004) and South Carolina (OC 474691) that may be more of an indication of the difficulty of spotting this species rather than truly disjunct records. Two other records—those for Delaware and Oklahoma—are of actual vagrants as both are at least 1,000 km away from where the species is thought to occur regularly. The Delaware record was of a ♀ specimen that Hal White found

Figure 14.18 Oklahoma's only record of *Gynacantha nervosa*, the Twilight Darner. Theodore H Hubbell collected this ♂ 1 mi NW of Page, Le Flore County, on 15 September 1935 (UMMZ specimen). Photo by ©Brenda D Smith.

Figure 14.19 The first and only record of *Gynacantha nervosa*, the Twilight Darner, in Oklahoma (Le Flore County; UMMZ) was found far from its normal range.

in the insect collection of the University of Delaware (White 2006[5]). That specimen was collected by Chuck Mason on 27 September 1975 in Newark, New Castle County. The Oklahoma record came in 1935, when TH Hubbell collected a ♂ ~2 km NW of Page, Le Flore County, on 15 September (UMMZ, OC 403468, Kormondy 1960; Figures 14.18 and 14.19). The specimen was not diagnosed for decades, and as such the species was not reported in a formal species list for the state until Beckemeyer's (2002a) checklist.

CORYPHAESCHNA INGENS (RAMBUR, 1842) – REGAL DARNER

The Regal Darner (*Coryphaeschna ingens*) now has two known populations in Oklahoma—Red Slough WMA, McCurtain County, and Boehler Seeps and Sandhills Preserve, Atoka County. A third location is conjectured to hold a population, and there are three records of vagrants.

The species has been seen in Oklahoma every year since it was first discovered in the state in 2008 (first seen on 1 July but not documented until 7 July, Red Slough WMA, McCurtain County, BAH and DA, 1♂, JCAC 24061, OC 282822, Arbour et al. 2009; Figure 14.20). Astonishingly,

Figure 14.20 A) Oklahoma's first documented record of *Coryphaeschna ingens*, the Regal Darner, ♂, Red Slough WMA, McCurtain County, 7 July 2008 (collected by Berlin A Heck, photographed by David Arbour, USDA Forest Service; JCAC 24061). This specimen looks awfully good ... especially for one shot out of the sky! B) ♂, Canton WMA, Dewey County, 21 July 2018 (©Michael A Patten, OC 485689), a record far to the northwest of where the species seems to be established in southeastern Oklahoma and a hint of how far the species may wander. C) ovipositing ♀, Red Slough WMA, McCurtain County, 25 July 2019 (USDA Forest Service; David Arbour, OC 500583), at the only site in Oklahoma where the species is encountered regularly and the only site with evidence of breeding. Notice the blue eyes of the females of this species.

BAH shot this specimen out of the air with his shotgun. That collecting method alone should go down in history, but even if he had not collected it the way he did, it was a great record for Oklahoma. Three of the four specimens from the state hail from Red Slough (1♂, 21 July 2009, DA, JCAC 24212, OC 314196; 1♀, 1 July 2014, SP 1305, OC 424070; 1♂, 8 May 2017, MAP, SP 2302); see below for details of the fourth of these specimens.

The species was first found outside of McCurtain County on 20 July 2014, when we saw 3♂ flying over the marshy area known as Boehler Lake at Boehler Seeps and Sandhills Preserve, Atoka County. BS-P photographed 1♂ (OC 424926). A ♂ and a ♀ were seen at Boehler Seeps again the following year (12 June 2015), missed the next, and then reported again in 2017 (2 July, 3♂, 1♀, MAP). It may be that Messer Bottoms in the Hugo WMA, Choctaw County, holds a third population for Oklahoma. At this point we cannot say so with certainty given the locality's single record (1♂, 25 June 2017, MAP), but given what appears to be suitable habitat for the species and its geographical position relative to Red Slough to the east and Boehler Seeps to the west, we predict that a population will eventually be documented there.

There are three vagrant records of *Coryphaeschna ingens* for Oklahoma. The first came on 4 July 2016, when MAP spotted a ♂ at Great Plains SP, Kiowa County, a location well outside the Gulf Coastal Plain, where the species is expected, by some 300 km (180 mi) west of the nearest known population. He snapped a couple of blurry photos, which in addition to getting good views of it in flight, persuaded him to submit the record as OC 449014. He had better luck on 9 June 2018, when he collected a ♂ at Sportsman Lake, Seminole County (SP 2620, OC 481654), but then the following month he was able to get only a photo of a ♂ he encountered at Canton Lake WMA, Dewey County (21 July 2018, OC 485689). As remarkable as these records are for Oklahoma, they are not the farthest west for the region. That distinction is held by a vagrant in the Texas panhandle (Lubbock County, Jerry Hatfield, 29 October 2012, OC 385434).

Regal Darners are known to fly in Oklahoma from 6 May until 18 September (early date is from 2012, 1♀, BAH and DA, OC 374781; late date is a DA sight record of 1U from 2018; both records are from Red Slough). In Texas it is known to fly from 31 March to 31 October (Abbott and Lasley 2019) and in Louisiana within that approximate timeframe (21 March–9 October, OC 462286, iNat 18309645). The species was first documented in Arkansas from nymphs collected in Drew County on 23 March 1970 (Harp and Rickett 1977, Arbour et al. 2009), yet an adult has

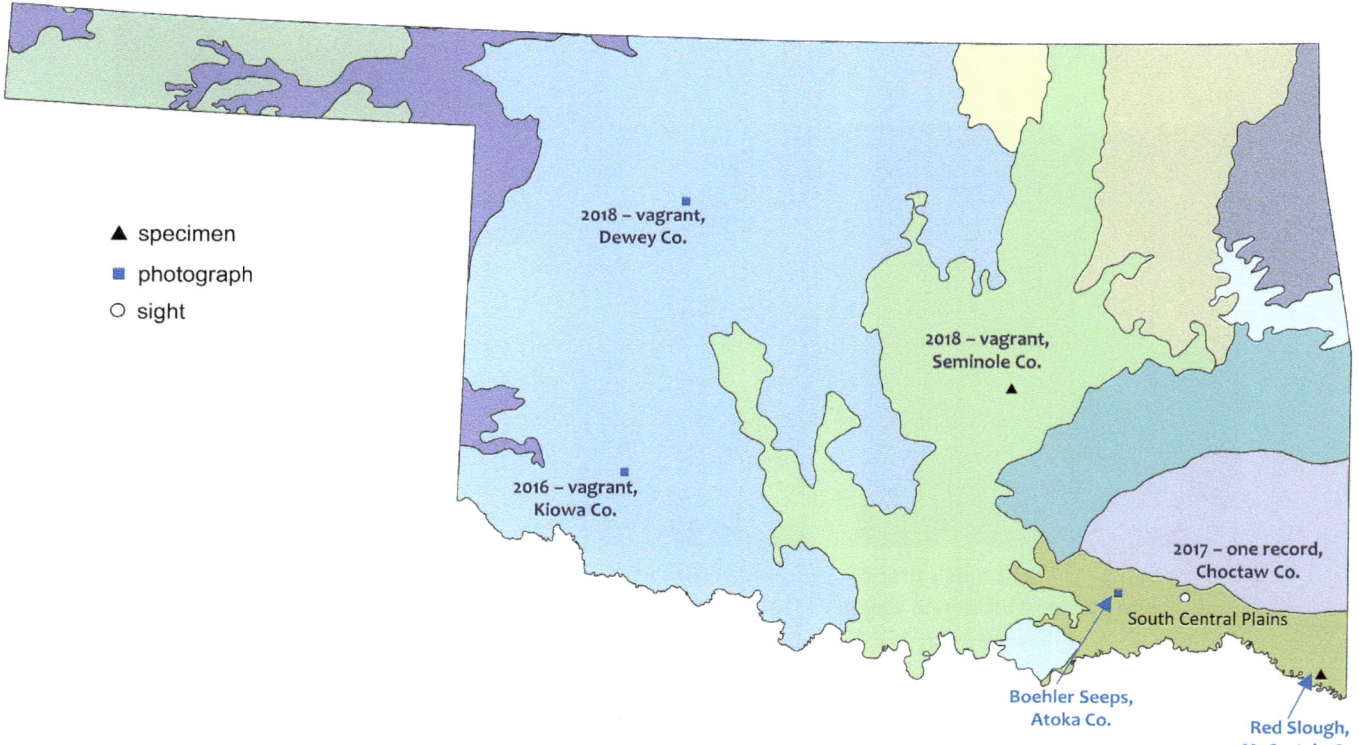

Figure 14.21 Documentation level of supporting records for *Coryphaeschna ingens*, the Regal Darner, in Oklahoma. The species has been encountered in six counties, but it has only been regularly found in one, McCurtain County. A second county, Atoka, has had multiple reports. Both of these counties lie within the South Central Plains, which is the only ecoregion in Oklahoma likely able to host a population of this species.

been documented only once (1♂, 20 May 2011, DeQueen, Sevier County, DA, OC 328040).

It is not known whether a small population has long existed in Oklahoma, or Arkansas for that matter, or if the species colonized recently. Like the Swamp Darner (*Epiaeschna heros*), the Regal Darner flies high, typically at or above treetop level, and thus is difficult to capture or even photograph well. (And even with the advent of modern field guides and more sophisticated field enthusiasts, these two large darner species can be confused in the field, especially the females). There is growing evidence that the species breeds in the region: one record of an immature ♀ (3 June 2015, OC 431303), a sighting of one pair on 2 August 2016, and two records of lone females ovipositing on water shield stems (*Brasenia schreberi*; 18 and 25 July 2019; all records from Red Slough by DA).[6] We expect that in the near future there will be clear evidence of successful, consistent breeding.

AESHNA CONSTRICTA SAY, 1839 – LANCE-TIPPED DARNER

There is one record of the Lance-tipped Darner (*Aeshna constricta*) for the state—1♂ photographed at Oxley Nature Center in Tulsa, Tulsa County, on 18 October 2004 by BC (OC 334055; Figure 14.22). The record went unknown until the photographer reviewed his photo again in 2011 and realized that he had a first state record. It is not surprising that this species was found in Oklahoma given that its known range dips down into southeastern and southcentral Kansas and adjacent Missouri (KS: Cherokee, Crawford, and Sedgwick County, Huggins et al. 1976, Donnelly 2004; MO: Jasper County, OC 136085).

In Kansas the species is known to fly between 1 August[7] and 13 October (SEMC 1315353, 1348931). The only published adult date that we know of for Arkansas was simply listed as May for Pulaski County in central Arkansas (Harp and Rickett 1977). In Missouri the species is known to fly from mid-June to mid-October (Sims 2012), a timeframe corresponding to its flight season elsewhere (Paulson 2009, 2011), so if it were to show again in Oklahoma, it can be expected anytime from summer to fall, and ought to be sought in the northeastern part of the state.

Aeshna constricta Say, 1839
Lance-tipped Darner
?
one record: 18 October 2004

Figure 14.22 Oklahoma's only record of *Aeshna constricta*, the Lance-tipped Darner, ♂, Oxley Nature Center, Tulsa, Tulsa County, 18 October 2004 (©Bill Carrell, OC 334055). The broad, even thoracic stripes and large blue spots on the abdomen eliminate *A. umbrosa*, the Shadow Darner, the only other species in the genus documented in Oklahoma.

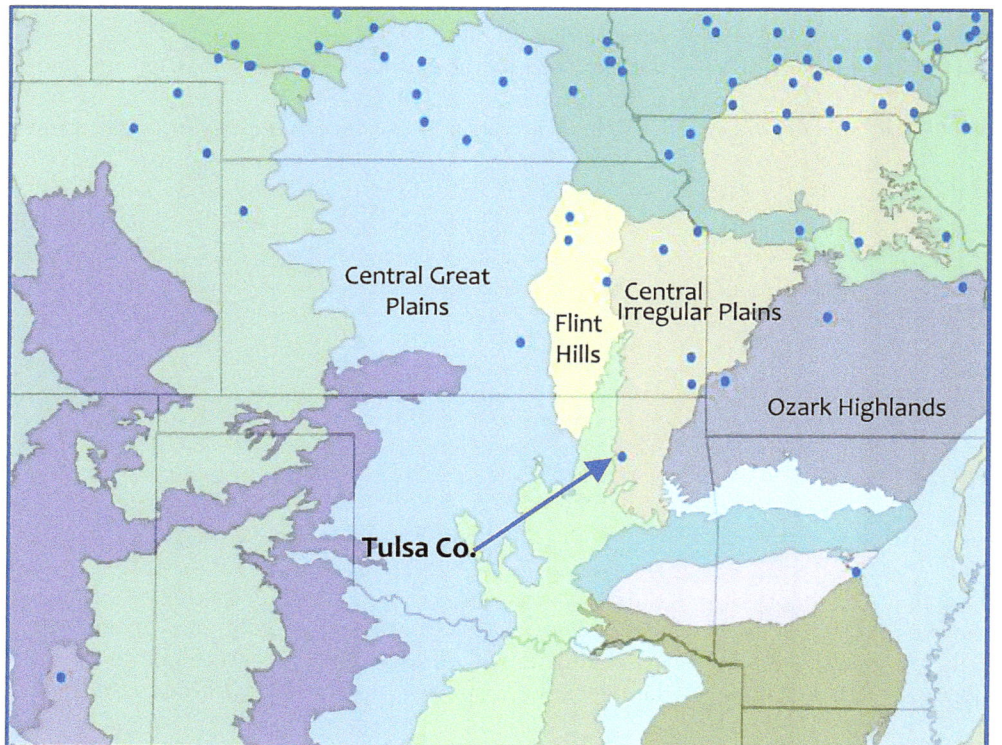

Central Great Plains

Central Irregular Plains

Flint Hills

Ozark Highlands

Tulsa Co.

Figure 14.23 Distribution of *Aeshna constricta*, the Lance-tipped Darner, showing Oklahoma's only record of the species (Tulsa County). Also note that, as indicated by shared ecoregions, the species could be found again in the Central Irregular Plains, but the Central Great Plains, Flint/Osage Hills, or the Ozark Highlands are possibilities, too.

AESHNA UMBROSA WALKER, 1908 – SHADOW DARNER

We agonized a great deal in our efforts to deduce the Shadow Darner's status in Oklahoma. Half of the agony was that the species is not well documented in the state—there are 17 confirmed records involving >30 adults. The other half of the agony stems from the (apparently high) potential of misidentifications—there are 22 reports of adults lacking documentation—and how that possibility made teasing out the species' phenology a challenge.

Until 2011, it was thought that the Shadow Darner was the only species of *Aeshna* in Oklahoma. We then realized that the Lance-tipped Darner (*Aeshna constricta*) occurs, at least as a vagrant but quite possibly regularly, in the northeastern corner of the state. And there is a chance that the Variable Darner (*A. interrupta*) and Paddle-tailed Darner (*A. palmata*) venture into the state at times. All four of these *Aeshna* species fly in at least parts of their ranges starting in the summer and ending in mid to late fall. To the uncertainty regarding congeners we add apparent confusion with other darners, especially the Springtime (*Basiaeschna janata*), Cyrano (*Nasiaeschna pentacantha*), and Blue-eyed (*Rhionaeschna multicolor*) Darners. Thus, the basis for the conundrum—are all reports of *A. umbrosa* actually *A. umbrosa*?

Initially (e.g., Smith-Patten et al. 2007) we drew a different, and we now believe erroneous, picture of the species' phenology in the state, one that we thought jibed with the phenology for the region, in that our impression was the species flew in Oklahoma from late spring or early summer to the fall. But once we examined records more closely, we noticed that there was a distinct seasonal break between the majority of undocumented reports and records documented with a photograph or specimen (*n* = 17; Table 14.1). *All documented records are from the fall*, with the bulk of them from mid-September to mid-October

The earliest of the documented records, in terms of seasonality, is a photograph that RC Anderson took on 26 August 2006 of a perched ♂ at Red Rock Canyon SP in Caddo County (OC 7299). Since that time there have been five other photos accepted as documentation for the species in the state (OC 323194, 409996, 438261, 461642, 473882). We know of 14–15 adult specimens, three of which are the EB Williamson specimens discussed below and two or three were collected by RD Bird in 1932. More than 70 years later, RJ Beckemeyer collected two of the 4♂ he encountered in Cimarron County in 1994. A much shorter time passed before the next specimen: in 2009, an OU student unwittingly caught a single ♂ at the Tallgrass Prairie Preserve, Osage County (SS Ansell, SP 1108), unwittingly because the specimen was presented unidentified as part of a collection for a class project. The specimen is oddly dark throughout much of the abdomen, although the dorsum of S10 is not fully black, and it lacks spots ventrally except on S8. It may be that the student treated this specimen with a chemical that caused discoloration, but if not, the odd patterning is of note. Other recent specimens have come from Kay, Harper, Cherokee, and Beaver Counties (SP 1031, 1431, 1792, 2224, 2522, 2713). The first of two Cherokee County records is of particular interest because not only is the location our easternmost locality but the record involved 4♂ and 3♀, including three pair (16 November 2016, 1♂ photographed in hand as OC 458375, 1♀ collected as SP 2224, Pipe Springs Hollow, Cookson WMA, C Farquhar; Figure 14.24), making it the highest count of the species in the state as well as the late flight date. The creek from which that recorded hailed is typical for the species in Oklahoma: shaded, shallow, narrow creeks with good flow and clear water, habitats that can be scarce (or not readily accessible) in the state. The species generally flies low over the water, often following each little bend and jag in the creek; some will forage, again flying low, in shaded clearings near such a creek.

The other 22 reports of *Aeshna umbrosa* for Oklahoma are essentially undocumented (Table 14.1). One record comes from a literature report by Bick (1991; Oklahoma County). Bick did not provide details (he did not even include the record in his notes), and we have been unable to find an associated specimen. Nonetheless, given the reliability of Bick we are inclined to consider the record valid, even if we are left without an idea of its seasonality. We labeled this record as probable as we did with two other records: a sighting of one adult at the Tallgrass Prairie Preserve, Osage County on 4 October 2003 (Smith-Patten et al. 2007) and a sight report by BC from 8 October 2017 at Doby Springs, Harper County, a well-documented location for the species. Seven other fall reports, dating from 5 September to 11 October, are possible records, although four were reported as likely, not confidently, the species (i.e., *A. umbrosa*, cf) and another as just *Aeshna*.

When we started to reconsider spring and summer reports, all of which fall between 13 May and 28 July, we did so by dismissing eight of our own sight records, one of which we previously published (1 June 2003, Mohawk Park, Tulsa County; Smith-Patten et al. 2007). Because we have no way to review these records given the lack of documentation we felt better playing it safe by assuming that we may have or, in some cases, probably misidentified those individuals. Throwing out our records left us with five spring and summer reports, each of them from good observers but none with good documentation. Three are sight records from KW, JWA, or both, of single individuals (29 May 2006 in Cimarron County, 19 June 2012, Okmulgee County, and 1 June 2012 in Tulsa County, respectively). Another report came from McCurtain County on 26 July 2010, when J Grzybowski photographed a ♂ he identified as a Shadow Darner (OC 321690). Unfortunately, the photograph is enough distant and out of focus to preclude identification, being either a species of *Aeshna* or *Rhionaeschna*. Finally,

Table 14.1 **Accepted, possible, and rejected records of *Aeshna umbrosa* from Oklahoma**

RECORD TYPE	DATA SOURCE	INDIVIDUALS	COUNTY	LOCALITY	DATE	COLLECTOR/ OBSERVER	GENERAL NOTES
Accepted — documented records							
specimen	UMMZ	3 (2♂, 1♀)	Caddo	12 km E of Hydro	19 Oct 1929	Williamson, EB	erroneously published as Custer Co.; photos of 1♂ and 1♀ as OC 403467
specimen	OMNH 2008	1 (1♂)	Cleveland	vic. Norman [presumed]	9 Oct 1932	Bird, RD	
specimen	OMNH 2010	1–2♂	Murray	Price Falls	15 Oct 1932	Bird, RD	Bick's note cards indicated there was 1♂, but OMNH has as 2♂; Bick and Bick 1957
specimen	RJB Collection	4 (4♂)	Cimarron	vic. Kenton	17 Sept 1994	RJB	Beckemeyer 1995; 2♂ collected
photo	OC 7299	1 (1♂)	Caddo	Red Rock Canyon SP, Rough Horsetail Trail	26 Aug 2006	Anderson, R	
specimen	SP 1108	1 (1♂)	Osage	TNC Tallgrass Prairie Preserve	10 Oct 2009	Ansell, SS	
photo	OC 323194	1 (1♀)	Pontotoc	TNC Pontotoc Ridge Preserve	23 Sept 2010	Alm, F	in hand
photo	OC 409996	1 (1♀)	Harper	Doby Springs Park	1 Sept 2013	BC	
specimen	SP 1031	1 (1♂)	Kay	Lake Ponca	15 Sept 2013	BS-P; MAP	along narrow outflow creek
photo	OC 461642	1 (1♂)	Harper	Doby Springs Park	7 Sept 2014	BC	teneral
specimen	SP 1431	1 (1♂)	Harper	Doby Springs Park	14 Sept 2014	BS-P; MAP	immature/somewhat teneral
specimen	SP 1792	1 (1♀)	Harper	Doby Springs Park	12 Sept 2015	MAP	
photo	OC 438261	1 (1♂)	Harper	Doby Springs Park	25 Oct 2015	BC	
specimen	SP 2224	7 (4♂, 3♀, incl. 3 pair)	Cherokee	Cookson WMA, Pipe Springs Hollow	16 Nov 2016	Farquhar, C	SP 2224, 1♀; 1♂, in hand
photo	OC 473882	5 (5♂)	Harper	Doby Springs Park	6 Oct 2017	MAP	3♂ in hand
specimen	SP 2522	1 (1♂)	Beaver	Beaver Dunes SP	8 Oct 2017	MAP	
specimen	SP 2713	1 (1♂)	Cherokee	Cookson WMA, Pipe Springs Hollow	5 Oct 2018	BS-P; Farquhar, C	
Probable record — documentation missing, seasonality unknown							
literature	Bick 1991	unknown	Oklahoma				we trust that Bick id'd correctly
Probable and possible records — undocumented, fall reports							
literature	Smith-Patten et al. 2007	1 (1U)	Osage	TNC Tallgrass Prairie Preserve	4 Oct 2003	BS-P; MAP	probable fall record, but lacks documentation
sight	BC	1 (1♂)	Harper	Doby Springs Park	8 Oct 2017	BC	probable fall record, but lacks documentation
sight	Fisher, J database	2 (2U)	Cimarron	Black Mesa SP	5 Sept 2006	Fisher, J	possible fall record, but lacks documentation
sight	Fisher, J database	3 (3U)	Cimarron	TNC Black Mesa Nature Preserve	6 Sept 2006	Harp, C; Nelson, J; Fisher, J	possible fall record, but lacks documentation
sight	BS-P/MAP field notes	1 (1♀)	Texas	Guymon, Sunset Lake	26 Sept 2010	MAP; BS-P	original id as *Aeshna umbrosa*, cf

(Continued)

Table 14.1 (*Continued*) Accepted, possible, and rejected records of *Aeshna umbrosa* from Oklahoma

RECORD TYPE	DATA SOURCE	INDIVIDUALS	COUNTY	LOCALITY	DATE	COLLECTOR/ OBSERVER	GENERAL NOTES
sight	BS-P/MAP field notes	1 (1U)	Atoka	Atoka WMA	11 Oct 2013	BS-P; MAP	original id as *Aeshna umbrosa*, cf
sight	BS-P field notes	1 (1U)	Harper	Doby Springs Park	28 Sept 2015	BS-P	original id as *Aeshna umbrosa*, cf
sight	MAP field notes	1 (1♂)	Beaver	Beaver Dunes SP	6 Oct 2017	MAP	MAP identified to genus only (returned to site on 8 October, captured SP 2522)
sight	MAP field notes	1 (1♂)	Lincoln	4 km SSE of Arlington, Deer Creek	23 Sept 2017	MAP	MAP identified to genus only
Rejected — undocumented, spring and summer reports							
literature	Smith-Patten et al. 2007	1 (1U)	Tulsa	Tulsa, Mohawk Park	1 June 2003	MAP; BS-P	
sight	Fisher, J database	1 (1U)	Cimarron	TNC Black Mesa Nature Preserve	29 May 2006	KW	
literature	Smith-Patten et al. 2007	1 (1U)	Osage	TNC Tallgrass Prairie Preserve, trail loop and Sand Creek	28 July 2007	BC	BC reconsidered record in 2013
sight	JWA	1 (1U)	Tulsa	Tulsa, Mohawk Park, Oxley Nature Center	1 June 2012	JWA; KW	
sight	JWA	1 (1U)	Okmulgee	Deep Fork NWR	19 June 2012	JWA	we feel we misled JWA and KW into identifying this as *Aeshna umbrosa*
sight	BS-P/MAP field notes	2 (2U)	McCurtain	Broken Bow	31 May 2003	MAP; BS-P	we have no confidence in this record
sight	BS-P/MAP field notes	1 (1♂)	Coal	Parker, Caney Boggy Creek	12 June 2004	MAP; BS-P	we have no confidence in this record
sight	BS-P/MAP field notes	1 (1♀)	Creek	Stroud Lake	21 June 2009	MAP; BS-P	original id as *Aeshna umbrosa*, cf
sight	BS-P/MAP field notes	2 (2U)	Dewey	Canton Lake WMA	13 May 2012	MAP; BS-P	we have no confidence in this record
sight	BS-P/MAP field notes	1 (1U)	Okmulgee	Okmulgee	26 May 2012	MAP; BS-P	we have no confidence in this record
sight	BS-P/MAP field notes	2 (2U)	Okmulgee	Deep Fork NWR	27 May 2012	MAP; BS-P	we have no confidence in this record
photo	OC 321690	1 (1U)	McCurtain	Beavers Bend SP	26 July 2010	Grzybowski, J	photo is too blurry for positive id

data source and other fields: BC = Bill Carrell, BS-P = Brenda D Smith-Patten, JWA = James W Arterburn, KW = Ken Williams, MAP = Michael A Patten, OC = Odonata Central, RJB = Roy J Beckemeyer, SP = Smith-Patten/Patten Collection, UMMZ = University of Michigan Museum of Zoology; individuals: incl. = including, U = undetermined/unsexed; locality: NWR = National Wildlife Refuge, SP = State Park, TNC = The Nature Conservancy, vic. = vicinity, WMA = Wildlife Management Area; general notes: id or id'd = identified, cf = compares favorably

we slung more egg onto our faces by retracting another published record (28 July 2007, Tallgrass Prairie Preserve, Osage County, BC; Smith-Patten et al. 2007).

Even now, after we have been admirably cautious, we may one day have to walk back our current assessment of the species' status in Oklahoma. It is true that currently all documented records are from the autumn, and it may be that the species moves southward into the state late in the season, just as do various meadowhawks (*Sympetrum* sp.), yet the possibility holds that there could be future spring or,

Figure 14.24 *Aeshna umbrosa*, the Shadow Darner. This species' status on the southern Great Plains is contentious. All documented records are from late summer or autumn, well after the peak flight seasons of most of our other darners. We expect that it "pushes" southward late in the year in the same way various *Sympetrum* meadowhawks do. A) ♂, Cookson WMA, Cherokee County, 16 November 2016 (©Colby Farquhar, OC 458375). B) ♀, Doby Springs Park, Harper County, 1 September 2013 (©Bill Carrell, OC 409996). Posited "pushes" aside, C) a teneral ♂ at Doby Springs on 7 September 2014 (©Bill Carrell, OC 461642) indicates the species breeds in Oklahoma.

at least, summer records accompanied by good documentation. Such records will support our suspicion that the species is, in addition to being present and regular in the fall, a spring and summer vagrant in the southern portions of its overall range, a hypothesis that would account for the lone Texas record (26 July, Randall County, in the TX panhandle, OC 363249, JCAC 37380) and may account for the early flight season for Chautauqua County, Kansas ("7 June–15 July", SEMC 1349022), which borders Oklahoma. Other records from Kansas for July are farther north in the state or, as with the next closest record to Oklahoma, has a date range that extends into August ("25 July–26 August", SEMC 1349021). In Missouri, the Shadow Darner is documented as early as 16 June, but no adult records have been reported from near Oklahoma (Trial 2005). Likewise, New Mexico has a flight season starting in May (Paulson 2009), but our

understanding is that those records are separated from Oklahoma by a considerable distance, both geographically and ecologically. Late dates for Kansas, Missouri, and New Mexico are all well into the fall (KS: 14 October, SEMC 1349026; MO: 8 November, Trial 2005; NM: 23 October, OC 457412).

Thus far we have scant, but compelling, evidence that the species breeds in Oklahoma. There are two records of tenerals (both in September 2014 at Doby Springs Park, Harper County: 1 teneral ♂, 7 September, BC, OC 461642, Figure 14.24; 1 slightly teneral ♂, 14 September, SP1431). Additionally, there are 5 records comprising 14 nymphs. Eleven were collected by WJ Matthews at springs in Cimarron (Negro Spring, 28 May 1982, *n* = 5; OMNH 1968, 1971, the latter was originally identified as *Anax*), McIntosh (High Spring, 19 May 1981, *n* = 2, OMNH 1942), and Washita Counties (Taylor Creek

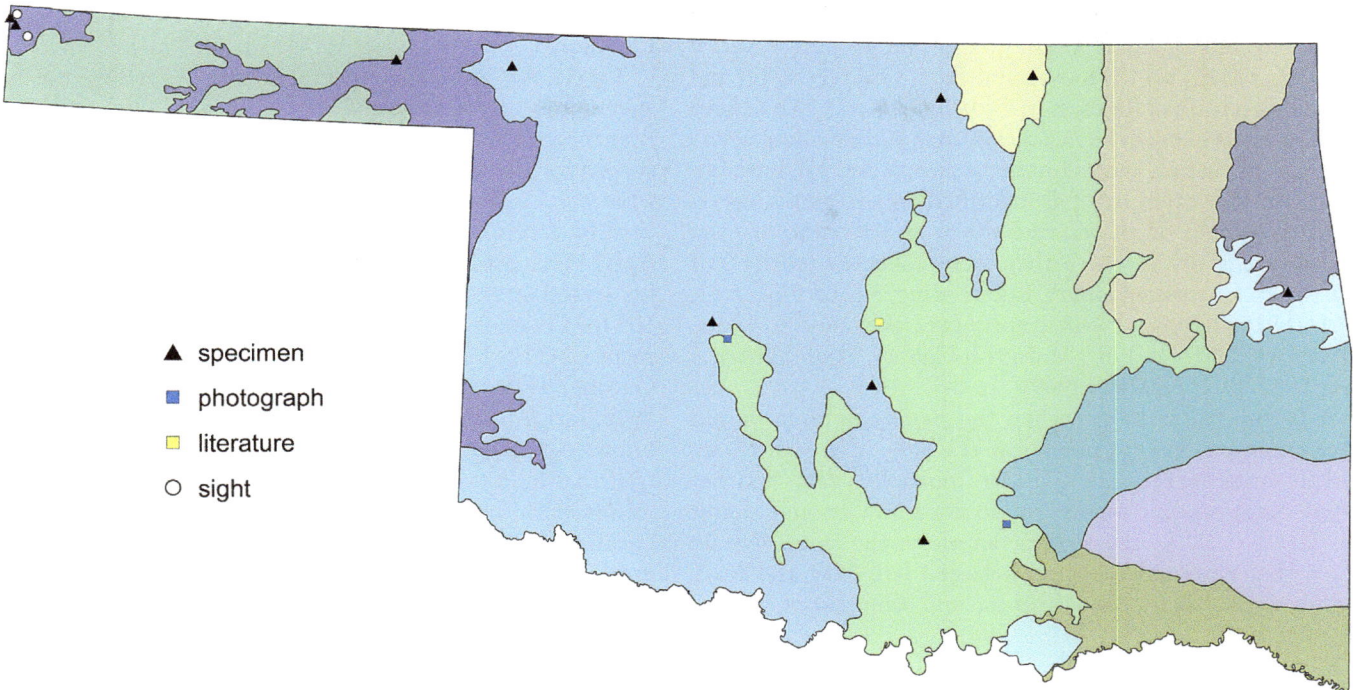

Figure 14.25 Documentation level of supporting records for *Aeshna umbrosa*, the Shadow Darner, in Oklahoma. Levels include specimen, archived photograph, literature reference, or sight record. Although sight records lack documentation, only those we feel are credible are plotted.

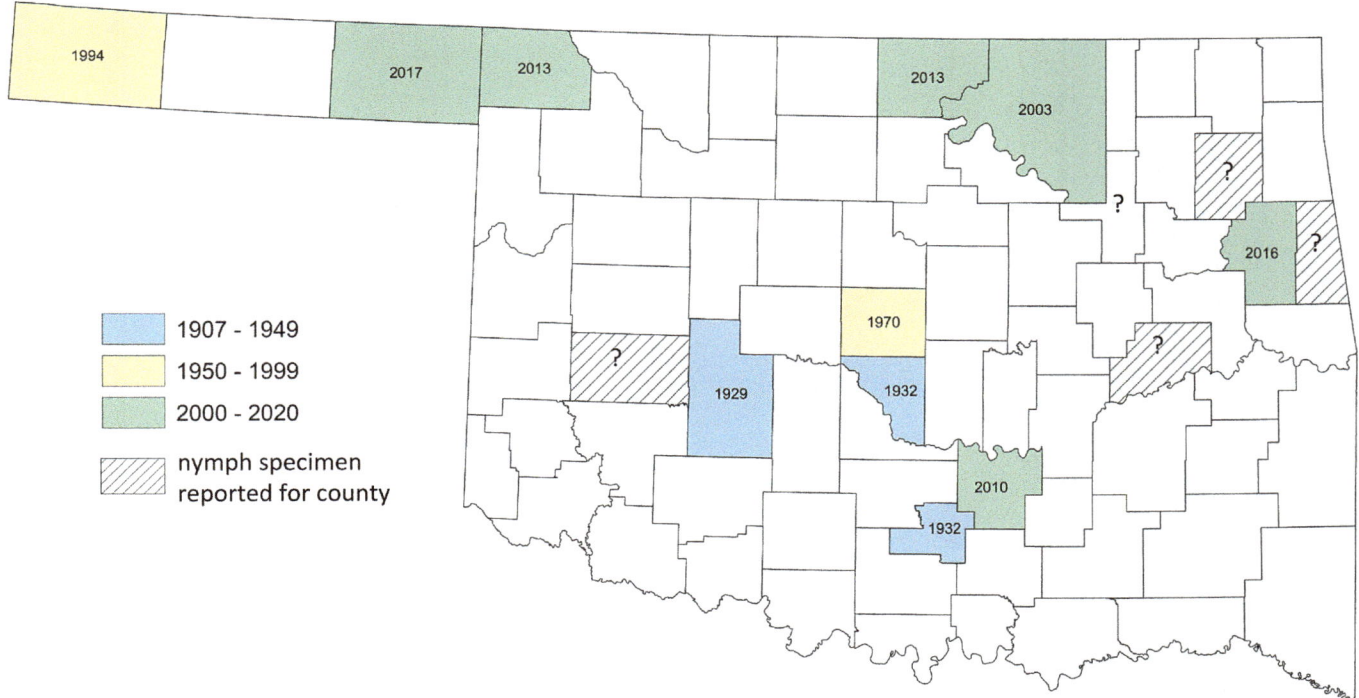

Figure 14.26 Counties in which *Aeshna umbrosa*, the Shadow Darner, are known to occur in Oklahoma. Year within the county is that in which the species was first reported. Hatched counties are those with only nymph specimens as support. We consider all nymphal identifications as tentative. There are two reports of *A. umbrosa* for Tulsa County, but given that *A. constricta*, the Lance-tipped Darner, could not be conclusively eliminated, that county will await documented records of the species.

Spring, 21 June 1981, $n = 4$, OMNH 1944). The other three were collected in Adair County (Chewey, Illinois River, 9 December 1983, $n = 1$, Hoover and Bart, OMNH 1742) and at an undetermined location in Oklahoma ($n = 2$). We believe these nymphs likely to be *Aeshna umbrosa*, yet definite species-level determination awaits further examination. We have not examined four other nymphs identified only to genus, specimens that could be *Aeshna umbrosa* or *A. constricta*, or even *A. interrupta*, given where they were collected (all by WJ Matthews: Cherokee County, Luck Spring, 3 June 1982, $n = 1$, OMNH 1966; Mayes County, spring at Lake Hudson, 3 June 1982, $n = 1$, OMNH 1967; McIntosh County, High Spring, 4 June 1982, $n = 2$, OMNH 1969).

It is impossible to say when the *Aeshna* nymphs would have emerged. We do not know the ages of any of them, and aeshnid larvae can live for multiple years (Corbet 1999). At this point we simply do not know enough about the larval cycle in the region, let alone the state. We do know that emergence occurs between late May and early October elsewhere, (in Virginia and Ontario, Landwer and Sites 2010), but those dates may not be applicable to Oklahoma.

Lastly, we offer an important aside regarding the first specimens of *Aeshna umbrosa* taken in Oklahoma. The first record was reported by Bird (1932) as the only county record for the species in the state, which he thought at the time was Custer County. That information was reported to Bird by Leonora K Gloyd of UMMZ in a letter dated 28 April 1932 (Bird, in a reply on 5 May, said he was finalizing his state list and that *Aeshna umbrosa* was species number 103 for Oklahoma!). Gloyd made the report based on specimens (2♂, 1♀, UMMZ) collected by EB Williamson on 19 October 1929 that were labeled as being taken "15 mi E of Weatherford, in Custer County, Okla." Had Williamson known Oklahoma geography better he might have realized that he was actually in Caddo County. His seemingly minor error was carried forward by Bick and Bick (1957), Donnelly (2004a), and our own records for many years until our 2013 visit to UMMZ clarified the situation (OC 403467).

RHIONAESCHNA MULTICOLOR (HAGEN, 1861) – BLUE-EYED DARNER

The Blue-eyed Darner is primarily a lentic species of open marshes often associated with ponds or small lakes. On the occasion it is found on a creek, it is at slow, marshy reaches. It has been reported in 32 counties throughout Oklahoma, although it is primarily a species of northwestern Oklahoma and the panhandle, from where >70% of records are known (*n* = 84).

We have flight records of the Blue-eyed Darner in Oklahoma as early as 2 April and as late as 29 November (early: 2008, Cimarron County: 6 mi E of Kenton, Hoot Owl Ranch, 1 mating pair, BAH, OC 281939; late: 2017, McCurtain County: Red Slough, 2–3 indiv., DA). Oklahoma's flight dates match well to those from New Mexico and Texas (NM: 30 March–November, OC, Paulson 2009; TX: 9 April–7 December, TTU-Z 31258, Abbott and Lasley 2019), but not so much with Colorado and Kansas (CO: 6 May–22 September, iNat 25429284, 16802720; KS: 18 May–11 September, iNat 10049523, SEMC 1348981, 1349007; MO has only one record, 30 April, Jackson County, Trial 2005).

The species' seasonality in Oklahoma is particularly interesting when one considers its geographical movements over the course of its flight season. Total records are roughly equally divided between those from spring and summer and those restricted to the fall. But when looking at the records in terms of distribution by season, we see that the spring/summer records are limited to 11 counties near the western edge of the state, whereas fall records have come from 23 counties, many of them much farther east. One of the two spring/summer exceptions is not terribly surprising, as it is only a couple of counties farther southeast than other seasonal records (El Reno Lake, Canadian County, 9 May 2012, 1♂). The real surprising spring/summer record is the anomalous ♀ captured and photographed on Rich Mountain, in Le Flore County, in southeastern Oklahoma near the Arkansas border (5 June 2011, TD Hibbitts, OC 329134). That record remains the only one known for the county, but it was the fourth for southeastern Oklahoma. Most other southeastern records are from McCurtain County from the fall (4–13 November 2008, OC 284348 and OC 284412/284438/JCAC 24023; 26 September 2009, OC 315222); the exception is a record of 5♂ and 1♀ (part of a tandem pair) at Lake Durant, Bryan County, on 22 October 2017 (MAP, 1♂, SP 2529). These records, taken along with other fall records from northeastern, north-central, and central Oklahoma (which constitute 51% of fall records), indicate that *Rhionaeschna multicolor* wanders eastward in the fall in Oklahoma (Figure 14.28). Fall wandering does not

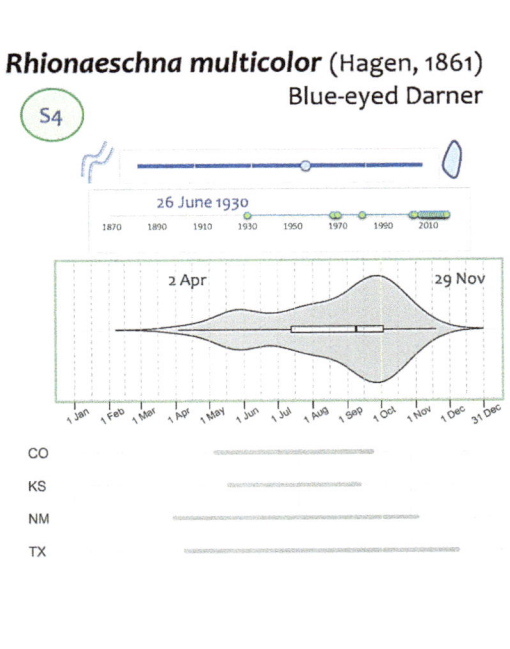

Rhionaeschna multicolor (Hagen, 1861)
Blue-eyed Darner

S4

26 June 1930

2 Apr   29 Nov

CO
KS
NM
TX

Figure 14.27 *Rhionaeschna multicolor,* the Blue-eyed Darner, A) ♂, Keystone Dam, Tulsa County, 31 October 2016 (©James W Arterburn, OC 457242). B) ♀, Rich Mountain, Le Flore County, 5 June 2011 (©Troy Hibbitts, OC 329134), a record that does not fit geographically or seasonally with patterns established elsewhere in the state.

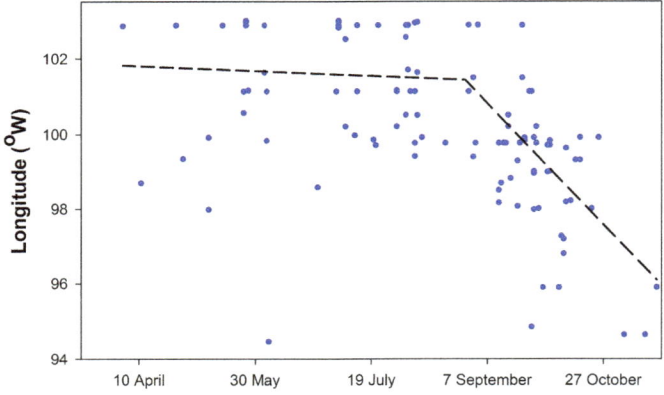

Figure 14.28 Relative abundance of *Rhionaeschna multicolor*, the Blue-eyed Darner, from across Oklahoma, plotted by longitude and Julian date. Piecewise regression indicates a break point corresponding approximately to 29 August, when occurrences correspond to lower longitudes (i.e., are farther east), essentially making the species a "fall wanderer."

Figure 14.29 *Rhionaeschna multicolor*, the Blue-eyed Darner, Drummond Flats WMA, Garfield County, Oklahoma, 20 September 2011, OC 333458. Photo by ©Jason R Heinen.

account for the lone summer Le Flore County record, however. Perhaps the species has simply been overlooked during the summer in eastern Oklahoma. At this point we are willing to discount that notion, given the extensive coverage the area has received by knowledgeable observers such as DA, BAH, BC, JWA, and ourselves.

We find it equally likely that RD Bird, GH Bick, and LE Hornuff did not overlook *Rhionaeschna multicolor* in their

respective eras. Bird reported the species only once, at the Glass (or Gloss) Mountains[8] in Major County, where he took a single ♀ on 26 June 1930 (OMNH 2001). That was the first record for the state. The species was not reported again for almost another four decades, until Hornuff collected 4♂ in 1968 (2 August, "Boise City, 15N, 3W," Cimarron County, FSCA), after which he and Bick had the species again in 1970 in Beaver (3 August, 1♂, 2.5 mi N of Beaver, FSCA), Cimarron (3–4 August, 1♂, Black Mesa SP, FSCA), and

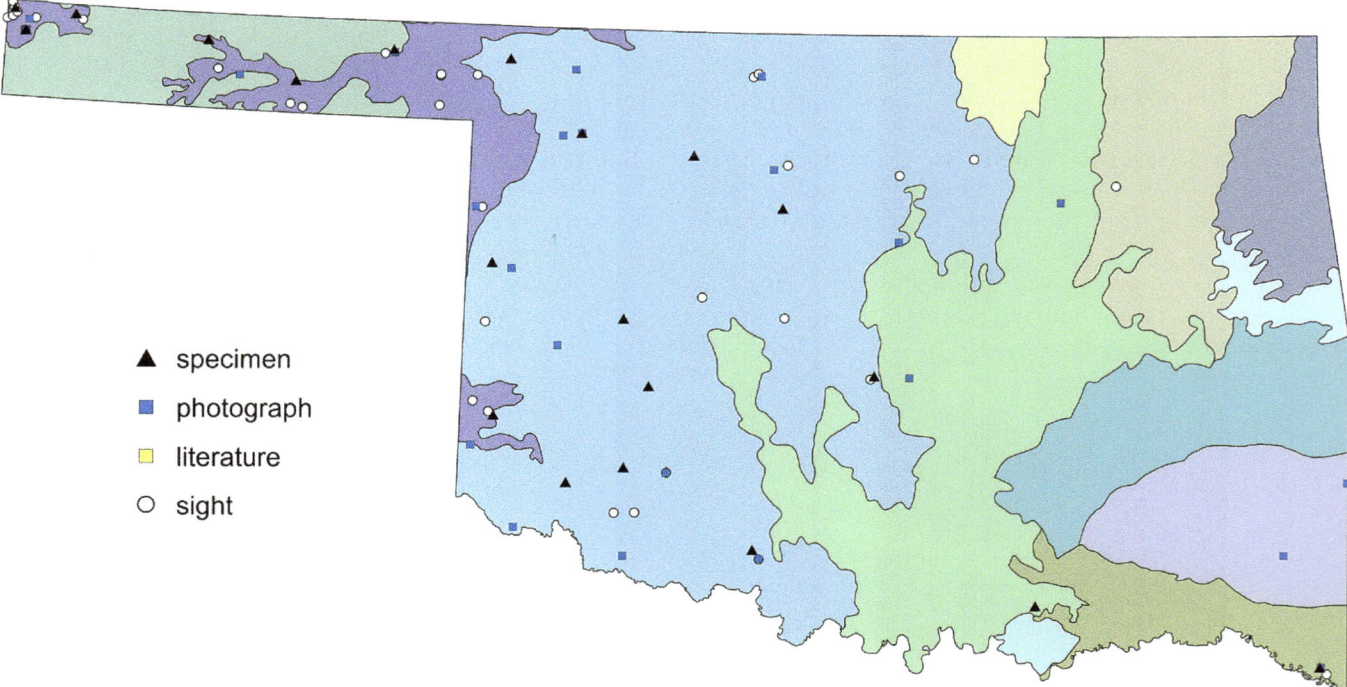

- ▲ specimen
- ■ photograph
- □ literature
- ○ sight

Figure 14.30 Documentation level of supporting records for *Rhionaeschna multicolor*, the Blue-eyed Darner, in Oklahoma. Levels include specimen, archived photograph, literature reference, or sight record. Although sight records lack documentation, only those we feel are credible are plotted.

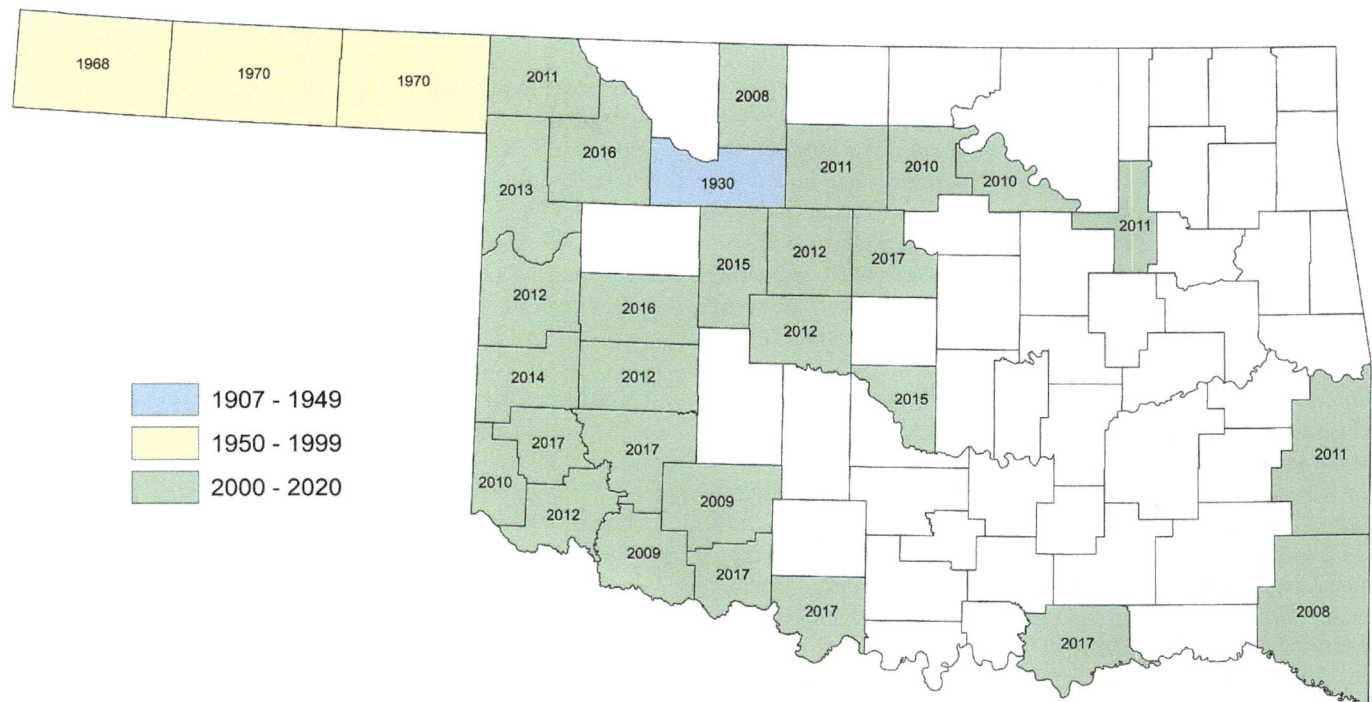

Figure 14.31 Counties in which *Rhionaeschna multicolor*, the Blue-eyed Darner, are known to occur in Oklahoma. Year within the county is that in which the species was first reported.

Texas (4 August, 3♂, 1♀, 7.1 mi W of Hough, 2♂ at FSCA) Counties. It makes sense that these three collectors missed the species in the fall because they rarely collected at that time of the year, but they all collected considerably during the summer. Yet a mere 4 encounters during the time (cumulatively about 25 years) they surveyed for Odonata in the state is dumbfounding. Perhaps this pattern provides anecdotal evidence for range expansion, in that Paulson (2011:202) remarked that *R. multicolor* is "such a successful species that it may move farther east," eventually bringing it into contact with the similar Spatterdock Darner (*R. mutata*).

There are only two other records between 1970 and 2003, both from 1981. The first was a nymph taken at a spring near Black Mesa in Cimarron County (5 May 1981, WJ

Matthews, OMNH 1945). This specimen was originally identified as *Anax*, but we re-identified it as *Rhionaeschna multicolor*. And on 3 June a Kansas Biological Survey team (P Liechti, DG Huggins, et al.) collected 12♂ and 1♀ (1 pair) at Black Mesa SP, Cimarron County.

We are responsible for the next two sightings, both in Cimarron County at Black Mesa SP (1 adult 26 April 2003 and 7 adults, including 1 pair, 15 May 2004). Since 2007, the species has been seen every year and appears to be becoming increasingly abundant (e.g., 18 ♂, 2♀ along Kiowa Creek near Slapout, Beaver County, 29 July 2012). If this species was in the state in any real numbers prior to the past couple of decades, then it is hard to believe that Bird, Bick, Hornuff, and possibly others, would not have recorded it more often.

ANAX LONGIPES HAGEN, 1861 – COMET DARNER

The Comet Darner is a species primarily of small to medium ponds embedded in pine or pine-oak woodlands. The ponds, often spring- or rain-fed, tend to be clear and so small (often only a few to tens of meters in diameter) that one would guess there is room enough only for one patrolling male. Yet in our experience if you capture that male another will soon appear to take his place. On one such instance, we would have bet there was only one male, but we captured him and a second one was there immediately. As we walked away from the pond, not even a minute later, we scared up a mating pair down in the knee-high grass that we had not seen earlier (Figure 14.32). For such large dragonflies, they certainly can seem as thick as thieves.

It is surprising, then, that in Oklahoma the first encounter with the species did not come until 29 June 1956 when the Bicks—*not* George, but Juanda and their daughters, Suzann and Patty—saw one fly over at the "Federal Fish Hatchery near Reagan" (Tishomingo Federal Fish Hatchery), Johnston County (Bick and Bick 1957). According to the Bicks, this sighting was the westernmost record of the species at the time. Given the considerable time Ortenburger, Bird, Pritchard, and other collectors devoted to surveying areas we currently find the species, one wonders if the Comet Darner has expanded its range westward into the state.

(We provide support for such a hypothesis purely on the basis of the venerable and indefatigable A Earl Pritchard, who found Southern Sprites, Fawn Darners, Burgundy Bluets, Double-ringed Pennants [a species he described to science], Ozark Emeralds, and on and on, but in all his years of field work miraculously failed to notice a single large darner with a conspicuous tomato-red abdomen).

But even during the Bick/Hornuff era, there were a meager 18 records. The Bick ladies had the species on George for seven years until he finally saw one in Marshall County at Blackland Farm, west of Kingston on 9 July 1963. George had various sight records between 1963 and 1970, with only one specimen taken: on 7 July 1968, at Cryerville, Marshall County, he recorded in his notes "1 adult sight/1 skin kept," although we have yet to find that exuvia. George's last sighting of the species, on 22 July 1970 (Little City, Marshall County, 1 indiv.), was also the last until 2003. As much coverage as the state received during this era, it is hard to understand how the Comet Darner was so rarely recorded if it was as common then as it is now. It may just be that its behavior and habitat choice are to blame. Lest one visits those small, forested ponds, one is unlikely to encounter the species.

Since our sighting of 2♂ at a small pond at the Atoka PHA, Atoka County, on 14 June 2003, there have been >80

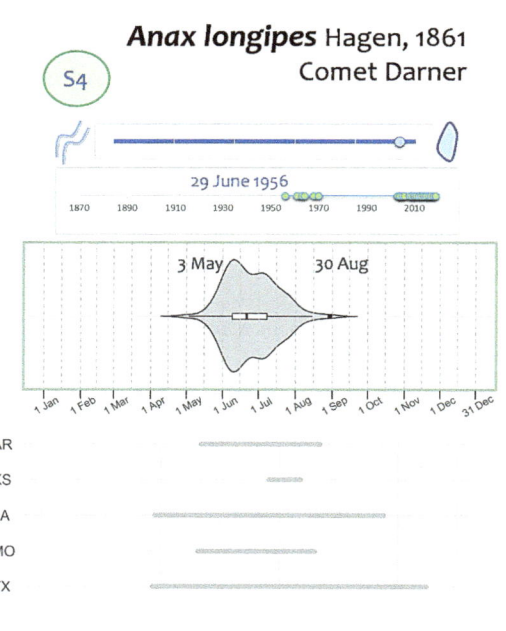

Figure 14.32 *Anax longipes*, the Comet Darner. In the southern Great Plains, this denizen of ponds embedded in woodland in eastern (especially) Oklahoma and the Cross Timbers is far less numerous than its congener. When seen, it is almost always in flight, such as this A) ♂ compulsively circling a pond at Pushmataha WMA, Pushmataha County, 10 June 2013 (©Brenda D Smith, OC 400668). It is more unusual to see the species perched, but B) this tandem pair at the same locale on the same date went so far as to choose the ground over a tree (©Brenda D Smith, OC 400669).

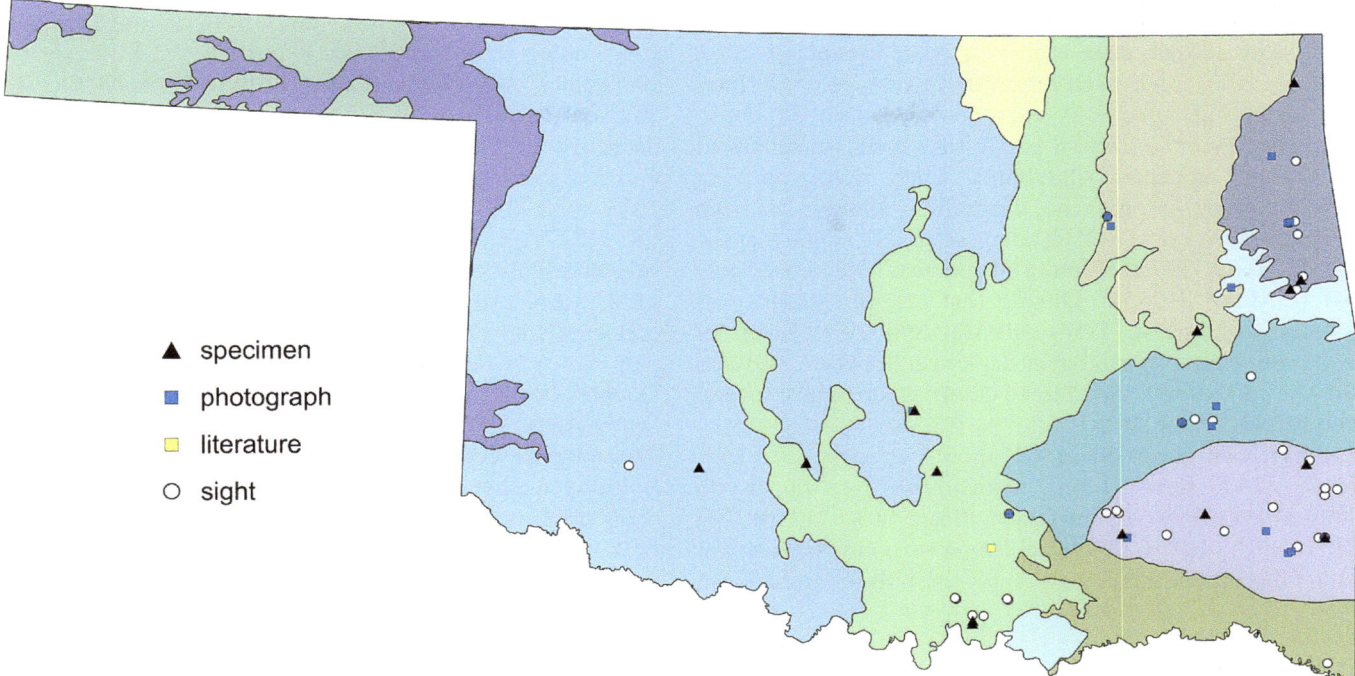

Figure 14.33 Documentation level of supporting records for *Anax longipes*, the Comet Darner, in Oklahoma. Levels include specimen, archived photograph, literature reference, or sight record. Although sight records lack documentation, only those we feel are credible are plotted.

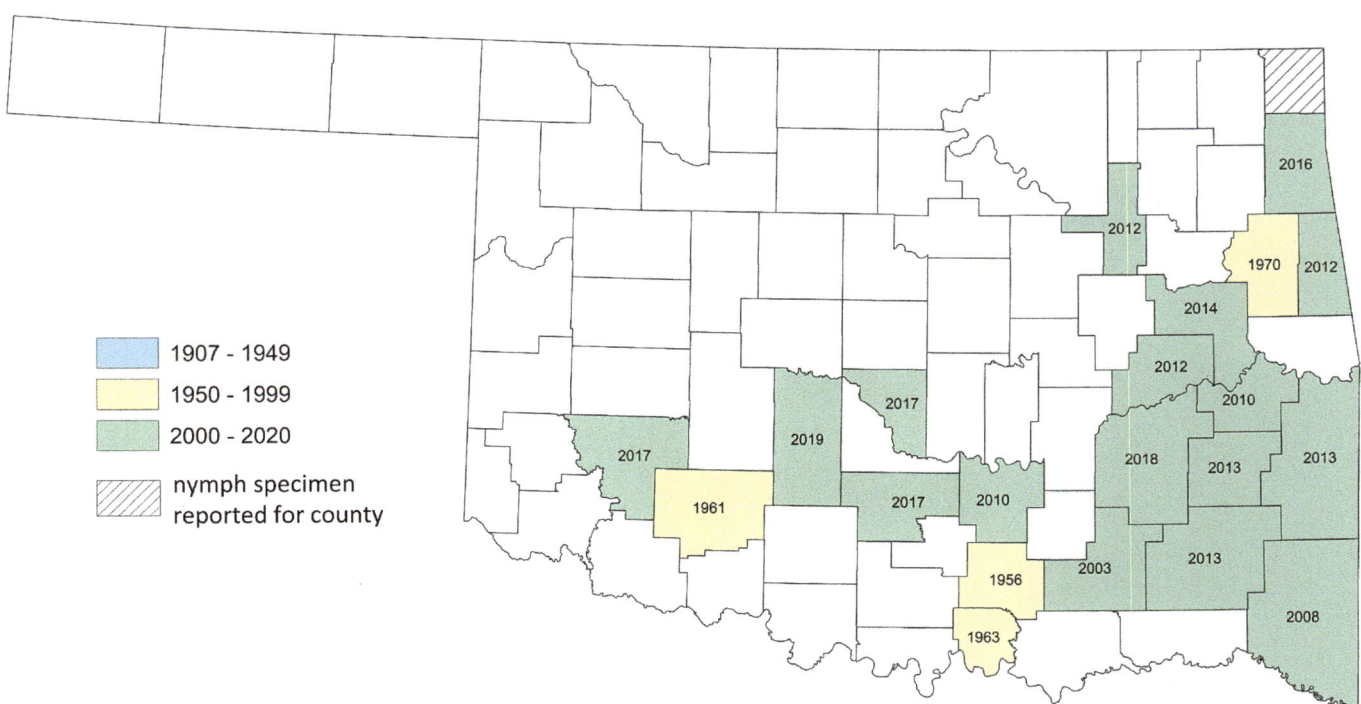

Figure 14.34 Counties in which *Anax longipes*, the Comet Darner, are known to occur in Oklahoma. Year within the county is that in which the species was first reported. The hatched county—Ottawa—only has a nymph specimen as support, but we are quite confident it was identified correctly.

reports of the Comet Darner in Oklahoma. It has been seen in the state at least once each year since except for 2007, and since 2013, it has been reported on average of 10 times each year. The species is currently found in 22 counties throughout the eastern part of the state, with various records west to central Oklahoma. There is one outlying western county—Comanche, naturally—where it has been recorded, albeit once: RW Cruden collected a single ♂ at the base of Mount Scott, Wichita Mountains WR, on 24 July 1961 (UMMZ 210064). During that era, the species was recorded in Johnston (1956) and Marshall (1970) Counties but has not been reported in either since. Between 2003 and 2018 the species was added to 16 counties, including several from central Oklahoma.

It has been recorded in Oklahoma as early as 3 May (BS-P, MAP sight record from 2015, 1♂, vic. Lynn Mountain, Le Flore County) and as late as 30 August (BC sight record, 1 indiv., JT Nickel Preserve, pond on Pumpkin Flats, Cherokee County), although the bulk of records fall between early June and mid-July. In Texas and Louisiana, it is known to fly both earlier and later (TX: 3 April–20 November, Abbott and Lasley 2019; LA: April–October, Paulson 2011) than in Oklahoma. The Oklahoma flight dates are most in line with those reported from Arkansas and Missouri (AR: 14 May–22 August, Harp and Harp 2003, iNat 31360552; MO: 11 May–18 August, Trial 2005, OC 403279), but are longer than those known for Kansas (10 July–6 August, SEMC). The late date for Kansas, 6 August 1982, is a record of note because among other *Anax longipes* collected that day, D. Huggins collected a mixed-species pair of a ♂ *A. longipes* and a ♀ *A. junius* (Common Green Darner) at the "reservoir pond" at Nelson Environmental Study Area, Jefferson County, Kansas (SEMC 1348953). To our knowledge, mixed pairings of any species with *A. longipes* have not been reported previously (although mixed pairs of *A. junius* with other *Anax* species have; Bick and Bick 1981).

ANAX JUNIUS (DRURY, 1773) – COMMON GREEN DARNER

The Common Green Darner was one of the first odonates recorded in what at the time was Indian Territory; i.e., just before Oklahoma became a state ("August 30 and September" 1907, F Collins, 2♂, 8♀, near Henryetta, Okmulgee County; Williamson 1914b). The species is common and widespread throughout the state, occurring during all eras of data collection and in all 77 counties. It is not picky about its selection of habitat, although it seldom appears around flowing water.

This species is often the first odonate that kicks off the field season (although *Ischnura posita*, the Fragile Forktail, also vies for that title), and is present almost all the year

Figure 14.35 *Anax junius*, the Common Green Darner, A), ♂, Greenleaf SP, Muskogee County, 5 October 2018 (©Bill Carrell, OC 491278). B) ♂, Neosho River 7 km SW of Bernice, Delaware County, 10 May 2014 (©Randy L Kelley, OC 422051), in a decidedly more typical view. C) ♀, Hackberry Flat WMA, Tillman County, 27 August 2007 (©Victor W Fazio III, OC 263055). D) At times later in the season, this super-common species strives to become even more super-common, as seen in this mass ovipositing at McGee Creek WMA, Atoka County, 25 September 2019 (©Bryan E Reynolds).

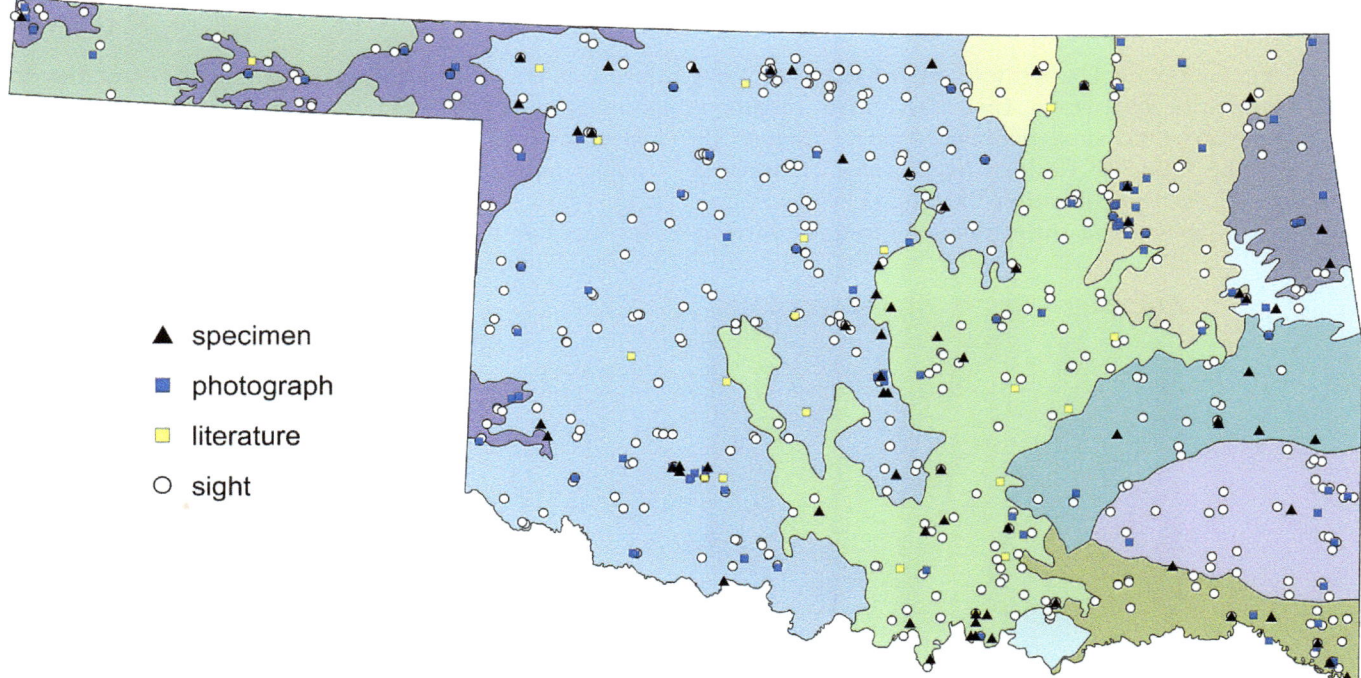

Figure 14.36 Documentation level of supporting records for *Anax junius*, the Common Green Darner, in Oklahoma. Levels include specimen, archived photograph, literature reference, or sight record. Although sight records lack documentation, only those we feel are credible are plotted.

through, at least in years with milder winters. This is also true of the species in Texas and Louisiana and is mostly true of New Mexico (TX: Abbott and Lasley 2019; LA: Mauffray 2014; NM: February–December, Paulson 2009). Elsewhere in the region the reported seasonality is more limited (AR: 30 March–14 November, OC 478819, 458265; KS: 29 March–5 October, SEMC 1315568, OC 491260, Allison 1921 indicated it flies until late October; MO: 14 March–26 November, Sims 2012/Enns, Trial 2005). The species occurrence in the state is bimodal, even if not as strikingly so as we had reported earlier (Smith-Patten et al. 2007), with peak abundance and occurrence in spring and, especially, autumn. In the fall the species is often seen in large numbers (100–250 at a time) along with gliders (*Tramea lacerata*, *T. onusta*, *Pantala flavescens*, *P. hymenaea*) when keying up for migration. They can also be seen ovipositing in large numbers (Figure 14.35).

NOTES

1. 13♂, 13♀ (EMEC 300371–300391, OSU, RWG).
2. 1♂, 2♀ at Nashoba (EMEC 300392–300393, OSU; the later specimen is labeled as 23 July, which seems to be in error, as it was likely collected on 23 June as were AEP's other specimens. 1♂ at Broken Bow, McCurtain County (EMEC 300394, 17 June). 1♂, 1♀ at Sallisaw, Sequoyah County (EMEC 300395–300396, 24 June).
3. Early odonatologists occasionally confused *Boyeria vinosa* with the phenotypically similar *B. grafiana*, the Ocellated Darner. Early confusion aside, careful work since has shown that *B.*

grafiana is confined to northeastern North America and thus exceedingly unlikely to occur on the Great Plains.
4. Note that Sims (2012) has late date as "9/5," which we assume was a typographical error for 9 May rather than 5 September.
5. White (2006) mentioned that he thought there was a record for southern Virginia. BS-P contacted him in September 2017 to ask if he knew the source of that record. He searched his records, finding that he made that reference based on Carle's (1982) map for *Gynacantha nervosa* that had a question mark in southern Virginia. White said that to his knowledge there is no basis for that record (*in litt*). Frank Carle later confirmed that the record was in error (*in litt*).
6. OMNH had two nymphs originally identified as *Coryphaeschna* but which BS-P re-identified as *Nasiaeschna pentacantha* (Comanche Co., Medicine Creek, 22 October 1977, coll. "LMS," OMNH 1760; Cherokee Co., Illinois River, 8 October 1977, OMNH 1761). Although we are comfortable with the re-id, we recommend they be re-examined by a nymphal expert.
7. There may be an earlier date for Kansas, from a ♀ specimen (PSU) with a label indicating it was collected on 4 May 1971 in Doniphan County. We are hesitant to accept this record because the handwriting on the label is unclear and May would be an extremely early time for the species (primarily flies throughout its range from July on, with but a handful of records from June, e.g., CMNH, OC, OSCU). Confounding the dubious nature of the specimen is that it was labeled, on its pin, as "det D Huggins 1975," but in a tray labeled "*Aeschna* [sic] *constricta*/det DG Huggins 1975." We were not able to confirm the identification except to say that it does appear to be a ♀ *Aeshna*.

8. There has been much confusion of whether this location should be the Glass or the Gloss Mountains. USGS favors Glass Mountains, indicating that Gloss Mountains is a variant name. Presumably, their use stems from "Glass" being the term that the General Land Office (GLO) used on its first survey that referenced the mountains (the first survey that included 22N township and 13W range, Indian Territory, dates to 18–24 March 1873, but the first survey notes to reference the mountains are from 10–16 June 1873; all surveys led by TH Barrett). "Glass" was also transcribed onto the first map (February 1874) as well as those since. However, the state of Oklahoma favors use of "Gloss," such as its use in Gloss Mountain State Park. We presume that use is based on the history presented by the Gloss Mountain Conservancy, who, on their historical marker at the state park indicated their version of the history: "In February 1873 the name Glass Mountains appeared on a map issued by the Federal General Land Office. Two years later the same office issued another map calling them the Gloss Mountains. Thus precipitating a conflict that continues to this day. The 1875 map resulted from a survey led by an engineer named TH Barrett. Historiographer James Cloud is of the opinion that a draftsman copied this map and misread the "a" for an "o". A persistent legend exists that a member of that first exploring party was British or Bostonian. This member awakened early one morning in the survey camp on the knoll located east of this point and saw the sun on the glistening clear crystals of selenite. In his long eastern dialect he exclaimed, 'Why, they look just like glaws.' The party's cartographer simply recorded what he thought he had heard. Indeed a passing error." Though that history is amusing, its accuracy is called into question by there being no record of a February 1873 or an 1875 map produced by the GLO (https://glorecords. blm.gov/search/). We abide by the USGS usage of "Glass Mountains" for reference to the mountains themselves but when specifically referring to the state park we comply with Oklahoma's usage of "Gloss Mountain," which also appears to be the preferred name used by locals.

## CHAPTER 15

GOMPHIDAE: CLUBTAILS

Oklahoma hosts a wide variety of species in the family Gomphidae, from the massive—the Dragonhunter (*Hagenius brevistylus*) is the hemisphere's largest clubtail—to the diminutive—the Interior Least Clubtail (*Stylogomphus sigmastylus*) is one of the continent's smallest clubtails. Species span the commonness–rarity scale, too, with some, such as the Common Sanddragon (*Progomphus obscurus*), Sulphur-tipped Clubtail (*Phanogomphus militaris*), Flag-tailed Spinyleg (*Dromogomphus spoliatus*), and Eastern Ringtail (*Erpetogomphus designatus*), widespread and found at times in large numbers (dozens or hundreds), whereas other species, such as the Two-striped Forceptail (*Aphylla williamsoni*), Bayou Clubtail (*Arigomphus maxwelli*), Banner Clubtail (*Hylogomphus apomyius*), and Cocoa Clubtail (*Gomphurus hybridus*), are so rare, habitat specific, and localized that they are not recorded annually in the state.

A key driver of this diversity is the extent to which species are lentic, lotic, or a mix of the two (Figure 15.1). Some species are strictly lotic; we would be shocked to find an Interior Least Clubtail away from a clear, rocky creek with moderate to strong flow, for example. Other species are strictly lentic; for instance, we have yet to encounter a Stillwater Clubtail (*Arigomphus lentulus*) away from a small pond, whether clear or muddy, embedded in forest or dense woodland. Perch location, in terms of both substrate and height above the ground, matters as well (Figure 15.1). Some species perch regularly in trees, with some, notably the Russet-tipped Clubtail (*Stylurus plagiatus*) and Dragonhunter, seldom seen perched elsewhere and seldom found lower than at least a meter above the ground (and sometimes found as high as four or more meters up). Other species perch routinely on the ground, such as the Common Sanddragon, a denizen of sandbars and mudflats or on rocks that jut from flowing waters. Rocks jutting from a cascading creek also are often favored by the Ozark Clubtail (*Gomphurus ozarkensis*) and Eastern Ringtail, whereas the Cobra Clubtail (*G. vastus*) can also be found rather regularly directly on dry bedrock patches in or near flowing water. Habitat selection in the sanddragon and ringtail is curious in that both species are strongly lotic and when at creeks or rivers seldom perch above ground level, yet each can be seen, at times by the dozens, obelisking on barbed-wire fences or roadside shrubs some distance from water, particularly in late summer and early autumn (Figure 15.2). We have seen both species behave in this manner atop trees five or six meters tall.

A phylogenetic analysis and associated taxonomic revision (Ware et al. 2017; Appendix A) increased the generic diversity of the Gomphidae in Oklahoma, in that the "garbage pail" genus of *Gomphus* was, for North American species, split into four … none of which is the genus *Gomphus*, a genus now restricted to Old World species (see Appendix A). Of the nine species in the state that were until recently placed in *Gomphus*, one, the Sulphur-tipped Clubtail, is by a comfortable margin the state's most widespread and numerous clubtail (Figure 15.3). The title of commonest species in *Gomphus sensu lato* is aided by the species' relatively long flight season; of the nine species, it is one of only two, the other being the Cobra Clubtail, that does not fly solely in spring and earliest summer (Figure 15.4A). Over the whole of the family, though, several species, including especially the Russet-tipped Clubtail, Flag-tailed Spinyleg, and Eastern Ringtail, have flight seasons shifted to later in the year, with the first of the three species bordering on being an autumn species in terms of median date of occurrence and peak abundance in Oklahoma (Figure 15.4B, C).

Several clubtail species that occur in Oklahoma, such as the Bayou and Banner Clubtails, are of conservation concern given their rarity *within* the state (Patten and Smith-Patten 2013b; see Chapter 7). Other species are of concern more because of their limited overall geographical ranges. The Ozark Clubtail is one such species. Although in recent years our knowledge of this species has greatly improved and we now know that it is more common and widespread than once thought, it still has a restricted range, confined to the four-state region of eastern Oklahoma, southeastern Kansas, Missouri, and western and northern Arkansas. Another gomphid of conservation concern found in that approximate limited geographical area, Westfall's Snaketail (*Ophiogomphus westfalli*), has yet to be recorded in Oklahoma, but we remain convinced that it will be found one day (see Appendix A). A little farther south, there is another species of concern, the Oklahoma Clubtail (*Phanogomphus oklahomensis*). Similar to the Ozark Clubtail, the Oklahoma Clubtail was once thought rare in Oklahoma, but our recent surveys have shown otherwise. Regardless, the species' restricted range, also in just four states, including eastern Oklahoma, Arkansas, northern Louisiana, and eastern Texas, makes it one that, like the snaketail and the Ozark Clubtail, should continue to be monitored, if for no other reason because of their intrinsic value as regional endemics.

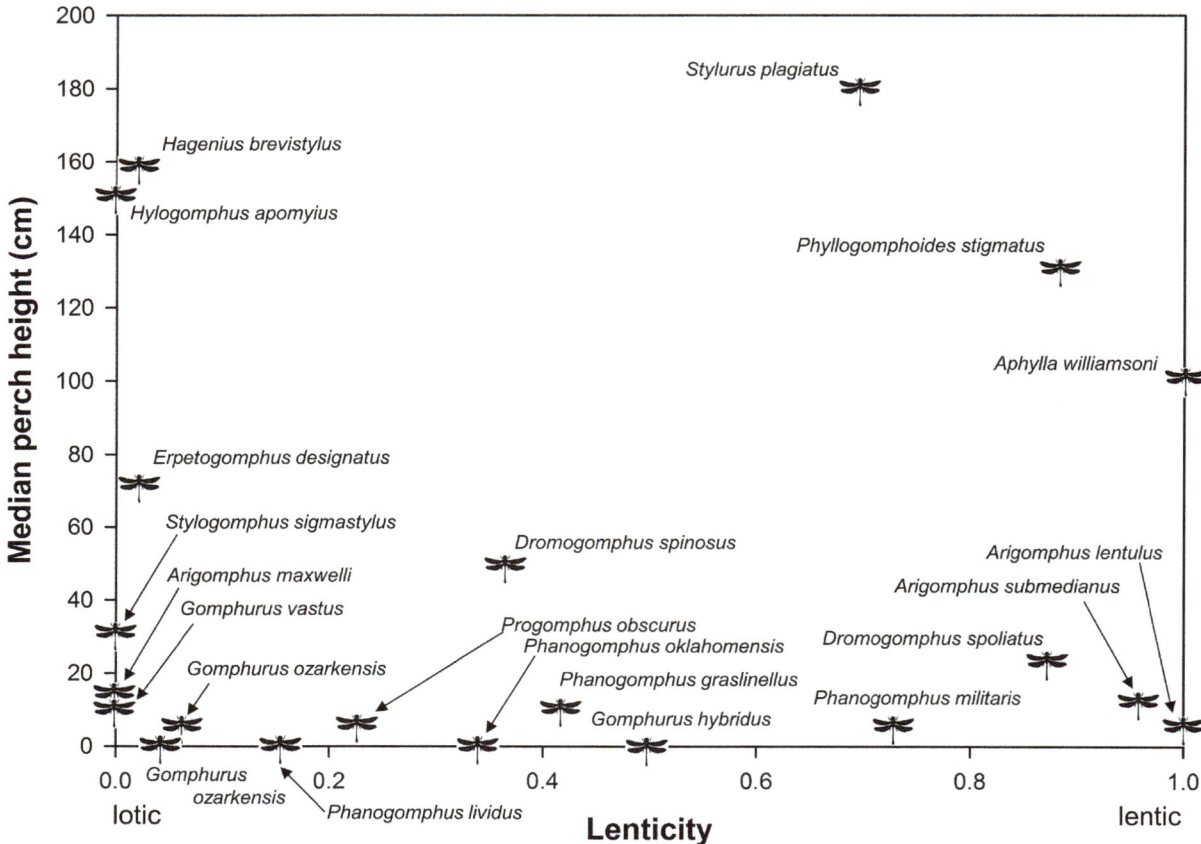

Figure 15.1 A scatter plot on two habitat axes of Oklahoma species in the family Gomphidae, the clubtails. The lenticity score was estimated directly from our field data as a proportion of individuals encountered in one habitat or the other. Median perch height, by contrast, is more subjective because we did not record such data directly; still, the "scores" were derived from our field observations in the state (as opposed to being culled from the literature).

Figure 15.2 Eastern Ringtail (*Erpetogomphus designatus*) obelisking on a barbed-wire fence. This behavior is common among various clubtail species.

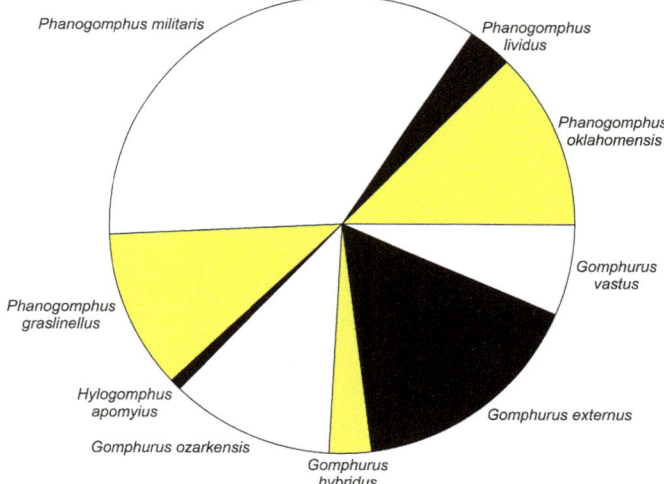

Figure 15.3 Relative commonness of nine clubtail species formerly merged into the genus *Gomphus*. The Sulphur-tipped Clubtail (*Phanogomphus militaris*) accounts for nearly 40% of occurrences, in line with its ubiquity (it has been recorded at least once in each of the Oklahoma's 77 counties).

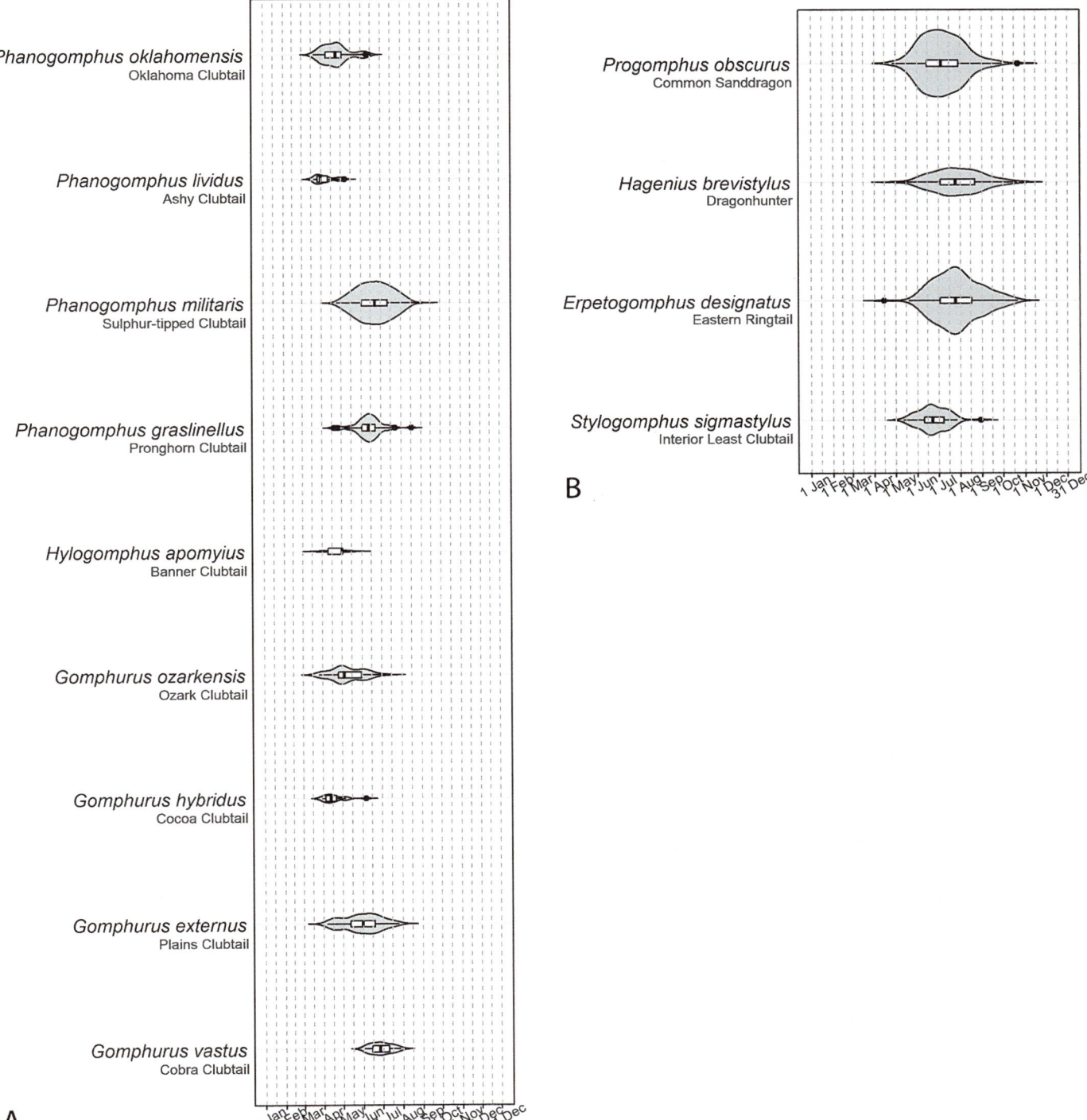

Figure 15.4 Flight seasons in Oklahoma of species in the family Gomphidae, the clubtails, broken out by A) species formerly merged in the genus *Gomphus*, B) all other clubtail species that are chiefly lotic, and C) other clubtail species that are chiefly lentic. Although *Arigomphus maxwelli* occupies more lotic settings, albeit with slow (almost imperceptibly) flow, it is grouped with its congeners.

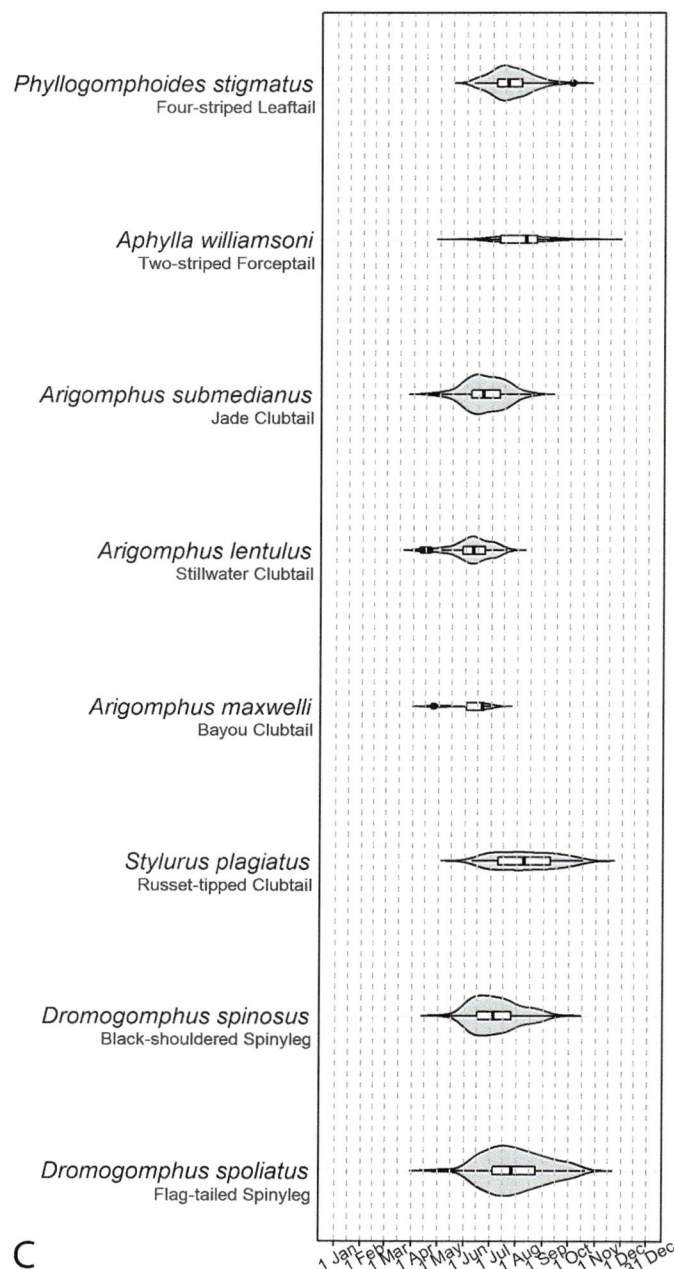

Figure 15.4 (*Continued*) Flight seasons in Oklahoma of species in the family Gomphidae, the clubtails, broken out by A) species formerly merged in the genus *Gomphus*, B) all other clubtail species that are chiefly lotic, and C) other clubtail species that are chiefly lentic. Although *Arigomphus maxwelli* occupies more lotic settings, albeit with slow (almost imperceptibly) flow, it is grouped with its congeners.

Figure 15.5 *It's a clubtail eat clubtail world.* This series of photos show how clubtails, like other odonates, can take prey as large or larger than themselves. A) ♀ *Gomphurus vastus*, the Cobra Clubtail (right), eating a ♂ *Dromogomphus spoliatus*, the Flag-tailed Spinyleg (left). B and C) The struggle, which began at 1:14 pm, continued until at least 1:18 pm when the last photo was taken. Record from JW Arterburn and K Williams, on 7 July 2011, at Kaw Lake, Kay County, Oklahoma. Photos by ©Ken Williams.

PROGOMPHUS OBSCURUS (RAMBUR, 1842) – COMMON SANDDRAGON

The Common Sanddragon is an uncommon to locally common lotic species of sandy, shallow creeks. It has been recorded across much of the state, being most abundant in the west. The species is typically reported singly or a few at a time, but it occasionally occurs in the double digits, including high counts of 70–80 individuals (65♂ and 5♀, including 4 tandem pairs, along the Salt Fork of the Red River 8 km SW of Vinson, Harmon County, 10 June 2017, MAP; 80U, including 3 pairs, 1♀ as SP1334, 13 July 2014, Spring Creek Lake, Roger Mills County).

The species was first recorded in adult form in Oklahoma on 5 July 1928 by Z Logsdon in the Wichita Mountains. This record is from GH Bick's notes in which he indicated that he examined the specimen, a ♂ that we have not relocated, which is not surprising given Bick's notation of "from LLE discarded." Fortunately, there remains an early-era adult specimen; actually, there are 86 early-era adult specimens, the oldest being a ♂ collected by RD Bird in Love County on 13 June 1929 (OMNH 859). Bird (1932) reported the species for 17 counties; it is now recorded in 68, with commonness increasing from east to west. The species has been reported regularly in the state in all eras of data collection.

This is a species for which we have many records of nymphs ($n = 162$; OBS, OMNH, UCO, UMMZ), far more than we have for any other odonate in the state. Prior to

Progomphus obscurus (Rambur, 1842)
Common Sanddragon

S5

Figure 15.6 *Progomphus obscurus*, the Common Sanddragon, A) ♂, Lake Stanley Draper, Cleveland County, 14 June 2016 (©Emily A Hjalmarson, OC 446686). B) ♀, 10 km SE of Idabel, McCurtain County, 30 June 2009 (©Berlin A Heck, OC 315590), in eastern Oklahoma where, despite the species' distribution across the eastern North America, it is notably less common and widespread than it is in western Oklahoma. Incidentally, the two individuals above belie the species' name. It is called a sanddragon because it indeed favors perching on sandbars and mudflats, evidenced by C) this ♂ on Little Beaver Creek 17 km SW of Duncan, Stephens County, 2 August 2014 (©Brenda D Smith, OC 425526), even if this ♂ is pale compared to more typical black and yellow males (see Figure 15.7).

Figure 15.7 *Progomphus obscurus*, the Common Sanddragon, abdomens. Note the difference in coloration of the more typical black and yellow pattern (A) versus a form with more brown on the club (B) that is encountered somewhat frequently in western Oklahoma. A) ♂, Buffalo Creek, Beckham County, Oklahoma, 10 August 2014. B) ♂, Little Beaver Creek, Stephens County, Oklahoma, 2 August 2014. Photos by ©Brenda D Smith.

the adult records mentioned above there are 14 nymph specimens that are technically the first records of the species in the state. Eight of these nymphs were collected in the Wichita Mountains in West Cache Creek, Comanche County (3♀ on 10 June 1926, AI Ortenburger, UMMZ Nymphs 3504 and 3♂ and 2♀ on 6 June 1927, UMMZ

Nymphs 2310), and 6U were collected 9 mi SE of Guymon in Coldwater Creek, Texas County, on 1 July 1926 by AI Ortenburger (UMMZ Nymphs 461). In addition to all of the nymphal records we know of for the species, there are four records of reared individuals. One record, from Bick's notes, indicated that LE Hornuff collected seven nymphs, from which he reared 5♂ and 2♀. Hornuff collected these nymphs in Cowan Creek, Willis, Marshall County, on 19 June 1955, but Bick did not indicate the larval stage or the emergence dates for these individuals. There is a ♂ specimen with this date and locality in the FSCA collection, but the specimen card does not mention it was a reared specimen. The other reared records are all from Bick, also from Cowan Creek in Willis, Marshall County. Bick reared 3♂ that he collected on 20 June 1955 that emerged on 28 June 1955, 1 July 1955, and 3 July 1955. Bad luck has it that Bick did not record the larval stage of these individuals either.

Somewhat surprising is that Oklahoma has the longest documented flight season for the species in the region, spanning from late April to late October (26 April 2012 and 21 October 2012, both from JWA, Fry Creek, Bixby, Tulsa County, OC 374590 and 382323, respectively; teneral records from 28 May–21 July). Both in Louisiana and Texas the species is known to fly starting a week or two before Oklahoma's early date, but it ends in late August/

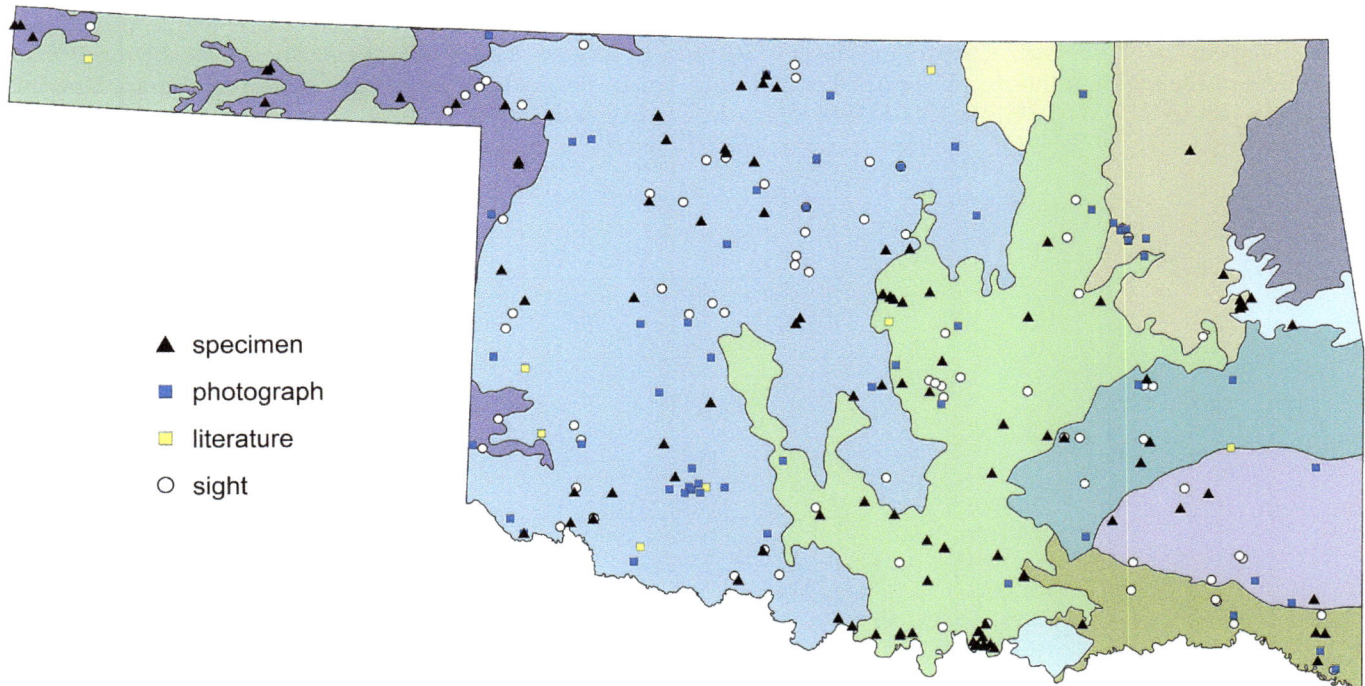

▲ specimen
■ photograph
□ literature
○ sight

Figure 15.8 Documentation level of supporting records for *Progomphus obscurus*, the Common Sanddragon, in Oklahoma. Levels include specimen, archived photograph, literature reference, or sight record. Although sight records lack documentation, only those we feel are credible are plotted.

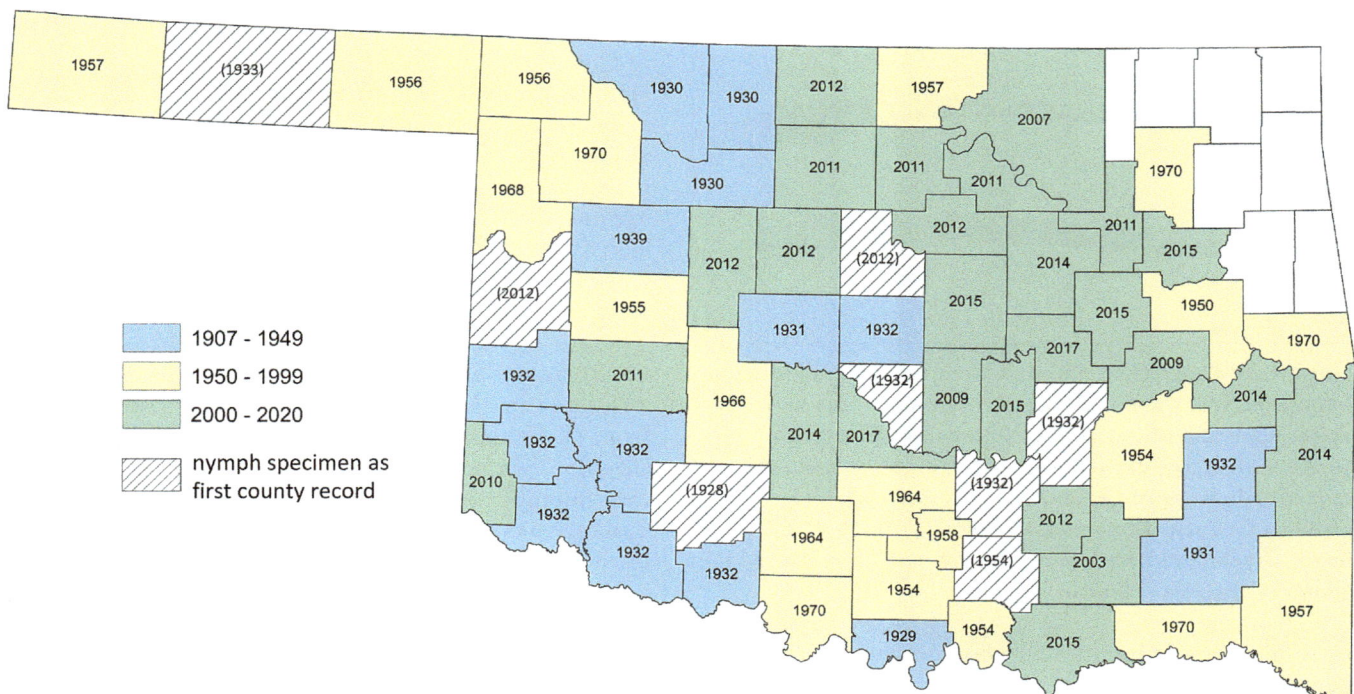

Figure 15.9 Counties in which *Progomphus obscurus*, the Common Sanddragon, are known to occur in Oklahoma. Year within the county is that in which the species was first reported. Hatched counties are those with a nymph specimen as the first documentation of the species in the county. When a parenthetical year is also presented, it indicates the year of the first adult record; no year indicates the county only has larval records as support. We did not examine all of the nymph specimens of this species, but we feel the identifications are likely good given the distinctiveness of the species at that life stage.

early September (LA: 11 April–25 August, Mauffray 2014; TX: 14 April–10 September, Abbott and Lasley 2019). Elsewhere in the region it flies from late April or May into July or August (AR: 9 May–20 July, OC 374854, Harp and Harp 2003; CO: 11–19 July, OC, limited records; KS: 30 April–13 August, SEMC 1326634, 1350172; MO: 19 May–13 August, Trial 2005). We suspect that the more limited date ranges in at least some of the region are a result of data deficiencies rather than a true indication of the species' phenology.

PHYLLOGOMPHOIDES STIGMATUS (SAY, 1839) – FOUR-STRIPED LEAFTAIL

The Four-striped Leaftail is generally uncommon in the state but it can be locally fairly common in southwestern Oklahoma. It is most often seen in single digits but it has twice been reported a dozen at a time (28 July 2011, Wichita Mountains WR, Comanche County, T Kompier, OC 333098; 13 July 2014, 10♂ [1 as SP 1333], 2♀, including one pair, Ellis County WMA, Ellis County). Its highest count in the state is our record of 20♂ and 6♀, including 3 pairs, on 19 June 2015 at Sandy Sanders WMA, Greer County. Paulson (2009:242) described the species' habitat as "slow-flowing streams and rivers and large open ponds." In Oklahoma we would consider it more of a lentic species, given its distinct preference for medium-sized lakes and medium to large ponds with clear (often crystal clear) water and abundant perches (usually upright sticks within the water). These water bodies are generally surrounded by grassland and shrubs, but can be quite devoid of shoreline vegetation.

RD Bird first recorded the species in Oklahoma when he collected 3♂ in the Wichita Mountains, Comanche County, on 29 June 1932 (OMNH 2346 and 2348). Other early records come from that same year in and around the Wichitas, which is curious. It is curious because the Bicks (Bick and Bick 1957) thought that their 19 July 1954 record of a single ♂ at "Sunset Lake" (likely Sun Set Pool) in the Wichitas was the first for the state, apparently being unaware of the 6 records—*of 15 ♂ specimens*—that the OU expedition collected in 1932.[1] (Granted, Bird did not report

the species but why did GHB not find the specimens when he examined the OMNH collection?) During the Bick and Hornuff era the species was recorded six times and was added to two additional counties (Marshall and Murray) between 1954 and 1970. Between 1983 and 1995 there were five more records of the species, with additions of three more counties (Carter, Love, and Major), including the first (Major) outside of southwestern and southcentral Oklahoma.

But it was not until the contemporary era that records of the species really increased and the number of counties grew fivefold. To the 17 records and 5 counties documented from 1932 to 1995, >100 records and 34 counties were added since 2003. This marked increase in records and distribution implies that the species has expanded its range in the state, particularly to the north and east, likely as a result of the proliferation of man-made cattle ponds and small- to medium-sized lakes. Indeed, in 2017 and 2018 the species was found in, for example, Choctaw (SP 2007, OC 453167), Atoka (OC 449016), and Pittsburg (OC 485121) Counties, well to the east of what was thought to be the species' geographic range in the state.

Within the United States, *Phyllogomphoides stigmatus* is restricted primarily to Texas and Oklahoma, with a slight range extension into New Mexico. It is known to occur only as far south as Coahuila and Nuevo León (Paulson 2009), making the species somewhat of a regional endemic. It is known to fly in Oklahoma from early June until early

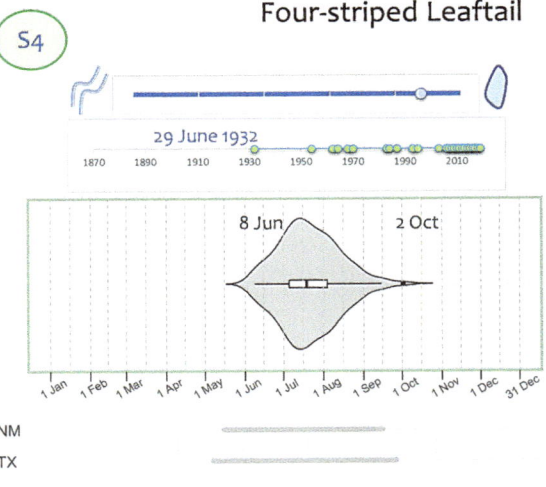

Figure 15.10 *Phyllogomphoides stigmatus,* **the Four-striped Leaftail, A)** ♂**, Doc Hollis Lake, Greer County, 3 July 2016 (©Bill Carrell, OC 448270). B)** ♀**, American Horse Lake, Blaine County, 8 July 2006 (©Randy C Anderson, OC 7306). There is not much to say about this species beyond "wow." It is one of our flashiest dragonflies.**

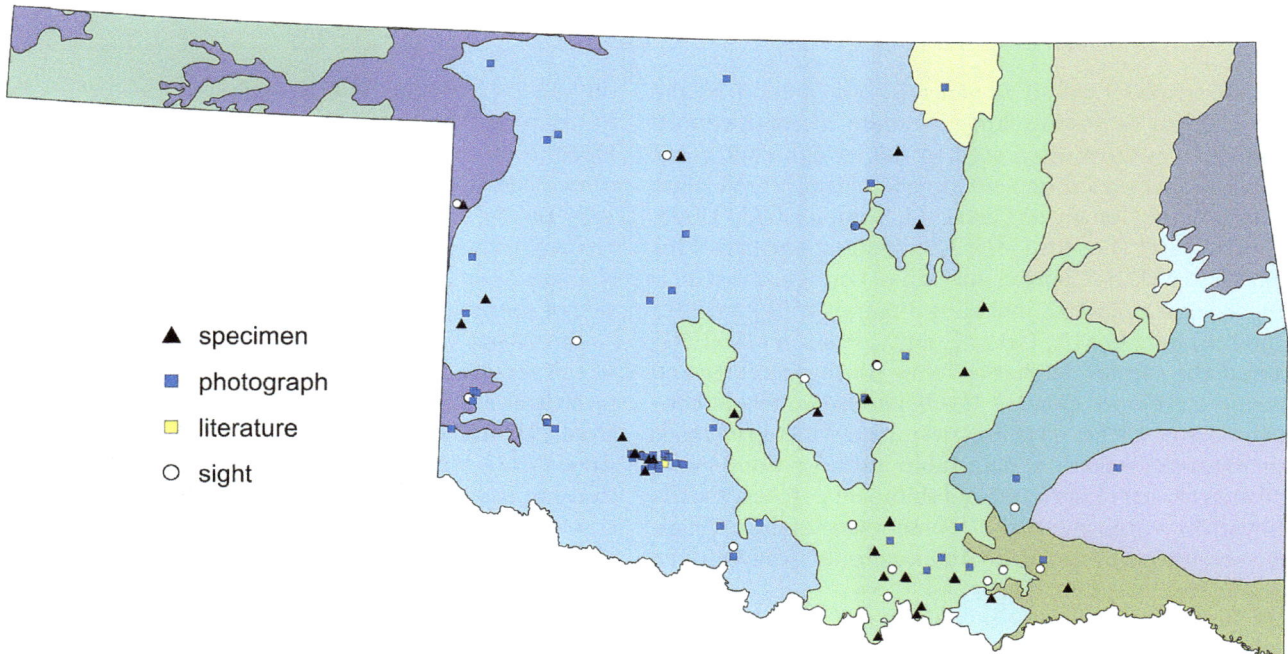

Figure 15.11 Documentation level of supporting records for *Phyllogomphoides stigmatus*, the Four-striped Leaftail, in Oklahoma. Levels include specimen, archived photograph, literature reference, or sight record. Although sight records lack documentation, only those we feel are credible are plotted.

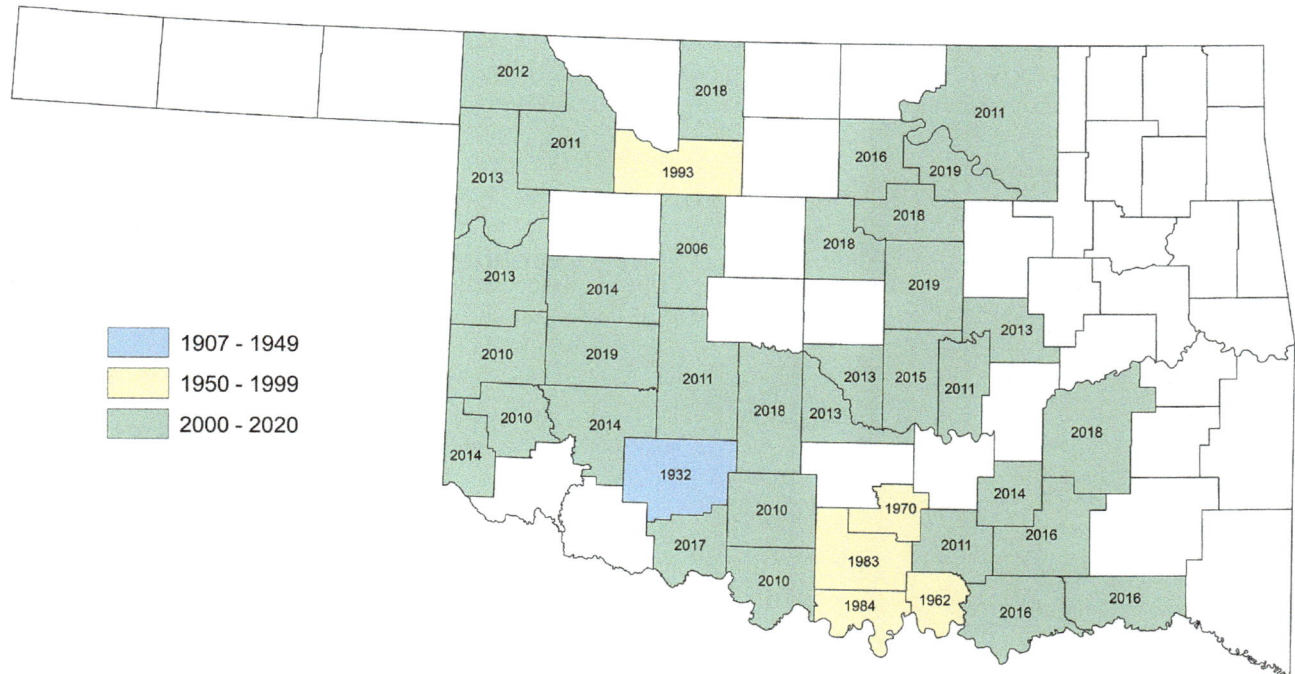

Figure 15.12 Counties in which *Phyllogomphoides stigmatus, the Four-striped Leaftail,* are known to occur in Oklahoma. Year within the county is that in which the species was first reported.

October, which is a month behind its start in New Mexico and Texas, but a slightly later ending than those two states (OK: 8 June 2012 and 2 October 2007, both records from Fort Sill MR, Comanche County, VWF, 1♂ each, OC 376026, 263479; NM: May–September, Paulson 2009; TX: 7 May–26 September; Abbott and Lasley 2019). We know of half a dozen reports of pairs, but there are no records of ovipositing or nymphs. Emergence data for the state consists of one record of a teneral (9 June 2007, 1♂ in hand, Fort Sill MR, Quanah Lake, Comanche County, VWF, OC 263577).

APHYLLA WILLIAMSONI (GLOYD, 1936) – TWO-STRIPED FORCEPTAIL

David Arbour holds the honor of having discovered the Two-striped Forceptail in the state as well as holding the records for early and late dates. His first record was on 28 August 2005[2] on a levee of Ward Lake near Red Slough WMA, McCurtain County (♂, OC 6542, JCAC 20789). This site and adjacent areas have hosted all but one of the subsequent records (approaching 30), the exception being in neighboring Choctaw County in 2011 (6 July, 1♂, Raymond Gary SP, BAH, OC 329467).

Aphylla williamsoni is a bit of an enigma in Oklahoma because of its irregularity of detection. It has been reported almost every year since its discovery in the state, but it was missed in 2007 and 2012–2014. And in adjacent northeastern Texas the species was not reported in 2007, or anywhere in the state for that matter, but it was in the northeast each of the latter years (OC, iNat, M Dillon, *in litt*). The year 2014 was particularly perplexing, because despite conducting regular surveys in southeastern Oklahoma, including one at Red Slough in mid-August, which is typically the high season for the species in the state, we did not encounter it. It was reported in northeastern Texas that year, including records in Red River (OC 320981) and Bowie (OC 426986) Counties, Texas, just across the Oklahoma border. The Oklahoma misses could be due to drought conditions in that part of the state in recent years, but really the species remains enigmatic. In subsequent years, the species was reported once in 2015 (11 August, DA and S Easley, sight record of 1 indiv.), four times in 2016 (5 July, DA, 1♂, OC 448397; 9 July, MAP, 6♂ [1♂ as SP 2164], OC 449015; 19 July, DA, 3♂, sight record; August, no day reported, DA, 2U, sight record), 11 times in 2017 (up to 5♂ at a time, 2 July–16 October; MAP, OC 466172, OC 471901, OC 473067), 3 times in 2018 (up to 4 indiv., 26 July–14 August, DA, two sight records, OC 486624), and multiple times in 2019.

Also of interest, from a regional perspective, is that *Aphylla williamsoni* has not been seen in the southwestern corner of Arkansas, where it was initially discovered in that state (2 September 2006, 1♂, Millwood Lake, Howard County, Mills 2007), since 9 July 2010 (specimen record, Mills, *in litt*) until C Mills reported 2–3♂ on 8 July 2017 at Bois D'Arc WMA, Hempstead County (OC 466311). It could be argued that southeastern Oklahoma and southwestern Arkansas are edge of range for this southeastern U.S. species and, hence, are not occupied annually or consistently. Regardless, we are concerned that the species may

blip off the state map and feel that Oklahoma has a responsibility to try and retain the species within its borders. Given the precarious status of *Aphylla williamsoni* in Oklahoma we designated it a "Watch List" species (an S3 species; Patten and Smith-Patten 2013b). We feel that conservation concern still is warranted, but because its status is difficult to fully discern at this juncture, we now relegate it to an "unknown status" category (Table 7.3).

The species is now known to fly from June to September in Oklahoma, with a decided peak in mid-summer, just as it does in Arkansas and Texas, but which is a much shorter season than that for Louisiana (OK: 8 June–16 October, OC 313201, DA sight record; AR: 16 June–9 September, OC 377112, 332981; LA: 14 April–2 November, Mauffray 2014; TX: 22 June–27 September, Abbott and Lasley 2019). Given the distance between the two known Oklahoma localities for the species—roughly 70 km (43 mi) apart—and the rather broad flight season now documented in the state, we agree with Arbour et al. (2009) that there is likely a small population in the southeastern corner of Oklahoma. That said, there has yet to be any documentation of breeding; all records are of adults (28♂, 2♀, 16U; total of 2♂ specimens) with no pairings or ovipositing documented, but we suspect that is just a matter of time.

In terms of individual variation, some Oklahoma records provide additional support of this species showing an extra thoracic stripe on its side, belying its name. That stripe ranges from just a hint to almost complete. It is found mid-thorax (interpleural region) in between the two stripes for which this species is named. Abbott (2015[3]) indicated that only females exhibit this variation whereas Paulson (2011) said that females were merely more likely to show it. We can offer little in regard to how often females exhibit the third stripe because of the fourteen Oklahoma records for which there are photographs or specimens and we can clearly see the thorax, only two are of females. What we can say is that of those 14 records, about half of the males are classic Two-striped Forceptails and the rest of the individuals have at least a hint of a stripe or a partial stripe up to a mostly full stripe (classic: 6♂; hint/partial: 4♂, 1♀; mostly full: 1♂, 1♀, 1U).[4] Perhaps we simply have a skewed sample, or perhaps the aberrant stripe is a little more common and is not as restricted to females as once thought; we suggest further study into the matter to determine the prevalence and sex ratio of the mark as well as to examine its spatial distribution.

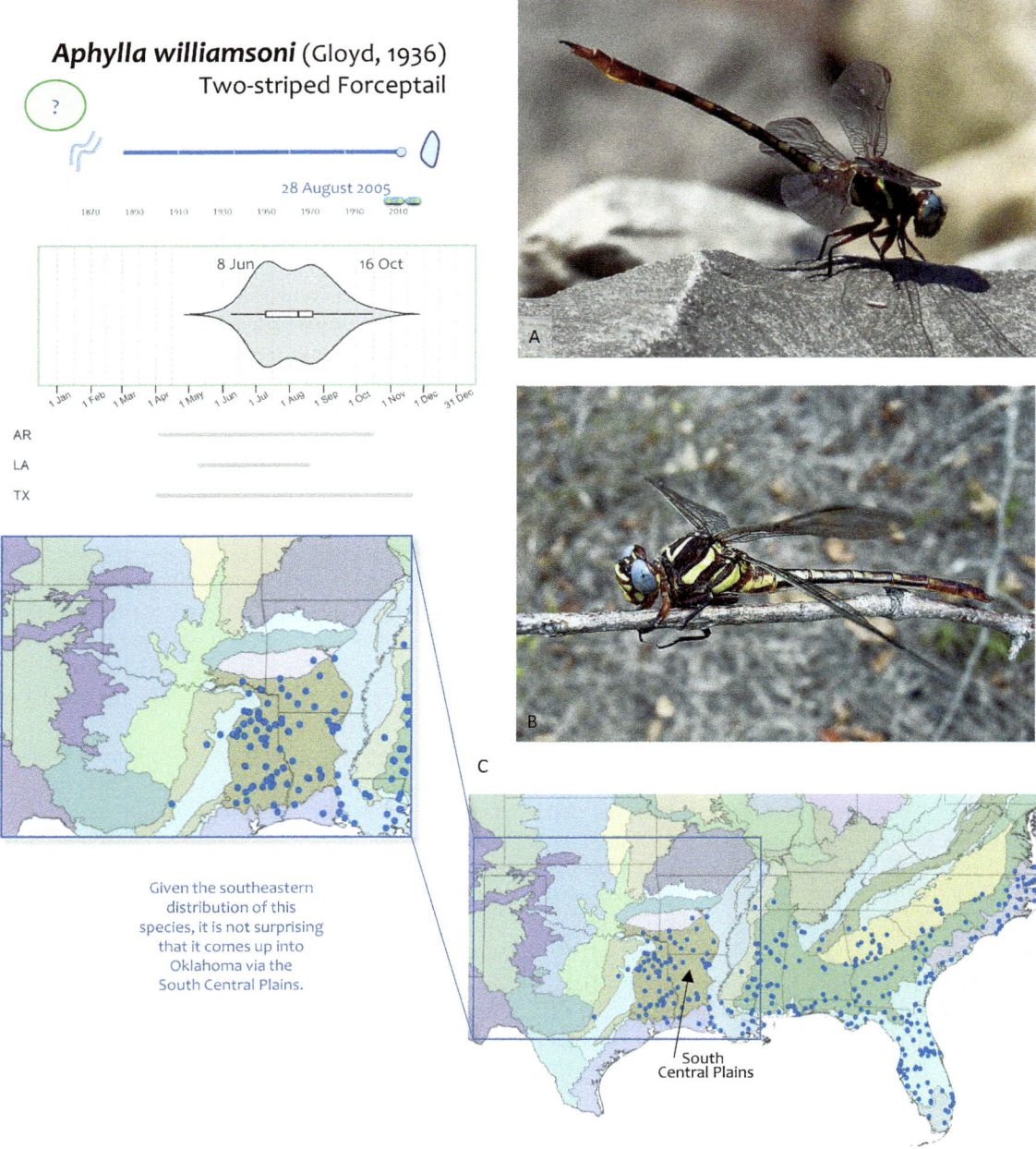

Figure 15.13 *Aphylla williamsoni*, the Two-striped Forceptail, A) ♂, Raymond Gary SP, Choctaw County, 6 July 2011 (©Berlin A Heck, OC 329467), the only individual recorded in Oklahoma away from Red Slough. B) ♀, Red Slough WMA, McCurtain County, 22 August 2006 (USDA Forest Service; David Arbour, OC 7333). C) Given the southeastern distribution of this species, it is not surprising that it comes up into Oklahoma via the South Central Plains.

ARIGOMPHUS SUBMEDIANUS (WILLIAMSON, 1914) – JADE CLUBTAIL

Of the three species of *Arigomphus* clubtails known from Oklahoma, *A. submedianus* is easily the most numerous and most widespread. Nevertheless, it is generally uncommon to occasionally fairly common in the eastern two-thirds of the state. It is a lentic species found around waterbodies of various sizes, including "large mud-bottomed lakes, sloughs, and canals" (Paulson 2009). Smaller lakes and ponds, whether muddy or clear, but especially if there are branches or rocks to serve as perches, and nearby fields seem to be favored in Oklahoma. Most records are of one to a few individuals at a time, but we also know of about half a dozen records that have double digit counts, the highest count being 25 individuals, reached twice in 2011 (10 June, 2♂, 23U, Pine Creek WMA, Turkey Creek Landing, McCurtain County, BAH, OC 328443 and 18 July, Tishomingo NWR, Johnston County, T Kompier, OC 332257).

The Jade Clubtail was first recorded in the state on 3–4 June 1907 from Wister, Le Flore County (3♂ [1 at MCZ], 2♀ Williamson 1914b). The OU expeditions recorded the species only on two days, once in Comanche County at the fish hatchery in Medicine Park (♂, 10 June 1932, WM Fisher, OMNH 2454) and once in Cotton County along the Red River (♀, 17 June 1932, RD Bird, OMNH 2464). Bird (1932) reported the species (as *Gomphus submedianus*) for Hughes, Le Flore, and Oklahoma. It is unclear on what he based the Hughes and Oklahoma County reports, but we since have obtained documentation for both counties (1♂, Dustin Lake, Hughes County, 29 July 2013, OC 402413; 1♂, Oklahoma City, Oklahoma County, 29 May 2017, E Isley, iNat 6473104).

There are not that many more records for the Bick/Hornuff era: a total of only six, one specimen for each of six counties (4♂, 2♀; FSCA; Atoka, Carter, Marshall,

Figure 15.14 *Arigomphus submedianus*, the Jade Clubtail, A) ♂, Sultan Park in Walters, Cotton County, 2 July 2012 (©Victor W Fazio III, OC 376808). B) ♀, Pretty Water Lake, Creek County, 17 July 2015 (©James W Arterburn, OC 434009). Of the three *Arigomphus* clubtails in Oklahoma, *A. submedianus* is by a wide margin the most numerous and widespread: it can be found across much of the Cross Timbers region and to an extent west and east of there. C) Although typically a percher on sticks over water, it will perch on lily pads like the one seen here (Kitchen Lake, Cleveland County, Oklahoma, 9 June 2016, ©Brenda D Smith).

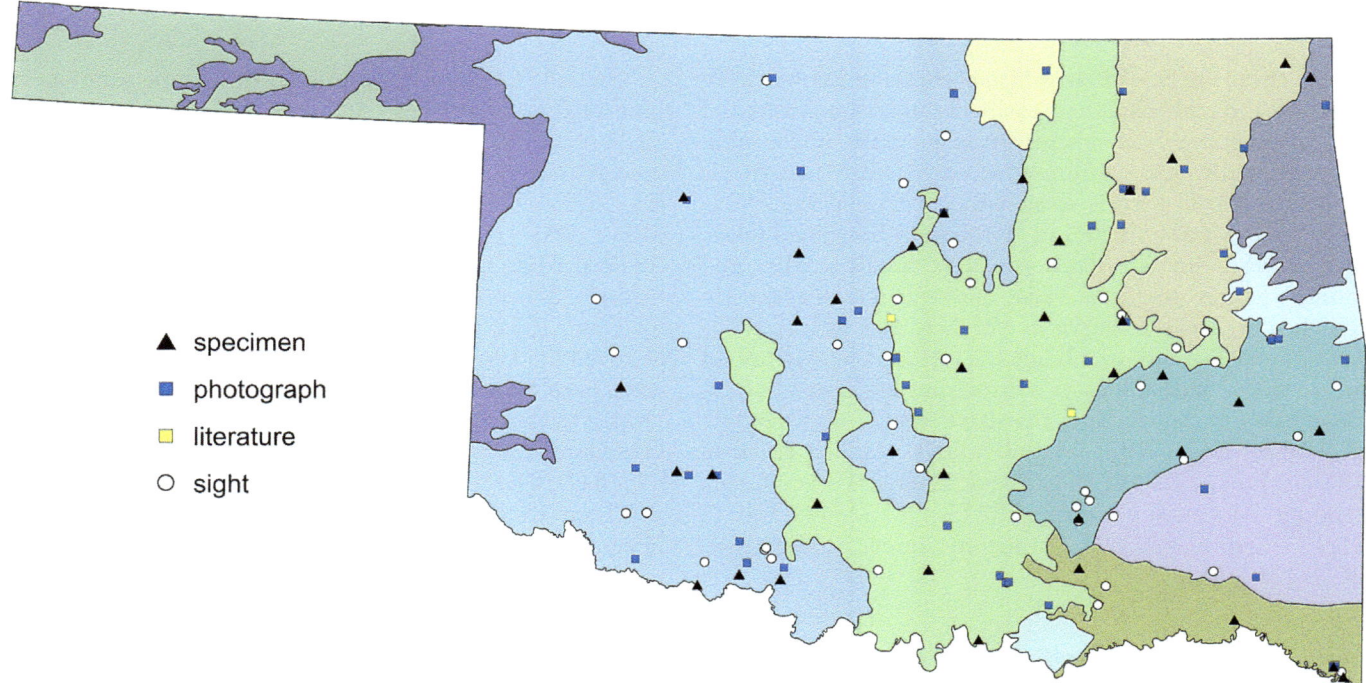

▲ specimen

■ photograph

□ literature

○ sight

Figure 15.15 Documentation level of supporting records for *Arigomphus submedianus*, the Jade Clubtail, in Oklahoma. Levels include specimen, archived photograph, literature reference, or sight record. Although sight records lack documentation, only those we feel are credible are plotted.

McCurtain, Ottawa, and Rogers). There are two records after that era and before the 2000s (one adult specimen, 4 June 1981, Oxley Nature Center, Tulsa County, B Ball, Beckemeyer 2002e; one nymph, discussed below). Between 2002 and 2019 there were >150 records added, bringing the total of counties the species is known in from to 55. Circumstantial evidence points to range expansion coincident with the thousands of man-made, muddy-bottomed cattle ponds and lakes that now populate the hydrological landscape of Oklahoma.

The species has a rather long flight season in Oklahoma, roughly equivalent to that of Texas in terms of number of days (121 for OK, 126 for TX). The early flight date is 18 April, from a 2012 photographic record from Red Slough WMA, McCurtain County (DA, OC 374458). The late date, also a photo record from Red Slough, is 19 August (2014, 1♀, BS-P, MAP, DA, M Dillon, OC 426162, SP 1395), but most records are from June and July. The date range corresponds to those in neighboring states, although both Kansas and Arkansas report only mid-season dates (May to June/July), but we imagine the ranges are an artifact of survey effort rather than a true indication of seasonality as, for example, the species flies from May to September in Nebraska (Paulson 2009) and March to August in Texas (Abbott and Lasley 2019), so it is doubtful that Kansas and Arkansas are intrinsically different. Emergence dates in Oklahoma, from seven teneral records from Coal, Comanche, Cotton, Rogers, and Ottawa Counties, date 25 April to 20 June. There is one nymph record positively identified as this species (UCO 5980; see Figure 15.16 for details).

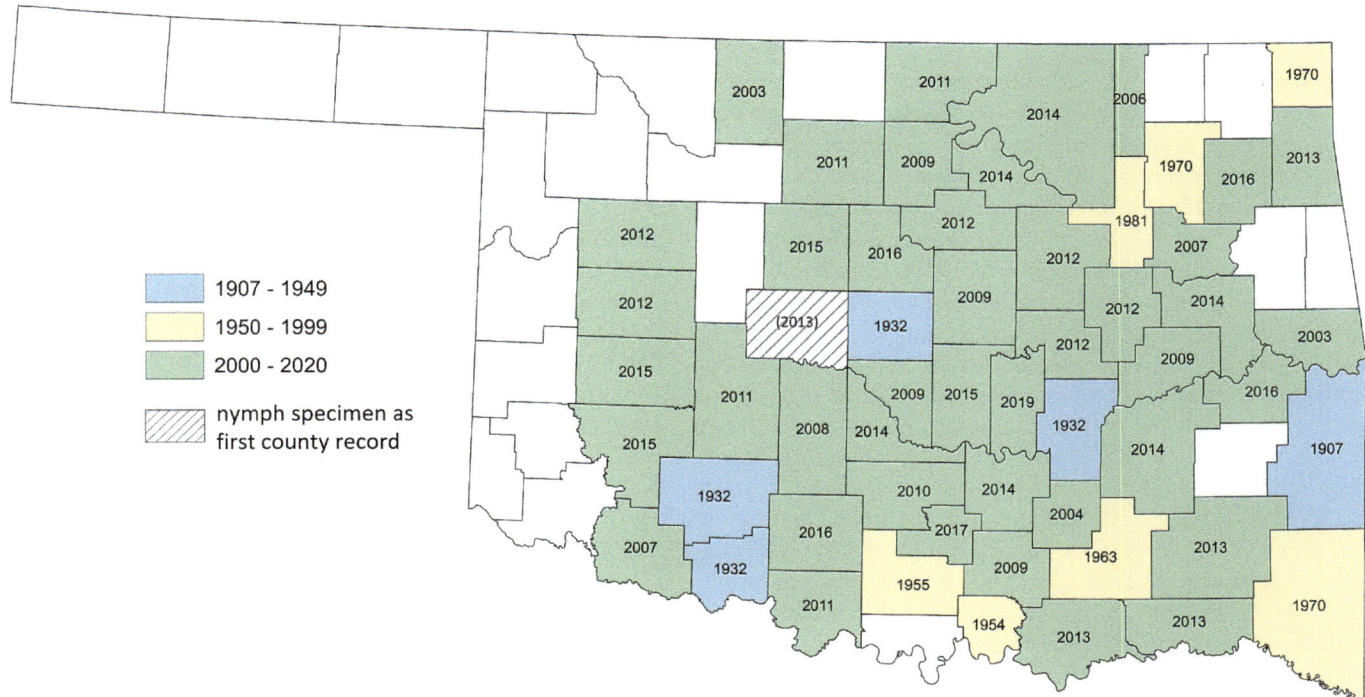

Figure 15.16 Counties in which *Arigomphus submedianus*, the Jade Clubtail, are known to occur in Oklahoma. Year within the county is that in which the species was first reported. The Jade Clubtail was first known from Canadian County (hatched) when a nymph, collected on 12 October 1994 in Hoggard Park, Piedmont, by D Bass, was re-identified as the species (UCO 5980, original id as *Dromogomphus spoliatus*, re-id by BS-P). The first adult record for the county come in 2013.

ARIGOMPHUS LENTULUS (NEEDHAM, 1902) – STILLWATER CLUBTAIL

The Stillwater Clubtail is a lentic species that shares some of the same habitats as the Jade Clubtail (*Arigomphus submedianus*), although it favors smaller ponds and pools, generally with clear water (but at times muddy) and nearly always with nearby shade. It is neither as common nor as widespread as its congener, but it has been reported rather consistently since it was first found in Oklahoma in 1929 (13 June, 1U, RD Bird, Love County, OMNH 828). Up to 1933, there were seven additional records from seven counties. Interestingly, Bird (1932, as *Gomphus subapicalis*) reported the species for only Johnston and Love Counties even though he personally collected the species in Hughes and Latimer Counties in 1931 (3♂, 10 mi E of Holdenville, Hughes County, 7 June, OMNH 2447, 2448, 2466; 4♂, Latimer County, 16 June, OMNH 2443, 2445). We understand why he did not report the species for Cleveland County even though he collected a nymph there in 1929 (Indian Springs, 24 November, OMNH 4228): that omission resulted from a misidentification of the specimen as *Ophiogomphus* (re-id, BS-P). The highest count we have on a given day and locality is 6♂ and 1♀, a record set in 1933 (28 May, Cleveland County, FSCA, OMNH), a total nearly equaled by 6♂ at Pushmataha WMA, Pushmataha County, 3 June 2018 (MAP). It was recorded between 1954 and 1994, in 8 counties, 26 times. From 2003 to 2019 there were about 60 records from 15 counties, such

that it is now known from 30 counties, all but 2—Comanche (always an exception, it seems) and Stephens—in the eastern half of the state.

The Stillwater Clubtail has a shorter flight season than the Jade Clubtail, as it is only known to fly from early April to mid-July (10 April 2012, 1U, Red Slough WMA, McCurtain County, DA, OC 374374; 18 July 1932, 1♂, 2 mi S of Marlow, Stephens County, RD Bird, OMNH 2462). Texas has the longest season in the region (TX: 21 March–12 August, Abbott and Lasley 2019). Date ranges for other states in the region appear to be truncated given the scarcity of available records (AR: 24 May–23 June, OC 282460, Bick 1959; KS: 14 May–20 June, OC 374967, SEMC; LA: 3 April–25 May, OC 479365, 463141; MO: 19 May–18 June, Trial 2005, Steve Hummel Collection). We know of three records of tenerals in Oklahoma, from Cleveland, Marshall, and McClain Counties, with a range of 15 April to 22 June. Only one ovipositing female has been reported (12 June 2016, James Collins WMA, Pittsburg County, MAP).

Of historical note, Bird (1934) inadvertently described the nymph of *Arigomphus lentulus* (OMNH 4250; Figure 15.18) in his paper titled "The emergence and nymph of *Gomphus militaris* (Odonata: Gomphinae)." Landwer and Sites (2003) cleared up this error, but problems still exist in

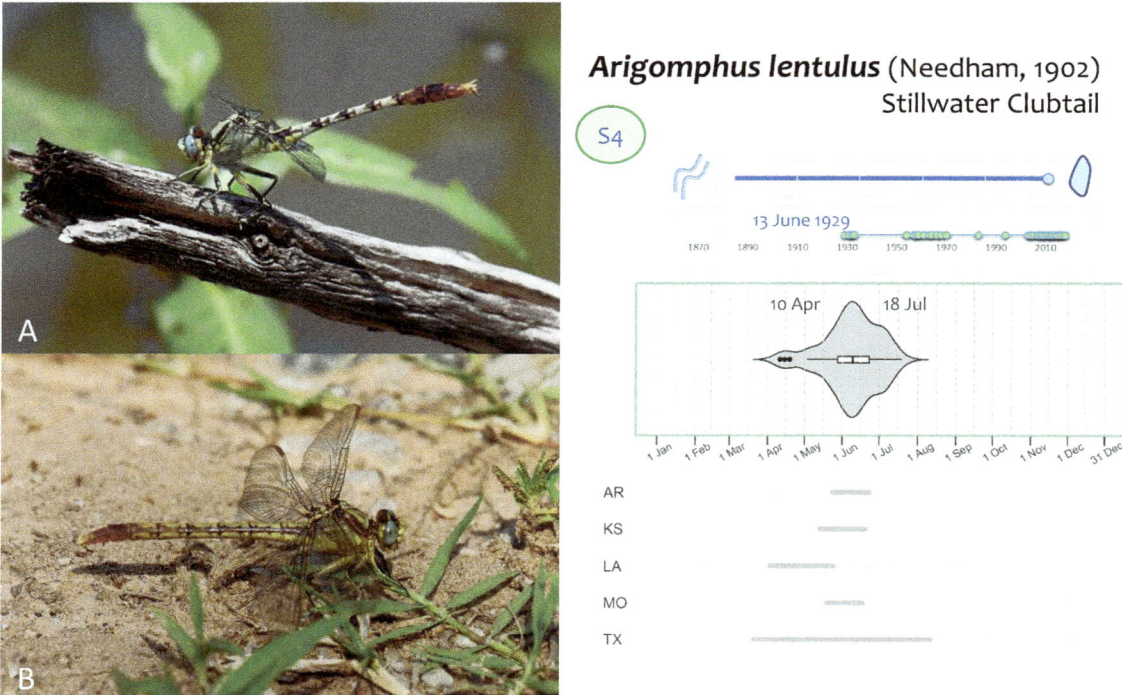

Figure 15.17 *Arigomphus lentulus*, the Stillwater Clubtail, A) ♂, Red Slough WMA, McCurtain County, 11 May 2008 (©Berlin A Heck, OC 281830). B) ♀, Deep Fork NWR, Okmulgee County, 11 June 2012 (©James W Arterburn, OC 375892). Unlike its commoner congener, *A. lentulus* is found almost exclusively at small, often fishless ponds embedded in woodland and forest in the eastern half of the state.

46   ENTOMOLOGICAL NEWS   [Feb., '34

Measurements. Total length, exclusive of antennae, 34 mm.; antennae 2; abdomen 22.5; hind wing 8; hind femur 7; width of abdominal segment 5, 9.4.

The specimen from which this description is made is deposited in the Museum of Zoology, University of Oklahoma.

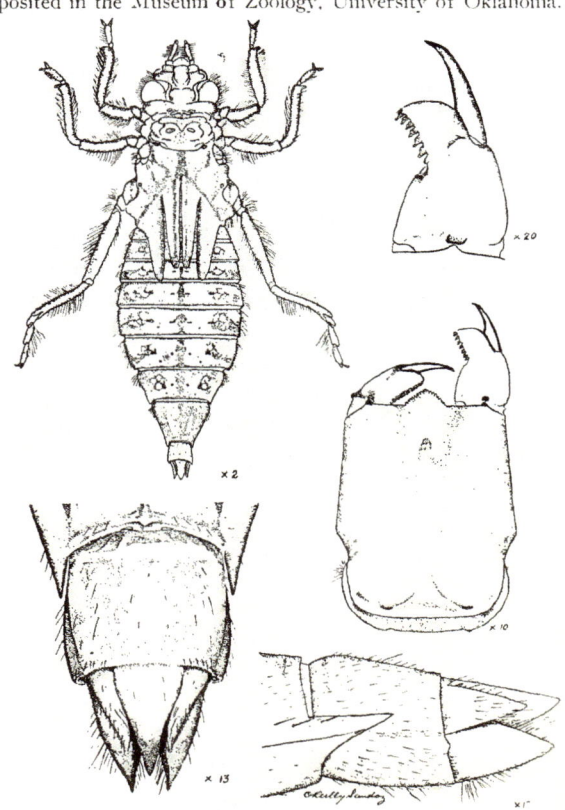

Female nymph last instar; left labial palpus; labium, ventral view; dorsal view of abdominal segment 10 and anal pyramid female nymph; lateral view of abdominal segment 10 and anal pyramid female nymph.

Figure 15.18 O'Reilly Sandoz's original drawing of what he and Ralph D Bird (1934) thought was the exuvia of a *Gomphus militaris* [now *Phanogomphus militaris*] nymph, but which was re-identified as *Arigomphus lentulus* by Landwer and Sites (2003). The specimen this drawing was based on is housed at the Sam Noble Museum of Natural History (OMNH 4250, from McClain County, 10 May 1932).

some dichotomous keys, although the proper characteristics are now reconciled in Needham et al. (2014) and Tennessen (2019). Bird (1934:44) described the emergence as beginning at 9:46 in the morning when

> the nymph was seen to crawl from deeper water to the shore. It was restless and seemed to find difficulty in getting a satisfactory place to emerge, sometimes coming out of the water completely and then going back. It finally came to rest at the very edge with only its back above water. Here it rested for ten minutes before there was twitching of the wing pads which caused the skin to split. A minute later the head appeared through the slit and in four minutes the thorax and first half of the abdomen.

Bird watched the nymph for 1 hour and 17 minutes before the teneral fully emerged, inflated, and flew away. Had he caught the specimen, he would have most certainly realized his error.

GH Bick recorded in his notes that he reared two specimens (1♂, 1♀) that he collected 8 mi N of Willis, Marshall County, in 1959. The dates of those records are 15 and 22 June, but it is unclear whether those are the emergence or collection dates. He indicated that the specimens were donated to FSCA, but we were unable to relocate them there.

Another item of note is that there is a pinned specimen in the OSU collection of a ♀ Stillwater Clubtail that was mislabeled as a Gray Sanddragon (*Progomphus borealis*). This specimen was collected, probably by an OSU student, in Nowata County at Coody's Bluff on 6 July 1983. And speaking of OSU, no, the Stillwater Clubtail is not named for Stillwater, Oklahoma, the home of Oklahoma State University. We do not know why the common name was chosen, but rumors (or wishful thinking) aside, there is no Oklahoma connection. The type locality for the species is in Illinois at Flora, 5 mi NE on the Little Wabash River in Clay County, and not even any of the type series is from Oklahoma. Alas, fame and fortune (ha!) cannot come of a clubtail for Stillwater, Oklahoma.

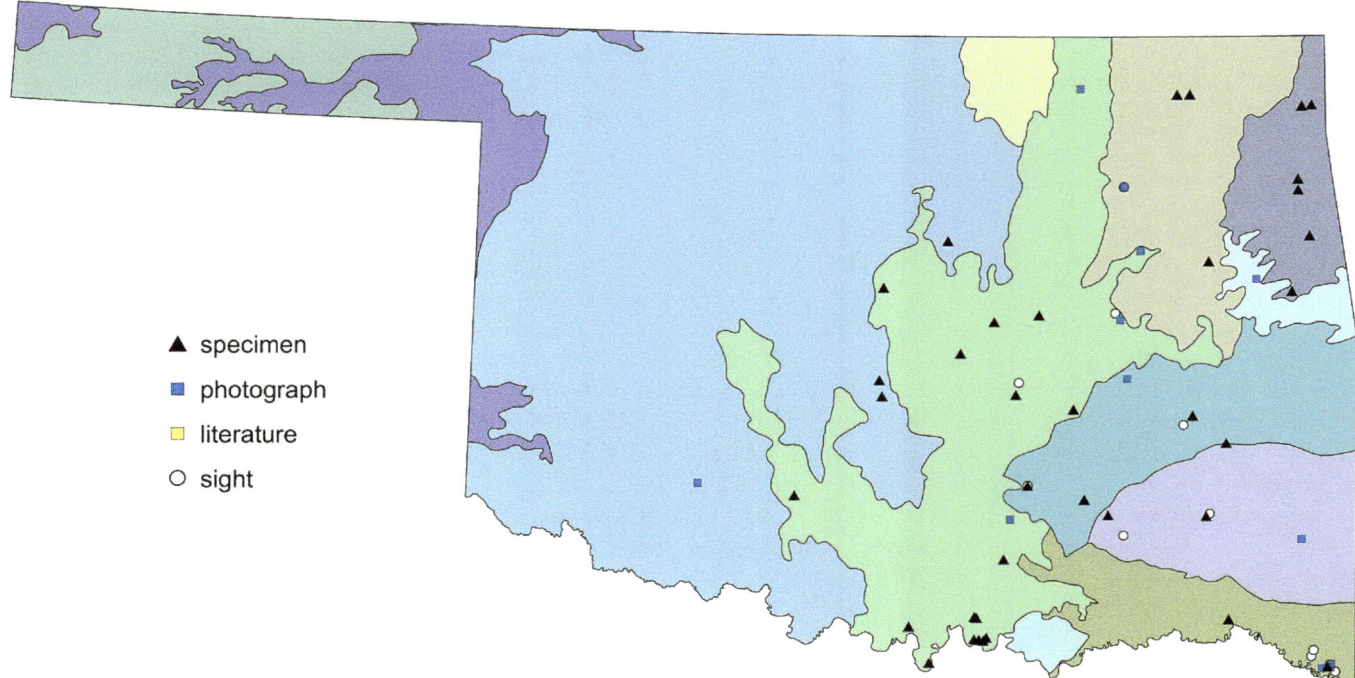

Figure 15.19 Documentation level of supporting records for *Arigomphus lentulus*, the Stillwater Clubtail, in Oklahoma. Levels include specimen, archived photograph, literature reference, or sight record. Although sight records lack documentation, only those we feel are credible are plotted.

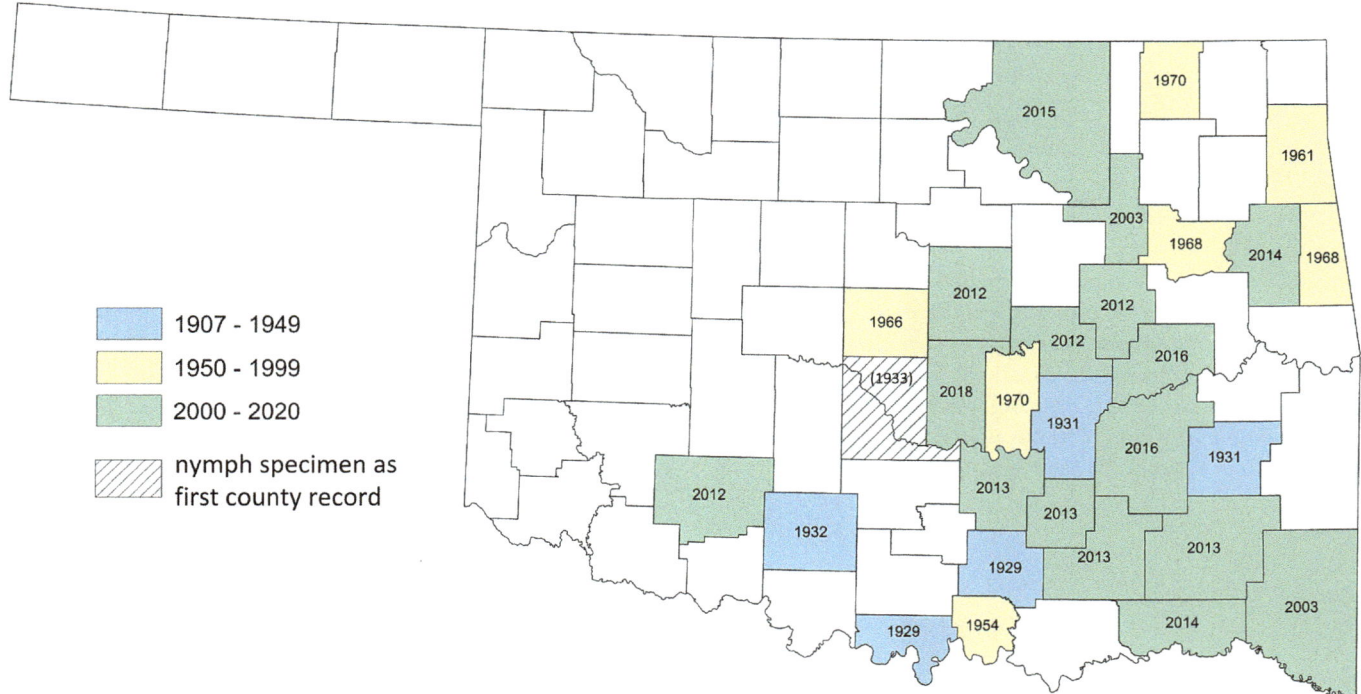

Figure 15.20 Counties in which *Arigomphus lentulus*, the Stillwater Clubtail, are known to occur in Oklahoma. Year within the county is that in which the species was first reported. Cleveland and McClain Counties are hatched because the first county records are based on nymph specimens. We have much confidence in these determinations. These specimens were identified as *A. lentulus* by BS-P, in the case of the Cleveland County record (see text above, OMNH 4228), and by BHP Landwer for McClain County (see text for details, OMNH 4250). There has not been another record for McClain County since the nymph was collected in 1932. However, a second record for Cleveland County came in 1933, this time of adults, which stands as the high count for the state (see text); it has not been reported in the county since.

ARIGOMPHUS MAXWELLI (FERGUSON, 1950) – BAYOU CLUBTAIL

The Bayou Clubtail has been reported six times in Oklahoma (Smith-Patten and Patten 2017a; OC, SP, photos on file, and one sight record). The first record was from 25 April 2002,[5] when DA caught a mating pair at Grassy Slough WMA, McCurtain County. He photographed the ♂ (OC 6584; Figure 15.21) and released it. He also reported two other individuals. Despite intensive, near-annual surveys of Grassy Slough and the nearby similar habitat of Red Slough, the species had not been found again in Oklahoma for a dozen years, until …

On a 1 July 2014 survey of Red Slough, DA and BS-P found a small population of Bayou Clubtails. The tale of finding this group of clubtails is one that readers who have co-discovered a good record can appreciate—the thrill of the first sighting, the slight childish irritation at sharing accolades, and then the hindsight of knowing all that matters is getting the bug! DA and BS-P began surveying a heavily wooded bayou on the east side of Teal Lake, while MAP. stayed in the clearing trying to catch a ♂ Regal Darner (*Coryphaeschna ingens*). The two started working

different parts of the bayou and, about 20 minutes later, DA yelled to BS-P to get her attention. BS-P answered back by saying she had spotted Bayou Clubtails and was trying to photograph and catch one. DA, obviously not able to hear BS-P, yelled "I've got a Bayou Clubtail!" Eventually the two met up where BS-P had spotted a couple of males and an ovipositing female. MAP was not far behind, so the 3 spent the next 30–40 minutes trying to catch 1 of the male clubtails. The males were furtive, barely sitting long enough for DA to get a quick photo and certainly not long enough to take a good swing with the net. Finally, one landed near enough to BS-P that she was able to slap the net down on it. Mud-splattered BS-P came up out of the creek triumphant with her equally mud-splattered Bayou Clubtail! By midday 4♂ and 2♀ were encountered. The 2♂ collected were the first specimens of the species for Oklahoma (OC 423827, SP1306–1307).

The individuals in 2014 were in a shaded hardwood woodland at pooled spots of a narrow, muddy, shallow-water stream. The streambanks were often steep, having drop-offs

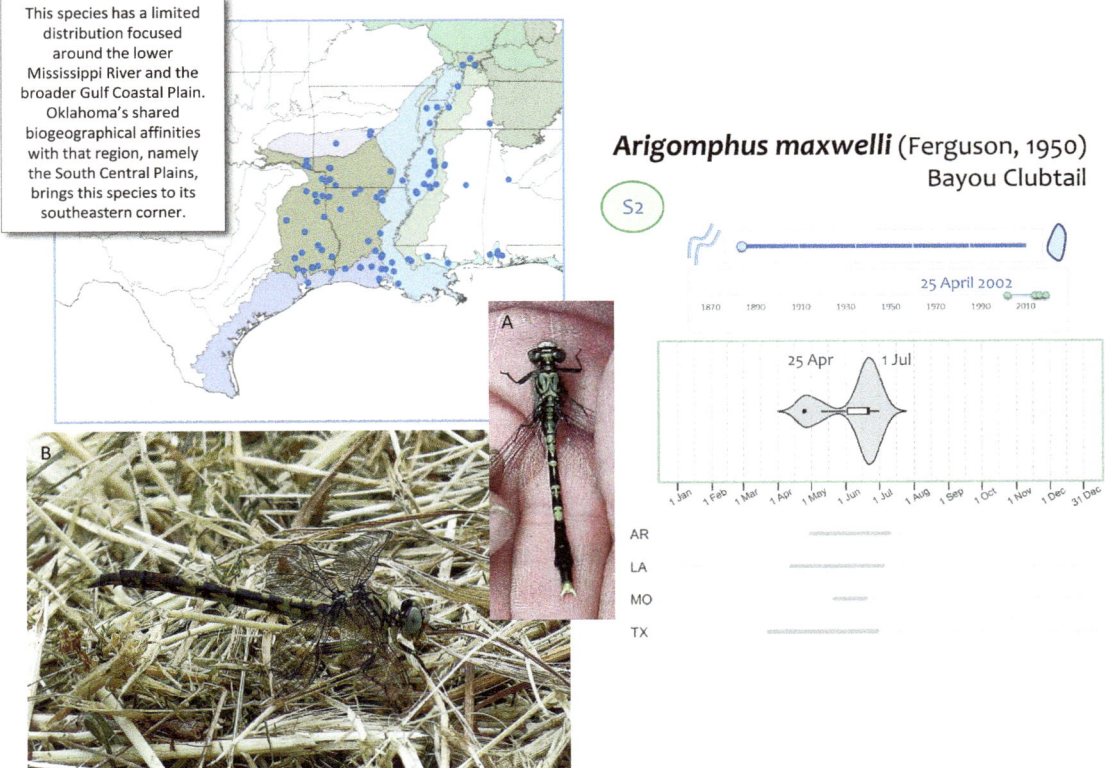

This species has a limited distribution focused around the lower Mississippi River and the broader Gulf Coastal Plain. Oklahoma's shared biogeographical affinities with that region, namely the South Central Plains, brings this species to its southeastern corner.

Arigomphus maxwelli (Ferguson, 1950)
Bayou Clubtail

Figure 15.21 *Arigomphus maxwelli*, the Bayou Clubtail A) ♂, Grassy Slough WMA, McCurtain County, 25 April 2002 (USDA Forest Service; David Arbour, OC 6584), supplying Oklahoma's first record. B) ♀, Red Slough WMA, McCurtain County, 9 May 2016 (©Sylvia Hanson, OC 444354). This clubtail is known only from three sites in southeastern McCurtain County, each of which supports sloughs and slow-moving creeks and backwaters that the species prefers. This species has a limited distribution focused around the lower Mississippi River and the broader Gulf Coastal Plain. Oklahoma's shared biogeographical affinities with that region, namely the South Central Plains, brings this species to its southeastern corner.

of 2 m in places, but there were some places at which one could walk to the water readily. The pools at which the Bayou Clubtails were found were in partial to full sun. The males patrolled the pools, often landing on the sandier parts of the lower stream banks. One of the females was seen ovipositing into shaded water at the base of a tree that was partially exposed in the stream bank. The males were skittish and, when disturbed, they inevitably flew high up into the trees. Of the two that were caught, one sat along the stream on the sand, while the other was perched in a shrub near the stream.

The species was missed in 2015 but was seen three times in 2016 at Red Slough. DA reported a sight record on 3 May of an immature ♂ along Push Creek on the north levee of Bittern Lake and a ♀ at the nearby bridge that crosses Push Creek. But it was not until 9 May when a photographic record was submitted after S Hanson and DA photographed a somewhat-teneral ♀ on Push Creek (OC 444354). A month later DA had the species again on 9 June, when he photographed a ♂ in the north parking lot off Mudline Road at the entrance to Teal Lake (OC 446030). No reports came in 2017 or 2018, but the species was encountered at three locations in Units 1 and 2 at LRNWR on 20–22 June 2019 (total of four adults; BS-P, DA, Paige Schmidt, Gary Murphy, BC; SP 2787, a ♀ captured by G Murphy; OC 497773). This species may be one that does not occur annually as adults, or it may occur in such low numbers as to go undetected in some years.

Obviously what we know of the seasonality in the state for this species is limited, but the date range of 25 April–1 July falls within the confines of the species' seasonality in Texas and Louisiana (TX: 23 March–28 June, OC 461556, Abbott and Lasley 2019; LA: 12 April–3 July, OC 374389, iNat 3626712). It is known in Arkansas from 30 April–9 July (OC 479483, iNat 14215136) and in Missouri from 21 May–18 June (Sims 2012).

Previously we treated the species as a vagrant to the state (Patten and Smith-Patten 2013b), despite the record of a mating pair. Given that the species was found again and that breeding behaviors were documented a second time, we think the designation of this species as a mere vagrant to the state may no longer be valid. When one considers the difficulty in detectability of this species, as well as its truncated flight season, it may be that it has simply been overlooked. The Bayou Clubtail is a cryptic species in behavior and in terms of the difficulty in accessing its preferred habitat, not unlike the Harlequin Darner (*Gomphaeschna furcillata*). As with that species, we suspect that, with time, the Bayou Clubtail will be found to be more common than once thought, but we have our suspicions that it may not range as far west in the South Central Plains ecoregion as the former species does. Until such time that the species can be found to be otherwise, we feel that the Bayou Clubtail ought to be considered an S2 species of conservation concern.

PHANOGOMPHUS OKLAHOMENSIS (PRITCHARD, 1935) – OKLAHOMA CLUBTAIL

Phanogomphus oklahomensis is one of two species of dragonflies AE Pritchard first found in Oklahoma and described to science. Pritchard first encountered this species on 25 April 1931 in, or likely north of, Wilburton, Latimer County (Pritchard 1935). The ♂ he collected that day became a paratype for the species; we have yet to locate this paratype, which is the case for another 26♂ and 7♀ paratypes he and/or CA Sooter collected (Figure 15.23). Fortunately, we do know the whereabouts of the holotype and allotype (safely housed at UMMZ) that were collected on 28 April 1934, the second time the species was recorded in the state (Table 15.1).

Of particular note is with a specimen at the Essig Museum of Entomology (EMEC 310258) that was mislabeled as the holotype, by an unknown hand, but was relabeled as a pseudotype[6] by RW Garrison (Figure 15.23). Note that the label, likely not the original, has a collection date of 29 April 1934, a day that Pritchard does not indicate he collected a specimen of *Phanogomphus oklahomensis* (Pritchard 1935); neither do we have any record of him being in the field that day. But we do know he was in the field the day before, as that is when he is known to have collected an *in copula* pair of *P. oklahomensis* he designated as the holotype and allotype (Pritchard 1935). One more data entry error of unknown origin worthy of note pertains to another EMEC specimen, this time a ♀ *P. oklahomensis* with a typewritten label, again likely not an original, shows a collection date of 12 June 1934 (EMEC 81372; Figure 15.23). Although Pritchard was collecting at or near Wilburton on that date, we find it odd that he never mentioned collecting this specimen, nor is it labeled as a paratype. Perhaps it is one of the missing paratypes collected on 12 May 1934, but we will probably never know.

The species was not recorded in the state again until 1959, when GH Bick collected one nymph and eight adults at Robbers Cave SP, Latimer County (18 April, 7♂, 1♀, Bick notes, FSCA, JCAC 22688); we have not found the nymph, but have found all adults but 1♂. Bick and LE Hornuff next found the species on 14 June 1970, also at Robbers Cave SP, where they collected 1♂ (FSCA). Those two records were the only ones for the Bick/Hornuff era, likely because they tended to survey from June onward, after the peak flight season for the species. There are three records between 1971 and 1992 that added another county, Creek County, to the list. Since 2007, there have been about 100 records and 17 counties added. Two counties, Marshall and Ottawa, have been reported but supporting details are partial or missing entirely. Marshall County was reported by

Figure 15.22 *Phanogomphus oklahomensis*, the Oklahoma Clubtail. Oklahoma's namesake dragonfly was discovered in and described from … wait for it … Oklahoma. Although the type locality is Pushmataha County in the southeastern part of the state, it is now known that the species ranges through much of eastern Oklahoma as well as into northeastern Texas, northwestern Louisiana, and western Arkansas. A) ♂, Haskell Lake, Muskogee County, 15 May 2014 (OC 422137). B) ♀, Bixhoma Lake, Wagoner County, 4 April 2012 (OC 374621). Photos by ©James W Arterburn.

Of the holotype, allotype, and 30♂ and 8♀ paratypes in the series, we have located but a few. Two of the missing paratypes may include two EMEC specimens (A), one of which (EMEC 310258) was mis-labeled as the holotype.

type	data from Pritchard (1935), verbatim	specimen(s) found?
holotype	Holotype male and allotype female taken in copulation, Fourche Maline Creek, 8 miles north of Wilburton, Oklahoma, April 28, 1934, A. E. Pritchard, collector	UMMZ
allotype	Holotype male and allotype female taken in copulation, Fourche Maline Creek, 8 miles north of Wilburton, Oklahoma, April 28, 1934, A. E. Pritchard, collector	UMMZ
paratype	24 males, 2 females same data as holotype (8 of these males collected by C. A. Sooter)	2♂, EMEC 81402, 310258 (?)
paratype	3 males, 3 females, Wilburton, May 12, 1934, (A. E. P.)	No, but EMEC 81372 (♀), indicated as 12 June 1934 could be a data entry error
paratype	3 females, Wilburton, June 9, 1934, (A. E. P.)	1♀, EMEC 81373
paratype	1 male, Wilburton, June 10, 1934, (A. E. P.)	1♂, EMEC 81401
paratype	1 male, Wilburton, April 25, 1931, (A. E. P.)	No
paratype	1 male, Idabel, Oklahoma, April 28, 1934, collected by C. A. Sooter	No

These specimens, EMEC 310258 (♂, left) and EMEC 81372 (♀, right) may not be associated with their original labels, which may explain why EMEC 310258 was labeled in error as the species' holotype (actual holotype is at UMMZ).

We believe these sample labels (B) are representative of AE Pritchard's handwriting as they are typical of what we have seen on many other specimens he collected. The writing on these labels does not resemble that pictured above (A).

Figure 15.23 Type series of *Phanogomphus oklahomensis*, the Oklahoma Clubtail.

Donnelly (2004a). The Ottawa County report comes from a specimen (1♂, id confirmed by BS-P and MAP, USNM 338751) that was collected in Commerce by Harwell but there is no date of collection on the specimen label,[7] and the locale is well to the north of the species' known geographical distribution.

Pritchard (1935:4) described the type locality as a "small stream in the sandstone hills … in a wooded area where the creek widened out." Pritchard continued to say that males "were found squatting on the ground in sunny spots along the banks, making only short journeys, and these mostly when disturbed" and that "exuviae were found several inches high on trees standing in the water, or along the banks close to the water." The ♂ that CA Sooter collected in Idabel was "found on a bank above a small muddy lake in which the species doubtlessly breeds" (Pritchard 1935:5). Abbott (2015:149) describe the habitat in Texas as, "sandy-bottomed ponds, lakes, and slow-moving streams that may or may not be wooded." In our experience in Oklahoma, the species is seen most often near small ponds, including

Table 15.1 Hindwing (HW) and total length (TL) measurements of *Phanogomphus oklahomensis* specimens examined from Oklahoma, Louisiana, and Texas (*n*=28; SP, USNM). Specimens tended to be smaller (see parenthetical numbers) than measurements presented by Needham et al. (2014) and Abbott (2015), neither of whom distinguished measurements by sex. Measurements are in mm.

	N =	HW MIN	HW MAX	HW MEDIAN	TL MIN	TL MAX	TL MEDIAN
♀	6	25.1 (−0.9)	28.3 (−1.7)	26.8 (−1.2)	39.6 (−4.4)	46.4 (−2.6)	45.3 (−1.2)
♂	22	23.2 (−2.8)	27.4 (−2.6)	25.3 (−2.7)	41.0 (−3.0)	49.3 (+0.3)	45.7 (−0.8)
♀♂	28	23.2 (−2.8)	28.3 (−1.7)	25.4 (−2.6)	39.6 (−4.4)	49.3 (+0.3)	45.5 (−1.0)
Needham et al. (2014)		26.0	30.0	28.0	46.0	49.0	47.5
Abbott (2015)		-	-	-	44.0	49.0	46.5

cattle ponds, in cross timbers or pineywoods with clear or muddy water and with or without lily pads or emergent vegetation, although we have found plenty of individuals at or near small creeks. Creeks have been described as rocky, 3–5 m wide, with moderate to fast-flowing clear water or they may be slow-moving creeks or sloughs, such as those at Grassy Slough WMA. McCurtain County. The species can be found in clearings in wooded areas and sometimes dirt or gravel roads, too, including at least once on an open dirt road with a water filled roadside ditch in a pine plantation. The species typically perches on the ground, usually within leaf litter or low vegetation but sometimes without cover, or on weedy or shrubby vegetation just above (<20 cm) above the ground.

The Oklahoma Clubtail flies in the state for about three months: from late March to mid-June (19 March 2012, 1♂, 10 km SE of Idabel, McCurtain County, BAH, OC 374127 and 14 June 1970, 1♂, Robbers Cave SP, Latimer County, GHB, LEH, FSCA). Previously it was thought the late date for Oklahoma was much later. However, that date was based on a specimen that BS-P determined was actually *Phanogomphus graslinellus*, the Pronghorn Clubtail (17 August 1994, 1♀, Stillwater, Payne County, Gates, USNM 388734; see that species account). Elsewhere in *P. oklahomensis'* range, it flies approximately as long as it does in Oklahoma (78–88 days; AR: 14 March–31 May, OC 374099, 356637, 364286; LA: 6 March–23 May, OC 440654, 367074; TX: 10 March–6 June, Abbott and Lasley 2019).[8]

Phanogomphus oklahomensis was "most abundant at the end of April" when Pritchard found it with *Epitheca cynosura* (Common Baskettail), *Didymops transversa* (Stream Cruiser), and *Enallagma divagans* (Turquoise Bluet), but it "was almost completely replaced by [*P.*] *graslinellus*" by early June. Pritchard was correct in his assessment; our data show that the species peaks in mid to late April and is done flying right around the time that *P. graslinellus* has its peak abundance for the season (Figure 15.4A).

Our highest counts for the species were of 50–80 individuals (25♂, 15♀, including 2 pairs [1 as SP 1511], and 10U tenerals, 10 km SE of Idabel, McCurtain County, 11 April 2015; 40 ♂, 30 ♀, 10U tenerals, Grassy Slough WMA, McCurtain County, 31 March 2018). Even so, just a few individuals at a time, whether adult or teneral, is more typical for counts, although our highest count for tenerals has been 15U (Grassy Slough WMA, McCurtain County, 20 March 2017, MAP, who also recorded 2♂ and 1♀ adults). Tenerals have been noted in Oklahoma between 26 March and 27 April (3♂, 8♀, 26U, from McCurtain and McIntosh Cos.). The only record of a nymph we know of is the 1959 record from Bick that was mentioned above. Nine pairs have been reported, but there are only two records indicating oviposition. The ovipositing females were both solitary. We saw one ♀ ovipositing on 25 April 2014 at Lake John Wells, Haskell County, where we noted one other ♀ and 5♂, including one brief pairing, and BS-P photographed a ♀ ovipositing at the lake shore of Dow Lake, Pittsburg County, on 28 April 2017, when no other individuals were noted.

yellow swirl

Figure 15.24 Note the telltale yellow swirl laterally on abdominal segment 8 of *Phanogomphus oklahomensis*, the Oklahoma Clubtail. Photos by ©Brenda D Smith.

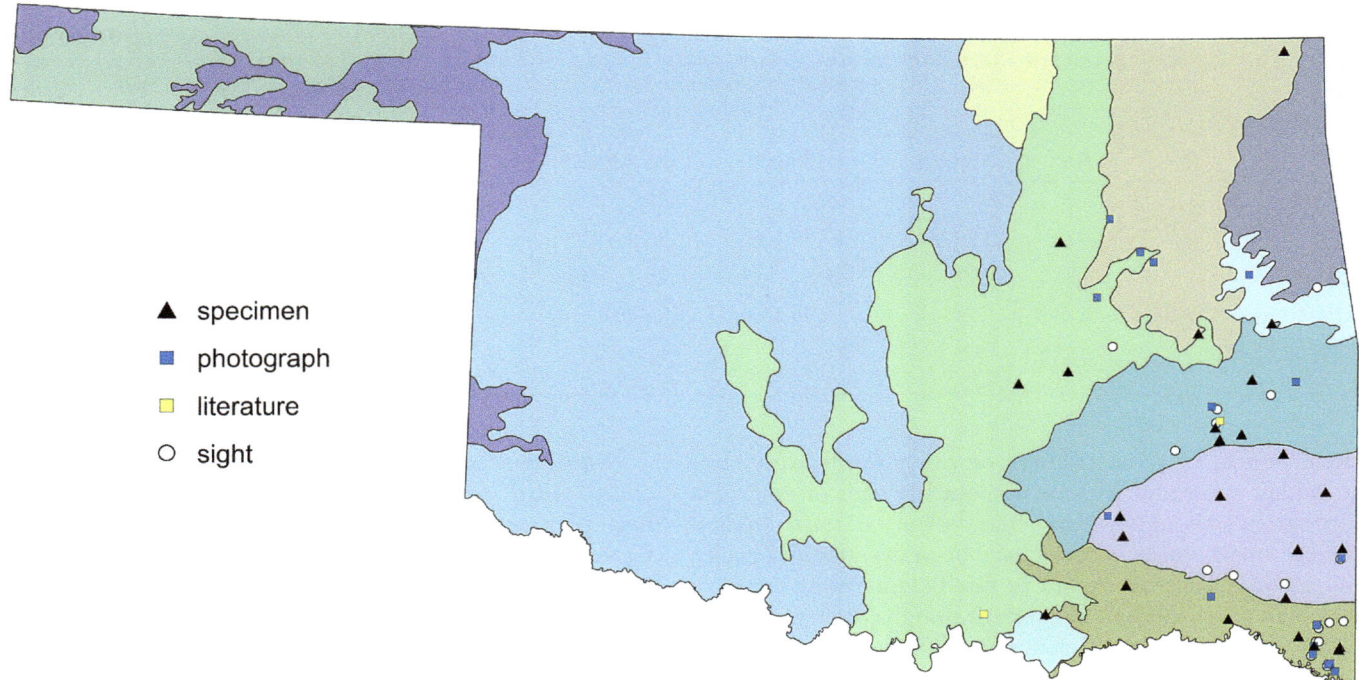

▲ specimen
■ photograph
□ literature
○ sight

Figure 15.25 Documentation level of supporting records for *Phanogomphus oklahomensis*, the Oklahoma Clubtail, in Oklahoma. Levels include specimen, archived photograph, literature reference, or sight record. Although sight records lack documentation, only those we feel are credible are plotted.

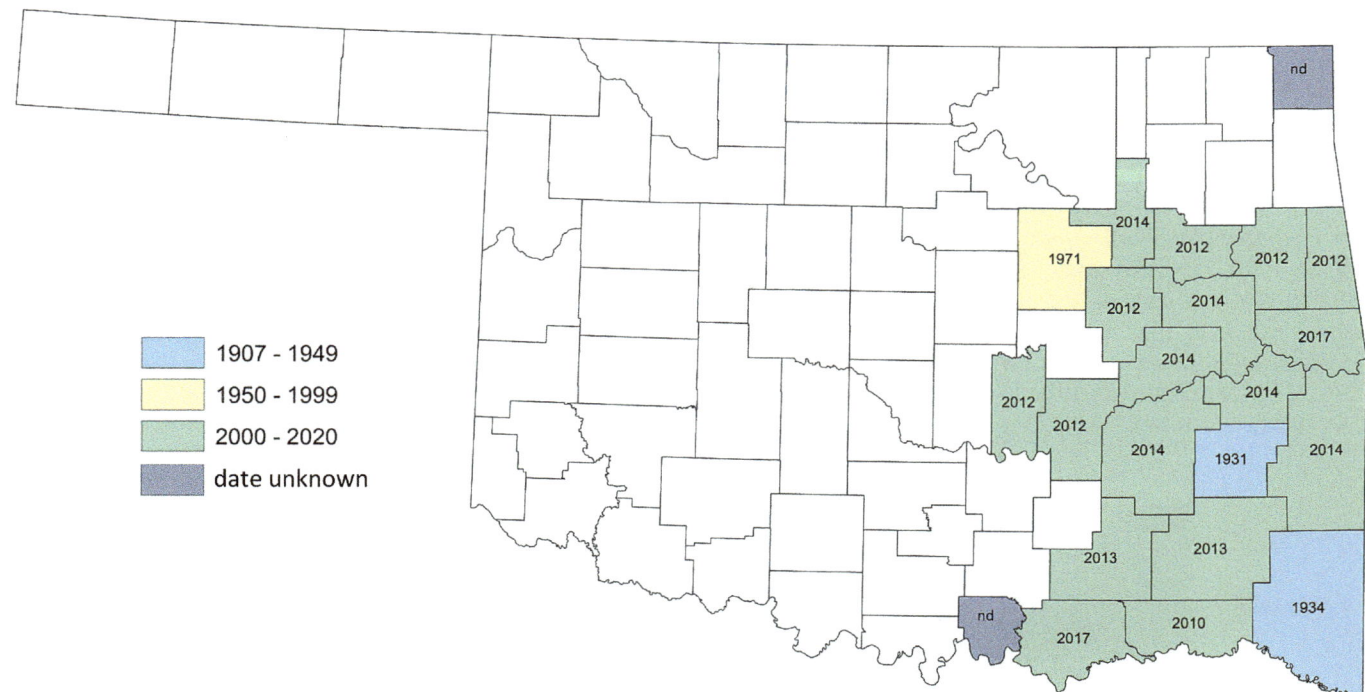

1907 – 1949
1950 – 1999
2000 – 2020
date unknown

Figure 15.26 Counties in which *Phanogomphus oklahomensis*, the Oklahoma Clubtail, are known to occur in Oklahoma. Year within the county is that in which the species was first reported. Two counties, Ottawa and Marshall (hatched), have adult specimens (JCAC and USNM) but no associated collection dates (nd = no date).

The Oklahoma Clubtail has been classified as a Tier II "species of greatest conservation need" by the Oklahoma Department of Wildlife Conservation (ODWC 2015). This designation seems to have been made chiefly on account of the species' rather small geographic range, as it occurs in only four states, from the Gulf coast of eastern Texas and western Louisiana north through central Arkansas and eastern Oklahoma. In our experience the species is locally fairly common in the east-central and southeastern portions of the state, as such we feel the Tier II status is not necessary as the species is apparently secure (Patten and Smith-Patten 2013b). We recommend an SRank of S4 or perhaps S3S4 to err on the side of caution.

PHANOGOMPHUS LIVIDUS (SÉLYS, 1854) – ASHY CLUBTAIL

There are just over 20 confirmed records, at 5 localities in 4 counties, of the Ashy Clubtail in Oklahoma. About two-thirds of those records are from one site, "Berlin's property," located about 10 km SE of Idabel, McCurtain County. To our knowledge, all encounters on the property have been within a scrubby field just west of a clear, shallow, slow-flowing creek with firm, sandy soil, where the species tends to alight (although sometimes it lands on short vegetation). The creek runs through an area that is a patchwork of mixed woodland, shrubs, grassland, and pasture. BAH first found the species in Oklahoma there, on 7 May 2010, when he photographed a ♀ (OC 318729). This record remains the late date for the species in the state. The early date for the species is 20 March, in 2017, when MAP and DA recorded a teneral ♀ at Grassy Slough WMA, McCurtain County (OC 461536). That was the second time the species was recorded at that locale, which is a primarily hardwood woodland with a network of slow-flowing, clear-water streams that occasionally flood out into a marshy area where we have recorded *Gomphaeschna furcillata*, the Harlequin Darner. The first

time *Phanogomphus lividus* was recorded at Grassy Slough was on 11 April 2014, when it became the second location for the species in Oklahoma (2♂, one as SP 1115). Later that same day we, along with BAH, also recorded 2♂ and 2♀ at "Berlin's property." The species has been reported from both properties up to 2018.

The third location for the species was discovered in 2015, when we captured a lone ♂ on the Hassell Lake side of the TNC Boehler Seeps Preserve, Atoka County (SP1536). That record was the first time the species was recorded outside McCurtain County. In 2017 MAP recorded the fourth and fifth localities, and the third and fourth counties, for the species in the state. On 26 March, there were 1♂ and 3♀ near the outflow of the South Fork of Holly Creek at Schooler Lake, Choctaw County. One ♀ was identified in hand and photographed (OC 461619) and another was noted as a teneral. The ♂ was collected as SP 2235. A couple of weeks later a single ♂ was collected at Belzoni, Pushmataha County (SP 2269). On 20 April 2019, MAP had 2♂ (1 verified in hand) and 2♀ (one as SP 2754, one as OC 494653) at Schooler Lake, suggesting presence of a small population there.

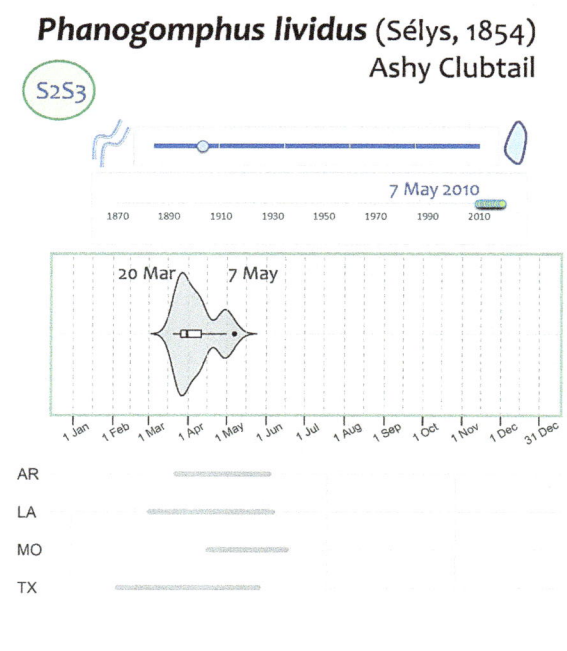

Figure 15.27 **A)** First record and the late flight date for Oklahoma of *Phanogomphus lividus*, the Ashy Clubtail. This ♀ was photographed on 7 May 2010, at Berlin Heck's property SE of Idabel, McCurtain County (©Berlin A Heck, OC 318729). **B)** ♂, at same location, 25 March 2012 (©James W Arterburn). In the state, the species was known solely from this locale until it was discovered at nearby Grassy Slough WMA, McCurtain County, in 2014. It has since been found at three other locales, only one of which, Schooler Lake, Choctaw County, appears to support a population.

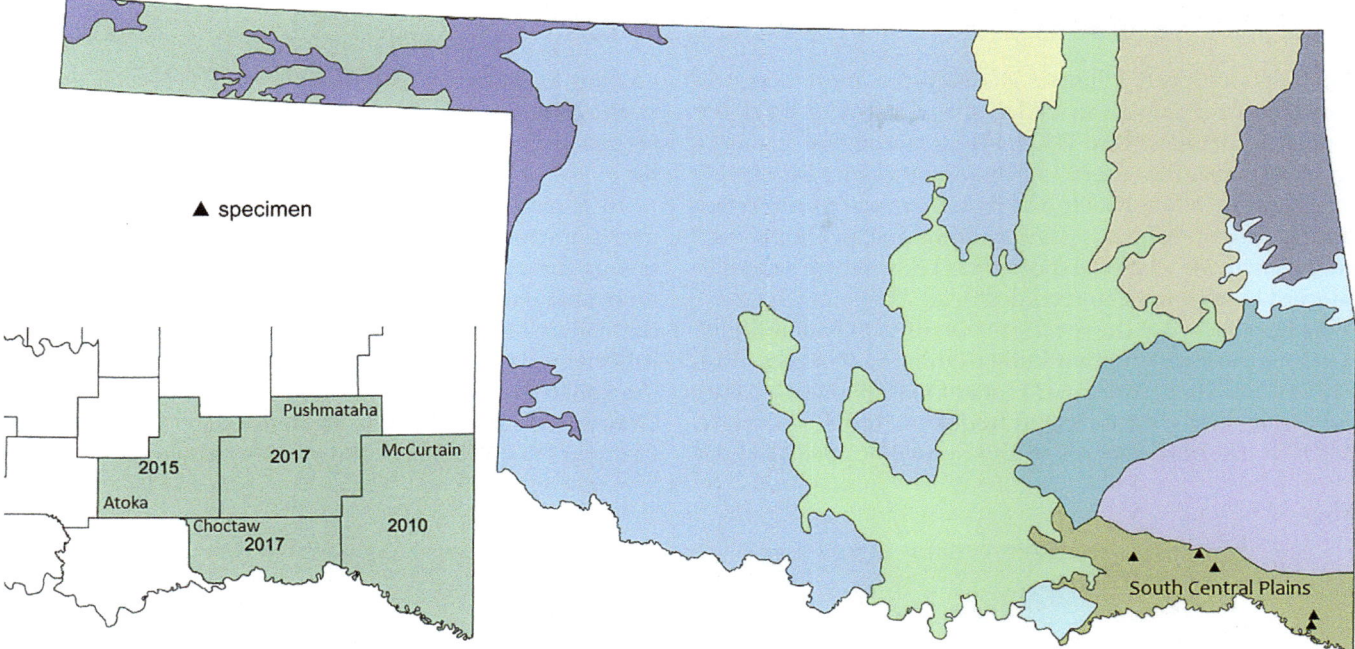

Figure 15.28 Documentation level of supporting records for *Phanogomphus lividus*, the Ashy Clubtail, in Oklahoma (above). All localities have specimen records. Note that the species only occurs within the South Central Plains ecoregion, where it has been documented in four counties (Atoka, Choctaw, McCurtain, and Pushmataha; left: years indicate first record of occurrence for that county).

All of the known localities are within the South Central Plains ecoregion (Figure 15.28).

Including the 4♂ specimens mentioned above, there are a total of 8 specimens of *Phanogomphus lividus* for the state. The first specimen was collected on 23 March 2012, when a ♂ was captured (DRP) and a ♀ was photographed (OC 374172) at "Berlin's property." Both were reported as tenerals, as were an additional 10 individuals (BAH). Two days after that record, 2♂ and 2♀ adults were reported from the site, followed the next day by a report of 1♀ (both JWA photographic records). We recorded 2♂ and 1♀, again at Berlin's, on 11 April 2015 when the only other male specimen was collected (SP 1510). We also captured the first female collected in the state, on 27 April 2013 (SP 527). That female was the only individual we encountered on that day.

The species tends to be seen a few at a time, but it has been reported multiple times in the 10–12 range, and the highest count is 20 individuals (14♂ [two in hand], 6♀, including 4 pair, Grassy Slough, 12 April 2018, BS-P, DA). On that day, BS-P and DA observed behavior that we had not noticed before. Instead of seeing individuals near the ground where we were used to seeing these small clubtails, some individuals were exhibiting an undulating flight (bobbing from about 1 m off the ground up to about 2 m while flying in a straight line. Some flew for 50+ meters, most for 15 or 20

m). At first, we were not sure what species of dragonfly was flying that way but after catching a few and watching some alight (all males) we realized that they were Ashy Clubtails. Initially we thought they were flying that way because of the windy conditions that day, but we later encountered others flying that way in sheltered areas as well. We do not know if this was feeding behavior or a mating display, nevertheless, it has been described as typical of the species elsewhere (GW Lasley, *pers comm*).

Records indicate that the species occasionally is fairly common (see above), but until additional sites are located in the state, its occurrence in Oklahoma is tenuous, although it likely does not warrant S1 status. We posit that the Ashy Clubtail is more widespread than data currently suggest, and it will likely be found elsewhere in the South Central Plains ecoregion. Its short flight period in Oklahoma hinders data gathering for this species, but the longer flight season in nearby states suggests that there is hope of extending the species' known phenology in Oklahoma (AR: 23 March–3 June, OC 440843, USNM 338505; LA: 2 March–6 June, OC, Mauffray 2014; MO: 17 April–17 June, OC 374445, Trial 2005; TX: 5 February–26 May, Abbott and Lasley 2019; the nymph record for Jackson County, Kansas, SEMC 1110818, is well out of range for the species and should be re-examined). We feel S2S3 status is currently sufficient for the species in Oklahoma.

PHANOGOMPHUS MILITARIS (HAGEN, 1858) – SULPHUR-TIPPED CLUBTAIL

The Sulphur-tipped Clubtail is easily the most common and widespread clubtail in Oklahoma, a status that has not changed in decades. Bird (1934:44) remarked that it was "a common species in western Oklahoma." Bick and Bick (1957) remarked that it "was widely distributed and was the most frequent species of the genus [*Gomphus senso lato*] in Oklahoma," a statement made when the species was known from a mere 17 counties! It has now been recorded in all 77.

We have recorded the species many times in double digits, including the state's highest known count of 40 adults (30♂, 10♀, 24 June 2015, American Horse Lake, Blaine County). The species also is encountered frequently in Oklahoma. In the OOP database, records of the Sulphur-tipped Clubtail account for almost a fifth of all gomphid records, and relative to its congeners [*Gomphus sensu lato*], *Phanogomphus militaris* is by far the most commonly encountered, accounting for 35.1% of all occurrences (*n* = 747; Figure 15.3).

In Kansas, Kennedy (1917:138) said that it was "found over the western half and along the southern tier of counties as far east as Montgomery county" and it was "the most abundantly represented of any of the Gomphines in the collection [SEMC]." He went on to comment, "either it is common over its habitat or else it is an easy mark for the collector." Allison (1921) did not report the species for Crawford County, which is along the Missouri border. Cringan (1979) reported the species, only in early June, in

Figure 15.29 *Phanogomphus militaris*, the Sulphur-tipped Clubtail, A) ♂, Fort Sill, Comanche County, 1 August 2010 (©Victor W Fazio III, OC 321415), with the abdomen tip extensively yellow. B) ♂, Collinsville City Lake, Rogers County, 14 June 2014 (©James W Arterburn, OC 432040), with the abdomen tip extensively black. C) ♀, Lake Lloyd Vincent, Ellis County, 14 June 2016 (©Greg W Lasley, OC 447959). Without dispute, this clubtail is our commonest in the southern Great Plains, including specifically in Oklahoma, where it has been recorded in all 77 counties, so much so that it is the "default" small clubtail.

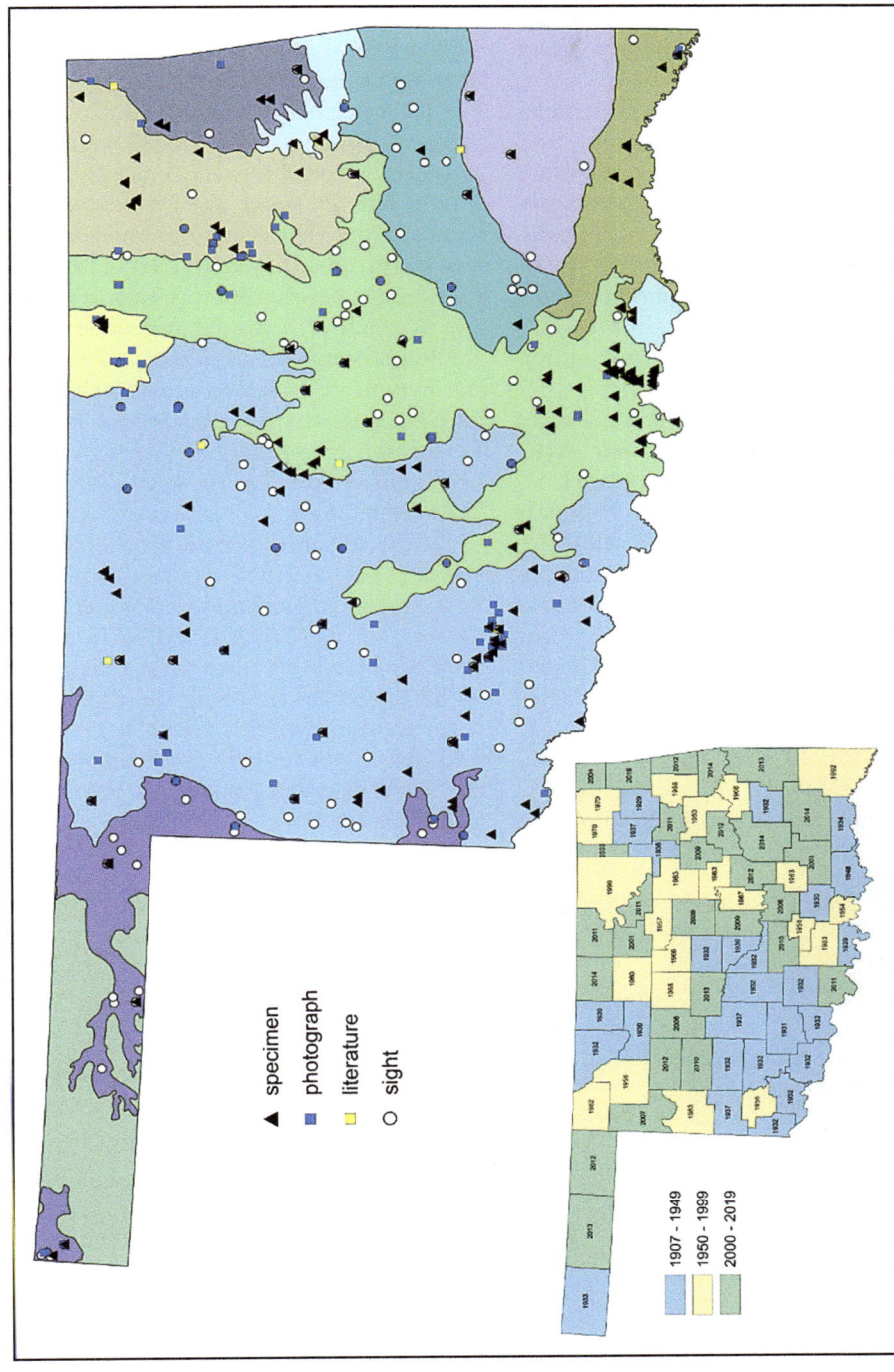

Figure 15.30 Documentation level of supporting records for *Phanogomphus militaris*, the Sulphur-tipped Clubtail, in Oklahoma (above). Note that this species is often recorded without supporting documentation. It has been reported from all 77 Oklahoma counties (left), with the earliest records mostly from the southwest.

Lyon County, which is roughly as far east as Montgomery County. Beckemeyer and Huggins (1997) indicated that it was common across Kansas. In Missouri, it is known to have expanded eastward in the state and is common as both adults and larvae in the prairie; it was the most frequently encountered gomphid as larvae in prairie ponds, where they are known to burrow into detritus and sediments along pond edges (Landwer and Sites 2003).

As adults, the species is found in a variety of Oklahoma habitats, from lakes to ponds, from rivers to creeks, and even in open grasslands or patrolling gravel roads. We often see individuals perched along lake shorelines, particularly on rocks or sitting directly on the ground, or occasionally perching on vegetation. Bird (1934:44) said that "its favorite habitat is about small muddy ponds and stagnant pools of streams." It has been reported consistently since he discovered it in the state in 1929 (13 June, 1♂, Love County, OMNH 849).

In Oklahoma the flight season is late April to early September (24 April 2003, 1 adult, Fort Sill MR, Comanche County, BC Kondratieff, Kondratieff et al. 2004; 9 September 2018, 1♂, 6 km N of Kenton, North Carrizo Creek, Cimarron County, BC, OC 490184). This flight season is, as one would expect, not as long as in Texas (3 March–21 November, OC 493415, Abbott and Lasley 2019). Its length is longer than the known ranges in Arkansas, Colorado, Kansas, Missouri, and New Mexico (AR: 13 April–20 July, OC 282474, 424826; CO: 27 May–29 August, OC 480567, 426686; KS: 21 April–12 August, SEMC 1349260, 1349301; MO: 29 April–17 August, OC 493538, Trial 2005; NM: 27 May–19 September, JCAC, OC 490705). We have teneral records dating from 6 May to 21 July, with a median date of emergence of 29 May ($n = 58$), whereas Bird (1934) reported peak emergence in 1932 as being in mid-May.

Landwer and Sites (2003) brought attention to the inadvertent error that Bird (1934) made when he described what he thought to be the nymph of *Phanogomphus militaris* (at the time, *Gomphus militaris*), but that turned out to be *Arigomphus lentulus*, the Stillwater Clubtail (Figure 15.18). For the nymphal description Bird used the exuviae of a ♀ gomphid (OMNH 4250) that he watched emerge on 10 May 1932. He watched this specimen for over an hour from the time it left the water until it flew away as a teneral. If he had captured the teneral, he undoubtedly would have realized his error. Bird's error caused much confusion over the years, as his description was used in various keys (Needham and Westfall 1955; Young and Bayer 1979; and Needham et al. 2000). Note that Landwer and Sites (2010:61) described the nymph of *P. militaris* as "superficially less similar to its congeners than to *Arigomphus* in general appearance." Fortunately, the mistaken identity of the exuviae has now been cleared up as has the key in Needham et al. (2014) and Tennessen (2019).

Lastly, we encountered an aberrant individual of note. We blithely identified a teneral ♀ gomphid that we saw at Kessler Atmospheric and Ecological Field Station, McClain County, on 18 May 2013 (SP 567), as *Gomphurus externus*, the Plains Clubtail, because of a distinct lateral black notch on S8 that extended along the edge of the segment. Later, when we examined this specimen closely in the lab, we realized that it was actually *Phanogomphus militaris* with extensive black laterally on S8—as the saying goes, *subgenital plates never lie.*

PHANOGOMPHUS GRASLINELLUS (WALSH, 1862) – PRONGHORN CLUBTAIL

The Pronghorn Clubtail is fairly common in Oklahoma. It is typically found along or near rocky rivers and streams, although it is encountered occasionally at ponds and lakes or along dirt roads or in scrubby fields. Such variety has been noted elsewhere (e.g., Abbott 2015, Paulson 2011). The variety of habitats inhabited by adults is not surprising given that larvae also are found in a variety of habitats (Landwer and Sites 2010).

Although we usually only record the species 1–6 adults at any one time, we have recorded it in double digits twice. Our highest count was of 30 individuals, all teneral, on 25 April 2014 at Lake John Wells, Haskell County, in the grass along the lakeshore (5♂, 25♀, we captured and released 3♂ and 4♀ and collected 1♂ and 1♀, SP 1138, 1139). Later that season we recorded 11 adults at the dam spillway at McGee Creek SP, Atoka County, on 6 June (8♂, 3♀, we captured and

released 1♂ and 1♀ and collected 3♂, SP 1257–1259). The males collected at McGee Creek SP all had varying amounts of yellow on the terminal segments of the abdomen, both laterally and dorsally. Individual variation in the species has been noted for many decades. Williamson (1914b:447) remarked that the specimen he collected as the first record for Oklahoma (5 June 1907, 1♀, Wister, Le Flore County) was "studied by Dr. Calvert and myself, and it seems safe to refer it to this species," indicating that he had some doubt, at least initially. He then goes on to discuss how in Indiana "considerable variation in marking is shown." His discussion focused on thoracic stripes, but we guess that he noticed variation in abdominal patterning, too, just as we have seen in Oklahoma. Kennedy (1917:137) said of the SEMC specimens he examined: "These specimens are very dark in color and at first glance would not be taken for [*Phanogomphus*]

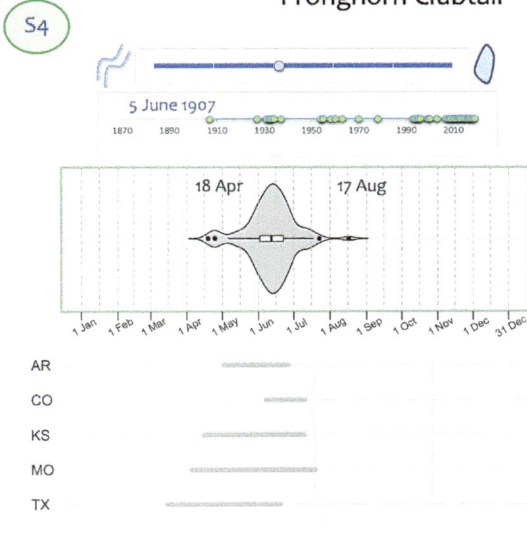

Phanogomphus graslinellus (Walsh, 1862)
Pronghorn Clubtail

S4

5 June 1907

18 Apr   17 Aug

Figure 15.31 *Phanogomphus graslinellus*, the Pronghorn Clubtail, A) ♂, Pushmataha WMA, Pushmataha County, 26 June 2013 (©James W Arterburn, OC 403320). B) ♀, Tallgrass Prairie Preserve, Osage County, 11 June 2016 (©Bill Carrell, OC 446490). The lotic species is uncommon in the southern Plains, yet in easternmost Oklahoma in June it tends to outnumber even its ubiquitous congener *P. militaris*.

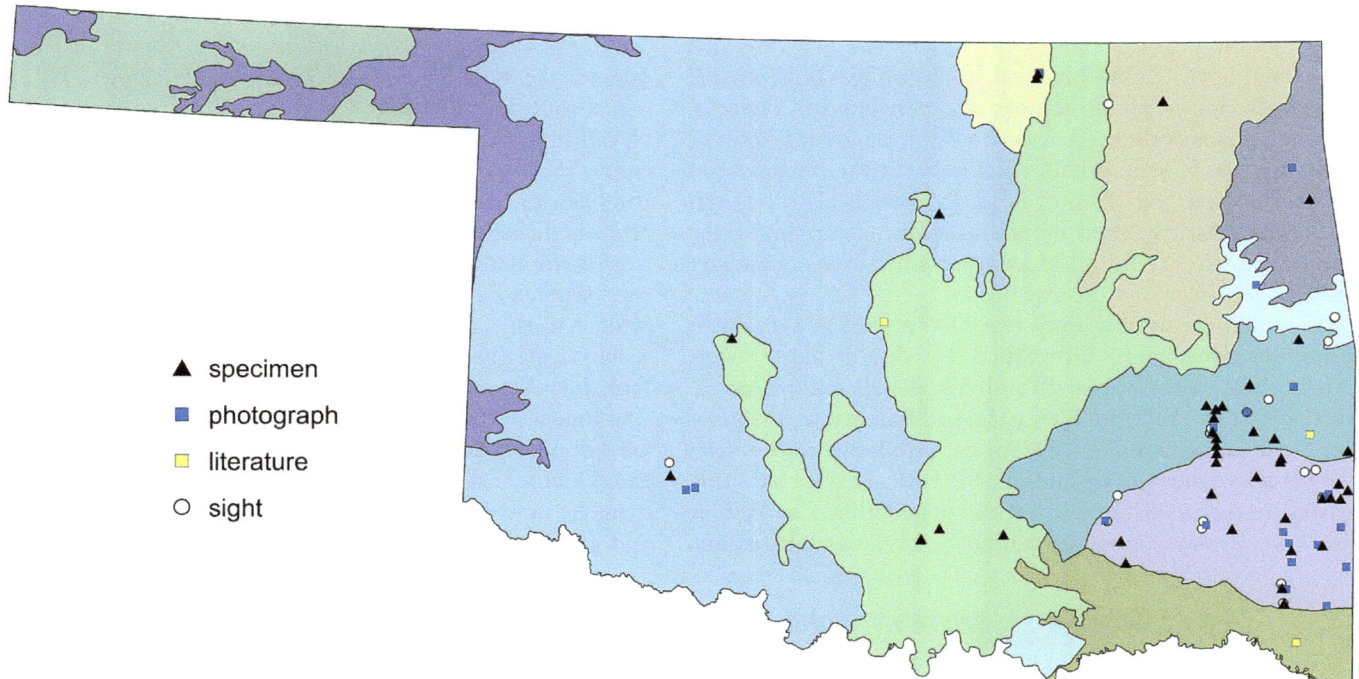

Figure 15.32 Documentation level of supporting records for *Phanogomphus graslinellus*, the Pronghorn Clubtail, in Oklahoma. Levels include specimen, archived photograph, literature reference, or sight record. Although sight records lack documentation, only those we feel are credible are plotted.

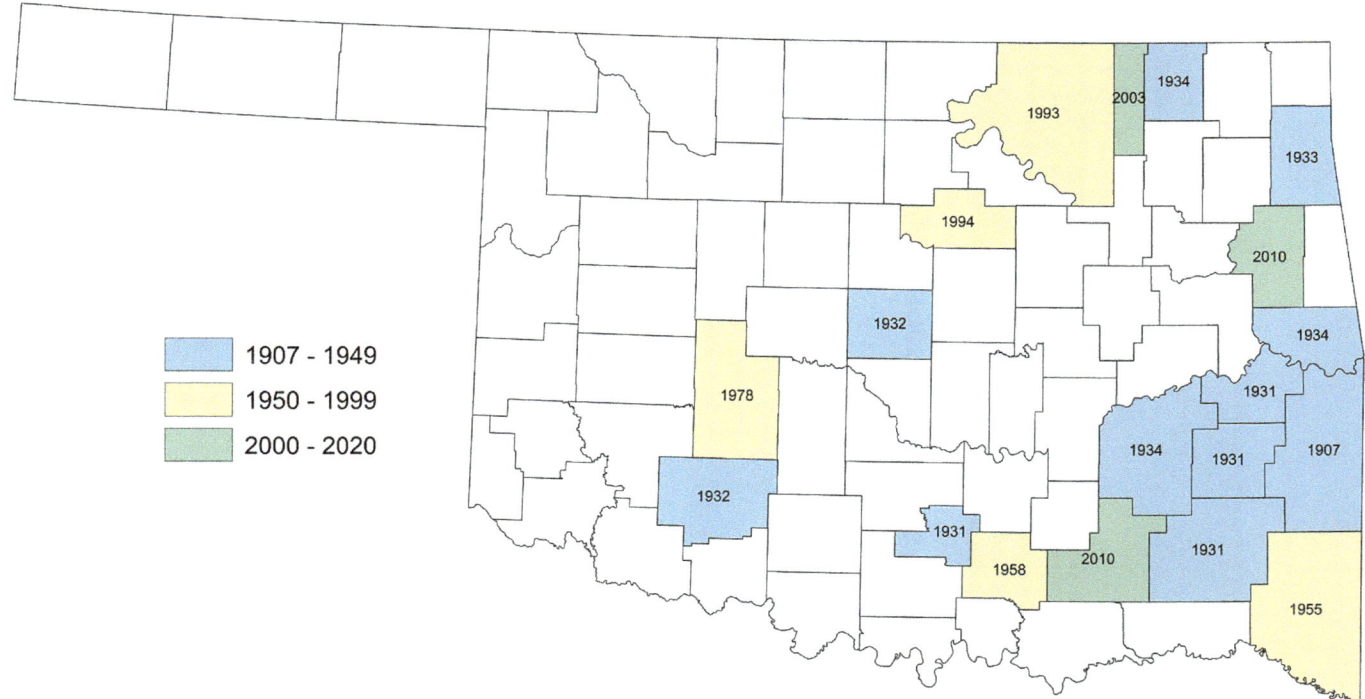

Figure 15.33 Counties in which *Phanogomphus graslinellus*, the Pronghorn Clubtail, are known to occur in Oklahoma. Year within the county is that in which the species was first reported.

graslinellus." Regarding variation, we urge caution about reliance on yellow on the hind femora as a strict characteristic of the Sulphur-tipped Clubtail (*P. militaris*) because we have encountered several ♀ Pronghorn Clubtails in Oklahoma with distinct yellow on the basal half of the hind femora or with a yellow stripe on the outer part of them.

The species has been reported more-or-less regularly since it was first discovered in the state. Bird (1932) reported it for 6 counties, but it was known from 12 by 1934. It is now known from 19 counties, primarily in the eastern third of the state, but with scattered records as far west as Comanche and Caddo Counties (Figure 15.33).

The early date for the species is of a teneral ♂ that was photographed by JWA along the Mountain Fork River, 8.5 km SSW of Smithville, McCurtain County, on 18 April 2012 (OC 374608). The latest record of a teneral we know of for the state is 7 June, but the overall late flight date for Oklahoma *and* the region is 17 August, which comes from a ♀ specimen originally identified as *Phanogomphus oklahomensis*, the Oklahoma Clubtail, that was collected in Stillwater, Payne County, in 1994 (coll. by Gates, USNM 388734, re-id BS-P by photo, confirmed by Oliver Flint, *in litt*). Pritchard (1935:5) noted that "by the first part of June, *Gomphus [Phanogomphus] oklahomensis* was almost completely replaced by [*P.*] *graslinellus*," suggesting that although the species coexist, there is species turnover by mid-season. This pattern holds this day, with *P. graslinellus* peaking in abundance in early to mid-June well after *P. oklahomensis*' peak season and right around the time that species stops flying (Figure 15.4A).

Pritchard made no mention of the beginning of the season in Oklahoma for the Pronghorn Clubtail, but we now know that it flies a month earlier than the records he had access to indicated (pre-1935 records, 14 May until 23 July, OMNH; now 18 April–17 August). Oklahoma's early date is not as early a start date as that known for Kansas, Missouri, and Texas, but it is earlier than Arkansas and the limited records from Colorado (AR: 2 May, OC 430664; CO: 7 June, OC 400406; KS: 15 March, SEMC 1349240; MO: 5 April, Trial 2005; TX: 15 March, Abbott and Lasley 2019). *Phanogomphus graslinellus* is known to stop flying between mid-June and mid-August elsewhere in the region (AR: 26 June, OC 401502; CO: limited records, 10 July, Steve Hummel Collection; KS: 10 July, SEMC 1349249; MO: 19 July, Trial 2005; TX: 18 June,[9] Abbott and Lasley 2019).

HYLOGOMPHUS APOMYIUS (DONNELLY, 1966) – BANNER CLUBTAIL

The tiny Banner Clubtail, known from a mere two locations in the state (Figures 15.34 and 15.35), is Oklahoma's only *Hylogomphus*. It was first documented in Oklahoma in 2010, when on 15 May BAH photographed 1♂ on the Glover River at State Highway 3, 10 km NE of Wright City (OC 318873). He returned on 18 and 25 May, but he did not see the species. He visited the site again the following year on 1 April, when he photographed a teneral ♂ (OC 327591), but on return visits (six in 2011 between 2 April and 11 July and four in 2012 between 26 March and 17 April) he did not encounter it again. We surveyed the site twice in 2014 (12 April and 4 May), neither time finding the species. On the latter visit we collected an aberrant pygmy *Gomphurus*

ozarkensis, the Ozark Clubtail (SP 1168; Figure 15.36), whose diminutive size, at a mere 39 mm, initially made us think we had captured a Banner Clubtail, but whose pattern and appendages did not jibe (Needham et al., 2014 has total lengths as 35–37 mm for *H. apomyius* and 50–53 mm for *G. ozarkensis*). We encountered 13♂ and 5♀ (2 imm ♀) Ozark Clubtails of more typical size at the same spot on the same day. Another visit to the location on 26 April 2015 did not turn up *H. apomyius*. Finally, on 2 April 2016, BS-P captured a single, still somewhat-teneral ♂ (SP 1843) with her baseball cap. No other Banner Clubtails, or other odonates for that matter, were seen that day. MAP returned to the site on 9 May with no luck, and EA Hjalmarson and

Figure 15.34 **A)** Oklahoma's first record and late flight date of *Hylogomphus apomyius*, the Banner Clubtail. Berlin Heck photographed this ♂ on the Glover River at State Highway 3, 10 km NE of Wright City, McCurtain County on 15 May 2010 (©Berlin A Heck, OC 318873). The species has been found at this site twice since ... but missed there dozens of others times, despite focused searches. **B)** ♂, McGee Creek SP, Atoka County, 2 May 2015 (©Brenda D Smith, OC 430739), on a typical perch on a low (shoulder height or lower) leaf, often over flowing water, like the scene pictured here (see center of photo). A few individuals at this locale on 2–3 May 2015 marked the only time the species has been found in the state away from the Glover River site. Nonetheless, the Glover River is the only place tenerals have been observed (**C**, teneral ♂, same location as photo A, 1 April 2011, ©Berlin A. Heck, OC 318873).

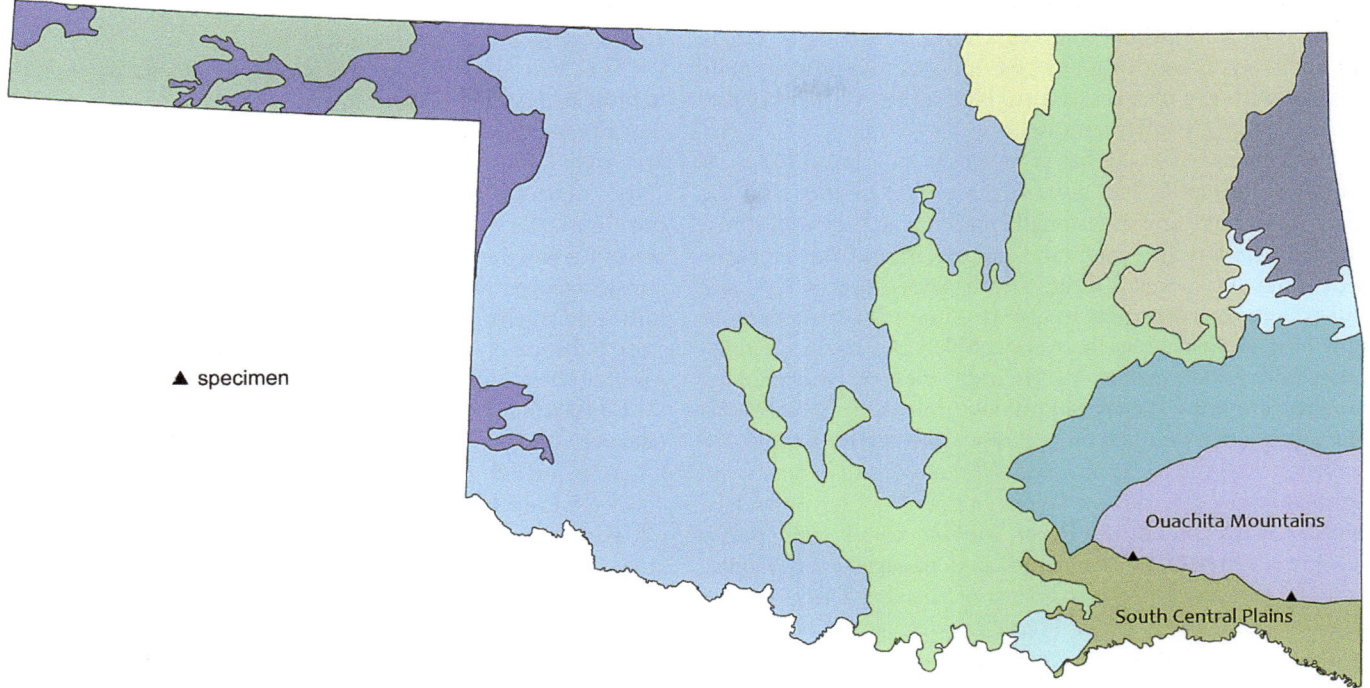

▲ specimen

Ouachita Mountains

South Central Plains

Figure 15.35 Documentation level of supporting records for *Hylogomphus apomyius*, the Banner Clubtail, in Oklahoma. Specimens have been taken at both locations known for this species in the state: McGee Creek State Park, Atoka County, and 10 km NE of Wright City, Glover River at State Highway 3, McCurtain County. The locations fall at the interface of two ecoregions, the South Central Plains and the Ouachita Mountains.

B Roberts visited the site on 25 June, again with no luck. All told we have 24 reported visits to this site since 1996, when SW Dunkle visited on 12 June, but Banner Clubtails have only been seen there three times. It could be argued that for Oklahoma the phenology of the species would preclude visits in June and July ($n = 7$). Banner Clubtails are known to fly in Louisiana, for example, from 25 March to 24 April (Mauffray 2014) and Texas from 9 March to 22 April (Abbott and Lasley 2019), although in the greater southeast they fly from March to 16 June (Beaton 2007, OC; until July in New Jersey, Paulson 2011); in Oklahoma we have records only from 1 April until 15 May. It is reasonable to suppose that it may fly in Oklahoma until late May, but probably not beyond. Nonetheless, eliminating the June and July visits still leaves us with 17 visits to the site between March and May with a mere three encounters. Surveys elsewhere on the Glover River and at nearby sites that appear to be suitable proved unproductive for finding the species. Not surprisingly the encounter rate at the Glover River and neighboring sites leaves us puzzled.

The Glover River site remained the only locale for Banner Clubtails in Oklahoma until early May 2015, when, on 1 May, we visited the dam spillway at McGee Creek SP late in the afternoon. There we spotted a lone ♂ perched on a mostly submerged bush in rapidly flowing, rather deep water. We managed to snap a few terrible photos of this ♂ and took a couple of swings at it with a net, but as the sun went down we realized our chances were up. We returned the next day to find 2♂ and 1♀, all skittish and perching mostly on submerged vegetation in deep, rushing water. BS-P managed to get a couple of passable photos (OC 430739) and then we eventually realized that at least one ♂ and the ♀ occasionally alighted at a nearby hillside seep, where MAP captured SP 1537 (Figure 15.36), the first specimen of the Banner Clubtail for Oklahoma. To our disappointment, this all took place just days before the state experienced severe flooding, enough so that the spillway was rendered inaccessible until the species' flight season was over. We were not able to survey the site in 2016, although Hjalmarson and Roberts visited on 26 June with no success, which is not surprising given that the flight season likely had passed. We made two surveys in 2017, on 25 and 29 May, with no sightings, and struck out again on 6 May 2018 and 20 April 2019.

The two known localities for this species fall within the interface of two ecoregions, that of the South Central Plains, or "pineywoods" region, and the Ouachita Mountains (Figure 15.35). Both regions have a mixture of hardwood and pine forests; although they were once blanketed with such forests but now are a mosaic of limited natural mixed forests, shortleaf pine plantations, and agricultural fields/ pastureland. The Glover River site, for example, is primarily surrounded by agricultural lands and pine plantations with few extant natural woodlands remaining. The McGee Creek site, however, is more-or-less the opposite, being primarily forested with the nearest pasture land about 1 km away and having no large timber stands.

The Glover River site lies at an elevation around 145 m. It is a fairly wide (50–75 m) river at the record locality, and the substrate is rocky and occasionally interspersed with vegetated sand bars. Submerged rocks are slippery, likely from presence of diatoms. The river is clear, mostly knee-deep or less, and generally flows moderately to quickly. There is little emergent vegetation and the shorelines are cobbled expanses or eroded vegetated/wooded banks <1 m high. Beyond the cobbled shoreline are sandy, vegetated hillocks, on which SP 1843 was captured when it perched on vegetation low to the ground.

Other than the rocky substrate and moderate to fast-flowing clear water, the McGee Creek site does not have much in common with the Glover River. McGee Creek right below the dam is 118 m in elevation and, at 45 m across, is narrower than the Glover, but the majority of the creek is narrower still, being 15–20 m wide. There are occasional patches of organic debris and submerged rocks are silty, with some algal growth. Most of the length of the creek has wooded shorelines, although it is not so at the record site. At that site there were scattered broadleaf emergent saplings, narrow mowed areas, bunch grasses, and medium-sized boulder riprap. We saw individuals perched on the emergent saplings but the individual that was captured was perched in a broadleaf shrub on a leaf about 1 m above the ground. That shrub was about 2.5 m tall and was at a seep at the top of the steep, riprap slope above the creek.

The two Oklahoma sites only superficially resemble habitat as described elsewhere in the species' range. Habitat descriptions include: "clean sand/gravel-bottomed streams and rivers, but can tolerate some silt" (Dunkle 2000:78); "clear streams and small rivers with sandy or gravelly bottoms" (Beaton 2007:192); and "small, clean woodland streams with acid [sic] water and sand bottom and accumulations of organic detritus" (Paulson 2011:249). Unlike that indicated in the latter description, water at both Oklahoma sites is more-or-less neutral (Glover River: 7.05–7.21 pH; McGee Creek: 7.32–7.50 pH).

The Banner Clubtail likely goes undetected in Oklahoma because of its short flight season, inconspicuous habits, and limited surveys at that time of year. Not too long ago we were inclined to call this species a mere vagrant to Oklahoma, but the two teneral records and the presence of at least one female have convinced us otherwise. At the least a small breeding population exists in Oklahoma. More surveys may show it to be a little more widespread, although likely it will remain uncommon as it is elsewhere (Paulson 2011). We categorized this species as S2 on the NatureServe ranking system (Patten and Smith-Patten 2013b) because, although it appears to have no major threats in the state at this time, its apparent rarity in the state as well as globally (G3G4), along with its low population density and specialized habitat, warrant remaining conservative about its conservation assessment.

It is easy to see why a pygmy *G. ozarkensis* taken at the Glover River (A) was initially mistaken for a *Hylogomphus apomyius*, the Banner Clubtail (B).

It is also easy to see why hybridization with *G. externus*, the Plains Clubtail (C), was initially proposed as an explanation for the extra yellow on the terminal segments that *G. ozarkensis howeryi* presents (D). Although occasionally nominate *G. ozarkensis* shows yellow dorsally on S9, it generally is but trace (E).

Hylogomphus apomyius, McGee Creek SP, Atoka County, Oklahoma, 2 May 2015, SP 1537

Gomphurus ozarkensis, Glover River, McCurtain County, Oklahoma, 4 May 2014, SP 1168

Gomphurus ozarkensis, Mountain Fork River, McCurtain County, Oklahoma, 28 April 2013, SP 531

Gomphurus ozarkensis howeryi, Salt Creek, Osage County, Oklahoma, 11 June 2016, SP 1942

Gomphus externus, Otter Creek, Logan County, Oklahoma, 31 May 2014, SP 1219

lower specimens are to scale

Figure 15.36 Variation in *Gomphurus ozarkensis*, the Ozark Clubtail, in comparison to other species.

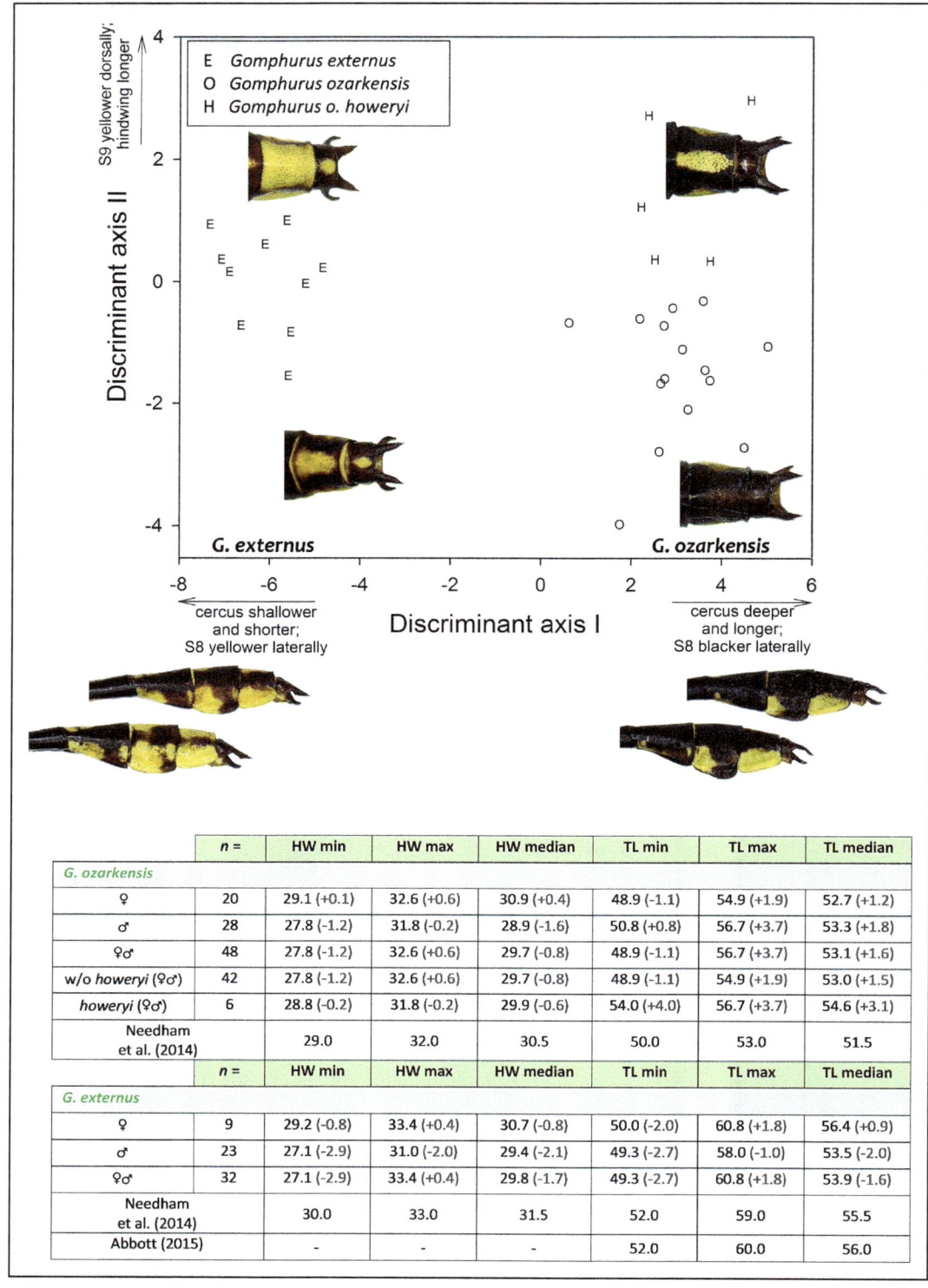

	n =	HW min	HW max	HW median	TL min	TL max	TL median
G. ozarkensis							
♀	20	29.1 (+0.1)	32.6 (+0.6)	30.9 (+0.4)	48.9 (−1.1)	54.9 (+1.9)	52.7 (+1.2)
♂	28	27.8 (−1.2)	31.8 (−0.2)	28.9 (−1.6)	50.8 (+0.8)	56.7 (+3.7)	53.3 (+1.8)
♀♂	48	27.8 (−1.2)	32.6 (+0.6)	29.7 (−0.8)	48.9 (−1.1)	56.7 (+3.7)	53.1 (+1.6)
w/o howeryi (♀♂)	42	27.8 (−1.2)	32.6 (+0.6)	29.7 (−0.8)	48.9 (−1.1)	54.9 (+1.9)	53.0 (+1.5)
howeryi (♀♂)	6	28.8 (−0.2)	31.8 (−0.2)	29.9 (−0.6)	54.0 (+4.0)	56.7 (+3.7)	54.6 (+3.1)
Needham et al. (2014)		29.0	32.0	30.5	50.0	53.0	51.5
	n =	HW min	HW max	HW median	TL min	TL max	TL median
G. externus							
♀	9	29.2 (−0.8)	33.4 (+0.4)	30.7 (−0.8)	50.0 (−2.0)	60.8 (+1.8)	56.4 (+0.9)
♂	23	27.1 (−2.9)	31.0 (−2.0)	29.4 (−2.1)	49.3 (−2.7)	58.0 (−1.0)	53.5 (−2.0)
♀♂	32	27.1 (−2.9)	33.4 (+0.4)	29.8 (−1.7)	49.3 (−2.7)	60.8 (+1.8)	53.9 (−1.6)
Needham et al. (2014)		30.0	33.0	31.5	52.0	59.0	55.5
Abbott (2015)		-	-	-	52.0	60.0	56.0

Figure 15.37 Hindwing (HW) and total length (TL) of *Gomphurus ozarkensis* and *G. externus*, the Ozark and Plains Clubtails, from Oklahoma and surrounding states. Note that *G. ozarkensis howeryi* is larger than measurements previously reported for the overall species. Also note that *G. externus* measured are both smaller and larger than previously reported. Discriminant analyses, including the one above, show that *G. externus* and *G. ozarkensis* are distinct taxa and that *G. ozarkensis howeryi* individuals fall broadly within *G. ozarkensis* but are taxonomically distinct as a subspecies.

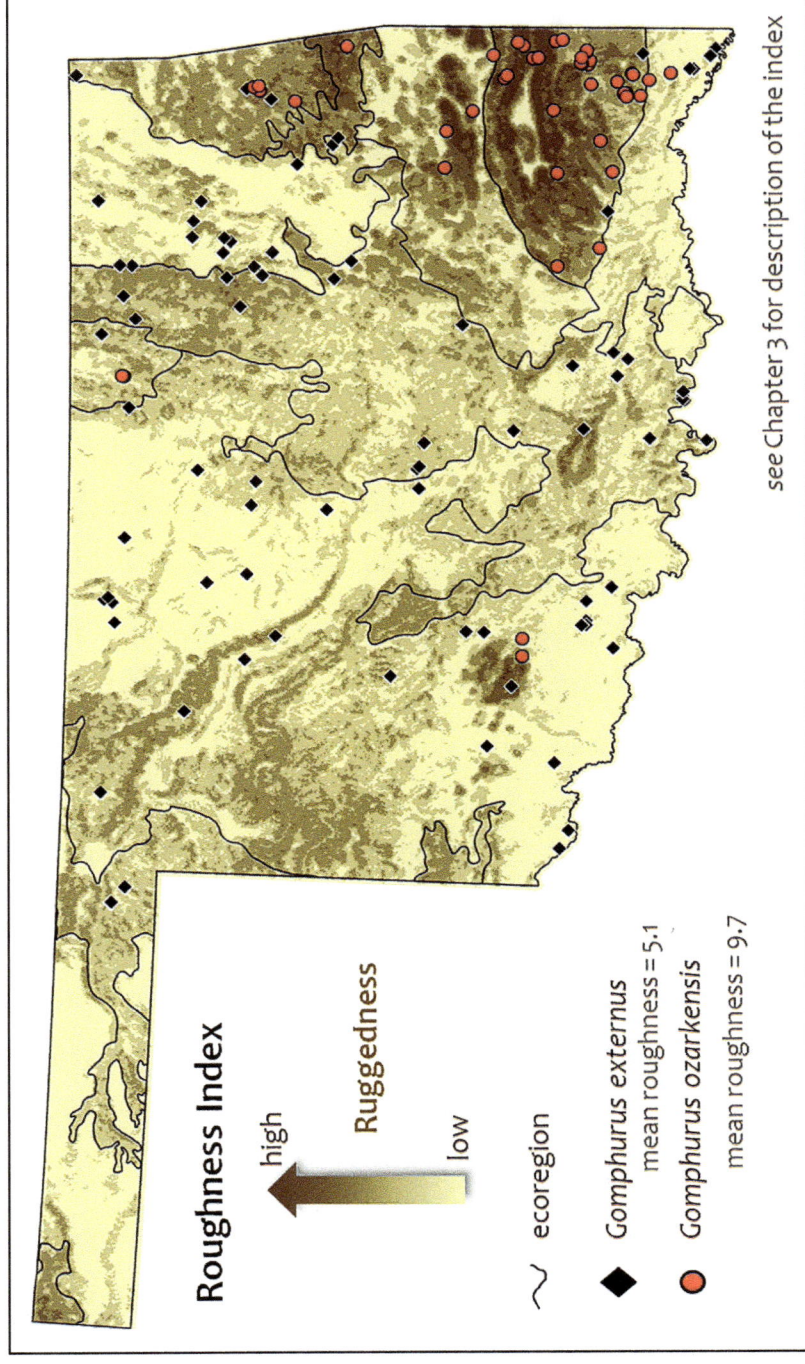

Figure 15.38 The geographic ranges of *Gomphurus externus* and *G. ozarkensis* barely overlap; topography may play a role in why. Elevation alone does not account for the division, but it may be that roughness, a measure of how quickly elevational changes occur in an area, contributes. In both cases, elevational range is approximately the same, but ruggedness differs markedly—*G. ozarkensis* occurs in areas twice as rugged as those where *G. externus* is found.

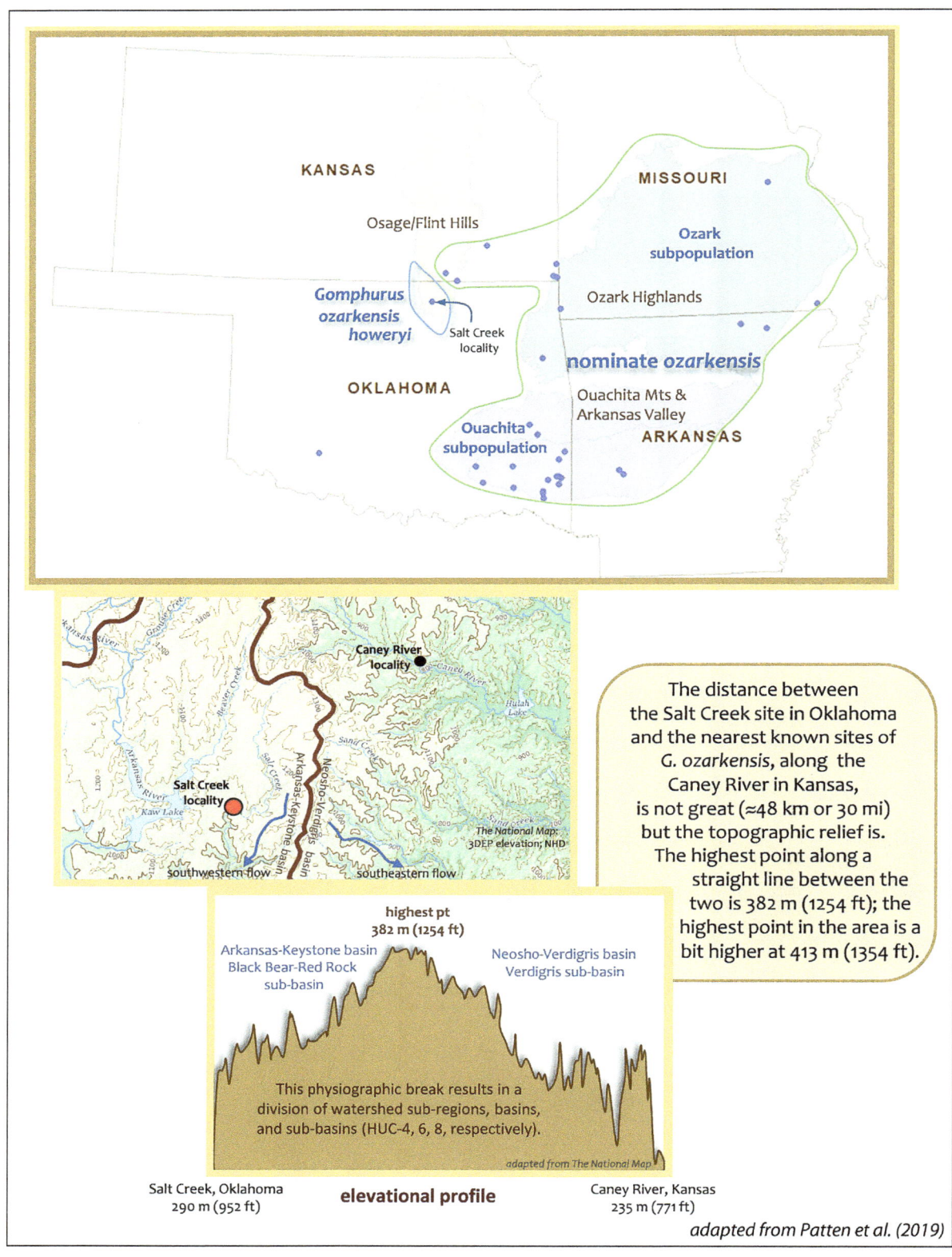

Figure 15.39 Geographical range of *Gomphurus ozarkensis*, showing the nominate and the *howeryi* subspecies' range. Adapted from Patten et al. (2019)

GOMPHURUS OZARKENSIS (WESTFALL, 1975) – OZARK CLUBTAIL

The Ozark Clubtail is a lotic species associated with clear, rocky creeks and rivers of moderate to large size and at modest elevation, in that the species is absent from low-elevation sites on the Gulf Coastal Plain or in valleys of the Arkansas and Illinois Rivers, where it is replaced by other species of *Gomphurus*. In our experience the species occasionally is found along narrow (<2 m) streams, too, and it even will occur around ponds embedded in piney-hardwood mixed forests that support breeding streams. It can be locally common in the Ouachita Mountains, especially in Le Flore and McCurtain Counties, where during peak season we have recorded the species as many as 35 adults at a single locale. It has also been reported from the Arkansas Ozarks in numbers we would characterize as locally common; for instance, Susanke and Harp's (1991) highest population estimate from their study of the species on the

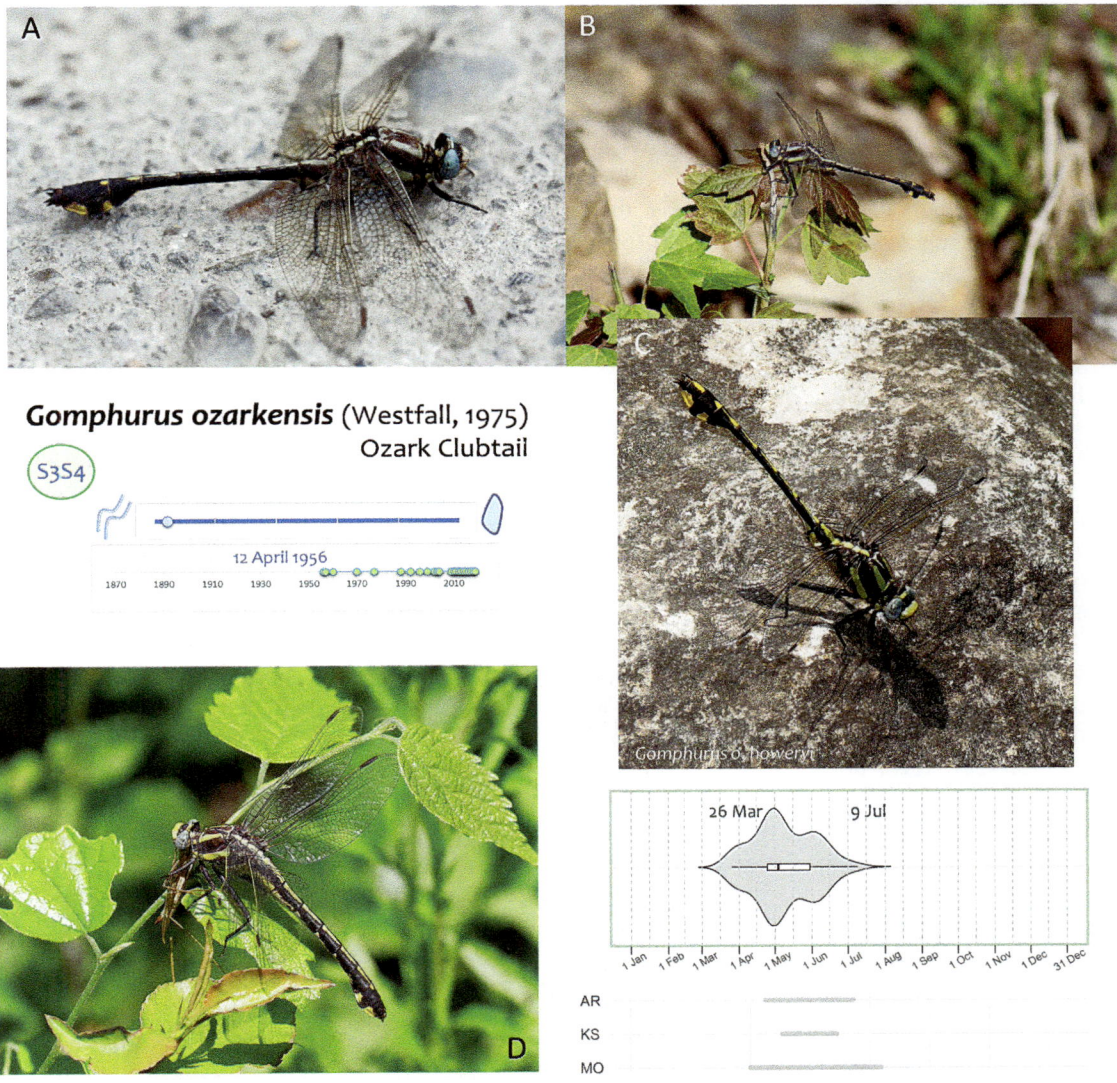

Figure 15.40 *Gomphurus ozarkensis*, the Ozark Clubtail, A) ♂, Mountain Fork River at The Narrows, McCurtain County, 17 June 2014 (©Terry Hibbitts, OC 486218). Note the mostly black dorsum of S9, a feature typical of individuals in the Ouachita Mountains. B) ♂, Mill Creek 3 km SSE of Quinton, Pittsburg County, 3 May 2014 (©Brenda D Smith, OC 422083), a ♂ with a spot of yellow on the dorsum of S9. C) ♂, Salt Creek 3 km NNE of Burbank, Osage County, 31 May 2011 (©Victor W Fazio III, OC 328214). Both males and females at this disjunct locale have the dorsum of S9 extensively yellow. D) ♀, Barren Fork Creek 3 km E of Park Hill, Cherokee County, 24 April 2012 (©James W Arterburn, OC 374596).

South Fork of the Spring River in eastern Fulton County was 63 individuals, also during what we now know of as peak season for the species.

In Oklahoma, the species has been reported from seven counties in the Ouachita Mountains region, but only from three counties (Adair, Cherokee, and Sequoyah) on the Ozark Plateau. Away from these areas, which constitute key highland regions in the core of the Ozark Clubtail's geographic range, the species has been recorded in the Wichita Mountains in southwestern Oklahoma—Boris Kondratieff and colleagues collected a pair on the Fort Sill MR, Comanche County, on 24 April 2002 (CSU; Zuellig et al. 2006), which we borrowed to confirm (♂, OC 381757; ♀, OC 381758; Figure 15.41). That pair's presence in the county would lead one to believe the species has a population there, but, mysteriously, the species has not been reported again from Comanche County even though many intensive surveys have been conducted there since Kondratieff's study. Another geographical outlier is on the western edge of the Osage Hills in western Osage County, but unlike Comanche County, this location appears to hold a population (*but see below*).

The species was first found in the state in McCurtain County by LE Hornuff in April 1956 (2♀, 1 and 1.5 mi SW of Hochatown, McCurtain County, 12[10] and 26 April, FSCA), two decades before it was described to science. GH Bick collected the species the following year, also in McCurtain County (♀, Little River at highway 70, 13 June 1957, FSCA). Hornuff then found the species in Pushmataha County in 1960 (♀, 3 mi W of Honobia, 9 July, FSCA). All of these specimens were originally identified as *Gomphurus externus* (Bick note cards). The misidentification is not surprising given that those specimens are females[11] and that *G. ozarkensis* had yet to be described. The latter reason explains why the ♂ Bick collected in 1970 (29 June, 8 mi E of Bethel, FSCA) went unidentified initially. It does not explain why it appears he did not re-identify those specimens or make a note card for *G. ozarkensis*. Nor does it explain why he did not report the species in his two updates for the state (Bick 1978, 1991). The only mention Bick (1983) made of the species for Oklahoma was a record from Pushmataha County

that was shown on a range map presented by Louton (1982), who did not specify what that record entailed (perhaps the 1♂ and 1♀ Louton collected 5.5 mi E of Cloudy on the Little River, on 27 May 1977 [KJT]).

More puzzling than the Bick era records is that AI Ortenburger, RD Bird, and AE Pritchard all failed to find the species in the state, despite extensive surveys in its range. Obviously, given that the species was yet to be described, they would have identified individuals as a species other than *Gomphurus ozarkensis*, likely *G. externus*, or left them undetermined to species. But we feel rather confident that we or others have re-examined extant specimens that were taken in the early and mid- eras (FSCA, OMNH, UMMZ, and USNM[12]) and have re-identified them, confirmed them as *G. externus*, or determined them not to be from locations likely to host both species,[13] given the extremely limited areas these two species overlap.[14] So why did those three eminent collectors not encounter *G. ozarkensis*?

The flight season in Oklahoma, lasting 105 days (from 26 March, see 2012 teneral record below, DRP, OC 374179; to 9 July, ♀, 5 km W of Honobia, Pushmataha County, LEH, FSCA), with a decided peak in April and May, exceeds that reported for Arkansas and Kansas but is similarly as long as Missouri's 109 days (AR: 23 April–5 July, OC; KS: 7 May–21 June, SEMC, CSU; MO: 10 April–28 July, Trial 2005). Of >85 records of the species for Oklahoma, there are four reports of tenerals, three from McCurtain County and one from Osage County. Three tenerals (2♂, 1♀; DRP) of a reported 20 were collected by BAH on 26 March 2012 at the Banner Clubtail (*Hylogomphus apomyius*) spot on the Glover River (see account above). At that same spot two years later we encountered ten *Gomphurus ozarkensis*, six of which were teneral (2♂, 4♀). And lastly for McCurtain County, we collected a ♀ teneral at "The Narrows" on the Mountain Fork of the Little River on 28 April 2013 (9 km SW of Smithville, SP 532; 2♂ adults also present). The Osage County record is of a ♂ we collected along Salt Creek where we encountered a total of four tenerals and no adults (4 km NNE of Burbank, on 11 May 2014, SP 1181). There are no confirmed nymphs of the species for the state.

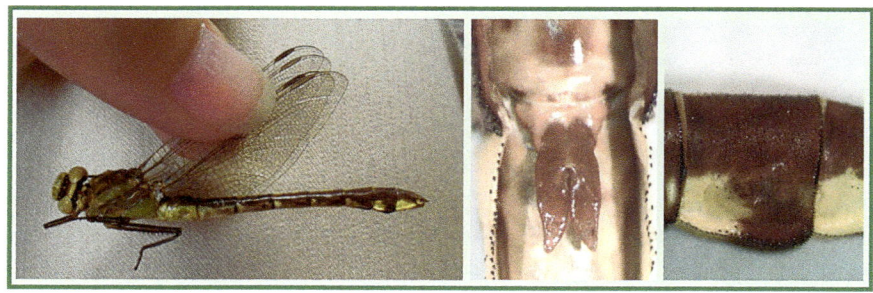

Figure 15.41 Westernmost record of *Gomphurus ozarkensis*, the Ozark Clubtail, within its entire range. This ♀, part of a pair, was collected on 24 April 2002 at Fort Sill Military Reserve, East Cache Creek, at South Boundary Road, Comanche County, Oklahoma. The specimen, along with its mate, is housed at the C. P. Gillette Museum of Arthropod Diversity, Colorado State University, Fort Collins, Colorado. This is the only record of the Ozark Clubtail for Comanche County.

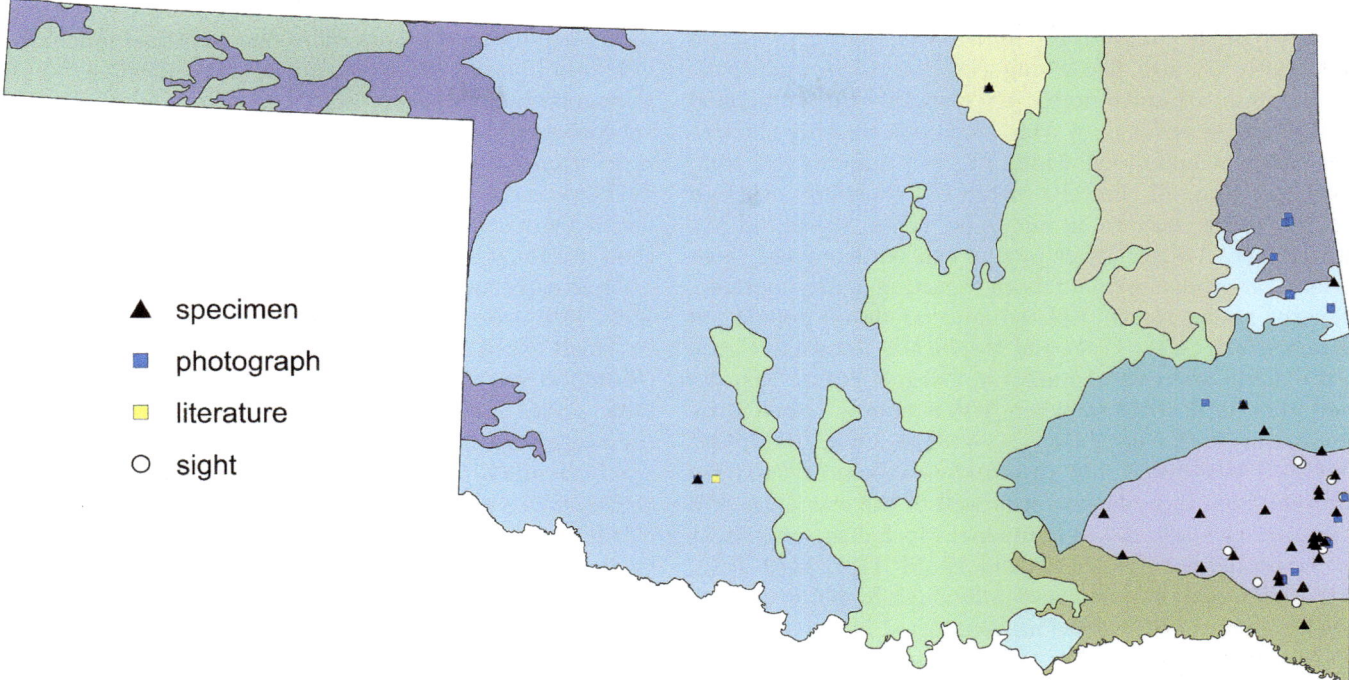

Figure 15.42 Documentation level of supporting records for *Gomphurus ozarkensis*, the Ozark Clubtail, in Oklahoma. Levels include specimen, archived photograph, literature reference, or sight record. Although sight records lack documentation, only those we feel are credible are plotted.

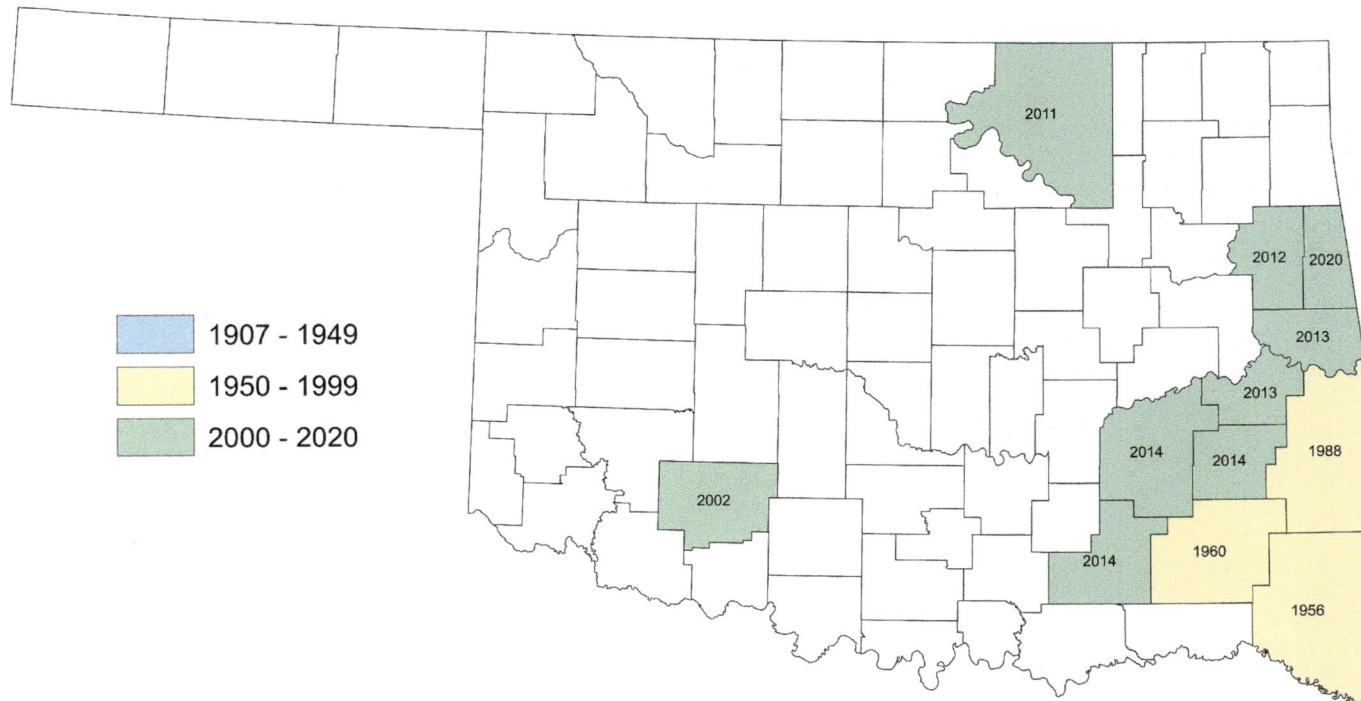

Figure 15.43 Counties in which *Gomphurus ozarkensis*, the Ozark Clubtail, are known to occur in Oklahoma. Year within the county is that in which the species was first reported. There are two disjunct locations, in Comanche and Osage Counties. The pair collected in Comanche County in 2002 is the only record of the species in the county. The Osage County population has proved to be a separate subspecies, *G. o. boweryi*.

We have noted much variation in this species, including both smaller and larger individuals than typical and those with atypical patterning (Figures 15.36 and 15.37. For example, we encountered a remarkably tiny individual on the Glover River on 4 May 2014; one roughly the size of a Banner Clubtail (SP 1168, TL: ~39 mm, ab: ~25 mm, HW: 26.2 mm; all genitalia match *G. ozarkensis*, although the abdominal pattern is not quite right; Figure 15.36). Whereas the Glover River pygmy can be dismissed as an oddball, another aspect of both variation and geography concerns an apparently isolated (and far-flung) population in the western Osage Hills, and specifically along Salt Creek to the north and east of Burbank, Osage County. Reports from this locale date to when VWF photographed various males 30–31 May 2011 (OC 328214, OC 328215, OC 329311). VWF noted that these individuals looked slightly different from those he had seen well to the southeast. We have since encountered 13 individuals, including tenerals, at the location; we collected 5♂ and 1♀ (SP 1181, 1210–1212, 1942–1943). Not only are these individuals larger (averaging *3–4 mm* longer) than previously reported by Needham et al. (2014) for the species but they are larger than elsewhere in the species' range (by 1.6 mm). They also consistently have the top of S9 extensively yellow as opposed to more typical black elsewhere (in southeastern counties, S9 typically is wholly black or has only a bit of yellow at its center, although a couple of individuals have been found that have S9 yellower; Figure 15.37). The ♂ paraprocts are near to those of typical *G. ozarkensis*, although they may splay a bit more, like those of *G. externus*. We quantified much of the variation in size and color pattern on the distal abdominal segments, and in doing so described this population as a new subspecies—Howery's Clubtail, *Gomphurus ozarkensis howeryi* Patten, Barnard, and Smith, 2019—a subspecies that, as far as we know, is confined to the Osage Hills of Oklahoma with a slight extension into the southwesternmost portion of the Flint Hills in Kansas (Figure 15.39).

In an early conservation assessment of the Ozark Clubtail, Bick (1983) considered it to be at risk, chiefly because of its small global range. As with the Oklahoma Clubtail (*Phanogomphus oklahomensis*), the Ozark Clubtail is confined to only four states, making it a regional endemic. Granted the range is small, consisting principally of the Ozark Plateau and Ouachita highlands of southeastern Kansas, southwestern Missouri, eastern Oklahoma, and western Arkansas, but it is found at many localities and when encountered, it can be seen a dozen at a time. Bick later (2003) concluded that the species was too common to be of compelling conservation concern, so he removed it from the national list of odonate species at risk. We agree with Bick's later assessment and we might be inclined to rank this species as S4 (= apparently secure) were it not for this its endemism and that *Gomphurus ozarkensis howeryi*, if ranked separately, would have S1 status (= critically imperiled). As such, we recommend the species remain one of some concern, as S3S4, in line with *Phanogomphus oklahomensis*.

GOMPHURUS HYBRIDUS (WILLIAMSON, 1902) – COCOA CLUBTAIL

The Cocoa Clubtail has been recorded about 20 times in the state, although all but 1 record come from only 3 sites in McCurtain County; the exception is a geographic outliner well to the west in Atoka County. The first time the species was reported for Oklahoma and most subsequent records have been from the Little River NWR. These records include the first (4 April 2007, 1♂, OC 281873), the early flight date (29 March 2012, 1♂, 9U, OC 374204), and the late flight date (6 June 2010, 1♀, OC 319632), all of which are BAH's records. Five of the state's six specimens are from the Little River NWR. Two specimens were taken on the west side of the refuge (17 April 2012, 1♂, DRP 24; 4 May 2014, 1♂, SP 1166), whereas three others were captured some 25 km away at Forked Lake on the east side of the refuge (3 April 2016, BS-P, 1♂ as SP 1848, 2♀ as SP1849, 1850; total of 4♂, 3♀, 3U at site that day). Elsewhere in McCurtain County it has been recorded once at Red Slough WMA (OC 318500), on 20 April 2010 by DA and once on the Glover River 10 km NE of Wright City, on 4 May 2014 (BS-P sight record).

The only record for *Gomphurus hybridus* outside of McCurtain County is a ♂ photographed and captured by BS-P in the Atoka County portion of TNC's Hottonia Bottoms Preserve (30 March 2017, OC 462591, SP 2241). This ♂ was the only individual of the species found despite surveying the preserve for most of the day. It was in low vegetation near a small stream (at most ½ m wide) in a wooded area <1 km from the Clear Boggy Creek. In our experience the species is most often found in swampy bottomlands near large creeks or medium-sized rivers (>15 m wide) that tend to be rather muddy.

The flight date range for Oklahoma of 29 March to 6 June is almost 70 days, which is the longest for the region; even so, almost all state records are from late March to early May. Texas and Louisiana are not far behind in length of the flight season (LA: 53 days, 14 March–6 May, OC 374098, 430676; TX: 59 days, 16 March–14 May, Abbott and Lasley 2019), but it is documented in Arkansas for a much shorter period (25 days, 7 April–2 May, OC374441, 374691). All records from Oklahoma are of adults, although the early

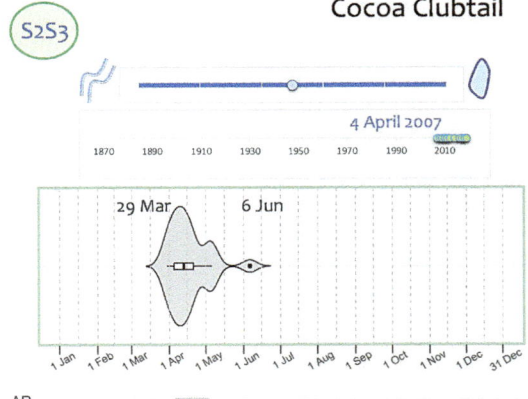

Gomphurus hybridus (Williamson, 1902)
Cocoa Clubtail

S2S3

4 April 2007

29 Mar   6 Jun

AR
LA
TX

Figure 15.44 A) Oklahoma's first state record of *Gomphurus hybridus*, the Cocoa Clubtail, ♂, Little River NWR, McCurtain County, Oklahoma, 4 April 2007 (©Berlin A Heck, OC 281873). This locale accounts for >95% of state records. B) ♀, Red Slough WMA, McCurtain County, 20 April 2010 (USDA Forest Service; David Arbour, OC 318500), the first of only two records of the species for this well-worked locale.

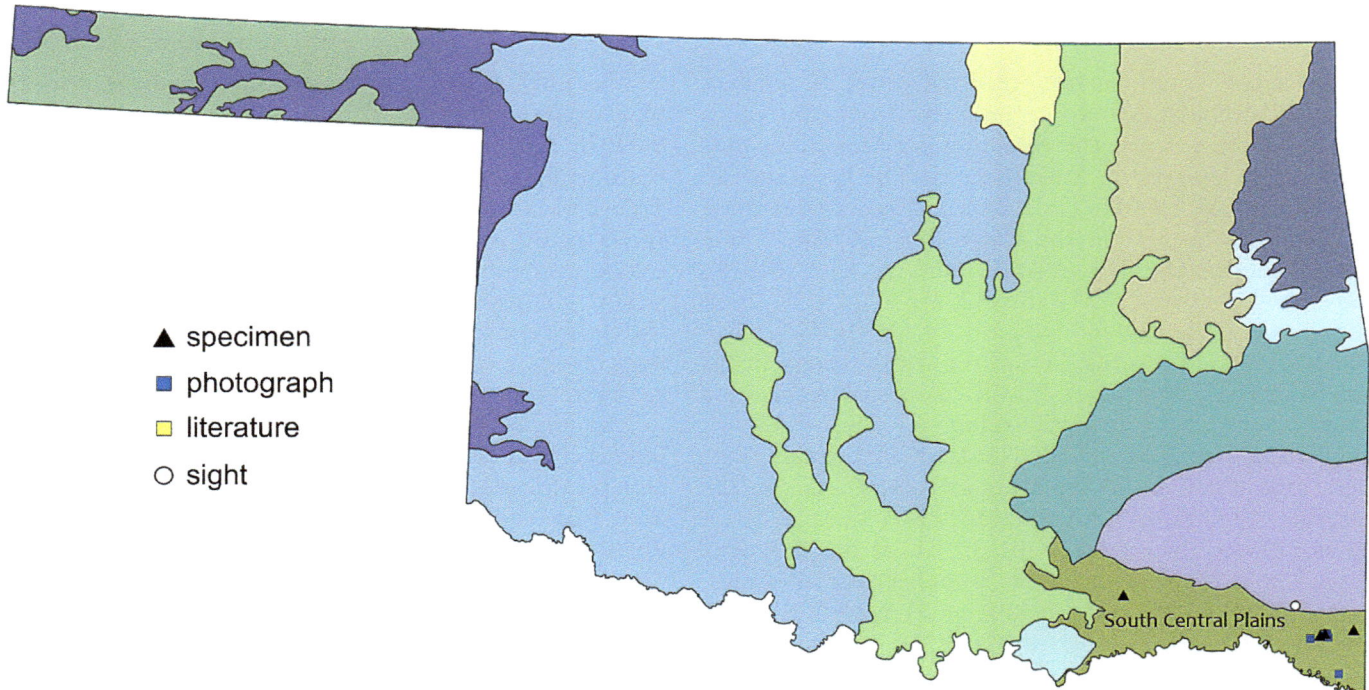

- ▲ specimen
- ■ photograph
- ☐ literature
- ○ sight

South Central Plains

Figure 15.45 Documentation level of supporting records for *Gomphurus hybridus*, the Cocoa Clubtail, in Oklahoma. Levels include specimens, archived photographs, and one sight record. Note that the species is essentially confined to the South Central Plains ecoregion.

flight date record mentioned tenerals may have been present. There are no records of nymphs, pairings, or ovipositing. The species typically has been encountered one or two at a time, but we have two records of five individuals and two of ten individuals (4♂, 1U, 6 May 2010, LRNWR, BAH and 5♂ [one in hand], 18 April 2018, LRNWR, BS-P; early date record, as above; Forked Lake record, as above).

Given the species' preference for muddy rivers in wooded bottomlands, a fairly common habitat in the southeastern corner of the state, we suspect that it has been overlooked in the South Central Plains ecoregion and will prove to be both more common and more widespread than it currently appears to be. But because Oklahoma is at the far edge of the species' range it likely will not be so common in the state as to warrant its removal from a list of species of conservation concern for Oklahoma. We recommend a refinement of our previous rank (Patten and Smith-Patten 2013b), from S2 to S2S3, to retain it as a species of conservation concern for the state, even though it appears generally not to be of conservation concern elsewhere.

GOMPHURUS EXTERNUS (HAGEN, 1858) – PLAINS CLUBTAIL

The Plains Clubtail lags behind the Sulphur-tipped Clubtail (*Phanogomphus militaris*) as Oklahoma's most widespread *Gomphus*—used here in the broad sense, including, in Oklahoma, the genera *Gomphurus*, *Phanogomphus*, and *Hylogomphus*, all of which were once considered just *Gomphus* (Figure 15.3; Appendix A). Admittedly, *Gomphurus externus* is not nearly as widespread (50 counties as opposed to 77), but relative to other *Gomphus sensu lato* species, it has a large range in the state. For example, other *Gomphus sensu lato* species, such as the Oklahoma (*P. oklahomensis*), Pronghorn (*P. graslinellus*), and Cobra (*Gomphurus vastus*) Clubtails have been found in 22 counties, for the former, and 19 counties for the latter two.

The species was first collected in Oklahoma by RD Bird in Murray County on Rock Creek on 18 April 1930 (3♂, 1♀, including a pair, OMNH 2297, 2301) and it has been recorded rather regularly since (but note that there has been confusion with this species and *Gomphurus ozarkensis*, which has made us skeptical about accepting records without supporting documentation[15]; see *G. ozarkensis* account for additional details). Bird (1932) reported *G. externus* from six counties: Alfalfa, Caddo, Cleveland, Jackson, Murray, and Pottawatomie. We have yet to determine the basis for the Pottawatomie record and it has not been reported from that county since, although there is no reason to think that it does not still occur there.

The Plains Clubtail, a lotic species that favors clear, shallow creeks and rivers with sandy or rocky substrate, tends to be uncommon, with but a few individuals found at a time. In Oklahoma it has an early flight date of 25 March and a late date of 26 July (25 March 2012, 1♀ teneral, Glover River, Highway 3, McCurtain County, JWA, photo on file, OOP; 26 July 1950, 1♀, OSU WCS, Muskogee County, GHB, FSCA). Oklahoma's seasonality is earlier than most but still comparable to neighboring states (AR: 28 April–17 to 19 May, OC 7646, Harp and Harp 2003; CO: limited records, 18 June–11 July, OC 367211, 367437; KS: 7 April–2 August,[16] SEMC; LA: limited records, 18–25 April, OC 421890, 430568; MO: 5 May–15 August, Sims 2012; NM: March–14 August, Paulson 2009, OC 315179; TX: 7 February–14 July, Abbott and Lasley 2019). We have no reports of nymphs for the state, but there are six records of tenerals, which date from 25 March to 31 May. The species exhibits a fair amount of morphological variation and body size in the region is both smaller and larger than previously reported (Figures 15.37 and 15.49).

Figure 15.46 *Gomphurus externus*, the Plains Clubtail, A) ♂, Bicentennial SP, Ottawa County, 8 June 2012 (©James W Arterburn, OC 375765). B) ♀, Redbud Valley Nature Preserve, Rogers County, 11 May 2016 (©Bill Carrell, OC 444355). Of the various species formerly lumped into the genus *Gomphus*, this lotic species is more widespread in the southern Great Plains than any but *Phanogomphus militaris*, the Sulphur-tipped Clubtail, even if it is never particularly common anywhere.

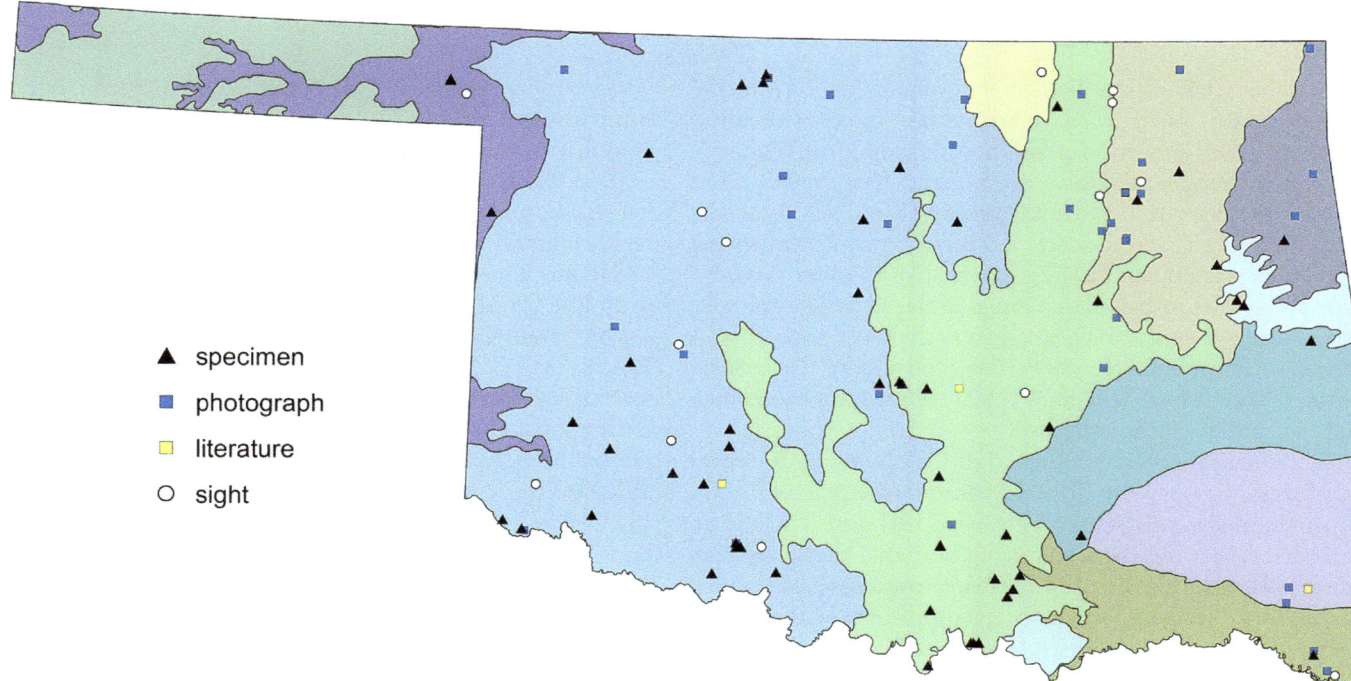

Figure 15.47 Documentation level of supporting records for *Gomphurus externus*, the Plains Clubtail, in Oklahoma. Levels include specimen, archived photograph, literature reference, or sight record. Although sight records lack documentation, only those we feel are credible are plotted.

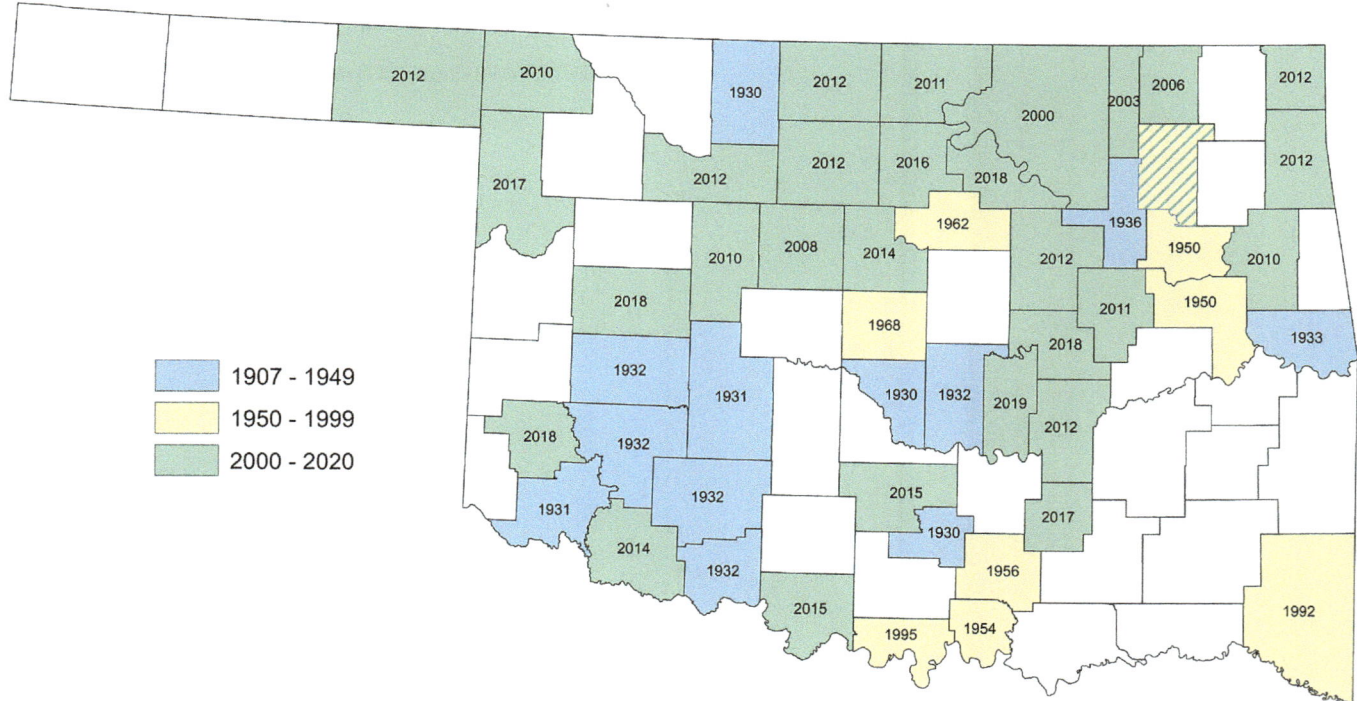

Figure 15.48 Counties in which *Gomphurus externus*, the Plains Clubtail, are known to occur in Oklahoma. Year within the county is that in which the species was first reported. Note that because there is some doubt about a 1956 record of this species for Rogers County, the county is labeled here as either from the mid-era or the current era (the second record for the county was not found until 2012). Also of note is that early on it was thought that the first records for the species in McCurtain and Pushmataha Counties were found in 1956 and 1960, respectively; however, re-identifications of those records proved that they were actually *G. ozarkensis*, the Ozark Clubtail.

13 km NNE of Crescent, Otter Creek, Logan County,
Oklahoma, SP1219

Lake Evans Chambers, Beaver County,
Oklahoma, SP1330

Figure 15.49 Individual variation in *Gomphurus externus*, the Plains Clubtail.

GOMPHURUS VASTUS (WALSH, 1862) – COBRA CLUBTAIL

For the longest time we felt that we must have had blinders on for this species, so much so that we missed out on a 2005 record (BC, OC 334214) at our old haunt of Mohawk Park, Tulsa County, as well as a 2006 record (KW, OC 7520) for Nowata County on Cedar Creek, 5 mi E of Lenapah (Smith-Patten et al. 2007). After we moved from the northeastern corner of the state to the center of the state, it appeared that our mistake was being rubbed in our faces because it seemed like everyone interested in odonates in Oklahoma saw the species except us. Our bruised egos were assuaged slightly by the knowledge that there were only three records of the species prior to our arrival in the state in 2003, and that the chief odonatologists for the state had about as much luck with the species as we had prior to 2013. GH Bick saw the species only once: a single ♂ 5 mi S of Bokchito, Bryan County, on 21 July 1954 (FSCA; the first record for the state). Interestingly, Bick never published that record (or the record that follows). His notes indicated that he received confirmation of his determination of this individual in 1966 from Leonora Gloyd and from Minter J Westfall in 1972 (Figure 15.51), yet the species did not appear in Bick and Bick (1957), prior to confirmation of the determination, or in Bick (1978, 1991), afterwards. LE Hornuff (with McDougal) also had the species only once, a ♂ and ♀ at Armstrong, Bryan County (9 July 1960, FSCA, JCAC 22768). Ortenburger, Bird, and Pritchard all missed it in the state. Currently, the Cobra Clubtail is documented in 17 counties in the eastern third of the state (we added it to 5 … finally breaking our jinx).

The early flight date for the species in Oklahoma is based on a sight record for Tulsa County at Fry Creek in

Figure 15.50 *Gomphurus vastus*, the Cobra Clubtail, A) ♂, Kaw Lake, Kay County, 15 July 2011 (©Victor W Fazio III, OC 330128). B) ♂, Muddy Boggy Creek 5 km ESE of Lehigh, Coal County, 25 June 2017 (©Michael A Patten, OC 465024), showing off the species' distinctive green eyes. C) ♀, Redbud Valley Nature Preserve, Rogers County, 1 July 2008 (©Ken Williams, OC 315346). In the southern Great Plains, this species is found exclusively on rivers and large creeks with emergent rocks for perches. Riprap at spillways of various major dams can provide ideal habitat.

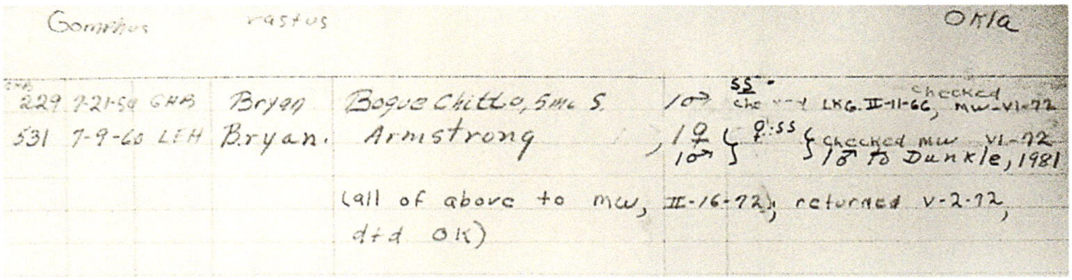

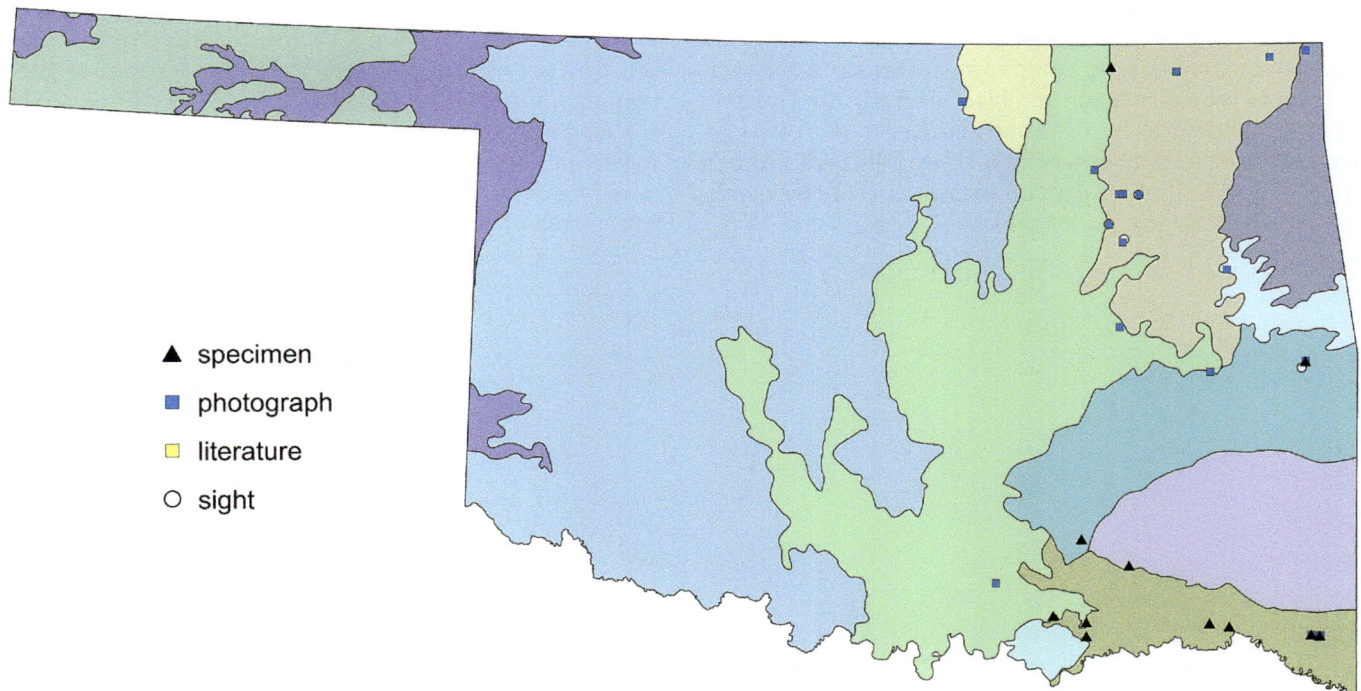

Figure 15.51 George H Bick's (GHB) note card for *Gomphurus vastus* (formerly *Gomphus vastus*), the Cobra Clubtail. In 1954, he collected the first known record of the species for Oklahoma, but mysteriously he never published it, nor did he publish Lothar E Hornuff's (LEH) record from 1960.

south Tulsa on 31 May 2012 (2U, JWA). The species' late date is a record on 28 July 2010 in Cherokee County along the Neosho River below the Fort Gibson Lake dam (1♂, JWA, OC 329039). The known phenology for the species in Oklahoma fits easily within the region's flight window of late March to mid-August (AR: 17 May–14 July, OC 375019, 448654; KS: 4 June–1August, OC 321042, SEMC; LA: 29 March–16 July, OC 461658, 320817; MO: 16 May–14 August, Trial 2005, TX: 1 April–29 July, OC 461926, Abbott and Lasley 2019). There is one report of a teneral, a female that was part of the highest count for the state: 14♂, 8♀, on 22 June 2014, along the Arkansas River below the Robert S Kerr Lake dam in Le Flore and Sequoyah Counties (river is the county line; 1 adult ♀ as SP1298 from Le Flore County;

OC 423379, 423380, one photo for each county). Other than that record, the species has been reported typically only 1–2 at a time.

In Oklahoma the species has been found in various habitats, including a farm pond in Nowata County, a gravel bar in the often muddy Little River in McCurtain County, a clear-water, moderate flow, bedrock creek below Raymond Gary Lake in Choctaw County, and a swift, clear-water, rocky creek with regular riffles in Atoka County. We most often encounter the species alighting directly on bedrock in more shallow creeks, or when waters are deeper and often fast-flowing, perched on riprap-type boulders or large rocks at or near the shoreline.

▲ specimen
■ photograph
□ literature
○ sight

Figure 15.52 Documentation level of supporting records for *Gomphurus vastus*, the Cobra Clubtail, in Oklahoma. Levels include specimen, archived photograph, literature reference, or sight record. Although sight records lack documentation, only those we feel are credible are plotted.

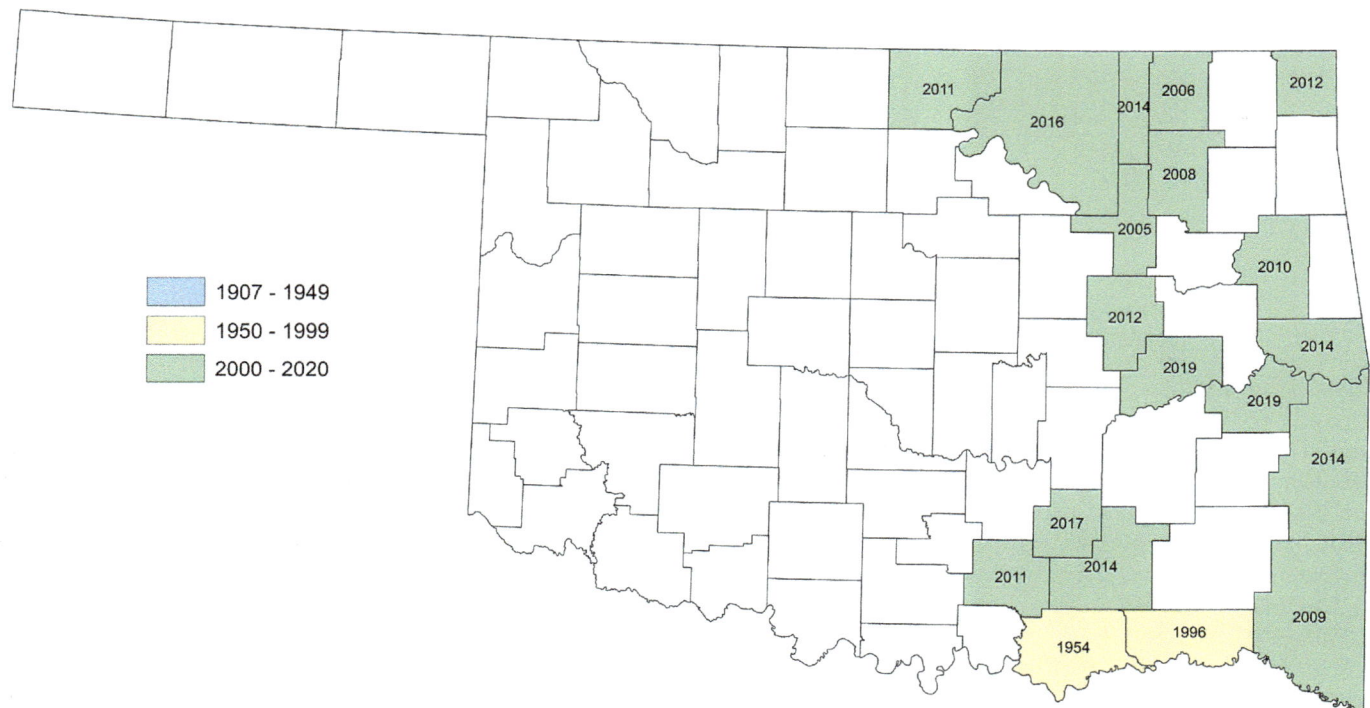

Figure 15.53 Counties in which *Gomphurus vastus*, the Cobra Clubtail, are known to occur in Oklahoma. Year within the county is that in which the species was first reported.

Like elsewhere, *Gomphurus vastus* exhibits considerable variation in its abdominal pattern in Oklahoma. The pale spot laterally on S8 can be anywhere from a barely perceptible dot to a square of yellow covering up to about a third of the anterior portion of the segment. The terminal appendages vary in base color from brown to black or in some cases brown laterally and black dorsally. In Texas the species is known to be browner and larger than its eastern and northern counterparts (Westfall 1974; Abbott 2015). We have limited data on overall size of the species in Oklahoma, but variation in base color on the terminal appendages suggests that Oklahoma lies in a transition zone between the more "typical" black form and the Texas brown form.

Behaviorally, JWA and KW documented that the species can take prey as large or larger as it is—a series of photos they took on 11 July 2011 at Kaw Lake, Kay County, show a ♀ *G. vastus* eating a ♂ *Dromogomphus spoliatus*, the Flag-tailed Spinyleg (*G. vastus*, 47–57 mm and *D. spoliatus*, 56–65 mm, Needham et al. 2014; Figure 15.5).

STYLURUS PLAGIATUS (SÉLYS, 1854) – RUSSET-TIPPED CLUBTAIL

The Russet-tipped Clubtail was long thought to be the only hanging clubtail in Oklahoma. The discovery of a 1932 record of a ♀ *Stylurus intricatus* (Brimstone Clubtail; Smith-Patten and Patten 2012) changed this notion. Still, the *Stylurus* expected regularly in the state is the Russet-tipped Clubtail, which is widespread albeit generally uncommon, although it becomes locally fairly common in late summer and autumn. The species typically is seen <5 at a time but it has occasionally been reported >10 individuals, including our highest count of 30 individuals on 9 August 2009 at Jim Hall Lake, Okmulgee County (1♂ as SP 76, 1♀ as SP 78). The encounter rate and number of individuals seen varies annually, but we cannot detect a specific pattern: encounters between 2009 and 2017 were as low as a handful to upwards of a dozen and number of individuals recorded within a given year varied as widely (6–48; Figure 15.55).

The species was first recorded in Oklahoma by F Collins between 2–6 August 1907 near Wister, Le Flore County (2♂, 2♀, Williamson 1914b). It was next recorded on 22 June 1932 by RD Bird in Kiowa County (5 mi W of Roosevelt, Elk Creek, 3♂, 2♀, including 1 pair, OMNH 2446, 2450, 2451, 2460), even if Bird (1932) reported it from only Le Flore County. From the Kiowa County record until 1941 it was added to an additional five counties— Cotton, Grady, Harper, Okmulgee, and Payne (notice the shocking absence of Comanche County from the list). The OU expeditions found the species in the county just to the west, but not in the magical place that is the Wichita Mountains area in Comanche County. Even though the species tends to be most numerous in the autumn, if the OU team found it elsewhere in the state during summer surveys, then they should have been able to find it in the Wichitas, *if it was there*. The dozen records between 1954 and 1999 added another 5 counties to the mix, but 2003–2019 saw a further 33,[17] bringing the overall total to 45 counties (Figure 15.57).

Figure 15.54 *Stylurus plagiatus*, the Russet-tipped Clubtail, A) ♂, Shawnee Twin Lakes, Pottawatomie County, 3 October 2006 (©John F Fisher, OC 7420). B) ♀, Ft. Gibson Dam, Cherokee County, 6 October 2017 (©James W Arterburn). No other "hanging clubtail" (genus *Stylurus*)—so named for their strong inclination to perch on vegetation (as seen above)—occurs in Oklahoma regularly. Moreover, unlike other species in the family, this one's flight season is shifted markedly to later in the year, such that its abundance peaks in late summer and autumn rather than in spring or early summer.

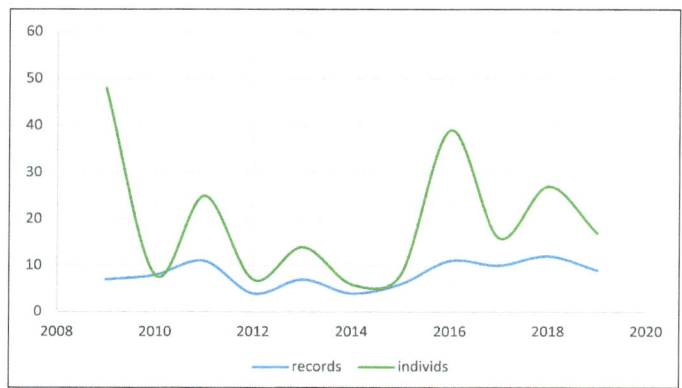

Figure 15.55 Annual variation in encounters and number of individuals reported of *Stylurus plagiatus*.

Since 2003 there has been not only a rather large increase in the number of counties in which *Stylurus plagiatus* has been recorded but there has also been a large increase in the number of records (~80%) in the state. We suspect that the increase in record numbers in the recent era is the result of a concomitant increase in surveys conducted in the autumn. Where it becomes trickier is to determine if the large increase in counties indicates a possible range expansion. Our suspicion is that the county total increased partly because of the increase in autumn surveys and partly because of increased habitat generalization. Paulson (2009:281) described the species' habitat as "slow-flowing rivers and streams down to fairly narrow ones." In his eastern guide he added "Common at some lakes in peninsular Florida." (Paulson 2011:282). Abbott (2015:167) described its habitat in Texas as "weedy rivers, streams, and lakes with moderate to little current." In Oklahoma, the species inhabits streams and lakes about equally. For those not accustomed to seeing the species patrolling over open water, it can be discombobulating, and we wonder if the species is occasionally overlooked when patrolling over open water, possibly attributed to a species more expected at a lake such as *Phanogomphus militaris*, the Sulphur-tipped Clubtail. It may be that when it expanded its range, *Stylurus plagiatus* came into the state via stream courses but has broadened its niche breadth to incorporate lentic habitat as well, making it not so much of a range expansion as a niche expansion.

The species tends to fly late in the season, especially relative to many other clubtails in the state. It is encountered most often from mid-July to mid-September yet has been recorded as early as 9 June[18] and as late as 16 October, a flight season that corresponds regionally (OK: 9 June 1934, 1U, Okmulgee County, CA Sooter, OSU and 16 October 2016, 4♂, 1♀, Hickory Creek WMA, Hickory Creek, Love County, MAP, OC 457237; AR: 17 June–7 September, OC 7647, 333384, JCAC 18969; KS: 15 July–2 August, SEMC; LA: 14 June–19 October, Mauffray 2014; MO: 15 July–18 September, Trial 2005; NM: June–4 November, Paulson 2009, OC 492131; TX: 1 April–24 November, Abbott and Lasley 2019, OC 492339). We know of no records of nymphs

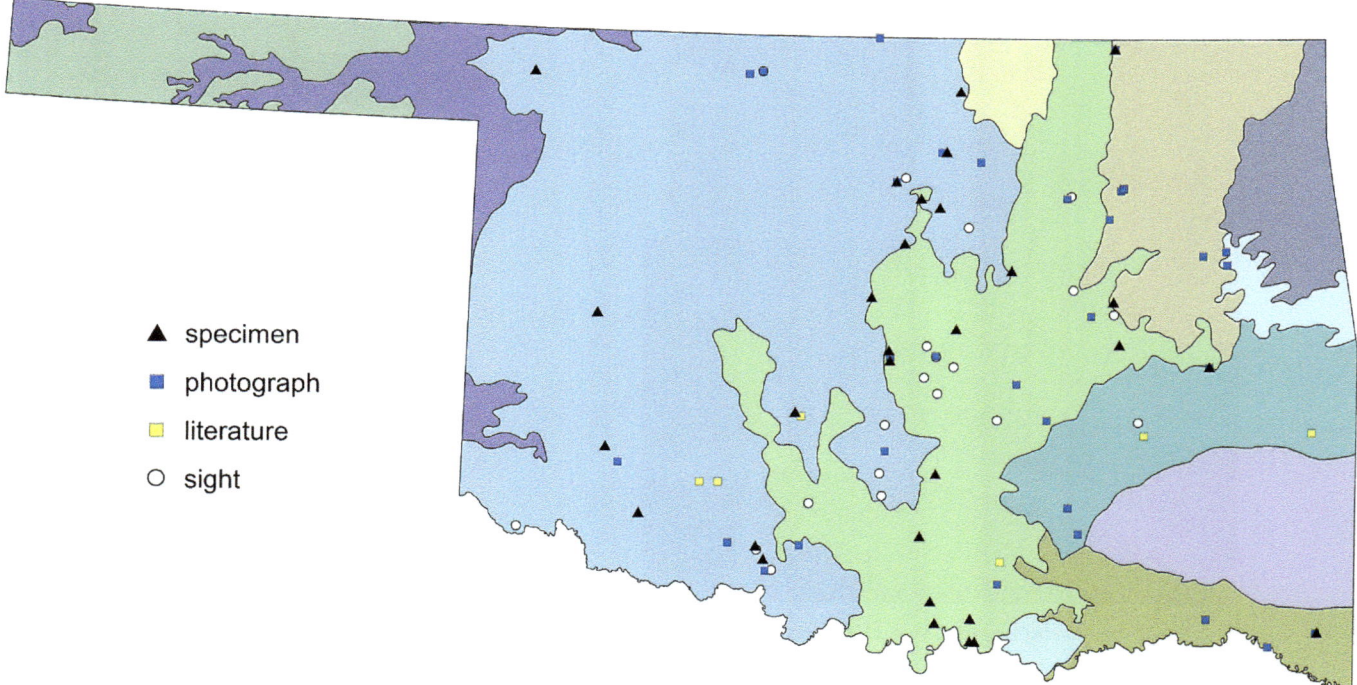

Figure 15.56 Documentation level of supporting records for *Stylurus plagiatus*, the Russet-tipped Clubtail, in Oklahoma. Levels include specimen, archived photograph, literature reference, or sight record. Although sight records lack documentation, only those we feel are credible are plotted.

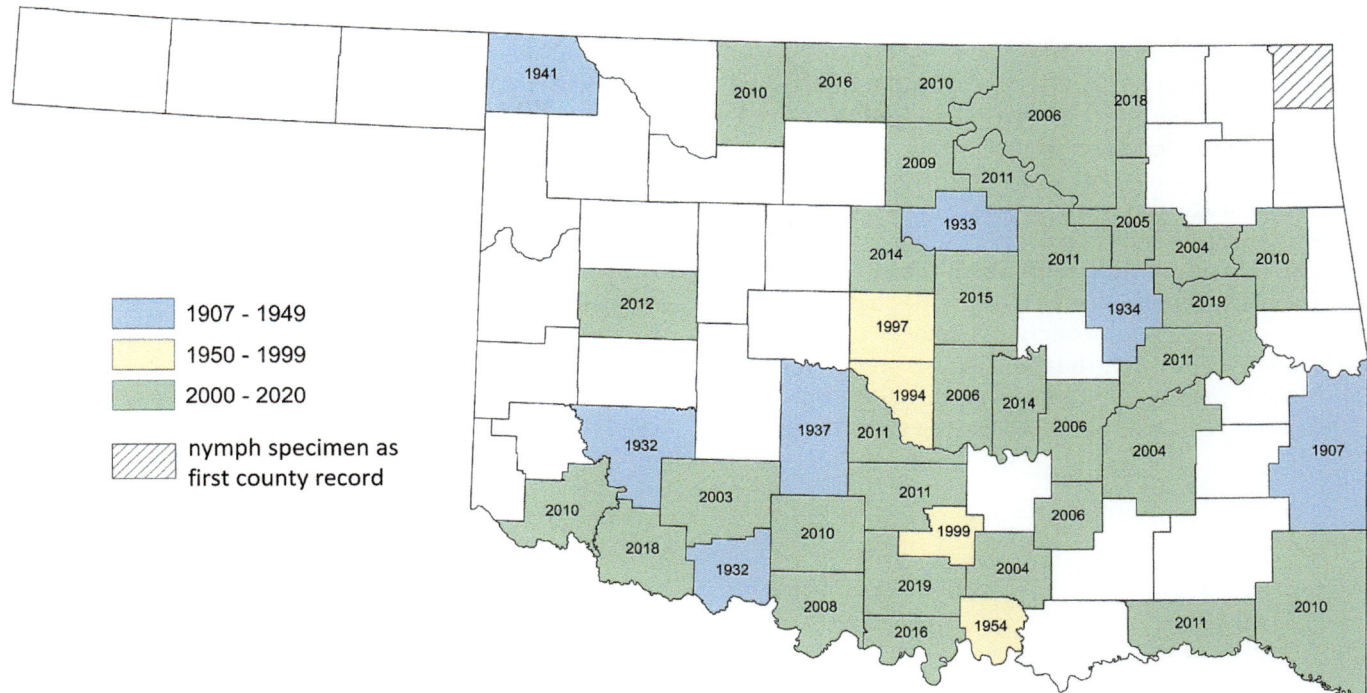

Figure 15.57 Counties in which *Stylurus plagiatus*, the Russet-tipped Clubtail, are known to occur in Oklahoma. Year within the county is that in which the species was first reported. We have not re-examined the two nymph specimens constituting the Ottawa County record. These nymphs were collected 4 mi N of Peoria, Newman Cemetery, Little Fivemile Creek, on 8 June 1982, by Gelhaus, Huggins, and Liechti (SEMC 1141957, 1141961) and are identified to the genus *Stylurus*. Our assumption has been that the genus was correctly determined and as such, the species would likely be *S. plagiatus*. However, it would behoove future researchers to confirm the determinations.

TL = 52.9 mm, ab = 39.2 mm, HW = 31.8 mm

Needham et al (2014): TL = 57–66 mm, ab = 40–50 mm, HW = 30–40 mm

Individuals are at same scale

Figure 15.58 The ♂ *Stylurus plagiatus*, the Russet-tipped Clubtail, on the left was collected along the shore of Kaw Lake, Kay County on 8 October 2010 (SP 209). Note the dark thorax, which obscures the lateral thoracic stripes, and the red on the thorax near the wing bases; all of which are abnormal for this species. This individual, too, is small for a *S. plagiatus* and its secondary genitalia are somewhat small and stubby, which we surmise is attributable to a developmental issue. The ♂ on the right is more typical of the species (Longmire Lake, Garvin County, Oklahoma, 30 September 2013, SP 1062).

for the state outside of the one nymph that GH Bick collected and reared (1♂, 19 June 1954, unsure if date was of collection or emergence, OU Biological Station, Marshall County, coll. from rocks at boat house, FSCA). There are four records of tenerals, from Cotton, Creek, Marshall, and Okmulgee Counties, that date between 4 July and 7 September. Likewise, there are only three observations of ovipositing, all by lone females.

We have noticed little variation in the phenotype of this species in Oklahoma, although it may be that the few individuals found in southwestern Oklahoma tend to have a little more yellow (and, thus, less russet) on the outer parts of the club (*pers obs*). Needham et al. (2014) and Paulson (2009, 2011) reported the species having a total length (TL) of 57–66 mm and a hindwing (HW) of 30–40 mm, whereas Abbott (2015) reported smaller individuals (53–66 mm) and Bailowitz et al. (2015) reported still smaller ones (50–55 mm). We also report small individuals, down to about 53 mm TL, but we see no geographic pattern to their distribution.[19] The most diminutive individual is also of note because it has a dark thorax (laterally) that obscures the thoracic stripes and there are small red patches showing near where the wing bases contact the thorax (Figure 15.58); to our knowledge, this aberrancy has not been reported previously. The hamules of this ♂ are not an exact match for the species but are likely within the range of individual variation. This ♂ also exhibited unusual behavior in that it was perched on the ground.

STYLURUS INTRICATUS (HAGEN, 1858) – BRIMSTONE CLUBTAIL

The Brimstone Clubtail is a lotic species of numerous disjunct populations throughout the southwestern United States and into the central U.S. and Canada (Paulson 2009). Considering one population is centered on Nebraska with a southward extension into northern and western Kansas, it is not surprising that AE Pritchard discovered one in Dunlap, Harper County, between 24 and 26 August 1932 (♀, OMNH 2413, OC 381939; Figures 15.59 and 15.60). But, the species remained unreported until BS-P discovered the fluid specimen in the OMNH collection (Smith-Patten and Patten 2012). Curiously, the specimen was labeled with the correct identification (*Gomphus intricatus*, as it was called then), but it was reported neither by Pritchard nor RD Bird, and it appears that GH Bick did not find the specimen when he perused the collection in his day. For years, we thought that Pritchard and Bird were not convinced of the identification given that the primary reference at the time (Needham and Heywood 1929; Figure 15.60) made it seem as if the subgenital plate of *Stylurus intricatus* could be confused with that of *S. olivaceus*, the Olive Clubtail. We now know that the Olive Clubtail does not occur anywhere near Oklahoma, but its range was not so clearly defined in the early 1930s. All of this sounded so reasonable. Some of it still is, but in 2018 additional details surfaced when BS-P received some of Pritchard's letters from 1932, courtesy of the Bentley Historical Library at the University of Michigan. It turns out that Pritchard was in no uncertain terms confident of the identification, so much so that as a 17-year-old kid he was bold enough to write a letter (dated 30 November 1932; Figure 15.60) to the esteemed

EB Williamson to proclaim that he had recently added "*G. intricatus*" to the Oklahoma state list. Williamson apparently was so taken with Pritchard's certitude that he mistook him for faculty at OSU (at the time Oklahoma Agricultural and Mechanical College). Subsequent letters hint that Williamson examined this specimen but there is no clear indication that he agreed with the id (the response letter is missing from the correspondence string). Perhaps something Williamson said in the missing letter discouraged Pritchard from pursuing a public announcement of the state record. Regrettably, why Pritchard's conviction did not translate into something beyond a handwritten specimen label remains a head-scratcher.

A much more frustrating part of the saga of *Stylurus intricatus* in Oklahoma is that despite many surveys in Harper and surrounding counties we have yet to have a chance meeting with this little bugger. Much like that missing Williamson letter, the status of this species in Oklahoma continues to escape us. But we hope that one day we will discover whether the Brimstone Clubtail is just a vagrant to Oklahoma or if there is a small population at the base of the panhandle.

Look for this species throughout the summer in northwestern Oklahoma, as it is known to fly from June to October throughout the region (CO: 7 July–24 August, UMMZ, CSU; MO: three records, 26 June–15 October, Trial 2005; NE: 17 June–11 September, FSCA, USNM; NM: 9 June–16 October, OC, iNat 20347475; TX: 8 June–5 October, Abbott and Lasley 2019). A report for Arkansas is in error (Harp and Rickett 1977, Harp 1983).

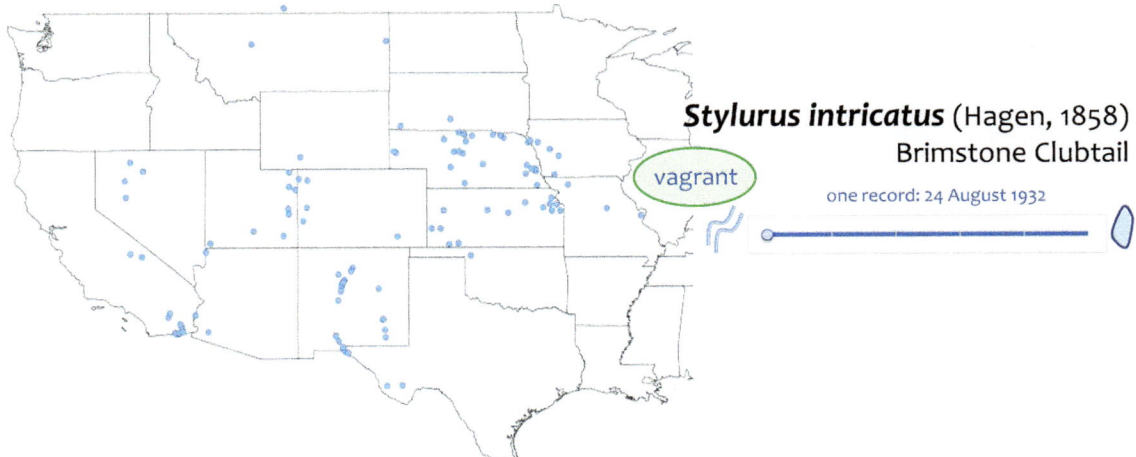

Figure 15.59 Oklahoma's only record of *Stylurus intricatus*, the Brimstone Clubtail, is a specimen collected in Harper County on 24 August 1932 (OMNH 2413). Look for the species to appear again anywhere in the panhandle.

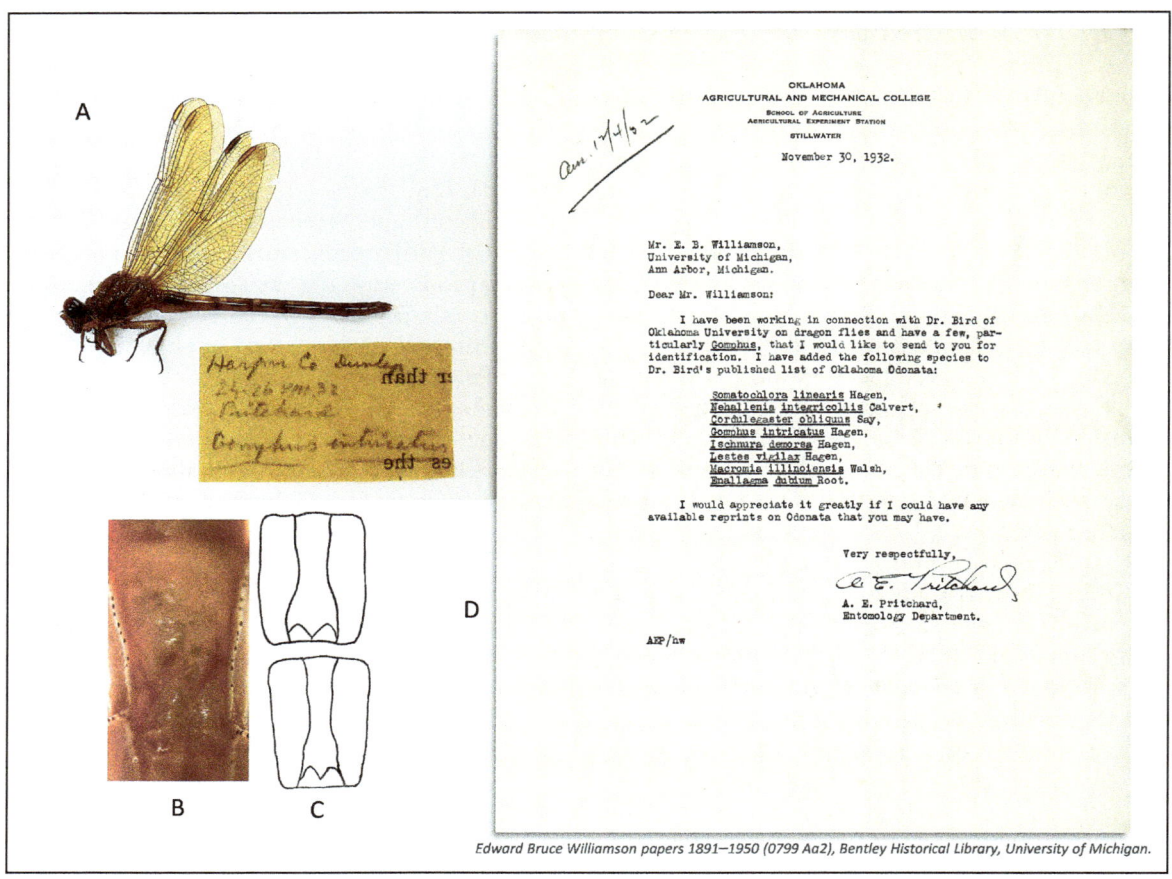

Edward Bruce Williamson papers 1891–1950 (0799 Aa2), Bentley Historical Library, University of Michigan.

Figure 15.60 Oklahoma's first and only record of *Stylurus intricatus*, the Brimstone Clubtail. This ♀, found in the fluid collection housed at the Sam Noble Oklahoma Museum of Natural History (OMNH 2413), is shown here with the original collecting label (A). At the time of her capture the only available reference that RD Bird and AE Pritchard would have had was Needham and Heywood (1929) to compare her subgenital plate (B) to illustrations of possible clubtail species (C, top is *S. intricatus*, bottom is *S. olivaceus*, the Olive Clubtail). The quality of these illustrations apparently did not dissuade Pritchard from claiming this to be a Brimstone Clubtail (D, letter to EB Williamson, 30 November 1932, Courtesy: Bentley Historical Library); nonetheless, he did not publish the record.

DROMOGOMPHUS SPINOSUS SÉLYS, 1854 – BLACK-SHOULDERED SPINYLEG

The Black-shouldered Spinyleg is uncommon to fairly common in the eastern third of the state, with scattered records farther west, including in Oklahoma's western outpost for various eastern species, the Wichita Mountains and the nearby Arbuckle Mountains. It is less common than—and not nearly as widespread as—its congener in the state, *Dromogomphus spoliatus*, the Flag-tailed Spinyleg: about 275 records in 37 counties versus about 475 in 71, respectively, and with contrasting high counts of 10 versus 35 (Figures 15.62 and 15.63). That said, the Black-shouldered Spinyleg often is the more numerous and frequently encountered in eastern woodland and forest, where it prefers lotic habitats (Bick and Bick 1957) typified by clear, shallow creeks with rocky substrate. Moreover, the species does not appear to be rare enough or have any pressing threats that warrant conservation concern in the state at this time. Clearing of forests and streamflow loss are the biggest concerns for this species.

Dromogomphus spinosus was first recorded in the state on 3–4 June 1907 by EB Williamson near Wister, Le Flore County (1♂, 1♀, both tenerals, UMMZ, Williamson 1914b). It has been reported regularly since. Bird (1932) reported the species from three counties: Haskell, Latimer, and Le Flore. It is unclear why he did not report it from Pushmataha County given his personal collection of the species there on 18 June 1931 (1♂, 7 mi NW of Tuskahoma, Buffalo Creek, OMNH 2173).

It is usually seen 1–5 individuals at a time, but there are a good number of records with higher counts up to 10 adults. There are three reports of ovipositioning, all of lone females. The species is known to fly as early as 8 May to as late as 19 September (8 May 2010, 1U, Cedar Lake, Le Flore County; 19 September 2009, 1U, Robbers Cave SP, Latimer County, BAH, OC 315571), which is an almost perfect match to the region (AR: 1 May–13 September, OC 374800, Harp and Harp 2003; KS: 26 May–4 August, SEMC 1349143, 1349145; LA: 23 May–11 September, OC 445405, Mauffray 2014; MO: 29 May–15 September, Trial 2005; TX: 8 May–23 August, Abbott and Lasley 2019, OC 502264).

Of note is the variation in the "black shoulder" of this species. Both sexes can have solid black at the "shoulder" (i.e., fused

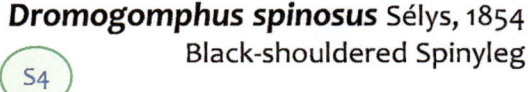

Dromogomphus spinosus Sélys, 1854
Black-shouldered Spinyleg

S4

3–4 June 1907

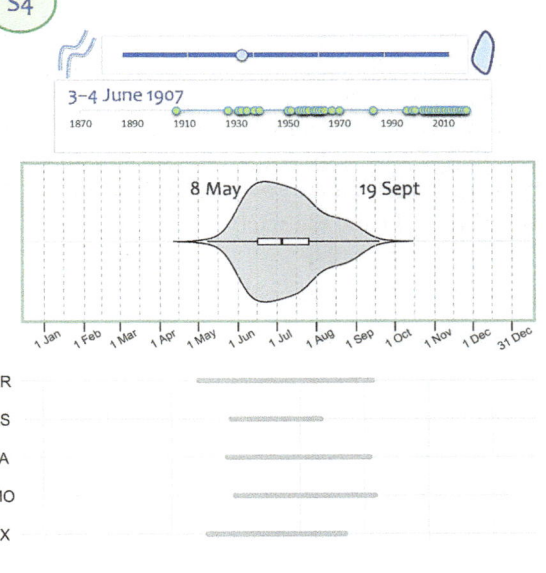

Figure 15.61 *Dromogomphus spinosus*, the Black-shouldered Spinyleg, A) ♂, Sardis Lake, Pushmataha County, 26 June 2013 (©James W Arterburn, OC 427974). B) ♀, Osage Hills SP, Osage County, 4 July 2015 (©Bill Carrell, OC 432686). In Oklahoma, apart from the "usual" isolated population in the Wichita Mountains, as well as one in the Arbuckles, this species essentially is confined to the eastern third of the state, where it can be locally common.

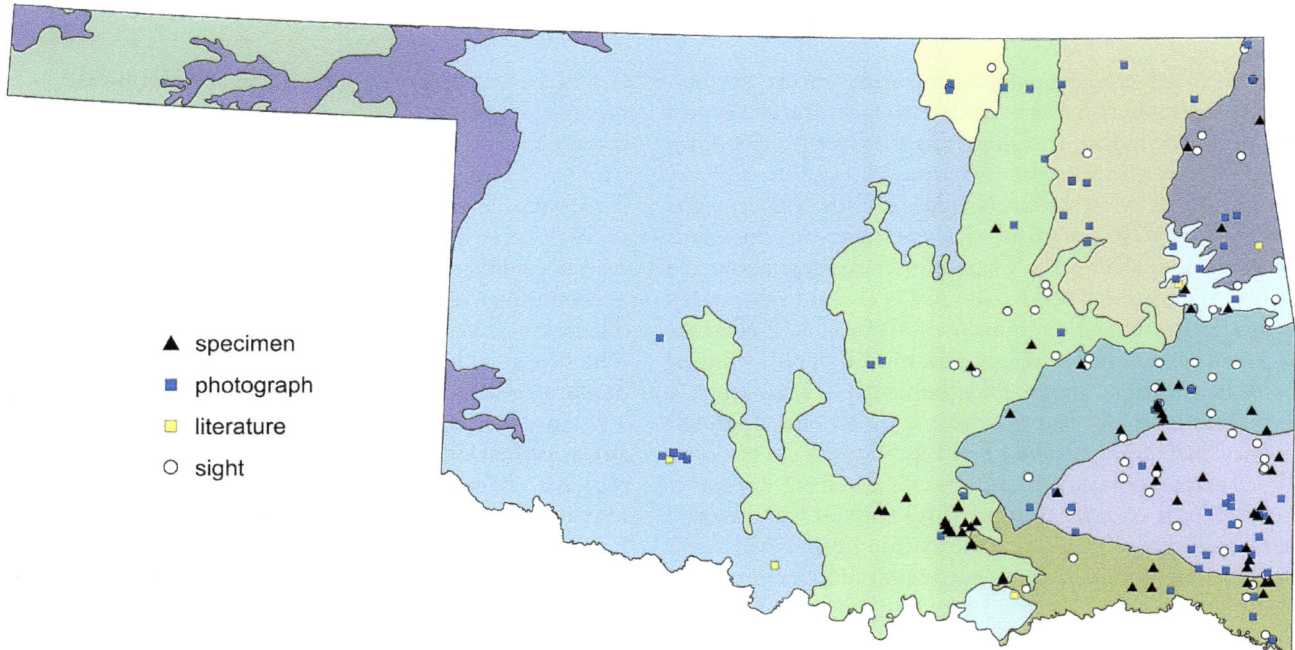

Figure 15.62 Documentation level of supporting records for *Dromogomphus spinosus*, the Black-shouldered Spinyleg, in Oklahoma. Levels include specimen, archived photograph, literature reference, or sight record. Although sight records lack documentation, only those we feel are credible are plotted.

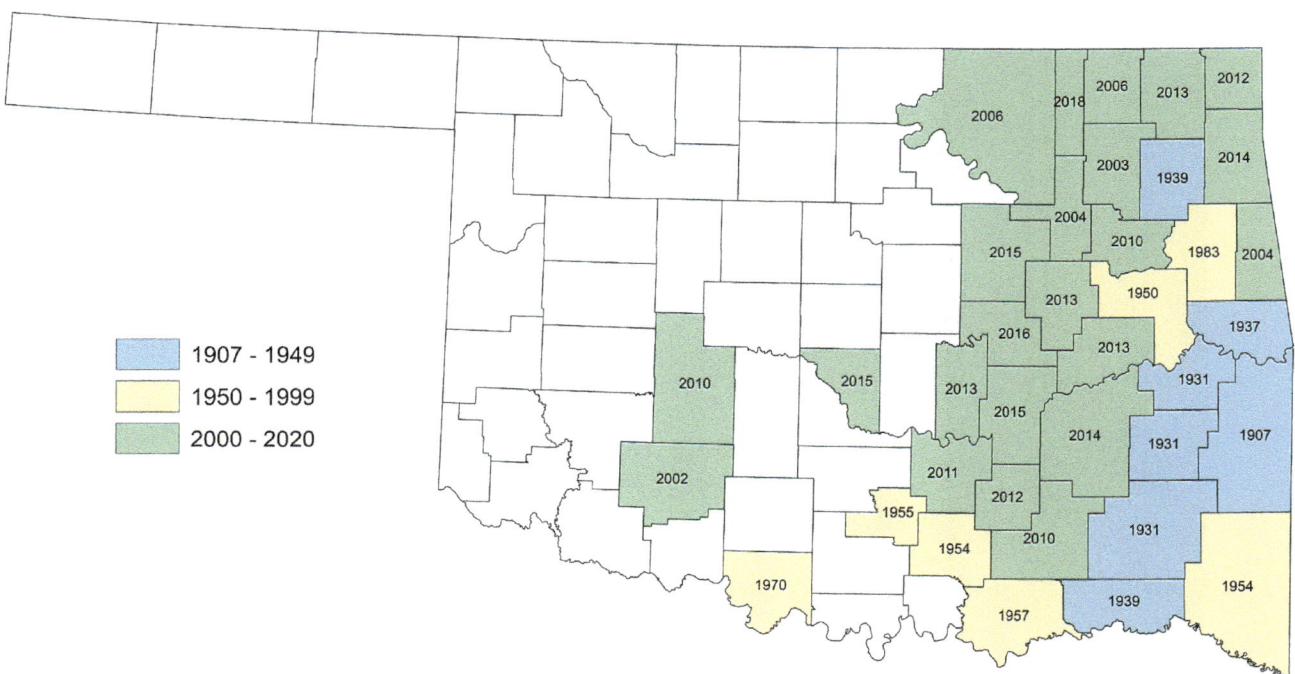

Figure 15.63 Counties in which *Dromogomphus spinosus*, the Black-shouldered Spinyleg, are known to occur in Oklahoma. Year within the county is that in which the species was first reported.

T1–2 thoracic stripes) or have a thin pale line forming a break or partial break between T1 and T2. We also have encountered a female with two pale lines in the shoulder, one at the suture line between T1 and T2 and one just below (SP 2905).

Such variation appears to be individual, as a review of regional records and a cursory review of photos from across the United States and Canada shows both fused and unfused individuals throughout the species' range and at the same locations.

DROMOGOMPHUS SPOLIATUS (HAGEN, 1858) – FLAG-TAILED SPINYLEG

The Flag-tailed Spinyleg is fairly common and widespread in Oklahoma, especially relative to its congener, *Dromogomphus spinosus*, the Black-shouldered Spinyleg (see species account). It occurs in a variety of habitats, both lotic and lentic, although it is partial to small lakes or impoundments with rocky shorelines (at least partly rocky, such as riprap against a dam). Bick and Bick (1957:3) said that it was often found "near lakes or ponds and not along wooded streams where *spinosus* was most frequent," a thumbnail description that fits our experience with the two species. The species has been recorded in numbers as high as 35 (7 July 2012, Lake Hall, Harmon County, SP 317), although counts of <10 are more typical.

The species was first noted in the state in 1931 when RD Bird collected 2♂ at Deans Lake, McClain County, on 4 July (OMNH 2204, 2221). Bird (1932) reported it from four counties: Comanche, Latimer, McClain, and Oklahoma. We have yet to discover the basis for the latter county record. By the late 1930s it was recorded in a total of ten counties. The species has been seen regularly since.

Bick and Bick (1957:3) noted that "unlike *spinosus*, [*spoliatus*] was not collected in several of the well studied [sic] southeastern counties." This statement, as shown by the dearth of early records, was most certainly true prior to the current era. In Bick's notes, where he indicated what he examined in the OSU collection, he mentioned that he looked at one of two specimens that we now know to have been collected in southeastern Oklahoma prior to the recent era. The specimen Bick examined was collected on 1 July 1937 in Grant, Choctaw County (1♀, Standish, RW Kaiser, OSU). Bick also noted "Leflore (?)" but made no other notes as to which OSU specimen that was. During our examinations of the collection we did not find a specimen that had the potential to match the Le Flore specimen although we did find the one from Choctaw County. The second specimen from southeastern Oklahoma that was collected in the 1930s is an OMNH specimen that Bick appears not to have examined, or at least he did not indicate in his notes he had looked at it. That specimen was collected by WM Fisher on 11 July 1931 (1♂, OMNH 2197).

Figure 15.64 *Dromogomphus spoliatus*, the Flag-tailed Spinyleg, A) ♂, Tishomingo NWR, Johnston County, 18 July 2011 (©Tom Kompier, OC 332255). B) ♀, Lake Overholser, Oklahoma County, 17 August 2012 (©James W Arterburn, OC 427969). It may be that only *Phanogomphus militaris*, the Sulphur-tipped Clubtail, is a commoner clubtail in the southern Great Plains, and it may be the clubtail that has benefited most from proliferation of man-made lakes.

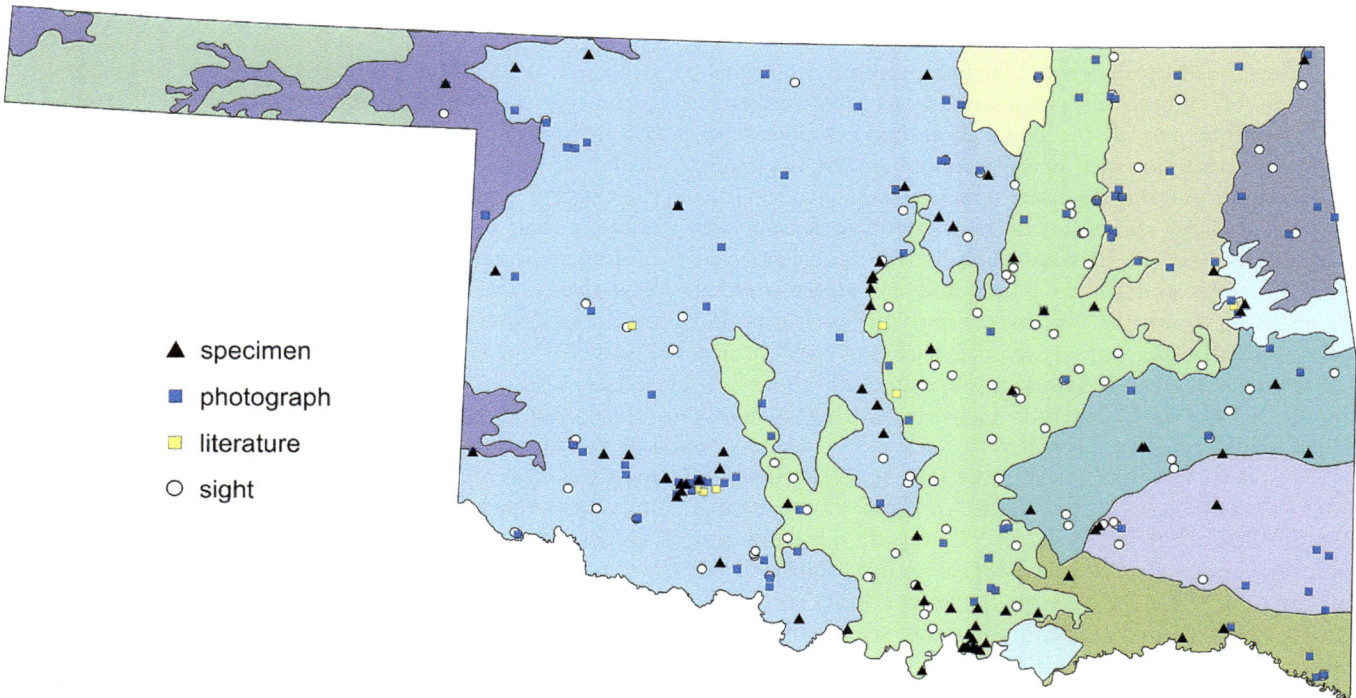

Figure 15.65 Documentation level of supporting records for *Dromogomphus spoliatus*, the Flag-tailed Spinyleg, in Oklahoma. Levels include specimen, archived photograph, literature reference, or sight record. Although sight records lack documentation, only those we feel are credible are plotted.

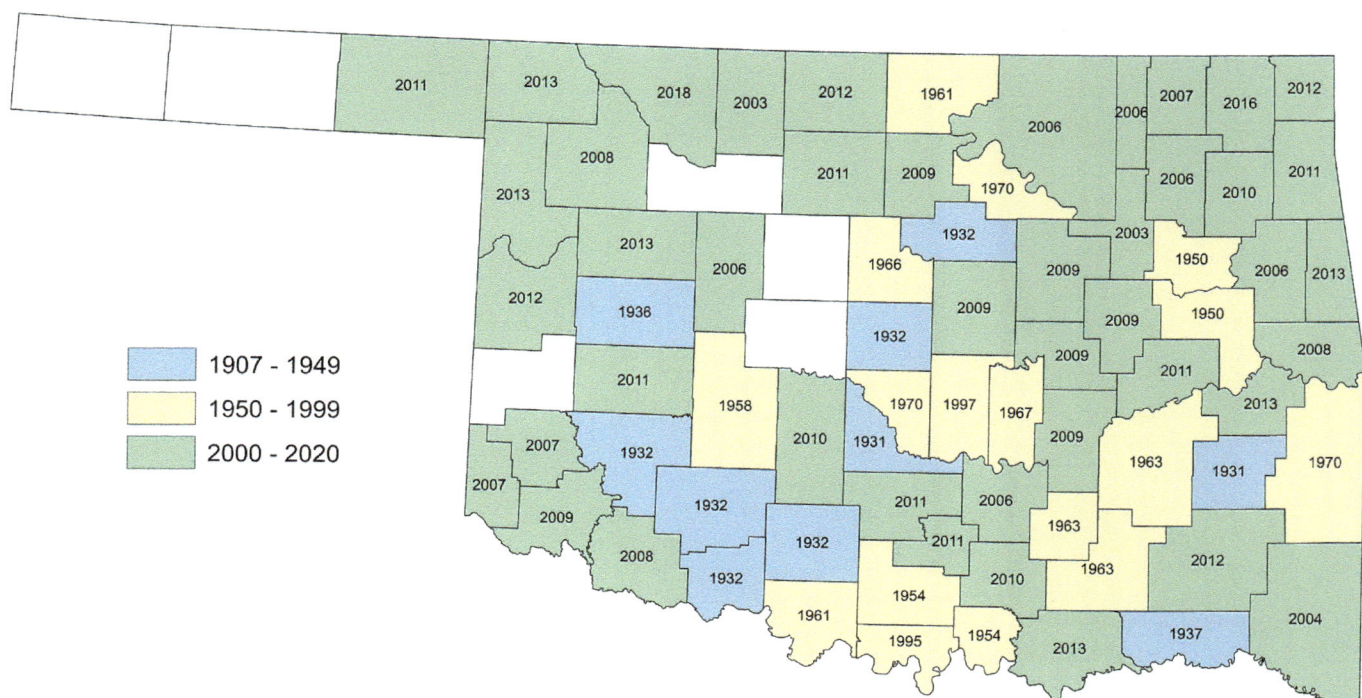

Figure 15.66 Counties in which *Dromogomphus spoliatus*, the Flag-tailed Spinyleg, are known to occur in Oklahoma. Year within the county is that in which the species was first reported.

Since Bick's time the species has been reported from southeastern Oklahoma with regularity, indicating that perhaps it has expanded into that part of the state in recent times. It may have expanded westward, too, if the Bicks' impression of it being "more frequent in central Oklahoma" was correct because in our experience, it is most common in the western third of the state.

Within the region the species flies the earliest in Arkansas and Texas and well into the fall regionally (OK: 31 May 2009,[20] 1U, 2 km NE of Sparks, Lincoln County, and 30 October 2016, 1♂, Healdton Municipal Lake, Carter County, MAP; AR: 18–21 May–24 October, Harp 2006, OC 382177; LA: 15 June–5 October, OC 464329, 473695; MO: 1 June–23 September, Trial 2005; NM: 17 July–14 September, JCAC 19897, OC 436703; TX: 30 March–29 November, Abbott and Lasley 2019). The season in Kansas is of particular note because of either a difference of opinion or a difference in the species' status and distribution between eras. Allison (1921) reported *Dromogomphus spoliatus* as common and occurring from mid-June to late October in Crawford County, whereas Cringan (1979; for Lyon County) reported the species as "infrequently observed" and only in mid-July. Stewart and Murphy (1968), in their study of dispersal of lentic odonates in southern Oklahoma, found 1♂ at a pond 0.72 km from where they had originally marked it, attesting to the species' dispersal ability. There are six records of ovipositing, four of which noted lone females.

HAGENIUS BREVISTYLUS SÉLYS, 1854 – DRAGONHUNTER

The Dragonhunter, as the name implies, is a fierce hunter of other dragonflies, including ones approaching its own size. It hunts other large insects, too, especially butterflies. It is the largest clubtail in North America and is the only species in the genus *Hagenius* (Paulson 2011). Its distribution across the eastern United States and southeastern Canada is perplexing because its closest relatives are species within *Sieboldius*, a genus that occurs in Japan, northern China, the Korean peninsula, and southeastern Asia west to India (Johnson 1972a; Needham et al. 2014).

As one would expect for a top odonate predator, the Dragonhunter is an uncommon species. There are a small handful of reports of up to eight adults at a time, but generally only one or two are encountered during a site visit. The strange flight style of this species—as if floating slowly down the river, seemingly slowed by the weight of its downturned, ball-like club—and its contrarian behavior—bold enough to fly right up to a person and snatch nearby insects, yet a tad skittish when approached—leaves one awestricken when a quick swing of a net produces only a quicker retreat

from a deceptively fast Dragonhunter. So, one wonders how F Collins collected ten adults in a three-day period in 1907 (reported as 10♂ by Williamson 1914b, although 1♂ and 1♀ are at UMMZ), when Oklahoma received its first record. Then again, he was a teenage boy at the time.

The second record of the species for the state came on 13 June 1931 when RD Bird collected 1♀ in Latimer County (OMNH 2468). This was followed four years later by the third record: five nymphs collected by Hubbs and Trautman on 13 September 1935 at Turkey Ford on the Elk River, Delaware County (UMMZ Nymphs 3584, id. by E Bright). These are the only records we have data for from the 1930s, despite Bird (1932) reporting the species from Delaware, Latimer, Le Flore, and McCurtain Counties. Clearly Bird would not have known of the 1935 Delaware County record in 1932, so there was an earlier record, but we do not know the basis for it. Likewise, the McCurtain County report from Bird (1932) remains a mystery because the earliest record we know of for the county is of two nymphs collected by LE Hornuff in June 1955 (Bick notes). Regardless

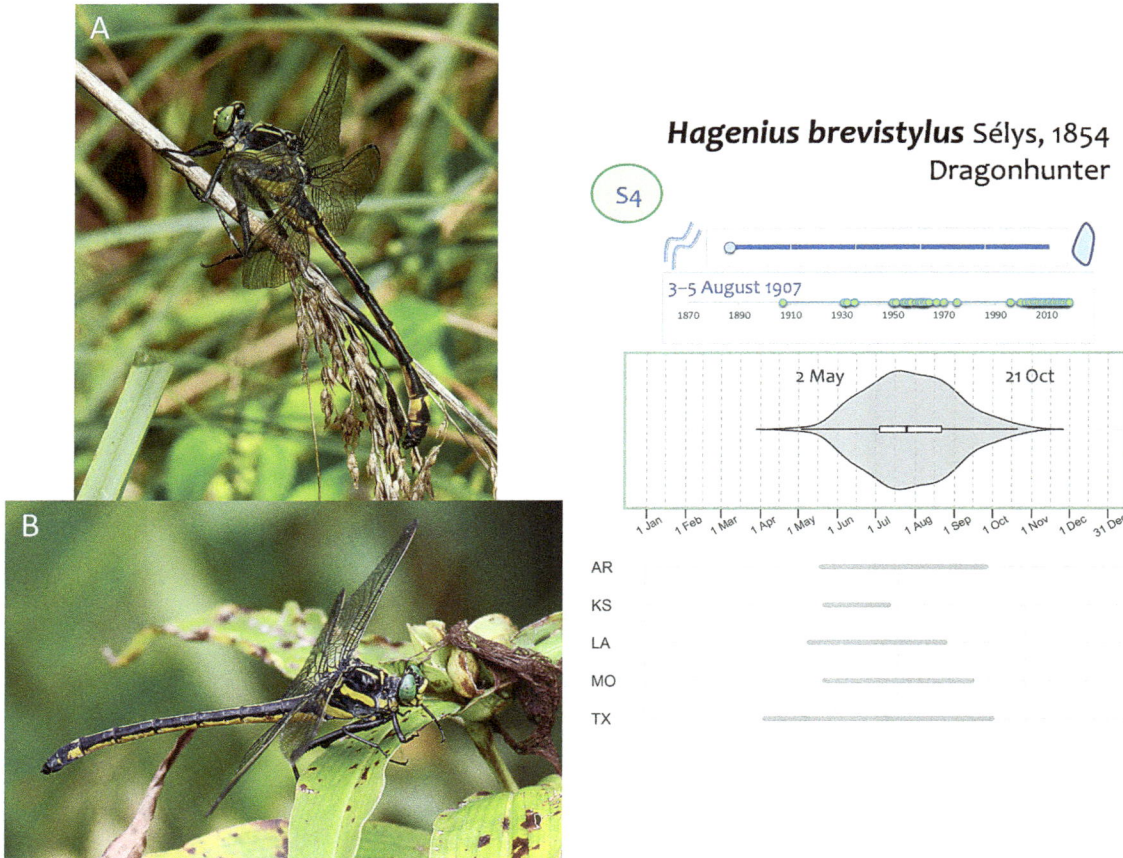

Figure 15.67 *Hagenius brevistylus*, the Dragonhunter, A) ♂, Medicine Creek on Fort Sill Military Reserve, Comanche County, 31 August 2009 (©Victor W Fazio III, OC 315010). A geographically isolated population occurs in these mountains (of course); there is one in the Arbuckle Mountains, too. Otherwise, it is restricted to shady streams primarily in the eastern half of the state. B) ♀, Beavers Bend SP, McCurtain County, 27 August 2006 (©Berlin A Heck, OC 316149).

of the basis of the records, if there truly were only a handful of records prior to the Bick/Hornuff era, one must wonder why. And we wonder why in Bick and Hornuff's time there were about 30 records but >115 thereafter, as we have little reason to think this species would have become more common and widespread in the state since the 1950s. Given various issues with streamflow in the state, we would have guessed the opposite—after all, this species borders on the strictly lotic, being found almost exclusively along shaded, slow-moving, shallow streams with sandy or rocky substrates. Atypical habitats include small ponds near streams and along sunny stretches of river.

There are little more than a handful of records outside of the eastern half of the state. One of these, from Garfield County, happens to also be the early date for the species (2 May 1960, 1♀, JF Reinert, USNM 391824). The rest of the outlying records are in the southwestern corner of the state. J Grzybowski photographed a probable ♀ Dragonhunter near Colony in Washita County on 23 June 2009 (OC 313434), but it was BC Kondratieff and colleagues who were the first to record the species in the southwest in, you guessed it, Comanche County. At Fort Sill MR they collected a nymph on 11 October 2002 and an adult on 1 July 2003 (Kondratieff et al 2004; Zuellig et al. 2006). The other Comanche County records are also from Fort Sill MR. VWF reported 4♂ from 7 August 2007, 25 June 2008, and 31 August 2009 (OC 263027, 313903, 315004, 315010). There appears to be no pattern to these wanderings, and it

may be that there is an outlying population in the Wichita Mountains in Comanche County (the first county record is of a nymph collected in 2002; CSU), although if that is the case then it is puzzling why the OU expeditions never encountered the species there.

The latest record we have for the species is 21 October from a 2010 photographic record of BAH's at Little River NWR, McCurtain County (2U, OC 323779). The Oklahoma late date is the latest for the region, being almost three weeks past the Texas late date of 1 October (Abbott and Lasley 2019). Conversely, the Oklahoma season starts about a month later than it does in Texas (5 April, Abbott and Lasley 2019), although the length of the seasons, around 170 days, is roughly the same. Other regional date ranges are shorter and within the confines of those of Oklahoma and Texas (AR: 18–21 May–26 September, Harp 2006, Harp and Rickett 1977; KS: limited records, 22 May–11 July, OC 448494, 433012; LA: 10 May–25 August, Mauffray 2014, USNM 339370; MO: 22 May–15 September, Trial 2005).

There are relatively many larval records of *Hagenius brevistylus* for Oklahoma (n = 13, of 23 individuals; UCO, UMMZ Nymphs, OMNH, Kondratieff et al. 2004, Bick notes), which likely attests to the ease to identify the species' distinctive larvae. We know of one teneral record: a freshly emerged ♀ at the Looney Unit of Ozark Plateau NWR, Delaware County, on 22 June 2016. We further know of six observations of ovipositing, four of which were indicated as being lone females, dating from 3 July to 25

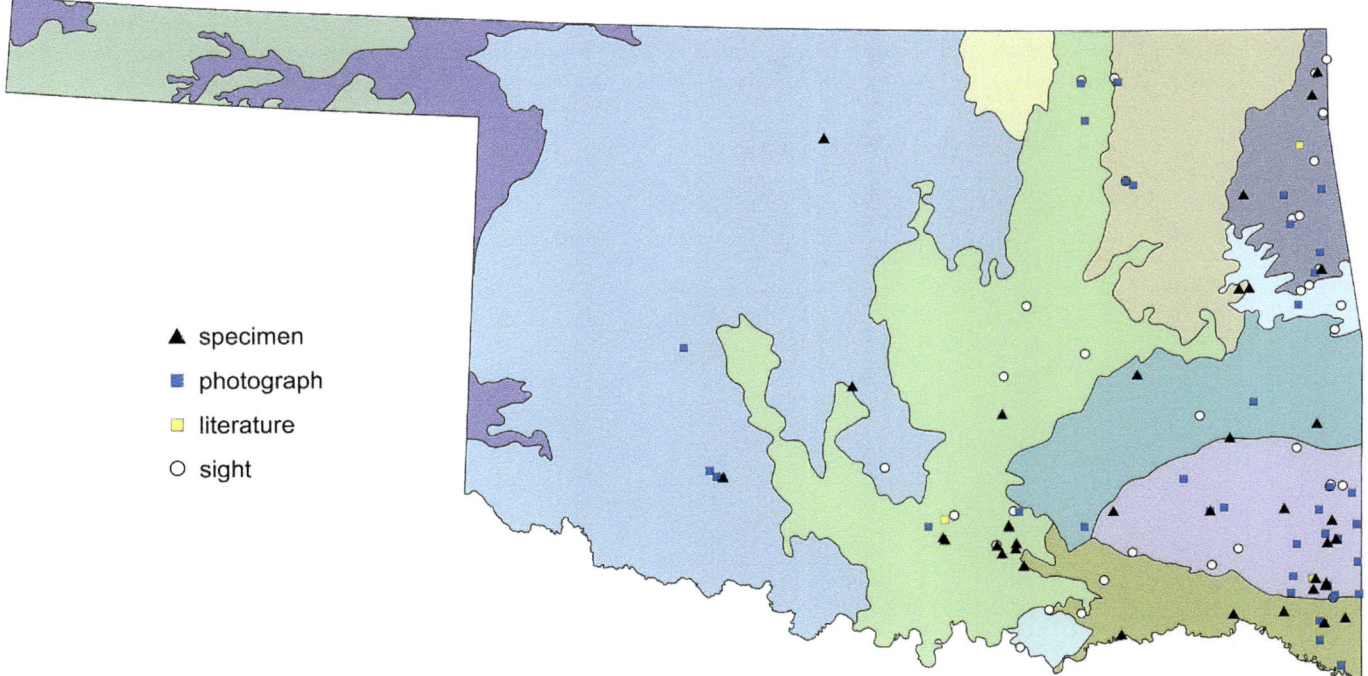

Figure 15.68 Documentation level of supporting records for *Hagenius brevistylus*, the Dragonhunter, in Oklahoma. Levels include specimen, archived photograph, literature reference, or sight record. Although sight records lack documentation, only those we feel are credible are plotted.

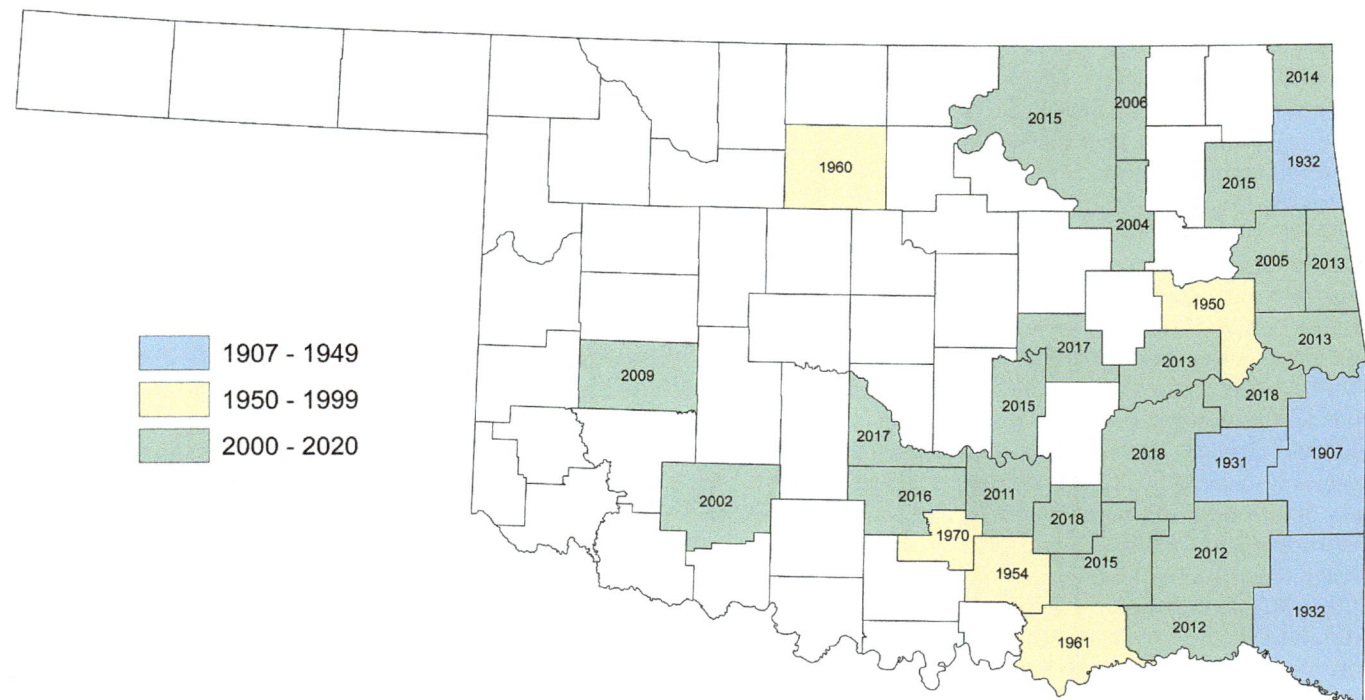

Figure 15.69 Counties in which *Hagenius brevistylus*, the Dragonhunter, are known to occur in Oklahoma. Year within the county is that in which the species was first reported. The first report of the species from Comanche County was a nymph collected at Fort Sill Military Reserve in 2002; the following year an adult was collected (both CSU specimens; Kondratieff et al. 2004).

September. Ovipositing behavior was described as "laying eggs in a creek, and then perched" on a branch (BAH, OC 283315, McCurtain County) and "ovipositing in the vegetation near the creek bank before flying over to the rocks" (BC, OC 315004, Tulsa County). In Comanche County, a ♀ was described as ovipositing in a quick-flowing stream about a foot from where she landed and was photographed. When she left that perch she was seized by a ♂, from whom she dangled for about two minutes while the pair flew around; she did not copulate with him (VWF, OC 437268).

ERPETOGOMPHUS DESIGNATUS HAGEN, 1858 – EASTERN RINGTAIL

The Eastern Ringtail is common or even abundant just about wherever there are clear, medium- to fast-flowing streams with exposed rocks, sandbars, or mudflats. It is far less common in the east yet decidedly more so to the west, until one reaches the central panhandle. To wit, the species has been found in all ecoregions in the state except the High Plains. We suspect it inhabits that ecoregion, too, but there are almost no publicly accessible streams in that part of the Oklahoma panhandle. It is typically encountered 1–5 individuals at a time, but occasionally >10, including four times ≥30 individuals. The highest count has been 35♂ and 1♀, including one pair, at Moneka Park, Lake Waurika, Jefferson County, on 26 July 2015 (MAP).

We have tallied >260 records for this species since it was first recorded in the state on 29 June 1930. On that date RD Bird collected a ♂ (OMNH 2317) and a ♀ (OMNH 2316) at Cleo Springs on Eagle Chief Creek in Major County. The second time it was collected was on 12 July 1931, by H. M. Smith, in Lawton, Comanche County (1♂, UMMZ). Bird (1932) reported those two counties. Shortly after that publication the species was recorded in an additional five

counties. During the Bick and Hornuff era it was added to another 18 and then 46 thereafter, for a total of 71 counties at present. It is unclear why, but seven of those counties have not had subsequent records: Le Flore (1950, only record), Muskogee (1950, only record), Pittsburg (1963, one other record, in 1954), Sequoyah (1970, only record), Stephens (1961, only record), Wagoner (1950, only record), and Woodward (1970, only record).

Many collectors have captured nymphs ($n = 36$) and collected exuviae ($n = 4$) of *Erpetogomphus designatus* across the state. GH Bick twice reared the nymphs. Although we do not know at what larval stage these specimens were taken, one appears to have been captured on 20 June 1955 near Willis, Marshall County, at a creek that Bick dubbed "Croceipennis Creek," which probably refers to Cowan Creek. It emerged as a ♂ (FSCA). The second individual was captured on 15 June 1955 at the same location; it emerged on 28 June 1955 as a ♀ (FSCA). Larval records, some only identified as *Erpetogomphus* and presumed to be *E. designatus*, are from five counties—Canadian, Comanche, Johnston, Oklahoma, and Roger Mills (OBS,

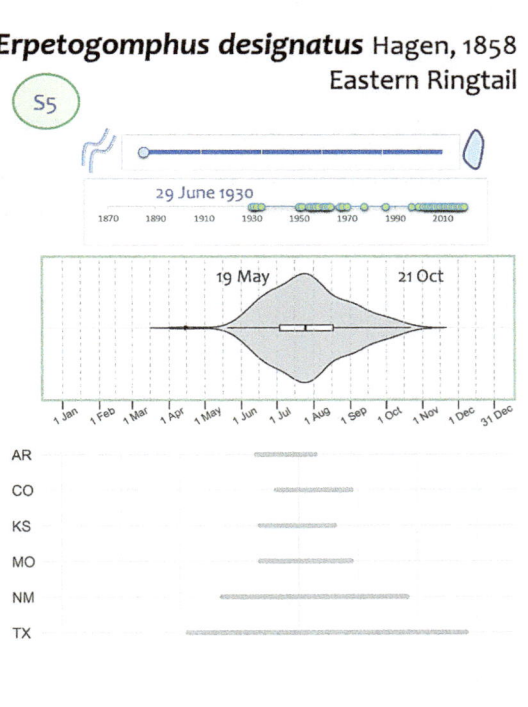

Figure 15.70 *Erpetogomphus designatus*, the Eastern Ringtail, A) ♂, Rock Island Park, Grant County, 19 May 2012 (©Jason R Heinen, OC 375050), the earliest noted in the state—typically, the species is not found until early June. B) ♀, 1 km E of Turner Falls, Murray County, 3 August 2014 (©Randy L Kelley, OC 425740). This clubtail provides yet another example of an ostensibly eastern species whose abundance across the southern Great Plains *increases* from east to west, so much so that it remains unknown from some counties in eastern Oklahoma and can be encountered in the dozens or even hundreds in western Oklahoma.

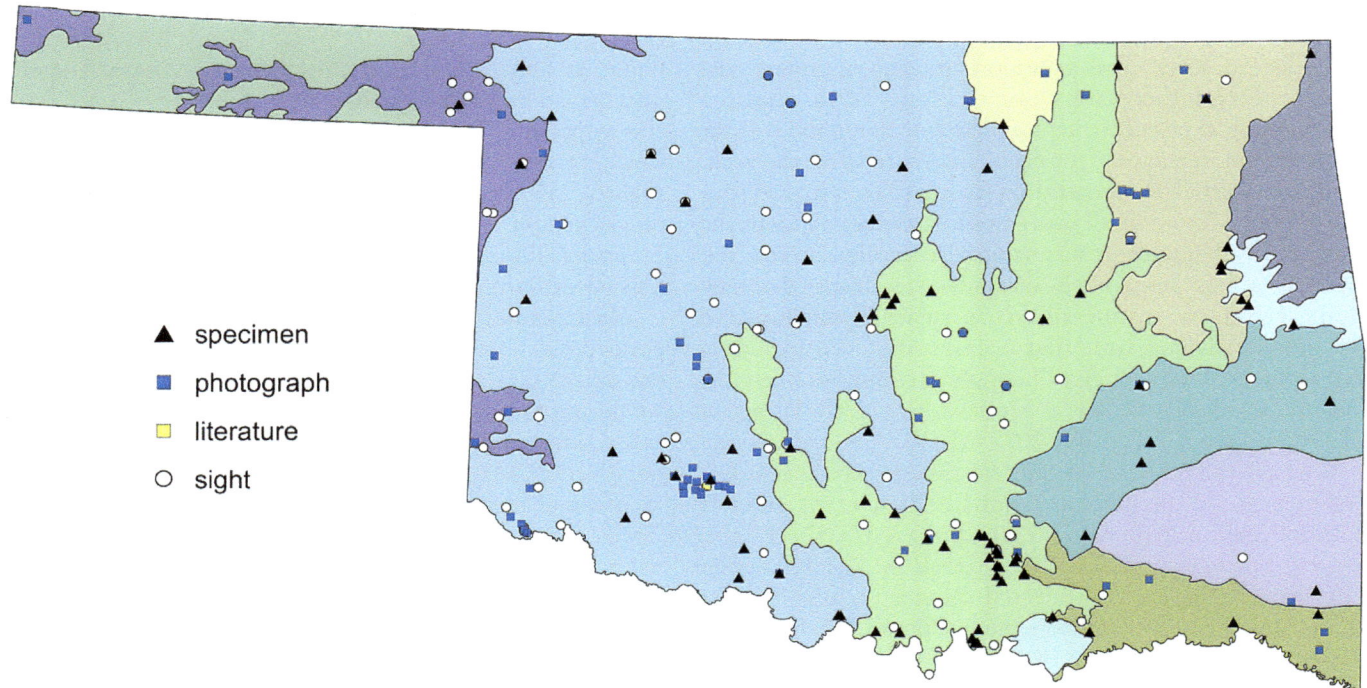

Figure 15.71 Documentation level of supporting records for *Erpetogomphus designatus*, the Eastern Ringtail, in Oklahoma. Levels include specimen, archived photograph, literature reference, or sight record. Although sight records lack documentation, only those we feel are credible are plotted.

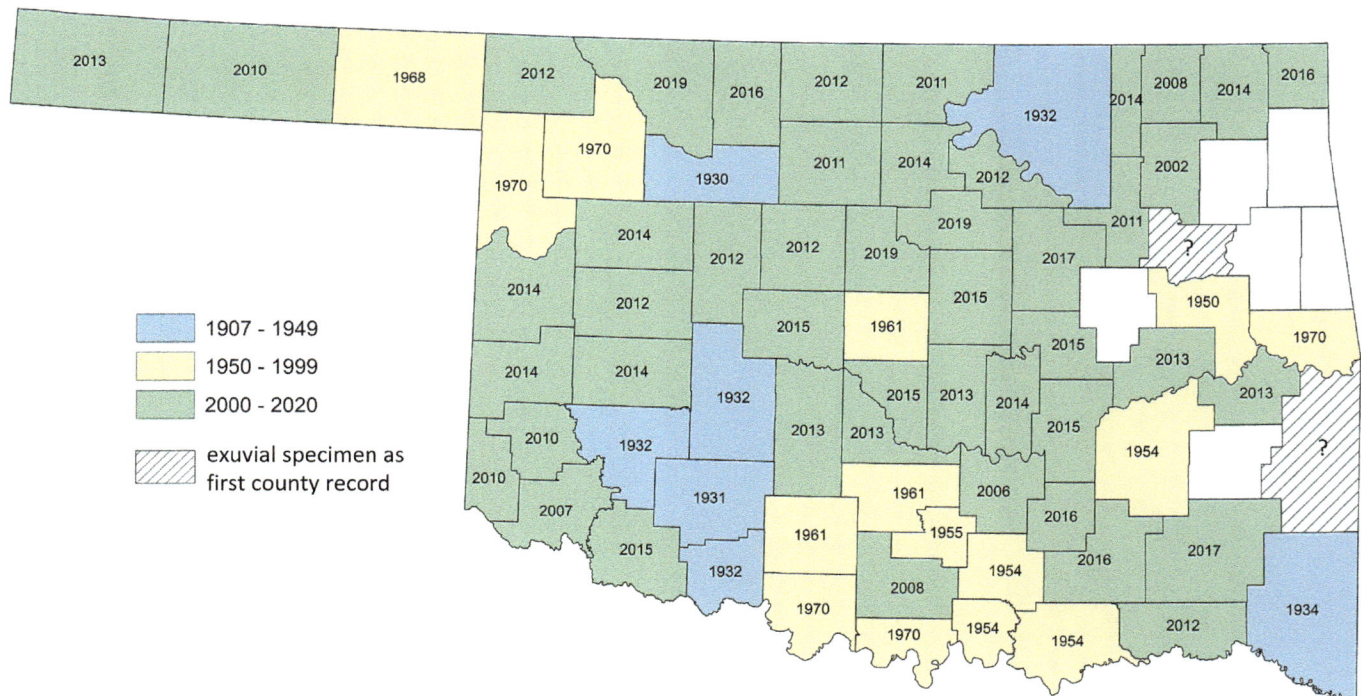

Figure 15.72 Counties in which *Erpetogomphus designatus*, the Eastern Ringtail, are known to occur in Oklahoma. Year within the county is that in which the species was first reported. Canadian and Roger Mills Counties may have earlier records, of nymphs, but they have not been confirmed. The only records for Le Flore and Wagoner Counties come from GH Bick from claimed exuviae taken in 1950. These specimens have not been re-located, so we cannot confirm their identifications.

OMNH, UCO, Bick notes). According to Bick's notes exuviae were collected 3–18 July 1950 from Wagoner and Le Flore Counties; Bick discarded one and we have yet to find the others.

The earliest documented (and unquestioned[21]) flight record for Oklahoma is a ♂ photographed by JR Heinen on 19 May 2012 at Rock Island Park, Jefferson, Grant County (OC 375050). The latest flight record for the state is from 21 October of the same year, when JWA photographed 1♂ in Tulsa County on Fry Creek near Bixby (OC 382324). Oklahoma's flight season is a touch earlier in marking the beginning of the season compared to Texas, but it ends much earlier than it does there (16 April–7 December, Abbott and Lasley 2019). Elsewhere in the region the species is known to start flying much later, from May to July, which may be due to limited records in some states, and it stops flying anywhere from mid-August to mid-October (AR: 13 June–2 August, OC 376031, Harp and Rickett 1977; CO: limited records, 30 June–1 September, OC 502793, 409849; KS: 16 June–18 August, SEMC, OC 502030; MO: limited records, 16 June–1 September, Sims 2012; NM: May–18 October, Paulson 2009, OC 334173).

STYLOGOMPHUS SIGMASTYLUS COOK AND LAUDERMILK, 2004 – INTERIOR LEAST CLUBTAIL

This diminutive species (almost as small as a third of the size of *Hagenius brevistylus*, the Dragonhunter, and comparable to many damselflies) is found in extreme eastern Oklahoma along the Arkansas border. In Oklahoma, the Interior Least Clubtail is a highland species[22] with two distinct populations, one northeastern and one southeastern. Recent encounters with the southeastern population of the Ouachita Highlands have made us wonder if there are biogeographical differences between it and the northeastern population of the Ozark and Boston Mountains. In both regions, it is a species of shallow, rocky, clear-water, woodland streams, but there are ostensible differences between the regions in width and flow. In northeastern Oklahoma, the species is found at moderate- to fast-flowing streams that are 10–15 m wide, occasionally wider, but rarely narrower, whereas in the Ouachitas the opposite appears to be true, flow is generally limited and width of streams is generally no more than 5 m, often just a couple of meters or less, and found less commonly on larger streams. Rangewide,

adults act similarly: they often alight on sunny or shaded rocks found within the stream, or they sit on flat concrete creek crossings on dirt roads and, just as often, up to head-height (or just above) on the leaves of trees or shrubs that edge the stream. Nymphs have been found at the same types of streams and have been reported at riffles (OMNH 1837, *n* = 1, Bidding Creek, Adair County) and "in gravel in riffles" (OMNH 1836, *n* = 3, Kiamichi River, Big Cedar, Le Flore County). A more detailed account comes from the Arkansas Ouachitas, where BS-P collected a nymph (19 April 2018, Fernwood Seep Natural Area, Polk County, SP 2553) at a seep complex with a fairly rocky, silty, moderately flowing, clear-water stream that was 50 cm–1 m wide. The nymph was buried in a patch of silt within <1 cm of flowing water. About 20 cm above the water was a dirt overhang that formed a tiny cave over the nymph. The site sits within a hardwood forest and contains many species of orchids as well as cinnamon ferns (*Osmundastrum cinnamomeum*) and Lescur's sphagnum (*Sphagnum lescurii*).

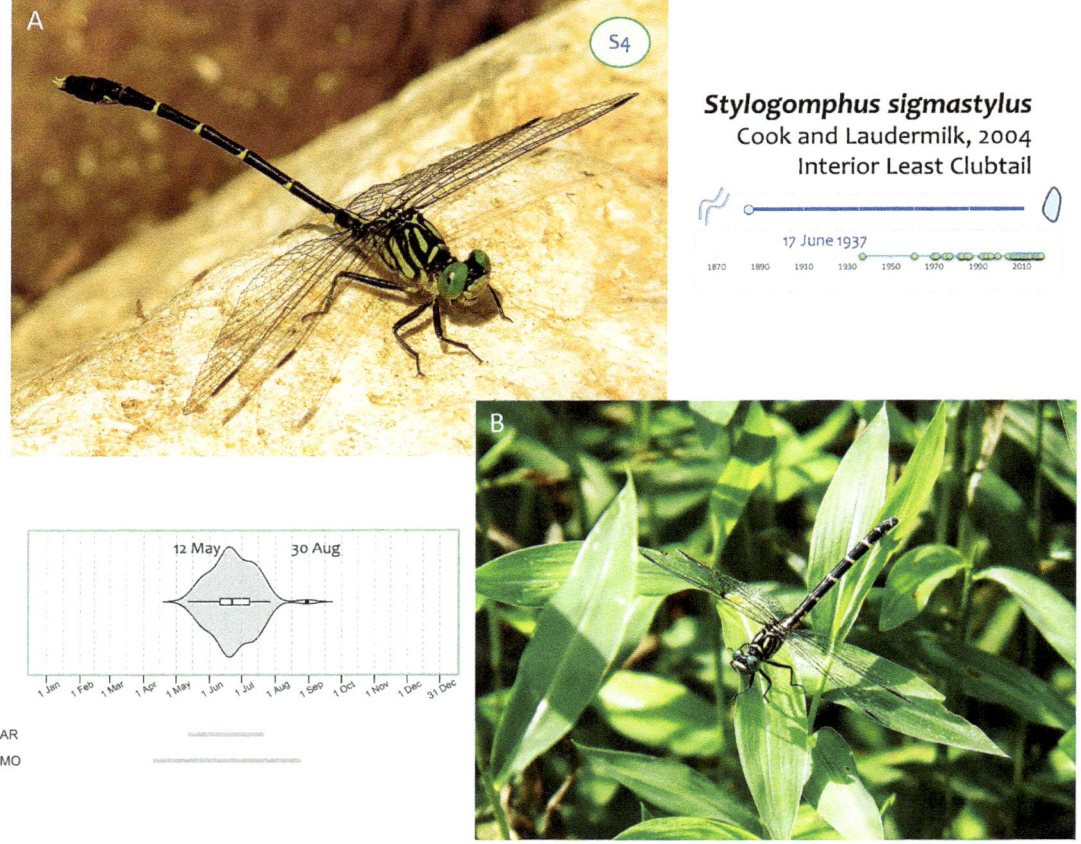

Figure 15.73 *Stylogomphus sigmastylus*, the Interior Least Clubtail, A) ♂, Snake Creek 3 km S of Locust Grove, Mayes County, 5 July 2014 (©Bill Carrell, OC 424318). Whereas males of this species will typically perch on rocks within streams, females especially tend to be found on vegetation, including that above head level. B) The ♀ pictured here was photographed on Caney Creek, 6 km NW of Stilwell, Adair County, Oklahoma, 24 June 2017 (©CA Ivy, OC 475228). This diminutive species, our smallest clubtail, is restricted to clear, rocky creeks and narrow rivers on the eastern fringe of the plains.

Typically, we see the species one to a handful at a time (92% of records are ≤5 indiv.), but occasionally there are reports of higher numbers. In northeastern Oklahoma, where the species has been reported >40 times, there have been four reports in the double digits, including our highest count of 20 (all ♂, 18 of which were tenerals) on 23 May 2014 on Lost Creek at Tecumseh Park in West Seneca, Ottawa County, and our next highest of 15 individuals at that same location on 22 June 2016 (10♂, 5♀ [1 as SP 1970], OC 447229). The highest count for the state also came from the northeast: JWA and KW sight record of 35 individuals on 10 July 2014 on Snake Creek, 4 km SSE of Locust Grove, Mayes County. In the Ouachitas, of the >30 reports, there have only been two with more than four individuals at a time: 8♂ at Big Creek, vic. Page, Le Flore County, 19 June 2016, MAP, and 19 individuals (see end of account for details) from Buffalo Creek, McCurtain County.

With few exceptions, the Interior Least Clubtail is confined to the "Ozark Uplift" (comprised of the Ozark Plateau and the Boston Mountains) and the Ouachita Highlands in Oklahoma, Arkansas, Kansas, and Missouri (Figure 15.75). The one exception for Oklahoma is a record of a teneral ♀ along Jackfork Creek 15 km SE of Ti, Pittsburg County, 2 June 2018 (MAP, SP 2612). Besides its geographic distance from other records, there is nothing about this site that is not within the scope of this species' habitat as it is known in the region. Two outlying records in Kansas, both of nymphs from Cherokee County,[23] are within the Central Irregular Plains. Falling outside of the more typical highland areas for the species may explain why those are the only two records for Kansas, being mere accidental occurrences. Cook and Laudermilk (2004:21) commented that the "largest populations occur in the Ozark Plateau of Arkansas and Missouri and Interior Low Plateaus of Kentucky and Tennessee." Prior to their study little research had been conducted to determine abundance of the species in Oklahoma. We believe that if they had access to data we now have they would have included the Oklahoma population in the above statement.

Until recently this species was considered the same as *Stylogomphus albistylus*. Because of that association, *S. sigmastylus* has also been called *Gomphus albistylus* Hagen, 1878, as it was initially described, and *Lanthus albistylus*, the genus it was moved to by Needham (1901). *Lanthus* is now used solely for the "Pygmy Clubtails," as opposed to *Stylogomphus* for the "Least Clubtails." Despite *S. albistylus* being moved to *Stylogomphus* 50 years ago (Chao 1954), researchers continued to use the old name of *Lanthus*, probably because Needham and Westfall (1955) did not recognize *Stylogomphus* (it was not recognized until Needham et al. 2000). The late recognition of the genus explains continued use of the old name *Lanthus* (e.g., Bick in his notes, Kormondy 1960), but it does not explain why some researchers used both names (Orth et al. 1982; Bass 1995), particularly given that neither of the American *Lanthus* species occurs anywhere near Oklahoma or the central United States (Needham and Westfall 1955; Paulson 2011). Regardless, *S. albistylus* and *S. sigmastylus* are parapatric[24] at their respective western and eastern extents,

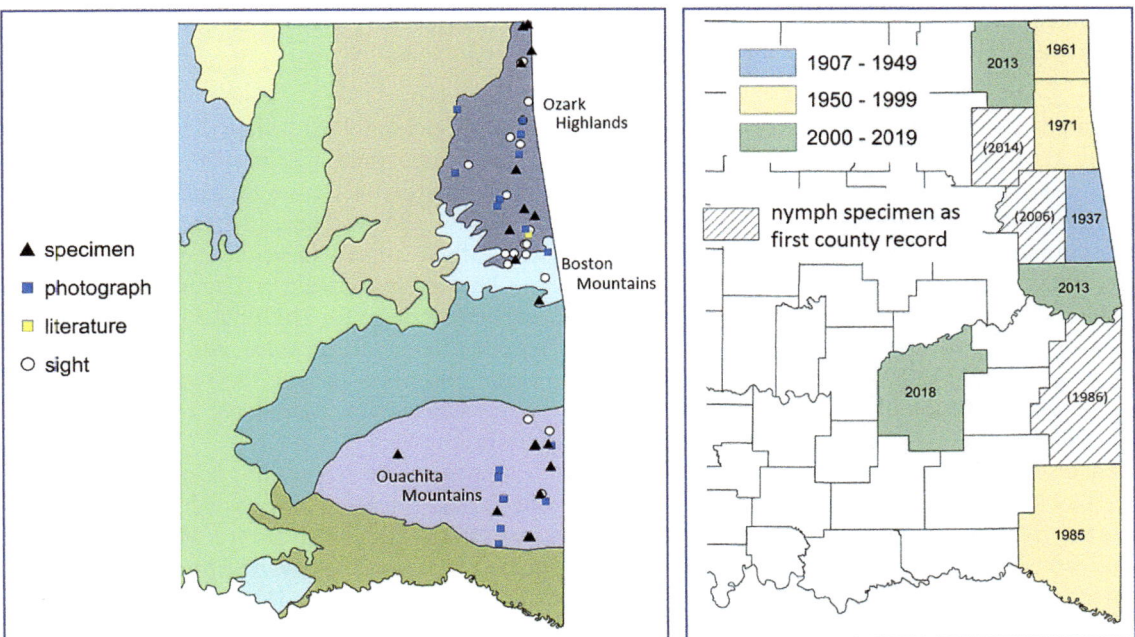

Figure 15.74 Documentation level of supporting records for *Stylogomphus sigmastylus*, the Interior Least Clubtail, in Oklahoma (right). The species is concentrated in the greater Ozark Plateau (Ozark Highlands and Boston Mountains) and the Ouachita Highlands. The species has been reported in ten counties in eastern Oklahoma since 1937, three of which were first reported as nymphs (left; parenthetical year is first adult record).

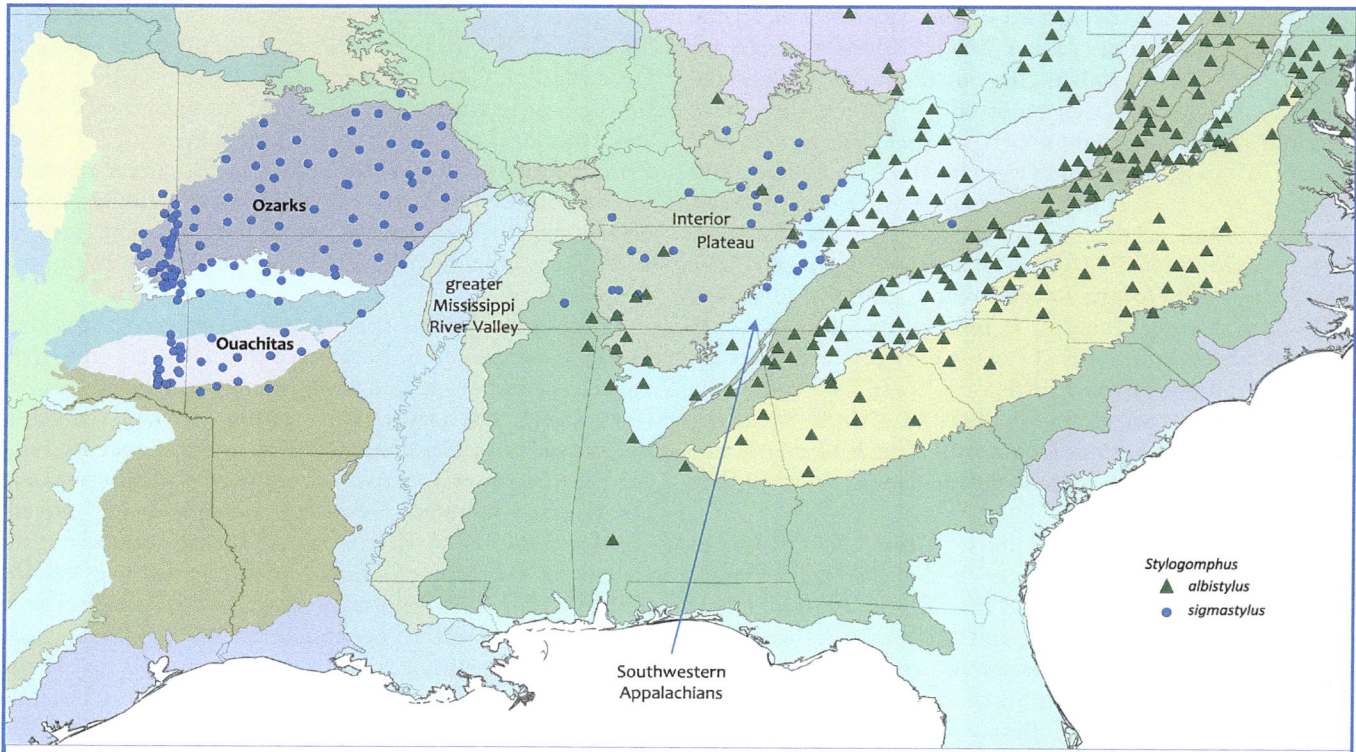

S. sigmastylus is known from all of the ecoregions in eastern Oklahoma and western Arkansas but it is concentrated in the Ozark and Ouachita Highlands. Notice, too, that it skips over the greater Mississippi River Valley to be found again farther east. The two species come into contact in the Southwestern Appalachians and the Interior Plateau of Kentucky and Tennessee, where hybridization is known to occur.

Figure 15.75 Distribution of *Stylogomphus albistylus* and *S. sigmastylus*, the Eastern and Interior Least Clubtails. *S. sigmastylus* is known from all of the ecoregions in eastern Oklahoma and western Arkansas but it is concentrated in the Ozark and Ouachita Highlands. Notice, too, that it skips over the greater Mississippi River Valley to be found again farther east. The two species come into contact in the Southwestern Appalachians and the Interior Plateau of Kentucky and Tennessee, where hybridization is known to occur.

with documented historical hybridization at two sites in Kentucky and Tennessee, with subsequent dominance of *S. sigmastylus* at those locales within the past three decades or so (Cook and Laudermilk 2004; Figure 15.75).

Another synonym issue with *Stylogomphus sigmastylus* regards a series of FSCA specimens that are labeled as *Stylogomphus divaricatus*, a *nomen nudum*. All but 2♀ (original locality of "Baron, Barren Fork Creek" coll. by LE Hornuff on 18 July 1970) designated as *S. divaricatus* were published as paratypes of *S. sigmastylus* (Cook and Laudermilk 2004), so we are confident that the name will remain naked (also, Carl Cook said that it would, *pers comm* and *in litt*). Cook and Laudermilk (2004) published an additional paratype, a ♂ from "3.2 km W of Stillwell [sic, Stilwell]" collected by WF and FA Blair on 17 June 1937. This specimen was published by Kormondy (1960, as a ♀), but we have not located it. The record makes for the first time the species was recorded in the state and makes one wonder why Bird et al. never reported it in the state despite working within its range. The species was also not published by Bick and Bick (1957). GH Bick did not encounter

it until 9 June 1961, when he collected 3♀ on Sycamore Creek in Ottawa County (FSCA, JCAC 40928, 41593). This was the only time Bick saw the species in Oklahoma. Likewise, Hornuff encountered the species on only a single day, albeit at two locations (Adair County: Christie and on Baron Fork Creek, at Baron, 18 July 1970). Combined, earlier records, for the first 35 years that we know of the species being in the state, add up to just a handful within three counties—Adair, Delaware, and Ottawa. In the 40 years subsequent, there have been >70 records and 7 other counties added.

Stylogomphus sigmastylus is documented as beginning to fly in Oklahoma as early as 12 May and has been seen flying as late as 30 August (12 May 2010, 1♂ teneral, Cedar Creek, 13 km WNW of Hochatown, BAH, OC 318824 and 30 August 2006, 1U, Tulley Hollow, TNC JT Nickel Preserve, Cherokee County, BC sight record). The early date matches that for Arkansas but it is a month later than that for Missouri and then the opposite is true for late flight dates, with Missouri matching Oklahoma, but Arkansas' season ending about a month earlier (AR: 14 May–20 July,

Harp 2007, Harp and Harp 2003, the later as *S. albistylus*; MO: 11 April–23 August, Trial 2005). We know of a handful of larval records and eight teneral records, the latter dating from 12 May to 5 July. One of those records was an encounter with 13 tenerals and 6 adults (27 May 2017, 11 km SSW of Smithville, Buffalo Creek, McCurtain County, tenerals: 1♂, 2♀ [one as SP 2348], 10U, adults: 2♂ [one found dead], 4♀, BS-P, W Boys, L Wishard). We noted ovipositing on two occasions, both times of lone females (21–22 June 2014, Adair and Sequoyah County).

NOTES

1. OMNH 2335, 2340, 2342, 2344, 2346–2351, 2357–2358.
2. Mis-reported by Mills (2007) as 30 August.
3. Prevalence of the extra thoracic stripe remains subjective and likely has a shifting baseline as more data are gathered: for example, in one part of his species account Abbott (2015:109) said that females "often" had a thin stripe but later in the account he said that the extra stripe is present "only rarely so … and only on females." Obviously, more research is needed.
4. Classic (no stripe or just a tiny dot dorsolaterally): 6♂ = OC 313201, 314747, 320410, 329467, 486624, SP 2164; hint/partial: 4♂, 1♀ = OC 283955, 283956, 315589, 448397, JCAC 20789; mostly full: 1♂, 1♀, 1U = OC 7333, 471901, 473067.
5. Misreported by Patten and Smith-Patten (2014b, Table 1) as 2005.
6. The term "pseudotype" is not a standard taxonomic term, as approved by the International Code of Zoological Nomenclature (ICZN). However, it was useful to RW Garrison's indication that the EMEC specimen was incorrectly labeled as a holotype. Likewise, we thought the term useful for our purposes in the species account.
7. Verbatim label data: Gomphus oklahomensis Pritchard ♂/Oklahoma: Commerce/Harwell/Robert H Gibbs, Jr. Collection/Collection U.S. National Museum//no 4
8. Abbott (2005) had a late flight date for *Phanogomphus* (at the time *Gomphus*) *oklahomensis* as 31 August for Texas. Abbott (2015) indicated the flight season extended into early September and Abbott and Lasley (2017) had 1 September as the late date for Texas. We pointed out that a fall flight date was incongruous with the species' regional phenology, as such Abbott determined that the September date was for a nymph specimen and that the latest date of an adult in Texas was 6 June (*in litt*, email dated 1 August 2018).
9. There is a specimen with a later date indicated for Texas but we feel it should be considered dubious given that it is so out of the flight season for the species in the region (USNM 338363, an adult ♀ collected in Bexar County, Texas, by JF Reinert, 8 September 1963). The identification and label data were verified by Dr. Oliver S Flint, Jr., on 4 May 2018 (*in litt*; original data label, as per Dr. Flint: "[printed] Bexar Co., Tex. [next line, printed] Coll: John F Reinert [next line, printed] 19 [in ink] 63 Sept. 8"), so we assume there was an error in label association or the wrong date was written on the label by the collector.
10. Bick's notes indicate that on 12 April Hornuff collected 3♀, all of which were designated *externus*. Because we have located only one of those females, we cannot say if the others, too, were actually *Gomphurus ozarkensis* or if they were originally identified correctly as *G. externus*.
11. Females of *Gomphurus externus* and *ozarkensis* have similar subgenital plates. It is understandable why earlier researchers, prior to *G. ozarkensis*' type description and subsequent morphological work, would have determined female *G. ozarkensis* specimens as *G. externus*. It is also plausible, given some overlap of characters between males of the two species (Figure 15.36, 15.37), that males would be confused early on.
12. Nonetheless, we suggest that all pre-1975 specimens of unidentified *Gomphurus* or those determined as *G. externus* from Oklahoma, Arkansas, Kansas, and Missouri be re-examined to ensure that they are not *G. ozarkensis*.
13. The one record that we have mixed feelings about is from Rogers County. In his notes, Bick report that Hornuff collected 1♀ at 6 mi NW of Claremore on 26 June 1956. We feel it is unlikely that *Gomphurus ozarkensis* occurs at that location, but because we have not located that specimen and cannot confirm its determination as *G. externus* we feel we should perhaps be cautious with it. If we were to consider that record dubious, then the date of the first record for Rogers County would change to 23 April 2012 (Collinsville City Lake, teneral ♀, JWA, OC 374603).
14. We were curious why the geographic ranges of *Gomphurus ozarkensis* and *G. externus* generally do not meet, so we examined various environmental variables (e.g., precipitation, temperature, and elevation) that might explain it. We discovered that the amount of topographic relief in their respective ranges appears to contribute to the division (Figure 15.38). On average, *G. ozarkensis* occurs in areas that are twice as rugged as where *G. externus* is found (mean roughness index: 9.7 for *ozarkensis*, 5.1 for *externus*; Figure 15.38).
15. For example, although BC is a talented field observer, we hesitate to accept a sight record he reported as an adult *Gomphurus externus* from Pumpkin Flats at the JT Nickel Preserve, Cherokee County, on 28 May 2005. We re-evaluated this record in August 2019 when we realized that there were but three records for the county. One of the other records was a specimen that we re-identified as *G. ozarkensis* (1♀, Sparrowhawk WMA, 4 May 2014, JT Bried, SP 1499). We feel the third record should stand as originally accepted by GW Lasley (1♂, 3 km NNE of Scraper, Illinois River, 26 May 2010, OC 329135) because it does indeed look like a *G. externus*. The peculiarity of that record is that despite rather extensive coverage of Cherokee County, it is the only verified record of *G. externus* in a part of the state one would expect to find *G. ozarkensis*, not *G. externus*, which leads us to wonder about the possibility it is a record of a vagrant instead of indicative of a population in the county.

One other record we consider dubious until proven otherwise is a specimen taken in Pushmataha County (1U, vic. Kellond, presumed to have been on Panther Creek, 27 April 1991, SR Moulton III, formerly JCAC/UTIC 36862, but the specimen appears to be lost, as per JC Abbott [*in litt*, 4 Oct 2019]). That record is from the foothills of the Kiamichi Mountains (part of the broader Ouachita Highlands), which we suspect does not host *Gomphurus externus*. We wonder if, instead, that specimen is actually *G. ozarkensis* or *Phanogomphus graslinellus*. If that specimen is correctly identified, it would stand as the only record of the species in Pushmataha County, a county that receives regular intensive and extensive survey coverage.

16. There is a ♀ specimen at PSU that is a *Gomphurus externus* (BS-P confirmed, 2018) but it is labeled as being collected "10-20-70," which is undoubtedly in error.

17. Donnelly (2004) added Johnston and Pittsburg Counties but we do not know the basis of those records. There has been a record subsequent for Johnston County but not for Pittsburg County; however, we have little reason to think the species does not regularly occur in both counties.

18. There is one adult male specimen (UCO 888) that we identified as *Stylurus plagiatus* that was labeled as being collected in Harper County in April (no day indicated) 1941 by Bickler. We have assumed that the collection month was actually August, as April is too early for the species in Oklahoma, especially given how far north in the state the specimen was collected.

19. Adults smaller than measurements provided by Needham et al. (2014) and Paulson (2009, 2011) are: **SP 209**, Kay County, ♂: TL = 52.9, HW = 31.8; **SP 129,** McIntosh County, ♀: TL = 53.7, HW = 32.8; **SP 77**, Noble County, ♂: TL = ~54, HW = 31.2; **SP 2445**, Stephens County, ♂: TL = 54.2, HW = 33.3; **SP 2664**, Tillman County, ♂: TL = 55, HW = 34.3; **SP 2216,** Love County, ♂: TL = ~56, HW = 32. All measurements are in mm.

20. We published a record for 30 April from 2006 of 6 individuals at Mohawk Park in Tulsa, Tulsa County (Smith-Patten et al. 2007). We have since decided that date is too early for the species' seasonality at that latitude, as such we rescind that record.

21. Two other records have earlier dates. However, MAP, and to some extent BS-P, have reservations about accepting those reports. The earliest report is a sighting of 1 individual by F Alm on 3 April 2012 at TNC's Oka' Yanahli Preserve, Johnston County. This record is more than a month and half earlier than any record in Oklahoma, and it would be quite early for the region as a whole. The second record is a ♀ specimen (UCO 889) that was collected by "Gardner" on 15 April 1961 at Lake Hefner, Oklahoma City, Oklahoma County. MAP feels the label is in error even if the species determination appears to be correct.

22. Reported from 72 localities falling within 125–465 m, with a mean elevation of 271 m.

23. SEMC 1141959 and 1141974, both were determined by D Huggins, so we have assumed they are correctly identified.

24. Parapatric species are two distinct species that come into contact where there is no known extrinsic barrier to gene flow, i.e., there is no physical, abiotic barrier preventing the mating of two species. In the case of *Stylogomphus albistylus* and *S. sigmastylus*, the lowlands of the Mississippi River Valley, which includes the ecoregions of the Mississippi Alluvial Plain, the Mississippi Valley Loess Plains, and the Interior River Valleys and Hills, acts superficially as a barrier between the two species except that the range of *S. sigmastylus* hops over the lowlands and continues farther east (Figure 15.75). In the eastern part of the range, arguably a disjunct population from the bulk of the species' range in Oklahoma, Kansas, Arkansas, and Missouri, it is parapatric with *S. albistylus*. The species are parapatric primarily within the Interior Plateau and the Southwestern Appalachians. Cook and Laudermilk (2004) indicated that at two locations, at Livingston, Rockcastle County, Kentucky, and on the Caney Fork River west of Crossville, Cumberland County, Tennessee, hybrids were collected historically, but recently *S. sigmastylus* has become dominate. If that is the case elsewhere where the two species meet, then there is no way to tell how much of a range shift the two species will undergo, presumably with *S. albistylus* losing ground to *S. sigmastylus*.

The division of *Stylogomphus sigmastylus* into what we presume are two genetically distinct populations argues for conservation measures specific to each population. Whereas it would appear that the eastern population has a chance to expand limitlessly, unless there is some extrinsic barrier to range expansion for *S. sigmastylus* that we have yet to detect, that population may not be as much of a conservation concern as its western counterpart that essentially has nowhere else to go, as it already spans the highland areas within the region. Moreover, we wonder if this western population warrants a distinct name, either as a subspecies or separate species, such as that suggested by Cook and Laudermilk's labeling of FSCA specimens as *S. divaricatus*. If that is the case, then that taxon would be of special conservation concern as a regional endemic.

## CHAPTER 16

CORDULEGASTRIDAE: SPIKETAILS

In Oklahoma, the spectacular spiketails, the family Cordulegastridae—a relatively small family (55 species worldwide, 11 in North America) of black-and-yellow dragonflies—haunt the edges of our perception. It is remarkable, for such large, showy species, to be so elusive. Part of the reason is seasonality: two of the three species known from Oklahoma, the Twin-spotted Spiketail (*Cordulegaster maculata*) and the Ouachita Spiketail (*C. talaria*) fly only for a short period in early spring (Figure 16.1). Part of the reason is rarity: as its English name implies, the Ouachita Spiketail occurs solely in the Ouachita Mountains of western Arkansas, from where the species was discovered and described to science in the early *twenty-first* century (Tennessen 2004), and southeastern Oklahoma, where there are only three records. Moreover, in Oklahoma, the Twin-spotted Spiketail is known from just two locations, both in southern McCurtain County. Part of the reason is habitat selection: all three species, including the only "widespread"

one in the state, the Arrowhead Spiketail (*C. obliqua*), breeds in seeps, seep-runs, and shallow, narrow, generally ephemeral streams embedded in woodland, slight waterways that males patrol quietly and inconspicuously and females frequent only to oviposit. Such habitats are distributed spottily in the landscape and may change from year to year, depending on the vagaries of rainfall and aquifer storage.

A taxonomic side note for Oklahoma's species of Cordulegastridae is that it has been argued that they belong in different genera (Carle 1983; Lohmann 1992), an idea that has been to some degree accepted elsewhere, but that has not yet caught on (impending?) in North America. If changes were accepted in North America, then *Cordulegaster obliqua* would become *Taeniogaster obliqua*; *C. maculata* would become *Pangaeagaster maculata*; and *C. sarracenia* and *C. talaria* would be *Zoraena sarracenia* and *Z. talaria*. Tennessen (2019) argued, based on nymph morphology, for acceptance of *Zoraena* as a distinct genus.

Cordulegaster maculata
Twin-spotted Spiketail

Cordulegaster obliqua
Arrowhead Spiketail

Cordulegaster talaria
Ouachita Spiketail

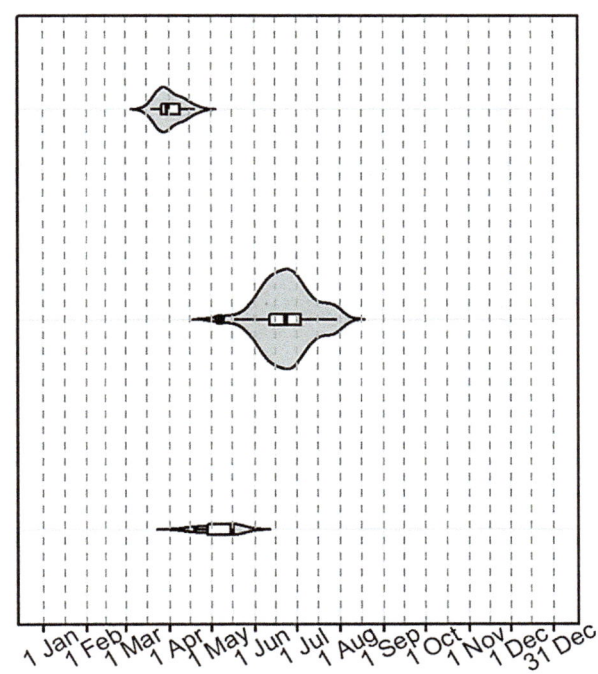

Figure 16.1 Seasonality in Oklahoma of the three species in the family Cordulegastridae. There are but two records, both in late April, of the Ouachita Spiketail (*Cordulegaster talaria*) for the state.

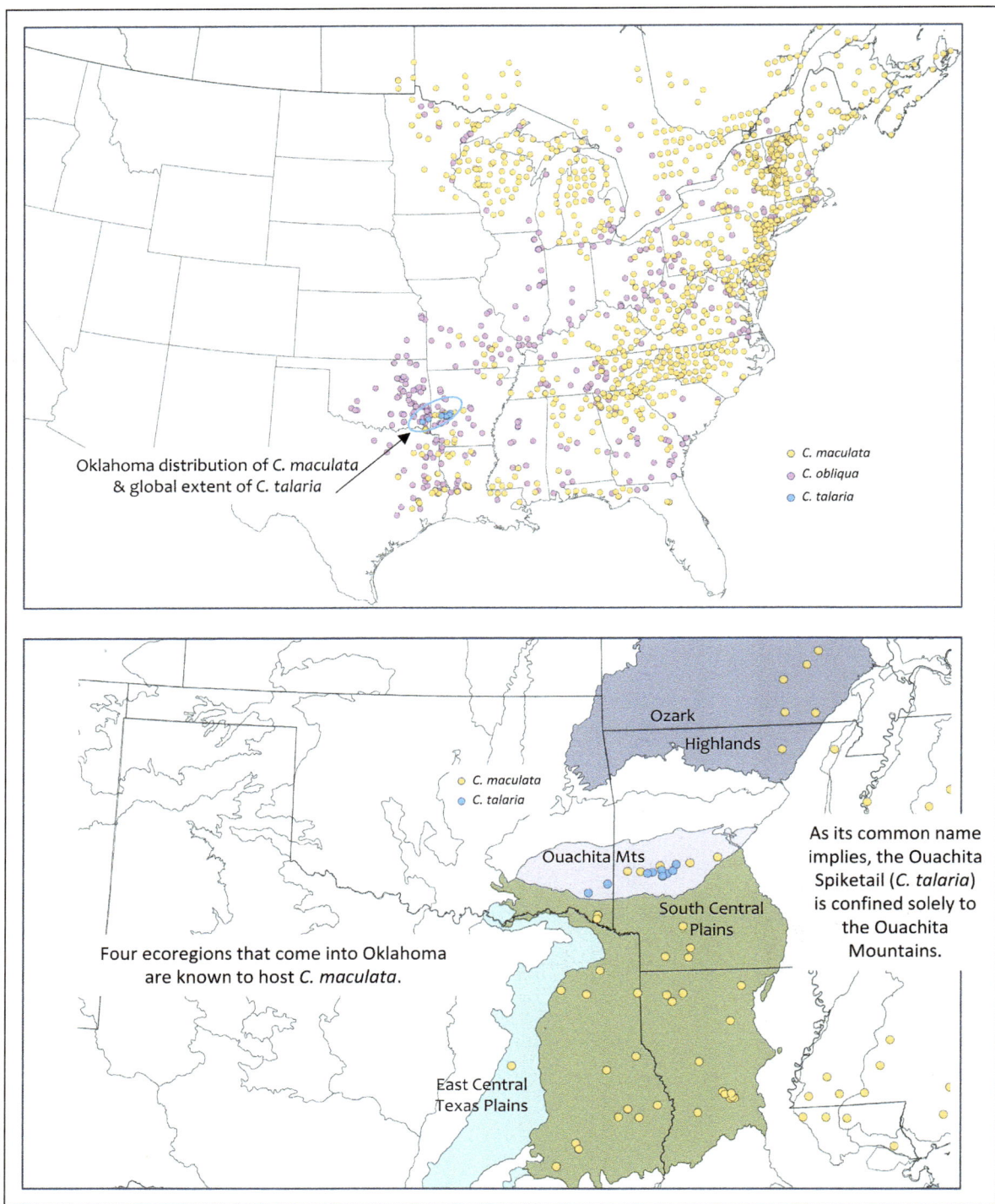

Figure 16.2 Distribution of the three *Cordulegaster* species known from Oklahoma. Although the distribution of *C. maculata* is extremely limited in Oklahoma, it has an overall distribution similar to the widespread *C. obliqua*. Conversely, *C. talaria* is not only limited in Oklahoma, but its entire distribution is comparatively minuscule. Four ecoregions that come into Oklahoma are known to host *C. maculata*.

CORDULEGASTER MACULATA SÉLYS, 1854 – TWIN-SPOTTED SPIKETAIL

The Twin-spotted Spiketail has been known from Oklahoma only from the past decade—from just eight confirmed records at two localities since 2010. The first record came from BA Heck's property southeast of Idabel, McCurtain County, when Heck photographed 1♂ on 20 April 2010 (OC 318493). The next year, on 3 April, Heck caught a pair, photographed them in hand (OC 327593), and let them go. In 2012, Heck reported 1♂ and 3U (OC 374173) on 23 March 2012, and then between 25–27 March, BA Heck, K Williams, and JW Arterburn reported a high count of 11 individuals (saying that they were "mostly females"). The species was not documented again until 2015, when we visited the site on 11 April and found 7♂ and 1♀. We examined 2♂ in hand and collected 2 others (SP 1507, 1508). Bad luck has it that the species has not been reported from Heck's property since, despite four reported visits.

In 2017, the species was found at Grassy Slough WMA, McCurtain County, when D Arbour and MAP collected three nymphs (SP 2226) on 20 March 2017. In 2018, we returned to the site on 19 March with Arbour and collected one exuvia (SP 2538), photographed 1♂ adult, and had in hand 4 early instar nymphs that were likely *Cordulegaster maculata*. MAP revisited the site on 31 March, when he encountered 4♀ (one taken as SP 2543). Arbour and/or BS-P visited the site at least five additional times that spring, but they did not report the species.

The 2 known localities have been rather well surveyed, with almost 40 visits reported from Heck's property during March and April (its flight season; see below) and close to 20 from Grassy Slough. Even if we separated out all of the days the species has been reported (i.e., treating 25–27 March 2012 as three records instead of one) that would mean it has been seen on 10 of 54 visits to those sites. That encounter rate is not great (especially if we were to also tally up all the visits made to suitable habitat in southeastern Oklahoma during the species' flight season). We suspect so few records is a matter of Oklahoma lying at the periphery of the species' geographic distribution. We would guess that it will continue to be more-or-less regularly reported from the state, and will be found eventually at other sites, but it will likely remain a rare species in Oklahoma.

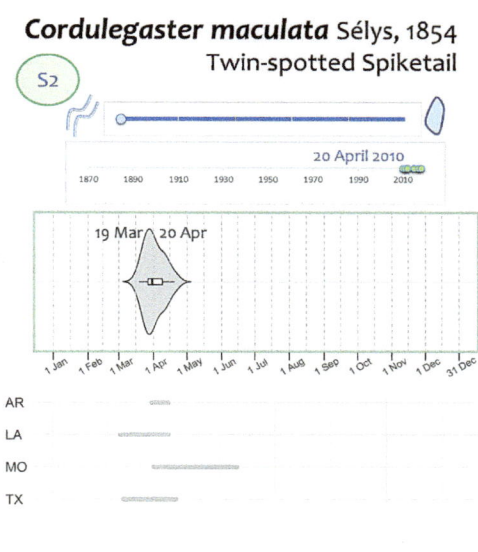

Figure 16.3 *Cordulegaster maculata*, the Twin-spotted Spiketail, A) ♂, Grassy Slough WMA, McCurtain County, 19 March 2018 (©Brenda D Smith, OC 478537). B) ♀, 10 km SE of Idabel, McCurtain County, 26 March 2012 (©Berlin A Heck, OC 374178), at the site of Oklahoma's first record (in 2010) and only records until 2017, when it was found at a second site. Its apparent scarcity is emphasized by its contracted flight season early in the year.

Cordulegaster maculata has been documented as flying in Oklahoma from 19 March until 20 April.[1] Texas and Louisiana have flight seasons that are not much longer (TX: 5 March–21 April, Abbott and Lasley 2019; LA: 2 March–14 April, Mauffray 2014). Missouri's season, however, extends into the summer (1 April–15 June, Trial 2005, Sims 2012), such as that reported for elsewhere in the United States and Canada, with New Jersey as the closest match (NJ: April–June; GA: March–June; KY: March–July; ME and OH: May–July; ON, QC, NS, and WI: May–August; FL: February–April; Paulson 2011). Fewer records have come from Arkansas, where the documented date range is 21 March to 14 April (iNat and OC).

CORDULEGASTER OBLIQUA (SAY, 1839) – ARROWHEAD SPIKETAIL

The Arrowhead Spiketail is generally uncommon in Oklahoma but has been reported fairly consistently since its discovery in the state. It was first found in the state on 25 June 1931 (2♂, Craig County, M Lasatey, OSU),[2] which is mid-season for this species in Oklahoma. Its early date is 7 May (OC 327881, GW Lasley, BAH) and its late date is 30 July (OC 331141, VWF; early and late records both from 2011, McCurtain County 11 km SSW of Smithville, Buffalo Creek).

Oklahoma's seasonality is consistent with Arkansas, Kansas, and Missouri (AR: 17 May–18 July, Harp and Harp 2003; KS: 13 May–25 July, SEMC 1349070, OC 321086; MO: 2 May–29 July, Trial 2005), although it starts rather late compared to Texas and Louisiana (TX: 31 March–19 July, Abbott and Lasley 2019; LA: 31 March–28 June, OC 313494, 478739).

The species was reported by Pritchard (1936) to fly late in the evening with *Somatochlora ozarkensis* (Ozark Emerald),

S. linearis (Mocha Emerald), and *S. tenebrosa* (Clamp-tipped Emerald) in Le Flore County in late June. Bick (1951:179) observed a ♀ "... in full sunlight while perched low on a small weed in a field ..." in Muskogee County in late July (FSCA). BA Heck observed the species "... patrolling on a small, clear, rocky, active stream (3-ft. across), flying up and down the stream about 1-foot above the water." J Fisher found one on the grill of his car, which is probably not the species' preferred habitat.

The species is primarily confined to the eastern half of the state, although there are a few records farther west. The Pontotoc Ridge Preserve, Pontotoc County, record of a ♂ (28 May 2008, D Wood, OC 421665) is likely the western extent of what can be considered the species' "normal" range in the state. But of course, Comanche County has two records—Wichita Mountains, 10 July 1932, RD Bird, ♂, OMNH 2099 and Fort Sill MR, 11 June 2002, BC Kondratieff, ♀, CSU and OC 381756 (Kondratieff et al. 2004, Zuellig et al. 2006).

Figure 16.4 *Cordulegaster obliqua*, the Arrowhead Spiketail, A) ♂, Lynn Mountain, Le Flore County, 27 June 2013 (©James W Arterburn). B) ♀, Cedar Creek 7 km E of Lenapah, Nowata County, 29 June 2013 (©Ken Williams, OC 401456). If one hopes to see a spiketail in the Southern Great Plains, pin those hopes on this spectacular species. It is common nowhere, and it is more than a little persnickety in its habitat preferences, but it is far, far more widespread in the region than its congeners.

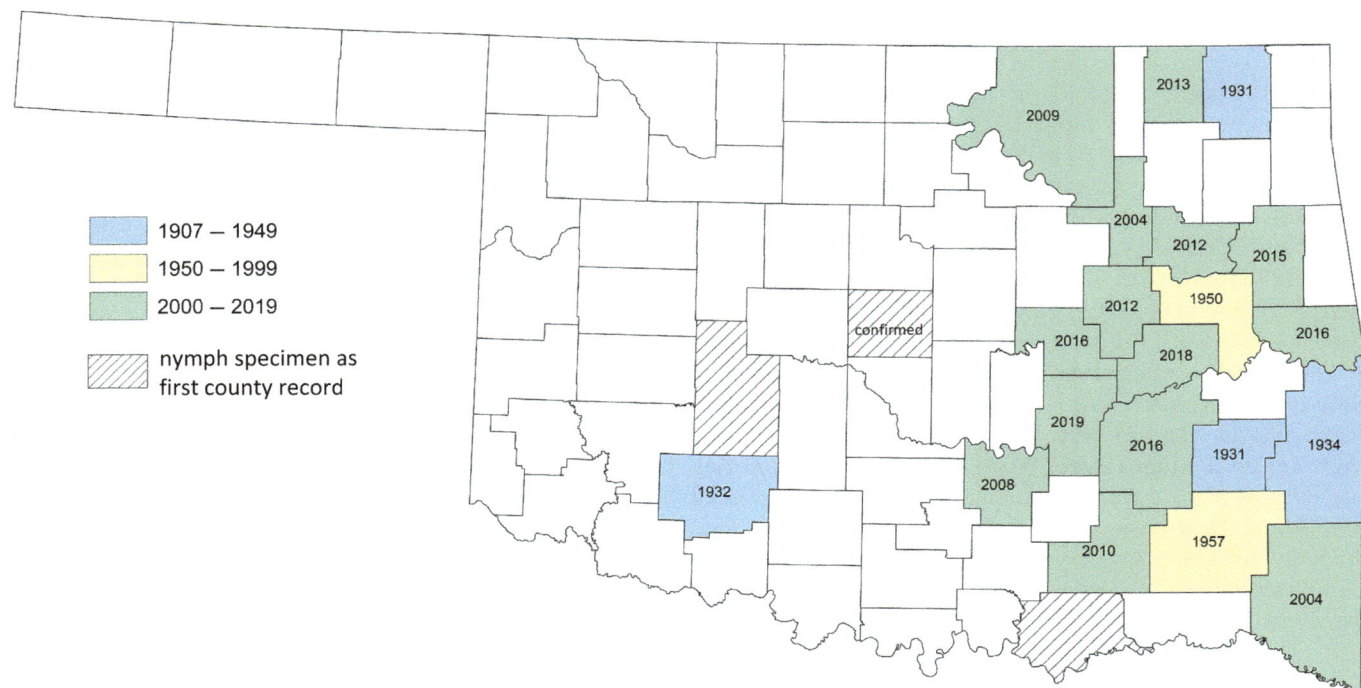

Figure 16.5 Counties in which *Cordulegaster obliqua*, the Arrowhead Spiketail, are known to occur in Oklahoma. Year within the county is that in which the species was first reported. Three counties (Bryan, Caddo, and Oklahoma; hatched) have only nymph specimens as documentation, only one of which has been confirmed.

These records could well be just vagrants, given the survey effort that has been dedicated to the Wichita Mountains Wildlife Refuge and the military base at Fort Sill, so we posit that if there was a population there, then more than two individuals would have been found by now.

We have two records outside of what we consider the species' range in the state. Both records are of nymphs. One is for Caddo County, Fort Cobb (1 indiv., 1 November 1975, D Dawson, OMNH 1766), but this specimen's determination has not been confirmed. The species identification of the second record has been confirmed (BS-P). That specimen was collected from Coffee Creek in Edmond, Oklahoma County (27 April 1991, D Bass, UCO 5183). It is possible then that a breeding population exists outside of the eastern half of the state, but we would not suggest holding one's breath on that.

There are two additional items, both regarding taxonomic issues, that warrant a brief mention. The first regards Caddo County nymph discussed above. This nymph was originally determined to be *Taeniogaster* Sélys, 1854, which at first blush sounds like a misidentification, but *Taeniogaster* is actually a synonym of *Cordulegaster* (see Cordulegastridae family discussion and hypothetical list). The other specimen posing some confusion in its designation is a ♂ from Latimer County that was collected by WM Fisher on 21 July 1931 (OMNH 2696). This specimen was determined by RD Bird as *Macromia annulata* and published as such (Bird 1932). GH Bick later re-identified the specimen as *C. obliqua* (actually as *C. obliquus* [sic]; Bick and Bick 1957). We are at a loss as to how this misidentification happened, but we hope mention of it will avoid any later confusion.

CORDULEGASTER TALARIA TENNESSEN, 2004 – OUACHITA SPIKETAIL

It could be said that the Ouachita Spiketail is the newest species to Oklahoma's dragonfly fauna, not that it was the most recently added to the state's species list, but because it was recognized as a species only a decade and a half ago (Tennessen 2004).[3] Its relative newness to science, its suspected low density and overall small population size, its short flight season (18 April–29 May), and its habitat specificity explain why there are a mere 17 records known for this species: 14 from Arkansas and 3 from Oklahoma (Table 16.1). Despite intensive and extensive adult and larval surveys conducted in Oklahoma and Arkansas in the past six years, only a handful of records have surfaced. To date, there are fewer than a dozen known localities: eight in two

counties (Garland and Montgomery[4]) in Arkansas and three in one county (McCurtain) in Oklahoma. That is relatively little profit returned on a huge time investment.[5]

Ken Tennessen was the first to catch what he suspected was a new species of *Cordulegaster* when he collected a ♀ *C. talaria* in Garland County (Tennessen 1990). He and others took the type series of specimens (16♂, 5♀) between 1990–1993 at or near a tributary to the Caddo River at Caddo Gap, Montgomery County (Table 16.1). The species has been seen sporadically in the state since (Table 16.1).

In Oklahoma, the Ouachita Spiketail was encountered the first time on 18 April 2011 (current early flight date) when the Heck family (Berlin, Pat, and Greta) startled a ♂

Cordulegaster talaria
Tennessen, 2004
Ouachita Spiketail

18 April 2011

18 Apr   29 May

dates are for
Oklahoma and Arkansas

Figure 16.6 *Cordulegaster talaria*, the Ouachita Spiketail, A) ♂, Cedar Creek 13 km WNW of Hochatown, McCurtain County, 18 April 2011 (©Berlin A Heck, OC 327732). B) ♀, Pine Mountain Spring, McCurtain County, 28 April 2018 (©Michael A Patten, OC 479444). These two records represented the first two for Oklahoma of this globally rare species that is restricted to the central Ouachita Mountains of southeastern Oklahoma and western Arkansas. It has since been found at a third locale in Oklahoma, but the total number of records known anywhere numbers fewer than 20.

Table 16.1 **All known records of *Cordulegaster talaria*, the Ouachita Spiketail.**

STATE	COUNTY	LOCALITY	ELEV	DATE	COLLECTOR	INDIVIDUALS	SOURCE
Arkansas	Garland	6 km NW of Crystal Springs, Red Branch	210 m	24 May 1990	KJT	1♀	OC 166999; Tennessen (2004)
Arkansas	Montgomery	type locality	250 m	26 May 1990	KJT	1♀; Allotype	FSCA; Tennessen (2004)
Arkansas	Montgomery	type locality	250 m	28–29 May 1990	KJT; JJD	7 (6♂, 1♀) and nymphs; Holotype and paratypes	JCAC 23733; FSCA; JJDC; OC 357172; Tennessen (2004); Tennessen, *in litt*
Arkansas	Montgomery	type locality	250 m	17 May 1992	KJT	6 (4♂, 2♀); paratypes	USNM 381990–381991; FSCA; Tennessen (2004)
Arkansas	Montgomery	type locality	250 m	23 May 1992	TEV	2♂; paratypes	ISM; Tennessen (2004)
Arkansas	Montgomery	type locality	250 m	26 May 1993	KJT	5 (4♂, 1♀); paratypes	FSCA; Tennessen (2004)
Arkansas	Montgomery	seep on logging road to Sharp Top Mountain	?	27 May 1993	KJT	nymphs	Tennessen, *in litt*
Arkansas	Montgomery	Forest Road 177, near Crystal Campground	320 m	23 July 1998	KJT	nymphs	Tennessen, *in litt*
Arkansas	Montgomery	17 km NE of Caddo Gap, Forest Road 476	240 m	18–19 May 2002	JCA	2♀	JCAC 37926–37927
Arkansas	Montgomery	7 km W of Black Springs, Caddo River	265 m	19 May 2006	DRP; GB; MT	3 (2♂, 1♀)	Johnson (2006); Paulson and Beaton field notes
Arkansas	Montgomery	spring seep on Caddo River, about 2 mi downstream of bridge at Caddo Gap	?	14 May 2007	GH	?	Harp (2007)
Oklahoma	McCurtain	13 km WNW of Hochatown, Cedar Ck.	220 m	18 Apr 2011	BAH; GH	1♂	OC 327732; Heck (2012)
Arkansas	Montgomery	type locality	220 m	11–12 May 2011	GWL; GLH; PH; DA*	4 (3♂, 1♀)**; 1♀ ovipositing	iNat 220628; OC 333477, JCAC 48828
Arkansas	Montgomery	type locality***	220 m	11 May 2013	DH; TH	1♂, 1♀	iNat 1269268 and 1269273
Oklahoma	McCurtain	Pine Mountain Spring	360 m	28 Apr 2018	MAP	1♀	SP 2561; OC 479444
Arkansas	Montgomery	18 km WSW of Norman	470 m	10 May 2018	BS-P; DA; WR	1♀ ovip; 13 nymphs?	BS-P field notes; rearing two possible *C. talaria* nymphs (SP 2575–2576)
Oklahoma	McCurtain	Arbour Seep	310 m	16–17 May 2019	BS-P; DA; MAP	3♂	SP 2737, OC 494904, OC 494956

Abbreviations: locality: Ck. = Creek; elev = elevation; collector: BAH = Berlin A Heck, BS-P = Brenda D Smith-Patten; DA = David Arbour, DH = Diana Hibbitts, DRP = Dennis R Paulson, GB = Giff Beaton, GH = Greta Heck, GLH = George L Harp, GWL = Greg W Lasley, JCA = John C Abbott, JJD = Jerrell J Daigle, KJT = Ken J Tennessen, MAP = Michael A Patten, MT = Mike Thomas, PH = Phoebe Harp, TEV = Timothy E Vogt, TH = Terry Hibbitts, WR = William Rainey; individuals: ovip = ovipositing, U. = undetermined; source: BS-P = Brenda D Smith-Patten, FSCA = Florida State Collection of Arthropods, ISM = Illinois State Museum, JCAC = John C Abbott Collection, JJDC = Jerrell J Daigle Collection, OC = Odonata Central, SP = Smith-Patten/Patten Collection, USNM = United States National Museum of Natural History/Smithsonian Institution

Notes: *Arbour was present only on 11 May 2013. He reported seeing one female ovipositing (*in litt*, email 12 November 2018). **Lasley reported seeing 3♂ and 1♀ on 11 May (iNat 220628; 1♂ as JCAC 48828) and at least 1♂ on 12 May (OC 333477; records *in litt*, email 12 November 2018). ***Coordinates from iNat plot at 10 km SW of Caddo Gap for the ♂ and at 6 km SW of Caddo Gap for the ♀, but Troy Hibbitts (*in litt*, email 13 November 2018) said that Diana and Terry Hibbitts were at the type locality.

from a blackberry bush (*Rubus* sp.) in the Ouachita NF, 13 km WNW of Hochatown[6] (OC 327732; Heck 2012; Figure 16.6). Unfortunately, they did not have an insect net with them at the time, so they were able only to photograph it. The species has not been re-found at the site despite multiple return visits. The second state record came in a much more circuitous manner. On 20 April 2018, BS-P and Wade Boys collected *Cordulegaster* nymphs at Pine Mountain Spring. Based on descriptions of *C. talaria* nymphs by JC Abbott, BS-P thought that she and Boys had found a breeding site for the species, so they reported it to MAP as such. MAP returned to the site on 28 April, when he managed to photograph and capture a ♀ (OC 479444, SP 2561). Joyful cheers were heard all around—"we found a breeding site in Oklahoma!" *Nope*. On 8 May, Greg Lasley, Mike Dillon, and BS-P made two visits, one in the morning and one in the afternoon. Neither visit was successful for adults or nymphs. But the death knell came a couple of weeks later when KJ Tennessen confirmed that two of the nymphs collected on the 20th were actually *C. obliqua*, the Arrowhead Spiketail. A third nymph, one that was kept alive until December 2019, also appears to be *C. obliqua*. Dreams dashed by a bizarre coincidence.

Surveys in Arkansas in 2018 produced slightly better results. On 19 April, BS-P and Boys surveyed at the species' type locality and then checked various seeps throughout the day while they made their way back towards Oklahoma.[7] The last stop of the day, at Fernwood Seep NA, Polk County, was the only one productive for *Cordulegaster*. No adults were seen but nine nymphs of various ages were found; however, the three nymphs that were collected turned out not to be *C. talaria* (*C. obliqua*, SP 2554; *C. maculata*, SP 2555, *C. maculata*, cf, SP 2556).[8] Better luck was had on 10 May, when BS-P, DA, Dillon, and William Rainey spent the day surveying at five seeps in the Ouachita NF, Polk and Montgomery Counties. Only one site was productive: an unassuming roadside trickle about 18 km WSW of Norman, Montgomery County, that they stopped at midday. Dip netting brought up 15 *Cordulegaster* nymphs[9] that squiggled out the gloppy mud into BS-P's hand. Two were taken for rearing (SP 2575 and 2576), but they have yet to be identified to species. While larval sampling, an adult ♀ *C. talaria* came in to oviposit, practically at BS-P's feet. She yelled for DA, who popped his head into the vegetation just long enough to see her (the ode, not BS-P) dip her abdomen a few times. Regrettably, DA did not have an aerial net in his hand, nor did BS-P, who instead had a handful of mud, which is a surprisingly inefficient way to capture adult dragonflies.

That seep was along a wide, graded dirt Forest Service road. It was obscured by shrubby vegetation, a look inside of which showed a seep run not much more than a trickle (¼–½ m wide, 1–2 cm deep, with a substrate of pea gravel and rich silt). Elsewhere, *Cordulegaster talaria* has been described as occurring at shaded seeps and associated narrow runs as well as in nearby fields. But few details have been given outside of Tennessen's (2004) type description.[10] Oklahoma's first record was at a locality described as "very rocky and is dense with trees, mostly hardwood (*Quercus* and *Carya* sp.) and pine (*Pinus* sp.). The slopes are steep, falling to a small stream that meanders through the area" (Heck 2012:9). A recent visit revealed a seep run along one of the roads leading up to Heck's site. Previously, we would not have considered such habitat suitable for the species but we now question if that small seep run is in fact the natal site of Heck's adult *C. talaria*. We have been rather focused on finding *C. talaria* at more pristine wooded seeps but we are now starting to think otherwise. Finding 3♂ at "Arbour Seep," a complex along Union Valley Trail (a wide, graded dirt road) in far eastern McCurtain County, on 16–17 May 2019, solidified the idea that, though a habitat specialist and likely sensitive to extreme disturbance, the species appears to be tolerant of dirt roads and nearby pastures (but see below). Records from throughout the species' range indicate that it is a highland species (relative to the area's topography) found at elevations of 210–470 m (689–1,542 ft).

The endemicity of this species—only found in the Ouachita Mountains of Arkansas and Oklahoma within the tiny range of <1,300 km[2]—*just 500 mi[2]*—should arguably warrant its protection. Recall, too, that it is a habitat specialist, a low-density species with a presumed small population, and one that, irrespective of years of intensive surveys, has few known occurrences. We anticipate that it will be found at additional sites in the Ouachitas (*with much more hard work*!) but the species will continue to be rare throughout its range and threatened, at least moderately, by human population growth and attendant land-use changes, as well as climate change. And let's also keep in mind that the species is undoubtedly subject to being roadkilled by logging trucks and other vehicles that speed along forest roads. But the greatest threat is pressure on seeps from water use and loss of quality, as well as habitat loss and degradation. It cannot be underestimated, too, that this species may have an "extinction lag" (*sensu* Kuussaari et al. 2009), which means that though it currently persists in areas disturbed by dirt roads, pastures, and pine plantations, the species could decline in the near future once the full impacts of habitat loss and degradation take effect. As such, the species' ranking as imperiled/critically imperiled by NatureServe (S1 in Arkansas and Oklahoma, N1N2 nationally, and G1G2 globally) is most certainly warranted.[11]

NOTES

1. There may be a later flight date from when we visited Heck's property late in the day on 27 April 2013. We saw what we thought were 2♀ but we ended up backtracking on the record because of the poor lighting and we were not able to obtain documentation. We still have a niggling feeling that we were too quick to dismiss the record, but perhaps it is best to play it safe than sorry.

2. AE Pritchard thought he had added the species to the state, as shown by his letter to EB Williamson, dated 30 November 1932 (Figure 15.60). No details were provided in the letter, nor have we found an associated specimen, so for now the OSU record will stand as the first state record.

3. One other species, *Stylogomphus sigmastylus*, the Interior Least Clubtail, was also recognized in 2004, but that was split from an existing species already known in the state as opposed to a species that had gone undetected previously.

4. The species was attributed to Clark County, Arkansas, in error, but has since been clarified. It appears that this error sprang from a miscommunication or a transcription error from an email, dated 7 January 1997, between George Harp and Bill Mauffray regarding an adult *Cordulegaster* that was thought to have been found by Ken Tennessen in Clark County, Arkansas. That record was published by Mauffray (1997, present still in the 2014 edition) as *C. erronea*, the Tiger Spiketail, but was later changed to *C. talaria* by Donnelly (2004a). When BS-P tried to track that record down in 2018, it became apparent that both Tennessen and Harp thought the record was a mistake. Therefore, it was agreed by all parties that the record should be considered in error (*in litt*, Harp, Mauffray, and Tennessen, emails dated 3–15 October 2018). On 25 October 2018, JC Abbott "unconfirmed" the Odonata Central record (OC 166998).

5. We know of at least 17 people, not counting ourselves, who have searched for the species. Some were lucky enough to find it (Table 16.1), but more often than not they came away with no records. For example, we conducted six years of targeted surveys for adults and BS-P and Wade Boys conducted larval sampling for two years, with little success.

6. Reported as "about 19.3 km northwest of Broken Bow … and 2.9 km south of the intersection of FS road 53000 and 53420, near the Cedar Creek crossing" (Heck 2012:9).

7. Note that there was an unexpected cold front that had come in, so temperatures hovered around 50°F in the morning. It was only 62°F by the end of the day.

8. SP 2554 and 2555 were identified by KJ Tennessen and JC Abbott.

9. Sizes ranged from barely 2 mm long to 28 mm.

10. Tennessen (2004:838) described the type locality: "The first-order tributary where most of the *C. talaria* specimens were collected originates at the base of a wooded hill approximately at the base of a wooded hill approximately 150 m from the Caddo River. The head of the seep branches from a densely shaded area, forming a small stream about 25–40 cm wide which runs through open pasture for most of its length before entering the wooded bank of the river. The dominant aquatic macrophyte in the open part of the stream was water starwort (*Callitriche heterophylla*). The shady seep comprises a small area, approximately 50 m², and is only 2.5–10 cm deep, with a substrate of organic ooze, mud, and dead sticks partly covered with moss…Cattle graze the pasture along the first-order stream and adjacent to the seep, although they apparently do not enter the actual seep probably because of the density of brush." Recent visits to the site indicate that the habitat has changed little since (including aerial images that show woodland density directly at the site as approximately the same, although the area in general has seen some loss; but we did not quantify density at or near the site).

11. The species does not qualify for such a ranking as per the last assessment by IUCN, which continues to consider it of Least Concern (Paulson 2018b; assessed 5 May 2016).

## CHAPTER 17

MACROMIIDAE: CRUISERS

One could say the dragonfly family Macromiidae, the cruisers, is under-represented in the New World. With 123 species worldwide, the family is not particularly large, but only 7%, just 9 species, are found in the United States. Similarly, a clear majority, 63% (78 of 123), of species in the family are in the genus *Macromia*, but a mere 7 of these (9%) occur in the United States. Oklahoma hosts 5 of those 7 *Macromia* with one other family member, *Didymops transversa*, the Stream Cruiser, thrown in for good measure.

In Oklahoma, the Stream Cruiser is both the earliest flier (Figure 17.1) and the runt of the litter (56–60 mm total length, but *Macromia* are 62–91 mm). It is also not easy to get a good look at one: it spends a great deal of time on the wing as it patrols lakes, rivers, or streams, flying fast and low over the water as it traces the contours of the shoreline. With its bulbous abdomen tip, it can be mistaken for a clubtail (Gomphidae), but once the flight style and shape are ingrained, one discovers that the species is not as scarce as would be assumed solely from chance encounters with perched individuals.

Until recently we would have said that the Stream Cruiser differs utterly from other Macromiidae species that occur in Oklahoma. With the addition of a fifth *Macromia* species to the state—*M. annulata*, the Bronzed River Cruiser, thus far confirmed just once, but perhaps with a regular presence in the far southwest—that is no longer the case because the two species are broadly similar in their brown and creamy white coloration.[1] The other *Macromia* deviate from that coloration by instead being black-and-yellow, although they are equally as large as *M. annulata*. Being large black-and-yellow dragonflies make these *Macromia* easy to mistake for a spiketail (*Cordulegaster*; although the latter have two stripes on the side of the thorax instead of one and generally, they are slower fliers). These *Macromia* share basic coloration and, to a disturbingly large extent, pattern and morphology. As such, it is too easy to mistake them for one another, rendering species diagnosis, shall we say for polite company, irksome.

Behavior helps only a little for species determination because all Macromiidae have a penchant for hurtling by along the shorelines of lakes and rivers. It is only the Royal River Cruiser (*Macromia taeniolata*) that can be said to be more lentic than lotic, because it is most often found at lakes than all the other Macromiidae. It can also be said that *Didymops transversa* forage low and tend to stay closer to water, whereas *Macromia*, in general, have a (maddening) fondness for foraging often well above head height (and net height!) in open fields and along quiet roads and power line cuts in eastern forests. It is not uncommon to find *Macromia* individuals swarming with other species of *Macromia*, as well as emeralds (*Somatochlora* sp.), saddlebags (*Tramea* sp.), and gliders (*Pantala* sp.).

There is enough individual morphological variation exhibited by each of Oklahoma's black-and-yellow species that we no longer accept most sight records, particularly of individuals seen (or photographed) only on the wing (Figure 17.2). Our own experience taught us that our initial identification of a river cruiser on the wing proved, once we had the individual in hand, too often to be the wrong one (we do not have a quantitative estimate of our error rate, but it was likely in the 20–25% range, which is much too high). Accounts herein thus are grounded exclusively in specimen records supplemented by clear, identifiable photographs, although we did sparingly include a few sight records if the individual was studied by a trusted observer as it perched (or in hand) and did not alter basic understanding (e.g., seasonality, habitat use, geographic distribution) derived from the specimen records.

High morphological variation has caused identification woes and long-standing taxonomic fits—species limits remain unclear, and it is possible that there are six, not five, species in the state, the sixth being undescribed to science. Or perhaps it is seven (or eight?), with a lumped species that ought to be split? Complexity is exemplified perfectly by the plight of the Allegheny River Cruiser (*M. alleghaniensis*), a species not thought to occur as far west as Oklahoma, until both an archived photograph (and accompanying specimen) that had been identified as another species was shown to be this one *and* a misidentified specimen collected decades earlier was re-identified (Patten and Smith-Patten 2016).

Taxonomic controversies swirl about the Swift River Cruiser (*M. illinoiensis*) complex, two taxa that, because of high levels of intermediacy in some areas (Donnelly and Tennessen 1994), generally are lumped into a single species. Even so, intermediacy is low to non-existent in other areas, suggesting to us an alternative hypothesis of mosaic hybrid zones between biological species, which would mean the Illinois River Cruiser (*M. [i.] illinoiensis*) and the Georgia River Cruiser (*M. [i.] georgina*) are species, not subspecies

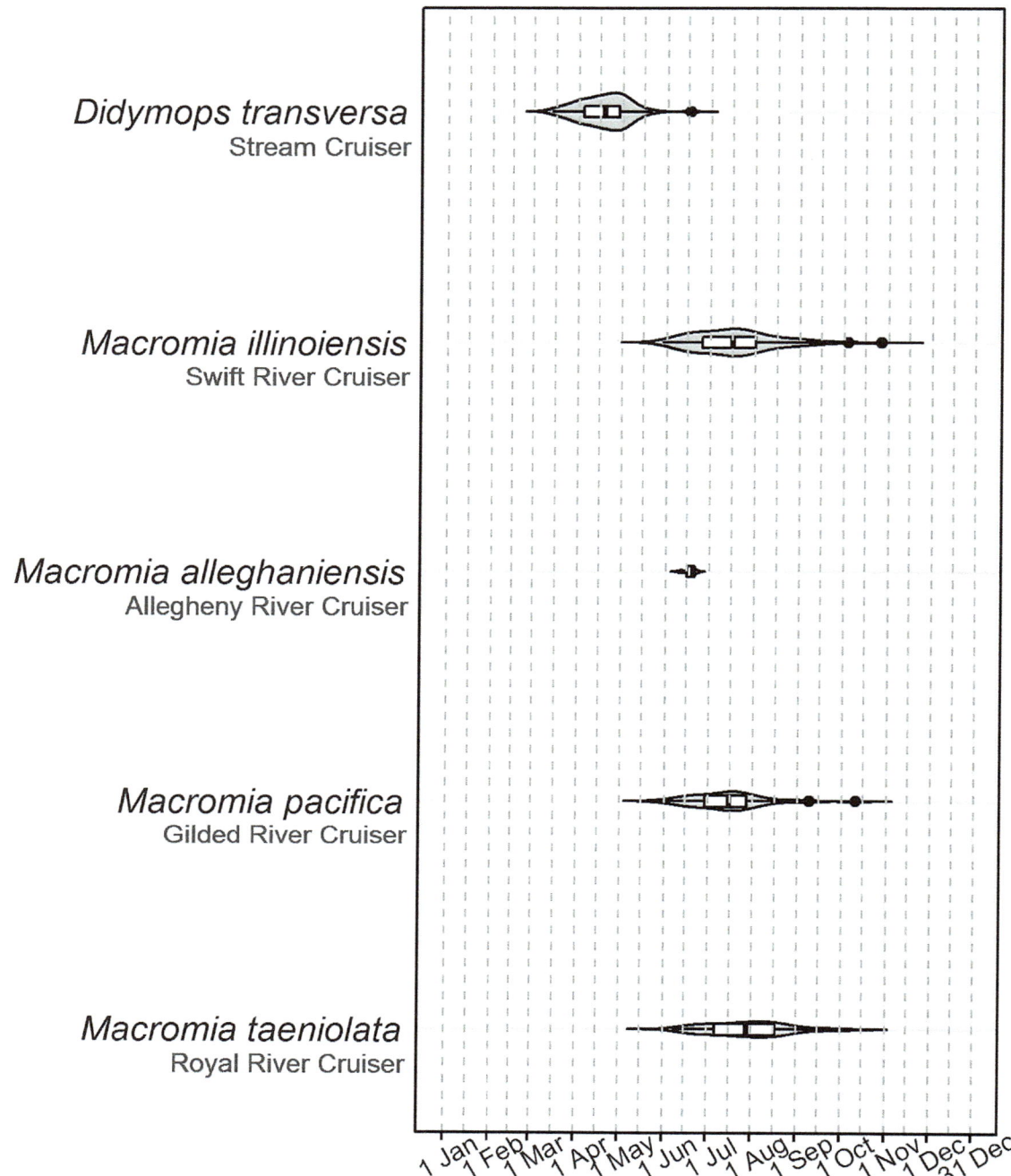

Figure 17.1 Seasonality in Oklahoma of the species in the Macromiidae. The Stream Cruiser (*Didymops transversa*) is one of the classic early season fliers in the state. The river cruisers (*Macromia* sp.), by contrast, fly chiefly in mid-summer, with one species, the Allegheny River Cruiser (*M. alleghaniensis*) recorded only three times. A sixth species, the Bronzed River Cruiser (*M. annulata*), has been recorded definitively just once, on 5 August.

Figure 17.2 An all-too-typical view of a river cruiser (*Macromia* sp.), in this case a ♂ at Stillwater, Payne County, on 11 July 2011. Identification to genus is a snap, but identification to species, even if an adult ♂, requires excellent views of perched individuals (fly-bys can have yellow appear extended due to movement) or in hand examination of the mesepisternal stripes, hamules, mesotibial keel, vertex, postfrons, pattern on S2 (dorsally and laterally), and pattern of yellow on the distal abdominal segments. Seldom are river cruisers kind enough to reveal these secrets, and so many—such as this one—must be left unidentified. Photograph by ©Jason R Heinen.

(Oklahoma records of the complex are of *georgina*, with one to three records of apparent *illinoiensis* and additional hybrids). The Royal River Cruiser (*M. taeniolata*) raises further taxonomic conundrums. A named taxon floating around, *M. wabashensis*, variously has been treated as a full species (Williamson 1909), a hybrid between the Royal

and Gilded (*M. pacifica*) River Cruiser (e.g., Abbott 2005), or a yellow color morph of the Royal River Cruiser (e.g., Needham et al. 2014). Many *wabashensis* or *wabashensis*-like individuals are known from Oklahoma. Even individuals that outwardly fit the Royal River Cruiser fall into two groups on the basis of certain morphological features, particularly the size and shape of the male's mesotibial keel. In Oklahoma, males with a small, leaf-like keel have been collected only on the Gulf Coastal Plain, whereas every male collected in the state's other ecoregions has a short, low, blade-like keel (see *M. taeniolata* account). Could the latter individuals be examples of the unnamed taxon that Cook (1994) reportedly discovered? More work is needed, but at this stage the geographic segregation in Oklahoma is compelling.

And we cannot close a short discourse on taxonomic questions in *Macromia* without acknowledging that, in most cases, the instance and extent of hybridization among species is unknown. As mentioned above, some argue that the Gilded and Royal River Cruiser hybridize enough to have created a stable phenotype that Williamson (1909) was so bold to name as a species. We, too, have wondered about hybridization in Oklahoma, never more so than regarding a ♂ we collected in 2013 in northwestern Oklahoma (SP 785, OC 425615). We fussed over this specimen for many months and vacillated repeatedly about its identification. Was it a Bronzed River Cruiser (*M. annulata*), a species then unrecorded in the state? Or was it an aberrant *georgina*-type Swift River Cruiser? Its hamules pointed toward the latter but most other characters pointed to the former (see the *M. annulata* species account for more discussion). Whereas it would be easy to place all faith in the hamules,[2] we cannot discount the vertex color and abdominal pattern; in short, we cannot eliminate a hybrid.

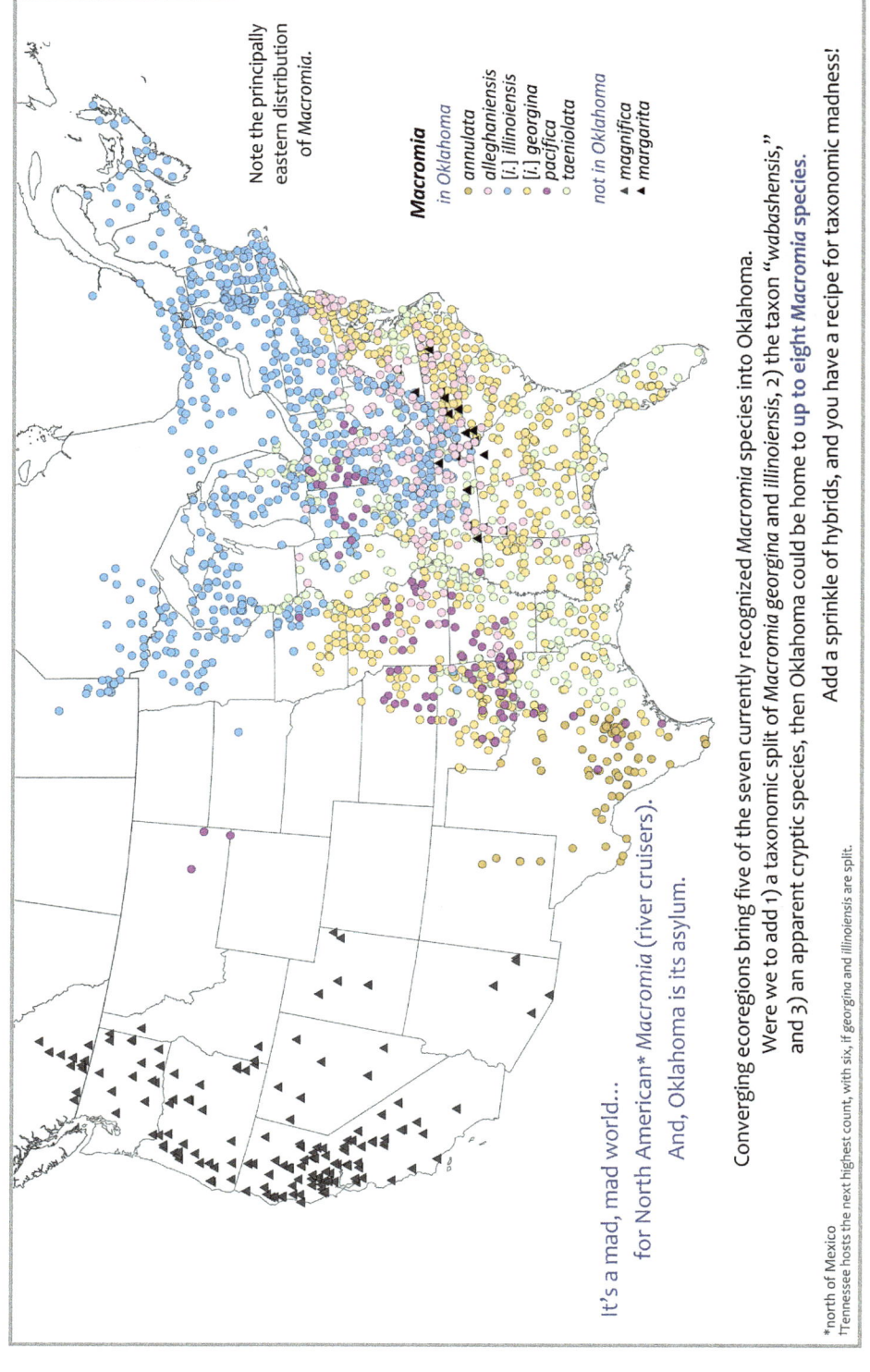

It's a mad, mad world…
for North American* *Macromia* (river cruisers).
And, Oklahoma is its asylum.

Note the principally eastern distribution of *Macromia*.

Macromia

in Oklahoma
- annulata
- alleghaniensis
- [i.] illinoiensis
- [i.] georgina
- pacifica
- taeniolata

not in Oklahoma
- ▲ magnifica
- ▲ margarita

Converging ecoregions bring five of the seven currently recognized *Macromia* species into Oklahoma. Were we to add 1) a taxonomic split of *Macromia georgina* and *illinoiensis*, 2) the taxon "wabashensis," and 3) an apparent cryptic species, then Oklahoma could be home to **up to eight *Macromia* species.**

Add a sprinkle of hybrids, and you have a recipe for taxonomic madness!

*north of Mexico
†Tennessee hosts the next highest count, with six, if *georgina* and *illinoiensis* are split.

Figure 17.3 **It's a mad, mad world … for North American**† ***Macromia*** **(river cruisers). And, Oklahoma is its asylum.**

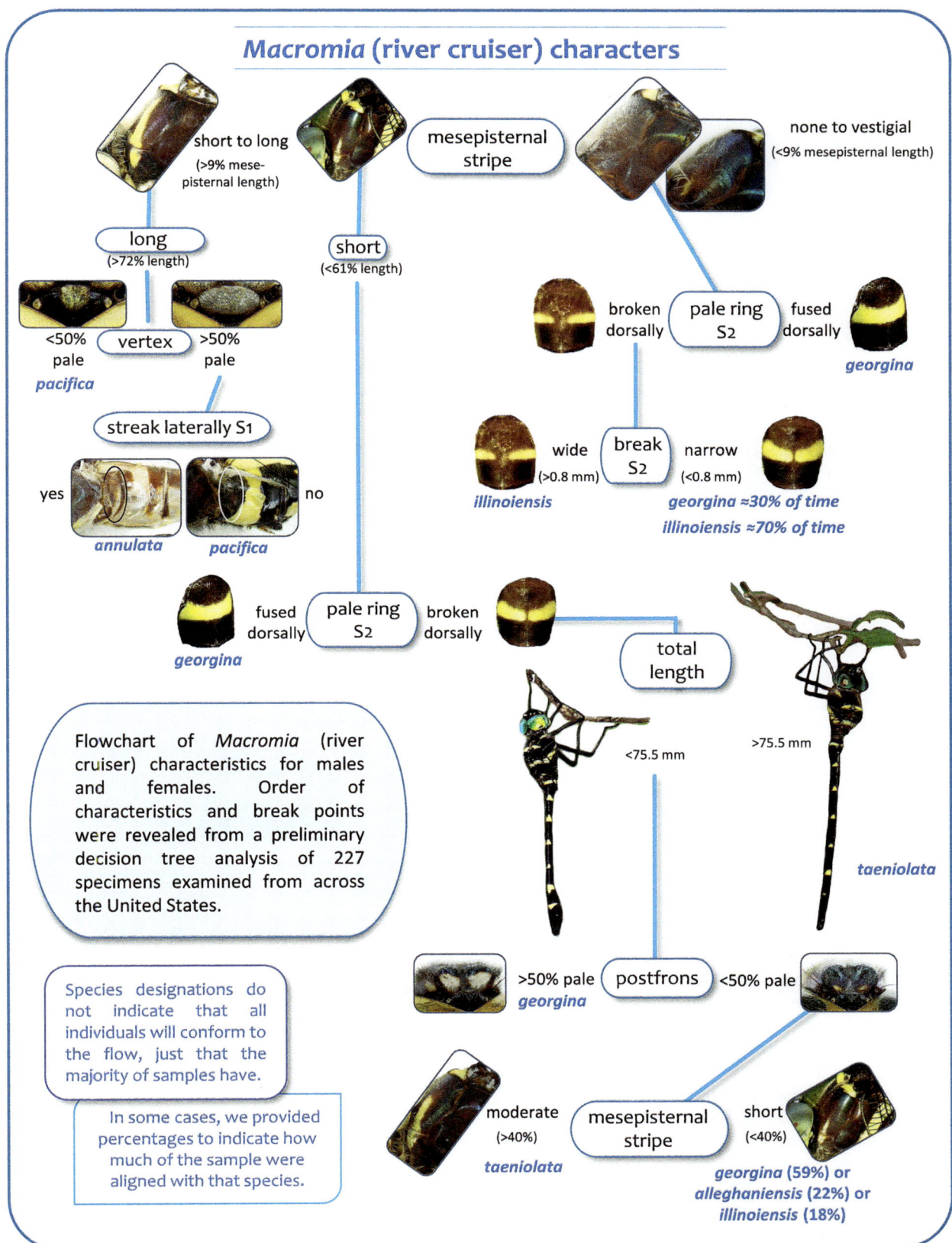

Figure 17.4 *Macromia* (river cruiser) characters and Clues, caveats, and conundrums of *Macromia* (river cruiser) identification.

Clues, caveats, and conundrums of *Macromia* (river cruiser) identification

Macromia do not always have straightforward species identifications

there is still debate as to if *M. illinoiensis* is a single species with two subspecies or is actually two separate species, in either case there are two distinct taxa: 1) *illinoiensis* and 2) *georgina*, the Illinois and Georgia River Cruisers.

Some useful characters are:

S5–6 yellow dorsally
though at times S6 can be almost entirely black

S5–6 all black or nearly so on *illinoiensis*

georgina

georgina exhibits much variability

S3 yellow wraps laterally to join with pale underneath segment

this S3 pale pattern* may be the only way to distinguish ♀ *georgina* and *alleghaniensis* in the region, but it remains to be seen if this character distinguishes the two species as well as it does in the southeastern US (Beaton 2007)

generally *illinoiensis* has no mesepisternal stripe

however, *georgina* can have a stripe that is near absent to nearly 40% of the mesepisternum

S2 entirely yellow laterally

S2 with black laterally at the auricle

the extent of yellow on the postfrons of *georgina* varies from a fifth to over half

rarely (<2% of instances), the vertex is partially pale

illinoiensis   *georgina*

black auricles can also be found on *M. magnifica*, the Western River Cruiser, and occasionally on *M. alleghaniensis*, the Allegheny River Cruiser

georgina generally has a wide break of the pale on S7 at the lateral carina

whereas the pale on S7 wraps entirely around with *M. alleghaniensis* but may not always be complete ventrally

one conundrum here... these two photos are actually of the same *M. alleghaniensis* ♂, not a *georgina* and *alleghaniensis*; nonetheless, generally the two species can be distinguished by a break along the lateral carina versus not

as much as 5% of individual *M. pacifica* have the vertex wholly black, defying the key in Needham et al. (2014)

Variation within what is called *M. taeniolata* is bewildering.

Various august researchers, from EB Williamson to Carl Cook, have opined that one or more cryptic taxa are included under this name. The former went so far as to provide a name for a yellower form (*wabashensis*), a form that some feel is a valid species, some feel represents only a color morph of *M. taeniolata*, and some feel is a hybrid between *M. taeniolata* and *M. pacifica*.

See the species account for *M. taeniolata* for more bewildering variation.

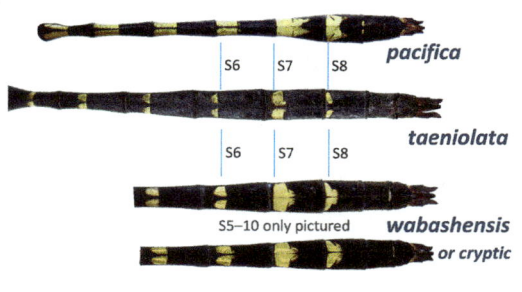

S6   S7   S8   *pacifica*

S6   S7   S8   *taeniolata*

S5–10 only pictured   *wabashensis* or cryptic

Note the differences in extent of yellow and shape of spots, especially on S6–8

Figure 17.4 (*Continued*) **Macromia** (river cruiser) characters and Clues, caveats, and conundrums of *Macromia* (river cruiser) identification.

DIDYMOPS TRANSVERSA (SAY, 1839) – STREAM CRUISER

The Stream Cruiser is one of Oklahoma's quintessential early season species: it flies from mid-March to mid-May[3] (13 March 1930, ♀, coll. by "Coats" at Norman, Cleveland County, OMNH 4361, which is also the first record of the species for Oklahoma, to 22 May 2013, 1 pair photographed by KW at Bixhoma Lake, Wagoner County, OC 492357). This flight season is similar to elsewhere in the region, although in Texas and Louisiana the species begins to fly earlier still (AR: 19 March–19 May, OC, Harp and Rickett 1977; KS: 22 April–9 May, SEMC, OC; LA: 22 February–10 May, OC; MO: 1 April–26 May, OC; TX: 8 February–2 May, Abbott and Lasley 2019).

Despite its English name, the Stream Cruiser is not restricted to lotic habitats; rather, it is numerous both at lake shores and along rivers and streams with slow current. It is largely confined to the east half of the state, with records thin even in the central mixed grass prairie, although, as is often the case, the Wichita Mountains, Comanche County, in the southwest are an exception. To wit, in the earliest part of its season, it can be common: RD Bird and students collected 18 adults in one day (26 April 1930; OMNH, UMMZ, USNM) and 15 on another (23 April 1932, OMNH) at the Wichitas. We routinely encounter half a dozen individuals at a time when in the field in April, and we conservatively have counted up to 14 on a given day across multiple localities. We have noted three occasions of ovipositing; all of which were lone females.

Fun fact: the Stream Cruiser can be confused easily with a large clubtail because of its clubbed abdomen and its habit of patrolling shorelines by repeatedly flying the same beat at roughly the same tempo, diverging only briefly to catch prey or fight with an intruder, yet none of the clubtails of comparable body size (e.g., the *Dromogomphus* spinylegs) fly so early in Oklahoma.

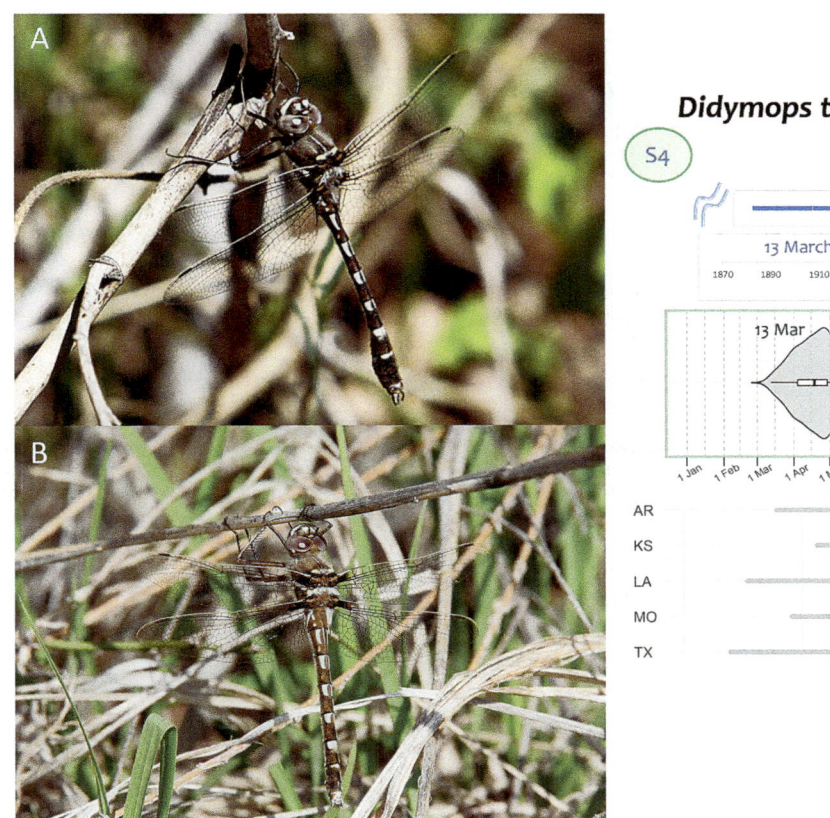

Figure 17.5 *Didymops transversa*, the Stream Cruiser, A) ♂, Little River NWR, McCurtain County, **24 March 2011** (©Berlin A Heck, OC 327524). B) ♀, Bixhoma Lake, Wagoner County, 28 March 2012 (©James W Arterburn, OC 375891). An "early season specialty" of the Southern Great Plains, this clubtail-esque species can be locally common during its brief flight season from late March to early May. Set its English name aside: it is as likely to cruise a lakeshore as it is a stream.

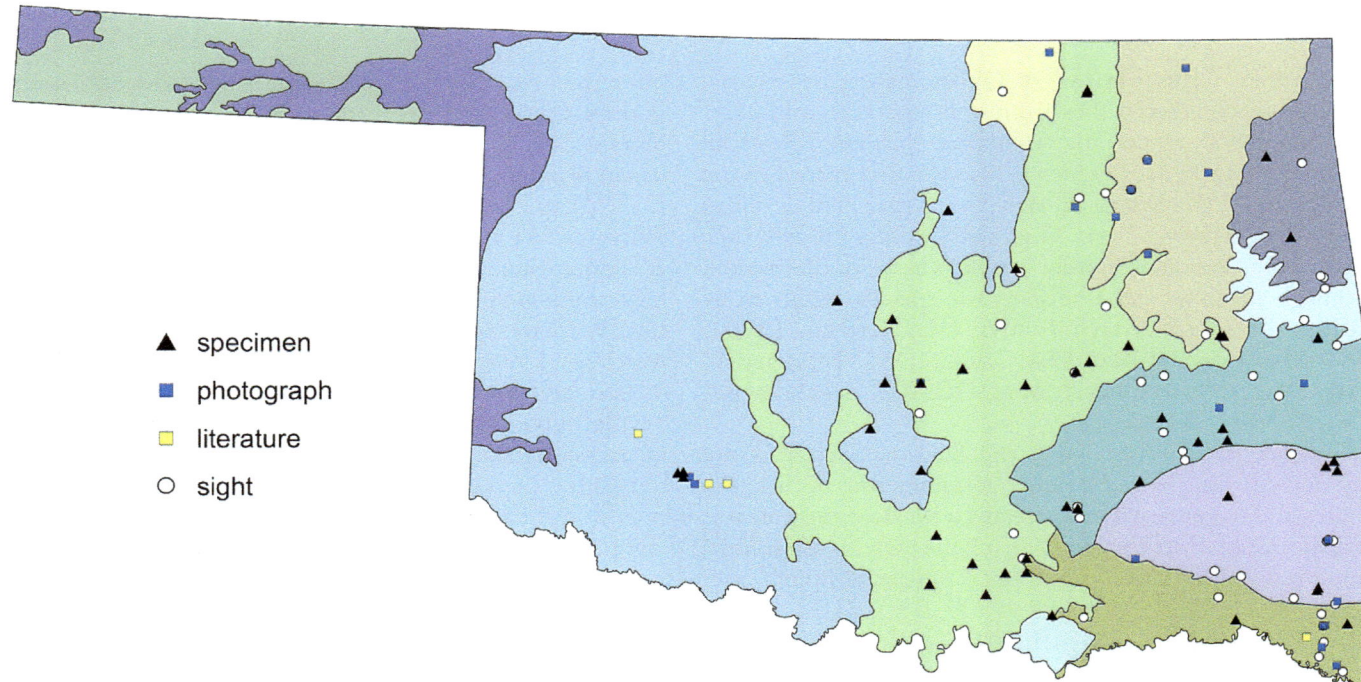

Figure 17.6 Documentation level of supporting records for *Didymops transversa*, the Stream Cruiser, in Oklahoma. Levels include specimen, archived photograph, literature reference, or sight record. Although sight records lack documentation, only those we feel are credible are plotted.

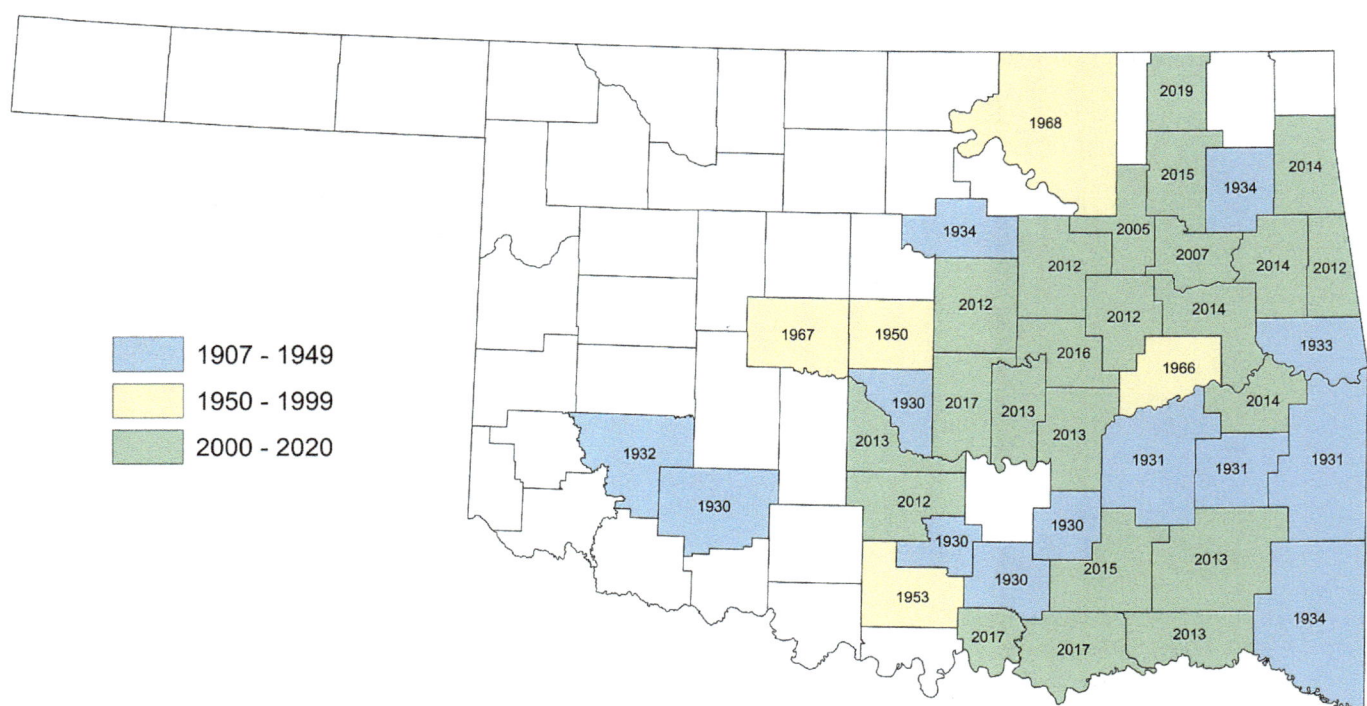

Figure 17.7 Counties in which *Didymops transversa*, the Stream Cruiser, are known to occur in Oklahoma. Year within the county is that in which the species was first reported.

MACROMIA ILLINOIENSIS WALSH, 1862 – SWIFT RIVER CRUISER

Macromia [illinoiensis] georgina (Selys, 1878) – Georgia River Cruiser

Virtually all of Oklahoma records of *M. illinoiensis sensu lato* (see below) are diagnosable as *M. [i.] georgina*, the Georgia River Cruiser, and some early publications (e.g., Bick and Bick 1957) refer to *Macromia georgina* as a species because, for many decades, it and the nominate were treated as species (Figure 17.8). A broad assessment of patterns of intermediacy (Donnelly and Tennessen 1994) led to a taxonomic lump that has been accepted generally, although that taxonomic treatment is not without controversy, enough so that we treat the taxa separately (Figure 17.9).

The Georgia River Cruiser is an uncommon to locally fairly common species in Oklahoma, where its status has changed little over the past 80-odd years (Table 17.1). Although the (sub)species may no longer be the most frequently encountered *Macromia* in the state, a distinction it held easily during the Bick/Hornuff era, it is the most widespread, having been documented in 48 counties compared to 35 for *M. taeniolata* (the Royal River Cruiser) and 22 for *M. pacifica* (the Gilded River Cruiser). Even that distinction may fall given the rapid, recent expansion of *M. taeniolata*

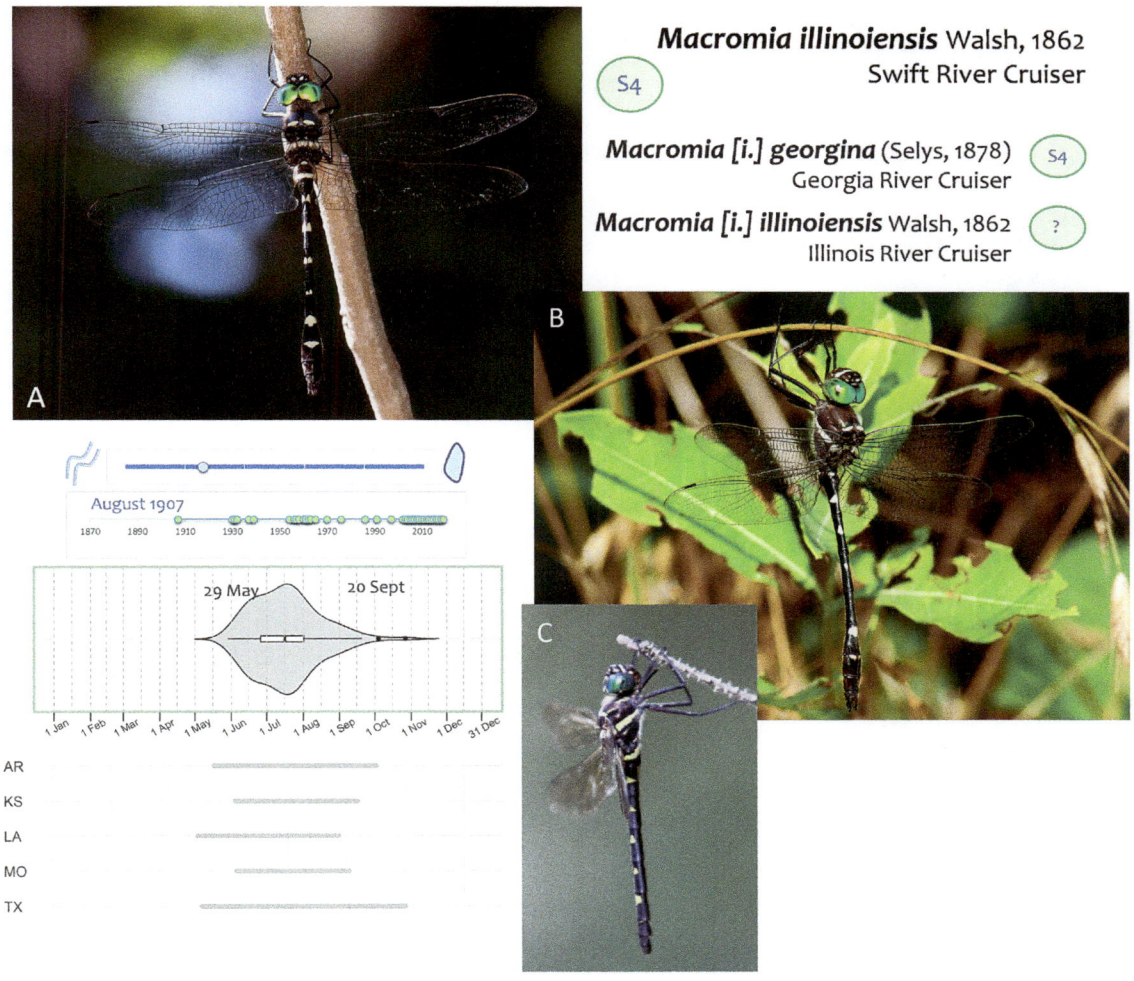

Macromia illinoiensis Walsh, 1862
Swift River Cruiser
S4

Macromia [i.] georgina (Selys, 1878)
Georgia River Cruiser
S4

Macromia [i.] illinoiensis Walsh, 1862
Illinois River Cruiser
?

August 1907

29 May — 20 Sept

AR
KS
LA
MO
TX

Figure 17.8 *Macromia illinoiensis*, the Swift River Cruiser. A) ♂, Stinchcomb Wildlife Preserve, Oklahoma County, 10 June 2006 (©Randy C Anderson, OC 7311), typical of the yellower southern subspecies (or species) *M. [i.] georgina*, the most widespread river cruiser in the Southern Great Plains and now the "default" species in lotic habitats. B) By contrast, this ♂ from Mohawk Park in Tulsa, Tulsa County, 29 May 2006 (©Randy C Anderson, OC 7313), is an apparent nominate *M. i. illinoiensis* or, if treated as a species, *M. illinoiensis sensu stricto*. C) ♀, Mountain Fork River at The Narrows, McCurtain County, 8 August 2007 (©Berlin A Heck, OC 281893).

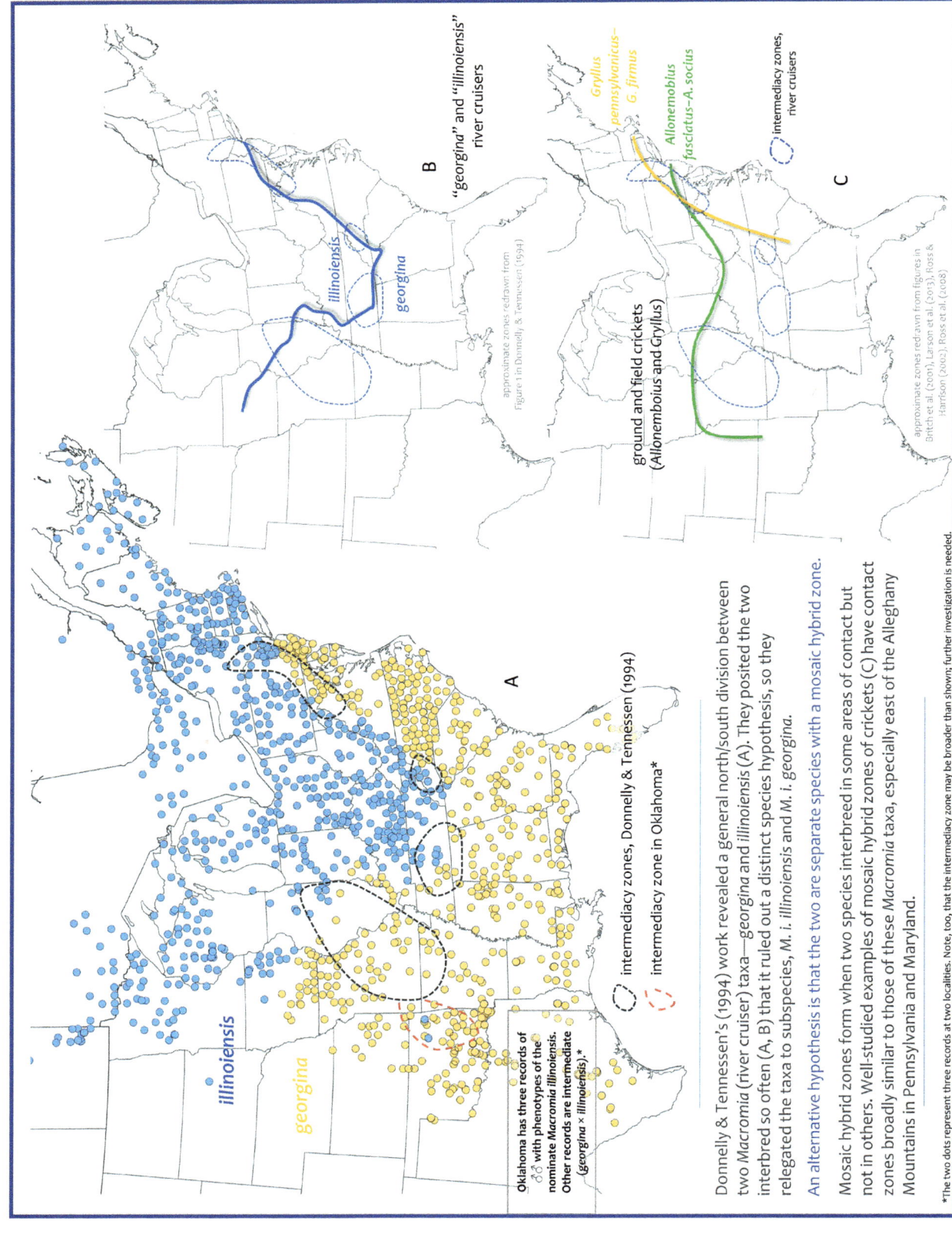

"georgina" and "illinoiensis"
river cruisers

illinoiensis

georgina

B

*approximate zones redrawn from
Figure 1 in Donnelly & Tennessen (1994)*

*Gryllus
pennsylvanicus–
G. firmus*

*Allonemobius
fasciatus–A. socius*

intermediacy zones,
river cruisers

**ground and field crickets
(Allonemobius and Gryllus)**

C

*approximate zones redrawn from figures in
Britch et al. (2001), Larson et al. (2013), Ross &
Harrison (2002), Ross et al. (2008)*

illinoiensis

georgina

A

**Oklahoma has three records of
♂♂ with phenotypes of the
nominate *Macromia illinoiensis*.
Other records are intermediate
(*georgina × illinoiensis*).***

intermediacy zones, Donnelly & Tennessen (1994)

intermediacy zone in Oklahoma*

Donnelly & Tennessen's (1994) work revealed a general north/south division between two *Macromia* (river cruiser) taxa—*georgina* and *illinoiensis* (A). They posited the two interbred so often (A, B) that it ruled out a distinct species hypothesis, so they relegated the taxa to subspecies, *M. i. illinoiensis* and *M. i. georgina*.

An alternative hypothesis is that the two are separate species with a mosaic hybrid zone.

Mosaic hybrid zones form when two species interbreed in some areas of contact but not in others. Well-studied examples of mosaic hybrid zones of crickets (C) have contact zones broadly similar to those of these *Macromia* taxa, especially east of the Alleghany Mountains in Pennsylvania and Maryland.

*The two dots represent three records at two localities. Note, too, that the intermediacy zone may be broader than shown; further investigation is needed.

Figure 17.9 Mosaic hybrid zones: River cruisers and crickets.

Table 17.1 **Lenticity and expected number of individuals encountered by era for the three *Macromia* species that occur regularly in Oklahoma. The lenticity scale ranges from purely lotic (= 0) to purely lentic (= 1). Number of individuals were standardized by the number of survey days in each of the three periods: early (before 1950), mid (1950–1999), and late (2000– present). Encounter rate for *M. illinoiensis*, a bit of a generalist that tilts toward lotic habitats, has not changed over time, whereas records of *M. pacifica*, a more-or-less strictly lotic species, have dropped markedly, especially after 1950. By contrast, records of *M. taeniolata*, the "expected" river cruiser in lentic habitats such as lakes and large ponds, have skyrocketed, with all but three records since the turn of the century.**

	LENTICITY	EARLY	MID	LATE	TREND
Macromia illinoiensis Walsh, 1862 – Swift River Cruiser	0.262	4.9	4.9	6.2	0.130
Macromia pacifica Hagen, 1861 – Gilded River Cruiser	0.071	5.9	2.1	2.0	−0.649↓
Macromia taeniolata Rambur, 1842 – Royal River Cruiser	0.609	0.2	0.1	6.7	20.186↑

Lenticity: light blue = lotic, darker blue = lentic; Trend: yellow = stable, orange = decline, green = increase

For methodological details, see associated endnotes for the similar table of *Argia* species.

in Oklahoma, but away from lakes and large ponds *M. [i.] georgina* remains the "default" river cruiser in the state, an unsurprising assertion given that Needham et al. (2014) described the (sub)species as North America's "most common and widespread *Macromia*." More often than not the Georgia River Cruiser is found along creeks and rivers, from muddy to clear and from deep to shallow, yet to an extent it does occur at lakes (e.g., a ♂ at Lake Holdenville, Hughes County, 16 August 2009, SP 66, and a ♂ at Wewoka Lake, Seminole County, 29 June 2019, SP 2790). Generally only one or several are encountered at a time, but occasionally it occurs in sizable numbers, sizable for a large-bodied dragonfly. For example, 16♂ were documented (3 as SP 2863–2865, 8 examined in hand, 5 seen well as they hung in trees) at Blue River Public Hunting and Fishing Area, Johnston County, 28 July 2019 (MAP), and that total sets aside another 15♂ *Macromia* whose identification could not be verified!

Records for Oklahoma date to the earliest days of field work in the state, with 7♂ and 2♀ found on the Poteau River, Latimer County, in early August 1907 (Williamson 1909, 1914b; 1♂, 2♀, UMMZ). Indeed, these specimens, collected by F. Collins, formed the basis for Williamson's (1909) type description of *Macromia australensis*, a name long since relegated to junior synonymy. Even so, Bird (1932) persisted in using this name. Per OMNH specimen labels, Bird identified various specimens as *M. alleghaniensis* as well, although we do not know why, other than confusion with extant dichotomous keys and diagnostic criteria. In our assessment (and GH Bick's), each of the specimens is typical *M. [i.] georgina*. The mischievous Bird also identified some *M. [i.] georgina* specimens as *M. wabashensis*—see the *M. taeniolata* account for details about that troublesome "taxon"— an error corrected by Bick and Bick (1957), with whom we agree after our examination of the specimens (OMNH[4]). AE Pritchard appeared confident enough in his ability to identify *M. illinoiensis* that he claimed to be the first to have recorded the species in the state (letter dated 30 November

1932, Figure 15.60), although presumably Williamson then informed him of Collins' 1907 collections.

In Oklahoma, the Georgia River Cruiser flies from late May to late September[5] (29 May 2017, McGee Creek SP, Atoka County, MAP; 20 September 1930, vic. Shawnee, South Canadian River, Pottawatomie County, EB Webster, 1♂, OMNH 1218).[6] Oklahoma runs in the middle of the pack for regional seasonality (AR: 17 May–2 October, USNM, OC; KS: 4 June–16 September, USNM, SEMC; LA: 3 May–31 August, JCAC, Mauffray 2014; MO: 5 June–9 September, OC, Trial 2005; TX: 7 May–27 October, Abbott and Lasley 2019). Ovipositing has been reported three times, all of which have been lone females, and one of the times the female was grabbed by a male and then carried into nearby trees.

Macromia [illinoiensis] illinoiensis
Walsh, 1862 – Illinois River Cruiser

As noted in the account for the Georgia River Cruiser, *M. [i.] georgina*, it and the Illinois River Cruiser, *M. [i.] illinoiensis*, were classified as separate species until Donnelly and Tennessen's (1994) examination of phenotypic intermediacy led to a taxonomic lump (see the *M. [i.] georgina* account for more). The status of the Illinois River Cruiser in Oklahoma is problematic, in that diagnosis can be a challenge without a specimen in hand. There are somewhere between one and four records for the state, with at least one for southeastern Kansas.[7] The firmest record for Oklahoma is of a ♂ collected at Spavinaw GMA, Delaware County, on 22 June 2016 (SP 1971). This individual has the auricles (flanges on the sides of S2) black, mesepisternal stripes lacking, yellow ring on S2 broken at the apex, mesotibial keel about ½ the length of the tibia, and extensive black on the abdomen (including small dots on S5 dorsally and a wholly black S6). Unfortunately, this ♂ has its hamules broken off, who knows how, meaning one phenotypic tidbit is not available for examination. Nevertheless,

nothing else about it suggests *M. [i.] georgina*, insofar as Cook (1994) and others treat the black auricles alone as diagnostic of *M. [i.] illinoiensis*, and this individual has a suite of characters typical of that taxon.

A ♂ collected at New Spiro Lake, Le Flore County, 10 August 2019 (SP 2894) likewise points toward *M. [i.] illinoiensis*: the auricles are black, the mesepisternal stripes are mere basal yellow dots, and the yellow ring on S2 is broken dorsally. It may show some intermediacy, though, because there are tiny yellow spots on S6—a feature more associated with *M. [i.] georgina* albeit one present on some ♂ *M. [i.] illinoiensis*, even those found squarely within territory solely inhabited by *M. [i.] illinoiensis* (e.g., Whiteshell Provincial Park, Manitoba, 25 June 2008, D Dodson, OC 459125; St. Louis County, Minnesota, 18 July 2011, GW Lasley, OC 332841).

The hamules and mesotibial keels, too, are somewhere between those typical of each taxon (per TW Donnelly's, *in litt*, illustrations). Males photographed at Mohawk Park in Tulsa, Tulsa County, on 29 May 2006 (RC Anderson, OC 7313) and at Spavinaw GMA on 19 July 2014 (BC, OC 424965) are extensively black and exhibit other characters that suggest *M. [i.] illinoiensis*, but various key features cannot be discerned well enough in the photographs to be certain (that said, were SP 7313 [Figure 17.8] from an area where only *M. [i.] illinoiensis* were thought to occur, the record would be accepted as such without a blink of an eye). If all four of the likely nominate records indeed are of the Illinois River Cruiser, then its flight season in Oklahoma extends from 29 May to 10 August and its occurrence is likely to be limited to the eastern fringe of the state.

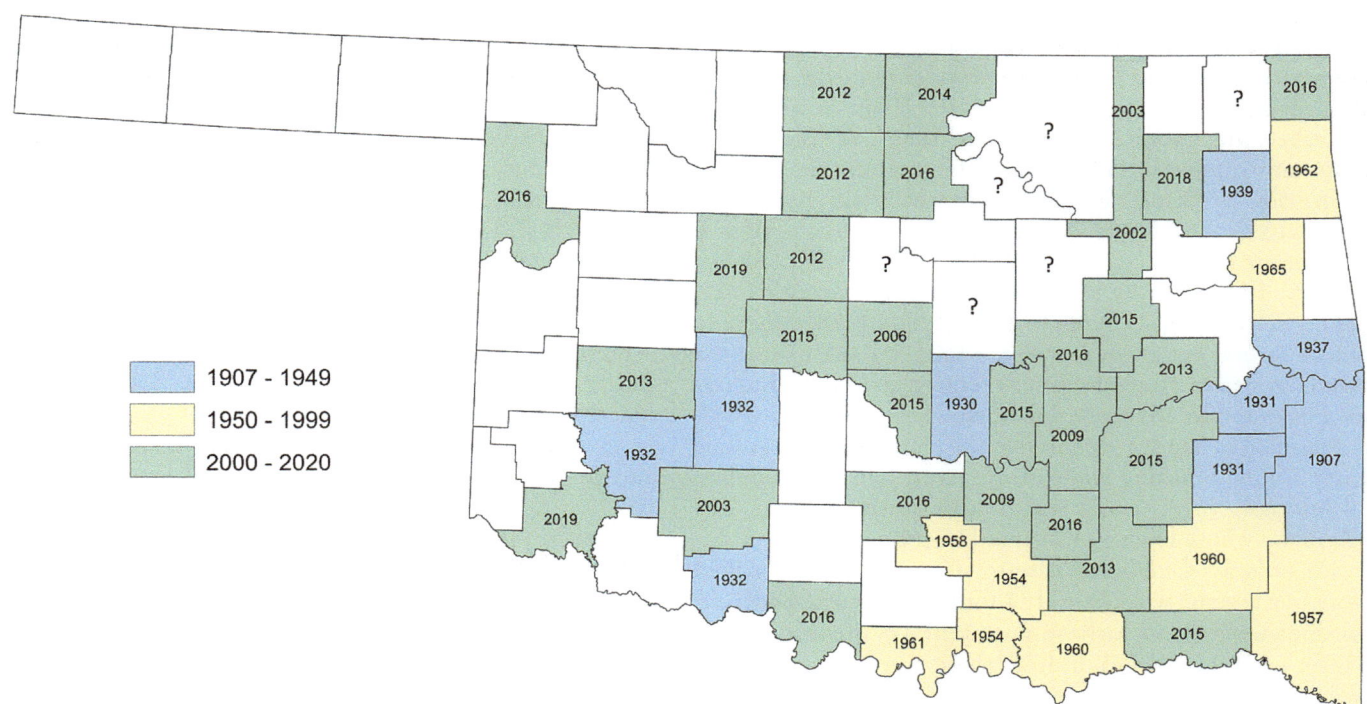

Figure 17.10 Counties in which *Macromia illinoiensis senso lato*, the Swift River Cruiser, are known to occur in Oklahoma. Distribution includes individuals identified as *M. [i.] georgina*, the Georgia River Cruiser, *M. [i.] illinoiensis*, the Illinois River Cruiser, and individuals considered intermediates. Year within the county is that in which the species was first reported. Counties with question marks are those from which only sight records have been reported.

MACROMIA ALLEGHANIENSIS WILLIAMSON, 1909 – ALLEGHENY RIVER CRUISER

Outwardly, the Allegheny River Cruiser can be confused easily with either taxon in the *Macromia illinoiensis* complex, enough so that not just the first but the *first and second* Oklahoma records of *M. alleghaniensis* languished undetected for years (Patten and Smith-Patten 2016).

Misidentification certainly played its part in Oklahoma's first record, yet that record is remarkable precisely because the misidentification was so egregious. On a routine visit to the entomology collection at OSU in 2015, BS-P examined a pinned specimen labeled as *Basiaeschna janata* (the Springtime Darner) that proved not only to be a *Macromia* rather than a darner but a ♂ *M. alleghaniensis*. This ♂ was collected "in flight" (presumably the insect, not the collector) at Jay, Delaware County, on 12 June 1962 (OC 434896). The second record (but the first to be discovered) of *M. alleghaniensis* in Oklahoma came when a ♂ had the misfortune to collide with BA Heck's windshield near Mountain Fork Park, McCurtain County, on 20 June 2012. He collected the specimen, which he sent to JC Abbott (JCAC 49444), and submitted photographs to Odonata Central under the name of *M. taeniolata* (the Royal River Cruiser; OC 376227). Abbott changed the id to *M. [illinoiensis] georgina* (Georgia River Cruiser). But in 2014, BS-P examined the record and determined it actually to be *M. alleghaniensis* because the photos showed that yellow encircled S7, the yellow ring on S2 was broken at its apex, and the mesepisternal stripe was vestigial. Later, Abbott confirmed the *re-re*-identification of the specimen.

Only the third Oklahoma record was recognized as *M. alleghaniensis* contemporaneous with collection:

MAP secured a ♂ in the Blackjack Mountains near the entrance to the northwestern unit of Honobia Creek WMA, Pushmataha County, on 19 June 2016, as it foraged over a dirt road in mixed pine-oak woodland amid a small mixed-species swarm that included 2♂ *Somatochlora ozarkensis*, the Ozark Emerald (SP 1980, OC 447527). Morphological similarity among *Macromia* reared its head when, as it perched in a roadside shrub, MAP took it to be *M. [illinoiensis] georgina*. Yet as soon as he had the individual in hand he noted the yellow ring on S7,[8] the vestigial yellow mesepisternal stripe (i.e., mere spots), the broken ring on S2, and, with aid of a hand lens, the shape of the hamules and was shocked to realize it was *M. alleghaniensis*.

If occurrence of the Allegheny River Cruiser in Oklahoma was not a surprise it was only because the venerable Sidney W Dunkle discovered the species at multiple sites in west-central Arkansas in 1984 (Harp and Harp 1996). Since, the species has been found several additional times in the same three-county region in the Ouachita Highlands, not far east of the Oklahoma state line (e.g., Harp and Harp 2009). Arkansas records fall between 30 May (JCAC 20132) and 18 July (Harp and Harp 2003), dates that bracket the three Oklahoma records, each of which falls in the narrow window of 12–20 June. In Missouri, the species is known to fly later, from 22 June until 16 August (iNat, SEMC). There is one possible record for Kansas of a ♂ on 31 July 1975, 1.5 mi W of Elgin, Big Caney River, Chautauqua County (SEMC 1367060, det. BS-P as cf).

We suspect the species is a scarce but regular component of Oklahoma's fauna, although confirming it as

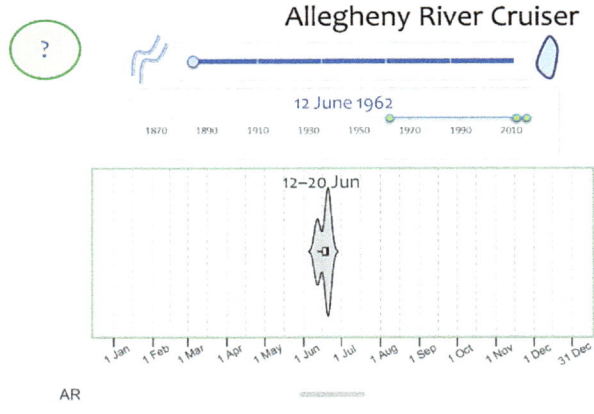

Macromia alleghaniensis Williamson, 1909
Allegheny River Cruiser

Figure 17.11 *Macromia alleghaniensis*, the Allegheny River Cruiser, ♂, Mountain Fork Park, McCurtain County, 20 June 2012 (©Berlin A Heck, OC 376227). This record had a tangled history, during which it traveled under several names. Once it was determined to be *M. alleghaniensis* it was thought to represent a first for Oklahoma, but not long after we found an old specimen (OSU) of the species.

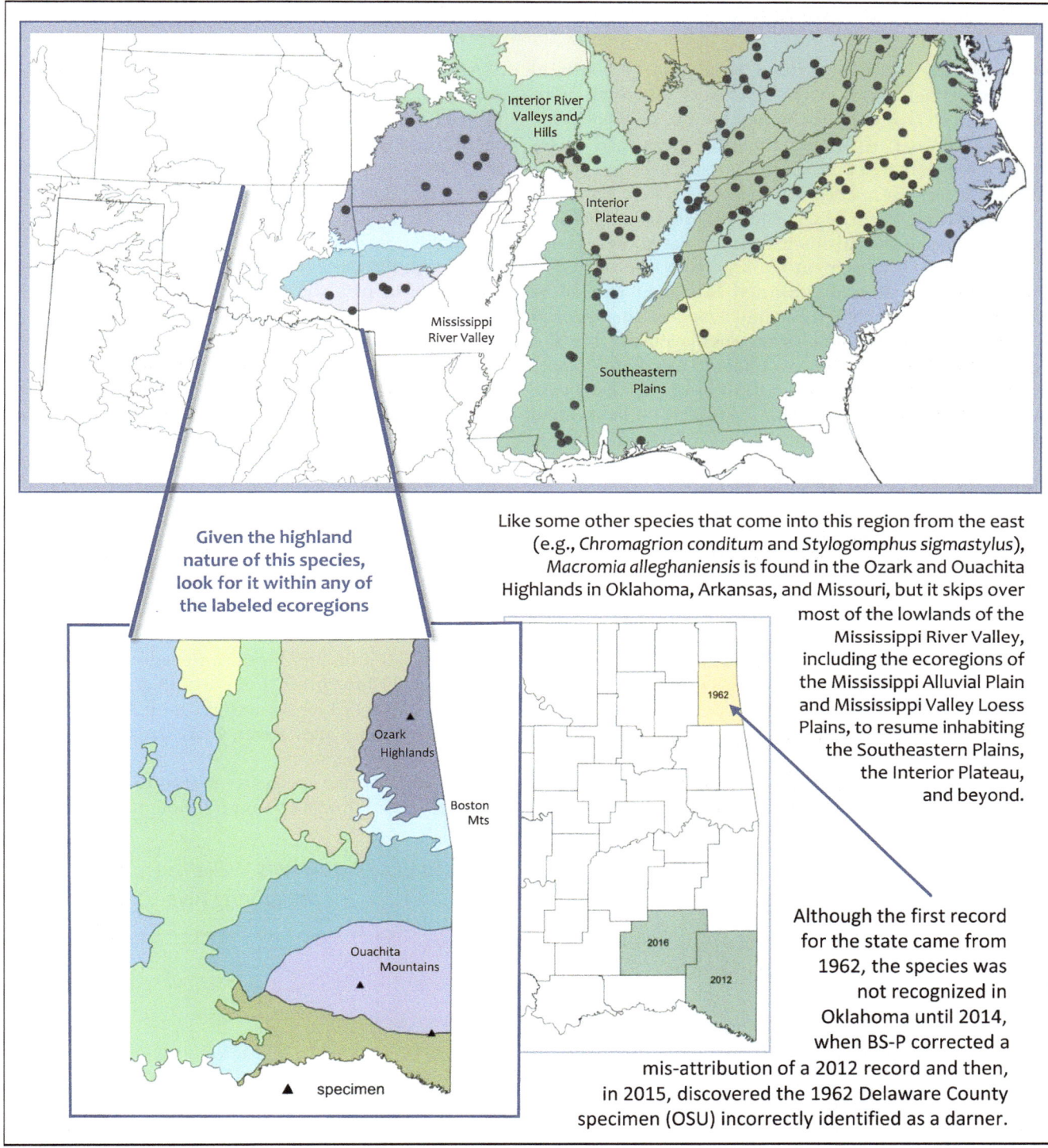

Given the highland nature of this species, look for it within any of the labeled ecoregions

Like some other species that come into this region from the east (e.g., *Chromagrion conditum* and *Stylogomphus sigmastylus*), *Macromia alleghaniensis* is found in the Ozark and Ouachita Highlands in Oklahoma, Arkansas, and Missouri, but it skips over most of the lowlands of the Mississippi River Valley, including the ecoregions of the Mississippi Alluvial Plain and Mississippi Valley Loess Plains, to resume inhabiting the Southeastern Plains, the Interior Plateau, and beyond.

Although the first record for the state came from 1962, the species was not recognized in Oklahoma until 2014, when BS-P corrected a mis-attribution of a 2012 record and then, in 2015, discovered the 1962 Delaware County specimen (OSU) incorrectly identified as a darner.

Figure 17.12 All accepted records in Oklahoma of *Macromia alleghaniensis*, the Allegheny River Cruiser, are supported by specimens (left). It is known from only three eastern Oklahoma counties (right).

such will be a daunting task in light of the difficulty to photograph a ♂ well or to capture one for in hand study. Diagnosis of a ♀ will be a greater challenge still and likely can be accomplished solely with careful examination of the subgenital plate and the extent of yellow laterally and ventrally on S3 (Beaton 2007, Needham et al. 2014). We ranked the species as S2S3 because of its apparent rarity in the state and limited geographic extent, but we acknowledge that current data do not allow us to fully discern its conservation status.

MACROMIA PACIFICA HAGEN, 1861 – GILDED RIVER CRUISER

Oh, how the mighty have fallen. If early collection records can serve as an index of abundance, then the Gilded River Cruiser once was the most numerous *Macromia* in Oklahoma (Table 17.1), being captured more often even than *M. [illinoiensis] georgina*, the Georgia River Cruiser, which generally is thought to be both the most frequently encountered and the most geographically widespread river cruiser in Oklahoma, although even that conventional wisdom may hold no longer.

The Gilded River Cruiser was discovered in Oklahoma on 5 July 1928, when Z Logsdon collected a ♂ in the Wichita Mountains, Comanche County (OMNH 881). In just 7 years of collecting in that early era, the species was reported 27 times from 8 counties, accounting for almost 40 individuals (Figure 17.14). And at least on three occasions, it was numerous, *for a river cruiser*. For example, on 28–29 June 1932 the Bird expedition collected 9 adults in the Wichita Mountains, Comanche County (2♂ on the 28th, OMNH 4191; 6♂ and 1♀, on the 29th, OMNH 766, 767, 4200, 4208, 4210). Since that early era, no more than three adults have been recorded at a time, and despite a vast increase in field effort in recent times *M. pacifica* has become easily the scarcest of Oklahoma's three regular *Macromia* (Table 17.1), so much so that years have elapsed during which we did not encounter the species in the state.

Our hunch is that steady loss of lotic habitat (see Chapter 6) drives the species' apparent population decline. The Gilded River Cruiser might be especially sensitive to habitat loss because it does not occur on just any river or stream;

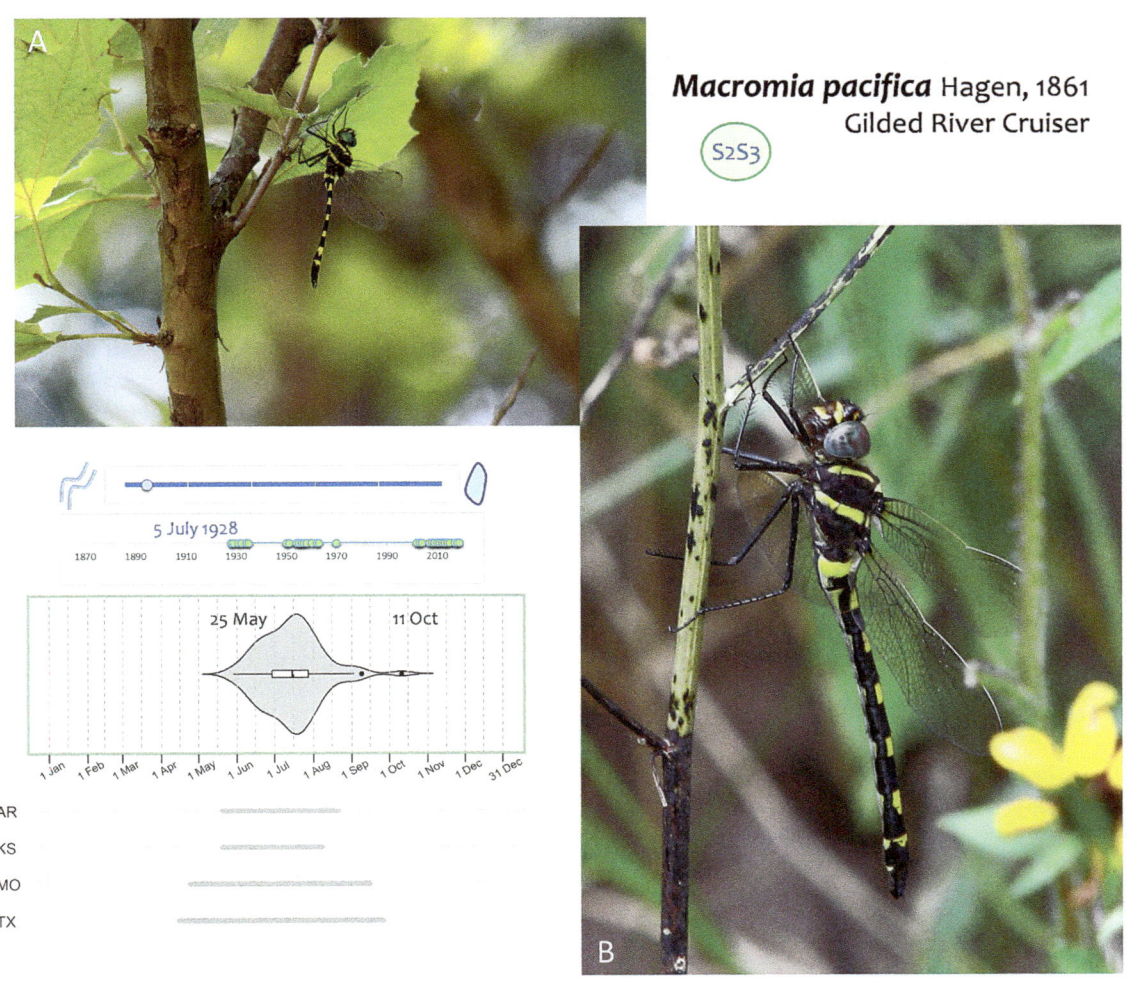

Macromia pacifica Hagen, 1861
Gilded River Cruiser
S2S3

Figure 17.13 *Macromia pacifica*, the Gilded River Cruiser, A) ♂, Mountain Lake, Carter County, 3 August 2013 (©Brenda D Smith, OC 402625, SP 913). B) ♀, Greenleaf Creek in Cherokee WMA, Cherokee County, 16 July 2009 (©Ken Williams, OC 315459). Historical data suggest this species was once the commonest river cruiser in Oklahoma, but now it is the rarest, by far, of the three (or four) species that occur regularly in the state.

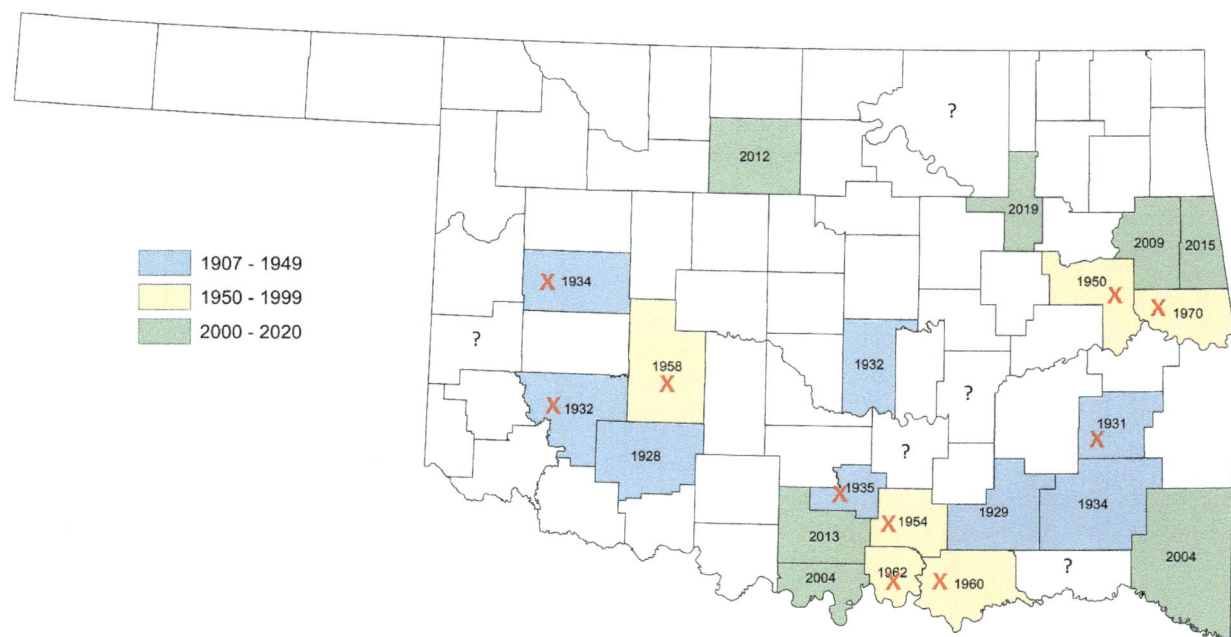

Figure 17.14 Counties in which *Macromia pacifica*, the Gilded River Cruiser, are known to occur in Oklahoma. Year within the county is that in which the species was first reported. There are 10 counties in which the species has not been reported from for approximately 50 years or more (up to about 90!) are marked with a red X. Five counties (marked with?) have unconfirmed reports.

rather, we have found it only on perennial streams that are relatively narrow (<15 m), shallow (<1 m), clear, and rocky. We have not detected a clear range retraction concomitant with the species' apparent population decline (Figure 17.15). It thus may be that suitable streams are harder to come by but are distributed evenly across the landscape, enough to allow the species to persist within the broad outline of its historical range even if now absent from many specific locations within that range; thus making for a rare, but rather widely distributed, species. All of this taken into consideration, we ranked the species as S2S3 because of concern of its continued presence in the state, while still acknowledging that data do not allow us to fully discern the conservation status of this species.

The Gilded River Cruiser is known to fly from late May to mid-October (25 May 2017, 1♂ on Cedar Creek, 15 km WNW of Hochatown, McCurtain County, BS-P, SP 2344; 11 October 2002, 1 indiv. on Medicine Creek, Fort Sill MR, Comanche County, Kondratieff et al. 2004). The flight season is shifted earlier in Missouri and Texas (MO: 24 April–15 September, USNM, Trial 2005; TX: 15 April–26 September, Abbott and Lasley 2019) and truncated in Arkansas and Kansas (AR: mid/late-May–21 August, Harp 2006, Johnson 2006, OC; KS: mid/late-May–8 August, SEMC).[9]

Macromia pacifica, M. taeniolata, and M. "wabashensis," the Gilded, Royal, and Wabash River Cruisers

Prior to 2000, *Macromia taeniolata* was known (or thought to be known) in Oklahoma from only three records (A), none of which can be said to inspire confidence.[†] The species is now known from the eastern half of the state (B).

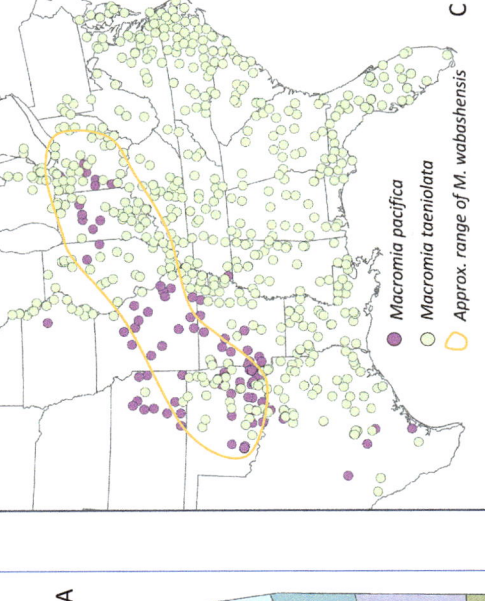

● *Macromia pacifica*
▲ *Macromia taeniolata*

Although there has been little loss in the overall geographical range of *M. pacifica* over time, there are many fewer records of the species in the state now. Concomitant, is a striking increase in the number and distribution of *M. taeniolata* records in Oklahoma.

We discuss these two species here, not because we think there is necessarily any direct relation to their population changes over time, rather it is because there is a long-held notion that *M. pacifica* and *M. taeniolata* hybridize to create *M. wabashensis*, a notion we would like to dispel.

[†]All records are purported female specimens, one of which is thought to be lost and the other two have yet to be examined fully using revised species characters.

What is "Macromia wabashensis?"

There are two prevailing hypotheses about *M. wabashensis*.
1) It is a "form" or "morph" of *M. taeniolata*.
2) It is a hybrid of *M. pacifica* and *M. taeniolata*.

● *Macromia pacifica*
○ *Macromia taeniolata*
◯ *Approx. range of M. wabashensis*

Problem: *M. wabashensis* has a distinct geographical range (C), which should not be the case for a "form" or "morph" because such a phenotype, by definition, should be distributed randomly across the species' range.

Problem: Although the geographical ranges of *M. pacifica* and *M. taeniolata* overlap, they rarely co-occur and when they do, they are in feeding swarms or in different microhabitats.

Problem: Upwards of 95%* of *M. taeniolata* in Oklahoma would be considered hybrids under this hypothesis, even though the putative parental species do not co-occur (*M. taeniolata* is distinctly lentic, whereas *M. pacifica* is distinctly lotic).

Problem: The phenotype of *M. wabashensis* is too stable to be considered a hybrid, which ought to show much intermediacy (mixture of characters); instead, it appears to have distinct characters that define it.

Alternative hypothesis: *M. wabashensis* is not a morph or hybrid, but its true status is complicated by a cryptic species (Fig. 17.19). In the end, it may prove to be part of a species complex with that cryptic species.

*by area

Figure 17.15 A look at *Macromia pacifica*, *M. taeniolata*, and *M. "wabashensis*," the Gilded, Royal, and Wabash River Cruisers.

MACROMIA TAENIOLATA RAMBUR, 1842 – ROYAL RIVER CRUISER

The Royal River Cruiser was unknown in Oklahoma during the heyday of the notorious Bird/Pritchard gang; despite their concerted efforts to hunt down the state's odonates, nary a one was collected (Bird 1932). Nor was the species reported by Bick and Bick (1957) when they updated the state checklist, although in their case a secret was hidden away in a major museum. The first record for Oklahoma was provided by a ♀ WF Blair collected on 20 June 1936 in Mayes County (UMMZ), a specimen that was not diagnosed for several decades (Kormondy 1960). The species was not recorded again until 5 July 1970, when LE Hornuff presumably collected a ♀ 1 mi SE of Dougherty, Murray County, a specimen we have not located but that Bick listed in his notes. A ♀ from Raymond Gary SP, Choctaw County, provided the only other pre-2000 record for the state (FSCA, RD Cuyler). It is curious that the only three early records were all of lone females. Why was that?

If Y2K had any effect, perhaps it was solely to push *M. taeniolata* into Oklahoma. The species is now encountered as often as, if not slightly more often than, *M. illinoiensis*, the Swift River Cruiser *senso lato* (Table 17.1). In under two decades the state has seen over 70 confirmed records of about 150 individuals, and it has now been recorded in a remarkable 33 counties, principally in the eastern half of the state. We expect that the vast change in water habitats in Oklahoma (see Chapter 6) have benefited this species enormously: lotic habitats have declined while lentic habitats have increased sharply. In our experience, the Royal River Cruiser is the default *Macromia* on lentic waters. If we see a *Macromia* patrolling the shoreline of a lake or circling a large pond, we expect it to be *M. taeniolata*, and once we have the critter in hand, we see that seldom are we wrong. Counts on lentic waters can be high, too, for a river cruiser. For example, estimates of 10 individuals have been provided twice (both BAH: Raymond Gary SP, Choctaw County, 22 June 2008, OC 282554 of 1♂; LRNWR, McCurtain County, 9 August 2008, 3 in hand, OC 283484 of 1♂), once of 9 individuals (Stroud Lake, at 2 spots, 1 in Creek County and 1 in Lincoln County, on 9 August 2015, MAP, 8♂ and 1♀, of which 2♂ were collected as SP 1747–1748 and 4♂ were in hand), and another of 7 (Osage Point, Copan Lake, Washington County, 18 August 2018, MAP, 6♂ and 1♀, of which 2♂ are SP 2696–2697, and 3♂ and 1♀ were in hand).

The flight season of this species in Oklahoma ranges from early June to early October (♀ on 8 June 2010, Little River NWR, McCurtain County, BAH, OC 319691; 2♂, 1♀, 3 October 2015, Raymond Gary SP, Choctaw County, MAP, 1♂ SP 1804). The length of Oklahoma's flight season is second only to that of Texas but roughly the same as in Louisiana, but in both instances, Oklahoma's is shifted a bit later in the year (OK: 117 days; LA: 116 days, 7 May–31 August, Mauffray 2014; TX: 146 days, 3 May–26 September,

Abbott and Lasley 2019). Records from Kansas are very limited (one adult ♀ confirmed by us, collected in Cherokee County, 22 August 1923, RH Beamer, SEMC 1367073) and some are perhaps dubious (primarily nymphal records). Records from Arkansas and Missouri are also limited and are likely not indicative of the species' full flight season in those states (AR: 20 June–17 September, BugGuide 131185, OC; MO: 17 July–7 September, Sims 2012). There are three ovipositing records for Oklahoma, all of which are of lone females.

We should feel grateful were the sole controversy about *M. taeniolata* the extent to which the species has expanded its range into the state (Figure 17.15). Yet taxonomy and systematics rear their heads, too, enough so that species limits remain a subject of (at times intense) debate. One source of disagreement is the status of *Macromia wabashensis* Williamson, 1909, a purportedly yellower version of a river cruiser otherwise much like *M. taeniolata*. Some (e.g., Abbott 2005) consider the taxon to represent a hybrid between *M. taeniolata* and *M. pacifica*,[10] a hypothesis we question (Figure 17.15) not for the least of which makes one wonder about the differences in body size between the two species that might make mating a challenge (total lengths: *pacifica* = 62–76 mm, *taeniolata* = 75–91 mm). Further, the two species are rarely encountered together. Others (e.g., Needham et al. 2014) treat the taxon as a "form" ("morph" or subspecies?) of *M. taeniolata* and in doing so emphasize that the taxon, as described, is much nearer that species than it is to the other putative parental species. Others still (e.g., Cook 1994) feel that *M. wabashensis* is best treated as a species, albeit a rare one. As outlined by Williamson (1909), Cook (1994), Glotzhober and McShaffrey (2002), and Rosche et al. (2008), subtle differences lie in coloration of the costal veins, shape of the hamules, and pattern of yellow dorsally on the abdomen, including what is now thought to be a key character—a moth-shaped pale spot on top of S8. Regardless of taxonomic treatment, river cruisers currently diagnosable as *wabashensis* or *wabashensis*-like have been recorded in Oklahoma a dozen or more times (perhaps many more, Figure 17.15, but see below regarding what may be a cryptic species). For example, males we collected at Atoka WMA, Atoka County, on 19 June 2013 (SP 715) and at Spavinaw GMA, Delaware County, on 22 June 2016 (SP 1972) have the vertex black, yellow ring on S2 broken at the apex, the telltale moth-shaped pale on S8, postfrons ¼–½ yellow, yellow mesepisternal stripe ½–⅔ the length of the thorax, and head of the hamules longer than wide. At least one female has a phenotype consistent with *wabashensis* (Pontotoc Ridge Preserve, Pontotoc County, 1 July 2010, F Alm, OC 320588).

A muddier issue is whether there is an undescribed species of *Macromia* in the Midwest (Cook 1994), including

Macromia taeniolata Rambur, 1842
Royal River Cruiser

S5

20 June 1936

8 Jun   3 Oct

AR
KS
LA
MO
TX

Figure 17.16 *Macromia taeniolata*, the Royal River Cruiser, A) ♂, Cherokee WMA, Cherokee County, 26 June 2009 (©James W Arterburn, OC 328971). B) ♂, Comanche Lake, Stephens County, 5 August 2010 (©Victor W Fazio III, OC 321637), how *Macromia* usually are seen. C) ♀, Shady Grove Creek 3 km SW of Shady Grove, McIntosh County, 15 July 2013 (©Brenda D Smith, OC 401756). The ♀ of any river cruiser may have her wings washed amber, but such color tends to be deeper and more extensive on *M. taeniolata*. Until the turn of the 21st century this species bordered on the unknown in Oklahoma, but since 2000 records have skyrocketed, so much so that it is encountered as often as *M. [illinoiensis] georgina*, the Georgia River Cruiser, and is the expected species in lentic habitats.

eastern Oklahoma.[11] In October 2017, BS-P visited Carl Cook's personal collection in Kentucky to examine the type series of this putative taxon. She could not reach a definitive taxonomic conclusion in so brief a visit, but she did see firsthand the characters Cook (1994) reported. A character on which he did not report, the mesotibial keel (present only on the ♂), may prove key. We have collected 29♂ and examined in hand an additional 14♂ ostensible *M. taeniolata* in Oklahoma, and in doing so we have noted a pattern: only those individuals from the South Central Plains ecoregion (part of the broad Gulf of Mexico plain) have a small, leaf-shaped keel (Figure 17.19) said to be typical of the species (TW Donnelly, *in litt*); everywhere else, the keel is low and short, rather like that of *M. alleghaniensis*. We are as yet unsure if these differences align with the characters Cook (1994) outlined for his new species, but we are sure that there are multiple taxa in Oklahoma that travel under the name *Macromia taeniolata*.

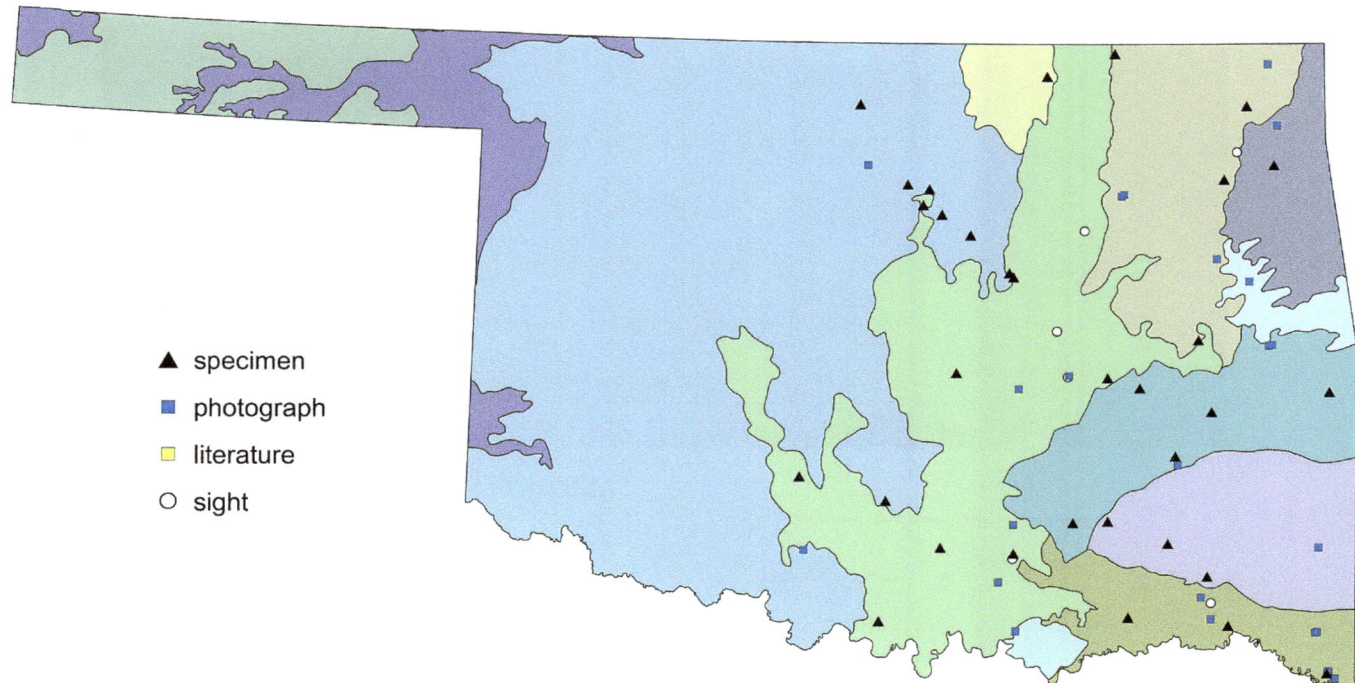

Figure 17.17 Documentation level of supporting records for *Macromia taeniolata*, the Royal River Cruiser, in Oklahoma. Levels include specimen, archived photograph, literature reference, or sight record. Although sight records lack documentation, only those we feel are credible are plotted.

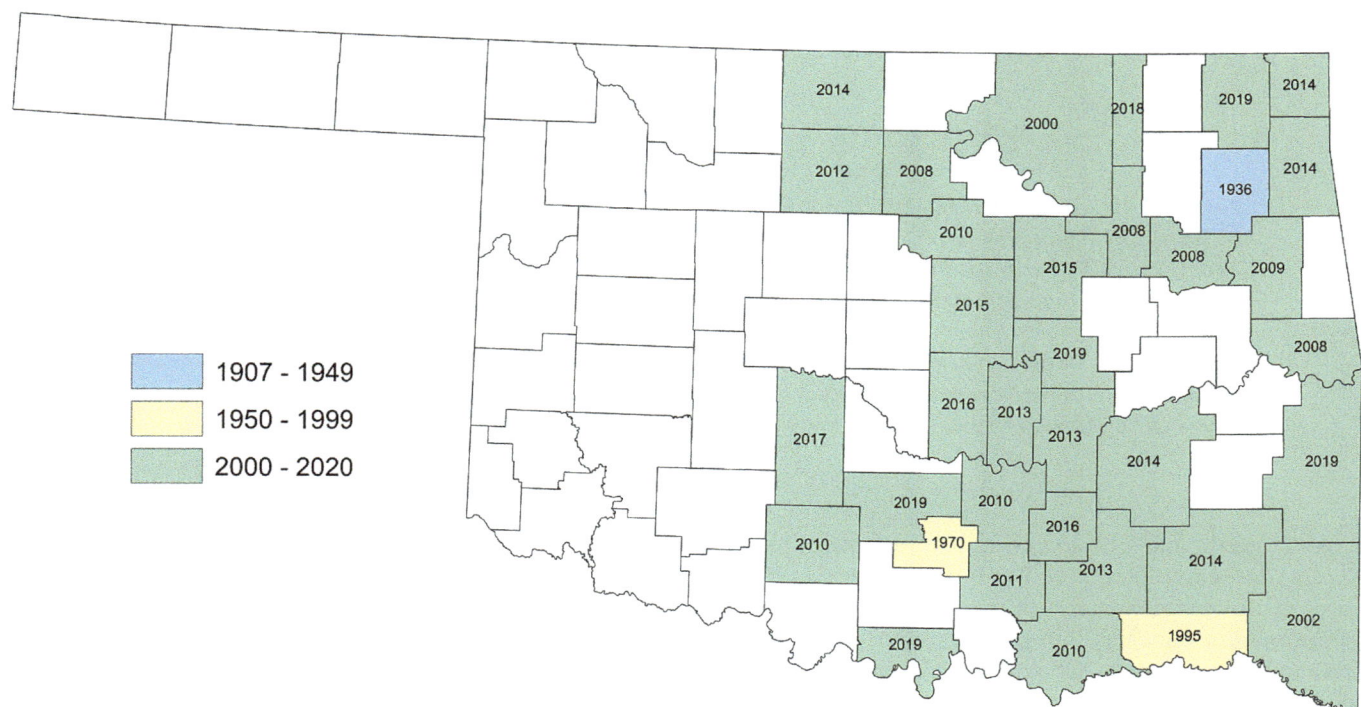

Figure 17.18 Counties in which *Macromia taeniolata*, the Royal River Cruiser, are known to occur in Oklahoma. Year within the county is that in which the species was first reported. Although the species was first recorded in the state in 1936, there are only three of 35 counties in which the species was recorded prior to 2000.

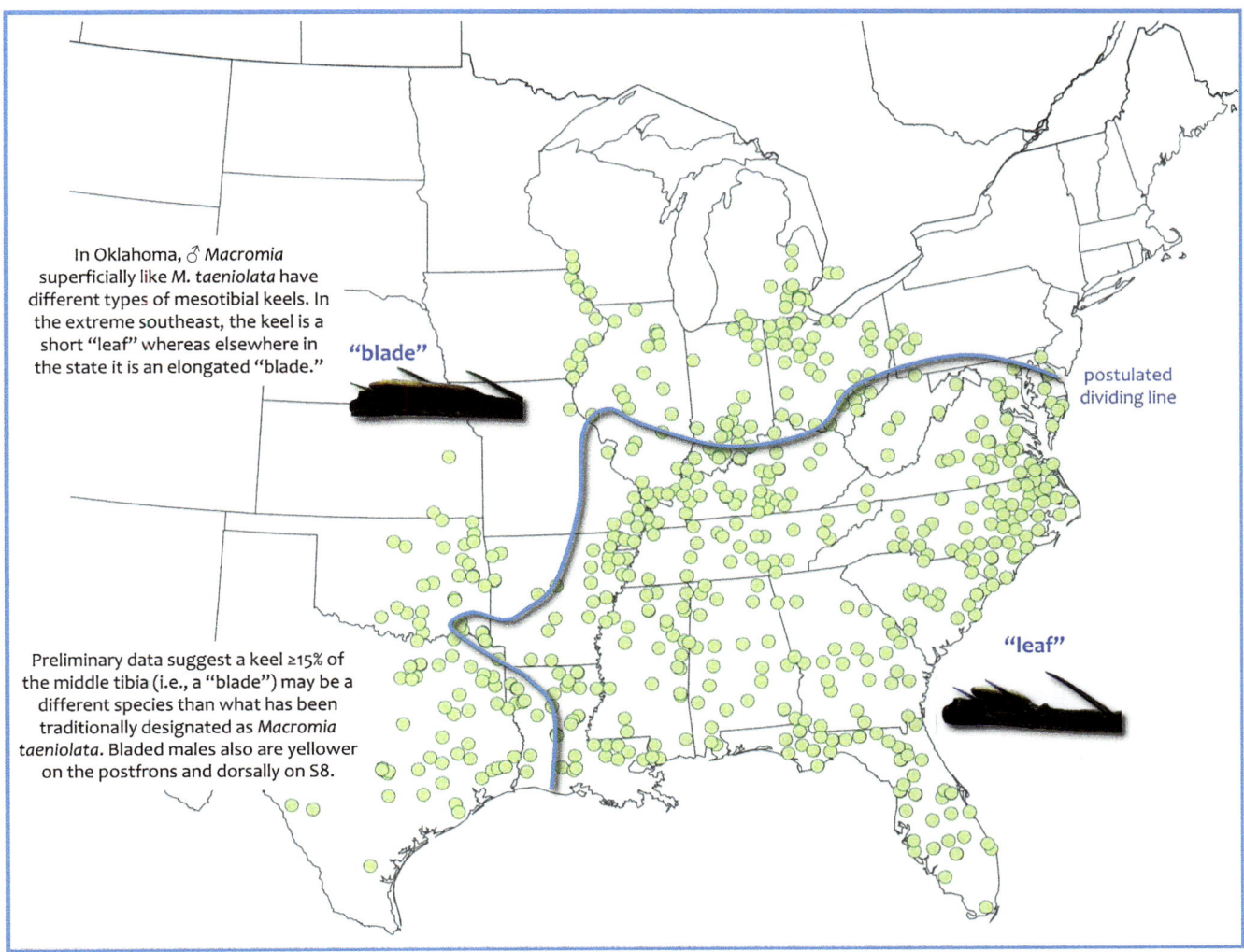

In Oklahoma, ♂ *Macromia* superficially like *M. taeniolata* have different types of mesotibial keels. In the extreme southeast, the keel is a short "leaf" whereas elsewhere in the state it is an elongated "blade."

"blade"

postulated dividing line

"leaf"

Preliminary data suggest a keel ≥15% of the middle tibia (i.e., a "blade") may be a different species than what has been traditionally designated as *Macromia taeniolata*. Bladed males also are yellower on the postfrons and dorsally on S8.

Figure 17.19 Mesotibial keels of *Macromia* presently considered *M. taeniolata*, the Royal River Cruiser.

MACROMIA ANNULATA HAGEN, 1861 – BRONZED RIVER CRUISER

The Bronzed River Cruiser is a muted, non-metallic, grayish brown-and-white (rather than black-and-yellow[12]) counterpart to the five species of river cruiser known from eastern North America. Prior to finding the species in Oklahoma, it was thought to range solely from the Hill Country of Texas and the Pecos River valley of New Mexico, south to Nuevo León, Mexico. In Texas, the species had been collected north to Jones and Palo Pinto Counties (Donnelly 2004b), not far north of the northern fringe of the Edwards Plateau. The status of the species in this under-worked part of Texas is unknown—perhaps it occurs regularly in small numbers in the region. That at least one reached Oklahoma, then, may or may not have been a surprise. A ♀ collected on Sandy Creek 7 km SSE of Eldorado, Jackson County, on 5 August 2018 (MAP, SP 2690, OC 487189; Figure 17.20) was in the "proper habitat" for the species and geographically was not an extreme outlier (<200 km to the north). Moreover, shortly after capture a ♂ river cruiser twice passed as it patrolled on long beats the north-south road that crosses the creek (MAP). This ♂ appeared, in quick views, to be gray/brown-and-white, suggesting strongly that a second *M. annulata* was present. Strengthen the argument for a small population in Oklahoma, at this same site, one or two males hurtled by, once each day, 8–9 June 2019 (MAP), and in each instance the ♂ was distinctly not black-and-yellow. Sandy Creek is one of several perennial creeks in southern Jackson County. Time will tell if a small population of this species occurs on those creeks, to say nothing of along the Red River, which forms the border with Texas only 3–4 km from this locale where the ♀ was collected.

In Oklahoma, there are four problematic records referred to this species: one a simple misidentification, the others make for troublesome identifications, but possible records, nonetheless. The first is OMNH 2696, a ♂ specimen collected by WM Fisher in Latimer County on 21 July 1931. Bird (1932) published this specimen as *Macromia annulata*, but it was re-identified by Bick as *Cordulegaster obliqua* (Bick and Bick 1957, as *C. obliquus*; confirmed by BS-P). The second record is a ♀ that was captured on the West Range of Fort Sill MR, where Medicine Creek crosses Punchbowl Road (Comanche County, 1 July 2003, BC Kondratieff, JP Schmidt, Owens; CSU, no catalog number assigned, but we assigned it as CSU_01072003-01 for our revision of the genus *Macromia* in North America, Smith-Patten et al., ms). The pale coloration on this individual is overall yellow, which is not the more typical white of the species (*viz* on the abdomen, not on the thorax[13]), but the yellow does not rule out *M. annulata*, at least for our current understanding of the species (one that is unquestionably understudied). The yellow spots on S3–4 are confluent dorsally, but broken on S5–6, and there is some yellow laterally on S1. This ♀ also has extensive mesepisternal stripes (6.5 mm), the postfrons almost entirely pale, and a vertex with yellow on most of the cone tips and across the "valley" between the cones. Its total length is 65.7 mm, which is too small for what is known of *M. annulata*'s size, but within the confines of *M. illinoiensis* and *M. pacifica*. The latter species is what this ♀ was originally identified as, although it was speculated that it was a possible hybrid with *M. annulata* (B. and I. Prather). MAP determined it to be a "typical" *M. pacifica*, although BS-P is inclined to agree with the Prathers.

An equally irksome specimen is the ♂ we captured at Lake Lloyd Vincent, Ellis County, on 9 July 2013 (SP 785; OC 425615, submitted and accepted as *Macromia annulata*, but later we unconfirmed the record). This ♂ is superficially a *M. annulata*, having the overall body pattern similar to *M. annulata*, including paired pale (yellow, as with the CSU specimen) spots on S3–6 that are mostly confluent dorsally,

Macromia annulata Hagen, 1861
Bronzed River Cruiser

Figure 17.20 The only documented record of *Macromia annulata*, the Bronzed River Cruiser, in Oklahoma is this ♀, captured on Sandy Creek 7 km SSE of Eldorado, Jackson County, on 5 August 2018 (©Michael A Patten, OC 487189). Sight records at this location indicate that the species may occur regularly in southern Jackson County in the southwestern corner of the state.

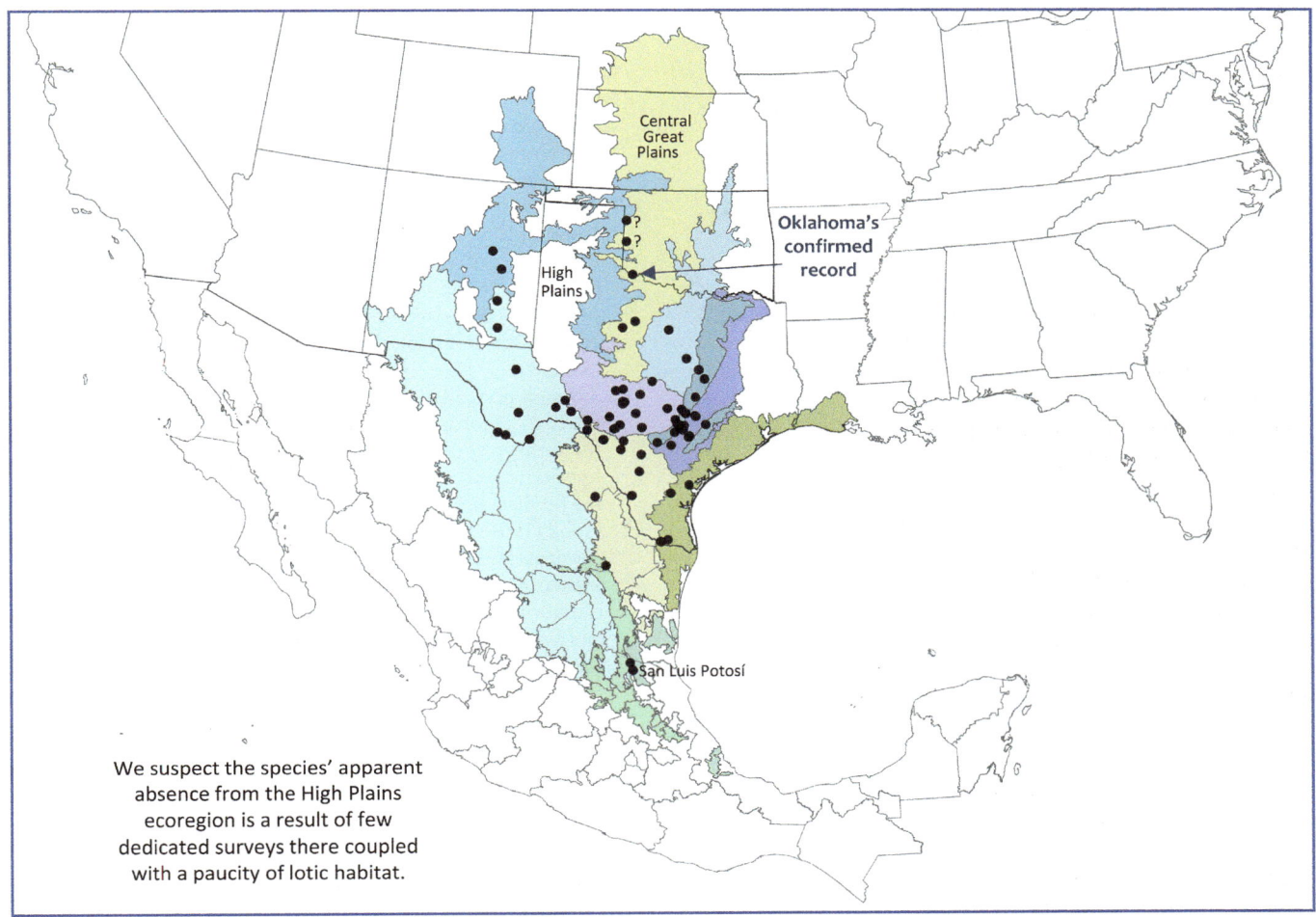

Central Great Plains

Oklahoma's confirmed record

High Plains

San Luis Potosí

We suspect the species' apparent absence from the High Plains ecoregion is a result of few dedicated surveys there coupled with a paucity of lotic habitat.

Figure 17.21 *Macromia annulata*, **the Bronzed River Cruiser, occurs from tropical dry forest in the Mexican state of San Luis Potosí north into the Southern Great Plains to at least southwesternmost Oklahoma. Two locations a bit farther north may harbor the species, too, but documentation is lacking, and a specimen from Ellis County (SP 785) may prove to be a hybrid.**

lateral streaking on S1, ring around S2 uninterrupted, elongated mesepisternal stripes (5.5 mm), almost entirely pale postfrons, and a largely pale vertex, but its hamules compare favorably to *M. illinoiensis georgina*, having a long hamulus head.[14] TW Donnelly examined the specimen and was inclined to identify it as "*georgina* with some *illinoiensis* intermediate character. It is not close to *annulata*." (*in litt*), whereas we feel that a putative *M. annulata* × *M. i. georgina* hybrid is more likely (Patten and Smith-Patten 2014b, 2016). For now, the specimen remains identified as *Macromia* sp., as does a second ♂ we saw flying around the same cove that day.

And finally, we saw a ♂ *Macromia* on West Buffalo Creek, 2 km SE of Sweetwater, Beckham County, on 3 June 2012, that we thought may have been *M. annulata*. Because we were not able to capture or photograph it, we left the record without a conclusive determination. But the latter three records indicate that there may be an actual population in Oklahoma of *M. annulata* or, at the very least, a hybrid population with *M. annulata* as a parental species.

Macromia annulata is best known from Texas, where it flies from 1 April until 11 October (Abbott and Lasley 2019). Limited records in New Mexico and Mexico fall much short of those dates but are undoubtedly not reflective of the species' flight there (NM: 14 June–22 July, JCAC, CSU, SWD; MX: 23 June–8 August, USNM).

NOTES

1. If the identifications are correct for records from Travis County, Texas, (e.g., OC 480156, 480158, 480316), then *Macromia annulata* can sometimes be black-and-yellow. Nevertheless, we feel hybrids cannot be ruled out for those records.

2. It is reasonable to assume that hamule shape is under quantitative genetic control—that it is the product of a suite of genes—yet truth be told we do not know if such an assumption is true and if it is whether many or few genes are involved, if any genes involved exhibit dominance, or if a gene complex tends to be inherited as a unit (i.e., it is tightly linked). If, say, the "neck" typical of *M. illinoiensis* hamules is the result of one or two dominant genes, then a hybrid of that species likely would exhibit a "neck."

3. Note that an Oklahoma specimen from FSCA was entered into the database as being collected on 16 June 1976 (♀, Cedar Lake, Le Flore County, JB Heppner), but we assume the month is a data entry error because the species is not known anywhere in the Southern Great Plains to extend its flight season past May.

4. OMNH 776, 777, 779, 780, 783, 784 from Kiowa, Haskell, and Latimer Counties on 21 June 1931, 11 July 1931, and 22 July 1932.

5. GH Bick's note cards have a record from 27 October 1957 of 1♂ at the confluence of the Little and Mountain Fork Rivers, McCurtain County. We hesitate to accept the record because of how late that date is relative to most elsewhere in the region; moreover, because that specimen is apparently lost, we cannot verify the species identification (and recall that diagnostics for *Macromia* have changed since Bick's time).

6. Definite *georgina* has been recorded 29 May–17 September, with the early date in a ♂ in 2017 at McGee Creek SP, Atoka County (MAP) and the late date a ♂ in 2016 at 5 km N of Tribbey, Little River, Pottawatomie County (SP 2204), whereas *illinoiensis sensu stricto* (including putative records) has been found 29 May–30 July.

7. An apparent *Macromia i. illinoiensis* (det. BS-P) from southeastern Kansas provided the late flight date for that state: 16 September 1976, Baxter Springs, below dam, Cherokee County, collected by "Huggins & Liechti" (SEMC 1367062).

8. Although seemingly extremely rare, it is worth mentioning that *georgina* can have a ring on S7 that wraps completely around the sides of the segment. We know of two such individuals, both males. One was encountered as part of our efforts to revise the taxonomy of North American *Macromia*. The CSU specimen (no CSU catalog number, but we assigned it as CSU_20072015 for our study) was collected about 7 km N of Robbins, on the Deep River, at North Howard Mill Road, Moore County, North Carolina, on 20 July 2015 by BC Kondratieff and D Leatherman. It was originally determined

as *M. illinoiensis* by B. and I. Prather in 2015. Later that year, we determined it as the *georgina* subspecies. BS-P discussed the matter with M Dobbs, G Beaton, and S Krotzer, asking if they had encountered this character on *georgina* in their fieldtime in the southeastern United States (*in litt*, 17–29 November 2016). Between the three of them, there was one record: 1♂ captured by Krotzer but that got away (data verbatim from Krotzer "29 June 2008; MISSISSIPPI, Stone County; Desoto National Forest, FS375B adjacent to Bluff Creek, east of MS 15 and S of MS 26, ca. 2 miles SE of Moore Crossing; N30.8439/W-88.9634.").

9. Arkansas' early date was not specific, i.e., 18–21 May (Harp 2006, Johnson 2006); neither was Kansas':12–28 May (SEMC).

10. Though not called *wabashensis*, there are, as of mid-January 2020, 41 records of *Macromia pacifica* × *taeniolata* on iNaturalist (*if* that is an indication of the acceptance of the hybrid notion). Also, see Garrison and von Ellenrieder (2019) for further discussion regarding the status of *wabashensis*.

11. Cook (1994) remarked that his "n. sp." is "known from Arkansas, Ohio, Texas and possibly Oklahoma [but] it is not presently known from Indiana." He thought that Oklahoma did not host *Macromia wabashensis*; if true, then Oklahoma's *wabashensis*-like individuals may be of his proposed taxon or part of a cryptic species complex.

12. Again, as mentioned in an earlier endnote, if all Odonata Central records for the species are determined correctly, then it can be black-and-yellow (see above).

13. Also note that the species was originally described (Hagen, 1861) as having the pale coloration yellow.

14. TW Donnelly (*in litt*, email dated 3 December 2014) described the hamule as "like *georgina*, but slightly shorter and stouter, like *georgina–illinoiensis* intergrades." Note that Donnelly considers *georgina* and *illinoiensis senus stricto* as subspecies of *M. illinoiensis*, as opposed to distinct species, as some researchers do.

## CHAPTER 18

CORDULIIDAE: EMERALDS

The Corduliidae is a notoriously difficult family. The difficulty lies in the cryptic nature of its species as well as in the ongoing struggle to confidently delimit species. There are 50 currently recognized species of corduliids known to occur in North America.[1] Fifteen are known from Oklahoma. Those species fall within four genera, although if one chooses to split *Epitheca* (baskettails) into two genera, then Oklahoma has five: *Helocordulia*, *Somatochlora*, *Neurocordulia*, *Epitheca* (*Tetragoneuria*), and *Epitheca* (*Epicordulia*).[2]

Two of Oklahoma's fifteen species of corduliids—the Texas Emerald (*Somatochlora margarita*) and the Cinnamon Shadowdragon (*Neurocordulia virginiensis*)—have been recorded only once in the state and, two others—Uhler's Sundragon (*Helocordulia uhleri*) and the Robust Baskettail (*Epitheca spinosa*)—have been recorded only twice. Accordingly, in some way the state hosts only 11 species, although each of these singleton or doubleton species occurs regularly near enough to Oklahoma that we expect each has a regular, albeit low-level, presence in the state.

Our conviction of regular occurrence is bolstered by the challenge to locate or identify most species of corduliids. Apart from the morphologically and behaviorally conspicuous Prince Baskettail (*Epitheca princeps*), each of the remaining 14 corduliid species recorded in the state is behaviorally cryptic, difficult to identify in the field (or even in hand for the small baskettails), or both. Each of the sundragons (*Helocordulia*), for example, has an extremely short flight season (Figure 18.1), lasting perhaps a month in early spring, each of the shadowdragons (*Neurocordulia*) is crepuscular,[3] and each of the emeralds (*Somatochlora*) sticks to shaded streams and creeks (except when they swarm, which is generally in the late afternoon and evening, in open, sunny glades, meadows, or roads, but even that behavior is ephemeral).

Below we provide summaries for each genus, including a detailed one for the small baskettails. We separated out the latter because they present a complex suite of issues, many stemming from profound questions of species limits and identification criteria. The small baskettails have been at times placed in their own genus, *Tetragoneuria* (e.g., Davis 1933; Hagen 1861; Martin 1906; Morse 1895; Needham and Heywood 1929; Needham and Westfall 1955), and at others treated as a subgenus (or subset) of *Epitheca* (e.g., May 1995; Needham et al. 2014; Walker 1966). Regardless of how *Tetragoneuria* is treated, it does comprise a useful division

in terms of discussing the smaller-bodied and early-flying *Epitheca* (relative to *E. princeps*, which has a total length of 58–78 mm versus all of the other North American *Epitheca* at 32–48 mm, as per Needham et al. 2014, and which starts flying, in Oklahoma, when the others have stopped or, at least,

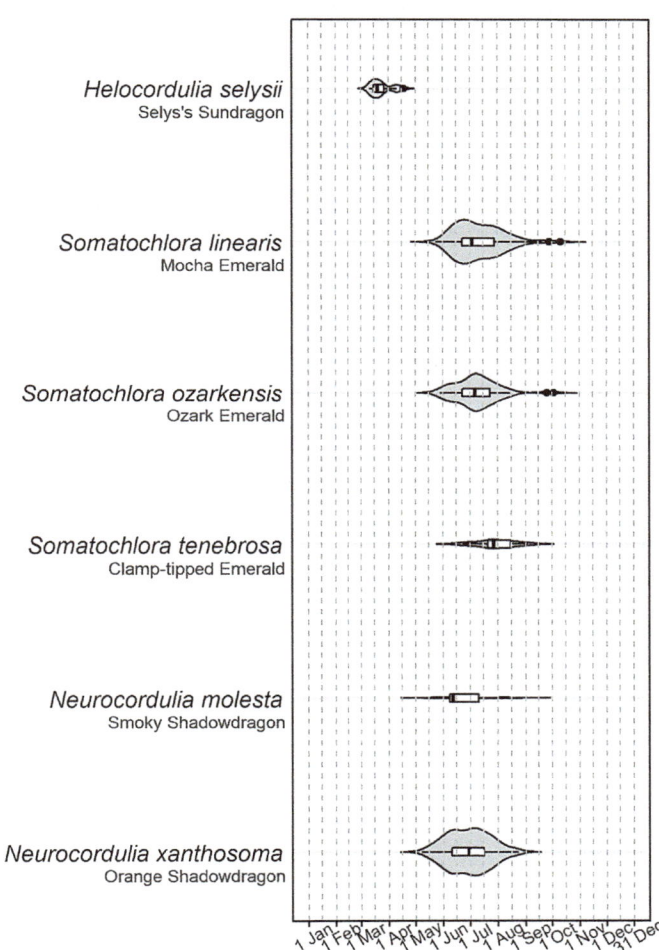

Figure 18.1 Seasonality in Oklahoma of species in the Corduliidae (see the separate introduction for the genus *Epitheca* for seasonality of the five regularly occurring baskettail species). Two additional species have been recorded once each: the Texas Emerald (*Somatochlora margarita*) on 22 June and the Cinnamon Shadowdragon (*Neurocordulia virginiensis*) on 18 June. Uhler's Sundragon (*Helocordulia uhleri*) has been recorded only twice.

have tapered off for the season). We do not present a separate summary for *Epicordulia*, which is also still generally considered a subgenus of *Epitheca* and which is represented by the single species of *Epitheca princeps*, the Prince Baskettail.

HELOCORDULIA

There are two species of *Helocordulia* sundragons in the world, both of which occur in Oklahoma. They very much superficially resemble the small baskettails, the *Tetragoneuria*-type *Epitheca*. These genera overlap extensively in size and general coloration, but if one looks closely the darker base color, the limited yellow patterning, and the clubbed abdomen should alert the observer that it is a *Helocordulia* not a baskettail. Nonetheless, we suspect that sundragons have been overlooked in Oklahoma, as elsewhere, by being mistaken for baskettails. That may be the case, for example, for the report of *Epitheca petechialis* for Latimer County (see that species account for details).

Aside from the single Uhler's Sundragon (*Helocordulia uhleri*) that Lothar E. Hornuff collected in 1956, the genus was unknown for Oklahoma until 2008, when Berlin A. Heck discovered a small population of Selys's Sundragon (*H. selysii*). Thereafter, that species was known solely from that small population, which seems to wax and wane, until a single male was photographed at a different location 90 km to the north. Uhler's Sundragon was not recorded again until 2020. We suspect that both species occur in small numbers in, at least, eastern McCurtain and Le Flore Counties in southeastern Oklahoma, with *H. uhleri* also in northeastern Oklahoma, but their short flight season and inconspicuous habits have conspired to limit our understanding of their status.

SOMATOCHLORA

That the common name of *Somatochlora*, the emeralds, is the same as the common name of the family Corduliidae, can cause some confusion, but the shared name simply indicates that *Somatochlora* is the most speciose genus of corduliids (just over half) in North America. Of those 26 species, Oklahoma can claim but 4; unsurprising given that the state lies at the periphery of where most *Somatochlora* species occur—the genus radiated in the eastern and northern parts of the United States and Canada.

The emeralds are sought-after jewels of eastern forests, none more so in Oklahoma than the Ozark Emerald (*Somatochlora ozarkensis*), a regional endemic found only in eastern (and chiefly southeastern) Oklahoma, extreme southeastern Kansas, southern Missouri, and western and central Arkansas. This species is found routinely in the Ouachita Mountains, with scattered records for the Ozark Plateau, and known or suspected breeding creeks number fewer than a couple of handfuls. The broadly similar Clamp-tipped Emerald (*S. tenebrosa*) occurs in this same geographic area, whereas the larger Mocha Emerald (*S. linearis*) is

more widespread but nonetheless encountered infrequently. Where they overlap, all three of these species can be found on the same creek. We find the single recent (2017) record of the Texas Emerald (*S. margarita*) intriguing because it hints to us the presence of a previously undetected, albeit likely small, breeding population in southeasternmost Oklahoma (and possibly southwesternmost Arkansas) within the South Central Plains ecoregion, where it is found in Texas and Louisiana.

There are two other *Somatochlora* that should be on the radar of Oklahoma dragonfly hunters, not for finding populations in the state but for the possibility of vagrant records. The northeastern Texas record of the Coppery Emerald (*S. georgiana*) makes that species a possible vagrant to Oklahoma, and the currently defined range of the Fine-lined Emerald (*S. filosa*) falls close enough to the southeastern corner of Oklahoma we would not be shocked if it is one day found in the state.

Before leaving *Somatochlora* we feel compelled to share a quote from EB Williamson taken from a letter to RD Bird, dated 14 November 1932. Williamson prefaced his statement with reference to the new *Somatochlora* RD Bird was to describe, i.e., *S. ozarkensis*, in comparison to *S. provocans*, *hineana*, and *tenebrosa*. He says of *Somatochlora* that:

> I believe no more elusive dragonflies occur in North America than these brutes. An active collector might miss them for years in a locality where he considered himself entirely informed. In my efforts to collect *hineana*, I made four trips to Ohio … accompanied on every trip by other collectors than myself, and the net result of all our work was one male and one female. As there is nothing else of interest there, and the locality is an unattractive, dirty, mosquito-infested region, we didn't get much fun out of those trips.

NEUROCORDULIA

Neurocordulia are generally lotic species, with most found at or near medium to large rivers. They are principally crepuscular (active at dusk or dawn), but they can occasionally be somewhat active during overcast days. Most of our encounters have been when we disturbed one (or more) during the day and roused it from its roost. Ordinarily they do not fly far (they also fly awkwardly in direct sunlight, much in the manner of teneral odonates), but their new perch can be challenging to find, typically buried well within dense thorny vegetation or rock crevices.

Five species of *Neurocordulia* have been claimed for Oklahoma, but we consider only three of those as valid. The Orange Shadowdragon, *N. xanthosoma*, is by far the state's most common, followed by the Smoky Shadowdragon, *N. molesta*. The former is the only *Neurocordulia* bearing two rows of cells between the margin of the hindwing and the toe of its anal loop. Some argue that the distinctiveness[4] of *N. xanthosoma* is cause for the species to be realigned with its original genus of *Platycordulia*.[5] *Neurocordulia molesta*, too, is distinctive morphologically, especially considering that it is the only species of *Neurocordulia* whose larvae have a pyramidal frontal horn. Adult males also have mesotrochanters that bear a diagnostic process. As such, some have considered separating *N. molesta* into its own genus of *Rostrocordulia*[6] (Needham and Westfall 1955:355). We cannot fully reconcile the mere half-dozen records of the Smoky Shadowdragon scattered through the eastern half of the state, other than to concede that it is overlooked. The Orange Shadowdragon is, in contrast, found routinely, we suspect both because it is by far the most numerous but also because, as far as is known, it tends to roost at lower height—shin to knee height, rather than head height or higher—and so it is flushed up more often than the other shadowdragons.

The third species of *Neurocordulia* known from Oklahoma is *N. virginiensis*, the Cinnamon Shadowdragon. The species has been documented once in the state, although the only specimen appears to be lost. The Umber Shadowdragon, *N. obsoleta*, was historically claimed for the state, but those records are now known to be *N. xanthosoma* or *Epitheca petechialis*, the Dot-winged Baskettail (see Appendix A). Lastly, the Stygian Shadowdragon, *N. yamaskanensis*, was published as occurring as larvae in the Wichita Mountains in southwestern Oklahoma. Although that record is undoubtedly a misidentification, the species could conceivably be found in northeasternmost Oklahoma given the rather close proximity of its westernmost range in southwestern Missouri and north-central Arkansas.

EPITHECA (TETRAGONEURIA)

Set aside female bluets (*Enallagma* sp.), forktails (*Ischnura* sp.), and dancers (*Argia* sp.). Speak not of river cruisers (*Macromia* sp.) of either sex. Say no more about certain meadowhawks (*Sympetrum* sp.) that appear in autumn or of certain pondhawks (*Erythemis* sp.) with disputed taxonomy that likely hybridize where ranges meet. Without question the most vexing and maddening identification problem in Oklahoma is presented by the small baskettails (*Epitheca* sp., subgenus *Tetragoneuria* Hagen 1861; Figure 18.2). Five species have been recorded in the state, only one of which, the Robust Baskettail (*E. spinosa*), causes no identification anxiety … *if* one is confronted with a male (and, fortunately, each of the two records for Oklahoma is of a male). The other four species overlap at their peripheries in Texas and Oklahoma (Figure 18.3).

Vexation is driven by an unholy combination of three interrelated factors: 1) a high degree of morphological similarity, 2) a bewildering degree of individual variation, and 3) an unknown propensity for hybridization and attendant preponderance of hybrids. Similarity in basic morphology among these species has long been a grievance of researchers who attempt to wade through the taxonomic confusion. It strikes us as a truism that no two experts agree on which characters are diagnostic and, hence, no two agree on the identity of any given specimen whose characters are not at the morphological extremes that place it well away from all but one species (Figure 18.4). (We have examined museum specimens identified as one species by the experienced collector, as another by an expert who assessed them years later, and as something else again by us when we attempted to establish uniform identification criteria. No wonder, then, photographs of *Tetragoneuria* often must go unidentified!) Even then, authorities differ markedly in what constitutes diagnosability within a species let alone between species (Figure 18.5). Muttkowski (1911:91), in speaking of all *Tetragoneuria*, nicely summed the situation:

> The close resemblance of the species, the generalized genitals and but little specialized anal appendages, and the variability of wing markings and wing venation are factors *which tend to make a mechanical separation of the species difficult, if not impossible on occasion.* [our emphasis]
>
> The genus *Tetragoneuria* has always, more or less, belonged to the category of "Splitters' and Lumpers' Paradise."

In a nutshell … no one, despite almost two hundred years of investigations,[7] has definitively determined the limits of species and subspecies within *Tetragoneuria*, and so species have come and gone, been downgraded to subspecific status (or "races") and risen back up to specific status. So the question remains, if we do not know where to draw the line to delimit species and subspecies, how can we define what constitutes a hybrid? And who is to say that individual variation does not account for the apparent intermediacy we think we see?

For now, our most straightforward path is to continue trying to determine specimens to species as currently defined (with the caveat that our future research and that of others may result in taxonomic changes), putting "classic" individuals into the various species' boxes (Figure 18.4) and leaving atypical or intermediate forms labeled as undetermined or as multiple species (e.g., *Epitheca* sp. or *E. costalis/cynosura*). As such, herein, we treat the Dot-winged (*Epitheca petechialis*) and Slender Baskettails (*E. costalis*) as species and individuals not clearly denoted are left undetermined. If the systematic hypothesis favored by some taxonomists is true that these taxa are western and eastern representatives of the same species (i.e., they are subspecies), then hybridization is expected where ranges meet, in our case along a north–south axis through central Oklahoma (Figure 18.6).

The extent of hybridization among other taxa, specifically between the Slender Baskettail and Common Baskettail (*Epitheca cynosura*) or between the Common Baskettail and the Mantled Baskettail (*E. semiaquea*), is unknown. Various specimens of intermediate morphology we have collected in Oklahoma suggest to us that these species hybridize at least rarely but perhaps uncommonly. Individuals of intermediate phenotype, and thus presumably of hybrid origin, between the Common and Slender Baskettails have been reported from the Midwest (Donnelly 2006b) and between the Common and Mantled Baskettails in the northeast (May 1995). As such, we certainly were not taken by surprise when we learned that Oklahoma's first record of the Mantled Baskettail was of a male in tandem with a female Common Baskettail (FSCA). Such mixed-species pairs and intermediate forms make for a strong argument not to heed the suggestion of females being "named by their association with the males" (Needham and Westfall 1955:366, Needham et al. 2014:365).

In the end, as tantalizing as evidence of hybridization may be, at base we are confronted with a dearth of direct evidence of interbreeding, leaving us ignorant of its extent until focused studies are conducted.[8] Consequently, we retreated to a conservative stance and base species accounts on precisely those specimens to which we allude above, those that fall clearly within the morphological space that absolutely (as far as known!) exclude other species. In other words, the species accounts are grounded in specimens for which we feel identification cannot be anything but certain,[9] regardless of authority followed. Our accounts are fleshed out, then, with additional records we feel are only likely to be of the particular species in question. Such specimens lend context and can point to areas where additional research is needed. Our seasonality plots (Figure 18.7), too,

A

Also, I am carrying out Mr. Williamson's wishes in writing a description of the new *Tetragoneuria* also from Florida. While searching through our unidentified material on the possibility of finding more specimens of the same species, I came upon six males of what I believe to be *T. petechialis* from the Wichita National Forest, collected by Dr. T. H. Hubbell. I am sending a specimen to Dr. Calvert for verification. When I receive a report on it I shall be glad to give you the complete data for your Oklahoma records.

B

I was interested in hearing of the report of *Tetragoneuria petechialis* from Wichita National Forest. I would certainly appreciate records of these specimens and would also like to borrow a specimen for comparison with a large series of specimens I have from the same region which I am calling *T. cynosura*. So far as I know we have four *Tetragoneuria* from Oklahoma. They are *T. cynosura* from Western central Oklahoma, *T. simulans* from Eastern Oklahoma, *T. williamsoni* from Eastern and Central Oklahoma, and the one record of *T. canis* from Eastern Oklahoma.

C

April 30, 1934.

Prof. R. D. Bird
Department of Zoology
University of Oklahoma
Norman, Oklahoma

Dear Prof. Bird:

In a letter last spring I promised to let you know what Dr. Calvert reported regarding the *Tetragoneuria petechialis* which I sent to him. His letter dated October 22, 1933 says "... I recently compared the specimen from Comanche Co., Oklahoma, which you sent me last spring as *T. petechialis*, with Muttkowski's and Davis' descriptions and with M's paratypes C1 and C4 and I confirm your identification."

D

25. *Tetragoneuria cynosura* Say
Beckham, Caddo, Cleveland, Comanche, Hughes, Latimer, LeFlore, McClain, McCurtain, Murray, Pittsburg, Pushmataha.

26. *Tetragoneuria williamsoni* Muttkowski
LeFlore (*Williamson).

27. *Tetragoneuria canis* McLachlan
Latimer (Hooper).

E

32. TETRAGONEURIA CYNOSURA (Say). Beckham, Caddo, Cleveland, Comanche, Hughes, Latimer, Le Flore, McClain, McCurtain, Murray, Pittsburg, Pushmataha (Bird, 1932). Muskogee (Bick, 1951). Bird (1932) did not differentiate *cynosura* and *petechialis* and there is some doubt that *petechialis* is a distinct species. Specimens (OU) from Beckham, Murray, and Comanche Counties determined as *cynosura* by Bird were determined as *petechialis* by both the writer and Dr. E. J. Kormondy. There are other specimens in the OU collection from Murray and Cleveland Counties which are definitely *cynosura*; some were undetermined by Bird, some were determined by him as *williamsoni* Muttkowski. Thus, of Bird's list of counties, I could verify material from only Cleveland and Murray.—Carter, Washita, Woodward.

*33. TETRAGONEURIA PETECHIALIS (Muttkowski). Beckham, Comanche, Murray (OU). Greer, Woodward. It is interesting to note that three males of *petechialis* and three of *cynosura* were collected by the authors at the same spot in Boiling Springs Park on June 2, 1956 and that there are specimens of the two species in the OU collection from Murray taken on May 14, 1932.

*34. TETRAGONEURIA SPINOSA (Hagen). A single male (OU) from Latimer determined as *canis* by Bird is probably the basis of his (1932) record of *canis* from Latimer. This specimen was determined as *spinosa* by both the writer and Dr. E. J. Kormondy.

35. TETRAGONEURIA WILLIAMSONI Muttkowski. Le Flore (Muttkowski, 1911; Williamson, 1914). Muttkowski's three type specimens and the Williamson record are from Wister.

Figure 18.2 Trying to make sense of *Epitheca* (subgenus *Tetragoneuria*) in Oklahoma during the early and middle eras of odonatology in the state. LK Gloyd and RD Bird corresponded about *Tetragoneuria*: A) letter from Gloyd to Bird, dated 5 April 1933, B) Bird's reply, dated 8 April 1933, and C) Gloyd's return letter on this topic, dated 30 April 1934. Bird (1932; panel D) reported three species of *Tetragoneuria* prior to his correspondence with Gloyd. Bick and Bick (1957:4–5; panel E) later indicated that there were four species in the state (notice the omission of *E. semiaquea*; *E. williamsoni* is now *E. costalis*). Bick and Bick re-identified some of Bird's specimens while acknowledging that there were ongoing taxonomic disputes.

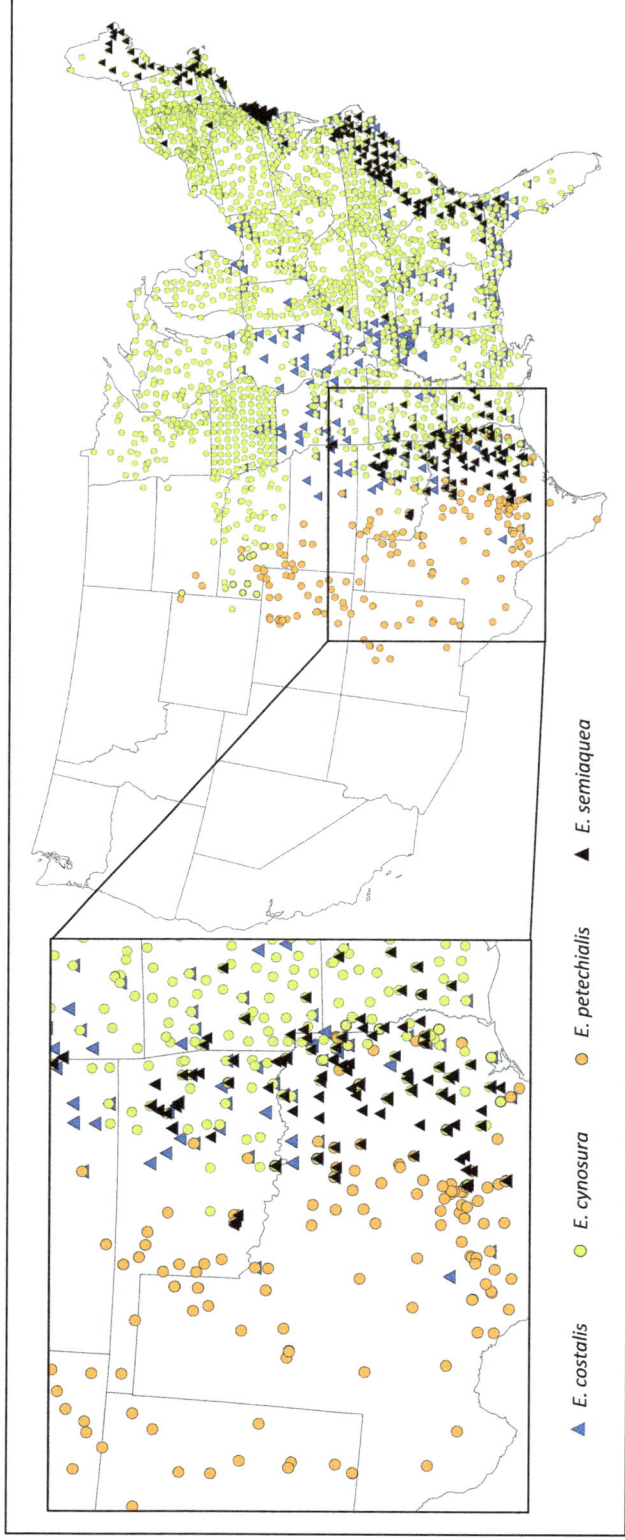

Figure 18.3 The geographic ranges of four species of *Epitheca* (subgenus *Tetragoneuria*) in the United States (right) meet in Texas and Oklahoma (left). Given the morphological similarities between these species—*E. costalis*, *E. cynosura*, *E. petechialis*, and *E. semiaquea*—it should be expected that at least some interbreeding would occur in the region.

▲ *E. costalis*   ● *E. cynosura*   ● *E. petechialis*   ▲ *E. semiaquea*

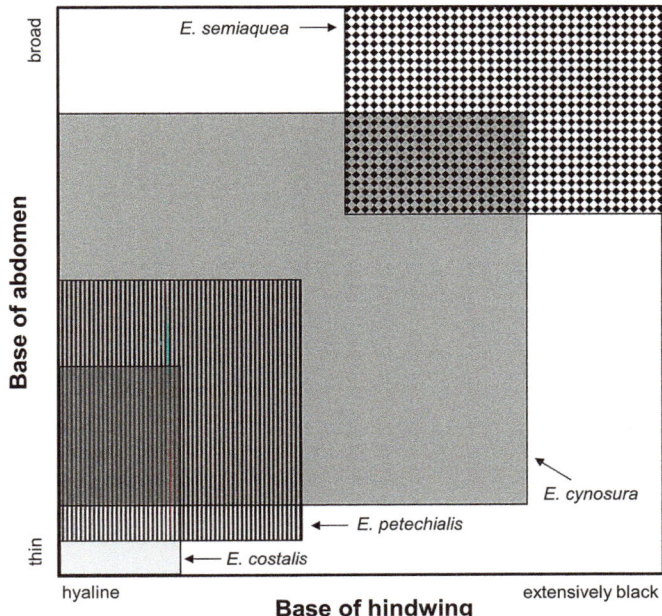

Figure 18.4 A schematic diagram of the range of variation within each of the four common small baskettail species (*Epitheca*, subgenus *Tetragoneuria*) that occur in Oklahoma. The wide range of individual variation in the Common Baskettail (*E. cynosura*) presents an especial problem.

were constructed from records that we feel are, at present, correct.

As far as it is known, ecological differentiation, such as habitat segregation, is unlikely to provide much of a clue: males of each species of small baskettail patrol short beats near the (invariably well-vegetated) edge of a pond, small lake, or slow-moving stream or creek. The beats tend to be below chest height and feature a good deal of hovering and frequent agonistic interactions. It may be that the Common Baskettail is more partial to shaded habitats, but that hypothesis awaits supporting data. Likewise, the Dot-winged Baskettail tends to be found more often in open habitats, but any such difference could be a function of the drier, more open habitats in western Oklahoma than any real difference in habitat preference. Swarms can be found anywhere there are clearings, from meadows to glades and from roads (especially gravel or dirt ones) to berms, near forest or woodland edge. In the eastern part of the state, we have noticed a tendency for the Mantled Baskettail and Slender Baskettail to co-occur and a tendency for the Common Baskettail to occur alone, but we know neither if such apparent tendencies hold in general nor what mechanism may drive the tendencies if they do hold.

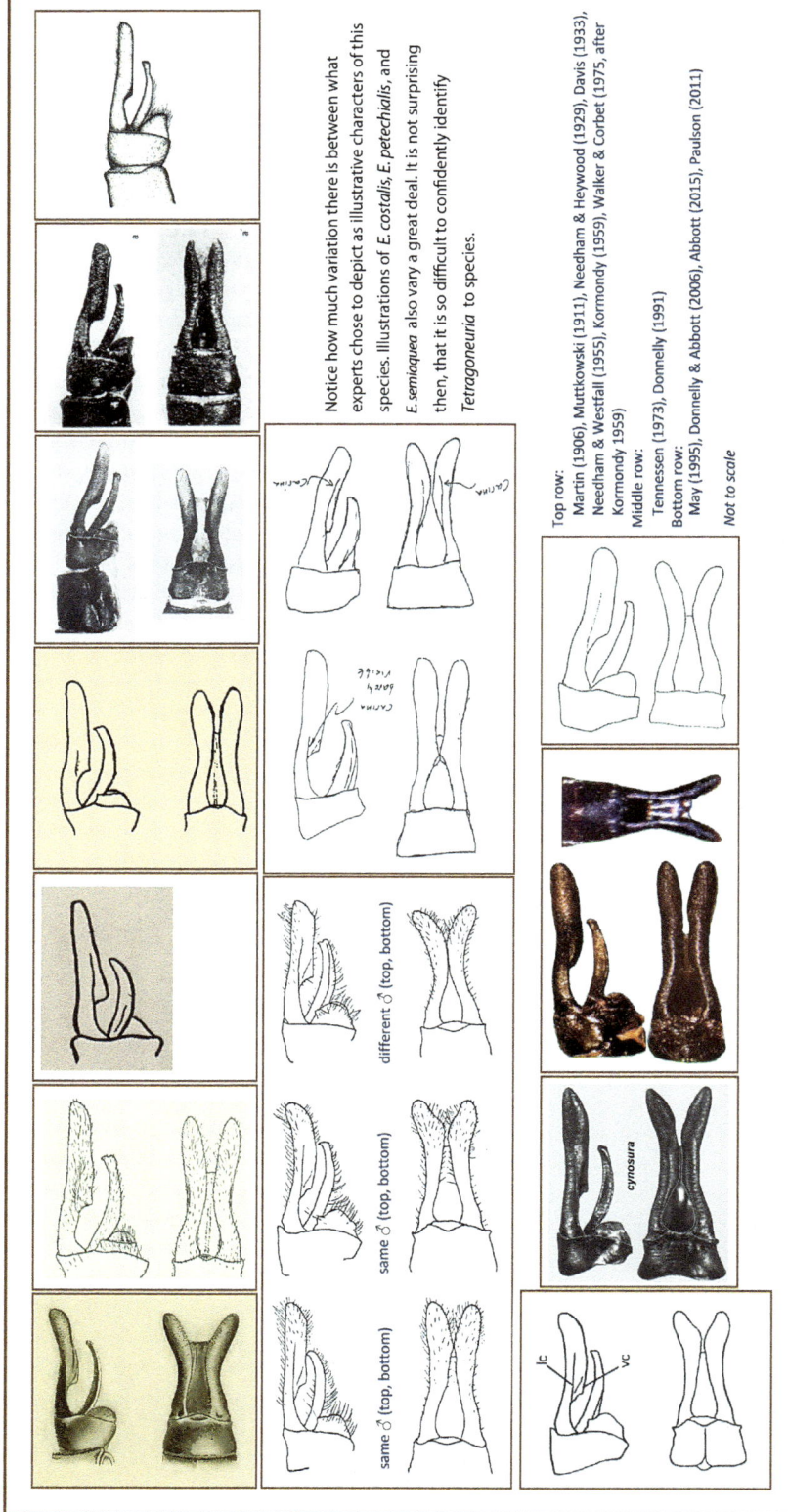

Notice how much variation there is between what experts chose to depict as illustrative characters of this species. Illustrations of *E. costalis*, *E. petechialis*, and *E. semiaquea* also vary a great deal. It is not surprising then, that it is so difficult to confidently identify *Tetragoneuria* to species.

Top row:
Martin (1906), Muttkowski (1911), Needham & Heywood (1929), Davis (1933), Needham & Westfall (1955), Kormondy (1959), Walker & Corbet (1975, after Kormondy 1959)
Middle row:
Tennessen (1973), Donnelly (1991)
Bottom row:
May (1995), Donnelly & Abbott (2006), Abbott (2015), Paulson (2011)

Not to scale

Images used with permission: New York Entomological Society, University of Michigan Museum of Zoology, University of Toronto Press, Kenneth J Tennessen, Thomas W Donnelly, Michael L May, University of Texas Press, and Princeton University Press.

Figure 18.5 Sample of illustrations of *Epitheca (Tetragoneuria) cynosura* terminal appendages."

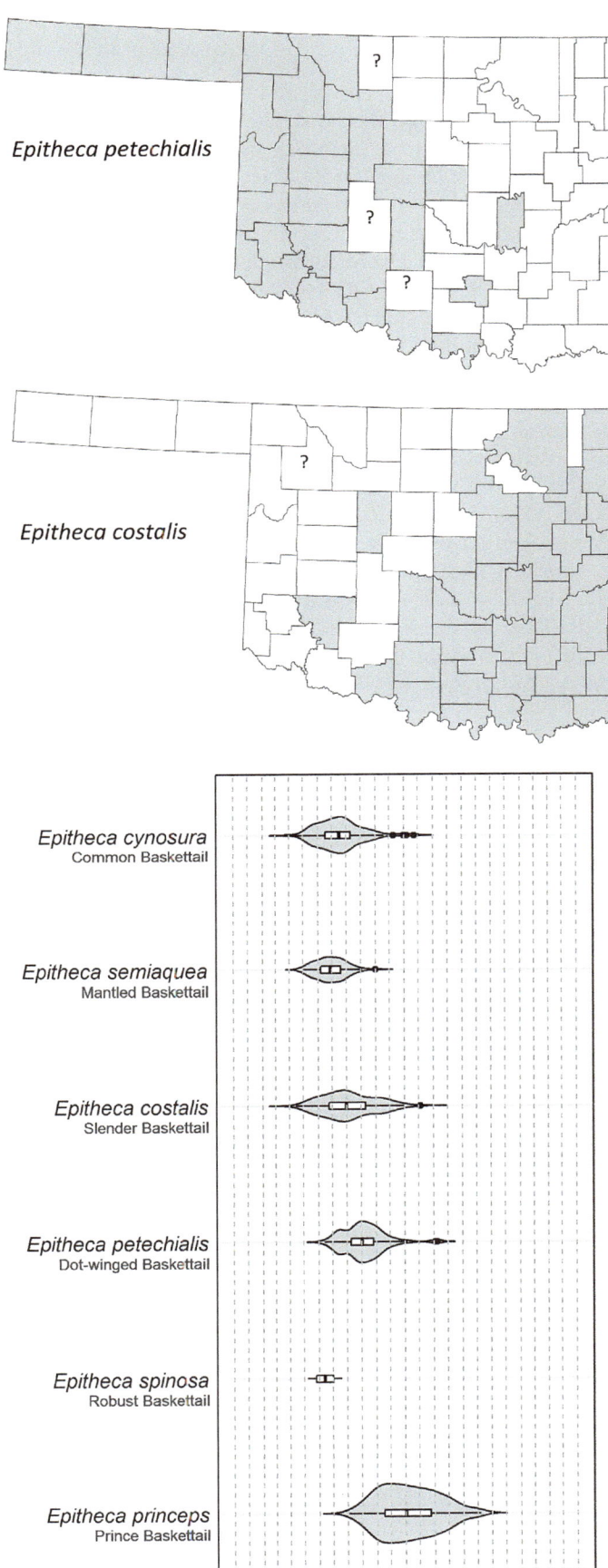

Figure 18.6 Comparative distributions, on a county-by-county basis, of the A) Dot-winged (*Epitheca petechialis*) and B) Slender (*E. costalis*) Baskettails in Oklahoma. A question mark signifies that the species has been attributed to a county but we have been unable to verify the identification. These species in effect are east–west replacements of one another, although no more so than, say, the superficially similar Comanche (*Libellula comanche*) and Spangled (*L. cyanea*) Skimmers (see Figure 19.47).

Figure 18.7 Seasonality of the baskettails (*Epitheca* sp.) that occur regularly in Oklahoma. Note that the distinctive Prince Baskettail (*E. princeps* of the subgenus *Epicordulia*) flies later than the small species (of the subgenus *Tetragoneuria*). Note, too, that the flight season of the Dot-winged Baskettail (*E. petechialis*) is shifted slightly later than that of the other small species, presumably because spring comes slightly later to western Oklahoma.

HELOCORDULIA SELYSII (HAGEN, 1878) – SELYS'S SUNDRAGON

Until recently Selys's Sundragon had only been documented from the seemingly magical spot that is nicknamed "Berlin's property," a privately owned (by Berlin and Pat Heck) plot of land located about 10 km SE of Idabel, McCurtain County. This is the same locale that for many years hosted the only known populations of *Phanogomphus lividus*, the Ashy Clubtail, and *Cordulegaster maculata*, the Twin-spotted Spiketail, until populations of each were found elsewhere. The shallow, clear water, sandy-bottomed perennial stream lined by trees that run through the property is the perfect habitat for Selys's Sundragon, and the other species too.

There are only 12 records of Selys's Sundragon for Oklahoma, all but three of which have been reported solely by BAH On 12 March 2012, other observers (DA, JWA, and KW) had the joy of seeing this species in Oklahoma when they found nine individuals, the majority of which were teneral (OC 374135, 374097, Figure 18.8; the following day BAH reported that none were seen). Another encounter with the species, on 19 March 2012, yielded 10 adults, but this time there were three tandem pairs, one photographed *in copula* (OC 374125[10]). The only other record of a pairing was the single *in copula* pair captured by MAP on 11 April

2014, which made for the second and third specimens for the state (SP 1118). Otherwise the species has been documented one at a time, including both the first time it was recorded in Oklahoma, on 21 March 2008 (♀, OC 281801), and the only time it has been recorded in Oklahoma away from the single site in McCurtain County (♂, Beech Creek Botanical Area, 15 km SE of Big Cedar, Le Flore County, 26 March 2017, MAP, OC 461622; Figure 18.8). The latter record was near a similar narrow, clear, shallow stream in deciduous woodland.

The flight season is short and early, which helps to explain the dearth of records. It has been seen only from 10 March to 17 April (OC 374088, 374610); phenology that is consistent with other areas nearby (Paulson 2009, 2011; AR 5 April–2 May, Harp and Harp 1996; LA: 1 March–10 April, OC, Mauffray 2014; TX: 23 Febuary–16 April, iNat, Abbott and Lasley 2019).

Its short flight season, its presumably low numbers, and the likelihood of it to be confused with the smaller baskettails (*Epitheca* species of the subgenus *Tetragoneuria*) make it difficult to assess its status in the state. Our guess is that Selys's Sundragon has been overlooked and is a bit more

Figure 18.8 *Helocordulia selysii*, Selys's Sundragon, A) ♂, Beech Creek Botanical Area, Le Flore County, 26 March 2017 (©Michael A Patten, OC 461622), the only to be recorded in Oklahoma away from a single site in McCurtain County where it has been found off-and-on beginning in 2008. B) ♀, 10 km SE of Idabel, McCurtain County, 12 March 2012 (©James W Arterburn), at said magic site.

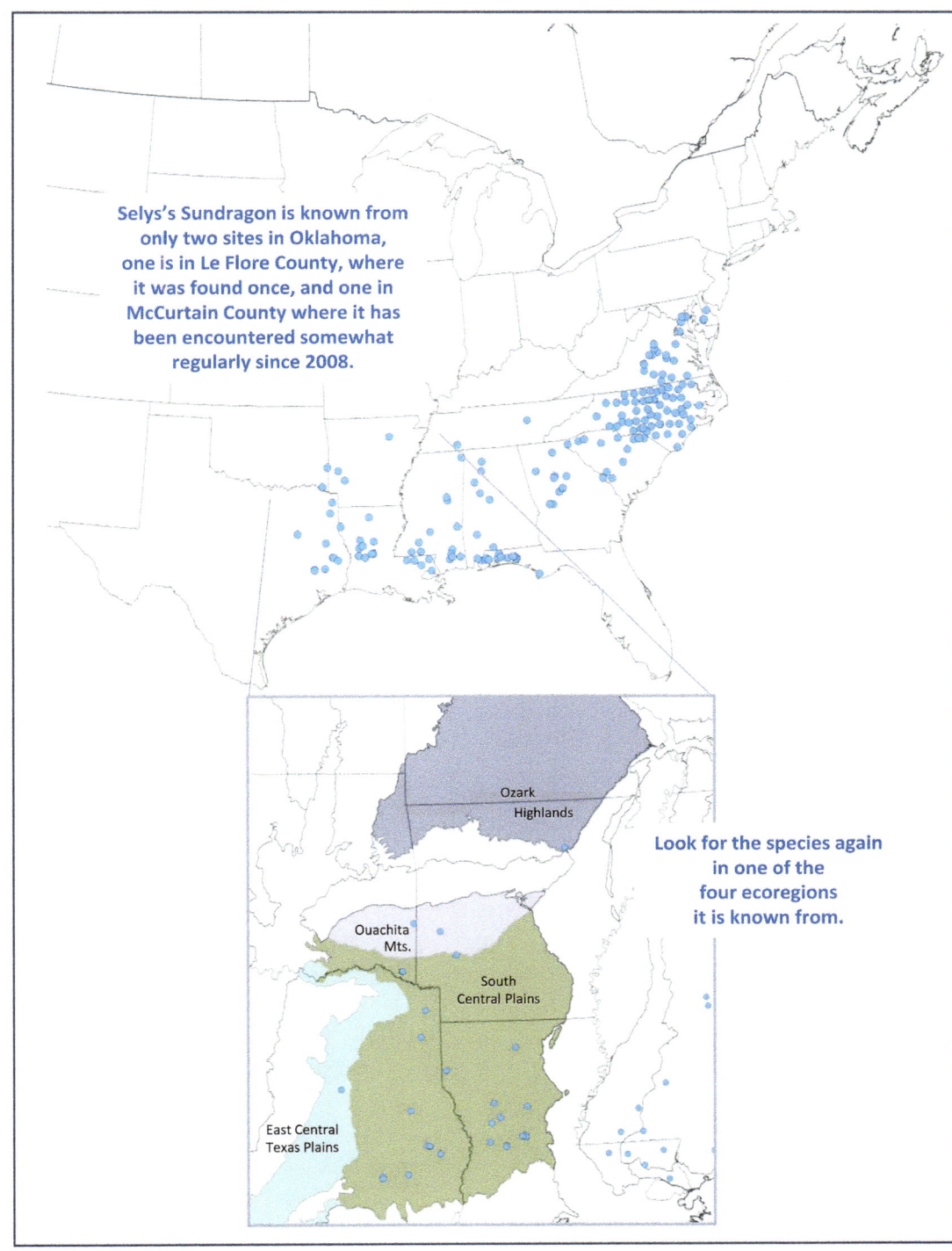

Selys's Sundragon is known from only two sites in Oklahoma, one is in Le Flore County, where it was found once, and one in McCurtain County where it has been encountered somewhat regularly since 2008.

Ozark Highlands

Look for the species again in one of the four ecoregions it is known from.

Ouachita Mts.

South Central Plains

East Central Texas Plains

Figure 18.9 The geographical range of *Helocordulia selysii*, Selys's Sundragon, barely reaches southeastern Oklahoma.

common and widespread than we have currently detected it to be. As such, we feel our earlier assessment of the species as S2 (Patten and Smith-Patten 2013b) is sufficient, but if the species is shown to be less common than we presume and threats to the species increase, an upgrade to S1 would be warranted.

One dubious record for the species is a report of larvae at the Wichita Mountains, Comanche County (Bass 1990; we have not been able to locate this specimen). Although we cannot entirely rule out the possibility that this species has bred in the Wichita Mountains, where there are plenty of odd records, it is nonetheless biogeographically unlikely (Figure 18.9).

HELOCORDULIA UHLERI (SÉLYS, 1871) – UHLER'S SUNDRAGON

The first record of Uhler's Sundragon for Oklahoma came on 12 April 1956 when LE Hornuff collected a single ♂ at Hochatown, McCurtain County (FSCA; Bick and Bick 1957 reported it as "Hochatown, Mountain Fork River"; Figure 18.10). The search was on for more than 60 years until Alex Harman beat us all to the punch by finding a mating pair in the Ozark Plateau on Cherokee Nation Tribal Trust land near Bell, Adair County, in 2020 (23 April, iNat 43016058; Figure 18.10).

Tennessen et al. (1995:62) described the species' habitat in Alabama as "small, upland forested streams usually where flow has been impeded." In Georgia it was described as "small sandy streams" (Beaton 2007:238) and throughout its range as "wide habitat choice, from small rocky woodland streams to large open rivers, usually with good current." These habitats are common throughout the Ouachita NF near the state's first record. Given that the area is adjacent to known localities just across the border in Arkansas, we see no reason the species could not be found again in the area; indeed, we expect that a small population exists in the Ouachitas of eastern Le Flore and McCurtain Counties. Still, we and others have lucklessly sought the species on numerous occasions. Similar habitat exists throughout the Ozark Plateau, so we are keeping our fingers crossed that the 2020 record comes from an established population.

Although known from relatively few records, the species flies in Arkansas from 19 March to 26 May (OC 312310, USNM 389575, 389576) and in Missouri from 7 May to 1 June (Trial 2005). Elsewhere it is known to fly from mid-February to late July (OC). There is one larval report for Oklahoma that is likely erroneous given its location, but we have not re-examined the specimen (Tallgrass Prairie Preserve, Osage County; Bass 1994).

Figure 18.10 There are but two records of *Helocordulia uhleri*, Uhleri's Sundragon, in Oklahoma. The first came on 12 April 1956 when LE Hornuff collected this ♂ in the Ouachita Mountains (A; FSCA). The second record, of a mating pair (B), was found six decades later, this time in the Boston Mountains of the Ozark Plateau (©Alex Harman, 23 April 2020, iNat 43016058). The status of this species in the state is unknown.

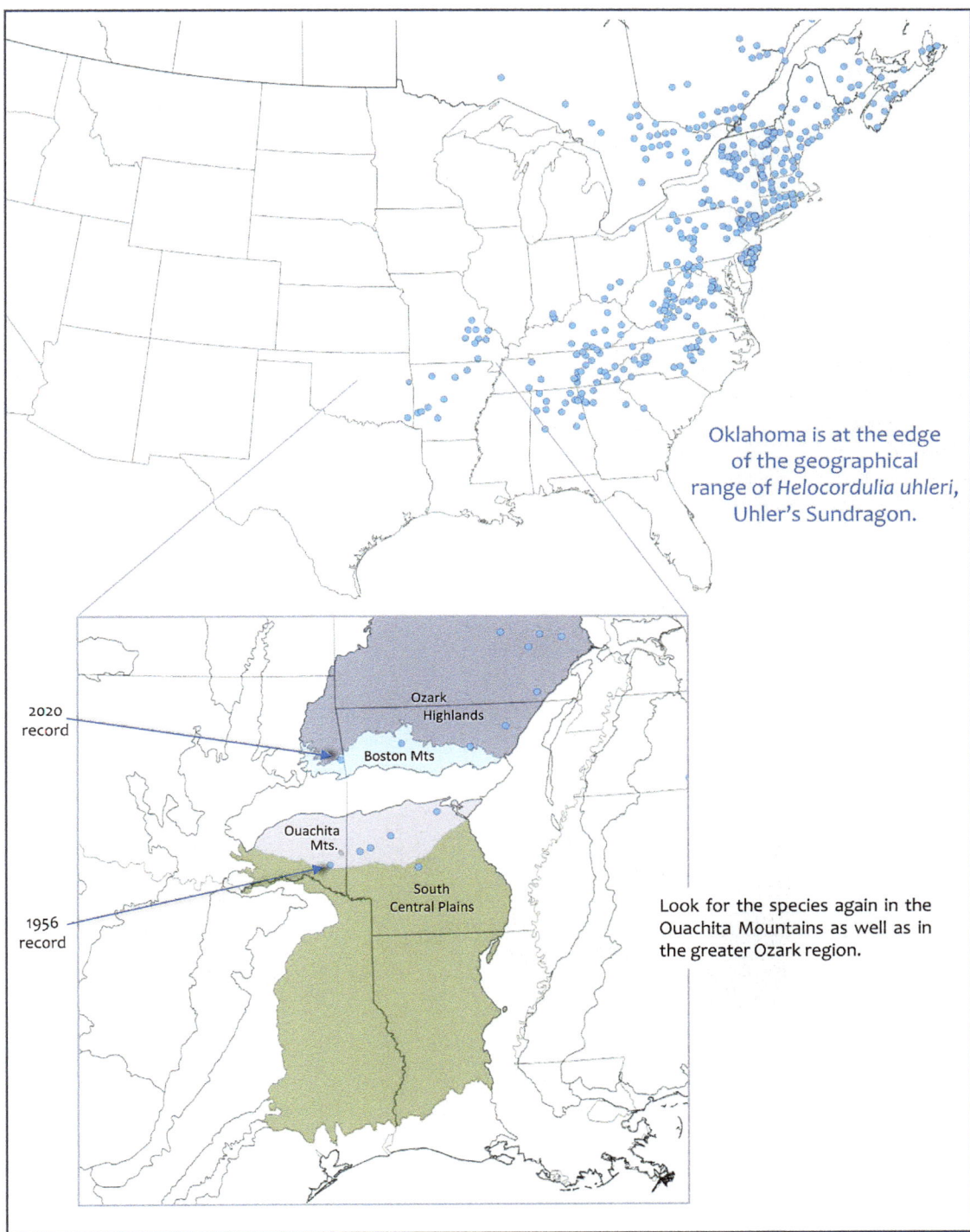

Oklahoma is at the edge of the geographical range of *Helocordulia uhleri*, Uhler's Sundragon.

2020 record

1956 record

Ozark Highlands

Boston Mts

Ouachita Mts.

South Central Plains

Look for the species again in the Ouachita Mountains as well as in the greater Ozark region.

Figure 18.11 The geographical range of *Helocordulia uhleri*, Uhler's Sundragon, reaches portions of Arkansas and Missouri that share ecoregions with Oklahoma. Look for the species again in the Ouachita Mountains as well as in the greater Ozark region.

SOMATOCHLORA LINEARIS (HAGEN, 1861) – MOCHA EMERALD

The Mocha Emerald is an (apparently) uncommon species of the eastern half of the state. We say "apparently" because detectability of emeralds is low; that is, a species may be numerous, but its habits make it difficult to detect, especially in large numbers. In this case, *S. linearis* usually is found one to half a dozen individuals at a time, but occasionally up to a dozen are encountered, and we have had the species once with a count of 25 in a swarm over a clearing adjacent to a mixed conifer-hardwood forest (near Tucker Knob, 8 km S of Kinta, Haskell County, 3 July 2016, MAP, 1♀ as SP 2009). In Oklahoma, we have found it near trickling beaver ponds in woodlands, at bayous, and small, shaded creeks in hardwood forests, including within cross timbers. Sometimes it is found in bottomlands, at spring and creek complexes, and foraging along hilltops with other *Somatochlora*. One would not expect any *Somatochlora* to be common in Oklahoma because the state lies at the periphery of the center of abundance and richness of the genus, which radiated in the eastern

Figure 18.12 *Somatochlora linearis*, the Mocha Emerald, A) ♂, Osage Hills SP, Osage County, 8 July 2018 (©Bill Carrell, OC 484488). B) ♀, 2 km N of Big Cedar, Le Flore County, 16 June 2010 (©Gary L Spicer, OC 319903). Were it a race, the outcome would not be close: this species is easily the most common and widespread small emerald in the Southern Great Plains. Even then it can be elusive given its penchant for shadowy, narrow creeks embedded in mature woodland.

and northern portions of the United States and Canada, yet the Mocha Emerald is certainly the most common of the state's *Somatochlora* species, the other three being the Ozark Emerald (*S. ozarkensis*), the Clamp-tipped Emerald (*S. tenebrosa*), and the Texas Emerald (*S. margarita*).

RD Bird (1932) technically found the first record of *Somatochlora linearis* for Oklahoma (♂, 11 June 1931, Latimer County[11]), but he misdiagnosed it as *S. ensigera*, the Plains Emerald, which is no wonder given the illustrations in Needham and Heywood (1929). Bick and Bick (1957) later re-identified the specimen as *S. linearis*. AE Pritchard somehow figured out how to identify the species correctly because he announced to EB Williamson, in a letter dated 30 November 1932 (Figure 15.60), that he had added the species to the Oklahoma state list. The pronouncement would have been made on the basis of the ♂ that Pritchard collected on 26 July 1932 in Pawnee County (OMNH 3018). In 1934, Pritchard collected and identified another 20 *S. linearis* from Latimer, Le Flore, McCurtain, and Pushmataha Counties (EMEC, OSU). Otherwise there is only one other record from the 1930s (13 June 1939, Eagletown, McCurtain County, RW Kaiser and WT Nailon, OSU). The species was reported equally as often in the middle era of odonate collection in Oklahoma but is now reported regularly. It is now documented in 21 counties, with sight records potentially of this species from another four (Noble, Okmulgee, Ottawa, and Sequoyah).

The early flight date in Oklahoma for this species was of a ♂ we caught and released at Mohawk Park, Tulsa, Tulsa County, on 24 May 2003 (Smith-Patten et al. 2007). The late date is of a ♀ ovipositing (although her ovipositor was broken off) along Clear Creek below the spillway at Clear Creek Lake, Stephens County, on 28 September 2019 (MAP; SP 2920). We now doubt one we reported for a later date (11 October 2003; Smith-Patten et al. 2007). This species' phenology in Oklahoma is broadly in line with surrounding states (AR: 19 May–28 August, OC 375134, 322772; KS: 28 April–12 August,[12] SEMC 1349081–2, OC 322220; LA: 18 May–1 September, OC 444878, 444944, 314916; MO: 12 June–28 August, Trial 2005; TX: 26 May–25 August, Abbott and Lasley 2019). Ovipositing has been observed twice, when a lone ♀ laid repeatedly in a shaded pool in a mostly dry, rock-strewn bed of Holly Creek at Three Rivers WMA, Pushmataha County, 25 July 2015 (MAP) and when a lone ♀ laid eggs on a mossy rock at a creek in the Ouachita NF, McCurtain County (for additional details, see 16 August 2019 record of ♂ *Somatochlora tenebrosa* in the habitat descriptions for that species). Two nymphs have been collected (13 June 1950, at OSU Wildlife Conservation Station, Muskogee County, Bick 1951, specimen not relocated; 1 April 2018, Crooked Branch Creek, about 500 m below dam, Le Flore County, MAP, SP 2547) and we know of three teneral records, all from early to mid-June.

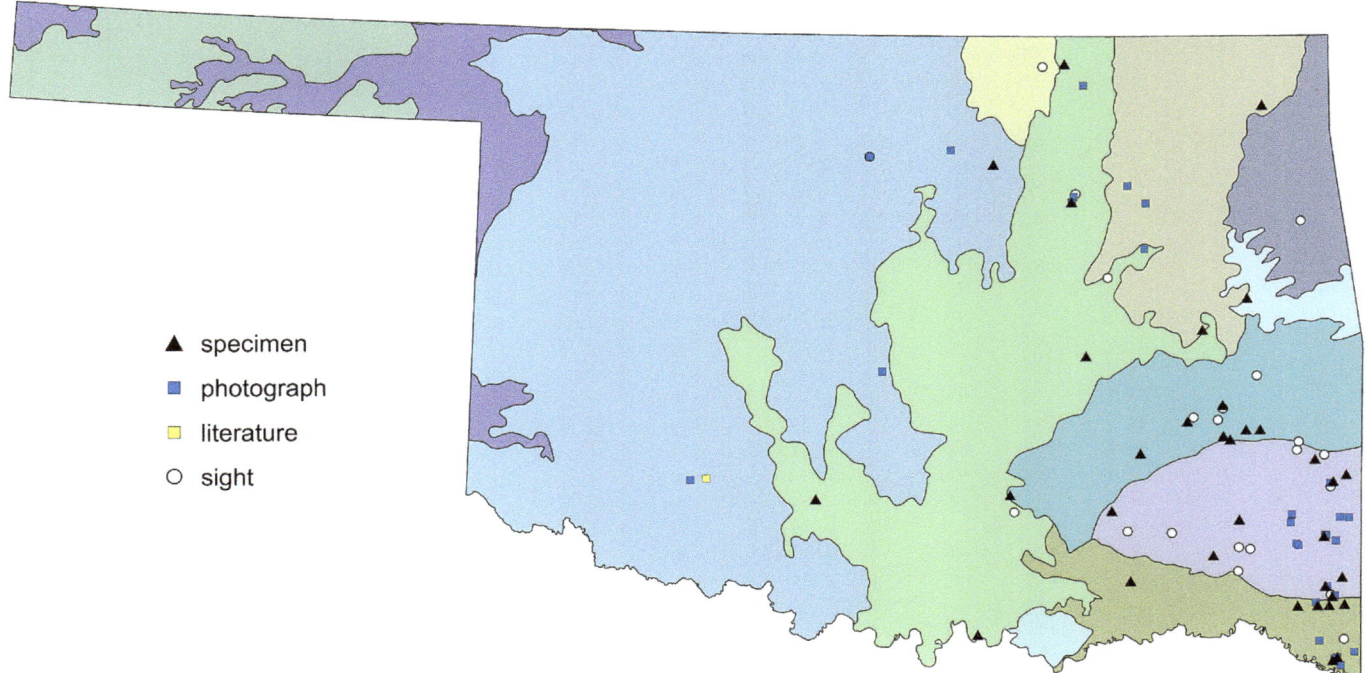

▲ specimen
■ photograph
□ literature
○ sight

Figure 18.13 Documentation level of supporting records for *Somatochlora linearis*, the Mocha Emerald, in Oklahoma. Levels include specimen, archived photograph, literature reference, or sight record. Although sight records lack documentation, only those we feel are credible are plotted.

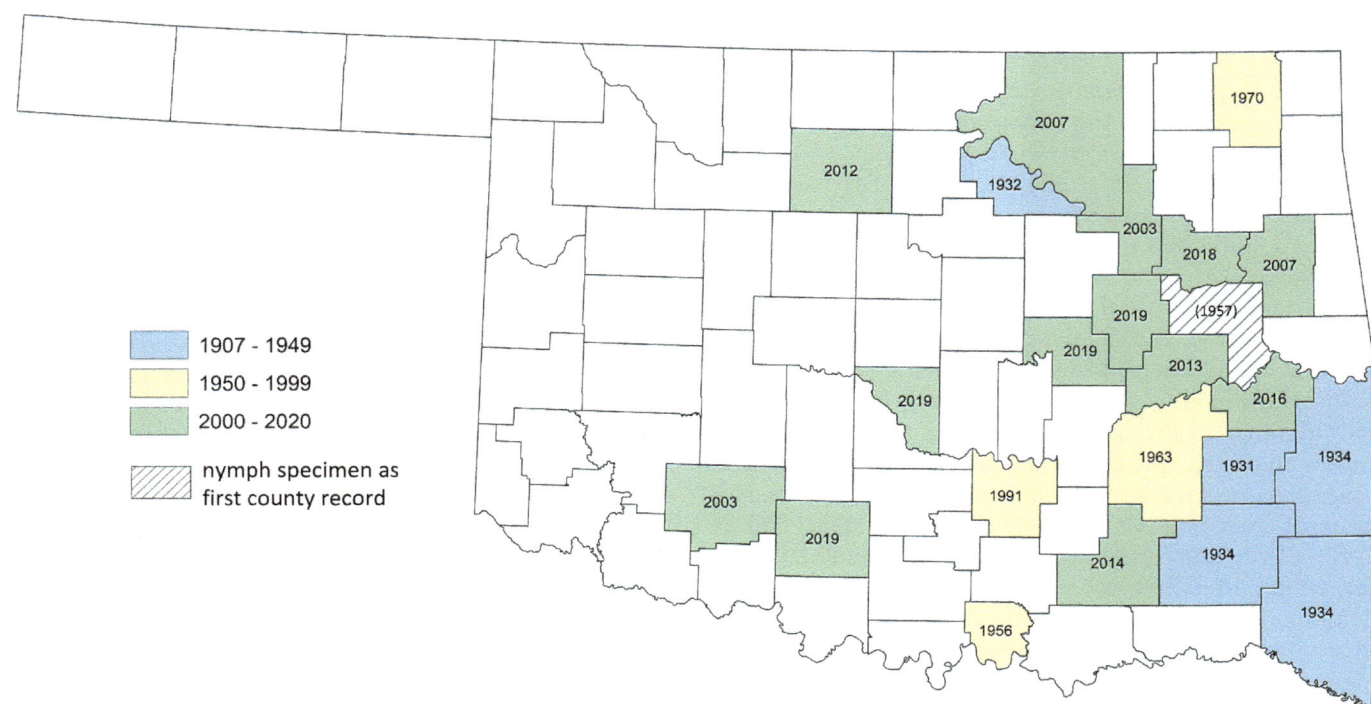

Figure 18.14 Counties in which *Somatochlora linearis*, the Mocha Emerald, are known to occur in Oklahoma. Year within the county is that in which the species was first reported. The first record for Muskogee County (hatched) was in 1950, represented by collection of a nymph; the first adult recorded for the county was in 1957.

SOMATOCHLORA OZARKENSIS BIRD, 1933 – OZARK EMERALD

Although its namesake is the Ozark Highlands, the Ozark Emerald has most often been encountered in the Ouachita Mountains, from where it was originally discovered and described (About 60% of records are directly from the Ouachitas (Figure 18.16). It rises to 86% if records from Latimer County, where the species was originally detected, but which technically lies in the Arkansas Valley just north of the Ouachitas, are included). Indeed, the species has been known from the Ouachitas since 1931, but was not confirmed in the Ozarks of Oklahoma until 2016!

It is from Oklahoma where the Ozark Emerald is best known. Prior to our work with the species, which began intensively in the spring of 2014, there were only about 45 reports of the Ozark Emerald throughout its entire geographical range—*there are now >130.* More than half of the known rangewide occurrences of the species and about half of the known locations fall within Oklahoma. The other three states[13] where this regional endemic is known to occur—Arkansas, Kansas, and Missouri (Figure 18.16)—fall short on comparable numbers. Arkansas lags behind, with 42 records reported from 27 locations,[14] and Missouri lags well behind, with only 20 records from about 15 localities.[15] There are eight reports from Kansas, but we have confidence in only one or two of those records, feeling that although we examined some of the specimens (primarily nymphs), all specimens should be re-examined in the future given some disputes in diagnosability of this species in early life stages.[16] Furthermore, we know of no recent records for Kansas, so its current status in that state is unknown. If it persists there, it must be within a tiny geographic range in the southeastern corner of the state.

Even if dubious records are excluded, the Ozark Emerald has been reported from at least eight ecoregions: Ouachita Mountains, Ozark Highlands, Boston Mountains, Arkansas Valley, Central Irregular Plains, Flint (or Osage) Hills, Cross Timbers, and the South Central Plains, in order of prevalence, with the first *three ecoregions harboring almost 80% of all known locations.* A ninth ecoregion, the Great Plains, is another possibility for the species. It has been reported twice in the Wichita Mountains of Comanche County, Oklahoma, well away from any other known population: on 20 June 2009, a ♂ was photographed in Ketch Canyon (OC 313402; Figure 18.15), and on 17 June 2011, a dead ♂ was discovered in a spider web at French Lake in the Wichita Mountains WR (OC 328764). The specimen was removed from the web to be sent to OMNH, but it never made it to the museum, so only photographic documentation of the species in the county exists. It is unclear whether the Comanche County records represent vagrants or if the species has an established disjunct population in the Wichita Mountains. Moreover, given that the Texas Emerald (*Somatochlora margarita*) since has been recorded in Oklahoma, on the basis of the archived photographs alone,

we cannot in good faith claim that the individuals found in Comanche County were unequivocally Ozark Emeralds.[17]

It is fitting that Oklahoma is best known for the Ozark Emerald because it was here, on 9 June 1931, that the species was first discovered, not just in the state but anywhere in the world (Table 18.1). Over the next two months Bird and his expedition collected another ten individuals that became part of the type series,[18] one of which became the species' holotype. An error of note regarding the type series is that there was a typo in the type description, indicating that "Cunneotubby Creek" was "northwest of Wilburton" whereas Cunneo Tubby (two words) Creek is north*east* of Wilburton. It is the Fourche Maline, a known current breeding site for the species, that is northwest of Wilburton. We therefore georeferenced the type locality as 2.5 mi (4 km) north*east* of Wilburton on Cunneo Tubby Creek.[19] In any case, Bird initially thought this series of specimens were of the Treetop Emerald (*Somotochlora provocans*); he later published the records as *Somatochlora* sp. (Bird 1932) and then realized he had discovered a species new to science (Bird 1933).

A Earl Pritchard collected the species again in 1934, this time in numbers as high as five adults in Latimer County, probably at Fourche Maline Creek (4♂, 1♀; EMEC, FSCA, OSU, USNM), and in Le Flore County at Page (1♂, 4♀; EMEC, FSCA). Pritchard is the only person, dead or alive, known indubitably to have encountered a nymph of this species in the wild. In his paper describing the nymph he said,

> Late one morning after a rain, a mud-encrusted naiad was taken from a tall sedge stem at the edge of a widened and quiet portion of the creek near Wilburton. Emergence followed several minutes later, and the imago turned out to be a female *ozarkensis*. [his description was] taken from this exuvia, and another taken several miles downstream.
>
> **(Pritchard 1936:100)**

The specimen upon which the description was made has not been relocated.

In addition to being the county from where the Ozark Emerald was described to science, Latimer County hosts the only breeding location that can be considered consistent for the species: within Robbers Cave WMA, along Fourche Maline Creek. Although the species has not been encountered on every visit to the site, it was reported there at least once each year between 2014 and 2018. In each of those years, breeding activity was witnessed, including the first time an ovipositing female was captured on video (BS-P, 19 July 2017; Appendix C). Numerous attempts have been made to collect nymphs at the site, but none have been successful. There are a small handful of sites that are considered semi-reliable, all of which are found in southeastern Oklahoma.

Somatochlora ozarkensis Bird, 1933
Ozark Emerald

Figure 18.15 *Somatochlora ozarkensis*, the Ozark Emerald, A) ♂, Keystone Dam, Tulsa County, 5 June 2016 (©Bill Carrell, OC 445702). B) After *S. margarita*, the Texas Emerald, was documented in Oklahoma in 2017, we realized we could not eliminate that possibility for the far-flung record of a ♂ in Ketch Canyon on Fort Sill Military Reserve, Comanche County, 20 June 2009 (©Victor W Fazio III, OC 313402). C) ♀, Beavers Bend SP, McCurtain County, 7 July 2011 (©Berlin A Heck, OC 329494), in the Ouachita Mountains, the heart of the species' distribution in Oklahoma. D) ♀, 15 km NE of Tahlequah, Cherokee County, 2 October 2016 (©CA Ivy, OC 456581), the latest flight date anywhere in the species' limited geographic range.

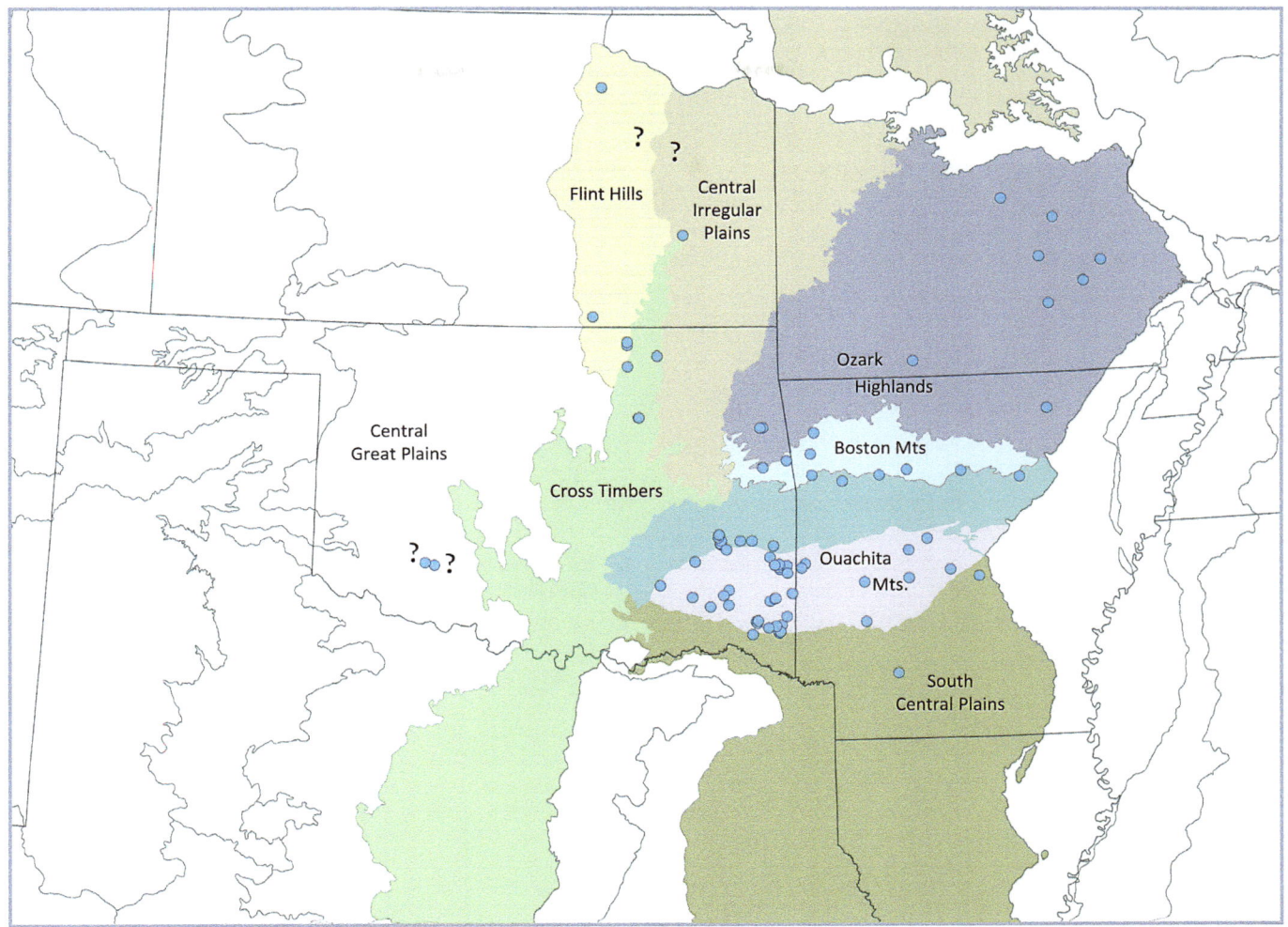

Figure 18.16 The Ozark Emerald (*Somatochlora ozarkensis*) is a regional endemic, found only in Arkansas, Kansas, Missouri, and Oklahoma. Some records, such as those from Comanche County, Oklahoma, and others in Kansas are best treated as dubious until further documentation can be obtained. Most records of the species have come from the Ouachita Mountains in Oklahoma.

For instance, in Le Flore County, feeding swarms have been encountered several times (see below), especially in the late afternoon, at Pipe Springs and along the Talimena Highway. In McCurtain County, the county that has the most known localities for the Ozark Emerald anywhere in the species' range, there are two areas that have proven somewhat consistent: a couple of access points along and near the stretch of the Mountain Fork River known as "The Narrows" and down river at and near Mountain Fork Park, including areas a little south of there, near Cooper Creek. Driving the Forest Service roads near the park near the close of the day has proven especially fruitful for finding feeding swarms, including those also harboring *Somatochlora linearis*, the Mocha Emerald, and the occasional *S. tenebrosa*, the Clamp-tipped Emerald (see below). Unfortunately, the vast majority of sites from where the Ozark Emerald has been reported have been one-offs, including at an unnamed creek 10 km SSW of Hartshorne, in Pittsburg County (14 July 2018,

MAP, 2♂, 1 as SP 2671, OC 485119), 10 km NNE of Moyers, Pushmataha County (14 July 2018, MAP, 1♂ and 1♀, the latter as SP 2674; OC 485123), Honobia Creek WMA near Cloudy, Pushmataha County (15 July 2018, MAP, 1 territorial ♂, OC 485127; 1 ovipositing ♀), and Crooked Branch Creek in the Well Hollow area, Ouachita NF, in Le Flore County (5 July 2014, 1♀, SP 1317), the last despite numerous subsequent visits. Just barely missing this list is Breadtown Creek at Atoka County WMA, in Atoka County, where on 13 July 2013 we encountered 2♂ and 3♀ (1♀ as SP 802) for the first and for a long-time-only county record. We could not re-find the species there or anywhere else in the county for nearly 6 years, until 2♂ were watched at length as they patrolled the dirt road near Breadtown Creek on 1 June 2019 (MAP).

We count ourselves lucky, at least relative to the misfortunes of GH Bick and LE Hornuff who conducted intensive surveys across eastern Oklahoma but never encountered

Table 18.1 The type series for *Somatochlora ozarkensis*, the Ozark Emerald, as presented by Bird (1933). The holotype and allotype are accounted for, residing at UMMZ, as are 2♂ paratypes. The other eight paratypes were to be distributed to ROM, ANSP, and OMNH but there are discrepancies. For example, it appears that the male that should have gone to ROM may still be at OMNH. The 2♂ gifted to ANSP have not be located and 1♀ remains unaccounted for.

	DATA FROM BIRD (1933), VERBATIM	SPECIMEN(S) FOUND?
holotype	Holotype male, along Cunneotubby Creek, two and one-half miles northwest [sic, northeast] of Wilburton, Latimer County, Oklahoma, July 14, 1931, Wilton Fisher, Collector	UMMZ
allotype	Allotype female, a teneral, same locality as the type, June 9, 1931, RD Bird, Collector	UMMZ
paratypes	Ten males, six adult and four tenerals, two females both teneral; all from the same locality as the types. The specimens are deposited as follows:	
	paratype males, Nos. 2 and 8, in the Museum of Zoology, University of Michigan	2♂, UMMZ; see also, 29 Nov 1932 Bird and Williamson correspondence
	paratype male No. 9, Dr. EM Walker, University of Toronto	No; ROM has no record of this specimen (Brad Hubley, *in litt*, 11 Nov 2014)
	paratype males, Nos. 4 and 5, Academy of Natural Sciences of Philadelphia, in care of Dr. PP Calvert	No, but as per a letter from Calvert, dated 5 Dec 1932, and a certificate in the OMNH archives, these were received by ANSP
	all others are in the Museum of Zoology, University of Oklahoma	4♂, 1♀, OMNH (1♂, 9 June), 1613 (1♀, 14 June), 1614 (1♂, 14 June), 1615 (1♂, 21 July), and 1616 (1♂, 21 July), all from 1931 except possibly 1616, which had the year entered as 1932

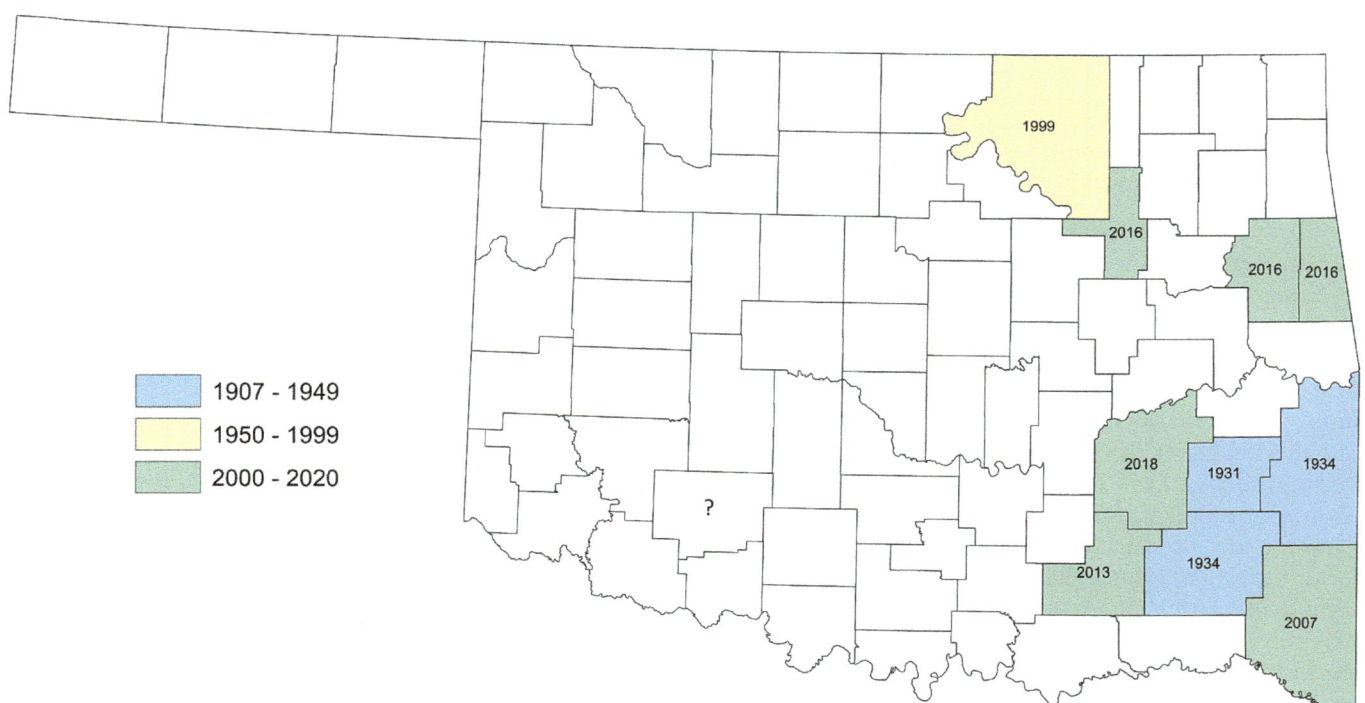

Figure 18.17 Counties in which *Somatochlora ozarkensis*, the Ozark Emerald, are known to occur in Oklahoma. Year within the county is that in which the species was first reported. Records from Comanche County lack conclusive documentation, so currently we treat them as dubious.

the Ozark Emerald. Well after his time surveying across Oklahoma, Bick (2003) reported the species from Osage County; presumably the basis of which was Vern LaGesse's 5♂ specimens taken at the Tallgrass Prairie Preserve in 1999 and 2000 (ISM 9831–9834 from 5 July 1999, ISM 14035 from 3 July 2000). Earlier we (Smith-Patten et al. 2007) questioned Bick's claim of the species in Osage County because the record had been confused with a possible nymph taken at a stock tank (Bick 1983; at present, an unsubstantiated Arkansas record). Furthermore, it was not until 2017 that LaGesse's specimens came to light. By that time, a photograph of a ♂ captured at Torpedo Switch in Osage County on 2 August 2006 had surfaced (OC 435703, Molly and Mark Ferguson). The species has not been reported in the county since, despite targeted surveys.

Although from the same county, the Osage County records fall within two ecoregions: the Flint/Osage Hills and the Cross Timbers. This is true, too, of the most credible records from Kansas.[20] In the Oklahoma Cross Timbers, it is a site in Tulsa County where BC recorded the species three times, that holds the most promise. On the southeastern side of Keystone Dam, where the Army Corps of Engineers maintains hiking and mountain biking trails, BC first encountered a teneral ♀ and an immature ♂ on 5 June 2016 (OC 445701–445702; Figure 18.15). On a return visit later in the summer, he documented an ovipositing ♀ (OC 453133, 14 August) and saw 1♂. Three years passed before the next record, of 1♂ on 6 July 2019 (OC 498552). The site has not produced records during all visits, but we surmise that it holds at least a small breeding population.

The long-awaited encounter with the Ozark Emerald in the Oklahoman Ozarks came for certain in 2016. The species has been known since at least the 1970s from the Ozarks (Trial 2005), but previous attempts to find it in the Oklahoman Ozarks fell flat, until, on 6 July 2016, Oklahoma picked itself up, dusted itself off, and said "let there be documentation." On that date, BS-P photographed and collected the species along Hastings Hollow at Cookson WMA, Cherokee County. That record stands as the highest count of adults at a breeding site in Oklahoma as well as from known records within the species' geographic range: 9 individuals, which included 6♂, 3 of which were in hand and released and 2 others that were collected (SP 1998–1999), and 3♀, 2 of which were in hand and released and 1 other that was collected (SP 2000). An earlier encounter, on 2 July 2015 at Hastings Hollow, with a ♀ *Somatochlora* was thought likely to be *S. ozarkensis*, but BS-P was unable to document it. She was elated in 2016 when she was able to document the species for the first time for the Oklahoman Ozarks. As icing on the cake, a day later she found another probable breeding population, this time on Eagle Pass Hollow Creek within the Ozark Plateau WMA in Adair County, where 2♂ and one ovipositing ♀ (Figure 18.18) were documented (OC 448782). These two sites lie at the interface of the Boston Mountains and the Ozark Highlands, so it could be said that

it was actually Jon and Cliff Ivy who are the true discoverers of the Ozark Emerald in the Oklahoman Ozarks. Their records, from the Bathtub Rocks at the TNC JT Nickel Preserve, Cherokee County, are unequivocally within the Ozarks (1♂, 24 September, at 11:17 am, OC 456397; 1♀, 2 October, at 12:42 pm, OC 456581; Figure 18.15). BS-P, in all her graciousness, is happy to share the discovery with her fellow bug nerds.

That last record, from 2 October, is the latest known flight date for the Ozark Emerald in Oklahoma and anywhere within its geographic range. The early flight date for the species anywhere is also from Oklahoma (28 May 2008, 2 teneral ♂, 13 km WNW of Hochatown, Cedar Creek, along FS 53000 Rd, McCurtain County, BAH, OC 282324). At just over four months long (127 days), the Oklahoma flight season is more than a month longer than that known from elsewhere (AR: 76 days long, 6 June–21 August, GL Harp specimens from Van Buren and Sharp County in 1981 and 1992, respectively, ASU; MO: 92 days, 16 [28] June–16 September, early date is sight record from 2014 from Troy Hibbitts, although the earliest documented record is 28 June 1996, Maries County, L Trial, EEM, and the late date is from 1999, TE Vogt specimens, Dent County, ISM 9779).

We know of 12 records of ovipositing, between 2 July and 15 August, and 7 teneral records, dating from 28 May to 22 July. Details of ovipositing have been noted only a few times; generally, females have laid their eggs directly onto damp pebbles, gravel, or moss where there is but a trickle of water (Figure 18.18; Appendix C). The only confirmed nymphal record taken in the wild is that reported by Pritchard (1936). An additional report from Oklahoma has yet to be confirmed. All other reports of nymphs from the wild are from Kansas and are considered dubious until they can be re-examined. One other record from Kansas is an exuvia and reared teneral, but the condition of the female was too poor for confident species identification. In two instances, eggs were harvested from adult females captured in the field. On 26 July 2017, L Wishard and W Boys captured an egg laden adult ♀ on Lee Creek at Devils Den SP, Washington County, Arkansas (OC 469451). Eggs were taken by Boys, who was able to rear them to nymphs for a couple of months before they died. The second instance was more successful: in mid-August 2010 TE Vogt took eggs from 2♀ he captured in Ketchum Hollow, Roaring River SP, Barry County, Missouri. Vogt sent the eggs to KJ Tennessen, who reared them (KJT; details in Tennessen, ms in prep.).

In the type description, Bird (1933:6) described the species' habitat: "The hills are covered by a heavy oak hickory forest and some pine (*Pinus echinata*). Along the creek there is considerable birch, sycamore, walnut, and willow." He elaborated, saying that the species was in the portion of the creek with "frequent rapids over rounded sandstone boulders bordered by a fringe of willows" (Bird 1933:6). Pritchard (1936:99) noted Bird's description was mostly

Figure 18.18 Ovipositing ♀ *Somatochlora ozarkensis*, the Ozark Emerald, on Eagle Pass Hollow Creek at the Ozark Plateau WMA, Adair County, 7 July 2016. The day prior to this record was the first time the species has been documented in the Ozarks of Oklahoma. Photos by ©Brenda D Smith.

expected, falling between 60 and 762 m (197–2,500 ft). We wonder if some of the derived numbers at the lower elevations are a result of measurement error from a few of the assigned geographic coordinates having less accuracy than is preferable. We have long expected that the species is more of a highlander (at least relatively speaking for the region), possibly venturing down into the foothills. Certainly, when records are classified into breeding, feeding, or records of both, then we see that the elevational ranges are 117–379 m, 140–762 m, and 227–303 m, respectively. As we long expected, >120 m (~395 ft) is probably more in line with this species' preferred elevational range. (It is the similar *S. margarita*, the Texas Emerald, that occurs at lower elevation.)

The resolute emerald hunter is far more likely to encounter the Ozark Emerald in late afternoon/evening feeding swarms. When encountered on creeks (almost always during the morning hours), numbers tend to be but a few at a time (≤6), whereas in the afternoon feeding swarms one can expect to encounter at least a half dozen but perhaps 10–25 adults in an area. It is likely those swarms contain other individuals but counts we and others have provided have been individuals confirmed by in hand examination, photograph, or as seen through binoculars (see Smith-Patten and Patten 2017b for details of some of those records). Of course, these swarms contain other emerald species or other dragonflies, particularly gliders (*Pantala* and *Tramea*), but sometimes Swamp Darners (*Epiaeschna heros*), river cruisers (*Macromia*), spiketails (*Cordulegaster*), and Orange Shadowdragons (*Neurocordulia xanthosoma*). With so much activity (≥50 adults), it is not surprising then that it can be difficult to determine exactly how many *Somatochlora ozarkensis* are present. And although seemingly more common in the afternoon, morning feeding swarms have also been reported, especially on overcast days. We most often encounter swarms on lightly traveled east–west roads that traverse forest pocked with clearings and glades and oftentimes pine plantations.[21]

Pritchard (1936:99) described some of these phenomena when he wrote that, at Fourche Maline Creek, Latimer County, Oklahoma, he found the species

> flying above fields in large clearings from just after daybreak into the early morning. Its flight was quite irregular, but not as extensive as that of *Somatochlora linearis* [Mocha Emerald] which was present in equally large numbers … Throughout the remainder of the day, *ozarkensis* was not encountered.
>
> **(Pritchard 1936:99–100)**

He went on to say that at Page, Le Flore County, he encountered the species "only in the late evening, often with regular beats across the road". On that evening, likely 23 June 1934, he had all three of Oklahoma's species of *Somatochlora*—the Mocha Emerald, Ozark Emerald, and Clamp-tipped Emerald (*S. tenebrosa*)—flying together. JJ Daigle and SW Dunkle also reported having all three species, but in a morning feeding swarm (JJD, 2♂, 1♀ *S. ozarkensis*, Pigeon

from "teneral specimens which were flushed from the willows along Cunneotubby Creek."

Preliminary results from our habitat assessments of Oklahoma creeks with records of the Ozark Emerald indicate that the species prefers small forested highland streams, including those that largely, or sometimes fully, dry by the end of the species' flight season. Other than that commonality, creeks vary radically (Figure 18.19). Substrate varies from bedrock exposure to streambeds full of gravel and boulders. Stream width tends to be about 0.5–5 m (~1.5–16 ft) but can be as wide as 12 m (~40 ft). Water depth can be just a trickle to >1 m (~3 ft), although it appears that males defend territory where the water is about 2 cm to <1 m (<1 inch to <3 ft) and females may have a preference of ovipositing where there is just a trickle of water over gravel or pebbles. The overall water regime of streams can be flowing or dried to the point where but a few pools exist along its length. Water temperature varies enormously (16.0–35.85°C [60.8–96.53°F]), as does pH (6.30–8.91). And streams can be found within hardwood or mixed forest. Elevation at sites across the species' geographic range was greater than

HABITAT

bare bedrock to gravel & boulders

0.5 – 5 m (12 m) wide

trickle – >1 m depth

flowing to series of pools

no emergents

choked with emergents

hardwood or mixed forest

WATER QUALITY
temp — 16.0 – 35.85°C (60.8 – 96.53°F)
pH — 6.30 – 8.91

Figure 18.19 Streams at which *Somatochlora ozarkensis*, the Ozark Emerald, have been found in Oklahoma. Much variety in habitat can be seen.

Roost Mountain, Montgomery County, Arkansas, 22 June 1984). MAP once had a comparable experience (*but with many more individuals*) on 24 July 2015, 10 km NE of Broken Bow, McCurtain County. On that late afternoon he estimated he had >80 *Somatochlora*: 2♀ *S. linearis* (1 in hand), 6 *S. tenebrosa* (2♂, 4♀; 1♂ and 2♀ as SP 1727–1729, and 1♀ captured and released), 25 *S. ozarkensis* (5♂, 20♀; 2♂, 9♀ in hand and released; 1♀ as SP 1726), and 50 *S. ozarkensis/tenebrosa*. Several *Anax junius*, *Macromia* sp., *Pantala* sp., and *Tramea* sp. were intermixed with the swarms.

Estimates of the species' overall geographic range have varied, from as little as 195,652 km² (75,542 mi²), if dubious records (from northernmost Kansas and the Comanche County, Oklahoma) are eliminated, to as high as 329,751 km² (127,318 mi²), if being particularly generous.[22] Those estimates may seem to encompass a large area, but when one considers how much of that range is actually, or potentially, used by the species given its apparent habitat limitations

(Boys et al., ms), then we become privy to one attribute of concern about the conservation of the Ozark Emerald. When we estimate the "area of suitable habitat currently occupied" by the species, or what is called the "area of occupancy (AOO)" (IUCN 2017:48), a criterion used by conservation organizations including NatureServe and IUCN, we find that the Ozark Emerald is currently known to occupy as little as 288–352 km² (111–136 mi²; Smith-Patten 2017a and re-estimates in 2018[23]) but most certainly no more than 2,000 km² (772 mi²; Abbott and Paulson 2017; see Chapter 7 for a discussion of how AOO factors into conservation assessments). Furthermore, if we examine range shifts in consideration of climate change scenarios, there will be further constriction of the species' range, including it potentially blipping out of the Ozarks entirely, being confined solely to the Ouachita and Boston Mountains (worst case scenario predicts the species to continue inhabiting only portions of those mountains; Boys et al., ms). The differences in overall range versus AOO (and potential habitat loss as a result of climate change) are akin to the species being found throughout an area the size of the state of Colorado versus confined to the city of Lubbock, Texas, or Kansas City, Kansas[24]—a contrast in size that must give pause to one's perspective on conservation and potential extinction risk of the Ozark Emerald.

Smith-Patten (2017a) indicated that "the threat level [to *Somatochlora ozarkensis*] is probably moderate but may be high"[25] and summarized those threat as:

> The main threat to this species appears to be loss or degradation of habitat. The region is a mosaic of National Forest lands, pine plantations, agriculture (farming and ranching), sand and gravel mines, coal mines, oil/gas extraction facilities, utility line corridors, and urban development. It is expected that most of these land-uses will continue and potentially expand, which may adversely affect *Somatochlora ozarkensis* by causing loss or degradation of habitat, including, but not limited to, pollution of waterways from pesticides, herbicides, fertilizers, and other chemicals. Possible expansion of roads, related to the above industries or other activities, may bring construction of additional culverts that degrade habitat and obstruct dispersal of eggs and nymphs. Climate changes, including drought and more erratic storms causing abnormal episodes of flooding, will undoubtedly affect this species in the long-term. However, given the amount of protected lands in the region, the species will likely experience limited risk of extirpation/extinction as long as federal and state land managers consider the needs of this species in their land management practices.

The Ozark Emerald was last reviewed globally for NatureServe by Smith-Patten (2017a), when it was ranked as vulnerable (G3). That rank roughly equates to the "Near Threatened" status attributed to the species by the IUCN Red List (Abbott 2007; Abbott and Paulson 2017; see Conservation chapter for comparison of ranks). The species has been designated as a Tier II Species of Greatest Conservation Need (SGCN) by the Oklahoma Department of Wildlife Conservation (ODWC 2005, 2015) for over a decade. Previously, we considered the species to be critically imperiled (S1) in Oklahoma given its restricted range and its presumed small population (Patten and Smith-Patten 2013b), but that assessment has since been revised to vulnerable (S3) due to supplemental data and by using NatureServe's current criteria (Smith-Patten 2017a). Smith-Patten (2017a, notes to NatureServe), although limited in jurisdictional authority to change conservation ranks in other states, recommended that Arkansas, Kansas, and Missouri also assign vulnerable (S3) status to the species there.[26] In short, *Somatochlora ozarkensis* is undoubtedly a species of conservation concern, but the level of concern is debatable as there is still much speculation and educated guessing that goes into its assessment. Even though huge strides made in recent years have increased our understanding of this species' distribution, ecology, and relative abundance, we have much to learn. Nonetheless, we do know the species is a regional endemic with an extremely limited area of occupancy. We also know that encounters are still relatively limited (generally only a handful a year) despite six years of intensive surveys. And, it appears that the overall population is relatively small and likely fluctuates between years, though quantitative analyses have yet to be made because we consider data still to be too deficient (Smith-Patten 2017a).

SOMATOCHLORA MARGARITA DONNELLY, 1962 – TEXAS EMERALD

Although there is but a single record of the Texas Emerald for Oklahoma, it is fair to say that the species' status in the state is unclear. It is possible that the ♀ DA collected at Red Slough WMA, McCurtain County, on 22 June 2017 (SP 2402, OC 464890; Figure 18.20) was a vagrant. In our view it is just as likely, if not more so, that the Red Slough area specifically and southeastern Oklahoma generally hosts a small population—the ♀ DA collected *just happened to be the first* one detected in the state. The small, striped emeralds are notoriously difficult to find, especially at their breeding sites, a difficulty that leads to an impression of rarity. They are notoriously difficult to capture as well, further elevating impressions of rarity. These species are detected much more frequently in feeding swarms; indeed, the Red Slough ♀ was caught on a heavily overcast day as she foraged at head height over the main north–south dirt road that cuts through Red Slough. We predict that, eventually, a breeding site or two will be found in southeasternmost Oklahoma at small creeks, perhaps especially those with sphagnum moss.[27] This area makes biogeographic sense, given that it is an extension of the South Central Plains ecoregion, where the Texas and Louisiana populations primarily are found. As such, we also predict that the species will eventually be found where that ecoregion comes into southern Arkansas.

We do not know if two records of ♂ striped emeralds from the Wichita Mountains, Comanche County, are of this species or of the similar Ozark Emerald (*Somatochlora ozarkensis*). Photographs (OC 313402, OC 328764) were submitted in support of the latter identification, but the identity of the first record is ambiguous—cercus shape cannot be seen well enough in the single photo—and at the time the Texas Emerald was not known from Oklahoma and likely was not considered as an alternative. Photographs for the second record, of a dead ♂ retrieved from a spider's web, are more suggestive of the Ozark Emerald, in that the superior appendage appears to be angled more sharply (see Donnelly 1962[28]), but the specimen was lost, so we cannot verify its identification. Regardless, either the Texas Emerald or the Ozark Emerald would be extraordinary in the Wichita Mountains, which lay in the Central Great Plains, far from the ecoregions either species is known to inhabit. More survey effort is needed to assess whether these individuals were vagrants or part of a small breeding population.

The Texas Emerald is known to fly in Louisiana and Texas from spring to late in the summer (LA: 2 April–2 August, JCAC, OC 434355; TX: 20 May–8 August, Abbott and Lasley 2019), so Oklahoma should expect seeing the species again during that general time frame. The species is of conservation concern—ranked as imperiled/vulnerable globally and nationally (G2G3, N2N3; Smith-Patten 2017b) and as imperiled in Louisiana and Texas (S2). If the species is regular in Oklahoma, then a rank of critically imperiled/imperiled (S1S2?) is appropriate (Smith-Patten 2017b).

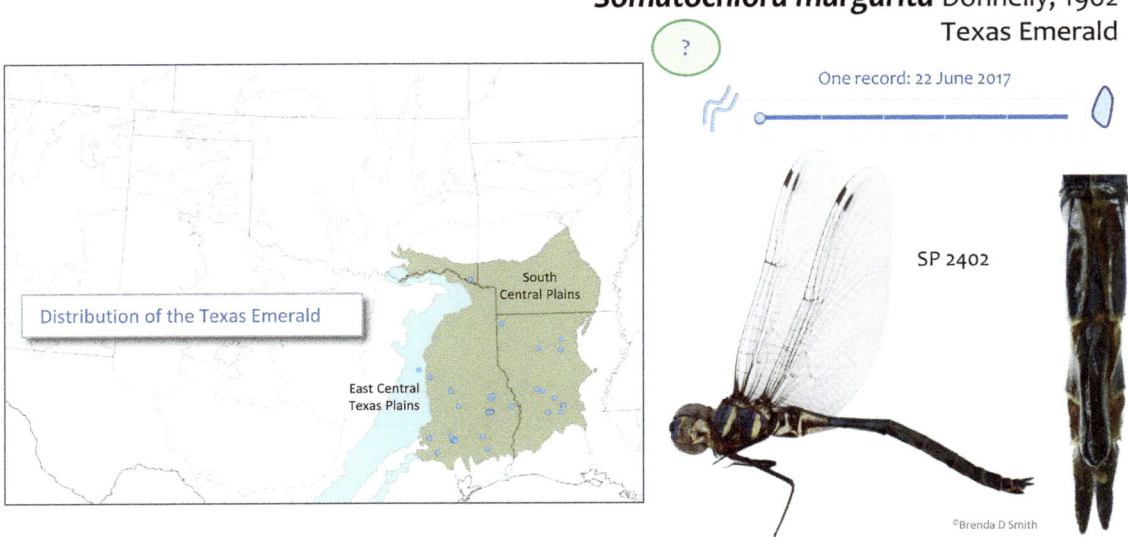

Somatochlora margarita Donnelly, 1962
Texas Emerald

One record: 22 June 2017

SP 2402

©Brenda D Smith

Figure 18.20 *Somatochlora margarita*, the Texas Emerald. This ♀ was captured at Red Slough WMA, McCurtain County, on 22 June 2017 (USDA Forest Service; David Arbour, OC 46489), providing the first and so far only record for Oklahoma, although the species may be a regular, albeit scarce, component of the odonate fauna in the Gulf-slope lowlands of southeastern Oklahoma (note that the species is primarily confined to the South Central Plains ecoregion of Oklahoma, Texas, and Louisiana).

SOMATOCHLORA TENEBROSA (SAY, 1839) – CLAMP-TIPPED EMERALD

The Clamp-tipped Emerald is the rarest of the three emeralds known to occur regularly in Oklahoma. We know of only about 15 records for the species in the state. It was first recorded in Oklahoma by AE Pritchard on 23 June 1934, when he captured a ♀ (EMEC 81619) that was flying around late in the evening with *Somatochlora ozarkensis*, the Ozark Emerald, *S. linearis*, the Mocha Emerald, and *Cordulegaster obliqua*, the Arrowhead Spiketail, at Page in Le Flore County (Pritchard 1936). Seventy-three years passed before the species was seen again in the state (1♂ in hand, 28 July 2007, DA, McCurtain County: near "The Narrows" of the Mountain Fork River, OC 262978). There were also reports of it in 2008 (*n*=4), 2011 (*n*=3), and 2015 (*n*=2). It was not reported in 2016–2018, but 2019 was a good year for the species (*n*=3).

Habitat elsewhere has been described similarly to where the species has been found in Oklahoma (except the description of streams with lots of leaf litter). When encountered in breeding habitat, it holds true to its name (rooted in *tenebrose*, meaning dark or gloomy), being a species of forested, and thus shady, streams. Those streams have intermittent riffles and pools and some dry out in mid-summer to the point that there are just a few scattered pools remaining. One example of a probable breeding location was nestled within hardwood forest at a spring complex at a moderate flowing, clear-water creek that was about 1 m wide and 2–4 cm deep. The portion of the creek where a patrolling ♂ was taken[29] (Ouachita NF about 10 km NNW of Hochatown, McCurtain County) had bedrock substrate and an abrupt, 1 m high, rocky slope on one side of the creek and a gentle sloping, less rocky side on the other, with herbaceous and mossy cover.

At another location, an ovipositing female was noted in a mostly shaded (hardwood woodland), shallow, spring-fed, clear water, perennial creek about 5 m wide, with little emergent vegetation and with bedrock substrate and occasional

Somatochlora tenebrosa (Say, 1839)
Clamp-tipped Emerald

S3

Figure 18.21 *Somatochlora tenebrosa*, the Clamp-tipped Emerald, A) ♂, Ozark Plateau NWR, Delaware County, 25 July 2011 (©James W Arterburn, OC 330565), the first to be found in Oklahoma in the Ozarks. B) ♂, Mountain Fork Park, McCurtain County, 3 July 2011 (©Greg W Lasley, OC 333432), in the Ouachita Mountains, where the species is much more numerous in Oklahoma. C) ♀, 5 km SE of Wright City, McCurtain County, 9 July 2008 (©Berlin A Heck, OC 283172).

pools about 3–30 cm deep. Another female was observed ovipositing in moss and mud in a shaded, mostly hardwood woodland spring run that was <50 cm wide and but a few cm deep that had clear water with moderate flow and a pebbly and rocky substrate. The former habitat description comes from Hastings Hollow Creek, at Cookson WMA, Cherokee County. On 4 September 2015, BS-P saw two females ovipositing, one in a shallow pool in the creek (SP 1781) and the other on a mossy rock in the creek (SP 1782). The latter female provided for much confusion initially because her ovipositor was broken off (maybe from ovipositing on rocks?), so she very much resembled a *Somatochlora ozarkensis* until closer inspection. The latter habitat description comes from Pine Mountain Spring, where one of two high counts have been noted (5♂, 1♀, 2 August 2019[30]; 1♂ as SP 2868; BS-P; possibly two other ovipositing females observed by BW Hoagland). The other high count, of 2♂ and 4♀, came from a record of the species foraging with 25 *S. ozarkensis* along dirt roads approximately 10 km NE of Broken Bow, McCurtain County, on 25 July 2015 (MAP; 1♂ as SP1727, 2♀ as SP1728–1729, 1♀ in hand). Otherwise, reports have been for only one individual at a time.

The species is only documented in five counties, three in the southeast and two in the northeast. The southeastern records include Pritchard's Le Flore County specimen, eight records in McCurtain County (DRP, JCAC, OC, SP), and one record that was published for Latimer County (Bick and Bick 1957). We do not know the basis for the Latimer County record because the Bicks did not indicate a source, nor is there a card for the species in Bick's notes. The two northeastern records are for Cherokee (see above) and Delaware Counties (Ozark Plateau NWR, 1♂, 25 July 2011, JWA, KW, OC 330565). All documented records are comprised of 6 photographic records and a total of 11 specimens (4♂ and 7♀). It would not be surprising that the species has been confused with the other Oklahoma emeralds, so an intensive study of the *Somatochlora* in the state may prove the species to be a little more common than once thought, but it will likely remain rare in Oklahoma given that the state is on the edge of the species' known range and does not appear to be common anywhere in that range.

Oklahoma has the second longest known flight season for the species in the region (OK: 23 June–4 September, see above; AR: 17–26 July, limited records; KS: about 11 June–19 July, SEMC[31]; MO: 25 June–2 October, Trial 2005; TX: 8 June–24 July, limited records, Abbott and Lasley 2019, OC). There are no known larval or teneral records for Oklahoma, nor have pairings been reported, but records of territorial males and ovipositing females indicate that there is a breeding population in the state.

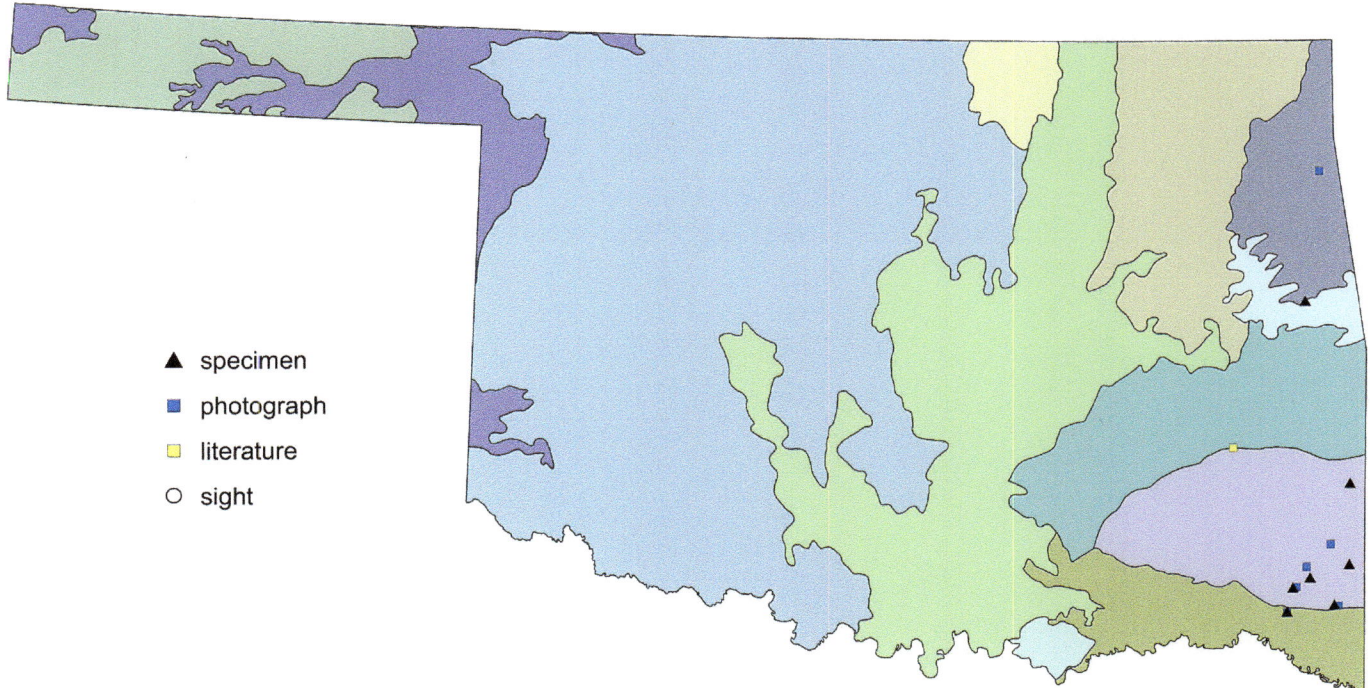

Figure 18.22 Documentation level of supporting records for *Somatochlora tenebrosa*, the Clamp-tipped Emerald, in Oklahoma. In this region, the species is one of the Ozark and Ouachita highlands.

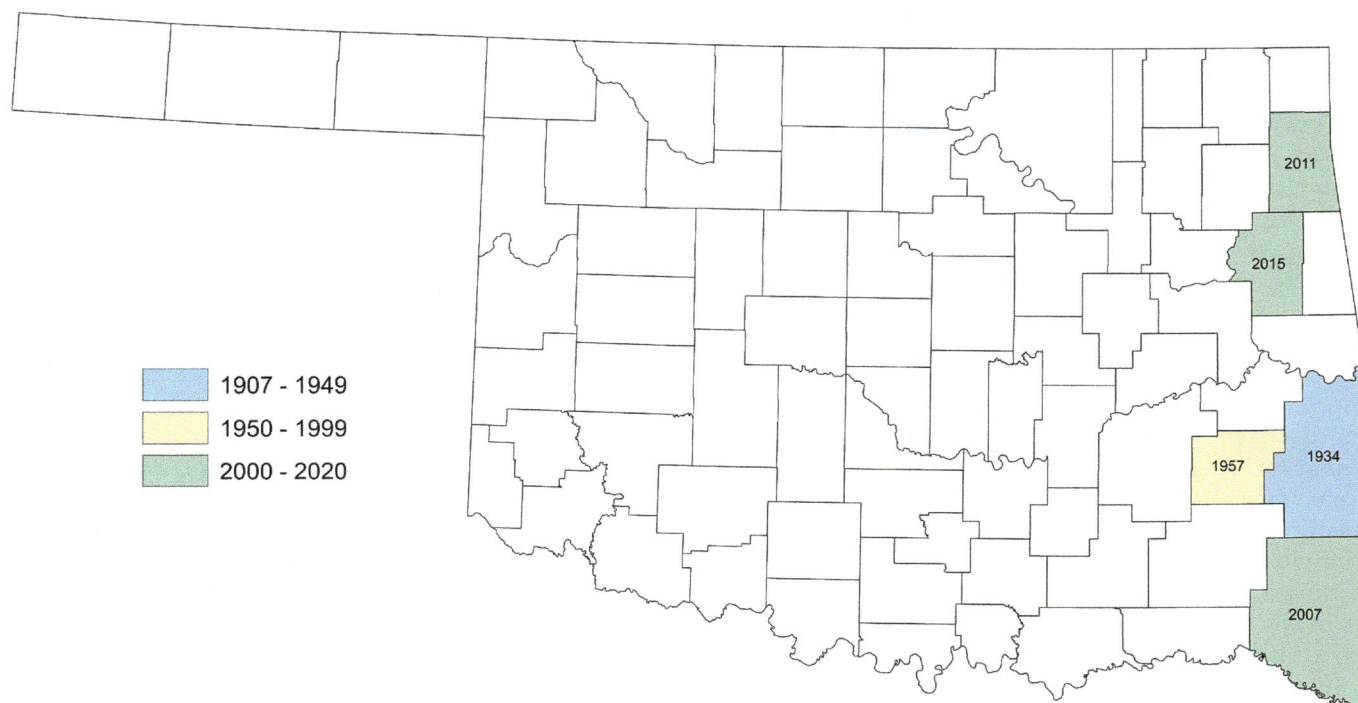

Figure 18.23 Counties in which *Somatochlora tenebrosa*, the Clamp-tipped Emerald, are known to occur in Oklahoma. Year within the county is that in which the species was first reported.

NEUROCORDULIA VIRGINIENSIS DAVIS, 1927 – CINNAMON SHADOWDRAGON

The Cinnamon Shadowdragon is known in Oklahoma from one record. A single ♀ was collected by John Stankavich on 18 June 1934 while he was collecting with AE Pritchard on the Mountain Fork River near Broken Bow, McCurtain County. Pritchard determined the specimen to be *Neurocordulia virginiensis* and then sent it to WT Davis, the species' authority, for confirmation. Davis compared it to the holotype, a female in his possession, and confirmed it as *N. virginiensis* (Byers 1937, Davis 1937; Figure 18.24). The specimen, which was apparently sent back to Pritchard but now appears to be lost, was originally published (Byers 1937) as being from the "Deep Fork River above Broken Bow, Oklahoma," but this was glaring error of the river's name and later was corrected by Davis (1937). Pritchard described the collection as such: "The specimen was at some rapids at a wide part of Mountain Fork River, at dusk, in company with *Platycordulia xanthosoma* Williamson." (Davis 1937:250). The Mountain Fork River, although dammed in the 1960s to create Broken Bow Lake, still fits the habitat description for this species, i.e., medium to large rivers within wooded areas (Paulson 2011), including still containing rapids in parts of its course.

To date the Oklahoma specimen remains the westernmost record for the species. The species has been reported multiple times from across the border in Arkansas, so it would be unsurprising to find that Oklahoma is part of its natural range, yet, there have been no additional records despite much survey effort in the region. Look for it in eastern Oklahoma, especially in the South Central Plains, but targeted surveys are needed to discover the true status and distribution in Oklahoma of this elusive, crepuscular species.

In Arkansas, the species is known to fly from 14–28 May (Harp 2007, Dunkle 1983). Elsewhere in the species' range it is known from March until late November (FL: March–June, GA: May–early August, Beaton 2007; AL: 28 November, OC 321607).

Neurocordulia virginiensis Davis, 1927
Cinnamon Shadowdragon

vagrant

one record: 18 June 1934

250   JOURNAL NEW YORK ENTOMOLOGICAL SOCIETY   [Vol. XLV

A SECOND RECORD OF THE DRAGONFLY NEUROCORDULIA VIRGINIENSIS

BY WILLIAM T. DAVIS, STATEN ISLAND, N. Y.

On June 21, 1919, a single female *Neurocordulia* of unknown species was collected close to the James River in Buckingham County, Virginia. No additional specimens were found though the locality was visited in subsequent years, so the insect was finally described in the "Bulletin of the Brooklyn Entomological Society," Vol. 22, June, 1927, under the name of *Neurocordulia virginiensis* Davis. Later it was again figured and compared with the type of *Neurocordulia clara* Muttkowski, in the JOURNAL, NEW YORK ENTOMOLOGICAL SOCIETY, Vol. 37, December, 1929.

On July 23, 1934, Mr. A. Earl Pritchard sent me a female *Neurocordulia* from his collection which he had correctly identified as *virginensis* and which was compared with the type from Virginia before its return to Mr. Pritchard. He has lately given me the data connected with the specimen and writes as follows: "The single female *virginensis* was taken near Broken Bow, Oklahoma, in the southeastern corner of the state, June 17, 1934, by John Standcavish. The specimen was at some rapids at a wide part of Mountain Fork River, at dusk, in company with *Platycordulia xanthosoma* Williamson. The flight of the crepuscular dragonflies is so irregular together with the dim light that selective collecting is impossible."

This is but the second specimen of this evidently rare dragonfly to be recorded in about seventeen years.

Material.—The only adult material of this species known to me is the female type (Davis, 1927) and the male allotype described above.[3]

NYMPH (Pl. II, Figs. 2 and 7; Pl. III, Fig. 3; and Pl. VIII, Fig. 2).—Description (supposition): body short and broad, depressed, abdomen arched dorsally. Head broadly convex above and on sides; eyes fairly prominent; a strongly convex shelf-like frontal ridge covered with a scurfy pubescence;

[3] Since this paper was written I have received a communication from Mr. A. Earl Pritchard in which he states: "You would undoubtedly be glad to know that *Neurocordulia virginiensis* occurs as far west as Oklahoma. A female was taken by John Stankavich on Deep Fork River above Broken Bow, Oklahoma, June 18, 1934. We were catching *Platycordulia* at the time. The specimen was compared with the type by Mr. Davis."

The finding of a specimen of *N. virginiensis* in Oklahoma extends the possible range of this species beyond the areas inhabited by *N. clara*, (Alabama) and Burmeister's *N. polysticta*, (Louisiana). The record constitutes the third for the capture of *N. virginiensis* and the second for the capture of a *Neurocordulia* west of the Mississippi River. (The first, *N. yamaskanensis*, was made by Williamson in 1932.)

Figure 18.24 Excerpt from Byers (1937:22, left) and whole article from Davis (1937, right) relating to the first and only record of *Neurocordulia virginiensis*, the Cinnamon Shadowdragon, from Oklahoma.

NEUROCORDULIA MOLESTA (WALSH, 1863) – SMOKY SHADOWDRAGON

We know of five records for Oklahoma of this crepuscular, elusive species (Figure 18.25). The first record for the state was a "last instar exuvium [sic]"[32] collected by GH and JC Bick on the Verdigris River near Okay, Wagoner County, on 3 July 1950 (Bick and Bick 1957:4; Figure 18.26). Note that the Bicks' published this record as Wagoner County, but the description of the locality in Bick's notes, i.e., the juncture of the Verdigris and Arkansas Rivers, is actually in Muskogee County. The river course, according to various historical topographic maps, has not changed, but it could be that if the Bicks were using the poorly scaled Oklahoma highways map of that era, then they thought the junction of the rivers laid in Wagoner County. We chose to leave this record as is, figuring that the Bicks knew which county they were in and only meant that they were near the junction, which seems reasonable given that they chose to publish the record as just on the Verdigris River. The Bicks collected a second individual, this time a nymph,[33] on 23 July 1954, on the Blue River, near Blue, Bryan County (Bick and Bick 1957; Figure 18.26). We have not been able to locate either of these specimens, but on the basis of these two records and one below, we reluctantly call *Neurocordulia molesta* a breeder in Oklahoma.

After the Bicks' time, the species was not recorded again until JC Criswell collected a ♀ in Bixby, Tulsa County, on 7 August 1984 (OSU; OC 398708; Figure 18.25). This specimen, the first adult for the state, went unknown until we found it in 2011 in a collection of unidentified student material at OSU. Around the same time we discovered this specimen, JWA and KW found and photographed a somewhat teneral ♂ at Oxley Nature Center in Tulsa, Tulsa County, on 5 June (OC 330413; Figure 18.25). In between those two records, SW Dunkle collected 1♂ on the Kiamichi River below Hugo Dam in Choctaw County, on 11 June 1996 (FSCA). A report for Oklahoma County is a transcription error of the county in place of the state (Donnelly 2004b [DMP]; OC 233922); DMP cited Bick and Bick (1957:4), which cited Needham and Westfall (1955:359), both of which reported the species for Oklahoma state.

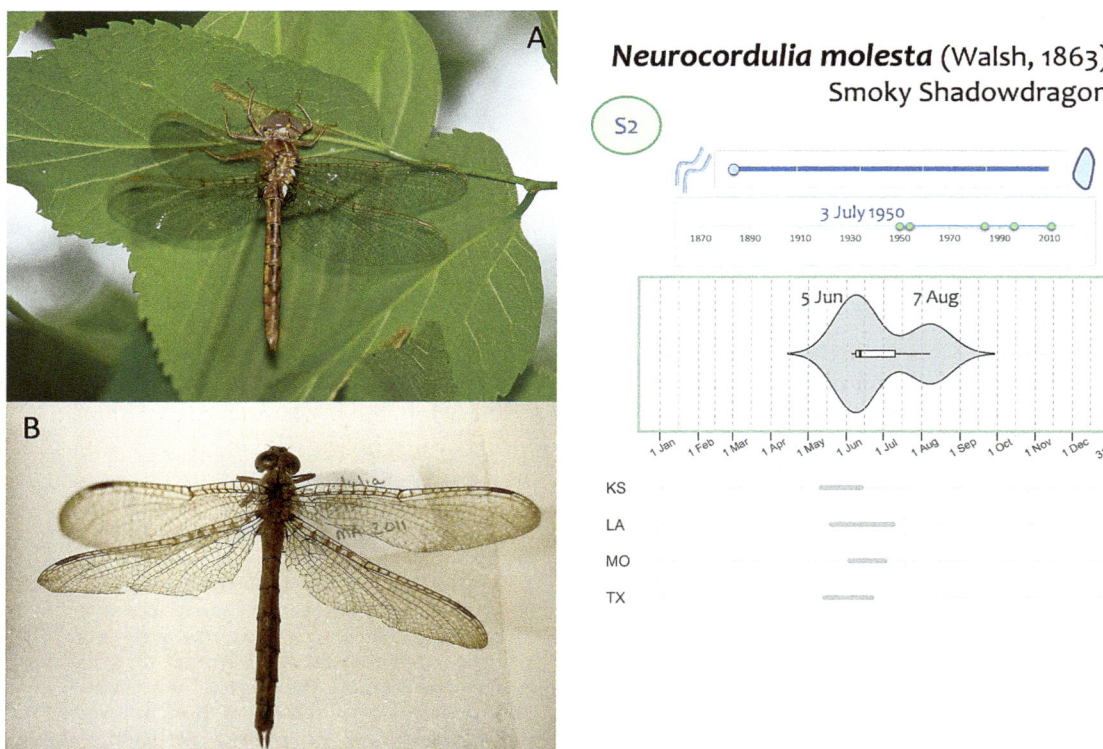

Figure 18.25 **A) One of Oklahoma's handful of records of *Neurocordulia molesta*, the Smoky Shadowdragon. This ♂ was found by James W Arterburn and Ken Williams on 5 June 2011 at Oxley Nature Center in Tulsa, Tulsa County (©James W Arterburn, OC 330413). B) But the first adult to be found in Oklahoma resided in an "unidentified Odonata" drawer at OSU for three decades (♀, Bixby, Tulsa County, 7 August 1984, JC Criswell, OSU, OC 398708). What to say about this shadowdragon? Apart from *Paltothemis lineatipes*, the Red Rock Skimmer, the status of no other dragonfly recorded in Oklahoma is as inexplicable. A mere handful of well-documented records are scattered across the eastern third of the state, but these records number just enough to convince us that the species likely occurs regularly. But where? And when?**

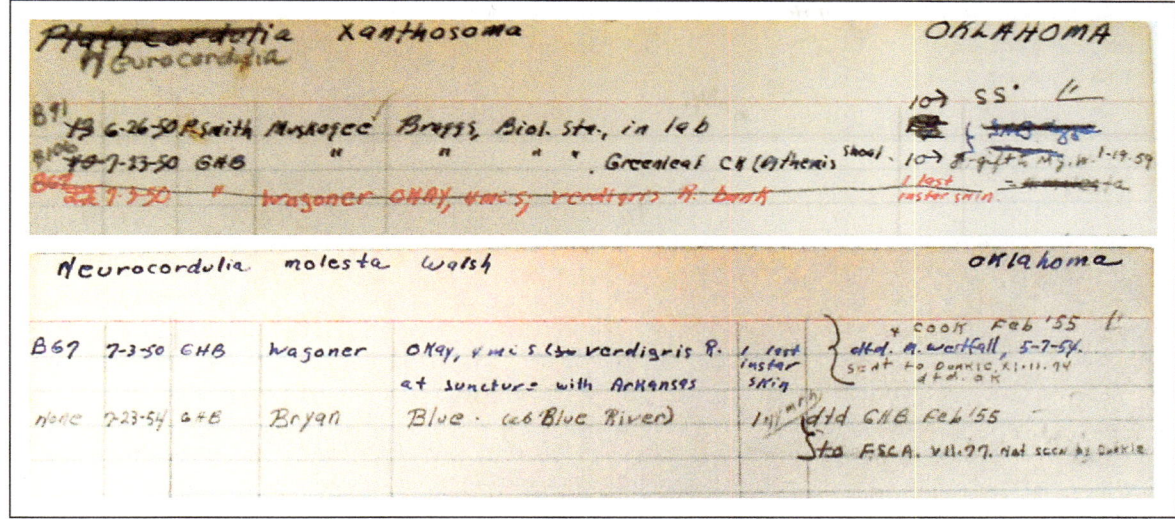

Figure 18.26 Two of GH Bick's note cards that appear to reference the same specimen, a "last instar skin" that he and JC Bick (Bick and Bick 1957) collected at 4 mi S of the Verdigris and Arkansas Rivers (actually in Muskogee County, but we retained the Bick's listing of it as being in Wagoner County) on 3 July 1950. These notes indicate that the specimen was originally identified as *Neurocordulia xanthosoma*, but later changed to *N. molesta*. This specimen has not been relocated, but presumably its determination is correct because it was examined by MJ Westfall (1954), Carl Cook (1955), and SW Dunkle (1974). Note, too, that the Bicks indicated they collected a nymph of *N. molesta* on 23 July 1954 on the Blue River, near Blue, Bryan County.

All but one of those localities make sense for the species as they are on large rivers with riparian woodland and sandy shores (suitable as per Paulson 2011). The only exception is Oxley Nature Park that, unlike Bixby, which is situated along the banks of the Arkansas River, is about 12 km from the nearest large river. The park is located near Bird Creek, but that creek does not seem to fit descriptions of the species' preferred habitat.

The limited adult records we have indicate that *Neurocordulia molesta* is a mid-summer flier in Oklahoma, as it primarily is elsewhere (OK: 5 June–7 August; KS: 11 May–13 June, OC 376396,[34] SEMC 1350206; LA: 20 May–9 July, Mauffray 2014; MO: 4 June–3 July, Trial 2005; TX: 15 May–22 June, Abbott and Lasley 2019, OC 501703). The mid-summer flight season makes it all the more surprising that the Bicks and LE Hornuff never encountered adults in Oklahoma given their intensive summer surveys in the state for more than two decades. We wonder why that is the case; of course, given the thousands of hours we have spent in the field, as well as the time others have spent, we wonder why we have not found it ourselves and why it has been encountered so rarely. Dunkle's experience with this species in Oklahoma may provide some insight: on the day he collected below Hugo Dam he captured a single ♂ *N. molesta* but 15 *N. xanthosoma* (FSCA, JCAC), a startling and impressive feat for these erratic fliers.

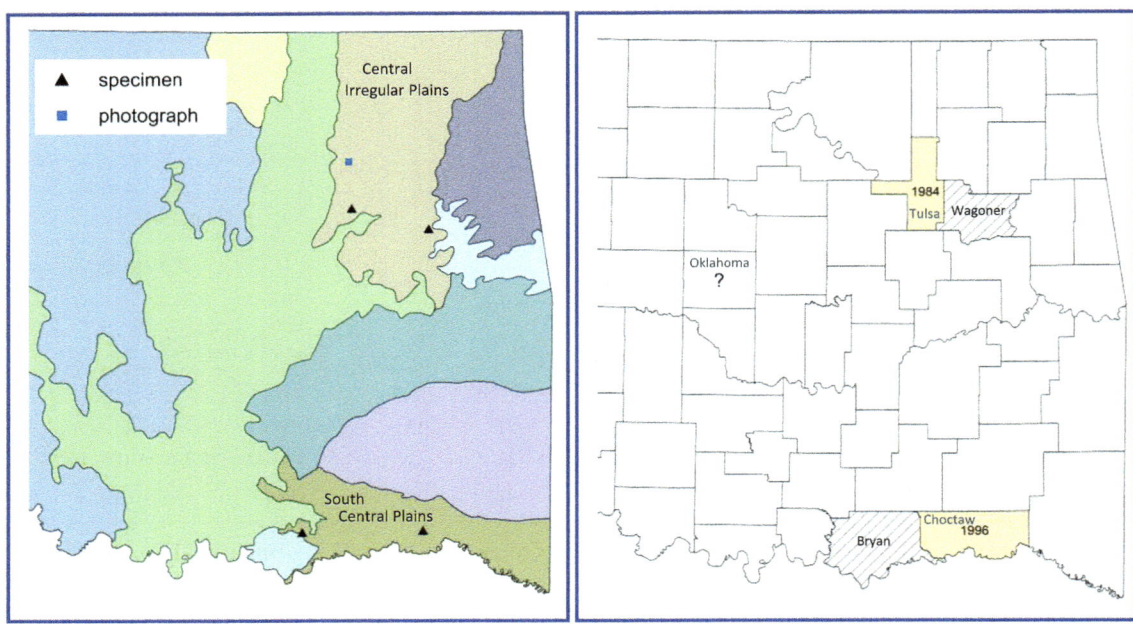

Figure 18.27 Documentation level of supporting records for *Neurocordulia molesta*, the Smoky Shadowdragon, in Oklahoma (left). The species has been confirmed only within the South Central and the Central Irregular Plains. Records of adults come from Choctaw and Tulsa Counties (right). Presence in two other counties, Bryan and Wagoner, is documented by nymphs (hatching) that GH Bick collected in the 1950s. No other records are known from those counties and the specimens have not been relocated. There is a literature record for Oklahoma County, the source of which is unknown and may be in error.

NEUROCORDULIA XANTHOSOMA (WILLIAMSON, 1908) – ORANGE SHADOWDRAGON

Shadowdragons are elusive and seldom seen without focused effort. Although three species are known to occur in Oklahoma, only the Orange Shadowdragon is encountered with any frequency. In comparison to encounter frequency of the other corduliids in the state, *Neurocordulia xanthosoma* is their rival, encountered roughly the same as *Somatochlora ozarkensis*, the Ozark Emerald, and *Epitheca semiaquea*, the Mantled Baskettail, but not quite as often as *E. cynosura* or *E. costalis*, the Common and Slender Baskettails, and being outright bested only by *E. princeps*, the Prince Baskettail, a species whose records well outnumber all of the other Oklahoma corduliids. And, even though often seen one or two at a time, *N. xanthosoma* can be encountered in Oklahoma in fairly large numbers. For example, we have recorded as many as 10 individuals and SW Dunkle recorded 15, which is the highest count known for the state (10♂, Lake Frederick, dam spillway, Tillman County, 8 July 2012, SP322, SP [RWG] 323, OC 399482; 15U, Kiamichi River below Hugo Lake Dam, Choctaw County, 11 June 1996, FSCA, JCAC). In McLennan County, Texas, Williams (1982:161) said "usually, but not every year, *N. xanthosoma* is abundant," so we ought not be too surprised by these numbers.

In Oklahoma, *Neurocordulia xanthosoma* is encountered most often in dense (often thorny) shrubbery along creeks or streams but occasionally in comparable habitat near the edge of small lakes. Creeks can be flowing or not, with clear to turbid water, forested or not, and with or without emergent vegetation.[35] Dam spillways are a favorite locale, especially when there are boulder-laden slopes that afford shelter along with pooled water. Williamson (1908:433–434), too, described the species at a spillway, although he made no mention of boulders. When speaking of the ♂ that he captured (4 June 1907, UMMZ) at "the overflow from the lake" that lay about a mile north of Wister, in what was then Choctaw Territory (now Le Flore County), he said:

A short distance from the lake, in passing some bushes which overhung the stream bed, I disturbed the only specimen I saw alive of *P.* [*Platycordulia*, now *Neurocordulia*] *xanthosoma*. This flew a short distance along the stream and alighted in a well concealed spot in bushes overhanging the water. Its flight, manner of alighting and position at rest suggested a teneral *Libellula*.

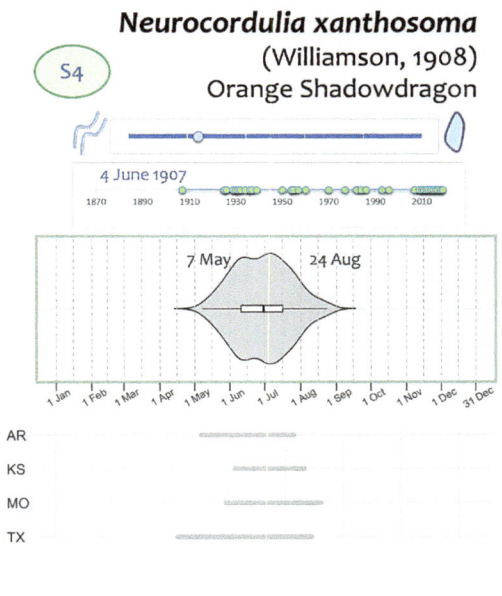

Figure 18.28 *Neurocordulia xanthosoma*, the Orange Shadowdragon. A) ♂, Osage Hills SP, Osage County, 20 July 2018 (©Bill Carrell). B) ♀, Coon Clear Lake, Latimer County, 16 June 2013 (©Brenda D Smith, OC 400690), a typical view, in that she perched within dense vegetation. Unlike its congeners, we have a good handle on this crepuscular species: it is fairly common and widespread on creeks, rivers, and even some lakeshores in the eastern half of the Southern Great Plains, with populations in southwestern Oklahoma, too.

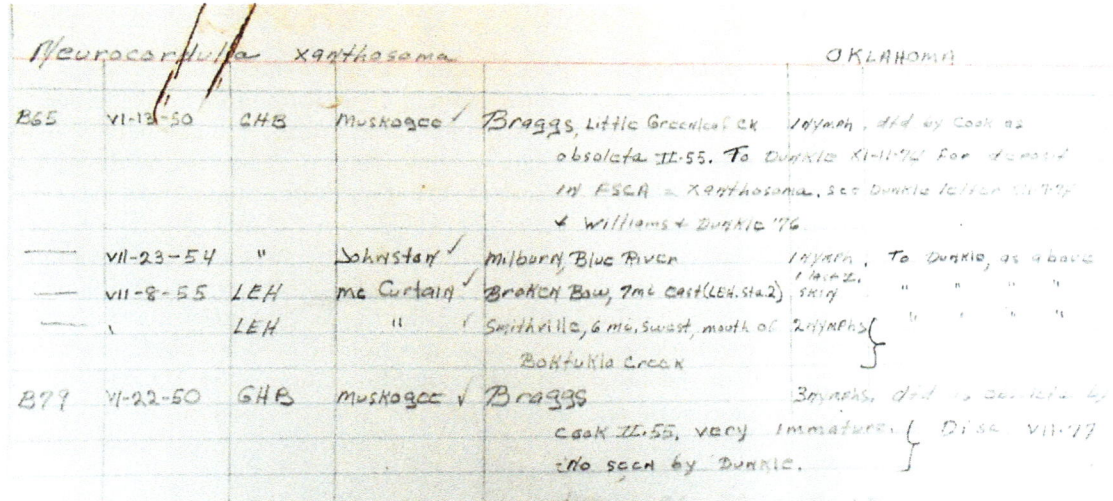

Figure 18.29 George H Bick's note card for *Neurocordulia xanthosoma* nymphs from Oklahoma. None of these nymphs have been relocated.

His description of flight after day-time disturbance matches our experience with the species. It can be difficult to re-find an individual once disturbed because it typically hangs well within dense shrubbery or flies into a rock crevice from which one has no hope of extricating a specimen.

As previously mentioned, the species was first collected in June 1907. Williamson's assistant, Frank Collins, later captured another ♂ on 2 August 1907 (along the Poteau River, near Wister, Le Flore County). Williamson (1908) used these specimens to describe the species, failing to designate one as the holotype. Byers (1937) also did not declare a holotype.[36] The UMMZ catalog, however, does list the early specimen as the holotype, though to our knowledge that has not been designated formally; leaving the Collins specimen to be a co-type.

Neurocordulia xanthosoma was reported by Ortenburger (1926b) and Bird (1932). We have tallied almost 30 records dating between 1926 and 1939, and almost as many between 1950 and 1996. It was during that second era that nymphs of *N. xanthosoma* were first collected. GH Bick collected a single nymph on 13 June 1950, during his first field season in the state, on Little Greenleaf Creek at the former Oklahoma State University Wildlife Conservation Station near Braggs, Muskogee County (Figure 18.29). He published this specimen as a new state record (Bick 1951) for *N. obsoleta*, the Umber Shadowdragon, a species unknown for Oklahoma. Curiously, his notes indicate that he had three other nymphs[37] that were identified as *N. obsoleta*, but these were not published in Bick (1951), nor in Bick and Bick (1957); perhaps the Bicks never felt confident of the identification because the larvae were early instars. At some point the record of the three nymphs found its way onto one of Bick's species note cards for *N. xanthosoma* (Figure 18.29; we did not find a note card for *N. obsoleta*). This record transfer must have occurred after Bick and Bick (1957) because

in that publication the Bicks included both *N. xanthosoma* and *N. obsoleta* as Oklahoma species. Moreover, Bick and Bick (1957:4) included in the *N. obsoleta* account new larval records that had been collected in 1954 and 1955 by Bick and LE Hornuff on the Mountain Fork River, McCurtain County, and on the Blue River, Johnston County.

In the decade-plus between 2007 and 2019, records of *Neurocordulia xanthosoma* increased for Oklahoma. The number of counties from which the species was known also increased. Roughly 80 records have been added (well over half of which are our own, hence our bafflement of not finding other *Neurocordulia* in the state!) from 26 counties—the species now is known from a total of 39 counties, with records concentrated in the eastern and southern parts of the state (Figure 18.31). An apparent increase in numbers may be an artifact of survey bias or maybe even just a fluke. We assumed it was the former, but we now wonder if possibly the species has become more common and possibly a little more widespread in the state. When considering the drastic changes in the hydrological landscape of Oklahoma, it is hard not to think that there have been effects on a species such as *N. xanthosoma*. We tend to think of *N. xanthosoma* as a lotic species, but as Williams (1976) suggested, it may be that only the adults are primarily lotic while larvae are primarily lentic. Williams found tenerals and exuviae[38] along lakeshores but almost always saw adults along streams, including most females ovipositing in streams. That would indicate adults mate at streams (either up or downstream), females oviposit at streams (presumably upstream of the lake), then eggs and larvae find their way into the lakes, emerge, and then tenerals fly to the streams to repeat the cycle. If this is indeed the case, then Oklahoma's hydrological changes would have benefited *N. xanthosoma* by providing lentic habitat for the larvae and lotic habitat immediately adjacent. We also suggest that the prevalence of boulder-lined steep slopes at the

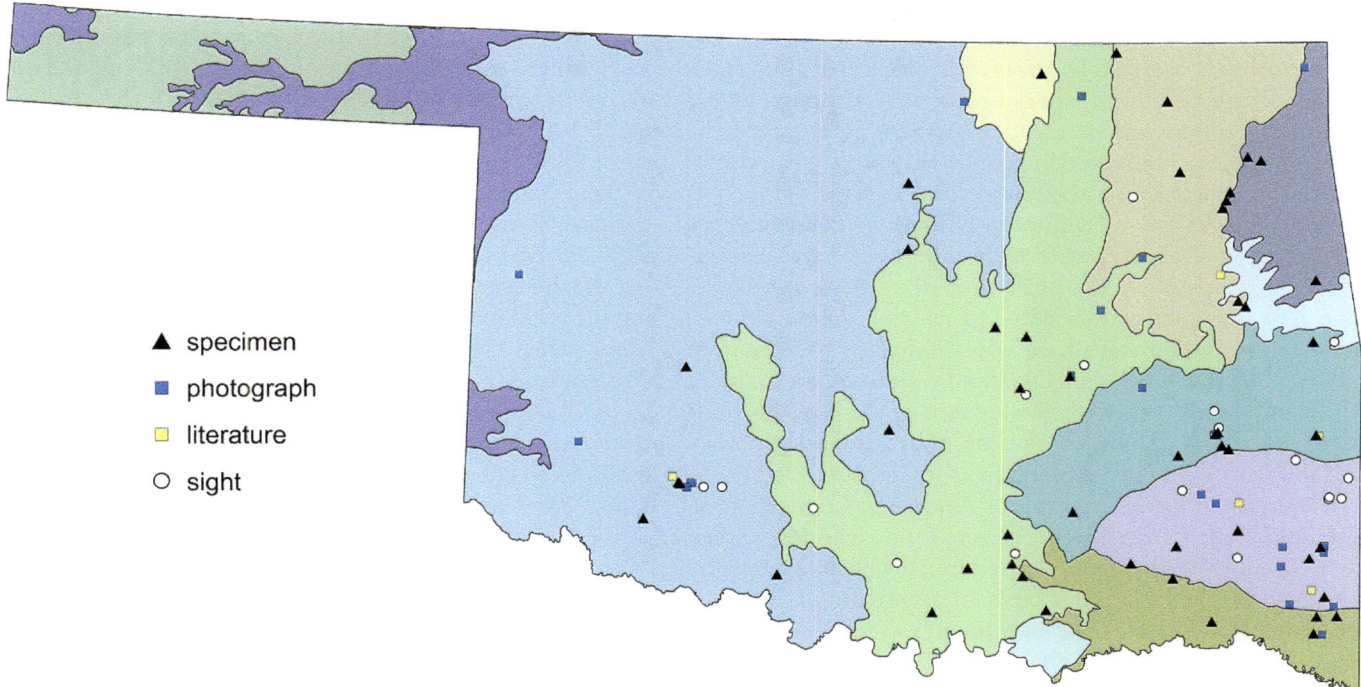

Figure 18.30 Documentation level of supporting records for *Neurocordulia xanthosoma*, the Orange Shadowdragon, in Oklahoma. Levels include specimen, archived photograph, literature reference, or sight record. Although sight records lack documentation, only those we feel are credible are plotted.

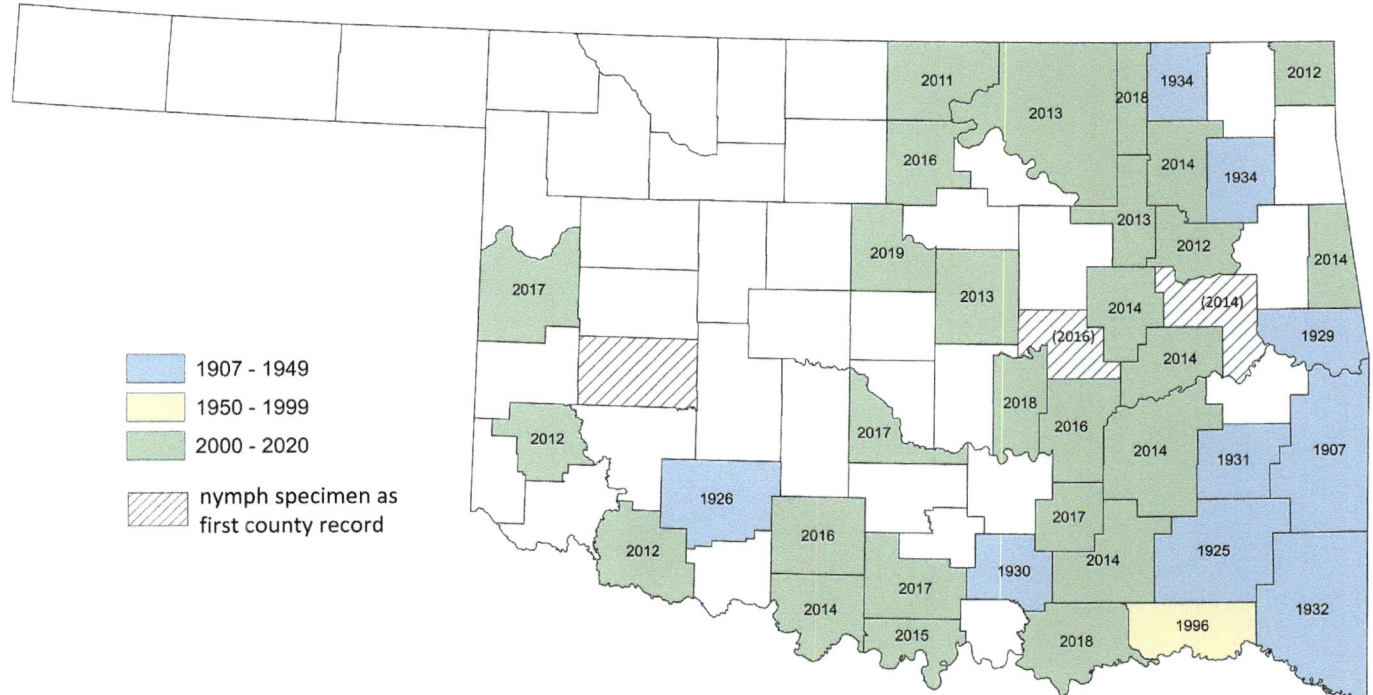

Figure 18.31 Counties in which *Neurocordulia xanthosoma*, the Orange Shadowdragon, are known to occur in Oklahoma. Year within the county is that in which the species was first reported. Muskogee, Okfuskee, and Washita Counties are hatched because the first documentation of the species for the counties came from nymphs. GH Bick also collected adults in Muskogee County the same year (1950) of his larval collections, but the species was not reported again until 2014. Okfuskee County also has since had a record of an adult (in 2016), but the single Washita County nymph remains the only record for that county.

dam spillways of these lakes has benefited the species. Again, Williams' study is enlightening because he opined that *N. xanthosoma* is strongly associated with vertical or steep banks, which befits our experience as well. Although he did not mention boulders on the banks where he observed the species, in Oklahoma we think there is a strong tie to natural rocky outcrops or their structural equivalent, such as boulder-laden slopes of spillways. Similar artificial outcrops are found at bridges over streams where we have encountered *N. xanthosoma*. We posit that all of these protective outcrops in conjunction with all of the impoundments now found in the state have contributed to the apparent increase of the species in Oklahoma. Because so many of those impoundments were created during the Bick/Hornuff years, a certain amount of lag time would explain why they did not detect the species as often as we do now.[39]

Regarding curious capture of *Neurocordulia xanthosoma*, GH Bick said that a ♀ was collected on a window screen and that a ♂ "was collected just after it had flown weakly and had perched low on a bush in a clearing at the edge of Greenleaf Creek. The specimen was collected about noon in almost full sunlight" (Bick 1951:179). In 2013 and 2014, J Fisher reported *N. xanthosoma* being attracted to blacklights. On three separate occasions at TNC's Tallgrass Prairie Preserve, Osage County, he found the species on his Lepidoptera collecting sheet. He noted a total of 6 individuals, all of which appeared at 21:00–21:30. In the case of 2♂ on 19 July 2013 and 1♂ on 3 July 2014, they stayed on the sheet until dawn. As for a ♂ and ♀ that came to the light on 29 June 2013, we will never know if they would have stayed because Fisher was kind enough to collect them for us (SP 1096, SP [RWG] 1097). A ♀ was also present on 19 July 2013, but she did not stay for long. It may be, then, that blacklighting is a good way to find *Neurocordulia* in general. That technique sure beats swinging wildly at spectral shadows that flit spastically along a river shore at dusk. Just saying.

In Oklahoma, we know of *Neurocordulia xanthosoma* flying from early May to late August, which is completely in line with elsewhere in the region (OK: 7 May 2016, Weleetka Lake, Okfuskee County, 1♀, MAP; 24 August 2014, 1 km S of Big Cedar, Kiamichi River, Le Flore County, 1♀; AR: 7 May–26 July, OC 374780, OC 331418; KS: 5 June–4 August, SEMC 1350215, SEMC 1350210; MO: 28 May–18 August, Trial 2005; TX: 16 April–10 August, Abbott and Lasley 2019). We know of a handful of teneral records dating from 14 May until 16 June.

EPITHECA CYNOSURA (SAY, 1839) – COMMON BASKETTAIL

Its English name aside, we have determined—painfully slowly, over the years—that although the Common Baskettail is, at times relatively common, it is not as common in Oklahoma as once thought. Previously (e.g., Smith-Patten et al. 2007) we had treated the Common Baskettail as the "default" species; i.e., we assumed it was the most numerous and widespread of Oklahoma's small baskettails (by which we mean *Tetragoneuria*-type *Epitheca*). Perhaps we were fooled by its name.

We cannot revisit many of the older records, but we stand by earlier assessments that at times this species outnumbers its congeners and, in our experience, when found in large swarms those swarms typically are not mixed—they hold only the Common Baskettail. Nonetheless, at this point in our understanding it is a near-certainty that it not the most common of the four small baskettail species that occur regularly in Oklahoma, especially if we restrict our dataset to specimens that can be diagnosed with a high level of certainty (that is, typologically bulletproof, with no intermediacy). All of this is to say that individuals that once would have been readily called Common Baskettails are, given our current understanding of the morphology of Oklahoma's

small baskettails, either better classified as *E. semiaquea*, the Mantled Baskettail, or intermediate toward that species or to *E. costalis*, the Slender Baskettail.

The first record for Oklahoma can be attributed to RD Bird, who collected a ♂ at Falls Creek, Murray County on 19 April 1930 (OMNH 329). Yet Bird himself, through no fault of his own given questions of species limits and diagnostic characters, muddied our understanding of the species' status even in the 1930s. Bick and Bick (1957:4–5) wrote that,

> Bird (1932) did not differentiate *cynosura* and *petechialis* and there is some doubt that *petechialis* is a distinct species. Specimens (OU) from Beckham, Murray, and Comanche Counties determined as *cynosura* by Bird were determined as *petechialis* by both the writer [senior author, GHB] and Dr. EJ Kormondy. There are other specimens in the OU collection from Murray and Cleveland Counties which are definitely *cynosura*; some were undetermined by Bird, some were determined by him as *williamsoni* Muttkowski. Thus, of Bird's list of counties, I [*sic*, we] could verify material from only Cleveland and Murray.

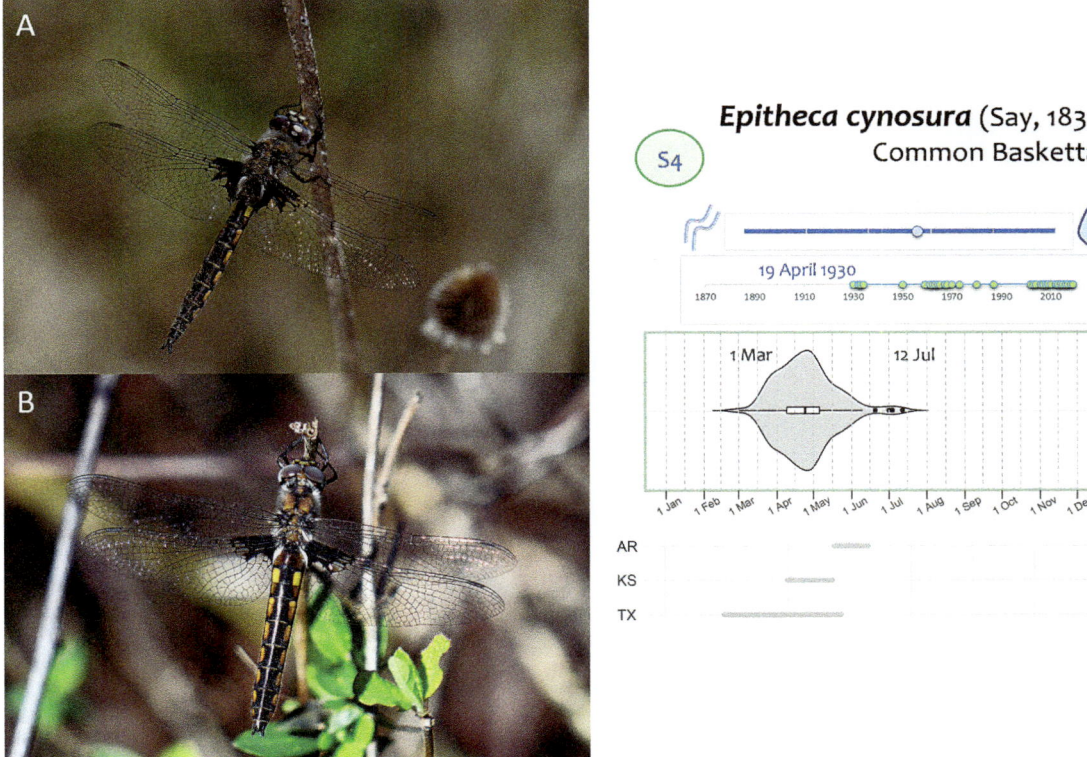

Figure 18.32 *Epitheca cynosura*, the Common Baskettail, A) ♂, Little River NWR, McCurtain County, 31 March 2018 (©Bryan E Reynolds). B) ♀, Red Slough WMA, McCurtain County, 1 March 2017 (USDA Forest Service; David Arbour, OC 461332). Please ignore its English name: we have determined that this species is often the least common small *Epitheca* encountered when baskettails seem to be everywhere in early spring.

How little has changed! And the Bicks themselves were not immune to the baskettail problem. Bick and Bick (1957) published a record of 3♂ *Epitheca cynosura* they collected from Boiling Springs SP, Woodward County, on 2 June 1956 (FSCA), but we re-identified these as *E. petechialis*, the Dot-winged Baskettail.

Our difficulties to determine the status and distribution of the Common Baskettail in Oklahoma stem from it being, by a wide margin, the most variable of the small baskettails to occur in the state. The base of the hindwings varies from hyaline (i.e., unmarked and clear) to extensively black, at times surpassing in extent the typical pattern on *Epitheca semiaquea* (see Figure 426 in Needham et al. 2014). The abdomen of the male can be short or long, nearly as wasp-waisted as that of *E. costalis*, or nearly as broad-based as that of *E. semiaquea*. Were we to cull our dataset to specimens that are unquestionably "perfect" Common Baskettails—the abdomen is long and slender, but not exceedingly so, and the base of the hindwing has a modest black patch—then one might draw the conclusions that the species is confined to southeastern Oklahoma: the bulk (~85%) of "classic" specimens come from Atoka, Choctaw, Le Flore, McCurtain, and Pushmataha Counties (i.e., those counties around the Ouachita Mountains). That said, we have found "classic" Common Baskettails north to the Ozark Plateau (e.g., Adair County, SP 161) and west to the Cross Timbers (e.g., Creek County, SP 479; Marshall County, SP 2262). In general, though, records away from the Ouachitas may have the wing hyaline more often (5 of 15 specimens vs. 6 of 30 specimens), but sample size is low

enough to be certain this apparent pattern is not a chance result. It is more likely that specimens away from the Ouachitas have the abdomen shorter with S3 less constricted, features that cloud diagnosis relative to *E. semiaquea*, if the hindwing is maculated. If it is hyaline, then diagnosis relative to *E. costalis* is clouded—we use cercus length, (*E. cynosura*: ♂ ≤ 3.0 mm, ♀ ≤ 2.1 mm), shape of the ventral surface of the cercus, if ♂ (sharply angled in both species but with a wider keel that lacks a protuberance in *E. cynosura*), and degree of constriction of S3 (more constricted on *E. costalis*). Some difficult to diagnose individuals may be hybrids (e.g., those with an abdomen of moderate length and width and hindwing maculation neither extensive nor reduced may be hybrid *E. cynosura* × *E. semiaquea*).

Bearing these caveats in mind, it appears that the Common Baskettail can be found in small numbers throughout the eastern two-thirds of the state (it is known from 38 counties; Figure 18.34), with occasional large swarms in the southeast. Likewise, the flight season ranges from the beginning of March, slightly earlier than that of *Epitheca semiaquea* or *E. costalis*, to early June, with a few reported as late as mid-July (1 March 2017, ♀, Red Slough WMA, McCurtain County, DA, OC 461332; 12 July 1959, ♀, Fairmont, Garfield County, J Reinert, FSCA[40]). As with some of the other small baskettail species, the flight season in Oklahoma is broader than that in neighboring states (AR: 17 March–14 June, Harp and Rickett 1977; KS: 9 April–16 May, OC 318391, 375008; TX: 17 Febuary–23 May, Abbott and Lasley 2019).

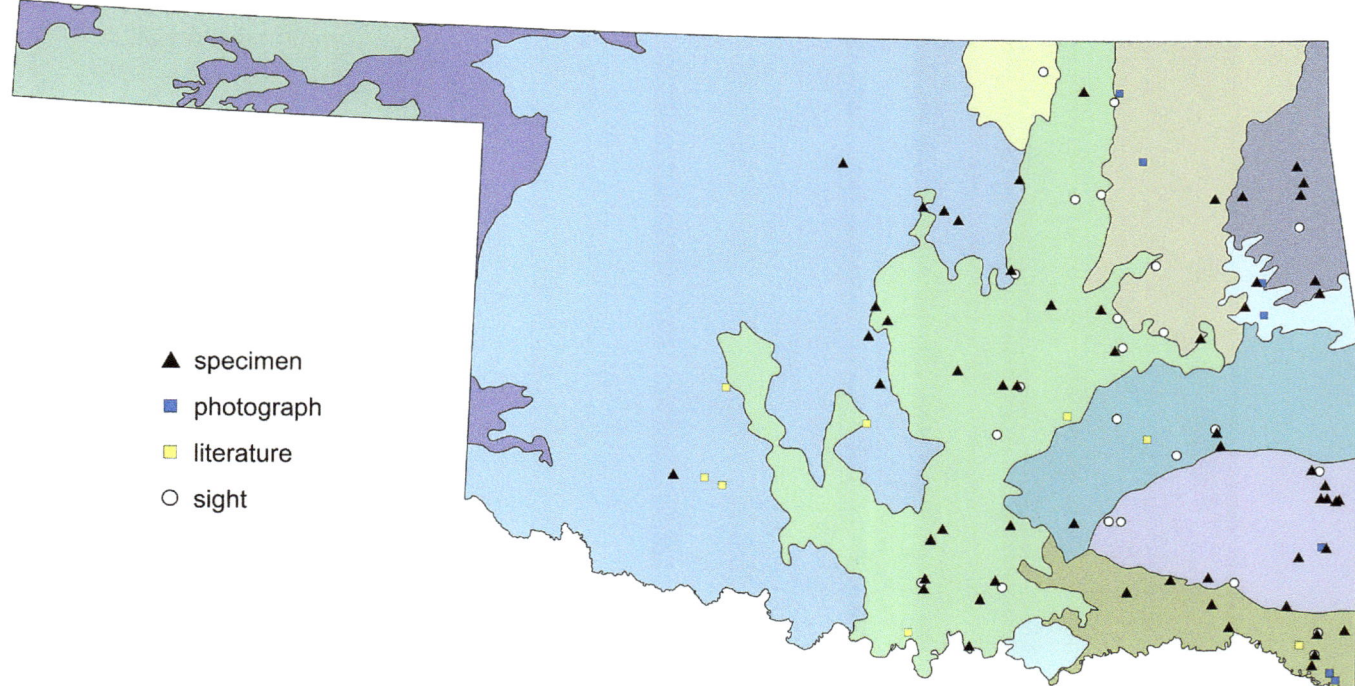

▲ specimen
■ photograph
□ literature
○ sight

Figure 18.33 Documentation level of supporting records for *Epitheca cynosura*, the Common Baskettail, in Oklahoma. Levels include specimen, archived photograph, literature reference, or sight record. Although sight records lack documentation, only those we feel are credible are plotted.

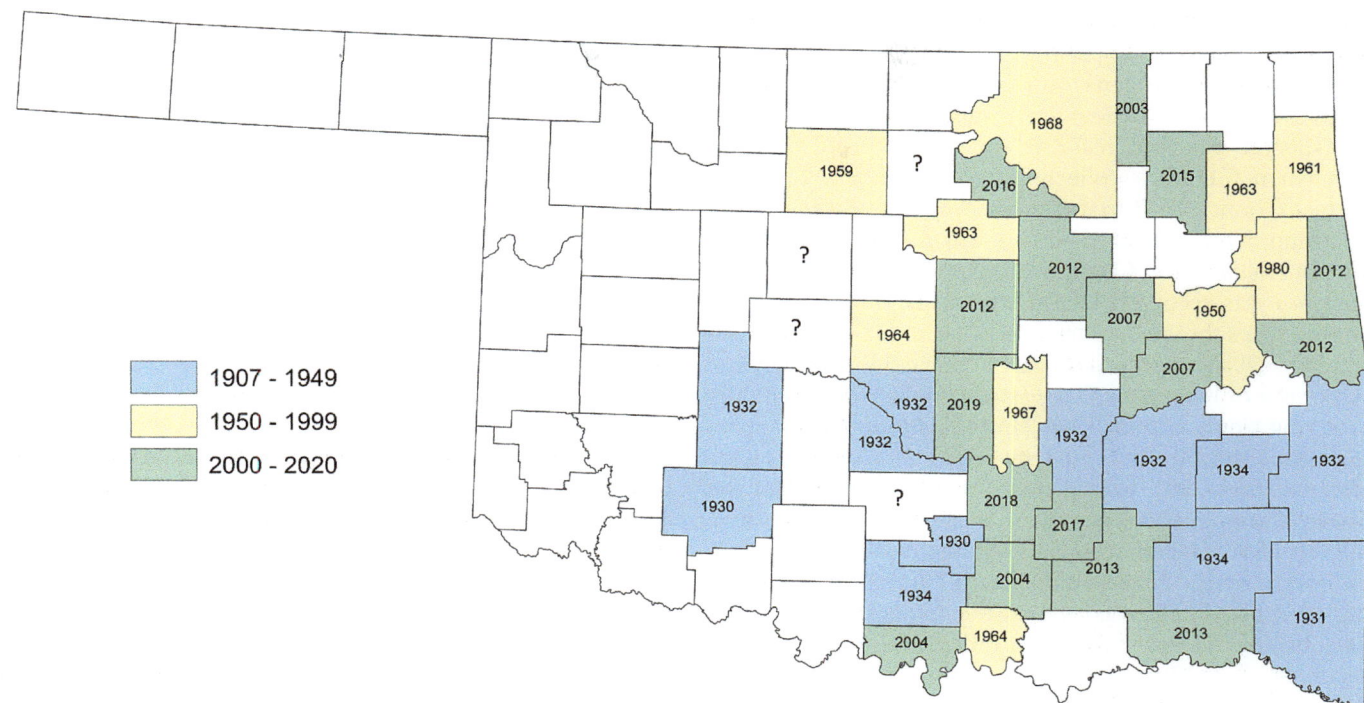

Figure 18.34 Counties in which *Epitheca cynosura*, the Common Baskettail, are known to occur in Oklahoma. Year within the county is that in which the species was first reported. Counties marked with a ?—Canadian, Garvin, Kingfisher, and Noble—denote inconclusive reports.

EPITHECA SEMIAQUEA (BURMEISTER, 1839) – MANTLED BASKETTAIL

The status in Oklahoma of the Mantled Baskettail has begun to come into focus only in the past half-decade. The species was first recorded in the state on 18 April 1959, when GH Bick collected a ♂ at Robbers Cave SP, Latimer County. As if to underscore inherent difficulties in the group, when captured, this ♂ was in tandem with a ♀ *Epitheca cynosura*, the Common Baskettail (FSCA). *Epitheca semiaquea* was recorded in the state only three additional times in the subsequent three decades, and it was not until the mid-1990s that records began to accrue. Even so, by the turn of the millennium, *E. semiaquea* was known from only three counties: Latimer, Love, and Oklahoma. The apparent commonness in Love County, where 21 specimens were collected 1994–1997, hinted that our understanding of the species' status in the state was inadequate. There are now >100 records spread across 33 counties, with roughly half of all occurrences in the southeastern "fourth" of the state, defined approximately as east of the Arbuckle Mountains and south of the Arkansas River valley, and thus including the whole of the Ouachita Mountains and the Gulf Coastal Plain (Figures 18.36 and 18.37).

Hence, *Epitheca semiaquea* is not as scarce or localized as it was previously thought—not long ago we classified the conservation status as S3 (Patten and Smith-Patten 2013b)—although we still do not know just how common and widespread it is. Mounting evidence suggests that in the southeastern quadrant of the state this species is the second most numerous small baskettail, behind *E. costalis*, the Slender Baskettail. Interestingly, these two species often co-occur to the point that feeding swarms in the southeast predictably consist of a mix of this duo to the exclusion of the *E. cynosura*. (By contrast, swarms of *E. cynosura* tend to hold that species alone.) Mounting evidence further suggests that the geographic distribution of *E. semiaquea* extends farther to the north and west than was thought previously, where it may be common at times (e.g., 8♂ and 5♀, many identified in hand and released, at Ardmore Regional Park, Carter County, 8 April 2017; MAP, 1♂ as SP 2558). Accordingly, it appears that the conservation status is better characterized as S4.

We do not know if this species has expanded its range in Oklahoma, yet we feel it more likely that its status was conflated with that of *Epitheca cynosura*. Not only can diagnosis be difficult when phenotype is at extremes (e.g., the minimal extent of the hindwing patch on *E. semiaquea* overlaps the maximal of *E. cynosura*; see Figure 426 in Needham et al. 2014), but the extent of hybridization of *Tetragoneuria* species is unknown. Some individuals

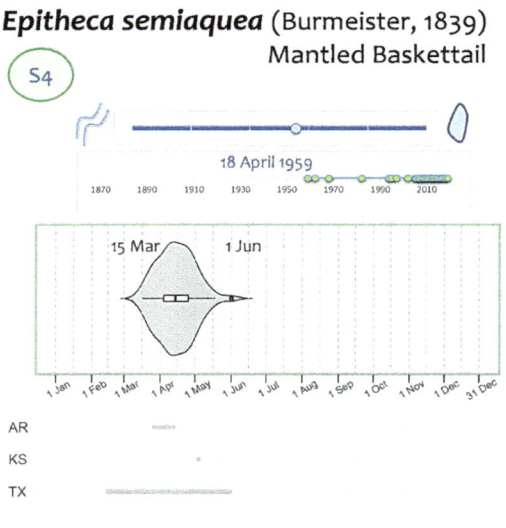

Figure 18.35 *Epitheca semiaquea*, the Mantled Baskettail, A) ♂, Salt Creek Recreation Area, Choctaw County, 30 March 2018 (©Bryan E Reynolds). B) ♀, Tenkiller WMA, Sequoyah County, 12 April 2016 (©James W Arterburn, OC 456434). For years we thought this species to be scarce and largely confined to the Ouachitas. We now know it is fairly common and much more widespread.

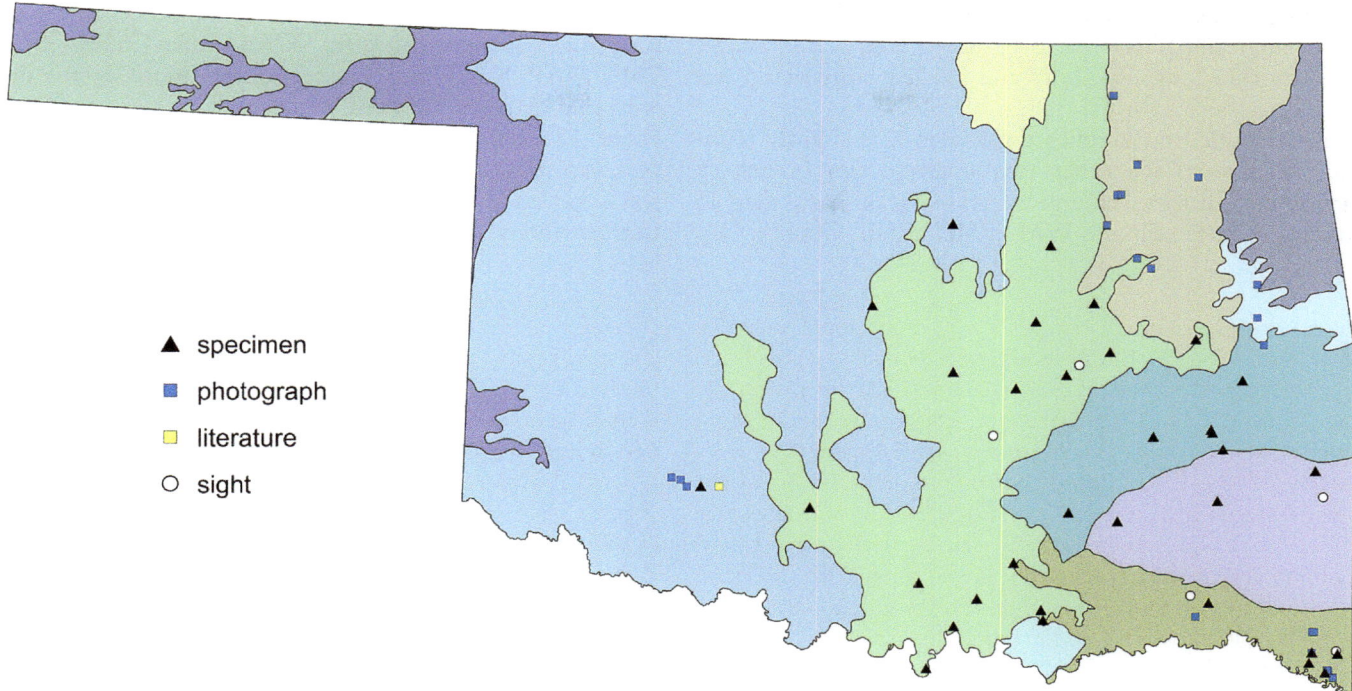

Figure 18.36 Documentation level of supporting records for *Epitheca semiaquea*, the Mantled Baskettail, in Oklahoma. Levels include specimen, archived photograph, literature reference, or sight record. Although sight records lack documentation, only those we feel are credible are plotted.

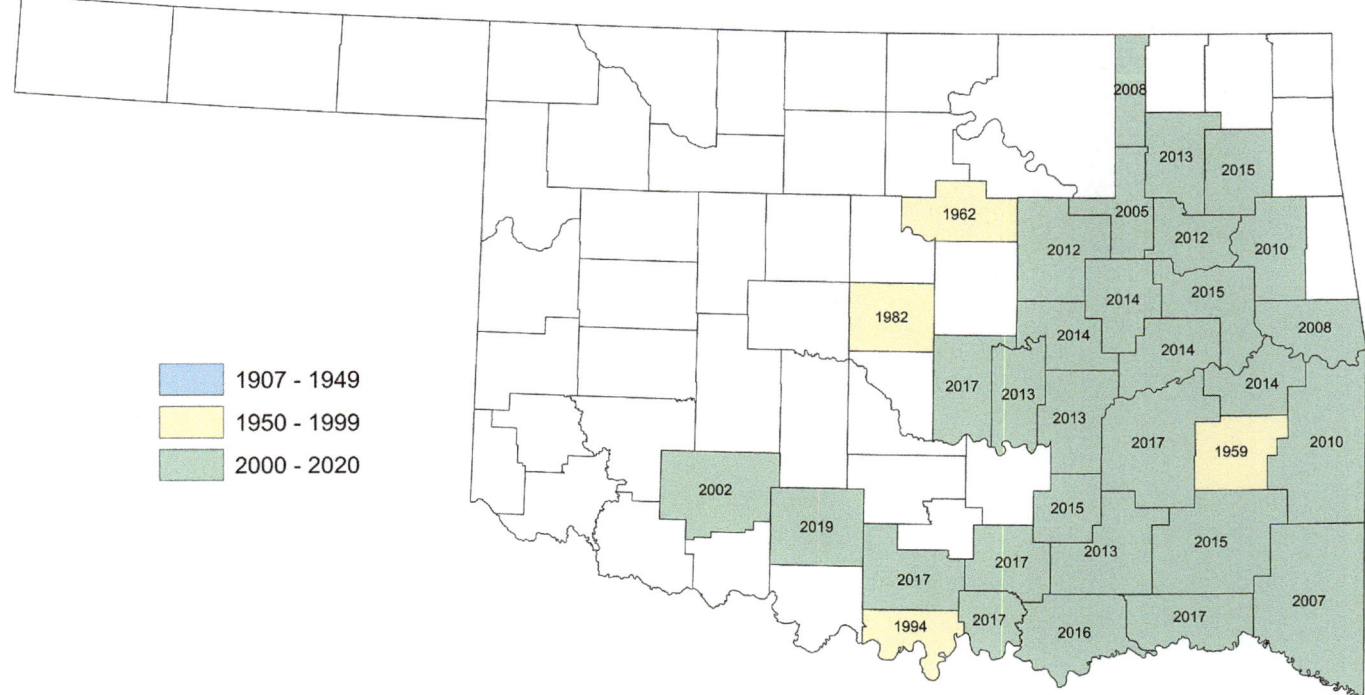

Figure 18.37 Counties in which *Epitheca semiaquea*, the Mantled Baskettail, are known to occur in Oklahoma. Year within the county is that in which the species was first reported.

collected in the state seem to have an "intermediate" hind-wing patch and problematic abdomen length and shape, being neither long nor short and neither constricted nor thick. Are these hybrids?

The Mantled Baskettail's flight season is slightly truncated relative to the other two small species in eastern Oklahoma. All three emerge in mid-March (early date 15 March 2012, Red Slough WMA, McCurtain County, DA, OC 374105), but *E. semiaquea* has been recorded beyond mid-May just once (♀ at Atoka WMA, Atoka County, 1 June 2019, MAP, SP 2779). This seasonality more-or-less jibes with that in Texas, although it is a touch longer at either end there (15 Febuary–31 May, Abbott and Lasley 2019). The one Kansas record (4 May, OC 422002) and the few records for Arkansas (27 March–12 April, OC 440905, 479110) fall in this window as well.

EPITHECA COSTALIS (SÉLYS, 1871) – SLENDER BASKETTAIL

The Slender Baskettail's occurrence in Oklahoma dates to the beginning of odonatology in the state: EB Williamson and colleagues collected several ♂ and ♀ specimens from Wister, Le Flore County, 3–4 June 1907 (UMMZ, Williamson 1914b), specimens that served as the basis for the type description of *Tetragoneuria williamsoni* (Muttkowski 1911), a name that proved to be a junior synonym of *Epitheca costalis*. Since that time hundreds of records have been amassed, such that the species is now known from 47 counties, chiefly across the eastern half of the state (Figures 18.39 and 18.40), although records are curiously sparse in the northeast and, as if to compensate, are scattered across the southwest.

Epitheca costalis is the most numerous and frequently encountered small baskettail in the eastern half of Oklahoma. It is not unusual to find swarms of individuals, such as 25 at Robbers Cave WMA, Latimer County, 9 June 2013 (1♂, SP 682) or 20 (of which 4♂ and 2♀ were identified in hand) at Lake Nanih Waiya WMA, Pushmataha County, 26 April 2015 (MAP). Swarms are formed solely of this species, but mixed-species swarms are common enough, with *E. semiaquea*, the Mantled Baskettail, the expected co-occurring species; we have seldom encountered *E. cynosura*, the Common Baskettail, flying with either of those species.

The flight season is long for a small baskettail (*Tetragoneuria*-type *Epitheca*): the first of the season are on the wing by the second week of March (early date 8 March 2017, a ♂ at Red Slough WMA, McCurtain County, DA, OC 461386), although the bulk appear a week or two later than *E. cynosura*. Unlike that species, stragglers have appeared as late as mid-July (late date 18 July 2013, Deep Fork WMA, Okmulgee County, JT Bried, SP 1486), although numbers are reduced greatly by mid-June. The documented flight season in Oklahoma is markedly longer than that in neighboring states (AR: 27 March–14 June, OC 493734, 482016; KS: possibly 11 March–27 May, SEMC[41]; TX: 12 March–22 May, OC 461447, Abbott and Lasley 2019).

Figure 18.38 *Epitheca costalis*, the Slender Baskettail, A) ♂, Tenkiller WMA, Sequoyah County, 12 April 2016 (©James W Arterburn, OC 456433). B) ♀, Bixhoma Lake, Wagoner County, 3 June 2018 (©James W Arterburn). This elegant little species appears to be the most numerous small *Epitheca* in the eastern half of the Southern Great Plains.

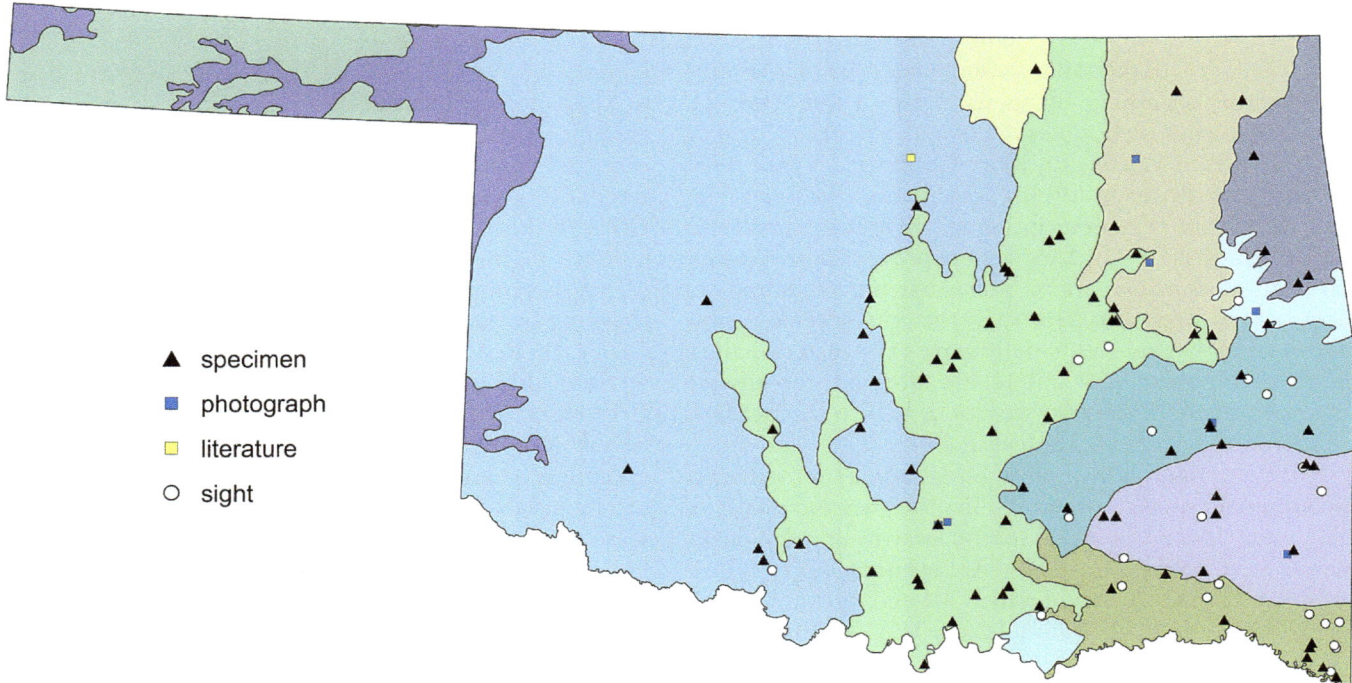

Figure 18.39 Documentation level of supporting records for *Epitheca costalis*, the Slender Baskettail, in Oklahoma. Levels include specimen, archived photograph, literature reference, or sight record. Although sight records lack documentation, only those we feel are credible are plotted.

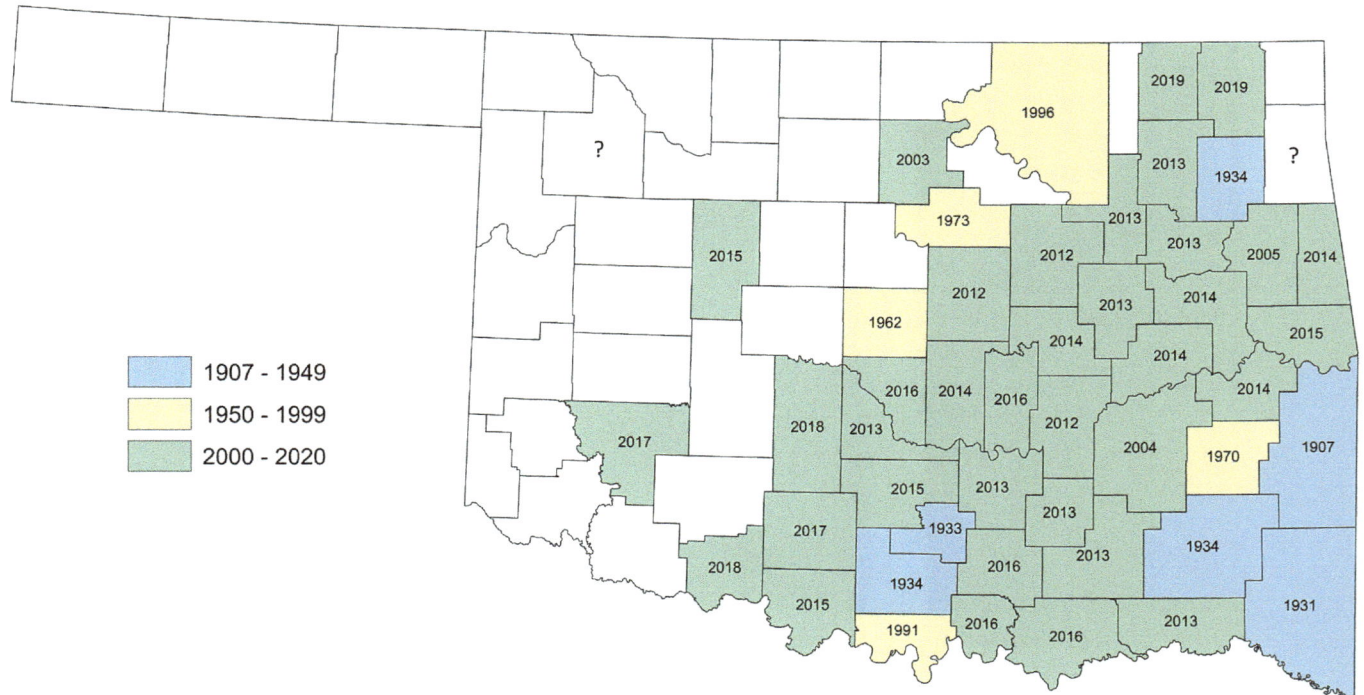

Figure 18.40 Counties in which *Epitheca costalis*, the Slender Baskettail, are known to occur in Oklahoma. Year within the county is that in which the species was first reported. Two counties—Delaware and Woodward—have reports that are not well documented.

Epitheca petechialis (Muttkowski, 1911)
Dot-winged Baskettail

S4

10 June 1926

6 Apr   5 Aug

KS
NM
TX

Figure 18.41 *Epitheca petechialis*, the Dot-winged Baskettail, A) ♂, Wichita Mountains WR, Comanche County, 26 May 2017 (©Eric Isley). B) ♀, Wichita Mountains WR, Comanche County, 10 May 2016 (©Bill Carrell, OC 444296). This species is the western counterpart to *E. costalis*, the Slender Baskettail. Its geographic distribution is shifted westerly enough that it is the only small baskettail expected in the western half of the Southern Great Plains.

EPITHECA PETECHIALIS (MUTTKOWSKI, 1911) – DOT-WINGED BASKETTAIL

The Dot-winged Baskettail is the only small baskettail expected in the western half of Oklahoma, where it has been documented in 29 counties (Figures 18.42 and 18.43). It was first recorded in Oklahoma when at least 5♂ were collected at "Boulder Camp" in what is now the Wichita Mountains WR, Comanche County, on 10 June 1926 (TH Hubbell, UMMZ, ANSP; reported in a letter, dated 5 April 1933, from LK Gloyd to RD Bird as 6♂ that she thought were *E. petechialis*). The species was encountered regularly thereafter, and it remains a common early to mid-season odonate in the western counties, including those in the panhandle.

Published reports or purported specimens east of central Oklahoma, where the species has been found occasionally, are problematic. Bick (1991) stated that the species was added to the Latimer County list, the source for which can be found on his note cards, where he stated that he collected a ♂ at Robbers Cave SP on 17 April 1959. We question the veracity of this claim, but we cannot re-examine the specimen, because remarkably, he noted that he discarded it. A careful read of his note card reveals that he thought he collected 8♀ (2 from nymphs he reared) and 1♂ the next day, although his notes further indicate that in 1972 he re-identified all (or most?) of these specimens as *Epitheca cynosura*, the Common Baskettail, a much more plausible determination. We do not know if he discarded the 17 April ♂ prior

to re-examining specimens in 1972, but if he did, then how did he know that ♂ specimen was identified correctly yet the other were not? Even if the venerable Bick recalled distinct dark crossbars at the base of the hindwing, how did he know, with the specimen unseen for perhaps as much as three decades, that a Selys's Sundragon (*Helocordulia selysii*) was eliminated? The sundragon was unknown from Oklahoma until 2008; it now is known from two sites, one each in McCurtain and Le Flore Counties, areas biogeographically comparable to Latimer County. Regardless, Bick's (1991) record was carried forward in various other publications.

With regard to potentially questionable occurrences, to Bick's record we add a ♂ we collected at Sportsman Lake, Seminole County, 12 May 2013 (SP 560), an individual that has a wing pattern typical of *E. petechialis* but the ventral surface of the cerci are intermediate toward that of *E. costalis*, the Slender Baskettail (i.e., between the smoothly rounded keel of *E. petechialis* and the sharply angular keel of *E. costalis*). Is this specimen in fact of a hybrid? If it is not, then it represents the easternmost record of *E. petechialis*. If it is a hybrid, then it and an unknown proportion of small baskettails in central Oklahoma may be best treated as unclassifiable intermediates. The extent of hybridization is unknown, and the vast bulk of specimens we have examined from the Southern

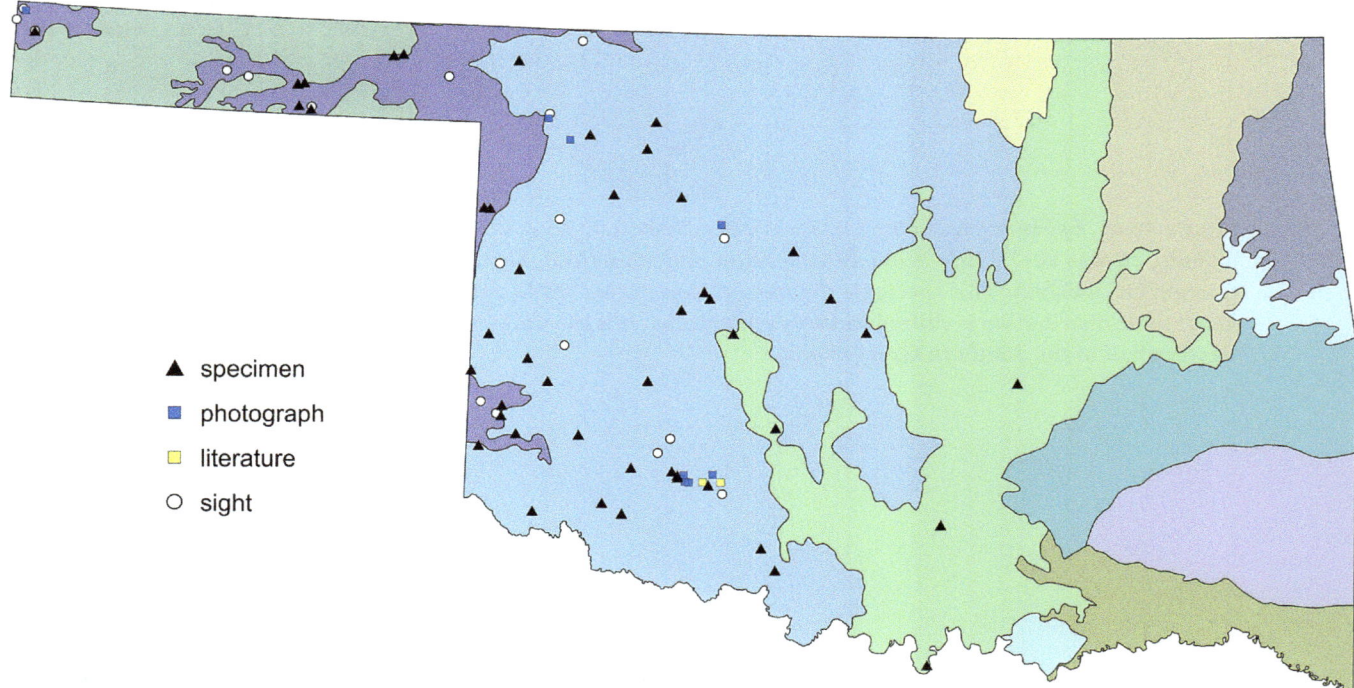

Figure 18.42 Documentation level of supporting records for *Epitheca petechialis*, the Dot-winged Baskettail, in Oklahoma. Levels include specimen, archived photograph, literature reference, or sight record. Although sight records lack documentation, only those we feel are credible are plotted.

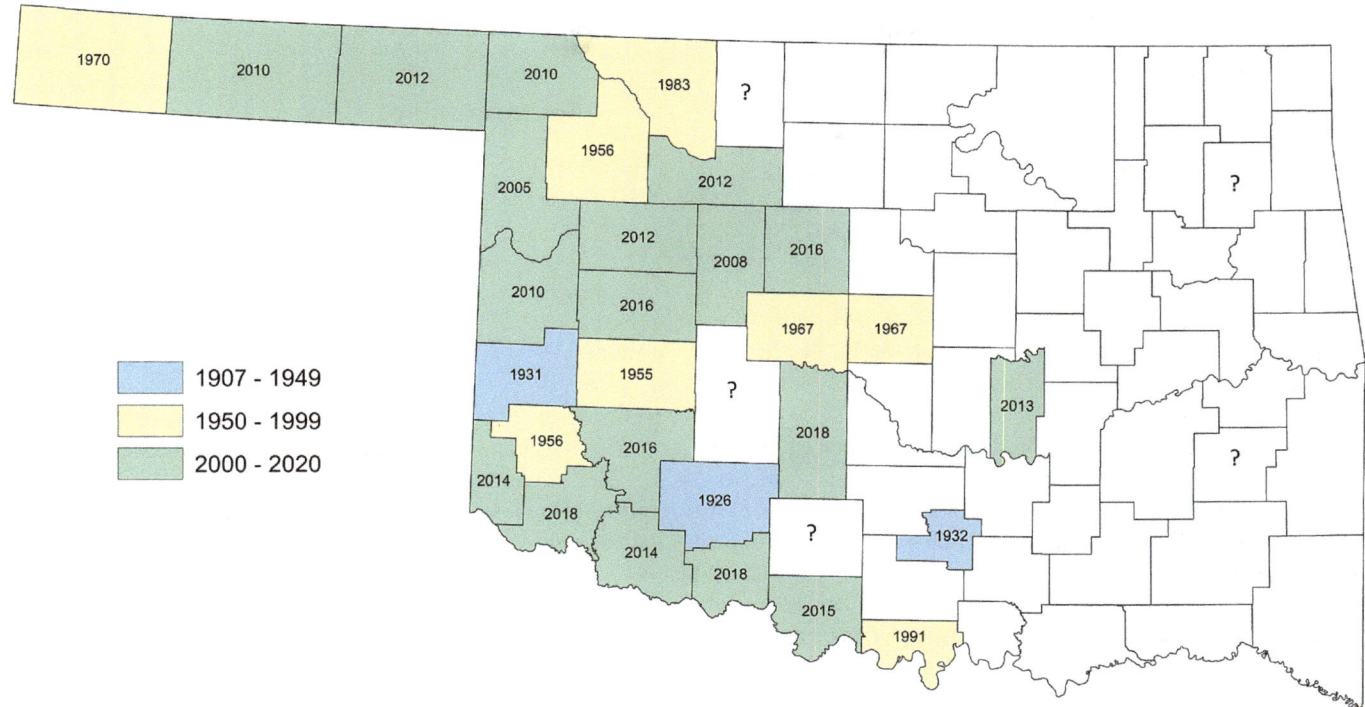

Figure 18.43 Counties in which *Epitheca petechialis*, the Dot-winged Baskettail, are known to occur in Oklahoma. Year within the county is that in which the species was first reported. Five counties, Alfalfa, Caddo, Latimer, Mayes, and Stephens, have reports that may be dubious.

Great Plains can be categorized readily as one species or the other, yet some authorities have argued that these two species ought to be lumped.

This species' flight season in Oklahoma extends from early April to, remarkably, for a small baskettail at this latitude, early August, although most cease to fly by the end of June (early date: 6 April 2016, American Horse Lake, Blaine

County, 1♂, SP 1858, and Doby Springs, Harper County, 2♂, one as SP 1856, both records BS-P, BW Hoagland; late date: 5 August 1970, Black Mesa SP, Cimarron County, 1♂, GHB, LEH, FSCA). Kansas has a much shorter known flight season but, no surprises, Texas has a much longer one (KS: 7 May–20 June, SEMC; TX: 9 Febuary–12 June, OC 459560, 447932).

EPITHECA SPINOSA (HAGEN, 1878) – ROBUST BASKETTAIL

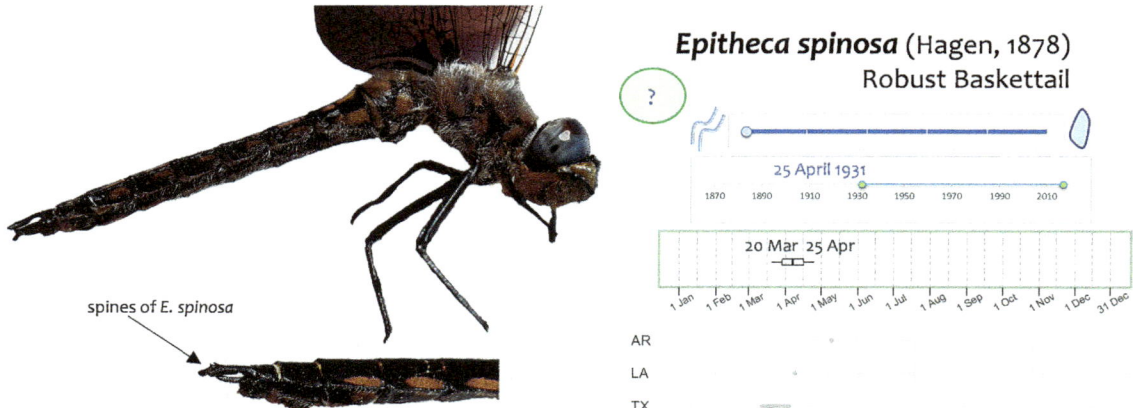

spines of *E. spinosa*

Epitheca spinosa (Hagen, 1878)
Robust Baskettail

25 April 1931

20 Mar 25 Apr

AR
LA
TX

Figure 18.44 Second record for Oklahoma of *Epitheca spinosa*, the Robust Baskettail (♂, Grassy Slough WMA, McCurtain County, 20 March 2017, Michael A Patten and David Arbour, OC 461537, SP 2227; photos: USDA Forest Service).

There are only two records of the Robust Baskettail for Oklahoma. The first came on 25 April 1931 when E Hooper collected a ♂ in an unknown location in Latimer County (OMNH 334). This specimen was originally identified and published as *Tetragoneuria canis* (Bird 1932) but was re-identified as *Epitheca spinosa* by GH Bick and EJ Kormondy (Bick and Bick 1957). The second record came 86 years later on 20 March 2017, when MAP and DA collected a single ♂ at Grassy Slough WMA, McCurtain County (OC 461537, SP 2227).

Because we do not know where in Latimer County the first specimen was collected, we can say nothing of the habitat. But the McCurtain County record—caught as the ♂ patrolled low in a shaded reach over a backwater part of the main creek that flows into the northeastern part of Grassy Slough—fits habitat descriptions from elsewhere. For example, Tennessen et al. (1995:62) said that in Alabama the species "occurs in wooded swamps with very little flow." Dunkle (2000:147) provided a similar description of "swamps with some water movement ... occasionally boggy ponds or lakes," as did Paulson (2011:396), "swampy woodland ponds."

The normal range of the species is odd in that it is primarily a species of the central eastern seaboard, especially the Chesapeake Bay area, but there are scattered records south and west of there, to Florida, Alabama, Mississippi, Louisiana, Texas, Arkansas, and Oklahoma. As Paulson (2011:396) said, "the sparseness of records south and west of North Carolina is puzzling." Either this species wanders a good deal or it occurs at low density but observers confuse it with other small *Epitheca* species. Given the difficulty, at least in the field, of telling apart some species of this genus we guess it is the latter option. Species confusion is a call for collecting more *Tetragoneuria*-type *Epitheca* until we can fully decipher species' limits, from a geographical standpoint in the case of *E. spinosa*, but also from the morphological

and genetic standpoints for the others. A specimen of a ♂ *Epitheca* photographed by BAH on 4 April 2009 at Little River NWR, McCurtain County, for example, would be much appreciated because we are unable to confirm the pictured species, which is either an *E. spinosa* or an especially broad *E. cynosura* (Figure 18.45).

Epitheca spinosa is known to fly throughout its range as early as 25 Febuary (OC 461289) and as late as June (Paulson 2011). It has been recorded in Texas 13 March–4 April (Abbott and Lasley 2019), from Louisiana once, on 9 April 1987 (Mauffray 2014), and from Arkansas, also once, on 10 May 1984 (Harp and Harp 1996).

Figure 18.45 A ♂ *Epitheca* (*Tetragoneuria*) photographed by Berlin A Heck at Little River NWR, McCurtain County, Oklahoma, on 4 April 2009. This individual remains unidentified to species (an *E. spinosa* or an *E. cynosura* with a fat abdomen?) because without a specimen we cannot confirm diagnostic characters.

EPITHECA PRINCEPS HAGEN, 1861 – PRINCE BASKETTAIL

The Prince Baskettail is a common and widespread species in Oklahoma, where it has been recorded many hundreds of times in 76 counties across the state; it is missing only from Cimarron County in the western panhandle, despite records to the south and north of this area. It is recorded regularly in Mountains (UMMZ). It has been a regularly reported species since, including at three different localities on its early flight date in Oklahoma (1 May 2016: 1♂, 1♀, Lake Durant, Bryan County, and 1♂, vic. Tishomingo NWR, Johnston County, and 3♂, Carter Lake, Marshall County; all records

Figure 18.46 *Epitheca princeps*, the Prince Baskettail, A) ♂, Sunset Lake in Guymon, Texas County, 10 August 2018 (©Tony Leukering, OC 487784), the westernmost record for the state in a typical view (i.e., in flight). The faint wing markings seem to be more prevalent along the western fringe of the Southern Great Plains. B) ♂, Red Slough WMA, McCurtain County, 7 May 2018 (©Greg W Lasley, OC 479567). Note the much heavier wing markings. C) ♀, Mohawk Park in Tulsa, Tulsa County, 14 July 2013 (©Bill Carrell, OC 403401).

numbers >10, including our highest count of 30 individuals, a number reached twice (1 July 2015, Horseshoe Lake, 7 km ESE of Verdigris, Rogers County, BS-P, BW Hoagland; 25 June 2016, American Horse Lake, Blaine County, MAP). It is encountered most often around ponds or lakes but can be found just about anywhere: pastures, prairie, and on or near creeks, particularly those with slow flow.

Epitheca princeps was first recorded in Oklahoma on 11 June 1926, when TH Hubbell captured 1♂ in the Wichita MAP). The species flies until 19 September in the state (from 2003, 1U, Lake George Fort Sill MR, CSU). The flight season in Oklahoma is in line with the region (AR: 21 April–21 September, Harp and Rickett 1977; KS: 15 May–15 August, SEMC[42]; LA: 30 April–19 August, iNat 5025324, Mauffray 2014; MO: 6 May–27 August, Trial 2005; TX: 10 March–27 September, Abbott and Lasley 2019). Occasionally pairs are reported, but ovipositing has only been reported once (lone ♀, Keechi Park, Cyril, Caddo County, 16 July 2017, MAP). We

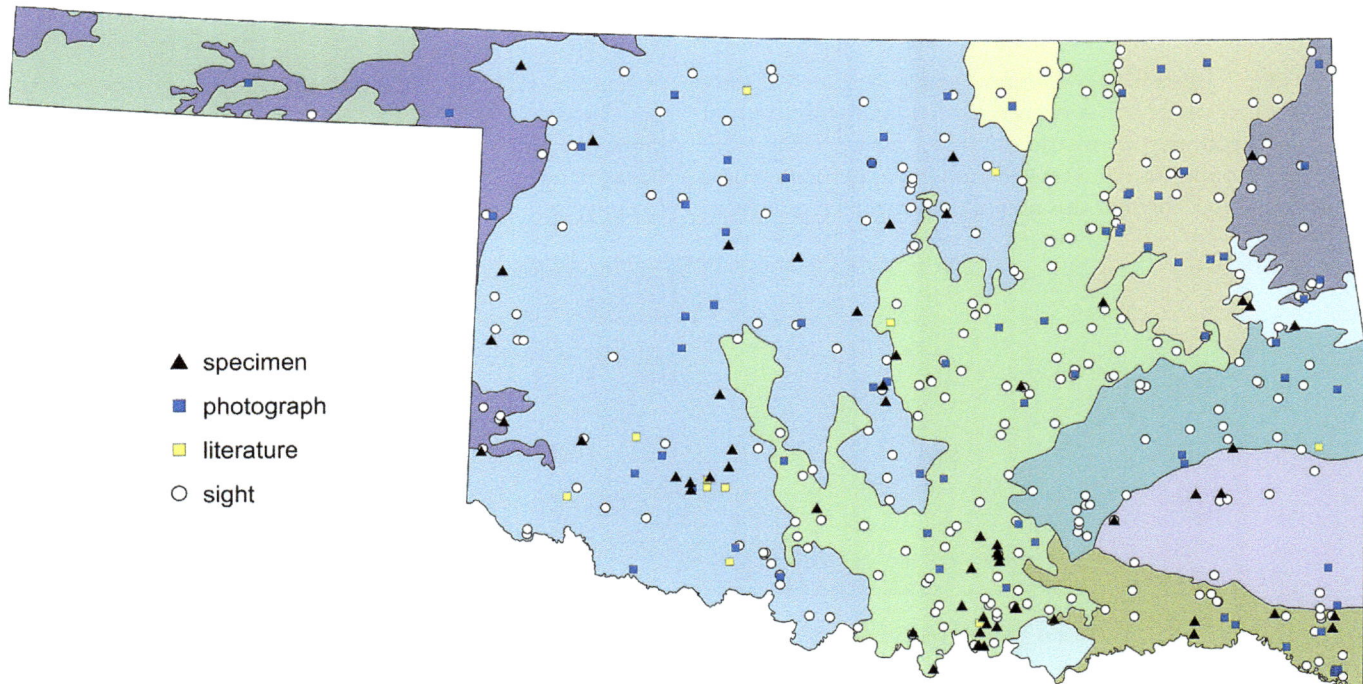

Figure 18.47 Documentation level of supporting records for *Epitheca princeps*, the Prince Baskettail, in Oklahoma. Levels include specimen, archived photograph, literature reference, or sight record. Although sight records lack documentation, only those we feel are credible are plotted.

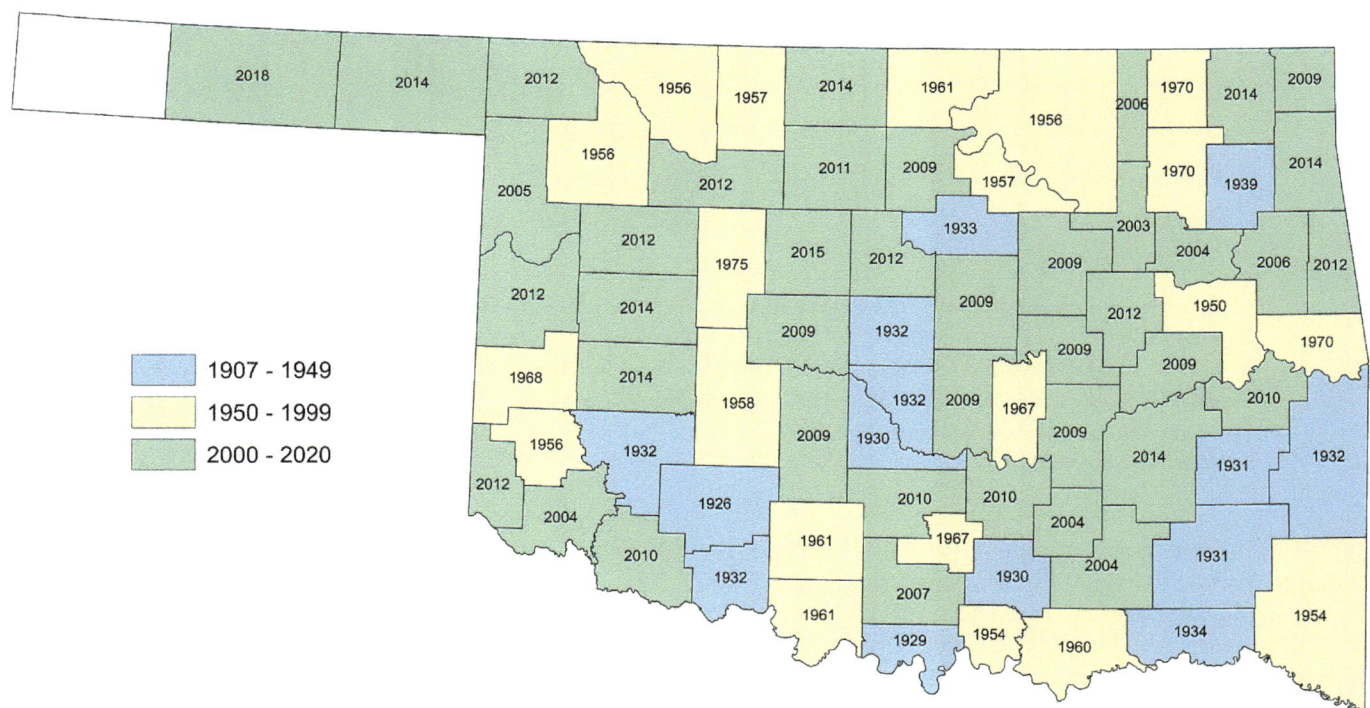

Figure 18.48 Counties in which *Epitheca princeps*, the Prince Baskettail, are known to occur in Oklahoma. Year within the county is that in which the species was first reported. Blaine County is hatched because its first record was of a nymph; its first adult record came in 2008.

Northwestern Oklahoma
SP 272 - Harper Co., Doby Springs, 2 June 2012

North-Central Oklahoma
SP 954 - Pawnee Co., Sooner Lake, 10 August 2013

East-Central Oklahoma
SP 884 - Okmulgee Co., Okmulgee WMA, 27 July 2013

Southwestern Oklahoma
SP 1919 - Stephens Co., Clear Creek Lake, 30 May 2016

Southeastern Oklahoma
SP 716 - Atoka Co., Atoka WMA, 19 June 2013

Figure 18.49 Variation in wing pattern of *Epitheca princeps*, the Prince Baskettail.

know of three larval records (OMNH, UCO) and four reports of tenerals, dating 11 May–24 July.

It is well known that the extent and saturation of wing markings varies greatly in the Prince Baskettail. There are hints of a geographic component in variation, but individual variation cannot be discounted. The large (HW 46–52 mm), heavily marked population from the southeastern United States has been named as subspecies *Epitheca princeps regina*, whereas the smaller (HW 38–42 mm), lightly marked population in the Northeast is nominate *E. p. princeps* (Needham et al. 2014). Subspecific classification of the population in the Southern Great Plains is unclear. As Paulson (2009:369) noted, generally individuals are smaller in the west and variation in size in the plains is minimal relative to eastern populations (TL 59–65 mm versus 59–75 mm), yet many individuals "with reduced markings [occur] in north Texas and farther north." Likewise, Abbott (2015:197): noted that "in Texas some individuals may have very faint markings." In Oklahoma, there may be an east-to-west gradient in reduced wing markings (Figure 18.49), although we have not studied this closely. Even so, the 3♂ we have collected in the state with exceedingly faint markings are

from the west: from Doby Springs, Harper County, 2 June 2012 (SP 272), from Clear Creek Lake, Stephens County, 30 May 2016 (SP 1919), and from Doc Hollis Lake, Greer County, 4 August 2018 (SP 2686). That said, some individuals in the broader Southern Great Plains, such as a ♂ photographed at Lake Charleston, Franklin County, Arkansas, 11 July 2017 (OC 469095), have the wing spots limited.

NOTES

1. Tennessen (2019).
2. Tennessen (2019) treated *Epicordulia* and *Tetragoneuria* as subgenera of *Epitheca*.
3. According to Corbet (1999:582–583), the technical term should be eocrepuscular, meaning "active during morning and evening twilight," rather than crepuscular, meaning "active during evening twilight." His use of eocrepuscular, although following A Haddow's use and introduction of the term (The Mosquitoes of Bwamba County, Uganda. II.—Biting Activity with special Reference to the Influence of Microclimate. *Bulletin of Entomological Research* 36[1]:33–73, cited as both 1945 and 1946), is not a largely abided term. Note, too, that basically every established dictionary defines "crepuscular"

in the way we use the term. As such, we retain the commonly used term crepuscular to connote flight during the early morning and evening hours.

4. The species is also distinctive in its "forewings each with 4–6 cell rows in trigonal interspace" (Needham et al. 2014:380). Before *Neurocordulia michaeli* was described as a new species, it was also thought that *N. xanthosoma* was the only neurocordulid with males having "cerci each with [a] ventromedial spine" (Needham et al. 2014:380).

5. However, Tennessen (2019) did not discuss *Platycordulia* as a possible separate genus.

6. *Rostrocordulia* is a *nomen nudum*, as per Garrison et al. (2006), which we presume resulted from Needham and Heywood (1929) publishing the name and saying that RS Hodges intended to write a description, but he failed to do so.

7. From Say (1839) onwards, including, but not limited to, Hagen (1861), Selys (1871, 1874), Morse (1895), Martin (1906), Muttkowski (1911, 1915), Needham and Heywood (1929), Davis (1933), Needham and Westfall (1955), Kormondy (1959), Walker (1966), Tennessen (1973, 1977), Donnelly (1991, 2003a, 2006b), May (1995), Paulson (2009, 2011), Abbott (2013, 2015), Needham et al. (2014), and our own foray into the taxonomic chaos through an ongoing morphological and genetic study of the genus.

8. We are currently undertaking such a study.

9. Species determinations were made primarily by MAP.

10. Published in error as OC 374215 in Patten and Smith-Patten (2013b).

11. Data from GH Bick's notes, which indicated this was an OMNH specimen. Unfortunately, although Bick wrote in his notes "OU collec returned," the specimen has not been relocated.

12. There are two specimens with possible later dates: "21 October 1974–10 June 1975" (SEMC 1349083) and "22 October 1974–11 June 1975" (SEMC 1349080).

13. Needham et al.'s (2014) report from Texas is an error.

14. The Dot Map Project (Donnelly 2004a,b,c) reported the species for Calhoun County, Arkansas (OC 247829), but that appears to be in error as no source has been found for that claim (*in litt*, John C Abbott, email dated 18 December 2017). As such, BS-P unconfirmed the OC record in November 2018.

15. Arkansas' records and localities increased much from 2017–2019 due to the work of Wade Boys. Missouri has also seen some increased survey effort in recent years, but it has not been as productive.

16. Also note that the source of the Dot Map Project (DMP; Donnelly 2004a,b,c) record for Riley County, Kansas, has not been discovered. Like the Calhoun County, Arkansas, DMP record, this one too may be in error (*in litt*, John C Abbott, email dated 18 December 2017).

17. The first record, especially; the second looks better for *Somatochlora ozarkensis* than *S. margarita*.

18. Note that Perkins (1980) listed the Cornell University Insect Collection (CUIC) as having two paratypes of *Somatochlora ozarkensis*. Jason J Dombroskie, CUIC collections manager, confirmed they have those specimens. However, it turned out that they were mislabeled as paratypes (*they were collected after the type description*); CUIC now has those specimens noted as not being true paratypes. Data are: Wilburton, Okla./ VI-10-1934/AE Pritchard//Somatochlora ozarkensis Bird/CU

Paratype No. 3020.1 ♂ AND Wilburton, Okla./VI-14-1934/ AE Pritchard//Somatochlora ozarkensis Bird/CU Paratype No. 3020.2 ♀

19. We have been unable to re-survey the type locality because it currently is private property, to which we have not been allowed access.

20. Kansas records of *Somatochlora ozarkensis* have long been problematic. Reports of the species fall within three ecoregions: Central Irregular Plains, Cross Timbers, and the Flint/Osage Hills. All three of the reports from the Central Irregular Plains are based on nymphal specimens (SEMC 1110492, 1110497, 11104508). Given the difficulty with identifying *S. ozarkensis* nymphs, we are inclined to consider these reports dubious. Of the two records from the Cross Timbers in Kansas, the most solid is an adult ♀ collected in Woodson County (Woodson State Fishing Lake [SFL], 15 June 1976, Huggins [probably DG] and PM Liechti, SEMC 1349085). A nymph (SEMC 1110469), identified as the species, was collected there on 18 July 1977, but we cannot verify the determination. A second site, near Chautauqua, Chautauqua County, produced a nymph that was reared (unclear if specimen date of 13 June 1984 is for collection or emergence). We could not confirm the id of the teneral ♀ because her ovipositor was bent, nor did we confirm the exuvia; however, we are inclined to think Huggins and Liechti made a correct determination. The Flint/Osage Hills also has two reports. The first, and the northernmost claim for the species, is from Riley County. It was reported by Donnelly (2004), the basis of which is unknown. The second, a ♂ photographed in poor lighting, is probably the species and was reported from Cowley County, just north of the Oklahoma border (Cowley SFL, 1 August 2010, TD Hibbitts, OC 322218).

21. Additional notes of feeding swarm behavior and composition in the region kindly were provided by TE Vogt (*in litt*). He described finding swarms (of 10–40 adults) along roads on broad ridgetops and often in young second growth. Encounters typically begin early evening but can be earlier on hotter days. Composition differed by ecoregions, with sites in the Missouri Ozarks having "almost exclusively *Somatochlora tenebrosa* with very few *S. ozarkensis* (0–3)" peppered in, whereas in the Arkansas Ouachitas, swarms consisted "predominantly of *S. linearis* and several *S. tenebrosa*, with a few *S. ozarkensis* (3–7)." He added that "very rarely were other odonates in the mix."

22. Estimates, some of which were published by Smith-Patten (2017a), were obtained through GIS analyses and from GeoCat (http://geocat.kew.org/).

23. Figured by GeoCat (http://geocat.kew.org/).

24. Mid-points of 195,652–329,751 km² and 288–352 km² equal 262,702 km² (i.e., about Colorado's size at 269,601 km²) and 320 km² (i.e., Lubbock, Texas, at 317 km² or Kansas City, Kansas, at 323 km²). Estimates, some of which were published by Smith-Patten (2017a), were obtained through GIS analyses and from GeoCat (http://geocat.kew.org/).

25. Medium to high threat as per NatureServe's threat assessment calculator.

26. As of February 2020, NatureServe has ranks as: Arkansas (S1), Kansas (S1), Missouri (S2S3), but note that their system has not yet been fully updated to reflect changes made by Smith-Patten (2017a), including Oklahoma still showing as "S4?".

27. The idea of the species being at sites with sphagnum moss is that of JC Abbott (*pers comm*).

28. The range of variation in cercus shape in either species is unknown given that the type description for the Texas Emerald (Donnelly 1962) compared only two, 1♂ and 1♀, Ozark Emeralds for diagnosis.

29. He began patrolling the spot within a few seconds after a ♀ Mocha Emerald, *Somatochlora linearis*, who had been ovipositing on moss and rocks nearby, flew away. Shortly after he began patrolling, another ♂ came in, tussled, and one ♂ flew off, leaving the one ♂ that was collected (SP 2905, taken at 10:30 am, 16 August 2019, 74° F, no wind or cloud cover; BS-P). A female was captured and photographed in the general area a couple of days later (19 August, J Burns, DA, OC 502135).

30. Interestingly, a survey at Pine Mountain Spring two weeks later, on 15 August, although well within the known flight season of the species, did not produce a record. Two males were encountered the next day along an unnamed creek in the Ouachita NF about 10 km NNW of Hochatown, McCurtain County (one as SP 2905; see text above, in endnote and species account, for details).

31. SEMC has a series of specimens (1349090–1349093, 1349098) that have date runs starting on 12 May and running into mid- to late-June, for example, "12 May–11 June 1982" and "12 May–21 June 1982." The only specimens for the species without date runs are those dating from 19 July (SEMC 1349086–1349088, 1349096–1349097); however, we chose to use 11 June as the possible early date because that is in line with regional dates.

32. According to GH Bick's notes, the exuvia was originally identified by MJ Westfall in 1954 (Figure 18.26). Presumably Westfall used the new larval key that had been prepared for the upcoming book, *A Manual of the dragonflies of North America (Anisoptera)*, which he and JG Needham were to publish the following year (published in 1955, copyrighted in 1954). That key placed *Neurocordulia molesta* as the first step, "Pyramidal horn on front of head" (p. 355). That first step still holds today (Needham et al. 2014) and so we ought to feel comfortable that Westfall identified the exuvia correctly (as did C Cook when he confirmed the identification in 1955). Westfall's determination is probably the reason Oklahoma was considered part of *N. molesta*'s range in the 1955 edition of the *Manual*. The last we know of this specimen was that Bick said he sent it to SW Dunkle in 1974 to be deposited at FSCA. We were not able to fully examine the FSCA fluid collection.

33. The nymph was determined by Bick in 1955 and then sent to FSCA in 1977 (Bick noted that it was "not seen by Dunkle"; Figure 18.26); we have not relocated it.

34. SEMC 1350202–3, 1350205 are listed as adults that were collected in early April 1976. Farther south and east, for example in Georgia, the species flies this early (Paulson 2011, OC), but April is suspiciously early for the species in this region. We suggest that these specimens be re-examined and the label data be confirmed before the dates are accepted. We therefore, included above the next earliest record (11 May).

35. *Neurocordulia xanthosoma*'s habitat has been described elsewhere in a variety of ways. Dunkle (2000:168) said: "small streams to large rivers with mud bottomed pools. Also lakes with clear to muddy water and wooded banks." Paulson (2011:384) said: "rivers with slow to moderate current in open or wooded landscape." Abbott (2015:215) said: "medium-sized turbid rivers and streams with strong currents." We have not seen the species on any waterway that we would call a "river," nor one with strong currents, but otherwise these descriptions fit various locales within Oklahoma that we have encountered the species. We generally consider *N. xanthosoma* (adults) to be a lotic species, but as with Dunkle's description, we have encountered it multiple times at small, clear-water lakes (e.g., Prague Lake, Lincoln County; Sportsman Lake, Seminole County; Langston Lake, Logan County; Perry Lake, Noble County) with at least some forested edge.

36. Also of note is that Byers (1937) indicated that a ♀ specimen collected by TH Hubbell in 1926 at the Wichita Mountains was collected on 10 July 1926, but that is likely an error, as Hubbell appears to only have been collecting in the Wichitas in June 1926.

37. Bick called these nymphs, which he collected on 22 June 1950 at "Braggs" (although likely the OSU field station), "very immature" (Figure 18.29). They were determined to be *Neurocordulia obsoleta* by C Cook in 1955. All were discarded by Bick in 1977. The nymph collected on 13 June 1950 was also determined by C Cook in 1955 as *N. obsoleta*.

38. Bick collected his "last instar exuvium" on 3 July 1950 on the banks of the Verdigris River at its juncture with the Arkansas River. Similarly to Williams (1976) describing exuviae being found on trees, Bick (1951:179) described his exuvia as "well attached to the trunk of a willow and was about 20 feet above the water level and four feet above the ground."

39. Bick seemed somewhat surprised by his encounters with the species. He said "in spite of the fact that my collections were made almost daily and at all possible times, including twilight and night collections, only three individuals of *Platycordulia* [*Neurocordulia xanthosoma*] were obtained" (Bick 1951:179).

40. There is an *Epitheca cynosura* specimen (we confirmed the id) at OSU that was collected in Stillwater, Payne County, by WR Duffer. The date on the label indicates it was collected on 27 August 1965. We decided not to include this specimen as part of Oklahoma's documented flight season because it is from so late in the year. We wonder if perhaps it was mislabeled and was actually collected in April, which would make *much* (*much*, *much*) more sense for the species' phenology.

41. We are willing to abide as early and late dates for Kansas the 11 March and 27 May SEMC records (1367145 and 1367153), both of which are exuvia, but the 23 February (SEMC 1349137, adult, and 1367151, exuvia) and the 19 September (SEMC 137177, exuvia) are too far off the mark to readily accept. We suggest that the SEMC specimens labeled as this species be re-examined for confirmation of date and species determination.

42. Early and late dates come from SEMC 1349100, 1349117–1349118. However, SEMC 1349101, has a date as "21 Apr–1 June 1978."

## CHAPTER 19

LIBELLULIDAE: SKIMMERS

Two taxonomic families dominate, in terms of species richness, the odonate fauna of Oklahoma, accounting for well over half of the species recorded: the damselfly family Coenagrionidae, the pond damsels (44 species, 25.4% of the state's richness), and the dragonfly family Libellulidae, the skimmers (58 species, 33.5% of the state's richness). Libellulidae also is diverse from the standpoint of the number of genera included in the family: 21 are recorded in Oklahoma, some represented by a single species and others by many (Figure 19.1). That diversity is well beyond that of the Gomphidae, the clubtails, which has the next highest count of genera in Oklahoma, with just 12.

Two genera dominate (numerically) Oklahoma's species richness in the Libellulidae: *Libellula*, the king skimmers, and *Sympetrum*, the meadowhawks. As such, we provide separate introductory summaries for *Libellula* and its near relatives, *Plathemis*, *Ladona*, and *Orthemis* (i.e., subfamily Libellulinae) and for *Sympetrum*. But first, we treat here all other species in the Libellulidae; even then, we do so with convenient groupings that may or may not correspond to named subfamilies (see Ware et al. 2007, Pilgrim and von Dohlen 2008[1]).

One such artificial grouping is of the pennants, artificial because inclusion of one species, the Marl Pennant (*Macrodiplax balteata*), a species with odd wing venation, does not fit phylogenetically into a grouping that otherwise contains *Perithemis* (the amberwings) and the *Brachymesia* and *Celithemis* pennants, but behaviorally the Marl Pennant fits nicely. As a group, artificial or otherwise, the pennants perch conspicuously, obelisk frequently, and often have flashy wing patterns. These species are characteristic of mid-summer (Figure 19.2), except for the aptly named *Celithemis verna* (taken from "vernal," as in the spring season). That species, otherwise known as the Double-ringed Pennant, was described to science from an Oklahoma type locality. Also of note with *Celithemis* in general is that there are some differences in perching height between the four Oklahoma species. All face the wind while perched atop vegetation (grasses, shrubs, cattails), but *C. verna* almost always perches below knee height whereas only *C. fasciata*

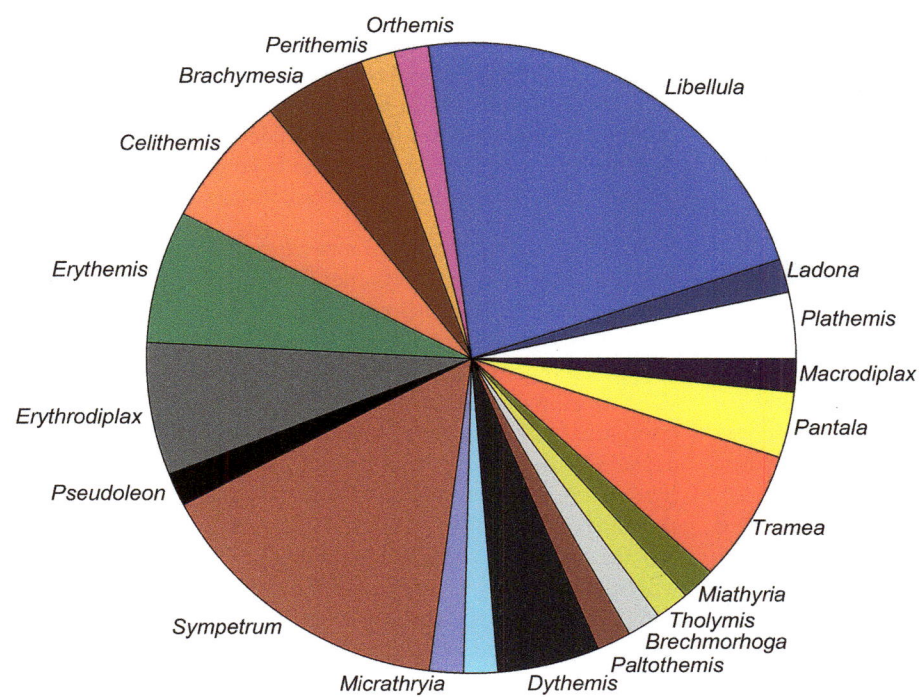

Figure 19.1 Breakdown of species richness by genus in Oklahoma for the Libellulidae.

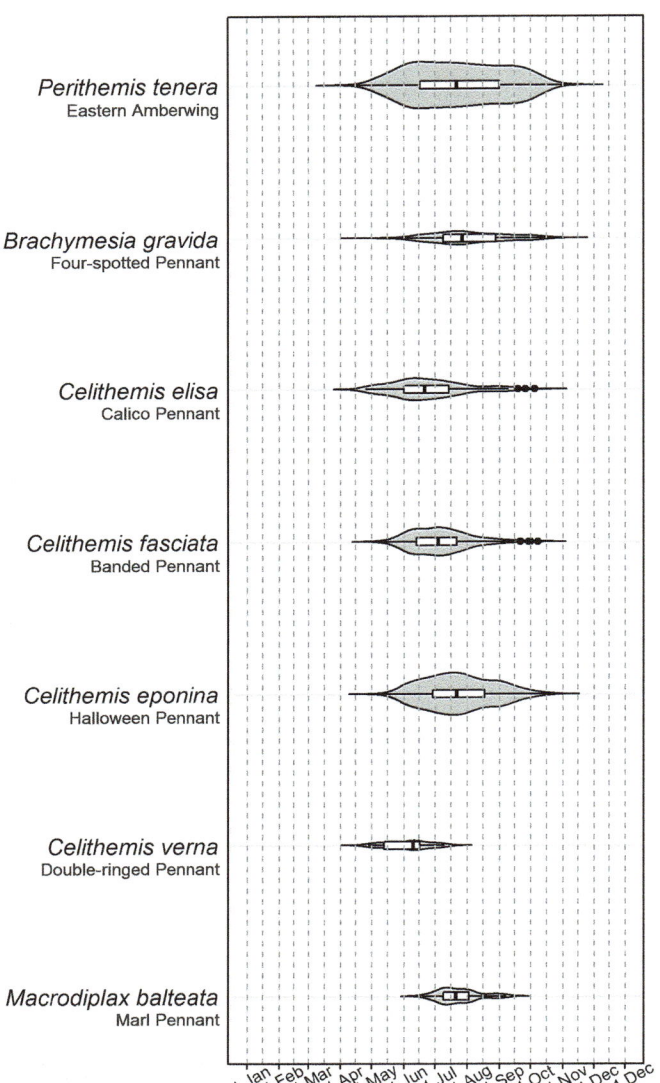

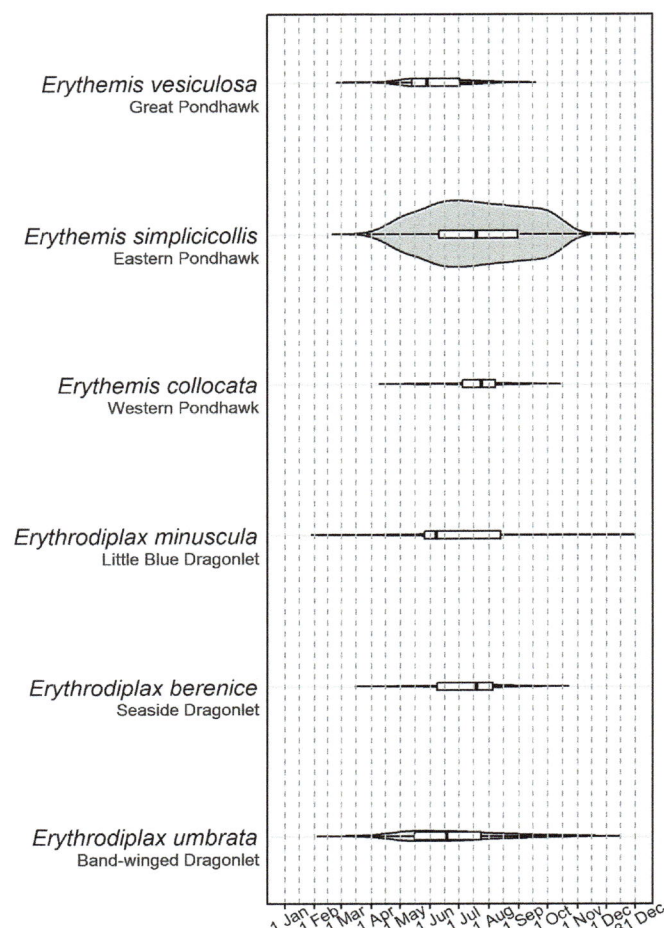

Figure 19.2 Seasonality in Oklahoma of the Eastern Amberwing (*Perithemis tenera*) and the various dragonflies known as pennants (*Celithemis, Brachymesia,* and *Macrodiplax*). Two additional species, the Red-tailed Pennant (*B. furcata*) and the Tawny Pennant (*B. herbida*) have been recorded, respectively, once or twice (4–5 October and, apparently, 23 July) and just once (1 October).

Figure 19.3 Seasonality in Oklahoma of the pondhawks (*Erythemis*) and dragonlets (*Erythrodiplax*). Two additional species, the Black Pondhawk (*Erythemis attala*) and Plateau Dragonlet (*Erythrodiplax basifusca*), have been recorded once each, on 6 October and 21 October, respectively. Another species that more-or-less fits in this group, *Pseudoleon superbus*, the Filagree Skimmer, has been recorded three times from 17 September–7 October.

and *C. eponina* perch in treetops, with *C. elisa* somewhere between those extremes. Fortunately, the two tree perchers have distinctive enough wings that even a photo from below will suffice for identification.

The pondhawks (*Erythemis*) are distributed chiefly in the Neotropics, but four species have reached Oklahoma, only one of which, the Eastern Pondhawk (*E. simplicicollis*), is common. Two others occur routinely, but are limited seasonally (Figure 19.3) or geographically. The dragonlets (*Erythrodiplax*), too, are chiefly a Neotropical group with four species recorded in the state, but only one, the boom-or-bust

Band-winged Dragonlet (*E. umbrata*), is recorded regularly. Another, the Seaside Dragonlet (*E. berenice*), was discovered in the state in 2014, although the site where it was initially found in southwestern Oklahoma supports a population.

The dashers (*Micrathyria* and *Pachydiplax*) and setwings (*Dythemis*), with clubskimmers (*Brechmorhoga*) thrown in for good measure, constitute another "group." Behaviorally, the dashers and setwings are united as sit-and-wait predators that perch with wings jutted forward and abdomens jutted skyward. The clubskimmers could not be more different. No skimmers in the United States act like clubskimmers, which relentlessly cruise riffles in rocky creeks, flying low and back-and-forth on short to medium beats, behavior that may bring to mind a cruiser (*Macromia* and *Didymops*) before it does a skimmer. Several species in this "group" are rather late-season fliers (Figure 19.4), and one, the ubiquitous Blue

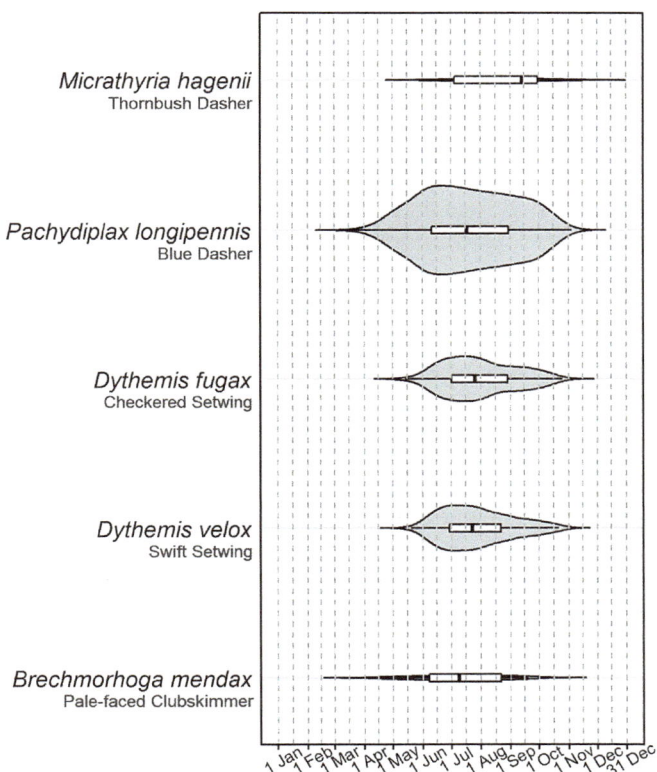

Figure 19.4 Seasonality in Oklahoma of the dashers (*Micrathyria* and *Pachydiplax*), setwings (*Dythemis*), and clubskimmers (*Brechmorhoga*). One other species, the Black Setwing (*D. nigrescens*), has occurred twice, on 9 May and 29 May.

Dasher (*Pachydiplax longipennis*), dominates relative abundance given that it vies with the Widow Skimmer (*Libellula luctuosa*) and Eastern Pondhawk for the title of most abundant libellulid in Oklahoma.

The gliders—taken in the broad sense (i.e., subfamily Trameinae)—form the final group. The seven species in this group, in three genera, seem to fly tirelessly, aided by the broadened triangular hindwing. Two species, the Hyacinth Glider (*Miathyria marcella*) and Striped Saddlebags (*Tramea calverti*), seem to be effectively late-season (and post-breeding?) wanderers from the south (Figure 19.5), with each occurring at times in sizable numbers (i.e., dozens), although since 2015 each has occurred in small numbers in spring

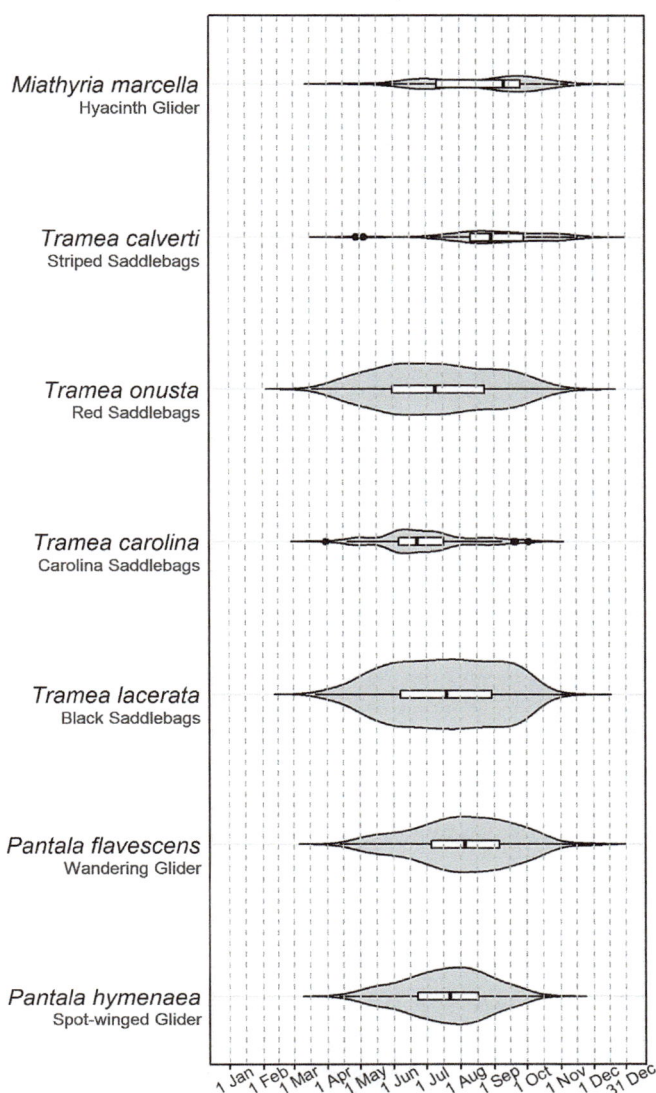

Figure 19.5 Seasonality in Oklahoma of the gliders, subfamily Trameinae (*Miathyria*, *Tramea*, and *Pantala*).

and summer, hinting at successful breeding and perhaps pointing toward range expansion northward. The other five species are common, with only the Carolina Saddlebags (*T. carolina*) other than widespread.

LIBELLULINAE (*LIBELLULA* AND NEAR RELATIVES)

In Oklahoma, the dragonfly subfamily Libellulinae is represented by 17 species in 4 genera, *Plathemis* (the whitetails; 2 species), *Ladona* (the corporals; one species), *Libellula* (the king skimmers; 13 species), and *Orthemis* (the tropical king skimmers; 1 species documented). Many of these species are among the most conspicuous and, thus, most frequently encountered odonates in the state. Two, the Common Whitetail (*Plathemis lydia*) and Widow Skimmer (*Libellula luctuosa*), are ubiquitous (Figure 19.6)—it is a rare summer survey that does not record at least of a few of each—to the point of being emblematic of the state's mid-summer dragonfly fauna.

Together, species in the Libellulinae occur nearly the whole of the odonate flight season in the state, from two quintessential early season species, the Blue Corporal (*Ladona deplanata*) and Painted Skimmer (*Libellula semifasciata*), to several of quintessential late-season species, such as the Flame Skimmer (*L. saturata*). Still, most species peak in mid-summer (Figures 19.7 and 19.8). Species also span the lotic–lentic habitat gradient, with some, such as the Spangled (*L. cyanea*) and Golden-winged Skimmers (*L. auripennis*), essentially restricted to ponds and small lakes, whereas others, such as the Desert Whitetail (*Plathemis subornata*) and the Neon Skimmer (*L. croceipennis*), essentially are restricted to streams and creeks (Figure 19.7).

Of note, two of the state's "lookalike" species, in that males of each are blue bodied with the pterostigma bright white, segregate themselves not only geographically—the Spangled Skimmer (*Libellula cyanea*) is an eastern species, whereas the Comanche Skimmer (*L. comanche*) is a western species (Figure 19.47)—but ecologically and seasonally, too, with the Spangled a lentic species that occurs almost two months earlier than the lotic Comanche Skimmer (Figures 19.7 and 19.8B). Curiously, another broadly similar species, the Yellow-sided Skimmer (*L. flavida*), fits nicely in between these two "lookalikes," even if it is nearly confined to the southeast, where it strongly favors seeps and seep-runs but occurs also in shallow marshes dominated by spikerush (*Eleocharis*) and sedge (*Carex*).

Other broad "lookalikes" are differentiated ecological, too. As noted above, the Neon Skimmer is strictly lotic in Oklahoma, whereas the Golden-winged Skimmer is nearly strictly lentic and the Flame Skimmer is somewhere in between[2] (Figure 19.7). Even if these species are not so differentiated seasonally (Figure 19.8C), they are more-or-less geographically, with a species of western (Flame), central (Neon), and eastern (Golden-winged; Figure 19.26) Oklahoma. The Painted Skimmer also is an orange species, but it has "picture wings," flies early, and is very nearly confined to shaded ponds in the Ouachita highlands. It is rare enough in the state, which is at the western edge of the species' geographic distribution, that years can go by without a record.

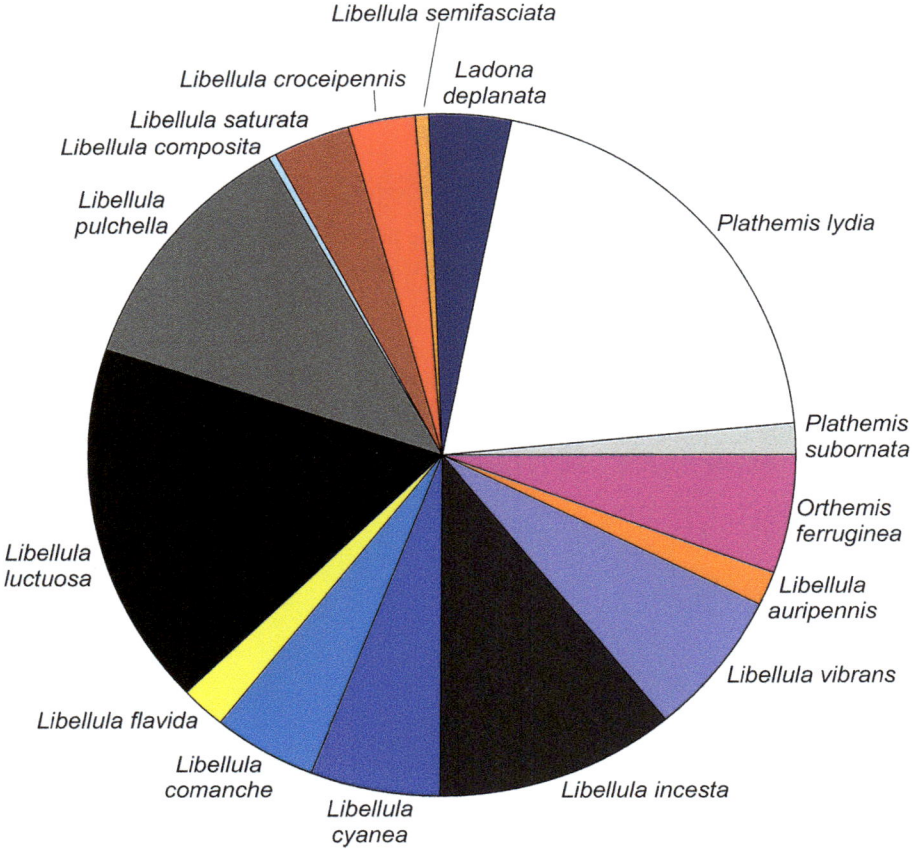

Figure 19.6 **Relative commonness in Oklahoma of the 16 regularly occurring skimmer species in the subfamily Libellulinae. Two species, the Common Whitetail (*Plathemis lydia*) and Widow Skimmer (*Libellula luctuosa*) are detected most often and are most widespread, with both being recorded in each of Oklahoma's 77 counties.**

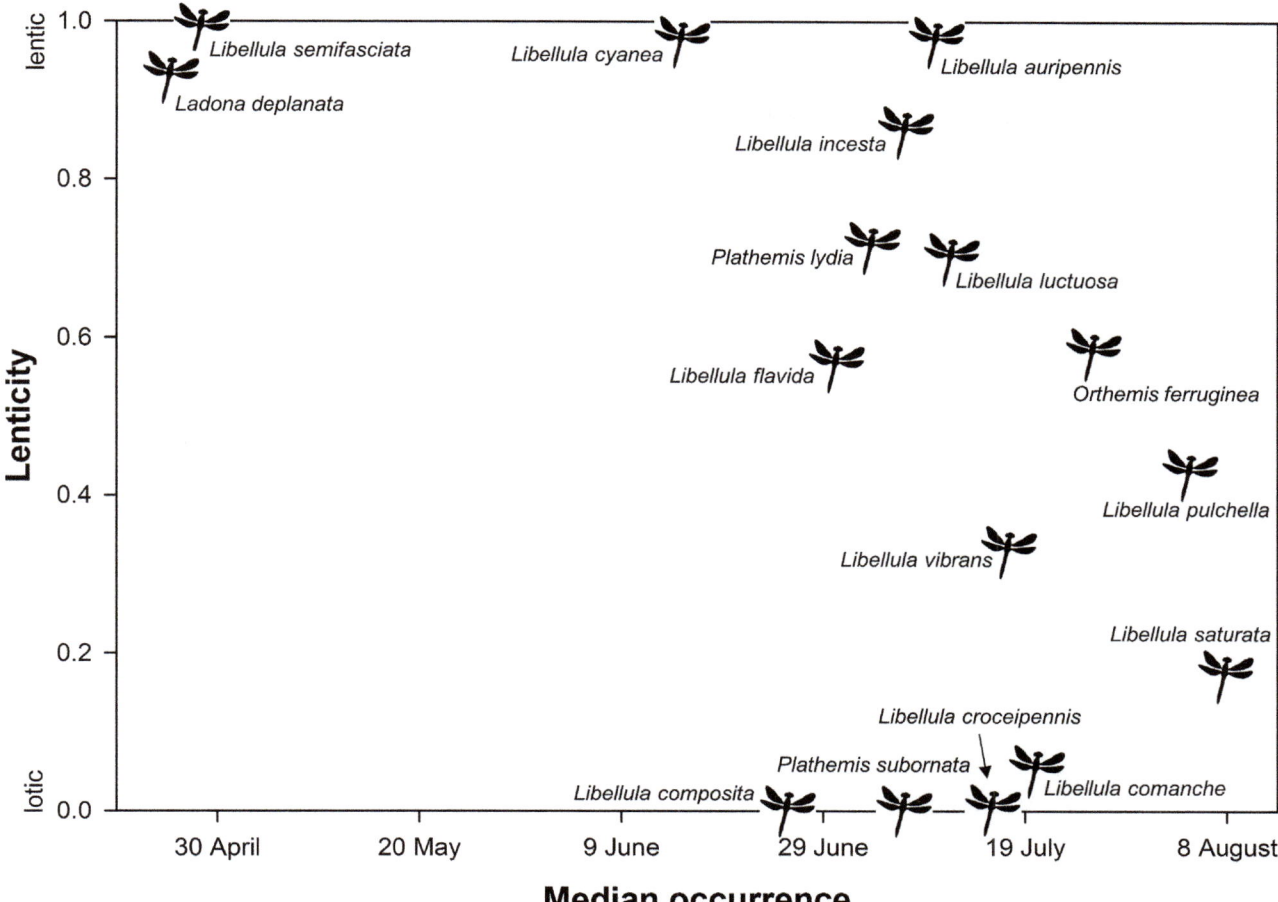

Figure 19.7 A scatter plot of the midpoint of seasonality against lenticity for 16 Oklahoma species in the subfamily Libellulinae (a 17th species, the Hoary Skimmer [*Libellula nodisticta*], is excluded because it has been recorded just once). Note that the two strictly early season species, the Blue Corporal (*Ladona deplanata*) and Painted Skimmer (*Libellula semifasciata*) are nearly strictly lentic, whereas several of the strictly or near-strictly lotic species occur much later in the season.

SYMPETRUM

Nine species of meadowhawks (*Sympetrum* sp.) have been recorded in Oklahoma, but only four of these species occur as regular breeders, which we infer by presence through the summer (Figure 19.9) and numerous records of tenerals or breeding activity. One species, the Band-winged Meadowhawk (*S. semicinctum*), is confined to westernmost Oklahoma, where most records come from the panhandle. The other three species are widespread, although the Blue-faced Meadowhawk (*S. ambiguum*) is chiefly an eastern species, with records and abundance decreasing from east to west.

The abundant Variegated Meadowhawk (*S. corruptum*), and the poorly named Autumn Meadowhawk (*S. vicinum*) exhibit a distinct bimodal occurrence pattern. The Variegated Meadowhawk appears to migrate through the state, in that peak numbers occur in spring (April and May) and autumn (especially, with the species abundant from mid-August to mid-October). The Autumn and Blue-faced Meadowhawks, by contrast, are found most commonly as adults in autumn (September and October), but there are

numerous records as early as May, typically of tenerals or immature individuals. It appears that these species emerge in early summer, "disappear" into the woods, and "reappear" later in the season to breed. Although this description sounds a bit tongue in cheek, there may be some truth in it if these species go into estivation or siccation, that is, decrease their activity or delay their adult development during the hotter times of the year by essentially resting in cooler places until the proper weather conditions set in. *S. vicinum* has been reported to do so (Corbet 1999:636) and we see no reason why *S. ambiguum* would not as well (see *S. ambiguum*'s species account for further discussion).

The remaining five species of *Sympetrum* recorded in the state are the true autumn meadowhawks (Figure 19.10) for Oklahoma: they migrate to the state only when winter weather hits areas farther north (or west), where these species normally occur. Two species, the Cherry-faced Meadowhawk (*S. internum*) and Saffron-winged Meadowhawk (*S. costiferum*), are northerly species that stage occasional southward incursions into the state. Such incursions have been known for

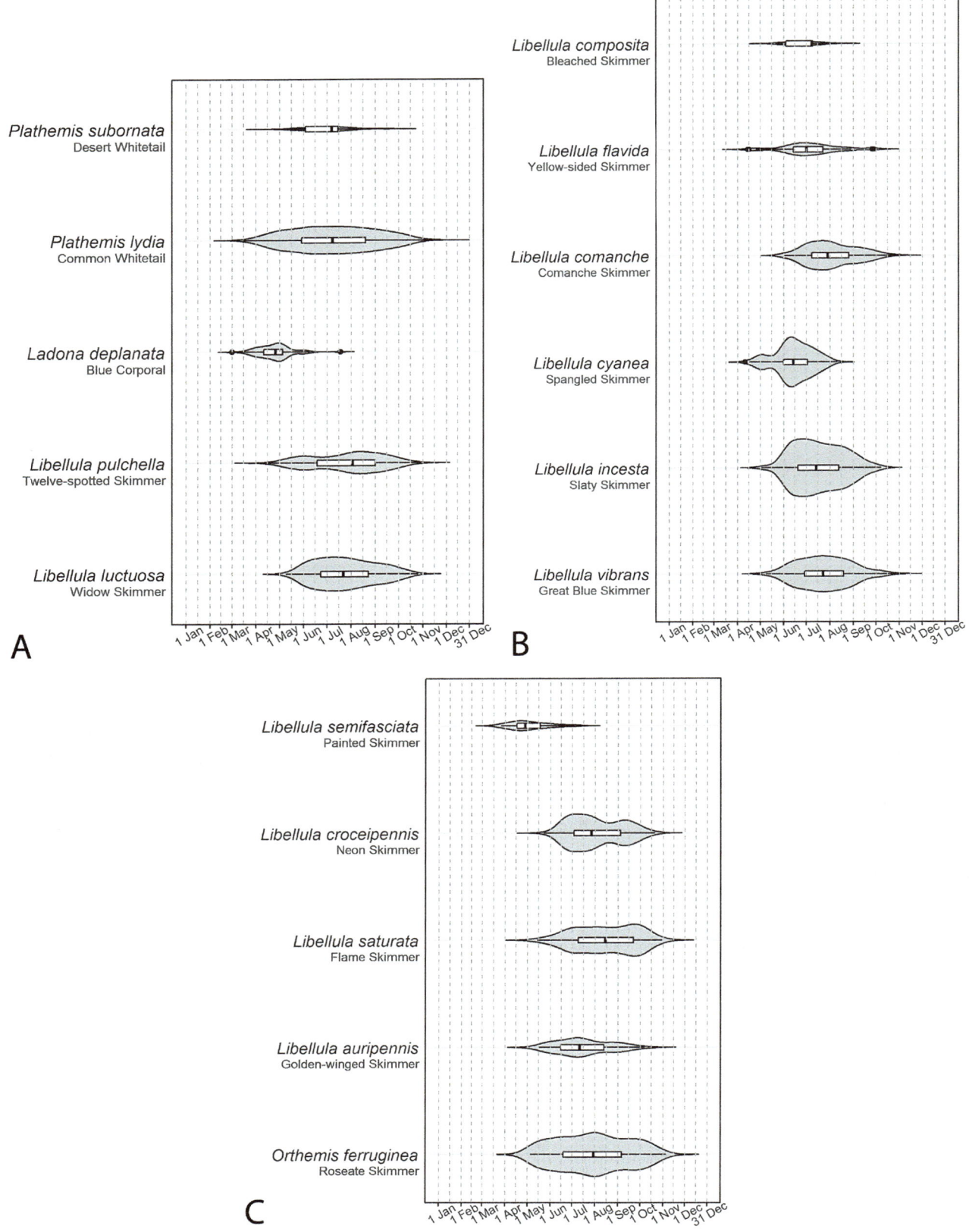

Figure 19.8 Flight seasons in Oklahoma of species in the subfamily Libellulinae, grouped by A) the whitetails (*Plathemis*), corporals (*Ladona*), and gray, "picture-winged" king skimmers (*Libellula*), B) the predominantly blue (or blackish, in one case) king skimmers, and C) the orange or reddish king skimmers and tropical king skimmers (*Orthemis*).

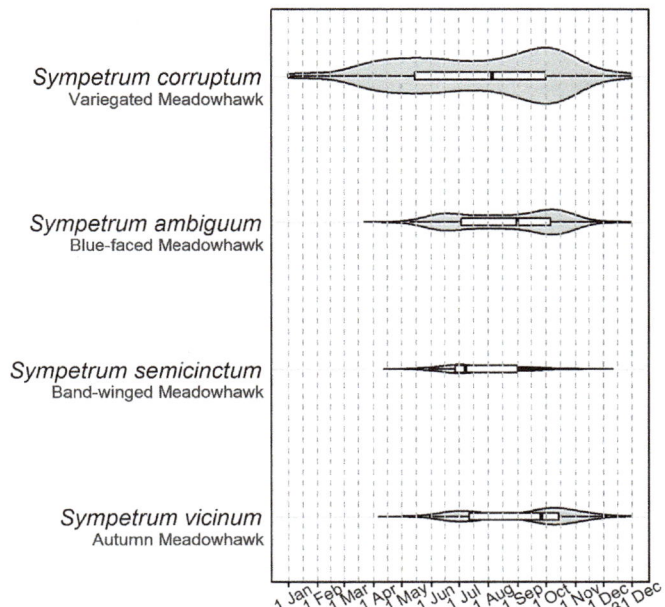

Figure 19.9 Seasonality of meadowhawks (*Sympetrum*) common in Oklahoma. Note the bimodal patterns evident in four of the species, patterns generated, we believe, by different processes.

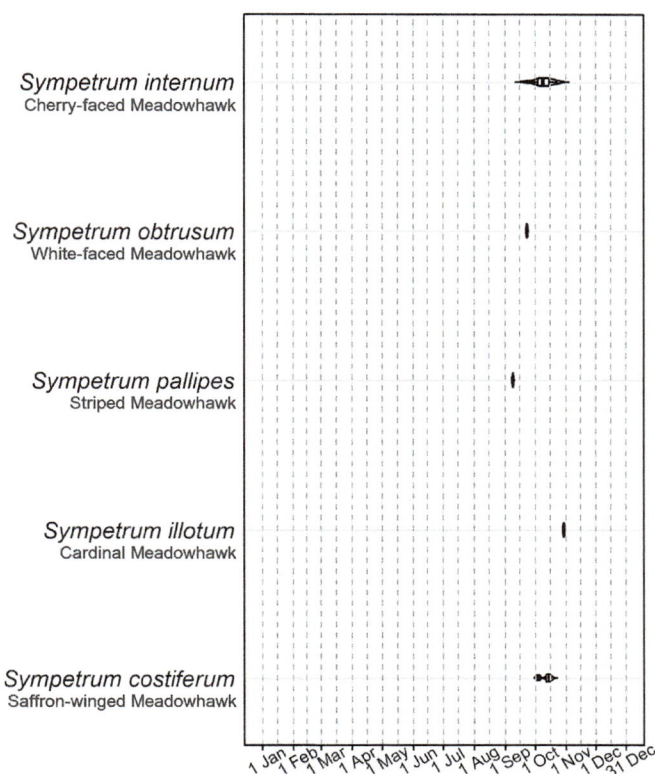

Figure 19.10 Seasonality of meadowhawks (*Sympetrum*) that are rare in Oklahoma, three of which have been recorded once each.

the Cherry-faced Meadowhawk since the 1950s (possibly as early as 1877), and recent "invasion years" of this species suggest to us that other northerly species co-occur with it. For instance, Oklahoma's only White-faced Meadowhawk (*S. obtrusum*) appeared during the 2012 Cherry-faced incursion, and multiple Saffron-winged Meadowhawks were detected during the 2012 and 2014 Cherry-faced incursions. Other species, such as the Ruby (*S. rubicundulum*) and Black (*S. danae*) Meadowhawks, could conceivably join future incursions. An understanding of climatic or ecological factors that drive southward incursions of the Cherry-faced Meadowhawk would help us understand factors that contribute to the appearance of other northerly meadowhawks.

Lastly, we cannot stress enough the importance in Oklahoma and the region of collecting meadowhawk specimens, at least of certain species. Species of the RuWhiChe [ru-WEE-chee] complex are an especially good argument in favor of collecting. RuWhiChes—the Ruby, White-faced, and Cherry-faced Meadowhawks—are closely related and phenotypically similar (Pilgrim and Von Dohlen 2007), so much so that, without specimens, reports often remain questionable at best, and unconfirmed at worst. To further complicate matters, these species are thought to interbreed (e.g., TW Donnelly 1997, *in litt*). The only way we will learn the extent of hybridization and what the status of RuWhiChes are in Oklahoma and the region is by having verifiable specimens at hand.

PLATHEMIS SUBORNATA HAGEN, 1861 – DESERT WHITETAIL

In Oklahoma, the Desert Whitetail is a species of clear, shallow, sandy-bottomed streams, a scarce and declining habitat in the relatively arid western part of Oklahoma. We know of 19 localities at which the species occurs, localities that fall primarily within the High Plains and Southwestern Tablelands ecoregions (Figure 19.12). The biogeographic regions it inhabits in the state make perfect sense for the species, because it is a species of the arid southwest whose range just extends into Oklahoma (Paulson 2009). It was thought to be confined to northwestern Oklahoma and the panhandle until we found a single ♂ at Lake Hall, Harmon County, in southwestern Oklahoma (19 June 2015, OC 432152). That record upped the number of counties the species has been recorded in to a mere seven (Figure 19.13).

The species was first discovered in the state on 8 July 1926 at 3 mi N of Kenton, Cimarron County (♂, UMMZ), by TH Hubbell, the first collector known to have explored the Oklahoma panhandle. Bird (1932) apparently was unaware of this collection because he did not list the species in the state's first checklist.

Since its discovery in the state the species has not been recorded often. AE Pritchard and W Chiles collected at least 27 specimens (EMEC, OSU, USNM; Bick and Bick

1957) when they were in the panhandle for roughly a month in June and July 1933. GH Bick and LE Hornuff had the species only twice—1♂, 1 August 1968, 3 mi W of Slapout, Beaver County (Hornuff only, FSCA), and 3♂ and 1♀ in the Black Mesa area, Cimarron County, on 5 August 1970 (2♂ at FSCA). The next record of the Desert Whitetail in Oklahoma came when WJ Matthews collected four nymphs at "Old Ranch Spring," Cimarron County, on 28 May 1982 (BS-P confirmed id, OMNH 1963).

Forty-two years after Bick and Hornuff collected their last adults we collected a somewhat teneral ♂ at Doby Springs, Harper County, on 12 May 2012 (SP 237). That teneral was the only Desert Whitetail we saw that day among 15 Common Whitetails (*Plathemis lydia*). We encountered the species again a few weeks later, this time at Lake Evans Chambers, Beaver County, on 2 June 2012. We assume that we hit peak emergence as we had 44 adults flying (40♂, 4♀; 1♂ SP 256) along with 22 Commons. Our next largest numbers seen flying were 25♂ at Doby Springs on 12 July 2014 and 25♂ and 1♀ the next day at Ellis County WMA (OC 424599, 424607, respectively). The numbers of Common Whitetails were a little more equal to the Desert Whitetails during

Figure 19.11 *Plathemis subornata*, the Desert Whitetail, A) ♂, Doby Springs, Harper County, 14 August 2013 (©Bill Carrell, OC 402815). B) ♀, Thurmond Ranch, Roger Mills County, 29 June 2017 (©Brenda D Smith, OC 465931). Unlike its pervasive congener, this species occurs in Oklahoma solely along small, clear creeks in the panhandle and in some of the westernmost counties.

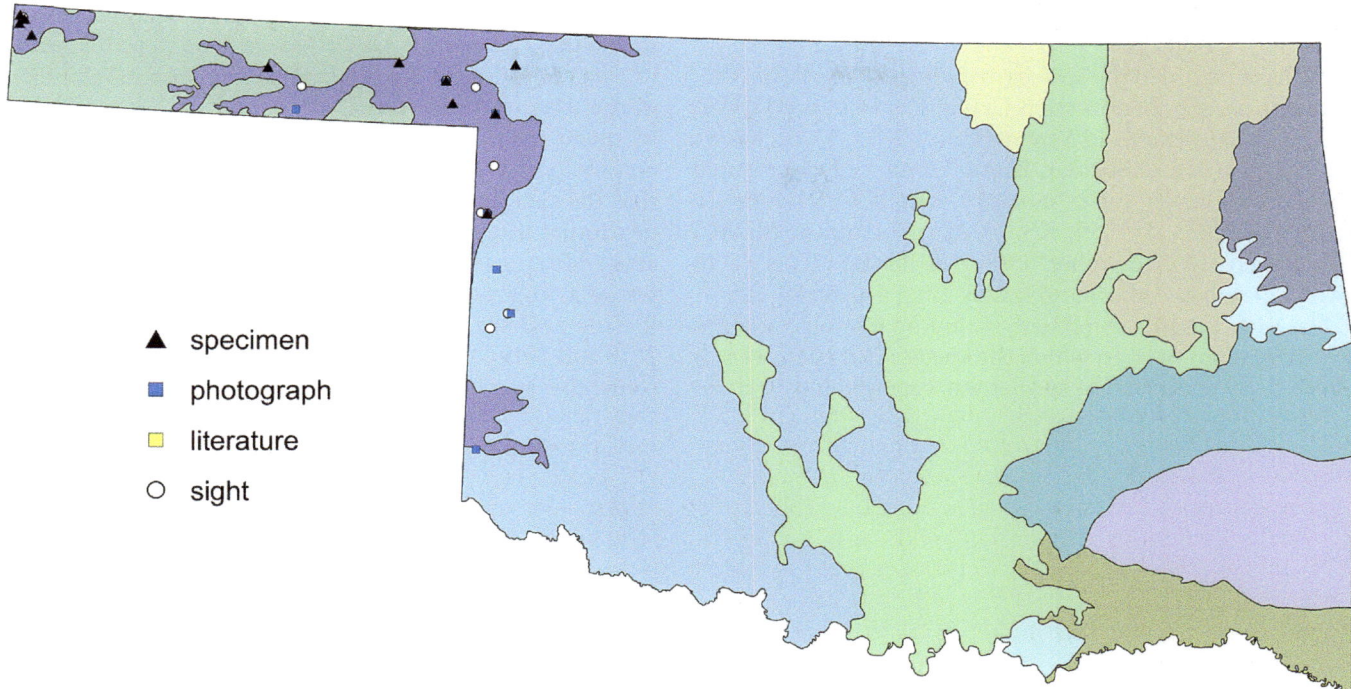

Figure 19.12 Documentation level of supporting records for *Plathemis subornata*, the Desert Whitetail, in Oklahoma. Levels include specimen, archived photograph, or sight record. Although sight records lack documentation, only those we feel are credible are plotted.

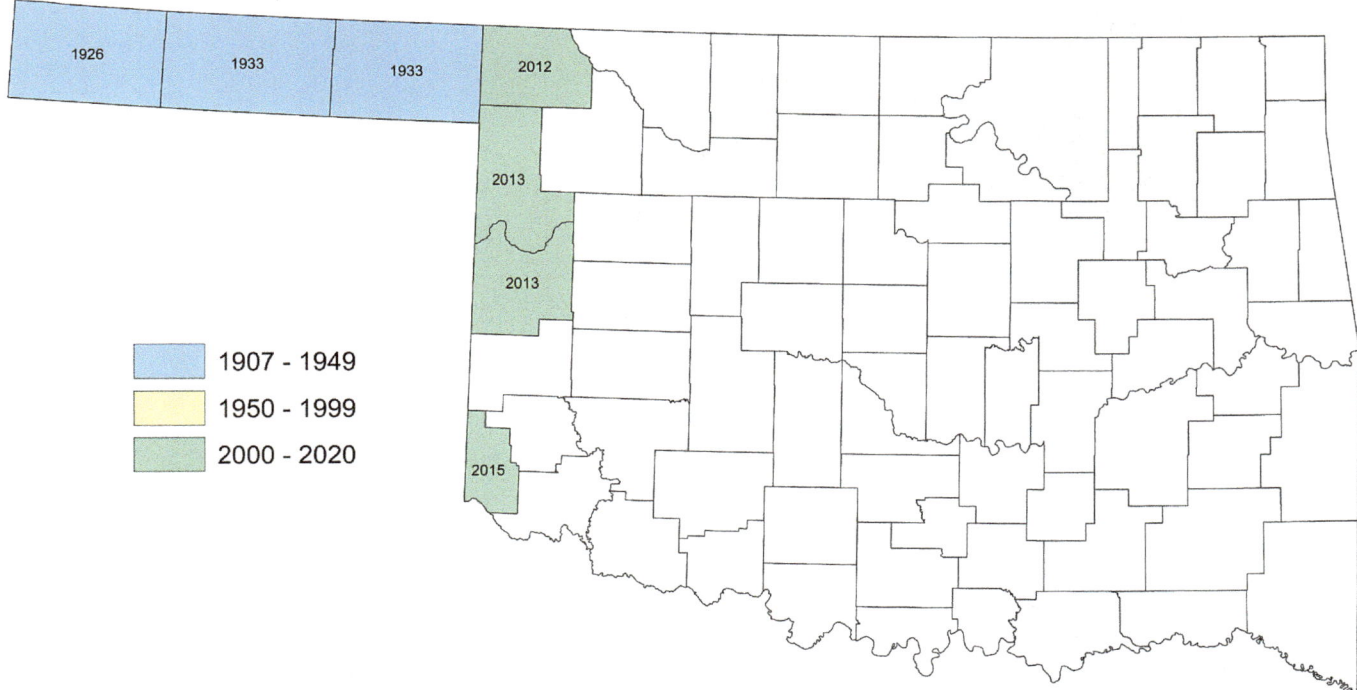

Figure 19.13 Counties in which *Plathemis subornata*, the Desert Whitetail, are known to occur in Oklahoma. Year within the county is that in which the species was first reported.

those days: on 12 July we noted 26 Common Whitetail, and on the 13th we had 18.

We have recorded the species as early as 24 April (in 2016, an immature ♂ at Commission Creek, Ellis County WMA, SP 1873) and as late as 15 September (in 2012, 3♂ on Kiowa Creek, 13 km NE of Slapout, Beaver County). Documented regional seasonality varies much by state (CO: 21 June–30 July, OC 482843, 501059; KS: 31 May–5 August, SEMC; NM: 6 May–13 September, OC 438030, 315127; TX: 26 April–3 October, Abbott and Lasley 2019, OC 504672).

The early records combined, including nymphs, account for a mere eight days on which the species had been seen or collected. By contrast, the species was encountered 30 times between 2012 and 2017, including on 8 separate occasions in just the 2 years of 2015–2016. It may be that this increase in records indicates an increase in the species' abundance in the state or it may simply be a result of our opportunity to spend more time in the panhandle than the esteemed entomologists who preceded us. The possibility still stands that the species is a relatively recent colonizer that was just beginning to take hold in the state in the earlier collecting eras and is now probably established. That said, the species' status at this point is difficult to discern because of conflicting evidence (such as there being no records of it from 2018 and 2019). We chose to error on the side of caution and refine this species' conservation rank to S2S3.

PLATHEMIS LYDIA (DRURY, 1773) – COMMON WHITETAIL

The Common Whitetail truly is a common species in Oklahoma. We have compiled >1,600 records for the species across all 77 Oklahoma counties. It is routinely seen 30 or 40 at a time. In the Bick/Hornuff era "*P.* [*Plathemis*] *lydia* was the most abundant and widely distributed odonate in Oklahoma." (Bick and Bick 1957:7). We imagine that the Bicks meant *P. lydia* was the most abundant and widespread anisopteran (i.e., excluding Zygoptera) in the state. Regardless, the species is the most common showy libellulid (of the Libellulinae) in the state (Figure 19.1), although it is not always so in a given part of Oklahoma, for example, *Libellula luctuosa*, the Widow Skimmer, is often the most common species of the Libellulinae in eastern Oklahoma.

The Common Whitetail was first collected in the state by F Collins in 1907 (2–6 August, 9♂, 11♀, near Wister, Le Flore County, 2♂, 4♀ at UMMZ; and 1♂, 2♀ at Henryetta,

Okmulgee County, no coll. date, specimens not relocated; Williamson 1914b). The species is known to fly from mid-March to mid-December (15 March 2012, 10 km SE of Idabel, McCurtain County, 1 imm ♂, BAH, OC 374113; 10 December 2016, Oxley Nature Center, Tulsa County, 1♂, BC, OC 458380) and it appears to fly straight through the season, by which we mean that we have accrued no additional evidence statewide to support a notion of bimodal seasonality that we had reported earlier for northeastern Oklahoma (Smith-Patten et al. 2007).

The Oklahoma early and late dates fit right into the date range for Texas of 7 February to 15 December (Abbott and Lasley 2019) and are similar to other neighboring states (AR: 19 March–19 November, OC 374132, Harp and Rickett 1977; CO: 16 May–2 October, OC 431075, 427337; KS: 17 April–2 October, early date is our unpublished sight record of 1 imm

Figure 19.14 *Plathemis lydia*, the Common Whitetail, A) ♂, Pontotoc Ridge Preserve, Pontotoc County, 15 June 2014 (©Bill Carrell, OC 423140). B) immature ♂, Lake Perry, Noble County, 21 May 2016 (©Bill Dobbins, OC 445135). C) ♀, Cedar Creek 7 km E of Lenapah, Nowata County, 2 July 2014 (©James W Arterburn, OC 424179).

♂ and 4♀ at Mined Lands WA, Cherokee County, in 2003, late date is SEMC 1365869; MO: 2 April–13 October, OC 374253, Trial 2005; NM: 4 May–27 September, OC 374711, 427211).

Bick and Bick (1958) found *Plathemis lydia* to be associated with a pond and spring (64 and 34, respectively, of 126 total adults) at their study area along Cowan Creek, Marshall County. Nymphs were most numerous at the pond, with only one at the spring, and a large number at an impoundment (117 and 51, respectively for pond and impoundment).

These findings indicate that the species is lentic, with little habitat specificity, which is congruent with our experience. Although they did not study the species throughout its full range in Oklahoma nor during its full flight season, they reported that abundance of *P. lydia* at their field site during their study peaked on 15 July (*n* = 38) and was fairly steady throughout July (*n* = 13–22 at non-peak times in the month). This peak is approximate to what we have found for the rest of the state.

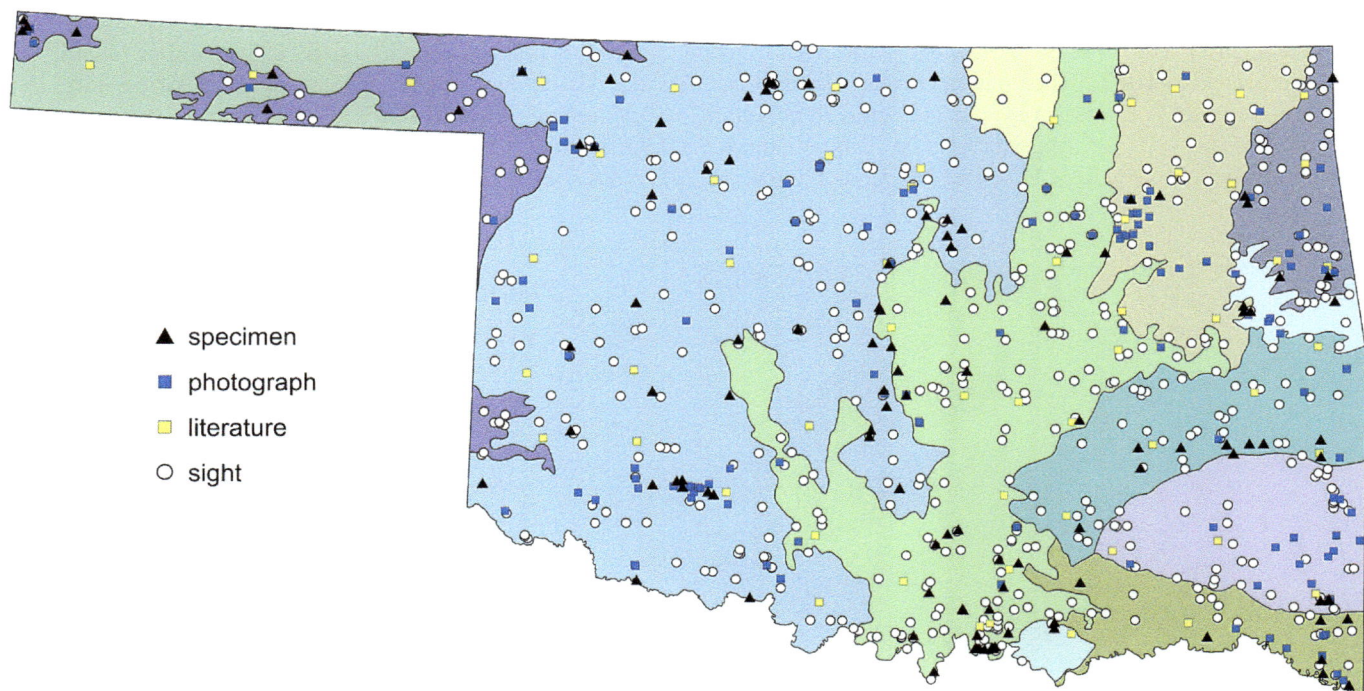

Figure 19.15 Documentation level of supporting records for *Plathemis lydia*, the Common Whitetail, in Oklahoma. Levels include specimen, archived photograph, literature reference, or sight record. Although sight records lack documentation, only those we feel are credible are plotted.

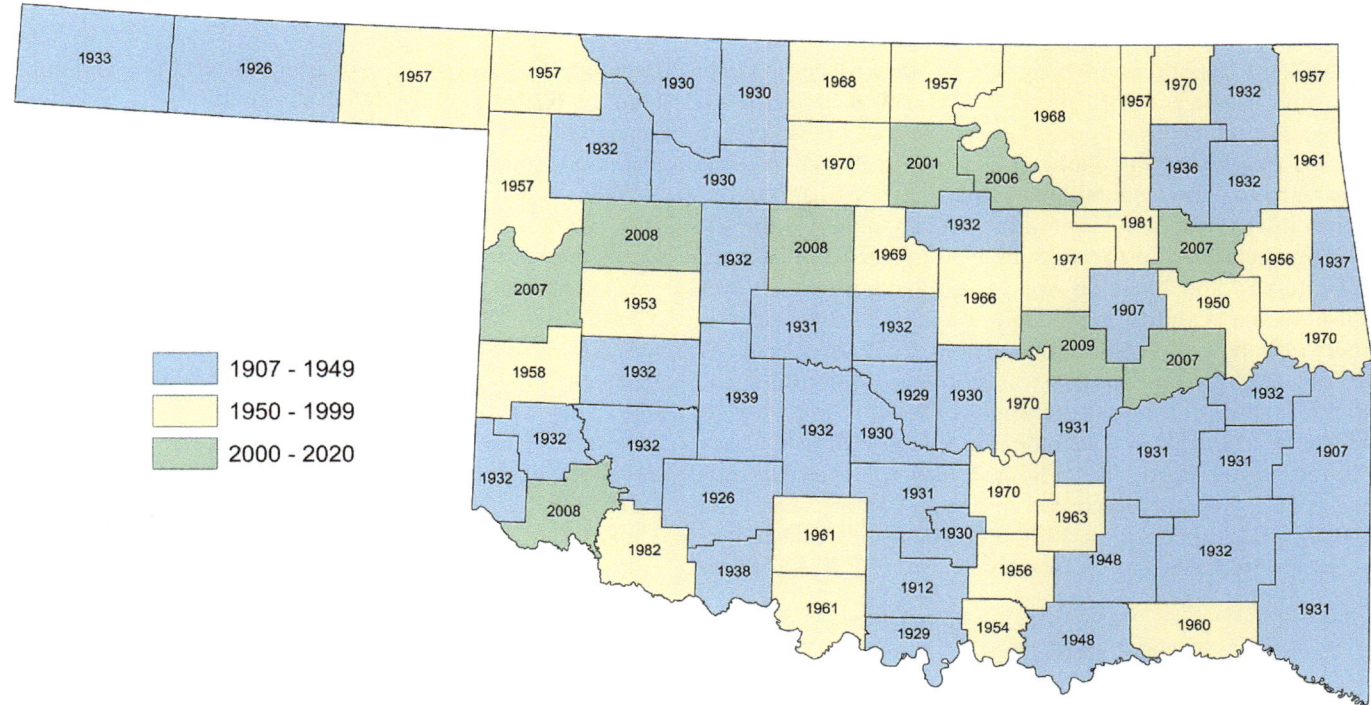

Figure 19.16 Counties in which *Plathemis lydia*, the Common Whitetail, are known to occur in Oklahoma. Year within the county is that in which the species was first reported.

LADONA DEPLANATA (RAMBUR, 1842) – BLUE CORPORAL

The Blue Corporal is an early season species that is fairly common in the eastern third of Oklahoma, where it has a particular affinity for ponds, lake shores, and clear creeks within moist oak woodland. There are scattered records westward to central Oklahoma and a small (regular?) population in the Wichita Mountains in Comanche County. It was first recorded in the state in 1934, when AE Pritchard collected 1♂ and 1♀ on 28 April at (probably near) Antlers, Pushmataha County (UMMZ, EMEC 81865); shortly thereafter he collected another ♂ and ♀ (12 May, EMEC 81864, 81866). The species was not seen again until the Bick/Hornuff era, but surprisingly only once and only by Hornuff, who collected a single ♂ at Osage Hills SP, Osage County, on 11 April 1968 (FSCA).[3] That was the only time the species has been recorded in Osage County. Just after the Bick/Hornuff era, in 1971, PD Harwood collected 1♀ at Heyburn Lake, Creek County (25 May, FSCA). There

were only two other records of adults (JCAC) and two of nymphs (OMNH 1854–1855) before 2002, but since it has been reported regularly and at times in sizable numbers (e.g., 100♂ and 3♀, including 1♀ ovipositing, at Three Rivers WMA, McCurtain County, 3 May 2015, and 90♂ and 30♀, including 3♀ ovipositing, at Schooler Lake, Choctaw County, 20 April 2019).

The species is restricted to spring and early summer (1 March 2017, 1♂, Red Slough WMA, McCurtain County, DA, OC 461331; 9 June[4] 2013, 2♀, Robbers Cave WMA, Latimer County, and 1♂ and 1♀ at Boney Ridge Lake, Le Flore County, 1♂ SP 675), which is the same throughout the region (AR: 16 March–27 May, OC 440791, JCAC 21696; KS: 6 April–3 June, OC 374302, 400318; MO: 4 April–9 June, CO 374283, Trial 2005; LA: 21 February–6 May, Mauffray 2014; TX: 15 February–27 May, OC 459615, 398590).

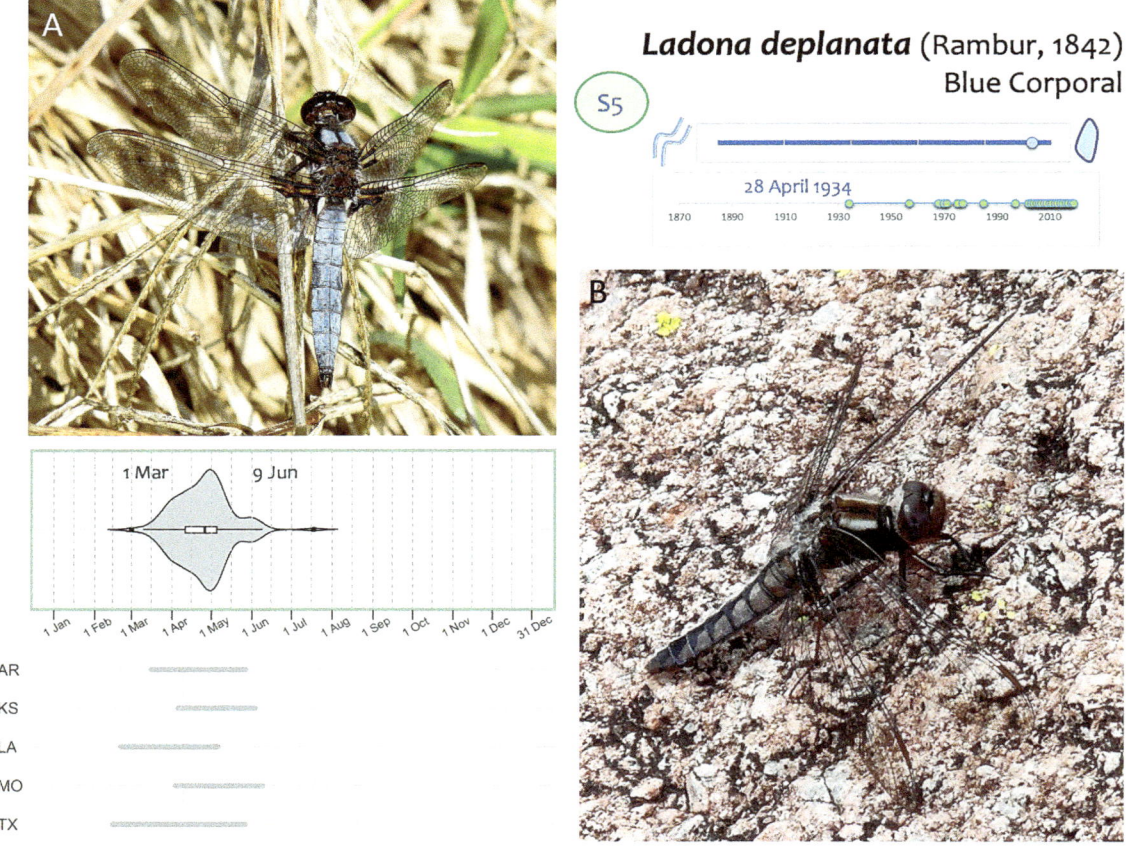

Figure 19.17 *Ladona deplanata*, the Blue Corporal, A) ♂, Mohawk Park in Tulsa, Tulsa County, 27 April 2017 (©Bill Carrell, OC 462277). B) ♀, Quanah Park Lake in the Wichita Mountains WR, Comanche County, 8 May 2007 (©Victor W Fazio III, OC 281950); this range hosts a geographically isolated population. This feisty little skimmer is one of the "early season specialties" of the Southern Great Plains.

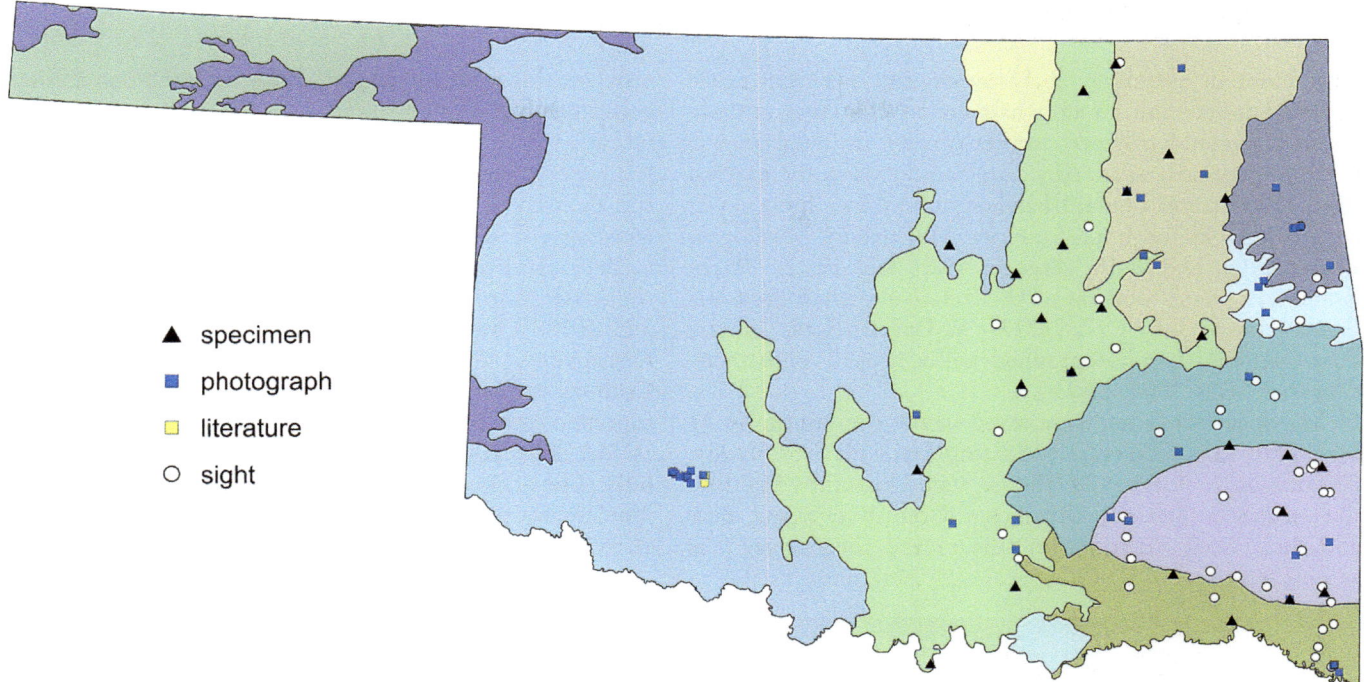

Figure 19.18 Documentation level of supporting records for *Ladona deplanata*, the Blue Corporal, in Oklahoma. Levels include specimen, archived photograph, literature reference, or sight record. Although sight records lack documentation, only those we feel are credible are plotted.

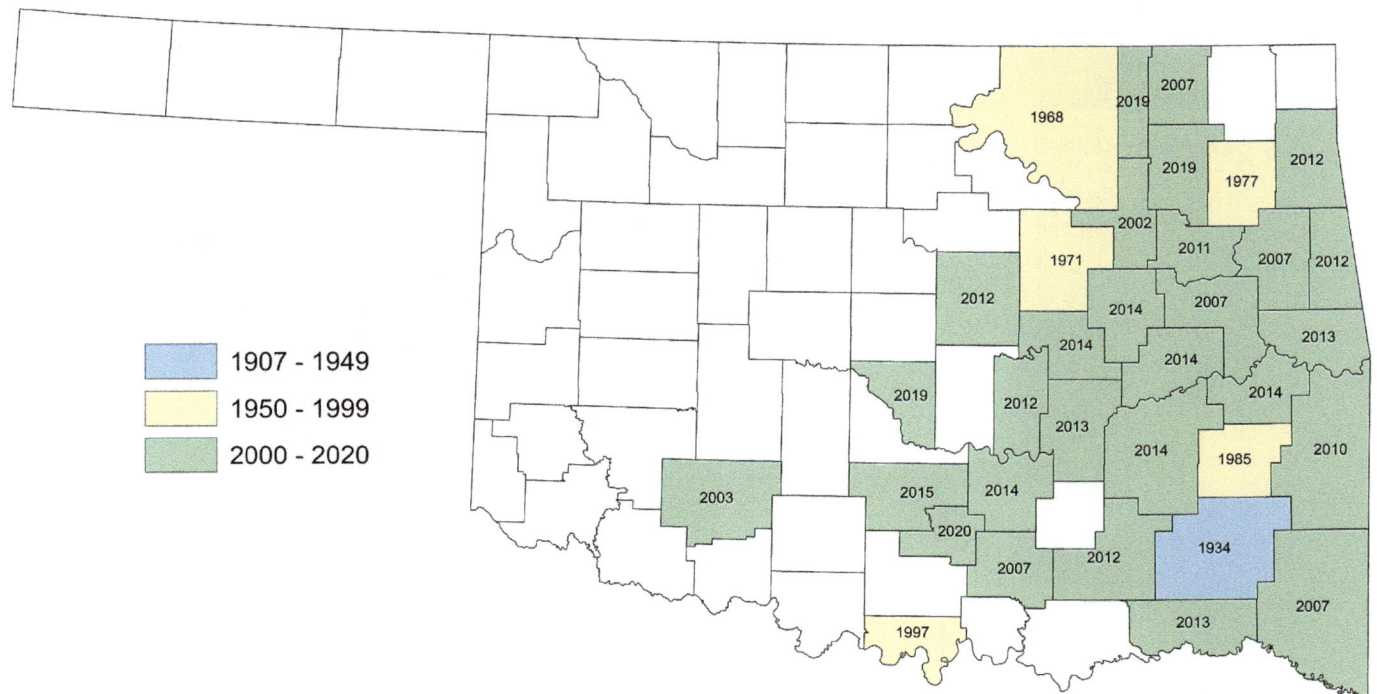

Figure 19.19 Counties in which *Ladona deplanata*, the Blue Corporal, are known to occur in Oklahoma. Year within the county is that in which the species was first reported.

LIBELLULA SEMIFASCIATA BURMEISTER, 1839 – PAINTED SKIMMER

The Painted Skimmer is a low-density species (never reported more than about a half dozen at a time, usually fewer) of mixed hardwood-conifer forests. In Oklahoma, it is found primarily at marshy, forested ponds in the eastern third of the state, where it has been found in a mere seven counties: Atoka, Latimer, Le Flore, McCurtain, Muskogee, Pushmataha, and Tulsa (Figures 19.21 and 19.22). There is but one record outside of this relatively small area, for Comanche County (OC 282273, at Canyon Lake on the Fort Sill MR, 1♀ photographed and a second adult mentioned, 25 May 2008, VWF).

The species was not discovered in the state until, on 23 June 1951, GH Bick collected a single ♂ 3 mi S of Braggs, probably at the former Oklahoma State University Wildlife Conservation Station, Muskogee County (FSCA, Bick and Bick 1957). Apart from this record the species was

only recorded one other time prior to 2007: one adult was collected in Latimer County by B Stark on 12 May 1968 (JCAC 37039). These two records provide two of the eight total specimens for the state and represent the only times the Painted Skimmer has been recorded in Latimer and Muskogee Counties.

Beginning in 2007, *Libellula semifasciata* began to be reported more frequently: >25 reports between 2007 and 2018. The first such report came when BAH photographed a ♀ at the Little River NWR, McCurtain County, on 7 April 2007 (OC 312537). The species was reported four other times that year: twice at Red Slough WMA (DA, BAH) and twice in Tulsa County (BC), the only times the species has been reported for that county. The next year brought four additional records for the state, one of which was the Comanche County record

Figure 19.20 *Libellula semifasciata*, the Painted Skimmer, A) ♂, 10 km SE of Idabel, McCurtain County, 15 April 2008 (©Berlin A Heck, OC 281998). B) ♀, Beech Creek Botanical Area, Le Flore County, 30 April 2018 (©Bryan E Reynolds, OC 479809). C) ♀, Fort Sill, Comanche County, 25 May 2008 (©Victor W Fazio III, OC 282273), one of few found in Oklahoma away from the Ouachita Mountains (and in this case well away). All told, there are but a couple of dozen records for Oklahoma, and it is scarce elsewhere in the Southern Great Plains, although the species' early flight season—effectively late March to early June—may contribute to the perception of scarcity.

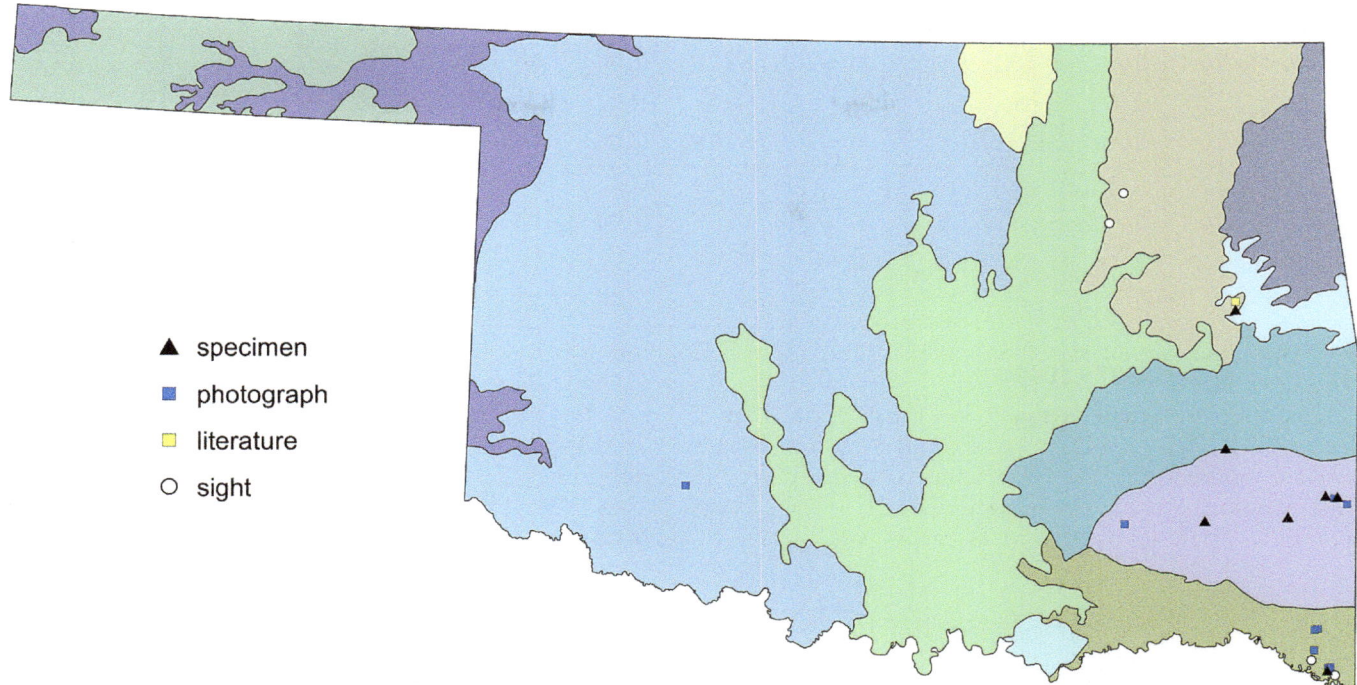

Figure 19.21 Documentation level of supporting records for *Libellula semifasciata*, the Painted Skimmer, in Oklahoma. Levels include specimen, archived photograph, literature reference, or sight record. Although sight records lack documentation, only those we feel are credible are plotted.

mentioned above and another, the only record for Atoka County (at Atoka PHA, 1♂ photographed by KW, 3 July, OC 315351). 2009 brought the first record for Le Flore County and another record for Red Slough. BAH submitted 2 records to OC in 2010 ... *and then reports dried up for 5 years*, until 3 May 2015, when we spotted 2♂ (1 as SP 1553) east of Big Cedar, Le Flore County, at a fairly large, partially open-water pond with a heavily vegetated marsh adjacent. Two years would elapse until the next record for the state, when the early flight date for Oklahoma was set (see below). In 2017 and 2018, *L. semifasciata* seems to have bounced back, with four and eight reports, respectively, and two more were added in 2019. The area around Big Cedar may be the only "go-to" spot for the species, although it has been reported almost as many times from Red Slough.

In Oklahoma, adults are known to be active from 26 March until 3 July (early: from 2017, 3♂ [1 as SP 2240], 1 ovipositing ♀, 7 km E of Big Cedar, Le Flore County, MAP; late: a 2008 photographic record from KW, Atoka WMA, Atoka County, OC 315351). Oklahoma's seasonality is more-or-less in line with nearby states (AR: 17 April–24 May, Harp and Rickett 1977; Harp 1983; LA: 6 March–30 May,[5] Mauffray 2014, OC; MO: 20 May–5 July, Trial 2005; TX: 18 March–14 June, Abbott and Lasley 2019).

In Texas, the species' habitat has been described as "marshy forest seepages, ponds, and ditches with ample emergent vegetation. Normally associated with woodlands." (Abbott 2015:329), which is similar to our encounters. The pond where the early date for Oklahoma was set is one that is just off the highway verge near a creek that floods behind several beaver dams (probably now inactive). The pond tends to be shallow around the edges with much *Eleocharis* (spikerush), and is somewhat open at its center, with much *Typha* (cattail). The surrounding area is fairly well forested with hardwood and mixed woodlands.

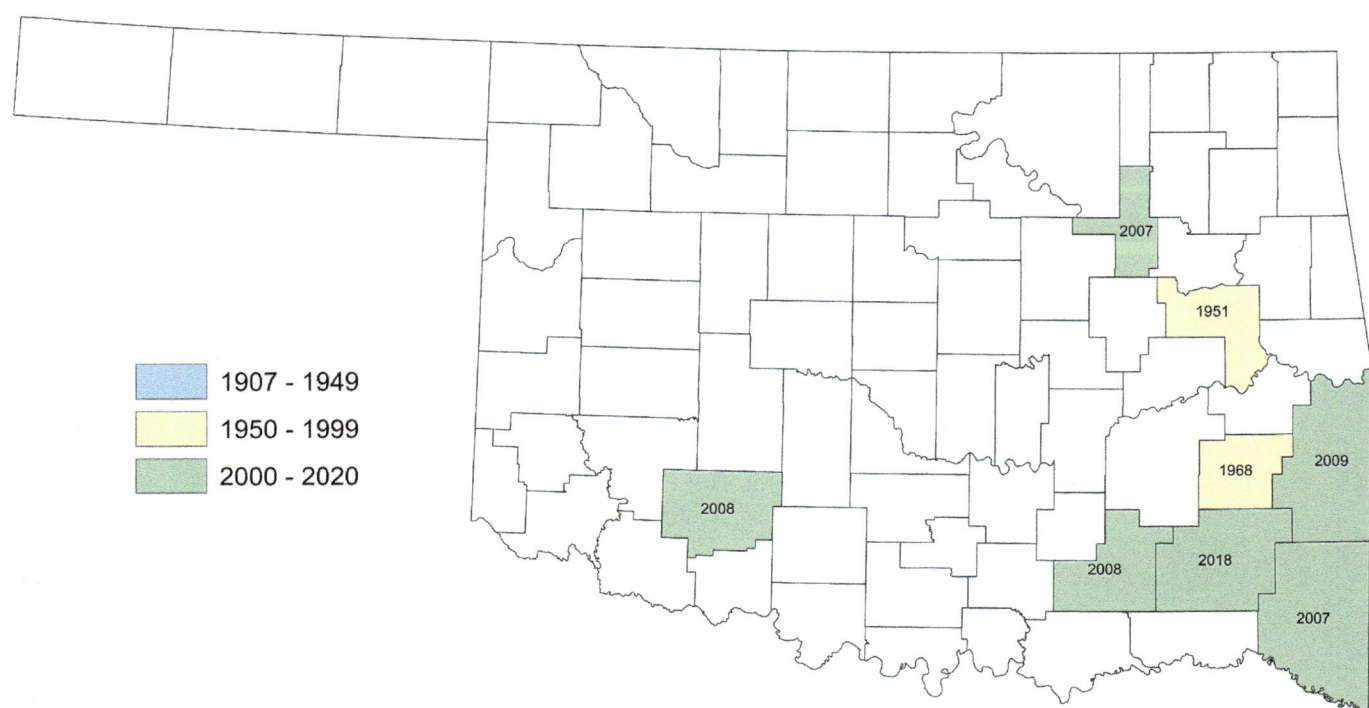

Figure 19.22 Counties in which *Libellula semifasciata*, the Painted Skimmer, are known to occur in Oklahoma. Year within the county is that in which the species was first reported.

LIBELLULA CROCEIPENNIS SÉLYS, 1868 – NEON SKIMMER

The Neon Skimmer is primarily found at small, flowing, clear-water streams, but it does tolerate developed areas, for example, city parks and suburban yards. It is an uncommon species in Oklahoma, where it tends to be seen in small numbers (1–4 at a time), although higher counts have been reported, such as 15 at Vensel Creek in Tulsa, Tulsa County, 7 July 2013 (JWA) and 11 (4♂, 7♀) below the dam at Purcell Lake, McClain County, 18 June 2017 (MAP, 1♀ SP 2398, 1♂, OC 464439).

The species was first found in the state on 6 July 1928 by Z Logsdon near Payne Springs on a tributary of Panther Creek in the Wichita Mountains area, Comanche County (♂, OMNH 929). It was recorded about another dozen times between 1928 and 1932. All of those records were confined to Oklahoma's southwest, except one for the south-central (Murray County; Bird 1932). All of the southwestern records are from the Wichita Mountains area, Comanche County, except two: Greer County, vic. Lake Altus, 1♂, 22 June 1932, OMNH 2490; Jackson County, Salt Fork of the Red River, 1♂, 18 June 1932, OMNH 2522. Oddly, Bird (1932) did not report the species for Jackson County. We know of 34 specimens that were collected during the OU expeditions era, 32 of which were collected in 1932, which would indicate an especially good year for the species in Oklahoma; 30 of the specimens collected in 1932 were taken at the Wichitas and

included 5 nymphs/exuviae, 1 teneral ♀ with exuvia, and 24 adults (all at OMNH except one housed at FSCA).

Between 2005 and 2019 the species was reported >50 times. Reports have come from across the entire stretch of what we now consider to be the species' range in Oklahoma, a diagonal swath through the central part of the state, starting with a wide base that encompasses the southwest and south-central, swooping northward through central Oklahoma, and into the north-central and northeastern parts of the state. This swath skips the southeast and northwest entirely, but oddly enough there is one record for Texas County in the panhandle (3 August 2014, 1♀, Optima Lake, BC, OC 445415), presumably of a vagrant (Figure 19.24). Also of interest is that the northernmost records have come in later years (Tulsa County in 2011, Texas County in 2014, and Kay County in 2015), perhaps indicating a range expansion, a plausible notion when we consider two lines of evidence: 1) early records and 2) the species' overall geographic range (Figure 19.25). Early records are confined to the southwestern part of the state: that, in itself, is not indicative of a range expansion, but what puzzles us is how RD Bird and his team of entomologists did not encounter the Neon Skimmer elsewhere, given that their team did a lot of collecting, for example, in Cleveland County, where the species was not recorded until 2008 and now occurs routinely during summer on and near

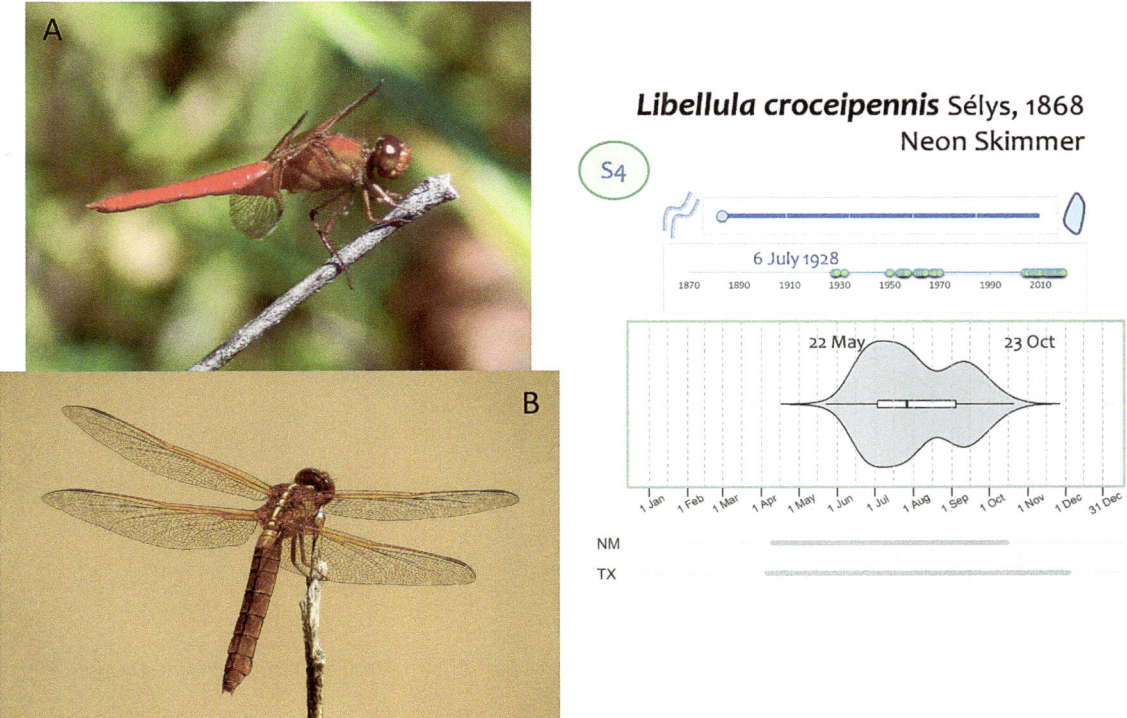

Libellula croceipennis Sélys, 1868
Neon Skimmer

S4

6 July 1928

22 May   23 Oct

Figure 19.23 *Libellula croceipennis*, the Neon Skimmer, A) ♂, Lake Ponca, Kay County, 7 October 2015 (©Brenda D Smith, OC 437736, SP 1811), nearly in Kansas, where the species is unrecorded. B) ♀, Optima Lake, Texas County, 3 August 2014 (©Bill Carrell, OC 445415), much farther west than the species is found typically in the Southern Great Plains.

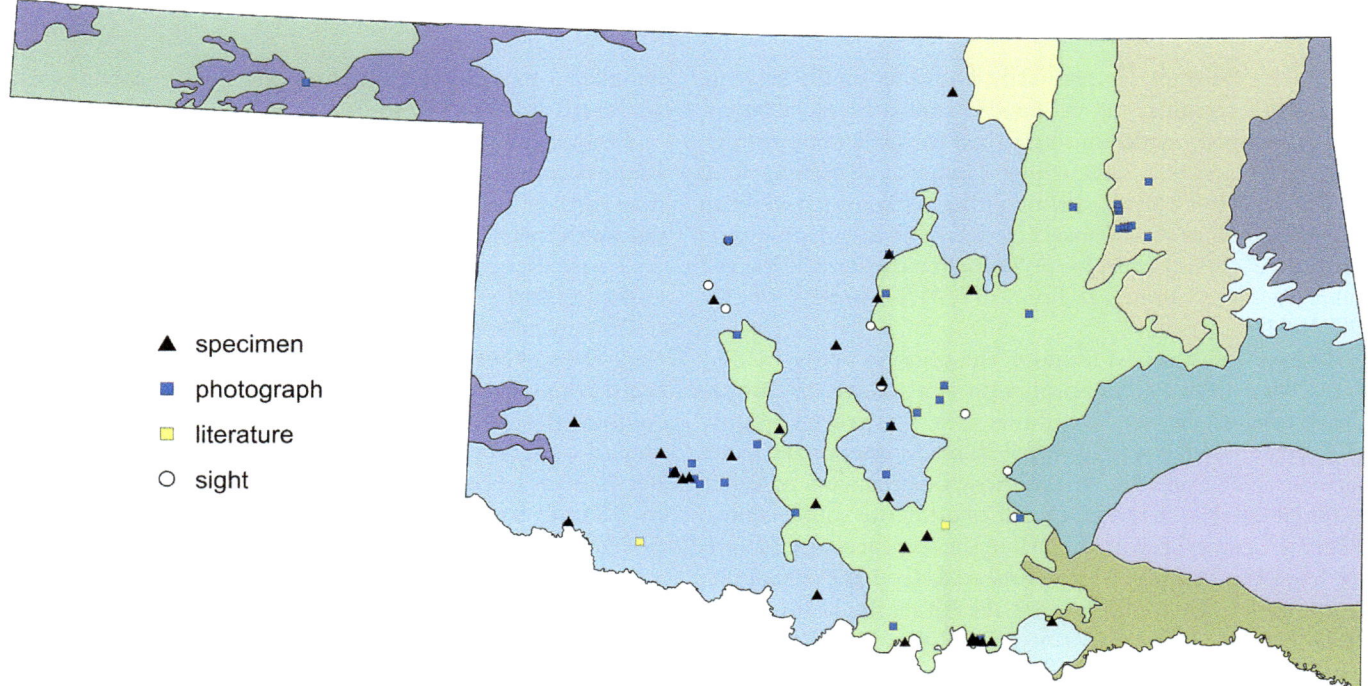

Figure 19.24 Documentation level of supporting records for *Libellula croceipennis*, the Neon Skimmer, in Oklahoma.

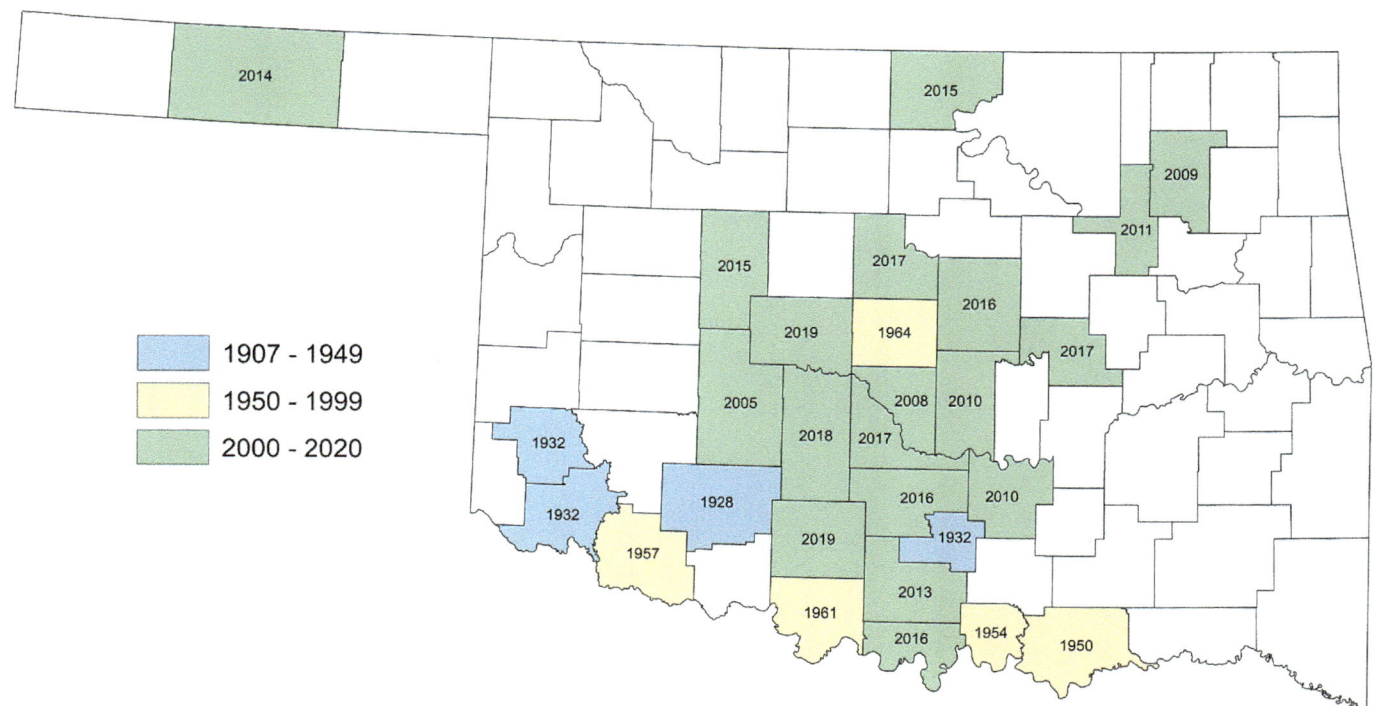

Figure 19.25 Counties in which *Libellula croceipennis*, the Neon Skimmer, are known to occur in Oklahoma. Year within the county is that in which the species was first reported.

the University of Oklahoma campus. It is hard to believe that none of those collectors, all of whom were based at the OU campus, encountered the species in that era *if it was present* on campus. It may be presumptuous, but we have faith that if one of the team saw a Neon Skimmer on campus, there would be a specimen (if BS-P can capture an adult ♂ on campus with her ball cap [SP 64/OMNH 9043, 1♂, 11 August 2009], then so could Bird's team).

Really *Libellula croceipennis* is more of a neotropical species that, relatively speaking, has a limited range in the United States. That "limited range" is particularly striking when compared to its equally stunning orange counterpart *L. saturata*, the Flame Skimmer. Even though the two species overlap to some degree in Oklahoma (Figure 19.26), the Flame Skimmer has a much broader distribution in the western U.S.. Moreover, the Flame Skimmer is less of a neotropical species in that although both species extend south throughout most of Mexico, only *L. croceipennis* reaches

Central America, south all the way to Costa Rica (Needham et al. 2014; also see *L. saturata* account). Given the time-frame and geographical spread of the Oklahoma records, the idea of northward expansion is one that may well prove itself in time. Further, although lotic species often seem to be disadvantaged in Oklahoma given various issues with water flow, if a species can adapt to urban environments, like we suspect *L. croceipennis* can—it does well around ditches and garden ponds—then it will likely continue to expand its range, albeit probably a little more slowly that *L. saturata* has been.

On a final note, the Neon Skimmer is a mid- to late-season flier, which is regionally consistent (OK: 22 May 2012, Fort Sill MR, Comanche County, VWF, OC 375184 and 23 October 2017, Roman Nose SP, Blaine County, S Johnson, OC 474512; NM: April–mid-October, Paulson 2009, OC 323785; TX: 5 April–3 December, Abbott and Lasley 2019).

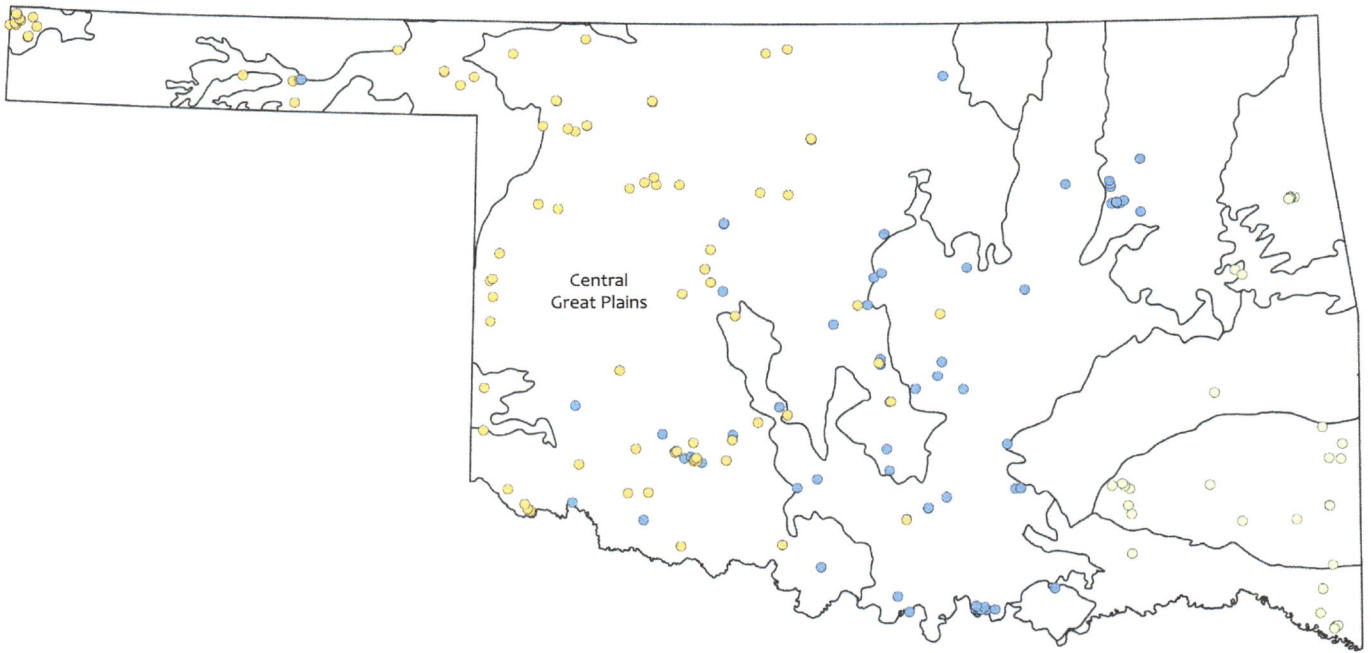

Figure 19.26 Distribution of three "lookalike" species, *Libellula croceipennis* (blue dots), *L. saturata* (orange dots), and *L. auripennis* (green dots), the Neon, Flame, and Golden-winged Skimmers. Outlines within the state are ecoregions. Notice that *L. croceipennis* and *L. saturata* overlap in the southern portion of the Central Great Plains. There is also one record of *L. croceipennis* in Texas County from 2014 (OC 445415). *L. auripennis* has a distinct range from the other two but in the future, there could be overlap.

LIBELLULA SATURATA UHLER, 1857 – FLAME SKIMMER

In Oklahoma, the Flame Skimmer is much like the Neon Skimmer (*Libellula croceipennis*) in its phenology and abundance. It is found usually only 2–4 at a time, rarely more, but it is known to be at least occasionally locally fairly common: for example, we recorded 8♂ and 1♀ on 21 September 2013 at Crystal Beach Park, Woodward, Woodward County.

Its seasonality in Oklahoma is from mid-May to late October (13 May 2016, Altus City Lake, Jackson County, 1 imm ♂, MAP; 30 October 2008, Lake Ellsworth dam, Comanche County, 1♂, VWF, OC 284890). This flight season is most similar to Colorado and Utah but is nowhere near as long as in Arizona, New Mexico, or Texas (CO: 7 June–7 September, OC 328368, 410101; AZ: 5 February–December, USNM 485304, Paulson 2009; NM March–November, Paulson 2009; TX: 29 March–18 November, Abbott and Lasley 2019; UT: 29 April–9 October, Myrup and Baumann 2016). It is difficult to say what the seasonality for Kansas is because there are so few records (19 June–30 August, OC), although that state's restricted seasonality may be simply a product of insufficient survey effort in its southwestern corner. We have no data about particulars of within-state seasonality vs. geographical distribution, but in Oklahoma the Flame Skimmer, somewhat like the Blue-eyed Darner (*Rhionaeschna multicolor*), tends to be found only in the panhandle and westernmost edge of the state in spring and early summer, but by autumn records fan eastward across western and into central Oklahoma (Figure 19.28).

Despite their similarity in abundance and phenology in the state, the Flame and Neon Skimmers differ in their biogeographical distributions (Figure 19.26). The Flame Skimmer is a species of Mexico and the southwestern United States. It comes into Oklahoma from the west (as opposed to the south, as does *L. croceipennis*) and extends into central Oklahoma. Although more widespread than *L. croceipennis*, *L. saturata* is primarily confined to the ecoregions of the Central Great Plains and Southwestern Tablelands. In contrast, *L. croceipennis* is found within the Cross Timbers and the Central Irregular Plains, with but one record in the Southwestern Tablelands (Texas County, see *L. croceipennis* account and Figure 19.24). The two species overlap within the Central Great Plains ecoregion, but *L. croceipennis* is more-or-less limited to the southern part of that ecoregion's extent in Oklahoma, whereas *L. saturata* is found throughout. Hence, although the two species' ranges overlap in southwestern Oklahoma, the Flame Skimmer is more-or-less currently bounded by the Cross Timbers region.

Unlike the Neon Skimmer, the Flame Skimmer has not been seen consistently since it was first discovered in the

Figure 19.27 *Libellula saturata*, the Flame Skimmer, A) ♂, Mountain Lake, Carter County, 5 October 2008 (©Victor W Fazio III, OC 284113). B) ♀, the "Jackson Salt Plains" 6 km S of Eldorado, Jackson County, 11 October 2013 (©Victor W Fazio III, OC 420224). Note the pale latitudinal line not far from the leading edge of the wing, a character not shared by ♀ *L. croceipennis*, which has these various veins concolorous (see Figure 19.23).

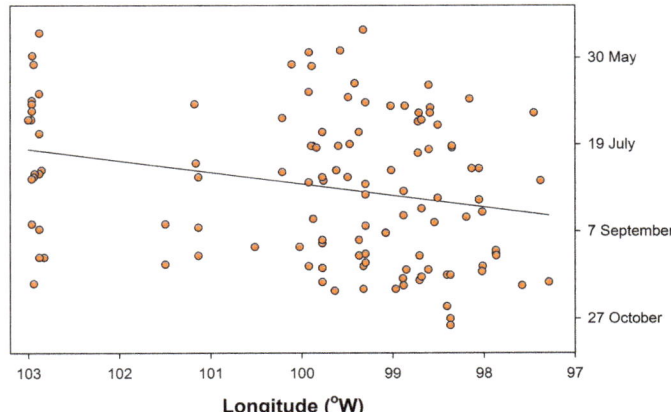

Figure 19.28 Eastward drift of Oklahoma records of *Libellula saturata*, the Flame Skimmer, as the flight season progresses. Note that longitude decreases as the season progresses—in effect, the species drifts eastward through its flight season.

state. The Flame Skimmer was first collected by RD Bird in the Wichita Mountains, Comanche County, on 30 June 1932 (2♂, OMNH 2702), but it was only recorded 6 other times in that era. All of the records of that era come from 3 counties—2 southwestern, Comanche and Tillman, and 1, Cimarron, in the panhandle (all are specimens at EMEC, OMNH, or OSU). Bird (1932) did not include *Libellula*

saturata in his *Dragonflies of Oklahoma*, so the first record must have been obtained shortly after that manuscript was completed. GH Bick and LE Hornuff encountered the species only two days of all of the many years they surveyed the state. Their records were limited to Cimarron County (four specimens at FSCA and one sight record).

Since 2004, there has been 10 times the number of reports than there were from all previous years, and 28 counties have been added to the species' range in Oklahoma. As we suggested for the Neon Skimmer, it is possible that earlier collectors failed to detect these showy, bright orange species, but we cannot abide that notion. We think a much more plausible explanation is that *L. saturata* has expanded its range eastward into Oklahoma (Figure 19.30). It is essentially a lentic species,[6] being found at ponds, lakes, and slow streams (Paulson 2009; Abbott 2015). We suspect this species has taken advantage of the lotic–lentic habitat turnover that occurred in Oklahoma post-Dust Bowl (see Chapter 6). This species can also be found well away from water, too, in that it is a desert-adapted species well suited to persist in the desiccated areas of western Oklahoma during the late summer and early autumn. As we have argued, being a lentic species in Oklahoma has its advantages; as such we expect that *L. saturata* will likely hop the apparent current barrier of the Cross Timbers and will continue to expand eastward, at least a little farther.

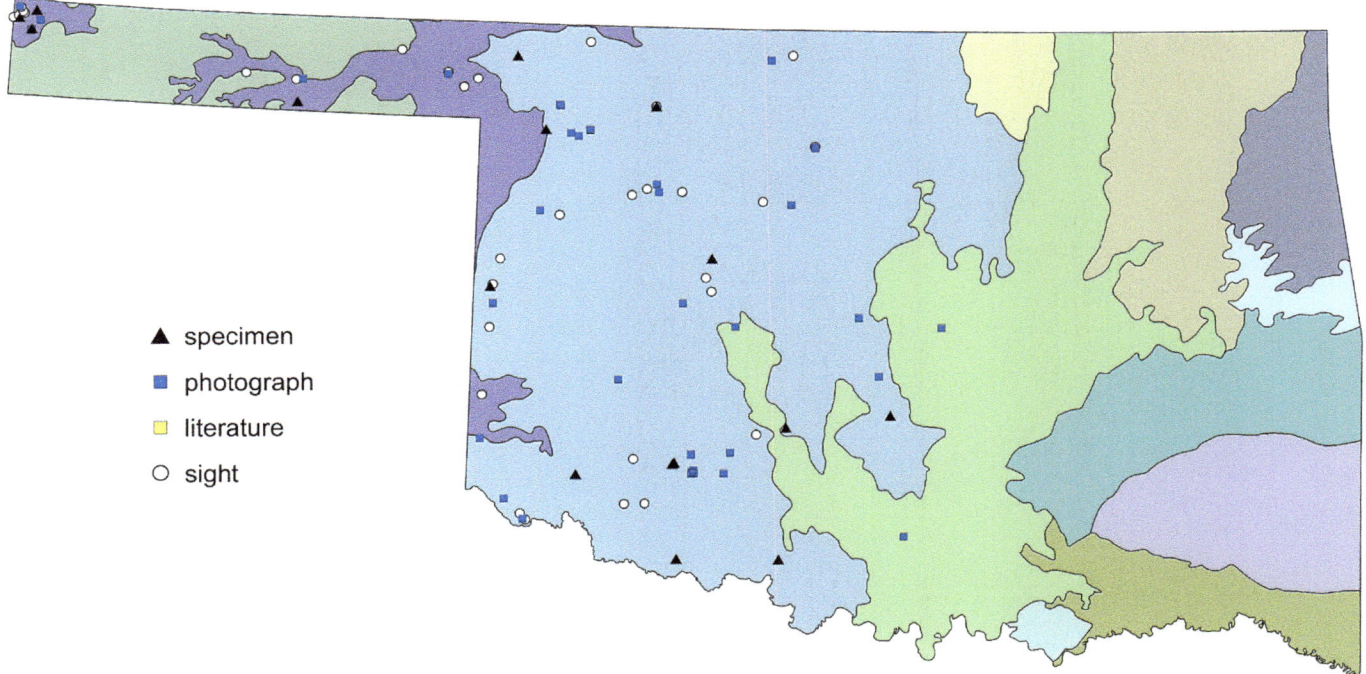

specimen
photograph
literature
sight

Figure 19.29 Documentation level of supporting records for *Libellula saturata*, the Flame Skimmer, in Oklahoma.

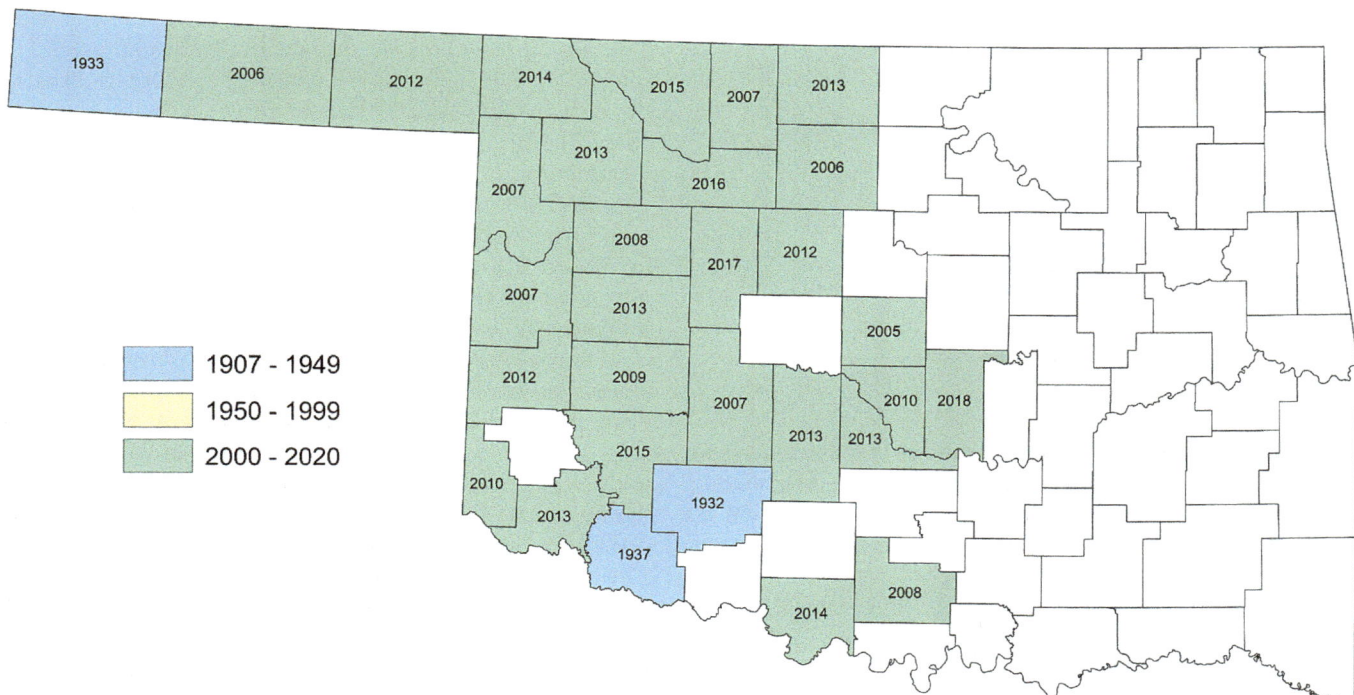

Figure 19.30 Counties in which *Libellula saturata*, the Flame Skimmer, are known to occur in Oklahoma. Year within the county is that in which the species was first reported.

LIBELLULA COMPOSITA (HAGEN, 1873) – BLEACHED SKIMMER

The Bleached Skimmer is known from just five locations in Oklahoma. It was added to the state list in 2011, when VWF found and photographed a lone ♂ in the central panhandle (Texas County: Optima Dam spillway, 5 August, OC 331112). We treated this as a vagrant record until 3 June 2012, when we bumped into a small population in Beckham County, 2 km SE of Sweetwater on West Buffalo Creek at the Buffalo Creek Lodge. How fortuitous it was that we were standing on a bridge above this inviting, crystal-clear, spring-fed creek with a *Typha* (cattail) marsh when the owner of the property, Jerry Alexander, drove up, asked us what we were doing, and then allowed us to survey his property. We were so grateful and remain so today. Before we gained access, we had already seen the county's first records of Flame Skimmer (*Libellula saturata*) and Band-winged Meadowhawk (*Sympetrum semicinctum*) and had glimpsed what we thought to be a ♂ Bleached Skimmer. As such, we were quite distressed that we were standing on the bridge instead of in the creek, so once granted permission we, of course, promptly plopped ourselves into the creek to try to catch a Bleached Skimmer! Once in the creek we saw 4♂ and 1♀ Bleached Skimmer, the latter part of a tandem pair; one of the males was so kind as to come home with us (SP 276, OC 375611).

In summer of the following year we came upon the third and fourth records for the state. On 8 July, we found an astonishing 18♂ at the same spillway as the state's first record. Despite spending more than an hour surveying the spot, we saw no females. The next day, at 13 km SW of Arnett, Ellis County, we stopped at a bridge on Red Bluff Creek, peered over the edge to see 3♂ and a ♀. The creek had no access point so the best we could do was to photograph a single ♂ (OC 401522) and one tandem pair (OC 401523), which is the only physical documentation of breeding behavior in the state. The creek here was shallow with a ponded area bordered by *Typha* and *Eleocharis* (spikerush) and surrounded by *Tamarix* (salt cedar) and *Salix* (willow).

Despite re-visits to the localities of the first four records it was not until 2016 when the species was documented again for Oklahoma. On 20 May 2016, MAP caught a teneral ♀ along Coldwater Creek at Optima NWR, Texas County (SP 1907, OC 444931). He saw the species again on 25 June 2016, this time a single ♂, also along Coldwater Creek not terribly far from the May record. And then on 2 June 2017, he encountered 2♂ at Shorb WMA, Texas County (MAP, OC 463901). Individuals at Coldwater Creek and Shorb were along a creek at spring-fed pools supporting *Eleocharis* and *Typha* and bordered by thickets of *Tamarix*. The other location at Optima NWR where *Libellula composita* has been recorded is at the dam spillway, which is about 4 km to the east, where a deep, permanent pool is bordered by boulder talus on the south side and a cattail marsh around the pool's northern edge. There is also a cattail marsh on the eastern edge where the pool shallows dramatically and becomes a

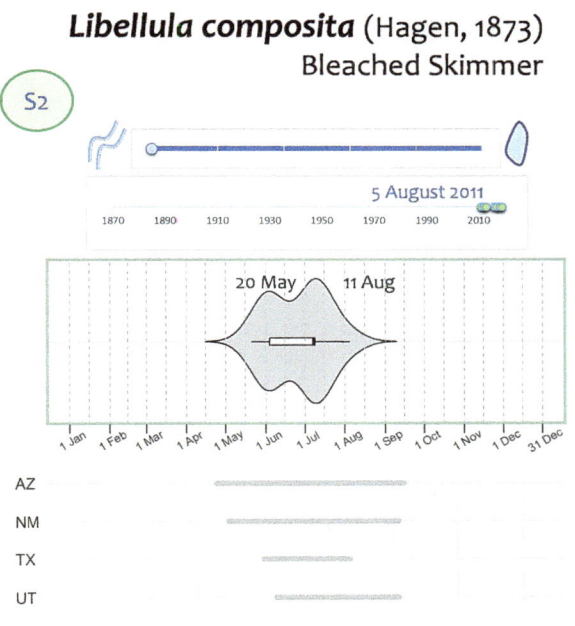

Libellula composita (Hagen, 1873)
Bleached Skimmer

S2

5 August 2011

20 May   11 Aug

AZ

NM

TX

UT

Figure 19.31 *Libellula composita*, the Bleached Skimmer, A) ♂, Red Bluff Creek 13 km SW of Arnett, Ellis County, 9 July 2013 (©Brenda D Smith, OC 401522). B) teneral ♀, Optima NWR, Texas County, 20 May 2016 (©Michael A Patten, OC 444931). It may be that this species is in the process of colonization, if patchily, of westernmost Oklahoma.

short reach of "creek" (really a linear pool). Although we have not conducted specific water or soil testing at Optima NWR nor at the other Oklahoma sites that support the species, the habitat at Optima NWR appears to generally corresponds to how Paulson (2009:385) describes the species' habitat elsewhere: "Alkaline marshes, lake borders, and springs … Common at some extremes of water chemistry where few other odonates are found …" Of interest in terms of the latter part of that statement is that varied numbers of co-occurring species have been reported with *L. composita* in the state—11 other species were noted during our 20 May 2016 survey and 17 during the 25 June 2016 survey at Optima, 30 species were recorded during our 3 June 2012 survey at the Buffalo Creek site, and 8 species were seen from where we were on a bridge about 10 m above Red Bluff Creek. Fewer co-occurring species have been noted at Oklahoma's fifth known location for *L. composita*, at Jackson Salt Plains, located 6 km S of Eldorado, Jackson County, which is also the sole locale in the state with a population of *Erythrodiplax berenice*, the Seaside Dragonlet (see that account). Apart from the bustling dragonlet population, the only odonates regular at the site are *Enallagma civile* (Familiar Bluet), *Ischnura barberi* (Desert Forktail), *Erpetogomphus designatus* (Eastern Ringtail), *Libellula comanche* (Comanche Skimmer), and *Macrodiplax balteata* (Marl Pennant).

Larsen (2008:19) stated that *Libellula composita* "is so closely associated with relict Pupfish populations in the desert southwest that it could be used as an indicator of relict or extinct Pupfish populations." He suggested that the Red River pupfish (*Cyprinodon rubrofluviatilis*) could be used in Oklahoma for such an exercise. Accordingly, we examined the distribution of Red River pupfish in Oklahoma, on the basis of specimen records, to determine overlap with *L. composita*. Two of the *L. composita* sites, from Beckham and Ellis Counties, are on or near rivers known to have Red River pupfish (North Fork of the Red River and the Canadian River, respectively), but the Texas County sites for *L. composita* are near the Beaver River, which to our knowledge, does not host the pupfish. One consideration with this analysis is that Red River pupfish are known to have been introduced into the first two rivers but, as yet, not to the Beaver River, so the non-native pupfish populations present an analytical hurdle for revealing overlap with *L. composita*. Nonetheless, the overlap of these species is an intriguing idea, although in-depth species distribution modeling is necessary to determine this notion's validity and its usefulness in finding populations of the odonate or the pupfish.

Even though we have limited records so far, the species' known flight season in Oklahoma, from 20 May to 11 August (late date is from 2018, EA Hjalmarson, B Roberts, SP 2693, from Jackson Salt Plains, Jackson County, where increasing numbers of records have come recently),[7] is consistent with that known for Texas as well as elsewhere in the species' range (TX: 30 May–5 August, Abbott and Lasley 2019; AZ: 24 April–September, OC 494292, Paulson 2009; NM: 3 May–11 September, OC 7642–7643, 333240; UT: 9 June–11 September, Myrup and Baumann 2016). Colorado

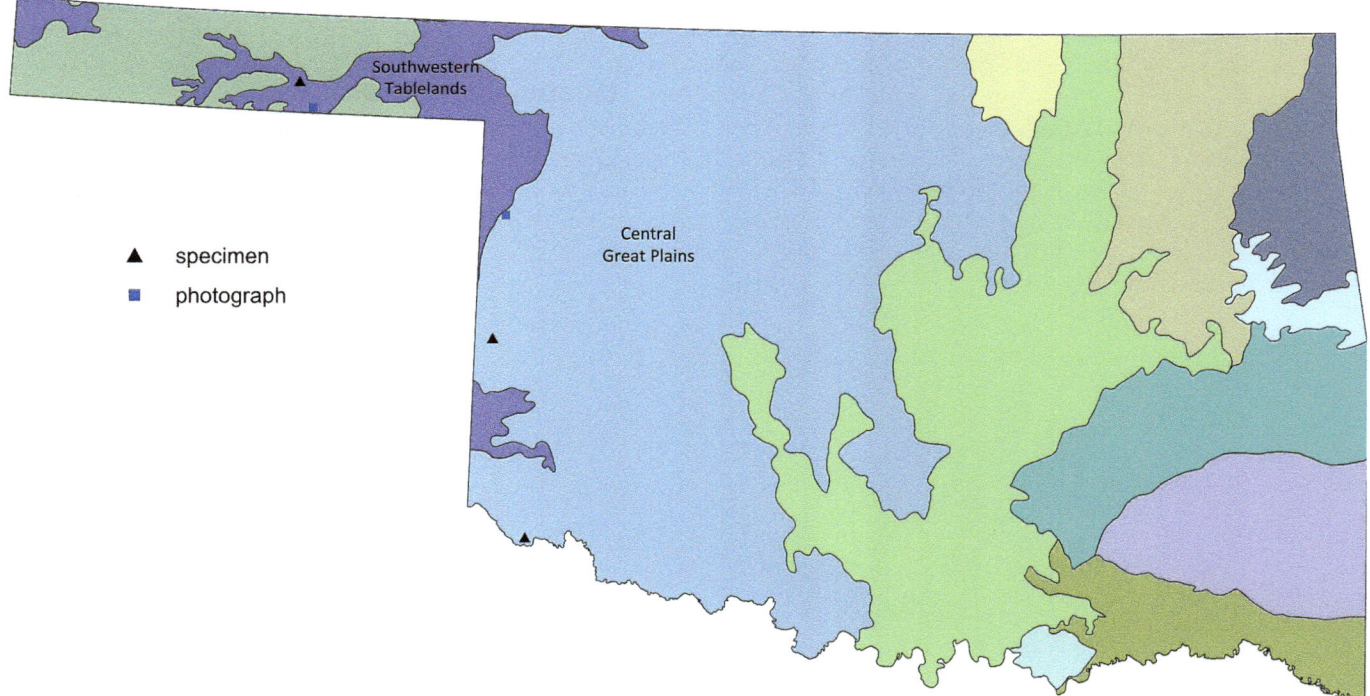

Figure 19.32 Documentation level of supporting records for *Libellula composita*, the Bleached Skimmer, in Oklahoma. Note that the species is confined to the Southwestern Tablelands and the western Central Great Plains.

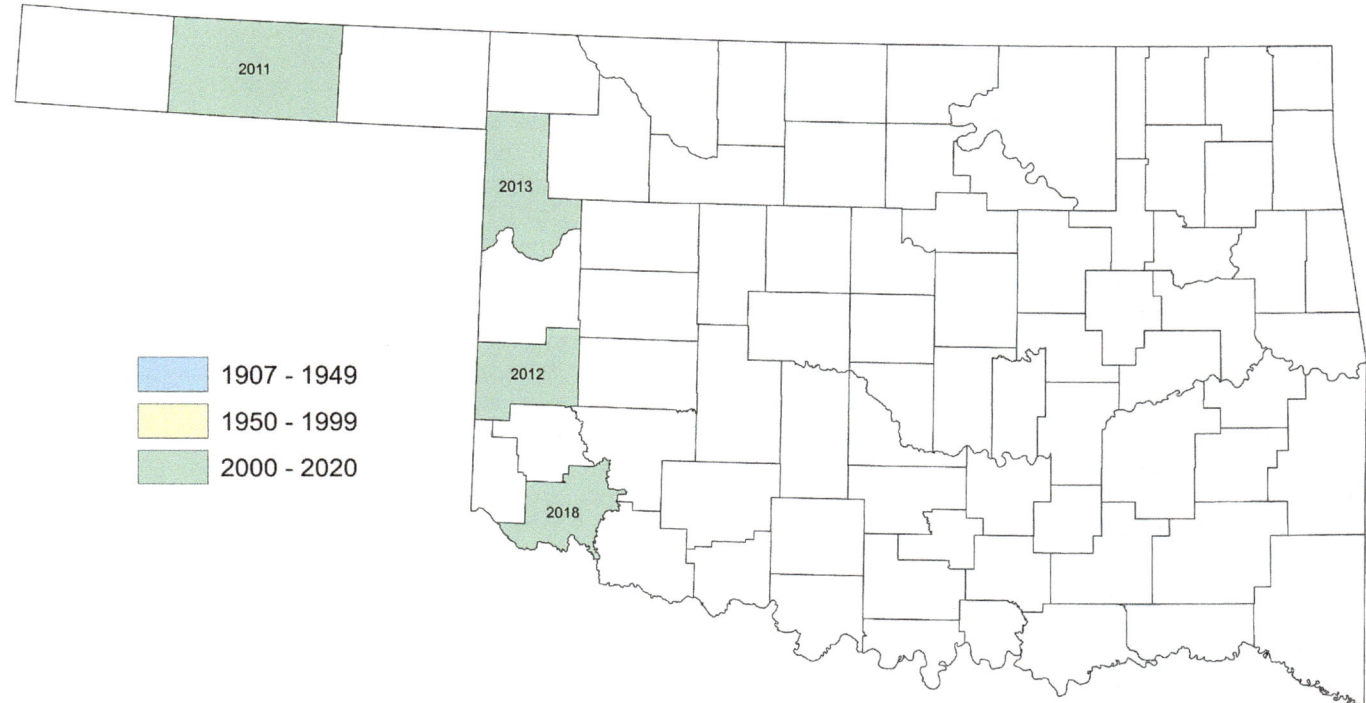

Figure 19.33 Counties in which *Libellula composita*, the Bleached Skimmer, are known to occur in Oklahoma. Year within the county is that in which the species was first reported.

has limited records, so its date range of 25 June–19 July (OC 330333, 330157) likely is not indicative of the flight season there. There are only three records for Kansas, from Meade (Beckemeyer and Huggins 1997), Stafford (OC), and Comanche (OC) Counties, with known dates from June.

Despite having around a dozen records in four counties in Oklahoma of the Bleached Skimmer, the numbers of adults, the mating pairs, ovipositing females, and a teneral suggests that there is a small population of the species in northwestern and southwestern Oklahoma. Thus far it has been found primarily in the Southwestern Tablelands ecoregion, especially within short-grass prairie. But it may be that the species is also colonizing the Southern Great Plains, as indicated by records from Oklahoma as well as

one from Kansas and another from Texas.[8] The Oklahoma population (or possibly more properly, subpopulations of a metapopulation) is far removed from the nearest known breeding populations in central New Mexico and southwestern Texas, but it is consistent with the species' spotty range of multiple disjunct populations (Paulson 2009). We expect that further surveys will prove the status of the species to be somewhat more common; nonetheless, we recommended this species be considered as an S2 (imperiled) species of conservation concern in Oklahoma because there is cause for concern that there will be continued loss of habitat, particularly of spring-fed creeks, in parts of the species' range within the state.

LIBELLULA NODISTICTA HAGEN, 1861 – HOARY SKIMMER

The Hoary Skimmer is a one-time vagrant to Oklahoma (Figure 19.34). GH Bick and LE Hornuff collected a single ♀ at Black Mesa SP, Cimarron County, during their August 1970 trip (Bick 1990, FSCA). The species has been reported in Colorado at least eight times (25 June–11 July, OC, iNat), three of which were in the southeastern corner of the state adjacent to the Oklahoma panhandle (Prather and Prather 2015; OC). The southeastern records are from Cottonwood Creek Canyon just across the Baca–Las Animas County line from one another (Baca County: 8 July 2006, Inez Prather, OC 502800, 1♀ photographed; 6 July 2011, GW Lasley, OC 330904, at least 2♂, 1 coll. as JCAC; Las Animas County: 3 July 2005, Bill Prather, OC 367274, CSU). As the skimmer flies, Cottonwood Creek Canyon is only about 19 km (12 mi) north of the Oklahoma border.

Elsewhere in the species' range the Hoary Skimmer is known to fly from early summer into the fall (AZ: 17 June–31 August, iNat 29772607, 31895285; NM: 11 June–13 August, OC 469370, 315149; UT: 29 May–6 October, BYU, UMNH 14227). This species ranges south to Venezuela. If a population were to establish in southeastern Colorado, then perhaps Oklahoma will get more records of the species. For now, we consider it a vagrant that stands a good chance to show up again in Oklahoma.

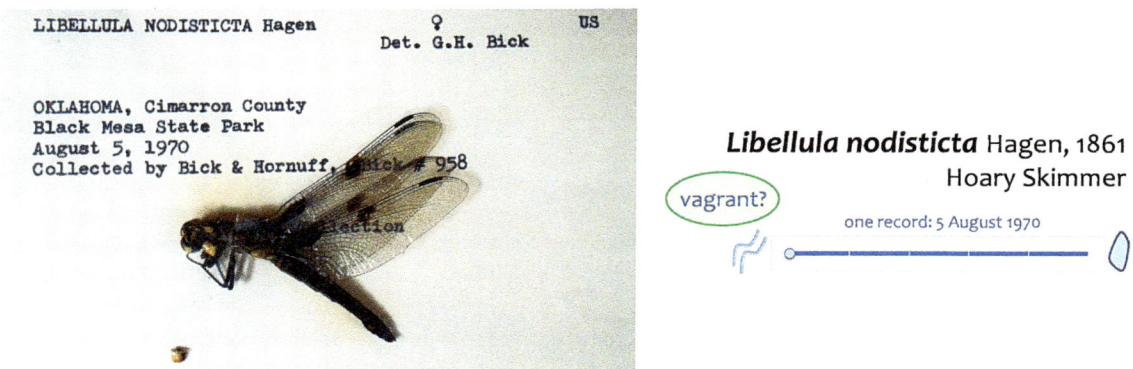

Figure 19.34 Oklahoma's only record of *Libellula nodisticta*, the Hoary Skimmer, is a ♀ from Black Mesa SP, Cimarron County, 5 August 1970 (GH Bick, LE Hornuff, FSCA). This site is not far, as the dragonfly flies, from a small population of the species that occurs in southeastern Colorado.

LIBELLULA PULCHELLA DRURY, 1773 – TWELVE-SPOTTED SKIMMER

The Twelve-spotted Skimmer is common and widespread throughout Oklahoma (recorded in all 77 counties!), although it is decidedly more numerous in the western half of the state, where in mid-season it has been recorded in numbers as high as 150–200 (19 August 2011, Lake Elmer, Kingfisher County; 5 August 2011, Optima Lake, Texas County, VWF, respectively). Even when not reaching numbers that high it is routinely counted in the double digits, fairly regularly >50 at a time. The species is decidedly most common in lotic habitats, with clear, shallow, flowing streams and creeks favored.

It was first recorded in the state by Williamson (3 June 1907, 1 teneral ♀, Wister, Le Flore County, UMMZ; Williamson 1914b) and has been recorded regularly and commonly since. The species has a long flight season in Oklahoma and the region. In Oklahoma it is known to fly from early April to late October (10 April 1965, Cleveland County portion of Stanley Draper Lake, 1♂, Yoachum, OMNH 1212; 29 October 1960, Bryan County, no location given, 2♂, LE Hornuff, FSCA). The species' seasonality in Oklahoma is consistent with neighboring states (AR: 12 May–21 October, OC 422102, Harp and Rickett 1977; CO: 6 June–6 October, USNM 485163, OC 427436; KS: 5 May–6 November, OC 318675, SEMC 1349870; MO: 2 May–20 October, Trial 2005; NM: 26 May–2 October, OC 375266, 333974; TX: 12 April–22 November, Abbott and Lasley 2019).

We have one specimen that documents hybridization between *Libellula pulchella* and *L. luctuosa*, the Widow Skimmer. This somewhat teneral immature ♂ (SP 1016; Figure 19.38) was captured at a drainage ditch, no wider that 1 m across with flowing water and emergent vegetation, in Legion Park in El Reno, Canadian County, on 14 September 2013. Although surprising to us at the time, this hybrid should not necessarily have been, as pairings of *L. pulchella* and *L. luctuosa* have been observed before (including by MAP, ♂ *L. pulchella* and ♀ *L. luctuosa*, 17 August 2019, Grand Lake SP, Mayes County). Calvert (1893) described a ♂ *L. pulchella* in copula with a ♀ *L. luctuosa* on 17 August 1893 in Pennsylvania. Wilson (1920) described seeing an identical pairing decades later. In Wilson's case, he captured the pair, took eggs from the

Figure 19.35 *Libellula pulchella*, the Twelve-spotted Skimmer, A) ♂, Pawnee Lake, Pawnee County, 16 August 2011 (©James W Arterburn, OC 427926). B) ♀, Winganon Flats at Lake Oologah, Rogers County, 9 August 2011 (©James W Arterburn, OC 427923). Despite its occurrence from coast to coast, in the Southern Great Plains this arresting king skimmer becomes rarer from west to east.

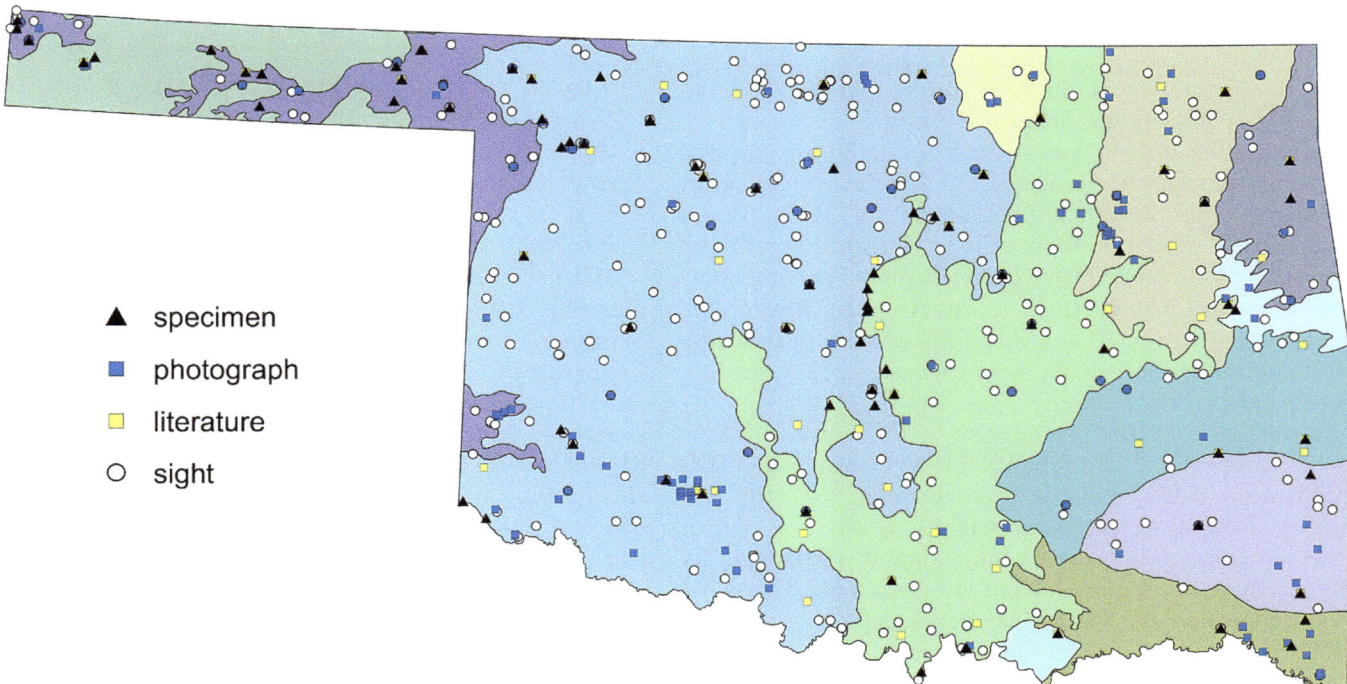

Figure 19.36 Documentation level of supporting records for *Libellula pulchella*, the Twelve-spotted Skimmer, in Oklahoma. Levels include specimen, archived photograph, literature reference, or sight record. Although sight records lack documentation, only those we feel are credible are plotted.

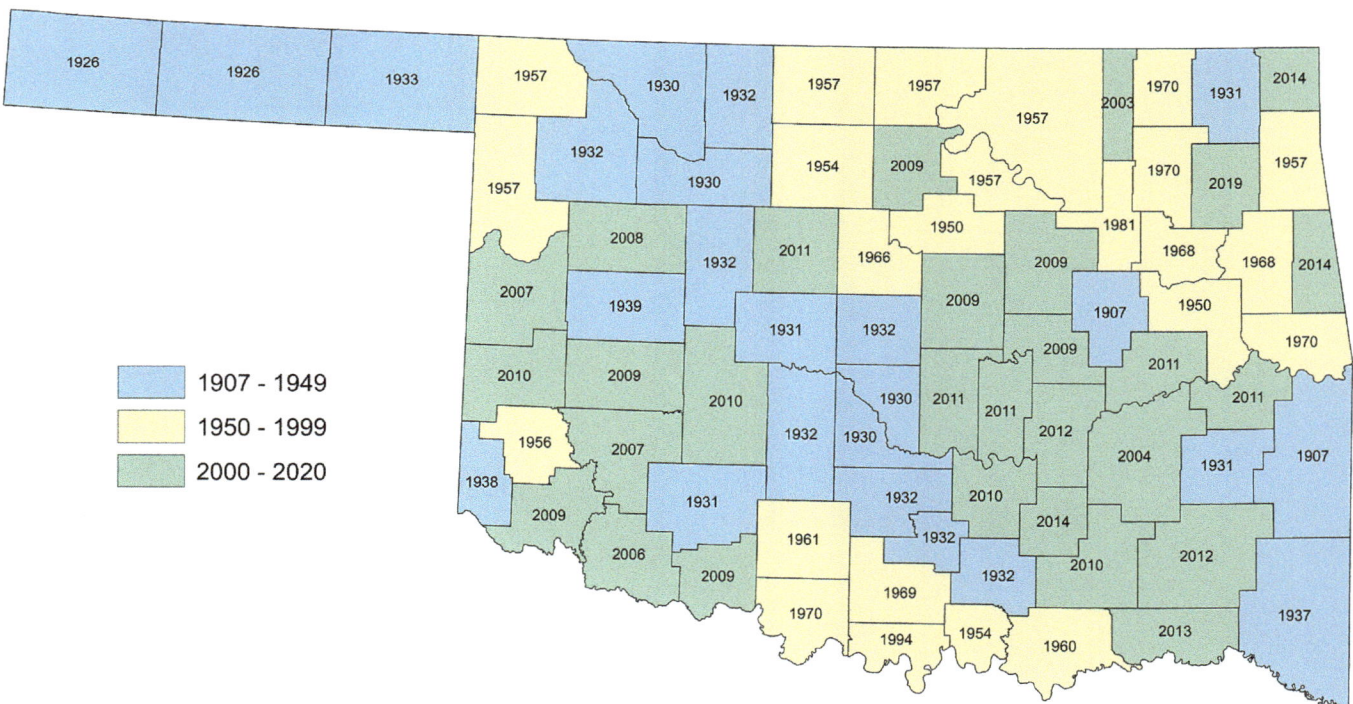

Figure 19.37 Counties in which *Libellula pulchella*, the Twelve-spotted Skimmer, are known to occur in Oklahoma. Year within the county is that in which the species was first reported.

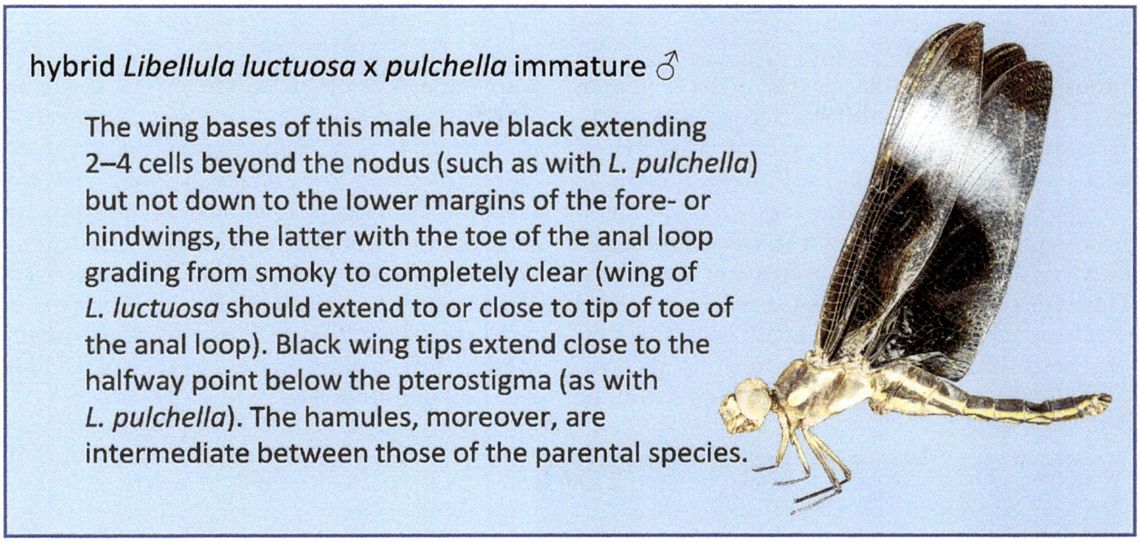

hybrid *Libellula luctuosa* x *pulchella* immature ♂

The wing bases of this male have black extending 2–4 cells beyond the nodus (such as with *L. pulchella*) but not down to the lower margins of the fore- or hindwings, the latter with the toe of the anal loop grading from smoky to completely clear (wing of *L. luctuosa* should extend to or close to tip of toe of the anal loop). Black wing tips extend close to the halfway point below the pterostigma (as with *L. pulchella*). The hamules, moreover, are intermediate between those of the parental species.

Figure 19.38 A hybrid *Libellula luctuosa × pulchella* immature ♂ (still somewhat teneral) was captured at a drainage ditch in Legion Park, El Reno, Canadian County, Oklahoma, on 14 September 2013 (SP 1016).

female, and reared the eggs. He described and illustrated what he thought to be hybrid nymphs of these two species (pp. 240–241). We claim no expertise with nymphs, so we yield here to Bick and Bick (1981:266)—"Because of the difficulty of confidently determining the early instars of many libellulids, particularly of the genus *Libellula*, one must wonder about the accuracy of the recognition of a hybrid based only on a second instar *Libellula* larva."—or any other authorities who care to judge Wilson's documentation for themselves.

LIBELLULA LUCTUOSA BURMEISTER, 1839 – WIDOW SKIMMER

The ubiquitous Widow Skimmer is the most common and widespread *Libellula* species in the state (Figure 19.1), especially in eastern Oklahoma, and probably is the third most numerous libellulid, behind the Eastern Pondhawk (*Erythemis simplicicollis*) and the Blue Dasher (*Pachydiplax longipennis*), although at times it vies for the first. The Bicks said that "*L. luctuosa* is one of the most frequent and widely distributed libellulines in Oklahoma but is not yet recorded from the Panhandle." (Bick and Bick 1957:6). We now have >1,200 records of this species, documenting its presence in all Oklahoma counties, including throughout the panhandle. It is often found in triple digits up to 100–250 at a time, and when it is not that abundant >50 at a time is routine. It is a habitat generalist, often being equally numerous around lake shores, at ponds, or along rivers and streams, although peak numbers tend to be found at lentic sites. Stewart and Murphy (1968), in their study of dispersal of lentic odonates in southern Oklahoma, found that of the 104 *L. luctuosa* adults they marked, 3♀ were re-sighted at the pond they were marked at and 4♀ and 4♂ had dispersed to other ponds, at a maximum distance of 0.53 km for females and 1.40 km for males.

Figure 19.39 *Libellula luctuosa*, the Widow Skimmer, A) ♂, Pontotoc Ridge Preserve, Pontotoc County, 15 June 2014 (©Bill Carrell, OC 423138). B) teneral ♂, 6 km SW of Ames, Major County, 10 July 2010 (©James W Arterburn, OC 427959). C) ♀, Commerce, Ottawa County, 2 July 2013 (©Randy L Kelley, OC 401109).

As with many other libellulids, the Widow Skimmer has a long flight season, extending from early May into the latter part of October (early and late date both from the Wichita Mountains WR, Comanche County: 8 May 2012, VWF, OC 374798; 26 October 2014, OC 427568). This flight season is consistent with the species' flight season in the region (AR: 14 May–15 October, Harp 2007; Harp and Rickett 1977; CO: 5 June–17 October, OC 334484, 437926; KS: 31 May–mid-October, OC 319429, Allison 1921, Cringan 1979; MO: 3 April–7 November, Trial 2005; NM: 26 May–November, OC 375269, Paulson 2009; TX: 4 April–31 October, Abbott and Lasley 2019). The Widow Skimmer was first collected by F Collins near Wister, Le Flore County, on 2 August 1907 (♀, UMMZ).

We know of one possible mixed-species pair and one definite hybrid involving *L. luctuosa* in Oklahoma. The possible mixed-species pair (and mixed-genera), that of a ♂ *L. luctuosa* and a ♀ *Erythemis simplicicollis* (Eastern Pondhawk), was collected by GH Bick on 3 June 1956 at Boiling Springs SP, Woodward County (both specimens at FSCA, ♀ as FSCA 66825). We consider it a possible mixed-species pair for two reasons. When we recorded data from the specimen cards we noted that this was a mixed pair, making the note "mixed tandem pair, with *Erythemis simplicicollis* ♀ FSCA #66825." We now question that notation because 1) GH Bick's notes do not mention this pairing and 2) neither do Bick and Bick (1981) in their paper discussing pairings across odonate families, genera, and species. It is possible that the Bicks simply forgot about this pairing, but we wonder if the more likely culprit is a transcription error from the FSCA specimens. The hybrid known from Oklahoma, involving *L. luctuosa* × *L. pulchella* (SP 1016), is discussed above in the latter species' account. Beyond hybridizing with *L. pulchella*, *L. luctuosa* is known to hybridize, at least in northern California, with *L. forensis*, the Eight-spotted Skimmer (Manolis and Bruun 2006), a species similar looking to *L. pulchella*.

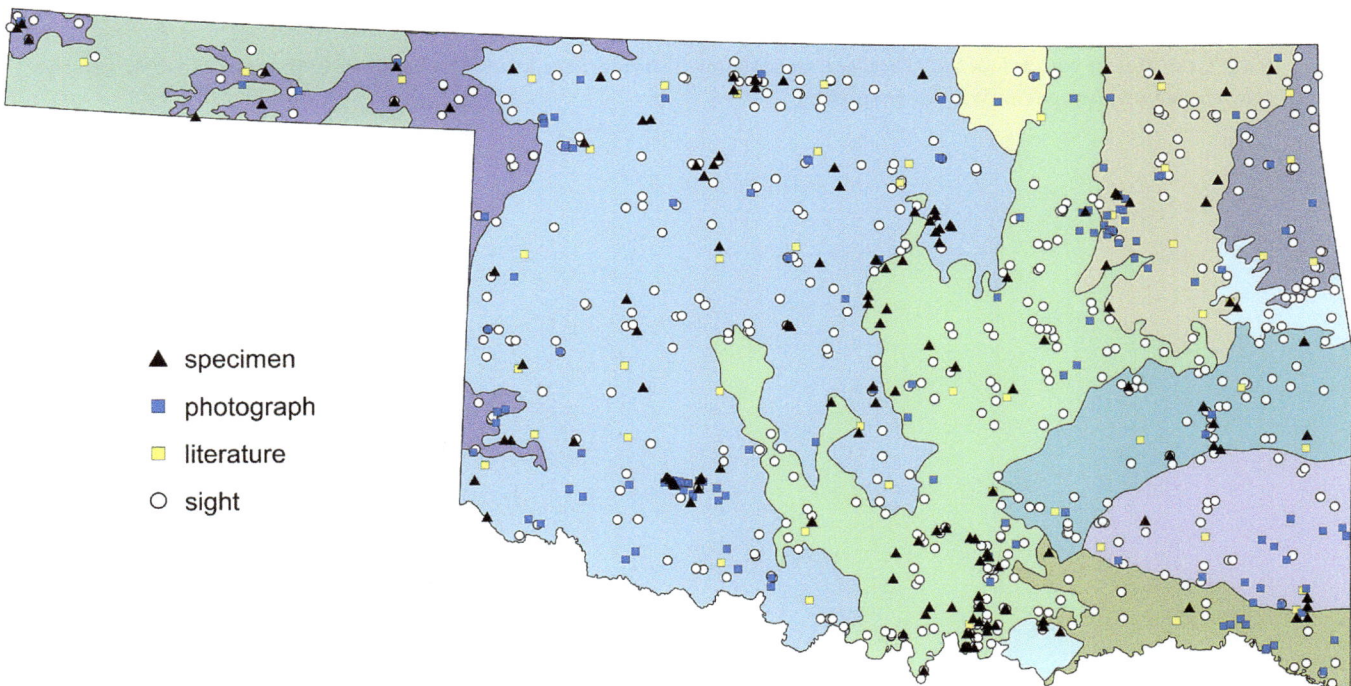

Figure 19.40 Documentation level of supporting records for *Libellula luctuosa*, the Widow Skimmer, in Oklahoma. Levels include specimen, archived photograph, literature reference, or sight record. Although sight records lack documentation, only those we feel are credible are plotted.

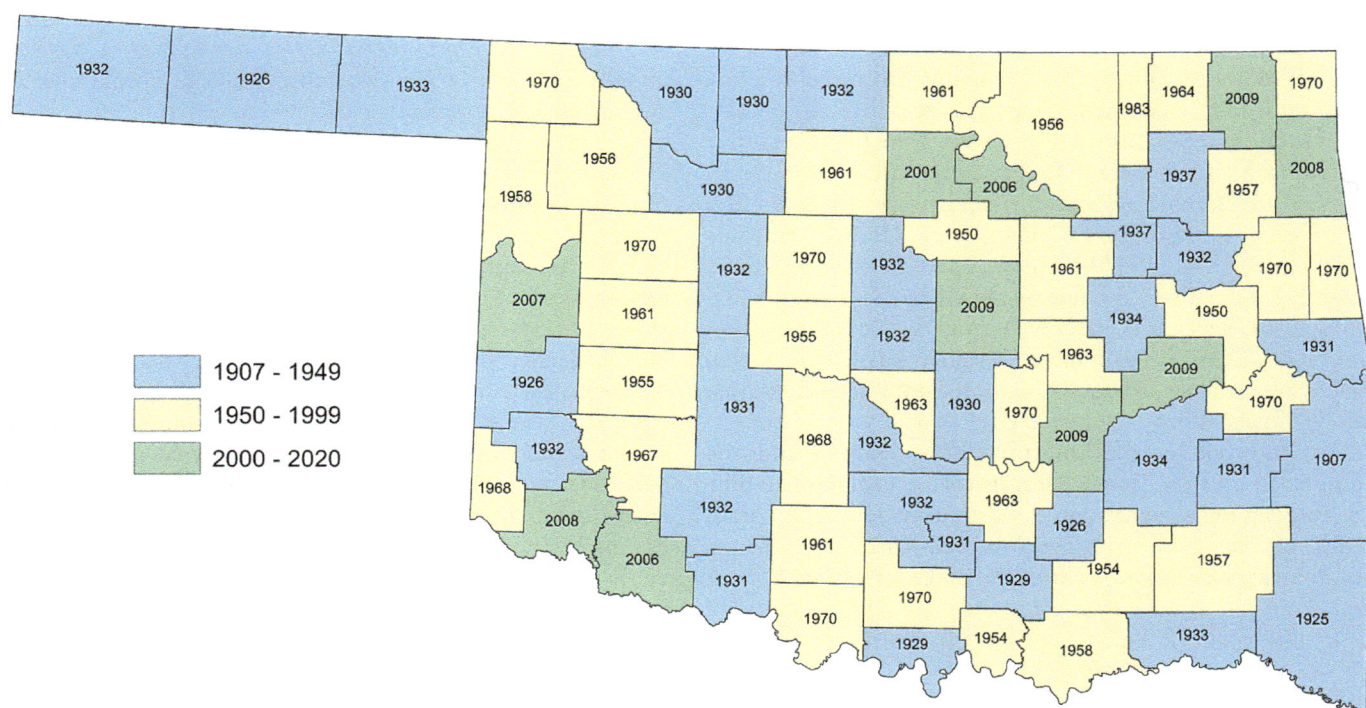

Figure 19.41 Counties in which *Libellula luctuosa*, the Widow Skimmer, are known to occur in Oklahoma. Year within the county is that in which the species was first reported.

LIBELLULA FLAVIDA RAMBUR, 1842 – YELLOW-SIDED SKIMMER

The Yellow-sided Skimmer is a low-density species of forests in the southeastern quadrant of the state. It has yet to be documented in the northeastern part of Oklahoma,[9] despite records that hug that corner. There are a few scattered records out to central and, surprisingly, western Oklahoma. Relative to the species' whole range, there are few records in the broader ecoregion of the Great Plains, with Oklahoma's westernmost record (1♀, Weatherford, Custer County, 8 July 1973, WA Drew, OSU) being the only one for the Central Great Plains level III ecoregion[10] (Figures 19.43 and 19.47; a report for Garfield County is now considered dubious; see below). There is but one record farther west for the species, a ♀ photographed in the Edwards Plateau, Kimble County, Texas (OC 377269).[11] Elsewhere the species is known to prefer "boggy ponds, seeps, and slow streams" (Paulson 2009), which is where it has generally been found in Oklahoma.

There are only six records of the species from the 1930s, one of which, from Latimer County, was the first record of the species in Oklahoma (8 June 1931, RD Bird, 1♂, 2♀, OMNH 2467, 2473, 2476), although Bird (1932) did not report it. The number of records from 1950–1996 more than doubled (n = 17) and have remained consistent since, more than doubling again between 2006 and 2019 (n = 45). We

caution that some earlier reports may be in error. For example, four specimens in the OSU collection—albeit unpublished, to our knowledge—were labeled as L. flavida. One ♀ from Grandfield, Tillman County, collected by Standish and RW Kaiser on 5 July 1937, was actually L. comanche, the Comanche Skimmer. We re-identified the three other specimens as L. cyanea, the Spangled Skimmer—1♀, Watts in Adair County, 10 June 1973, coll. by J Pickle; 1♂, also Watts, 14 July 1972, coll. by DR Molnar; and 1♀, Beaver's Bend SP, McCurtain County, 13 June 1972, coll. unknown.

Generally, the species is found only one or two at a time, but it has occasion to be in larger numbers, reaching its highest known count in the state twice. At a site in McCurtain County nicknamed "Bee Seep," in honor of the senior author, nine adults were recorded (8♂, 1♀, including 1♀ ovipositing while a male guarded her and another ♂ taken as SP 2872, all on 1 August 2019, 21 km N of Eagletown in the Ouachita NF, BS-P). GH Bick had an earlier record: 6♂ and 3♀ on 22 July 1951 near Braggs, likely at the former Oklahoma State University Wildlife Conservation Station, Muskogee County (1♀ at UMMZ, Bick and Bick 1957). Bick also recorded seven adults (5♂ [2 at FSCA, 1 at UMMZ], 2♀ [1 at FSCA]) at the same spot on 23 June 1951, but oddly, he

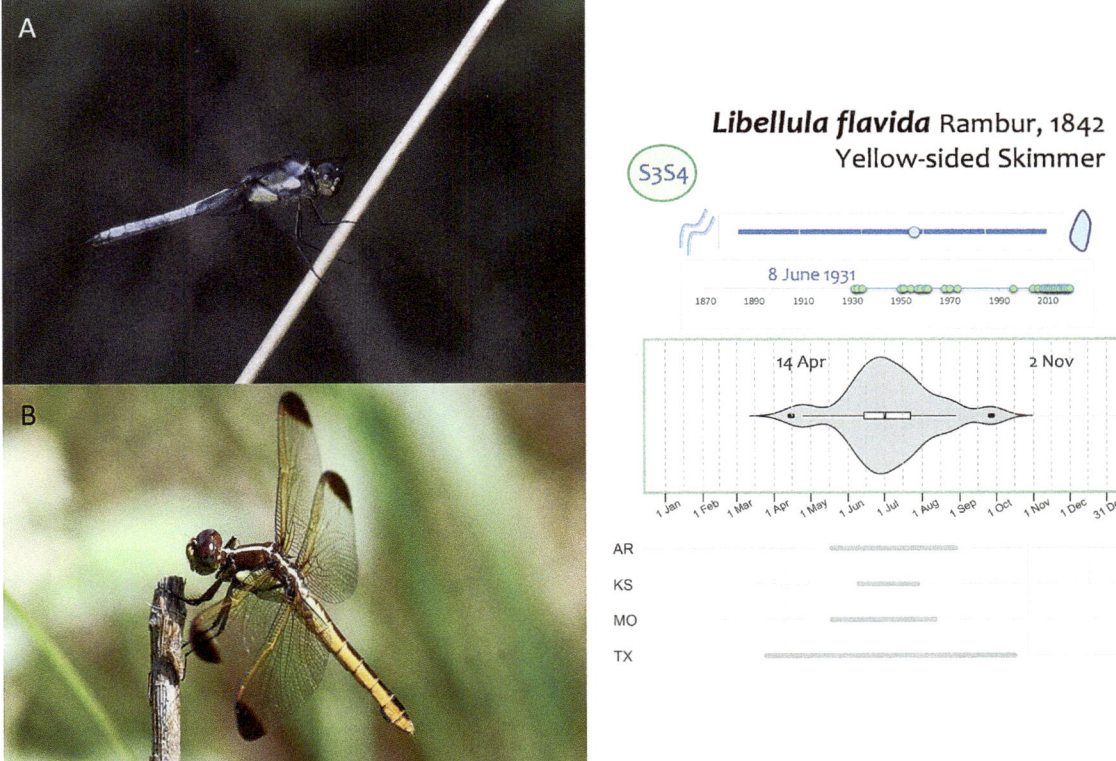

Figure 19.42 *Libellula flavida*, the Yellow-sided Skimmer, A) ♂, Lexington WMA, Cleveland County, 13 July 2008 (©Bryan E Reynolds, OC 382122), well to the west of the species' established range. In Oklahoma, this species is largely restricted to the Ouachita region, where it favors seeps and similar habitats. B) ♀, Beavers Bend SP, McCurtain County, 29 August 2006 (©Berlin A Heck, OC 281886).

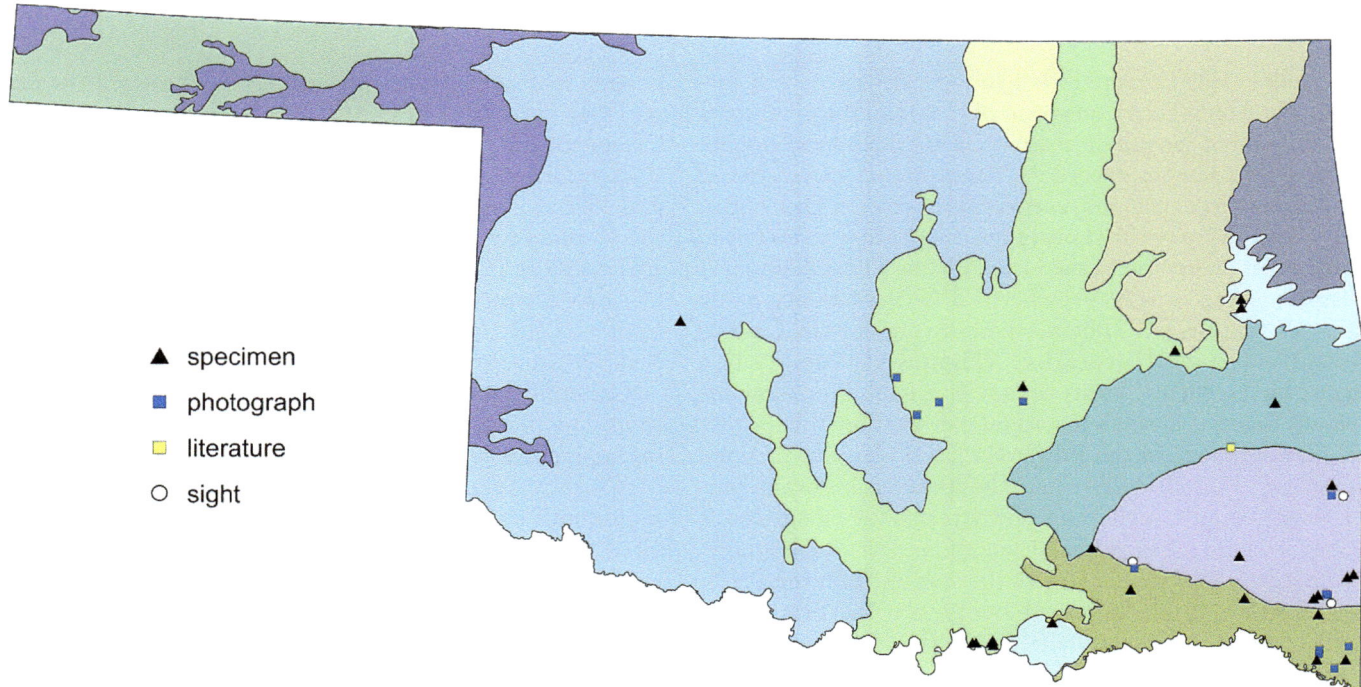

Figure 19.43 Documentation level of supporting records for *Libellula flavida*, the Yellow-sided Skimmer, in Oklahoma. Levels include specimen, archived photograph, literature reference, or sight record. Although sight records lack documentation, only those we feel are credible are plotted.

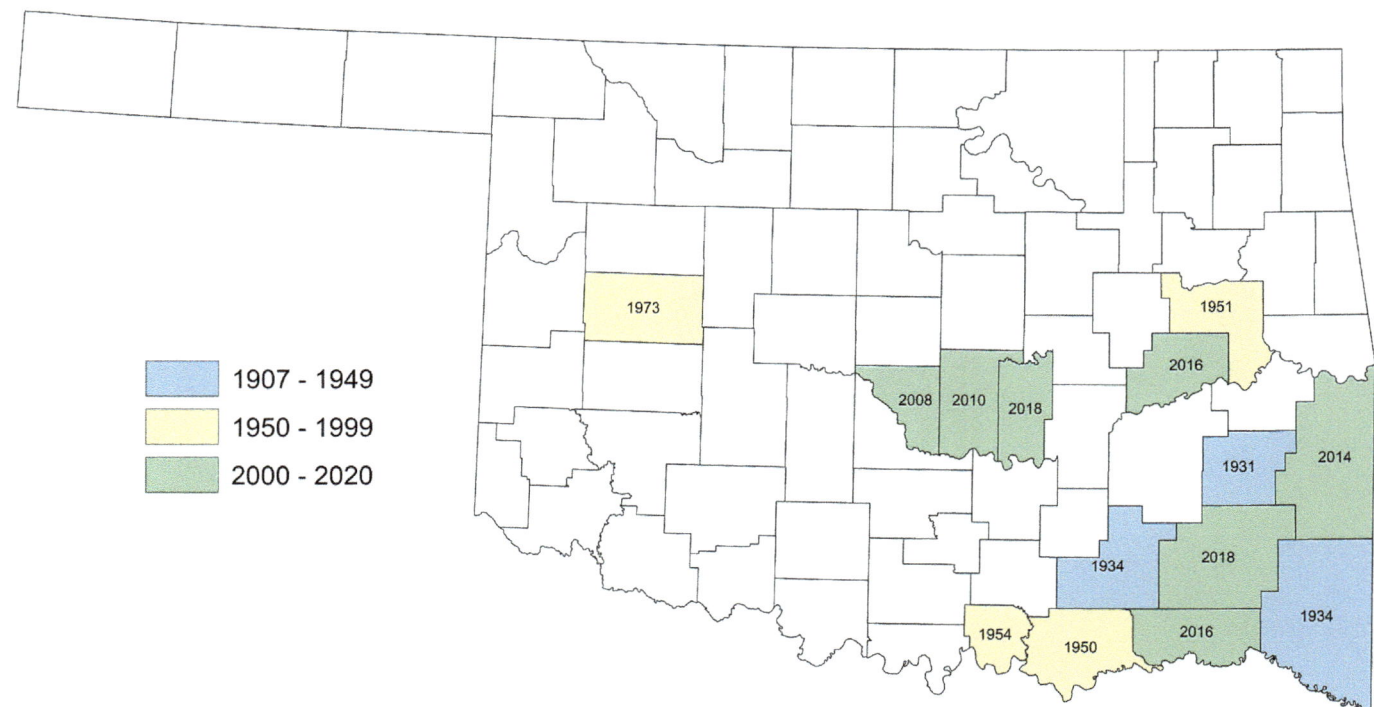

Figure 19.44 Counties in which *Libellula flavida*, the Yellow-sided Skimmer, are known to occur in Oklahoma. Year within the county is that in which the species was first reported.

did not publish the record in Bick and Bick (1957). There is one other record of such high numbers: RW Cruden collected 3♂ and 4♀ at "Happy Hollow" in Marshall County, on 30 June 1961 (UMMZ 210473–210479). Although it undoubtedly breeds in the state we know of no nymphs reported, but there are two teneral records (19 June 2016, Boehler Seeps, Atoka County, 1♂, MAP; 28 April 2017, 7 km E of Big Cedar, Le Flore County, 1♂, BS-P) as well as several records of tandem pairs, including from as far west as Seminole and McIntosh Counties.

One other record that ought to be mentioned is a sighting of the species for Garfield County, which we reported unofficially in one of our Oklahoma Odonata Project updates (5 June 2012). That record consisted of 1♂ flying along a stream at Drummond Flats WMA on 21 April 2012. Although it is one of our records and we are still rather confident it was of *L. flavida*, we decided to rule on the side of caution and discard it, chiefly because without documentation we did not want to stand by a record that was so far to the north and west of the species' known range in Oklahoma. The date, by contrast, did not raise a red flag because the Yellow-sided Skimmer is one of the earliest fliers in the genus *Libellula* in Oklahoma. It is known to fly in Oklahoma from mid-April to early November (14 April 2010, 10 km SE of Idabel, McCurtain County, 1 imm ♂, BAH, OC 318423; 2 November 2017, Red Slough WMA, McCurtain County, 1♀, DA, OC 474601). Oklahoma's documented flight season is longer than known from neighboring states (AR: 18 May–28 August, JCAC 37916, Harp and Rickett 1977; KS: possibly as early as 12 May because SEMC 1350453 has date as "12 May–9 June," late date is 28 July, which is based on USNM 31195; MO: 18 May–11 August, Trial 2005; TX: 25 March–16 October, Abbott and Lasley 2019). Females and immature males of this species have been confused with *L. cyanea*, the Spangled Skimmer, and *L. comanche*, the Comanche Skimmer, especially when the latter species shows an aberrant wing pattern (see accounts for those species).

LIBELLULA COMANCHE CALVERT, 1907 – COMANCHE SKIMMER

The Comanche Skimmer is a western species that was first recorded in the state on 3 July 1930, when a single ♂ was collected in Waynoka (probably on Dog Creek), Woods County (OMNH 924). This was the only county of occurrence that Bird (1932) reported in the first checklist for the state. During the 1930s the species was recorded 4 other times—3 times in the Wichita Mountains, Comanche County, by RD Bird in 1932 (12 specimens, OMNH and FSCA), and once at Grandfield, Tillman County, (1♀, misidentified as *L. flavida*, coll. by "Standish" and RW Kaiser, 5 July 1937, OSU). Records increased in the passing decades, and in recent years the species has become locally fairly common, including in areas known to have been surveyed by earlier collectors. It is now recorded across the western "half" of Oklahoma and increasingly eastward along the southern border (Figure 19.46; easternmost records are from along the Red River below Denison Dam, Bryan County, of 2♂ and 2♀, 17 June 2018, MAP, 1♂ as SP 2629 and 5♂ at Willis Spring in Bromide, Coal County, 28 July 2019, MAP, 1♂ as SP 2866).

In earlier eras it was almost unheard of to record the species in more than single digits, yet we have had the species multiple times in double digits. For example, we had 25♂ and 1♀ on West Buffalo Creek, 2 km SE of Sweetwater, Beckham County, on 3 June 2012 (1♂, SP 275), and 40♂ and 10♀ at Roman Nose SP,

Blaine County, on 4 August 2013. This apparent increase in records and number of individuals may indicate that the species has expanded its range and increased in abundance over the years. Recent records in the eastern half of the state support such supposition. If true, why? We find the Comanche Skimmer almost exclusively along clear creeks and streams with good flow, and the species is known elsewhere to be one of "springs, seeps, and pools in clear streams" (Paulson 2009), habitats not abundant (and at increasing threat) in the western half of the state but decidedly commoner farther east. And, two or three putative hybrids involving the Comanche×Great Blue Skimmer (*Libellula comanche×vibrans*) attests to increasing contact between those species (see *L. vibrans* account for more details).

Two female *Libellula comanche* that present a mix of characters of several species have made us scratch our heads in confusion, again to consider the possibility of hybridization involving the Comanche Skimmer, this time with two other species *L. flavida* and *L. cyanea*, the Yellow-sided and Spangled Skimmers. The first ♀ was the sole *L. comanche* (and one of two *Libellula* species, the other being *L. luctuosa*) at a site in southwestern Oklahoma on the day she was collected (27 June 2015, Clear Creek Lake, Stephens County, SP 1699). She was still somewhat teneral but the dark basal streaks and the extensive brown and amber wash along the

Figure 19.45 *Libellula comanche*, the Comanche Skimmer, A) ♂, Boiling Springs SP, Woodward County, 20 June 2015 (©Bill Carrell, OC 432100). B) ♀, West Buffalo Creek 2 km SE of Sweetwater, Beckham County, 20 July 2013 (©Brenda D Smith, OC 402129). This species is a common sight along clear creeks in western Oklahoma during the heart of summer.

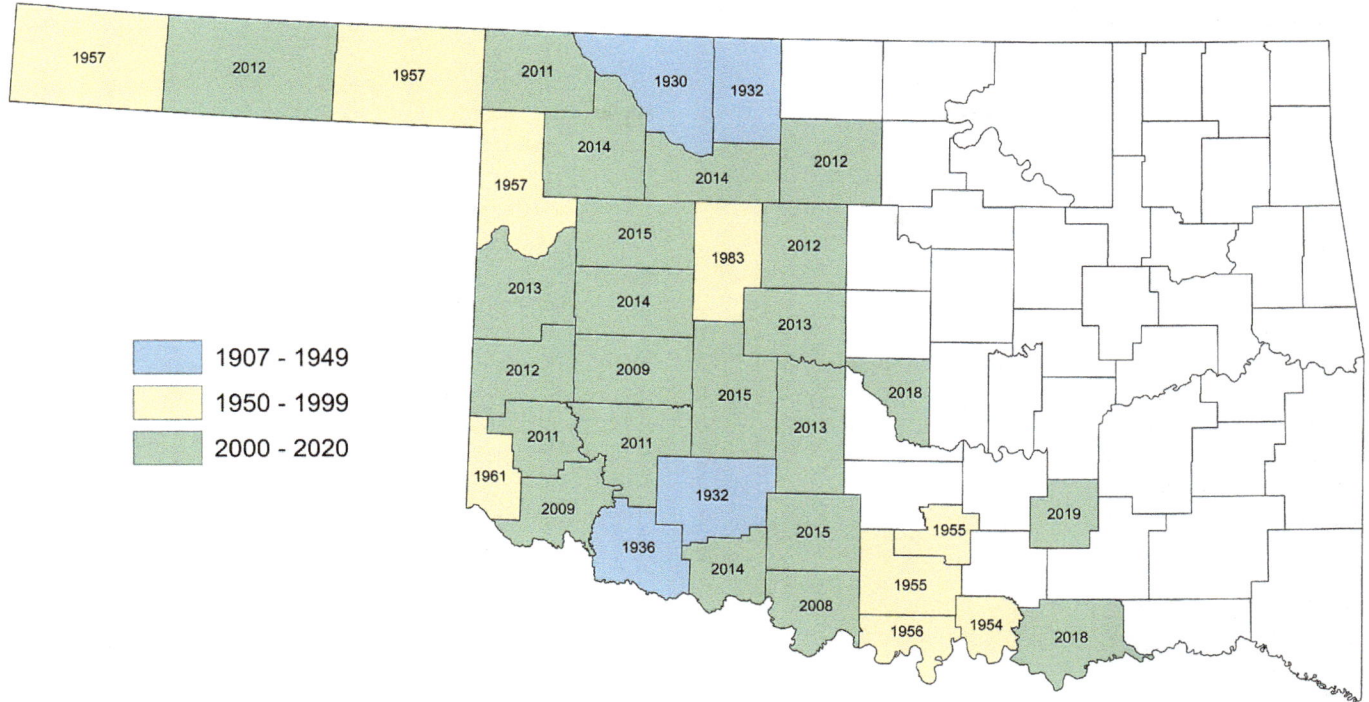

Figure 19.46 Counties in which *Libellula comanche*, the Comanche Skimmer, are known to occur in Oklahoma. Year within the county is that in which the species was first reported.

leading edge of the wing (i.e., within the cell membranes between the costa and the radial/median veins) immediately brought her to MAP's attention. These characters, along with the pterostigma being elongated and about two-thirds yellow, pointed toward *L. flavida*, but the dark costa, subcosta, and crossveins (all of those veins are yellow in *L. flavida* females) and her diminutive size eliminated that species. Her total length, at a mere 44.7 mm, put her in the size range of *L. cyanea* (*L. cyanea* = 41–46 mm, *L. flavida* = 48–51 mm, Needham et al. 2014). However, the extensive amber wash in the wing (although *L. cyanea* often has a wash of color along the front edge of the wing, it typically is more muted, not always along the entirety of the front edge, nor does it venture far from the costa) and the length and color of the pterostigma ought to eliminate *L. cyanea*. Her tan face does not eliminate either species (nor *L. comanche*). Although not able to reconcile various conflicting characters, we resigned ourselves to a label of "aberrant *L. comanche*," admittedly (and ambivalently) allowing known geographic range to be the deciding factor. Four years later, MAP captured a

near-identical ♀, this time in Blaine County >125 km to the northwest (American Horse Lake, 3 August 2019, SP 2878). This ♀ made us ask if these specimens were actually aberrant *L. flavida* or hybrids, even if the parental species were uncertain.[12] After much investigation and head scratching, we circled back around to our initial determination of aberrant *L. comanche* because we learned the range of phenotypic variation in females of this species greatly exceeded what has been reported in the literature (Figure 19.47).[13]

The Comanche Skimmer is known to fly from late May to late October (26 May 2018, Crowder Lake SP, Washita County, 2♂, MAP; 29 October 2014, Lake Elmer, Kingfisher County, 1♀, BS-P, SP 1484), a truncated season compared to Texas but longer than that known for New Mexico (TX: 27 April–1 November, Abbott and Lasley 2019; NM: May–6 October, Paulson 2009, OC 284274). We know of only three records of the Comanche Skimmer for Kansas (Comanche County, 22 June 1983, "young adult" ♂, SEMC 1142167; Meade County, Beckemeyer and Huggins 1997; Seward County, 25 July 2019, OC 500224).

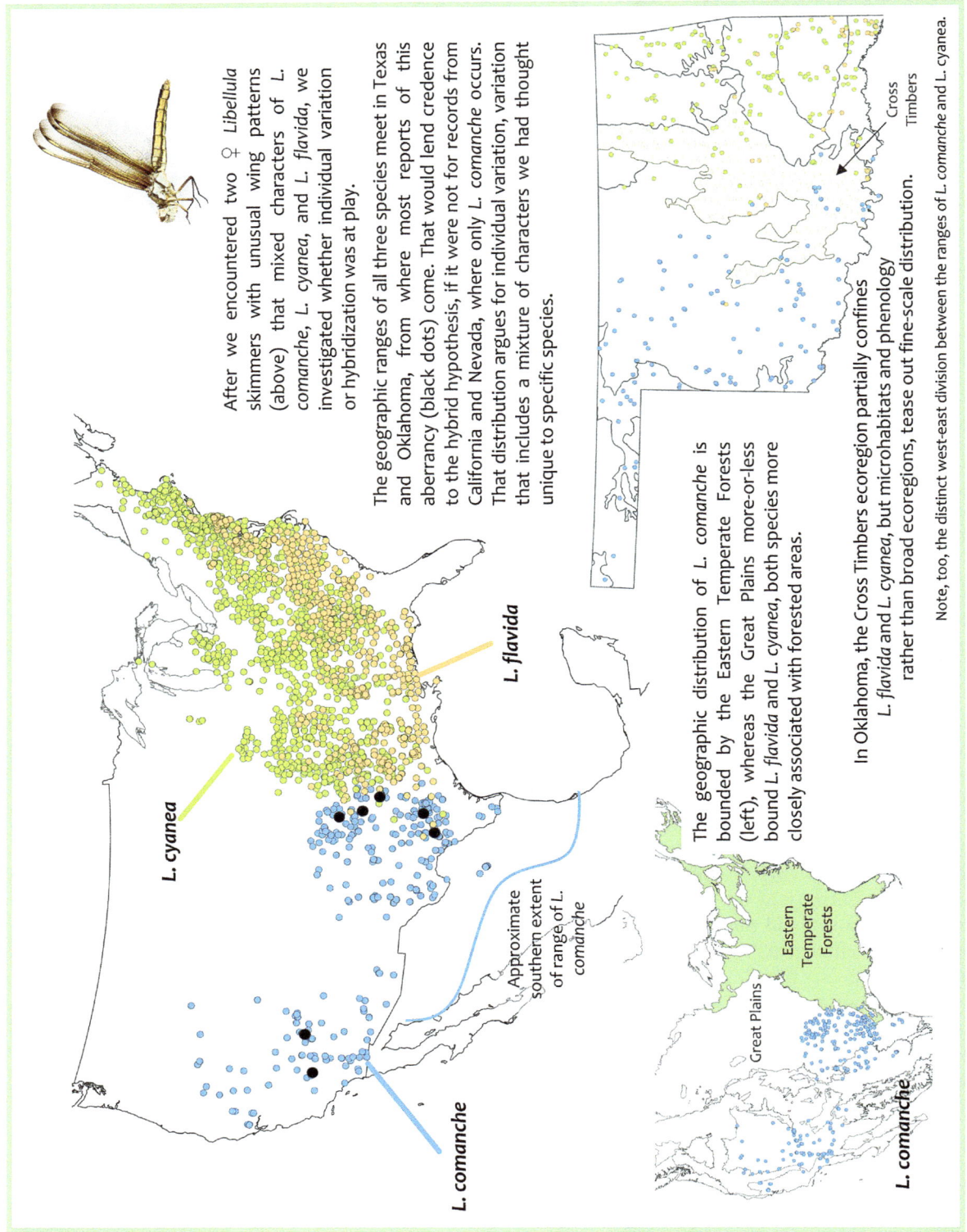

After we encountered two ♀ *Libellula* skimmers with unusual wing patterns (above) that mixed characters of *L. comanche*, *L. cyanea*, and *L. flavida*, we investigated whether individual variation or hybridization was at play.

The geographic ranges of all three species meet in Texas and Oklahoma, from where most reports of this aberrancy (black dots) come. That would lend credence to the hybrid hypothesis, if it were not for records from California and Nevada, where only *L. comanche* occurs. That distribution argues for individual variation, variation that includes a mixture of characters we had thought unique to specific species.

The geographic distribution of *L. comanche* is bounded by the Eastern Temperate Forests (left), whereas the Great Plains more-or-less bound *L. flavida* and *L. cyanea*, both species more closely associated with forested areas.

In Oklahoma, the Cross Timbers ecoregion partially confines *L. flavida* and *L. cyanea*, but microhabitats and phenology rather than broad ecoregions, tease out fine-scale distribution.

Note, too, the distinct west-east division between the ranges of *L. comanche* and *L. cyanea*.

Figure 19.47 Variation or hybridization? Part I.

How to identify
♀ *Libellula comanche*

There is more variation in the ♀♀ of this species than previously described.

A major discovery was that wings can be hyaline (clear; A), subhyaline (B), or have a wash of color across the entirety of the front edge of the wing (right, C).

The "face" may be white, yellowish, tan, brown, gray, dusky, or a combination. Vertex color, too, can be white, black, or any color in between.

S8 flare, rounded, sometimes squared

Yellow on sides of abdomen can be wide or more restricted.

Dark color gradually rises up above the lateral carina, with the most on S2–3.
Also on L. cyanea and L. flavida, although the color generally is smudged on L. flavida.

This ♀ *Libellula comanche* caused much consternation, but she proved to be useful in investigating how to distinguish similar species.

SP 1699

Pterostigma length and color varies.

The side of the thorax in younger ♀♀ has much pale, which is broken by a thin dark line, much like the pattern of *L. flavida*. The pattern is obscured in old females (A), which become pruinose and look like ♂♂. Occasionally, the pale will be broken more widely ventrolaterally (C).

basal streaks

It was thought that only *L. cyanea* and *L. flavida* had basal streaks in the wings (e.g., key in Needham et al. 2014). We now know that is in error. *L. comanche* can have basal streaks in the wings

A

B

C

Photos ©Brenda D Smith

For specimen data, see Appendix F, Table 5.

Figure 19.47 (*Continued*) **Variation or hybridization? Part I.**

While investigating possible hybrids, researchers often discover other taxonomic issues that, with some luck and perseverance, refine our understanding of all species involved.

Such was the case with the aberrant ♀ *Libellula comanche* (above).

Many of the characters previously used to distinguish ♀ *L. comanche*, *L. cyanea*, and *L. flavida* are now known to overlap. These include overall size, vertex and "face" color, pterostigma size and color, basal streaks on wings, lateral color on the abdomen, the lateroventral flare of S8, and, to some extent, the wing coloration.

There is one character that holds to distinguish *L. cyanea* from *L. flavida* and *L. comanche*—

the pale oval that projects toward the front leg

also note the wide dark area that cleanly breaks the pale laterally

sometimes oval is pointed (above), or partially flattened posteriorly, but always rounded and smooth anteriorly and does not reach up to base of wings

©Brenda D Smith

L. cyanea shows some variation in wing pattern.

Wings can be subhyaline, with little color outside of the basal streaks, or have an amber or brown wash almost through the entirety of the wing's front edge. Generally, the color in the costal area is postnodal but sometimes it bleeds antenodally.

Regardless of extent, color is often spottily distributed.

Extent of color on wingtips varies somewhat, too.

©Bill Farrell

©Bryan E Reynolds

©James W Arterburn

©Greg W Lasley

L. flavida shows limited variation in wing pattern.

L. flavida is the only species of the three with "dipped-in-ink" wingtips (left) and that always has yellow veins (*L. comanche* purportedly can have yellowish crossveins, as per Needham et al. 2014, but we have noted only dark veins in that species and in *L. cyanea*, which is consistent with descriptions of the latter species).

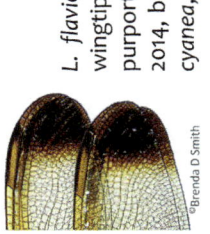

©Brenda D Smith

For photo data, *see* Appendix F, Table 5.

Figure 19.47 (*Continued*) **Variation or hybridization? Part II.**

LIBELLULA CYANEA FABRICIUS, 1775 – SPANGLED SKIMMER

The Spangled Skimmer is the eastern counterpart to the Comanche Skimmer (*Libellula comanche*; Figure 19.47). As with the Comanche Skimmer, the Spangled Skimmer can be locally fairly common with 10–25 individuals often reported at a time, and up to 100 in mid-season (Atoka WMA, Atoka County, 14 June 2003). Bick and Bick (1957:6) reported the species as "… common in Muskogee [County] but was not collected during three summers of intensive effort in Marshall County." GH Bick eventually did find the species in Marshall County, but only a single ♂ (1.5 mi SE of Enos, 17 July 1967, OMNH 217). The species has not been reported in the county since.

The Spangled Skimmer is now recorded in 37 counties—a sharp increase over the one county, Le Flore, to which both Williamson (1914b) and Bird (1932) attributed the species. Each of these esteemed odonatologists only encountered the species once, which is not shocking for Williamson given that he was in the state but a few days in 1907 (nonetheless he recorded 50 adults, 16 of which he collected!).[14] Bird on the other hand, despite living in Oklahoma for years, only ever collected a single ♂ (21 June 1931, OMNH 2444). It may be that neither of these collectors envisioned a day when the Spangled Skimmer would be known across the eastern "half" of Oklahoma (Figure 19.47).

Figure 19.48 *Libellula cyanea*, the Spangled Skimmer, A) ♂, Prague Lake, Lincoln County, 24 July 2010 (©Victor W Fazio III, OC 321122). B) tandem "pair," Sportsman Lake, Seminole County, 29 June 2014 (©Brenda D Smith, OC 423904). The wry quotes are to draw your attention to the ♀: this is a mixed-species pair, a ♂ *L. cyanea* and a ♀ *L. incesta*, the Slaty Skimmer. C) ♀, Haskell Lake, Muskogee County, 15 May 2014 (©James W Arterburn, OC 422141). Unlike its "lookalike" congener, *L. comanche*, the Comanche Skimmer, this species flies chiefly in the spring, not summer.

The species appears, like the Comanche Skimmer, to have increased in abundance and in its range. But unlike that species, there is a logical reason—the increase in muddy-bottomed ponds and lakes in Oklahoma post-Dust Bowl. Paulson (2011:414) described this species as one of "lakes, ponds, and marshes, primarily those with silty/muddy bottoms, in open or woodland." These habitats are exactly the types that became more ubiquitous in the eastern part of the state after Oklahoma undertook its mission to build reservoirs in an attempt to avoid another Dust Bowl. Although we have found that the species avoids the supersized reservoirs and similar large lakes in favor of marshy ponds or small lakes, it is decidedly more lentic than is the Comanche Skimmer.

The Spangled Skimmer flies earlier, too, by almost two months compared to the Comanche Skimmer. It is known to be on the wing in Oklahoma from 10 April to 11 August (both from McCurtain County, early: 2012, Red Slough WMA, BAH, DA, OC 374375; late: 2010, 8 km S of Smithville, Ouachita NF 28000 Road ponds, BAH, OC 322453). Its flight season in Oklahoma is documented to start a bit earlier than in nearby states, but its season overall is completely in line with the region (AR: 22 April–6 August, OC 375041, SEMC; KS: 14 May–8 August, OC 374969,

SEMC; LA: 16 April–28 July, OC 479246, Mauffray 2014; MO: 8 May–3 August, BugGuide 1065729, UMMZ; TX: 23 April–2 August, Abbott and Lasley 2019). There are reports of later dates into October,[15] which is out of line with this species' known phenology in the region. We wonder if these reports were in fact misattributions of *Libellula flavida*, the Yellow-sided Skimmer, which flies much later and is occasionally confused with the Spangled Skimmer, especially when reports involve immature males or females.

We know of one mixed-species pairing in Oklahoma involving *Libellula cyanea*. BS-P photographed a ♂ *L. cyanea* in copula with a ♀ *L. incesta* (Slaty Skimmer) at Sportsman Lake, Seminole County, on 29 June 2014 (OC 423872, 423904). The ♂ *L. cyanea* had some difficulty keeping a hold on the female during copulation. The ♀ *L. incesta* oviposited afterwards. To our knowledge this multi-species pairing has not been reported before (Figure 19.48). Bick and Bick (1981) only reported pairings of ♀ *L. cyanea* with *L. auripennis* (Golden-winged Skimmer, as *L. jesseana*, the Purple Skimmer, which is now a species confined to Florida) and with *L. flavida* (Yellow-sided Skimmer). While investigating putative hybrids involving this species, we discovered that some characters once thought unique to *L. cyanea* females were not so (Figure 19.47).

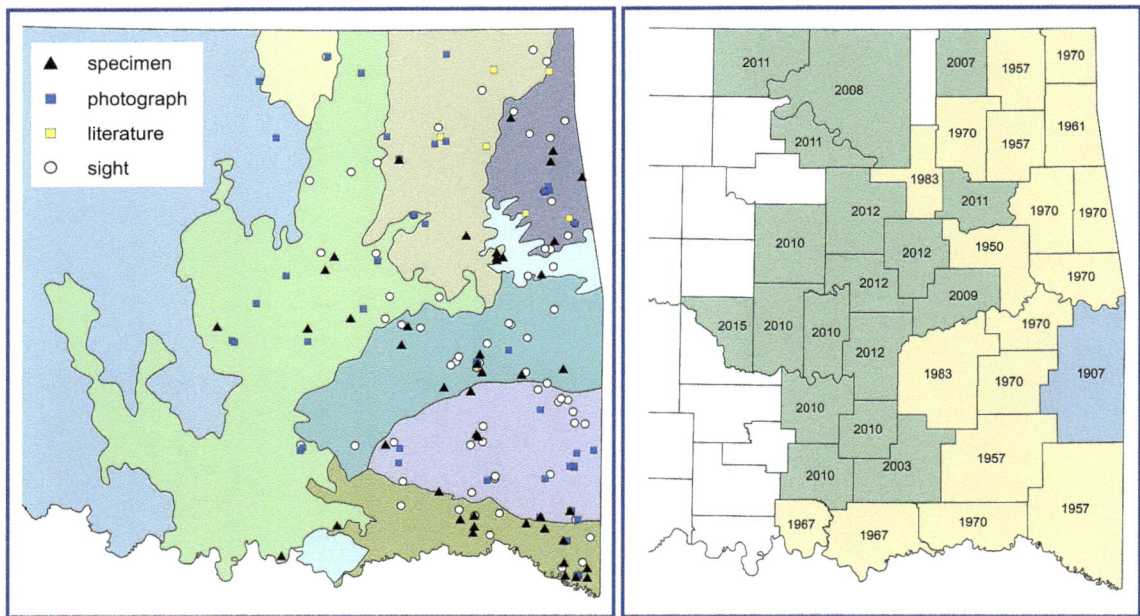

Figure 19.49 Documentation level of supporting records for *Libellula cyanea*, the Spangled Skimmer, in Oklahoma (left). Although not yet documented in the East Central Texas Plains, the species occurs in all of the other eastern ecoregions in the state. It has also been found in almost all of the eastern counties (right); its absence in some counties, for example, Washington County, is undoubtedly an artifact of survey effort during this species' flight season. Year within the county is that in which the species was first reported.

LIBELLULA INCESTA HAGEN, 1861 – SLATY SKIMMER

The Slaty Skimmer is a common species of principally lentic waters, although it can be numerous along slow-moving creeks and rivers. Observers elsewhere have described the species' habitat as creeks, muddy or rocky, and sloughs. In our experience in Oklahoma the species does tend toward sloughs, backwaters, boggy ponds, and sometimes creeks, but we have not taken note of the substrate. These spots are always at least near, if not directly in, deciduous woodland, where the species tends to be close to shade. Waters range from clear to muddy.

In Oklahoma the species is especially common in the eastern half of the state. There are scattered records farther west and, of course, a population in Comanche County. During the Bick/Hornuff era the species was known from 22 counties (7 recorded from the 1930s, five of which were known when Bird 1932 was published). The Bicks characterized the species as "widely distributed in eastern Oklahoma but … never abundant." (Bick and Bick 1957:6). We now have records of the species in 61 counties, and in our experience, although the species is more often seen in numbers around 20, it can be abundant at times. For example, we had 85 individuals at Boehler Seeps, Atoka County, on 20 July 2014 and 95 individuals at Red Slough WMA, McCurtain County, on 26 August 2015 (BS-P, MAP, DA, M Dillon). Ten years prior, we noted 125 at Red Slough, which is the largest count that we know of for the state. Williamson (1914b) reported a close second when F Collins had 115 (46♂, 69♀) near Wister, Le Flore County, on 2 August 1907, the first time the species was encountered in the state. That record was anomalous for the era. All other records for the species in the state with counts more than just a handful of individuals are from post-2000. It is possible that *L. incesta* has expanded its range and has become more abundant with the proliferation of lentic habitat in the state.

Libellula incesta Hagen, 1861
Slaty Skimmer

S5

2 August 1907

1870   1890   1910   1930   1950   1970   1990   2010

2 May   8 Oct

1 Jan 1 Feb 1 Mar 1 Apr 1 May 1 Jun 1 Jul 1 Aug 1 Sep 1 Oct 1 Nov 1 Dec 31 Dec

AR
KS
MO
TX

Figure 19.50 *Libellula incesta*, the Slaty Skimmer, A) ♂, Bull Creek in Vinita, Craig County, 7 July 2013 (©Bill Carrell, OC 401661). B) ♀, Medicine Park, Comanche County, 30 July 2012 (©Victor W Fazio III, OC 378252). Among the king skimmers (genus *Libellula*), only the omnipresent *L. luctuosa*, the Widow Skimmer, is more numerous and widespread than this species is.

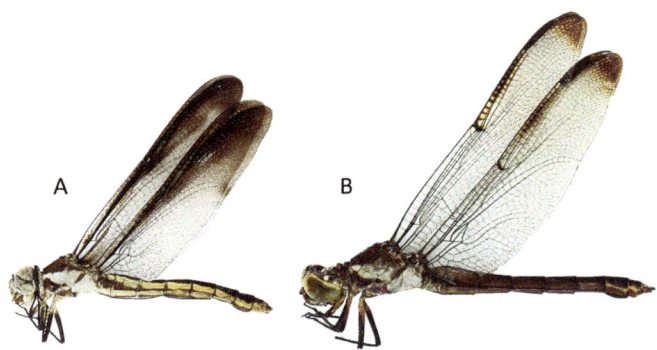

Figure 19.51 A) This somewhat teneral female *Libellula incesta*, the Slaty Skimmer, is unquestionably aberrant in both her size (about 10 mm smaller than typical) and wing pattern (not clear-winged as a typical *L. incesta*; B) typical *L. incesta*; SP 375, Lake Ozzie Cobb, Pushmataha County, 1 September 2012). She was captured at Schooler Lake, Choctaw County, on 2 July 2017 (MAP, SP 2438).

The species is known to fly from 2 May until 8 October (early: 2015, Lake Ozzie Cobb, Pushmataha County, 1 imm ♂, OC 430734; late: 2009, 2.5 mi NW of Mt. Herman, McCurtain County, 1♂, BAH, OC 315574). In northeastern Oklahoma it peaks in late June but tends not to be as abundant as the Great Blue Skimmer

(*Libellula vibrans*) in bottomlands (Smith-Patten et al. 2007). Elsewhere in Oklahoma, the species also peaks in late June, but it is typically the most numerous or second (behind *L. luctuosa*, the Widow Skimmer) most numerous *Libellula* encountered in the eastern part of the state. Oklahoma's flight season is similar to dates known for neighboring states (AR: 14 May–13 October, OC 374964, Harp and Rickett 1977; KS: 29 May–30 September, OC 375383, 315542; MO: 6 May–30 September, Trial 2005; TX: 1 April–15 September, OC 479439, Abbott and Lasley 2019).

We know of one mixed-species pairing with the Slaty Skimmer, which was with the Spangled Skimmer (*Libellula cyanea*; see that species for details). A decidedly aberrant female (SP 2438; Figure 19.51) and an unusual male (SP 2855) are also of note.[16] Finally, Needham et al. (2014) reported total length (TL) of *L. incesta* as 50–52 mm, abdomen length (ab) as 32–34 mm, and hindwing length (HW) as 36–42 mm. Others, for example, Curry (2001), Beaton (2007), and Rosche et al. (2008), have reported a wider range in total length, 47–54 mm, 45–56 mm, and 47–55.1 mm,[17] respectively. We also report a wider range: TL ♂ 48.0–55.0 mm, ♀ ≈50.0–50.8 mm; ab ♂ 31.5–36.3 mm, ♀ 33.4–34.6 mm; HW ♂ 36.3–42.0 mm, ♀ 41.2–42.0 mm.[18]

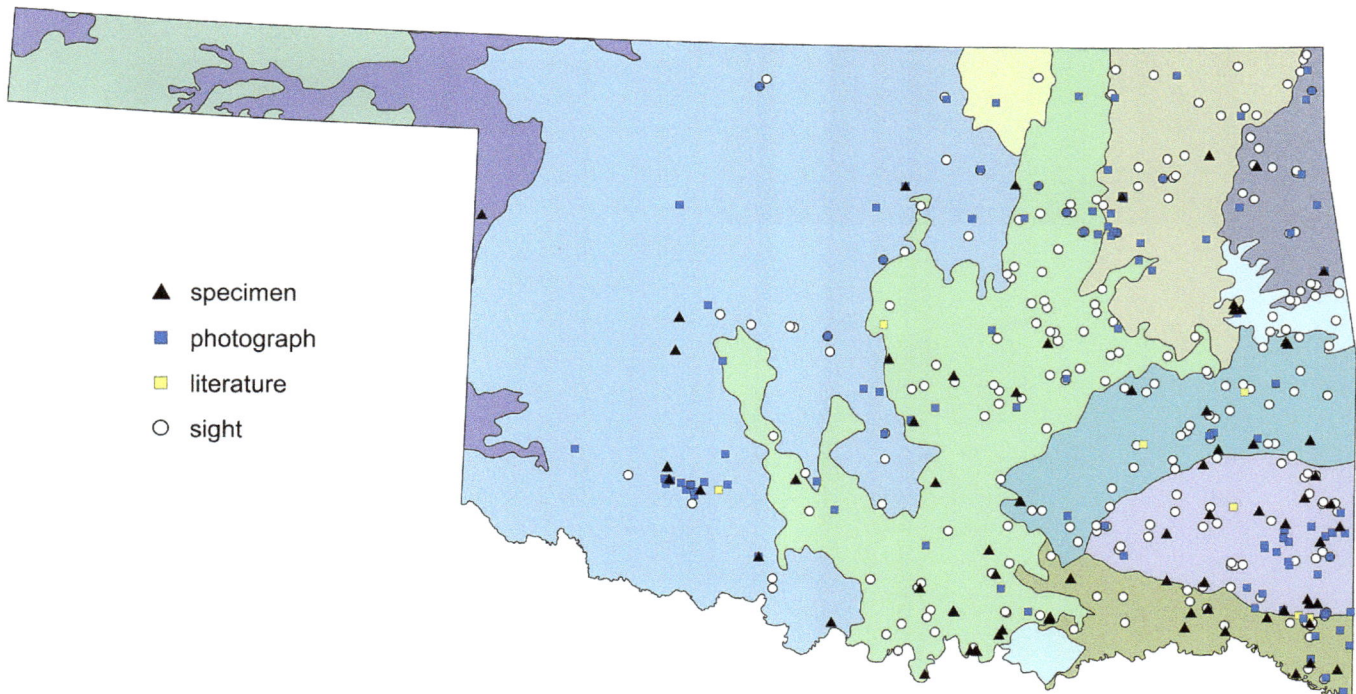

▲ specimen
■ photograph
□ literature
○ sight

Figure 19.52 Documentation level of supporting records for *Libellula incesta*, the Slaty Skimmer, in Oklahoma. Levels include specimen, archived photograph, literature reference, or sight record. Although sight records lack documentation, only those we feel are credible are plotted.

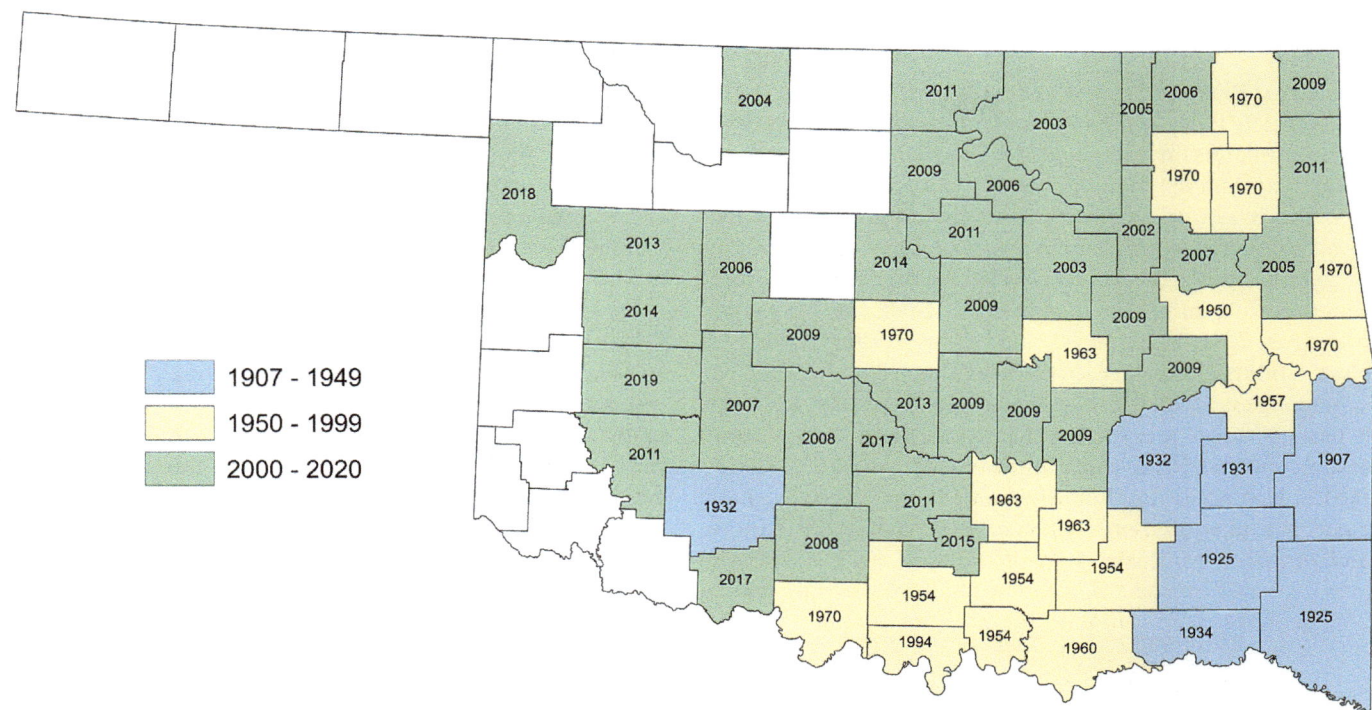

Figure 19.53 Counties in which *Libellula incesta*, the Slaty Skimmer, are known to occur in Oklahoma. Year within the county is that in which the species was first reported.

LIBELLULA VIBRANS FABRICIUS, 1793 – GREAT BLUE SKIMMER

The Great Blue Skimmer is generally uncommon but sometimes locally fairly common, particularly in wooded swamps or bottomlands in the eastern half of the state, where it frequents shaded, stagnant, muddy pools. It is commonly recorded in counts in the lower teens, roughly from a half to a fourth the typical abundance of the Slaty Skimmer (*L. incesta*), but it is recorded occasionally in larger numbers (e.g., 68 adults at Mohawk Park, Tulsa, Tulsa County, 16 August 2015, BC, and 60♂ and 5♀ at Wister WMA, Le Flore County, 11 August 2019, MAP).

Its range is much like that of the Slaty Skimmer, although it has not been recorded at nearly as many sites despite being reported from three more counties (*n* = 63). In the earliest years of collection, *Libellula vibrans*, a species of the Eastern Temperate Forests, was known from only Le Flore, McCurtain, and Pushmataha Counties. As is often the case with eastern species, the first record of the Great Blue

Skimmer in Oklahoma is from Williamson (1914b; 1♂, 8♀, near Wister, Le Flore County, F Collins, 2 August 1907). Bird (1932) reported only this Le Flore County record, but there were actually two records from McCurtain County at the time of his publication. The first was published by Ortenburger (1926b) as a Bar-winged Skimmer (*L. axilena*, misspelled as *axillena*). We re-identified this specimen (♀, 4 mi W of Broken Bow, McCurtain County, 8 July 1925, OMNH 922) as *L. vibrans*. Bird (1932:54) must have been suspicious of this record because, although also publishing it as *Libellula axillena* [sic], he said "One female identified by Williamson but with a query." The second record is a ♀ collected on 10 June 1931 by CC Deonier, which we came across at the OSU collection. Two years after Bird's publication, AEP collected the species for the first time in Pushmataha County, in May and June 1934 (11♂, 3♀, EMEC 82079–82091, OSU).

Figure 19.54 *Libellula vibrans*, the Great Blue Skimmer, A) ♂, Cat Creek in Claremore, Rogers County, 29 June 2014 (©Bill Carrell, OC 423930). B) ♀, Kaw Lake, Kay County, 7 July 2011 (©James W Arterburn, OC 427951). C) ♀, Roman Nose SP, Blaine County, 5 August 2012 (©Brenda D Smith, OC 381329). We thought we had found Blaine County's first *L. vibrans*, but an *Argiope* orb-weaver spider found it first.

In the Bick–Hornuff era, *Libellula vibrans* began to be known from outside southeastern Oklahoma, the first instances being when GH Bick reported it for Muskogee County in the summers of 1950–1951. Since, the species has been creeping toward northeastern and western Oklahoma (Figure 19.57; westward expansion appears to be the case, too, for Texas). In examining range shifts with this species, Marshall and Comanche Counties were instructive. Recall that the Bicks intensively surveyed at and near the OU Biological Station in Marshall County for all but five summers from 1952 until 1970. We have great faith that had *L. vibrans* been more than a vagrant to the area, the Bicks would have recorded it. Rather, it was reported only once, from the OU Biological Station on 23 June 1963, not by the Bicks, but by W McCarley (OMNH 159). The Bicks never reported the species from Marshall County and only twice in its neighbor, Johnston County (1961 and 1970, one adult each encounter). It has since been reported multiple times from that county, as it has from Love County, to the west, starting in the mid-1990s; however, the 1963 record for Marshall County stands as the only one from there. If it were not for the species' history in Comanche County, it would be easy to pass off the south-central records as mere happenstance. Comanche County received much survey effort during the OU expeditions era, as well as other times (surveyed on at least 370 days!), yet its first record of *L. vibrans* did not come until 2006 (28 July, Fort Sill MR, 1♂, VWF). If the species were resident in the county prior to that, then it would have somehow eluded the >50 surveyors (including us) during the 175 visits made to the county prior to the first record. Although not surveyed as intensively, other counties have similar histories. For example, despite surveys by Bick, Hornuff, and others, *L. vibrans* was not reported in Woodward County until 2011 (20 August, Boiling Springs SP, 5♂, 2♀, 10U, OC 331908, 331910, VWF). This species' wanderlust (e.g., to Weld County, Colorado, OC 482443) at play with landscape changes, such as the creation of impoundments and woody encroachment into areas of former prairie (e.g., westward spread of red cedars and the human tendency to favor trees over grasslands), it is no wonder *L. vibrans* is now found so much farther west than it was once known.

Westward range expansion of *Libellula vibrans* has brought it into contact with *L. comanche*, the Comanche Skimmer, and while *L. vibrans* has fanned out across Oklahoma, *L. comanche* has inched its way eastward in the southern part of the state (Figures 19.55 and 19.57). These range shifts have produced some interesting results, including at least two or three putative hybrids and other "intermediate" specimens (Figure 19.55). The first hybrid is a ♂ we collected on 21 September 2013 at Doby Springs, Harper County (SP 1036), which was one of 2♂ that day that we identified in the field as *L. vibrans*, albeit on the small side (TL: 54.4 mm; Needham et al. 2014 has TL = 56–63 mm), yet given the intermediate genitalia, we have determined it to be *L. comanche* × *vibrans* (Figure 19.55). A second, equally convincing hybrid, was taken at the same location on 12 July 2014 (♂, SP 1329) and provides for an excellent example of how photographic records can be problematic, given that we submitted as documentation for the species in Harper County a photo of that ♂ (OC 424597, now "unconfirmed"), which superficially appears to be *L. vibrans*, but which, upon a close examination of the specimen, is a hybrid. Predating these two records is a ♂ we collected at Roman Nose SP, Blaine County, on 4 August 2013 (SP 920), an individual as large as *L. vibrans* but with its hamules intermediate or nearer to those of *L. comanche*. Also of note are a ♂ taken in Coal County (Willis Spring, Bromide, 29 July 2019, SP 2867, MAP), in association with 5♂ *L. comanche* (one as SP 2866), that has hamules of *L. vibrans* but is small and has largely unmarked wings, and a ♂ taken in Roger Mills County (Spring Creek Lake, 11 August 2018, SP 2692, MAP) that has hamules of *L. comanche* but basal streaks on the wings and dark pterostigma. Interestingly, Bick and Bick (1981) did not present interspecific pairings involving *L. vibrans* or *L. comanche* (with each other or other species).

Libellula vibrans is known to fly in Oklahoma from 8 May to 30 October (early: 2017, Red Slough WMA, McCurtain County, 1♀, J Trahan, OC 462777; late: 2016, Ardmore Regional Park, Carter County, 1♂, MAP, OC 457239), but its known seasonality varies some throughout the region (AR: 2 May–12 October, OC 374689, Harp and Rickett 1977; KS: 23 May–2 September, OC 319183, 381497, Allison 1921 reported it flies until late October; LA: 19 March–7 October, OC 493548, 491542; MO: 8 June–30 September, Sims 2012; TX: 1 April–16 September, Abbott and Lasley 2019).

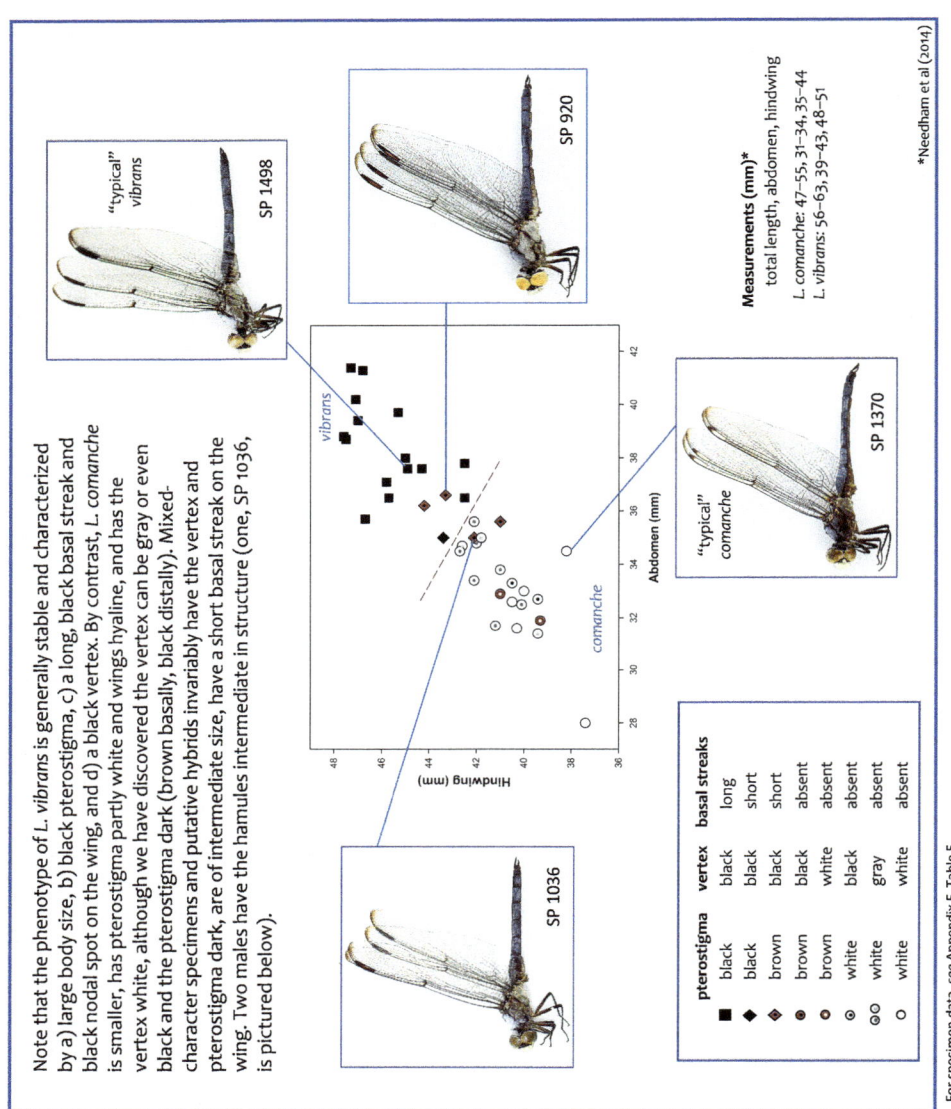

Note that the phenotype of *L. vibrans* is generally stable and characterized by a) large body size, b) black pterostigma, c) a long, black basal streak and black nodal spot on the wing, and d) a black vertex. By contrast, *L. comanche* is smaller, has pterostigma partly white and wings hyaline, and has the vertex white, although we have discovered the vertex can be gray or even black and the pterostigma dark (brown basally, black distally). Mixed-character specimens and putative hybrids invariably have the vertex and pterostigma dark, are of intermediate size, have a short basal streak on the wing. Two males have the hamules intermediate in structure (one, SP 1036, is pictured below).

pterostigma	vertex	basal streaks
black	black	long
black	black	short
brown	black	short
brown	black	absent
brown	white	absent
white	black	absent
white	gray	absent
white	white	absent

*Needham et al (2014)

Measurements (mm)*
total length, abdomen, hindwing
L. comanche: 47–55, 31–34, 35–44
L. vibrans: 56–63, 39–43, 48–51

For specimen data, *see* Appendix F, Table 5.

Figure 19.55 Coded scatter plot of *Libellula comanche* (Comanche Skimmer) and *L. vibrans* (Great Blue Skimmer) specimens from Oklahoma. Several specimens (brown diamonds) are intermediate and may be hybrids between these two species, interbreeding that was previously unknown, likely because the species' respective geographic ranges were until recently strictly allopatric (i.e., geographically isolated).

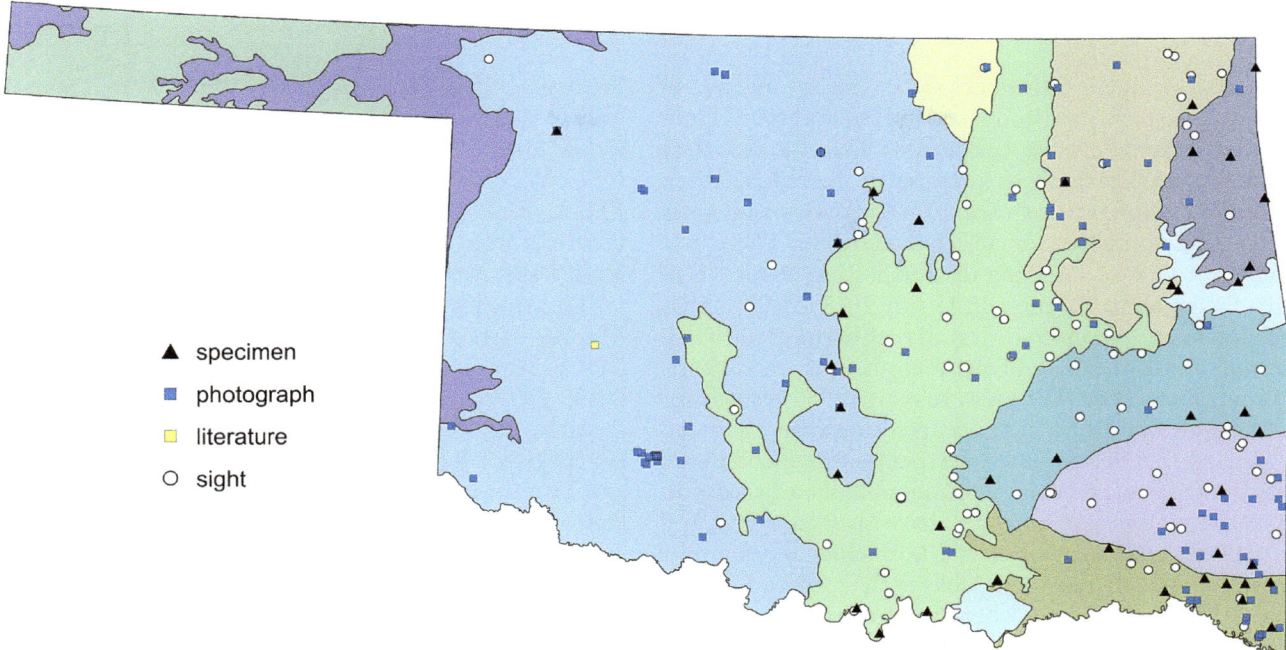

Figure 19.56 Documentation level of supporting records for *Libellula vibrans*, the Great Blue Skimmer, in Oklahoma. Levels include specimen, archived photograph, literature reference, or sight record. Although sight records lack documentation, only those we feel are credible are plotted.

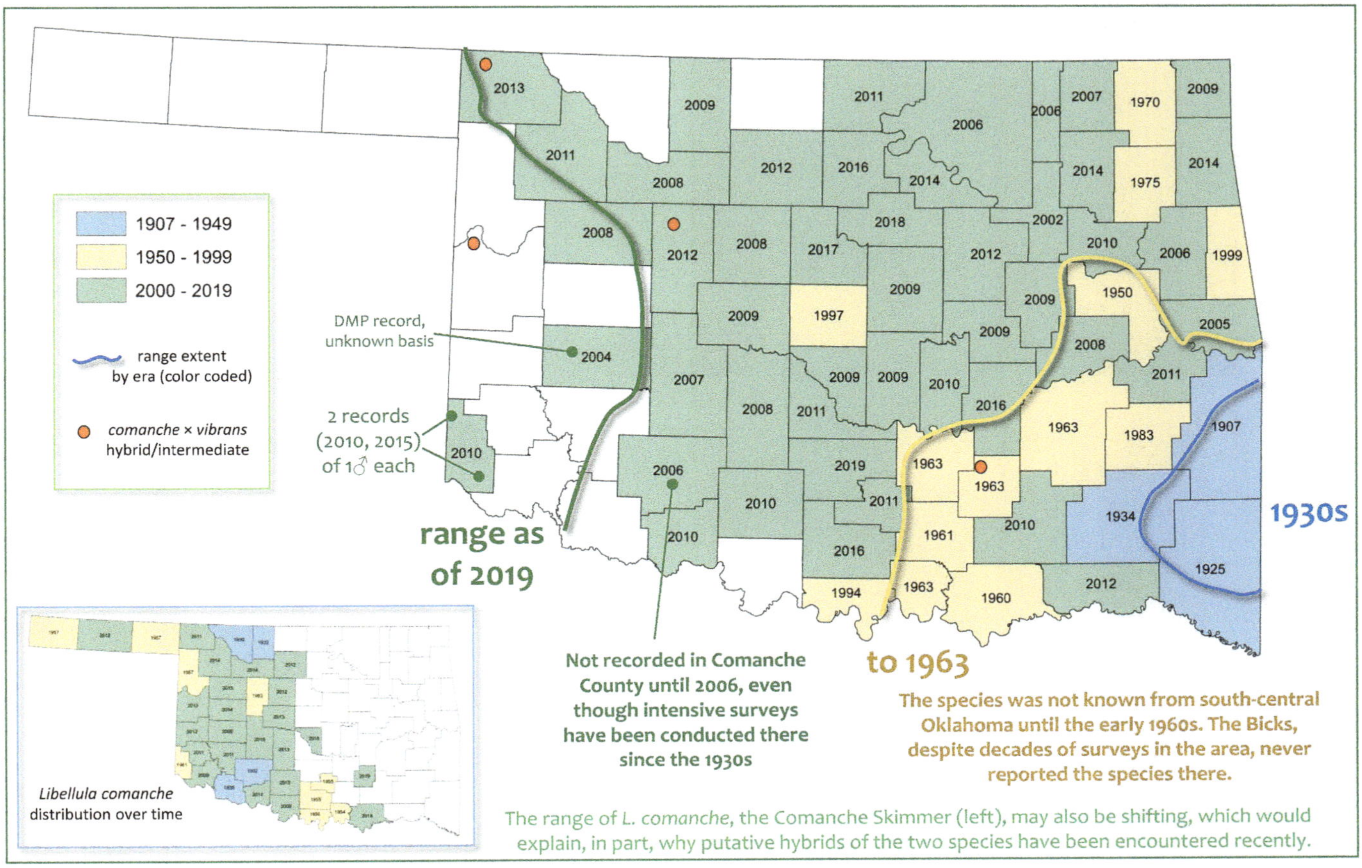

- ▲ specimen
- ■ photograph
- ▢ literature
- ○ sight

1907 - 1949
1950 - 1999
2000 - 2019

range extent
by era (color coded)

comanche × vibrans
hybrid/intermediate

DMP record,
unknown basis

2 records
(2010, 2015)
of 1♂ each

**range as
of 2019**

Libellula comanche
distribution over time

Not recorded in Comanche
County until 2006, even
though intensive surveys
have been conducted there
since the 1930s

to 1963

1930s

The species was not known from south-central
Oklahoma until the early 1960s. The Bicks,
despite decades of surveys in the area, never
reported the species there.

The range of *L. comanche*, the Comanche Skimmer (left), may also be shifting, which would
explain, in part, why putative hybrids of the two species have been encountered recently.

Figure 19.57 The geographical range of *Libellula vibrans*, the Great Blue Skimmer, has shifted over time, from four known localities in southeastern Oklahoma prior to 1935, to now with a range that covers most of Oklahoma. The same appears to be true elsewhere in the region.

LIBELLULA AURIPENNIS BURMEISTER, 1839 – GOLDEN-WINGED SKIMMER

The Golden-winged Skimmer is a scarce species of Oklahoma's eastern hardwood and mixed forests (Figure 19.26). It is a lentic species that can be found at a variety of ponds and seasonal pools (pristine to degraded), often embedded in coniferous or mixed forest, with open water and vegetated shorelines or adjacent marshes. There is a single lotic record, of a ♂ along the Little River 8 km NE of Cloudy, Pushmataha County, on 15 July 2018 (MAP). It has been found in only seven counties, all within the eastern half of the state, with the bulk of the records in the southeastern corner of the state. We know of roughly 60 records for Oklahoma, all but 3 of which occurred between 2006 and 2019. During that decade-plus, the species was encountered at least once per year. Although undoubtedly a breeder in the state, no nymphs are known to have been collected, and only 2 tenerals have been reported (1♀, Boehler Seeps, Atoka County, 2 July 2017, and 1♀, Atoka PHA, Atoka County, 1 June 2019), both at sites where tandem pairs have been reported (20 July 2014 and 2 June 2018, respectively).

The first time *Libellula auripennis* was recorded in Oklahoma was when TH Hubbell captured a single ♂ 1 mi NW of Page in the Ouachita NF, Le Flore County, on 15 September 1935 (UMMZ, Kormondy 1960). This was the only record for the 1930s. The 1950s did not fare any better, as GH Bick reported it only once (1♂, 10 July 1951, former OSU Wildlife Conservation Station, Muskogee County, FSCA, Bick and Bick 1957). The only other non-2000s record is from J Nelson and HC Reed of a single ♂ from Kulli Lake, McCurtain County, on 29 June 1983 (OSU).

The species is known to fly in Oklahoma from mid-May to early October (early: 17 May 2007, Kulli Lake, McCurtain County, 1♂, BAH, OC 314721; late: 7 October 2014, JT Nickel Preserve, Pumpkin Flats, Cherokee County, 1♂, JWA, OC 427389). Oklahoma's flight season begins later than it does in Louisiana and Texas but earlier than in Arkansas; it extends later than all of these states (AR: 31 May–14 September, OC 315135, Harp and Rickett 1977; LA: 19 April–9 September, OC 374784, Mauffray 2014; TX: 1 April–8 September, Abbott and Lasley 2019). In Missouri the species appears to have been collected twice, so the date range there spans just two days (6–7 August, Trial 2005, Sims 2012). The species was not reported for Kansas by Kennedy (1917), Montgomery (1967), Cringan (1979), or Beckemeyer and Huggins

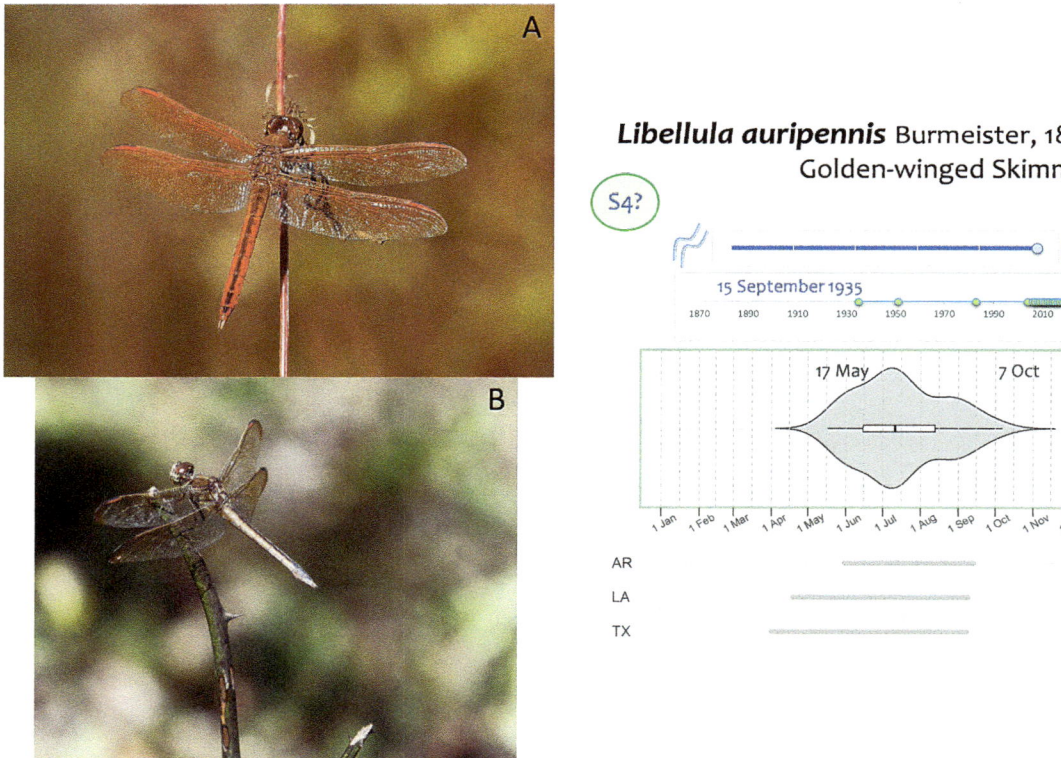

Figure 19.58 *Libellula auripennis*, the Golden-winged Skimmer, A) ♂, JT Nickel Preserve, Cherokee County, 7 October 2014 (©James W Arterburn, OC 427389). In Oklahoma, this species is seldom found away from the Ouachita Mountains, but it does occur regularly at this site in the northeast. B) ♀, 8 km S of Smithville, McCurtain County, 26 September 2009 (©Berlin A Heck, OC 315575).

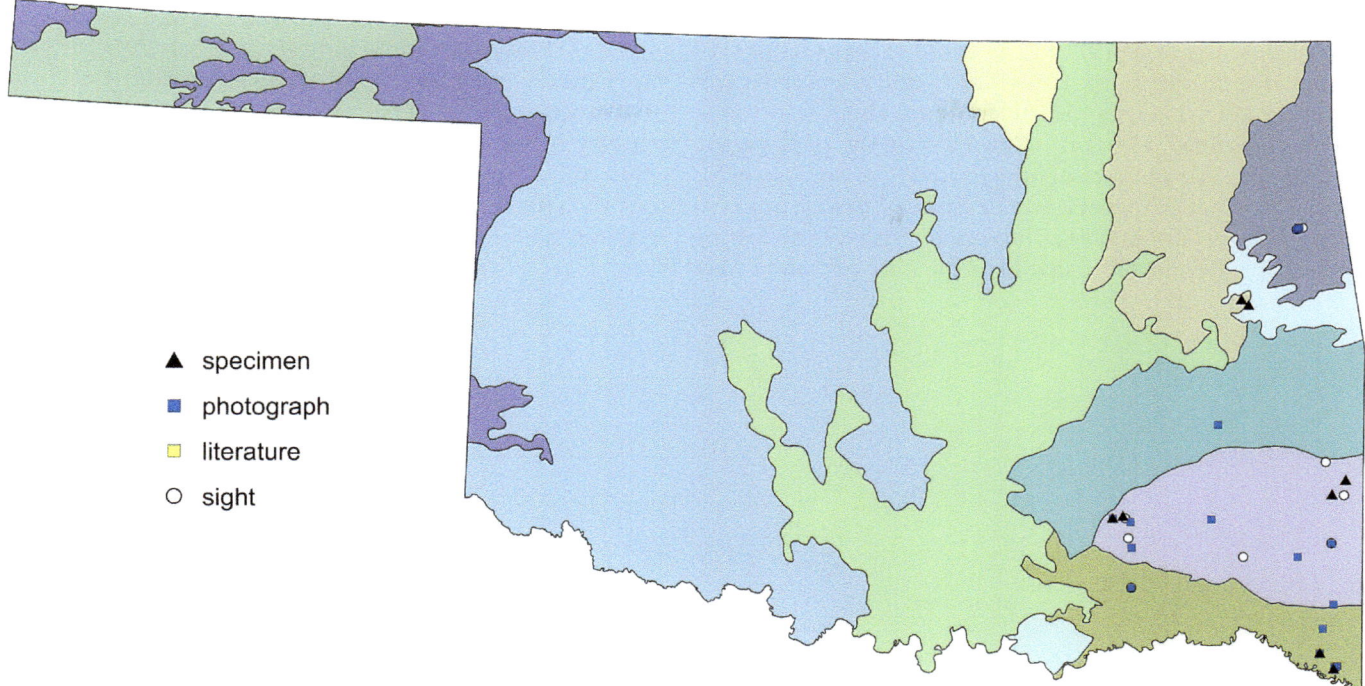

Figure 19.59 Documentation level of supporting records for *Libellula auripennis*, the Golden-winged Skimmer, in Oklahoma. Levels include specimen, archived photograph, literature reference, or sight record. Although sight records lack documentation, only those we feel are credible are plotted.

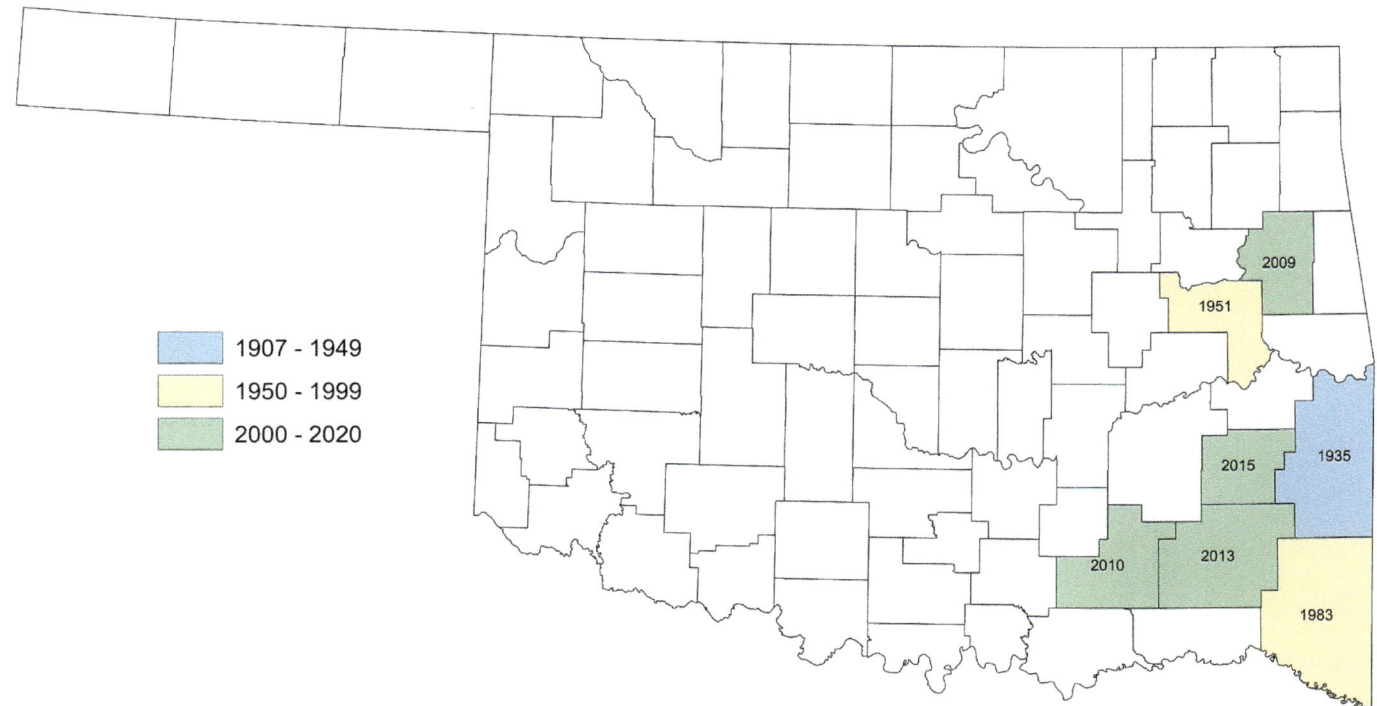

Figure 19.60 Counties in which *Libellula auripennis*, the Golden-winged Skimmer, are known to occur in Oklahoma. Year within the county is that in which the species was first reported.

(1997), nor are there any known specimens from the state; nonetheless, Allison (1921:51) reported the species at sites in and around Pittsburg, Crawford County, where he collected in 1913–1915, as "Fairly common about still water. July to October." He also indicated that the peak season was August. We admit to being puzzled here because in his species account Allison clearly describes *Libellula auripennis*, so it seems unlikely that he simply mistook something else for it.[19] But how is it that he is the only person to have ever recorded this species in Kansas, *in good numbers no less?* If valid, Allison's records indicate that *Libellula auripennis* has experienced a southward range retraction.

Our broadening perspective of the species' habitat and threats in the state have made us refine its conservation rank, thinking it is probably of less concern than we previously thought (Patten and Smith-Patten 2013b); we now suggest an "S4?" rank.

ORTHEMIS FERRUGINEA (FABRICIUS, 1775) – ROSEATE SKIMMER

The Roseate Skimmer is an uncommon but widespread species in Oklahoma. It is usually only seen a few at a time, but there are a dozen records of 6–15 individuals at a time, including when BS-P recorded 12♂ and 3♀ (3 pair) at Lake Hall, Harmon County, on 9 May 2014, and had the same (12♂, 3♀, but only one pair) at Moneka Park, Lake Waurika, Jefferson County, on the following day.

In Oklahoma the species is found most often at mud-bottomed ponds filled with muddy water, usually with little to no emergent vegetation to speak of, although typically there are at least a few bare vertical sticks used as perches. These muddy ponds often are drawn down to the extent that they attract few other odonates. Paulson (2009:404) described the species' habitat as open marshes, ponds, and ditches of various sizes, preferring "mud bottoms for larval habitat" and that "open water seems necessary." He also said the species likes "scummy stock tanks." All of these habitats have become more common since the Dust Bowl era and have

likely led to what appears to be the species' expansion in the state.

This is a species that we believe was first recorded in Oklahoma on 17 June 1931, when RD Bird collected 3♂ in Latimer County at "C Pool" and "Nepro," two localities that we have not been able to pinpoint (OMNH 2836, 2847, 2850). We are not certain that was the first state record because although we have located only three records from the 1930s, Bird (1932) reported the species for 10 counties (Comanche, Garvin, Haskell, Hughes, Latimer, McClain, Oklahoma, Pittsburg, Pushmataha, and Washita). He cited AE Pritchard as the source for the Oklahoma County record but did not indicate the provenance of the other records. We have not found documentation of the others, except for OMNH Latimer County specimens. The other two records we have for the 1930s come after Bird's publication—9 June 1934, Wilburton, Latimer County, 1♂, AEP, EMEC 82167 and 12 June

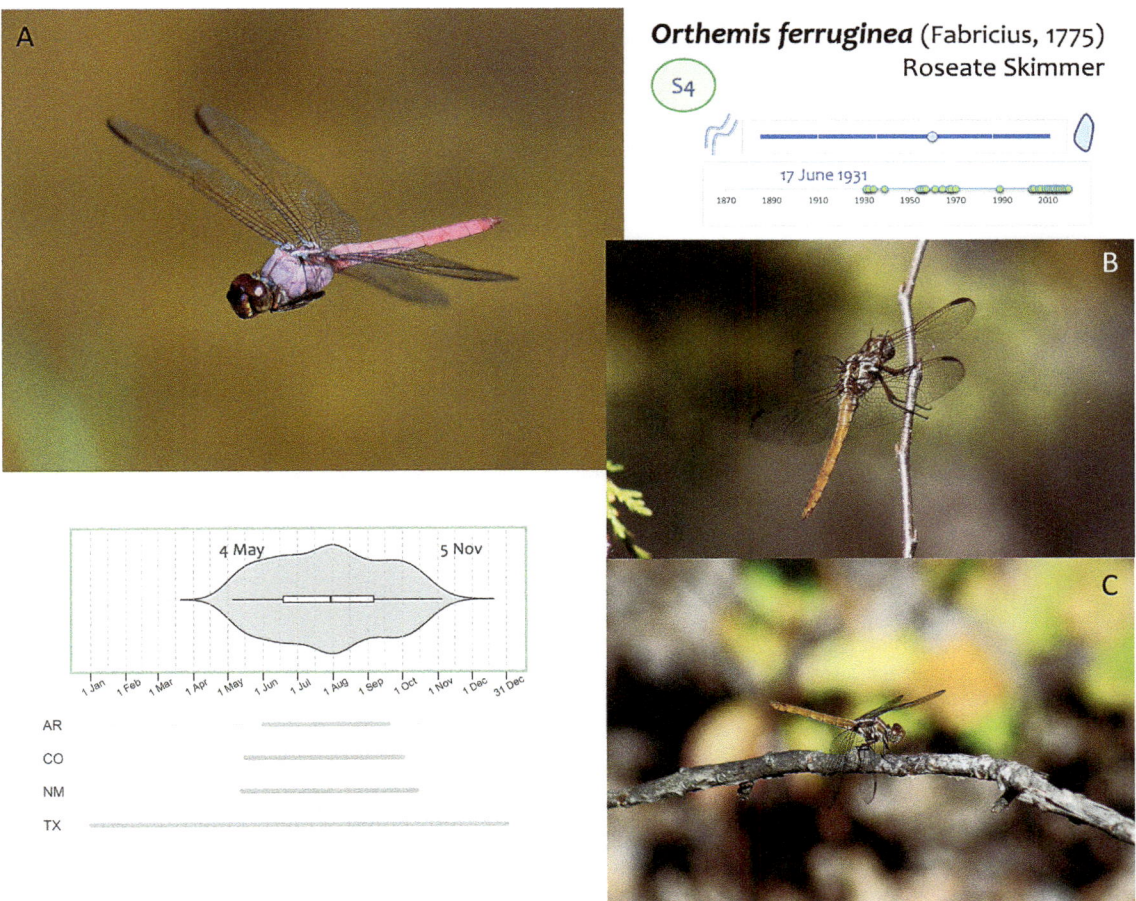

Figure 19.61 *Orthemis ferruginea*, the Roseate Skimmer, A) ♂, Turkey Creek Reservoir, Garfield County, 15 August 2012 (©Jason R Heinen, OC 379427). B) immature ♂, Great Plains SP, Kiowa County, 5 November 2017 (©Michael A Patten, OC 474668), which provided Oklahoma's late flight date. C) ♀, Bullfrog Creek 3 km ENE of Pink, Pottawatomie County, 29 September 2016 (©Brenda D Smith, OC 456478).

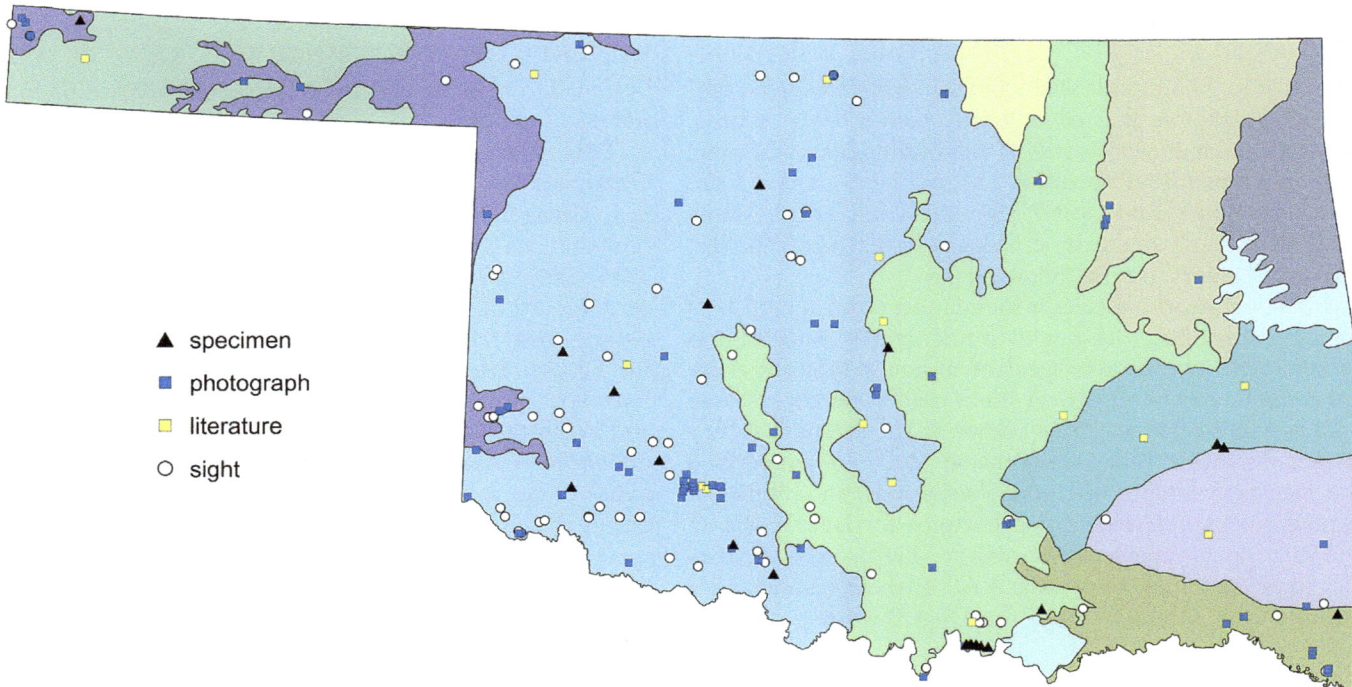

Figure 19.62 Documentation level of supporting records for *Orthemis ferruginea*, the Roseate Skimmer, in Oklahoma. Levels include specimen, archived photograph, literature reference, or sight record. Although sight records lack documentation, only those we feel are credible are plotted.

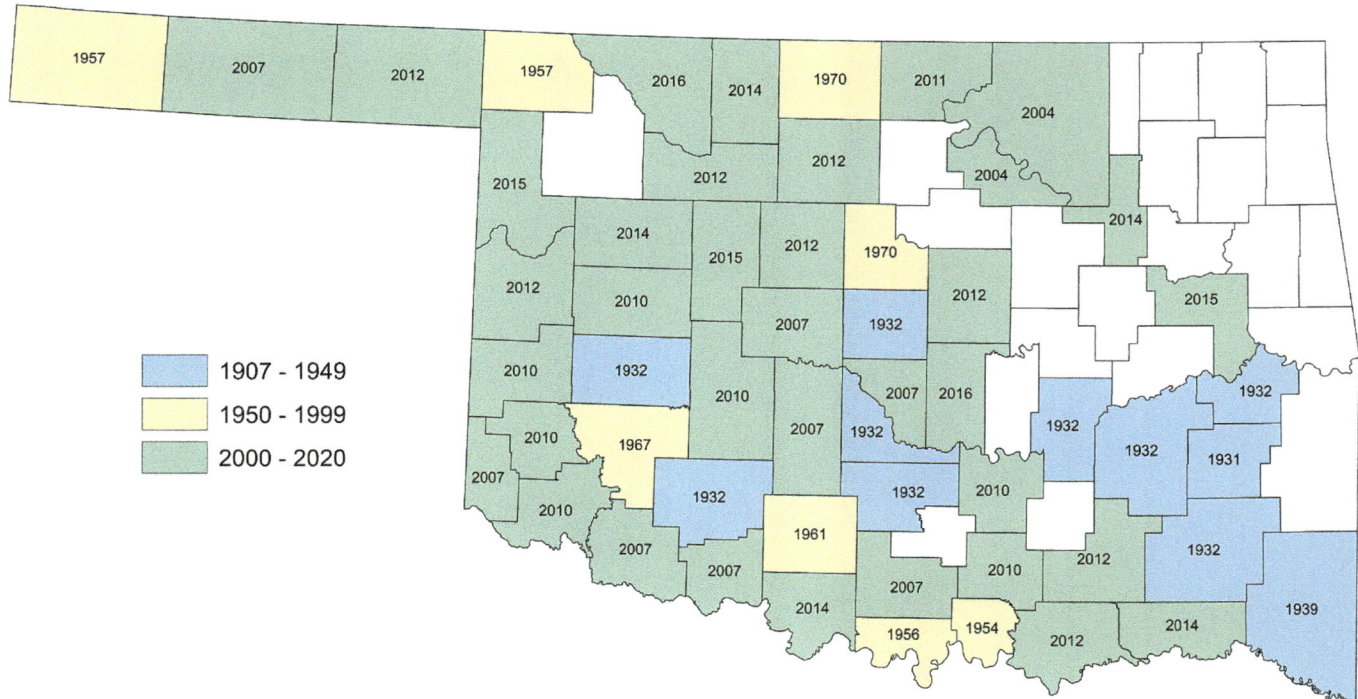

Figure 19.63 Counties in which *Orthemis ferruginea*, the Roseate Skimmer, are known to occur in Oklahoma. Year within the county is that in which the species was first reported.

1939, Eagleton, McCurtain County, 1♂, RW Kaiser, WT Nailon, OSU. It may be that Bird reported sight records for various counties, given the species' distinctive coloration and conspicuousness.

In the Bick/Hornuff era, the species seems to have become regular and was documented in an additional eight counties, as far as the opposite corner of the state (Cimarron County, at the tip of the panhandle). But it was not until the recent decade that records of the species' numbers really burgeoned. Of the roughly 175 records of the species, about 145 are from 2004–2019. During that span it was added to another 36 counties. It now is an expected species each year between early May and early November (4 May 2012, 1♂, Purcell Lake, McClain County; 5 November 2017, imm ♂, Great Plains SP, Kiowa County, MAP, OC 474668; Figure 19.61). Oklahoma's flight season is the longest of all its neighbors except that of Texas (AR: June–18 September, Harp and Rickett 1977, OC 6620; CO: 16 May–1 October, OC 431073, 437549; NM: 13 May–13 October, OC 329222, 284281; TX: year round, Abbott and Lasley 2019).

Bick and Bick (1958) described oviposition of the Roseate Skimmer at Cowan Creek, Marshall County. They observed one pair on 15 July 1955 at a muddy, turbid impoundment where the water was 5 cm (2 in) deep. The female hovered over the water no higher than about 25 cm (10 in), then dipped her abdomen into the water 5–10 times, hovered, dipped again 5–10 times, then moved to another spot. The male of this pair continued to circle about 60 cm (2 ft) above the female during oviposition, but other females they witnessed ovipositing were lone. This description mirrors observations we have made of the species elsewhere in Oklahoma.

PERITHEMIS TENERA (SAY, 1839) – EASTERN AMBERWING

The Eastern Amberwing (jokingly referred to as the Pugnacious Amberwing) is a strongly lentic species. It can be found along shorelines and within vegetation at all types of lentic habitat, including, as Bick and Bick (1958) found at their study site along Cowan Creek, Marshall County, at muddy impoundments. All of the 24 adults they encountered during their study were at the deep and muddy impoundment that had high turbidity and little vegetation. We have also found the species in such places, including along turbid, nearly still streams, as well as at clear-water wetlands and everything in between. It occurs in all 77 Oklahoma counties, with no apparent geographic variation in abundance, apart from being generally scarce in the panhandle.

This species is common enough that one often dons blinders when in its presence. But one never knows what might be in store for a watchful eye. A marvelous example is a perplexing encounter BS-P had with the species. Not one in favor of perpetuating traditional gender roles, nonetheless she took a double take when she noticed what looked like a ♂ Eastern Amberwing ovipositing while a ♀ mate guarded. At first thinking perhaps she had been taken by heat stress, but then realizing it was a rather cool day, she suddenly doubted her ability to differentiate males from females, so she took a swing with her net, capturing both individuals. Sure enough, she saw what she saw … a ♂ "ovipositing" and a ♀ "mate guarding," reversed behaviors we were unaware this species exhibited.[20]

Figure 19.64 *Perithemis tenera*, the Eastern Amberwing, A) ♂, Lexington WMA, Cleveland County, 1 June 2017 (©Kate Goodenough, OC 464376). B) A danger of being little: a *Diogmites* robber fly ("hanging thief") snacks on a ♂ at Brandt Park in Norman, Cleveland County, 19 July 2017 (©Emily A Hjalmarson, OC 467902). C) ♀, Tallgrass Prairie Preserve, Osage County, 11 June 2014 (©James W Arterburn).

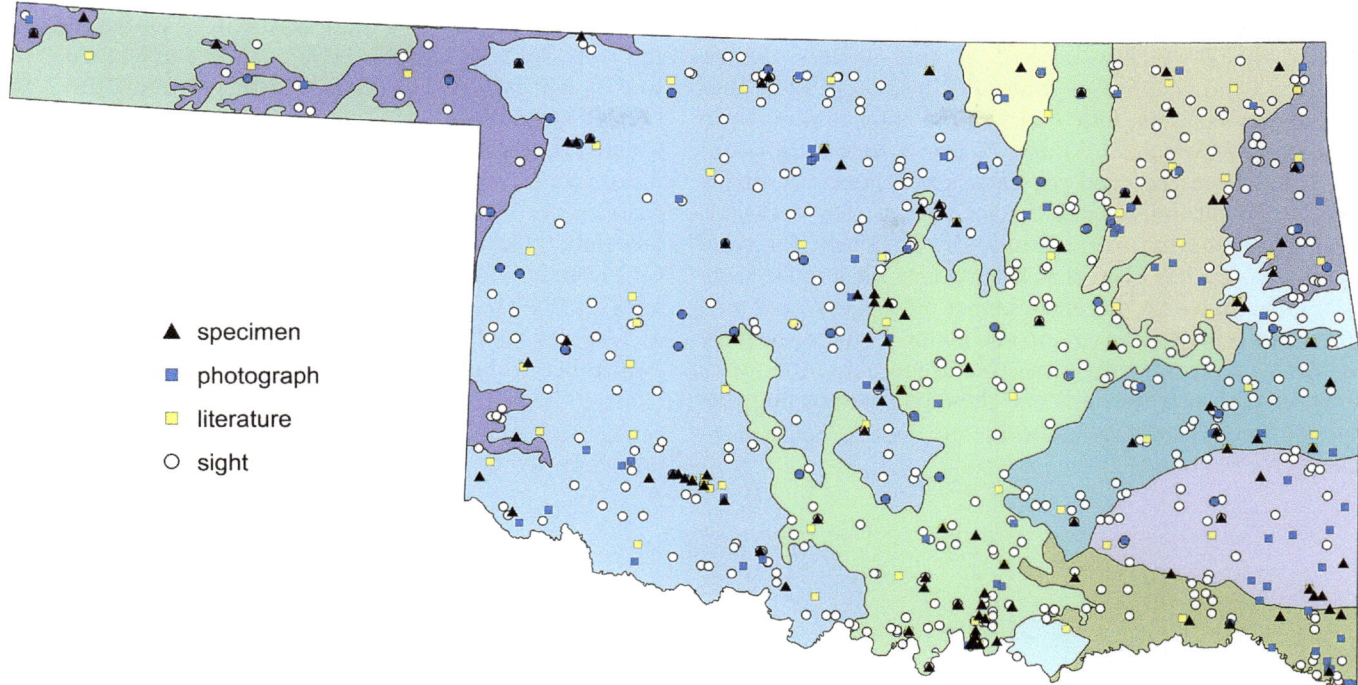

Figure 19.65 Documentation level of supporting records for *Perithemis tenera*, the Eastern Amberwing, in Oklahoma. Levels include specimen, archived photograph, literature reference, or sight record. Although sight records lack documentation, only those we feel are credible are plotted.

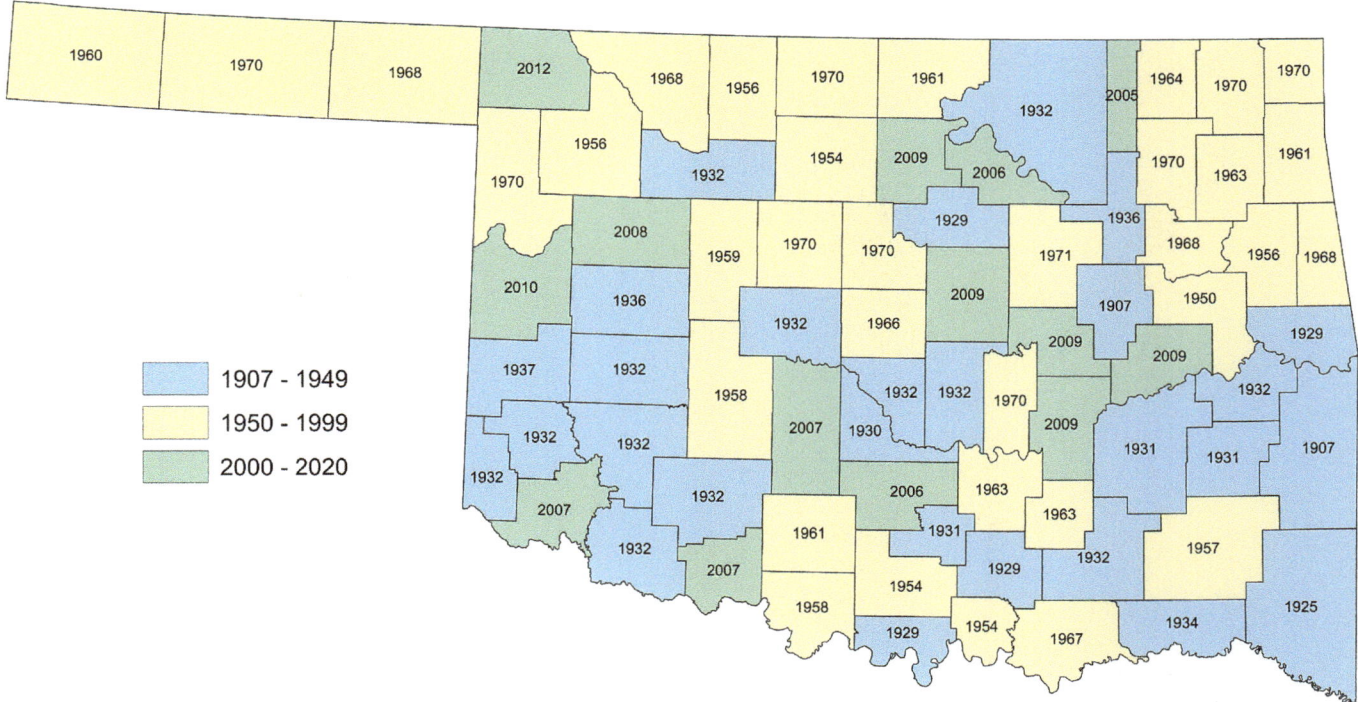

Figure 19.66 Counties in which *Perithemis tenera*, the Eastern Amberwing, are known to occur in Oklahoma. Year within the county is that in which the species was first reported.

Perithemis tenera was first recorded in Oklahoma in 1907 when EB Williamson had 9♂ and 3♀ near Wister, Le Flore County, on 3–4 June (Williamson 1914b). Bird (1932) reported it as both *Perithemis domitia* (unknown in the state; see Appendix A) and *P. tenera*, across 22 counties. It has been recorded regularly since, generally in the range of 20–30 individuals, but on at least four occasions it has been reported in the hundreds. Curiously, for years our impression has been that this species likes to sleep in: often enough, we do not encounter it on surveys until well after many other species are up and about. Perhaps the daily tussles this species incites with other dragonflies (often with much, much larger species!) tire these pugnacious pugilists so much so to prevent early rising.

The species has a long flight season in Oklahoma, extending from 10 April to 6 November (early and late date from Red Slough WMA, McCurtain County: 10 April 2012, DA, OC 374373; 6 November 2009, BAH, OC 315792). This seasonality corresponds fairly well with Arkansas, Missouri, and Texas, but it is markedly longer than the known date ranges for Colorado, Kansas, and New Mexico (AR: 2 April–2 December, Harp and Rickett 1977; CO: 23 June–7 October, OC 423455, 437917, 437918; KS: 14 May–27 September, OC 374972, 504407; MO: 14 May–30 October, Trial 2005; NM: 31 May–9 October, OC 375414, 317505; TX: 23 March–28 November, Abbott and Lasley 2019).

BRACHYMESIA FURCATA (HAGEN, 1861) – RED-TAILED PENNANT

The genus *Brachymesia* is tiny: it contains a mere three species, each of them largely Neotropical yet each of them ranging northward into the southern United States. Remarkably, all three species have been found in Oklahoma, albeit two have appeared solely as vagrants. In the Southern Great Plains, one species, *B. furcata*, the stunning Red-tailed Pennant, normally does not occur north of central Texas, but inexplicably found its way north to Sequoyah SP, Cherokee County, 4–5 October 2019. Bill Carrell added the species to the state list when he discovered 2♂ during the opening hours of the Oklahoma Biological Survey's annual Bioblitz! (OC 504675). The next morning, he and MAP counted, as carefully as they could given that the sewage pond the pennants frequented was fenced off, 6♂ (OC 504796), one of which MAP collected (SP 2925). Over the course of the day they surveyed another half dozen small sewage ponds in the park, which occupies a triangular peninsula that juts south into Fort Gibson Lake. Each of these ponds looked outwardly much like the discovery pond, yet none supported pennants.

This species has a curious history in Oklahoma and would have appeared in this book even had BC not struck the motherload. On 23 July 2012, Doug Danforth snapped a blurry photo of an adult ♂ at Rush Lake in the Wichita Mountains WR, Comanche County. He submitted the photo a week later (OC 378215). Danforth said that the individual "was perched on a half submerged tree branch about twenty feet out from shore. It had a brown thorax, red abdomen with upturned cerci, swollen segment two and a small dark basal hind wing patch." After much discussion with GW Lasley about photo quality (blurry) and observer experience (extensive), we were confident as we could be that Danforth correctly identified the species but agreed the documentation was insufficient for a first state record. With the glory of hindsight, we feel it is sufficient for a first county record, and so we place the species on the Comanche list yet treat BC's record as the state's first.

Prior to its occurrence in Oklahoma, the northernmost records on the Southern Great Plains were from Collin and Lubbock Counties, Texas, just <200 km from Oklahoma. Other vagrant records sprinkled across the southeastern United States emphasize that the species wanders north its geographic range.

Brachymesia furcata (Hagen, 1861)
Red-tailed Pennant

vagrant

first confirmed: 4 October 2019

Figure 19.67 *Brachymesia furcata*, the Red-tailed Pennant. Oklahoma's first well-documented record was of a ♂ at Sequoyah SP, Cherokee County, 4 October 2019 (©Bill Carrell, OC 504675). Remarkably, a second ♂ was observed that day at the same small sewage pond in the park, and even more remarkably, at least 6♂ were counted at the pond on 5 October 2019, one of which was kind enough to be caught for posterity (SP 2925).

BRACHYMESIA HERBIDA (GUNDLACH, 1889) – TAWNY PENNANT

The Tawny Pennant[21] is a neotropical species that has strayed north to Oklahoma only once, when a ♂ was studied at length (roughly four hours!), averted all efforts at capture, and was photographed distantly but definitively at a clear, spring-fed pond at Waurika WMA, Cotton County, on 1 October 2017 (MAP, OC 473523). This occurrence did not come as a complete surprise because the Tawny Pennant was on our "short list" of species likely to appear one day in Oklahoma given a historical record of a ♂ collected in Cass County, Nebraska, on 2 August 1914 (USNM) and three recent (2008–2015) records for Chaves County, New Mexico (OC 284261, OC 317350, and OC 437000), to say nothing of scattered records for Texas north of the species' usual range. The Waurika WMA individual was furtive and was hassled repeatedly by various larger ♂ Four-spotted Pennants (*Brachymesia gravida*) present contemporaneously.

Brachymesia herbida (Gundlach, 1889)
Tawny Pennant

vagrant

one record: 1 October 2017

Figure 19.68 *Brachymesia herbida*, the Tawny Pennant, ♂, Waurika WMA, Cotton County, 1 October 2017 (©Michael A Patten, OC 473523), Oklahoma's sole record. Forgive the less-than-ideal photograph—it was heroic … and the best available given that this ungrateful little bug refused to be caught or even approached closely.

BRACHYMESIA GRAVIDA (CALVERT, 1890) – FOUR-SPOTTED PENNANT

The Four-spotted Pennant is an uncommon and unpredictably irruptive species of clear, vegetated ponds and (usually small) lakes along the southern fringe of the state, with scattered records farther north. At times the species is fairly common, such as when we recorded 32 adults on one day in McCurtain County (30 at Grassy Slough and 2 at Red Slough, 22 June 2003). Our highest record was far outnumbered by an encounter that VWF had with the species in which he estimated there were 300 adults flying at the Cache sewage ponds, Comanche County (23 July 2009, OC 314277). More typical are counts of one to several. The species' occurrence varies by year, with some years having few sightings and others having many, as was the case in 2013, 2016, and 2018, when, respectively, 59 adults from 7 counties, 63 adults from 6 counties, and 81 adults from 9 counties were reported. By contrast, in 2012 we recorded the species only twice—a single ♂ and a single ♀ (7 July, Harmon County, Lake Hall and 1 September, Coal County,

Coalgate, SP 366)—and we received only four reports of the species in 2014 and five in 2015. Numbers by visit varies, too. For example, MAP visited Jasper "Jap" Beaver Lake,[22] Jefferson County, on 22 June 2019 and encountered one immature male. Just over a month later, BS-P visited the same local and counted 24♂ (one as SP 2889, 7 August) during a quick survey. If there are any sites that one could call "reliable" for the species in Oklahoma, this location may be one given that it has been encountered there four times in three separate years, which can also be said of Keechi Park, Cyril, Caddo County. Lake Louis Burtschi in Grady County, Waurika WMA in Cotton County, and Red Slough WMA in McCurtain County may be "reliable" locations for the species as well, with the latter site holding the record of 14 reports in ten years (2003–2017).

The species has a spotty early history, too. No one prior to 1955 recorded it in the state. GH Bick obtained the first record, of 2♂ at the Ardmore Rod and Gun Club lake,

Figure 19.69 *Brachymesia gravida*, the Four-spotted Pennant, A) ♂, Doby Springs, Harper County, 7 August 2015 (©Michael A Patten, OC 434751), the northernmost ever found on the Great Plains. B) ♀, Mountain Park WMA, Kiowa County, 7 September 2007 (©Victor W Fazio III, OC 263006). C) ♀, Lake Elmer, Kingfisher County, 23 August 2014 (©Bill Carrell, OC 426263). Older females such as this one can darken to the point of appearing to be males.

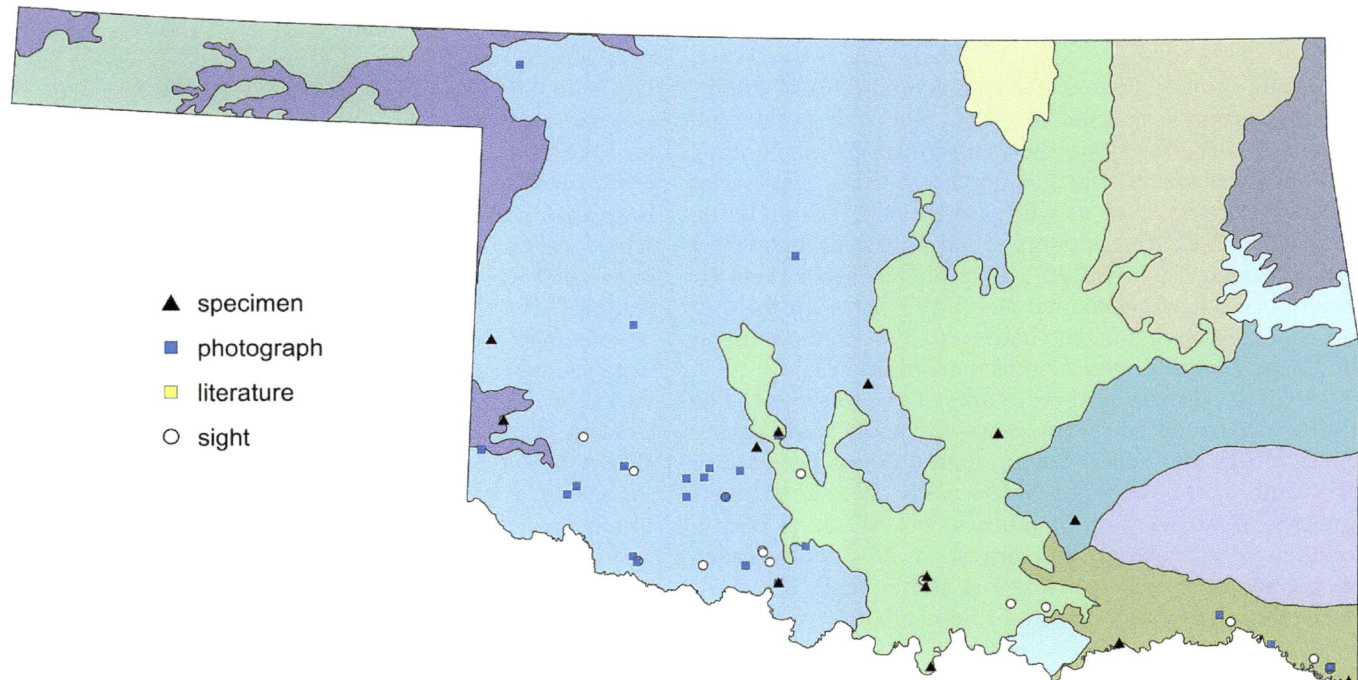

Figure 19.70 Documentation level of supporting records for *Brachymesia gravida*, the Four-spotted Pennant, in Oklahoma. Levels include specimen, archived photograph, or sight record. Although sight records lack documentation, only those we feel are credible are plotted.

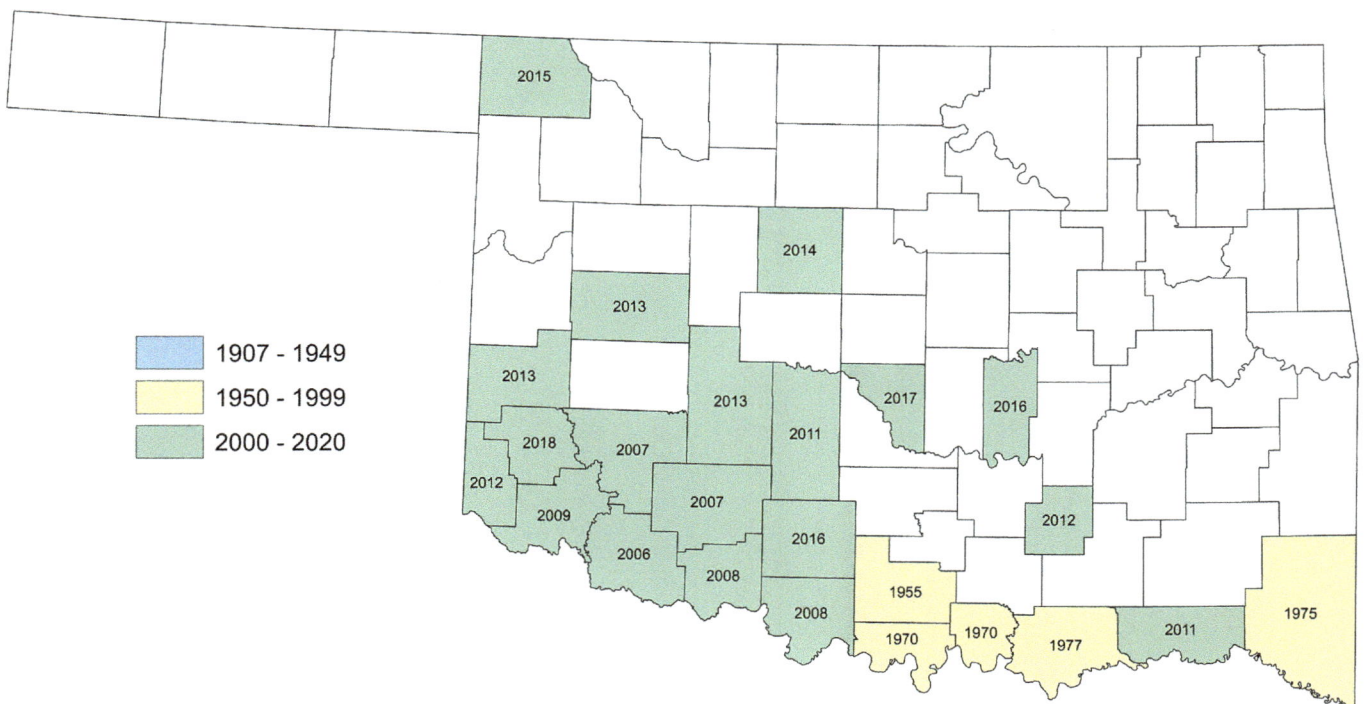

Figure 19.71 Counties in which *Brachymesia gravida*, the Four-spotted Pennant, are known to occur in Oklahoma. Year within the county is that in which the species was first reported.

Carter County, 7 July 1955 (FSCA). It was recorded two other times by Bick and Hornuff in July 1970 (Love and Marshall County, FSCA) and then twice as nymphs in 1975 and 1977 (Bryan and McCurtain County, OMNH 1887, 1888). Those five records over a 22-year period are in stark contrast to the >75 records in the 17 years from 2003 to 2019. The species has been seen just about every year since 2003, and in that span it was added to 18 counties, bringing the county total to 23 for the state. The number of records in the latter era, along with its distribution, leads us to conclude that the species has expanded its range into the state (Figure 19.71).

Oklahoma's flight season is more-or-less what one would expect in this region, although all nearby states fall short of Texas (OK: both from Red Slough WMA, McCurtain County, 8 May 2007, 1♀, BAH, OC 314712 and 19 October 2009, 1♂, DA, OC 315535; AR: 31 May–30 September, OC 445310, 456891; LA: 24 April–24 October, OC 430649, Mauffray 2014; NM: April–7 October, Paulson 2009, OC 284276; TX: 20 March–15 November, Abbott and Lasley 2019).

CELITHEMIS ELISA (HAGEN, 1861) – CALICO PENNANT

The Calico Pennant is a generally uncommon to locally fairly common species found primarily in the eastern half of Oklahoma, although as is the case for many eastern species, Comanche County, in the southwestern part of the state, harbors a small, geographically isolated population. The species generally is seen fewer than 10 at a time, but we have recorded as many as 90 (John Wells Lake, Haskell County, 31 May 2010). In Oklahoma it tends to be found at or near clear (often spring-fed and fishless) ponds or small coves of clear-water lakes with emergent vegetation and surrounding grassland. Pritchard (1935) found the species to be abundant on 10 June in Quinton, Pittsburg County, at a fairly large open lake with many cattails (*Typha*). *Celithemis elisa* perched with *C. fasciata*, the Banded Pennant, "facing the wind from the tips of high reeds with the abdomen down and their wings seemingly poised to hold their balance." (Pritchard 1935:8).

This species provides another example of a lentic species whose geographic range has expanded with the dramatic increase of lentic habitat in Oklahoma in the wake of the Dust Bowl. When the state had few lakes and ponds, the species was only record three times, but records increased markedly since post-Dust Bowl damming. It was first collected in Oklahoma on 17 June 1931 by RD Bird in Latimer County (at "C Pool" and "Nepro," 11 adults, including 4 pairs, OMNH 2134, 2146–2148). The other two early records are from 1934 when AE Pritchard and CA Sooter collected in the east-central part of the state (10 June, 3♂, AEP, Quinton, Pittsburg County, EMEC 81652–81653, OSU; 8 July, AEP and CA Sooter, Okmulgee County, ♂ at OSU, ♀ EMEC 81651). Bick and Hornuff recorded the species regularly, a pattern that has held to the present. There are now records for 46 counties. Three of those counties were well covered by the OU expeditions, and yet the species was not recorded in them until recent decades: Comanche (2006), Le Flore (2011), and McCurtain (1983). Those late detections argue for either a range expansion or that the species has become more common, or both.

Celithemis elisa is known to fly in Oklahoma from late April to early October, which is perfectly in line with regional dates (OK: 25 April 2015, 1♀, McGee Creek SP, Atoka County, MAP sight record and 5 October 2002, 1♂, Oxley Nature Center, Tulsa County, BC, OC 334187;

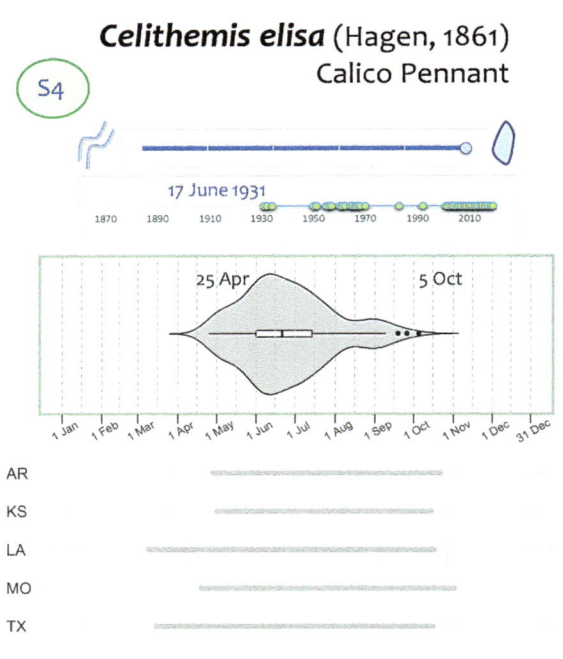

Figure 19.72 *Celithemis elisa*, the Calico Pennant, A) ♂, JT Nickel Preserve, Cherokee County, 9 June 2013 (©Bill Carrell, OC 401578). B) ♀, Bixhoma Lake, Wagoner County, 26 April 2012 (©James W Arterburn, OC 374589). Of the three more widespread *Celithemis* pennants on the Southern Great Plains, *C. elisa* is the most finicky, preferring clear (often spring-fed and fishless) ponds in wooded habitats.

AR: 28 April–22 October, OC 7649, 457097; KS: 1 May–mid-October, OC 374665, Allison 1921; LA: 9 March–17 October, Mauffray 2014, OC 334144; MO: 19 April–2 November, Sims 2012; TX: 15 March–16 October, OC 479041, Abbott and Lasley 2019). We have no records of nymphs, but we do have two records of tenerals: 1♀, Ellis County WMA, Commission Creek, Ellis County, 28 May 2016, MAP, and 12 individuals at Three Rivers WMA, McCurtain County, 3 May 2015.

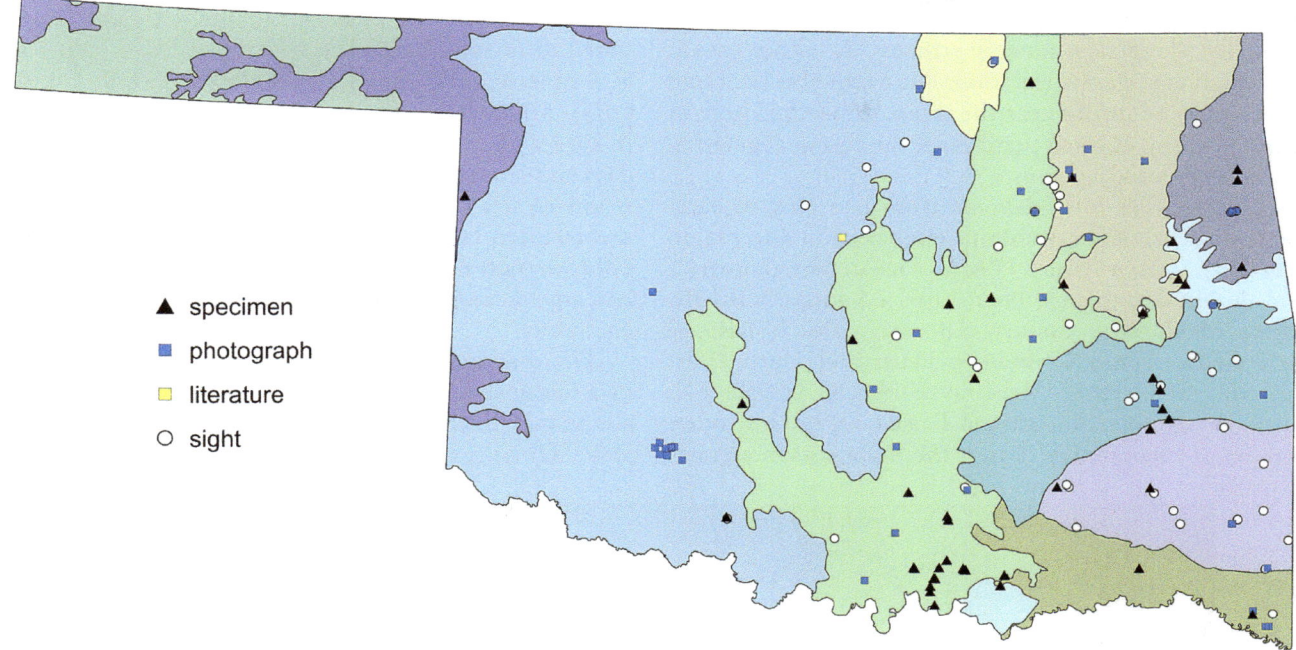

Figure 19.73 Documentation level of supporting records for *Celithemis elisa*, the Calico Pennant, in Oklahoma. Levels include specimen, archived photograph, literature reference, or sight record. Although sight records lack documentation, only those we feel are credible are plotted.

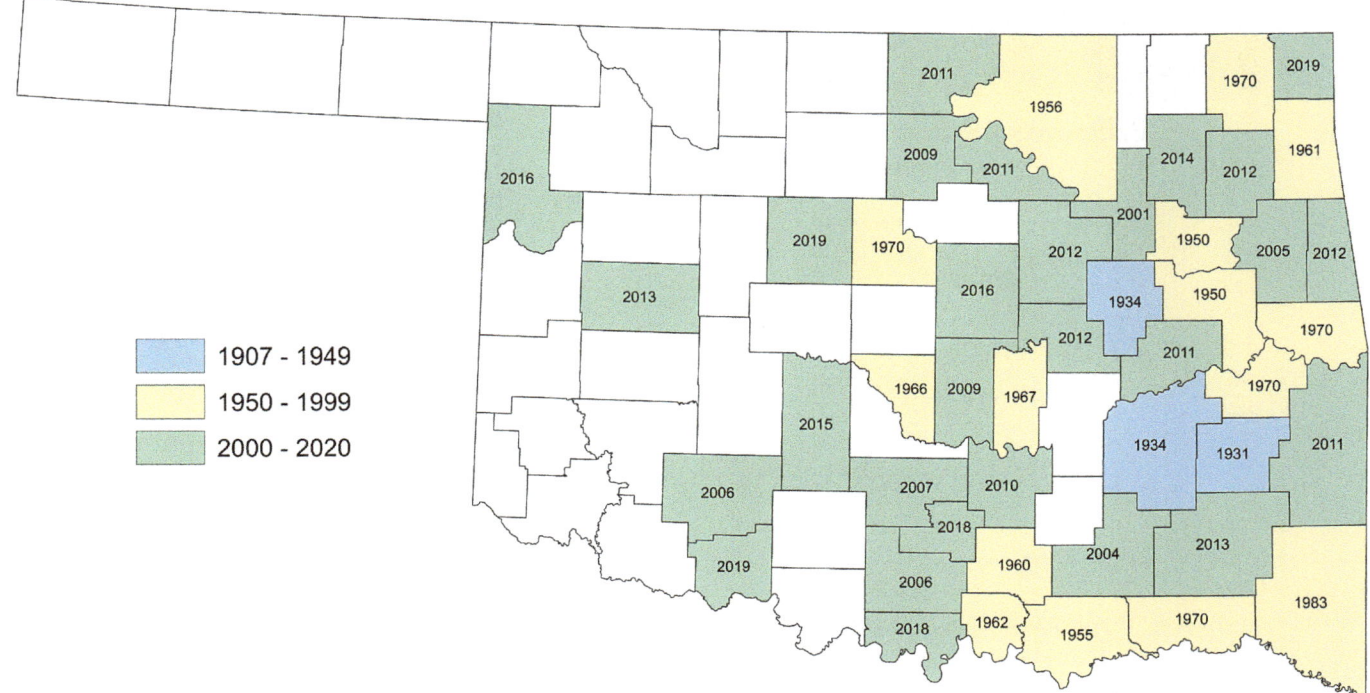

Figure 19.74 Counties in which *Celithemis elisa*, the Calico Pennant, are known to occur in Oklahoma. Year within the county is that in which the species was first reported.

CELITHEMIS FASCIATA KIRBY, 1889 – BANDED PENNANT

The Banded Pennant has a curious distribution in Oklahoma. It is reasonably common in southeastern Oklahoma, but there are also populations (possibly disjunct) in the northeast, southwest, and even along the west-central parts of the state. It is found at clear ponds and lakes of various sizes and, occasionally, along slow-flowing streams. In either case it prefers much emergent vegetation, but it can also be found in grass, reeds, or shrubs near the shore, or perched high in dead trees nearby. It often perches on the tips of vegetation to face into the wind.

The species flies in Oklahoma from late May to early October, which is comparable to elsewhere in the region (9 May 2017, Three Rivers WMA, McCurtain County, 1 teneral ♂, MAP and 8 October 2009, 2.5 miles NW Mt. Herman, McCurtain County, 1♂, OC 315573; AR: 14 May–17 September, OC 444439, Harp and Harp 2003; LA: 17 May–31 October, Mauffray 2014, OC 323958; TX: 29 April–19 October, Abbott and Lasley 2019). How often the species is encountered during the flight season appears to vary widely. This variation may simply be an artifact of how much people managed to get to the field in a given year, but since 2003 the records have varied from 1 to 16. It is an uncommon to fairly common species, generally ten or fewer adults seen at any one time, although often it is noted in double digits. We have recorded as many as 120 at a time in early June (Vian Lake, Sequoyah Co, 6 June 2015). Abundance also seems to vary by year; for example, in 2012 we personally encountered the species on 5 separate days and saw a total of 29 individuals, encountering between 1 and 12 at a time. In 2013 we saw the species on 16 days and counted 140, with 1 to 45 individuals at a time. And in 2015, we noted it on 14 days, counted 229 individuals, and had anywhere from a single individual to 120 at any given encounter.

The Banded Pennant provides yet another example of a lentic species that has likely spread and increased in abundance in the wake of the creation of lakes across the state. During the OU expeditions it was recorded only

Figure 19.75 *Celithemis fasciata*, the Banded Pennant, A) ♂, Rogers State University in Claremore, Rogers County, 29 June 2014 (©Bill Carrell, OC 423917). B) ♂, Sooner Lake, Pawnee County, 10 August 2013 (©Brenda D Smith, OC 403428). Note limited wing markings; this species is renowned for its individual variation. C) ♀, Pontotoc Ridge Preserve, Pontotoc County, 14 June 2014 (©Bill Carrell, OC 423137).

four times, including the first time it was collected in the state (12 July 1925, likely by AI Ortenburger, 0.5 mi W of Sawyer along the Kiamichi River, Choctaw County, 1♂, OMNH 912; sources of other records are: EMEC 81667, OMNH 161, OSU, USNM 293097, Bick notes). Bird (1932) reported the species in Choctaw and Latimer Counties, the latter under the junior synonym of *C. monomelaena* (see Appendix A for record details). Pritchard added the species to Pittsburg two years after Bird's paper. Between 1950 and 1994, the species was reported about 25 times from 12 counties (added to 10 counties). Since 2002, we have received approximately 160 reports from 42 counties. During the 2000s, the Banded Pennant was added to 34 counties, bringing the total of counties the species has been recorded in to 49 (although we do not know the basis for the Johnston County record reported by Donnelly 2004b). Again, it strikes us as odd that the OU expeditions did not record the species in a handful of counties they surveyed intensively, counties in which the species was first recorded much later: Cleveland (2008), Comanche (1989), Le Flore (2013), and McCurtain (1983).

On a final note for the species, we know of one nymph record: 21 April 1989, D Bass collected two nymphs at the Wichita Mountains WR, West Cache Creek, Forty Foot Hole, Comanche County (UCO).[23] We have noted tenerals numerous times, as singletons up to 20 individuals at a time. The first time we encountered 20 tenerals was at Deep Fork WMA, Okfuskee County, on 28 June 2015, and the second time was at Coon Clear Lake, Latimer County, on 16 June 2013, when we also noted 25 adults.

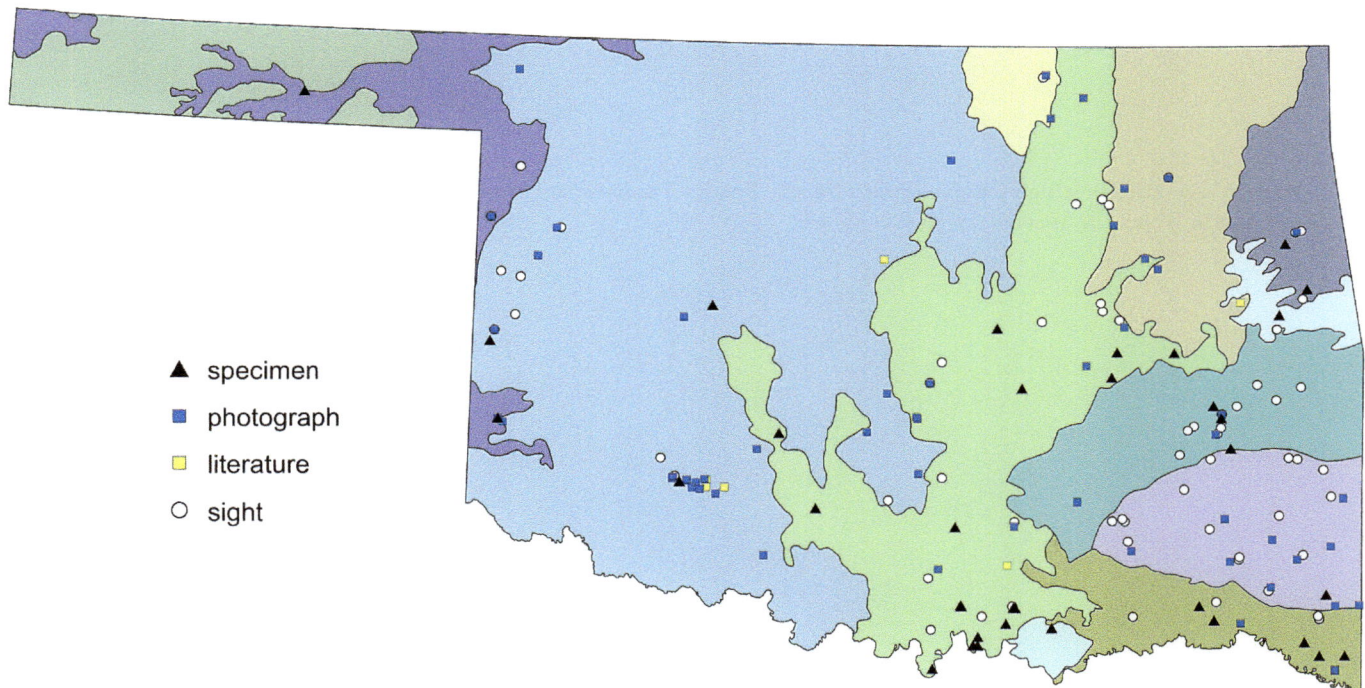

▲ specimen
■ photograph
□ literature
○ sight

Figure 19.76 Documentation level of supporting records for *Celithemis fasciata*, the Banded Pennant, in Oklahoma. Levels include specimen, archived photograph, literature reference, or sight record. Although sight records lack documentation, only those we feel are credible are plotted.

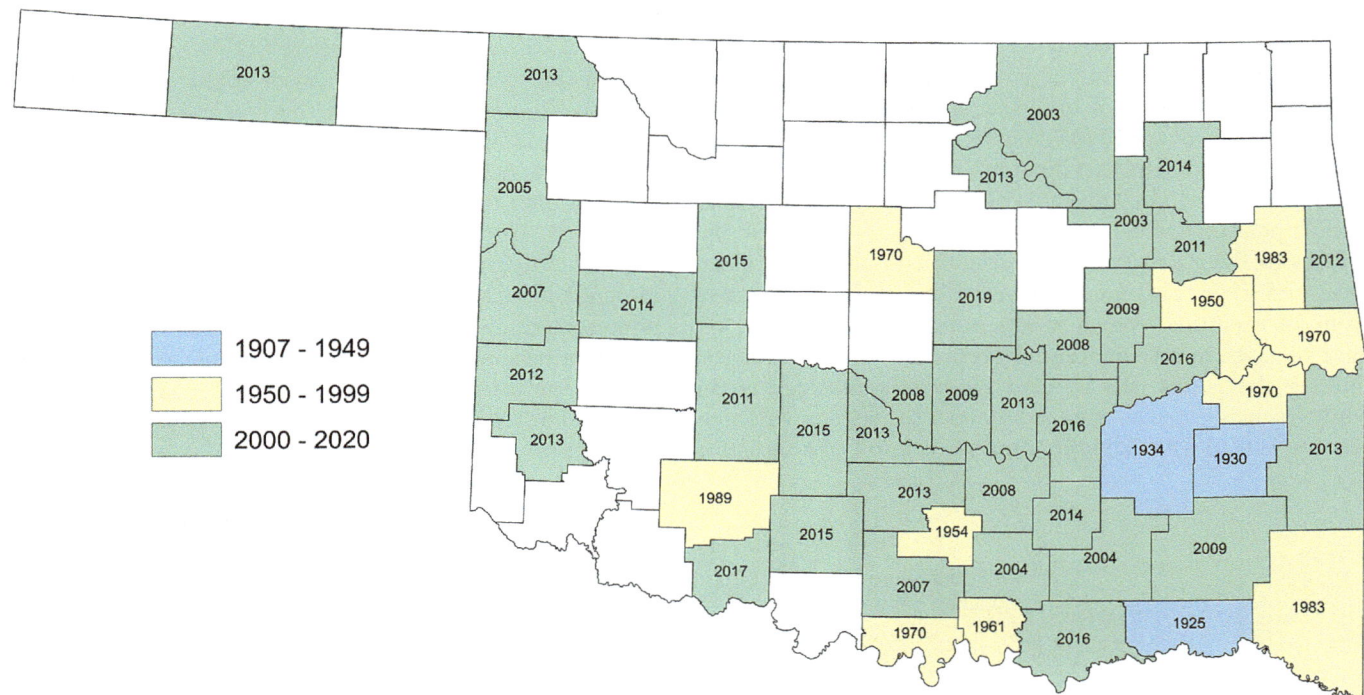

Figure 19.77 Counties in which *Celithemis fasciata*, the Banded Pennant, are known to occur in Oklahoma. Year within the county is that in which the species was first reported.

CELITHEMIS EPONINA (DRURY, 1773) – HALLOWEEN PENNANT

The Halloween Pennant is easily the most common and widespread *Celithemis* species in Oklahoma. It is found in similar habitats as the Banded and Calico Pennants (*C. fasciata* and *C. elisa*), but it perches in grasslands or fields much more than the other two, often perching at the tips of grasses or rushes (*Juncus* sp), looking much like the orangey-brown inflorescence (Figure 19.78).

As with *Celithemis fasciata* and *C. elisa*, *C. eponina* appears to have become more common and widespread since the Dust Bowl era. It was first reported for Oklahoma on 2 July 1925 from "1 mile west of the Arkansas-Oklahoma line, on the

Red River" (Ortenburger 1926b:221). Bird (1932) reported only two counties for the species—Love and McCurtain. It is unclear why he did not report it for Comanche given that O Sandoz collected 2♂ on 21 July 1932 at Ketch Lake in the Wichita Mountains WR and Bird determined them to be *C. eponina* (OMNH 2141). It appears that Bird reported other records collected in the summer of 1932, so perhaps it was simply an oversight. Pritchard collected the species again in Comanche County in 1934, as well as in Choctaw and Pittsburg Counties (5 adults from June–July, EMEC 81656, 81659–81660, OSU). The species was collected later in the

Figure 19.78 *Celithemis eponina*, the Halloween Pennant, A) ♂, 11 km ENE of Prue, Osage County, 9 June 2015 (©James W Arterburn, OC 432495). B) ♀, 1 km E of Turner Falls, Murray County, 3 August 2014 (©Randy L Kelley, OC 425742). C) decorating the vegetation at Brown, Bryan County, 18 July 2011 (©Tom Kompier, OC 332266), and underlining its status are (easily) our most common *Celithemis*.

decade two other times, both times by non-OU collectors: Carter (1939, OSU) and Tulsa (1937, UMMZ) Counties. It was recorded much later in some counties that the OU expeditions covered heavily—Cleveland (first recorded in 2008), Latimer (2009), and Le Flore (2014, but perhaps a bit earlier).[24]

Records increased dramatically in GH Bick and LE Hornuff's time, and they have continued to increase since (approaching 500 records between 2000 and 2019). The species is now known from all 77 Oklahoma counties, and is counted routinely in the 10–20 range, although there are a handful of times it has been counted >100, all of which are from July, which is the month the species peaks in abundance. The highest counts include: MAP estimated 150 individuals at Lake Frederick, Tillman County, on 24 July 2016, and we had 120 in the Wichitas, Comanche County, on 8 July 2012. Interestingly, it appears that Kansas has a similar history. Kennedy (1917) reported the species for two counties (Douglas and Pratt), Allison (1921) said that he captured it once during his 1913–1915 study, and Cringan (1979) said that although the species flew throughout August, he encountered it "infrequently." SEMC has only 5 early, i.e., 1911–1924, adult specimens, but almost 80 collected 1971–1980. We doubt such an apparent increase in distribution and abundance is simply a result of expanded survey effort, unless our trust in the ability of earlier surveyors is misplaced. The species' dispersal ability, as indicated by a study conducted in Marshall County (Stewart and Murphy 1968), is an element likely at play here. Two of the adult *Celithemis eponina*, 1♂ and 1♀, dispersed 0.53 km; a distance that conjures an image of leap-frogging from pond to pond as the number of ponds increased across the landscape over time post-Dust Bowl.

The Halloween Pennant's flight season in Oklahoma is from early May to late October, which is most similar to Arkansas and Missouri (OK: 7 May 2017, Red Slough WMA, McCurtain County, J. Donnell, 1♂, OC 462996 and 30 October 2016, Healdton Municipal Lake, Carter County, 1♂, MAP; AR: 2 May–13 October, OC 374685, Harp and Rickett 1977; CO: 8 June–1 September, OC 409903, CSU; KS: 9 June–17 September, OC 485925, SEMC; LA: 20 April–1 October, USNM 292702, BYUObserv 6177; MO: 11 May–30 October, Sims 2012; TX: 8 April–17 November, Abbott and Lasley 2019). Pritchard (1935:9) said that *C. eponina* was "making its debut … in the first part of June" as *C. verna* (Double-ringed Pennant) made "its last appearance of the season." His early observation remains generally correct.

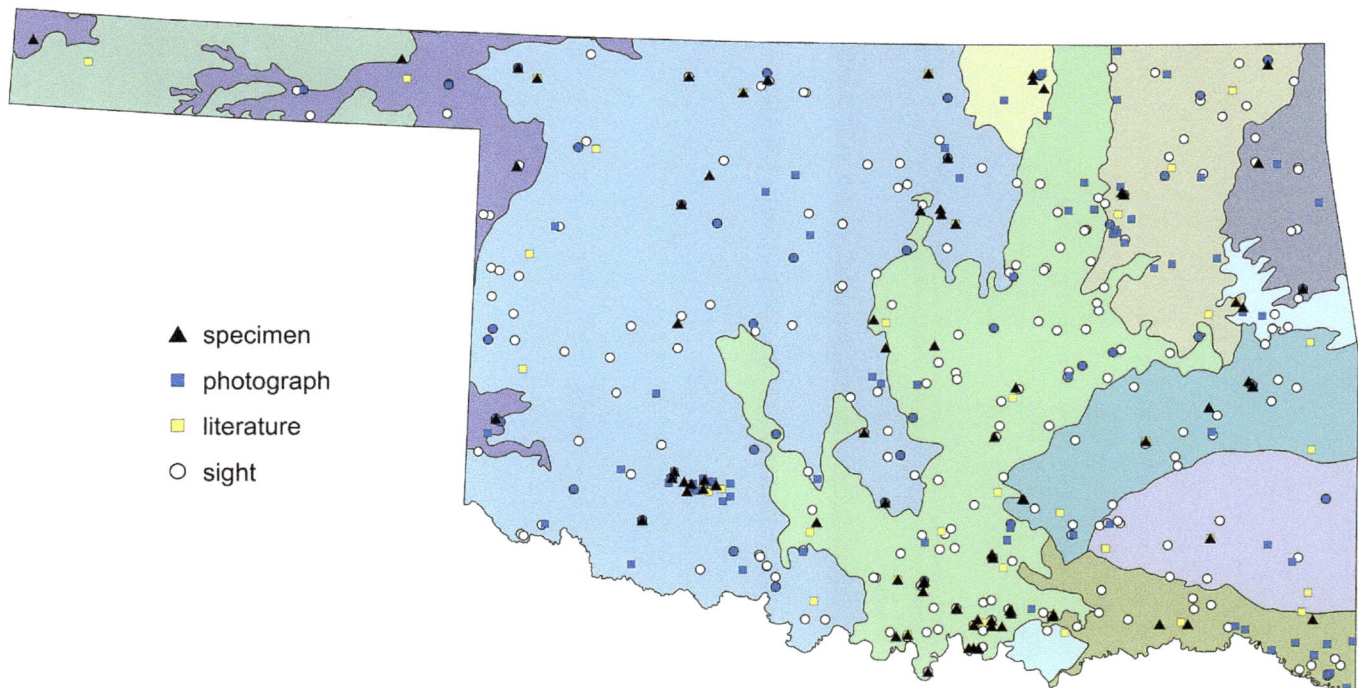

Figure 19.79 Documentation level of supporting records for *Celithemis eponina*, the Halloween Pennant, in Oklahoma. Levels include specimen, archived photograph, literature reference, or sight record. Although sight records lack documentation, only those we feel are credible are plotted.

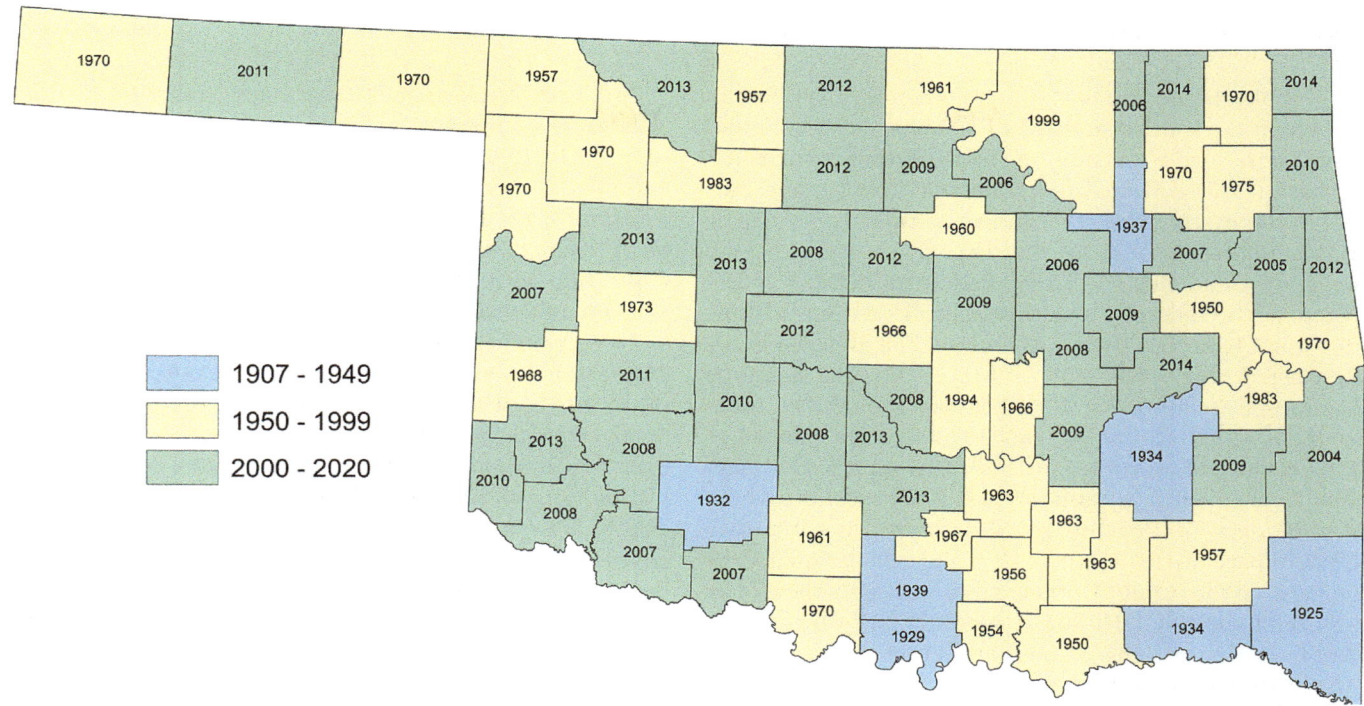

Figure 19.80 Counties in which *Celithemis eponina*, the Halloween Pennant, are known to occur in Oklahoma. Year within the county is that in which the species was first reported.

CELITHEMIS VERNA PRITCHARD, 1935 – DOUBLE-RINGED PENNANT

The Double-ringed Pennant is more common elsewhere in its range in the southeastern United States, but it nonetheless was described to science on the basis of specimens AE Pritchard (1935) collected in Oklahoma. Pritchard discovered, in the literal sense, the species just north of Antlers, Pushmataha County, on 15 May 1932 (♀, paratype, but we have not located this specimen; Pritchard 1935). He designated as the holotype a ♂ that he collected on 10 June 1934 near Quinton, Pittsburg County. The allotype is a ♀ he collected near Antlers on 12 May 1934. The holotype and allotype are housed at UMMZ, even though two other specimens (EMEC 309046 and 309047) were labeled as the holotype and allotype until RW Garrison found them and re-labeled them as pseudotypes. Both specimens have the same date and locality as the allotype; as such, they are likely part of the 7♂ and 15♀ paratypes designated by Pritchard (1935). Only 2 other paratypes with these data have been located (CUIC 3021.1 and 3021.2),[25] which leaves at least 18 paratypes unaccounted from that series. The rest of the paratypes designated by Pritchard were also collected by him from near Antlers—1♀, 28 April 1934, probably is EMEC 81675; 1♀, 15 May 1932, specimen not located; and 1♂, 1♀ (pair), 24 June 1934, specimens not located.

It is remarkable that Pritchard collected 7♂ and 16♀ in one day, as anyone who has tried to collect the species knows that it is furtive and skittish. As Needham et al. (2014:456) explained "The species may be locally common but difficult to capture, for it often keeps well out from land near the outer fringe of shore vegetation and is very alert and active." Collecting so many individuals attests, yet again, to how amazing of a collector Pritchard was.

In Oklahoma, *Celithemis verna* is a species of clear ponds and seasonal pools with much emergent vegetation extending from the shoreline to a few meters out but often with open water in the middle. These ponds and pools are typically found in pine or pine-oak woodlands, including pine plantations, or near mixed conifer-hardwood stands. We also have found the species at a few small lakes with back coves reminiscent of ponds just described. Although it has been recorded in 12 counties, in general it is restricted to the southeast (broadly the Ouachita Highlands), in that it is found with some regularity only in Atoka, Le Flore, McCurtain, and Pushmataha Counties (Figures 19.82 and Figure 19.83).

Pritchard (1935:8) said that the type locality was "a rather large open lake whose edges and bays were overgrown

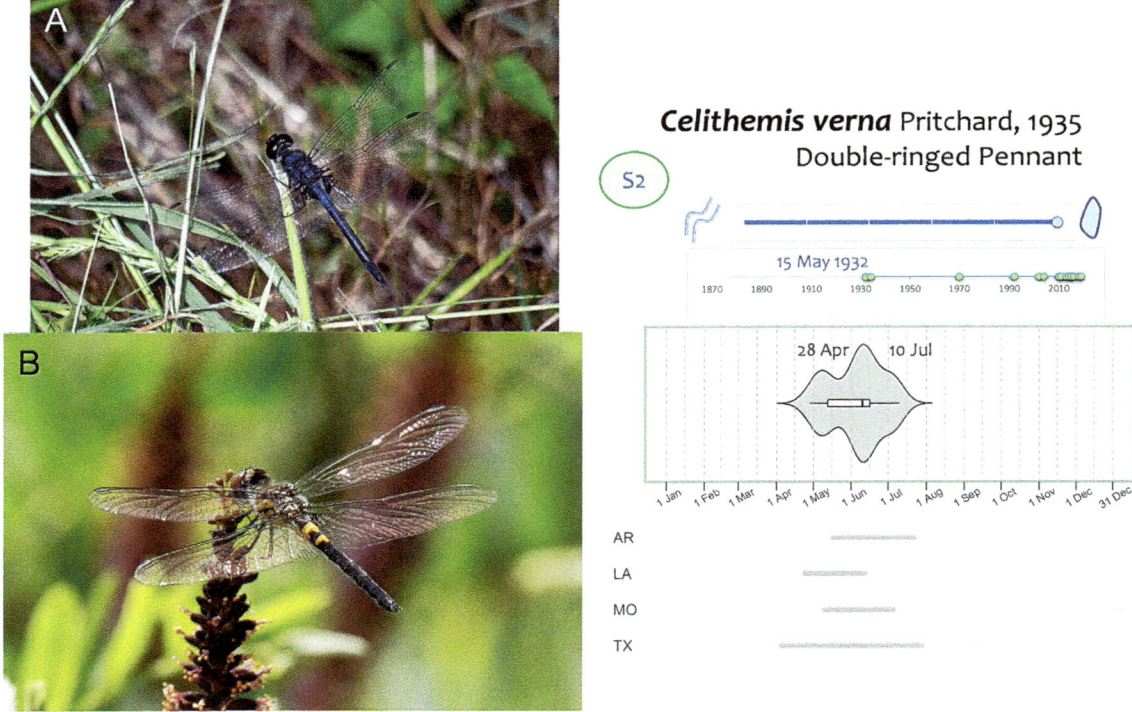

Figure 19.81 *Celithemis verna*, the Double-ringed Pennant, A) ♂, 8 km S of Smithville, McCurtain County, 4 July 2010 (USDA Forest Service; David Arbour, OC 320467). B) ♀, Bixhoma Lake, Wagoner County, 9 May 2014 (©James W Arterburn, OC 422055), showing off the double rings that give the species its English name and representing one of the more striking geographic outliers in Oklahoma.

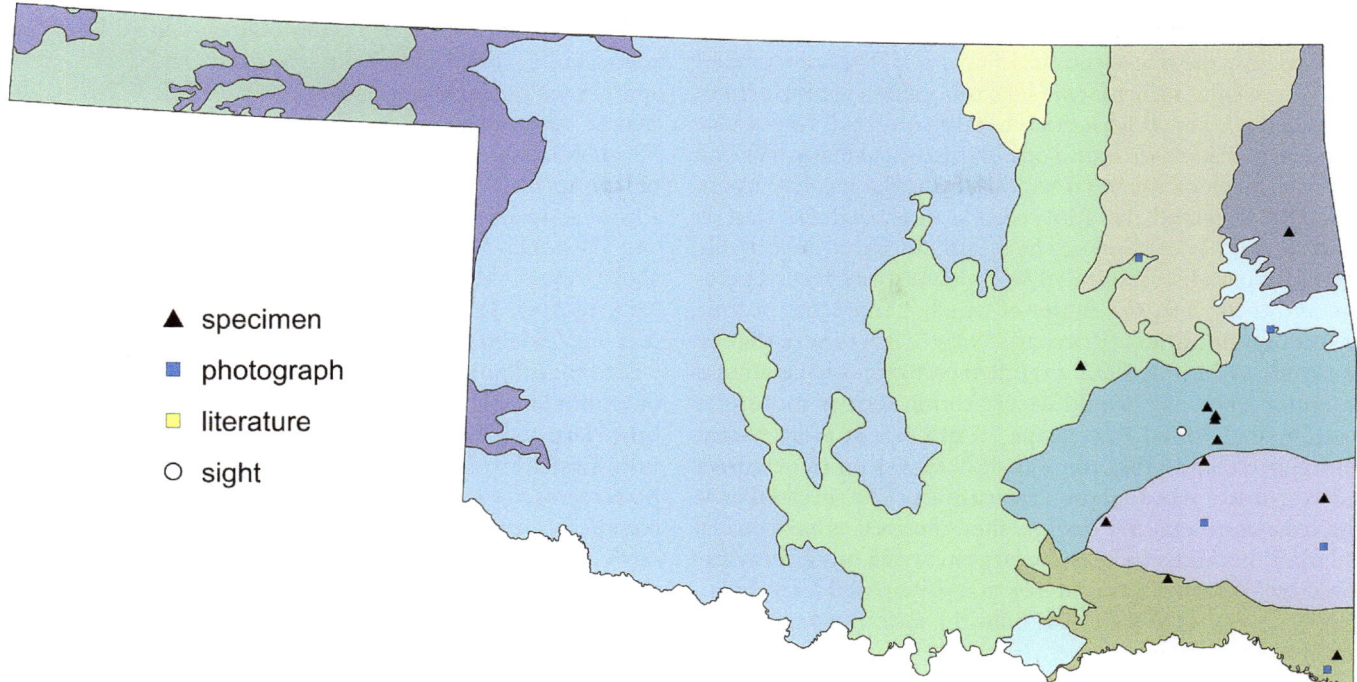

Figure 19.82 Documentation level of supporting records for *Celithemis verna*, the Double-ringed Pennant, in Oklahoma. Levels include specimen, archived photograph, literature reference, or sight record. Although sight records lack documentation, only those we feel are credible are plotted.

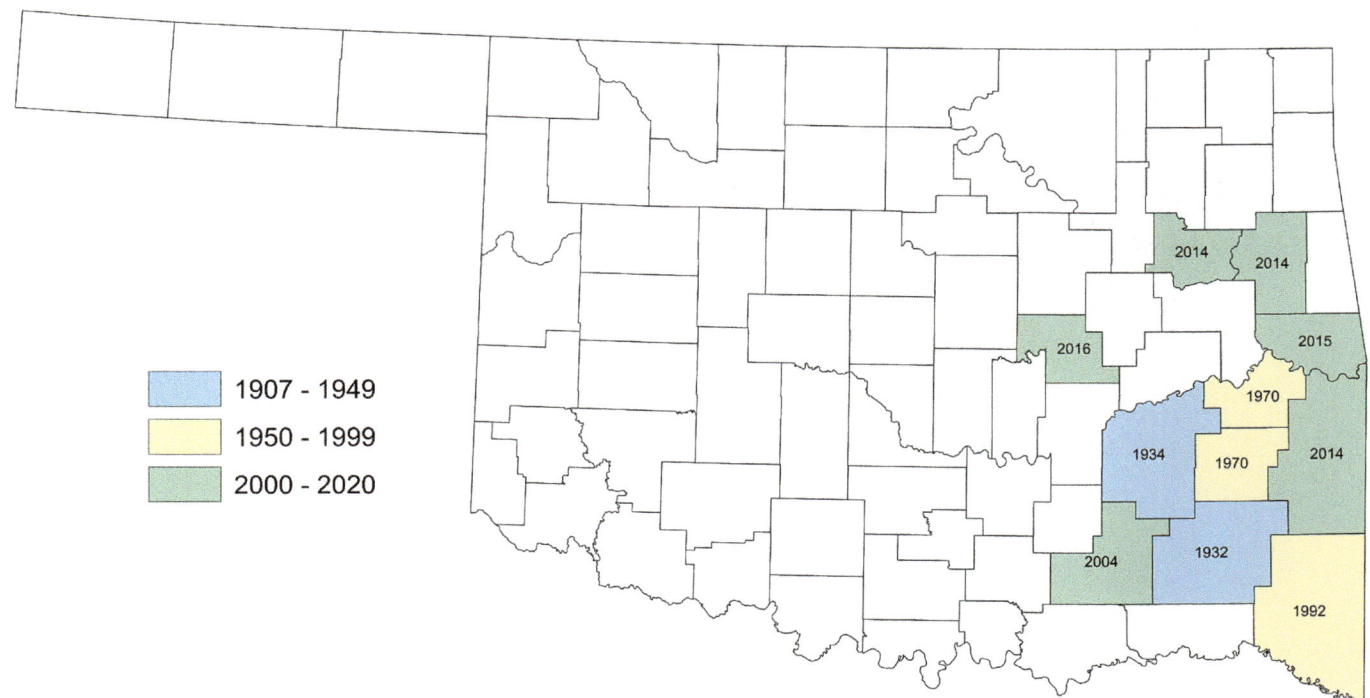

Figure 19.83 Counties in which *Celithemis verna*, the Double-ringed Pennant, are known to occur in Oklahoma. Year within the county is that in which the species was first reported.

with cat-tails." He went on to say that "Several day's search revealed only the one specimen of *verna* ..." We are stumped as to where this "rather large" lake is because currently there are no large lakes (at least not what we could call large) near Quinton, nor have we found one on historical maps. We had better luck relocating what we think was the locality where Pritchard collected the allotype. He described the spot as "a small clear lake close to the Kiamichi River, a veritable dragonfly paradise concealed by a dense growth of vegetation and trees; only the center of which is free from prolific masses of water lilies" (Pritchard 1935:9). It was here that he captured, on 24 June 1932, a tandem pair ovipositing among the water lilies. We would describe this spot as more of a medium-sized pond of perhaps 70 m × 275 m at its widest points. It could be that the water there was more extensive in 1934, or it is just an issue of semantics. The small lakes at which we or others have documented the species have been a bit larger. Bixhoma Lake is the largest, at 525 m × 775 m, followed by Weleetka Lake, at 425 m × 700 m, and Vian Lake, at 340 m × 360 m. Ponds that hold the species are about 50 m × 75 m or smaller.

Pritchard mentioned that there were many *Celithemis elisa* (Calico Pennant) and *C. fasciata* (Banded Pennant) present at the type locality. At the allotype locality, he indicated he had *Lestes vigilax* (Swamp Spreadwing), *L. inaequalis* (Elegant Spreadwing), *Enallagma dubium* (Burgundy Bluet), *E. daeckii* (Attenuated Bluet), *Nehalennia integricollis* (Southern Sprite), *Ischnura kellicotti* (Lilypad Forktail), and *Ladona deplanata* (Blue Corporal). We have encountered *C. verna* alongside *C. eponina*, *C. fasciata*, *Libellula cyanea* (Spangled Skimmer), *L. incesta* (Slaty Skimmer), *L. semifasciata* (Painted Skimmer), *Ladona deplanata*, *Sympetrum ambiguum* (Blue-faced Meadowhawk), *Enallagma divagans* (Turquoise Bluet), *E. geminatum* (Skimming Bluet), *E. signatum* (Orange Bluet), *E. traviatum* (Slender Bluet), *E. vesperum* (Vesper Bluet), *Ischnura posita* (Fragile Forktail), *I. hastata* (Citrine Forktail), and *Nehalennia integricollis*.

Pritchard said of his holotype collection on 10 June that it was when "*C. eponina* was making its debut, and *C. verna* its last appearance of the season." (Pritchard 1935:9). We now know that *C. verna*, although certainly having a short flight season, flies a little longer than Pritchard thought. His early flight date of late April still stands, which is as early as in Louisiana and almost as early as Texas, but somewhat earlier than Arkansas and Missouri (OK: 28 April 1934 ♀ paratype mentioned above; AR: 17 May, Harp 2007; LA: 24 April, Mauffray 2014; MO: 10 May, Trial 2005; TX: 5 April, OC 493925). The late date for Oklahoma has been extended to mid-July, which is bested by Arkansas and Texas (OK: 10 July 2010, 1♂, 8 km S of Smithville, McCurtain County, BAH, OC 320828; AR: 22 July, Harp and Harp 2003; LA: 12 June, Mauffray 2014; MO: 5 July, Trial 2005; TX: 28 July, OC 500503). Ovipositioning has been reported twice (10 and 16 June 2013), but no details were recorded except that it was done by lone females.

Pritchard (1935) found the species emerging from among water lilies "at daybreak" on 12 May 1934 (EMEC 309046). Early morning emergence and rarely seeing the species during the day made him think it might be crepuscular. Such behavior might account for there being so few records for the state (about 30 records across 12 counties, only 1 of which is of a teneral), but we have encountered it at various times of the day, and because we are often out and about just after dawn and around dusk—and all times in between!—we are skeptical of the crepuscular hypothesis.

One dubious record was reported from the Wichita Mountains, Comanche County, of nymphs (Bass 1990). This was likely just a misidentification of one of the common *Celithemis* (*C. elisa*, *eponina*, or *fasciata*; record likely relates to two nymphs at UCO re-identified by BS-P as *C. fasciata*, cf.). The species has been counted 1–3 adults at a time except for the 22 individuals reported by Pritchard (see above) and once by us of 25 adults (12 June 2004, Atoka WMA, Atoka County).

LIBELLULIDAE: SKIMMERS 563

ERYTHEMIS ATTALA (SÉLYS, 1857) – BLACK PONDHAWK

The Black Pondhawk is a tropical species whose range barely extends into the Rio Grande valley of southernmost Texas, where the species is scarce with just a couple of established populations (Abbott 2005, 2015). A few have been found in Texas north of the Rio Grande valley, notably on the western edge of the Edwards Plateau (OC 319911, OC 480982), and a ♀ found her way to southeastern New Mexico at Last Chance Canyon, Eddy County, on 3 September 2013 (OC 381571). By themselves these records failed to hint that the Black Pondhawk could reach Oklahoma, yet a ♂ collected 2 km SSE of Old Bingham, Ellmore County, Alabama, on 2 May 1989 (Krotzer and Krotzer 1992) not only provided the first record for the United States but hinted (strongly) that this species was capable of long-distance vagrancy. Even so, Oklahoma's single record, of a ♀ along the creek in Beaver Dunes SP, Beaver County, 6 October 2017 (SP 2514, OC 473884, MAP), came as a surprise.

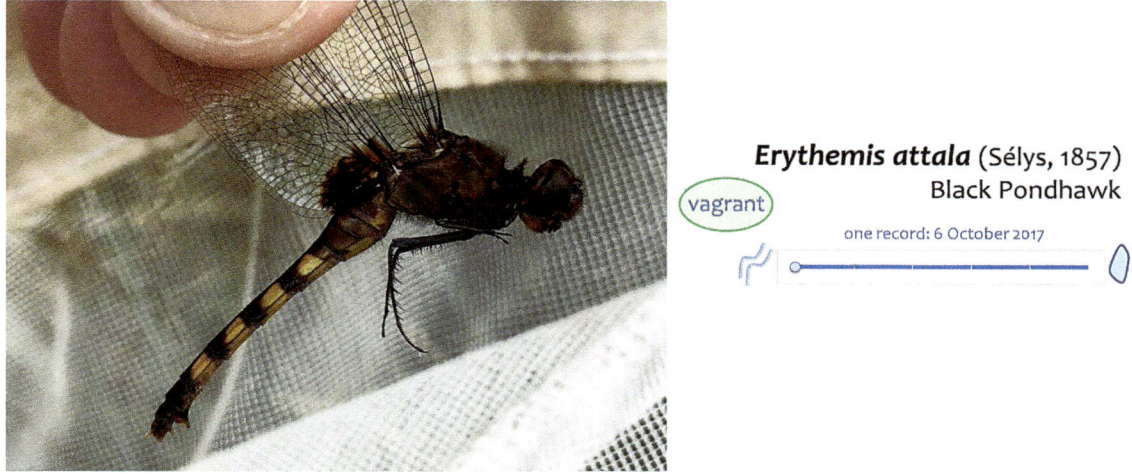

Erythemis attala (Sélys, 1857)
Black Pondhawk
vagrant
one record: 6 October 2017

Figure 19.84 *Erythemis attala*, Black Pondhawk. A wayward ♀ at Beaver Dunes SP, Beaver County, on 6 October 2017 (©Michael A Patten, OC 473884) represents the sole record for the Southern Great Plains of this tropical species.

ERYTHEMIS VESICULOSA (FABRICIUS, 1775) – GREAT PONDHAWK

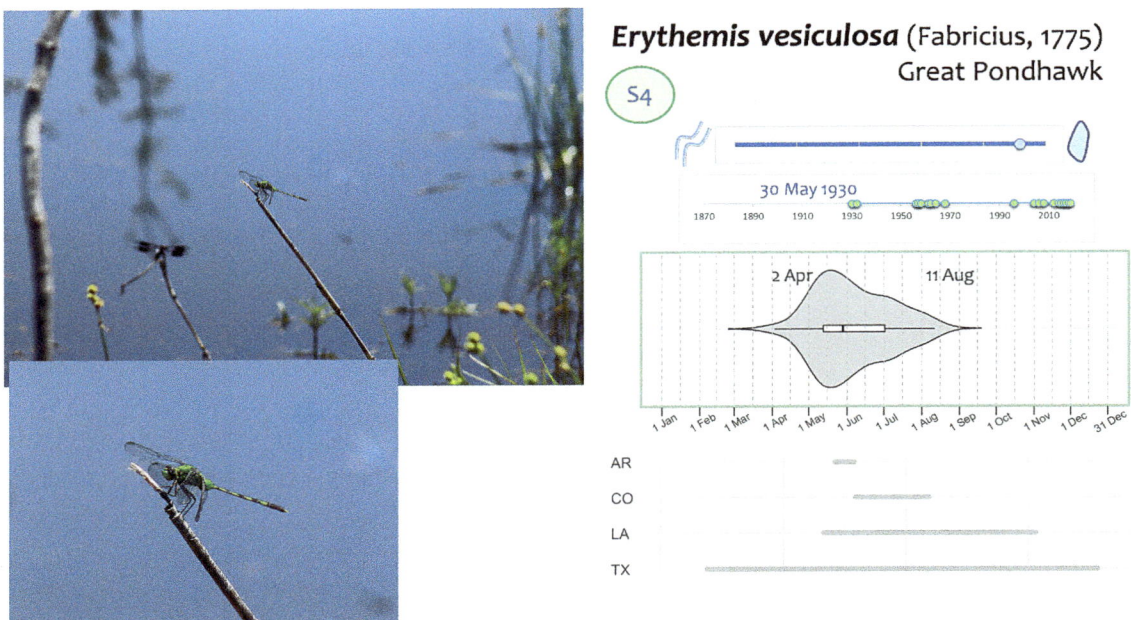

Figure 19.85 *Erythemis vesiculosa*, the Great Pondhawk, ♂, pond at Lake Frederick, Tillman County, 28 May 2016 (©Brenda D Smith, OC 445299); see Figure 19.86 for a ♀. This species ranges across the Neotropics, but it also ranges northward into the southernmost United States. Oddly, it flies through much of the year in Texas, but in Oklahoma is appears only in spring and early summer.

The Great Pondhawk is a tropical and subtropical species whose range extends from Argentina north into the Southern Great Plains. It is well established in Texas, but its status in Louisiana, Arkansas, and Oklahoma is unsettled, even if the species is annual in each state. We considered it a "putative vagrant" when we first ranked its status for Oklahoma (Patten and Smith-Patten 2013b). Two dozen more records were added since, bringing the overall total to 55 (of ~75 adults) in 22 counties scattershot across the state, although a preponderance of records are from southwestern Oklahoma and Red Slough in McCurtain County in the southeast. We now lean toward calling the species a rare but regular visitor to the state, in that we have yet to find larvae, exuviae, or tenerals, although 2♀ were seen ovipositing at Waurika WMA, Cotton County, on 5 May 2018 (MAP). When the species is seen it is almost always singly, but we have a handful of records of 2–3 individuals. The species has been found at well-vegetated habitats, most often at small clear ponds, or rarely along clear flowing streams and creeks.

The first record for Oklahoma goes all the way back to 1930, when RD Bird collected one at Dean's Slough, McClain County (30 May, 1♂, OMNH 921, in Bird 1932 as *Lepthemis vesiculosa*, although possibly originally considered *Mesothemis plebija* [sic, *plebeja*], see Appendix A). It was only recorded one other time in the 1930s: AE Pritchard collected a ♀ in Cherokee County on 11 August 1932 (OMNH

2706), which is also the late date for the species. The early date in Oklahoma is 2 April (from 1956, Cleveland County, locality and coll. unknown; data from GHB notes that said the specimen was from the OU student collection, but we have not found this specimen; record reported by Bick and Bick 1957 and Donnelly 2004b). The Great Pondhawk was recorded another nine times before the 2000s. Starting in 2006 there have been 45 additional records. Whether the

Figure 19.86 Comparison of *Erythemis vesiculosa*, the Great Pondhawk, to *E. simplicicollis*, the Eastern Pondhawk (11 km SW of Wakita, Crooked Creek, Grant County, 28 July 2012, SP 345 and 344, respectively, ©Brenda D Smith).

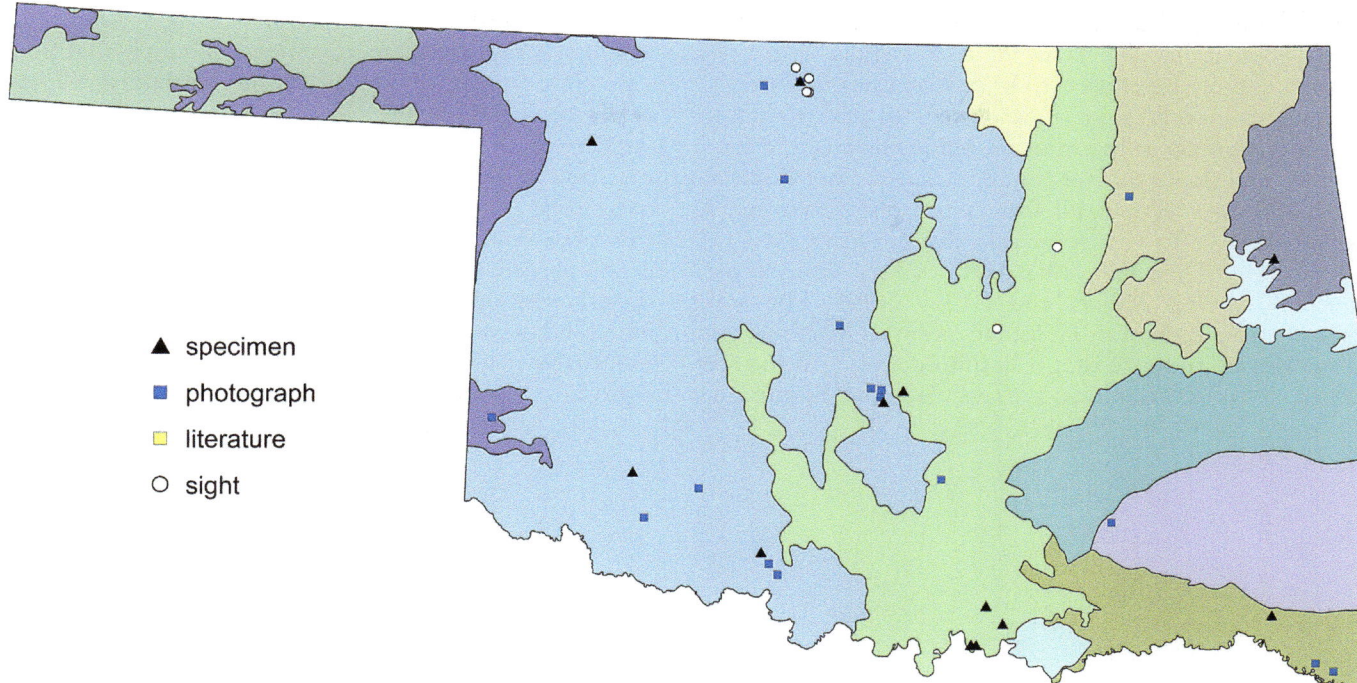

Figure 19.87 Documentation level of supporting records for *Erythemis vesiculosa*, the Great Pondhawk, in Oklahoma. Levels include specimen, archived photograph, literature reference, or sight record. Although sight records lack documentation, only those we feel are credible are plotted.

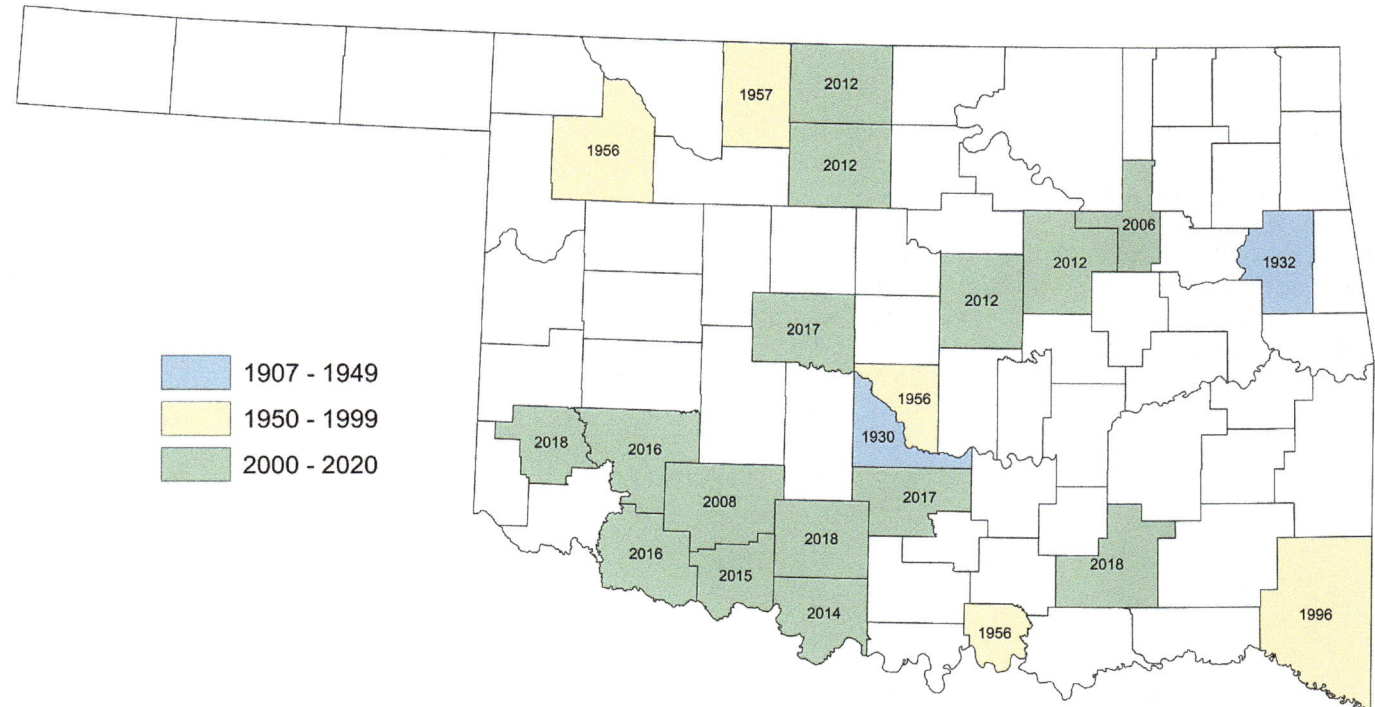

Figure 19.88 Counties in which *Erythemis vesiculosa*, the Great Pondhawk, are known to occur in Oklahoma.

increase in records means that the species now has an established population, has become more of a regular visitor, or is a result of more coverage across the state is unknown.

The species flies in Texas close to year round, in Louisiana for about five months, but is known from fewer records and thus a shorter time in Arkansas and Colorado (AR: 6 known records, all from southwestern corner, 22 May–7 June, OC; CO: 4 records, 8 June–7Aug, OC; LA: 13 May–12 November, iNat, OC; TX: 7 February–23 December, OC, Abbott and Lasley 2019). The scant records from Arkansas, Colorado, Kansas, Missouri, and New Mexico indicate that it is simply a vagrant to those states, as it is to Oklahoma. We speculate that, like another tropical species, *Tramea calverti* (Striped Saddlebags), which is a vagrant to southwestern Oklahoma (although a possible breeder in southeastern Oklahoma), vagrants of *Erythemis vesiculosa* come north when they emerge farther south in high numbers and population pressure and weather patterns push individuals away from natal areas. As odonates are want to do, they will breed when the opportunity arises, but we presume that they do not breed with any regularity in areas of vagrancy; nonetheless, opportunistic breeding in conjunction with increased movement northward fueled by climate change eventually may introduce breeding populations into the region.

ERYTHEMIS SIMPLICICOLLIS (SAY, 1839) – EASTERN PONDHAWK

The Eastern Pondhawk is wicked common—it is the most ubiquitous and widespread libellulid in the state. If you go afield in Oklahoma in summer to hunt odonates, it is a virtual certainty you will see a good many. We have accumulated over 1,500 records in the OOP database, many of which have adult counts of 50–300. Our highest estimation was 500 individuals on one day at Red Slough WMA, McCurtain County (26 August 2015). It can be found at just about any wetland habitat, and it is not unusual to see the species well away from water. Its status in the western panhandle is clouded by the presence of small numbers there, perhaps not annually, of the Western Pondhawk (*Erythemis collocata*), and likely interbreeding of the two.[26]

Erythemis simplicicollis was first recorded by E B Williamson on 3 June 1907 at Wister, Le Flore County (♀, UMMZ). By the OU expeditions it had been reported from 20 counties (Bird 1932, as *Mesothemis simplicicollis*), and by 1941 it was known from 23. It has now been recorded in all 77 Oklahoma counties. It flies much of the year, from late March to mid-December (early and late date from Red Slough: 26 March 2012, 1♀, DA, OC 374180 and 10 December 2016, 1♀, BC, OC 458381). This 8½ month season is closest to Missouri (9 months) in its duration, but not in its timing. We suspect that eventually the species will be recorded year-round, as it is in Texas, and that the seasons in other neighboring states will also be extended from their currently somewhat truncated seasons (AR: 9 April–5 October, OC 374354, 456887; CO: 10 June–1 September, OC 432952, 409898; KS: 8 May–30 November, OC 374810, SEMC 1315525; MO:1 May–probably October/November[27]; NM: May–October, Paulson 2009; TX: all year, Abbott and Lasley 2019).

The Eastern Pondhawk shows much phenotypic variation. We believe some of the variation results from hybridization or intergradation with the Western Pondhawk (see that species account for more details). Teasing out that taxonomic headache will take much effort. Variation in abdomen shape, specifically the wasp-waisted variants (Figure 19.90) noted in southeastern Oklahoma, will take directed research to elucidate evolutionary forces behind that morph. Outside of interbreeding between the Eastern and Western Pondhawks, we know of one possible mixed-species/genera pair involving *Erythemis simplicicollis*—see, remarkably, the *Libellula luctuosa* (Widow Skimmer) species account, above.

Erythemis simplicicollis (Say, 1839)
Eastern Pondhawk

Figure 19.89 *Erythemis simplicicollis*, the Eastern Pondhawk, A) ♂, Snake Creek 3 km S of Locust Grove, Mayes County, 5 July 2014 (©Bill Carrell, OC 424314). Note the thin abdomen and grayish cerci. B) ♂ Black Mesa SP, Cimarron County, 2 August 2014 (©Bill Carrell, OC 425629). Note the broader abdomen and white cerci. Individual variation is marked, but in this region the species hybridizes with *E. collocata*, too. C) immature ♂ Moneka Park at Lake Waurika, Jefferson County, 17 May 2014 (©Brenda D Smith, OC 422162).

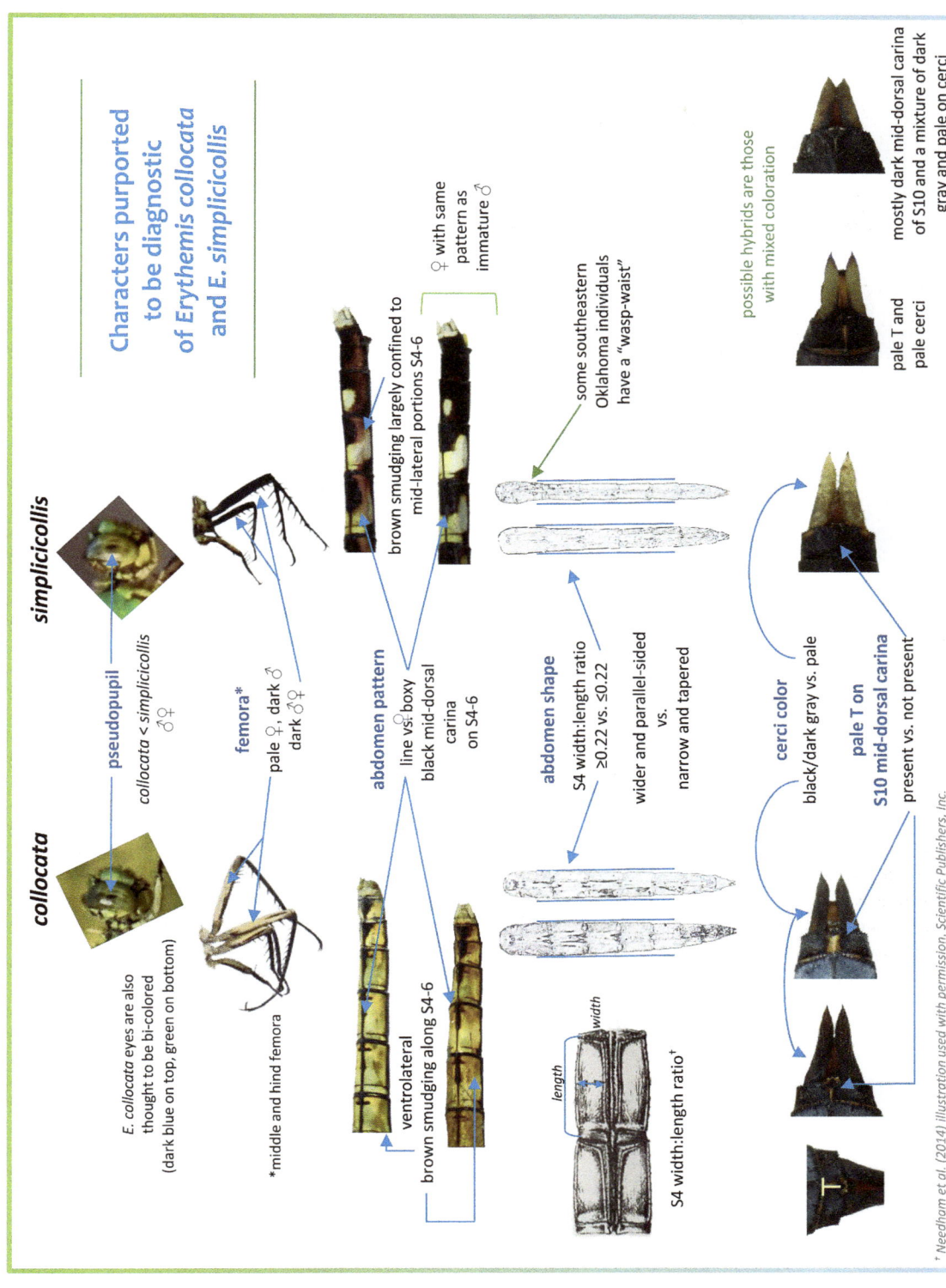

Figure 19.90 Characters purported to be diagnostic of *Erythemis collocata* and *E. simplicicollis.*

† Needham et al. (2014) Illustration used with permission, Scientific Publishers, Inc.

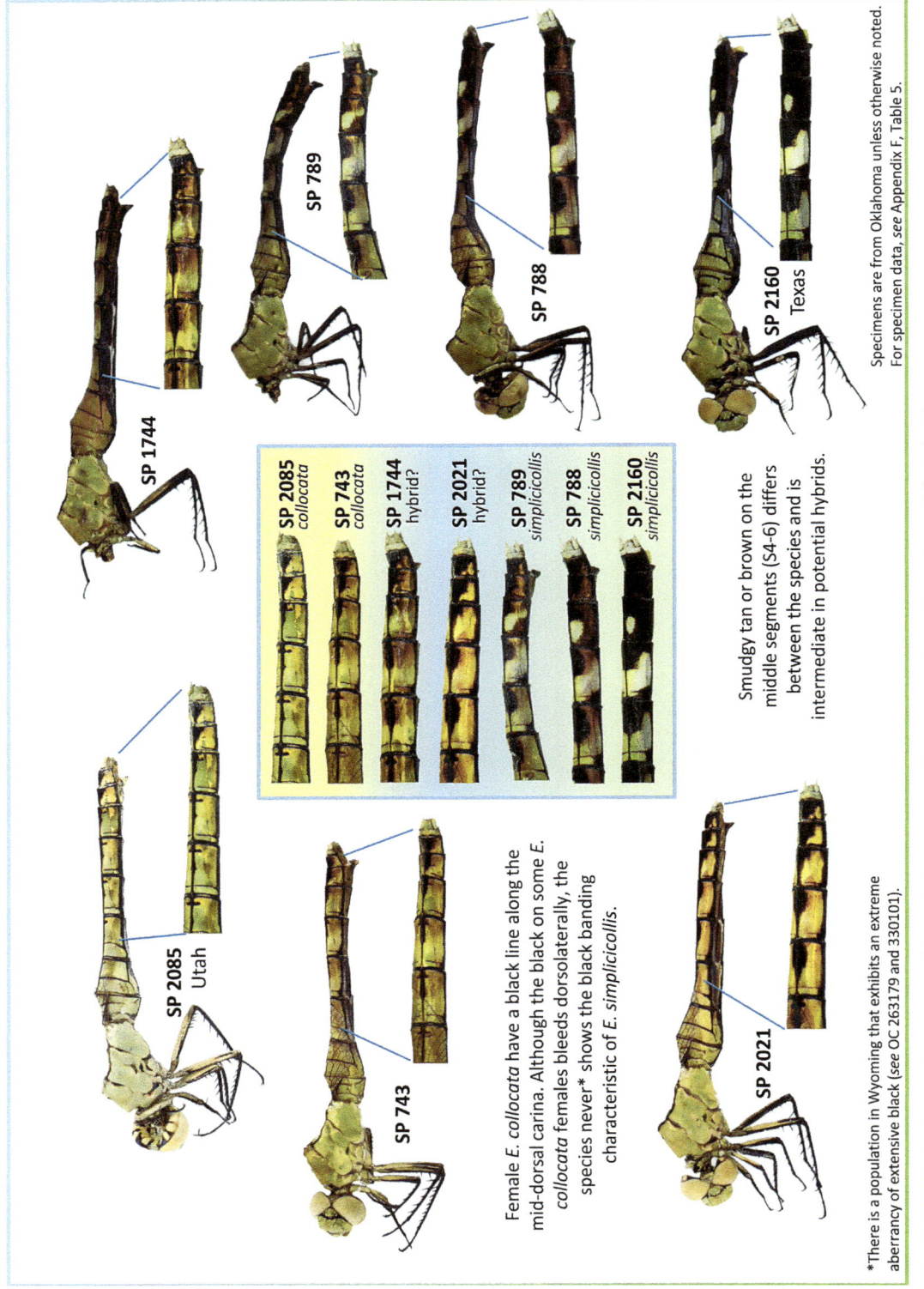

Figure 19.91 Abdominal patterns of female *Erythemis collocata*, *simplicicollis*, and possible hybrids.

Female *E. collocata* have a black line along the mid-dorsal carina. Although the black on some *E. collocata* females bleeds dorsolaterally, the species never* shows the black banding characteristic of *E. simplicicollis*.

Smudgy tan or brown on the middle segments (S4–6) differs between the species and is intermediate in potential hybrids.

Specimens are from Oklahoma unless otherwise noted. For specimen data, *see* Appendix F, Table 5.

SP 2085 *collocata*
SP 743 *collocata*
SP 1744 hybrid?
SP 2021 hybrid?
SP 789 *simplicicollis*
SP 788 *simplicicollis*
SP 2160 *simplicicollis*

SP 1744
SP 789
SP 788
SP 2160 Texas
SP 2085 Utah
SP 743
SP 2021

*There is a population in Wyoming that exhibits an extreme aberrancy of extensive black (*see* OC 263179 and 330101).

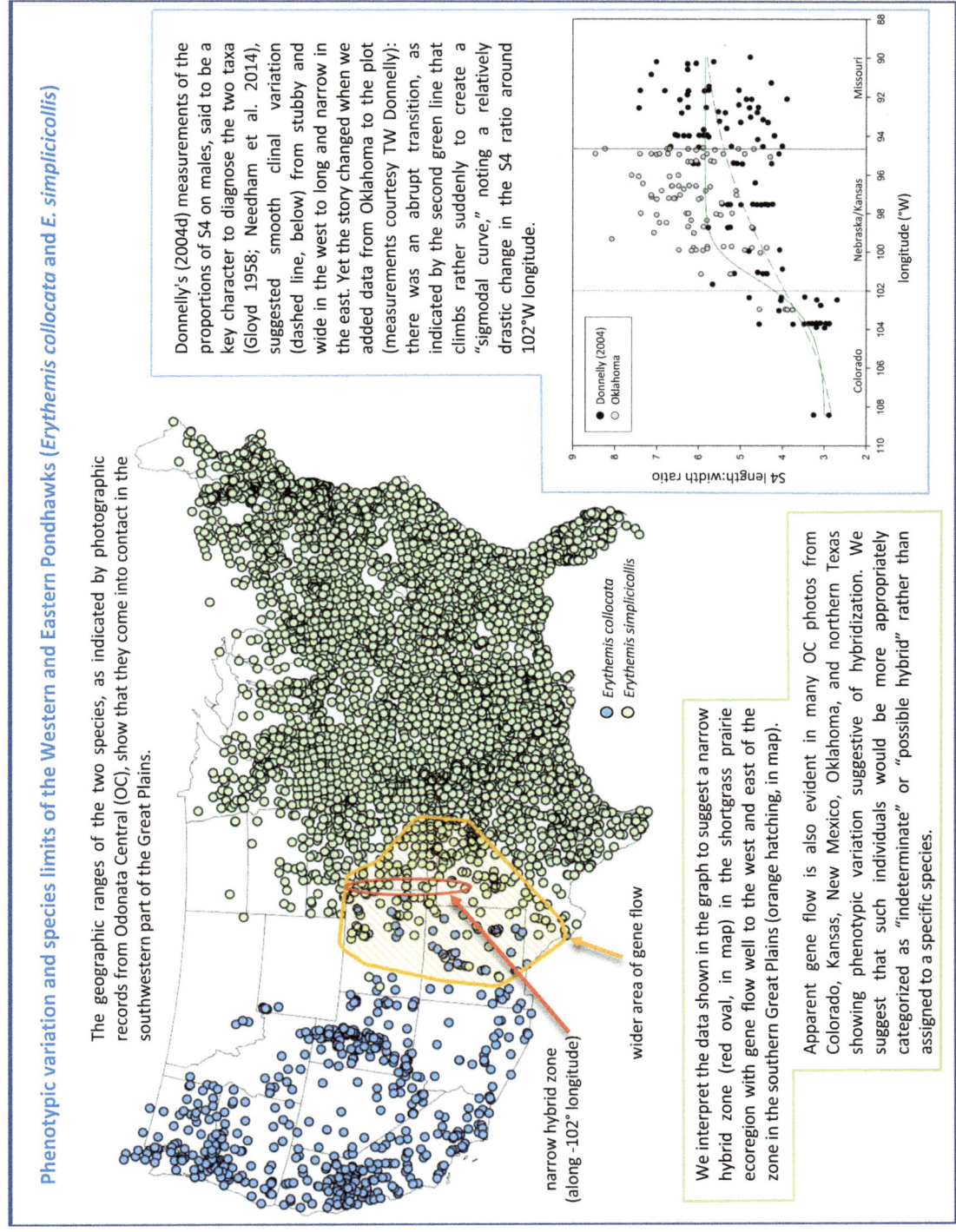

Phenotypic variation and species limits of the Western and Eastern Pondhawks (*Erythemis collocata* and *E. simplicicollis*)

The geographic ranges of the two species, as indicated by photographic records from Odonata Central (OC), show that they come into contact in the southwestern part of the Great Plains.

Donnelly's (2004d) measurements of the proportions of S4 on males, said to be a key character to diagnose the two taxa (Gloyd 1958; Needham et al. 2014), suggested smooth clinal variation (dashed line, below) from stubby and wide in the west to long and narrow in the east. Yet the story changed when we added data from Oklahoma to the plot (measurements courtesy TW Donnelly): there was an abrupt transition, as indicated by the second green line that climbs rather suddenly to create a "sigmodal curve," noting a relatively drastic change in the S4 ratio around 102°W longitude.

● *Erythemis collocata*
○ *Erythemis simplicicollis*

narrow hybrid zone
(along -102° longitude)

wider area of gene flow

We interpret the data shown in the graph to suggest a narrow hybrid zone (red oval, in map) in the shortgrass prairie ecoregion with gene flow well to the west and east of the zone in the southern Great Plains (orange hatching, in map).

Apparent gene flow is also evident in many OC photos from Colorado, Kansas, New Mexico, Oklahoma, and northern Texas showing phenotypic variation suggestive of hybridization. We suggest that such individuals would be more appropriately categorized as "indeterminate" or "possible hybrid" rather than assigned to a specific species.

Figure 19.92 Phenotypic variation and species limits of the Western and Eastern Pondhawks (*Erythemis collocata* and *E. simplicicollis*).

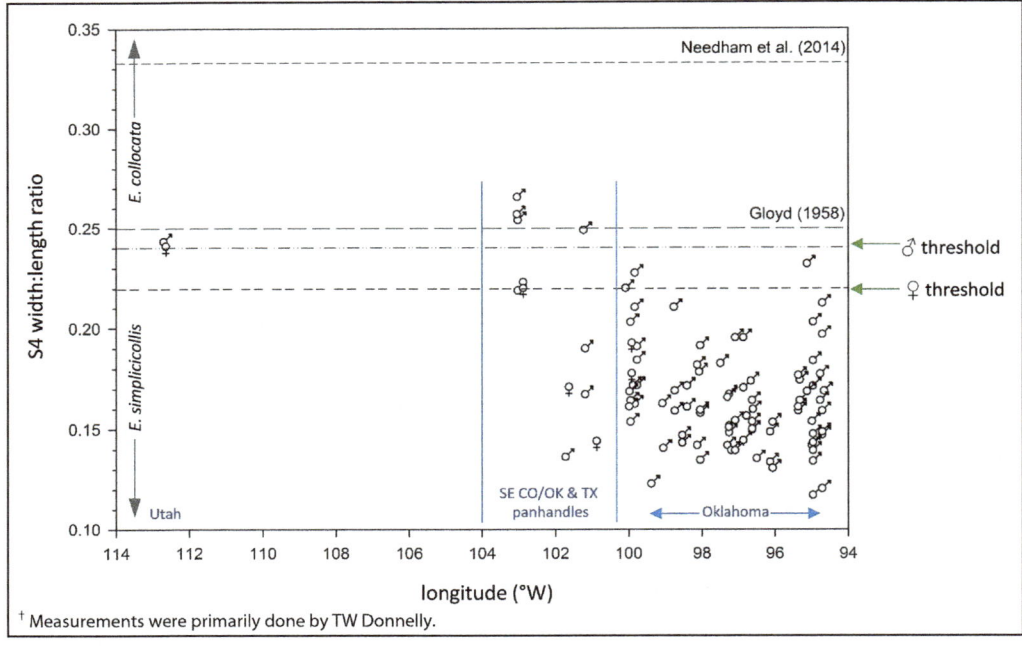

Figure 19.93 **Width:length ratio of abdominal segment 4 of a sampling of *Erythemis* specimens in the SP Collection.**[†] **Gloyd (1958:12) suggested that *E. collocata* could be distinguished from *E. simplicicollis*, by the latter having an S4 ratio of about ⅙ (0.17) and the former between ¼ and ⅓ (0.25–0.33; the lower end of that range is indicated in the graph). Needham et al. (2014) indicated the threshold should be > ⅓, which would exclude *E. collocata* from Utah. Specimens from the SP collection indicate the threshold for *E. collocata* ought to be around 0.22 to 0.24.**

ERYTHEMIS COLLOCATA (HAGEN, 1861) – WESTERN PONDHAWK

There are nine records of the Western Pondhawk from Oklahoma (Table 19.1). Eight records are from the Black Mesa area in northwestern Cimarron County, where it was first collected in the state by LE Hornuff and GH Bick on 5 August 1970 (Figure 19.94). This ♂ specimen went unreported until we found it at FSCA and identified it as *Erythemis collocata* (Smith-Patten and Patten 2013b; identification confirmed by B Mauffray). We examined this specimen and four others that Bick and Hornuff collected on 4–5 August 1970 at Black Mesa because Bick's notes for these specimens were annotated with "collocata?" We identified with confidence only this one specimen; the others were not in good condition. Not even two months after returning from our visit to FSCA in 2013 we found 1♀ Western Pondhawk at Lake Etling, Black Mesa SP, on 5 July (SP 743, OC 401514) and collected 1♂ (SP 746) on North Carrizo Creek, 7 km N of Kenton later that day. That fall, BC encountered 2♂ and 1♀ on 2 September on South Carrizo Creek at Black Mesa SP (OC 410019). There is one record

of the species farther east, for Texas County, where a ♂ was collected along Coldwater Creek in Optima NWR on 20 May 2016 (MAP, SP 1908). This individual has mixed dark gray and pale cerci, some pale on the mid-dorsal carina of S10, a wide, parallel-sided abdomen, an S4 width–length ratio of 0.25, and a hindwing–abdomen ratio of 0.908; characters predominantly indicative of *Erythemis collocata*. Other records (Table 19.1) in the eastern panhandle are suggestive of *E. collocata* or are possible hybrid Western × Eastern (*E. collocata × simplicicollis*) Pondhawks given their intermediate characters.

The Eastern and Western Pondhawks are distinguished by abdomen pattern and shape as well as by color of the cerci, femora, and eyes (e.g., Gloyd 1958; Paulson 2009, 2011; Needham et al. 2014; Figures 19.90 and 19.91). Hybrids are expected along the western edge of the Great Plains (including Oklahoma; Smith-Patten and Patten 2013b), where there may be, in Paulson's (2009) words, a "hybrid swarm." Accordingly, species limits of the Western

Table 19.1 **There are nine confirmed records of *Erythemis collocata*, the Western Pondhawk, from Oklahoma. Five other records are currently unconfirmed because they may be hybrids or they lack sufficient supporting evidence, or both. There are also at least 10 SP specimens and >50 Odonata Central records that are possible hybrids.**

COUNTY	LOCALITY	DATE	COLLECTOR	SOURCE	INDIVIDUALS	NOTES
Confirmed records						
Cimarron	Black Mesa SP	5 Aug 1970	GHB; LEH	FSCA	1♂	Smith-Patten and Patten (2013b)
Cimarron	Black Mesa SP, Lake Carl Etling	5 July 2013	BS-P; MAP	SP 743	1♀	Smith-Patten and Patten (2013b)
Cimarron	7 km N of Kenton, North Carrizo Ck.	5 July 2013	BS-P; MAP	SP 746	1♂	
Cimarron	4 km NNE of Kenton, Cimarron R.	7 Aug 2015	MAP	field notes	1♀	
Cimarron	Black Mesa SP, South Carrizo Ck.	12 July 2016	BS-P	SP 2021	2♀	another ♀ in hand (photo)
Cimarron	Black Mesa SP	18 July 2018	Schwartz, B	iNat 14800702	1♀	
Cimarron	Black Mesa SP, South Carrizo Ck.	2 Sept 2018	BC	OC 410019	3 (2♂, 1♀)	
Texas	Optima NWR, Coldwater Ck.	20 May 2016	MAP	SP 1908	1♂	
Cimarron	7 km N of Kenton, North Carrizo Ck.	25 May 2019	MAP	SP 2772	1♀	
Unconfirmed records/possible hybrids						
Cimarron	3 km E of Kenton	30 Sept 2008	BAH	OC 283762	1♂	photo not conclusive
Cimarron	3 km NE of Kenton, Cimarron R.	22 Sept 2013	BS-P; MAP	field notes	1♀	
Cimarron	Black Mesa SP	20 Aug 2017	BC	OC 470944	1♀	
Cimarron	6 km NW of Kenton, North Carrizo Ck.	9 Sept 2018	BC	OC 490818	1♂	probable hybrid
Texas	12 km N of Goodwell, Beaver R.	8 Aug 2015	MAP	SP 1744 (♀)	2 (1♂, 1♀)	♂ identified in hand but no photos or written description; ♀ an apparent hybrid

Ck. = Creek, NWR = National Wildlife Refuge, R. = River, SP = State Park

BAH = Berlin A Heck, BC = Bill Carrell, BS-P = Brenda D Smith-Patten, GHB = George H Bick, LEH = Lothar E Hornuff, MAP = Michael A Patten

FSCA = Florida State Collection of Arthropods, iNat = iNaturalist; OC = Odonata Central, SP = Smith-Patten/Patten Collection

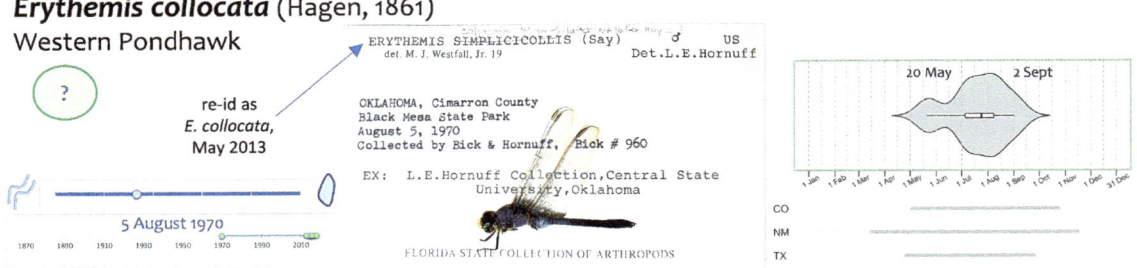

Figure 19.94 *Erythemis collocata*, the Western Pondhawk, ♂, Black Mesa SP, Cimarron County, 5 August 1970 (FSCA, OC 400673), the first record for Oklahoma, even if it was not determined as such until 2013.

and Eastern Pondhawks are complex and, in some circles, contentious. The taxa generally are classified as biological species (Paulson 2018a), a view not held universally. Donnelly (2004d), for example, presented morphometric data that showed a smooth cline in a key measure of abdomen shape—length–width ratio of S4—from Missouri west to Colorado. However, when we added Oklahoma specimens to his plot, which Donnelly was kind enough to measure for us (thus ensuring a uniform standard), we instead see an abrupt east–west transition from thinner to thicker abdomens at roughly 102°W longitude (roughly the border between Cimarron and Texas Counties in the Oklahoma panhandle and north along the western border of Kansas; Figure 19.92). We interpret this result as support for at most a narrow hybrid zone, akin to the situation with many closely related species of birds that hybridize on the western Great Plains, most of which have followed a common taxonomic pattern of first being described as species, later being lumped when hybridization was discovered, and finally being re-split when evidence of assortative mating and phylogenetic history was amassed. We stand by our assessment (Smith-Patten and Patten 2013b) that "'at worst' the taxa are subspecies, in that they are diagnosably distinct (by the standard 75% rule; Patten and Unitt 2002) and occupy separate geographic ranges but are not reproductively isolated." Most important, whether classified as a species or as a subspecies, the taxon *collocata* has been documented in Oklahoma.

Yet even this story is not so simple. Setting aside the abrupt transition in abdomen shape (and abdomen length:hindwing ratio), it may be that gene flow extends westward and eastward away from the contact zone (Figure 19.92). We reviewed Odonata Central records of both species from Colorado, Kansas, New Mexico, northern Texas, and Oklahoma; we concluded that a non-trivial number had intermediate morphology, such as gray cerci (i.e., neither black nor white) on males or indeterminate abdomen color and pattern on females (Figure 19.92). A rangewide

assessment is needed to elucidate whether apparent intermediacy is more geographically widespread. Further investigation is needed, too, as to where the threshold lies with using the S4 width–length ratio to diagnose species (Figure 19.93). Whereas Gloyd (1958:12) said that *Erythemis simplicicollis* has a ratio of "about one-sixth" and *E. collocata* of "between one-fourth and one-third," Needham et al. (2014:465–466) said that *E. simplicicollis* is "about ¼" and *E. collocata* is "more than ⅓." With the latter's criteria, none of our specimens would be diagnosable as *E. collocata*, even those from Utah, which is well within the species' accepted geographic range (Figure 19.94). The Utah specimens do not fare any better with Gloyd's (1958) criteria, but specimens from southeastern Colorado and the Oklahoma panhandle, areas generally considered as territory held by *E. simplicicollis*, would be diagnosable as *E. collocata*. We therefore recommend a preliminary revision of the S4 width–length ratio threshold—shifting it to 0.22, above which would diagnose *E. collocata* (with females at the lower end and males probably around 0.24 and higher) and below which would be *E. simplicicollis*. Obviously, this threshold cannot be used as the sole criterion for species diagnosis where the two species come into contact. Individuals falling near the threshold would need to be closely examined for other characteristics to determine if it is clearly one species or the other or is potentially a hybrid.

The few well-documented records of the Western Pondhawk for Oklahoma have occurred from 20 May to 2 September (Table 19.1). This flight season is narrower than that reported for the neighboring states of Colorado (3 May–24 October, iNat, OC), New Mexico (March–November, Paulson 2009), or Texas (26 April–25 September, Abbott and Lasley 2019, OC). We did not provide a conservation rank for this species previously, instead we called it an apparent vagrant (Patten and Smith-Patten 2013b); we still refrain from providing a rank until more research can be conducted with this taxon.

ERYTHRODIPLAX BASIFUSCA (CALVERT, 1895) – PLATEAU DRAGONLET

There is but one record of this species for the state. On 21 October 2014, while at Lake Hall, Harmon County, in the southwestern corner of Oklahoma, we encountered a single ♂ at the marshy, clear-water outflow from the dam. MAP came upon this male shortly after arriving at the stream. He was positioned in such a way that he was able to look down, almost right over this individual, for a minute or more. While studying this individual, he called for BS-P, who ran upstream through the adjacent woods, unfortunately arriving at a spot downstream of the ♂ that, depending on the limited area in which she could maneuver around, ranged from somewhat to considerably backlit. She snapped several photographs from a strongly backlit spot, and several more from a spot less back-lit (OC 427500; Figure 19.95). As she moved to a better angle, and just as MAP was ready to attempt to capture the ♂, it bolted, never to be refound. Before it bolted, it perched atop dead reed stalks about 20–30 cm above the clear, flowing creek (Figure 19.95). It was attacked multiple times by Blue Dashers (*Pachydiplax longipennis*), but when it perched it would throw its wings forward.

Previously this species was lumped into the *Erythrodiplax connata* species complex. Paulson (2003) determined that there were multiple species (most of which are neotropical, including: *abjecta, atroterminata, basifusca, bromeliicola, cauca, cleopatra, fusca, ines, justiniana, media, melanorubra, minuscula*, and *paraguayensis*) under the one name. As such *E. connata* is now the name reserved for a species in southern South America. Three of the allospecies[28] occur in the United States. The Little Blue Dragonlet (*E. minuscula*) is found in the eastern U.S., mostly confined to the southeast, but with records north up to New York and west to central Texas. The Plateau Dragonlet (*E. basifusca*) generally occurs in western Mexico and the southwestern United States, including as far east as south-central Texas. The Red-faced Dragonlet (*E. fusca*) ranges up from Mexico into southwestern and south-central Texas. Generally, these species have little overlap with their geographic ranges, yet they present an identification challenge where they might co-occur or if a vagrant was discovered in the other's normal range.

The first character noted about the Oklahoma record was that its frons was a shiny midnight blue, which immediately eliminated a more expected species, the Blue Dasher (*Pachydiplax longipennis*) as well as *E. fusca* (has a red frons). In life, the eyes were fuscous, without a trace of blue or red, and the thorax was a dark velvety reddish-brown, which further eliminated *E. fusca* as well as *E. minuscula*. The latter species was also excluded because it does not exhibit a crisp division between the dark color of abdominal segments 1–2 and the pruinosity of S3 and beyond, as both *E. basifusca* and *E. fusca* show. Black on the terminal segments help to distinguish *E. fusca* from *E. basifusca*, with the former (and *E. minuscula*) having black on S8–10 whereas *E. basifusca* has it on S7 as well. Although not necessarily of diagnostic value in this case, it should be noted that the wings were clear (it is our understanding that *E. fusca* can sometimes have greatly reduced markings, but all of the other characters excluded that species). And lastly, we were not as concerned about cerci color because it varies too much within and between species (see *E. minuscula* species account for further discussion); however, we note that although two of the photos taken show paler cerci than they were in life, we attributed the "paleness" to an artifact of backlighting, as the cerci, although not as dark as we expected them to be, were mouse gray in life, a color that fits nicely with *E. basifusca*.

one record: 21 October 2014

Erythrodiplax basifusca
(Calvert, 1895)
Plateau Dragonlet

vagrant

Figure 19.95 Oklahoma's only record of *Erythrodiplax basifusca*, the Plateau Dragonlet (left, close up view; right, view of habitat). This ♂ was photographed on 21 October 2014 at Lake Hall, Harmon County (OC 427500). Photos by ©Brenda D Smith.

ERYTHRODIPLAX MINUSCULA (RAMBUR, 1842) – LITTLE BLUE DRAGONLET

The Little Blue Dragonlet is a perplexing species in Oklahoma—we cannot assess whether it is a resident breeder, albeit a scarce one, or it just wanders occasionally into the state. The records to date account for only about 20 reported encounters from four counties (Comanche, Marshall, McCurtain, and Le Flore). So relatively few records make it difficult to fully assess this species' status in Oklahoma. We are currently hesitant to put it on a watch list (S3, "vulnerable"), but we are also not inclined to consider it "apparently secure" (S4), so we have chosen to rank it S3S4 with a recommendation of further surveys conducted to determine more fully this species' geographic distribution, breeding status, and conservation needs.

Bird (1932) was the first to report the species: he collected a single ♂ on 26 April 1930 at the Wichita Mountains WR in Comanche County (Bird 1932; this specimen remains a bone of contention[29]). GH Bick examined this specimen, confirmed it was an *Erythrodiplax minuscula*, and said he returned it to OMNH (Figure 19.97), yet we have been unable to find it.[30] That record remains the only one for Comanche County and one of two records away from southeastern Oklahoma. It is also the early flight date for the species in Oklahoma.

The second record for the species is from Cedar Lake, Le Flore County, where TW Donnelly collected four ♂ (FSCA) on 26 August 1954. Bick was the next to see the species when he collected a single ♂ near McMillian, Marshall County, in south-central Oklahoma on 27 July 1968 (FSCA). McCurtain County received its first record on 12 June 1985, when DC Arnold collected a single adult

Figure 19.96 *Erythrodiplax minuscula*, the Little Blue Dragonlet. This tiny skimmer is found in the Southern Great Plains rarely, with almost all records in the Ouachita region. A and B) ♂♂, Three Rivers WMA, McCurtain County, 26 May 2008 (©Berlin A Heck, OC 282305). Females have been seen (and collected) in Oklahoma, but none has been photographed in the field. C) This ♀ was photographed in Texas by ©Greg W Lasley (Bastrop State Park, Bastrop County, 14 July 2010).

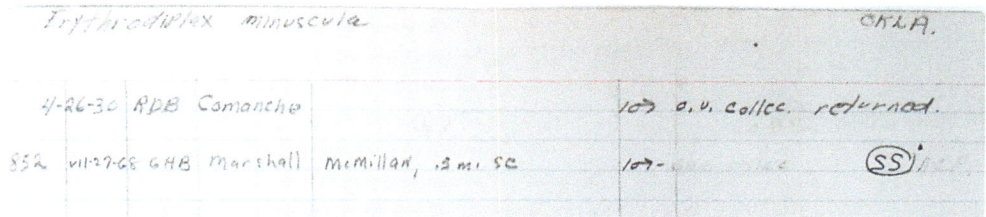

Figure 19.97 George H Bick's note card for *Erythrodiplax minuscula*, the Little Blue Dragonlet. Note that he examined a specimen collected by RD in 1930. Bick collected the species once, in Marshall County, 1968.

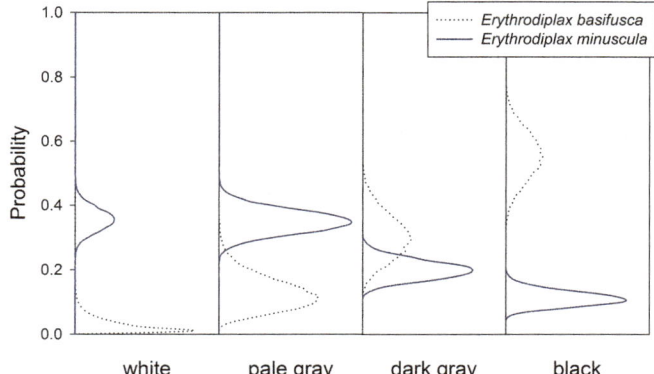

Figure 19.98 Variation in cercus color of ♂ *Erythrodiplax minuscula* and *basifusca*, the Little Blue and Plateau Dragonlets. Individuals with dark gray cerci, especially, are nearly equi-probable between the two species. By contrast, it is much more probable that an individual with white cerci is *E. minuscula* ($p \sim 0.35$ vs. $p < 0.05$), whereas it is much more probable that an individual with black cerci is *E. basifusca* ($p \sim 0.6$ vs. $p \sim 0.1$).

at Beavers Bend SP (OSU); this record accounted for the fourth time the species was recorded in the state. The species was not reported again until 2007, when BAH found 1♂ on his property 10 km SE of Idabel, McCurtain County, on 9 August 2007 (OC 263205). He saw the species there again in early May of the following year (OC 282149), and later that month at a small pond in the Ouachita NF, McCurtain County (OC 282305); again, these records were of single males. In 2009 the species was reported multiple times by BAH, this time all at Red Slough WMA's Teal Lake: 6 days between 6 September (JCAC 40146) and 11 November, with his highest count being 8U (OC 315705, 315793). That record accounts for the late date for the species in Oklahoma. BAH reported the species only once in 2010 (OC318599). There were no additional documented until we found an ovipositing ♀ on 6 June 2015 (SP 1632) and 2♂ the next day (SP 1641, OC 431510) at a well-vegetated pond at Big Cedar, Le Flore County. The species was not seen again for two years until 27 May 2017, when, at a small pond in mixed forest, BS-P found 2♂, 8 km S of Smithville, Ouachita NF 28000 Road, McCurtain

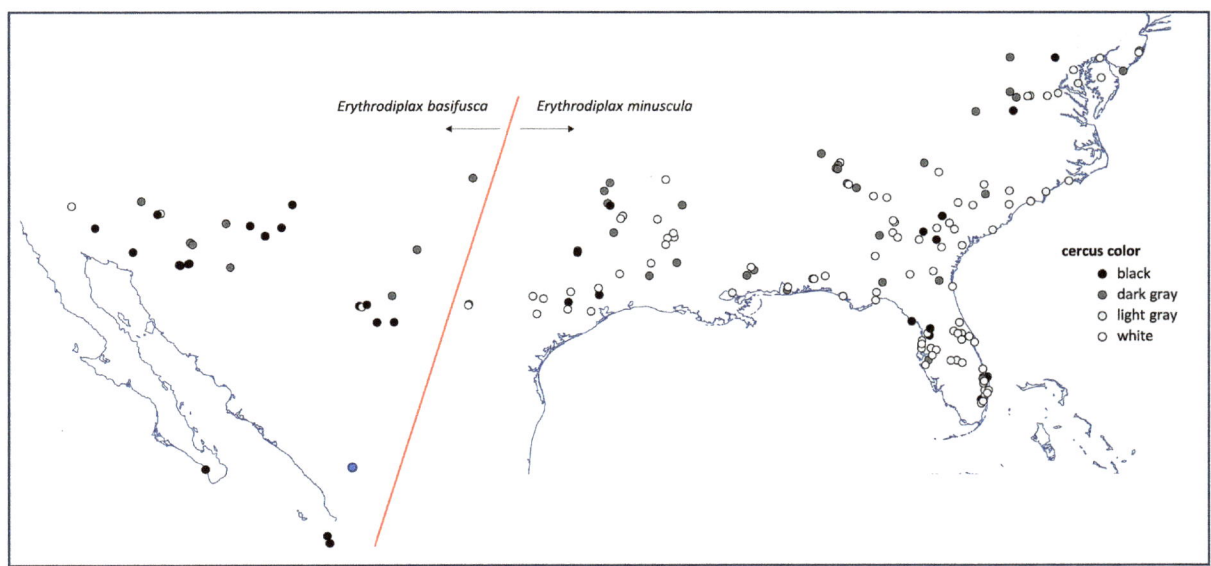

Figure 19.99 Cercus color of ♂ *Erythrodiplax minuscula* and *basifusca*, the Little Blue and Plateau Dragonlets, plotted against longitude and shown on a map of the United States. Some overlap in cercus color occurs regionally; yet, there is steep clinal variation in the color, with darker cerci becoming less prevalent farther east (see Figure 19.98).

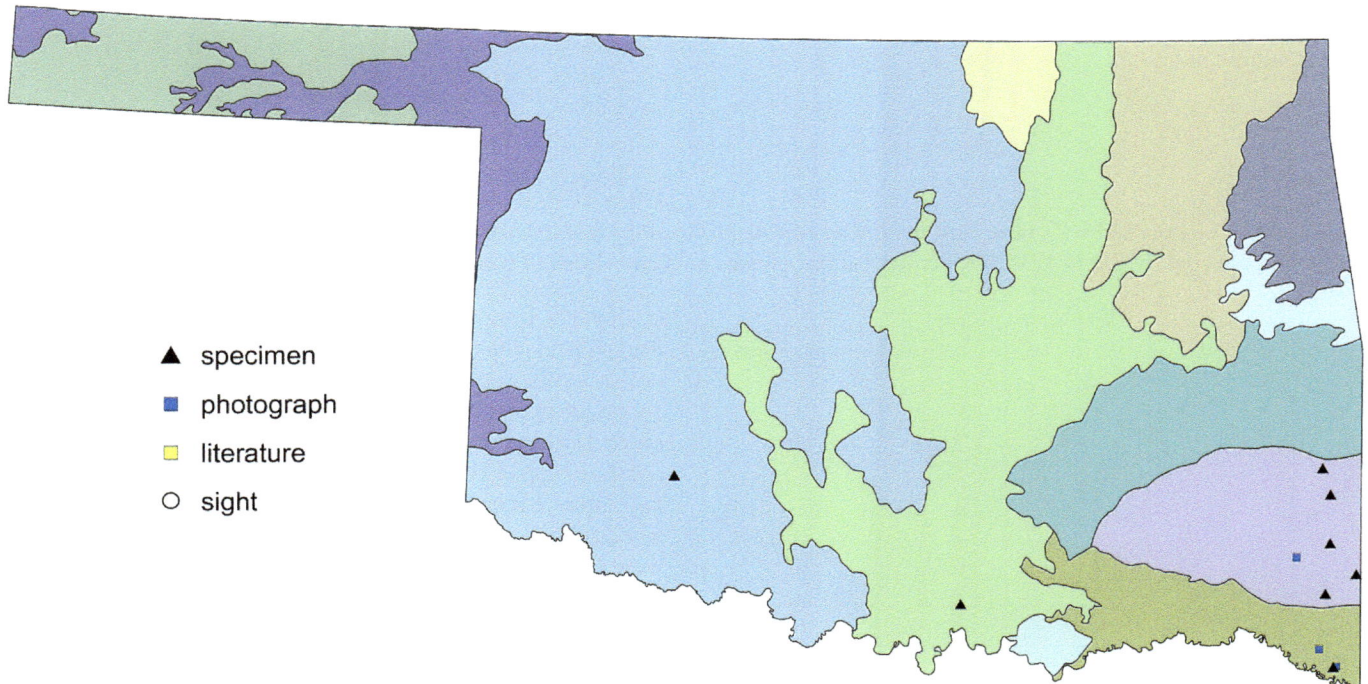

Figure 19.100 Documentation level of supporting records for *Erythrodiplax minuscula*, the Little Blue Dragonlet, in Oklahoma. Levels include specimen, archived photograph, literature reference, or sight record. Although sight records lack documentation, only those we feel are credible are plotted. Some dispute exists as to the identification of the Comanche County specimen (see discussion above).

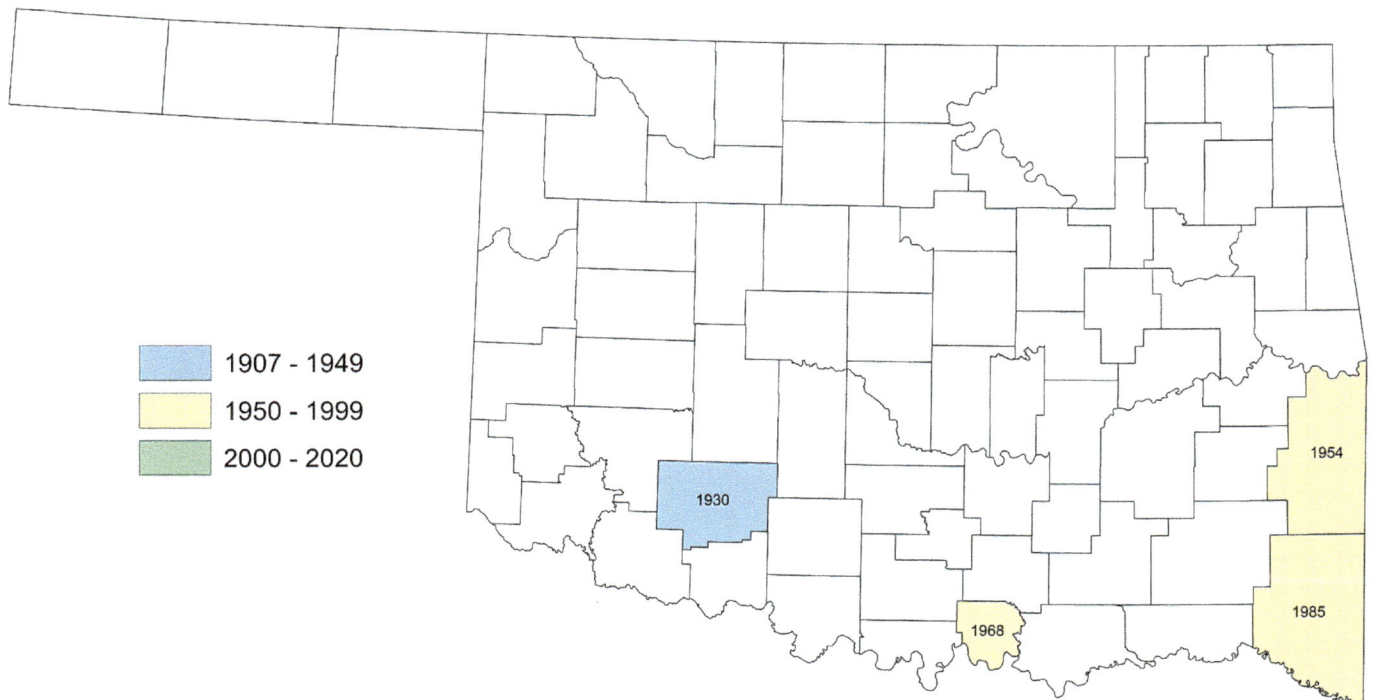

Figure 19.101 Counties in which *Erythrodiplax minuscula*, the Little Blue Dragonlet, are known to occur in Oklahoma. Some dispute exists as to the identification of the Comanche County specimen (see discussion above).

County (1♂, SP 2349; photo, possibly of same ♂). Also in 2017, a ♂ was photographed in hand at Big Cedar on 3 July (MAP, OC 466248). No records were reported in 2018, yet it was found twice in 2019, at a shallow, fishless quarry pond 23 km NNE of Eagletown (McCurtain County, 1♂, 24 May, BS-P, DA, M Dillon, SP 2747; 1♀, 2 June, examined in hand, MAP). We wonder to what extent this species is confused with the superficially similar Blue Dasher (*Pachydiplax longipennis*), obscuring the species' status in the state; however, we trust that if it were terribly common our encounter rate would be markedly higher.

The known flight dates from Oklahoma extend longer into the season than dates known for Arkansas but they are shy a month on either side compared to Louisiana and Texas (AR: 23 April–5 September, OC; LA: 27 March–9 December, USNM 296735, Mauffray 2014; TX: 26 February–19 December, Abbott and Lasley 2019). Flight season in the region generally matches that known for elsewhere in the species' range (Paulson 2011).

It has been thought that a good distinguishing character between *Erythrodiplax minuscula* and *E. basifusca* is cercus color, with the former being white, the latter black (Paulson 2003, Abbott 2005, Needham et al. 2014). Preliminary results from our review of Odonata Central records of these species from across the United States shows 1) that this distinction holds generally, 2) that there are many individuals that have intermediate coloration (pale or dark gray cerci), 3) and that individuals with dark gray cerci are nearly equally probable to be of either species (Figure 19.98).[31] Further, we found that cercus color varies clinally in *E. minuscula* such that males in the southeastern United States—excepting central Florida, for reasons unknown—typically have the cerci gleaming white whereas many at the western edge of the range, such as in southeastern Oklahoma, southern Arkansas, and eastern Texas have the cerci black or dark gray, much the same as on a typical *E. basifusca* (Figure 19.99).[32] Consequently, we urge observers not to rely on cercus color to distinguish these two species.

ERYTHRODIPLAX BERENICE (DRURY, 1773) – SEASIDE DRAGONLET

Few dragonflies tolerate saltwater, but the Seaside Dragonlet is one of them. Its distribution is chiefly coastal, yet populations occur along the Pecos River drainage inland to central New Mexico (Dunkle 1975). This species was discovered in Oklahoma when we found a ♂ at the "Jackson Salt Flats,"[33] a saline marsh 6 km S of Eldorado, Jackson County, 3 August 2014 (Patten and Smith-Patten 2014a; SP 1377, OC 425536). At this same location the following year, 7♂ and 2♀ were found, including a tandem pair (MAP, OC 433554), implying that a small population occurs at this locale, an implication borne out by a remarkable 32♂ (one SP 2457) and 13♀ (one SP 2458) there on 16 July 2017 (MAP) and around a dozen adults reported during visits in 2018 and 2019 (MAP, BC).

Away from Jackson Salt Plains, a ♂ was photographed at the city lake in Altus, Jackson County, on 24 July 2016 (MAP, OC 451280) and, astonishingly, a ♂ was photographed at Brandt Park in Norman, Cleveland County, on 30 May 2017 (EA Hjalmarson, OC 463299). A probable, additional record of the species could not be documented: a small, wholly black ♂ dragonfly observed for several minutes at Waurika WMA, Cotton County, 18 June 2016 (MAP) likely was this species, but an adult ♂ Black Meadowhawk (*Sympetrum danae*), a species yet unrecorded in Oklahoma, could not be eliminated with certainty.

In Oklahoma, the species has been documented from 6 May until 1 September (both records from 2018, MAP: 6 May, 2♂, 1♀, OC 479584; 1 September, 4♂, 5♀, [1♀ as SP 2703]). In New Mexico the species is known to fly from 27 May until October (JCAC, Paulson 2009), which is just shy of Texas's span, even though that state's seasonality includes coastal populations (2 April–11 November, Abbott and Lasley 2019).

Figure 19.102 *Erythrodiplax berenice*, the Seaside Dragonlet. A) ♂, the "Jackson Salt Flats" 6 km S of Eldorado, Jackson, 3 June 2018 (©Bill Carrell, OC 480999), site of the sole population in Oklahoma, discovered in 2014. B) ♂ Brandt Park in Norman, Cleveland County, 30 May 2017 (©Emily A Hjalmarson, OC 463299), a vagrant far from suitable habitat. C) This ♀ at the "Jackson Salt Flats" on 1 September 2018 (©Michael A Patten, OC 489663) provided a late date for the state.

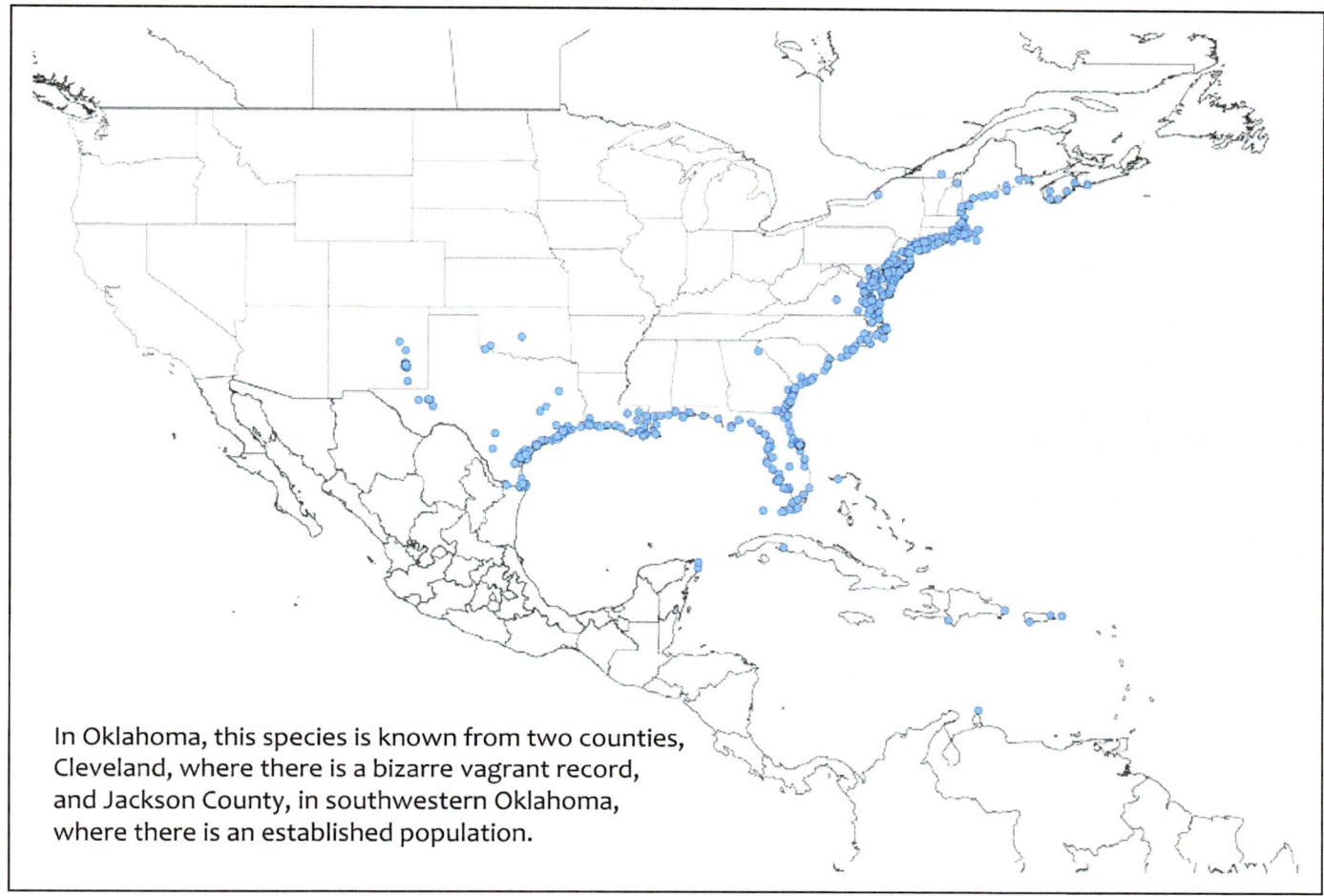

In Oklahoma, this species is known from two counties, Cleveland, where there is a bizarre vagrant record, and Jackson County, in southwestern Oklahoma, where there is an established population.

Figure 19.103 The curious distribution of the not-always-seaside Seaside Dragonlet (*Erythrodiplax berenice*), with its inland populations in the Southern Great Plains.

ERYTHRODIPLAX UMBRATA (LINNAEUS, 1758) – BAND-WINGED DRAGONLET

The Band-winged Dragonlet is most numerous in the southern part of Oklahoma, but it is recorded spottily across much of the state, and its numbers vary markedly from year-to-year. It can be found at small (often tiny and ephemeral) ponds and seasonal wetlands that support considerable fringing vegetation, typically spikerush (*Eleocharis*). It is generally uncommon, normally seen one or two at a time, although once five were collected (1 May 1930, Dean's Slough, McClain County, OMNH 916–918, USNM 297945–297946). In more recent years it was reported in larger numbers, including 14 adults at Moneka Park, Lake Waurika, Jefferson County (10 May 2014, 12♂, one as SP 1177, 2♀, including one pair, BS-P) and a high count of "at least 24 individuals" at Keystone Dam, Tulsa County (4 July 2014, OC 424303, BC). In

most years it has been reported only once or twice, but in some there are booms, such as when the species was reported thirteen times in 2012 (Figure 19.105). But it is 2014 that was the most remarkable year—42 reports of 113 adults! Since 2002 the species was added to 35 counties, bringing the overall total to 47[34] (Figure 19.107).

Erythrodiplax umbrata was first recorded in the state on 8 July 1925, when it was collected at 4 mi W of Broken Bow, McCurtain County (♀, FSCA). Bird (1932) only reported the species for Cleveland, Cotton, Latimer, McClain, and Woods Counties. Although others reported it in 1961 and 1964, GH Bick and LE Hornuff apparently never saw the species in the state. Those reports and specimens came from Marshall and Comanche Counties (23 June–24 July, FSCA, UMMZ 210531–210532). It is curious that neither Bick nor

Figure 19.104 *Erythrodiplax umbrata*, the Band-winged Dragonlet, A) ♂, Pawpaw Bottoms 9 km SSE of Muldrow, Sequoyah County, 18 May 2016 (©CA Ivy, OC 444881). B) immature ♂, 10 km SE of Idabel, McCurtain County, 21 August 2009 (©Berlin A Heck, OC 330851). C) ♀, Tulsa, Tulsa County, 3 June 2014 (©Bill Carrell, OC 422550). Perhaps no other Oklahoma dragonfly typifies better a "boom-or-bust" species: in some years there are many spread across the state, whereas in other years it borders on absent.

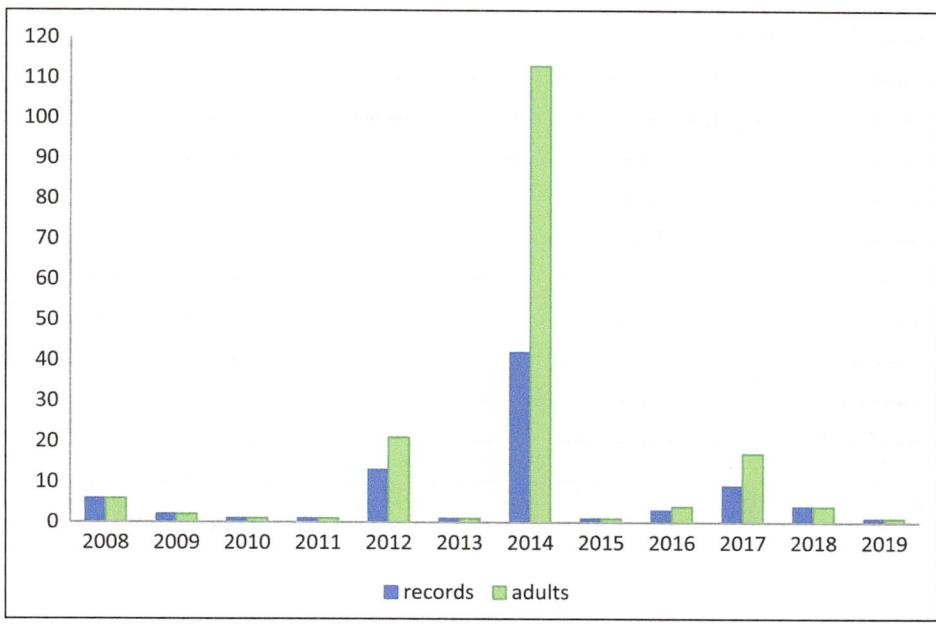

Figure 19.105 Annual variation in records and relative abundance of *Erythrodiplax umbrata*, the Band-winged Dragonlet, 2008–2019.

Hornuff nor the OU expeditions found the species in those counties.

Its flight season is from late March to late October, which outside of Texas, where it flies year round (Abbott and Lasley 2019), is the longest flight season in the region (OK: 27 March 2008, BAH, Red Slough WMA, McCurtain County, OC 330853 and 24 October 2006, J Fisher and J Nelson, Bryan County). The Oklahoma season starts well before nearby states and ends well after all but Louisiana, which it basically ties (AR: 17–19 May–8 September, Harp and Harp 2003, JCAC 23493; KS: few records, 11 July–29 August, USNM 297943, OC 426540; LA: 13 June–17 November, Mauffray 2014, OC 463619; NM: 13 August–7 September, OC 426004, 381665). We know of no records of nymphs, but there are four reports of tenerals (3 June 2014, 4U, Tulsa, Tulsa County, BC; 4 July 2014, 1♀, Okmulgee SP, Okmulgee County; 4 July 2016, 1♂, 1♀, Great Plains SP, Kiowa County, MAP; 8 July 2017, 1♂ being eaten by *Erythemis simplicicollis*, Longmire Lake, Garvin County, MAP), so we believe the species is a resident breeder in the state, albeit a sporadic one.

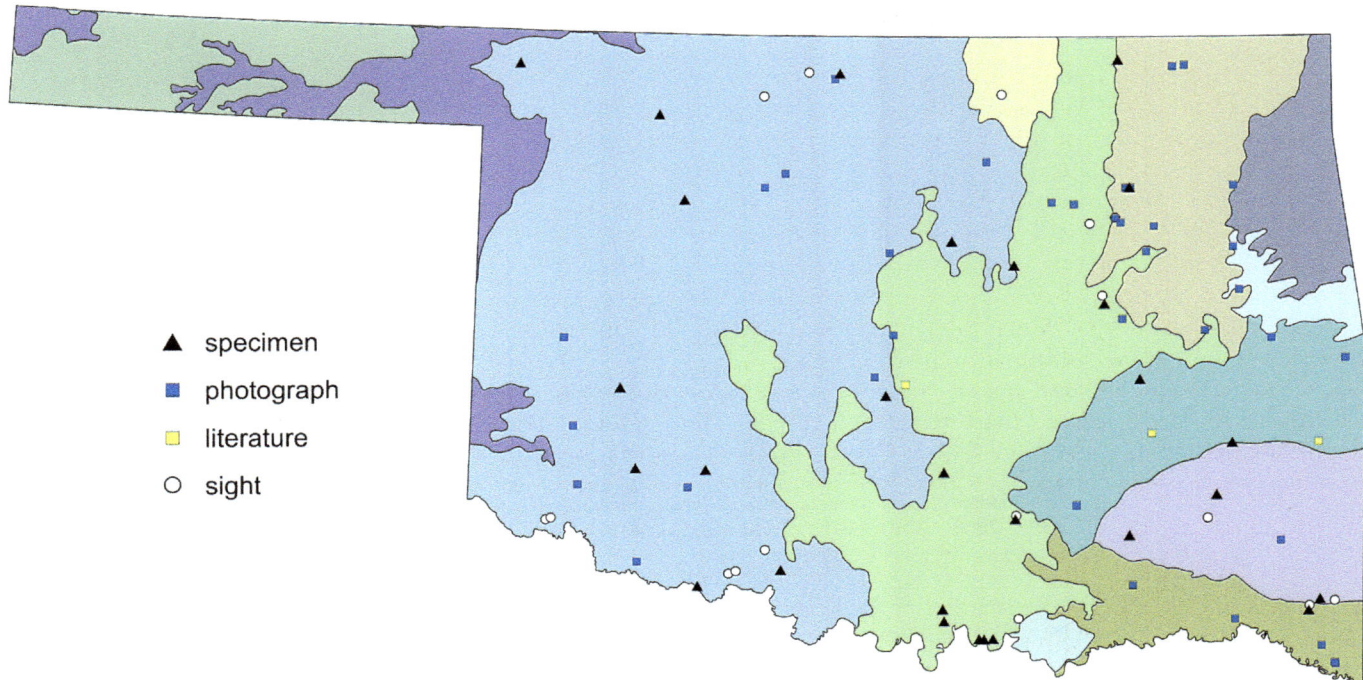

Figure 19.106 Documentation level of supporting records for *Erythrodiplax umbrata*, the Band-winged Dragonlet, in Oklahoma. Levels include specimen, archived photograph, literature reference, or sight record. Although sight records lack documentation, only those we feel are credible are plotted.

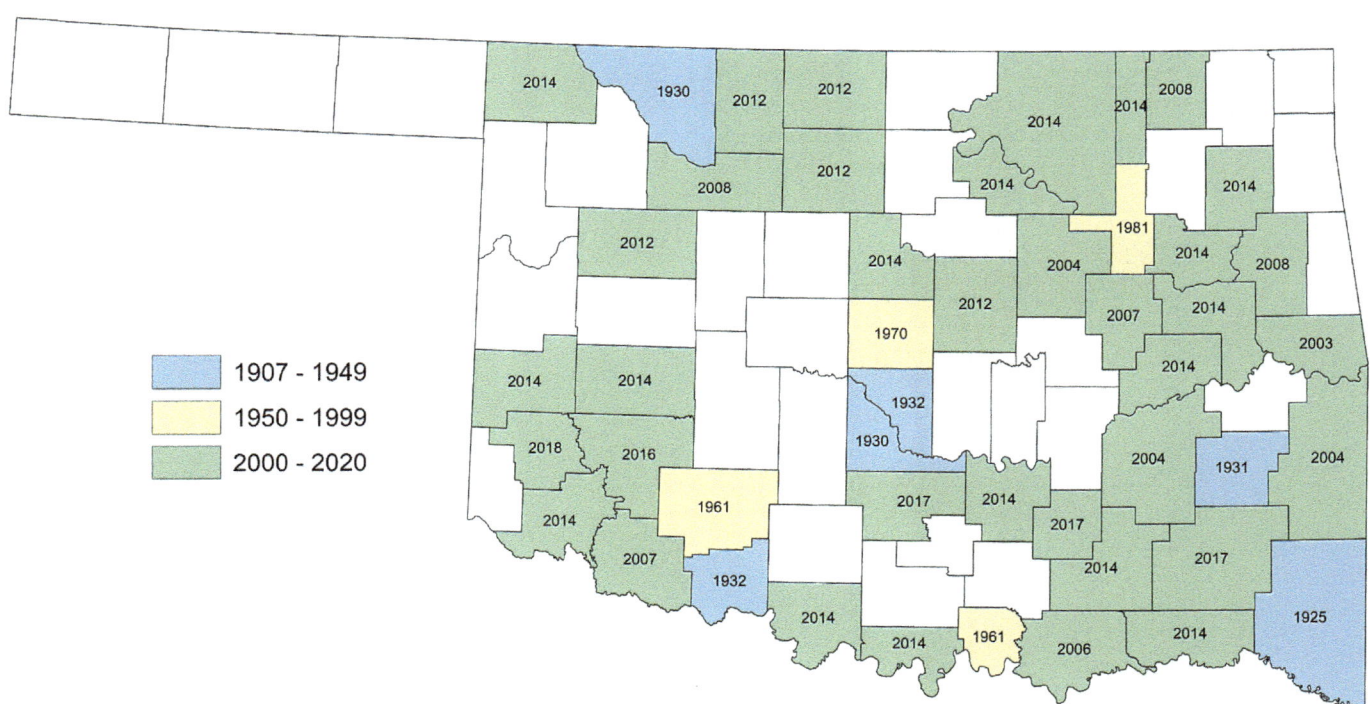

Figure 19.107 Counties in which *Erythrodiplax umbrata*, the Band-winged Dragonlet, are known to occur in Oklahoma.

PSEUDOLEON SUPERBUS (HAGEN, 1861) – FILAGREE SKIMMER

This spectacular tropical species with a penchant for wandering has graced Oklahoma with its presence three times. The first record was of a ♂ that frequented Beaver Creek at Moneka Park, Lake Waurika, Jefferson County, 30 September–5 October 2008 (VWF, OC 284013, 284120). The second was of a ♀ at the Cimarron River 4 km NNE of Kenton, Cimarron County, 7 October 2017 (MAP, SP 2521, OC 473886). And the third was on 17 September 2018 at Black Mesa SP, Cimarron County, when BC photographed an immature ♂ (OC 490587).

The species is found regularly in New Mexico, where it is known to fly from late March to early October (26 March–6 October, USNM 333648–333649, OC 504737). Although found year-round elsewhere in Texas, it has only been reported from the northern Texas panhandle once (Wolfe Creek SP, Ochiltree County, 14 July 2009, JCAC 24224) and multiple times at its base (May–June, OC). There are a growing number of records in east-central Colorado (El Paso County, 2–20 July, OC) that may constitute a breeding population.[35] Future records in Oklahoma's southwestern corner up into the panhandle are expected.

One report of larvae for Oklahoma was re-identified as *Erythemis simplicicollis* (Eastern Pondhawk; UCO 5191). An additional report of a nymph from the Tallgrass Prairie Preserve, Osage County (Bass 1994), is likely erroneous given the nearest breeding populations are in southwestern Texas and east-central Colorado.

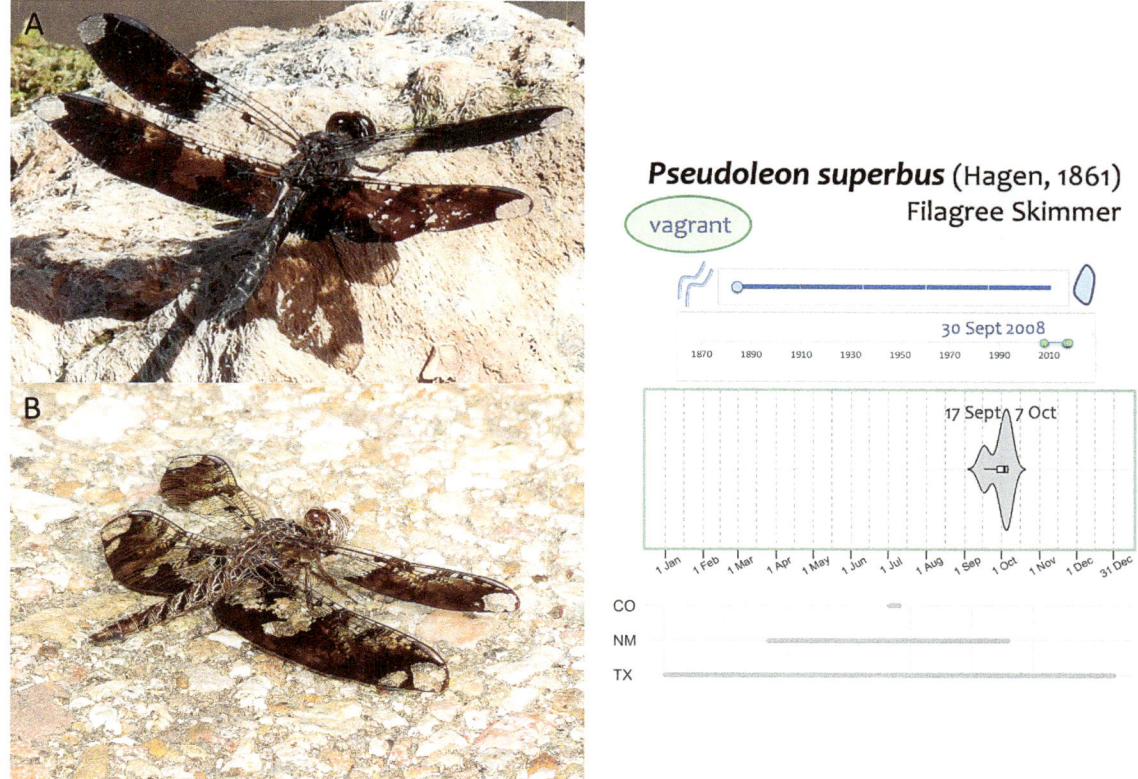

Figure 19.108 *Pseudoleon superbus*, the Filagree Skimmer, A) ♂, Moneka Park at Lake Waurika, Jefferson County, 5 October 2008 (©Victor W Fazio III, OC 284120). This individual had been present since 30 September 2008 and represented Oklahoma's first record. It has since been recorded twice more, one of which B) was an immature ♂ at Black Mesa SP, Cimarron County, on 17 September 2018 (©Bill Carrell, OC 490587).

SYMPETRUM CORRUPTUM (HAGEN, 1861) – VARIEGATED MEADOWHAWK

The Variegated Meadowhawk is a common to abundant species in a variety of habitats in Oklahoma, habitats that include suburban yards, parking lots, and roadside verges. It is typical to record the species 50–100 at a time, with even higher numbers in the western half of the state (e.g., 500 reported from Black Mesa area, Cimarron County, 22 September 2013, and 500 tenerals from Fort Supply Lake, Woodward County, 20 June 2015, BC). It was first recorded in the state on 22 September 1907, when F Collins collected 5♂ (2 at UMMZ, Williamson 1914b) at Henryetta, Okmulgee County. It has been seen regularly through all eras of data collection in the state.

It has been found in all but one county in Oklahoma, Adair, in the northeastern corner of the state. It is found year-round, just as it is in New Mexico and Texas (Paulson 2009, Abbott and Lasley 2019), and is both a resident breeder and a migrant, with spring and fall migrations through the state, which gives it a bimodal distribution in terms of its abundance and seasonality (Smith-Patten et al. 2007). In other neighboring states the species flies most of the year, but ranges are still truncated compared to those known for Oklahoma, New Mexico, and Texas (AR: 4 March–10 December, OC 461349–461351, 492451; CO: 17 April–3 October, OC 374492, 429156; KS: 18 March–28

Figure 19.109 *Sympetrum corruptum,* the Variegated Meadowhawk, A) ♂, Fobb Bottom WMA, Marshall County, 30 September 2016 (©Bill Carrell, OC 456619). B) ♀, Lake Perry, Noble County, 11 October 2015 (©Bill Carrell, OC 437822). It is likely that a sum of records or numbers of all other *Sympetrum* meadowhawks known from the Southern Great Plains would lag behind sums for *S. corruptum.* C) Mixed-species pairing of ♀ andromorph *Sympetrum ambiguum,* the Blue-faced Meadowhawk (left), and a ♂ *S. corruptum* (right; Oxley Nature Center, Tulsa, Tulsa County, 26 October 2019, ©Bill Carrell).

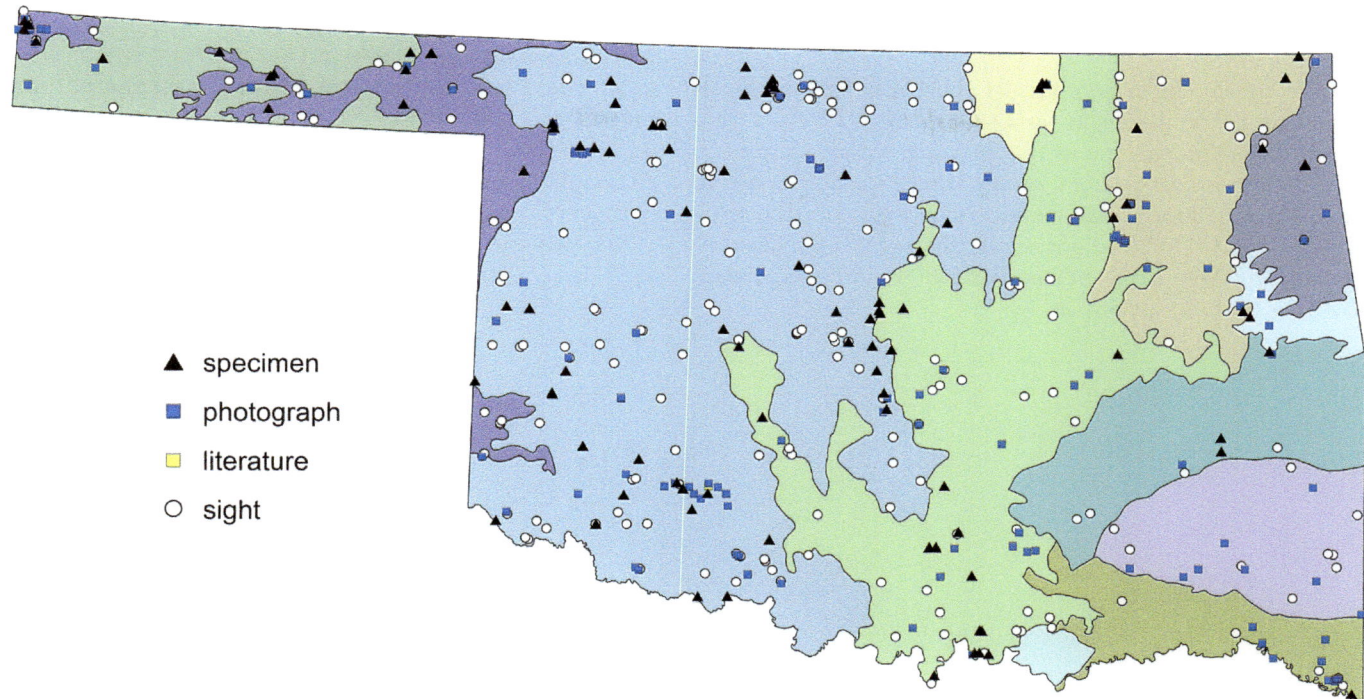

Figure 19.110 Documentation level of supporting records for *Sympetrum corruptum*, the Variegated Meadowhawk, in Oklahoma. Levels include specimen, archived photograph, literature reference, or sight record. Although sight records lack documentation, only those we feel are credible are plotted.

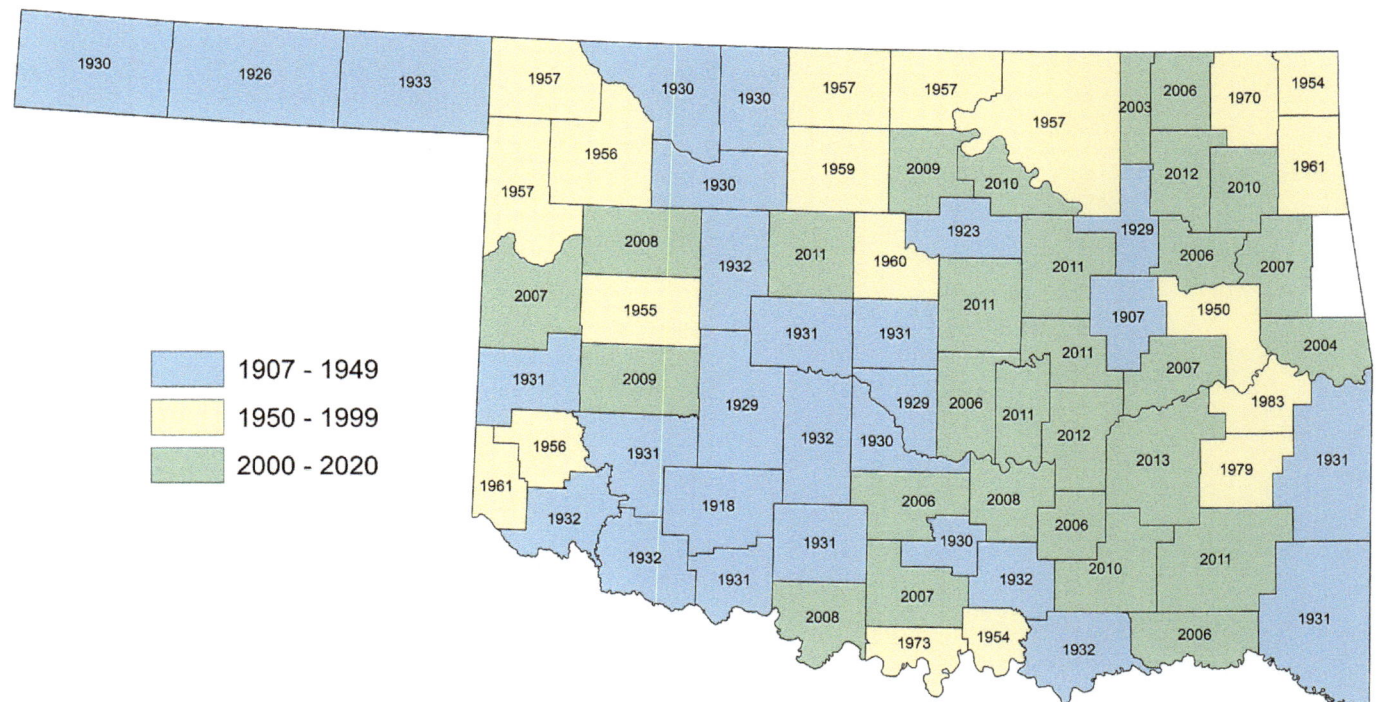

Figure 19.111 Counties in which *Sympetrum corruptum*, the Variegated Meadowhawk, are known to occur in Oklahoma.

October, OC 374122, 315585; MO: 16 April–26 November, Sims 2012).

Two specimens of note are 2♂ (UMMZ) that were collected by EB Williamson on 19 October 1929. As with a handful of other specimens that he collected at "15 miles E of Weatherford, in Custer County," these specimens had to be re-cataloged for Caddo County, where the location actually lies. Bird (1932) did not report the species for Custer[36] but the Dot Map Project (Donnelly 2004b) did, perhaps on the basis of these specimens.

One mixed-species pairing with this species is known. A ♂ *Sympetrum corruptum* and a ♀ andromorph *S. ambiguum*, the Blue-faced Meadowhawk, was photographed by BC at Oxley Nature Center, Tulsa, Tulsa County, on 26 October 2019 (Figure 19.109). To our knowledge, although mixed-species pairings between *Sympetrum* have been reported relatively often (e.g., Bick and Bick 1981), this specific combination has not.

SYMPETRUM ILLOTUM (HAGEN, 1861) – CARDINAL MEADOWHAWK

This southern and western species is a vagrant to Oklahoma, where it has been recorded only once: an adult ♂ was photographed at Fort Sill MR, Comanche County, 30 October 2007 (VWF, OC 263515; Figure 19.112). The records nearest to Oklahoma are from eastern New Mexico and southwestern Texas (NM: we know of three records with dates, all of which are of single ♂, 19 May–28 August, OC 430916, 436335, other records may be dubious, *pers comm*, JN Stuart; TX: 20 April–13 September, Abbott and Lasley 2019). The nearest documented breeding population we know of is in southeastern Arizona and adjacent Sonora (Bailowitz et al. 2015). To our knowledge there were no major storms around this time of year that would have driven the species into the state, so its presence in Oklahoma on that day remains a mystery.

Figure 19.112 *Sympetrum illotum*, the Cardinal Meadowhawk, ♂, Fort Sill, Comanche County, 30 October 2007 (©Victor W Fazio III, OC 263515), the only recorded anywhere in the Southern Great Plains.

SYMPETRUM AMBIGUUM (RAMBUR, 1842) – BLUE-FACED MEADOWHAWK

The Blue-faced Meadowhawk is a denizen of broadleaf (especially oak and oak–hickory) forests in the eastern half of the state, although there are scattered records throughout the rest of the Oklahoma "mainland" (i.e., *sans* panhandle). It seems to be fairly strongly associated with wooded areas and is a particular fan of steep-sided, slow-moving creeks or seasonally flooded woodlands, both with little light penetration. Allison (1921:54) reported the species to be common in Kansas and said it occurred at "Still water, swampy ground and meadows."

It is usually seen in small numbers in Oklahoma, but in the fall, particularly in September and October, the species can be common. For example, BC reported 100 individuals, including 5 pairs, at Bixhoma Lake, Wagoner County, on 18 October 2015 (OC 437963), and we recorded 75 individuals, including 2 pairs, on 2 October 2004 at Mohawk Park in Tulsa, Tulsa County (Smith-Patten et al. 2007).

It was first recorded in Oklahoma at Waynoka, Woods County, on 2 July 1930 by RD Bird (1♂, 1♀, OMNH 309), although Bird (1932) did not report this record. Instead, he reported the species from one county, Le Flore, which was a record that he cited as originating from AI Ortenburger. Our tally of counties during the early 1930s is six (Choctaw, Le Flore, McCurtain, Pittsburg, Pushmataha, Woods). It has been reported fairly consistently since, with a substantial increase in record numbers in recent times (9 records 1930–1934, 23 in 1948–1994 from an additional 14 counties, and about 190 in 2002–2019 from an additional 39 counties). 2015 was a particularly good year, with 33 records, followed by 22 records in 2016. *Sympetrum ambiguum* is now reported from 61 counties across the state (>220 reports), with records as far west as the base of the panhandle (♂ at Beaver Dunes SP, Beaver County, 6 October 2017; MAP, SP 2515, OC 473883).

Figure 19.113 *Sympetrum ambiguum*, the Blue-faced Meadowhawk. A) ♂, Chickasaw NRA, Murray County, 5 September 2016 (©Emily A Hjalmarson, OC 454784). B) ♀, Tallgrass Prairie Preserve, Osage County, 15 June 2012 (©Bryan E Reynolds, OC 434882). C) ♀ andromorph, Tulsa, Tulsa County, 20 October 2013 (©Bill Carrell, OC 411316). This stunning meadowhawk, one of our more colorful dragonflies, may be common in moist eastern woodlands in autumn.

In the southeastern part of the state it has been reported from early May to mid-December (both records from Red Slough, McCurtain County: 8 May 2016, 1♀, MAP sight and 15 December 2015, 1U, DA sight). These dates are also the seasonal date range for the species in general in Oklahoma, which although being the latest date for the region, it is the third earliest behind Arkansas and Texas (AR: 23 April–17 November, OC 374543, Harp and Rickett 1977; KS: 15 June–mid-November, SEMC 1366676, Allison 1921; LA: 27 June–27 November, Mauffray 2014; MO: 21 May–14 November, Sims 2012; TX: 25 April–5 December, Abbott and Lasley 2019, OC 439506).

Earlier we classified the species as a "classic late-season emergent" in northeastern Oklahoma (Smith-Patten et al. 2007:11). We mischaracterized the species by making a bad assumption. At the time we had no emergence data, so we assumed that the striking increase in abundance in October was an indication that the species emerged then. Now, there are 15 teneral records, none of which support the idea of late-season emergence because all teneral records, accounting for 58 individuals, fall between 12 May and 25 June. Additionally, all records of immature individuals also fall earlier in the year between 26 May and 1 August. There are records of adults during that time frame as well, but to compare numbers, we see that tenerals and immature individuals greatly outnumber adults: 8 records of 32 adults versus 26 records of 79 teneral and immatures before 15 June. Interesting, too, is that although there are adult records earlier in the season, all records of mating are from October (n = 12, 2–27 October).

There remains a mystifying lull in records between peak emergence, such as a dozen or more tenerals at different sites in Atoka County, on 2 June 2014, or twenty tenerals at Langston Lake, Logan County, on 5 June 2016 (MAP), and peak counts of adults, which tend to be recorded in October. (There are plenty of records from July, August, and September, but there tend not to be more than one or two individuals recorded at a time during those months.) We posit two hypotheses that may or may not be mutually exclusive. It is possible that *Sympetrum ambiguum* is a species, like others in the genus (Corbet 1999), that estivate or siccatate during the hot summer months. If either of those behavioral or physiological phenomena do occur, i.e., reduction of activity, as with estivation, or suspension of development to full adulthood, as with siccatation, then the species would essentially wait out the heat in a hidden location as pre-productive adults and then would "re-emerge" as mature adults when temperatures cool in the autumn.[37] Alternatively, we may see an uptick in adult numbers in the autumn because individuals migrate into or through the state. We would not be surprised to learn that the late-season uptick can be attributed to a combination of these two hypotheses. In any case, we now doubt a hypothesis of mass emergences in the autumn, emergences that elude our detection year after year.

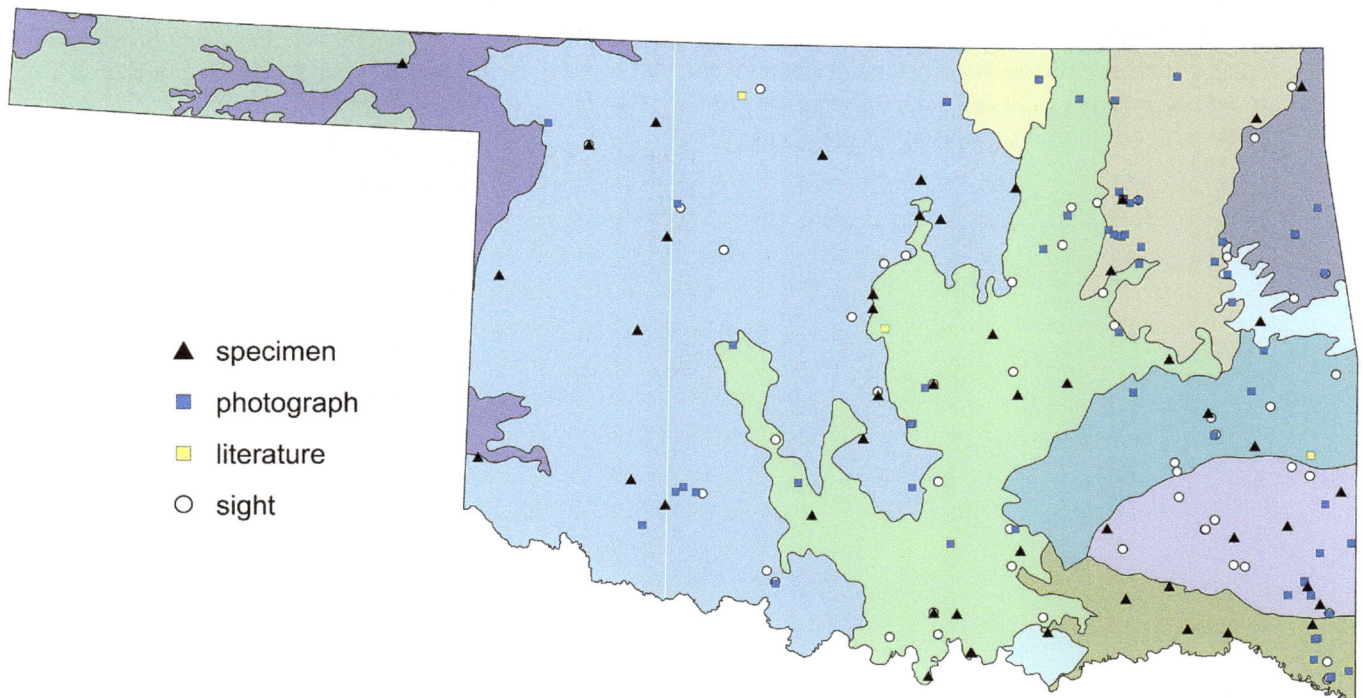

Figure 19.114 Documentation level of supporting records for *Sympetrum ambiguum*, the Blue-faced Meadowhawk, in Oklahoma. Levels include specimen, archived photograph, literature reference, or sight record. Although sight records lack documentation, only those we feel are credible are plotted.

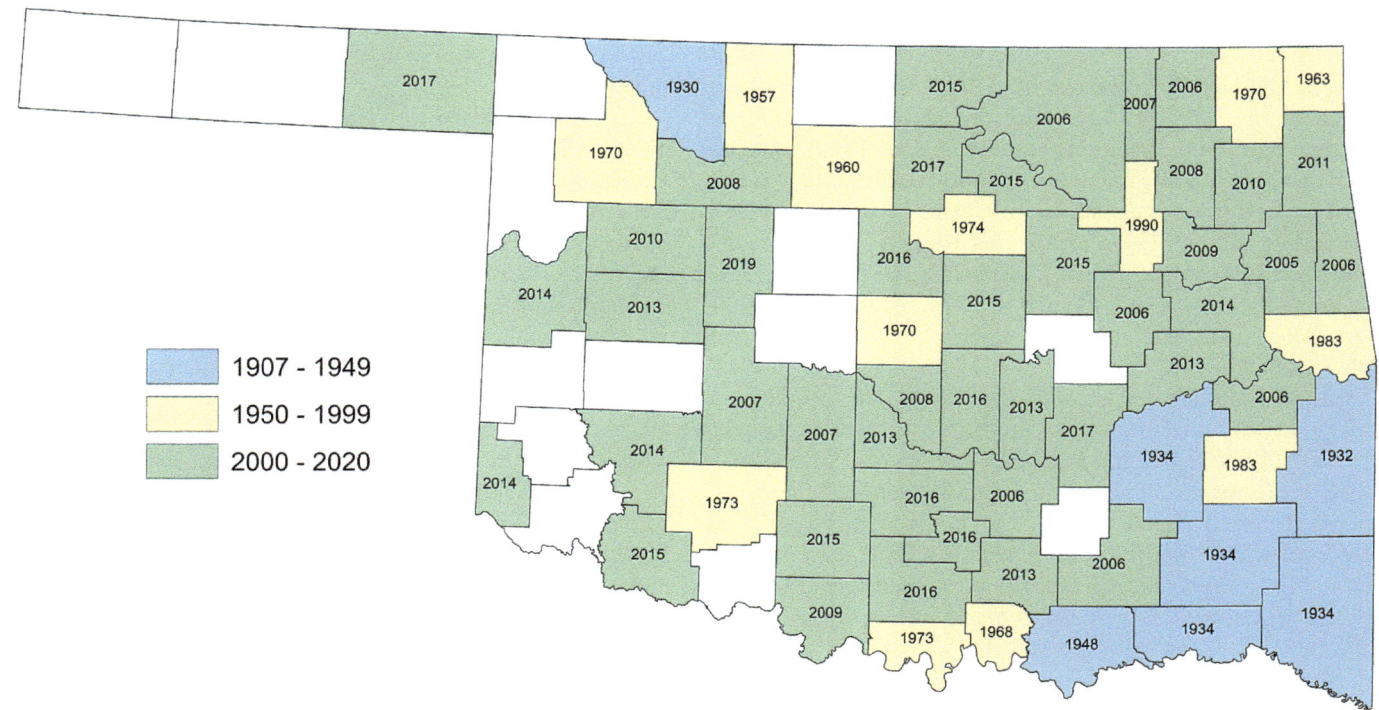

Figure 19.115 Counties in which *Sympetrum ambiguum*, the Blue-faced Meadowhawk, are known to occur in Oklahoma.

Lastly, we know of half a dozen records of females with the abdomen as bright red as on a male. One has the red limited to the top of S3 (OC 330929, possibly some red on S4, too), but all of the others have red extending the length of their abdomens dorsally and dorsolaterally (OC 262958, 263281, 411316 [Figure 19.113], 434878, and 436306). Four of these females also have the face green-blue. Andromorph females are rare elsewhere (Paulson 2011), which is the same for Oklahoma: the 6♀ identified as andromorph-types are out of about 660 individuals in the OOP database. However, we cannot discount how easily mistaken for males these females are and so andromorphs may be undercounted; nevertheless, in our experience, they are encountered rarely. We know of one instance where a ♀ andromorph was mistaken for a ♀ *Sympetrum corruptum*, the Variegated Meadowhawk, by a ♂ *S. corruptum* (see that species' account and Figure 19.109).

SYMPETRUM OBTRUSUM (HAGEN, 1867) – WHITE-FACED MEADOWHAWK

There is one documented record for this species in Oklahoma: we collected a single ♂ at Lake Elmer, Kingfisher County, 22 September 2012 (Smith-Patten and Patten 2013a; SP 431, OC 381935; Figure 19.116). This ♂ was with a handful of red meadowhawks of the RuWhiChe [ru-WEE-chee] complex, the closely related and phenotypically similar Ruby (*Sympetrum rubicundulum*), White-faced, and Cherry-faced (*S. internum*) Meadowhawks (Pilgrim and Von Dohlen 2007).

A second ♂ we captured (SP 430) had a yellowish face and what we felt were atypical hamules. We posited that it was a hybrid *S. obtrusum*×*S. internum* (Smith-Patten and Patten 2013a). We feel that given the shape of the hamules (Figure 19.117), which in our opinion are intermediate between the two species, the possibility of this specimen being a hybrid cannot be excluded; however, after an examination of the specimen, TW Donnelly (*in litt*) identified it as *S. obtrusum*. The

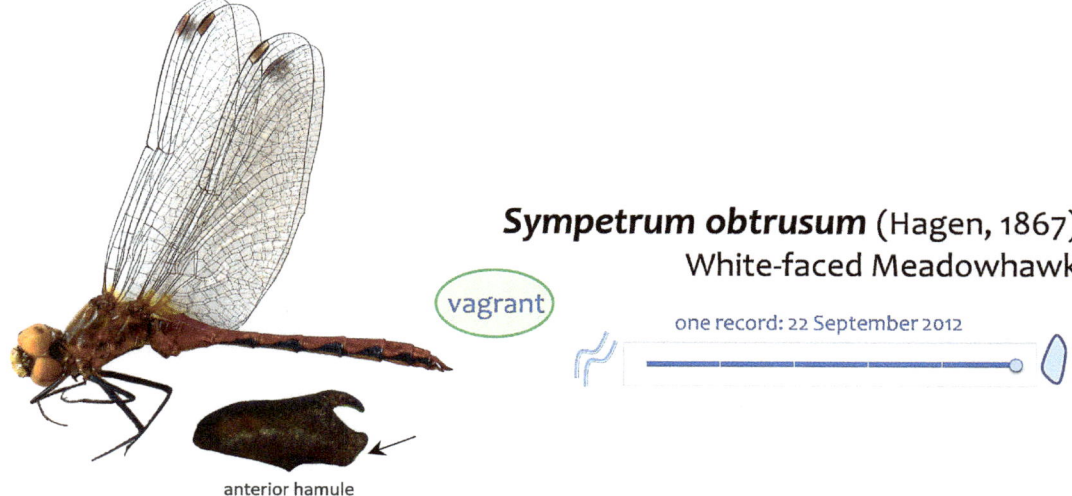

Sympetrum obtrusum (Hagen, 1867)
White-faced Meadowhawk

vagrant

one record: 22 September 2012

Figure 19.116 Oklahoma's first record of *Sympetrum obtrusum*, the White-faced Meadowhawk. This ♂ was collected on 22 September 2012 at Lake Elmer, Kingfisher County (SP 431). Arrow points to the wide, squared lobe distinctive of this species.

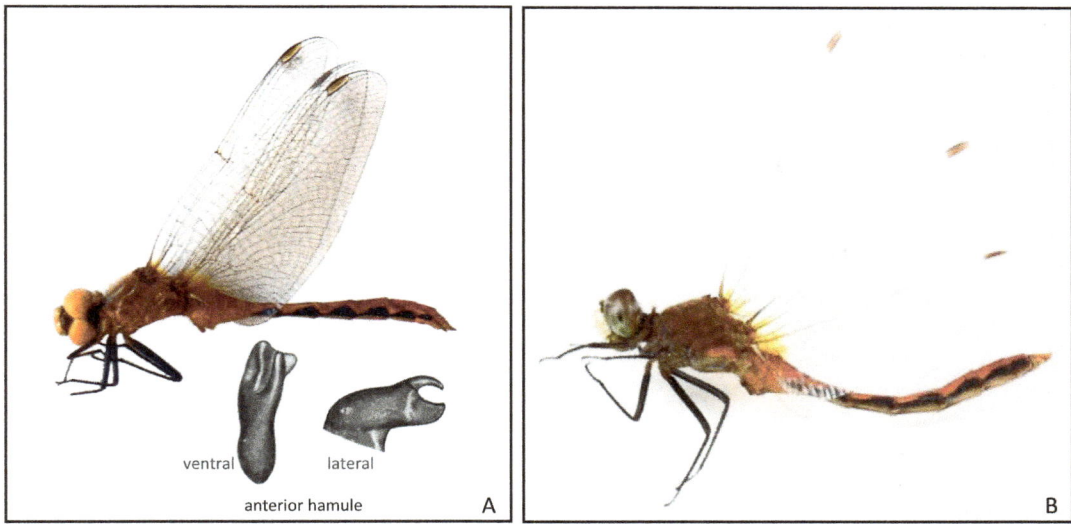

ventral   lateral
anterior hamule   A

B

Figure 19.117 Dubious records of *Sympetrum obtrusum*, the White-face Meadowhawk, in Oklahoma. A) This ♂ was collected on 22 September 2012 at Lake Elmer, Kingfisher County (SP 430; hamule illustrations by ©Thomas W Donnelly). There is debate regarding hamule shape. Is this *S. obtrusum* or a hybrid *S. obtrusum*×*S. internum*? B) Identification of a ♂ meadowhawk of this persuasion at Texhoma, Texas County, 24 September 2011 (©Kurt F Schaefer, OC 434895) could not be established because hamule shape is unknown and because the specimen was discarded after being photographed in harsh lighting.

Lake Elmer record is the southernmost for the species in the Great Plains by at least 240 km, the next nearest being from Stafford and Crawford Counties, Kansas (Beckemeyer 1995, Beckemeyer and Todd 1996, OC).

A possible second record of the species for Oklahoma is of a ♂ RuWhiChe from Texhoma, Texas County, 24 September 2011 photographed in captivity under harsh artificial light (KF Schaefer, OC 434895; Figure 19.117). The specimen was not retained. This ♂ has an apparently whitish face, but it is difficult to determine to what extent the strong light and white background has bleached out the color. It may be that the face is slightly yellow, which could indicate an immature *Sympetrum internum* or *S. rubicundulum* (although the hyaline wings would seem to eliminate individuals in the western part of that species' geographical range). We expect *S. obtrusum* to occur in Oklahoma again one day, but it likely will appear only during years when *S. internum* stages a "southward irruption" (see that species' account).

SYMPETRUM PALLIPES (HAGEN, 1874) – STRIPED MEADOWHAWK

The sole record for Oklahoma is of a ♂ photographed at Black Mesa SP, Cimarron County, 9 September 2014 (BC, OC 426895; Figure 19.118). The geographically nearest records to the state are from northeastern New Mexico (♂ at Sugarite Canyon SP, Colfax County, 23 July 2012, JN Stuart, OC 377946) and southeastern Colorado (♀ at Lamar Community College Woods, Prowers County, 22 June 2013, B Maynard, OC 400841), 135 km west and 140 km north, respectively. The sole record for Texas, in the southern panhandle (♂ at Llano Estacado Audubon Trail, Lubbock County, 2 September 2009, J Hatfield, OC 314925), indicates a possibility that the species could occur in southwestern Oklahoma someday.

Sympetrum pallipes flies from late June to late October in nearby Colorado and New Mexico (CO: 21 June–20 October, OC 465126, 438573; NM: 30 June–19 October, OC 376716, 427470). These dates suggest that the species could occur in Oklahoma again anytime from summer to fall.

Sympetrum pallipes (Hagen, 1874)
Striped Meadowhawk

vagrant

one record: 9 September 2014

Figure 19.118 *Sympetrum pallipes*, the Striped Meadowhawk. A) ♂, Black Mesa SP, Cimarron County, 9 September 2014 (©Bill Carrell, OC 426895), Oklahoma's sole record of this species.

SYMPETRUM INTERNUM MONTGOMERY, 1943 – CHERRY-FACED MEADOWHAWK

The Cherry-faced Meadowhawk is an irregular autumn visitor to Oklahoma (alternatively, its scarcity masks its presence in the state every autumn). The species appears to irrupt southward into the state, thus far at indeterminate intervals. From the limited data for the state ($n = 23$ records), we know that when it appears it tends to be seen one to a few individuals at a time, although higher numbers can occur, such as when we encountered 5♂ and 2♀ at Legion Park in Blackwell, Kay County, on 15 October 2014 (3♂ as SP 1455–1457, 1♀ as SP 1458) or when "The Baumgartners" (likely Frederick and Marguerite) encountered 6♂ and 5♀ in Stillwater, Payne County, on 8 October 1950 (GH Bick notes). The last record is the highest count known for Oklahoma.

The species has been documented to fly in the state from August until late October (late date 23 October 2011, 3♂, 7.5 km E of Lenapah, Nowata County, KW, in DRP). The early flight date comes from the state's oldest specimen of any species (USNM 487037)—from August 1877. That specimen, collected by Charles Valentine Riley, a founder of modern entomology, has the vague locality of "Red River" and a difficult to read second line on the collecting label (see

Figure 2.1 and endnote about this specimen in Chapter 2). The specimen has long been attributed to Indian Territory, which at the time, spanned the entirety of the length of the Red River that we now consider to be the southern border of Oklahoma. This record would be the southernmost for the species. We have long been wary of this record, thinking that perhaps the specimen actually was collected along the Red River of the North, which forms the border between Minnesota and North Dakota and extends into Canada. Confusion between the two Red Rivers is long standing, and Riley was known to have traveled extensively from the southern Plains north into Canada. We have begrudgingly accepted this record, partly on the basis of geography and partly on the extralimital distribution of another northern species known to occasionally wander nearly as far south. Documented records for *Sympetrum internum* in Oklahoma are only approximately 130 km on a bee line away from the Red River and only 300 km at its farthest point. Such distances can easily be traversed by a vagrant, such as *S. costiferum* (the Saffron-winged Meadowhawk) that, at its farthest south record in Oklahoma, is >500 km out of its "normal" autumn range. If it is possible for *S. costiferum* to stray

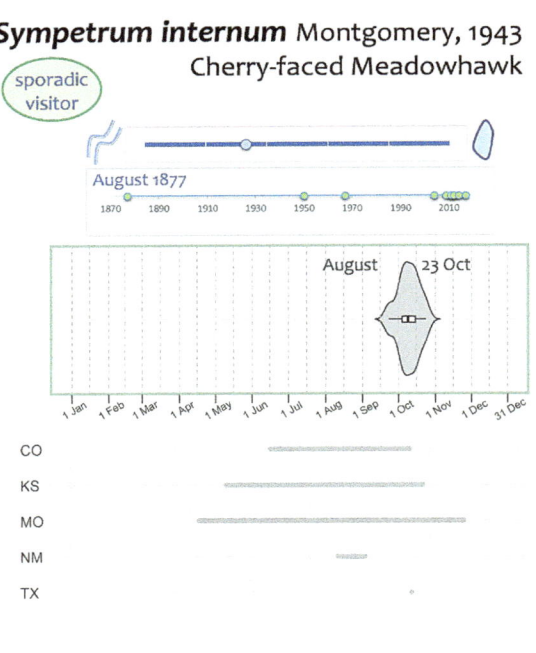

Figure 19.119 *Sympetrum internum*, the Cherry-faced Meadowhawk. A) ♂, Woodward Park in Tulsa, Tulsa County, 6 October 2014 (©James W Arterburn, OC 427347). B) ♂, 6 km NNE of Tonkawa, Kay County, 15 October 2012 (©Jason R Heinen, OC 382108). This species "pushes" southward into Oklahoma once every 2 to 20 years. How is that for uncertainty?

so far, then maybe it is not such a long shot for *S. internum* to do so. That said, an August occurrence is odd. The next earliest record for Oklahoma is 22 September (2012, 2♂ at Lake Elmer, Kingfisher County, one as SP 429; OC 381934), a date that jibes better with a pattern of autumnal irruptions into the *southern* Great Plains.

Regardless of the ultimate provenance or validity of that first record, Oklahoma's flight season falls within the range-wide species' phenology of June to October (Paulson 2009, 2011). It also falls within the regionwide pattern of primarily autumn visitation. For example, even though Kansas and Missouri have records from earlier in the year, the vast majority of their records are autumnal (KS: 10 May–21 October, PSU, OC; MO: 17 April–24 November, Sims 2012). That is also the case for the limited records from New Mexico (11 August–3 September, OC), but not necessarily the case for Colorado (16 June–10 October, with subequal summer and fall records, OC), although the latter might be more of an elevational, rather than a latitudinal, variation. It is known in Texas only from one collected by GL Wise in Deaf Smith County, in the panhandle, on 12 October 1968 (FSCA). The general latitudinal pattern of northern species dropping southward as colder temperatures start to set in the north makes perfect sense for our region given that it lies on the southern fringe of the species' overall range.

Records of the Cherry-faced Meadowhawk in the state are from 12 counties, and all are of adults. At least one record from each county, except one (Alfalfa County, see below), has been documented by photograph or specimen. We know of about 30 specimens that have been taken in the state, from 8 counties, but only 15 specimens are known to be extant, 10 of which are in the SP collection (9♂, 1♀, SP 429, 502–506, 1455–1458), FSCA and USNM have one each, and DRP has 3♂. An additional specimen was mistakenly attributed to the state (EMEC 300207, 1♀, verbatim collecting label as per EMEC is "Itasca OKLA: VII-4-37 AE Pritchard"), but seeing as there is no Itasca in Oklahoma, the specimen was likely collected at Itasca, Minnesota, where Pritchard was during July 1937. The remaining 16 specimens, from Cleveland and Payne Counties, appear to have all been discarded or lost. Fortunately, they were examined and confirmed by GH Bick beforehand. The Cleveland and Payne County records were between 8–15 October 1950 (coll. by: the Baumgartners, Coats, FE Stuart, HT Russell, and HS Hervey; Bick and Bick 1957, GHB notes). According to Bick's notes the only other record he knew of was of a ♀ collected in Oklahoma City, Oklahoma County, by L Alleman on 24 September 1967 (FSCA).

Thirty-seven years passed before the species was reported again in Oklahoma, which happened to be our 2004 sight record for Alfalfa County (1♂, 23 October 2004, Salt Plains NWR). We had just moved to Oklahoma from New Hampshire, so we did not even think to catch this individual, still being in the habit of not doing so with obvious

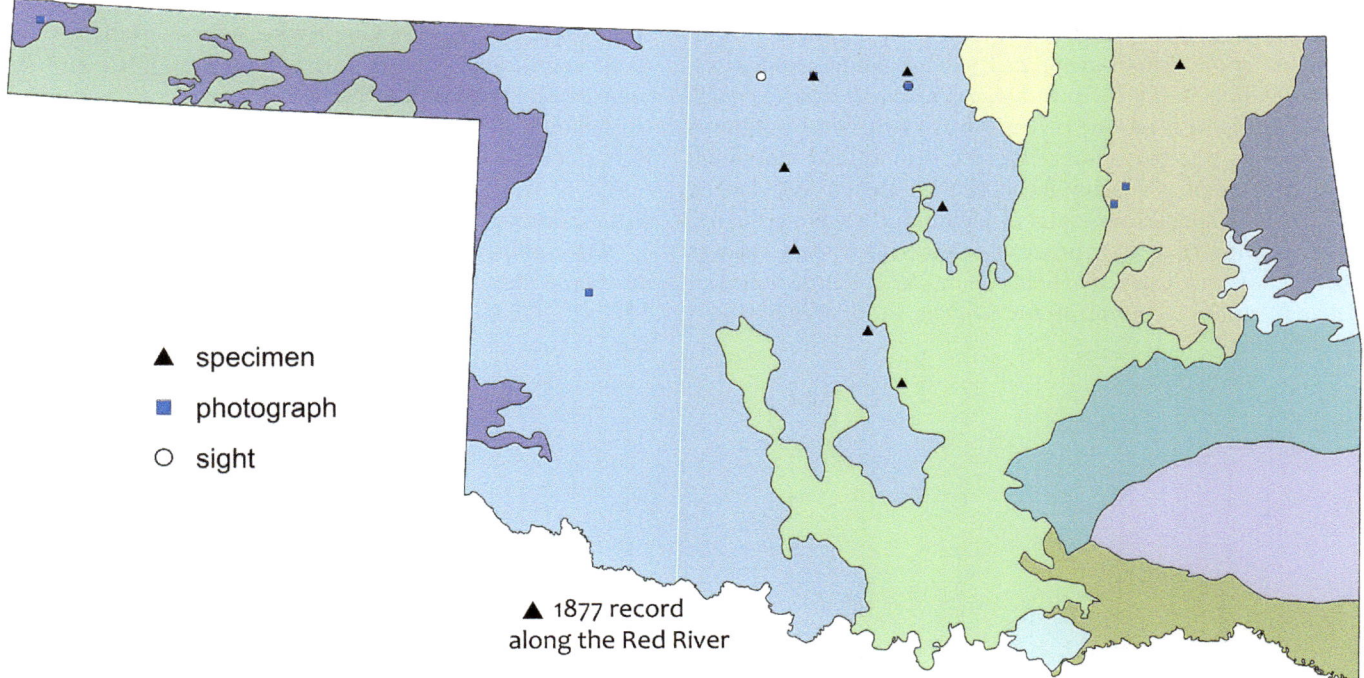

▲ specimen

■ photograph

○ sight

▲ 1877 record along the Red River

Figure 19.120 Documentation level of supporting records for *Sympetrum internum*, the Cherry-faced Meadowhawk, in Oklahoma. Levels include specimen, archived photograph, literature reference, or sight record. Although sight records lack documentation, only those we feel are credible are plotted. It is not known where the 1877 specimen record came from except that it was apparently along the Red River, which makes for the southern border of Oklahoma.

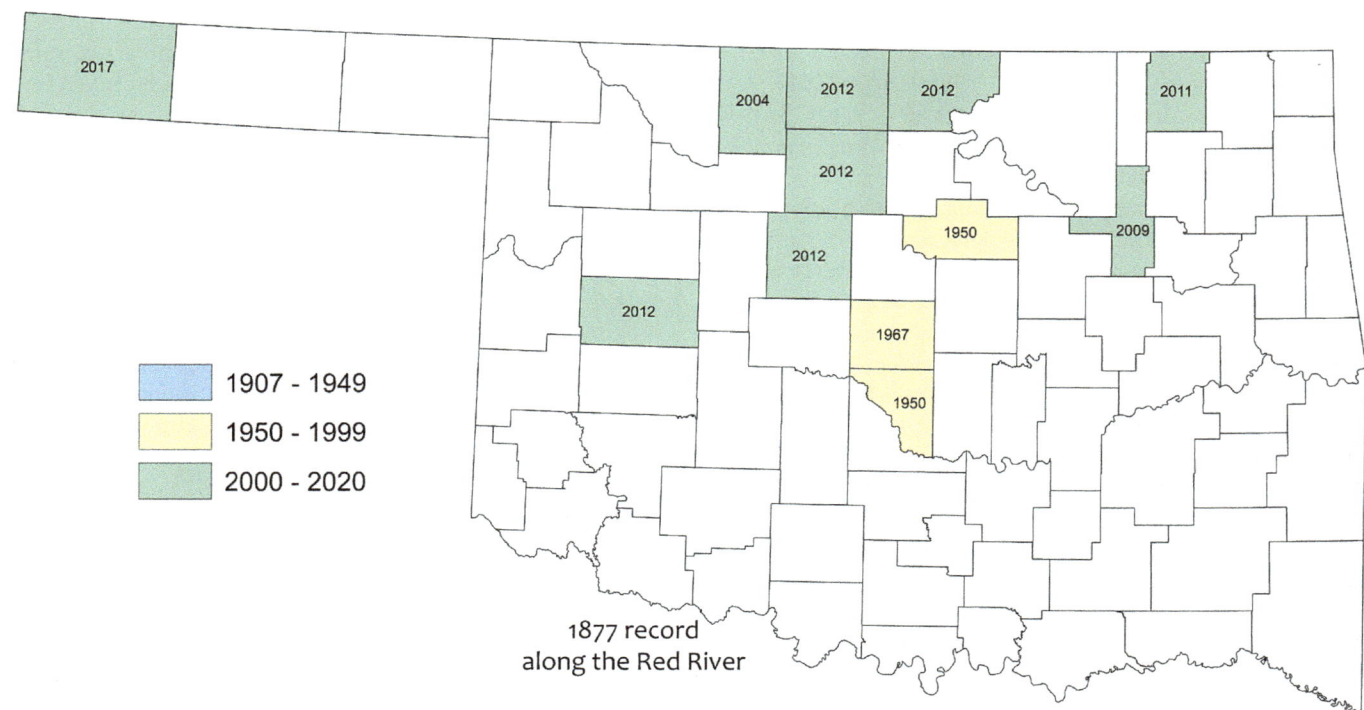

Figure 19.121 Counties in which *Sympetrum internum*, the Cherry-faced Meadowhawk, are known to occur in Oklahoma.

Cherry-faced Meadowhawks. The magnitude of that error did not hit us until eight years had passed with no other records. Our first chance to take a specimen was on 22 September 2012 at Lake Elmer, Kingfisher County, when we came upon a small group of RuWhiChe meadowhawks, one of which was a definite *Sympetrum internum* (SP 429, OC 381934), at least one other ♂ was identified by sight as that species (the frons was deep red), and a third individual may have been a Cherry-faced (Smith-Patten and Patten 2013b). The species was seen elsewhere in the state during 2012, making that year the most productive to date—it was recorded from Custer, Garfield, Grant, Kay, Kingfisher, and Tulsa Counties, accounting for 8 of the 14 records between 2004 and 2017.

There is one unconfirmed record of *Sympetrum internum* in Oklahoma. A series of photographs was taken on 19 October 2011 and submitted to Odonata Central (374661) from the same Nowata County location as the DRP specimens mentioned above. The submitters, JWA and KW, upon realizing that their photos were not sufficient for distinguishing the species from the other RuWhiChe types, went out a few days later and collected the 3♂ that currently reside in the DRP. Nonetheless the photographic record will remain unconfirmed.

And finally, there is one record of a possible *S. internum × obtrusum* hybrid that we collected on 22 September 2012 at Lake Elmer. This specimen (SP 430) is discussed in the *S. obtrusum* account.

SYMPETRUM COSTIFERUM (HAGEN, 1861) – SAFFRON-WINGED MEADOWHAWK

We thought we had discovered the first state record of the Saffron-winged Meadowhawk on 15 October 2014 when we found 3♂ at Lake Ponca in Kay County (Patten and Smith-Patten 2015). We collected two (SP 1453, 1454) and managed to snap decent photographs of one of those males (OC 427449; Figure 19.122). We also saw 1♂ at Legion Park in Blackwell, Kay County, later in the day, but we were not able to document it.

A few days after returning home, we realized that we already had specimens of the species in the SP Collection. How was this possible? Well, two years earlier, JR Heinen collected three *Sympetrum* specimens, two of which he photographed and submitted to Odonata Central as the Autumn Meadowhawk (*S. vicinum*). Because these submissions were confirmed (before we took over as vetters for the state), we blithely accessioned them into the SP Collection as Autumn Meadowhawks without examining them closely. Our 2014 records spurred us to compare those specimens to vouchers in the SP Collection, at which point we caught the errors. Heinen's records are: 1♂ and 1♀ (SP 507 and SP 508, respectively) from Three Lakes, Grant County, 11 October 2012 (OC 382090 is a photograph of the ♂; Figure 19.122), and 1♂ (SP 509) from Legion Park, Blackwell, Kay County, 15 October 2012 (OC 382107). To rub more salt in our wounds, we discovered a further oversight years later: on 4 October 2014, MAP photographed but could not capture a ♂ *Sympetrum* at Spring Creek Lake, Roger Mills County (OC 427331), that for lack of precedence for *S. costiferum*, he identified as *S. internum*, the Cherry-faced Meadowhawk, even though certain key aspects of the phenotype did not match that species, especially the pattern of black on the abdomen but also the extent of soft amber in the costal area of wing. Fast forward a half-decade later to the realization that the ♂ was another participant of the *S. costiferum* "invasion" of October 2014.

To recover a bit from our embarrassment, MAP was contemporaneously cognizant of the identification of the southernmost record of the species in its overall range. While on a routine survey of the city lake in Altus, Jackson County, a few days after our "first state record" he collected 1♂ (18 October 2014, OC 427492, SP 1459). So not only were the northern Oklahoma records extralimital to the species' more "normal" range by at least a few hundred kilometers, but the Jackson County record is another 300 km farther south.

Sympetrum costiferum is a northern species found across the northern United States and well into Canada (Paulson 2009, 2011). There are relatively few records for Colorado, Kansas, and Missouri, although the species is much more regular in those states than it appears to be in Oklahoma.

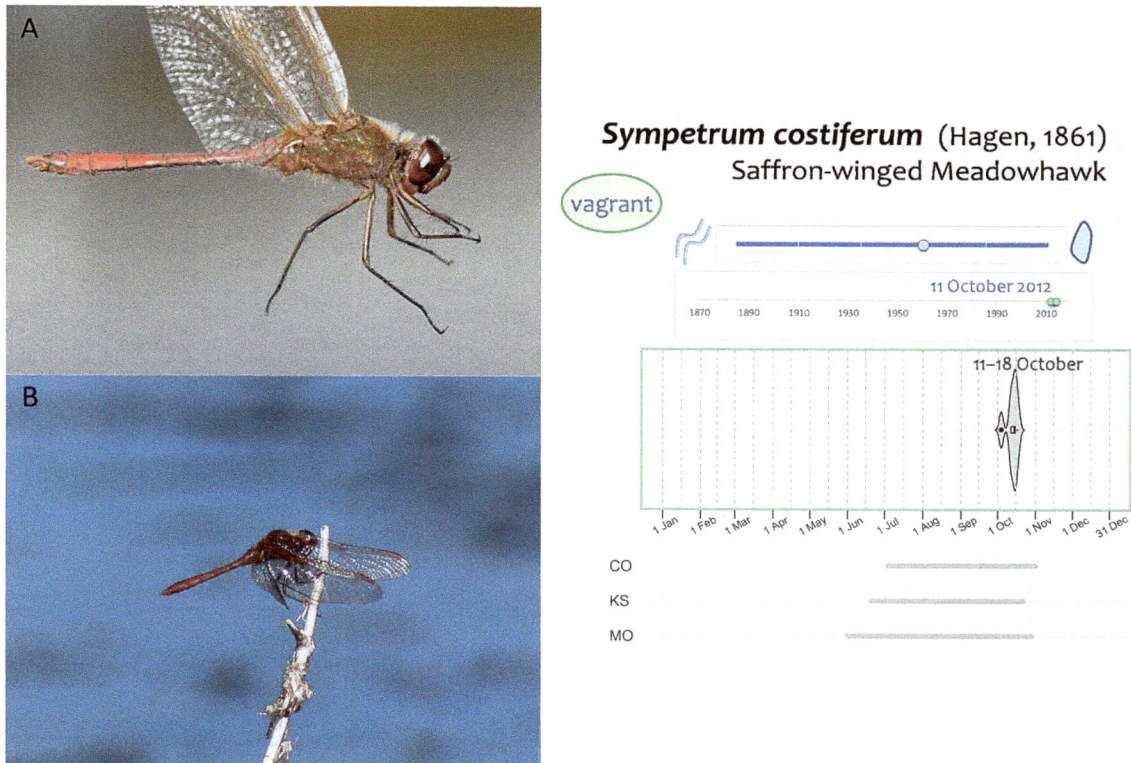

Figure 19.122 *Sympetrum costiferum*, the Saffron-winged Meadowhawk. A) ♂, Three Lakes, Grant County, 11 October 2012 (©Jason R Heinen, OC 382090, SP 507), Oklahoma's first record of the species, albeit not determined as such until two years later. B) ♂, Lake Ponca, Kay County, 15 October 2014 (©Brenda D Smith, OC 427449), one of three present that we thought at the time to be the state's first.

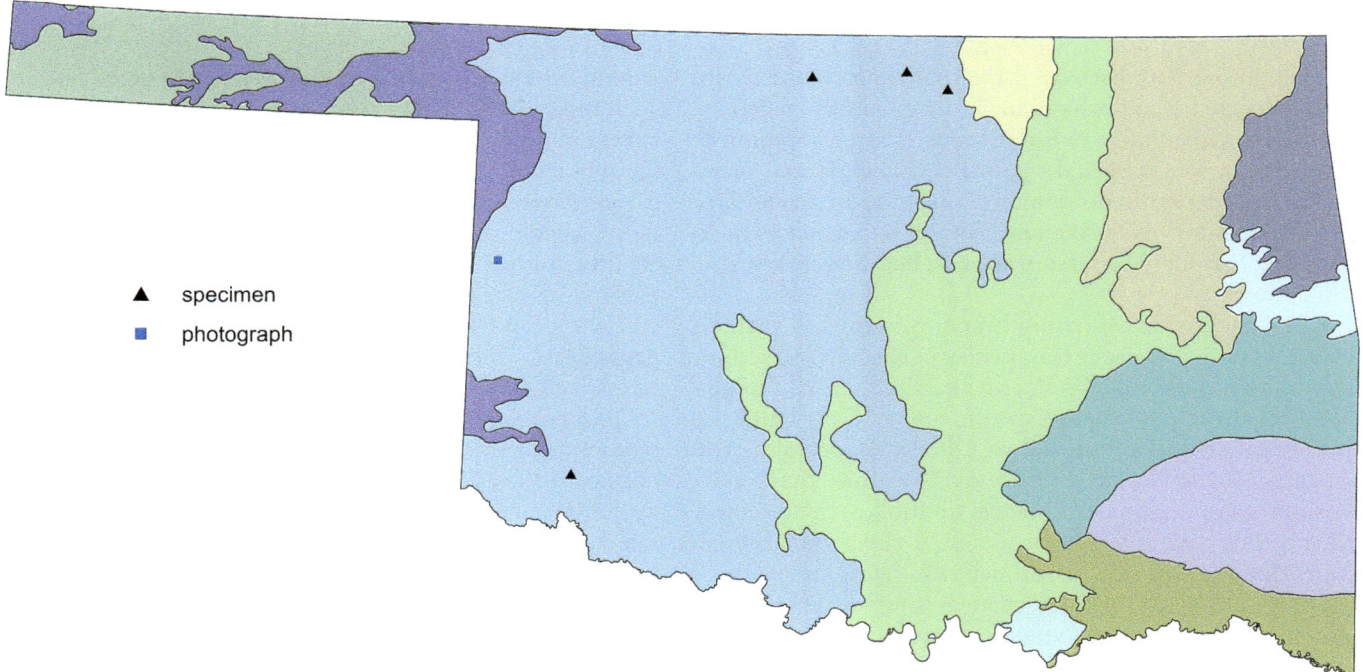

Figure 19.123 Documentation level of supporting records for *Sympetrum costiferum*, the Saffron-winged Meadowhawk, in Oklahoma.

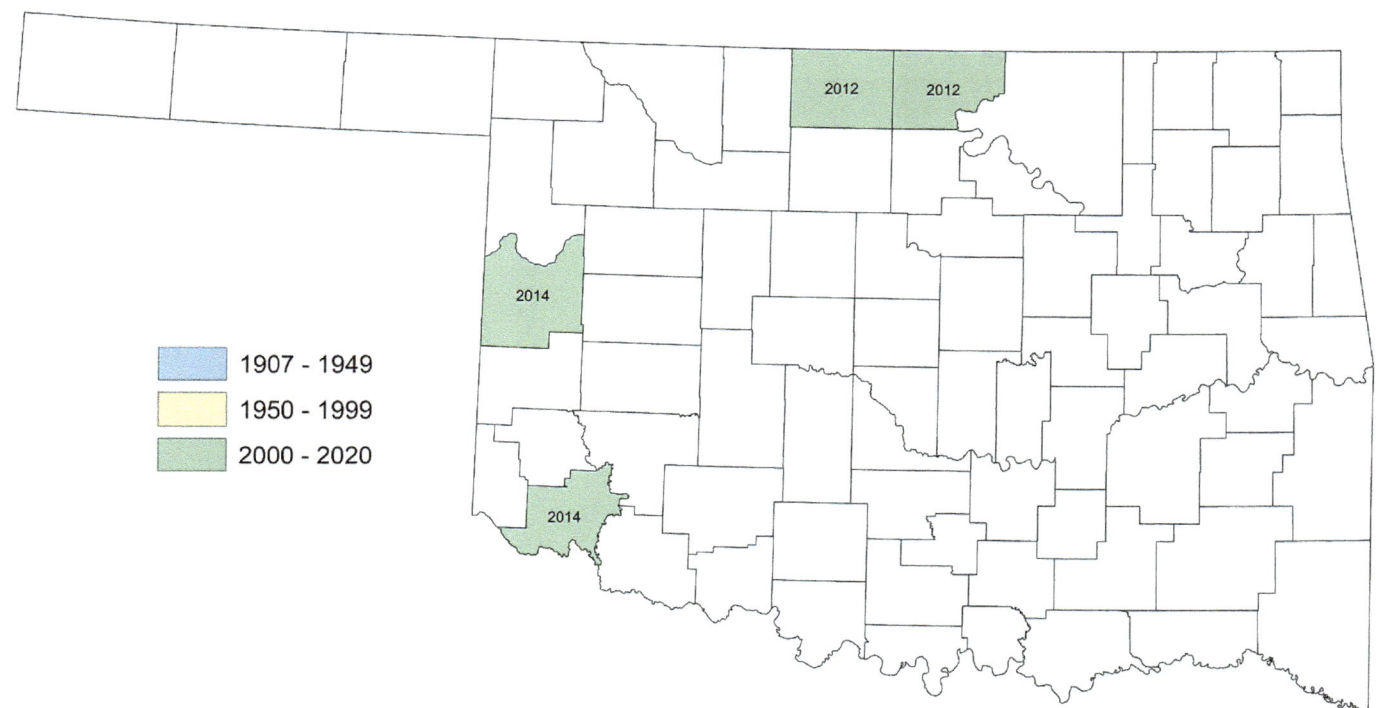

Figure 19.124 Counties in which *Sympetrum costiferum*, the Saffron-winged Meadowhawk, are known to occur in Oklahoma.

In those states it is known to fly from the summer into the late fall (CO: 4 July–1 November, FMNH 0003 007 621, OC; KS: possibly as early as 14 June–22 October, SEMC 1366999–1367002, specimens labeled as "14–24 June 1978," OC 382164; MO: 1 June–29 October, Trial 2005). Previously it was thought there was one record for New Mexico, from Eddy County (Evans 1995), but that was apparently a mis-identified ♀ *S. vicinum* specimen from CSU. In the decision notes for OC 252757, JN Stuart wrote, "Bill Prather (*pers comm*, 24 September 2013) notes that in the CSU collection "there is a specimen (in bad shape) labelled *Sympetrum costiferum* from Eddy County, NM that has an obvious *S. vicinum* subgenital plate remaining." A verified record of this species for Eddy County and for New Mexico may be lacking." (Patten and Smith-Patten 2015). With ongoing climate change and the predicted resultant poleward shifts of many species' geographic ranges, Oklahoma may be unlikely to host this species in the future.

SYMPETRUM SEMICINCTUM (SAY, 1839) – BAND-WINGED MEADOWHAWK

The Band-winged Meadowhawk is a species of the western fringe of Oklahoma, especially, its panhandle, although there is one dubious report for central Oklahoma (Logan County, 1♂, 4 July 1930, EB Webster, OMNH 1175)[38]. The species is uncommon in the state and is not regularly reported. We have fewer than 40 records and it is generally found one or two at a time, but it can be locally fairly common (e.g., 8♂, 3♀, Coldwater Creek, Optima NWR, Texas County, 25 June 2016, MAP and 3♂, 4♀, Lake Evans Chambers, Beaver County, 8 July 2013). It favors deciduous sheltered and shaded locales around ponds and along slow-moving creeks.

The species was first recorded in the state on 1 July 1930 by RD Bird at Waynoka, Woods County (1♂, OMNH 1182, 1♀, USNM 314182). He collected additional specimens until 5 July at the same locality (7♂, 2♀, OMNH 1176–1181, USNM 314181). Bird (1932) reported only Woods County for the species, apparently unaware that EB Webster had collected the ♂ in Logan County. The species was also recorded during the OU expedition era in Cimarron County, by AE Pritchard, in 1933 (at Kenton: 3♂, 30 June, EMEC 300272–300273, 314186; at Boise City: 4♂, 1♀, 9 July, EMEC 300274–300275, 314184–314185, OSU

and 1♀, 17 July, USNM 314183). GH Bick never reported seeing the species in the state, but LE Hornuff had it once (only record for that era) on 10 August 1973 at Lake Carl Etling, Cimarron County (1♂, FSCA). Bick and Bick (1957) reported that record as *Sympetrum occidentale fasciatum*, an older name for the taxon. Although there has been some dispute regarding *fasciatum* as a valid subspecies (Needham et al. 2014, Pilgrim and von Dohlen 2007), we follow Paulson's (2009, 2011) view that there of four subspecies of this species, with *fasciatum* being the one found in Oklahoma. Since 2007, the species has been recorded fairly regularly in the state, accounting for 27 records.

Regionally, *Sympetrum semicinctum* flies from early summer to late fall, a span consistent with the dates known for Oklahoma (OK: 28 May 2016, Ellis County WMA, Ellis County, 1 adult ♂, 1 imm ♂, MAP and 25 October 2015, Black Mesa SP, Cimarron County, 1♂, BC, OC 438263; CO: 7 June–8 November, OC 328375, 427662; KS: 31 May–late November, SEMC 1326815–1326816, Allison 1921; MO: 11 July–22 September, Trial 2005, as *S. occidentale fasciatum*; NM: 3 May–9 November, CSU as *S. occidentale*, OC 427828; TX: 25 May–9 September, JCAC 38293, OC 503443).

Figure 19.125 *Sympetrum semicinctum*, the Band-winged Meadowhawk, A) ♂, Black Mesa SP, Cimarron County, 3 September 2017 (©Bill Carrell). B) ♀, Lake Lloyd Vincent, Ellis County, 14 June 2015 (©Greg W Lasley, OC 447955). In Oklahoma, this striking species is restricted to the panhandle and some of its westernmost counties.

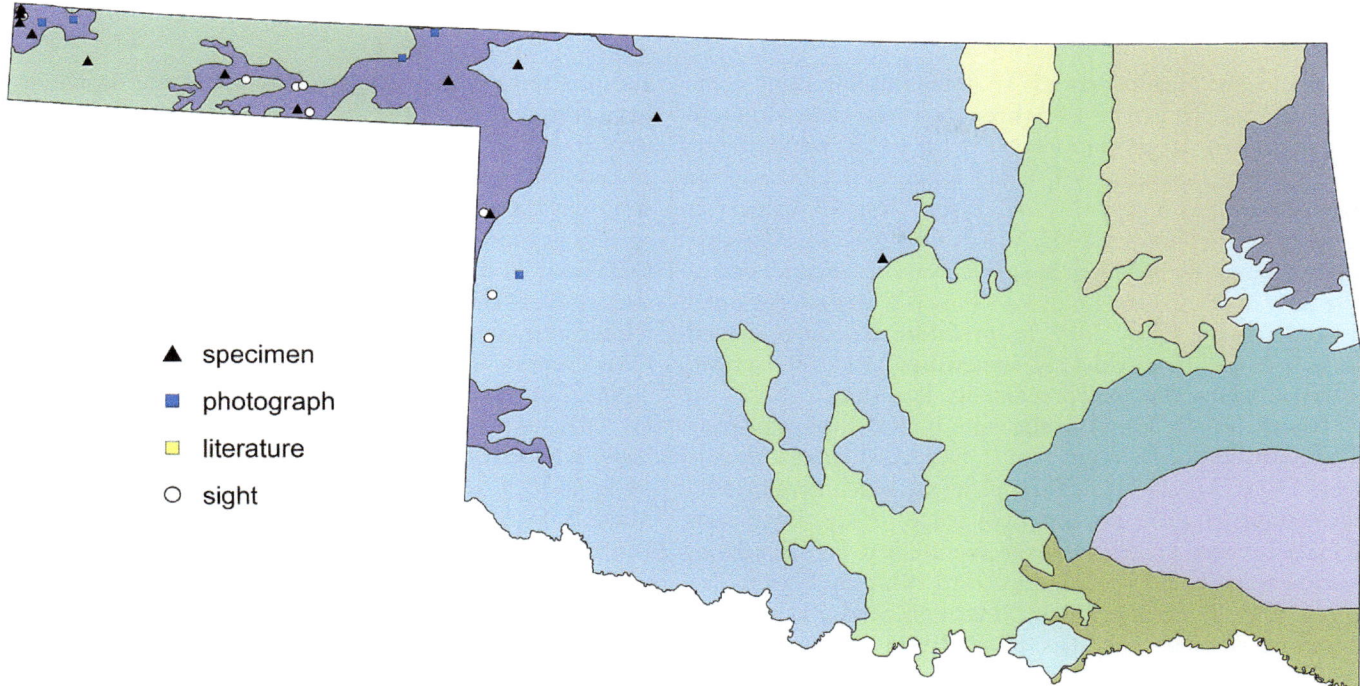

Figure 19.126 Documentation level of supporting records for *Sympetrum semicinctum*, the Band-winged Meadowhawk, in Oklahoma. Levels include specimen, archived photograph, literature reference, or sight record. Although sight records lack documentation, only those we feel are credible are plotted.

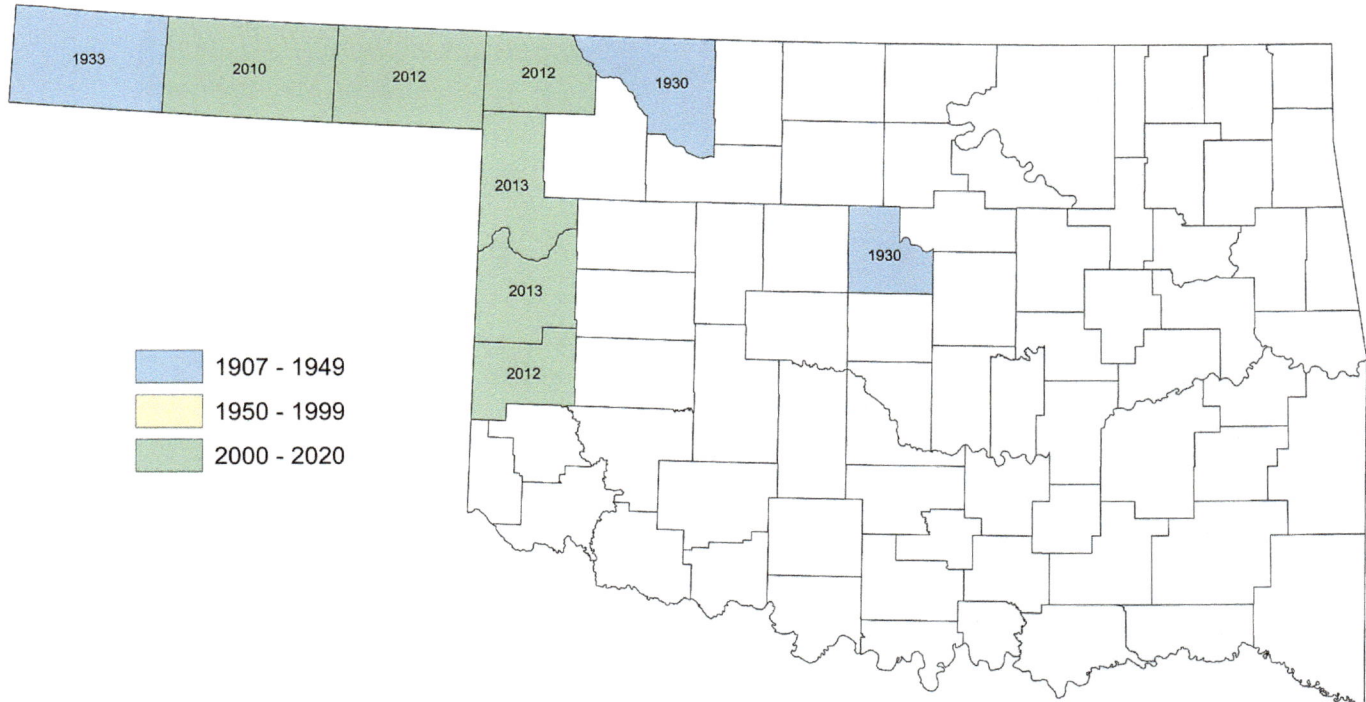

Figure 19.127 Counties in which *Sympetrum semicinctum*, the Band-winged Meadowhawk, are known to occur in Oklahoma.

SYMPETRUM VICINUM (HAGEN, 1861) – AUTUMN MEADOWHAWK

The Autumn Meadowhawk is an uncommon to fairly common species that can be seen throughout the state, although usually not in large numbers. It tends to be found a couple at a time, sometimes 4–5, but on at least 1 occasion it was recorded as high as "100+" (Black Mesa SP, Cimarron County, 22 September 2007, VWF, OC 263126). The next highest count we have for the species was on the other end of the state at Big Cedar, Le Flore County, where we recorded 15 tenerals on 5 July 2014. Its preferred habitat is shaded locales in deciduous thickets, sometimes far (>100 m) from water; a habitat that may preclude its detection.

It was initially found in the state by RD Bird at or near Waynoka, Woods County, on 2 July 1930, when he collected 1♀ (OMNH 1183). Perhaps he never felt comfortable identifying the (on present knowledge, easily diagnosed) female because he reported the species just for Cleveland, Latimer, and Le Flore Counties (Bird 1932). By 1934, it also was known from Comanche, Harmon, and Pushmataha Counties. During the Bick/Hornuff era the species was seen as regularly as it is currently.

Despite the currently accepted American English common name for this species, it is seen in Oklahoma from late spring/early summer into the winter. Summer and autumn records are fairly equally split, and although we have limited records of tenerals ($n=7$), all are from late May to mid-July, not the autumn (suggesting siccatation, as known elsewhere with *S. vicinum* [Corbet 1999:363], and as discussed with *S. ambiguum*, above). In our region the species is generally seen from May until November, but in Oklahoma and Missouri the late dates fall in December, and in Texas the species has been recorded throughout the winter into January (OK: 29 May 2017, Coalgate Reservoir, Coal County, 1♀, MAP, SP 2361 and 5 December 2015, TNC Tallgrass Prairie Preserve, Osage County, 1♂, BC, OC 439344; AR: 21 June–22 November, OC 438954, 438953; CO: 31 July–14 November, OC 330945, 438741; KS: 14 June–30 November, SEMC 1366587, 1326825; MO: 25 May–23 December, Sims 2012; NM: May–13 November, Paulson 2009, OC 438221; TX: 22 May–22 January, Abbott and Lasley 2019). Elsewhere the species is also reported outside of autumn (e.g., May: FL, GA, OH; June: BC, KY, NJ, NS, ON, WI; July: IA, ME, NE, QC, WA; Paulson 2009, 2011). Even if abundance tends to be slightly higher in the autumn in Oklahoma, use of a former common name, the Yellow-legged Meadowhawk, may be more appropriate.

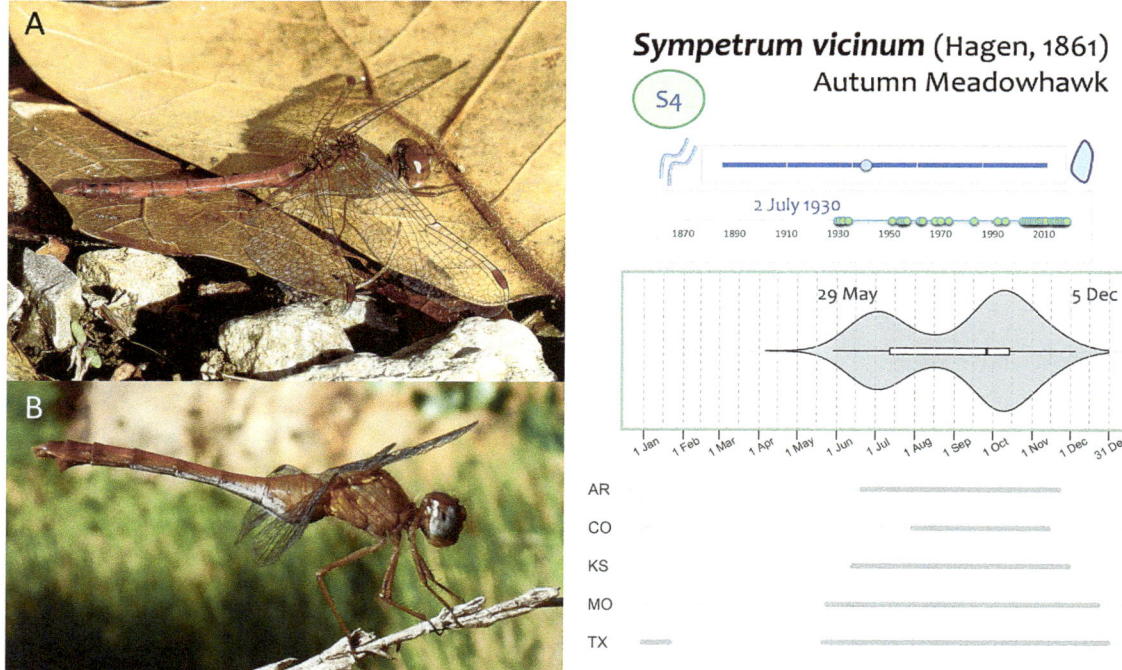

Figure 19.128 *Sympetrum vicinum*, the Autumn Meadowhawk. A) ♂, Tallgrass Prairie Preserve, Osage County, 5 December 2015 (©Bill Carrell, OC 439344). B) ♀, Lake Carl Etling, Cimarron County, 22 September 2007 (©Victor W Fazio III, OC 263126). True to its current name—the species used to be known as the Yellow-legged Meadowhawk—this meadowhawk is found chiefly September–November.

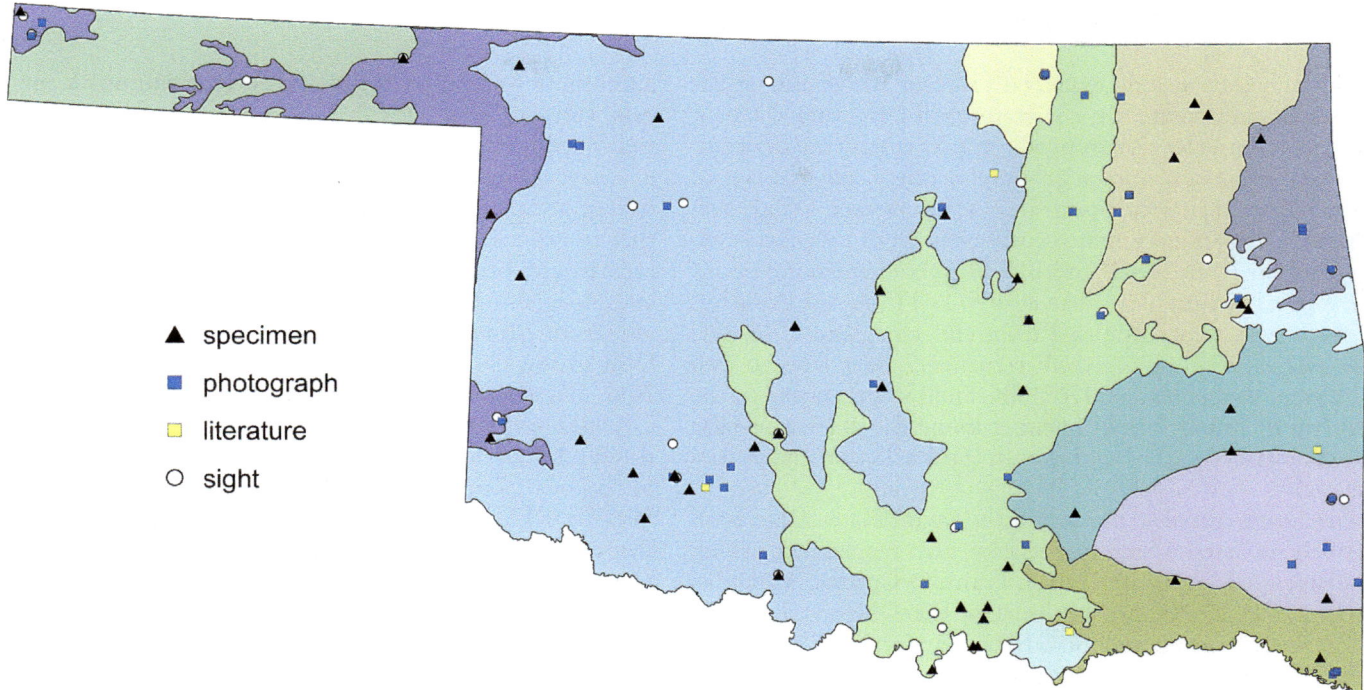

Figure 19.129 Documentation level of supporting records for *Sympetrum vicinum*, the Autumn Meadowhawk, in Oklahoma. Levels include specimen, archived photograph, literature reference, or sight record. Although sight records lack documentation, only those we feel are credible are plotted.

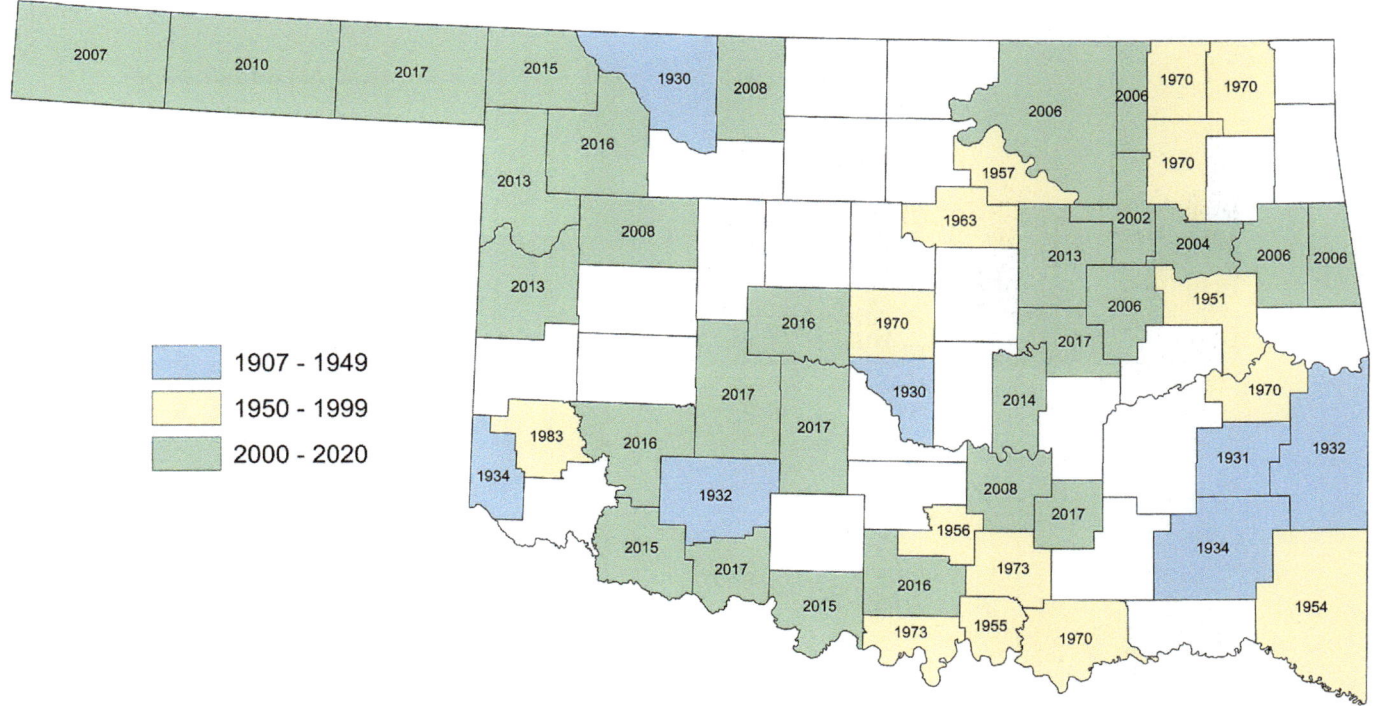

Figure 19.130 Counties in which *Sympetrum vicinum*, the Autumn Meadowhawk, are known to occur in Oklahoma.

MICRATHYRIA HAGENII KIRBY, 1890 – THORNBUSH DASHER

This species was long expected to occur in Oklahoma but was not documented in the state until 2015, and where it was first found was far from where it was predicted to occur. Given the species' known geographic range, the first record was expected in south-central or southwestern Oklahoma; instead, Oklahoma's first record came from the northeast, where, on 3 July, BS-P and BW Hoagland encountered 8♂ at a pond at Pumpkin Flats at TNC's JT Nickel Preserve, Cherokee County (Smith-Patten and Hoagland 2015, OC 432602, SP695, 1696). MAP returned 2 days later to find 7♂ (retaining one as SP 1700). BC and JWA returned to the site on 14 and 31 July, but neither found the species, nor has anyone else since at that location (Smith-Patten and Patten 2017a).

There is one older record for the Ozark Plateau: Houston (1970) reported a single specimen, sex unspecified, which he collected on 8 June 1968 in Franklin County, Arkansas, some 100 km (63 mi) southeast of the Oklahoma record. The record was carried forward by Harp and Rickett (1977), but there is currently some doubt regarding its veracity. The specimen has not been found, and the author cannot be located by us nor by George Harp (*pers comm*). We suggest

continuing to treat that record with caution until it can be fully verified. That said, the Oklahoma record, even though well documented, is still some 650 km (400 mi) northeast of where the species breeds regularly in Texas and 400 km (250 mi) from the northernmost Texas record. It was posited that severe weather may have been the culprit behind this record and other long-distance dispersals (Smith-Patten and Hoagland 2015), including the lone Tennessee record, also from 2015 (Owl's Hill Nature Sanctuary, Brentwood, Williamson County, Robert and Andrea English, 14 July 2015, OC 433072, English and English 2015).

Oklahoma's second record came later that same year, when 1♂ and 1♀ were found at a pond at Great Plains SP, Kiowa County, on 13 September (MAP, OC 436455, SP 1793). The species was spotted there four other times that autumn: 1♀ on 18 September (OC 436913), 3♂ on 27 September (OC 437267), 2♂ and 1♀ on 10 October, and 1♂ on 11 October (OC 459327). On the 10 October visit MAP observed an ovipositing female. Despite annual surveys of Great Plains SP since, the species was detected only twice, a single ♂ each time (26 October 2017, SP 2532; 14 September 2019; both records MAP). This pattern is

Figure 19.131 *Micrathyria hagenii*, the Thornbush Dasher. A) ♂, JT Nickel Preserve, Cherokee County, 3 July 2015 (©Brenda D Smith, OC 432602), one of eight present that represented Oklahoma's first record (Smith-Patten and Hoagland 2015). B) ♀, Great Plains SP, Kiowa County, 18 September 2015 (©Brenda D Smith, OC 436913), at the site that has hosted every subsequent record for the state.

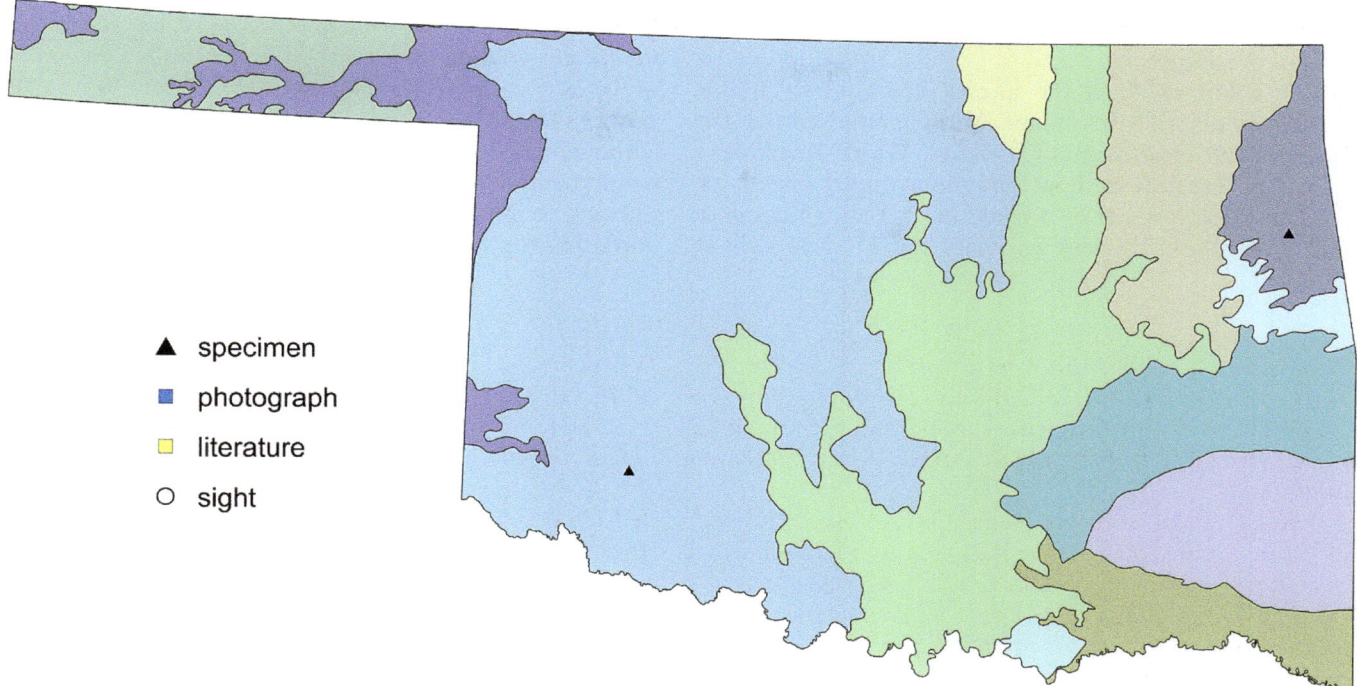

Figure 19.132 The two localities in Oklahoma known for *Micrathyria hagenii*, the Thornbush Dasher, are both supported by specimen records.

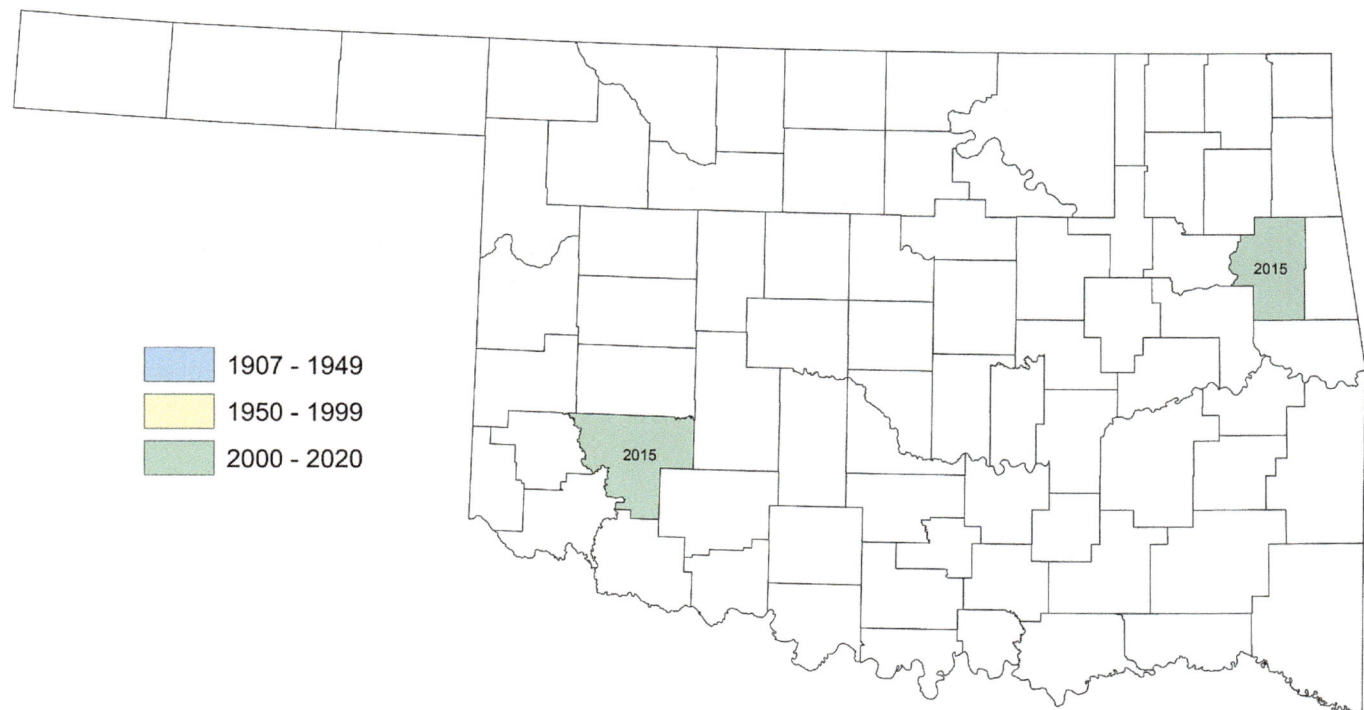

Figure 19.133 *Micrathyria hagenii*, the Thornbush Dasher, is known from two counties in Oklahoma, Cherokee in the northeast and Kiowa in the southwest.

confounding and leaves us unable to discern the species' status in the state.

Little is known about the species' life history. Dunkle (1976), from his study in Sinaloa, Mexico, reported that the egg stage lasts a minimum of 7–8 days. We are unaware of any studies that indicate how long the nymphal stage lasts. If it is multiple years, then it could be that a population of larval Thornbush Dashers are maturing in that pond at Great Plains SP (or elsewhere) as we write, but at this point we are not holding our breath. The species is known to fly in Texas from mid-March to late December, so its emergence in Oklahoma may be expected within that timeframe (although likely not as late as December). Both the Cherokee County and Kiowa County records have come from small, isolated, manmade ponds with clear water and extensive bordering vegetation.

PACHYDIPLAX LONGIPENNIS (BURMEISTER, 1839) – BLUE DASHER

We consider the Blue Dasher to be the second most common and widespread libellulid in the state. Even though it only has the fourth highest total of records (>1,300), it is found in all 77 counties, and it vies with the Eastern Pondhawk (*Erythemis simplicicollis*) for the record of high counts, which often enough reach triple digits.

The species was first recorded by EB Williamson in Wister, Le Flore County, on 3 June 1907 (♂, UMMZ). It is strongly associated with ponds but can be found almost anywhere, from the shores of large lakes to along rivers and creeks. In a study by GH Bick and JC Bick (1958), almost 50 adults occupied a pond they described as "partially shaded" and "heavily vegetated, clear, shallow, [and] mud-bottomed." Elsewhere in Oklahoma observers have described habitats as flooded areas (including those by the side of the road) and creeks 2–10 m wide with much vegetation (emergent and surrounding), with slow-flowing water and sometimes small riffles present.

Pachydiplax, a monospecific genus, flies for much of the year in Oklahoma (early date of 17 March, in 2011, VWF, Wichita Mountains WR, Comanche County, OC 330147; late date of 5 November, in 2007, M Dreiling, Bartlesville, Washington County, OC 263864). This flight season is similar to elsewhere in the region, although it flies all year in Texas (AR: 9 April–14 November, OC 374352, 458264; CO: 16 May–24 October, OC 375000, 438575; KS: 2 April–30 November, SEMC; MO: 20 April–20 November, Sims 2012; NM: March–October, Paulson 2009). With continued climate change in the region, it is likely that this species will one day be recorded year-round.

Figure 19.134 *Pachydiplax longipennis*, the Blue Dasher, A) ♂, Cedar Creek 7 km E of Lenapah, Nowata County, 23 August 2014 (©James W Arterburn, OC 424176). B) immature ♂, Enid, Garfield County, 27 May 2016 (©Bill Dobbins, OC 445665). C) ♀, Salt Plains NWR, Alfalfa County, 1 August 2014 (©Bill Carrell, OC 425622). Is this our most numerous and widespread dragonfly? If it is not, it is in the running.

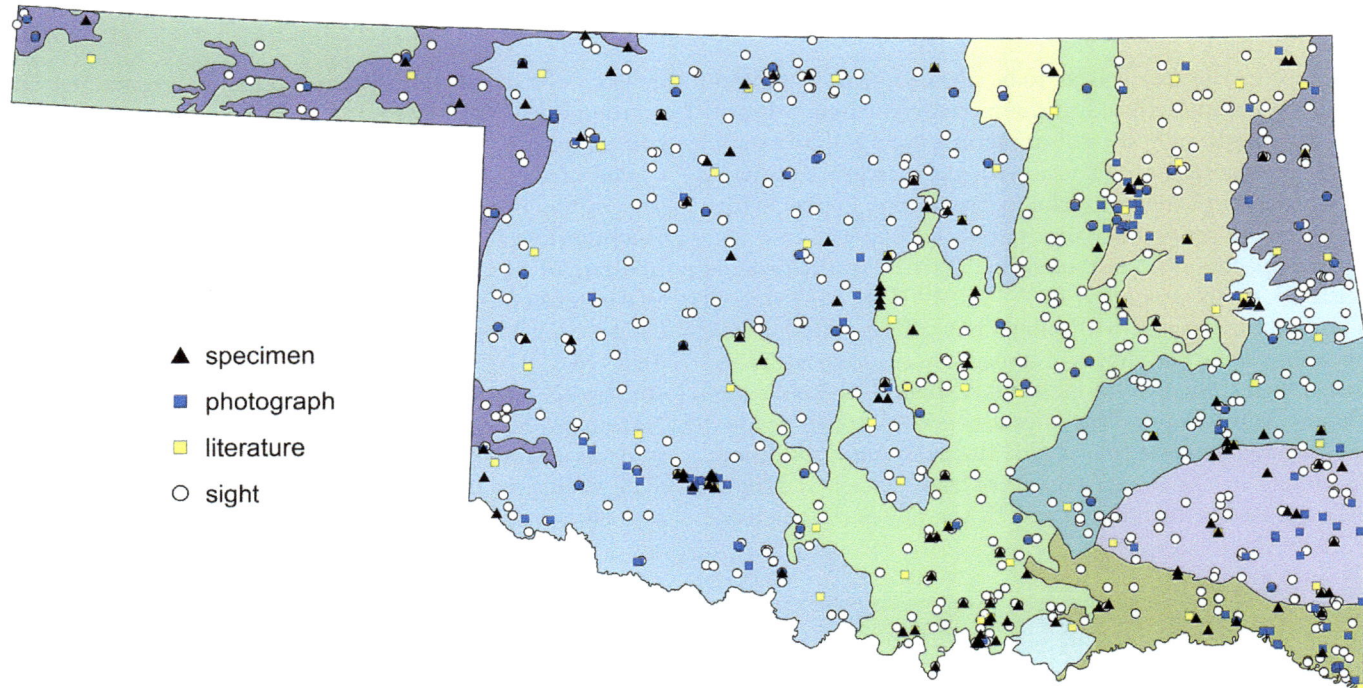

Figure 19.135 Documentation level of supporting records for *Pachydiplax longipennis*, the Blue Dasher, in Oklahoma. Levels include specimen, archived photograph, literature reference, or sight record. Although sight records lack documentation, only those we feel are credible are plotted.

DYTHEMIS FUGAX HAGEN, 1861 – CHECKERED SETWING

The Checkered Setwing is a species that was once thought to be chiefly of the western half of the state, but now we know it occurs across most of the state, albeit more commonly in the western half. Bird (1932) reported it from four counties (Blaine, Comanche, Murray, Oklahoma), although we have tracked down early era records only from Comanche County (*n* = 8, all from in or near Wichita Mountains WR). One of those records, from 28 June 1931 at Medicine Park, was the first for the state (2♂, AE Pritchard, OMNH 2236). Between 1954 and 1996, the species was added to another 12 counties from approximately 40 records. And since 2001, 44 other counties were added, bringing the total to 61 (although we do not know the basis for the Noble County, record reported in Donnelly 2004b). There have been >200 records between 2001 and 2019.

It is uncommon to locally fairly common, regularly occurring in the teens and twenties, and on at least three occasions much higher. The higher numbers include our records of 50 individuals (including 3 pair, 2 km SE of Sweetwater, West Buffalo Creek, Beckham County, 20 July 2013), 60 individuals (including 3 pair, 2 km SW of Shady Grove, McIntosh County, 21 June 2014), and 70♂, 5♀ (including 2 pair, Sandy Sanders WMA, Greer County, 10 August 2014). Observers have described the species' habitat in Oklahoma as spring-fed lakes and ponds; farm ponds with little emergent vegetation but small trees and shrubs along some of the margins;

in low shrubbery at the end of a pasture; in mixed grass/herbaceous cover; in a sand plum (presumably *Prunus angustifolia*) patch within an otherwise open grassland; and in a prairie upland area near a stream and beaver wetland. In our experience this species is found exclusively where water is clear, favoring spring-fed or spring-like ponds, and it tends toward the lentic end of the spectrum.

Its flight season in Oklahoma is mid-May to late October (14 May 2017, Lake Hall, Harmon County, 3♂, MAP; 26 October 2017, Great Plains SP, Kiowa County, 1♀, MAP), which is longer than dates for Kansas, much shorter than those for Texas, a later start than Arkansas, but pretty close to New Mexico (AR: 28 April–7 September, OC 374623, 333380; KS: 7 June–22 September, OC 375641, USNM 293500; NM: May–8 October, Paulson 2009, OC 491409; TX: 1 April–26 December, Abbott and Lasley 2019). There are few records for the species in Colorado and Missouri, so those date spans are not indicative of the region (CO: 8 July–14 August, OC 377258, 488549; MO: 19 June–1 August, OC 376179, 402432). We have five teneral records for Oklahoma spanning from 4 June to 27 August. A study of odonate dispersal in Oklahoma included 1♀ *Dythemis fugax* that was found 0.59 km from the pond at which she was marked and 1♂ that dispersed 1.62 km, which was the greatest distance between study ponds (Stewart and Murphy 1968).

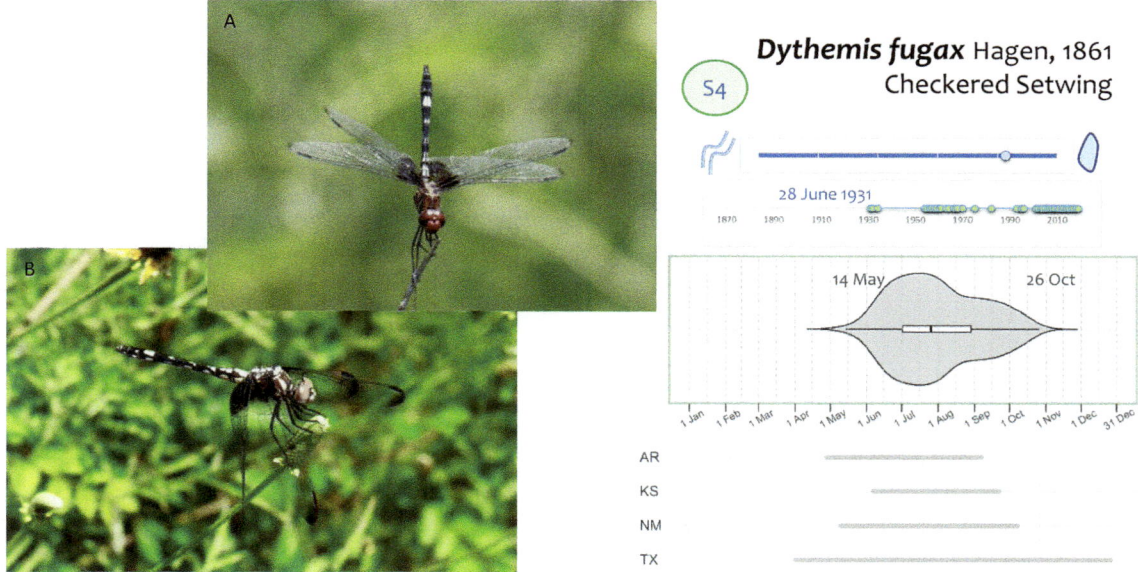

Figure 19.136 *Dythemis fugax*, the Checkered Setwing, A) ♂, Tallgrass Prairie Preserve, Osage County, 24 June 2011 (©James W Arterburn, OC 329235). B) ♀, Oklahoma City, Oklahoma County, 10 June 2019 (©J Harrell Johnson, iNat 32921660). Of the two *Dythemis* setwings that occur regularly in the Southern Great Plains, this species is the more western one. It favors clear-water ponds and lakes.

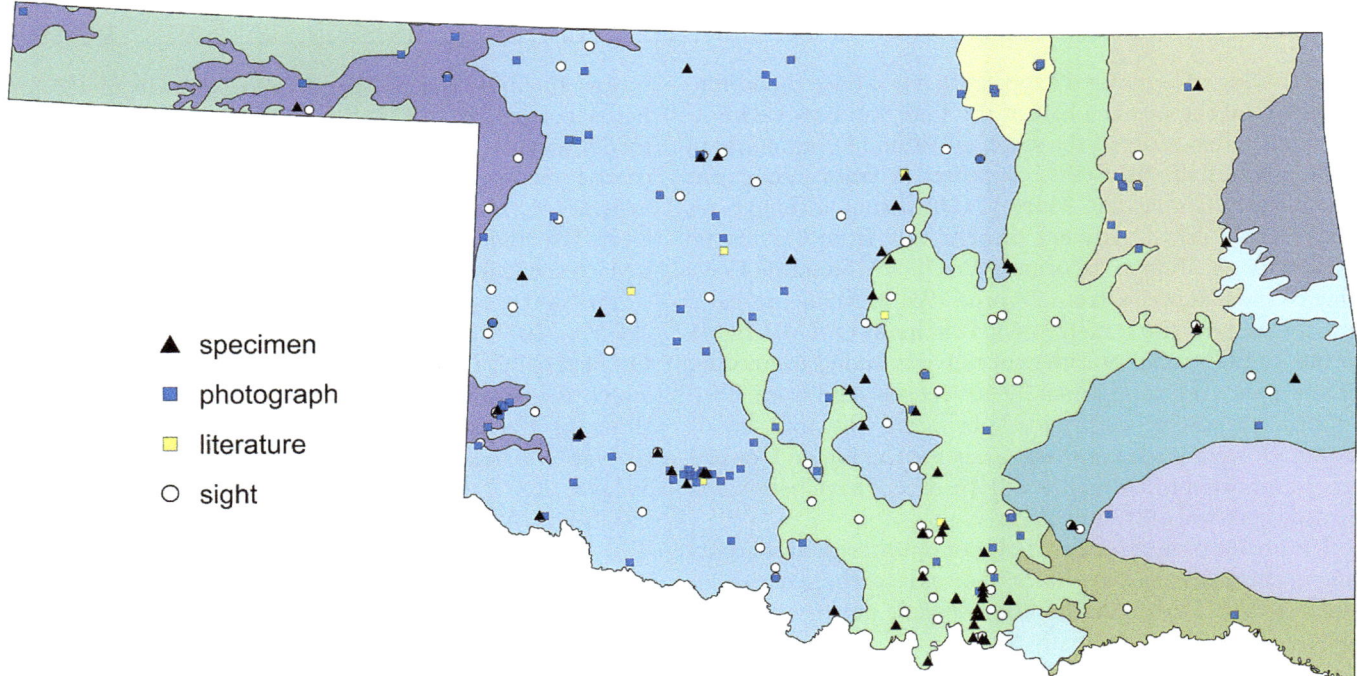

Figure 19.137 Documentation level of supporting records for *Dythemis fugax*, the Checkered Setwing, in Oklahoma. Levels include specimen, archived photograph, literature reference, or sight record. Although sight records lack documentation, only those we feel are credible are plotted.

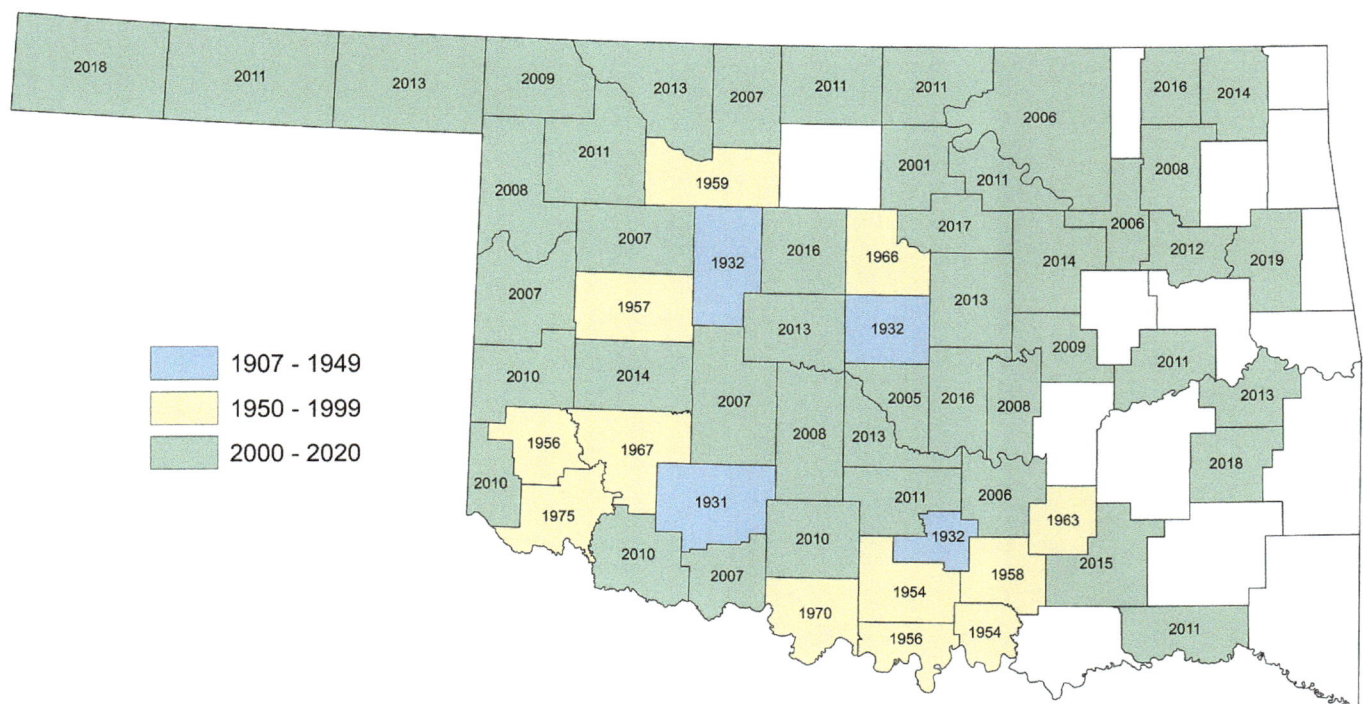

Figure 19.138 Counties in which *Dythemis fugax*, the Checkered Setwing, are known to occur in Oklahoma.

DYTHEMIS NIGRESCENS CALVERT, 1899 – BLACK SETWING

This species has been found in Oklahoma twice. The first encounter was on 9 May 2014 when BS-P found a single ♂ at Altus City Lake, Jackson County (OC 422099; Smith-Patten 2014; Figure 19.139). One could have predicted, which we did, the species to come into Oklahoma at some point, but the more likely bet probably would have been in south-central Oklahoma given that the nearest record to Oklahoma at the time was one from Fannin County, Texas (13 October 2007, Jessica Womack, OC 375274), which is adjacent to Bryan County, Oklahoma, just 25 km (~15 mi) south of the border.

We predicted, and still do, that eventually the Black Setwing will be found more consistently along the southern edge of the state. That said, the second state record came from the far panhandle—an unexpected record given its distance north (416 km, or ~260 mi, from Altus and 322 km, ~200 mi, from the nearest Texas panhandle record, OC 464692, in Briscoe County). This northernmost record for the species was of a ♂ Robert Sanders photographed along North Carrizo Creek near Black Mesa, Cimarron County, on 29 May 2018 (OC 480702; Figure 19.139).

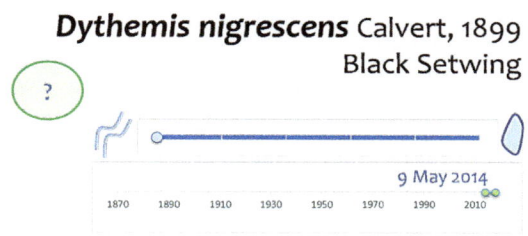

Dythemis nigrescens Calvert, 1899
Black Setwing

9 May 2014

1870 1890 1910 1930 1950 1970 1990 2010

Figure 19.139 *Dythemis nigrescens*, the Black Setwing, A) ♂, Altus, Jackson County, 9 May 2014 (©Brenda D Smith, OC 422099), the first to be recorded in Oklahoma (Smith-Patten 2014). B) ♂, North Carrizo Creek 7 km N of Kenton, Cimarron County, 29 May 2018 (©Robert Sanders, OC 480702), the second and only other to be recorded in the state.

DYTHEMIS VELOX HAGEN, 1861 – SWIFT SETWING

The Swift Setwing is uncommon to locally fairly common in Oklahoma. It is usually found in the handfuls, but we have recorded it as many as 30 at once (25♂, 5♀, including 2 pair, Memorial Park, Duncan, Stephens County, 31 July 2016, 1♂ as SP 2188). It has been recorded fairly regularly across eras of surveys in the state, and is now known from 64 counties (Figure 19.142), with this species the expected *Dythemis* is the eastern third of the state (by contrast, *D. fugax* is the expected species in the western third of the state; the middle is up for grabs). McCurtain County hosted the state's first record (19 June 1925, 2 mi N of Broken Bow, Yanubbe Creek, Ortenburger 1926b).

In Oklahoma and Texas, the species begins to fly in mid to late March, but everywhere else, including in Colorado, Kansas, and Missouri, which are states with limited records, it begins to fly in May or June (AR: 2 June, OC 481334; CO: 11 June, OC 325624; KS: 16 June, OC 319902, 319896; LA: 30 May, OC 313363; MO: 4 June, Trial 2005; NM: May, Paulson 2009; OK: 20 May 2017, Ardmore Regional Park, Carter County, 1♂, MAP, OC 463132; TX: 13 March, OC 494348). Texas is an outlier for late date, and outside two of the states with limited dates, the species stops flying in the region somewhere from early September to late October

(OK: 30 October 2016, 3♂, Ardmore Regional Park, Carter County, MAP; AR: 5 September, Harp and Rickett 1977; CO: 3 August, OC 367417; KS: 18 July, OC 320874; LA: 8 September, Mauffray 2014; MO: 15 September, Trial 2005; NM: October, Paulson 2009; TX: 27 December, Abbott and Lasley 2019).

Bick and Bick (1958) found the species at their study site along Cowan Creek, Marshall County, primarily at an impoundment, but also at a pond (4 and 1, respectively, of 5 total adults). The impoundment was described as deep and muddy with high turbidity and little aquatic vegetation but with a bottom of organic materials, whereas the pond was "heavily vegetated, clear, shallow, mud-bottomed" and was partially shaded. Paulson (2009:490) described the species' habitat as "streams and rivers with slow to moderate current, less often at pond and lake shores. Usually wooded or shrubby banks." Some Oklahoma observers have described the species' habitat as small, rocky, clear-water streams with some emergents, or ponds in oak woodland. Our experience is more in line with Paulson's description, in that this species tends to be more numerous in lotic habitats than in lentic ones, a pattern opposite that of its common congener in the state, the Checkered Setwing (*Dythemis fugax*).

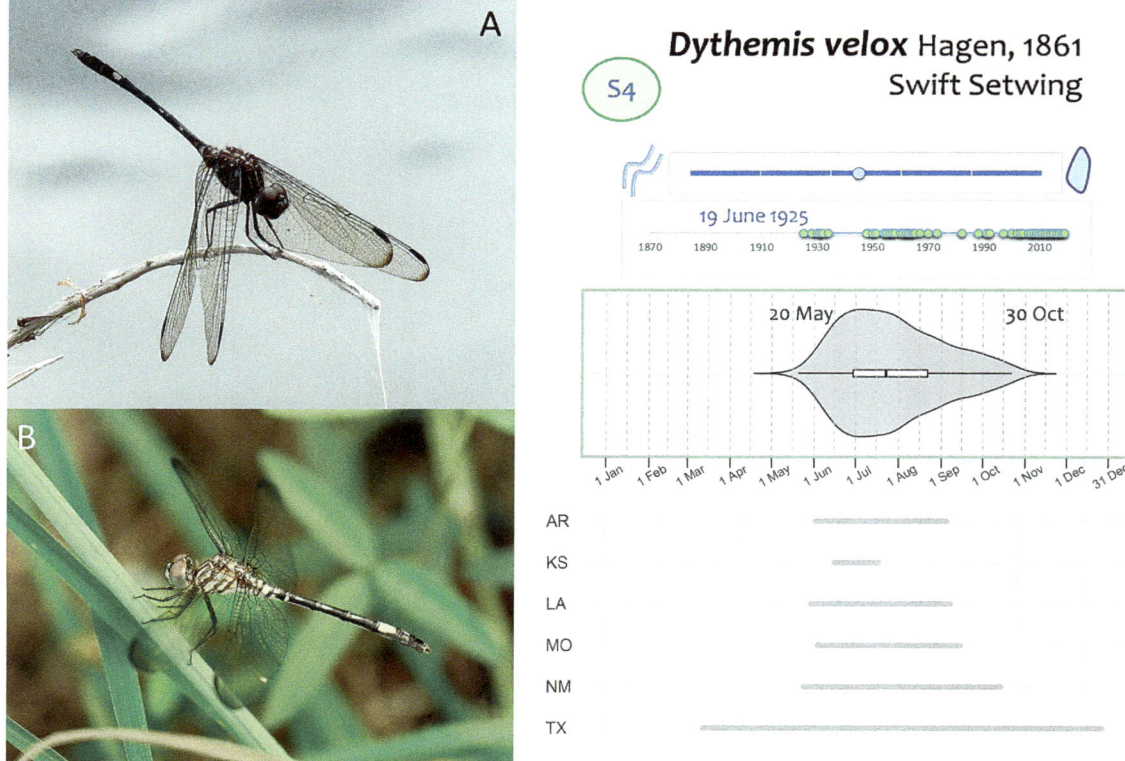

Figure 19.140 *Dythemis velox*, the Swift Setwing, A) ♂, Little Elk Creek 1 km W of Rocky, Washita County, 8 September 2009 (©Victor W Fazio III, OC 315092). B) ♀, Pretty Water Lake, Creek County, 17 July 2015 (©James W Arterburn, OC 434016). This species is the other setwing that occurs regularly in the Southern Great Plains, where it has a more eastern distribution and where it favors lotic habitats.

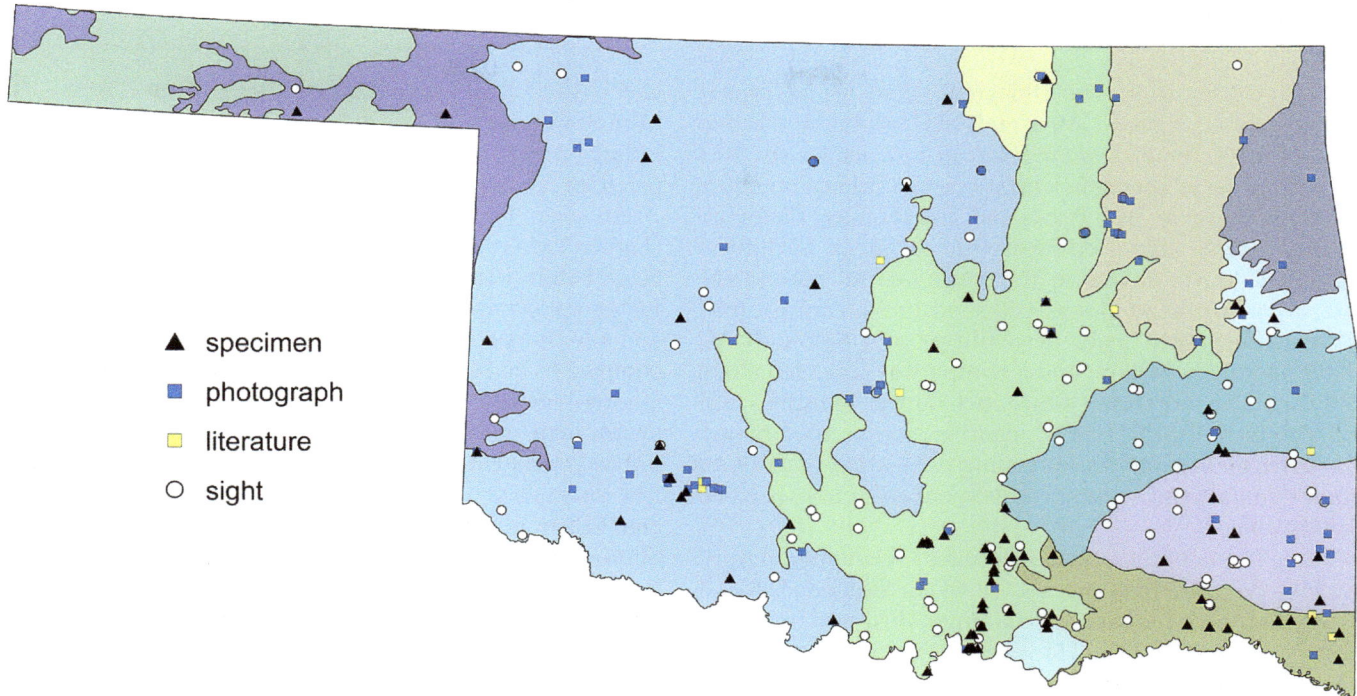

Figure 19.141 Documentation level of supporting records for *Dythemis velox*, the Swift Setwing, in Oklahoma. Levels include specimen, archived photograph, literature reference, or sight record. Although sight records lack documentation, only those we feel are credible are plotted.

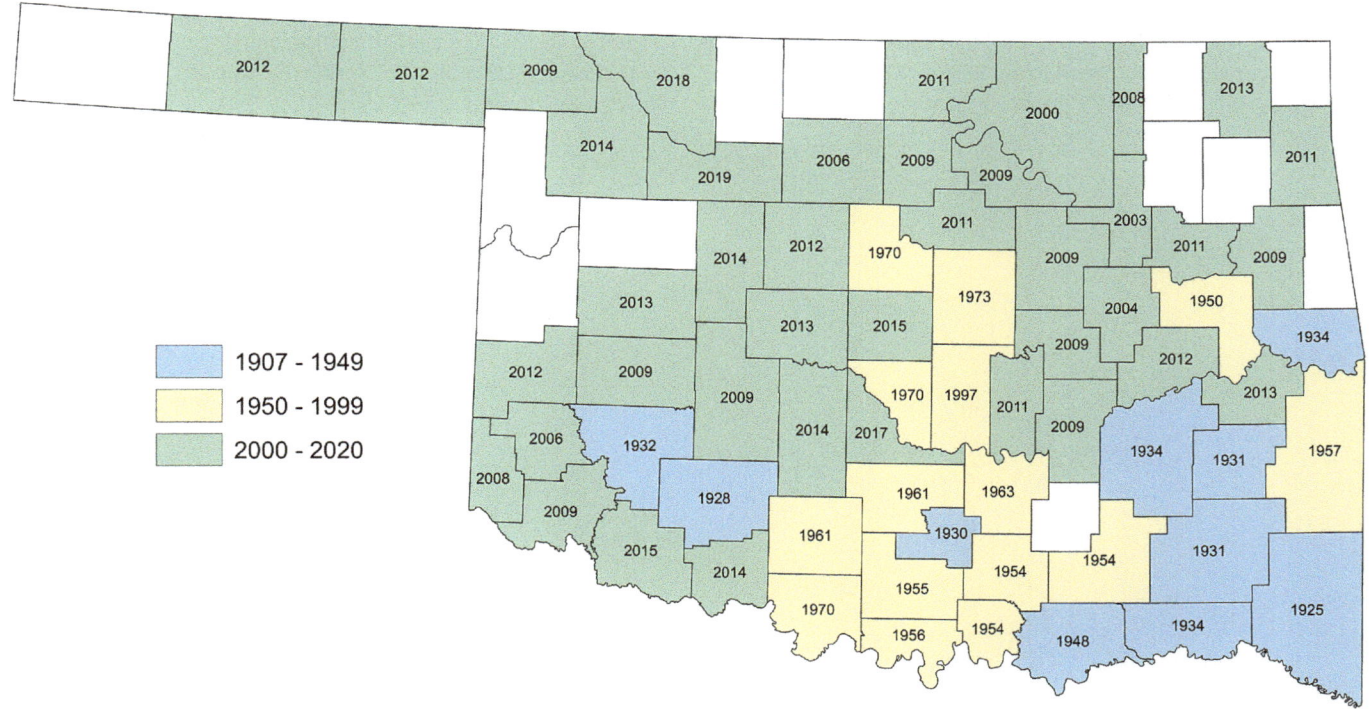

Figure 19.142 Counties in which *Dythemis velox*, the Swift Setwing, are known to occur in Oklahoma.

PALTOTHEMIS LINEATIPES KARSCH, 1890 – RED ROCK SKIMMER

Perhaps no other species on Oklahoma's state list has caused us more consternation. We simply do not know what to make of it. In the end, we are forced to consider the Red Rock Skimmer a "data deficient" species in the state because we do not know its true status and distribution. There are but two records supported by extant specimens, both from the Arbuckle Mountains in the south-central part of the state, where an adult ♂ was collected in Murray County, 11 September 1968 (coll. by: Milliger and Solon, JCAC 18898; Donnelly 2004b), and a nymph was collected along Mill Creek in northern Johnston County, 18 April 1976 (W Magdych; OMNH 1874; identification verified by us). The latter record implies a breeding population in the state rather than just two vagrant records, but there have been no records since.

There is, evidently, a third record for Oklahoma. In April and May 1933 RD Bird and LK Gloyd corresponded at least three times about an adult ♀ skimmer Bird was unable to identify (Figure 19.143). He sent the specimen to Gloyd, then at UMMZ, with an explanation that he could not determine which species of *Libellula* it was. She responded that the specimen was in fact a *Paltothemis lineatipes*! Nowhere in their correspondence does either mention the location or date of collection, and we have been unable to locate the specimen despite extensive searches in the OMNH collection (Bird acknowledged, in his 9 May 1933 letter, that he received it back from Gloyd and that it was species number 121 for the state!). Moreover, Bird never published any details about the record, and knowledge of it was lost by the time Bick and Bick (1957) updated Bird's (1932) Oklahoma checklist. If Bird's specimen also came from the Arbuckle Mountains, then there may have been a geographically isolated population of the species that has since been extirpated, or it is so small it has evaded further detection.

An even bigger mystery lies in a set of Bird's notes (Figure 19.143). He indicated that in June and July 1931 (two years before his correspondence with Gloyd) he and WM Fisher collected a series of 5♂ and 2♀ *Paltothemis lineatipes* and saw one pair (and perhaps 8 other individuals). We are completely bewildered by these notes because 1) the notes are for Latimer County, in southeastern Oklahoma, which makes little geographical sense for the species and 2) Bird did not claim the species for Oklahoma until Gloyd identified an adult ♀ in 1933 (see above). We presume that Bird and Fisher mistook some other species for the Red Rock Skimmer because nothing collected those days could have been confused with *Paltothemis lineatipes*, at least as far as our imaginations can stretch.[39]

The range of *Paltothemis lineatipes* has retracted southward elsewhere—for example, it formerly occurred north into southern Oregon but no longer occurs there (Kerst and Gordon 2011). This species is a specialist of arid, rocky streams, habitats not well surveyed since the earlier eras of Oklahoma odonatology. Because we have not been able to determine if the Arbuckles support a population—there is little public access to suitable habitat—or if the range retracted to outside the state, we recommended this species be listed as an S1 (critically imperiled) species of conservation concern (Patten and Smith-Patten 2013b) until its status can be further assessed.

Paltothemis lineatipes Karsch, 1890
Red Rock Skimmer

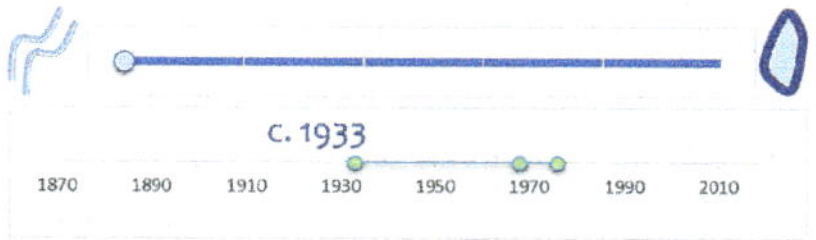

c. 1933

1870 1890 1910 1930 1950 1970 1990 2010

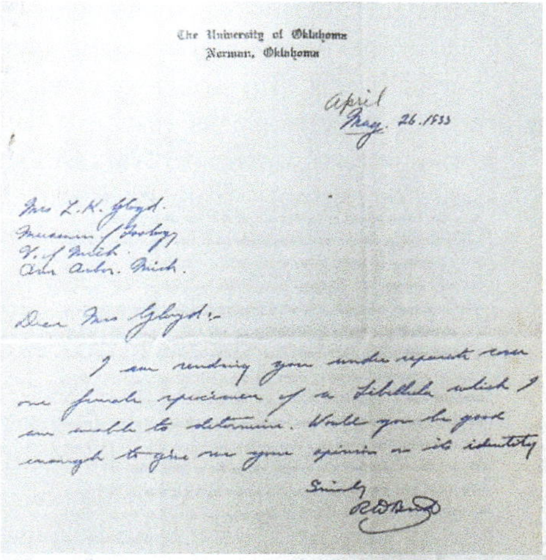

A) Correspondence between LK Gloyd and RD Bird, dated April and May 1933. Bird sent an adult ♀ "*Libellula*" to Gloyd on 26 April (top left). She responded on 2 May (top right) saying that it was, in fact, a *Paltothemis lineatipes*. On 9 May (left), Bird acknowledged receiving the specimen back. That specimen has not been re-located. Neither party mentioned when or where that specimen was collected (for the record, we do not think it was one of the specimens relating to Bird's notes below).

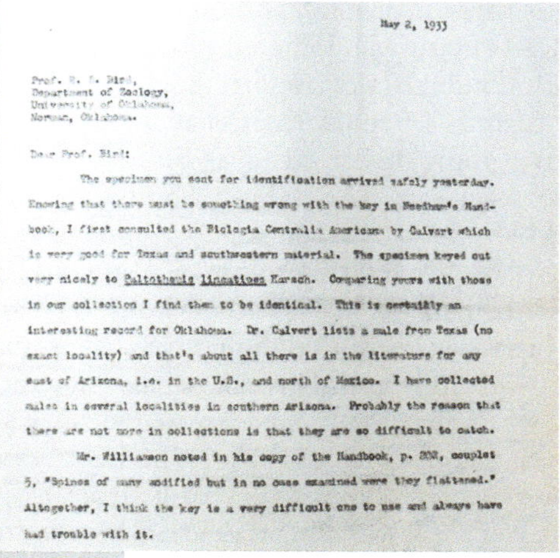

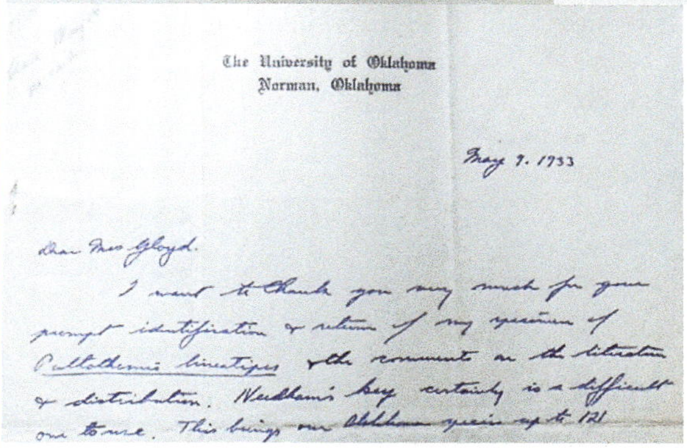

B) RD Bird's notes indicated that he and WM Fisher ("RDB" and "WF") collected *Paltothemis lineatipes*, the Red Rock Skimmer, in June and July 1931 in Latimer County, Oklahoma. We assume that Bird and Fisher re-thought the determinations and that those specimens (none of which have been re-located) were not this species.

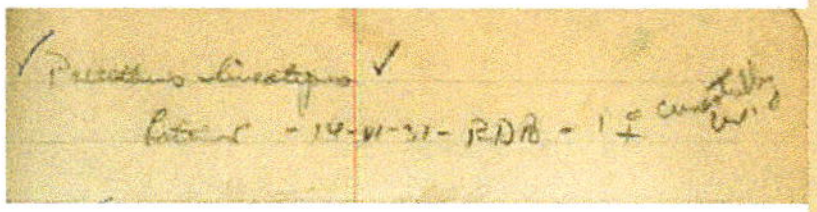

Figure 19.143 Early history of *Paltothemis lineaptipes*, the Red Rock Skimmer, in Oklahoma.

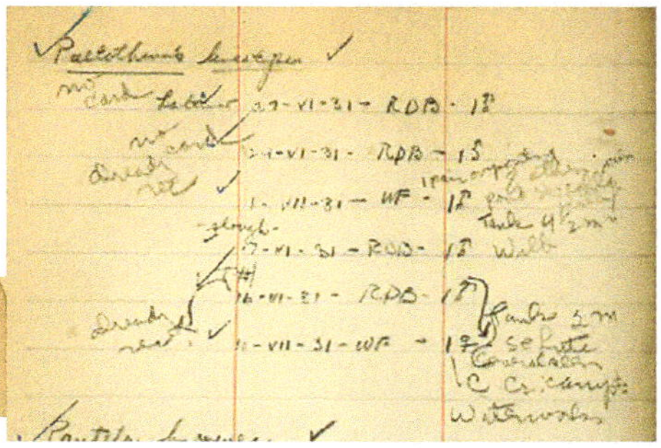

BRECHMORHOGA MENDAX (HAGEN, 1861) – PALE-FACED CLUBSKIMMER

As of yet we have had little luck with predicting where and when we might encounter the Pale-faced Clubskimmer in Oklahoma. There are relatively few confirmed records for the state (*n* = 34, plus three probable), and those records are dispersed widely (12–14 counties). When it is seen, typically fewer than three individuals are reported. Nonetheless, we do have three records of more than that, the highest count of which was 6 individuals: 6♂ (2 as SP 1770–1771, and 2 in hand and released; MAP) along Main Creek at Major County WMA, Major County, on 29 August 2015, on the heels of a ♀ photographed there on 15 August (OC 435075). As well as the high count, 2015 turned out to be an impressive year for the species in Oklahoma because it was reported six times from three counties. An equally impressive year was 2017, when there were six identifiable records from four counties, three of which the species had not been recorded in previously (three additional records, from Ellis and Osage County, were left as "*mendax*, cf"). There were but three reports of the species in 2018, although two new

counties were added, and one record in 2019. Overall, the species is rare in Oklahoma, which is not too surprising given that it is more a species of the southwestern United States and northwestern and central Mexico (Paulson 2009).

We initially thought of this species as primarily one of southwestern and south-central Oklahoma, but various records have made us re-think this. Recent records from northwestern, north-central, and southeastern parts of the state, for example, from Beckham, Ellis, Major, Osage, and Atoka Counties, now make previously outlying Oklahoma records and those in Kansas and Arkansas (Figure 19.145) make sense. Previously, some early records, such as those reported by Bick and Bick (1957 and GH Bick's note cards), were somewhat perplexing. They reported that they examined specimens (OSU) from Cimarron and Payne Counties as well as from Comanche County. We mention the Comanche County record not because it was in an unexpected part of the state but because that specimen appears to be lost and because the species was not seen in the county

Brechmorhoga mendax (Hagen, 1861)
Pale-faced Clubskimmer

S2S3

18 April 1930

18 Apr   20 Sept

Figure 19.144 *Brechmorhoga mendax*, the Pale-faced Clubskimmer, A) ♂, Sandy Creek 7 km SSE of Eldorado, Jackson County, 3 June 2018 (©Bill Carrell, OC 481006). B) ♂, McGee Creek SP, Atoka County, 29 May 2017 (©Michael A Patten, OC 463291), in a far more typical view of the species, males of which compulsively fly a beat low over riffles in a clear stream. C) ♀, Major County WMA, Major County, 15 August 2015 (©Michael A Patten, OC 435075).

again until 2003, when between 12 June and 20 September, Kondratieff et al. (2004) had five individuals on Medicine Creek, Fort Sill MR. Since then there have been two other records: 1♀ at Wichita Mountains WR on 15 September 2007 (SP 199); and 1 individual on Medicine Bluff Creek, Fort Sill MR, 31 August 2009 (VWF, OC 315007). The specimens examined by the Bicks for Cimarron and Payne Counties also cannot be relocated. Those records, too, initially appeared to be outliers. We now know that is not the case. Although not reported again specifically for Payne County, we now have a record for nearby Osage County (Osage Hills SP, 24 June 2017, BC) of a *Brechmorhoga* that in all likelihood is *B. mendax*. As for Cimarron County, on 3 June 2015 BS-P photographed a single ♂ flying around

riffles on the Cimarron River, 4 km NNE of Kenton (OC 431499). We were unable to catch that individual, but MAP returned to the site on 8 August and caught one of 3♂ he saw (SP 1742). He also encountered 1♀ that was part of a pair with one of those males. BC encountered one individual on his visit to the site on 7 September.

Brechmorhoga mendax was recorded only twice in the 1930s—18 April 1930, RD Bird, on Rock Creek, Murray County (1♂, OMNH 3017), which is both the first time the species was recorded in the state and the early date; and 22 April 1933 when AE Pritchard and Bird had 2♂ (OMNH 3522) and 1♀ (OMNH 3027) in Murray County. Between 1954 and 1991 there were 11 records, 8 of which involved GH Bick and/or LE Hornuff (in 3 counties: Johnston,

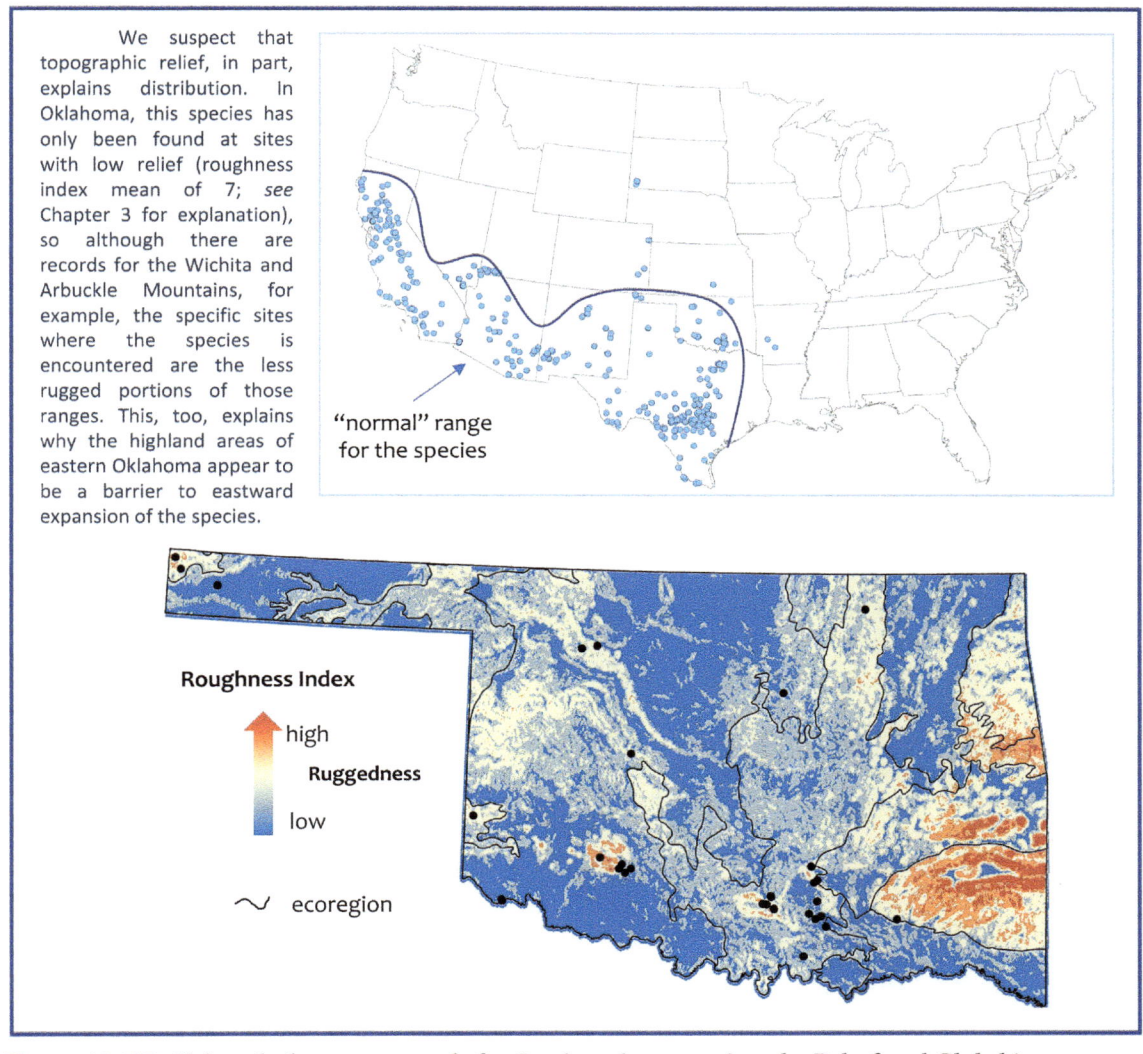

Figure 19.145 Although there are records for *Brechmorhoga mendax*, the Pale-faced Clubskimmer, as far east as Arkansas and north to South Dakota, Oklahoma appears to be the biogeographical limit for this species' range in the western United States.

Murray, and Pontotoc). Since 2003, the species (or at least the genus) has been reported >20 times, including the late date for the state of 20 September 2003 (see above).

The species has been reported in its overall range to fly as early as March and as late as December, and in New Mexico it is known from May to October (Arizona/Sonora: late March to mid-December, Bailowitz et al. 2015; NM: Paulson 2009; TX: 17 March–3 December, Abbott and Lasley 2019). Although Oklahoma has two records from April and four from May, the vast majority of records are from early June to mid-September. Despite there being scattered records in Arkansas (*n* = 2), Colorado (*n* = 5), Kansas (*n* = 1; Figure 19.145), and even as far north as South Dakota (*n* = 2 from 1980, OC 325363, and 2010, OC 322073), it is likely that Oklahoma currently represents the northern extent of the species' normal range in the central United States. To our knowledge no nymphs have been collected in Oklahoma. All known records are of adults, with the only indication of breeding being the one pair mentioned above. Nevertheless, the species is probably a rare breeding resident in the state.

Bailowitz et al. (2015:300) indicated that the species in Arizona and Sonora "occupies a bewildering variety of habitats, but, whether over a stream or a city street, individuals patrol in the same relentless, straight, back and forth manner." In Oklahoma it is a lotic species encountered at locations similar to how Paulson (2009:501) described the species' habitat: "shallow rocky streams with riffles and pools, some current. Banks wooded or open but must have sun." Indeed, we have yet to encounter a male away from a riffle on shallow, clear stream.

Morphological variation of this species is of interest. The species was named for its pale "face," a term that generally refers collectively to the labrum, postclypeus, and front portion of the frons, but some also include the post-frons and vertex. There are slightly differing descriptions of the distinctiveness of this region of the body (regrettably, authors use a variety of terms not always clearly defined)—nonetheless, there is general agreement that the Pale-faced Clubskimmer is indeed pale "faced." For example, Abbott (2015) said the face is pale; Paulson (2009) said the face is light brown; Needham et al. (2014), in their species account, said that the face *and* vertex are yellow and in their key said, "vertex and postfrons mostly pale" as opposed to the "vertex and postfrons dark, usually metallic blue" (p. 443, key step to distinguish *Brechmorhoga mendax* from *B. vivax* and *B. praecox*, but *B. pertinax* also has a dark vertex and postfrons). Bailowitz et al. (2015) also emphasized a pale labrum[40] and frons, as well as a pale vertex lacking dark metallic. However, Paulson (2009:500) said that, "In Texas, [a] small percentage ... have dark metallic frons," a peculiarity with which Abbott (2015) concurred.[41] We offer further support of atypical individuals: in Oklahoma's case, there are but two adults that do not have entirely pale faces (typically yellowish labrum and frons with a gray

postclypeus). One is a female (SP 199, Comanche County) that has a brownish non-metallic labrum and a frons of a shimmery gray. The other is a ♂ (SP 2479, Beckham County) with smudged discoloration in the dorsolateral portion of the frons but not so much that we characterize it as "dark." What is more interesting about that ♂ is that he has a postfrons that is roughly half dark metallic and a vertex that is entirely so (Figure 19.146), which by Bailowitz et al.'s (2015) characterization of the species would eliminate *B. mendax*. Another ♂, collected in Atoka County (SP 2364), has a postfrons about a third dark metallic with a vertex almost entirely so. These two males are the darkest and only metallic adults yet noted from the state. All

Figure 19.146 Variation in *Brechmorhoga mendax*, the Pale-faced Clubskimmer. The postfrons and vertex of individuals varies from a "classic" looking *B. mendax* with both entirely pale (A: SP 1742, ♂, Cimarron County) to the postfrons with some dark and the vertex brown with dark cone tips (B: SP 1770, ♂, Major County; not metallic) to the postfrons half dark metallic and the vertex entirely so (C: SP 2479, ♂, Blaine County). Also shown are lateral portion of abdominal segment 2 and the hamules.

other adult specimens (all ♂) vary from having entirely pale vertices to those that are gray or brown and from entirely pale postfrons to those with just under a quarter dark (Figure 19.146). All of these males have the hamules of *B. mendax*, so we do not question the species-level identification; we do, however, wonder why individuals in this region diverge from the typical postfrons and vertex coloration. We see no clear pattern of geographic distribution of pale, intermediate, and dark individuals, so it appears that a study of geographical or individual variation of *B. mendax* in Oklahoma and Texas is in order.

THOLYMIS CITRINA HAGEN, 1867—EVENING SKIMMER

Oklahoma has a single record of this tropical skimmer. On 20 August 2006, DA caught a lone ♀ at Red Slough WMA, McCurtain County. He first spotted it "around 7 AM" on a cloudy morning when it was "flying low to the water with a quick bouncing flight … over a large pool in a drying up canal" in open woodland (Arbour 2007 and *pers comm*, Arbour et al. 2009; JCAC 39785, OC 7269, 7322).

The Oklahoma record was at the time only the seventh for the United States. The first record for the U.S. was from Hidalgo County, Texas, where it was collected on 16 September 1950 (FSCA, Barber and Elia 1994). It was next found at the Stock Island Botanical Garden, Monroe County, Florida, where it was photographed by V Elia (Barber and Elia 1994). The next two records for the U.S. also came in Florida, both reported by DR Paulson when he visited the state in 2000. He also encountered the species at the Stock Island Botanical Garden, where he said

it was common, and on that same day, 7 January, noted "one or more were seen at Southeast Point on Big Pine Key" in Monroe County (Paulson 2001:64). The other two records that preceded the Oklahoma specimen were from Texas (22 October 2003, Borregos Creek, Santa Gertrudis Division of King Ranch, Kleburg County, T Langschied, JCAC 20725; 2 November 2004, San Ygnacio, Zapata County, M Reid, OC 281520). Since 2007 there have been 22 other records submitted to Odonata Central for Texas and two for Florida. There is also a single vagrant record in southeastern Arizona (12 June 2007, Bailowitz et al. 2015).

We had considered a second record for Oklahoma a long shot until one was photographed not far south of the state line at Hagerman NWR, Grayson County, Texas, on 17 October 2015 (R Cantu, OC 438055), implying occasional wanderers to the north and with it the possibility of another Oklahoma record.

Figure 19.147 *Tholymis citrina*, the Evening Skimmer, ♀, Red Slough WMA, McCurtain County, 20 August 2006 (USDA Forest Service; David Arbour), the sole individual to be recorded in Oklahoma and one of few north of the Rio Grande Valley of southernmost Texas.

MIATHYRIA MARCELLA (SÉLYS, 1856) – HYACINTH GLIDER

The Hyacinth Glider is a tropical species that ranges south from Argentina up into the southeastern United States, with vagrant records north on the east coast to Maryland and Virginia (OC). It is found especially where water hyacinth (*Eichhornia crassipes*) or water lettuce (*Pistia stratioles*), both invasive species in the U.S., are abundant in ponds and lakes. Neither of those plant species are naturalized in Oklahoma (BW Hoagland, *pers comm*). If the species does have a population in Oklahoma it is safe to say that it must coexist with plants other than those more typical invasives.

In the southeastern U.S. it is known primarily from near the coast, although it ventures inland as far as northern Alabama, northern Texas, Arkansas, and even Kansas. In Arkansas it has been reported 3 times, in 3 counties, the most recent of which was 12 adults at Lake Columbia, Columbia County, in southwestern Arkansas (29 June 2017, Devin Moon, OC 465431). In Kansas it is known from one record (1♂, 27 September 2008, Ninnescah Field Station, Sedgwick County, R Beckemeyer, OC 283980).

For years we had thought of the species as an occasional visitor to Oklahoma, but we now think that it may be attempting to establish itself as a breeder, at least in the southeastern corner of the state, where the first record appeared: DA photographed an immature ♂ on the east end of Otter Lake at Red Slough WMA on 10 October 2007 (OC 263169). The next day he caught the state's first specimen (OC 263237, JCAC 23556), again an immature ♂, but this time at the northern parking lot on Mudline Road that is less than 1 km to the east of Otter Lake. DA photographed another immature ♂ at the end of October that year (28 October, OC 263487). The next year he photographed the first record of a ♀ in the state (8 October, OC 284142), which was the only time the species was reported in 2008. In 2009, DA and BAH reported it on four days between 14 October and 3 November. Three of those times were at Red Slough (2♂, 2♀), but on 25 October, BAH photographed a ♀ at his property 10 km SE of Idabel (OC 315546).

Our perception of the species' seasonality in the state was blown away in 2017, when DA reported it on 2 May.

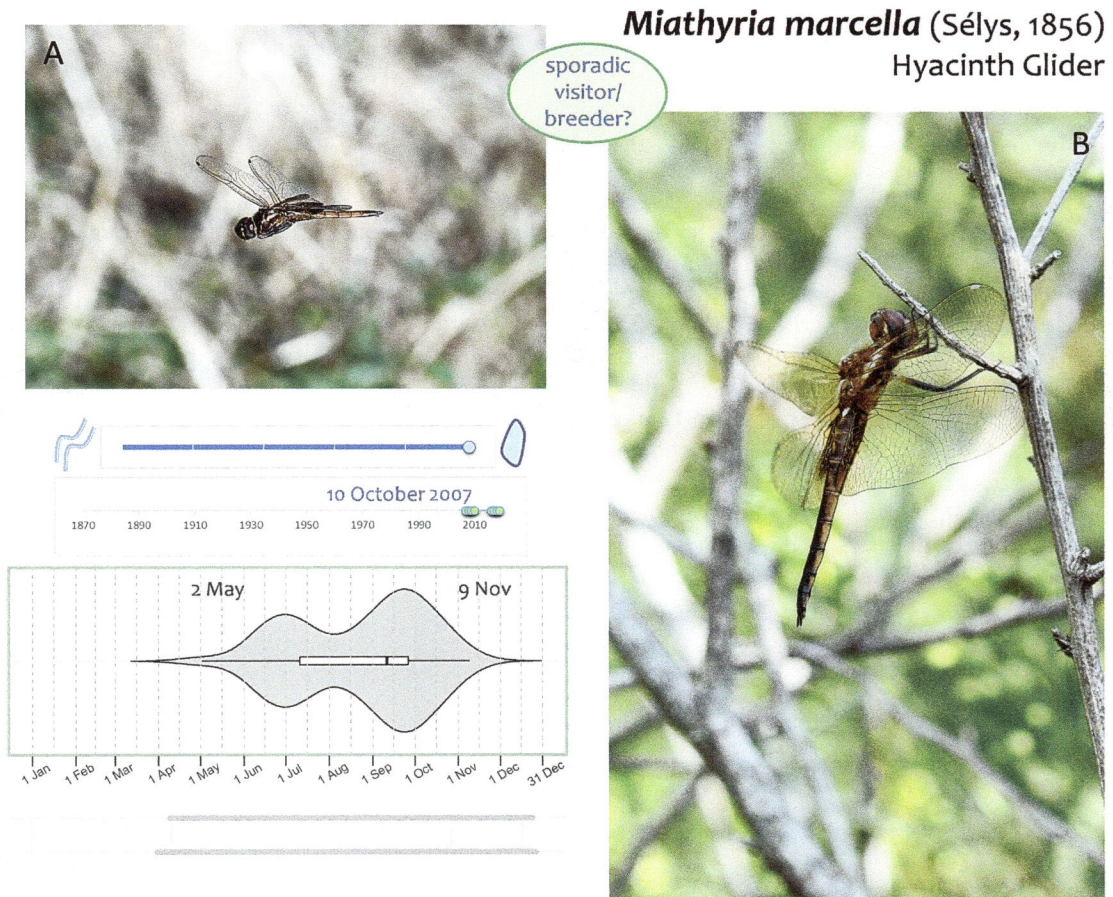

Figure 19.148 *Miathyria marcella*, the Hyacinth Glider, A) ♂, Red Slough WMA, McCurtain County, 9 November 2017 (USDA Forest Service; David Arbour, OC 474717). B) ♀, Red Slough WMA, McCurtain County, 25 September 2017 (USDA Forest Service; David Arbour, OC 473639). All but four records for Oklahoma are from this locale.

Although not able to get documentation he did briefly capture a single ♂ and got a look at it before it wriggled out of his net and hurtled away. About a month later he happened upon a mini invasion of about 13 adults and then 2 days later, on 15 June 2017, he and MAP counted 30♂ and 3♀, including 1 tandem pair. It was on that day, nearing ten years since the first specimen was collected, that the state's second and third specimens were taken (SP 2395, 2396). A total of 35 reports of the Hyacinth Glider came from Red Slough in 2017, including the high count for the state (40 adults on 19 September, DA) and the last report of the season, on 9 November (OC 474717), which stands as the current late date. Double digits persisted into October (e.g., 24 on 26 September, DA; last double-digit count was 11U on 10 October, DA, OC 473941). Two mating pairs were reported, but no ovipositing was observed. In 2017, the species also was reported from Broken Bow, McCurtain County (9 October, 1♂ in hand, OC 473940), and, stunningly, 4 km W of Nuyaka, Okmulgee County (23 September, 1♂, MAP, SP 2509). Breeding may have failed: there have been but a couple of reports (each of a lone adult, 2 and 9 October 2019, Red Slough, DA) since.[42]

There are two other counties in the state with records of *Miathyria marcella*: Pontotoc and Kiowa. On 15 September 2010, Franklin Alm captured a ♀ at TNC's Pontotoc Ridge Preserve, Pontotoc County, (OC 323053), as that county's first and only record of *M. marcella*. On 3 July 2016, a much more shocking report was received by BC saying that he found a ♂ at Great Plains SP, Kiowa County (OC 449669). We had some hesitation with the record given that it was the only one that was outside of autumn and that he was unable to get good documentation, managing only to procure a blurry photo. Nevertheless, we accept the record based on his skill level in the field and his description of the individual, an obvious adult ♂. He said that he "Observed [it] off and on for over an hour, often interacting with Red and Black Saddlebags. Eyes and face dark, thorax dark violet or purple, abdomen yellow orange, dark hindwing bases." We also thought the record credible because this was not the only out of range species recorded at the locality in early July. The next day, MAP visited the site and found southwestern Oklahoma's first *Tramea calverti* (Striped Saddlebags, ♂, SP 2015). He also

saw a ♂ *Coryphaeschna ingens* (Regal Darner, OC 449014) that he managed to only get lousy photos of but watched on and off for about 30 minutes. The darner was well out of its "normal" range. The *Tramea* was more-or-less so, as it had been recorded closer to Kiowa County before but, similarly to the *M. marcella*, previously only in autumn. Consequently, the July *Tramea* and *Miathyria* records were both out of range and season. It may be that weather played a role in the displacement of these species given the storms that rolled through the area in late June 2016 (Oklahoma Climatological Survey 2016). The storms may also explain why *M. marcella* appeared at Red Slough in 2016, after not being seen there since 2009, and then had a boom year in 2017.

In Louisiana the Hyacinth Glider is known from early April to late November and in Texas it is present from late March to late December (LA: 9 April–24 November, iNat 5631336, Mauffray 2014; TX: 31 March–26 December, Abbott and Lasley 2019). We now know that the species appears in Oklahoma not just in the fall but in the summer, too.

There are a few problematic reports for Oklahoma. The species was reported in larval form as rare in creeks of the Wichita Mountains, Comanche County (Bass 1990, as *Miatheria* [sic]). Despite the Kiowa County record (see above) and the fact that the Wichita Mountains have many disjunct populations of a variety of plants and animals, we consider the report dubious. Given the amount of survey effort in the area with only one adult reported, we doubt the species breeds there, although failed attempts producing an occasional nymph cannot be fully ruled out. Another larval record was reported for the Tallgrass Prairie Preserve, Osage County (Bass 1994), but we assume this was a misidentification. Unfortunately, neither of these larval records have extant specimens to be re-examined. There was also a nymph (UCO 5173) reported from Tinker Creek, Arcadia Lake, Oklahoma County, from 19 September 1986, collected by D Bass that was originally identified as *Miathyria* but that we re-identified it as *Libellula luctuosa* (Widow Skimmer). If the species breeds in the state, we surmise it would be in the southern part of the state not in northern or central Oklahoma (Figures 19.149 and 19.150).

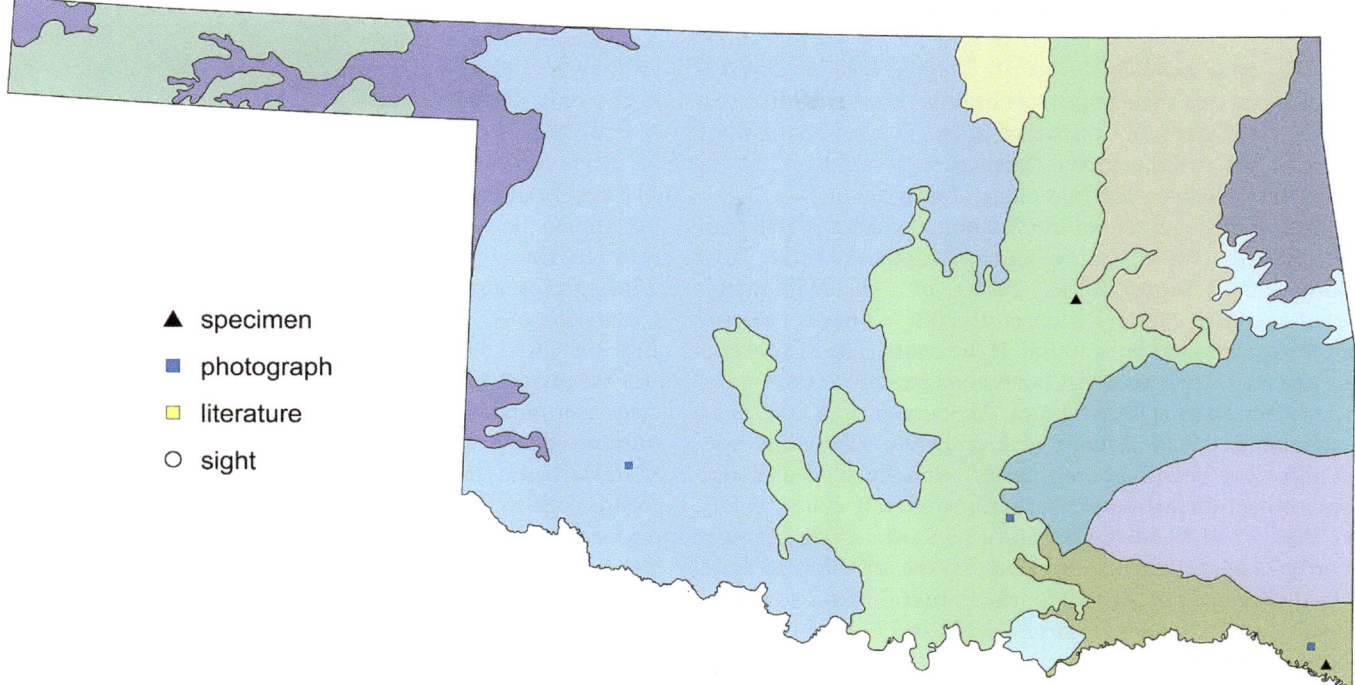

Figure 19.149 Documentation level of supporting records for *Miathyria marcella*, the Hyacinth Glider, in Oklahoma. Levels include specimen, archived photograph, literature reference, or sight record. Although sight records lack documentation, only those we feel are credible are plotted.

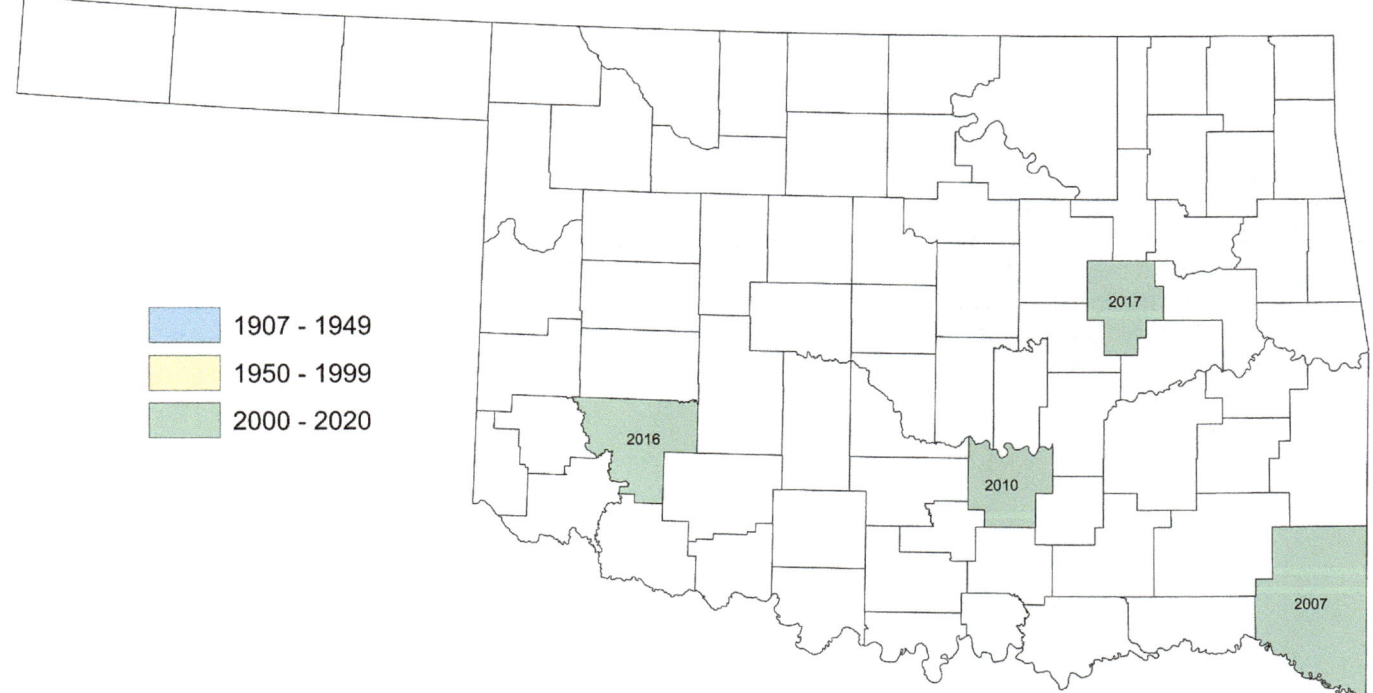

Figure 19.150 Counties in Oklahoma in which *Miathyria marcella*, the Hyacinth Glider, are known to occur.

TRAMEA CALVERTI MUTTKOWSKI, 1910 – STRIPED SADDLEBAGS

Like *Miathyria marcella* (Hyacinth Glider), *Tramea calverti* is a tropical species that ranges south to Argentina. Unlike that species, *T. calverti* is not restricted to the southeastern United States; rather, it has a spotty distribution across much of the southern and eastern portions of the country.

It was first documented in Oklahoma when a ♂ was photographed at Oxley Nature Center, Mohawk Park, Tulsa County, on 19 August 2006 (J Fisher, BC, OC 7350; misreported as a ♀ in Smith-Patten et al. 2007). That is the only time the species has been found in the county. Just five weeks after the state's first record, Oklahoma's second record turned up at Otter Lake at Red Slough, McCurtain County, on 26 September 2006 (1♀, Arbour et al. 2009, OC 7398). The species since has been recorded nearly annually at Red Slough from mid-summer through late autumn (16 July 2008, 1♀, DA, OC 283175 and 18 November 2015, 1♂, Lotus Lake, DA sight record). The late date for Red Slough is the same for the state.

Only 9 records of the >50 for the state come from locations away from Red Slough (Figures 19.152 and Figure 19.153). A record for Love County, from along the Red River (♀, Love Valley WMA, 20 September 2014, SP 1434), is the only one away from a clear-water pond or small lake. Kiowa County is the only county besides McCurtain that has more than one record. The first for that county came on 4 July 2016 (♂, SP 2015) and the second, and only other, on 12 May 2018 (♂, MAP sight); both were at Great Plains SP. The early date for Oklahoma came on 28 April 2017 when BC photographed a ♂ at Hackberry Flat WMA, Tillman County (OC 462290). The species also has been documented in northwestern Oklahoma (2♂, 5 km SW of Lookout, Woods County, 22 July 2018, OC 485691).

Our current knowledge of the flight season in the state indicates that our initial assessment of this species as an autumn vagrant (Patten and Smith-Patten 2013b) was a miscalculation. We based that assessment in part on there being only one record each from July and August, meaning that eight of the ten occurrences between 2006 and 2011 were autumn records. We have since obtained 6 additional July records and 16 for August, which has weakened the argument of this species being an autumn vagrant to Oklahoma. Its documented seasonality in the state now suggests that it is probably regular, albeit rare, throughout the entire season it is known for elsewhere in the

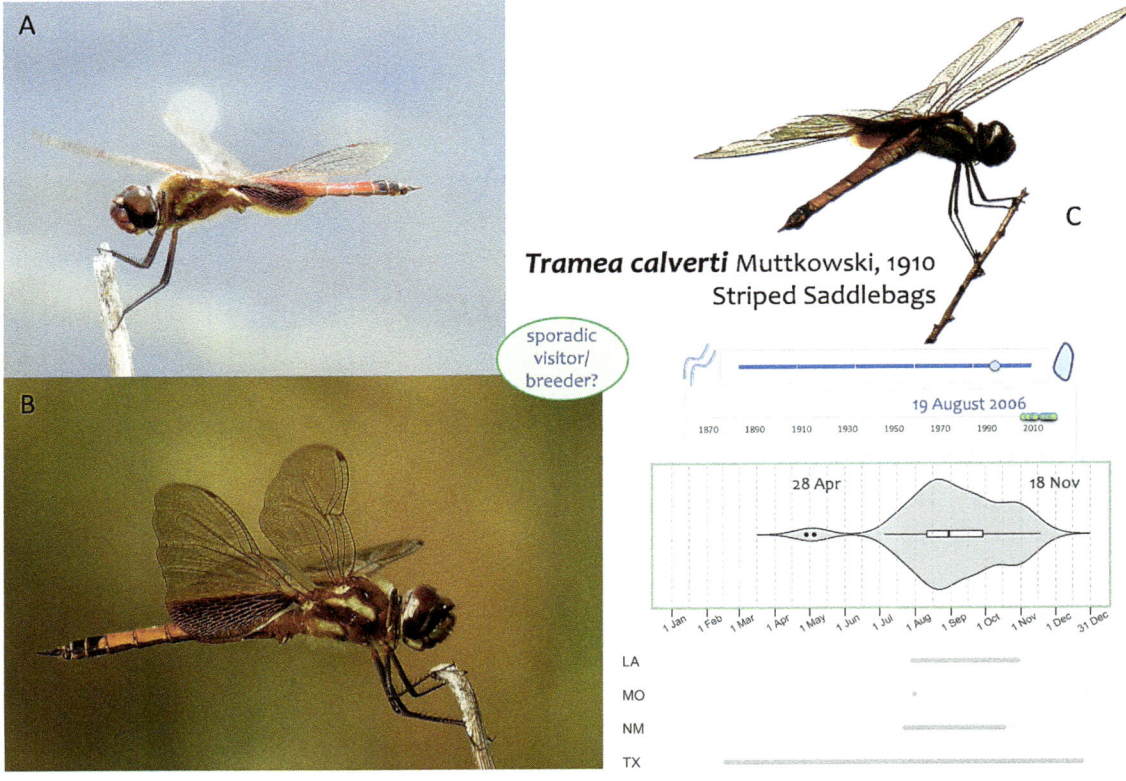

Figure 19.151 *Tramea calverti*, the Striped Saddlebags, A) ♂, Hackberry Flat WMA, Tillman County, 28 April 2017 (©Bill Carrell, OC 462290). Curiously, the few records for southwestern Oklahoma are from spring or early summer (late April to early July). B) immature ♂, Mohawk Park in Tulsa, Tulsa County, 19 August 2006 (©John F Fisher, OC 7350), the first of the species to be found in Oklahoma. C) ♀, Love Valley WMA, Love County, 20 September 2014 (©Brenda D Smith, OC 427081, SP 1434). Records for eastern Oklahoma, which greatly outnumber those for the western part of the state, exclusively fall between mid-summer and early autumn (late July to early November).

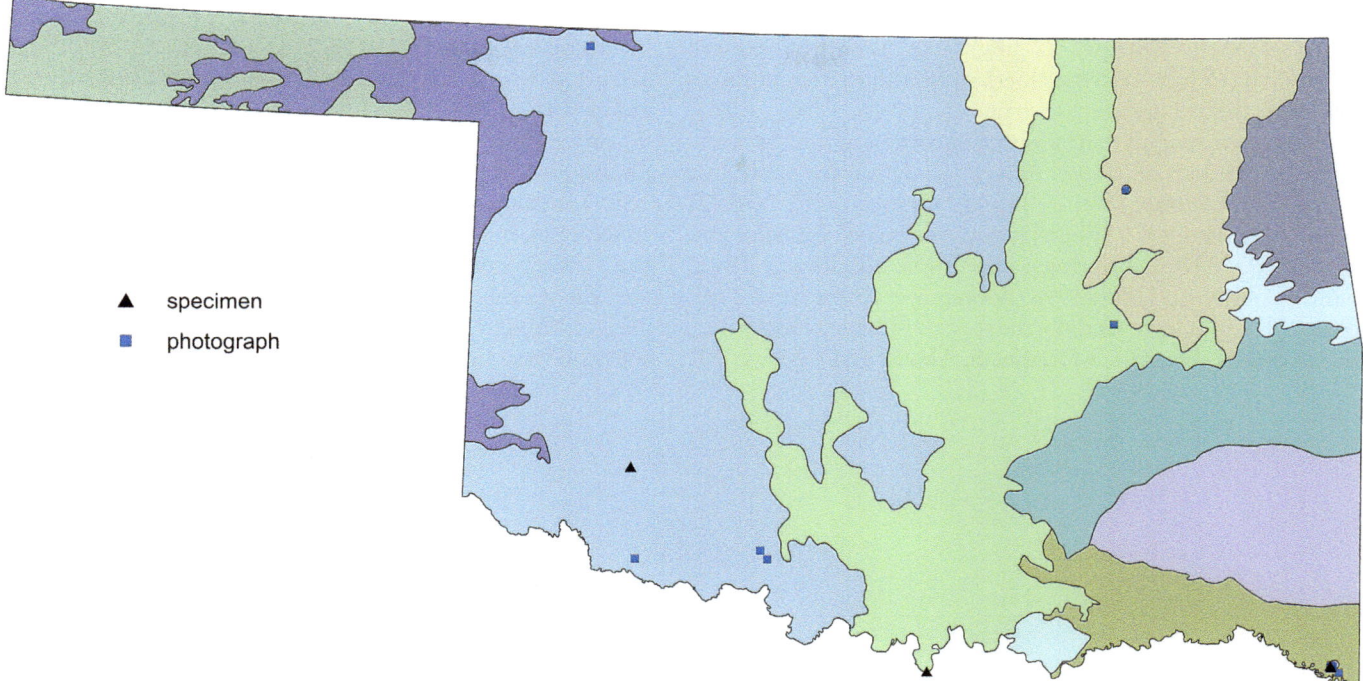

Figure 19.152 Documentation level of supporting records for *Tramea calverti*, the Striped Saddlebags, in Oklahoma.

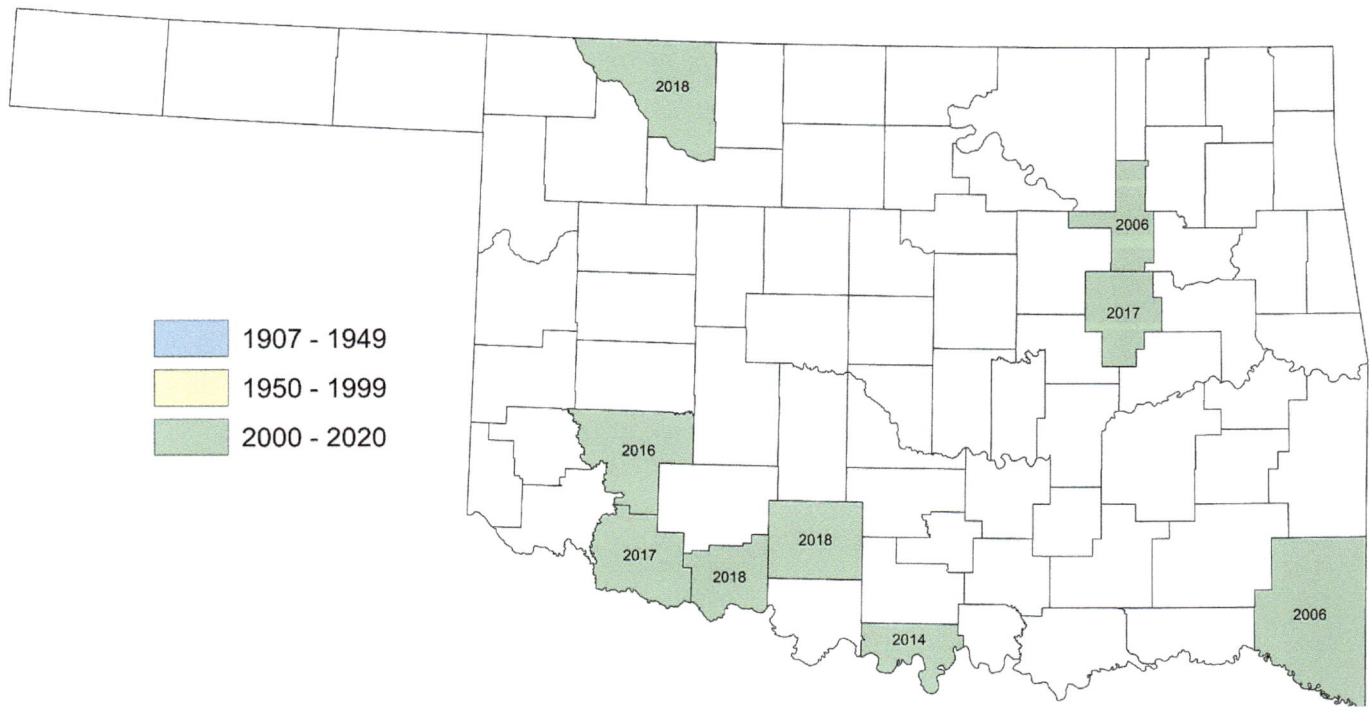

Figure 19.153 Counties in which *Tramea calverti*, the Striped Saddlebags, are known to occur in Oklahoma.

region. Curiously, recent records (*n* = 6) for western Oklahoma differ seasonally (28 April–22 July) from those from eastern Oklahoma (16 July–18 November), suggesting different forces drives northward drift into the state.

There are roughly half a dozen records each in Arkansas and New Mexico of *Tramea calverti*, all dating from late July to mid/late October (AR: 29 July–29 October, OC; NM: 23 July–17 October, OC). There is only one record for Missouri, from the McAllister Springs area at the Blackwater River, Saline County, on 31 July 1998 (Beckemeyer 1998a, Trial 2005, Sims 2012). Not unexpectedly the species flies for a longer time in Texas (17 February–23 December, Abbott and Lasley 2019).

Typically the species is reported as single individuals but there are a handful of records with counts >3 individuals, all of which are from Red Slough, including when we, along with DA and M Dillon, recorded the high count of 8♂ on 19 August 2014 (1♂ as SP 1396). The species has been photographed many times (>20 known to us), but there are only 4 specimens for the state (3♂ and 1♀: JCAC 233810, SP 1396, 1434, and 2015), all of which are adults, as are all of the state's records. As such, there is currently no evidence that the species breeds in Oklahoma (no reports of pairings, even), although it may be in the process of spreading northward, so breeding may occur in the future.

TRAMEA ONUSTA HAGEN, 1861 – RED SADDLEBAGS

The Red Saddlebags is a widespread species, recorded in all 77 Oklahoma counties, and it is fairly common, typically recorded in the range of 5–10 individuals at a time. We have recorded it as numerous as 150 adults (Salt Plains NWR, Alfalfa County, 6 June 2004). It occurs in a range of habitats, although it is far more numerous in lentic settings or foraging (or migrating) over fields and lawns. It was first found in the state on 8 July 1925 on Lukfata Creek 6 km (4 mi) northwest of Broken Bow, McCurtain County (Ortenburger 1926b) and has been reported consistently since. The early (18 March) and late (10 November) dates for the species in Oklahoma are also from McCurtain County, albeit from Red Slough (both DA, early from 2013, OC 398553; late from 2009, OC 315680). The Oklahoma date range is not as long as the species' year-round presence in Texas, being closer to its known seasonality elsewhere in the region (AR: 10 April–9 October, OC 422280, Harp and Rickett 1977; KS: 17 April–mid-October, early date is BS-P and MAP unpublished 2004 sight record of 1U, Mined Lands WA, Cherokee County, late date is from Allison 1921; MO: 25 April–12 October, USNM, Trial 2005, Sims 2012; NM: March–October, Paulson 2009; TX: Abbott and Lasley 2019).

Figure 19.154 *Tramea onusta*, the Red Saddlebags, A) ♂, 3 km W of Gene Autry, Carter County, 17 July 2013 (©Ken Williams, OC 403374). B) ♀, Cimarron River 16 km NW of Knowles, Beaver County, 8 August 2015 (©Bill Carrell, OC 434870).

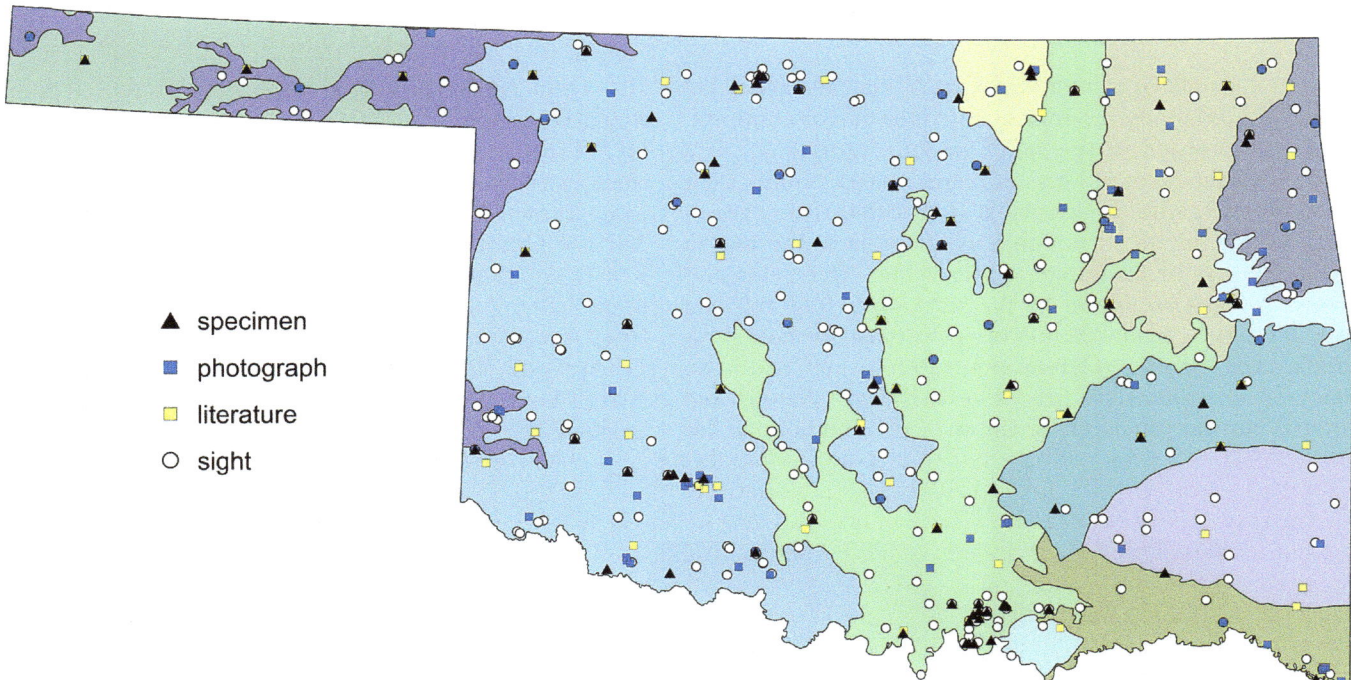

Figure 19.155 Documentation level of supporting records for *Tramea onusta*, the Red Saddlebags, in Oklahoma. Levels include specimen, archived photograph, literature reference, or sight record. Although sight records lack documentation, only those we feel are credible are plotted.

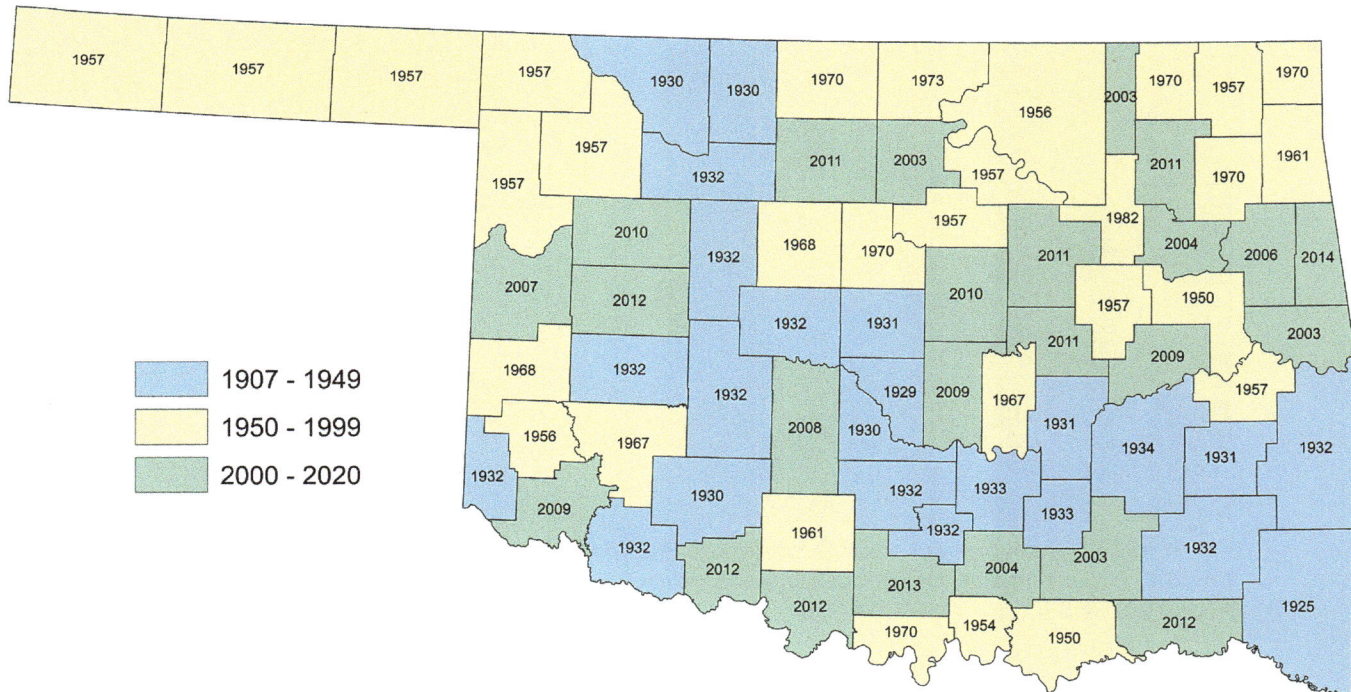

Figure 19.156 Counties in which *Tramea onusta*, the Red Saddlebags, are known to occur in Oklahoma.

TRAMEA CAROLINA (LINNAEUS, 1763) – CAROLINA SADDLEBAGS

Tramea carolina is much rarer and range-restricted than the other red *Tramea* in the state (*T. onusta*, Red Saddlebags)—the Carolina Saddlebags has >100 records from about 20 counties, whereas the Red Saddlebags has roughly 650 records spread across all 77 Oklahoma counties. Even so, the Carolina Saddlebags can be locally common in the eastern third of the state, although, counts cannot always be taken at face value because confirmation of identification often is possible only in hand (see below). Nevertheless, we place confidence in some records, such as the highest reported total, of 37 individuals: "14 tenerals kicked up in grasses around several ponds on the preserve. In addition I had 23 adults on the preserve" (JT Nickle Preserve, Cherokee County, 13 June 2014, JWA, OC 423043). Our highest count for the species is 30 adults in Le Flore County (5♂, 5♀, 20U, including 5 pairs, Big Cedar, 5 July 2014), with the next highest of 15 adults (12♂ [1 as SP 695], 3♀, including 2 pairs, Pushmataha WMA, Pushmataha County, 10 June 2013). In the northeast it has been reported as high as 7–8 individuals in Cherokee County (5♂ [1 as SP 2001, 1 in hand], 2♀ [1 as SP 2002], including 2 pairs, Cookson WMA, 6 July 2016, C Farquhar, BS-P; 8U, JT Nickel Preserve, 6 June 2009, BC).

Tramea carolina has a spotty distribution across the eastern half of the state. Difficulties with proper identification, specifically distinguishing it from *T. onusta*, especially on the wing and from photos with problematic lighting, likely masks its true status in the state. The 30+ records in the OOP database that are labeled as "*carolina/onusta*," "*onusta*, cf," or "*carolina*, cf".are indicative of the overall difficulty of confident identification. Many times is the case that one needs to examine an individual under magnification in order to confirm the species and even then, there can be problems. One such example is a specimen reported by Bick and Bick (1957:8); they wrote:

> The only Oklahoma record (Bick 1951) of this eastern species was based on a single early instar nymph. The presence of *carolina* in Oklahoma is now substantiated by a male which I examined (OAM [now OSU]) from Payne County collected on August 29, 1938.
>
> **(Bick and Bick 1957:8)**

We have not relocated the nymph he collected in June 1950 in Muskogee County (Bick note card), which he never seemed terribly confident about, but we did find the adult ♂

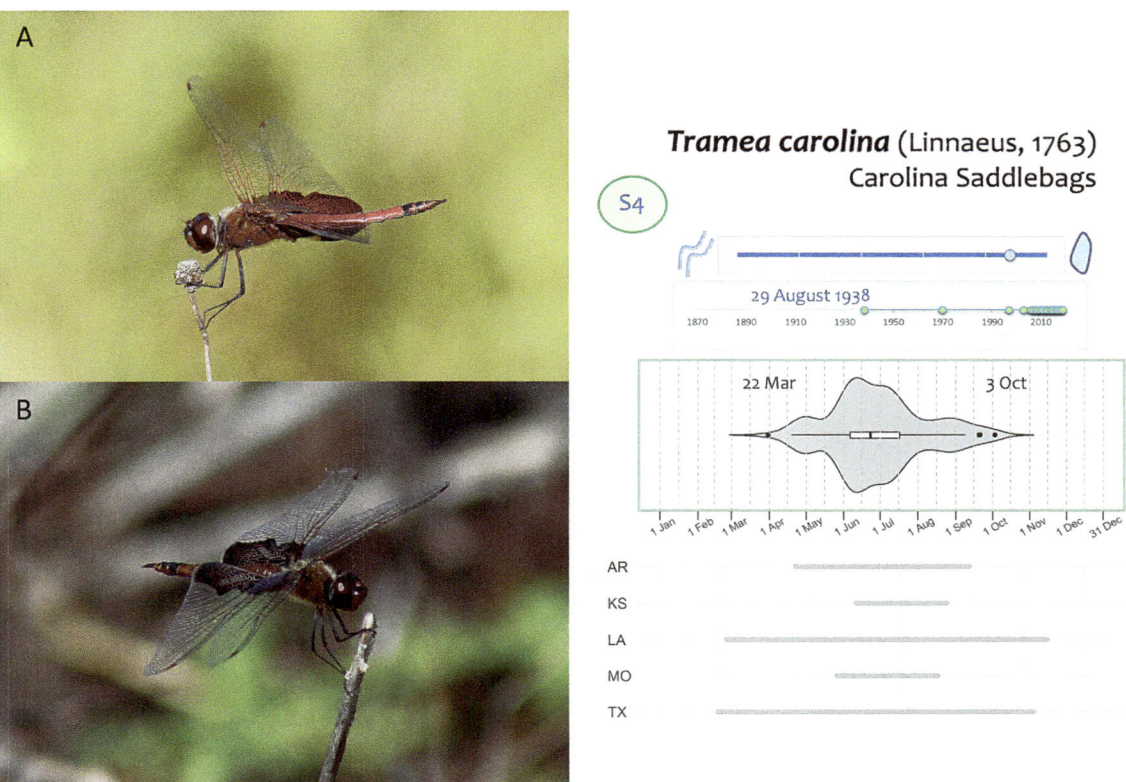

Figure 19.157 *Tramea carolina*, the Carolina Saddlebags, A) ♂, 3 km NNW of Braggs, Muskogee County, 3 June 2014 (©James W Arterburn, OC 422551). B) ♀, Grassy Slough WMA, McCurtain County, 31 March 2018 (©Michael A Patten, OC 479443). Of the two red-colored *Tramea* species that occur commonly in Oklahoma, *T. carolina* is easily the rarer, although in wooded highlands in the eastern third of the state it is the commoner.

he mentioned. Many years ago, during a quick visit to the OSU collection, MAP came across the specimen and re-identified it as *T. onusta*, on the basis of the extent of red in the hindwing, which extends distally well short of the anal loop. Hindwing aside, that identification niggled at us for years, until BS-P re-examined the ♂ and determined that it, in fact, was *T. carolina* on the basis of its hamules and metallic purple frons. Yet she, too, was bothered by the hindwing pattern and so contacted DR Paulson to ask if the specimen could be *T. carolina* or was perhaps a hybrid *T. carolina × T. onusta*. Paulson (*in litt*) examined photos BS-P sent and confirmed the identification as *T. carolina*. In doing so, he noted that the hindwing pattern can vary too much to rely on it, in that *T. carolina* occasionally has as little red in the hindwing as is present on a typical *T. onusta*, although there is no evidence the opposite can be true (i.e., *T. onusta* never has red on the hindwing that extends distally past the anal loop).

Tramea carolina was not reported again for Oklahoma until 1970, on 17 July, when LE Hornuff collected 1♂ and 1♀ at 5.8 mi N of Vian, Sequoyah County (FSCA; reported in error as 2♂ in Bick's notes). And then another 27 years went by, until 17 April 1997, when SW Dunkle and JC Abbott collected 5 adults near Thackerville, Love County (JCAC 24765, 26966; FSCA). We are baffled by this detection rate because now the species is regularly detected in the highlands of eastern Oklahoma (about 100 records between 2003 and 2019), and now scattered records extend westward through the Cross Timbers to Grady County in central Oklahoma (3♂, Grady County WMA, 22 June 2019, MAP, 1♂ as SP 2842) (Figures 19.158 and 19.159).

Tramea carolina is known to fly in Oklahoma from late March to early October (both from Red Slough WMA, McCurtain County: 22 March 2018, 1U, DA, OC 478579; 3 October 2009, ♂, BAH, OC 315618). Oklahoma's documented seasonality, at about six months long, is almost three months shorter than that for Louisiana and Texas but it is over a month longer than in Arkansas (AR: 22 April–12 September, OC 6950, 455401; KS: 11 June–25 August, OC 319781, 315528; LA: 24 February–15 November, Mauffray 2014; MO: 26 May–17 Aug, OC 480384, Trial 2005; TX: 17 February–4 November, Abbott and Lasley 2019).

One problematic record comes from Bass (1990). He reported occasional *Tramea carolina* nymphs found in macroinvertebrate samples from the Wichita Mountains, Comanche County. Of the samples we were able to examine from those collections, none was labeled as *T. carolina*, and other nymphs were not *Tramea* or could only be identified to Libellulidae. Although possible in light of the odd distributions supported by the Wichita Mountains, we think that *T. lacerata* (Black Saddlebags) or *T. onusta* would be much more likely than *T. carolina* to breed there.

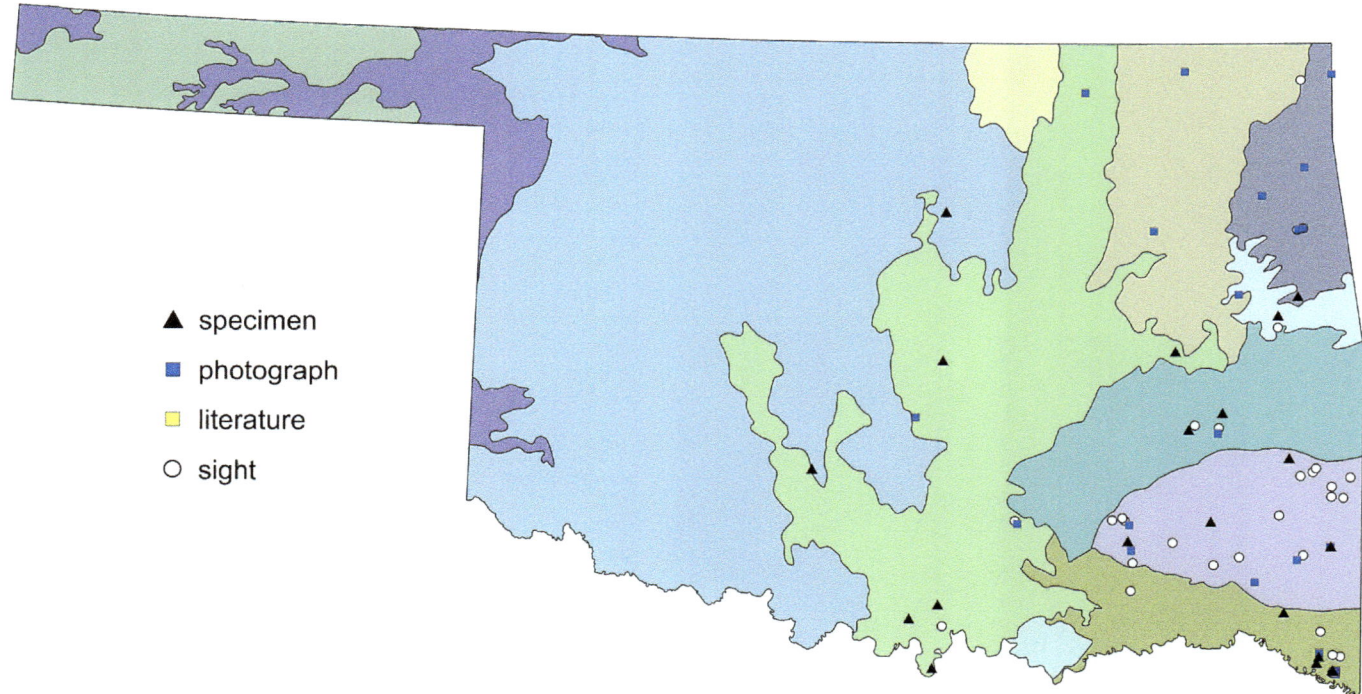

▲ specimen
■ photograph
□ literature
○ sight

Figure 19.158 Documentation level of supporting records for *Tramea carolina*, the Carolina Saddlebags, in Oklahoma. Levels include specimen, archived photograph, or sight record. Although sight records lack documentation, only those we feel are credible are plotted.

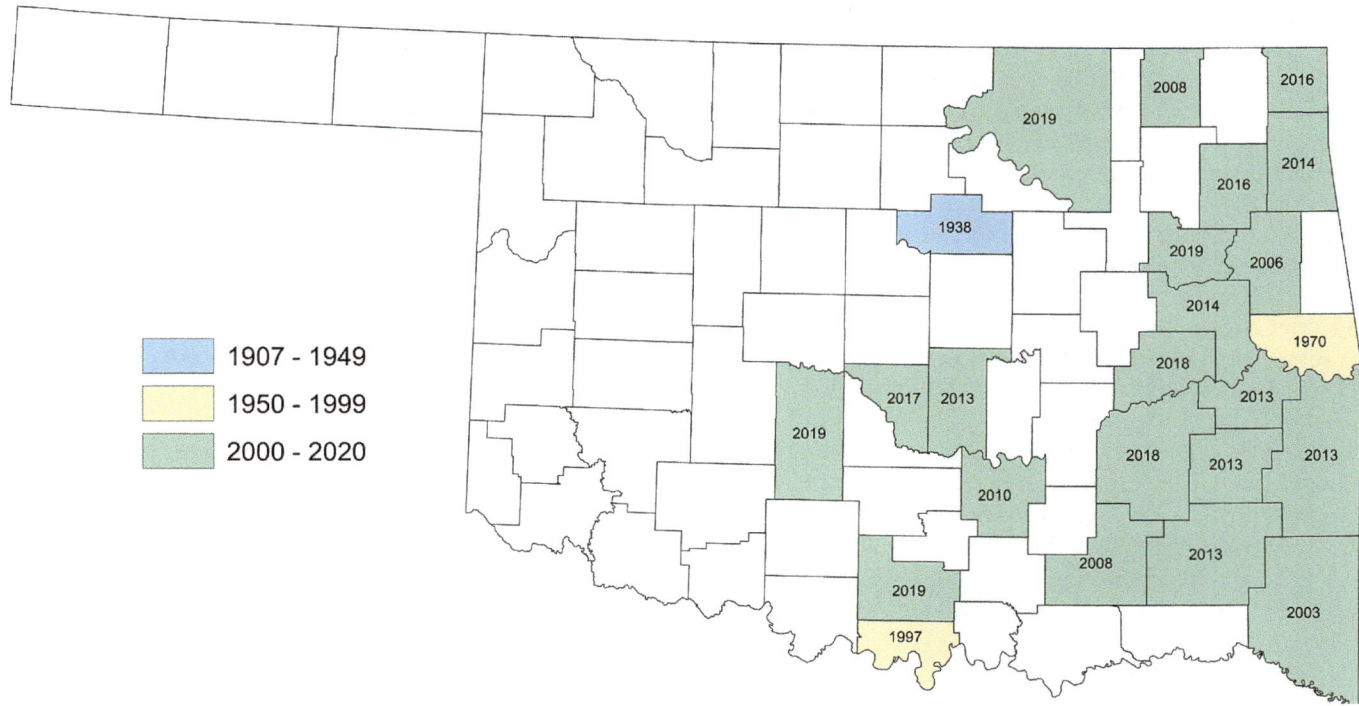

Figure 19.159 Counties in which *Tramea carolina*, the Carolina Saddlebags, are known to occur in Oklahoma.

TRAMEA LACERATA HAGEN, 1861 – BLACK SADDLEBAGS

The Black Saddlebags has been recorded from all Oklahoma counties, where it can be seen in just about any habitat, although it is commonest near lakes and ponds (Figure 19.160). It is routinely seen in double digits, and single-locality counts have exceeded 100 individuals on four occasions. It was first recorded in Oklahoma on the same day and at the same place as the first record of the Red Saddlebags (*Tramea onusta*), on Lukfata Creek 6 km (4 mi) northwest of Broken Bow, McCurtain County, on 8 July 1925 (1♂, OMNH 1192; Ortenburger 1926b). It has been reported regularly through all eras of survey effort.

The species flies most of the year regionwide, including through the winter in Texas (OK: 21 March 2017, Red Slough WMA, McCurtain County, 1U, DA and 10 November 2009, Red Slough, BAH, OC 315794; AR: 25 April–24 October, early from unpublished BS-P and MAP 2004 sight record of 4U, Pea Ridge National Military Park, Benton County, late date is from OC 457095; KS: 20 April–6 November, early from unpublished BS-P and MAP 2003 sight record of 1U, Mined Lands WA, Cherokee County; MO: 17 April–16 November, Trial 2005, Sims 2012; NM: March–October, Paulson 2009; TX: 22 March–13 January, Abbott and Lasley 2019).

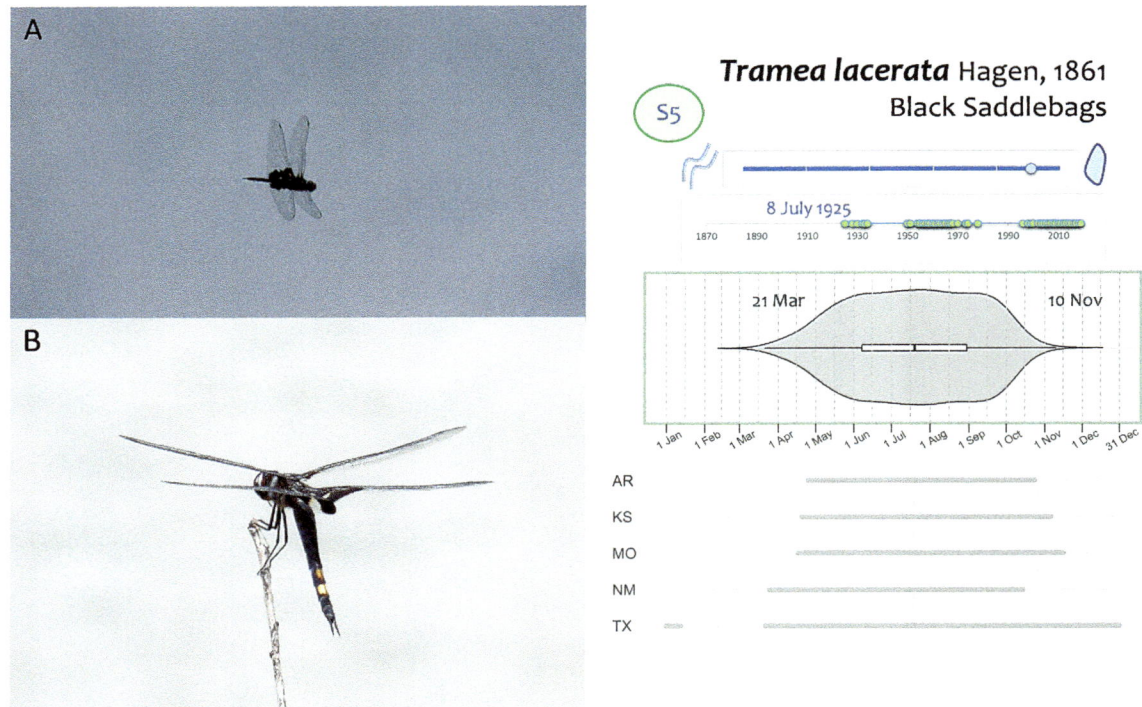

Figure 19.160 *Tramea lacerata*, the Black Saddlebags, A) ♂, Stroud Lake, Lincoln County, 4 October 2013 (©Brenda D Smith, OC 410773), seen here, as species in the genus are encountered more often than not: on the wing and difficult to determine to species. B) ♀, Lake Perry, Noble County, 17 July 2011 (©Victor W Fazio III, OC 330142). When perched, *Tramea* tend to jut their abdomen downward. Note, too, the yellow patch on the abdomen typical of females.

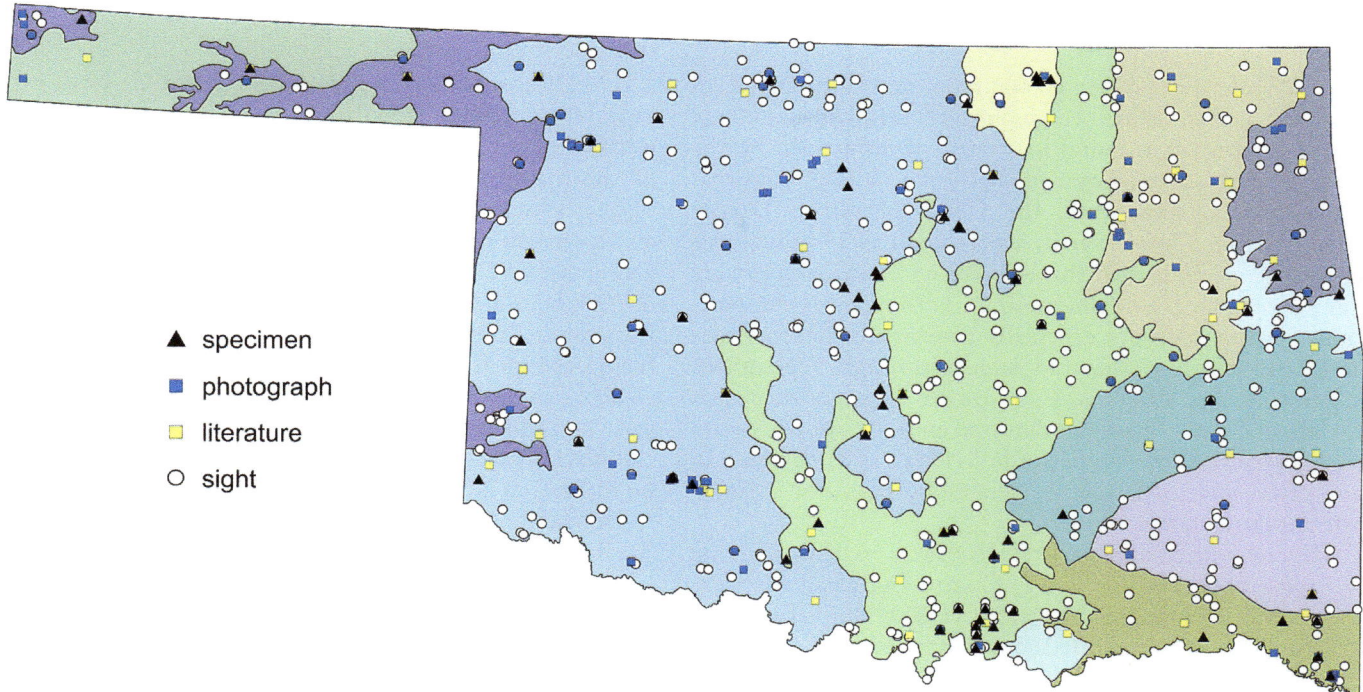

Figure 19.161 Documentation level of supporting records for *Tramea lacerata*, the Black Saddlebags, in Oklahoma. Levels include specimen, archived photograph, literature reference, or sight record. Although sight records lack documentation, only those we feel are credible are plotted.

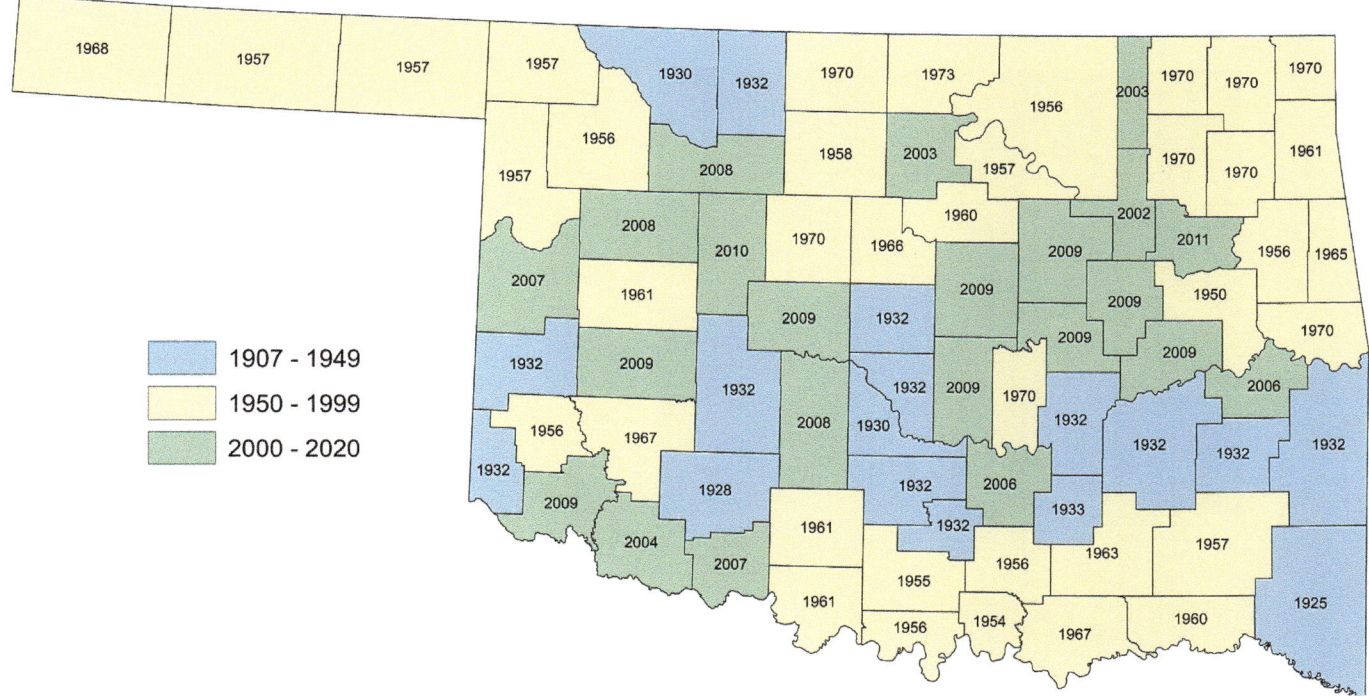

Figure 19.162 Counties in which *Tramea lacerata*, the Black Saddlebags, are known to occur in Oklahoma.

PANTALA FLAVESCENS (FABRICIUS, 1798) – WANDERING GLIDER

The Wandering Glider is common and widespread in Oklahoma, where it has been reported from all 77 Oklahoma counties. It is often found in swarms accompanied by the Spot-winged Glider (*P. hymenaea*), Black Saddlebags (*Tramea lacerata*), and, sometimes, the Common Green Darner (*Anax junius*). These swarms are as small as 25 individuals or as high as 300, with the Wandering Glider comprising anywhere from 5% to almost 100% of the species composition. Such swarms are especially common in late summer and early autumn (i.e., mid-August through September), suggesting that all of these species are migratory through Oklahoma, but we have not detected an association between seasonality and the number of individuals of *P. flavescens* in any given swarm. Likewise, geography does not seem to affect swarm composition.

Pantala flavescens was first recorded in the state on 30 September 1907, when F Collins collected 2♂ near Henryetta, Okmulgee County (UMMZ, Williamson 1914b). It has been recorded regularly since, with an apparent increase in frequency in the recent era. Bick and Bick (1957:8) said that the species "was neither as frequent nor as abundant as *P. hymenea* [sic]." The opposite now holds, in that we have compiled and personally recorded the Wandering Glider both more often and in larger numbers than its congener (Figure 19.164). We have no reason to doubt the Bicks' conclusion, which jibes with Kennedy's (1917:143) assessment for Kansas, where he noted the Wandering Glider in only one county in the far southwestern part of the state but noted four scattered records of the Spot-winged Glider. Allison (1921), in his study in southeastern Kansas, indicated that both species were common, but he gave no indication of relative abundance. Cringan's (1979:18) assessment mirrors ours for present-day Oklahoma: in a year-round study in southeastern Kansas, he found that the Wandering Glider was regular from late June to mid-September, whereas the Spot-winged Glider was only recorded for a short time in late June and was "infrequently observed." Huggins et al. (1976) indicated that both species were found across Kansas but gave no indication of relative abundance.

In Oklahoma, the Wandering Glider flies from mid-April to late November, which is closest to the date range known for Louisiana; but outside of Texas, where the species flies year-round, the seasonality elsewhere in the region does not diverge much from Oklahoma's (OK: 12 April 2014, 1♂, Little River NWR, and 30 November 2016, 1U, Red Slough, DA, both McCurtain County; AR: 7 May–1 November, Harp and Rickett 1977; CO: 26 May–14 September, OC 376552, 425551; KS: 21 May–late October, SEMC, Allison 1921; LA: 13 April–5 December, Mauffray 2014; MO: 21 May–12 November, Sims 2012; NM: May–October, Paulson 2009; TX: Abbott and Lasley 2019).

Figure 19.163 *Pantala flavescens*, the Wandering Glider, A) ♂, Fort Sill, Comanche County, 29 August 2011 (©Victor W Fazio III, OC 332314). B) ♀, Sunset Lake in Guymon, Texas County, 18 July 2017 (©Tony Leukering, OC 467411), in a far more typical in-flight view. They are called gliders for a reason.

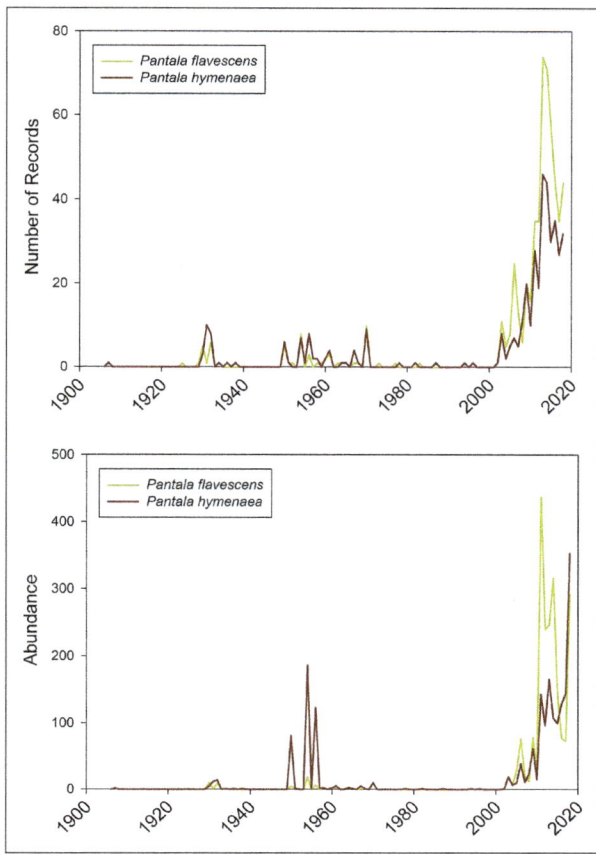

Figure 19.164 Changes in record numbers and abundance *Pantala flavescens* and *P. hymenaea*, the Wandering and Spot-winged Gliders, through time.

In Indiana and Montreal, Quebec, *Pantala flavescens* nymphs are known to die over the winter, and so those populations are maintained by migrating adults coming back into those areas (Landwer and Sites 2010). It is not known if this is also the case with the species in Oklahoma because, remarkably, no nymphs have been collected or at least positively identified as such. The latest date of emergence known for Oklahoma is 22 September (in 2013, 2 unsexed tenerals, Black Mesa area, Cimarron County).

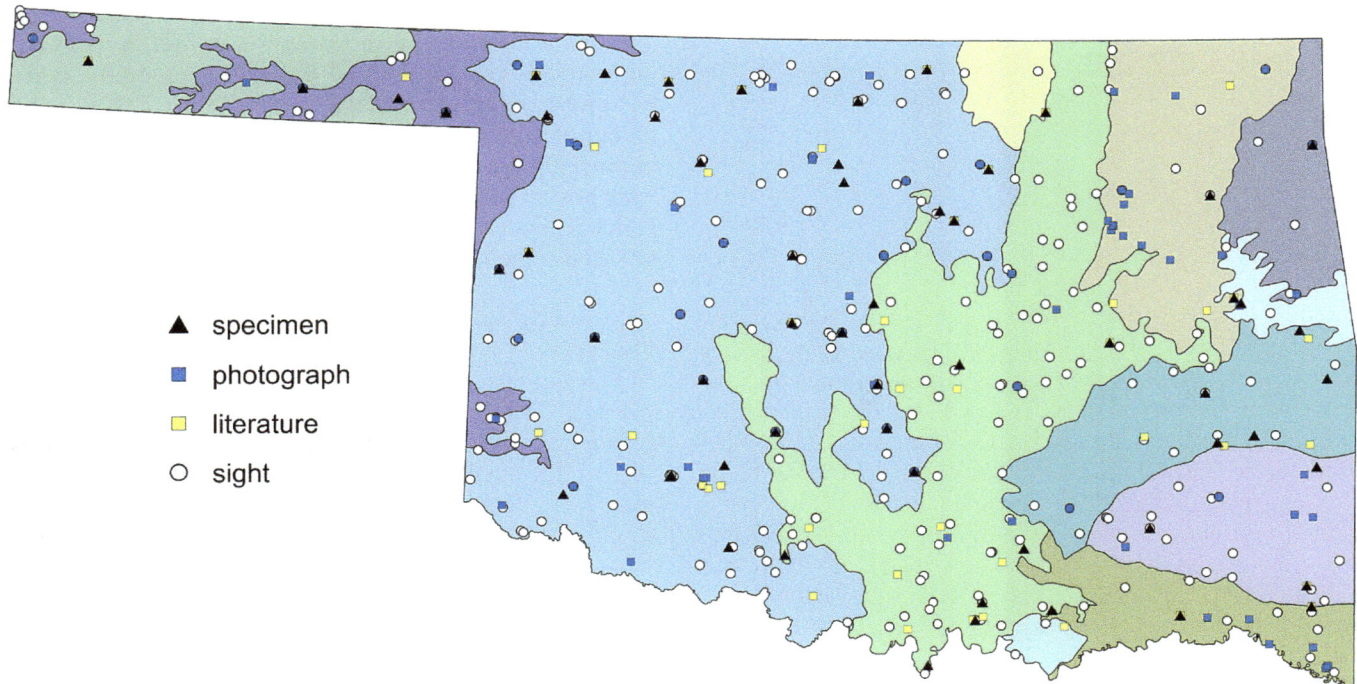

Figure 19.165 Documentation level of supporting records for *Pantala flavescens*, the Wandering Glider, in Oklahoma. Levels include specimen, archived photograph, literature reference, or sight record. Although sight records lack documentation, only those we feel are credible are plotted.

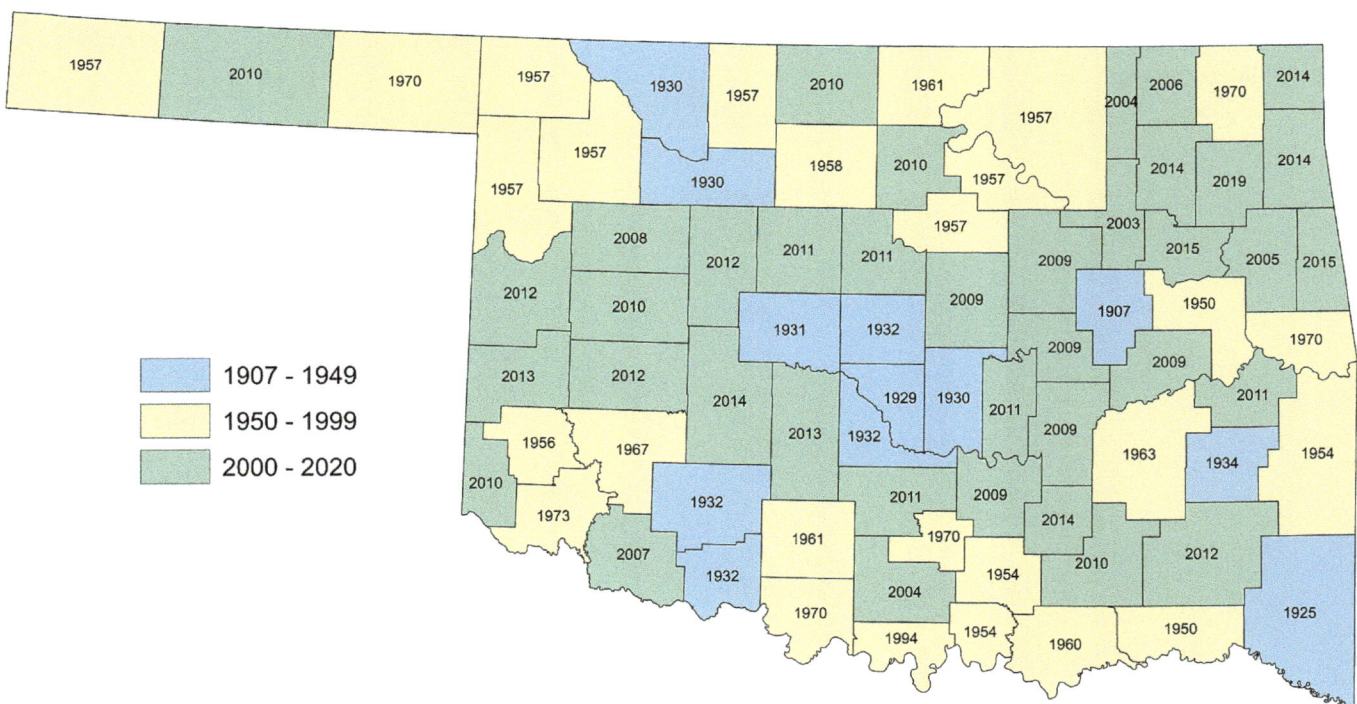

Figure 19.166 Counties in which *Pantala flavescens*, the Wandering Glider, are known to occur in Oklahoma.

PANTALA HYMENAEA (SAY, 1839) – SPOT-WINGED GLIDER

In Oklahoma, *Pantala hymenaea* is as widespread, but not as common, as *P. flavescens* (Wandering Glider; see that species for discussion of its and *P. hymenaea*'s status in the region). Its status may have changed over the past half-century, in that Bick and Bick (1957) remarked that *P. hymenaea* "was the only odonate frequently observed in swarms," a statement no longer true (Figure 19.164). It may be that the Bicks' observations were biased because they were in Oklahoma only during mid-summer, not during the periods of migration. If the Bicks' had conducted surveys in the state during the late summer and early autumn, then we venture they would have noticed that the Spot-winged Glider often swarms alongside the Wandering Glider, Black Saddlebags (*Tramea lacerata*), and, often, Common Green Darner (*Anax junius*). When present in swarms, *P. hymenaea* can number as few as 10 to more than 200 individuals, comprising anywhere from 20% to 90% of a given swarm. The highest count of the species that we have estimated in a swarm was at Lake Overholser, where on 8 August 2009 we had a 150 on the Oklahoma County side of the lake and 230 on the Canadian County side. These swarms were made up almost entirely of *P. hymenaea* (Oklahoma County swarm had 10 each of *P. flavescens* and *Tramea lacerata*; Canadian County swarm had no *P. flavescens* and 50 *T. lacerata*). As with the Wandering Glider, we cannot see a pattern in how many individuals there are in any given swarm—sometimes there are more Spot-winged Gliders in the swarm, other times there are more Wandering Gliders. We believe that these swarms indicate that the species is migratory through Oklahoma.

The Spot-winged Glider was first recorded in the state when F Collins collected a ♀ near Henryetta, Okmulgee County, on 29 September 1907 (UMMZ; Williamson 1914b). Its phenology spans most of the typical odonate season in Oklahoma, from mid-April to late October (10 April 2012, DA, Red Slough WMA, McCurtain County, OC 374371; 22 October 2017, Norman, Cleveland County, 1♂, EA Hjalmarson, OC 474274). Within the remainder of the region, excluding the species' year-round presence in Texas, Oklahoma's flight season begins the earliest.

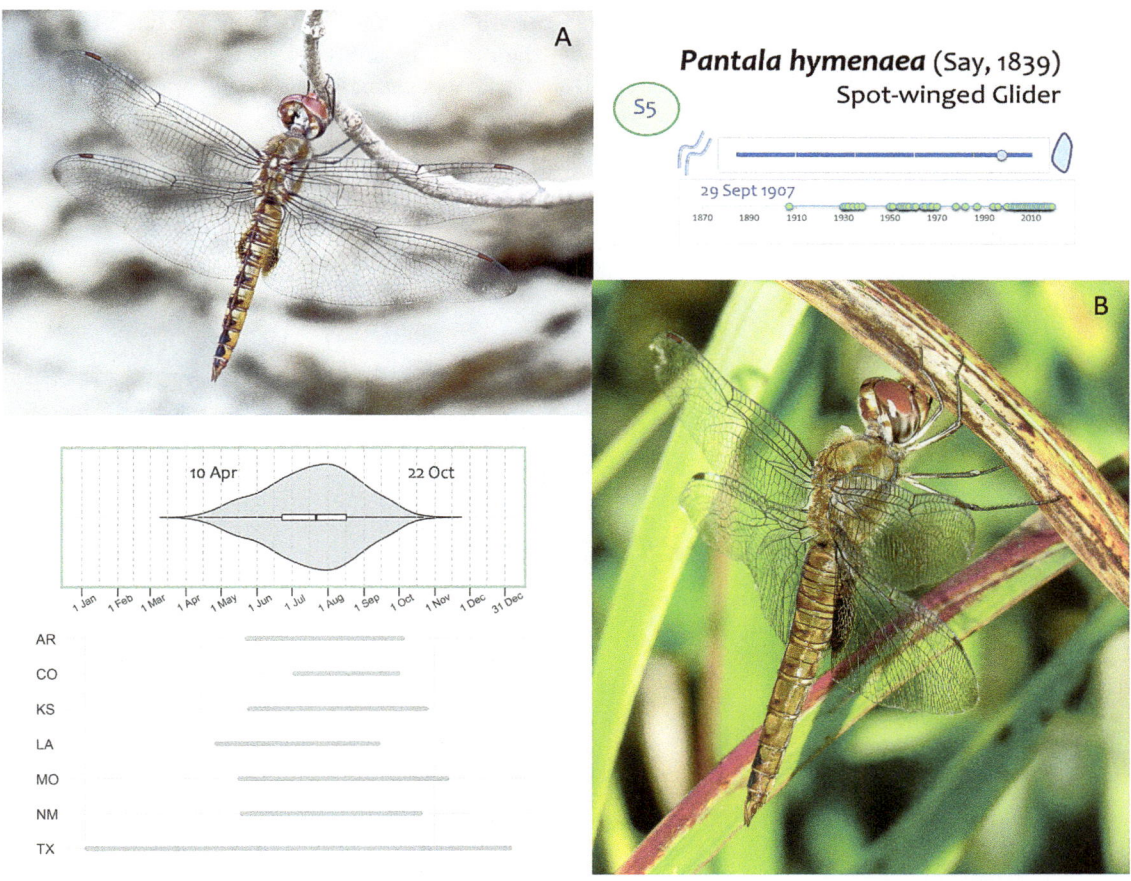

Figure 19.167 *Pantala hymenaea*, the Spot-winged Glider, A) ♂, Kaw Lake, Kay County, 10 July 2016 (©Bill Carrell, OC 449043). B) ♀, Fobb Bottom WMA, Marshall County, 30 September 2016 (©Bill Carrell, OC 456620).

And although the flight seasons of the two *Pantala* gliders begin concurrently, the Spot-winged ceases to fly earlier in Oklahoma and regionwide, with the latest date being in early November from Missouri (AR: 18–21 May–31 September, Harp 2006, Harp and Rickett 1977; CO: 29 June–26 September, OC 423855, 334507; KS: 21 May–late October, SEMC, Allison 1921; LA: 23 April–9 September, OC 462248, Mauffray 2014; MO: 13 May–7 November, Trial 2005; NM: May–October, Paulson 2009; TX: Abbott and Lasley 2019).

We have one teneral record for the species in Oklahoma that is also one of two accounts we have that indicate the species has diapause eggs, has a remarkably short larval life span, or both. In July 2017, Thomas Parr, an aquatic biologist with the Oklahoma Biological Survey, asked BS-P to identify a nymph that he had taken from a mesocosm tank (an outdoor experimental natural environment replication tank). She identified the specimen as a probable *Pantala hymenaea*. Later, on 13 July, Parr sent BS-P photos of a recently emerged *P. hymenaea* (on file, OOP). Parr (*in litt*) said that the mesocosm tanks had been completely dry between October 2016 and late May 2017. He began flushing them that May, finally filling the tanks on 7 June. In mid-June, he noted odonate larvae (unspecified). This time frame indicates that, whether new eggs were laid or there were already eggs in diapause within the gravel, eggs hatched, nymphs developed, and emergence happened in just over a

month's time. This account is reminiscent of one from 1950 reported by Bick (1951:179–180). He reasoned that the species must have "a remarkably short nymphal period" because a concrete swimming pool where he was working was filled partially on 26 June, flushed out, then filled on 29 June, and by 27 July he collected "numerous apparently mature nymphs." By 1 August "numerous last instar exuviae were taken clinging to the concrete sides about three feet above the water level." These anecdotal data suggest that the Spot-winged Glider has an even shorter developmental period than the <2 months for the Wandering Glider (*P. flavescens*; Bick 1951; Landwer and Sites 2010). Only a handful of *P. hymenaea* nymphs have been collected in Oklahoma. Two were reared by GH Bick, but he did not indicate length of larval period. The other specimens were collected between 20 June and 1 August. Bick's collection of 16 exuviae on 1 August (Bick 1951) is the latest date we have documentation for of emergence but we doubt that is truly the latest emergence for the species in Oklahoma.

On a final tragic note, both species of *Pantala* will oviposit on surfaces not conducive to successful breeding, such as sidewalks and car roofs (e.g., video taken by BS-P of ovipositing on a car around 10:00 am on 24 July 2018 in an urban neighborhood in Norman, Cleveland County; later in the morning, she witnessed another ♀ ovipositing on a car roof in a parking lot with what appeared to be a male guarding her; Appendix C).

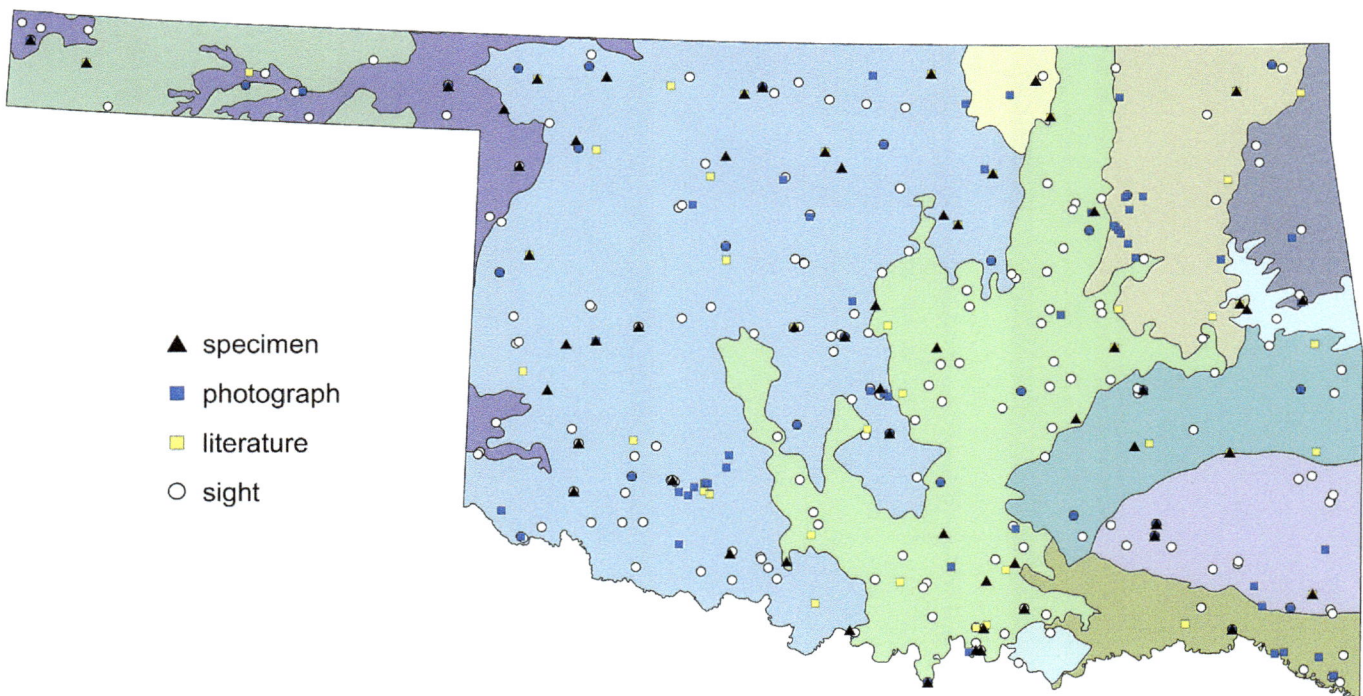

▲ specimen

■ photograph

▢ literature

○ sight

Figure 19.168 Documentation level of supporting records for *Pantala hymenaea*, the Spot-winged Glider, in Oklahoma. Levels include specimen, archived photograph, literature reference, or sight record. Although sight records lack documentation, only those we feel are credible are plotted.

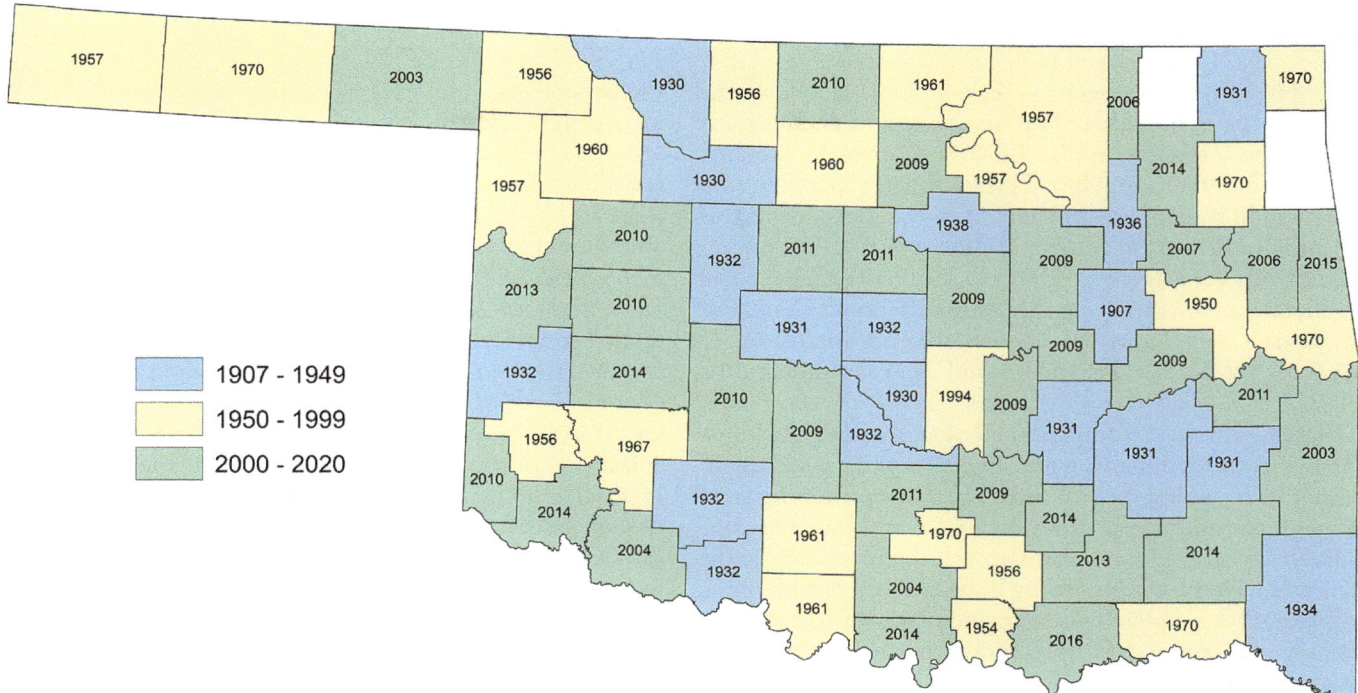

Figure 19.169 Counties in which *Pantala hymenaea*, the Spot-winged Glider, are known to occur in Oklahoma.

MACRODIPLAX BALTEATA (HAGEN, 1861) – MARL PENNANT

The Marl Pennant appears to have colonized Oklahoma chiefly in the past decade. The first record of an adult was reported by VWF from Fort Sill MR, Comanche County, on 12 July 2009 (OC 313904). This lone ♂ was "perched atop a 1.4-m high stalk of grass meters from the shoreline of this very small (30×50 m) muddy shallow pond." The species was not reported again in 2009. In 2010 VWF again had the species, on 2 occasions, this time in Jackson County at the Jackson Salt Plains. He encountered 3♂ and 6♀ on 25 July (OC 321133–321134) and 1♂ and 5♀ on 9 August (OC 321897).

Either 2011 was an exceptional year for the species or VWF was exceptionally determined to and proficient in finding it: he added it to 13 more counties across the state (through the southwest, the northwest, and into the panhandle). Of the 149 immatures and adults reported for the year, all but 8 were reported by him; the others were by T Kompier (OC 333101). In 2012 the numbers declined somewhat ($n = 98$), and it was only added to one other county. That year we had 10♂ and 6♀ (3 pairs) at Altus City Lake, Jackson County, where BS-P captured the first specimen for

Oklahoma (1♂, 7 July, SP 314). Beginning in 2013 records have trended downward, although in that year the species was found in 3 new counties, bringing the total number of counties with documented records to 19. We know of at least 30 localities at which the species has been recorded, and we currently have >70 records for the state (Figures 19.171 and 19.172).

Records of *Macrodiplax balteata* for Oklahoma have tended to involve only a few individuals at a time, although counts have reached double digits multiple times, peaking at 47 and 57 adults at a time (both VWF records from Altus City Lake, Jackson County, 31 July 2012, OC 378253; 19♂, 33♀, 5U, 9 July 2011, OC 329576–329577). The latter record included two ovipositing pairs. We have one other record of ovipositing: a single ♀ and a pair ovipositing at Artesian Beach Park in Gage, Ellis County, 13 July 2014. There was a total of 12♂ and 5♀, including 2 pairs, on that day at the park. Other indications of breeding for the state include one teneral record and one nymphal specimen. The teneral record is also from Artesian Beach Park, but from the year prior, on 9 July. We also counted 2♂ and 4♀ adults on that day.

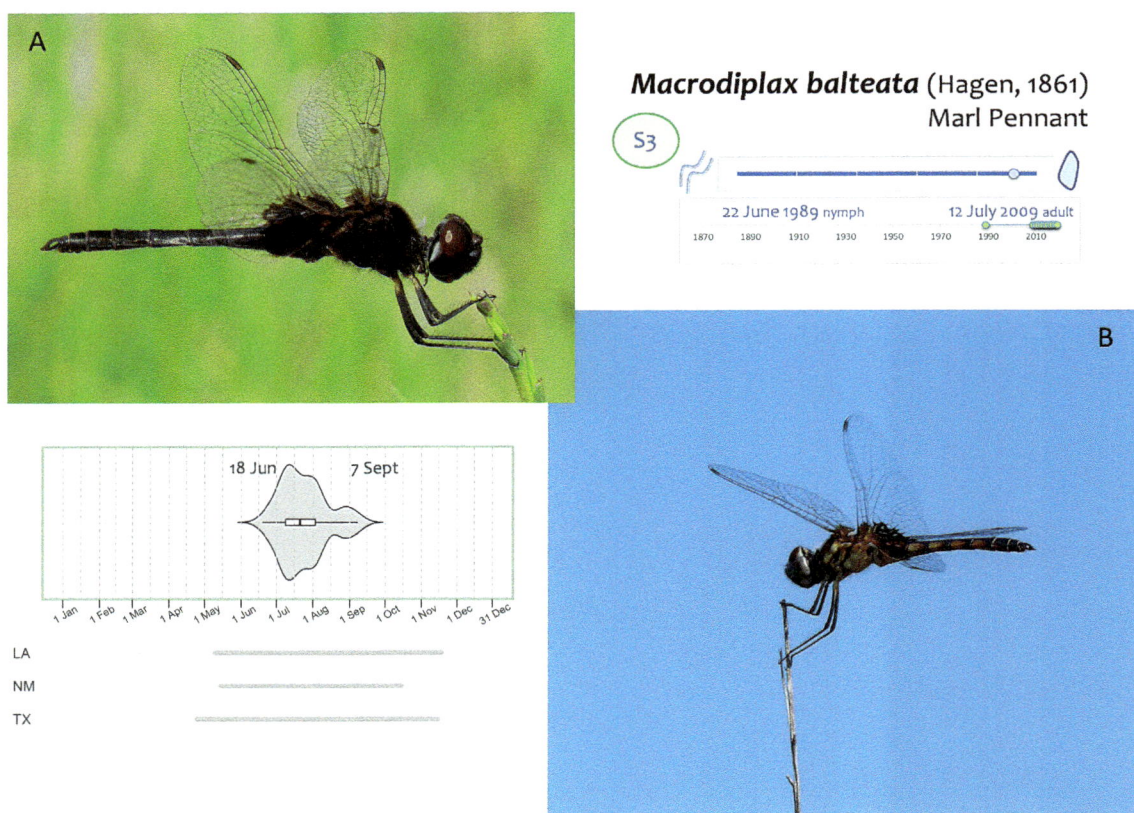

Figure 19.170 *Macrodiplax balteata*, the Marl Pennant, A) ♂, Crystal Beach in Woodward, Woodward County, 4 August 2013 (©Bill Carrell, OC 402800). B) ♀, Fort Sill, Comanche County, 12 July 2009 (©Victor W Fazio III, OC 313904). At the time this record was thought to be Oklahoma's first, but a nymph from two decades early has since surfaced. This ♀ nonetheless provided the first record of an adult; more important, it alerted us to the species' occurrence in the state.

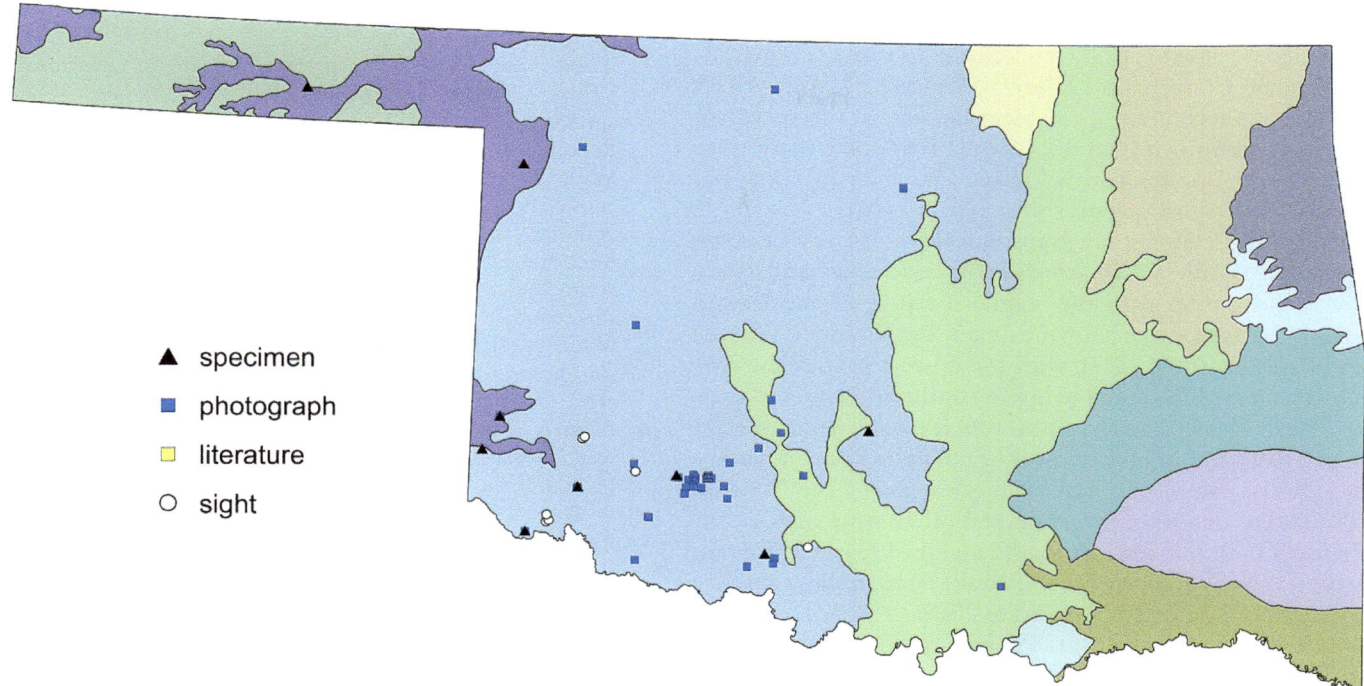

Figure 19.171 Documentation level of supporting records for *Macrodiplax balteata*, the Marl Pennant, in Oklahoma.

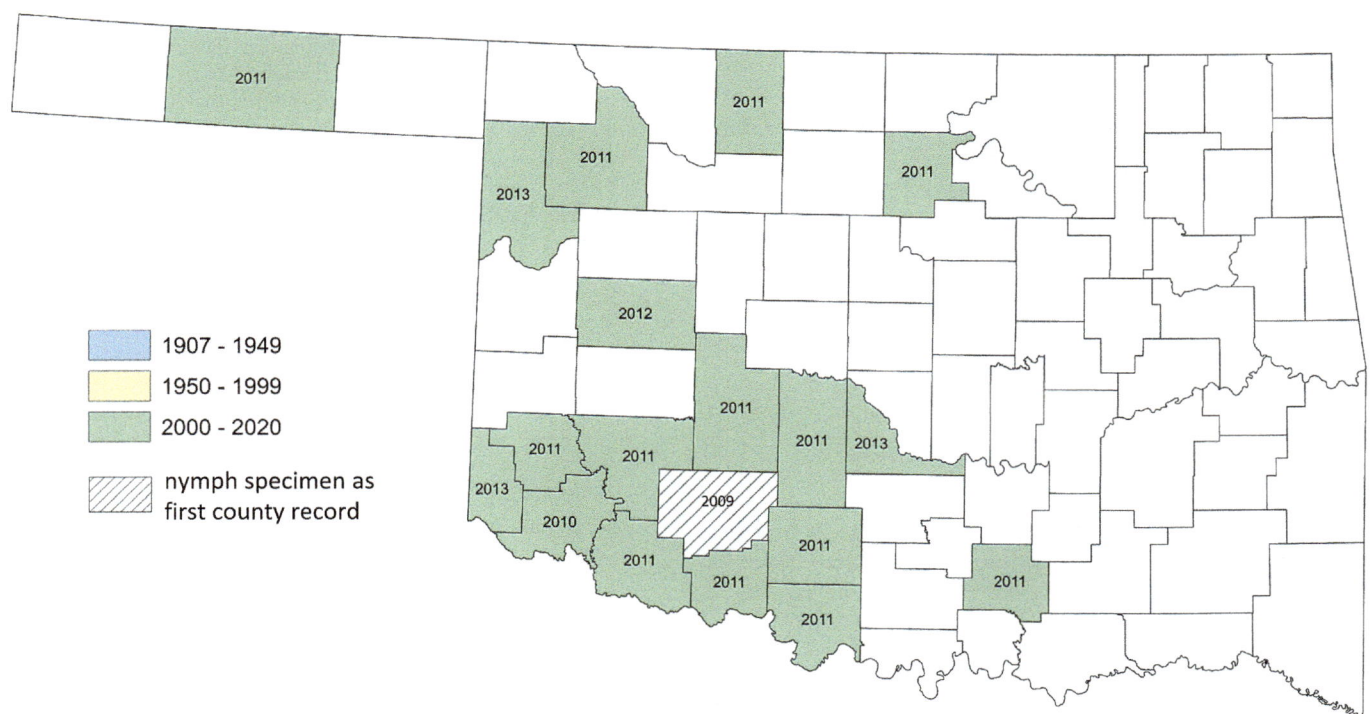

Figure 19.172 Counties in which *Macrodiplax balteata*, the Marl Pennant, are known to occur in Oklahoma. The first record of the species for Comanche County came in 1989 when a nymph was collected (UCO 5180). The first adult record for the county came 20 years later.

We had considered dubious the identification of a nymph collected on West Cache Creek in the Wichita Mountains, Comanche County, on 22 June 1989 (coll: "KWF," UCO 5180), but we confirmed the identification of the specimen when we examined it in 2015. Until that confirmation we thought the species had colonized the state in 2009, but clearly the first occurrences were much earlier.

The Marl Pennant is a Neotropical species, with a range extending south to Venezuela, that has been principally previously confined to the southern fringe of the United States. Its range was shown as such by Donnelly (2004b) and Paulson (2009, 2011), with the nearest part of the species' range to Oklahoma being in southern Texas and eastern New Mexico. In the same timeframe that the species began to appear in Oklahoma it began to be found in northern Texas. Given the extensive coverage of the Wichita Mountains area since the 1930s and other areas such as Salt Plains NWR, Jackson Salt Plains, and the Quartz Mountain area, it is hard to believe that if adults were in the state they had not been detected previous to 2009.

We have found the Marl Pennant at lakes and ponds of various sizes, multiples of which have been spring-fed. A distinct majority of occurrences are from localities with a saline or alkaline chemistry, enough so that the Desert Forktail (*Ischnura barberi*) often co-occurs at the same sites. The pennant perches conspicuously atop bare trees and shrubs, usually near shorelines but occasionally well away from shore or from water. Males will patrol over water, acting much like a small baskettail (*Tetragoneuria*-type *Epitheca*), the flight season of which usually has ended before the first pennants appear. Even though *M. balteata* is reported as early as May elsewhere in our region, in Oklahoma the species is known to fly from mid-June to early September (18 June 2016, 1♂, Comanche Lake, Stephens County, MAP and 7 September 2012, 1♂, 11U, Altus City Lake, Jackson County, VWF, OC 381602). In Texas the species flies from late April to mid-November (Abbott and Lasley 2019). In New Mexico and Louisiana, it begins to fly in May and is reported until October and November, respectively (NM: Paulson 2009; LA: 10 May–18 November, Mauffray 2014).

NOTES

1. Much work still needs to be done to tease out taxonomic breaks within Libellulidae, particularly at the subfamily level. Consequently, named subfamilies do not necessarily have the morphological or phylogenetic support that is ideal to sustain their use. That lack of support, and sometimes conflicting results, lead Needham et al. (2014, and earlier editions), for example, to not discuss subfamilies in any great length.

2. See the species account for the Flame Skimmer (*Libellula saturata*), particularly the endnote (number six, below), for a discussion why this species is generally considered a lentic species even though it plots closer to the lotic end of the spectrum on the lenticity scale.

3. The specimen card indicated that it was collected on 4 November, but we assumed this was a misreading of the original label (no longer associated with the specimen) that probably had the date as 4/11.

4. There is one report of the species flying later in Oklahoma. The report comes from Beckemeyer (2002e) and regards a specimen from Oxley Nature Center in Tulsa, Tulsa County. Beckemeyer reported the collection date as 18–19 July 2002. We have long suspected that the date was in error, being from June rather than July. Unfortunately, when BS-P visited the collection she learned that the specimen had been destroyed by dermestid beetles so it, and its label, was thrown away. We continue to feel the July date is in error, so we have set aside this report.

5. Mauffray (2014) reported 25 August as the late date for the species in Louisiana but we believe that date to be in error.

6. *Libellula saturata* falls toward the lotic end of the lenticity scale; however, it should be noted that although often found on creeks, the species really does tend to stay nearer to pools or slow-moving portions of the creek. That is not to say that it will not be found where water is flowing, it just tends to stay in areas that mimic lentic habitat. We suspect that *L. saturata*, as a desert-adapted species, has evolved to cope with limited water availability, thereby using flowing streams when that is what is available but being most accustomed to lentic conditions brought on by seasonal drying.

Libellula saturata's rating as more of a lotic species is an issue that we are working to remedy with the lenticity calculations. The scale is not so refined, as yet, to be able to distinguish between a creek that flows and a creek that is pooled and essentially not flowing; basically, it currently cannot differentiate linearity (e.g., creek width) and flow rate, so species occurring on the latter are still labeled as lotic. For the most part, it seems that flaw does not present a problem when determining the lenticity of most species but we have a handful of species for which we have discovered this issue. For consistency, we chose not to make manual corrections; instead we left the scores as they were calculated by the algorithm. Eventually, we will refine the scaling to better capture types of lotic habitat rather than using a blanket classification.

7. In 2019, a ♂ was photographed in hand at Jackson Salt Plains on 8 June (MAP, OC 496475), another ♂ and a ♀ were photographed there on 13 and 14 July, respectively (BC, OC 499470, 499474), and 2♀ were watched as they oviposited within 1 m of each other on 19 July (MAP).

8. The Stafford County, Kansas, record comes from an alkaline salt marsh at Quivira NWR where "a number of these odes [were] flying on a cool cloudy day" on 24 June 2017 (OC 465067; Dave Rogers). A more recent record (19 June 2018) for Kansas came from Comanche County ("above Big Gyp Cave and stream running out of it"; OC 483101; Daniel Johnson), although that location falls within the Southwestern Tablelands ecoregion. We do not know the basis of the Dot Map Project (Donnelly 2004b) record for Jack County, Texas, the coordinates of which from Odonata Central plot as a centroid to the county and thus within the Cross Timbers ecoregion. However, other parts of the county fall within the Great Plains ecoregions, so we place that record broadly with the Great Plains.

9. Although see the *Libellula cyanea* account regarding possible confusion between these two species.

10. A lone ♀, photographed by VWF on 2 August 2010 at 4 km SW of Cyril, Caddo County (OC 321494), was long thought to be another record of the species for the Central Great Plains ecoregion. However, when we first reviewed the record in 2019, we determined that it was not *L. flavida* (species could not be determined from photos, but it is definitely not *L. flavida*).

11. OC 313341, a female from Kerr County, also in the Edwards Plateau of Texas, is actually an aberrant *Libellula comanche* (see that account), not *L. flavida* as it was originally submitted and accepted as by JC Abbott.

12. Interestingly, DR Paulson (*in litt*) remarked after viewing photographs of the specimen, "Michael, as I think you're implying, you might well consider it a putative hybrid between *cyanea* and *flavida*, as it is somewhat intermediate. Maybe you can't rule out *comanche* entirely, either, as I think either *comanche* or *cyanea* could have mated with *flavida* to produce your specimen. Why are they so darned similar? But how to know if it is a hybrid and between which species?"

13. Along with the Oklahoma records (SP 1699, 2878), we consider the following OC records to be of aberrant *Libellula comanche* females that have an amber or brown wash along the entirety of the front of the wing (i.e., approximate area between the costal vein and radial/medial veins from the base of the wing to the pterostigma). From California: OC 5243 (Kern County). From Nevada: OC 314191, 471426, 480960 (Clark County). From Texas: OC 481022 (Denton County); OC 313341 (Kerr County); OC 328167, 401596, 445418, 445629 (Williamson County).

14. Williamson recorded 46♂ and 4♀ on 3–4 June 1907 near Wister. He collected 13♂ and 3♀, all of which are at UMMZ.

15. Allison (1921:52, 58) reported *Libellula cyanea* for Kansas from "June on" and, in his seasonality graph, showed it flying until late October. Harp and Rickett (1977) reported a late date of 23 October for adults in Arkansas. Trial (2005:130) reported *L. cyanea* adults in Missouri for 14–31 May, 1–30 June, 1–29 July, 1–31 August, 14 September, and 15 October. We suspect reports after early, perhaps mid-August, are in error. Interestingly, dates reported by Trial (2005:131; 18 May–11 August) for *L. flavida* are more in line with the seasonality of *L. cyanea*, making us wonder if the date charts were transposed in that publication. In Oklahoma, previous late dates have been expunged. A sight record from the JT Nickel Preserve, Cherokee County, for 30 August (4U, BC, from 2006) is now considered dubious. And, we reported a record of *L. cyanea* for 24 August, but later realized we mis-identified a female *L. flavida* (OC 426357).

16. Measurements of SP 2438 are given with Needham et al.'s (2014) ranges presented parenthetically; all are in mm. TL: 40 (50–52), ab: 27 (32–34), HW: 34.5 (36–42). SP 2855 is not aberrant in its measurements. Also note that the wings of SP 1311, an immature male, suggests that in maturity it would have exhibited a similar pattern to that of SP 2855.

17. Rosche et al. (2008) reported 1.85–2.17 in (47–55.1 mm) for males and 1.85–2.0 in (47–50.8 mm) for females.

18. Specimens in SP collection from Oklahoma and Texas (*n* = 11, two of which are ♀).

19. We have been unable to find Allison's specimens; visits to PSU and SEMC, for example, have not turned them up.

20. The ♀ (SP 2870) also occasionally harassed the ♂ (SP 2869), so much so that he flew off a couple of times, but then he returned to "oviposit" again (manmade pond near Pine Mountain Spring, McCurtain County, 1 August 2019).

21. Account authored solely by MAP.

22. This lake is officially called "Lake Jap Beaver." We chose to modify it in the text because we find it to be an unfortunate (and tone deaf) name, in much need of context. The lake is named after Jewell Jasper "Jap" Beaver (1922–1964), a long-time game warden in the area (Oklahoma Wildlife Conservation Commission meeting notes, 2 March 1964, p. 9; *Waurika News Journal*, installment of "Woods and Water" column by Mike Gaines, 25 May 2018). We know not when the nickname "Jap" was bestowed upon Mr. Beaver but we hope it was prior to the derogatory use of the term (it is a shame that he did not go by "Jasp"). Nonetheless, without context, we understand why USGS chose to rename it, albeit in error, to "Japanese Beaver Lake." Google incorrectly calls Lake Jap Beaver "Waurika Lake," presumably because that was the original name of the lake (1953–1964) prior to the opening, in 1977, of the large reservoir to the north that now wields that name.

23. These nymphs were examined by BS-P and thought to be *Celithemis fasciata*, cf. We have presumed these to be the basis for the record of *C. verna* reported by Bass (1990).

24. Documented in 2014, but reported by Donnelly (2004b), although the basis for that record is unknown.

25. There is another specimen at CUIC that is labeled as having been determined by AE Pritchard but its other label just has "Oklahoma" on it (i.e., no location, date, or collector indicated). This may be a paratype that has lost its corresponding data.

26. As such, many records labeled as "*simplicicollis*" (in Odonata Central and museums, for example) may actually be *E. collocata* × *E. simplicicollis* or *E. simplicicollis* × *E. collocata* hybrids.

27. Trial (2005) and Sims (2012) reported the species as adults occuring from 1 May through the winter to 23 January. We have a feeling that the winter reports are actually of nymphs, as the species' persistence as adults during the winter at such a latitude would be extraordinary. We therefore consider the winter adult reports dubious until documentation can be had.

28. Allospecies form by way of allopatric speciation, also called vicariant or geographic speciation, in which a species is broken into subpopulations because a geographical barrier arises (e.g., continental separation, mountains form, or a river changes course) that causes separation of individuals. That separation excludes gene flow between subpopulations and eventually results in those subpopulations becoming distinct genetically and thus forming a new species.

29. The acceptance of Bird's record is a point on which the authors disagree. MAP feels it ought to be set aside as indeterminate because *Erythrodiplax basifusca*, the Plateau Dragonlet, cannot be eliminated. His argument is that "we ought to bear in mind that at the time there was no such species, meaning it was natural to assume, post taxonomic split, that it was *E. minuscula* because that is what was known elsewhere in Oklahoma." He sees no reason an *Erythrodiplax* in the Wichita Mountains was not *E. basifusca*, and without a specimen we cannot say one way or another. Further, we do not know when Bick examined the specimen, other than it was sometime prior to 1968 (Figure 19.97). And, we do not know the age of the specimen: if this ♂ was an adult, then coloration

alone would lend confidence that a correct determination was made, but if this ♂ was an immature, then it could easily have been confused with *E. basifusca* (see color plates in Abbott 2005). He is also concerned about accepting the record because "despite a ton of survey effort" in the county there has yet to be another record of *E. minuscula*. In summary, he thinks it is clear then, that *E. basifusca* cannot be eliminated and that this "seasonally early and geographically far-flung" record is less than ideal.

BS-P notes that Bird's record is three days earlier than the next earliest record of 29 April and that April is well within the species' seasonality elsewhere. She acknowledges that the geographical outlier issue is trickier because that is left up to happenstance and forces we do not fully understand. Yes, Comanche County makes for an outlying record from *Erythrodiplax minuscula*'s more "normal" distribution in southeastern Oklahoma, but we ought to remember that the nearest record of *E. minuscula* is from only 178 km to the southeast where Bick collected it in 1968 in Marshall County. The nearest *E. basifusca* record (reported once for Oklahoma) is 112 km to the west. Neither of those distances are much for most odonates to traverse, so distance, in the case could be a toss-up between *E. minuscula* and *E. basifusca*. But a vagrant *E. minuscula* in Comanche County would be just one of a good number of eastern species that have been recorded once, such as *Anax longipes* collected in 1961, a pair of *Gomphurus ozarkensis* collected in 2002, a nymph and adult of *Hagenius brevistylus* collected in 2002, and two adult *Libellula semifasciata* reported in 2008. And then there are disjunct populations of eastern species in the Wichita Mountains area, such as *Dromogomphus spinosus* and *Celithemis elisa*. Comanche County also holds disjunct records of western species, such as *Enallagma praevarum*, *Ischnura demorsa*, *I. perparva*, *Argia leonorae*, and *Amphiagrion abbreviatum*. The Wichita Mountains area has many species of plants and animals that are disjunct, as one-off records or as populations, so it is not unrealistic in the slightest to think that *E. minuscula* made its way there.

In terms of species identification, BS-P has chosen to trust that because both Bird and Bick examined the specimen and that they came to the same determination; the identification is probably correct and ought to stand as an accepted record. But a process of elimination is arguably worthwhile. She feels that *Erythrodiplax umbrata* or *E. berenice* would have been correctly eliminated. The coloration, at least as a fresh specimen, probably would have immediately eliminated those species, but one could argue that by Bick's time, especially if the specimen was in fluid, which was Bird's preferred preservation technique, the coloration could have been obscured. Nonetheless, BS-P trusts that with such an unusual specimen both Bird and Bick would have taken the time to key the specimen. The taxonomic keys in Needham and Heywood (1929) and Needham and Westfall (1955) would have made Bird and Bick readily dispense with *E. umbrata* and *E. berenice*. That leaves the fourth *Erythrodiplax* known for Oklahoma, *E. basifusca*. It was described by 1930 but it was not included in Needham and Heywood (1929); therefore Bird would have needed to track down the type description, which he probably would not have thought to do, given that the species was then called *Trithemis basifusca* and it was published in an article called *The Odonata of Baja California, Mexico*

(Calvert 1895). Nevertheless, Bick would have had a way to readily diagnose *E. basifusca* under one of its former names, *E. connata* (Needham and Westfall 1955). Consequently, unless both Bird and Bick completely bollixed the identification, say, misidentifying the genus, then by eliminating other *Erythrodiplax* and putting some faith in Bird's and Bick's keying ability, alas, BS-P came to the decision that although lost, this outlier *E. minuscula* must be just that.

We hope one day that this specimen will be relocated and prove one of us wrong; for now, it is up to the reader to decide the merits of Bird's record.

30. We even examined all the specimens collected on 26 April 1930 and all of the Blue Dashers, *Pachydiplax longipennis*, we could find in the collection, in case it was subsequently mislabeled. OMNH does not have any *Erythrodiplax* from Comanche County.

31. MAP built a Bayesian model to estimate multinomial probabilities of each color class for each species. He calculated estimated probabilities, which yielded a posterior distribution that he then plotted for each color class.

32. Data on cercus color of ♂ *Erythrodiplax minuscula* ($n = 164$) and *E. basifusca* ($n = 36$) were assessed on photographs archived in Odonata Central (years 2005–2018, OC 6211–490103). Color was scored in four categories: white, pale gray, dark gray, and black. An ordinary least squares regression line was fit purely to show the strength and direction of the pattern— we were interested only in a visual depiction of effect size (i.e., because of potential violations of statistical assumptions, test statistics were not interpreted).

33. Also called "Jackson Salt Plains," as it was called during the OU Expedition Era.

34. We do not know the basis of reports for Le Flore and Pittsburg Counties reported by Donnelly (2004b).

35. Odonata Central records from 2014–2017: OC 424649, 1♂, 3 July 2014; OC 432933, 1♀, 10 July 2015; OC 459601, 2♂, 1♀ ovipositing, 5 and 7 July 2016; OC 465621, 1♂, 2 July 2017; OC 468417, 1♂, 20 July 2017. Prather and Prather (2015:48) said of OC 424649 that "Searches by several individuals in the following days failed to locate this or other individuals. Although, the habitat was similar to habitats we have seen for this species in Arizona, this site is 300 km north of available reported records for the Filigree Skimmer (Odonata Central Abbott 2006-2014)." In his comments for some of his records, Eric Eaton indicated that the stream is slated for improvements in 2018, which he thinks may negatively impact the population. However, there were two more reports of the species in July 2018 (OC).

36. We assume that Bird (1932) did not report the species for Custer County because the letter from LK Gloyd, dated 28 April 1932, from which he obtained other erroneous Custer County data (from EB Williamson's label errors) omitted *Sympetrum corruptum*.

37. If this scenario is true, then adult females would lay eggs that either could be in diapause throughout the winter, hatch in spring, and grow rapidly to emerge in summer or the species overwinters as nymphs, delays development during the winter, grows rapidly in spring, and emerges in summer.

38. Incorrectly reported in Patten and Smith-Patten (2013b) as FSCA.

39. Anisopterans collected in Latimer County by RD Bird and WM Fisher in 1931 on 14, 16, 17, 24, and 27 June 1931 and

1 and 11 July (OMNH): *Nasiaeschna pentacantha, Arigomphus lentulus, Dromogomphus spinosus, D. spoliatus, Phanogomphus graslinellus, Macromia illinoiensis, M. pacifica, Neurocordulia xanthosoma, Somatochlora ozarkensis, Celithemis elisa, Dythemis velox, Erythemis simplicicollis, Erythrodiplax umbrata, Libellula incesta, L. luctuosa, L. pulchella, Orthemis ferruginea, Pachydiplax longipennis, Pantala hymenaea, Perithemis tenera,* and *Tramea onusta.* Not only can we not fathom any of these species being mistaken for *Paltothemis lineatipes,* but none of these species were collected on all of the days that Bird reported in his notes, nor do numbers of individuals collected match. All signs point to Bird and Fisher recanting their original determination but failing to annotate as such in Bird's field notes.

40. Needham et al. (2014:443), in their key, also use the labrum to distinguish species. The first key step includes "labrum mostly dark, usually somewhat metallic" versus "usually mostly pale, not metallic." The first description leads to *Brechmorhoga pertinax* and *B. tepeaca* and the second refers to the other *Brechmorhoga* (*mendax, vivax,* and *praecox*).

41. When questioned about the extent of dark metallic on the frons and vertex of *Brechmorhoga mendax* from Texas, JC Abbott examined a series of *B. mendax* and *B. pertinax* from his collection. His response was, "Many specimens in Texas have the top of the frons dark metallic blue (vertex as well). In a few specimens it does extend down on to the front of the frons, but rarely to the extent that you see in *pertinax*. When it does come on to the front of the frons, the sides generally remain pale. In all my *pertinax* specimens, the area where the dark and pale colors come together is more sharply defined than in *mendax*." (JC Abbott, *in litt,* 4 February 2019). In a recent collection of three specimens we took in Texas, one has dark metallic on the vertex and postfrons but the other two have more extensive dark metallic similar to that described above (SP 2800, 2826, 2827).

42. As of 10 October 2019.

SYNONYMS, ERRORS, AND DUBIOUS RECORDS

Herein, we provide taxonomic synonyms (i.e., older scientific names) for Oklahoma Odonata that have been used in publications or written on specimen labels. By no means do we provide a full list of synonyms (see Garrison and von Ellenrieder 2019); rather, we include only those taxa we encountered during museum visits and while poring over the literature. It is important to note, too, that the synonyms presented here are specific to Oklahoma and should not necessarily be applied generally. For instance, *Enallagma pollutum*, in regards to Oklahoma, is a synonym for *E. vesperum*, but that is not the case elsewhere, where *E. pollutum*, the Florida Bluet, is a valid application.

In addition, we report here typographical errors (or "write-o's") encountered in publications and on specimen labels. We provide these errors not to be critical but rather in the hopes of aiding those who are searching for a given taxon but do not realize there are misspellings that may cause one to overlook a specimen reference. And finally, we included species that we feel are misattributions or dubious records. Many of the records have been re-identified through examination of a specimen or were synonymized with another scientific name. Others have no extant documentation, making it impossible to re-examine them, or we felt it unwise to assume the error was simply due to synonymy. We deem such records as dubious.

Species are listed below in alphabetical order by the scientific name used in whatever source(s) we encountered that name. The original scientific name is listed in bold and italics, followed by its common name (if there is one), and then by an equals sign (=) and the current accepted common name for that species. Sometimes the name that follows the equals sign is a synonym, such as when *Rhionaeschna multicolor* follows **Aeshna multicolor**. Other times, such as when *Nasiaeschna pentacantha* follows **Aeshna eremita**, it is not a synonym, but was instead a re-identification. In the case of a synonym, we indicate "syn.," and in the case of a re-identification, we indicate "re-id.,". When a species was split from the older name, we indicated "sep.," to designate that separation. When a species is considered a misidentification but we have not necessarily determined what the re-identification should be, then we labeled the species with "mis-id."

For each of the original scientific names, we generally included the primary sources in which the Oklahoma-specific reference was made, as well as any pertinent details. Nevertheless, because it can generally be assumed that when a name was used in Bird (1932), there is an associated specimen(s) at OMNH and that the same is probably also true for Ortenburger (1926b), we felt citing those publications was sufficient for sourcing the name used, rather than necessitating a list of OMNH specimens. We made a similar assumption for Bick and Bick (1957), except that their specimens are much more dispersed and, as such, it would take too much space to list out all associated specimens.

Aeshna eremita – Lake Darner = re-id., *Nasiaeschna pentacantha* – Cyrano Darner

A nymph collected in Oklahoma County (Tinker Creek, Arcadia Lake, 19 September 1986, D Bass, UCO 5171), was a misidentification as *Aeshna eremita*. We re-identified the specimen as *Nasiaeschna pentacantha*. The nearest confirmed record of *A. eremita* to Oklahoma is in central Colorado.

Aeshna multicolor = syn., *Rhionaeschna multicolor* – Blue-eyed Darner

The older name of *Aeshna multicolor* was used by Bird (1932) and Bick and Bick (1957). The genus name *Rhionaeschna* Förster, 1909, although old, was not accepted for use with this species until fairly recently (von Ellenrieder 2003).

Agrion maculata/maculatum = syn., *Calopteryx maculata* – Ebony Jewelwing

These are older names for *Calopteryx maculata*. Two UMMZ specimen are labeled as *Agrion maculata* and *A. maculatum* was used by Ortenburger (1926b), Bird (1932), Bick (1951), and Bick and Bick (1957).

***Amphiagrion* intermediate**, n. sp.? = syn., *Amphiagrion abbreviatum* – Western Red Damsel

See *Amphiagrion saucium*.

Amphiagrion mesonum, n. sp.? = syn., *Amphiagrion abbreviatum* – Western Red Damsel

See *Amphiagrion saucium*.

Amphiagrion saucium – Eastern Red Damsel = sep., *Amphiagrion abbreviatum* – Western Red Damsel

Species limits in *Amphiagrion* are debatable. The names intermediate (sometimes even italicized, as if a scientific name), *mesonum*, and *saucium* have been applied to Oklahoma specimens, all of which appear to refer to *A. abbreviatum*. See that species account for details.

Anax juneus = typo, *Anax junius* – Common Green Darner

Four specimens (OMNH 2044, 2047, 2049, 2077) were mislabeled with this typographical error.

Anomalagrion hastatum = syn., *Ischnura hastata* – Citrine Forktail

This is an older name for *Ischnura hastata*. It was used by Bird (1932), Bick (1951), and Bick and Bick (1957).

Argia agrioides – California Dancer = re-id, multiple species

Bird (1932) identified some Oklahoma specimens as *A. agrioides*, but those specimens were re-identified as *A. immunda*, the Kiowa Dancer, and *A. nahuana*, the Aztec Dancer. Bick and Bick (1957) used the name *A. agrioides nahuana*, a treatment prior to universal recognition of a species-level split.

Argia intruda = syn., *A. moesta* – Powdered Dancer

Once thought to be a different species from *A. moesta*, it was eventually subsumed. Bird (1932) used both names, probably because both were used by Needham and Heywood (1929). In a letter to EB Williamson, dated 8 December 1932, Bird wrote "I have never been able to satisfactorily separate *moesta* from your *intruda*. There seems to be a complete intergradation."

Argia putrida = syn., *A. moesta* – Powdered Dancer

Older name for *Argia moesta*, used in Ortenburger (1926b). It was also mentioned in Bick and Bick (1957) as the name used by Ortenburger. It was not used by Bird (1932), probably because by that time it was considered a variant of *A. moesta* (Needham and Heywood 1929).

Argia violacea – Violet Dancer = syn., *A. fumipennis* – Variable Dancer

At various times *Argia violacea* has been considered its own taxon, as by Needham and Heywood (1929; and by Ortenburger 1926b, Bird 1932, Bick 1951, and Bick and Bick 1957; Paulson 2004 used the common name Violet Dancer), but more often it is classified as a subspecies of *Argia fumipennis* (see Gloyd 1968a).

Argia vivida – Vivid Dancer = sep., *A. plana* – Springwater Dancer

Argia plana was considered a variety of *A. vivida* from the time it was described (Calvert 1902:96), and at the time Bird (1932) considered the species, *A vivida* var. *plana* was still in use (Needham and Heywood 1929), although inexplicably he chose to use the naked name *A. vivida*. By the Bick/Hornuff era, some still considered *plana* as a variety, or in some cases, as a subspecies. Bick and Bick (1957) included *Argia vivida*

plana and mentioned that thought that *plana* deserved species status. A year later, Gloyd (1958) raised the rank of *plana* to a species, at which it has stood ever since.

Arigomphus pallidus – Gray-green Clubtail = re-id, multiple species

There are two reports of this species for Oklahoma. The first comes from GH Bick's species card for "*Gomphus (Arigomphus) submedianus*," on which he indicated that he collected a ♀ on 5 July 1954 at Willis, Marshall County, that he initially identified as *pallidus* but he re-identified as *submedianus* in August 1970. A larval *Arigomphus pallidus* was reported by Bass (1994) from the Tallgrass Prairie Preserve, Osage County, but it almost certainly was a misidentification of a *A. lentulus* or *A. submedianus*. We have not been able to relocate either of the specimens on which these records were based. The closest to Oklahoma that *A. pallidus* gets to Oklahoma is western Kentucky, as it is not known west of the Mississippi River (Paulson 2011).

Boyeria grafiana – Ocellated Darner = mis-id.

In a letter dated 20 October 1932 to EB Williamson, RD Bird indicated that he had recorded *Boyeria grafiana* for the first time in Oklahoma. The first *Boyeria vinosa* for the state was not found until two years later, so we doubt he misdiagnosed the species. It may be reaching, but because Needham and Heywood (1929:128) described *B. grafiana* as "A handsome brown species with bluish face and side spots …," perhaps Bird initially mistook the ♂ *Aeshna umbrosa* he collected in Norman, Cleveland County, on 9 October 1932 (OMNH 2008) as *B. grafiana*.

Calopteryx maculatum = syn., *Calopteryx maculata* – Ebony Jewelwing

See *Agrion maculata/maculatum* entry above.

Cannacria gravida = syn., *Brachymesia gravida* – Four-spotted Pennant

Bick and Bick (1957) used *Cannacria gravida* because that was the accepted name at the time (Needham and Westfall 1955).

Celithemis monomelaena = syn., *C. fasciata* – Banded Pennant

Bird (1932) used this junior synonym for *Celithemis fasciata*. Bick and Bick (1957) mentioned *C. monomelaena* only to say that they examined the specimen on which they thought Bird's record was based and re-identified it as *C. fasciata*. Bick's notes indicated that when he examined the specimen in 1954, it was "somewhat intermediate between *mono.* and *fasciata*." Bick's notes showed that the specimen he examined was a ♂ from Latimer County that he said Bird collected on 14 July 1931. He also said that MJ Westfall concurred with Bick's re-identification of the specimen as *C. fasciata*. OMNH 161 is probably this specimen, although the OMNH catalog lists the collection year as 1930 and the collector as WM Fisher. It is unclear why Bird chose to determine the specimen as *C. monomelaena* over *C. fasciata*, as both were accepted as species at the time (Needham and Heywood 1929), but neither was known for the region.

Chromagrion* sp. – Aurora Damsel = mid-id., *Ischnura

One nymph (UCO 5255, Stroud Lake, Creek County, 23 July 1997) was originally identified as *Chromagrion* sp. but it was later re-identified as *Ischnura* sp. by BS-P.

***Coenagrion* sp. = re-id, multiple species**

We know of 40 specimens that were originally labeled as *Coenagrion* (UCO 5273, 5285, 7006, 11688, 12547, 12577, 12628, 12652, 12678), from Canadian, Johnston, and Oklahoma Counties. All have been re-identified as definitely or probably *Argia* (*n* = 3), *Enallagma* (*n* = 35), or *Ischnura* (*n* = 2). In addition to the specimens, there are some *Coenagrion* reported in the macroinvertebrate literature. Bass and Potts (2001) reported nymphs of this genus from Boehler Seeps, Atoka County. Bass (1990) reported it at the Wichita Mountains, Comanche County, and Bass (1994) from the Tallgrass Prairie Preserve, Osage County. None of these localities is close to the geographic ranges of any of the three *Coenagrion* ("Eurasian Bluets") species in North America; if a *Coenagrion* were to occur in Oklahoma, the likeliest would be *C. resolutum*, the Taiga Bluet, which would be found in the Oklahoma panhandle. It is likely that the published records are, akin to the specimens we re-identified, *Argia*, *Enallagma*, or *Ischnura*. Unfortunately, specimens that serve as the basis for those reports have not been relocated.

***Cordulegaster obliquus* = syn., *Cordulegaster obliqua* – Arrowhead Spiketail**

Cordulegaster obliquus was used by Bick (1951) and Bick and Bick (1957). The name change is a reconciliation between the gender of the genus and species.

Cordulia* sp. = re-id, probably *Epitheca

Cordulia is the genus of the American Emerald (*Cordulia shurtleffii*) that occurs no closer to Oklahoma than central Colorado (DuBois 2010, Paulson 2009). The genus was reported in larval form by Harrel (1969) and Bass (1994). If the species were to occur in Oklahoma it likely would be as an adult vagrant to the panhandle rather than as nymphs in central or north-central Oklahoma. We assume these specimens, none of which we have relocated, are likely misidentifications of *Epitheca*.

***Enallagma clausum* – Alkali Bluet**

Paulson (2009) and Abbott (2011) presumed that *Enallagma clausum* occurs in the Oklahoma panhandle, so despite not having any records for the state they mapped the far western Oklahoma panhandle as part of the species' range. Despite much survey effort, we have yet to find *E. clausum* in the state. We do agree with Paulson and Abbott that given records of the species just to the north in Colorado, to the west in New Mexico, and to the south, at one site, in the Texas panhandle, that the species ought to occur in the Oklahoma panhandle. A specimen will be needed to document this species for the state list.

***Enallagma laurenti* = syn., *E. vesperum* – Vesper Bluet**

See *Enallagma vesperum* species account.

***Enallagma piscinarium* = syn., *E. geminatum* – Skimming Bluet**

Bick and Bick (1957:12) said, "Bird (1932) records this species from Latimer County. I [GHB] could not find specimens of it in the OU collection. Montgomery (1942) states that records of *piscinarium* from Oklahoma are almost certainly in error." We have not found any specimens labelled as *E. piscinarium*.

***Enallagma pollutum* – Florida Bluet = syn., *E. vesperum* – Vesper Bluet**

See *Enallagma vesperum* species account. This synonymy is Oklahoma specific.

Enallagma westfalli* = syn., *Enallagma traviatum westfalli

Donnelly (1964) described this taxon as a separate species but later relegated it to subspecific status, as *Enallagma traviatum westfalli* under the subgenus *Teleallagma*, and indicated that this is the subspecies found in Oklahoma (Donnelly 1973).

***Epicordulia princeps* = syn., *Epitheca princeps* – Prince Baskettail**

Until Walker (1966) recommended relegating *Epicordulia* to subgeneric status, it was the genus for the Prince Baskettail, so it is not surprising that Bird (1932) and Bick (1951) used *Epicordulia* instead of *Epitheca* for *E. princeps*.

***Erythrodiplax connata minuscula* = sep., *E. minuscula* – Little Blue Dragonlet**

Bick and Bick (1957) used this name even though Bird (1932) used *E. minuscula*. Additionally, both Needham and Heywood (1929) and Needham and Westfall (1955) considered *E. connata* and *E. minuscula* as separate species. We suspect that the Bicks felt the separation of species was not warranted, thus agreeing with Borror (1942). It is now widely accepted that *E. minuscula* is not a subspecies of *E. connata* (Paulson 2003; Needham et al. 2014).

***Gomphoides stigmata/stigmatus* = syn., *Phyllogomphoides stigmatus* – Four-striped Leaftail**

Bird et al. determined 16 specimens as *Gomphoides stigmata* (EMEC 80227; OMNH 2335, 2340, 2342, 2344, 2346–2351, 2357–2358) and the slight spelling variant of *G. stigmatus* is found in Bick and Bick (1957). *Phyllogomphoides stigmatus* was originally described as *Aeshna stigmatus* Say, 1839, then it was changed to

Progomphus stigmatus Sélys, 1854, *Gomphoides stigmata* Hagen in Sélys, 1858, and finally to *Phyllogomphoides stigmatus* (Gloyd 1973), as it has been since. The different specific name endings used by Bird and the Bicks are explained by what was used in the manuals at the time, i.e., Needham and Heywood (1929) used the name *G. stigmata* and Needham and Westfall (1955) used *G. stigmatus*.

Gomphus spp. = syn., multiple species

Gomphus has historically been a "garbage can" genus, i.e., a polyphyletic genus comprised of numerous species that are not closely related evolutionarily but have broadly similar morphology. Various genera have been split from *Gomphus*, including *Arigomphus* and *Stylurus*. Most recently, Ware et al. (2017) relegated *Gomphus* to the Old World and split the remaining North American *Gomphus sensu lato* into four genera (by elevating subgenera): *Gomphurus*, *Hylogomphus*, *Phanogomphus*, and *Stenogomphurus*, the latter does not occur in Oklahoma. See Table A.1.

Gomphus albistylus = syn. and sep., *Stylogomphus sigmastylus* – Interior Least Clubtail

See *Stylogomphus sigmastylus* species account.

Gomphus intricatus = syn., *Stylurus intricatus* – Brimstone Clubtail

Older name used in Needham and Heywood (1929). See *Stylurus intricatus* species account.

Gomphus lentulus = syn., *Arigomphus lentulus* – Stillwater Clubtail

Gomphus was the original genus for this species. Bird (1932) called the species *G. subapicalis*; whereas Bick and Bick (1957) called it *G. (Arigomphus) lentulus*.

Gomphus pallidus

See *Arigomphus pallidus* above.

Gomphus plagiatus = syn., *Stylurus plagiatus* – Russet-tipped Clubtail

Like many species of gomphids, *Stylurus plagiatus* was originally described as part of the genus

Gomphus. Bird (1932) used the name *G. plagiatus*, even though at the time *Stylurus* was considered its subgenus (Needham and Heywood 1929); as such, Bird could have published the name as Bick and Bick (1957) did, i.e., "*Gomphus (Stylurus) plagiatus.*"

Gomphus subapicalis = syn., *Arigomphus lentulus* – Stillwater Clubtail

Gomphus was the original genus for *Arigomphus lentulus*. Although *G. lentulus* and *G. subapicalis* were both recognized as species during Bird's time (Needham and Heywood 1929), *G. subapicalis* eventually became a junior synonym. In Bird's era, *G. lentulus* was known only from Illinois, whereas *G. subapicalis* was known from Texas; perhaps explaining why Bird (1932) published determinations of *G. subapicalis*.

Gomphus submedianus = syn., *Arigomphus submedianus* – Jade Clubtail

Gomphus was the original genus for this species. Bird (1932) used *G. submedianus* and Bick and Bick (1957) called the species *G. (Arigomphus) submedianus*.

Helocordulia selsii = typo, *Helocordulia selysii* – Selys Sundragon

Typographic error of *Helocordulia selysii* (Bass 1990). We assume that the larval specimen this record was based on was a misidentification of another corduliid. We have not been able to locate any specimens in the UCO collection labelled as *Helocordulia*.

Hetaerina tricolor = syn., *Hetaerina titia* – Smoky Rubyspot

In RD Bird's era, *Hetaerina tricolor* and *H. titia* were separate species (Needham and Heywood 1929). At other times *H. tricolor* has been considered a form of *H. titia*. By the time Bick and Bick (1957) summarized Oklahoma Odonata, they treated *H. tricolor* as a junior synonym of *H. titia*. The Murray County specimen Bird (1932) published as *H. tricolor* apparently was not examined by GH Bick, as evidenced by the record

Table A.1 Taxonomic changes of North American species once placed in the genus **Gomphus**, which is now restricted to the Old World (Ware et al. 2017). All common names were retained.

FORMER NAME	CURRENT NAME
Gomphus apomyius Donnelly 1966	*Hylogomphus apomyius* (Donnelly 1966) – Banner Clubtail
Gomphus externus Hagen, 1858	*Gomphurus externus* (Hagen, 1858) – Plains Clubtail
Gomphus graslinellus Walsh, 1862	*Phanogomphus graslinellus* (Walsh, 1862) – Pronghorn Clubtail
Gomphus hybridus Williamson, 1902	*Gomphurus hybridus* (Williamson, 1902) – Cocoa Clubtail
Gomphus lividus Selys, 1854	*Phanogomphus lividus* (Selys, 1854) – Ashy Clubtail
Gomphus militaris Hagen, 1858	*Phanogomphus militaris* (Hagen, 1858) – Sulphur-tipped Clubtail
Gomphus oklahomensis Pritchard, 1935	*Phanogomphus oklahomensis* (Pritchard, 1935) – Oklahoma Clubtail
Gomphus ozarkensis Westfall, 1975	*Gomphurus ozarkensis* (Westfall, 1975) – Ozark Clubtail
Gomphus vastus Walsh, 1862	*Gomphurus vastus* (Walsh, 1862) – Cobra Clubtail

being omitted from his species' notes, nor have we relocated it.

Hyponeura lugens = syn., *Argia lugens* – Sooty Dancer

Argia lugens was originally described as a species of *Agrion* and later as *Hyponeura*. *Hyponeura* was synonymized with *Argia* (Gloyd 1968b) after Bick and Bick (1957) published AE Pritchard's 29 June 1933 record of 2♂ from Kenton, Cimarron County. The Bicks indicated those specimens were housed at OSU, but we have only found 1♂, which is an EMEC specimen (#300741). Additional reports of *Hyponeura* are from three nymphs (OMNH 1671–1672, UCO 5238) from McCurtain, Pushmataha, and Oklahoma Counties. BS-P identified those specimens as *Argia* sp.

Ischnura utahensis = syn., *Ischnura barberi* – Desert Forktail

This old name was used by Bird (1932) for an Alfalfa County record. In addition, Bird published records of *Ischnura barberi* for Alfalfa, Jackson, Woods, and Woodward Counties. Needham and Heywood (1929:353) included both *I. barberi* and *I. utahensis* in separate accounts and in their key, but they said that *I. utahensis* was "Possibly, a synonym of *I. barberi*." In his notes, Bick made no mention of examining Bird's *I. utahensis*, nor did Bick and Bick (1957). We have not found any specimens labeled as *I. utahensis*.

Lanthus – a pygmy clubtail = syn., in part, *Stylogomphus*, least clubtail

See the *Stylogomphus sigmastylus* species account.

Lanthus albistylus = syn. and sep., *Stylogomphus sigmastylus* – Interior Least Clubtail

See the *Stylogomphus sigmastylus* species account.

Lepthemis vesiculosa = syn., *Erythemis vesiculosa* – Great Pondhawk

This junior synonym of *Erythemis vesiculosa* was used by Bird (1932) and Bick and Bick (1957). Both names were used by various authors until Pinto et al. (2012) petitioned the International Commission on Zoological Nomenclature to declare *Erythemis* a senior synonym of *Lepthemis* despite the latter having page precedence.

Lestes congener – Spotted Spreadwing = re-id., *Lestes alacer* – Plateau Spreadwing

Lestes congener, had long been treated as a vagrant to Oklahoma until we removed it from the state list (Patten and Smith-Patten 2013a). It was added to the state list on the basis of a ♂ collected on 10 April 1966 by Clarice Kerfoot (mistakenly published as collected by LE Hornuff) at Woodward, Woodward County (FSCA). GH Bick (1978) published his original identification of *L. congener*, but he later privately retracted his determination (Figure A.1). In 1985, Bick re-identified the specimen as *L. alacer*, the Plateau Spreadwing, which, upon our examination of it in July 2011, we confirmed. Yet because Bick never corrected his error in print, the erroneous claim of *L. congener* was carried forward by Donnelly (2004c), Paulson (2009), and Abbott (2011). Presently, there is no documentation of *L. congener* in Oklahoma, although records for southeastern Colorado (CSU, OC), northeastern New Mexico (OC), and southwestern Missouri (Trial 2005) suggest that it may occur in the state, particularly in the western panhandle. One final note about this record is that because it is the only claim of the species in Oklahoma that we have been able to find, we assumed that the Dot Map Project record (OC 211917) for Creek County was a data entry error and actually relates to the Woodward County claim. As such, we, with JC Abbott's blessing, unconfirmed the Odonata Central record in October 2012.

Lestes disjunctus – Northern Spreadwing = syn., in part, *Lestes australis* – Southern Spreadwing

Surprisingly, Bird (1932) did not include *Lestes disjunctus* in his species list for Oklahoma; instead he identified his specimens as *L. forcipatus* and *L. forficula* (see below). Later, Bird's specimens were re-identified as *L. disjunctus* (Figure A.2) and then *L. australis* (others as *L. alacer*). In Bird's era, *L. disjunctus* included the subspecies *L. disjunctus australis*, which is the name Bick and Bick (1957) used. Now, *L. australis* is considered a separate species and is the one known from Oklahoma. Outside of the above reports, there has been one report of *L. disjunctus* for Oklahoma (OC 283840), but that report is unconfirmed because no specimen was taken for confirmation.

The range of *Lestes disjunctus* was once considered to come right up to the Oklahoma border and is still pictured as such in some sources (e.g., Odonata

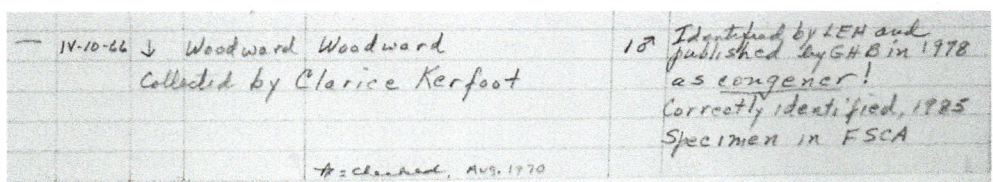

Figure A.1 GH Bick's notes indicating his retraction of his earlier report of *Lestes congener, the Spotted Spreadwing,* for Oklahoma. This specimen was re-identified as *L. alacer,* the Plateau Spreadwing.

Central shows Dot Map Project [DMP; Donnelly 2004c] records for Kansas, recent and DMP records for Colorado, and near border records in New Mexico) but others, such as Paulson (2009, 2011), have shifted the range westward and northward. Kansas DMP records are in need of re-examination (Figure 11.14) but records from Colorado and New Mexico indicate that the species could occur in the Oklahoma panhandle. Photographic records cannot be accepted because careful examination of the cerci and hamules are necessary for species determination of males, and females are trickier still.

Lestes dryas – Emerald Spreadwing = mis-id., probably *Lestes vigilax* – Swamp Spreadwing

Bick and Bick (1957) said that *Lestes dryas* was "Recorded from Oklahoma by Walker (1953)." Walker (1953:103) indeed included Oklahoma as part of this species' range, but we know not why. We suppose it may have been a mix up with *Lestes vigilax*, the Swamp Spreadwing, a species that occurs in Oklahoma but was not noted to occur by Walker. Beckemeyer (2002a) carried forward the error, but Donnelly (2004c) must have realized the mistake and so it was not published again. *Lestes dryas* does occur close to the Oklahoma panhandle, so a watchful eye there could produce a vagrant record.

Lestes forcipatus – Sweetflag Spreadwing = re-id., multiple species

In 2013, Oklahoma's only confirmed record of *Lestes forcipatus* was found (see species account). Until then it was thought that all records of *L. forcipatus* were actually attributable to *L. australis*, the Southern Spreadwing. OMNH has a long series of specimens that were identified as *L. forcipatus* by the RD Bird expedition (OMNH 67–70, 74–100, 1336–1355, 3737–3739, 3741–3761, 3763–3379, 3781–3785, 3790, 3796–3797, 3799–3802, 3806–3807, 3813, 3820–3822, 3824–3825, 3827, 3829, 3833–3834) that account for 187♂, 89♀, and 2U, and includes 14 pairs. These specimens appear to be those that LK Gloyd re-identified as *L. disjunctus* (Figure A.2). Bick and Bick (1957) examined and further identified 66♂ and 29♀ as *L. disjunctus australis*. We re-examined, re-identified, or

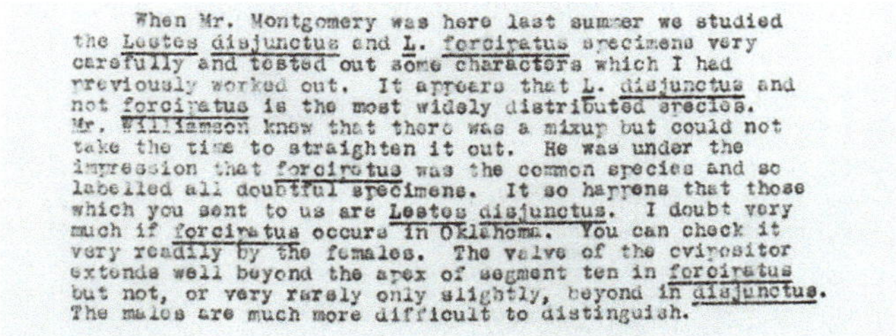

re-assigned all to *L. australis*. Specimens cataloged as OMNH 3780 were originally labeled as *L. forcipatus* or *L. forficula*, the Rainpool Spreadwing, but we re-identified them as *L. alacer* (see *L. forficula* below).

Bird (1932) included 19 counties under his heading of *Lestes forcipatus*. He did not include *L. disjunctus*, a species certainly known at the time (Needham and Heywood 1929), in his list of odonates for Oklahoma (Ortenburger 1926b did not include any *Lestes*). At the time, neither species was well described (*L. australis* Walker, 1952, had yet to be) and either range description could have fit the state. It may also have been that Bird knew of Kennedy's (1917) paper for Kansas that indicated *L. forcipatus*, not *L. disjunctus*, occurred there; records have subsequently been re-assigned to *L. australis* for Kansas, too. Nonetheless, all but perhaps the Bird expedition collections in Latimer County can rather safely be called *L. australis*. Given that the Latimer County specimens are in fluid currently and that differentiating between *L. australis* and *L. forcipatus* depends upon miniscule differences in shape and measurements, we feel that trying to identify those specimens with 100% certainty is out of the question given that some shrinkage and distortion undoubtedly has occurred.

We suggest that if older specimens or records for Oklahoma are found in the future with the identification of *Lestes forcipatus*, then it is safe to assume they are actually *L. australis* (of course, examination of specimens is always preferable to assumption). The only caveat is if those specimens or records are from extreme eastern Oklahoma where the range of *L. australis* and vagrant records of *L. forcipatus* meet.

Lestes forficula – Rainpool Spreadwing = re-id., multiple species

There are four records for Oklahoma that were originally attributed to *Lestes forficula* by RD Bird. OMNH 3780, consisting of 2♂ and 5♀ specimens, were identified as *L. forcipatus* or *L. forficula* (original labels with these data: Lestes/forficula/Det. RD Bird 1932; Exuvia of/this sp. Or/forcipatus/Det. RD Bird 1932; Wichita Nat'l Forest/9 VI 1932 Okla./RD Bird). It appears that GH Bick, despite examining other "*L. forficula*" from OMNH, did not examine these. We re-identified the specimens as *L. alacer*. On that same collection date and locality, Bird captured a ♂ and a ♀, probably a pair (OMNH 3808), that Bick did examine. He left that record on his note card for *L. forficula* (Figure A.3). At the bottom of that card he indicated that "All Bird material on hand (Aug. '70) seem greatly distorted …" so it may be that he felt he could not confidently identify OMNH 3808. We have not examined those specimens. Bick included another Comanche County record, of 2♀ collected on 23 July 1932. He also did not indicate a re-identification. These

When Mr. Montgomery was here last summer we studied the Lestes disjunctus and L. forcipatus specimens very carefully and tested out some characters which I had previously worked out. It appears that L. disjunctus and not forcipatus is the most widely distributed species. Mr. Williamson knew that there was a mixup but could not take the time to straighten it out. He was under the impression that forcipatus was the common species and so labelled all doubtful specimens. It so happens that those which you sent to us are Lestes disjunctus. I doubt very much if forcipatus occurs in Oklahoma. You can check it very readily by the females. The valve of the ovipositor extends well beyond the apex of segment ten in forcipatus but not, or very rarely only slightly, beyond in disjunctus. The males are much more difficult to distinguish.

Figure A.2 Letter from LK Gloyd to RD Bird, 30 April 1934, regarding re-identification of Bird's *Lestes forcipatus*, the Sweetflag Spreadwing, specimens as *L. disjunctus*.

specimens match 2♀ *L. alacer* at FSCA. The specimens from the single Cotton County record, of 5♂ and 1♀ collected on 19 July 1932, listed on Bick's *L. forficula* card likewise appear to now reside at FSCA. The final record originally reported to be *L. forficula* is OMNH 3803, a single ♂ from Stephens County, taken on 18 July 1932, probably on Little Beaver Creek (label indicated "Beaver Cr." and Bird's field notes had "Beaver Creek close to Hulen" as locality). According to Bick's *L. disjunctus australis* species' card, he re-identified this specimen as *L. australis* (Figure A.4), which is what we had independently confirmed it as. Bird (1932) did not publish any of these records, nor did he include the species in his list of Oklahoma Odonata. Bick and Bick (1957) did not discuss re-identification of the specimens they examined.

None of the above dismisses the possibility of *Lestes forficula* finding its way to Oklahoma. Although not high on our list of expected species, we will not be surprised if one someday turns up in the state, especially in the Red River valley.

Libellula axilena/axillena – Bar-winged Skimmer = re-id., *Libellula vibrans* – Great Blue Skimmer

Ortenburger (1926b) was the first to present a claim of *Libellula axilena* in Oklahoma. He indicated there was a specimen taken on 8 July 1925 on Lukfata Creek 4 mi NW of Broken Bow, McCurtain County.

Lestes forficula OKLA
— 7-9-32 RDB Comanche called forficula by RDB 1♂ 1♀ O.U. collection - returned
— 7-19-32 " Cotton " " " " 5♂ 1♀ O.U. collection - KEPT papered by us - GHB series
— 7-23-32 " Comanche " " " " 2♀ " " " " "
These were studied August 1970, all males check more closely with alacer in BCA than with forficula. This collection, called forficula by Bird, differs from his alacer substantially only in color! forficula - more black on thorax than alacer, but the former (studied in 1970) were all collected

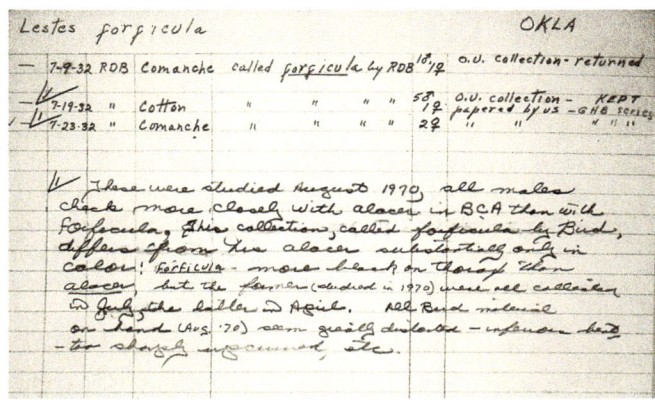

Figure A.3 GH Bick's note card for *Lestes forficula*, the Rainpool Spreadwing. Reports of this species, from RD Bird in 1932, have been re-identified as *L. alacer*, the Plateau Spreadwing, or have remained undetermined.

Lestes disjunctus australis OKLA
— 7-18-32 RDB Stephens called forficula by RDB 1♂ O.U. collection - returned

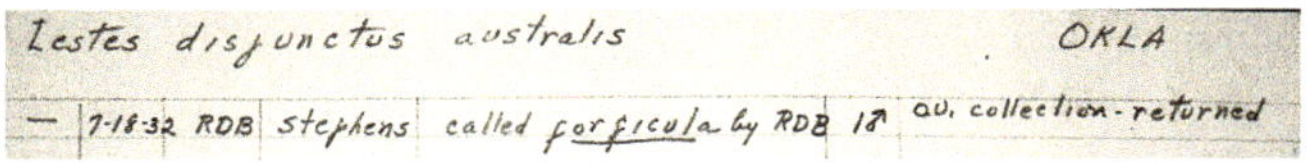

Figure A.4 GH Bick's notes regarding a report of *Lestes forficula*, the Rainpool Spreadwing, from Stephens County, Oklahoma, from 1932. Bick re-identified this specimen as *L. australis*, the Southern Spreadwing.

That record appears to match OMNH 922, a single ♀ that MAP re-identified as *Libellula vibrans.* Bird (1932:54) carried the record forward as "McCurtain (*Ortenburger). One female identified by Williamson but with a query." Note that both Ortenburger and Bird misspelled the specific name as *axillena.* Bick and Bick (1957:12) included the species in their "Synonyms and doubtful records" section and noted, "Ortenburger (1926b) records this species from Latimer [*sic*] County. Bird (1932) states that this was based on a single female determined by Williamson with query. I doubt that the species occurs in Oklahoma." We disagree with the Bicks; the species may occur, at least occasionally, in the southeastern corner of the state given the relative proximity (~105 km, or <70 mi) of *L. axilena* records in northeastern Texas and southern Arkansas. Moreover, two records of males for Larimer County, Colorado (Loveland, 6 May 2012, CSU specimen; Spruce Gulch 7 km W of Carter Lake, 12 July 2018, OC 485315) are well west of the species' geographic range and suggest to us that *L. axilena* might be encountered as a vagrant anywhere in Oklahoma.

Macromia australensis = syn., *Macromia illinoiensis georgina* – Georgia River Cruiser

Williamson (1909, 1914b) described *Macromia australensis* as a separate species. He based his description on a ♂ and ♀ Frank Collins collected 3 August 1907 along the Poteau River, at Wister, Indian Territory (now Le Flore County, Oklahoma; UMMZ). Bird (1932) carried forward Williamson's record and reported the species also for Latimer County (OMNH 782; Figure A.5). Bick and Bick (1957; and MJ Westfall) re-determined those specimens (and OMNH 773, from Caddo County) to be *M. georgina*, a taxon now generally treated as the subspecies *M. illinoiensis georgina.*

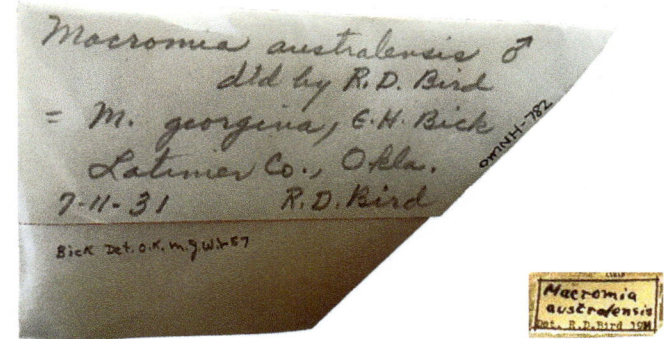

Figure A.5 Specimen envelope with GH Bick's handwriting and re-determination as *Macromia illinoiensis georgina* (right). Original specimen label from RD Bird (lower right, enlarged), 11 July 1931, determination as *M. australensis* (OMNH 782).

Macromia georgina = syn., *Macromia illinoiensis georgina*
– Georgia River Cruiser

    See *Macromia australensis* (above) and the *Macromia illinoiensis senso lato* species account.

Macromia wabashensis – Wabash River Cruiser

    Macromia wabashensis was described as a species by Williamson (1909); since, it has been relegated to being either a subspecies (or "form") or color morph of *M. taeniolata*, the Royal River Cruiser (Needham et al. 2014) or a hybrid of *M. taeniolata* and *M. pacifica*, the Gilded River Cruiser (e.g., Donnelly 2004b; Abbott 2005; Paulson 2011). Bird (1932) reported *M. wabashensis* from Haskell and Latimer Counties. Those specimens, 4♂ (OMNH 776–777, 783–784), were re-determined as *M. georgina* by GH Bick and MJ Westfall (Bick and Bick 1957). Two additional OMNH (779–780) specimens, 2♂ from Kiowa County, were originally identified as *M. wabashensis* by Bird, but later re-examined and re-identified by Bick and Westfall as *M. georgina*.

    See *Macromia taeniolata* species account for further discussion of records we feel are attributable to actual *M. taeniolata wabashensis*.

Macrothemis sp. – a sylph = mis-id.

    Harrel (1969) reported larval *Macrothemis* for the Otter Creek drainage basin in Garfield and Logan Counties. None of the four *Macrothemis* species (sylphs) known from the United States occurs closer to Oklahoma than central Texas, so it is unlikely that larval stages of *Macrothemis* occur in the state.

Mesothemis plebija = typo and re-id, *Erythemis vesiculosa*
– Great Pondhawk

    In a letter dated 20 October 1932 to EB Williamson, RD Bird said that he had added *Mesothemis plebija* [sic] to the state list. We presume Bird referred to the ♂ *Erythemis vesiculosa* he took at Deans Lake, McClain County (OMNH 921). We imagine that Bird thought the specimen looked like a large *E. simplicicollis*, the Eastern Pondhawk, which was then called *Mesothemis simplicicollis*, the Green Jacket. The other *Mesothemis* at the time was *M. plebeja* (Needham and Heywood 1929; now *Erythemis plebeja*, the Pin-tailed Pondhawk). Although the coloration description did not fit entirely, the description of the abdomen as slender may have persuaded Bird to call it *M. plebeja* rather than to flip the page to *Lepthemis vesiculosa* (now *E. vesiculosa*) and consider a species classified under a different genus. This said, *E. plebeja* is known to come within ~95 km (~60 mi) of Oklahoma's southern border, so it is plausible that the state will gain a record one day.

Mesothemis simplicicollis = syn., *Erythemis simplicicollis* –
Eastern Pondhawk

    Synonym used by Bird (1932). The species was put into the genus *Erythemis* by Needham and Westfall (1955), and thus by Bick and Bick (1957).

Miatheria [sic, *Miathyria*] = mis-id.?

    We assume a reference to larval "*Miatheria*" referred to *Miathyria marcella*, the Hyacinth Glider. See that account from problematic reports.

Nannothemis sp. – Elfin Skimmer = mis-id.

    Bass and Potts (2001) reported larval *Nannothemis* at Boehler Seeps, Atoka County. *Nannothemis* is the monotypic genus of the Elfin Skimmer (*N. bella*), which occurs no closer to Oklahoma than central Kentucky. We have not relocated this specimen for re-identification, but we are confident it is not *N. bella*.

Neurocordulia obsoleta – Umber Shadowdragon = re-id., multiple species

    GH Bick and LE Hornuff collected seven nymphs and one exuviae in the 1950s that were originally determined to be *Neurocordulia obsoleta* but were subsequently identified by SW Dunkle and others as *N. xanthosoma*. These specimens were published in Bick (1951) and Bick and Bick (1957) as *N. obsoleta*; none have been relocated. One specimen, an adult ♂ from Comanche County (OMNH 888), was identified by RD Bird as *N. obsoleta*, but BS-P re-identified it as *Epitheca petechialis*, the Dot-winged Baskettail.

Neurocordulia yamaskanensis – Stygian Shadowdragon = re-id., probably *Neurocordulia xanthosoma* – Orange Shadowdragon

    Although we feel that an adult *Neurocordulia yamaskanensis* being found in Oklahoma is definitely possible (see hypothetical/expected species list), breeding in the state is unlikely except perhaps in the far northeastern corner. We consider the report of this species in larval form from the Wichita Mountains, Comanche County (Bass 1990), as a misidentification of, probably, *Neurocordulia xanthosoma*, the Orange Shadowdragon. These two species are one key step apart in the reference we assume was used (Needham et al. 2000). We have not been able to locate the specimen in the UCO collection.

Ophiogomphus sp. = re-id., multiple species

    Nymphs identified as *Ophiogomphus* were found in two collections; all were re-identified. OMNH 4228, a single nymph collected at Indian Springs, Cleveland County, 24 November 1929, by RD Bird, was re-identified as *Arigomphus lentulus*, cf. Three nymphs from Pennington Creek, near Reagan, Johnston County, collected on 24 June 2002 (UCO 5224), were also re-identified: one as *Dromogomphus spinosus*, the Black-shouldered Spinyleg, and two as *Erpetogomphus designatus*, the Eastern Ringtail. Also, Harrel (1969) reported one nymph in the Otter Creek drainage basin of northcentral Oklahoma in Logan and Garfield Counties. All of these records are out of range for the only expected *Ophiogomphus* species in Oklahoma, *O. westfalli*, Westfall's Snaketail, which would be found only in northeastern or east-central Oklahoma (see hypothetical/expected species section, below).

Pangaeagaster = syn., *Cordulegaster*

See Cordulegastridae introduction for details. We know of no records from Oklahoma labeled with this genus.

Perithemis domitia – Slough Amberwing = sep., *Perithemis tenera* – Eastern Amberwing

When Bird (1932) published this species for Oklahoma, it was considered to be the only one in the U.S. (Needham and Heywood 1929). *Perithemis tenera* was recognized as a distinct species by Needham and Westfall (1955) and was published as such by Bick and Bick (1957).

Platycordulia xanthosoma = syn., *Neurocordulia xanthosoma* – Orange Shadowdragon

Older name used by Ortenburger (1926b), Bird (1932), and Bick (1951).

Progomphus borealis – Gray Sanddragon = re-id., multiple species

There are two OSU pinned specimens that were labeled as *Progomphus borealis*, but we re-identified them. One specimen was a ♂ collected at Elmer, Jackson County, on 6 July 1937 by Standish and RW Kaiser; we determined it as *P. obscurus*, the Common Sanddragon. The second specimen was a ♀ collected at Coody's Bluff, Nowata County, on 6 July 1983; we determined it as *Arigomphus lentulus*, the Stillwater Clubtail. Although these records were not identified correctly, we expect the species to occur in southwestern Oklahoma given multiple records for Caprock Canyons SP, Briscoe County, Texas (OC 400260, 463974), only 100 km (62 mi) west of the state line.

Progomphus obscuratus = typo, *Progomphus obscurus* – Common Sanddragon

OMNH 859 was labeled as *Progomphus obscuratus*. It was probably just a "write-o" in the field. The ♂ *Progomphus obscurus* was collected by RD Bird on 13 June 1929 in Love County.

Somatochlora ensigera – Plains Emerald = re-id., *Somatochlora linearis* – Mocha Emerald

We are perplexed by this re-identification. Bird (1932) published *Somatochlora ensigera* for Latimer County. According to GH Bick's notes, he examined the specimen Bird's record was based on (1♂, Latimer County, 11 June 1931, RD Bird) and re-identified it as *S. linearis*. What is perplexing about Bird's original identification is that *S. ensigera* and *S. linearis* have reasonably different terminal appendages, differences clearly illustrated by Needham and Heywood (1929:186), which Bird absolutely had at his disposal. Granted, the early illustrations were not perfect, but Bird should have been able to distinguish the two species readily. We would very much like to see Bird's

specimen in hopes of understanding why he chose *S. ensigera* over *S. linearis*, but, alas, although Bick said he returned the specimen to OMNH, it has not been relocated. Even though Bird's record of *S. ensigera* was dispelled, there is an ever so slight chance the species could one day show in the Oklahoma panhandle. We are not confident enough about that possibility, so we did not include it in the expected species list below.

Somatochlora provocans – Treetop Emerald = re-id., *Somatochlora ozarkensis* – Ozark Emerald

Of the *Somatochlora* specimens Bird and his team collected in the summer of 1931 on Cunneo Tubby Creek north of Wilburton, Latimer County, he said, "In the field they were identified as *provocans*, but on closer examination they proved to be new" (Bird 1933:1). Those specimens were used to describe *Somatochlora ozarkensis* Bird, 1933.

Stylogomphus divaricatus = syn., *Stylogomphus sigmastylus* – Interior Least Clubtail

See the *Stylogomphus sigmastylus* species account for discussion of this taxon.

Sympetrum occidentale fasciatum = syn., *Sympetrum semicinctum fasciatum* – Band-winged Meadowhawk

Used by Bick and Bick (1957). The specific name has changed to *semicinctum* but the subspecific name stays the same. See *Sympetrum semicinctum* species account for further discussion.

Taeniogaster = syn., *Cordulegaster*

See Cordulegastridae introduction for details. We know of one record from Oklahoma originally labeled as *Taeniogaster* (OMNH 1766, a nymph from Fort Cobb, Caddo County, 1 November 1975, D Dawson) but it is assumed that the identification is *Cordulegaster obliqua*.

Tarnetrum corruptum = syn., Sympetrum corruptum – Variegated Meadowhawk

This species was originally described as *Mesothemis corruptum* Hagen, 1861, but was later transferred into *Sympetrum* (Needham and Heywood 1929), then split out as *Tarnetrum* (Needham and Westfall 1955), which is why Bick and Bick (1957) published it as such. Eventually it was rejoined as *S. corruptum* (Needham et al. 2000, 2014).

Teleallagma daecki = syn., *Enallagma daeckii* – Attenuated Bluet

Used by Bick and Bick (1957).

Telleallagma sp. [sic? *Teleallagma*] = typo and mis-id.?

Harrel (1969) reported a nymph of this genus in the Otter Creek drainage basin of Garfield and Logan Counties. The genus or subgenus of *Teleallagma* most specifically has related to what is now *Enallagma daeckii*, the Attenuated Bluet. But it has also been applied, on occasion to other species, including, *E. traviatum* and *E. westfalli* (as a separate species or as a subspecies of *E. traviatum*; e.g., Donnelly 1964, 1973). If Harrel's specimen referred to *E. daeckii*, then we highly doubt the identification is correct, but if it referred to the latter species then it may be correct. Unfortunately, we have not relocated the specimen for re-examination.

Tetragoneuria canis – Beaverpond Baskettail = mis-id., *Epitheca spinosa* – Robust Baskettail

Used by Bird (1932). Relates to OMNH 334, re-identified as *Epitheca spinosa* (Bick and Bick 1957; see species account).

Tetragoneuria cynosura = syn., *Epitheca cynosura* – Common Baskettail

Used by Bird (1932), Bick (1951), and Bick and Bick (1957).

Tetragoneuria petechialis = syn., *Epitheca petechialis* – Dot-winged Baskettail

Used by Bick and Bick (1957).

Tetragoneuria spinosa = syn., *Epitheca spinosa* – Robust Baskettail

Used by Bick and Bick (1957).

Tetragoneuria williamsoni = syn., *Epitheca costalis* – Slender Baskettail

Used by Williamson (1914b), Bird (1932), and Bick and Bick (1957). *Tetragoneuria williamsoni* Muttkowski 1911, later became a junior synonym of *Epitheca costalis* (Sélys, 1871). Muttkowski (1911:122–123) used the three specimens Williamson collected on 3–4 June 1907 as the holotype, allotype, and only paratype for the species' description. In that same publication just a few pages later Muttkowski discussed *costalis*. It appears that he personally was not familiar with that species given that he indicated the male was unknown and, for the female, he provided a transcription of a letter he received of its description. It is unsurprising, then, that *T. williamsoni* was later subsumed by *E. costalis*.

Zoraena = syn., *Cordulegaster*

Tennessen (2019) argued that given nymph morphology, *Zoraena* ought to be elevated to generic status. For Oklahoma, this change would affect *Cordulegaster talaria* (and *C. sarracenia*, if that species is found one day in Oklahoma, a possibility given the shared ecoregions of Oklahoma and the species' currently defined geographic range). See Cordulegastridae introduction for details.

HYPOTHETICAL/EXPECTED SPECIES

There are 41 species we feel reasonably can be "expected" in Oklahoma in the future, as vagrants or as yet undiscovered populations. There are five records we consider to be hypothetical (†), meaning that they lack sufficient documentation—either a verifiable specimen or identifiable photograph—to be accepted as a first state record. We will continue to consider these species as hypothetical until solid documentation can be obtained. Species are listed here taxonomically by family, following the order as presented by Paulson (2009, 2011).

We do not intend this list to be exhaustive; we merely included species that more-or-less lie near the periphery of the state. Vagrants, on the other hand, are unfathomable and often mind-blowing, so we cannot profess to know what nature has in store for Oklahoma. Many tropical species are known to wander considerable distances northward during the autumn—for examples, see the species accounts for *Argia tezpi*, *Gynacantha nervosa*, *Erythemis attala*, *Erythrodiplax basifusca*, and *Tholymis citrina*, to name but a few. Consequently, we urge you to have southern species (even those that barely come into the United States) on your radar in the autumn, just in case (and *please* have a net handy). At that time of year, too, keep your eye to the north when cold fronts dip down, as they can bring northern species into the state. Also keep an eye on hurricanes and tropical storms, which can bring in species from great distances (those storms do not have to pass over Oklahoma; winds at the peripheries of storms can bring critters with them).

Zygoptera
Calopterygidae
Calopteryx dimidiata – Sparkling Jewelwing

This species could show up in the southeastern corner of the state given that it has been recorded <140 km (85 mi) to the south in Gregg County, Texas (Donnelly 2004c).

Lestidae
Lestes forficula – Rainpool Spreadwing

A recent record from Rusk County, Texas (OC 472867) of this species suggests it could make its way farther north another ~100 km (~60 mi) to reach the southeastern corner of Oklahoma.

Lestes congener – Spotted Spreadwing

See entry in *Synonyms, errors, and dubious records* section above.

Lestes disjunctus – Northern Spreadwing

See entry in *Synonyms, errors, and dubious records* section above.

Lestes dryas – Emerald Spreadwing

See entry in *Synonyms, errors, and dubious records* section above.

Coenagrionidae
Coenagrion resolutum – Taiga Bluet

Records <125 km (~75 mi) west of the Oklahoma panhandle suggest that the species could be found, at least as a vagrant, in the state (see OC 284246 and OC 326904 for discussion).

Enallagma anna – River Bluet

Records of this species from eastern Colorado and New Mexico suggest that it could be found in the Oklahoma panhandle within the Southwestern Tablelands ecoregion.

Enallagma clausum – Alkali Bluet

This species is long overdue for Oklahoma. See entry in *Synonyms, errors, and dubious records* section above.

Enallagma annexum – Northern Bluet

Records purported to be this species occur within ~110 km (70 mi) of the Oklahoma panhandle. Diagnosability issues with this species (especially with *E. boreale*, see below) make it such that photographic records cannot be accepted, so collection of a specimen will be necessary.

Enallagma boreale – Boreal Bluet

Records purported to be this species occur within ~45 km (<30 mi) of the Oklahoma panhandle. As with *E. annexum*, the Northern Bluet, its confusion with other species necessitates a verifiable specimen in order to add it to the official Oklahoma state list.

Hesperagrion heterodoxum – Painted Damsel

Although reports of this species north to Colorado (Donnelly 2004c, Paulson 2009) have been found to be erroneous (Prather and Prather 2015), the recent record from Lubbock County, Texas, (OC 473521, 22 Sep 2017) indicates that this species' geographic range may be bulging eastward, perhaps one day as far as southwestern Oklahoma.

Argia hinei – Lavender Dancer

Records from the Texas panhandle indicate that this species could be found in southwestern Oklahoma or its panhandle, likely within the Southwestern Tablelands or High Plains ecoregions. This species is easily confused with *Argia fumipennis*, the Variable Dancer, and because we know of one specimen (SP 1611) with overlapping characters of *A. fumipennis* and *A. hinei*, a verifiable specimen of *A. hinei* will be necessary to add this species to the official Oklahoma state list.

Argia vivida – Vivid Dancer

Records within ~100 km (~60 mi) of the Oklahoma panhandle in Colorado and New Mexico suggest that the species could find its way to the state. This species is extremely easy to confuse with *Argia plana*, the Springwater Dancer, so a specimen is a must to document its occurrence in the state.

Argia emma – Emma's Dancer

Records from southeastern Colorado (Las Animas and Otero Counties) that are <90 km (~55 mi) away indicate that this arid Canyonlands species could occur in the Oklahoma panhandle.

Aeshnidae

Gomphaeschna antilope – Taper-tailed Darner

Though there are recent records from the South Central Plains ecoregion of central and northern Louisiana, an ecoregion shared by Oklahoma, this species may be a long shot for the state, but it should nonetheless be on the radar. It may be that the species is bounded by sub-ecoregions farther south (especially the Southern Tertiary Uplands), but we still urge caution in blithely identifying an encountered *Gomphaeschna* as *G. furcillata* without close inspection (Shields and Petranka 2018).

Aeshna interrupta – Variable Darner

If DMP records in Arkansas and Kansas are correctly identified, then this species could occur in eastern Oklahoma. On the other side of the state, records from Las Animas County, Colorado, and Colfax County, New Mexico, being <125 km (~75 mi) from the Oklahoma panhandle, suggest *Aeshna interrupta* could occur there as well. If it is found in the state, it would likely be as a fall vagrant.

Aeshna palmata – Paddle-tailed Darner

As with *Aeshna interrupta*, this species occurs within <125 km (~75 mi) from the Oklahoma panhandle, so it could occur in the state as a fall vagrant.

†*Aeshna persephone* – Persephone's Darner

One was seen rather well by BS-P and fairly well by MAP on 3 August 2014 near Jackson Salt Plains (i.e., southeast of Eldorado), Jackson County. We were not able to obtain documentation. Vagrant records for this species indicate it wanders considerable distances on occasion, even if reports from Colorado are in error (Prather and Prather 2015).

Rhionaeschna psilus – Turquoise-tipped Darner

Perhaps but a long shot for Oklahoma, we feel that given a Texas panhandle record (<200 km, or <125 mi, away from southwestern Oklahoma; Lubbock County, OC 263216), this species ought to be kept on Oklahoma's radar as a potential vagrant.

†*Anax amazili* – Amazon Darner

One ♀ was seen well by MAP on 26 October 2014 at French Lake in the Wichita Mountains WR, Comanche County, but he was not able to get documentation.

†*Anax walsinghami* – Giant Darner

One ♂ was reported by BC on 14 May 2017 from Black Mesa SP, Cimarron County. He was unable to obtain documentation.

Gomphidae

Arigomphus villosipes – Unicorn Clubtail

This species has multiple long-distance vagrant records, including one from nearby Pike County, Arkansas (10 June 1978, pond at 4 mi W of Glenwood, C Cook; Harp 1983; site is 78 km, or <50 mi east of Oklahoma). A record in eastern Oklahoma would not be shocking and could occur anytime between May to perhaps August.

Phanogomphus quadricolor – Rapids Clubtail

Records in Arkansas that are within about 50 km (~30 mi) of the Oklahoma border suggest that this species could show up in southeastern and east-central Oklahoma, likely in the Athens Plateau ecoregion (level IV) in the vicinity of Smithville and Watson. It could be found between May and July and, during that species' early flight season, could easily be confused with the Oklahoma Clubtail, *Phanogomphus oklahomensis*, if not inspected closely; as such photographic documentation cannot be accepted to include this species on the official state list.

Gomphurus modestus – Gulf Coast Clubtail

Arkansas records of this species that are within 35 km (22 mi) of Oklahoma indicate that its presence in southeastern Oklahoma is a definite possibility (Howard and Sevier County records came from Harp and Rickett 1985, but we do not know the source of the Clark County record). It should be sought out at medium to large muddy or sandy rivers and creeks between May and July, perhaps especially the Red and Little Rivers or some of the nearby creeks in eastern McCurtain County. It can be confused with *G. ozarkensis* and *G. vastus*, the Ozark and Cobra Clubtails, so a specimen will be necessary to verify this species in the state.

Stylurus laurae – Laura's Clubtail

Records from northeastern Texas and southwestern Arkansas indicate this species could appear in southeastern Oklahoma.

Ophiogomphus severus – Pale Snaketail

Records of this species from western Kansas, eastern Colorado, and New Mexico range close enough to the Oklahoma panhandle to make it a possibility for the state.

Ophiogomphus westfalli – Westfall's Snaketail

We will go out on a limb and say that this species is a part of Oklahoma's odonate fauna. There are records, of both adults and nymphs, in Arkansas, Missouri, and southeastern Kansas, that embrace the northeastern corner of Oklahoma, so it would be astonishing if the species does not occur here. It is known from the Elk River basin (Harp and Trial 2001),

of which Oklahoma's Elk River and Buffalo Creek, for example, are part, although hydrology are now altered because of damming for Grand Lake O' the Cherokees. Surveys for this species in northeastern Oklahoma have been impeded greatly by lack of public access to streams. Records south to Montgomery and Saline Counties in west-central and central Arkansas suggest that the species could be found anywhere in extreme eastern Oklahoma, but it is likely to be much rarer farther south. The species flies from May to July and should be sought along clear, rocky stream courses similar to where *Stylogomphus sigmastylus*, the Interior Least Clubtail, is found.

Progomphus borealis – Gray Sanddragon

See entry in *Synonyms, errors, and dubious records* section above.

Phyllogomphoides albrighti – Five-striped Leaftail

Records in northern Texas indicate that this species should, especially with increased climate change, occur in southwestern and south-central Oklahoma (currently, Oklahoma only shares the Cross Timbers ecoregion with this species). *Phyllogomphoides albrighti* can be difficult to distinguish from *P. stigmatus*, the Four-striped Leaftail, so close inspection is a must and a verifiable specimen is preferred to include the species on the state list of odonates.

Aphylla protracta – Narrow-striped Forceptail

The geographic range of this species is thought to approach the southeastern corner of Oklahoma, so it ought to one day be found at mud-bottom ponds or lakes within the East Central Texas Plains or perhaps within the South Central Plains.

Cordulegastridae

Cordulegaster dorsalis – Pacific Spiketail

The multiple disjunct populations of this species along with recent records from Colorado and New Mexico that are as close as ~100 km (~60 mi) to the Oklahoma panhandle hint that this species might make it to the state, at least as a vagrant.

Corduliidae

Somatochlora georgiana – Coppery Emerald

Even though the sole record of this species for Texas, from Titus County, is from the 1950s (Abbott 2015), the county lies only 15 km (10 mi) south of the Oklahoma border, so a vagrant record of the enigmatic (and apparently rare) species for Oklahoma would not be out of the question.

Somatochlora filosa – Fine-lined Emerald

A recent record of a ♂ from Marion County, Texas (OC 488887), about 60 km

(<40 mi) southeast of the southeastern corner of Oklahoma, has made the possibility of an Oklahoma record of this species even more of a reality. Look for it from mid-summer into the fall at "small sandy forest streams and tiny, tannin-stained seeps" (Paulson 2011:366).

Neurocordulia yamaskanensis – Stygian Shadowdragon

Records from Arkansas and Missouri within the Ozark Highlands from ~130 km (~80 mi) away from the northeastern corner of Oklahoma indicate that this is a real possibility for an additional to the Oklahoma state list.

Libellulidae

Libellula quadrimaculata – Four-spotted Skimmer

There are records of this species from nearby New Mexico and Colorado, the closest of which is ~140 km (88 mi) away from the Oklahoma panhandle, making a vagrant record of this species a possibility.

Libellula forensis – Eight-spotted Skimmer

This species occurs in Colorado and New Mexico as close as 120 km (<75 mi) from the Oklahoma panhandle. We have long wondered if it simply has been overlooked given its similarity to other species of *Libellula* and *Plathemis*. It is most easily distinguished from the expected *L. pulchella*, the Twelve-spotted Skimmer, by having clear wingtips (the Twelve-spotted has dark wingtips, accounting for the additional four "spots" of this species' common name). The clear wingtips is a character of both sexes of *L. forensis* and is a mark that also distinguishes it from the similar female *Plathemis lydia*, the Common Whitetail. But study closely the differing wing pattern of female *P. subornata*, the Desert Whitetail, because they have clear wingtips but the jagged bands in the middle of the wing, including a second set, should readily distinguish the two species. Look for *L. forensis* anywhere in the Oklahoma panhandle where these other species are found. It is also known to occur with *L. saturata*, the Flame Skimmer, and *L. composita*, the Bleached Skimmer. And, it has hybridized with *L. luctuosa*, the Widow Skimmer (Manolis and Bruun 2006).

Libellula axilena – Bar-winged Skimmer

See entry in *Synonyms, errors, and dubious records* section above. Due to its similarity to *Libellula vibrans*, the Great Blue Skimmer, and *L. incesta*, the Slaty Skimmer, a specimen will be necessary to substantiate a claim of the species in the state.

Libellula needhami – Needham's Skimmer

This species has long been expected in Oklahoma, but with continued climate change moving species northward, it is almost guaranteed to pop up in the state. There are records in Arkansas, Louisiana, and Texas that are relatively close to Oklahoma's southeastern corner, the nearest of which are a mere 120 km (<75 mi) away.

†*Orthemis discolor* – Carmine Skimmer

A furtive adult ♂ seen well at Great Plains SP, Kiowa County, 10 October 2015 (MAP) obliged with neither specimen nor photograph. The red frons was compared directly to the purple frons of multiple ♂ *Orthemis ferruginea*, the Roseate Skimmer, present with it.

Orthemis schmidti

Ken Williams churned up quite the controversy when he photographed a ♂ *Orthemis* at a pool along the salt flats at Sandpiper Trail, Salt Plains NWR, Alfalfa County, on 8 September 2007 (OC 262983; Figure A.6). He submitted the record as *O. discolor*, the Carmine Skimmer, at the time the only other *Orthemis* species one would have considered for Oklahoma. However, this ♂ bears a striking resemblance to *O. schmidti*, a species of South and Middle America that is known to range no farther north than Nayarit, Mexico (OC 385454) and northern Belize.[1] Granted, there is a pending record submitted as *O. schmidti* from Miami-Dade County (OC 492214), and JJ Daigle has said that the species occurs in Florida;[2] however, it is unclear how much confusion still exists between this species and *O. macrostigma* and potentially other *Orthemis* in the region (Meurgey and Daigle 2007; Paulson 2011).

Figure A.6 A ♂ *Orthemis* from Salt Plains NWR, Alfalfa County, Oklahoma, that was submitted as *O. discolor*, but which remains unidentified because of the possibility of a vagrant record of *O. schmidti* or a hybrid *O. discolor × O. ferruginea*. **Photo by ©Ken Williams (OC 262983).**

Regardless, the nearest Florida record is no closer than the others, so it remains that the range of *O. schmidti* may be no nearer to Alfalfa County than ~1,900 km (1,200 mi)!

In his decision comments for the record, JC Abbott said, "Photo doesn't look like the expected *O. ferruginea*, but also doesn't quite match up with *O. discolor*. Actually looks closest to *O. schmidti*. We need a specimen to determine for sure what this is." Others have offered opinions, such as Daigle, who feels the photo is of *O. schmidti* (*pers comm*, 2016). We have waffled on this record many a time. The extreme distance this individual would have traveled seemingly rules *O. schmidti* out as a possibility, but the fact remains that we do not have a perfect understanding of the species' geographic range. Equally as great a distance to travel would be a vagrant of one of the *Orthemis* from the Antilles or Florida that is undescribed (or under-described); again, a flight of at least 1,900 km. Alternatively, this ♂ could be a hybrid *O. ferruginea* × *O. discolor*. That idea borders still on the implausible because it entails a vagrant hybrid traveling >725 km (~450 mi) north of where the two species are known to co-occur (even the nearest vagrant *O. discolor* record, from Lubbock County, Texas, is about 475 km away). The last option we can provide is an exceptionally aberrant *O. ferruginea*; a notion we are inclined to reject. Needless to say, this record will remain unconfirmed. The take home message … anything this out of the ordinary demands a specimen is taken.

Celithemis amanda – Amanda's Pennant

Multiple records of this species from south-central Arkansas (Calhoun, Ouachita, and Union County; about 170 km, or 105 mi, to the east of southeastern Oklahoma) indicate that a population exists there. Additional records from northeastern Texas make it possible that this species one day could be found in Oklahoma.

Celithemis ornata – Ornate Pennant

Records from Arkansas, Louisiana, and Texas, within a similar proximity to southeastern Oklahoma as *Celithemis amanda*'s current range is (see above), makes this a possible species to one day be added to the Oklahoma state list.

Erythemis plebeja – Pin-tailed Pondhawk

See entry for *Mesothemis plebija* [sic, *Mesothemis plebeja*] in *Synonyms, errors, and*

dubious records section above. This species is overdue in Oklahoma.

Sympetrum rubicundulum – Ruby Meadowhawk

There is little reason that this species could not show up in Oklahoma along with the other northern *Sympetrum*, such as *S. costiferum*, *S. internum*, and *S. obtrusum*, which are known to come to the state during southward pushes in the autumn. Look for this species in northeastern and north-central Oklahoma when strong autumnal winds blow from the north.

Sympetrum danae – Black Meadowhawk

Records of this species are as close as ~125 km (~75 mi) west of the Oklahoma panhandle. Like its congener, *Sympetrum pallipes*, the Striped Meadowhawk, the Black Meadowhawk could visit Oklahoma one day, likely also in the autumn.

†*Tramea insularis* – Antillean Saddlebags

A frustratingly skittish ♂ was watched for several hours at Great Plains SP, Kiowa County, on 30 September 2018 (MAP), but all attempts to secure a specimen or usable photographs were futile. It was seen in direct comparison to many (>15) *Tramea onusta*, the Red Saddlebags, but differed somewhat in color (deeper red) and had narrow saddles (as on *T. calverti*, the Striped Saddlebags). It was similar to *T. onusta* in the unmarked thorax and limited black atop distant abdominal segments. Key to identification was the dark violet frons.

SPECIES NO LONGER EXPECTED

There would have been a time that we would have expected *Ischnura cervula*, the Pacific Forktail, to find its way into Oklahoma. However, recently records from Colorado and New Mexico were re-identified as *I. damula*, the Plains Forktail (B Prather, *in litt*, 8 June 2017; JN Stuart, *in litt*, 22 Aug 2017). As such *I. cervula*'s range no longer comes as near the Oklahoma panhandle as was once thought.

NOTES

1. See "*Orthemis* species images," www.pugetsound.edu/academics/academic-resources/slater-museum/biodiversity-resources/dragonflies/image-collection/orthemis-species-images/
2. Daigle said he has had males and females in Florida that compare to the holotype for *Orthemis schmidti* (www.odonatacentral.org/index.php/PageAction.get/name/O_schmidti; copy also on file, OOP).

We obtained data for the Oklahoma Odonata Project database and for the book from various sources including online and personal databases and from numerous museum and personal collections. Those sources are listed below along with their codens (acronym), abbreviations, or, in some cases, their URL.

ANSP: Academy of Natural Sciences, Philadelphia, Pennsylvania, USA

ASU: Arkansas State University, Jonesboro, Arkansas, USA

Bick note cards: George H. Bick's species notes (large index cards), housed at FSCA/IORI

CASENT: California Academy of Sciences, Entomology Collection, San Francisco, California, USA

CCC: Carl Cook Collection, Center, Kentucky, USA [transferred to FSCA in 2019]

CMNH: Cleveland Museum of Natural History, Invertebrate Zoology Collection, Cleveland, Ohio, USA

CSU: Colorado State University, C. P. Gillette Museum of Arthropod Diversity, Department of Bioagricultural Sciences and Pest Management, Fort Collins, Colorado, USA

CUIC: Cornell Insect Collection, Ithaca, New York, USA

DMP: Dot Map Project – Donnelly (2004a, b, c)

DRP: Dennis R. Paulson's personal collection, Washington, USA

EEM: Enns Entomological Museum, University of Missouri, Columbia, Missouri, USA

EMEC: Essig Museum of Entomology Collection, Berkeley, California, USA

FMNH: The Field Museum of Natural History, Chicago, Illinois, USA

FSCA: Florida State Collection of Arthropods, Gainesville, Florida, USA

Fisher, J database: see J Fisher database, below

GBIF: Global Biodiversity Information Facility, www.gbif.org

iDigBio: www.idigbio.org/portal/search

iNat: iNaturalist, www.inaturalist.org

INHS: Illinois Natural History Survey, Champaign, Illinois, USA

IORI: International Odonata Research Institute, Gainesville, Florida, USA (see FSCA, as preferred acronym)

ISM: Illinois State Museum, Springfield, Illinois, USA

JCAC[1]:   John C. Abbott personal collection (currently housed at the University of Alabama)

J Fisher database: John Fisher's personal database of records reported to him; data now housed with the Oklahoma Biodiversity Information System

KJT: Kenneth J. Tennessen personal collection (likely at FSCA/IORI by publication)

MCZ: Museum of Comparative Zoology, Harvard University, Cambridge, Massachusetts, USA

OBS: Oklahoma Biological Survey, University of Oklahoma, Norman, Oklahoma, USA

OC: Odonata Central, www.odonatacentral.org

OMNH: Sam Noble Oklahoma Museum of Natural History, Recent Invertebrates, University of Oklahoma, Norman, Oklahoma, USA

ONC: Oxley Nature Center, Tulsa, Oklahoma, USA

OSU[2]:   K.C. Emerson Entomology Museum, Oklahoma State University, Stillwater, Oklahoma, USA [also used to refer just to the university]

OSUC: CA Triplehorn Insect Collection, Ohio State University, Columbus, Ohio, USA

PSU: Pittsburg State University, Pittsburg, Kansas, USA

QMOR: Ouellet-Robert entomological collection, Université de Montréal, Quebec, Canada

RWG: Rosser W Garrison personal collection, California, USA

SCAN: Symbiota Collections of Arthropods Network portal, http://scan-bugs.org/portal/collections/index.php

SEMC: University of Kansas, Snow Entomological Museum, Lawrence, Kansas, USA

SESC: Southeastern State College (now Southeastern Oklahoma State University[3]), Durant, Oklahoma, USA

SP: Smith-Patten/Patten personal collection, currently housed at the Oklahoma Biological Survey, Norman, Oklahoma, USA

TAMU: Texas A & M University Insect Collection, College Station, Texas, USA

TTU: Texas Tech University, Lubbock, Texas, USA

TWD: Thomas W. Donnelly personal collection

UCBME: University of California, Davis, RM Bohart Museum of Entomology, Davis, California, USA

UCO: University of Central Oklahoma, Edmond, Oklahoma, USA

UMMZ: University of Michigan, Museum of Zoology, Ann Arbor, Michigan, USA

UNSM: University of Nebraska State Museum, Lincoln, Nebraska, USA

USNM: United States National Museum of Natural History/Smithsonian Institution, Washington, D. C., USA

YPM: Yale University Peabody Museum, Stanford, Connecticut, USA

YUHO: Yucca House National Monument (associated with CSU)

In addition to the above specimen collections and data sources, we obtained Oklahoma-specific records from many published papers and reports (e.g., Abbott 1996; Ahrens 1938; Bass 1990; Beckemeyer 1995, 1998b, 2002a,c,e,f, 2004; Bick 1951, 1978, 1991; Bick and Bick 1957, 1961, 1965; Bird 1932, 1934; Byers 1937; Davis 1937; Donnelly 2004a,b,c ["Dot Map Project"]; Johnson 1973, 1974; Kondratieff et al. 2004; Kormondy 1960; Landwer and Sites 2003; Mills 2007; Muttkowski 1911; Ortenburger 1926b; Perkins 1980; Pritchard 1935, 1936; Reece and McIntyre 2008; Smith-Patten et al. 2007; Walker 1952, 1953; Williamson 1908, 1909, 1912a,b, 1914a,b; Zuellig et al. 2006). Steve Hummel also provided specimen data from his personal collection (no coden assigned).

NOTES

1. Formerly as UTIC – University of Texas Insect Collection, Brackenridge Field Laboratory, Austin, Texas, USA. Note that there is some confusion for some specimens as to whether they are actually part of JCAC or UTIC, so there may be a couple of instances where we miscited a specimen as JCAC that remained with UTIC when JCA departed that institution.
2. Also cited as OSEC (Evenhuis 2017), but we chose to use OSU instead.
3. Norris (2009).

OKLAHOMA ODONATA PROJECT

The Oklahoma Odonata Project's purpose is to document the diversity of the state's odonates (dragonflies and damselflies). Our field surveys, along with efforts to find all extant specimens for the state, produced an extensive database of odonate records. This database also contains data gathered from field notes of other researchers and photographic and sight records. Together these sources add up to >55,000 records dating back to 1877 and accounting for >250,000 individual odonates. With these data we have investigated many aspects of odonate ecology including biogeographical affinities, conservation status, and the effects of land-use changes.

The OOP is archived as part of the Oklahoma Biodiversity Information System (OBIS) of the Oklahoma Natural Heritage Inventory.

Webpages for the OOP (biosurvey.ou.edu/smith/Oklahoma_Odonata.html) have a variety of helpful resources. We regularly update these resources and will continue to add to and improve them. Resources include project history and updates, a species checklist for Oklahoma (.pdf or .doc), seasonality information (dates and basic chart), a listing of species by county and documentation level, a species richness map by county, and information about species of conservation concern. We also provide an Oklahoma gazetteer of localities at which data have been collected and a dragonfly finding guide.

Videos of interest are also posted on the OOP webpages, including a female glider (*Pantala sp.*) ovipositing on a car, a behavior not conducive to successful breeding, and the first ever video of a female Ozark Emerald (*Somatochlora ozarkensis*) ovipositing.

FURTHER READING, IDENTIFICATION GUIDES, AND ONLINE RESOURCES

We do not provide here a complete list of books, field guides, and other resources about odonates, but rather we hope these will get you started down the path of learning more about the wonderful world of dragonflies. In addition, we included some other resources that may be useful to you.

General

Dragonflies and damselflies: A natural history, by Dennis R. Paulson, 2019. Princeton University Press, Princeton, New Jersey.

Dragonflies and damselflies: Model organisms for ecological and evolutionary research, edited by Alex Córdoba-Aguilar, 2008. Oxford University Press, Oxford, UK.

Dragonflies of the world, by Jill Silsby, 2011. Smithsonian Institution Press, Washington, D. C.

Field Guides

Both of Dennis Paulson's guides are absolutely indispensable for Oklahoma; however, keep in mind that the maps have become somewhat outdated given the number of records generated since their publication. The field guides for Texas (Abbott 2011, 2015) are also useful for identification of odonates in Oklahoma, but they do not always reflect current distributional data. There are also many field guides from across the United States and Canada that are helpful with identifications.

Dragonflies and damselflies of the West, by Dennis R. Paulson, 2009. Princeton University Press, Princeton, New Jersey.

Dragonflies and damselflies of the East, by Dennis R. Paulson, 2011. Princeton University Press, Princeton, New Jersey.

Technical References

General:

Dragonflies: Behavior and ecology of Odonata, by Philip S. Corbet, 1999. Cornell University Press, Ithaca, New York.

Keys – Damselflies:

Damselflies of North America, by Minter J. Westfall, Jr., Michael L. May, 2006. Revised Edition. Scientific Publishers, Gainesville, Florida.

Damselfly genera of the New World: An illustrated and annotated key to the Zygoptera, by Rosser W. Garrison, Natalia von Ellenrieder, Jerry A. Louton, 2010. John Hopkins University Press, Baltimore, Maryland.

Keys – Dragonflies:

Dragonflies of North America: The Odonata (Anisoptera) fauna of Canada, the continental United States, northern Mexico and the Greater Antilles. 3rd edition, by James G. Needham, Minter J. Westfall, Jr., Michael L. May, 2014. Scientific Publishers, Gainesville, Florida.

Dragonfly genera of the New World: An illustrated and annotated key to the Anisoptera, by Rosser W. Garrison, Natalia von Ellenrieder, Jerry A. Louton, 2006. John Hopkins University Press, Baltimore, Maryland.

Dragonfly nymphs of North America: an identification guide, by Kenneth J. Tennessen, 2019. Springer International Publishing, Switzerland.

Online Resources

Booksellers and Equipment:
    BioQuip
        https://www.bioquip.com/
    NHBS
        https://www.nhbs.com/
    Pemberley Books
        https://www.pemberleybooks.com/
Record Submission:
    Odonata Central
        https://www.odonatacentral.org/

Societies and Odonate Groups

Dragonfly Society of the Americas
    https://www.dragonflysocietyamericas.org/
International Odonata Research Institute
    http://www.iodonata.net/[links to a wide variety of groups]
Worldwide Dragonfly Association
    https://worlddragonfly.org/

A CHECKLIST OF OKLAHOMA ODONATA (DRAGONFLIES AND DAMSELFLIES)

Adapted from Smith-Patten & Patten (2020) in alphabetical order by family and scientific name

State total: 176 species

ZYGOPTERA – DAMSELFLIES (n = 58)
Calopterygidae – Broad-winged Damsels (*n* = 3)
- [ ] *Calopteryx maculata* – Ebony Jewelwing
- [ ] *Hetaerina americana* – American Rubyspot
- [ ] *Hetaerina titia* – Smoky Rubyspot

Lestidae – Spreadwings (*n* = 10)
- [ ] *Archilestes grandis* – Great Spreadwing
- [ ] *Lestes alacer* – Plateau Spreadwing
- [ ] *Lestes australis* – Southern Spreadwing
- [ ] *Lestes eurinus* – Amber-winged Spreadwing
- [ ] *Lestes forcipatus* – Sweetflag Spreadwing
- [ ] *Lestes inaequalis* – Elegant Spreadwing
- [ ] *Lestes rectangularis* – Slender Spreadwing
- [ ] *Lestes sigma* – Chalky Spreadwing
- [ ] *Lestes unguiculatus* – Lyre-tipped Spreadwing
- [ ] *Lestes vigilax* – Swamp Spreadwing

Coenagrionidae – Pond Damsels (*n* = 45)
- [ ] *Amphiagrion abbreviatum* – Western Red Damsel
- [ ] *Argia alberta* – Paiute Dancer
- [ ] *Argia apicalis* – Blue-fronted Dancer
- [ ] *Argia bipunctulata* – Seepage Dancer
- [ ] *Argia fumipennis* – Variable Dancer
- [ ] *Argia immunda* – Kiowa Dancer
- [ ] *Argia leonorae* – Leonora's Dancer
- [ ] *Argia lugens* – Sooty Dancer
- [ ] *Argia moesta* – Powdered Dancer
- [ ] *Argia nahuana* – Aztec Dancer
- [ ] *Argia plana* – Springwater Dancer
- [ ] *Argia sedula* – Blue-ringed Dancer
- [ ] *Argia tezpi* – Tezpi Dancer
- [ ] *Argia tibialis* – Blue-tipped Dancer
- [ ] *Argia translata* – Dusky Dancer
- [ ] *Chromagrion conditum* – Aurora Damsel
- [ ] *Enallagma antennatum* – Rainbow Bluet
- [ ] *Enallagma aspersum* – Azure Bluet
- [ ] *Enallagma basidens* – Double-striped Bluet
- [ ] *Enallagma carunculatum* – Tule Bluet
- [ ] *Enallagma civile* – Familiar Bluet
- [ ] *Enallagma daeckii* – Attenuated Bluet
- [ ] *Enallagma divagans* – Turquoise Bluet
- [ ] *Enallagma doubledayi* – Atlantic Bluet
- [ ] *Enallagma dubium* – Burgundy Bluet
- [ ] *Enallagma exsulans* – Stream Bluet
- [ ] *Enallagma geminatum* – Skimming Bluet
- [ ] *Enallagma praevarum* – Arroyo Bluet
- [ ] *Enallagma signatum* – Orange Bluet
- [ ] *Enallagma traviatum* – Slender Bluet
- [ ] *Enallagma vesperum* – Vesper Bluet
- [ ] *Ischnura barberi* – Desert Forktail
- [ ] *Ischnura damula* – Plains Forktail
- [ ] *Ischnura demorsa* – Mexican Forktail
- [ ] *Ischnura denticollis* – Black-fronted Forktail
- [ ] *Ischnura hastata* – Citrine Forktail
- [ ] *Ischnura kellicotti* – Lilypad Forktail
- [ ] *Ischnura perparva* – Western Forktail
- [ ] *Ischnura posita* – Fragile Forktail
- [ ] *Ischnura ramburii* – Rambur's Forktail
- [ ] *Ischnura verticalis* – Eastern Forktail
- [ ] *Nehalennia gracilis* – Sphagnum Sprite
- [ ] *Nehalennia integricollis* – Southern Sprite
- [ ] *Telebasis byersi* – Duckweed Firetail
- [ ] *Telebasis salva* – Desert Firetail

ANISOPTERA – DRAGONFLIES (n = 118)
Petaluridae – Petaltails (*n* = 1)
- [ ] *Tachopteryx thoreyi* – Gray Petaltail

Aeshnidae – Darners (*n* = 12)
- [ ] *Aeshna constricta* – Lance-tipped Darner
- [ ] *Aeshna umbrosa* – Shadow Darner
- [ ] *Anax junius* – Common Green Darner
- [ ] *Anax longipes* – Comet Darner
- [ ] *Basiaeschna janata* – Springtime Darner
- [ ] *Boyeria vinosa* – Fawn Darner
- [ ] *Coryphaeschna ingens* – Regal Darner
- [ ] *Epiaeschna heros* – Swamp Darner
- [ ] *Gomphaeschna furcillata* – Harlequin Darner
- [ ] *Gynacantha nervosa* – Twilight Darner
- [ ] *Nasiaeschna pentacantha* – Cyrano Darner
- [ ] *Rhionaeschna multicolor* – Blue-eyed Darner

Gomphidae – Clubtails (*n* = 22)
- [ ] *Aphylla williamsoni* – Two-striped Forceptail
- [ ] *Arigomphus lentulus* – Stillwater Clubtail
- [ ] *Arigomphus maxwelli* – Bayou Clubtail
- [ ] *Arigomphus submedianus* – Jade Clubtail
- [ ] *Dromogomphus spinosus* – Black-shouldered Spinyleg
- [ ] *Dromogomphus spoliatus* – Flag-tailed Spinyleg
- [ ] *Erpetogomphus designatus* – Eastern Ringtail
- [ ] *Gomphurus externus* – Plains Clubtail
- [ ] *Gomphurus hybridus* – Cocoa Clubtail
- [ ] *Gomphurus ozarkensis* – Ozark Clubtail
- [ ] *Gomphurus vastus* – Cobra Clubtail
- [ ] *Hagenius brevistylus* – Dragonhunter
- [ ] *Hylogomphus apomyius* – Banner Clubtail
- [ ] *Phanogomphus graslinellus* – Pronghorn Clubtail
- [ ] *Phanogomphus lividus* – Ashy Clubtail
- [ ] *Phanogomphus militaris* – Sulphur-tipped Clubtail
- [ ] *Phanogomphus oklahomensis* – Oklahoma Clubtail
- [ ] *Phyllogomphoides stigmatus* – Four-striped Leaftail
- [ ] *Progomphus obscurus* – Common Sanddragon
- [ ] *Stylogomphus sigmastylus* – Interior Least Clubtail
- [ ] *Stylurus intricatus* – Brimstone Clubtail
- [ ] *Stylurus plagiatus* – Russet-tipped Clubtail

Cordulegastridae – Spiketails (*n* = 3)
- ☐ *Cordulegaster maculata* – Twin-spotted Spiketail
- ☐ *Cordulegaster obliqua* – Arrowhead Spiketail
- ☐ *Cordulegaster talaria* – Ouachita Spiketail

Macromiidae – Cruisers (*n* = 6)
- ☐ *Didymops transversa* – Stream Cruiser
- ☐ *Macromia alleghaniensis* – Allegheny River Cruiser
- ☐ *Macromia illinoiensis* – Swift River Cruiser
- ☐ *Macromia pacifica* – Gilded River Cruiser
- ☐ *Macromia taeniolata* – Royal River Cruiser
- ☐ *Macromia annulata* – Bronzed River Cruiser

Corduliidae – Emeralds (*n* = 15)
- ☐ *Epitheca costalis* – Slender Baskettail
- ☐ *Epitheca cynosura* – Common Baskettail
- ☐ *Epitheca petechialis* – Dot-winged Baskettail
- ☐ *Epitheca princeps* – Prince Baskettail
- ☐ *Epitheca semiaquea* – Mantled Baskettail
- ☐ *Epitheca spinosa* – Robust Baskettail
- ☐ *Helocordulia selysii* – Selys's Sundragon
- ☐ *Helocordulia uhleri* – Uhler's Sundragon
- ☐ *Neurocordulia molesta* – Smoky Shadowdragon
- ☐ *Neurocordulia virginiensis* – Cinnamon Shadowdragon
- ☐ *Neurocordulia xanthosoma* – Orange Shadowdragon
- ☐ *Somatochlora linearis* – Mocha Emerald
- ☐ *Somatochlora ozarkensis* – Ozark Emerald
- ☐ *Somatochlora margarita* – Texas Emerald
- ☐ *Somatochlora tenebrosa* – Clamp-tipped Emerald

Libellulidae – Skimmers (*n* = 59)
- ☐ *Brachymesia furcata* – Red-tailed Pennant
- ☐ *Brachymesia gravida* – Four-spotted Pennant
- ☐ *Brachymesia herbida* – Tawny Pennant
- ☐ *Brechmorhoga mendax* – Pale-faced Clubskimmer
- ☐ *Celithemis elisa* – Calico Pennant
- ☐ *Celithemis eponina* (Drury, 1773) – Halloween Pennant
- ☐ *Celithemis fasciata* – Banded Pennant
- ☐ *Celithemis verna* – Double-ringed Pennant
- ☐ *Dythemis fugax* – Checkered Setwing
- ☐ *Dythemis nigrescens* – Black Setwing
- ☐ *Dythemis velox* – Swift Setwing
- ☐ *Erythemis attala* – Black Pondhawk
- ☐ *Erythemis collocata* – Western Pondhawk
- ☐ *Erythemis simplicicollis* – Eastern Pondhawk
- ☐ *Erythemis vesiculosa* – Great Pondhawk
- ☐ *Erythrodiplax berenice* – Seaside Dragonlet
- ☐ *Erythrodiplax basifusca* – Plateau Dragonlet
- ☐ *Erythrodiplax minuscula* – Little Blue Dragonlet
- ☐ *Erythrodiplax umbrata* – Band-winged Dragonlet
- ☐ *Ladona deplanata* – Blue Corporal
- ☐ *Libellula auripennis* – Golden-winged Skimmer
- ☐ *Libellula comanche* – Comanche Skimmer
- ☐ *Libellula composita* – Bleached Skimmer
- ☐ *Libellula croceipennis* – Neon Skimmer
- ☐ *Libellula cyanea* – Spangled Skimmer
- ☐ *Libellula flavida* – Yellow-sided Skimmer
- ☐ *Libellula incesta* – Slaty Skimmer
- ☐ *Libellula luctuosa* – Widow Skimmer
- ☐ *Libellula nodisticta* – Hoary Skimmer
- ☐ *Libellula pulchella* – Twelve-spotted Skimmer
- ☐ *Libellula saturata* – Flame Skimmer
- ☐ *Libellula semifasciata* – Painted Skimmer
- ☐ *Libellula vibrans* – Great Blue Skimmer
- ☐ *Macrodiplax balteata* – Marl Pennant
- ☐ *Miathyria marcella* – Hyacinth Glider
- ☐ *Micrathyria hagenii* – Thornbush Dasher
- ☐ *Orthemis ferruginea* – Roseate Skimmer
- ☐ *Pachydiplax longipennis* – Blue Dasher
- ☐ *Paltothemis lineatipes* – Red Rock Skimmer
- ☐ *Pantala flavescens* – Wandering Glider
- ☐ *Pantala hymenaea* – Spot-winged Glider
- ☐ *Perithemis tenera* – Eastern Amberwing
- ☐ *Plathemis lydia* – Common Whitetail
- ☐ *Plathemis subornata* – Desert Whitetail
- ☐ *Pseudoleon superbus* – Filigree Skimmer
- ☐ *Sympetrum ambiguum* – Blue-faced Meadowhawk
- ☐ *Sympetrum corruptum* – Variegated Meadowhawk
- ☐ *Sympetrum illotum* – Cardinal Meadowhawk
- ☐ *Sympetrum internum* – Cherry-faced Meadowhawk
- ☐ *Sympetrum obtrusum* – White-faced Meadowhawk
- ☐ *Sympetrum costiferum* – Saffron-winged Meadowhawk
- ☐ *Sympetrum pallipes* – Striped Meadowhawk
- ☐ *Sympetrum semicinctum* – Band-winged Meadowhawk
- ☐ *Sympetrum vicinum* – Autumn Meadowhawk
- ☐ *Tholymis citrina* – Evening Skimmer
- ☐ *Tramea calverti* – Striped Saddlebags
- ☐ *Tramea carolina* – Carolina Saddlebags
- ☐ *Tramea lacerata* – Black Saddlebags
- ☐ *Tramea onusta* – Red Saddlebags

Oklahoma Odonata species color-coded by era (blue = early era; peach = OU Expeditions era, 1925–1949; yellow = Merry Trio era, 1950–1999; blue = 2000–2020), as provided for by published records and lists, as well as by photographs and specimens. Note the discrepancies between the total number of species thought to occur in the state by era, versus the number that we now know to have occurred during that era. Discrepancies are highlighted by color-coded daggers (†) indicating that the species was either not known as a record during that era or was not conclusively identified at the time. Such species include the *Progomphus obscurus* that AI Ortenburger collected just after his 1926 publication, the *Argia leonorae* that was collected by RD Bird but the species had not been described at the time, and the *Stylurus intricatus* is an example of a species that was identified by AE Pritchard at the time but was not published until Smith-Patten & Patten (2012). Another good example is the *Gynacantha nervosa* that was collected in 1935 but it went unreported until Kormondy (1960). The Oklahoma state list now stands at 176 species.

Legend:
- ● = known previously in the state
- ● = new to the era
- ● (any color) = unknown in that era
- † (any color) = unknown in that era*

Taxon	Williamson 1914[a]	others	Ortenburger 1926[b]	Bird 1932[c], 1933	Pritchard[d]	others	total for era	Bick 1951[e]	Bick & Bick 1957[f]	Bick 1978, 1991[g]	total for era	Beckemeyer[h]/Abbott[i]	2003–2019	current state list	notes (subscripts refer to citations to left)
	1	2	3	4	5	6	7	8	9	10	11	12	13	14	
ODONATA: ZYGOPTERA – DAMSELFLIES															
Calopterygidae - Broad-winged Damsels															
Calopteryx maculata – Ebony Jewelwing	●		●	●			●	●	●	●	●	●	●	●	as Agrion maculatum [b,c,e,f]
Hetaerina americana – American Rubyspot			●	●	●		●	●	●	●	●	●	●	●	
Hetaerina titia – Smoky Rubyspot	●			●	●		●	●	●	●	●	●	●	●	also as Hetaerina tricolor [c]
Lestidae - Spreadwings															
Archilestes grandis – Great Spreadwing	●			●	●		●		●	●	●	●	●	●	
Lestes alacer – Plateau Spreadwing				●	●		●		●	●	●	●	●	●	
Lestes australis – Southern Spreadwing				●	●		●		●		●	●	●	●	L. forcipatus [c], L. disjunctus australis [f,g,h], L. disjunctus [i]
Lestes eurinus – Amber-winged Spreadwing													●	●	
Lestes forcipatus – Sweetflag Spreadwing													●	●	
Lestes inaequalis – Elegant Spreadwing	●			●	●		●		●	●	●	●	●	●	as L. inequalis [a,c]
Lestes rectangularis – Slender Spreadwing				●	●		●		●	●	●	●	●	●	
Lestes sigma – Chalky Spreadwing										●		●	●	●	
Lestes unguiculatus – Lyre-tipped Spreadwing					●		●		●		●	●	●	●	
Lestes vigilax – Swamp Spreadwing					●		●		●		●	●	●	●	
Coenagrionidae - Pond Damsels															
Amphiagrion abbreviatum – Western Red Damsel					●		●		●	●	●	●	●	●	as A. saucium [c]; as A. sp. [h]
Argia alberta – Paiute Dancer				●	●		●	●	●	●	●	●	●	●	
Argia apicalis – Blue-fronted Dancer	●			●	●		●		●	●	●	●	●	●	
Argia bipunctulata – Seepage Dancer				●			●		●	●	●	●	●	●	
Argia fumipennis – Variable Dancer	●		●	●	●		●	●	●	●	●	●	●	●	as A. violacea [a,b,c,e,f]; as A. fumipennis violacea [g,h]
Argia immunda – Kiowa Dancer				●			●		●	●	●	●	●	●	also as A. agrioides [c]
Argia leonorae – Leonora's Dancer				†			†	†	†	†	†	●	●	●	
Argia lugens – Sooty Dancer					●		●		●	●	●	●	●	●	as Hyponeura lugens [f]

(Continued)

column #	1	2	3	4	5	6	7	8	9	10	11	12	13	14	
Argia moesta - Powdered Dancer	●						●	●	●	●	●	●	●	●	also as *A. intruda* [a,c]; *A. putrida* [b]
Argia nahuana - Aztec Dancer			●	●			●		●	●	●	●	●	●	as *A. agrioides* [c]; as *A. agrioides nahuana* [f]; as *A. nehuana* [g] [sic]
Argia plana - Springwater Dancer	●		●	●	●		●	●	●	●	●	●	●	●	as *A. vivida* [c]; as *A. v. plana* [f]
Argia sedula - Blue-ringed Dancer			●	●			●	●	●	●	●	●	●	●	
Argia tezpi - Tezpi Dancer													●	●	
Argia tibialis - Blue-tipped Dancer	●		●	●	●		●		●		●	●	●	●	
Argia translata - Dusky Dancer				●			●	●	●	●	●	●	●	●	
Chromagrion conditum - Aurora Damsel					●		●		●		●	●	●	●	
Enallagma antennatum - Rainbow Bluet				●			●		●		●	●	●	●	
Enallagma aspersum - Azure Bluet				●	●		●		●		●	●	●	●	
Enallagma basidens - Double-striped Bluet				●	●		●		●	●	●	●	●	●	
Enallagma carunculatum - Tule Bluet				●			●		●	●	●	●	●	●	
Enallagma civile - Familiar Bluet	●				●		●	●	●	●	●	●	●	●	
Enallagma daeckii - Attenuated Bluet	●				●		●				●		●	●	as *Teleallagma daecki* [f]
Enallagma divagans - Turquoise Bluet	●			●	●		●		●		●	●	●	●	
Enallagma doubledayi - Atlantic Bluet													●	●	
Enallagma dubium - Burgundy Bluet					●		●		●		●	●	●	●	
Enallagma exsulans - Stream Bluet	●			●	●		●	●	●		●	●	●	●	
Enallagma geminatum - Skimming Bluet	●			●	●		●		●		●	●	●	●	as *E. piscinarium* [c]
Enallagma praevarum - Arroyo Bluet				●			●		●		●	●	●	●	
Enallagma signatum - Orange Bluet	●			●	●		●		●		●	●	●	●	
Enallagma traviatum - Slender Bluet	●			●	●		●		●		●	●	●	●	
Enallagma vesperum - Vesper Bluet	●			●	●		●		●	●	●	●	●	●	*E. pollutum* [a,c]. *E. laurenti* [g,c]
Ischnura barberi - Desert Forktail				●			●		●		●	●	●	●	also as *I. utahensis* [c]
Ischnura damula - Plains Forktail	●				●						●	●	●	●	
Ischnura demorsa - Mexican Forktail								●	●	●	●	●	●	●	as *Anomalagrion hastatum* [a,c,e,f]
Ischnura denticollis - Black-fronted Forktail				●			●		●		●	●	●	●	
Ischnura hastata - Citrine Forktail	●						●		●	●	●	●	●	●	
Ischnura kellicotti - Lilypad Forktail	●			●	●		●		●		●	●	●	●	
Ischnura perparva - Western Forktail											●	●	●	●	missed by Abbott (2005)
Ischnura posita - Fragile Forktail	●			●			●		●	●	●	●	●	●	
Ischnura ramburii - Rambur's Forktail				●					●		●	●	●	●	
Ischnura verticalis - Eastern Forktail				●	●		●	●	●		●	●	●	●	

(Continued)

column #	1	2	3	4	5	6	7	8	9	10	11	12	13	14	Notes
Nehalennia gracilis - Sphagnum Sprite													●	●	
Nehalennia integricollis - Southern Sprite					●				●		●	●	●	●	
Telebasis byersi - Duckweed Firetail													●	●	
Telebasis salva - Desert Firetail				●			●		●	●	●	●	●	●	
ODONATA: ANISOPTERA – DRAGONFLIES															
Petaluridae - Petaltails															
Tachopteryx thoreyi - Gray Petaltail	●					●	●		●	●	●	●	●	●	
Aeshnidae – Darners															
Aeshna constricta - Lance-tipped Darner													●	●	as Aeschna [sic] umbrosa [c]
Aeshna umbrosa - Shadow Darner				●	●		●		●			●	●	●	
Anax junius - Common Green Darner				●	●		●		●	●	●	●	●	●	
Anax longipes - Comet Darner					●					●	●	●	●	●	
Basiaeschna janata - Springtime Darner				●	●		●		●	●	●	●	●	●	
Boyeria vinosa - Fawn Darner					●		●		●	●	●	●	●	●	
Coryphaeschna ingens - Regal Darner													●	●	
Epiaeschna heros - Swamp Darner	●			●			●		●	●	●	●	●	●	
Gomphaeschna furcillata - Harlequin Darner			●								●	●	●	●	
Gynacantha nervosa - Twilight Darner						+	+	+	+	+	+	●	●	●	
Nasiaeschna pentacantha - Cyrano Darner				●	●		●	●	●	●	●	●	●	●	
Rhionaeschna multicolor - Blue-eyed Darner				●			●		●		●	●	●	●	Aeschna [sic] multicolor [c]; Aeshna multicolor [e,i]
Gomphidae - Clubtails															
Aphylla williamsoni - Two-striped Forceptail													●	●	
Arigomphus lentulus - Stillwater Clubtail							●				●	●	●	●	Gomphus subapicalis [c]
Arigomphus maxwelli - Bayou Clubtail													●	●	
Arigomphus submedianus - Jade Clubtail	●			●			●		●	●	●	●	●	●	
Dromogomphus spinosus - Black-shouldered Spinyleg	●			●			●	●	●	●	●	●	●	●	Gomphus submedianus [a,c]
Dromogomphus spoliatus - Flag-tailed Spinyleg							●	●	●	●	●	●	●	●	
Erpetogomphus designatus - Eastern Ringtail							●	●	+	+	+	●	●	●	
Gomphurus externus - Plains Clubtail				●	●		●				●	●	●	●	
Gomphurus hybridus - Cocoa Clubtail													●	●	
Gomphurus ozarkensis - Ozark Clubtail								+	+	+	+	●	●	●	
Gomphurus vastus - Cobra Clubtail								+	+	+	+	●	●	●	
Hagenius brevistylus - Dragonhunter	●			●			●		●	●	●	●	●	●	
Hylogomphus apomyius - Banner Clubtail													●	●	

(Continued)

Species	1	2	3	4	5	6	7	8	9	10	11	12	13	14	Notes
column #															
Phanogomphus graslinellus - Pronghorn Clubtail	●			●	●		●		●	●	●	●	●	●	
Phanogomphus lividus – Ashy Clubtail											●	●	●	●	
P. militaris - Sulphur-tipped Clubtail				●	●		●	●	●		●	●	●	●	
P. oklahomensis - Oklahoma Clubtail				●	●		●		●		●	●	●	●	
Phyllogomphoides stigmatus - Four-striped Leaftail						●	●	●	●	●	●	●	●	●	as *Gomphoides stigmatus*[f]
Progomphus obscurus - Common Sanddragon			†	●	●		†				●	●	●	●	
Stylogomphus sigmastylus - Interior Least Clubtail						●	●				●	●	●	●	as *S. albistylus*[g,h,i]
Stylurus intricatus - Brimstone Clubtail					†		†	†	†	†	†		●	●	actually AE Pritchard knew he had documented the species but it was not published until Smith-Patten & Patten (2012)
Stylurus plagiatus - Russet-tipped Clubtail	●			●			●		●	●	●	●	●	●	as *Gomphus plagiatus*[a,c]
Cordulegastridae – Spiketails															
Cordulegaster maculata - Twin-spotted Spiketail												●	●	●	
Cordulegaster obliqua - Arrowhead Spiketail				●	●		●	●	●		●	●	●	●	
Cordulegaster talaria - Ouachita Spiketail				●	●		●			●	●	●	●	●	as *Macromia annulata*[c]
Macromiidae – Cruisers															
Didymops transversa - Stream Cruiser				●			●		●		●	●	●	●	
Macromia alleghaniensis - Allegheny River Cruiser								†	†	†	†	●	●	●	
Macromia annulata - Bronzed River Cruiser				●	●		●	●			●	●	●	●	
Macromia illinoiensis - Swift River Cruiser	●			●	●		●		●		●	●	●	●	as *M. australensis*[a,c]; as *M. georgina*[f,g]
Macromia pacifica - Gilded River Cruiser				●	●		●	●	●		●	●	●	●	
Macromia taeniolata - Royal River Cruiser				●	●		●	●	●		●	●	●	●	as *M. wabashensis*[c]
Corduliidae – Emeralds															
Epitheca costalis - Slender Baskettail	●			●	●		●		●		●	●	●	●	as *Tetragoneuria williamsoni*[a,c,f]
Epitheca cynosura - Common Baskettail				●	●		●	●	●		●	●	●	●	as *Tetragoneuria cynosura*[c,e,f,g]; as *T. cynosura simulans*[d]
Epitheca petechialis - Dot-winged Baskettail				●	●		●				●	●	●	●	as *Tetragoneuria petechialis*[f,g]
Epitheca princeps - Prince Baskettail				●	●		●	●	●	●	●	●	●	●	as *Epicordulia princeps*[c,e,f,g]

(Continued)

column #	1	2	3	4	5	6	7	8	9	10	11	12	13	14	
Epitheca semiaquea - Mantled Baskettail								†	†	†	†	•	•	•	
Epitheca spinosa - Robust Baskettail				•			•		•		•	•	•	•	as *Tetragoneuria canis*[c]; as *T. spinosa*[f]
Helocordulia selysii - Selys's Sundragon												•	•	•	
Helocordulia uhleri - Uhler's Sundragon									•		•	•		•	
Neurocordulia molesta - Smoky Shadowdragon									•		•	•	•	•	
Neurocordulia virginiensis - Cinnamon Shadowdragon						•	•					•	•	•	
Neurocordulia xanthosoma - Orange Shadowdragon	•			•	•		•	•	•		•	•	•	•	as *Platycordulia xanthosoma*[a,b,c,e], as *N. obsoleta*[e,f]
Somatochlora linearis - Mocha Emerald				•	•		•	•	•	•	•	•	•	•	as *S. ensigera*[c]
Somatochlora ozarkensis - Ozark Emerald				•	•		•		•		•	•	•	•	as *Somatochlora sp*[c]
Somatochlora tenebrosa - Clamp-tipped Emerald					•		•		•		•	•	•	•	
Somatochlora margarita - Texas Emerald													•	•	
Libellulidae - Skimmers															
Brachymesia furcata - Red-tailed Pennant													•	•	
Brachymesia gravida - Four-spotted Pennant									•	•	•	•	•	•	as *Cannacria gravida*[f]
Brachymesia herbida - Tawny Pennant													•	•	
Brechmorhoga mendax - Pale-faced Clubskimmer				•							•	•	•	•	
Celithemis elisa - Calico Pennant				•	•		•	•	•	•	•	•	•	•	
Celithemis eponina - Halloween Pennant			•	•	•		•	•	•	•	•	•	•	•	
Celithemis fasciata - Banded Pennant			•	•	•		•	•	•	•	•	•	•	•	
Celithemis verna - Double-ringed Pennant				•	•		•		•		•	•	•	•	
Dythemis fugax - Checkered Setwing				•	•		•		•		•	•	•	•	as *Dythemis fujax* [sic][c]
Dythemis nigrescens - Black Setwing													•	•	
Dythemis velox - Swift Setwing			•	•	•		•		•		•	•	•	•	
Erythemis attala - Black Pondhawk													•	•	
Erythemis collocata - Western Pondhawk								†	†	†	†		•	•	
Erythemis simplicicollis - Eastern Pondhawk	•		•	•	•		•	•	•	•	•	•	•	•	as *E. semplicicollis* [sic][b], as *Mesothemis simplicicollis*[c]
Erythemis vesiculosa - Great Pondhawk				•	•		•		•		•	•	•	•	as *Lepthemis vesiculosa*[c,f]
Erythrodiplax berenice - Seaside Dragonlet													•	•	
Erythrodiplax basifusca - Plateau Dragonlet													•	•	
Erythrodiplax minuscula - Little Blue Dragonlet				•			•		•		•	•	•	•	as *E. connata minuscula*[d]

(Continued)

column #	1	2	3	4	5	6	7	8	9	10	11	12	13	14	
Erythrodiplax umbrata - Band-winged Dragonlet							•				•	•	•	•	
Ladona deplanata - Blue Corporal				•	•		•				•	•	•	•	as Libellula deplanata[g]
Libellula auripennis - Golden-winged Skimmer						•	•				•	•	•	•	
Libellula comanche - Comanche Skimmer				•			•		•		•	•	•	•	
Libellula composita - Bleached Skimmer													•	•	
Libellula croceipennis - Neon Skimmer	•						•		•		•	•	•	•	
Libellula cyanea - Spangled Skimmer			•	•			•	•			•	•	•	•	
Libellula flavida - Yellow-sided Skimmer					•		•		•		•	•	•	•	
Libellula incesta - Slaty Skimmer	•		•		•		•	•	•		•	•	•	•	
Libellula luctuosa - Widow Skimmer	•		•		•		•	•	•	•	•	•	•	•	
Libellula nodisticta - Hoary Skimmer									•	•	•	•	•	•	
Libellula pulchella - Twelve-spotted Skimmer	•		•	•	•		•		•		•	•	•	•	
Libellula saturata - Flame Skimmer					•		•				•	•	•	•	
Libellula semifasciata - Painted Skimmer							•		•	•	•	•	•	•	as Libellula axillena[b,c] [sic, spelling and id]
Libellula vibrans - Great Blue Skimmer	•		•	•	•		•			•	•	•	•	•	
Macrodiplax balteata - Marl Pennant								†	†	†	†		•	•	
Miathyria marcella - Hyacinth Glider													•	•	
Micrathyria hagenii - Thornbush Dasher													•	•	
Orthemis ferruginea - Roseate Skimmer			•	•	•		•		•		•	•	•	•	
Pachydiplax longipennis - Blue Dasher	•		•	•	•		•	•	•		•	•	•	•	
Paltothemis lineatipes - Red Rock Skimmer				†			†				†	•	•	†	species not reported by Beckemeyer (2002a)
Pantala flavescens - Wandering Glider	•		•	•	•		•	•	•		•	•	•	•	
Pantala hymenaea - Spot-winged Glider	•		•	•	•		•	•	•		•	•	•	•	
Perithemis tenera - Eastern Amberwing	•		•	•	•		•		•		•	•	•	•	also as P. domitia[c]
Plathemis lydia - Common Whitetail	•		•	•	•		•	•	•		•	•	•	•	as Libellula lydia[g]
Plathemis subornata - Desert Whitetail							•					•	•	•	
Pseudoleon superbus - Filigree Skimmer											•	•	•	•	
Sympetrum ambiguum - Blue-faced Meadowhawk	•				•		•		•		•	•	•	•	
Sympetrum corruptum - Variegated Meadowhawk			•	•	•		•		•		•	•	•	•	as Tarnetrum corruptum[f]
Sympetrum illotum - Cardinal Meadowhawk													•	•	
Sympetrum internum - Cherry-faced Meadowhawk		•				†	†		•				•	•	
Sympetrum obtrusum - White-faced Meadowhawk													•	•	

(Continued)

column #	1	2	3	4	5	6	7	8	9	10	11	12	13	14
Sympetrum costiferum - Saffron-winged Meadowhawk													●	●
Sympetrum pallipes - Striped Meadowhawk													●	●
Sympetrum semicinctum - Band-winged Meadowhawk				●	●		●		●		●	●	●	●
Sympetrum vicinum - Autumn Meadowhawk				●	●		●		●	●	●	●	●	●
Tholymis citrina - Evening Skimmer													●	●
Tramea calverti - Striped Saddlebags													●	●
Tramea carolina - Carolina Saddlebags			●					●	●	●	●	●	●	●
Tramea lacerata - Black Saddlebags			●	●			●	●	●	●	●	●	●	●
Tramea onusta - Red Saddlebags				●	●		●	●	●	●	●	●	●	●
total Zygopterans	17	0	6	40	26	0	46	14	46	36	49	51	57	58
total Anisopterans	22	1	15	58	50	6	71	34	75	61	80	86	114	118
total odonates	39	1	21	98	76	6	117	48	121	97	129	137	171	176

Note: *S. occidentale fasciatum* [f,h] (*Sympetrum semicinctum* - Band-winged Meadowhawk)

	1877–1924	OU Expeditions era	Merry Trio era	2000–2020
total new for era		83	16	37
state list by era - known at time	40	117	129	135 (in 2003)
unknown at the time	0	6	10	4
state list actual	**40**	**123**	**139**	**176**

First state records of Oklahoma Odonata, indicating county where the species was found, year discovered, collector/observer, and documentation.

#	FAMILY	SPECIES	AGE	COUNTY	YEAR	COLLECTOR/OBSERVER	DOCUMENTATION
1	Libellulidae	*Sympetrum internum*	a	unknown	1877	Riley, CV	USNM
2	Aeshnidae	*Anax junius*	a	Okmulgee	1907	Collins, F	Williamson 1914b
3	Aeshnidae	*Epiaeschna heros*	a	Le Flore	1907	Collins, F	Williamson 1914b
4	Calopterygidae	*Hetaerina americana*	a	Le Flore	1907	Collins, F	Williamson 1914b
5	Calopterygidae	*Hetaerina titia*	a	Le Flore	1907	Collins, F	Williamson 1912a, 1914b
6	Gomphidae	*Hagenius brevistylus*	a	Le Flore	1907	Collins, F	UMMZ
7	Gomphidae	*Stylurus plagiatus*	a	Le Flore	1907	Collins, F	Williamson 1914b
8	Libellulidae	*Libellula incesta*	a	Le Flore	1907	Collins, F	Williamson 1914b
9	Libellulidae	*Libellula luctuosa*	a	Le Flore	1907	Collins, F	UMMZ
10	Libellulidae	*Libellula vibrans*	a	Le Flore	1907	Collins, F	Williamson 1914b
11	Libellulidae	*Pantala flavescens*	a	Okmulgee	1907	Collins, F	UMMZ
12	Libellulidae	*Pantala hymenaea*	a	Okmulgee	1907	Collins, F	UMMZ, Williamson 1914b
13	Libellulidae	*Plathemis lydia*	a	Le Flore	1907	Collins, F	UMMZ
14	Libellulidae	*Sympetrum corruptum*	a	Okmulgee	1907	Collins, F	UMMZ
15	Macromiidae	*Macromia illinoiensis*	a	Le Flore	1907	Collins, F	UMMZ, Williamson 1909
16	Coenagrionidae	*Argia fumipennis*	a	Le Flore	1907	Williamson, EB; Collins, F	Williamson 1914b
17	Coenagrionidae	*Argia moesta*	a	Le Flore	1907	Williamson, EB; Collins, F	UMMZ, Williamson 1912b, 1914b
18	Coenagrionidae	*Argia tibialis*	a	Le Flore	1907	Williamson, EB; Collins, F	UMMZ
19	Coenagrionidae	*Enallagma civile*	a	Le Flore	1907	Williamson, EB; Collins, F	UMMZ, Williamson 1914b
20	Coenagrionidae	*Enallagma divagans*	a	Le Flore	1907	Williamson, EB; Collins, F	UMMZ, Williamson 1914b
21	Coenagrionidae	*Enallagma exsulans*	a	Le Flore	1907	Williamson, EB; Collins, F	UMMZ
22	Coenagrionidae	*Enallagma geminatum*	a	Le Flore	1907	Williamson, EB; Collins, F	UMMZ, Williamson 1914b
23	Coenagrionidae	*Enallagma signatum*	a	Le Flore	1907	Williamson, EB; Collins, F	UMMZ
24	Coenagrionidae	*Enallagma traviatum*	a	Le Flore	1907	Williamson, EB; Collins, F	UMMZ
25	Coenagrionidae	*Enallagma vesperum*	a	Le Flore	1907	Williamson, EB; Collins, F	UMMZ, Williamson 1914b
26	Coenagrionidae	*Ischnura hastata*	a	Le Flore	1907	Williamson, EB; Collins, F	UMMZ
27	Coenagrionidae	*Ischnura kellicotti*	a	Le Flore	1907	Williamson, EB; Collins, F	Williamson 1914b
28	Coenagrionidae	*Ischnura posita*	a	Le Flore	1907	Williamson, EB; Collins, F	UMMZ, Williamson 1914b
29	Corduliidae	*Epitheca costalis*	a	Le Flore	1907	Williamson, EB; Collins, F	UMMZ, Muttkowski 1911

(Continued)

#	FAMILY	SPECIES	AGE	COUNTY	YEAR	COLLECTOR/OBSERVER	DOCUMENTATION
30	Corduliidae	*Neurocordulia xanthosoma*	a	Le Flore	1907	Williamson, EB; Collins, F	UMMZ, Williamson 1908, 1914b
31	Gomphidae	*Dromogomphus spinosus*	a	Le Flore	1907	Williamson, EB; Collins, F	UMMZ, Williamson 1914b
32	Gomphidae	*Phanogomphus graslinellus*	a	Le Flore	1907	Williamson, EB; Collins, F	Williamson 1914b
33	Lestidae	*Lestes inaequalis*	a	Le Flore	1907	Williamson, EB; Collins, F	UMMZ
34	Libellulidae	*Libellula cyanea*	a	Le Flore	1907	Williamson, EB; Collins, F	UMMZ
35	Libellulidae	*Libellula pulchella*	a	Le Flore	1907	Williamson, EB; Collins, F	UMMZ, Williamson 1914b
36	Libellulidae	*Pachydiplax longipennis*	a	Le Flore	1907	Williamson, EB; Collins, F	UMMZ
37	Libellulidae	*Perithemis tenera*	a	Le Flore	1907	Williamson, EB; Collins, F	Williamson 1914b
38	Coenagrionidae	*Argia apicalis*	a	Le Flore	1907	Williamson, EB; Collins, F	UMMZ
39	Libellulidae	*Erythemis simplicicollis*	a	Le Flore	1907	Williamson, EB; Collins, F	UMMZ
40	Gomphidae	*Arigomphus submedianus*	a	Le Flore	1907	Williamson, EB; Collins, F	Williamson 1914a,b
41	Gomphidae	*Gomphurus externus*	a	Cleveland	1921?	Bird, RD	OMNH
42	Lestidae	*Calopteryx maculata*	a	McCurtain	1925	Ortenburger expedition?	OMNH, Ortenburger 1926b
43	Coenagrionidae	*Argia sedula*	a?	McCurtain	1925	Ortenburger expedition?	Ortenburger 1926b
44	Libellulidae	*Celithemis eponina*	a?	McCurtain	1925	Ortenburger expedition?	Ortenburger 1926b
45	Libellulidae	*Celithemis fasciata*	a	Choctaw	1925	Ortenburger expedition?	OMNH, Ortenburger 1926b
46	Libellulidae	*Dythemis velox*	a?	McCurtain	1925	Ortenburger expedition?	Ortenburger 1926b
47	Libellulidae	*Erythrodiplax umbrata*	a	McCurtain	1925	Ortenburger expedition?	FSCA
48	Libellulidae	*Tramea lacerata*	a	McCurtain	1925	Ortenburger expedition?	OMNH, Ortenburger 1926b
49	Libellulidae	*Tramea onusta*	a?	McCurtain	1925	Ortenburger expedition?	Ortenburger 1926b
50	Coenagrionidae	*Enallagma basidens*	a	Comanche	1926	Hubbell, TH	UMMZ
51	Coenagrionidae	*Enallagma carunculatum*	a	Cimarron	1926	Hubbell, TH	UMMZ
52	Coenagrionidae	*Ischnura damula*	a	Cimarron	1926	Hubbell, TH	UMMZ
53	Coenagrionidae	*Ischnura verticalis*	a	Texas	1926	Hubbell, TH	UMMZ
54	Corduliidae	*Epitheca petechialis*	a	Comanche	1926	Hubbell, TH	ANSP, UMMZ
55	Corduliidae	*Epitheca princeps*	a	Comanche	1926	Hubbell, TH	UMMZ
56	Libellulidae	*Plathemis subornata*	a	Cimarron	1926	Hubbell, TH	UMMZ
57	Gomphidae	*Progomphus obscurus*	n	Comanche	1926	Ortenburger, Al, et al.	UMMZ
58	Lestidae	*Lestes australis*	a	Cleveland	1927	Ortenburger expedition?	OMNH
59	Coenagrionidae	*Argia plana*	a	Comanche	1928	Logsdon, Z	OMNH
60	Libellulidae	*Libellula croceipennis*	a	Comanche	1928	Logsdon, Z	FSCA, OMNH
61	Macromiidae	*Macromia pacifica*	a	Comanche	1928	Logsdon, Z	OMNH
62	Aeshnidae	*Nasiaeschna pentacantha*	a	Love	1929	Bird, RD	OMNH
63	Gomphidae	*Phanogomphus militaris*	a	Love	1929	Bird, RD	OMNH
64	Lestidae	*Archilestes grandis*	a	McClain	1929	Bird, RD	OMNH
65	Coenagrionidae	*Argia translata*	a	Murray	1929	Bird, RD	OMNH
66	Gomphidae	*Arigomphus lentulus*	a	Love	1929	Bird, RD	OMNH

(Continued)

#	FAMILY	SPECIES	AGE	COUNTY	YEAR	COLLECTOR/OBSERVER	DOCUMENTATION
67	Aeshnidae	*Aeshna umbrosa*	a	Caddo	1929	Williamson, EB	UMMZ
68	Coenagrionidae	*Argia nahuana*	a	Caddo	1929	Williamson, EB	UMMZ
69	Coenagrionidae	*Telebasis salva*	a	Caddo	1929	Williamson, EB	UMMZ
70	Lestidae	*Lestes unguiculatus*	a	Lincoln	1929	Williamson, EB	UMMZ
71	Coenagrionidae	*Enallagma praevarum*	a	Cimarron	1930	Bird expedition?	Bick and Bick 1957
72	Coenagrionidae	*Ischnura barberi*	a	Alfalfa	1930	Bird expedition?	USNM
73	Libellulidae	*Libellula comanche*	a	Woods	1930	Bird expedition?	OMNH
74	Aeshnidae	*Basiaeschna janata*	a	Murray	1930	Bird, RD	OMNH
75	Aeshnidae	*Rhionaeschna multicolor*	a	Major	1930	Bird, RD	OMNH
76	Coenagrionidae	*Argia alberta*	a	Alfalfa	1930	Bird, RD	USNM
77	Coenagrionidae	*Argia immunda*	a	Murray	1930	Bird, RD	Bick note cards
78	Coenagrionidae	*Enallagma antennatum*	a	Alfalfa	1930	Bird, RD	OMNH, Bick and Bick 1957
79	Coenagrionidae	*Ischnura denticollis*	a	McClain	1930	Bird, RD	OMNH
80	Corduliidae	*Epitheca cynosura*	a	Murray	1930	Bird, RD	OMNH
81	Gomphidae	*Erpetogomphus designatus*	a	Major	1930	Bird, RD	OMNH
82	Lestidae	*Lestes alacer*	a	McClain	1930	Bird, RD	OMNH, USNM
83	Lestidae	*Lestes rectangularis*	a	Cleveland	1930	Bird, RD	OMNH
84	Libellulidae	*Brechmorhoga mendax*	a	Murray	1930	Bird, RD	OMNH
85	Libellulidae	*Erythemis vesiculosa*	a	McClain	1930	Bird, RD	OMNH
86	Libellulidae	*Erythrodiplax minuscula*	a	Comanche	1930	Bird, RD	Bick note cards
87	Libellulidae	*Sympetrum ambiguum*	a	Woods	1930	Bird, RD	OMNH
88	Libellulidae	*Sympetrum semicinctum*	a	Woods	1930	Bird, RD	OMNH, USNM
89	Libellulidae	*Sympetrum vicinum*	a	Woods	1930	Bird, RD	OMNH
90	Macromiidae	*Didymops transversa*	a	Cleveland	1930	Coats	OMNH
91	Coenagrionidae	*Argia bipunctulata*	a + t	Latimer	1931	Bird, RD	OMNH
92	Coenagrionidae	*Enallagma aspersum*	a	Latimer	1931	Bird, RD	OMNH
93	Corduliidae	*Somatochlora linearis*	a	Latimer	1931	Bird, RD	Bick note cards
94	Corduliidae	*Somatochlora ozarkensis*	a	Latimer	1931	Bird, RD; Fisher, WM	OMNH, UMMZ
95	Gomphidae	*Dromogomphus spoliatus*	a	McClain	1931	Bird, RD	OMNH
96	Libellulidae	*Celithemis elisa*	a	Latimer	1931	Bird, RD	OMNH
97	Libellulidae	*Libellula flavida*	a	Latimer	1931	Bird, RD	OMNH
98	Libellulidae	*Orthemis ferruginea*	a	Latimer	1931	Bird, RD	OMNH
99	Corduliidae	*Epitheca spinosa*	a	Latimer	1931	Hooper, E	OMNH
100	Coenagrionidae	*Amphiagrion abbreviatum*	a	Alfalfa	1931	Kaiser, RW	UMMZ
101	Cordulegastridae	*Cordulegaster obliqua*	a	Craig	1931	Lasatey, M	OSU
102	Gomphidae	*Phanogomphus oklahomensis*	a	Latimer	1931	Pritchard, AE	Pritchard 1935
103	Libellulidae	*Dythemis fugax*	a	Comanche	1931	Pritchard, AE	OMNH

(Continued)

#	FAMILY	SPECIES	AGE	COUNTY	YEAR	COLLECTOR/OBSERVER	DOCUMENTATION
104	Coenagrionidae	*Argia leonorae*	a	Comanche	1932	Bird, RD	OMNH
105	Gomphidae	*Phyllogomphoides stigmatus*	a	Comanche	1932	Bird, RD	OMNH
106	Libellulidae	*Libellula saturata*	a	Comanche	1932	Bird, RD	OMNH
107	Gomphidae	*Stylurus intricatus*	a	Harper	1932	Pritchard, AE	OMNH, OC
108	Lestidae	*Lestes vigilax*	a	Pushmataha	1932	Pritchard, AE	OMNH
109	Libellulidae	*Celithemis verna*	a	Pushmataha	1932	Pritchard, AE	Pritchard 1935
110	Coenagrionidae	*Argia lugens*	a	Cimarron	1933	Pritchard, AE	EMEC, FSCA
111	Libellulidae	*Paltothemis lineatipes*	a	unknown	1933?	Bird, RD	OMNH (lost?)
112	Aeshnidae	*Boyeria vinosa*	a	Pushmataha	1934	Pritchard, AE	EMEC, RWG, OSU
113	Coenagrionidae	*Enallagma daeckii*	a	Pushmataha	1934	Pritchard, AE	EMEC
114	Coenagrionidae	*Enallagma dubium*	a	Pushmataha	1934	Pritchard, AE	Pritchard 1935
115	Coenagrionidae	*Nehalennia integricollis*	a	Pushmataha	1934	Pritchard, AE	EMEC
116	Corduliidae	*Somatochlora tenebrosa*	a	Le Flore	1934	Pritchard, AE	EMEC, Pritchard 1936
117	Libellulidae	*Ladona deplanata*	a	Pushmataha	1934	Pritchard, AE	EMEC, UMMZ
118	Corduliidae	*Neurocordulia virginiensis*	a	McCurtain	1934	Stankavich, J	Byers 1937, Davis 1937
119	Aeshnidae	*Gynacantha nervosa*	a	Le Flore	1935	Hubbell, TH	UMMZ, OC, Kormondy 1960
120	Libellulidae	*Libellula auripennis*	a	Le Flore	1935	Hubbell, TH	UMMZ, Kormondy 1960
121	Macromiidae	*Macromia taeniolata*	a	Mayes	1936	Blair, WF	UMMZ, Kormondy 1960
122	Gomphidae	*Stylogomphus sigmastylus*	a	Adair	1937	Blair, WF; Blair, FA	Kormondy 1960
123	Petaluridae	*Tachopteryx thoreyi*	a	Latimer	1948	unknown	Bick note card
124	Corduliidae	*Neurocordulia molesta*	e	Wagoner	1950	Bick, GH; Bick, JC	Bick and Bick 1957
125	Libellulidae	*Libellula semifasciata*	a	Muskogee	1951	Bick, GH	FSCA, Bick and Bick 1957
126	Coenagrionidae	*Ischnura ramburii*	a	Marshall	1954	Bick, GH	FSCA, Bick and Bick 1957
127	Gomphidae	*Gomphurus vastus*	a	Bryan	1954	Bick, GH	FSCA
128	Libellulidae	*Brachymesia gravida*	a	Carter	1955	Bick, GH	FSCA, Bick and Bick 1957
129	Aeshnidae	*Anax longipes*	a	Johnston	1956	Bick, JC; Bick, S; Bick, P	Bick note cards
130	Coenagrionidae	*Ischnura demorsa*	a	Harper and Woodward	1956	Bick, GH*	FSCA, Bick and Bick 1957
131	Corduliidae	*Helocordulia uhleri*	a	McCurtain	1956	Hornuff, LE	FSCA, Bick and Bick 1957
132	Gomphidae	*Gomphurus ozarkensis*	a	McCurtain	1956	Hornuff, LE	FSCA
133	Corduliidae	*Epitheca semiaquea*	a	Latimer	1959	Bick, GH	FSCA
134	Macromiidae	*Macromia alleghaniensis*	a	Delaware	1962	unknown	OSU
135	Lestidae	*Lestes sigma*	a	Marshall	1968	Bick, GH	Bick 1978
136	Libellulidae	*Erythemis collocata*	a	Cimarron	1970	Bick, GH; Hornuff, LE	FSCA, OC
137	Libellulidae	*Libellula nodisticta*	a	Cimarron	1970	Bick, GH; Hornuff, LE	FSCA
138	Libellulidae	*Tramea carolina*	a	Sequoyah	1970	Hornuff, LE	FSCA
139	Libellulidae	*Macrodiplax balteata*	n	Comanche	1989	F, KW	UCO
140	Gomphidae	*Arigomphus maxwelli*	a	McCurtain	2002	Arbour, D	OC
141	Coenagrionidae	*Ischnura perparva*	a	Comanche	2003	Kondratieff, BC; Schmidt, JP; Zuellig, RE	CSU, OC, Kondratieff, et al. 2004

(Continued)

#	FAMILY	SPECIES	AGE	COUNTY	YEAR	COLLECTOR/OBSERVER	DOCUMENTATION
142	Aeshnidae	*Aeshna constricta*	a	Tulsa	2004	Carrell, B	OC
143	Gomphidae	*Aphylla williamsoni*	a	McCurtain	2005	Arbour, D	JCAC, OC
144	Libellulidae	*Tholymis citrina*	a	McCurtain	2006	Arbour, D	JCAC, OC
145	Libellulidae	*Tramea calverti*	a	Tulsa	2006	Fisher, J; Carrell, B	OC
146	Libellulidae	*Miathyria marcella*	a	McCurtain	2007	Arbour, D	OC
147	Libellulidae	*Sympetrum illotum*	a	Comanche	2007	Fazio, VW, III	OC
148	Gomphidae	*Gomphurus hybridus*	a	McCurtain	2007	Heck, BA	OC
149	Aeshnidae	*Coryphaeschna ingens*	a	McCurtain	2008	Arbour, D; Heck, BA	JCAC, OC
150	Libellulidae	*Pseudoleon superbus*	a	Jefferson	2008	Fazio, VW, III	OC
151	Aeshnidae	*Gomphaeschna furcillata*	a	McCurtain	2008	Heck, BA	OC
152	Corduliidae	*Helocordulia selysii*	a	McCurtain	2008	Heck, BA	OC
153	Coenagrionidae	*Telebasis byersi*	a	McCurtain	2010	Arbour, D	JCAC, OC
154	Cordulegastridae	*Cordulegaster maculata*	a	McCurtain	2010	Heck, BA	OC
155	Gomphidae	*Hylogomphus apomyius*	a	McCurtain	2010	Heck, BA	OC
156	Gomphidae	*Phanogomphus lividus*	a	McCurtain	2010	Heck, BA	OC
157	Libellulidae	*Libellula composita*	a	Texas	2011	Fazio, VW, III	OC
158	Cordulegastridae	*Cordulegaster talaria*	a	McCurtain	2011	Heck, BA; Heck, GL	OC
159	Libellulidae	*Sympetrum costiferum*	a	Grant	2012	Heinen, JR	SP, OC
160	Coenagrionidae	*Enallagma doubledayi*	a	Atoka	2012	Smith-Patten, BD; Patten, MA	SP
161	Libellulidae	*Sympetrum obtrusum*	a	Kingfisher	2012	Smith-Patten, BD; Patten, MA	SP, OC
162	Lestidae	*Lestes eurinus*	a	Pushmataha	2013	Smith-Patten, BD; Patten, MA	SP, OC
163	Lestidae	*Lestes forcipatus*	a	Le Flore	2013	Smith-Patten, BD; Patten, MA	SP, OC
164	Libellulidae	*Sympetrum pallipes*	a	Cimarron	2014	Carrell, B	OC
165	Libellulidae	*Erythrodiplax basifusca*	a	Harmon	2014	Patten, MA; Smith-Patten, BD	OC
166	Libellulidae	*Erythrodiplax berenice*	a	Jackson	2014	Patten, MA; Smith-Patten, BD	SP, OC
167	Libellulidae	*Dythemis nigrescens*	a	Jackson	2014	Smith-Patten, BD	OC
168	Coenagrionidae	*Nehalennia gracilis*	a	Atoka	2014	Smith-Patten, BD; Tucker, J	SP, OC
169	Libellulidae	*Micrathyria hagenii*	a	Cherokee	2015	Smith-Patten, BD; Hoagland, BW	SP, OC
170	Coenagrionidae	*Argia tezpi*	a	Harmon	2017	Patten, MA	SP, OC
171	Corduliidae	*Somatochlora margarita*	a	McCurtain	2017	Arbour, D	SP, OC
172	Libellulidae	*Brachymesia herbida*	a	Cotton	2017	Patten, MA	OC
173	Libellulidae	*Erythemis attala*	a	Beaver	2017	Patten, MA	SP, OC
174	Macromiidae	*Macromia annulata*	a	Jackson	2018	Patten, MA	SP, OC
175	Coenagrionidae	*Chromagrion conditum*	a	McCurtain	2019	Smith-Patten, BD; Arbour, D	SP, OC
176	Libellulidae	*Brachymesia furcata*	a	Cherokee	2019	Carrell, B	SP, OC

* possibly actually AE Pritchard in 1932
  age: a = adult, e = exuviae, n = nymph, t = teneral

Table F.1 **Data references for specimens and photographs included in Coenagrionidae Figures 12.12, 12.37, 12.81, 12.82, 12.84, and 12.88.**

FIGURE, PART	SOURCE	SPECIES AND SEX	DATA	NOTES
Figure 12.12				
top	SP 1610	*Ischnura damula* ♂	Oklahoma: Cimarron Co.: 10 km N of Kenton, North Carrizo Creek, 3 June 2015, BS-P	thorax dorsum
middle left	SP 1345	*Ischnura denticollis* ♀	Oklahoma: Washita Co.: Little Elk Creek near Rocky Lake, 24 July 2014	thorax dorsum
middle right	SP 1528	*Ischnura denticollis* ♀	Oklahoma: Jackson Co.: Altus city lake, 19 Apr 2015, BS-P	thorax dorsum
bottom, left	SP 2378	*Ischnura verticalis* ♂	Oklahoma: Cimarron Co.: 4km NNE of Kenton, Cimarron River, 3 June 2017, MAP	thorax dorsum
bottom, right	SP 1868	*Ischnura verticalis* ♂	Oklahoma: Beaver Co.: Beaver River WMA, 22 Apr 2016, MAP	thorax dorsum
bottom row, left to right	GW Lasley	*Ischnura damula* ♀	Texas: Dallam Co.: Buffalo Springs, Rita Blanca National Grasslands, 31 Aug 2004, GW Lasley	
	GW Lasley	*Ischnura denticollis* ♂	Texas: Dallam Co.: Buffalo Springs, Rita Blanca National Grasslands, 31 Aug 2004, GW Lasley	
	GW Lasley	*Ischnura posita* ♂	Arkansas: Sevier Co.: Pond Creek NWR, 8 Sept 2007, GW Lasley	♂ from pair in copula
	GW Lasley	*Ischnura verticalis* ♂	Texas: Dallam Co.: Buffalo Springs, Rita Blanca National Grasslands, 31 Aug 2004, GW Lasley	
Figure 12.37				
	SP 2016	*Enallagma basidens* ♀	Oklahoma: Ellis Co.: Ellis Co. WMA, 11 July 2016, BS-P	
	SP 2611	*Enallagma divagans* ♀	Oklahoma: Pittsburg Co.: 15 km SE of Ti, Jackfork Creek, 2 June 2018, MAP	♀ from pair
	SP 630	*Enallagma exsulans* ♀	Oklahoma: Blaine Co.: 15 km WSW of Greenfield, unnamed creek, 1 June 2013	
	SP 1202	*Enallagma exsulans* ♀	Oklahoma: Delaware Co.: Lake Eucha, 22 May 2014, BS-P, MAP	
	SP 1768	*Enallagma exsulans* ♀	Oklahoma: Major Co.: Major Co. WMA, 29 Aug 2015, MAP	♀ from pair
	SP 2568	*Enallagma exsulans* ♀	Oklahoma: McCurtain Co.: Grassy Slough WMA, 7 May 2018, BS-P	
	SP 1011	*Enallagma signatum* ♀	Oklahoma: Canadian Co.: Mustang, Wildhorse Park, 14 Sept 2013	
Figure 12.81				
A	SP 2609	*Ischnura verticalis* ♂	Oklahoma: Pottawatomie Co.: Shawnee, Oklahoma Baptist University, 28 May 2018, MAP	thorax dorsum

(Continued)

Table F.1 (*Continued*) **Data references for specimens and photographs included in Coenagrionidae Figures 12.12, 12.37, 12.81, 12.82, 12.84, and 12.88.**

FIGURE, PART	SOURCE	SPECIES AND SEX	DATA	NOTES
B	SP 104	*Ischnura verticalis* ♂	Oklahoma: Texas Co.: Optima Lake dam spillway, 26 Sept 2010	thorax dorsum
C	SP 1594	*Ischnura verticalis* ♂	Oklahoma: Texas Co.: Optima Lake dam spillway, 2 June 2015, BS-P	thorax dorsum
D	SP 621	*Ischnura verticalis* ♂	Oklahoma: Cimarron Co.: Conrad, 27 May 2013	thorax dorsum
E	SP 1595	*Ischnura verticalis* ♂	Oklahoma: Texas Co.: Optima Lake dam spillway, 2 June 2015, BS-P	thorax dorsum
F	SP 611	*Ischnura verticalis* ♂	Oklahoma: Cimarron Co.: 2 km W of Kenton, Carrizozo Creek, 26 May 2013	thorax dorsum
G	SP 1868	*Ischnura verticalis* ♂	Oklahoma: Beaver Co.: Beaver River WMA, 22 Apr 2016, MAP	thorax dorsum
H	SP 420	*Ischnura verticalis* ♂	Oklahoma: Harper Co.: Doby Springs Park, 15 Sept 2012	thorax dorsum
I	SP 2378	*Ischnura verticalis* ♂	Oklahoma: Cimarron Co.: 4km NNE of Kenton, Cimarron River, 3 June 2017, MAP	thorax dorsum
J	SP 747	*Ischnura verticalis* ♂	Colorado: Baca Co.: North Carrizo Creek, 5 July 2013	thorax dorsum
K	SP 1373	*Ischnura verticalis* ♂	Oklahoma: Tillman Co.: 3 km W of Manitou, 3 Aug 2014	thorax dorsum
Figure 12.82				
A	OC 425308	*Ischnura verticalis* ♂	Oklahoma: Tulsa Co.: Tulsa, Turkey Mountain Urban Wilderness Area, 25 July 2014, BC	overview
B	OC 401675	*Ischnura verticalis* ♂	Oklahoma: Craig Co.: at Co. roads 240 and 4460, 7 July 2013, BC	overview and close-up of head and thorax
C	OC 282024	*Ischnura verticalis* ♂	Oklahoma: Comanche Co.: Wichita Mountains Wildlife Refuge, French Lake, 17 Apr 2008, VWF	overview
Figure 12.84				
A	SP 195	*Ischnura verticalis* ♀	Oklahoma: Kiowa Co.: Mountain Park WMA, 11 June 2010, CM Curry	thorax dorsum
B	SP 1179	*Ischnura verticalis* ♀	Oklahoma: Noble Co.: 3 km SSE of Perry, Perry Lake, 11 May 2014	lateral body, S7–10 dorsum, S7–10 lateral
C	SP 1922	*Ischnura verticalis* ♀	Oklahoma: Greer Co.: Doc Hollis Lake, 4 June 2016, MAP	lateral body, S7–10 dorsum, S7–10 lateral
D	SP 588	*Ischnura verticalis* ♀	Oklahoma: Cimarron Co.: 7 km N of Kenton, North Carrizo Creek, 26 May 2013	lateral body, S7–10 dorsum, S7–10 lateral
E	SP 1614	*Ischnura verticalis* ♀	Oklahoma: Texas Co.: 12 km N of Goodwell, Beaver River, 3 June 2015	lateral body, S7–10 dorsum, S7–10 lateral
Figure 12.88				
A, L	SP 1014	*Ischnura denticollis* ♀	Oklahoma: Canadian Co.: El Reno, Legion Park, 14 Sept 2013	thorax dorsum, prothorax
B	SP 1430	*Ischnura denticollis* ♀	Oklahoma: Harper Co.: Doby Springs Park, 14 Sept 2014	thorax dorsum
C, G, I	SP 415	*Ischnura denticollis* ♀	Oklahoma: Harper Co.: Doby Springs Park, 14 Sept 2012	thorax dorsum, prothorax
D	SP 1331	*Ischnura denticollis* ♀	Oklahoma: Ellis Co.: Ellis Co. WMA, 13 July 2014	thorax dorsum
E	SP 1528	*Ischnura denticollis* ♀	Oklahoma: Jackson Co.: Altus city lake, 19 Apr 2015, BS-P	thorax dorsum

(Continued)

Table F.1 (*Continued*) **Data references for specimens and photographs included in Coenagrionidae Figures 12.12, 12.37, 12.81, 12.82, 12.84, and 12.88.**

FIGURE, PART	SOURCE	SPECIES AND SEX	DATA	NOTES
F, H	SP 1345	*Ischnura denticollis* ♀	Oklahoma: Washita Co.: Little Elk Creek near Rocky Lake, 24 July 2014	thorax dorsum, prothorax
J	SP 2019	*Ischnura denticollis* ♀	Oklahoma: Ellis Co.: Ellis Co. WMA, 11 July 2016, BS-P	prothorax
K	SP 140	*Ischnura denticollis* ♀	Oklahoma: Garfield Co., Drummond Flats WMA, 13 Apr 2012	prothorax
M	SP 1586	*Ischnura denticollis* ♀	Oklahoma: Harper Co.: Doby Springs Park, 18 May 2015, BS-P	prothorax
N	SP 1336	*Ischnura denticollis* ♀	Oklahoma: Roger Mills Co.: Skipout Lake, 13 July 2014	prothorax

source: SP = Smith-Patten/Patten Collection; data: BC = Bill Carrell, BS-P = Brenda D. Smith-Patten, MAP = Michael A. Patten, NWR = National Wildlife Refuge, VWF = Victor W Fazio III, WMA = Wildlife Management Area

Table F.2 *Ischnura verticalis* records, Smith-Patten/Patten data (SP) for the Oklahoma panhandle and adjacent areas: Cimarron, Texas, Beaver, and Harper Counties, Oklahoma, and Baca County, Colorado (the latter shares North Carrizo Creek with Oklahoma). Number of "typical" and aberrant males encountered are indicated, as well as specimens of each. Total number of *I. verticalis* encountered is noted, including indication of sex and pairs noted. When unsexed (U) individuals were included in counts, we treated males as minimum counts (*). Individuals captured, examined, and released are labeled as "in hand." Bear in mind that some aberrant individuals are not easily diagnosed without microscopic examination, so number of aberrant individuals, particularly for females, may be undercounted.

LOCATION	TYPICAL ♂	ABERRANT ♂	TOTAL *I. VERTICALIS* SEX RATIO	SPECIMEN(S) TYPICAL	SPECIMEN(S) ABERRANT	DATE	NOTES
Baca Co., Colorado	0	1					
North Carrizo Creek	0	1	**2** (1♂, 1♀)		SP 747 ♂	5 July 2013	
Cimarron Co., Oklahoma	134*	11					
Black Mesa area	43*	5	**138** (48♂, 26♀, 64U)	SP 614, 617-618 ♂	SP 588 ♀, 610–611, 615–616, 619 ♂	26 May 2013[a]	SP 610 and 615 donated to RWG
	23	0	**33** (23♂, 10♀)			5 July 2013[b]	4♂ in hand
	10	0	**22** (10♂, 12♀)			22 Sept 2013[c]	6♂ in hand
	12*	3	**100** (15♂, 1♀, 84U)	SP 1600 ♀	SP 1597–1599 ♂	3 June 2015[d]	12♂ in hand
	4	0	**9** (4♂, 5♀)			8 Aug 2015[e]	2♂ in hand
	38	1	**72** (39♂, 33♀)		SP 2378 ♂	3 June 2017[f]	2♂ in hand
	3	0	**7** (3♂, 4♀)			7 Oct 2017[g]	2♂ in hand
Conrad	1*	2	**30** (3♂, 27U)	SP 622 ♂	SP 620–621 ♂	26 May 2013	SP 620 donated to RWG

(*Continued*)

Table F.2 (*Continued*) *Ischnura verticalis* records, Smith-Patten/Patten data (SP) for the Oklahoma panhandle and adjacent areas: Cimarron, Texas, Beaver, and Harper Counties, Oklahoma, and Baca County, Colorado (the latter shares North Carrizo Creek with Oklahoma). Number of "typical" and aberrant males encountered are indicated, as well as specimens of each. Total number of *I. verticalis* encountered is noted, including indication of sex and pairs noted. When unsexed (U) individuals were included in counts, we treated males as minimum counts (*). Individuals captured, examined, and released are labeled as "in hand." Bear in mind that some aberrant individuals are not easily diagnosed without microscopic examination, so number of aberrant individuals, particularly for females, may be undercounted.

LOCATION	TYPICAL ♂	ABERRANT ♂	TOTAL I. VERTICALIS SEX RATIO	SPECIMEN(S) TYPICAL	SPECIMEN(S) ABERRANT	DATE	NOTES
Texas Co., Oklahoma	**272***	**7**					
12 km N of Goodwell, Beaver R.	15	0	**26** (15♂, 11♀)		SP 1614 ♀ andromorph	3 June 2015	6♂ in hand
	1	0	**1** (1♂)			8 Aug 2015	
Guymon, Sunset Lake	1	0	**1** (1♂)			8 July 2013	
	1	0	**1** (1♂)			2 June 2017	
Optima Lake	5	1	**18** (6♂, 12♀)		SP 106 ♂	25 Sept 2010	
	23	2	**65** (25♂, 40♀)	SP 103 ♂, 108 ♀	SP 104–105 ♂	26 Sept 2010	
	12	0	**17** (12♂, 5♀)			29 July 2012	2♂ in hand
	8	0	**14** (8♂, 6♀)			25 May 2013	1♂ in hand
	7	0	**9** (7♂, 2♀)			4 July 2013	
	1	0	**2** (1♂, 1♀)			8 July 2013	
	58	2	**100** (60♂, 40♀)		SP 1594–1595 ♂	2 June 2015	6♂ and 1♀ in hand
	79	1	**130** (80♂, 50♀)		SP 1616 ♂	4 June 2015	2♂ in hand
	2	0	**2** (2♂)			7 Aug 2015	1♂ in hand
	3	0	**7** (3♂, 4♀)			23 April 2016	1♂ in hand
	4	0	**7** (4♂, 3♀)			20 May 2016	2♂ in hand
	2	0	**2** (2♂)			25 June 2016	
Range, Palo Duro Ck.	2	0	**4** (2♂, 2♀)			27 May 2013	
Schultz WMA	1	0	**5** (1♂, 4♀)			29 July 2012	1♂ in hand
	6*	0	**50** (6♂, 2♀, 42U)			27 May 2013	6♂ in hand; 2 pair noted
Shorb WMA	1	0	**2** (1♂, 1♀)	SP 1854 ♂		5 April 2016	
	8	0	**10** (8♂, 2♀)			20 May 2016	4♂ in hand
	10	0	**16** (10♂, 6♀)			2 June 2017	2 pair noted
	3	0	**8** (3♂, 5♀)			3 Sept 2017	1♂ in hand
2 km S of Straight	19	1	**50** (20♂, 30♀)		SP 609 ♂	25 May 2013	
Beaver Co., Oklahoma	**55**	**1**					
Beaver Dunes SP	1	0	**1** (1♂)			15 Sept 2012	
	5	0	**7** (5♂, 2♀)			25 May 2013	1♂ in hand

(*Continued*)

Table F.2 (*Continued*) ***Ischnura verticalis*** records, Smith-Patten/Patten data (SP) for the Oklahoma panhandle and adjacent areas: Cimarron, Texas, Beaver, and Harper Counties, Oklahoma, and Baca County, Colorado (the latter shares North Carrizo Creek with Oklahoma). Number of "typical" and aberrant males encountered are indicated, as well as specimens of each. Total number of *I. verticalis* encountered is noted, including indication of sex and pairs noted. When unsexed (U) individuals were included in counts, we treated males as minimum counts (*). Individuals captured, examined, and released are labeled as "in hand." Bear in mind that some aberrant individuals are not easily diagnosed without microscopic examination, so number of aberrant individuals, particularly for females, may be undercounted.

LOCATION	TYPICAL ♂	ABERRANT ♂	TOTAL *I. VERTICALIS* SEX RATIO	SPECIMEN(S) TYPICAL	SPECIMEN(S) ABERRANT	DATE	NOTES
	3	0	**3** (3♂)			2 June 2015	3♂ in hand
	2	0	**2** (2♂)	SP 1888 ♂		10 May 2016	1♂ in hand
Beaver River WMA	9	1	**14** (10♂, 4♀)		SP 1868 ♂	22 April 2016	3♂ and 1♀ in hand
	3	0	**3** (3♂)			10 May 2016	3♂ in hand
Kiowa Ck., Logan	1	0	**1** (1♂)			8 July 2013	1♂ in hand
Kiowa Ck., 7 km NE of Logan	3	0	**3** (3♂)			15 Sept 2012	
Kiowa Ck., 5 km N of Slapout	5	0	**10** (5♂, 5♀)	SP 360 ♂		29 July 2012	
Kiowa Ck., 13 km NE of Slapout	8	0	**14** (8♂, 6♀)			15 Sept 2012	4♂ in hand
	1	0	**1** (1♂)			4 July 2013	
Lake Evans Chambers	6	0	**8** (6♂, 2♀)			2 June 2012	3♂ in hand
	6	0	**11** (6♂, 5♀)			15 Sept 2012	1♂ in hand
	1	0	**1** (1♂)			4 July 2013	1♂ in hand
	0	0	**1** (1♀)	SP 1581 ♀		17 May 2015	
	1	0	**4** (1♂, 3♀)	SP 1885 ♂ teneral		10 May 2016	
Harper Co., Oklahoma	**104***	**4**					
Doby Springs	12	0	**24** (12♂, 12♀)			12 May 2012	3 individuals in hand
	4	0	**7** (4♂, 3♀)			2 June 2012	1♂ in hand
	1	0	**1** (1♂)			28 July 2012	
	5*	0	**30** (5♂, 25U)			14 Sept 2012	5♂ in hand
	36	4	**55** (40♂, 15♀)		SP 417–420 ♂	15 Sept 2012	
	25	0	**35** (25♂, 10♀)			25 May 2013	2 pair noted
	5	0	**8** (5♂, 3♀)			4 July 2013	
	2	0	**3** (2♂, 1♀)			26 Aug 2013	1 individual in hand
	0	0	**2** (2♀)			12 July 2014	
	2	0	**2** (2♂)			18 May 2015	2♂ in hand
	1	0	**1** (1♂)			2 June 2015	1♂ in hand
	7	0	**11** (7♂, 4♀)			21 May 2016	3♂ and 2♀ in hand

(*Continued*)

Table F.2 (*Continued*) *Ischnura verticalis* records, Smith-Patten/Patten data (SP) for the Oklahoma panhandle and adjacent areas: Cimarron, Texas, Beaver, and Harper Counties, Oklahoma, and Baca County, Colorado (the latter shares North Carrizo Creek with Oklahoma). Number of "typical" and aberrant males encountered are indicated, as well as specimens of each. Total number of *I. verticalis* encountered is noted, including indication of sex and pairs noted. When unsexed (U) individuals were included in counts, we treated males as minimum counts (*). Individuals captured, examined, and released are labeled as "in hand." Bear in mind that some aberrant individuals are not easily diagnosed without microscopic examination, so number of aberrant individuals, particularly for females, may be undercounted.

LOCATION	TYPICAL ♂	ABERRANT ♂	TOTAL *I. VERTICALIS* SEX RATIO	SPECIMEN(S) TYPICAL	SPECIMEN(S) ABERRANT	DATE	NOTES
	4	0	5 (4♂, 1♀)			21 May 2017	1♂ in hand; 1 pair noted

a   localities: Carrizozo Creek at 2 km W of Kenton, North Carrizo Creek at 7 km N of Kenton, and Black Mesa State Park.

b   localities: Carrizozo Creek at 2 km W of Kenton, North Carrizo Creek at 7 km N of Kenton, Cimarron River at 1 km NNE of Kenton, Cimarron River at 13 km E of Kenton, and Black Mesa State Park.

c   localities: not specified.

d   localities: Black Mesa State Park and vicinity.

e   locality: North Carrizo Creek at 7 km N of Kenton.

f   localities: Cimarron River at 4 km NNW of Kenton and Black Mesa State Park.

g   locality: Black Mesa State Park.

Table F.3 Records of aberrant *Ischnura verticalis* and other *Ischnura* species recorded on same day and location (= collection data point). Records were included here only if full species counts were made and specimens were collected; additional records of aberrant individuals are discussed in the *Ischnura* genus summary and the *I. verticalis* species account. Specimen numbers for non-*verticalis* species are not provided here.

COLLECTION DATA POINT (COUNTY PT#)	TOTAL VERTICALIS	ABERRANT VERTICALIS	DAMULA	PERPARVA	DEMORSA	DEM/PERP/VERT	DENTICOLLIS	POSITA	HASTATA	BARBERI
Beaver pt 1	14 (10♂, 4♀)	1♂, SP 1868					110 (75♂, 35♀)	3♂		1♀
Cimarron pt 2	99 (4♂, 96U)	4♂, SP 1597–1599	3 (1♂, 2♀)	1♂	4♂		2♂			
Cimarron pt 3	72 (39♂, 33♀)	1♂, SP 2378		2♂	4♂	45 (25♂, 20♀)	11 (8♂, 3♀)			
Cimarron pt 4	138 (48♂, 26♀, 64U)	8♂ SP 610–611, 614–619	4 (3♂, 1♀)	1♂		9♂	4 (2♂, 2♀)			
Cimarron pt 5	30 (3♂, 27U)	3♂, SP 620–622	18 (10♂, 8♀)				50(25♂, 25♀)			
Custer pt 6	2♂	2♂, SP 1341						12U	2♂	
Harper pt 7	55 (40♂, 15♀)	4♂, SP 417–420					4 (3♂, 1♀)	12 (8♂, 4♀)		
Texas pt 8	100 (60♂, 40♀)	2♂, SP 1594–1595					10 (7♂, 3♀)			
Texas pt 9	130 (80♂, 50♀)	1♂, SP 1616					1♂	2♂		4♂
Texas pt 10	50 (20♂, 30♀)	1♂, SP 609								

Notes: dem = *I. demorsa*, perp = *I. perparva*, vert = *I. verticalis*; SP 610, 615, and 620 were donated to RWG.

Collection data points:

Beaver pt 1 = Beaver Co.: Beaver River WMA, 22 April 2016, MAP

Cimarron pt 2 = Cimarron Co.: Black Mesa area (localities not specified), 3 June 2015, BS-P, MAP; *I. verticalis*: 12♂ in hand.

Cimarron pt 3 = Cimarron Co.: Black Mesa area (Cimarron River at 4 km NNW of Kenton and Lake Carl Etling), 3 June 2017, MAP; *I. verticalis*: 2♂ in hand.

Cimarron pt 4 = Cimarron Co.: Black Mesa area (Carrizozo Creek at 2 km W of Kenton, North Carrizo Creek at 7 km N of Kenton, Black Mesa State Park), 26 May 2013, BS-P, MAP

Cimarron pt 5 = Cimarron Co.: Conrad, 26 May 2013, BS-P, MAP

Custer pt 6 = Custer Co.: Weatherford, Rader Park, 24 July 2014, BS-P, MAP

Harper pt 7 = Harper Co.: Doby Springs Park, 15 Sept 2012, BS-P, MAP

Texas pt 8 = Texas Co.: Optima Lake, 2 June 2015, BS-P, MAP; *I. verticalis*: 6♂ and 1♀ in hand.

Texas pt 9 = Texas Co.: Optima Lake, 4 June 2015, BS-P, MAP; *I. verticalis*: 2♂ in hand.

Texas pt 10 = Texas Co.: 2 km S of Straight, 25 May 2013, BS-P, MAP

Table F.4 Data used to construct aberrant male Ischnura verticalis maps (Figure 12.81 and 12.83). Data are primarily from the Smith-Patten/Patten collection (SP) specimens and Odonata Central (OC) records. Additionally, we included six OC records here that are potentially aberrant individuals (?) but we were hesitant to conclusively categorize them as such because of the possibility that the constriction may be a result of blurriness or an odd angle of the photograph. The potential records were not part of the data used for the map. The "unequal" field indicates that the antehumerals differ from side to side.

LOCATION	TOTAL INDIVIDUALS	CONSTRICTED	BROKEN	EXCLAMATION POINTS	DOTS, ELONGATED	DOTS, DISTINCT	UNEQUAL	NOTES
Colorado	4	0	1	2	1	0	1	
Baca Co.	1				1		1	SP 747; elongated dots on left side, splattered pattern on right side; counted as elongated dots
Kiowa Co.	1			1				OC 480626
Kit Carson Co.	1			1				OC 329761
Lincoln Co.	1		1					OC 328653
Nebraska	1	0	1	0	0	0	0	
Nemaha Co.	1		1					JCAC
Oklahoma	37	11-18	8	9	6	3	2	
Beaver Co., Beaver Dunes SP	1			1				OC 479843
Beaver Co., Beaver R WMA	1			1				SP 1868
Beckham Co., Elk City	1	1						SP 1422
Cimarron Co., vic. Kenton	5			2		3		SP [RWG] 610, SP 611 and 2378; Hjalmarson photos, on file with OOP
Cimarron Co., Black Mesa SP	7	1	3	1	2		1	SP [RWG] 615, SP 616, 619, and 1597–1599, OC 383424; SP 1597 broken left side, full/slightly broken right side; SP 1599 full outline but spotty color
Cimarron Co., Conrad	2	1		1				SP [RWG] 620, SP 621
Comanche Co., Fort Sill MR		?						OC 315017
Comanche Co., Wichita Mts WR	1				1			OC 282024; observer noted "several males present," but he did not say others were aberrant
Craig Co., vic. Vinita	1	1						OC 401675
Custer Co., Weatherford	1		1					SP 1341
Ellis Co., Ellis Co. WMA	1	1						SP 2017
Grady Co., Lake Louis Burtschi	1	1						SP 862
Harper Co., Doby Springs	5		1	2	2			SP 417–420, OC 381715
Rogers Co., Claremore		?						OC 425897
Texas Co., vic. Straight	1			1				SP 609

(Continued)

Table F.4 (*Continued*) Data used to construct aberrant male *Ischnura verticalis* maps (Figure 12.81 and 12.83). Data are primarily from the Smith-Patten/Patten collection (SP) specimens and Odonata Central (OC) records. Additionally, we included six OC records here that are potentially aberrant individuals (?) but we were hesitant to conclusively categorize them as such because of the possibility that the constriction may be a result of blurriness or an odd angle of the photograph. The potential records were not part of the data used for the map. The "unequal" field indicates that the antehumerals differ from side to side.

LOCATION	TOTAL INDIVIDUALS	CONSTRICTED	BROKEN	EXCLAMATION POINTS	DOTS, ELONGATED	DOTS, DISTINCT	UNEQUAL	NOTES
Texas Co., Optima Lake	6	3	2		1			SP 104–106, 1594–1595, 1616
Tillman Co., vic. Manitou	1		1				1	SP 1373; narrows centrally and has slight break on left side, full right side
Tulsa Co., Tulsa	1	1, 3?						OC 409706, 423133, 424319, 425308
Woodward Co., Boiling Springs SP	1	1						OC 331903
Woodward Co., Ft Supply Lake	1	?						OC 331968
Texas	**4**	0	2	2	0	0	0	
Donley Co.	1		1					OC 282653
Ochiltree Co.	1		1					JCAC
Randall Co.	1			1				OC 312386
Randall Co.	1			1				OC 471736
totals	**46**	**11-18**	**12**	**13**	**7**	**3**	**3**	

JCAC = John C. Abbott Collection, OC = Odonata Central, SP = Smith-Patten/Patten Collection

Table F.5 **Data references for specimens and photographs included in Libellulidae Figures 19.47, 19.55, and 19.91.**

FIGURE, PART	SOURCE	SPECIES AND SEX	DATA
Figure 19.47			
top right corner, two pages	SP 1699	*Libellula comanche* ♀	Oklahoma: Stephens Co.: Clear Creek Lake, 27 June 2015, MAP
Part II, page 1, image A	SP 1484	*Libellula comanche* ♀	Oklahoma: Kingfisher Co.: Lake Elmer, 29 Oct 2014, BS-P
Part II, page 1, image B	SP 2484	*Libellula comanche* ♀	Oklahoma: Texas Co.: Shorb WMA, 3 Sept 2017, MAP
Part II, page 1, image C	SP 1428	*Libellula comanche* ♀	Oklahoma: Major Co.: Hoyle Creek, Ames, 9 Sept 2014
Part II, page 2, blue box, lateral view	BD Smith	*Libellula cyanea* ♀	Missouri: Reynolds Co.: Kay Branch fen, 21 June 2016
Part II, page 2, right, top	B Carrell	*Libellula cyanea* ♀	Oklahoma: Tulsa Co.: Oxley Nature Center, Tulsa, 25 June 2016
Part II, page 2, right, middle	JW Arterburn, OC 329291	*Libellula cyanea* ♀	Oklahoma: Wagoner Co.: Bixhoma Lake, 17 June 2011
Part II, page 2, right, middle inset	BE Reynolds	*Libellula cyanea* ♀	Oklahoma: Cherokee Co.: J. T. Nickel Preserve, 29 May 2014
Part II, page 2, right, bottom	GW Lasley	*Libellula flavida* ♀	No details
Figure 19.55			
	SP 920	*Libellula comanche × vibrans* ♂	Oklahoma: Blaine Co.: Roman Nose SP, 4 Aug 2013
	SP 1036	*Libellula comanche × vibrans* ♂	Oklahoma: Harper Co.: Doby Springs Park, 21 Sept 2013
	SP 1370	*Libellula comanche* ♂	Oklahoma: Cotton Co.: 10 km SE of Temple, 2 Aug 2014
	SP 1498	*Libellula vibrans* ♂	Oklahoma: Adair Co.: Sallisaw Creek, 8 km SW of Stilwell; 12 Aug 2014, JT Bried
Figure 19.91			
	SP 743	*Erythemis collocata* ♀	Oklahoma: Cimarron Co.: Black Mesa SP, 5 July 2013
	SP 788	*Erythemis simplicicollis* ♀	Oklahoma: Ellis Co.: Ellis Co. WMA, 9 July 2013
	SP 789	*Erythemis simplicicollis* ♀	Oklahoma: Ellis Co.: Ellis Co. WMA, 9 July 2013
	SP 1744	*Erythemis collocata × simplicicollis?* ♀	Oklahoma: Texas Co.: Beaver River, 12 km N of Goodwell, 8 Aug 2015, MAP
	SP 2021	*Erythemis collocata × simplicicollis?* ♀	Oklahoma: Cimarron Co.: South Carrizo Creek, Black Mesa State Park, 12 July 2016, BS-P
	SP 2085	*Erythemis collocata* ♀	Utah: Tooele Co.: Big Spring, 17 July 2016, BS-P
	SP 2160	*Erythemis simplicicollis* ♀	Texas: Gray Co.: Lake McClellan, McClellan Creek National Grassland, 24 July 2016, BS-P

source: OC = Odonata Central, SP = Smith-Patten/Patten Collection; data: BS-P = Brenda D. Smith-Patten, Co. = County, MAP = Michael A. Patten, SP = State Park, WMA = Wildlife Management Area

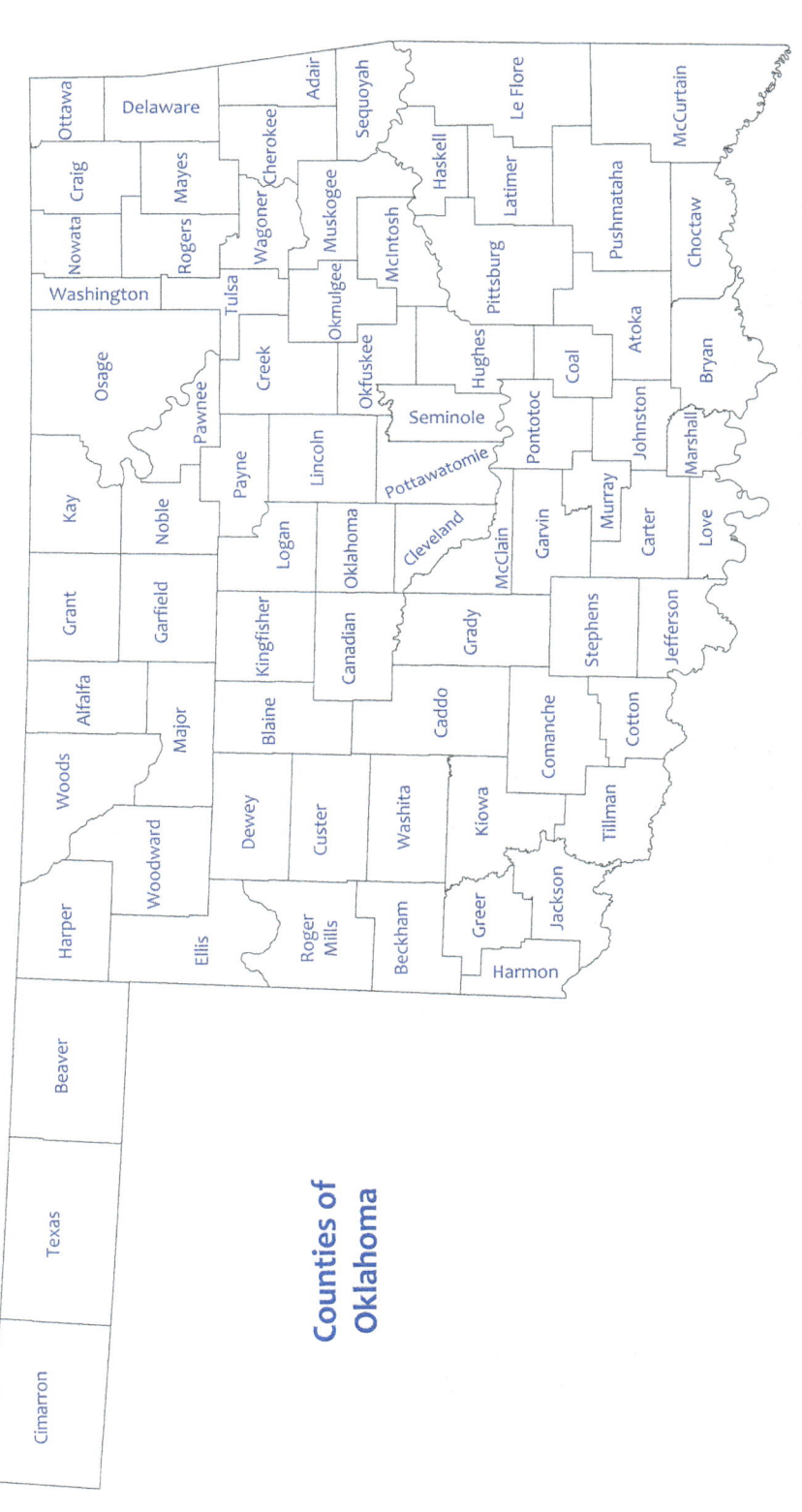

Counties of Oklahoma

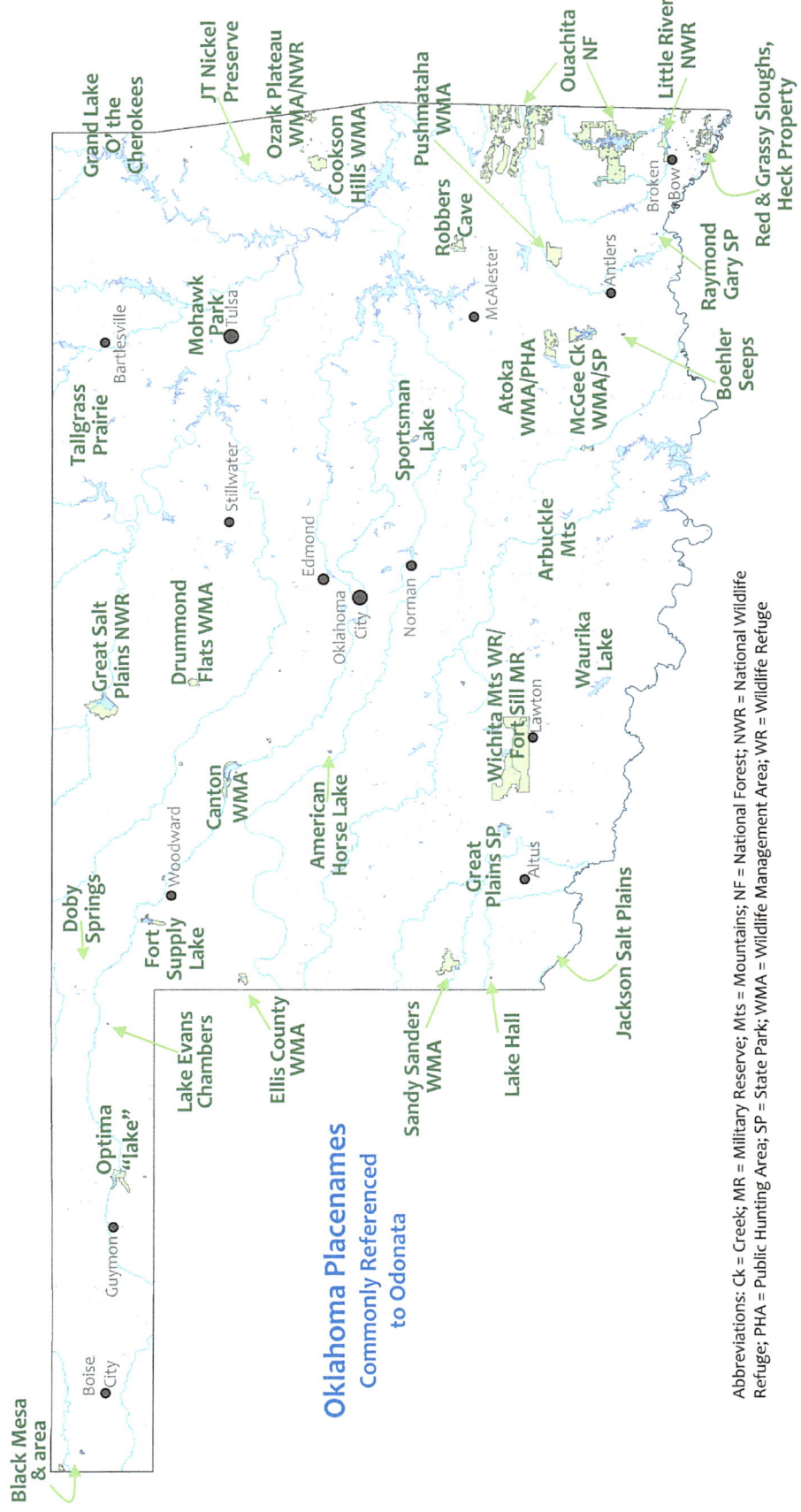

Oklahoma Placenames
Commonly Referenced to Odonata

Black Mesa & area

Optima "lake"

Doby Springs

Lake Evans Chambers

Ellis County WMA

Sandy Sanders WMA

Lake Hall

Jackson Salt Plains

Great Plains SP

Waurika Lake

Wichita Mts WR/ Fort Sill MR

American Horse Lake

Canton WMA

Fort Supply Lake

Great Salt Plains NWR

Drummond Flats WMA

Tallgrass Prairie

Mohawk Park

Arbuckle Mts

Sportsman Lake

Atoka WMA/PHA

McGee Ck WMA/SP

Boehler Seeps

Robbers Cave

Pushmataha WMA

Cookson Hills WMA

Ozark Plateau WMA/NWR

JT Nickel Preserve

Grand Lake O' the Cherokees

Ouachita NF

Little River NWR

Red & Grassy Sloughs, Heck Property

Raymond Gary SP

Boise City · Guymon · Woodward · Altus · Lawton · Oklahoma City · Edmond · Norman · Stillwater · Bartlesville · Tulsa · McAlester · Antlers · Broken Bow

Abbreviations: Ck = Creek; MR = Military Reserve; Mts = Mountains; NF = National Forest; NWR = National Wildlife Refuge; PHA = Public Hunting Area; SP = State Park; WMA = Wildlife Management Area; WR = Wildlife Refuge

Abbott, J. C. 1996. New and interesting records from Texas and Oklahoma. *Argia* 8(4):14–15.

Abbott, J. C. 2001. Distributions of dragonflies and damselflies (Odonata) in Texas. *Transactions of the American Entomological Society* 127:199–228.

Abbott, J. C. 2005. *Dragonflies and damselflies of Texas and the south-central United States*. Princeton University Press, Princeton, New Jersey.

Abbott, J. C. 2007. Somatochlora ozarkensis. *IUCN 2012. IUCN Red list of threatened species*. Version 2012.2. www.iucnredlist. org. Accessed 31 May 2013.

Abbott, J. C. 2011. *Damselflies of Texas: A field guide*. University of Texas Press, Austin, Texas.

Abbott, J. C. 2013. Identification of male *Epitheca* (*Tetragoneuria*) in Texas. *Copy on file*.

Abbott, J. C. 2015. *Dragonflies of Texas: A field guide*. University of Texas Press, Austin, Texas.

Abbott, J. C., G. W. Lasley. 2017. Seasonality of Texas dragonflies and damselflies. Unpublished ms., 23 March 2017. *Copy on file*.

Abbott, J. C., G. W. Lasley. 2019. Seasonality of Texas dragonflies and damselflies. Unpublished ms., 23 May 2019. *Copy on file*.

Abbott, J. C., D. R. Paulson. 2017. *Somatochlora ozarkensis*. The IUCN Red List of Threatened Species 2017: e.T20345A80697604. doi:10.2305/IUCN.UK.2017-3.RLT S.T20345A80697604.en. Assessment 5 May 2016. Accessed 12 November 2018.

Adamski J. C., J. C. Petersen, D. A. Freiwald, J. V. Davis. 1995. *Environmental and hydrologic setting of the Ozark Plateaus study unit, Arkansas, Kansas, Missouri, and Oklahoma*. U.S. Geological Survey, Water-Resources Investigations Report 94-4022. Little Rock, Arkansas.

Ahrens, C. 1935. A new record of *Archilestes grandis* (Odonata: Agrionidae *sensu* Selysii). *Entomological News* 46:183.

Ahrens, C. 1938. A list of dragonflies taken during the summer of 1936 in western United States (Odonata). *Entomological News* 49(8):9–16.

Allison, V. C. 1921. Some dragon flies [sic] of southeastern Kansas. *Transactions of the Kansas Academy of Science* 30:45–58.

Anonymous. 2005. *George H. Bick obituary*. Drennan & Ford Funeral Home and Crematory, Port Angeles, Washington.

Arbenz, J. K. 1968. Structural geology of the potato hills, Ouachita Mountains, Oklahoma. In: *A guidebook to the geology of the Western Arkoma Basin and Ouachita Mountains, Oklahoma*. Cline, L. M., editor. Oklahoma City Geological Society, Oklahoma City, Oklahoma, pp. 109–121.

Arbour, D. 2007. *Tholymis citrina* (Evening Skimmer) found in Oklahoma. *Argia* 18(4):29.

Arbour, D., B. Heck, R. Bastarache. 2009. Recent notable odonate records from the Red Slough Wetlands, Oklahoma. *Proceedings of the Oklahoma Academy of Science* 89:75–76.

Bailey, R. G. 1996. *Ecosystem Geography*. Springer-Verlag, New York.

Bailowitz, R., D. Danforth, S. Upson. 2015. *A field guide to the damselflies & dragonflies of Arizona and Sonora*. Nova Granada Publications, Tucson, Arizona.

Barber, B., V. Elia. 1994. *Tholymis citrina*: A recent record from Florida and an historical record from Texas. *Argia* 5(4):10–11.

Barker, B. M., W. C. Jameson. 1975. *Platt National Park: Environment and ecology*. University of Oklahoma Press, Norman, Oklahoma.

Barnard, A. A., O. M. Fincke, M. A. McPeek, J. P. Masly. 2017. Mechanical and tactile incompatibilities cause reproductive isolation between two young damselfly species. *Evolution* 71(10): 2410–2427.

Basara, J. B., J. N. Maybourn, C. M. Peirano, J. E. Tate, P. J. Brown, J. D. Hoey, B. R. Smith. 2013. Drought and associated impacts in the Great Plains of the United States—a review. *International Journal of Geosciences* 4:72–81.

Bass, D. 1990. A survey of aquatic invertebrates from Wichita Mountain streams. *Proceedings of the Oklahoma Academy of Science* 70:35–36.

Bass, D. 1994. Community structure and distribution patterns of aquatic macroinvertebrates in a tall grass prairie stream ecosystem. *Proceedings of the Oklahoma Academy of Science* 74:3–10.

Bass, D. 1995. Species composition of aquatic macroinvertebrates and environmental conditions in Cucumber Creek. *Proceedings of the Oklahoma Academy of Science* 75:39–44.

Bass, D., C. Potts. 2001. Invertebrate community composition and physiochemical conditions of Boehler Lake, Atoka County, Oklahoma. *Proceedings of the Oklahoma Academy of Science* 81:21–29.

Beaton, G. 2007. *Dragonflies & damselflies of Georgia and the southeast*. The University of Georgia Press, Athens, Georgia.

Bechly, G. 1996. Morphologische Untersuchungen am Flügelgeäder der rezenten Libellen und deren Stammgruppenvertreter (Insecta; Pterygota; Odonata), unter besonderer Berücksichtigung der Phylogenetischen Systematik und des Grundplanes der *Odonata. *Petalura* 2:402.

Bechly, G. 2008. Phylogenetic systematics of Odonata. https://be chly.lima-city.de/phylosys.htm. Accessed 1 November 2018.

Bechly, G., C. Brauckmann, W. Zessin, E. Gröning. 2001. New results concerning the morphology of the most ancient dragonflies (Insecta: Odonatoptera) from the Namurian of Hagen-Vorhalle (Germany). *Journal of Zoological Systematics and Evolutionary Research* 39:209–226.

Beckemeyer, R. [J]. 1995. Some county records for Kansas and Oklahoma. *Argia* 7(3):28–29.

Beckemeyer, R. [J]. 1998a. *Tramea calverti* collected in Missouri. *Argia* 10(4):13.

Beckemeyer, R. [J]. 1998b. Some miscellaneous Odonata collected in the Midwest in 1998. *Argia* 10(4):26–27.

Beckemeyer, R. J. 2000. The Permian insect fossils of Elmo, Kansas. *The Kansas School Naturalist* 46(1):1–16.

Beckemeyer, R. J. 2002a. A checklist of dragonflies and damselflies (Odonata) of Oklahoma. 19 January 2002, *copy on file*.

Beckemeyer, R. J. 2002b. Checklist of Kansas Odonata (dragonflies and damselflies) with data sources, notes, and historical records. 25 January 2002. www.gpnc.org/images/pdf/CheckOdos.pdf. *Copy on file*.

Beckemeyer, R. [J]. 2002c. Some Great Plains Odonata records for 2000 and 2001. *Argia* 13(4):7–8.

Beckemeyer, R. [J]. 2002d. George H. Bick: Honorary member, the dragonfly society of the Americas. *Argia* 14(3):4–5.

Beckemeyer, R. [J]. 2002e. Some Odonata records for the Oxley Nature Center, Tulsa County, Oklahoma. *Argia* 14(3):12–13.

Beckemeyer, R. J. 2002f. Odonata in the Great Plains States: Patterns of distribution and diversity. *Bulletin of American Odonatology* 6(3):49–99.

Beckemeyer, R. [J]. 2004. Some Odonata records for the Midwest and West for 2003. *Argia* 16(1):26.

Beckemeyer, R. J. 2006. Hind wing fragments of *Meganeuropsis* (Protodonata: Meganeuridae) from the Lower Permian of Noble County, Oklahoma. *Bulletin of American Odonatology* 9(3/4):85–89.

Beckemeyer, R. J. 2012. Robin J. Tillard in Kansas: Travels of an early-twentieth-century palaeoentomologist. *Journal of the Kansas Entomological Society* 85(4):353–373.

Beckemeyer, R. J. 2013. Rediscovered photographs of a 1927 visit by Frank Morton Carpenter to the Elmo, Kansas, fossil insect beds. In: *The Carboniferous-Permian Transition*. Lucas, S. G., W. A. DiMichele, J. E. Barrick, J. W. Schneider, J. A. Spielmann, editors. New Mexico Museum of Natural History and Science Bulletin 60:12–15.

Beckemeyer, R. J., J. D. Hall. 2007. The entomofauna of the Lower Permian fossil insect beds of Kansas and Oklahoma, USA. *African Invertebrates* 48(1):23–39.

Beckemeyer, R. J., D. G. Huggins. 1997. Checklist of the Kansas dragonflies. *The Kansas School Naturalist* 43(2), 16 pp.

Beckemeyer, R. J., D. G. Huggins. 1998. Checklist of the Kansas damselflies. *The Kansas School Naturalist* 44(1), 16 pp.

Beckemeyer, R. [J]., R. Todd. 1996. Additions to Kansas Odonata records for 1996. *Argia* 8(4):13–14.

Bick, G. H. 1951. Notes on Oklahoma dragonflies. *Journal of the Tennessee Academy of Science* 26:178–180.

Bick, G. H. 1957. The Odonata of Louisiana. *Tulane Studies in Zoology* 5(5):71–135.

Bick, G. H. 1959. Additional dragonflies (Odonata) from Arkansas. *The Southwestern Naturalist* 4:131–133.

Bick, G. H. 1963. Reproductive behavior in *Enallagma civile* and *Argia apicalis*. *Proceedings of the North Central Branch of the Entomological Society of America* 18:110–111.

Bick, G. H. 1978. New state records of United States Odonata. *Notulae Odonatologicae* 1:17–36.

Bick, G. H.. 1983. Odonata at risk in [the] conterminous United States and Canada. *Odonatologica* 12:209–226.

Bick, G. H. 1985. Obituary: Lothar E. Hornuff, Jr. *Odonatologica* 14:257–260.

Bick, G. H. 1990. Unpublished records in [the] Florida State Collection of Arthropods. *Argia* 2(1):3–4.

Bick, G. H. 1991. Oklahoma revisited: Unpublished records. *Argia* 3(1):1–4.

Bick, G. H. 1996. Looking back. *Argia* 8(2):22–26.

Bick, G. H. 2003. At-risk Odonata of [the] conterminous United States. *Bulletin of American Odonatology* 7:41–56.

Bick, G. H., J. C. Bick. 1957. The Odonata of Oklahoma. *The Southwestern Naturalist* 2:1–18.

Bick, G. H., J. C. Bick. 1958. The ecology of the Odonata at a small creek in southern Oklahoma. *Journal of the Tennessee Academy of Science* 33:240–251.

Bick, G. H., J. C. Bick. 1961. An adult population of *Lestes disjunctus* Walker (Odonata: Lestidae). *The Southwestern Naturalist* 6(3–4):111–137.

Bick, G. H., J. C. Bick. 1963. Behavior and population structure of the damselfly, *Enallagma civile* (Hagen) (Odonata: Coenagrionidae). *The Southwestern Naturalist* 8:57–84.

Bick, G. H., J. C. Bick. 1965. Demography and behavior of the damselfly, *Argia apicalis* (Say), (Odonata: Coenagrionidae). *Ecology* 46(4):461–472.

Bick, G. H., J. C. Bick. 1970. Oviposition in *Archilestes grandis* (Rambur) (Odonata: Lestidae). *Entomological News* 81(7):157–163.

Bick, G. H., J. C. Bick. 1972. Substrate utilization during reproduction by *Argia plana* Calvert and *Argia moesta* (Hagen) (Odonata: Coenagrionidae). *Odonatologica* 1(1):3–9.

Bick, G. H., J. C. Bick. 1981. Heterospecific pairing among Odonata. *Odonatologica* 10(4):259–270.

Bick, G. H., J. C. Bick, L. E. Hornuff. 1977. An annotated list of the Odonata of the Dakotas. *Florida Entomologist* 60:149–165.

Bick, G. H., L. E. Hornuff. 1971. Survey of the Odonata of Wyoming with emphasis on their altitudinal distribution. *Jackson Hole Research Station Annual Report* 1971:11–13. http://repository.uwyo.edu/jhrs_reports/vol1971/iss1/7.

Bick, G. H., L. E. Hornuff. 1972. Odonata collected in Wyoming, South Dakota, and Nebraska. *Proceedings of the Entomological Society of Washington* 74:1–8.

Bick, G. H., L. E. Hornuff. 1974. New records of Odonata from Montana and Colorado. *Proceedings of the Entomological Society of Washington* 76:90–93.

Bick, S. 2006. Memories of my daddy, George Bick. *Argia* 18(1):5–6.

Bird, C. D. 1972. Ralph Durham Bird. *Canadian Field Naturalist* 86:393–399.

Bird, R. D. 1931. The nymph of *Enallagma basidens* Calvert (Odonata: Agrionidae). *Entomological News* 42:276–277.

Bird, R. D. 1932. Dragonflies of Oklahoma. *Publications of the University of Oklahoma Biological Survey* 4:50–57.

Bird, R. D. 1933. *Somatochlora ozarkensis*, a new species from Oklahoma. *Occasional Papers of the Museum of Zoology, University of Michigan* 261:1–7.

Bird, R. D. 1934. The emergence and nymph of *Gomphus militaris* (Odonata, Gomphinae). *Entomological News* 45:44–46.

Blair, W. F. 1938. Ecological relationships of the mammals of the Bird Creek region, northeastern Oklahoma. *American Midland Naturalist* 20(3):473–526.

Blair, W. F., T. H. Hubbell. 1938. The biotic districts of Oklahoma. *American Midland Naturalist* 20(2):425–454.

Blome, C. D., D. J. Lidke, R. R. Wahl, J. A. Golab. 2013. *Geologic map of Chickasaw National Recreation Area*, Murray County, Oklahoma. U.S. Geological Survey Scientific Investigations Map 3258, 28 p., 1 sheet, scale 1:24,000, http://pubs.usgs.gov/sim/3258/

Borror, D. J. 1942. *A revision of the libelluline genus Erythrodiplax (Odonata)*. Ohio State University, Columbus, Ohio.

Brauckmann, C., W. Zessin. 1989. Neue Meganeuridae aus dem Namurium von Hagen-Vorhalle (BRD) und die Phylogenie der Meganisoptera [New Meganeuridae from the Namurian of Hagen-Vorhalle and the phylogeny of the Meganisoptera]. *Deutsche Entomologische Zeitschrift* 36(1–3):177–215.

Bried, J. T., A. M. Dillon, B. J. Hager, M. A. Patten, B. Luttbeg. 2015. Criteria to infer local species residency in standardized adult dragonfly surveys. *Freshwater Science* 34:1105–1113.

Brikowski, T. H. 2008. Doomed reservoirs in Kansas, USA? Climate change and groundwater mining on the Great Plains lead to unsustainable surface water storage. *Journal of Hydrology* 354:90–101.

Britch, C., M. L. Cain, D. J. Howard. 2001. Spatio-temporal dynamics of the *Allonemobius fasciatus–A. socius* mosaic hybrid zone: A 14-year perspective. *Molecular Ecology* 10:627–638.

Brooks, R. L. 2009. Prehistoric native peoples. In: *The encyclopedia of Oklahoma history and culture*. Everett, D., L. D. Wilson, L. O'Dell, J. D. May, editors. Oklahoma Historical Society, Oklahoma City, Oklahoma.

Bruner, W. E. 1931. The vegetation of Oklahoma. *Ecological Monographs* 1(2):99–188.

Buck, P. 1964. Relationships of the woody vegetation of the Wichita Mountains Wildlife Refuge to geologic formations and soil types. *Ecology* 45(2):336–344.

Bybee, S., Q. Hansen, S. Büsse, H. M. Cahill Wightman, M. A. Branham. 2015. For consistency's sake: The precise use of larva, nymph and naiad within Insecta. *Systematic Entomology* 40:667–670.

Byers, C. F. 1927. Key to the North American species of *Enallagma*, with a description of a new species (Odonata: Zygoptera). *Transactions of the American Entomological Society* 53(3):249–260.

Byers, C. F. 1937. A review of the dragonflies of the genera *Neurocordulia* and *Platycordulia*. *Miscellaneous Publications of the Museum of Zoology* 36:1–36. University of Michigan.

Calvert, P. P. 1893. Pairing of different species. *Entomological News* 4:268.

Calvert, P. P. 1895. The Odonata of Baja California, Mexico. *Proceedings of the California Academy of Science* 4:463–558.

Calvert, P. P. 1902. *Odonata, in Biologia Centrali Americana: Insecta Neuroptera*. R. H. Porter & Dulau Co., London, pp. 73–128.

Calvert, P. P. 1919. Gundlach's work on the Odonata of Cuba: A critical study. *Transactions of the American Entomological Society* 45(4):335–396.

Cannings, R. A. 1989. *Enallagma basidens* Calvert, a dragonfly new to Canada, with notes on the expansion of its range in North America (Zygoptera: Coenagrionidae). *Notulae Odonatologicae* 3(4):49–64.

Carle, F. L. 1982. A contribution to the knowledge of the Odonata. PhD dissertation, Virginia Polytechnic Institute and State University, Blacksburg, Virginia. https://vtechworks.lib.vt.edu/handle/10919/37497

Carle, F. L. 1983. A new *Zoraena* (Odonata: Cordulegastridae) from eastern North America, with a key to the adult Cordulegastridae of America. *Annals of the Entomological Society of America* 76:61–68.

Carpenter, C. C. 1990. The Ortenburger field notebooks. *Proceedings of the Oklahoma Academy of Science* 70:45–50.

Carpenter, C. C. 2000. Early Oklahoma naturalists and collectors. *Occasional Papers of the Sam Noble Oklahoma Museum of Natural History* 6:1–46.

Carpenter, F. M. 1931. The Lower Permian insects of Kansas. Part 2. The orders Paleodictyoptera, Protodonata, and Odonata. *American Journal of Science* (Fifth Series) 21:97–139.

Carpenter, F. M. 1933. The Lower Permian insects of Kansas. Part 6. Delopteridae, Protelytroptera, Plectoptera and a new collection of Protodonata, Odonata, Megasecoptera, Homoptera, and Psocoptera. *Proceedings of the American Academy of Arts and Sciences* 68:411–504.

Carpenter, F. M. 1939. The Lower Permian insects of Kansas. Part 8. Additional Megasecoptera, Protodonata, Odonata, Homoptera, Psocoptera, Protelytroptera, Plectoptera, and Protoperlaria. *Proceedings of the American Academy of Arts and Sciences* 73:29–70, plates 1–2.

Carpenter, F. M. 1943. The Lower Permian insects of Kansas. Part 9. The orders Neuroptera, Raphidiodea, Caloneurodea and Protorthoptera (Probnisidae), with additional Protodonata and Megasecoptera. *Proceedings of the American Academy of Arts and Sciences* 75(2):55–84.

Carpenter, F. M. 1947. Lower Permian insects from Oklahoma. Part 1. Introduction and the orders Megasecoptera, Protodonata, and Odonata. *Proceedings of the American Academy of Arts and Sciences* 76:25–54.

Carpenter, F. M. 1960. Studies on North American Carboniferous insects: 1. The Protodonata. *Psyche* 67:98–110.

Carpenter, F. M. 1992. *Treatise on invertebrate paleontology, Part R, Arthropoda 4, vols 3, 4 (Hexapoda)*. Geological Society of America and University of Kansas Press, Boulder, Colorado.

Chao, A. 1987. Estimating the population size for capture-recapture data with unequal catchability. *Biometrics* 43:783–791.

Chao, H. F. 1954. Classification of Chinese dragonflies (Odonata) of the family Gomphidae (Odonata), Part II. *Acta Entomologica Sinica* 4:23–84.

Clark, L. G. 2011. Survey of the vascular flora of the Boehler Seeps and Sandhills Preserve. *Oklahoma Native Plant Record* 11:4–21.

Cook, C. 1994. *The status of a rare dragonfly Macromia wabashensis (Wabash Belted Skimmer) in Indiana and Ohio*. Report submitted to the U.S. Fish and Wildlife Service and Ohio Historical Society, 36 pp.

Cook, C, E. L. Laudermilk. 2004. *Stylogomphus sigmastylus* sp. nov., a new North American dragonfly previously confused with *S. albistylus* (Odonata: Gomphidae). *International Journal of Odonatology* 7(1):3–24.

Corbet, P. S. 1999. *Dragonflies: Behavior and ecology of Odonata*. Cornell University Press, Ithaca, New York.

Corbet, P. S. 2002. Stadia and growth ratios of Odonata: A review. *International Journal of Odonatology* 5(1):45–73.

Costa, G. C., C. Wolfe, D. B. Shepard, J. P. Caldwell, L. J. Vitt. 2008. Detecting the influence of climatic variables on species distributions: A test using GIS niche-based models along a steep longitudinal environmental gradient. *Journal of Biogeography* 35(4):637–646.

Costigan, K. H., M. D. Daniels. 2012. Damming the prairie: Human alteration of Great Plains river regimes. *Journal of Hydrology* 444–445:90–99.

Cotterman, C. W. 1931. *Archilestes* in Ohio (Odonata, Agrionidae). *Entomological News* 42:64.

Covich, A. P., M. A. Palmer, T. A. Crowl. 1999. The role of benthic invertebrate species in freshwater ecosystems: Zoobenthic species influence energy flows and nutrient cycling. *BioScience* 49:119–127.

Cowardin, L. M., V. Carter, F. C. Golet, E. T. Laroe. 1979. *Classification of wetlands and deepwater habitats of the United States*. Biological Service Program, U.S. Fish and Wildlife Service, Washington, D. C., Publication # FWS/OBS-79/31.

Cringan, M. S. 1979. Dragonflies and damselflies of McKinney Marsh. *The Emporia State Research Studies* 27(3):5–19.

Cunfer, G. 2005. *On the great plains: Agriculture and environment*. Texas A & M University Press, College Station, Texas.

Curry, J. R. 2001. *Dragonflies of Indiana*. Indiana Academy of Science, Indianapolis, Indiana.

Curtis, N. M., W. E. Ham, K. S. Johnson. 2008. Geomorphic provinces of Oklahoma. In: *Earth sciences and mineral resources of Oklahoma*. Johnson, K. S., K. V. Luza, editors. Oklahoma Geological Survey, Norman, Oklahoma.

Daigle, J. J., E. Pilgrim. 2014. *Amphiagrion* (Red Damsel) update. *Argia* 26(1):19.

Danks, H. V. 2007. How aquatic insects live in cold climates. *Canadian Entomologist* 139:443–471.

Davis, W. T. 1933. Dragonflies of the genus *Tetragoneuria*. *Bulletin of the Brooklyn Entomological Society* 28:87–104.

Davis, W. T. 1937. Second record of the dragonfly *Neurocordulia virginiensis*. *Journal of the New York Entomological Society* 45:250.

de Borhegyi, S. F. 1957. A survey of Oklahoma museums: 1893–1957. *Chronicles of Oklahoma* 35(2):204–227.

DeForest, D. K., K. V. Brix, W. J. Adams. 2007. Assessing metal bioaccumulation in aquatic environments: The inverse relationship between bioaccumulation factors, trophic transfer factors and exposure concentration. *Aquatic Toxicology* 84(2):236–246.

Denning, D. G., W. W. Allen. 1965. Arthur Earl Pritchard. *Journal of Economic Entomology* 58:807–808.

Diamond, D. D., L. F. Elliott. 2015. *Oklahoma ecological systems mapping interpretive booklet: Methods, short type descriptions, and summary results*. Oklahoma Department of Wildlife Conservation, Oklahoma City, Oklahoma.

Dice, L. R. 1943. *The biotic provinces of North America*. University of Michigan Press, Ann Arbor, Michigan.

Doerr, A. H. 1962. A preliminary survey of plant succession on coal mine spoil in Oklahoma. *Journal of Geography* 61(7): 301–309.

Donnelly, T. W. 1961. The Odonata of Washington, D. C., and vicinity. *Proceedings of the Entomological Society of Washington* 63:1–13.

Donnelly, T. W. 1962. *Somatochlora margarita*, a new species of dragonfly from eastern Texas. *Proceedings of the Entomological Society of Washington* 64:235–240.

Donnelly, T. W. 1964. *Enallagma westfalli*, a new damselfly from eastern Texas, with remarks on the genus *Teleallagma* Kennedy. *Proceedings of the Entomological Society of Washington* 66(2): 103–109.

Donnelly, T. W. 1973. Status of *Enallagma traviatum* and *westfalli* (Odonata, Coenagrionidae). *Proceedings of the Entomological Society of Washington* 75(3):298–302.

Donnelly, T. W. 1991. Taxonomic problems (?) with *Tetragoneuria*. *Argia* 4(1):11–14.

Donnelly, T. W. 1997. The hunt of red *Sympetrum*. *Argia* 9(3): 17–18.

Donnelly, T. W. 2003a. Problems with *Tetragoneuria*. *Argia* 14(4):10–11.

Donnelly, T. W. 2003b. *Lestes disjunctus, forcipatus*, and *australis*: A confusing complex of North American Damselflies. *Argia* 15(3):10–13.

*Donnelly, T. W. 2004a. Distribution of North American Odonata. Part I: Aeshnidae, Petaluridae, Gomphidae, Cordulegastridae. *Bulletin of American Odonatology* 7:61–90.*

*Donnelly, T. W. 2004b. Distribution of North American Odonata. Part II: Macromiidae, Corduliidae, Libellulidae. *Bulletin of American Odonatology* 8:1–32.

*Donnelly, T. W. 2004c. Distribution of North American Odonata. Part III: Calopterygidae, Lestidae, Coenagrionidae, Protoneuridae, Platystictidae. *Bulletin of American Odonatology* 8:33–99.

Donnelly, [T. W.] N. 2004d. *Erythemis simplicicollis* and *collocata*—subspecies? *Argia* 15(4):11–13.

Donnelly, [T. W.] N. 2006a. George Bick, 1914–2005. *Argia* 18(1):2–5.

Donnelly, T. W. 2006b. Rediscovery of intergrade *Epitheca* in southern Indiana. *Argia* 18(2):12–13.

Donnelly, T. W., J. C. Abbott 2006. *Request for help with baskettails (Epitheca)*. Posted, 12 April 2006, to Odonata Central. *Copy on file.*

Donnelly, T. W., K. J. Tennessen. 1994. *Macromia illinoiensis* and *georgina*: A study of their variation and apparent subspecific relationship (Odonata: Corduliidae). *Bulletin of American Odonatology* 2:27–61.

Droke, J. A. 2009. Sam Noble Oklahoma museum of natural history. In: *The encyclopedia of Oklahoma history and culture*. Everett, D., L. D. Wilson, L. O'Dell, J. D. May, editors. Oklahoma Historical Society, Oklahoma City, Oklahoma.

DSA [Dragonfly Society of the Americas]. 1999. Juanda C. Bick: 1919–1999. *Argia* 11(4):2–4.

DuBois, R. 2010. *Dragonflies and Damselflies of the Rocky Mountains*. Kollath-Stensaas Publishing, Duluth, Minnesota.

Duck, L. G. 1943. *A game type map of Oklahoma*. Oklahoma Game and Fish Department, Oklahoma City, Oklahoma.

Duck, L. G., J. B. Fletcher. 1943. *Survey of the game and furbearing animals of Oklahoma*. Oklahoma Game and Fish Department [now Oklahoma Department of Wildlife Conservation], Oklahoma City, Oklahoma.

Dunbar, C. O., A. A. Baker, G. A. Cooper, P. B. King, E. D. McKee, A. K. Miller, R. C. Moore, N. D. Newell, A. S. Romer, E. H. Sellards, J. W. Skinner, H. D. Thomas, H. E. Wheeler. 1960. Correlation of the permian formations of North America. *Bulletin of the American Geological Society of America* 71:1763–1808, *plate 1*.

Dunkle, S. W. 1975. New records of North American anisopterous dragonflies. *Florida Entomologist* 58:117–119.

* Donnelly (2004a, b, c) are essentially the same publication that was printed in three installments. The series has become known as the Dot Map Project (DMP), in reference to the maps presented in the publication. Although we cite Donnelly (2004a, b, c) throughout the book, generally we are actually making reference to the DMP's source data, which is archived by J. C. Abbott (county-level data are also part of the Odonata Central database, but those records were regrettably striped of the information indicating where Donnelly obtained those data).

Dunkle, S. W. 1976. Notes on the Anisoptera fauna near Mazatlan, Mexico, including dry to wet seasonal changes. *Odonatologica* 5(3):207–212.

Dunkle, S, W. 1983. New records of North American Odonata. *Entomological News* 94(4):136–138.

Dunkle, S. W. 2000. *Dragonflies through binoculars: A field guide to dragonflies of North America.* Oxford University Press, Oxford, UK.

Dunkle, S., R. Novelo, R. Cannings, D. Paulson, B. Mauffray. 2006. Some tributes to George. *Argia* 18(1):6–7.

Egan, T. 2006. *The worst hard time: The untold story of those who survived the great American Dust Bowl.* Houghton Mifflin, Boston, Massachusetts.

Ellzey, K. D. 2004. *Enallagma doubledayi* population in Kisatchie National Forest, Natchitoches Parish, Louisiana. *Argia* 16(3):24–25.

Engel, M. S. 1998. *Megatypus parvus* spec. nov., a new giant dragonfly from the Lower Permian of Kansas (Protodonata: Meganeuridae). *Odonatologica* 27(3):361–364.

English, R., English, A. 2015. Thornbush Dasher (*Micrathyria hagenii*), a new record for Williamson County, Tennessee. *Argia* 27(3):19.

EPA (Environmental Protection Agency). 1996. *Superfund record of decision: National Zinc corporation.* Bartlesville, Oklahoma, 12/13/1994. PB95-964206, EPA/ROD/R06-95/098, June 1996. Environmental Protection Agency, Washington, D. C.

Epstein, H. E., W. K. Lauenroth, I. C. Burke, D. P. Coffin. 1996. Ecological responses of dominant grasses along two climatic gradients in the Great Plains of the United States. *Journal of Vegetation Science* 7:777–788.

Evans, M. A. 1995. Checklist of the Odonata of New Mexico with additions to the Colorado checklist. *Proceedings of the Denver Museum of Natural History* series 3(8):1–6.

Evenhuis, N. L. 2017. The insect and spider collections of the world website. http://hbs.bishopmuseum.org/codens/. Accessed 24 June 2017.

Everett, D. 2009a. Indian territory. In: *The encyclopedia of Oklahoma history and culture.* Everett, D., L. D. Wilson, L. O'Dell, J. D. May, editors. Oklahoma Historical Society, Oklahoma City, Oklahoma.

Everett, D. 2009b. Land openings. In: *The encyclopedia of Oklahoma history and culture.* Everett, D., L. D. Wilson, L. O'Dell, J. D. May, editors. Oklahoma Historical Society, Oklahoma City, Oklahoma.

Everett, D. 2009c. Tri-State lead and Zinc district. In: *The encyclopedia of Oklahoma history and culture.* Everett, D., L. D. Wilson, L. O'Dell, J. D. May, editors. Oklahoma Historical Society, Oklahoma City, Oklahoma.

Fairchild, R. W., R. L. Hanson, R. E. Davis. 1990. *Hydrology of the Arbuckle Mountains Area, southcentral Oklahoma.* Oklahoma Geological Survey, Circular 91.

Ferguson, A. 1944. The nymph of *Enallagma basidens* Calvert (Odonata: Coenagrionidae). *Field and Laboratory* 12(1):19.

Ferris, C. D. 1951. The discovery of *Archilestes grandis* in eastern Pennsylvania. *Entomological News* 62:304.

Fisher, W. M. 1937. *An ecological study of Oklahoma ants.* MS Thesis, University of Oklahoma, Norman, Oklahoma.

Fixico, D. 2009. American Indians. In: *The encyclopedia of Oklahoma history and culture.* Everett, D., L. D. Wilson, L. O'Dell, J. D. May, editors. Oklahoma Historical Society, Oklahoma City, Oklahoma.

Fuller, R. Sr. 2011. A baseline of diversity. *Outdoor Oklahoma Magazine* Jan/Feb(2011):6.

Garbrecht, J., M. Van Liew, G. O. Brown. 2004. Trends in precipitation, streamflow, and evapotranspiration in the Great Plains of the United States. *Journal of Hydrologic Engineering* 9:360–367.

Garrison, R. W. 1991. A synonymic list of the New World Odonata. *Argia* 3(2):1–30.

Garrison, R. W. 1994. A synopsis of the genus *Argia* of the United States with keys and descriptions of new species, *Argia sabino, A. leonorae,* and *A. pima* (Odonata: Coenagrionidae). *Transactions of the American Entomological Society* 120(4):287–368.

Garrison, R. W., N. von Ellenrieder. 2019. A synonymic list of the New World Odonata. Revised version, 18 October 2019. *Copy on file.*

Garrison, R. W., N. von Ellenrieder, J. A. Louton. 2006. *Dragonfly genera of the New World: An illustrated and annotated key to the Anisoptera.* John Hopkins University Press, Baltimore, Maryland.

Garrison, R. W., N. von Ellenrieder, J. A. Louton. 2010. *Damselfly genera of the New World: An illustrated and annotated key to the Zygoptera.* John Hopkins University Press, Baltimore, Maryland.

GTSF [Geological Time Scale Foundation]. 2016. Stratigraphic Chart—GTS2016: Phanerozoic and Precambrian Chronostratigraphy. https://engineering.purdue.edu/Stratigraphy/charts/chart.html. Accessed 1 November 2018.

George, P., S. R. Wood. 1943. The railroads of Oklahoma. *The Railway and Locomotive Historical Society Bulletin* 60:7–79.

Gleick, P. H. 2003. Water use. *Annual Review of Environment and Resources* 28:275–314.

Gloyd, L. K. 1958. The dragonfly fauna of the Big Bend region of Trans-Pecos Texas. *Occasional Papers of the Museum of Zoology, University of Michigan* 593:1–23, pl. 8.

Gloyd, L. K. 1968a. The union of *Argia fumipennis* (Burmeister, 1839) with *Argia violacea* (Hagen, 1861), and the recognition of three subspecies (Odonata). *Occasional Papers of the Museum of Zoology, University of Michigan* 658:1–6.

Gloyd, L. K. 1968b. The synonmy of *Diargia* and *Hyponeura* with the genus *Argia* (Odonata: Coenargrionidae: Argiinae). *The Michigan Entomologist* 1(8):271–274.

Gloyd, L. K. 1973. The status of the generic names *Gomphoides, Negomphoides, Progomphus,* and *Ammogomphus* (Odonata: Gomphidae). *Occasional Papers of the Museum of Zoology, University of Michigan* 668:1–7.

Gloyd, L. K. 1980. The taxonomic status of the genera *Superlestes* and *Cyptolestes* Williamson, 1921 (Odonata: Lestidae). *Occasional Papers of the Museum of Zoology, University of Michigan* 694:1–3.

Glotzhober, B., D. McShaffrey, editors. 2002. *Dragonflies and Damselflies of Ohio.* Ohio Biological Survey, Columbus, Ohio.

Gollehon, N., B. Winston. 2013. *Groundwater irrigation and water withdrawals: The Ogallala aquifer initiative.* Economic Series REAP Report 1, August 2013. U.S. Department of Agriculture, Washington, D. C.

Gould, C. N. 1900. The Oklahoma salt plains. *Transactions of the Annual Meetings of the Kansas Academy of Sciences* 17:181–184.

Gould, C. N. 1926. Oklahoma—An example of arrested development. *Economic Geography* 2(3):426–450.

Gould, C. N. 1932. Beginning of the geological work in Oklahoma. *Chronicles of Oklahoma* 10(2):196–203.

Green, D. E. 1977a. Beginnings of wheat culture in Oklahoma. In: *Rural Oklahoma*. Green, D. E., editor. Oklahoma Historical Society, Oklahoma City, Oklahoma.

Green, D. E. 1977b. King cotton in Oklahoma. In: *Rural Oklahoma*. Green, D. E., editor. Oklahoma Historical Society, Oklahoma City, Oklahoma.

Gregoire, S., J. Gregoire. 2005. Mass emergence of *Lestes unguiculatus* in a small pond in central New York. *Argia* 16(4):9–10.

Gregoire, S., J. Gregoire. 2006. Update on mass emergence of *Lestes unguiculatus* in Central New York. *Argia* 18(1):14.

Grimaldi, D., M. S. Engel. 2005. *Evolution of the Insects*. Cambridge University Press, Cambridge, England, xv + 755 pp.

Hagen, H. A. 1861. Synopsis of the Neuroptera of North America, with a list of the South American species. *Smithsonian Miscellaneous Collections* 4:1–347.

Hall, J. D. 2004. Depositional facies and diagenesis of the Carlton Member (Kansas) and the Midco Member (Oklahoma) of the Wellington Formation (Permian, Sumner Group). MS Thesis, Wichita State University, Wichita, Kansas, viii + 112 pp.

Hall, J. D., R. J. Beckemeyer, W. J. May. 2016. Stratigraphy, sedimentology, and paleontology of the Midco Member, Wellington Formation (Permian-Sumner Group) of Oklahoma. *Shale Shaker* 66(6):306–316.

Handlirsch, A. 1937. Neue Untersuchungen über die fossilen Insekten: Mit Ergänzungen und Nachträgen sowie Ausblicken auf phylogenetische, palaeogeographische und allgemein biologische Probleme. I. Teil. *Annalen Des Naturhistorischen Museums in Wien* 48:1–140.

Harp, G. L. 1983. New and unusual records of Arkansas Anisoptera, United States. *Notulae odonatologicae* 2(2):26–27.

Harp, G. L. 2005. *Ischnura kellicotti* and lily pads. *Argia* 16(4):11.

Harp, G. L. 2006. Odonata collected/identified during the 2006 SE regional meeting [18–21 May, Ouachita Mountains, Arkansas]. Unpublished report, *copy on file*.

Harp, G. L. 2007. Dragonflies recorded for the Ouachita National Forest and adjacent areas, 14–17 May 2007. Report submitted to the U.S. Forest Service, Ouachita National Forest. *Copy on file*.

Harp, G. L., P. A. Harp. 1996. Previously unpublished Odonata records for Arkansas, Kentucky and Texas. *Notulae Odonatologicae* 4(8):125–136.

Harp, G. L., P. A. Harp. 2003. Dragonflies (Odonata) of the Ouachita National Forest. *Journal of the Arkansas Academy of Science* 57:68–75.

Harp, G. L., P. A. Harp. 2009. Dragonflies (Odonata) observed during 18–20 June 2009, Ouachita National Forest and nearby sites. Unpublished report, *copy on file*.

Harp, G. L., J. D. Rickett. 1977. The dragonflies (Anisoptera) of Arkansas. *Arkansas Academy of Science Proceedings* 31:50–54.

Harp, G. L., J. D. Rickett. 1985. Further distributional records for Arkansas Anisoptera. *Arkansas Academy of Science Proceedings* 39:131–135.

Harp, G. L., L. Trial. 2001. Distribution and status of *Ophiogomphus westfalli* (Odonata: Gomphidae) in Missouri and Arkansas. *Journal of the Arkansas Academy of Science* 55(7):43–50.

Harrel, R. C. 1969. Benthic macroinvertebrates of the Otter Creek drainage basin, northcentral, Oklahoma. *The Southwestern Naturalist* 14:231–248.

Hart, D. L., R. E. Davis. 1981. *Geohydrology of the Antlers aquifer (Cretaceous), southeastern Oklahoma*. Oklahoma Geological Survey, Circular 81.

Hart, D. L., G. L. Hoffman, R. L. Goematt. 1976. *Geohydrology of the Oklahoma Panhandle, Beaver, Cimarron, and Texas counties*. U.S. Geological Survey Water Resources Investigation, 25–72. U.S. Geological Survey, Washington, D. C.

Hasik, A., J. T. Bried. 2017. *Lestes alacer* (Plateau Spreadwing), new for Arkansas. *Argia* 29(3):16–17.

Haukos, D. A., L. M. Smith. 1992. *Ecology of playa lakes*. Waterfowl Management Handbook Leaflet 13.3.7. U.S. Fish and Wildlife Service, Washington, D. C.

Heck, B. 2012. Ouachita Spiketail (*Cordulegaster talaria*), new for Oklahoma. *Argia* 24(3):9–10.

Hill, M. C. 1946. The all-Negro communities of Oklahoma: The natural history of a social movement: Part I. *Journal of Negro History* 31(3):254–268.

Hintze, J. L., R. D., Nelson. 1998. Violin Plots: A box plot-density trace synergism. *The American Statistician* 52(2):181–184.

Hoagland, B. [W]. 2000. The vegetation of Oklahoma: A classification for landscape mapping and conservation. *The Southwestern Naturalist* 45(4):385–420.

Hoagland, B. W. 2002. A classification and analysis of western Oklahoma wetlands. *Proceedings of the Oklahoma Academy of Sciences* 82:5–14.

Hoagland, B. W. 2006. Rivers, lakes, and reservoirs. In: *Historical atlas of Oklahoma*, 4th edition. Goins, C. R., D. Goble, editors. University of Oklahoma Press, Norman, Oklahoma.

Hoagland, B. W., S. L. Collins. 1997. Heterogeneity in shortgrass prairie vegetation: the role of playa lakes. *Journal of Vegetation Science* 8(2):277–286.

Hoagland, B. W., L. R. Sorrels, S. M. Glenn. 1996. Woody species composition of floodplain forests of the Little River, McCurtain and LeFlore Counties, Oklahoma. *Proceedings of the Oklahoma Academy of Science* 76:23–26.

Hoekstra, A. Y., M. M. Mekonnen. 2012. The water footprint of humanity. *Proceedings of the National Academy of Sciences of the United States of America* 109:3232–3237.

Hofsommer, D. L., editor. 1977. *Railroads in Oklahoma*. Oklahoma Historical Society, Oklahoma City, Oklahoma.

Hollon, W. E. 1961. History of the museum. In: *The University of Oklahoma Museum of Natural History, Case Study*. Prepared by the University of Oklahoma Foundation, Inc., and the University of Oklahoma Stovall Museum, Norman, Oklahoma, pp. 5–25.

Hornuff, L. E. 1950. A further study of *Aphylla williamsoni* Gloyd (Odonata, Aeschnidae [sic]). *Proceedings of the Louisiana Academy of Sciences* 14:39–44.

Hornuff, L. E. 1979. To Dr. George H. Bick on his 65th birthday. *Odonatologica* 8:155–158.

Houston, J. 1970. Notes on the habitat and distribution of the Odonata of Franklin County, Arkansas. *Arkansas Academy of Science Proceedings* 24:69–73.

Huggins, D. G. 1978. Redescription of the nymph of *Enallagma basidens* Calvert (Odonata: Coenagrionidae). *Journal of the Kansas Entomological Society* 51:222–227.

Huggins, D. G., P.M. Liechti, D.W. Roubik. 1976. Species accounts for certain aquatic macroinvertebrates from Kansas (Odonata, Hemiptera, Coleoptera and Spaheriidae). *Technical Publications of the State Biological Survey of Kansas* 1:13–77.

Huguet, A., A. Nel, X. Martinez-Delclos, G. Bechly, R. Martins-Neto. 2002. Preliminary phylogenetic analysis of the Protanisoptera (Insecta: Odonatoptera). *Geobios* 35:537–560.

Illinois DNR [Department of Natural Resources]. 1998. *Spoon River Area Assessment*, vol. 3. Living Resources. Unpublished report, *copy on file*.

IUCN [International Union for Conservation]. 2017. IUCN Standards and Petitions Subcommittee guidelines for using the IUCN Red List categories and criteria. Version 13. www.iucnredlist.org/documents/RedListGuidelines.pdf.

Johnson, C. 1972a. Tandem linkage, sperm translocation, and copulation in the dragonfly, *Hagenius brevistylus* (Odonata: Gomphidae). *The American Midland Naturalist* 88(1):131–149.

Johnson, C. 1972b. The damselflies (Zygoptera) of Texas. *Bulletin of the Florida State Museum: Biological Sciences* 16(2):55–128.

Johnson, C. 1973. Distributional patterns and their interpretation in *Hetaerina* (Odonata: Calopterygidae). *Florida Entomologist* 56(1):24–42.

Johnson, C. 1974. Taxonomic keys and distributional patterns for Nearctic species of *Calopteryx* damselflies. *Florida Entomologist* 57(3):231–248.

Johnson, H. L. 2008. *Climate of Oklahoma*. In: *Earth sciences and mineral resources of Oklahoma*. Johnson, K. S., K. V. Luza, editors. Oklahoma Geological Survey, Norman, Oklahoma.

Johnson, H. L., C. E. Duchon. 1995. *Atlas of Oklahoma climate*. University of Oklahoma Press, Norman, Oklahoma.

Johnson, J. 2006. Anything you Caddo, I Caddo better. *Argia* 18(2):8–9.

Johnson, K. S., C. C. Branson, N. M. Curtis, Jr., W. E. Ham, W. E. Harrison, M. V. Marcher, J. F. Roberts. 1972. *Geology and earth resources of Oklahoma: An atlas of maps and cross sections*. Oklahoma Geological Survey, Educational Publication #1. http://ogs.ou.edu/docs/educationalpublications.EP1.pdf

Karl, T. R., J. M. Melillo, T. C. Peterson, S. J. Hassol. 2009. *Global climate change impacts in the United States*. Cambridge University Press, Cambridge, UK.

Keddy, P. 2010. *Wetland ecology*, 2nd edition. Cambridge University Press, Cambridge, UK.

Kennedy, C. H. 1917. The Odonata of Kansas with reference to their distribution. *University of Kansas, Department of Entomology Bulletin* 11:129–145, 7 pls.

Kerst, C., S. Gordon. 2011. *Dragonflies and damselflies of Oregon: A field guide*. Oregon State University Press, Corvallis, Oregon.

Klots, E. B. 1944. Notes on the Protodonata and Protozygoptera of the Lower Permian of Kansas. *American Museum Novitates* 1260:7.

Kondratieff, B. C., P. A. Opler, M.C. Garhart, J. Schmidt. 2004. *Survey of selected insect taxa of Fort Sill, Comanche County, Oklahoma. Dragonflies (Odonata), stoneflies (Plecoptera) and selected moths (Lepidoptera)*. CP Gillette Museum of Arthropod Diversity, Department of Bioagricultural Sciences and Pest Management, Colorado State University, Fort Collins, Colorado, 91 pp.

Kormondy, E. J. 1959. The systematics of *Tetragoneuria* based on ecological, life history, and morphological evidence (Odonata: Corduliidae). *Miscellaneous Publications of the Museum of Zoology* 107:1–79. University of Michigan.

Kormondy, E. J. 1960. New North American records of anisopterous Odonata. *Entomological News* 71:121–135.

Krotzer, R. S., M. J. Krotzer. 1992. Significant new records of Odonata from the southeastern United States. *Notulae Odonatologicae* 3:168–169.

Kustu, M. D., Y. Fan, A. Robock. 2010. Large-scale water cycle perturbation due to irrigation pumping in the U.S. High Plains: A synthesis of observed streamflow changes. *Journal of Hydrology* 390:222–244.

Kuussaari, M., R. Bommarco, R. K. Heikkinen, A. Helm, J. Krauss, R. Lindborg, E. Öckinger, M. Pärtel, J. Pino, F. Rodà, C. Stefanescu, T. Teder, M. Zobel, I. Steffan-Dewenter. 2009. Extinction debt: A challenge for biodiversity conservation. *Trends in Ecology and Evolution* 24(10):564–571.

Lam, E. 2004. *Damselflies of the northeast: A guide to the species of eastern Canada & the northeastern United States*. Biodiversity Books, Forest Hills, New York.

Landwer, B. H. P., R. W. Sites. 2003. Redescription of the larva of *Gomphus militaris* Hagen (Odonata: Gomphidae), with distributional and life history notes. *Proceedings of the Entomological Society of Washington* 105:304–311.

Landwer, B. H. P., R. W. Sites. 2010. The larval Odonata of ponds in the prairie region of Missouri. *Transactions of the American Entomological Society* 136:1–105.

Larsen, R. R. 2008. Notes on the fragile habitat, distribution, and ecology of the Bleached Skimmer (*Libellula composita*). *Argia* 20(3):19–20.

Larson, E. L., C. Guilherme Becker, E. R. Bondra, R. G. Harrison. 2013. Structure of a mosaic hybrid zone between the field crickets *Gryllus firmus* and *G. pennsylvanicus*. *Ecology and Evolution* 3(4):985–1002.

Lasley, G. W., J. C. Abbott. 2009. Two new damselflies for Texas. *Argia* 21(3):17–18.

Little, E. L. 1938. The vegetation of the Caddo County canyons, Oklahoma. *Ecology* 20(1):1–10.

Little, E. L. 1980. Baldcypress (*Taxodium distichum*) in Oklahoma. *Proceedings of the Oklahoma Academy of Science* 60:105–107.

Lohmann, H. 1992. Revision der Cordulegastridae. I. Entwurf einer neuen Klassifizierung der Familie (Odonata: Anisoptera). *Opuscula Fluminea Zoologicae* 96:1–18.

Louton, J. A. 1982. Lotic dragonfly (Anisoptera: Odonata) nymphs of the Southeastern United States: Identification, distribution and historical biogeography. PhD dissertation, University of Tennessee, Knoxville, Tennessee.

Luckey, R. R., M. F. Becker. 1999. *Hydrogeology, water use, and simulation of flow in the High Plains aquifer in northwestern Oklahoma, southeastern Colorado, southwestern Kansas, northeastern New Mexico, and northwestern Texas*. Water-Resources Investigations Report 99–4104. U.S. Geological Survey and Oklahoma Water Resources Board, Oklahoma City, Oklahoma.

Manolis, T., R. Bruun. 2006. A hybrid *Libellula* (*forensis* × *luctuosa*) from northern California. *Argia* 18(3):8–9.

Martin, R. 1906. Cordulines. In: *Catalogue Systematique et Descriptif des Collections Zoologiques du Baron Edmond de Selys Longchamps*. Fasc. 17, Hayez, Impr. des Academies, Brussels, pp. 1–94.

Mauffray, B. 2014. *The dragonflies and damselflies (Odonata) of Louisiana*. http://iodonata.updog.co/index_files/lalist.html. Accessed 19 May 2014.

May, J. D. 2009. Bartlesville. In: *The encyclopedia of Oklahoma history and culture*. Everett, D., L. D. Wilson, L. O'Dell, J. D. May, editors. Oklahoma Historical Society, Oklahoma City, Oklahoma.

May, M. L. 1995. The subgenus *Tetragoneuria* (Anisoptera: Corduliidae: *Epitheca*) in New Jersey. *Bulletin of American Odonatology* 2(4):63–74.

May, M. L. 2019. Odonata: Who they are and what they have done for us lately: Classification and ecosystem services of dragonflies. *Insects* 10(3):62, 17. doi:10.3390/insects10030062.

May, M. L., S. W. Dunkle. 2007. *Damselflies of North America: Color supplement*. Scientific Publishers, Gainesville, Florida.

McGuire, V. L. 2009. *Water-level changes in the High Plains aquifer, predevelopment to 2007, 2005–06, and 2006–07*. Scientific Investigations Report 2009-5019, U.S. Geological Survey, 9 pp. https://pubs.usgs.gov/sir/2009/5019.

McGuire, V. L. 2017. *Water-level and recoverable water in storage changes, High Plains aquifer, predevelopment to 2015 and 2013–15*. Scientific Investigations Report 2017–5040, U.S. Geological Survey, 14 pp. doi:10.3133/sir20175040.

McKnight, D. T., J. Tucker, D. B. Ligon. 2012. Western Chicken Turtles (*Deirochelys reticularia miaria*) at Boehler Seeps and Sandhills Preserve, Oklahoma. *Proceedings of the Oklahoma Academy of Science* 92:47–50.

Menning, M., A. S. Alekseev, B. I. Chuvashov, V. I. Davydov, F.-X. Devuyst, H. C. Forke, T. A. Grunt, L. Hance, P. H. Heckel, N. G. Izokh, Y.-G. Jin, P. J. Jones, G. V. Kotlyar, H. W. Kozur, T. I. Nemyrovska, J. W. Schneider, X.-D. Wang, K. Weddige, D. Weyer, D. M. Work. 2006. Global time scale and regional stratigraphic reference scales of Central and West Europe, East Europe, Tethys, South China, and North America as used in the Devonian-Carboniferous-Permian Correlation Chart 2003 (DCP 2003). *Palaeogeography Palaeoclimatology Palaeoecology* 240:1–2, 318–372.

Meurgey, F., J. J. Daigle. 2007. New status for *Orthemis macrostigma* (Rambur, 1842) from the Lesser Antilles (Anisoptera: Libellulidae). *Odonatologica* 36(1):71–78.

Mills, C. 2007. *Aphylla williamsoni* (Two-striped Forceptail) new for Arkansas. *Argia* 18(4):34.

Miser, H. D. 1929. *Structure of the Ouachita Mountains of Oklahoma and Arkansas*. Oklahoma Geological Survey, Bulletin 50.

Montgomery, B. E. 1942. The distribution and relative seasonal abundance of the Indiana species of *Enallagma* (Odonata: Agrionidae). *Proceedings of the Indiana Academy of Science* 51:273–278.

Montgomery, B. E. 1967. Geographical distribution of the Odonata of the North Central states. *Proceedings of the North Central Branch, Entomological Society of America* 22:121–129.

Moore, C. A. 1953. John Willis Stovall—Vertebrate paleontologist. *Proceeding of the Oklahoma Academy of Science* 34:243–244.

Morris, J. W., C. R. Goins, E. C. McReynolds. 1986. *Historical Atlas of Oklahoma*, 2nd edition. University of Oklahoma Press, Norman, Oklahoma.

Morse, A. P. 1895. New North American Odonata. *Psyche* 7:207–211.

Moskowitz, D. P., D. M. Bell. 1998. *Archilestes grandis* (Great Spreadwing) in central New Jersey, with notes on water quality. *Bulletin of American Odonatology* 5(3):49–54.

Moss, S. P. 1992. Oviposition site selection in *Enallagma civile* (Hagen) and the consequences of aggregating behaviour (Zygoptera: Coenagrionidae). *Odonatologica* 21(2):153–164.

Muttkowski, R. A. 1911. Studies in *Tetragoneuria*. *Bulletin of the Wisconsin Natural History Society* 9:91–134.

Muttkowski, R. A. 1915. Studies in *Tetragoneuria* (Odonata), II. *Bulletin of the Wisconsin Natural History Society* 13:49–61.

Myrup, A. R., R. W. Baumann. 2016. The dragonflies and damselflies (Odonata) of Utah. *Monographs of the Western North American Naturalist* 9:1–114.

Needham, J. G. 1901. Odonata. In: Aquatic insects in the Adirondacks. *Bulletin of the New York State Museum* 47:381–612.

Needham, J. G., H. B. Heywood. 1929. *A handbook of the dragonflies of North America*. Charles C. Thomas, Springfield, Illinois.

Needham, J. G., M. J. Westfall. 1955. *A manual of the dragonflies of North America (Anisoptera): Including the Greater Antilles and the provinces of the Mexican border*. University of California Press, Berkeley and Los Angeles, California.

Needham, J. G., M. J. Westfall, W. L. May. 2000. *Dragonflies of North America*. Scientific Publishers, Gainesville, Florida.

Needham, J. G., M. J. Westfall, Jr., M. L. May. 2014. *Dragonflies of North America: The Odonata (Anisoptera) fauna of Canada, the continental United States, northern Mexico and the Greater Antilles*, 3rd edition. Scientific Publishers, Gainesville, Florida.

Nel, A., G. Bechly, J. Prokop, O. Béthoux, G. Fleck. 2012. Systematics and evolution of Paleozoic and Mesozoic damselfly-like Odonatoptera of the 'Protozygpteron' grade. *Journal of Paleontology* 86(1):81–104.

Nel, A., G. Fleck, R. Garrouste, G. Gand, J. Lapeyrie, S. M. Bybee, J. Prokop. 2009. Revision of Permo-Carboniferous griffenflies (Insecta: Odonatoptera: Meganisoptera) based upon new species and redescription of selected poorly known taxa from Eurasia. *Palaeontographica, Abteilung A, Paläozoologie-Stratigraphie* 289(4–5):89–121.

Norris, L. D. 2009. Southeastern Oklahoma State University. In: *Encyclopedia of Oklahoma history and culture*. Everett, D., editor. Oklahoma Historical Society, Oklahoma City, Oklahoma.

O'Brien, M. 1999. Changes to the 1999 Michigan Odonata list. *Williamsonia* 3(2):4–5.

O'Brien, M. 2008. Odonatological history in Michigan — 1875–1996. *The Great Lakes Entomologist* 41(1/2):1–11.

O'Dell, L. 2007. Wichita Mountains National Wildlife Refuge. In: *Encyclopedia of Oklahoma history and culture*. Everett, D., editor. Oklahoma Historical Society, Oklahoma City, Oklahoma.

Oklahoma Climatological Survey. 2016. *Oklahoma monthly climate summary, June 2016*. Oklahoma Climatological Survey, Norman, Oklahoma.

Oklahoma Climatological Survey. 2019. *Climate facts by county*. Oklahoma Climatological Survey, Norman, Oklahoma.

ODWC [Oklahoma Department of Wildlife Conservation]. 2005. *Oklahoma comprehensive wildlife conservation strategy*. Oklahoma Department of Wildlife Conservation, Oklahoma City, Oklahoma.

ODWC [Oklahoma Department of Wildlife Conservation]. 2015. *Oklahoma comprehensive wildlife conservation strategy*. Oklahoma Department of Wildlife Conservation, Oklahoma City, Oklahoma.

OKWRB [Oklahoma Water Resources Board]. 2011. *Oklahoma comprehensive water plan, supplemental report: Agricultural issues and recommendations*. June 2011. OKWRB, Oklahoma City, Oklahoma.

OKWRB [Oklahoma Water Resources Board]. 2012. *Oklahoma comprehensive water plan, executive report*, February 2012. Oklahoma Water Resources Board, www.owrb.ok.gov/supply/ocwp/ocwp.php.

OKWRB [Oklahoma Water Resources Board]. 2018. *Water facts.* Oklahoma Water Resources Board. www.owrb.ok.gov/util/waterfact.php. Accessed 30 March 2018.

Omernik, J.M., G.E. Griffith. 2014. Ecoregions of the conterminous United States: Evolution of a hierarchical spatial framework. *Environmental Management* 54(6):1249–1266.

Ortenburger, A. I. 1926a. The University of Oklahoma Museum of Zoology expedition of 1925. *Proceedings of the Oklahoma Academy of Science* 6:85–88.

Ortenburger, A. I. 1926b. Odonata collected in southeastern Oklahoma during the summer of 1925. *Proceedings of the Oklahoma Academy of Science* 6:219–221.

Ortenburger, A. I. 1926c. Exploring Oklahoma's Unknown: Rare, almost extinct specimens of animals, reptiles, insects found in the Wichita Mountains. *Outdoor Oklahoma* 2(10):6–7, 30–31.

Orth, D. J., R. N. Jones, O. E. Maughan. 1982. Species composition and relative abundance of benthic macroinvertebrates in Glover Creek, southeast Oklahoma. *Proceedings of the Oklahoma Academy of Science* 62:18–21.

Parmesan, C. 2006. Ecological and evolutionary responses to recent climate change. *Annual Review of Ecology, Evolution, and Systematics* 37:637–669.

Patten, M. A., A. A. Barnard, B. D. Smith. 2019. Geographic variation in a restricted-range endemic dragonfly *Gomphurus ozarkensis* (Odonata: Gomphidae), with description of a new subspecies. *Journal of Insect Biodiversity* 13(2):15–26.

Patten, M. A., J. T. Bried, B. D. Smith-Patten. 2015. Survey data matter: Predicted niche of adult vs. confirmed breeding Odonata. *Freshwater Science* 34:1114–1122.

Patten, M. A., E. A. Hjalmarson, B. D. Smith-Patten, J. T. Bried. 2019. Use of opportunistic occurrence data to determine breeding status of Odonata. *Ecological Indicators* 106:105460.

Patten, M. A., B. D. Smith-Patten. 2011. "As if" philosophy—Conservation biology's real hope. *BioScience* 61:425–426.

Patten, M. A., B. D. Smith-Patten. 2012. First record of the Atlantic Bluet (*Enallagma doubledayi*) for Oklahoma. *Argia* 24(4):16–17.

Patten, M. A., B. D. Smith-Patten. 2013a. Two new species, *Lestes eurinus* Say and *L. forcipatus* Rambur, for Oklahoma, with comments on other vagrant *Lestes* recorded in the state (Zygoptera: Lestidae). *Notulae Odonatologicae* 8:29–32.

Patten, M. A., B. D. Smith-Patten. 2013b. Odonata species of special concern for Oklahoma, USA. *International Journal of Odonatology* 16:327–350.

Patten, M. A., B. D. Smith-Patten. 2014a. First Record of the Seaside Dragonlet (*Erythrodiplax berenice*) for Oklahoma. *Argia* 26(3):17–18.

Patten, M. A., B. D. Smith-Patten. 2014b. The Oklahoma Odonata project: Progress and trends. *Argia* 26(4):19–25.

Patten, M. A., B. D. Smith-Patten. 2015. The Saffron-winged Meadowhawk (*Sympetrum costiferum*) in Oklahoma. *Argia* 27(1):10–11.

Patten, M. A., B. D. Smith-Patten. 2016. The Allegheny River Cruiser (*Macromia alleghaniensis*) in Oklahoma. *Argia* 28(3):12–14.

Patten, M. A., P. Unitt. 2002. Diagnosability versus mean differences of Sage Sparrow subspecies. *Auk* 119:26–35.

Paulson, D. R. 1974. Reproductive isolation in damselflies. *Systematic Biology* 23(1):40–49.

Paulson, D. R. 2001. Recent Odonata records from southern Florida-effects of global warming? *International Journal of Odonatology* 4(1):57–69.

Paulson, D. R. 2003. Comments on the *Erythrodiplax connata* (Burmeister, 1839) group, with the elevation of *E. fusca* (Rambur, 1842), *E. minuscula* (Rambur, 1842), and *E. basifusca* (Calvert, 1895) to full species (Anisoptera: Libellulidae). *Bulletin of American Odonatology* 6:101–110.

Paulson, D. [R]. 2004. New common names for some North American Odonata. *Argia* 16(3):29–30.

Paulson, D. R. 2009. *Dragonflies and damselflies of the West.* Princeton University Press, Princeton, New Jersey.

Paulson, D. R. 2011. *Dragonflies and damselflies of the East.* Princeton University Press, Princeton, New Jersey.

Paulson, D. R. 2018a. The Odonata of North America, including Mexico, Central America and the West Indies. *Bulletin of American Odonatology* 12(4):35–46.

Paulson, D. R. 2018b. *Cordulegaster talaria.* The IUCN Red List of Threatened Species 2018: e.T165047A80689957. doi:10.2305/IUCN.UK.2018-1.RLTS.T165047A80689957.en. Accessed 20 December 2018.

Paulson, D. R. 2019. *Dragonflies and damselflies: A natural history.* Princeton University Press, Princeton, New Jersey.

Penfound, W. T. 1953. Plant communities of Oklahoma lakes. *Ecology* 34:561–583.

Perkins, P. D. 1980. North American insect status review, Final report to the Office of Endangered Species, U.S. Fish and Wildlife Service. (no date on report but others have cited it as 1980).

Pilgrim, E.M., C. D. Von Dohlen. 2007. Molecular and morphological study of species-level questions within the Dragonfly genus *Sympetrum* (Odonata: Libellulidae). *Annals of the Entomological Society of America* 100(5):688–702.

Pilgrim, E. M., C. D. Von Dohlen. 2008. Phylogeny of the Sympetrinae (Odonata: Libellulidae): Further evidence of the homoplasious nature of wing venation. *Systematic Entomology* 33(1):159–174.

Pinto, A. P., R. W. Garrison, D. R. Paulson, T. W. Donnelly, M. L. May. 2012. Case 3584 *Erythemis* Hagen, 1861: Proposed precedence over *Lepthemis* Hagen, 1861 (Insecta, Odonata). *The Bulletin of Zoological Nomenclature* 69(2):92–100.

Prather, I., B. Prather. 2015. The dragonflies and damselflies (Odonata) of Colorado: An updated annotated checklist. *Insects of Western North America*. Contributions of the Department of Bioagricultural Sciences and Pest Management, CP Gillette Museum of Arthropod Diversity, Colorado State University. http://hdl.handle.net/10217/88633.

Pritchard, A. E. 1935. Two new dragonflies from Oklahoma. *Occasional Papers of the Museum of Zoology, University of Michigan* 319:1–10, 1 pl.

Pritchard, A. E. 1936. Notes on *Somatochlora ozarkensis* Bird (Odonata: Libellulidae: Corduliinae). *Entomological News* 47:99–101.

Raasch, G. O. 1946. *The Wellington Formation in Oklahoma.* PhD dissertation, University of Wisconsin, Madison, Wisconsin, 157 p.

Raney, H. 2002. Red Slough Odonata, March-September 2002. Unpublished report submitted to David Arbour, Oklahoma Department of Wildlife Conservation. *Copy on file.*

Reece, B. A., N. E. McIntyre. 2008. Dragonfly (Odonata: Anisoptera) holdings of the Museum of Texas Tech University. *Occasional Papers of the Museum of Texas Tech University, No. 279.*

Reese, L. 2009. Freedmen. In: *The encyclopedia of Oklahoma history and culture*. Everett, D., L. D. Wilson, L. O'Dell, J. D. May, editors. Oklahoma Historical Society, Oklahoma City, Oklahoma.

Rehn, A. C. 2003. Phylogenetic analysis of higher-level relationships of Odonata. *Systematic Entomology* 28:181–239.

Richards, A. 1929. The tentative program of the University of Oklahoma Biological Survey. *Publications of the University of Oklahoma Biological Survey* 1(1):9–13.

Riffell, S. K. 1999. Road mortality of dragonflies (Odonata) in a Great Lakes coastal wetland. *The Great Lakes Entomologist* 32:63–73.

Rogers, C. M. 1953. The vegetation of the Mesa de Maya region of Colorado, New Mexico, and Oklahoma. *Lloydia* 16:257–291.

Rosche, L. O., J. M. Semroc, L. K. Gilbert. 2008. *Dragonflies and damselflies of northeast Ohio*, 2nd Edition. Cleveland Museum of Natural History, Cleveland, Ohio.

Rosenberg, N. J., D. J. Epstein, D. Wang, L. Vail, R. Srinivasan, J. G. Arnold. 1999. Possible impacts of global warming on the hydrology of the Ogallala aquifer region. *Climatic Change* 42:677–692.

Ross, C. L., R. G. Harrison. 2002. A fine-scale spatial analysis of the mosaic hybrid zone between *Gryllus firmus* and *Gryllus pennsylvanicus*. *Evolution* 56(11):2296–2312.

Ross, C. L., J. H. Benedix, Jr., C. Garcia, K. Lambeth, R. Perry, V. Selwyn, D. J. Howard. 2008. Scale-independent criteria and scale-dependent agents determining the structure of a ground cricket mosaic hybrid zone (*Allonemobius socius–Allonemobius fasciatus*). *Biological Journal of the Linnean Society* 94(4):777–796.

Rothrock, E. P., A. C. Noe. 1925. *Geology of Cimarron County, Oklahoma*. Oklahoma Geological Survey, Bulletin 34.

Sargent, C. S. 1884. *Report on the forests of North America (exclusive of Mexico)*, Tenth Census, vol. 9. U.S. Government Printing Office, Washington, D. C.

Sawin, R. S., E. K. Franseen, R. R. West, G. A. Ludvigson, W. L. Watney. 2008. Clarification and changes in Permian stratigraphic nomenclature in Kansas. *Current Research in Earth Sciences* 254(2):1–4.

Say, T. 1839. Description of new North American neuropterous insects and observations on some already described by (the late) Th. Say. *Journal of the Academy of Natural Sciences of Philadelphia* 8:9–46.

Selys-Longchamps, M.-E. de. 1871. Synopsis des Cordulines. *Bulletin de l'Académie Royale des Sciences de Belgique* 31:238–316.

Selys-Longchamps, M.-E. de. 1874. Additions au Synopsis des Cordulines. *Bulletin de l'Académie Royale des Sciences de Belgique* 37:16–34.

Sellards, E. H. 1906. Types of Permian insects. *American Journal of Science* (series 4) 22:249–258.

Sellards, E. H. 1909. ART. IX. Types of Permian insects. *American Journal of Science (1880–1910)* 27(158):151.

Shantz, H. L., R. Zon. 1923. *Natural vegetation*. Map part of the Atlas of American Agriculture. U.S. Department of Agriculture, Washington, D. C. (map scale 1:8,000,000).

Shields, M. A., J. Petranka. 2018. A new field mark for differentiating between females of *Gomphaeschna antilope* (Taper-tailed Darner) and *Gomphaeschna furcillata* (Harlequin Darner). *Argia* 30(2):15–17.

Sims, G. G. 2012. A seasonality of Missouri odonates. Unpublished report, on file with the Oklahoma Odonata Project.

Smith, M. M. 1981. Beyond the Borderlands: Mexican labor in the central plains, 1900–1930. *Great Plains Quarterly* 1(4):239–251.

Smith-Patten, B. D. 2014. First record of Black Setwing (*Dythemis nigrescens*) for Oklahoma. *Argia* 26(2):13–14.

Smith-Patten, B. D. 2017a. Ozark Emerald (*Somatochlora ozarkensis*), species account for Nature Serve Explorer. http:// explorer.natureserve.org.

Smith-Patten, B. D. 2017b. Texas Emerald (*Somatochlora margarita*), species account for Nature Serve Explorer. http:// explorer.natureserve.org.

Smith-Patten, B. D. 2019. Odonate species richness in the United States and Canada. *Argia* 31(1):12–17.

Smith-Patten, B. D., E. S. Bridge, D. J. Hough, J. F. Kelly, M. A. Patten. 2015. Is extinction forever? *Public Understanding of Science* 24:481–495.

Smith-Patten, B. D., B. W. Hoagland. 2015. First record of the Thornbush Dasher (*Micrathyria hagenii*) for Oklahoma. *Argia* 27(3):25–28.

Smith-Patten, B. D., M. A. Patten. 2008. Diversity, seasonality, and context of mammalian roadkills in the southern Great Plains. *Environmental Management* 41:844–852.

Smith-Patten, B. D., M. A. Patten. 2010. Broken antehumeral stripes in a male *Enallagma civile* (Familiar Bluet). *Argia* 22(3):20–21.

Smith-Patten, B. D., M. A. Patten. 2012. *Stylurus intricatus*, a new old record for Oklahoma. *Argia* 24(3):10–11.

Smith-Patten, B. D., M. A. Patten. 2013a. First record of the White-faced Meadowhawk (*Sympetrum obtrusum*) for Oklahoma, and a review of the status of the Cherry-faced Meadowhawk (*S. internum*) in the state. *Argia* 25(2):12–13.

Smith-Patten, B. D., M. A. Patten. 2013b. First records of the Western Pondhawk (*Erythemis collocata*) for Oklahoma. *Argia* 25(3):19–21.

Smith-Patten, B. D., M. A. Patten. 2016. Update on the Oklahoma Odonata project: Zygopterans and mixed species pairs. *Argia* 28(4):35–41.

Smith-Patten, B. D., M. A. Patten. 2017a. Update on the Oklahoma Odonata project: Anisopterans. *Argia* 29(1):1–10.

Smith-Patten, B. D., M. A. Patten. 2017b. The Ozark Emerald (*Somatochlora ozarkensis*): Status, distribution, and ecological notes from Oklahoma. *Argia* 29(1):17–21.

Smith-Patten, B. D., M. A. Patten. 2020. *A checklist of Oklahoma Odonata (Dragonflies and damselflies)*. Oklahoma Biological Survey, Norman, Oklahoma.

Smith-Patten, B. D., M. A. Patten, M. J. Dreiling, J. Fisher. 2007. Phenology and new county records of Odonata of northeastern Oklahoma. *Publications of the Oklahoma Biological Survey* 8:1–13.

Smith-Patten, B. D., J. A. Tucker. 2014. First record of the Sphagnum Sprite (*Nehalennia gracilis*) for Oklahoma. *Argia* 26(4):26–27.

Soluk, D. A., K. Moss. 2003. Roadway and exuvial surveys of the Hine's emerald dragonfly (*Somatochlora hineana*) in Door County, Wisconsin. Part II: Roadway fatalities. *Aquatic Ecology Technical Report 03/09*. Natural History Survey, Champaign, Illinois.

Soluk, D. A., D. S. Zercher, A. M. Worthington. 2011. Influence of roadways on patterns of mortality and flight behavior of adult dragonflies near wetland areas. *Biological Conservation* 144(5):1638–1643.

Sophocleous, M. 2010. Review: Groundwater management practices, challenges, and innovations in the High Plains aquifer, USA—lessons and recommended actions. *Hydrogeology Journal* 18:559–575.

Späth, H.-J., G. L. Thompson, H. Eisenhart, editors. 1998. *Oklahoma resources for economic development.* Oklahoma Geological Survey Special Publication, Oklahoma Geological Survey, Norman, Oklahoma, 98–4, 266.

Stevenson, R. D., W. A. Haber, R. A. Morris. 2003. Electronic field guides and user communities in the eco-informatics revolution. *Conservation Ecology* 7(1):3.

Stewart, K. W., R. Murphy, Jr. 1968. Notes on inter-pond dispersion of some marked adult dragonflies in Oklahoma. *The Texas Journal of Science* 20(2):177–182.

Stovall, J. W., W. Langston. 1950. *Acrocanthosaurus atokensis*, a new genus and species of Lower Cretaceous Theropoda from Oklahoma. *American Midland Naturalist* 43(3):696–728.

Suneson, N. H. 1997. *The geology of the eastern Arbuckle Mountains in Pontotoc and Johnston Counties, Oklahoma: An introduction and field-trip guide.* Oklahoma Geological Survey, Norman, Oklahoma.

Susanke, G. R., G. L. Harp. 1991. Selected biological aspects of *Gomphurus ozarkensis* (Westfall) (Anisoptera: Gomphidae). *Advances in Odonatology* 5:143–151.

Svoboda, M., D. Lecomte, M. Hayes, R. Heim, K. Gleason, J. Angel, B. Ripper, R. Tinker, M. Palecki, D. Stooksbury, D. Miskus, S. Stephens. 2002. The drought monitor. *Bulletin of the American Meteorological Society* 83(8):1181–1190.

Tasch, P. 1961. Paleolimnology: Part 2. Harvey and Sedgwick Counties, Kansas: Stratigraphy and biota. *Journal of Paleontology* 35(4):836–865.

Tasch, P. 1963. Paleolimnology: Part 3. Marion and Dickinson Counties, Kansas, with additional sections in Harvey and Sedgwick counties: Stratigraphy and biota. *Journal of Paleontology* 37(6):1233–1251.

Tasch, P., J. R. Zimmerman. 1959. New Permian insects discovered in Kansas and Oklahoma. *Science* 130:1656.

Tasch, P., J. R. Zimmerman. 1962. The *Asthenohymen-Delopterum* bed—a new Leonardian insect horizon in the Wellington of Kansas and Oklahoma. *Journal of Paleontology* 36(6):1319–1333.

Tennessen, K. J. 1973. *A preliminary report on the systematics of Tetragoneuria (Odonata: Corduliidae) in the southeastern United States.* MS thesis, University of Florida, Gainesville, Florida.

Tennessen, K. J. 1977. Rediscovery of *Epitheca costalis* (Odonata: Corduliidae). *Annals of the Entomological Society of America* 70:267–273.

Tennessen, K. J. 1990. New species of *Cordulegaster* discovered in Arkansas. *Argia* 2:14.

Tennessen, K. J. 2004. *Cordulegaster talaria*, n. sp. (Odonata: Cordulegastridae) from west-central Arkansas. *Proceedings of the Entomological Society of Washington* 106:830–839.

Tennessen, K. J. 2019. *Dragonfly nymphs of North America: An identification guide.* Springer International Publishing, Switzerland.

Tennessen, K. J., J. D. Harper, R.S. Krotzer. 1995. The distribution of Odonata in Alabama. *Bulletin of American Odonatology* 3(3):49–74.

Tillyard, R. J. 1925. Kansas Permian insects. 5. The orders Protodonata and Odonata. *American Journal of Science (V)* 10:41–73.

Tillyard, R. J. 1926. Kansas Permian Insects. Part 6. Additions to the orders Protohymenoptera and Odonata. *American Journal of Science (V)* 11:58–73.

Tillyard, R. J. 1937. Kansas Permian Insects. Part 17. The order Megasecoptera and additions to the Palaeodictyoptera, Odonata, Protoperlaria, Copeognatha and Neuroptera. *American Journal of Science (V)* 33:81–110.

Tolson, A. L. 1970. Black towns of Oklahoma. *The Black Scholar* 1(6):18–22.

Transeau, E. N. 1903. On the geographic distribution and ecological relations of the bog plant societies of northern North America. *Botanical Gazette* 36(6):401–420.

Trewartha, G. T. 1968. *An introduction to climate*, 4th edition. McGraw-Hill, New York, NY.

Trial, L. 2005. *Atlas of Missouri Odonata.* Missouri Department of Conservation, Columbia, Missouri.

Trottier, R. 1973. Influence of temperature and humidity on the emergence behaviour of *Anax junius* (Odonata: Aeshnidae). *The Canadian Entomologist* 105(7):975–984.

UNEP [United Nations Environmental Programme]. 2008. *Vital water graphics—an overview of the state of the world's fresh and marine waters*, 2nd edition. UNEP, Nairobi.

U.S. Census Bureau. 1907. *Population of Oklahoma and Indian Territories, 1907.* Department of Commerce and Labor, Bureau of the Census, Bulletin 89, Government Printing Office, Washington, D. C.

van der Fels-Klerx, H. J., L. Camenzuli, M. K. van der Lee, D. G. A. B. Oonincx. 2016. Uptake of Cadmium, Lead and Arsenic by *Tenebrio molitor* and *Hermetia illucens* from contaminated substrates. *PLoS ONE* 11(11):e0166186. doi:10.1371/journal.pone.0166186.

von Ellenrieder, N. 2003. A synopsis of the Neotropical species of 'Aeshna' Fabricius: The genus *Rhionaeschna* Förster (Odonata: Aeshnidae). *Tijdschrift voor Entomologie* 146:67–207.

Wahl, K. L., T. L. Wahl. 1988. Effects of regional ground-water level declines on streamflow in the Oklahoma panhandle. *Proceedings of Symposium on Water-Use Data for Water Resources Management*, American Water Resources Association, Tucson, Arizona, 239–249.

Wahlgreen, H. F. 1941. Climate of Oklahoma (Oklahoma: Climatic Survey). In: *Climate and man, yearbook of agriculture, 1941.* U.S. Department of Agriculture, Washington, D. C., pp. 1065–1074.

Walker, E. M. 1952. The *Lestes disjunctus* and *forcipatus* complex (Odonata: Lestidae). *Transactions of the American Entomological Society* 78:59–74.

Walker, E. M. 1953. *The Odonata of Canada and Alaska, volume one: Part I: General, Part II: The Zygoptera—Damselflies.* University of Toronto, Toronto.

Walker, E. M. 1966. On the generic status of *Tetragoneuria* and *Epicordulia* (Odonata: Corduliidae). *The Canadian Entomologist* 98(9):897–902.

Walker, E. M., P. S. Corbet. 1975. *The Odonata of Canada and Alaska, volume three: Part III: The Anisoptera—three families.* University of Toronto Press, Toronto.

Ware, J., M. May, K. Kjer. 2007. Phylogeny of the higher Libelluloidea (Anisoptera: Odonata): An exploration of the most speciose superfamily of dragonflies. *Molecular Phylogenetics and Evolution* 45(1):289–310.

Ware, J. L., E. Pilgrim, M. L. May, T. W. Donnelly, K. Tennessen. 2017. Phylogenetic relationships of North American Gomphidae and their close relatives. *Systematic Entomology* 42(2):347–358.

West, R. R., K. B. Miller, W. L. Watney. 2010. The Permian System in Kansas. *Kansas Geological Survey Bulletin* 257:vi + 82. Kansas Geological Survey, Lawrence, Kansas.

Westfall, M. J., Jr. 1974. A critical study of *Gomphus modestus* Needham, 1942, with notes on related species (Anisoptera: Gomphidae). *Odonatologica* 3(1):63–73.

Westfall, M. J., Jr. 1975. A new species of *Gomphus* [*ozarkensis*] from Arkansas (Odonata: Gomphidae) [Insects]. *Florida Entomologist* 3(1):63–73.

Westfall, M. J., Jr. 1990. Descriptions of larvae of *Argia munda* Calvert, *A. plana* Calvert, *A. tarascana* Calvert and *A. tonto* Calvert (Zygoptera: Coenagrionidae). *Odonalologica* 19(1):61–70.

Westfall, M. J., Jr., M. L. May. 1996. *Damselflies of North America*. Scientific Publishers, Gainesville, Florida.

Westfall, M. J., Jr., M. L. May. 2006. *Damselflies of North America*, Revised Edition. Scientific Publishers, Gainesville, Florida.

White, H. 2006. *Gynacantha nervosa* from Delaware! *Argia* 18(2):13.

Williams, C. E. 1976. *Neurocordulia* (*Platycordulia*) *xanthosoma* (Williamson) in Texas (Odonata: Libellulidae: Corduliinae). *Great Lakes Entomologist* 9(1):63–73.

Williams, C. E. 1982. The dragonflies of McLennan County, Central Texas, United States. *Notulae Odonatologicae* 1(10):160–161.

Williamson, E. B. 1908. A new dragonfly (Odonata) belonging to the Cordulinae [sic], and a revision of the classification of the subfamily. *Entomological News* 19:428–434.

Williamson, E. B. 1909. The North American dragonflies (Odonata) of the genus *Macromia*. *Proceedings of the United States National Museum* 37:369–398.

Williamson, E. B. 1912a. *Hetaerina titia* and *tricolor* (Dragonflies—Odonata). *Entomological News* 23(3):98–101.

Williamson, E. B. 1912b. The dragonfly *Argia moesta* and a new species. *Entomological News* 23(5):196–203.

Williamson, E. B. 1914a. *Gomphus pallidus* and two new related species (Odonata). *Entomological News* 25(2):49–58, 2 pls.

Williamson, E. B. 1914b. Dragonflies (Odonata) collected in Texas and Oklahoma. *Entomological News* 25(9 & 10):411–415, 444–455.

Williamson, E. B. 1923. Notes on American species of *Triacanthagyna* and *Gynacantha*. *Miscellaneous Publications of the Museum of Zoology* 9:1–80, 7 pls. University of Michigan.

Williamson, E. B. 1931. *Archilestes grandis* (Ramb.) in Ohio (Odonata: Agrionidae). *Entomological News* 42:63–64.

Williamson, E. B. 1932. Dragonflies collected in Missouri. *Occasional Papers of the Museum of Zoology, University of Michigan* 240:1–40.

Williamson, E. B., P. P. Calvert. 1906. Copulation of Odonata. *Entomological News* 17:143–148.

Wilson, C. B. 1920. Dragonflies and damselflies in relation to pondfish culture: With list of those found near Fairport, Iowa. *Bulletin of the Bureau of Fisheries* 36(882):84.

Woods, A. J., J. M. Omernik, D. R. Butler, J. G. Ford, J. E. Henley, B. W. Hoagland, D. S. Arndt, B. C. Moran. 2005. *Level III and IV Ecoregions of Oklahoma* (color poster with map, descriptive text, summary tables, and photographs). U.S. Geological Survey, Reston, Virginia (map scale 1:1,250,000).

Young, W. C., C. W. Bayer. 1979. The dragonfly nymphs (Odonata: Anisoptera) of the Guadalupe River Basin, Texas. *The Texas Journal of Science* 31:85–97.

Zuellig, R. E., B. C. Kondratieff, J. P. Schmidt, R. S. Durfee, D. E. Ruiter, I. E. Prather. 2006. An annotated list of aquatic insects of Fort Sill, Oklahoma, excluding Diptera with notes on several new state records. *Journal of the Kansas Entomological Society* 79:34–54.

We indexed all pages for collectors/observers except when that person (e.g., Ralph Bird, George Bick, Lothar Hornuff, Bill Carrell, David Arbour) had inordinate occurrences that would result in dozens of cited pages. Instead, we included a sampling of pages appended with "*etc.*" to denote there are others. Citations of some terms, such as ecoregion, ovipositing, population trends, roads, seasonality, and streams, were limited but were not indicated as such with "*etc.*" Page numbers followed by "*f*", "*t*", or "*n*" refer to figures, tables, or notes, respectively. For geographical entries, see the Geographical Index.

We provide this separate index because 1) we feel geography is crucial to understanding organisms and 2) because we have found it frustrating when books have hidden records in them. What we mean by the second part of this statement is that when a book ostensibly is about one country, state, or area but it contains records outside of that jurisdiction, those records can be overlooked given the difficulty posed in finding them. A fine example of this is *The Odonata of Canada and Alaska*. By the title, one would not think to look there for Oklahoma records, but indeed it has some, at least one of which is the only source for that record.

To avoid saturating the index, we were selective with indexing general geographical statements and we mostly shied away from referencing seasonality (see the phenology section in the account for a species of interest). We indexed localities especially when the discussion provides species-specific context or when there are details of a record (e.g., *Ischnura damula* ♀, Texas: Dallam County: Buffalo Springs, Rita Blanca National Grasslands, 31 Aug 2004, GW Lasley; indexed under the species and county). In some cases, we referenced a locality for historical or other reasons of interest, not just for odonate-related topics.

National and subnational names are in bold for easier reference. More detailed jurisdictions are provided for U.S. states (under "counties" or "parishes," as appropriate). We provided only select localities for Oklahoma.

Note: Page numbers followed by "*f*", "*t*", or "*n*" refer to figures, tables, or notes, respectively.